To convert from	To	Multiply by
Mass moment of inertia		
slug · ft^2	kilogram · $meter^2$ ($kg \cdot m^2$)	1.355 818
lbf · s^2 · in	kilogram · $meter^2$ ($kg \cdot m^2$)	1.129 848 E − 01
Mass per unit area		
oz/ft^2	kilogram per $meter^2$ (kg/m^2)	3.051 517 E − 01
lb/ft^2	kilogram per $meter^2$ (kg/m^2)	4.882 428
Mass per unit volume (Includes **Density**)		
g/cm^3	kilogram per $meter^3$ (kg/m^3)	1000*
oz(avoirdupois)/in^3	kilogram per $meter^3$ (kg/m^3)	1729.994
lb/ft^3	kilogram per $meter^3$ (kg/m^3)	16.01846
lb/in^3	kilogram per $meter^3$ (kg/m^3)	2.767 990 E + 04
aluminum #	kilogram per $meter^3$ (kg/m^3)	2775
cast iron #	kilogram per $meter^3$ (kg/m^3)	7200
reinforced plastic #	kilogram per $meter^3$ (kg/m^3)	1450
steel #	kilogram per $meter^3$ (kg/m^3)	8000
Power·[1 W = 1 J/s = 1 N · m/s]		
Btu (international table)/h	watt (W)	2.930 711 E − 01
Btu (international table)/s	watt (W)	1055.056
erg/s	watt (W)	1.000 000* E − 07
ft · lbf/h	watt (W)	3.766 161 E − 04
ft · lbf/min	watt (W)	2.259 697 E − 02
ft · lbf/s	watt (W)	1.355 818
horsepower (550 ft · lbf/s)	watt (W)	745.6999
Pressure or stress (force per unit area) [1 Pa = 1 N/m^2, 1 MPa = 1 N/mm^2]		
atmosphere, standard	pascal (Pa)	1.013 25 E + 05
$dyne/cm^2$	pascal (Pa)	1.000 000* E − 01
kgf/m^2	pascal (Pa)	9.806 650*
kgf/mm^2	pascal (Pa)	9.806 650* E + 06
lbf/ft^2	pascal (Pa)	47.880 26
lbf/in^2 (psi)	pascal (Pa)	6894.757
Time		
day (mean solar)	second (s)	8.640 000* E + 04
hour (mean solar)	second (s)	3600*
year (365 days)	second (s)	3.153 600 E + 07
year (sidereal)	second (s)	3.155 815 E + 07
Torque		
dyne · cm	newton-meter (N · m)	1.000 000* E − 07
kgf · m	newton-meter (N · m)	9.806 650*
ozf · in	newton-meter (N · m)	7.061 552 E − 03
lbf · in	newton-meter (N · m)	1.129 848 E − 01
lbf · ft	newton-meter (N · m)	1.355 818

(*Continued*)

Kinematics and Dynamics of Machinery

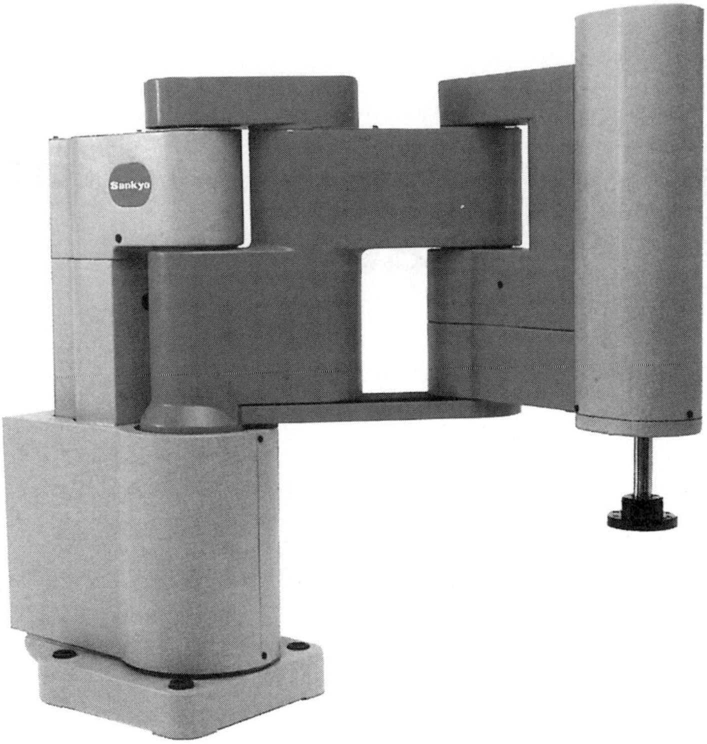

A robot arm for industrial applications. The robot may be taught a sequence of motions for performing a given task. (Source: Sankyo Robotics.)

Kinematics and Dynamics of Machinery

THIRD EDITION

Charles E. Wilson
New Jersey Institute of Technology
and
J. Peter Sadler
The University of Kentucky

Prentice Hall

Pearson Education, Inc.
Upper Saddle River, New Jersey 07458

Library of Congress Cataloging-in-Publication Data

Wilson, Charles E.
Kinematics and dynamics of machinery / Charles E. Wilson, J. Peter Sadler.--3rd ed.
p. cm.
ISBN 0-201-35099-8
1. Machinery, Kinematics of. 2. Machinery, Dynamics of. I. Sadler, J. Peter. II. Title.

TJ175. W758 2002
621.8'11--dc21 2002035453

Vice President and Editorial Director, ECS: *Marcia J. Horton*
Acquisitions Editor: *Laura Fischer*
Editorial Assistant: *Erin Katchmar*
Vice President and Director of Production and Manufacturing, ESM: *David W. Riccardi*
Executive Managing Editor: *Vince O'Brien*
Managing Editor: *David A. George*
Production Editor: *Scott Disanno*
Director of Creative Services: *Paul Belfanti*
Art and Cover Director: *Jayne Conte*
Art Editor: *Greg Dulles*
Manufacturing Manager: *Trudy Pisciotti*
Manufacturing Buyer: *Lisa McDowell*
Marketing Manager: *Holly Stark*
Cover Illustration: *ABB Flexible Automation*

Pearson Education, Inc.
Upper Saddle River, NJ 07458

Printed in the United States of America
10 9 8 7 6 5 4 3 2 1

ISBN 0-201-35099-8

Pearson Education Ltd., *London*
Pearson Education Australia Pty. Ltd., *Sydney*
Pearson Education Singapore, Pte. Ltd.
Pearson Education North Asia Ltd., *Hong Kong*
Pearson Education Canada, Inc., *Toronto*
Pearson Educación de Mexico, S.A. de C.V.
Pearson Education—Japan, *Tokyo*
Pearson Education Malaysia, Pte. Ltd.
Pearson Education, Inc., *Upper Saddle River, New Jersey*

There is no expedient to which a man will not resort to avoid the real labor of thinking.

SIR JOSHUA REYNOLDS

It is the authors' sincere hope that users of this book will be encouraged to think and to explore designs and methods not explicitly described herein. Although worked-out examples have been used extensively to illustrate applications to engineering problems, the intent is to build a capacity for solving new problems that one may encounter in the future.

Dedication

To the memory of Walter Michels, who inspired a generation of students to achievements beyond their own expectations.

Contents

Preface

What Abilities Define an Engineer?

Part of the answer is given by the *program outcomes and assessment* criteria of the Accreditation Board for Engineering and Technology[1]. Engineering programs must demonstrate that their graduates have specific abilities. These include:

- an ability to apply knowledge of mathematics, science, and engineering
- an ability to analyze and interpret data
- an ability to design a system, component, or process to meet desired needs
- an ability to identify, formulate, and solve engineering problems
- an ability to communicate effectively
- an ability to use techniques, skills, and modern engineering tools necessary for engineering practice.

Goals of the Text

A course in the kinematics and dynamics of machinery provides many opportunities to develop the abilities listed above. This text is designed to help foster development and application of those skills. One goal is to develop the ability of students to formulate and solve problems in the kinematics and dynamics of machinery. Engineering tools used to achieve this goal include motion simulation software and general-purpose mathematical software. These tools relieve the designer of repetitive tasks and provide a powerful means of communicating results through graphs and animation simulations. An equally important goal is the development of an understanding of the implications of computed results. That is, what do the results mean; how can we improve the design? Knowledge gained in previous courses is reinforced when applied to problems in the kinematics and dynamics of machinery. For example, matrix methods become meaningful when applied to equations describing velocities and accelerations in a spatial linkage. The skills learned and sharpened in studying the kinematics and dynamics of machinery are carried forward, even to unrelated courses and to engineering practice.

Scope

The coverage of this text includes mechanisms and machines, basic concepts; motion in machinery; velocity and acceleration analysis of mechanisms; design and analysis of

[1] Accreditation Board for Engineering and Technology, *Engineering Criteria 2000*, Third Edition, December 1997.

cams, gears, and drive trains; static and dynamic force analysis; synthesis; and an introduction to robotic manipulators. Practical applications are considered throughout the text. Example problems and homework problems involve engineering design and provide a basis for design courses to follow. Analytical and graphical vector methods are illustrated, as well as complex number methods.

The text illustrates the use of motion simulation software, mathematics software, and user-written programs to solve problem and to present the results in plotted or tabulated form.

There are also many problems that can be solved by "hand calculations," however, using only a scientific calculator and/or simple drafting tools. The latter group may be useful as short practice problems and examination problems when laptop computers are unavailable.

What's New in the Third Edition?

The text was updated throughout. A few of the changes and additions include:

- A list of "Concepts you will learn and apply ...." for each chapter.
- Chapter summaries
- Review and discussion items for each chapter
- Thorough revision of the material on cam design, including application of step and interval functions, and higher order polynomials for cam design.
- Practical design implications of results
- Gear train diagnostics based on noise and vibration frequencies
- Updates, example problems and homework using motion simulation software
- Updates example problems and homework problems using mathematics software throughout the text.
- Other updates for clarity and brevity
- Suggestions for "working smart," particularly with computers. Emphasis on computer-aided matrix solutions where appropriate
- Interpretation and assessment of results. Is it logical? Does it check? Let's compare results with another solution. What does it mean? What does the graph show?
- Practical design implications of results

Course Development

Professors who regularly teach the kinematics and dynamics of machinery will know what topics suit their students best. This note is for instructors who have not taught the course recently.

Most of the topics in this text can be covered in a three-credit-hour course given to engineering students who have completed a course in the statics and dynamics of rigid bodies. But a single course designed around the entire book would be likely to have insufficient depth. Instructors may decide to cover the parts of the text that they deem essential, and then select additional topics and solution methods according to goals set for their students. For example, either analytical vector methods or complex number methods may be used as a basis for writing computer programs to solve planar linkages. However, if analysis of spatial linkages is to follow analysis of planar linkages, then vector methods might be used for both.

For courses built around the use of motion simulation software and mathematics software, graphical methods are likely to be de-emphasized. For example, the velocity polygon might be used only to spot check a detailed analytical velocity analysis of a planar linkage. In a course concerning the kinematics and dynamics of machinery, uniformity of course content is not essential. Differences in emphasis and methods among university engineering departments may strengthen the "gene pool" of future engineers.

We have attempted to provide sufficient rigor and advanced material to challenge the student and provide a basis for further study. Student creativity may be fostered by the demands of the task, particularly if a few homework problems are expanded into open-ended design-type projects. Additional projects requiring creativity may be suggested by articles in technical publications or by an instructor's current research and consulting.

Disclaimer

The kinematics and dynamics of machinery and the design of mechanisms involve modeling of physical systems. Relationships so developed have limits of applicability. The user of this text is urged to interpret the results of calculations, rather than simply obtain problem solutions. It is the reader's responsibility to assess formulas and methods to determine their applicability to a particular situation. Although the publisher, the reviewers, and the authors have made every effort to ensure accuracy, errors invariably creep in. Suggestions and corrections are most welcome.

About the Authors

Charles E. Wilson is a Professor with the Department of Mechanical Engineering, New Jersey Institute of Technology. He received the B.S. and M.S. degrees in mechanical engineering from the Newark College of Engineering, the M.S. in engineering mechanics from New York University, and the Ph.D. degree in mechanical engineering from Brooklyn Polytechnic Institute. He is a licensed professional engineer, and has been awarded fellowships by the National Aeronautics and Space Administration, Department of Energy, and National Science Foundation.

Dr. Wilson has published papers in a number of journals and transactions. Textbooks he has authored and co-authored are widely used in the United States and Canada. English language versions are also published in Britain, Taiwan, India, and the Philippines, and translations are published in Korea and Mexico.

Dr. Wilson served as a U.S. Air Force electronics and armament officer, and as an engineer and consultant for a number of companies. He is often called on to investigate functional and design problems in vehicles, machinery, and consumer products. He has investigated and given expert testimony on auto, truck, bus, and ambulance accidents, and accidents involving elevators, hydraulic presses, welds, playground equipment, garden equipment, and truck-mounted machinery.

J. Peter Sadler is a Professor with the Department of Mechanical Engineering, University of Kentucky. He has previously held faculty positions at the State University of New York at Buffalo and the University of North Dakota. He received the B.S.M.E, M.S.M.E., and Ph.D. degrees from Rensselaer Polytechnic Institute.

Dr. Sadler is a registered professional engineer and a member of many technical societies. He served as Editor for dynamics for the *Journal of Mechanism and Machine Theory* and Associate Editor of the *Journal of Applied Mechanics and Robotics*.

Dr. Sadler holds a U.S. patent related to predicting optimum machining coditions. His industrial projects and research include kinematics and dynamics, robotics, computer aided design, engineering optimization, and "lean" manufacturing.

Acknowledgments

We wish to express our appreciation to all who helped with this book. Users of earlier editions and manuscript reviewers made many worthwhile suggestions based on their extensive teaching and engineering experience. Their expert analysis resulted in many changes in the text. Those sharing their expertise included: Leo Maier of Ohio Northern University, Joseph M. Mansour of Case Western Reserve University, Charles Mallory North, Jr. of Rose-Hulman Institute of Technology, G.K. Ananthasuresh of the University of Pennsylvania, Koorosh Naghshineh of Western Michigan University, Charles C. Adams of Dordt College, Noah Manring of the University of Missouri—Columbia, Melvin R. Corley of Louisiana Tech University, Dan R. Marghitu of Auburn University, Huh Jing Ying of the University of South Florida, Gary H. McDonald of the University of Tennessee at Chattanooga, Ara Arabyan of the University of Arizona, Saeed B. Nicu of California Polytechnic State University, Ferdinand Freudenstein of Columbia University, Robert Williams of Control Data Corporation, Kenneth Waldron of Ohio State University, and William Park of Pennsylvania State University. We also wish to thank our students and colleagues for their suggestions and comments and to thank the companies that provided photographs and illustrations. Thanks also to our editors and others who saw this work through to completion.

Symbols

Vectors and matrices are shown in **boldface**, scalar magnitudes in lightface.

$\boldsymbol{A}^{-1}$	Inverse of matrix $\boldsymbol{A}$
$\boldsymbol{A} \cdot \boldsymbol{B}$	Dot (scalar) product of vectors $\boldsymbol{A}$ and $\boldsymbol{B}$
$\boldsymbol{A} \times \boldsymbol{B}$	Cross (vector) product of vectors $\boldsymbol{A}$ and $\boldsymbol{B}$
a	Gear tooth addendum
$a, \boldsymbol{a}$	Acceleration
$a^c, \boldsymbol{a}^c$	Coriolis acceleration
$a^n, \boldsymbol{a}^n$	Normal acceleration
$a^t, \boldsymbol{a}^t$	Tangential acceleration
bc	Velocity of C relative to B (velocity difference)
C	Cylinder pair; planet carrier
$\boldsymbol{C}$	Force couple
$\boldsymbol{C}_i$	Inertia couple or inertia torque
c	Center distance
CAD	Computer-aided design
D	Determinant
d	Diameter of pitch circle
d_b	Diameter of base circle
DF	Degrees of freedom
e	Instantaneous efficiency; cam-follower offset; piston offset; eccentricity
$e^{j\theta}$	Polar form of a complexnumber
$\boldsymbol{F}$	Force
F_a	Axial or thrust gear tooth force component
$\boldsymbol{F}_e$	External force
$\boldsymbol{F}_i$	Inertia force
$\boldsymbol{F}_{ij}$	Force exerted by a member i on member j
$\boldsymbol{F}_n$	Normal gear tooth force
F_r	Radial gear tooth force component
F_s	Shaking force
F_t	Tangential gear tooth force component
f	frequency
f_i	Joint connectivity
G	Center of mass
H	Helix pair
h	Cam follower lift
hp	Horsepower
I	Mass moment of inertia
$\boldsymbol{i}, \boldsymbol{j}, \boldsymbol{k}$	Cartesian unit vectors
j	Cam follower jerk; $\sqrt{-1}$, the imaginary unit used to represent quantities on the complex plane
L	Link length, sound level
L_d	Length of diagonal (of linkage polygon)
l	Lead of worm
l_i	Length of link i
$\boldsymbol{M}_s$	Shaking moment
m	Mass; module; slope; meters
m^n	Normal module
N	Number of gear teeth; newtons
$\boldsymbol{N}$	Normal force
n	Rotational speed (revolutions per minute)
n_c	Number of constraints
n_J	Number of joints
n_L	Number of links
O_1	Fixed bearing on link 1
ob	Absolute velocity of point B
P	Prism pair; planet gear; power; diametral pitch
$\boldsymbol{P}$	Piston force
P^n	Normal diametral pitch
${}^i\{P\}$	Position vector in frame i
p	Transverse circular pitch, pressure
p_b	Base pitch
p^n	Normal circular pitch
p_w	Axial pitch of worm
R	Revolute pair; ring gear; length of crank
$\boldsymbol{R}$	Position vector
${}^i_j[R]$	Rotational transformation matrix
r	Radius of pitch circle
$\boldsymbol{r}$	Position vector; vector representing a link
$\dot{\boldsymbol{r}}$	Derivative of $\boldsymbol{r}$ with respect to time

r^*	Train value (speed ratio) for a planetary train relative to the carrier
r_a	Length of cam-follower arm; radius of addendum circle
r_b	Base circle radius; radius of back cone element
r_c	Center distance between cam and follower pivots
r_f	Radius of cam-follower roller; radius of friction circle
r_m	Mean pitch radius
r_v	Velocity ratio
$\boldsymbol{r}^u$	Unit vector
r_x	x component of vector $\boldsymbol{r}$
S	Sphere pair; sun gear
$\boldsymbol{s}, s$	Displacement
s	Seconds
s_n	Axial spacing of engine cranks and cylinders; joint offset along axis n
$\boldsymbol{T}, T$	Torque
$\boldsymbol{T}_e$	External torque
${}^i_j[T]$	Transformation matrix from frame j to frame i
t	Time; gear tooth thickness
U	Universal joint
$\boldsymbol{v}$	Velocity
v_p	Pitch line velocity
W	Work; watts
w	Gear tooth width; weight
x, y, z	Cartesian coordinates
$\times$	Cross product
$\alpha, \boldsymbol{\alpha}$	Angular acceleration
α	Cam rotation angle; angle of approach
β	Angle of recess
Γ	Pitch angle
γ	Cam follower rotation; pitch angle; mass density
$\delta\phi, \delta x$	Virtual displacements
θ	Angular position of link; cam angle; angle of action; connecting rod angle
θ_i	Angular position of link i
θ_n	Angular spacing of engine cylinders; joint angle about axis n
λ	Lead angle of worm
μ	Coefficient of sliding friction
ρ	Radius of curvature
ρ_p	Radius of curvature of pitch curve
Σ	Angle between shafts
τ_i	Link twist of member i
Φ	Heaviside step function
ϕ	Transmission angle; pressure angle; transverse pressure angle; friction angle
ϕ_i	Angular position of link i
ϕ^n	Normal pressure angle
ψ	Involute angle; helix angle
ψ_n	Angular spacing of engine cranks
$\omega, \boldsymbol{\omega}$	Angular velocity

What You Will Learn and Apply in the Study of the Kinematics and Dynamics of Machinery

The following is a partial list of the knowledge and skills you will acquire or enhance. In many cases, you will be applying mathematics and scientific principles that you learned previously.

- Effective computer use and software selection
- Application of animation software to linkage design
- Application of mathematics software to mechanism design
- Computer-aided solutions to engineering problems using vector and matrix equations
- Mobility of planar and spatial linkages
- Determination of motion characteristics of linkages
- Design to avoid binding and interference
- Design and selection of mechanisms for specific applications

- Analytical vector methods applied to linkage design
- Complex number methods applied to linkage design
- Analytical and graphical methods for finding linkage velocities
- Analytical and graphical methods for finding linkage accelerations
- Design and analysis of cams
- Design and analysis of spur gears
- Design and analysis of helical, worm, and bevel gears
- Arrangement of gears to produce desired input–output speed ratios
- Design of planetary speed changers
- Analysis of static forces in linkages and gear and cam mechanisms
- Analysis of dynamic forces in linkages
- Balancing of rotors and reciprocating machines
- Synthesis of linkages to produce predetermined motion
- Design and analysis of simple robotic manipulators
- Critical thinking applied to mechanism design. Critical thinking involves identification of a problem, gathering of data, objective analysis, and an attempt at solving the problem by a scientific process. This skill should be honed throughout an engineer's education and practice.
- Engineering creativity. The text and problems are designed to foster creativity, but this goal depends almost entirely on the student (with encouragement from an instructor).

CHAPTER 1

Mechanisms and Machines: Basic Concepts

In this chapter, you will learn

- The terms we use to describe mechanisms
- The degrees of freedom of mechanisms
- How to design crank-rocker, double-rocker, drag link, and other mechanisms
- How to design linkages to prevent jamming and interference
- How to optimize a mechanism design
- Computer animation of linkages to check the validity of a design
- The design of quick-return mechanisms
- The design and selection of mechanisms for special applications
- Numerical solutions
- Other basic concepts.

Kinematics and dynamics are vital components of machine design. An understanding of the kinematics and dynamics of machinery is important to the design of

- Production machinery, including robots and other programmable machines
- Consumer goods and office machines
- Aircraft and surface transportation vehicles
- Agricultural and construction machinery
- Many other items considered essential to modern living

1.1 INTRODUCTION

Kinematics and dynamics of machinery involve the design of machines on the basis of their motion requirements. A combination of interrelated parts having definite motions and capable of performing useful work may be called a **machine**. A **mechanism** is a component of a machine consisting of two or more bodies arranged so that the motion of one compels the motion of the others. The design of an automotive power train (a type of machine) is concerned with several mechanisms, including slider-crank linkages, cam and follower linkages, and gear trains. Many mechanisms undergo **planar motion**, motion in a single plane or in a set of parallel planes. The more general case, **spatial motion**, applies to mechanisms in which the motion must be described in three dimensions.

Kinematics is the study of motion in mechanisms without reference to the forces that act on the mechanism. **Dynamics** is the study of the motion of individual bodies and mechanisms under the influence of forces and torques. The study of forces and torques in stationary systems (and systems with negligible inertial effects) is called **statics**.

Synthesis is a procedure by which a product (a mechanism, for example) is developed to satisfy a set of performance requirements. If a product configuration is tentatively specified and then examined to determine whether the performance requirements are met, the process is called **analysis**. The design of mechanisms involves both synthesis and analysis.

The design process begins with the recognition of a need. A set of requirements is then listed. Creativity and inventiveness are key to selecting the connectivity and form of a mechanism or machine to satisfy the need. The designer may use formal synthesis procedures in which specifications and corresponding decision sets and design variables must be identified. The designer then prepares an adequacy assessment procedure, formulating a figure of merit and an optimization strategy.

Detailed analyses of displacements, velocities, and accelerations are usually required. This part of the design process is followed by an analysis of forces and torques. The design process may continue long after the first models have been produced and may include redesigns of components that affect velocities, accelerations, forces, and torques. To compete successfully from year to year, most manufacturers must continuously modify their products and their methods of production. Increases in the production rate, upgrading of product performance, redesign for cost and weight reduction, and motion analysis of new product lines are frequently required. Success may hinge on the accuracy of the kinematic and dynamic analysis.

1.2 TOOLS AVAILABLE TO THE DESIGNER OF LINKAGES AND OTHER MECHANISMS

A designer will ordinarily begin the design process by making various design decisions based on his or her experience and creative ability. These decisions may be verified and modified through analytical, graphical, numerical, and empirical methods. If a linkage is to be analyzed in only one position, graphical vector methods may provide the quickest solution. Complex-number methods are convenient for analyzing planar linkages. Analytical vector methods are used for solving planar and spatial linkages. While a calculator is adequate for solving a linkage problem for a single position, it is worthwhile to write a computer program when a solution is required over a range of values. Computer solutions are also effective for analysis and synthesis when it is necessary to evaluate several alternatives.

HINTS FOR EFFECTIVE COMPUTER USE

Software Selection

Every year, personal computers get faster and can store more information. New and more powerful versions of software appear as well. In selecting new software for designing and analyzing linkages and other mechanisms, some of the following considerations may be relevant:

- **Animation.** Motion simulation software (e.g., Working Model™) allows the designer to "build" and analyze linkage simulations. The linkages can be animated to verify design criteria and compute velocities, accelerations, and forces.
- **Plotting routines.** Clearly labeled plots showing the position, velocity, and acceleration of linkages help the designer gain an insight into the motion of a mechanism and give clues leading to an improved design.
- **Equation form.** Mathematics software (e.g., Mathcad™) shows subscripts, superscripts, upper- and lowercase Greek and Roman symbols, and built-up equations. In this form, equations are readable and easier to debug. Equations embedded in typical programming languages are somewhat less readable. In spreadsheets, equations are usually hidden, but tabular data are clearly displayed. An important spreadsheet feature is that when one cell is changed, all related cells are updated to reflect the change.
- **Computational features.** Mathematics software that includes equation solvers, numerical differentiation routines, and routines to manipulate complex numbers, vectors, and matrices can be a significant time-saver in working with mechanisms.
- **Trigonometric functions.** Linkage solutions require direct and inverse trigonometric functions. Software that offers a two-argument arctangent function (ANGLE or ARCTAN_2) is preferable.
- **Symbolic capabilities.** Symbolic equation solvers, symbolic integration and differentiation, and symbolic matrix operations are useful software features.
- **User experience.** Familiarity with a particular type of spreadsheet, programming language, or mathematics package may govern one's selection of software. For example, a person skilled in using a particular programming language (BASIC, FORTRAN, C++, etc.) may find it inefficient to switch to an unfamiliar type of mathematics software.
- **Educational considerations.** Educational goals sometimes override other considerations. Kinematics problems can be solved by means of spreadsheets (Lotus 1–2–3™, Excel™, etc.) if the course goals include learning to deal with spreadsheets.
- **Hardware limitations.** Software packages indicate minimum hardware requirements. The performance of a given package will be unsatisfactory with inadequate random-access memory or inadequate hard-disk space.
- **Presentation form.** Engineers must be prepared to present their work to others. Desirable software features include a word-processor-like text capability, a

cut-and-paste capability, and the ability to mix calculations, graphs, tables, and comments. Such features allow the engineer to "work smart" by completing the report while performing design and analysis calculations.

Identifying a Need or a Problem

Most academic problems are clearly defined. By contrast, typical real-world problems are ill-defined and require many assumptions. A computer cannot identify a real need or problem for you. Try to ask the right questions; identify the right problem before beginning detailed work. A correct solution to the wrong problem is of little value.

Programming

Some of the following suggestions may apply, whether a programming language, mathematics software, or a spreadsheet is selected to aid in kinematics design:

- Begin with a simple program. Test it with known data if possible. Then build on the program to solve the required problem.
- Be generous with titles and comments in your program. Identify variables. Note the limitations of the program.
- Output intermediate results so that you can check and debug the program. Do these results look reasonable? For example, has the length of a rigid link changed? Spot-check computer results by using independent calculations. Try to write a self-verifying program.
- Let the computer serve you; avoid wasting time making unnecessary improvements in your programs (unless improving programming skills is an educational goal).
- Make personal quick-reference cards. Include notes on the best utilization of software for your most common tasks.
- Interpret your results. Do "what if" analysis. What if the link length is changed? What if the angular velocity is increased? Computer software cannot replace creativity and interpretation of results. It reduces the time spent on repetitive tasks, leaving more time for important tasks related to the design of linkages.

Using Motion Simulation Software to Produce Coupler Curves

We sometimes need a mechanism with an output link that rotates through a limited range (oscillates) as the input crank rotates at constant speed. Or we may want more complicated output motion. Figure 1.1 illustrates the use of Working Model™ motion simulation software to describe a four-bar linkage. The moving links are the drive crank O_1B, the coupler BFGCDE, and the driven crank CO_3. A motor is located at the fixed bearing O_1, and a fixed bearing at O_3 supports the driven crank. The drive crank and the driven crank are joined to the coupler at bearings B and C.

As the drive crank rotates continuously, bearing B, which joins it to the coupler, traces a circle. The driven crank oscillates, and bearing C traces a circular arc. Near the

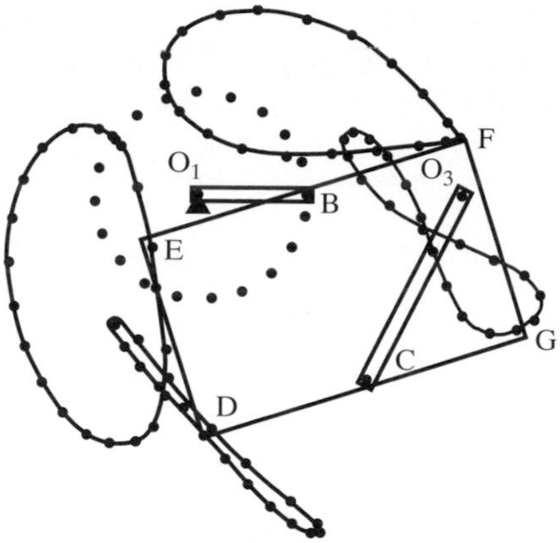

FIGURE 1.1 Coupler curves.

corners of the coupler, points D, E, F, and G trace various figures called **coupler curves**. Point G, for example, produces a figure eight. We might join another link at that point to produce some needed motion.

It is difficult to design a linkage to produce a specified motion pattern. However, we may be successful if we use ingenuity and test our designs with motion simulation software. If we cannot produce the desired motion with coupler curves, then we will try cams, numerically controlled systems, or computer-controlled robots for the task.

1.3 SYSTEMS OF UNITS

Any appropriate set of units may be used in the study of kinematics and dynamics of machinery, as long as consistency is maintained. We invite errors when we fail to check the consistency of units. Preferred systems are the International System of Units, or Systâme International (SI), a modernized version of the meter–kilogram–second (mks) system; and the customary U.S. inch–pound–second system. The following basic and derived units are suggested:

SI (m – kg – s) SYSTEM

Quantity	Unit	Symbol	Relationship
Acceleration			m/s^2
Energy and work	joule	J	$N \cdot m$
Force	newton	N	$kg \cdot m/s^2$
Length	meter	m	
Mass	kilogram	kg	$N \cdot s^2/m$
Mass moment of inertia			$kg \cdot m^2$
Power	watt	W	J/s or $N \cdot m/s$
Pressure and stress	pascal	Pa	N/m^2
Torque and moment			$N \cdot m$ ($N \cdot m/rad$)
Velocity			m/s

CUSTOMARY U.S. (in – lb_f – s) SYSTEM

Quantity	Unit	Symbol	Relationship
Acceleration			in/s^2
Energy and work			lb · in
Force	pound	lb or lb_f	
Length	inch	in	
Mass			$lb \cdot s^2/in$
Mass moment of inertia			$lb \cdot s^2 \cdot in$
Power	horsepower	hp	
Pressure and stress		psi	lb/in^2
Torque and moment			lb · in (lb · in/rad)
Velocity			in/s

COMMON TO BOTH SI AND CUSTOMARY U.S. SYSTEMS

Quantity	Unit	Symbol	Relationship
Angular acceleration			rad/s^2
Angular velocity			rad/s
Frequency	hertz	Hz	(cycles)/s
Plane angle	radian	rad	
Time	second	s	

SI prefixes may be used to eliminate nonsignificant digits and leading zeros. The following are in most common use:

Multiplication Factor	Prefix	Symbol
$1{,}000{,}000 = 10^6$	mega-	M
$1000 = 10^3$	kilo-	k
$0.01 = 10^{-2}$	centi-	c
$0.001 = 10^{-3}$	milli-	m
$0.000001 = 10^{-6}$	micro-	μ

Although prefixes representing powers of 1000 are preferred, the *centi-* prefix is also used (e.g., centimeters, cubic centimeters).

It is generally most convenient to perform calculations by using scientific notation (powers of 10) or engineering notation (10^{-6}, 10^{-3}, 10^3, 10^6, etc.). A suitable unit and prefix should be chosen to express the results of calculations so that the numerical value falls between 0.1 and 1000, where convenient. An exception to this suggestion is engineering drawings, in which, for consistency, linear dimensions are expressed in millimeters (mm). When a number of values are tabulated or discussed, consistent units and prefixes are preferred (e.g., a velocity range given as 0.09 m/s to 1100 m/s would be preferred over the same range expressed as 90 mm/s to 1.1 km/s).

The advantage of SI as a coherent system may be lost if it is used together with units from other systems. However, convenience and common usage suggest the use of the degree (and decimal parts of the degree) for the measurement of plane angles. But one should always be careful when using degrees. An angle that stands alone in an equation will be in radians, and the argument of the tangent, sine, and cosine functions must be in radians for use in most software and programming languages. Obviously, time expressed in minutes, hours, and days will often be more practical than the use of seconds in some applications (e.g., the velocity of a vehicle is commonly expressed in kilometers per hour, (km/h), and the kilowatt-hour (kW · h) is used as a measure of energy).

Like the pound (lb), the kilogram (kg) is sometimes used as a unit of force as well as a unit of mass. The accepted SI force unit, however, is the newton (N). Torque may be expressed in newton-meters (N · m). Although 1 N · m equals 1 joule (J), the term *joule* should be reserved for work and energy.

Conversion Factors

The following are few of the conversion factors that are useful in the kinematics and dynamics of machinery (an extensive list of conversion factors is given inside the front and back covers):

$$1\ g \text{ (gravitational constant)} = 386.09 \text{ in/s}^2 = 9.80665 \text{ m/s}^2$$
$$1 \text{ horsepower (hp)} = 6600 \text{ lb} \cdot \text{in / s} = 745.7 \text{ W}$$
$$1 \text{ in} = 25.4 \text{ mm};\ 1 \text{ m} = 39.37 \text{ in}$$
$$1 \text{ lb} = 4.4482 \text{ N (force)}$$
$$1 \text{ lb} = 0.45359 \text{ kg (mass)}$$
$$1 \text{ mi/h} = 0.44704 \text{ m/s}$$
$$1 \text{ psi} = 6894.8 \text{ Pa};\ 1 \text{ MPa} = 145.04 \text{ psi}$$
$$1 \text{ rad} = 180^\circ / \pi = 57.2958^\circ$$
$$1 \text{ revolution per minute (rev/min)} = \pi \text{ rad / 30 s} = 0.10472 \text{ rad/s}$$

1.4 TERMINOLOGY AND DEFINITIONS

Many of the basic linkage configurations have been incorporated into machines designed centuries ago, and the terms we use to describe them have changed over the years. Thus, definitions and terminology are not consistent throughout the technical literature. In most cases, however, the meanings will be clear from the context of the descriptive matter. A few terms of particular interest to the study of kinematics and dynamics of machines are defined next.

Link

A *link* is one of the rigid bodies or members joined together to form a kinematic chain. The term *rigid link*, or sometimes simply *link*, is an idealization used in the study of mechanisms that does not take into account small deflections due to strains in machine members. A perfectly rigid or inextensible link can exist only as a model of a real machine member. For typical machine parts, the maximum changes in dimension are on the order of only one-thousandth of the part length.

Frame

The fixed or stationary link in a mechanism is called the *frame*. When there is no link that is actually fixed, we may consider one as being fixed and determine the motion of the other links relative to it. In an automotive engine, for example, the engine block is considered the frame, even though the automobile may be moving.

Joint or Kinematic Pair

The connections between links that permit constrained relative motion are called *joints*. The joint between a crank and connecting rod, for instance, may be called a *revolute joint* or a *pin joint*. The revolute joint has one degree of freedom, in that, if one element is fixed, the revolute joint allows the other only to rotate in a plane. (Degrees of freedom are discussed in more detail in a section that follows.) A *sphere joint* (*ball joint*) has three degrees of freedom, thus allowing relative motion in three angular directions. A number of common joint types are idealized in Figure 1.2. Some of the practical joints that they represent are made up of several elements. Examples include the universal joint, ball and roller bearings that are represented by the revolute joint, ball slides represented by the spline joint, and ball screws represented by the helix.

Lower and Higher Pairs

Connections between rigid bodies consist of lower and higher pairs of elements. Theoretically, the two elements of a *lower pair* are in surface contact with one another, while the two elements of a *higher pair* are in point or line contact (if we disregard deflections). Lower pairs include revolutes or pin connections—for example, a shaft in a bearing or the wrist pin joining a piston and connecting rod. Both elements joined by the pin may be considered to have the same motion at the pin center if clearance is neglected. Other basic lower pairs include the sphere, cylinder, prism, helix, and plane (Figure 1.2). Waldron (1972) shows that these six are the only basic lower pairs possible.

Examples of higher pairs include a pair of gears or a disk cam and follower. The Hook-type universal joint is a combination of two lower pairs. A Bendix–Weiss type of constant-velocity universal joint includes higher pairs. (See illustrations later in the chapter.)

From a design standpoint, lower pairs are desirable, since the load at the joint and the resultant wear are spread over the contact surface. Thus, geometric changes or failure due to high contact stresses or excessive wear may be prevented. In practice, we may utilize a ball or roller bearing as a revolute pair to reduce friction; However, the advantages of contact over a large surface are sacrificed.

Closed-Loop Kinematic Chains

A kinematic chain is an assembly of links and pairs (joints). Each link in a closed-loop kinematic chain is connected to two or more other links. Consider, for example, the slider-crank mechanism, a component of the vertical compressor shown in Figure 1.3. Bearings (represented by a revolute joint) connect the casing (frame) and

Type of joint (pair)	Lower pair (L) or higher pair (H)	Symbol	Degrees-of-freedom (connectivity) of the joint in a spatial linkage	Schematic representation	Possible configuration	Descriptive example
Revolute	*L*	*R*	1 θ	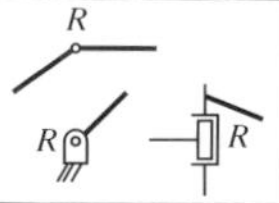	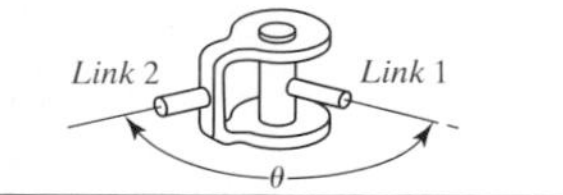	A pin joint that permits rotation only
Prism	*L*	*P*	1 x	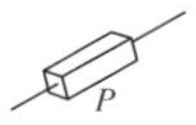	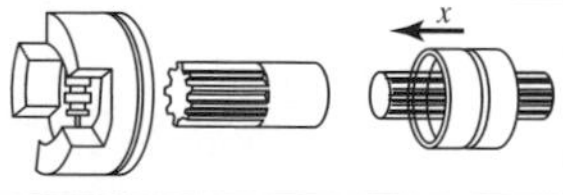	A straight spline that permits sliding only
Helix	*L*	*H*	1 x or θ	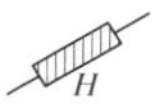	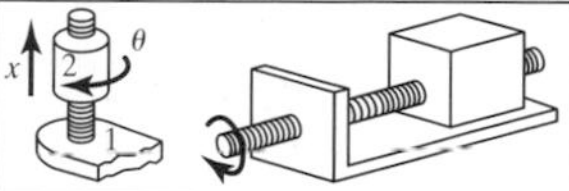	Power screw or helical spline
Cylinder	*L*	*C*	2 x, θ	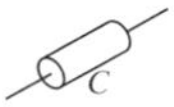	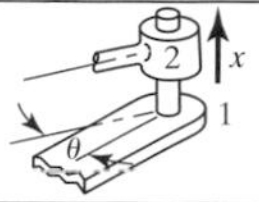	A sleeve that permits both rotation and sliding
Sphere	*L*	*S*	3 θ, ϕ, γ	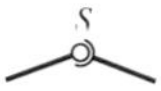	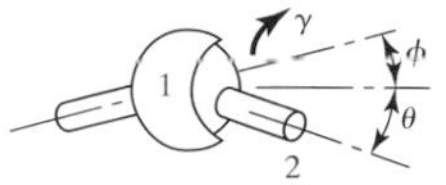	A ball (and socket) joint permitting rotation in three angular directions
Plane	*L*	P_L	3 x, y, θ	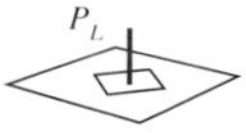	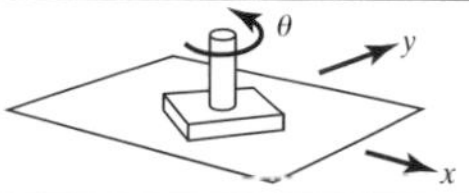	A surface restraint permitting rotation and motion parallel to the plane of the surface
Universal joint		*U*	2 θ, ϕ	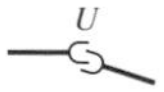	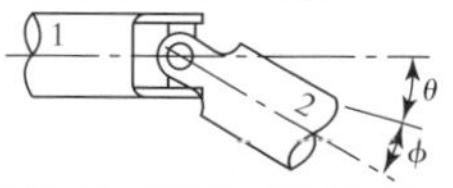	The Hooke-type universal joint that combines two revolute pairs
Gear pair	*H*	*G*	2 (rolling and sliding)	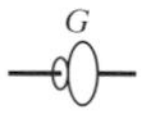	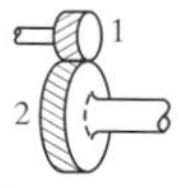	Spur gears, helical gears, and other gears
Cam pair	*H*	*Cam*	2 (rolling and sliding)			Disk cam and follower

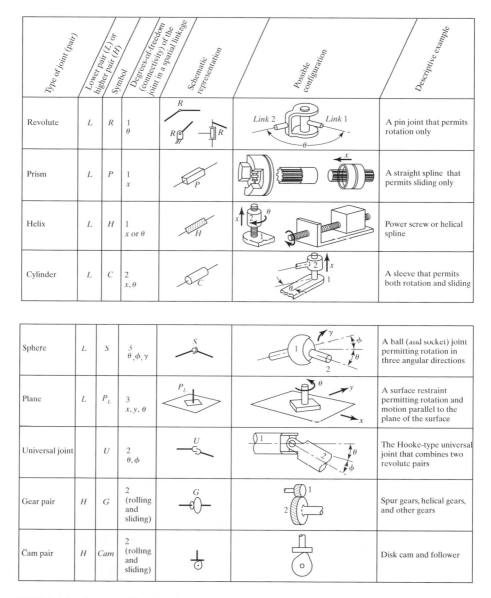

FIGURE 1.2 Common linkage joints (pairs).

crank; the crankpin (another revolute joint) connects the crank and connecting rod; the connecting rod and crosshead are joined at the wrist pin (a third revolute joint); finally, the piston and cylinder (frame) constitute a sliding pair (cylinder pair), closing the loop.

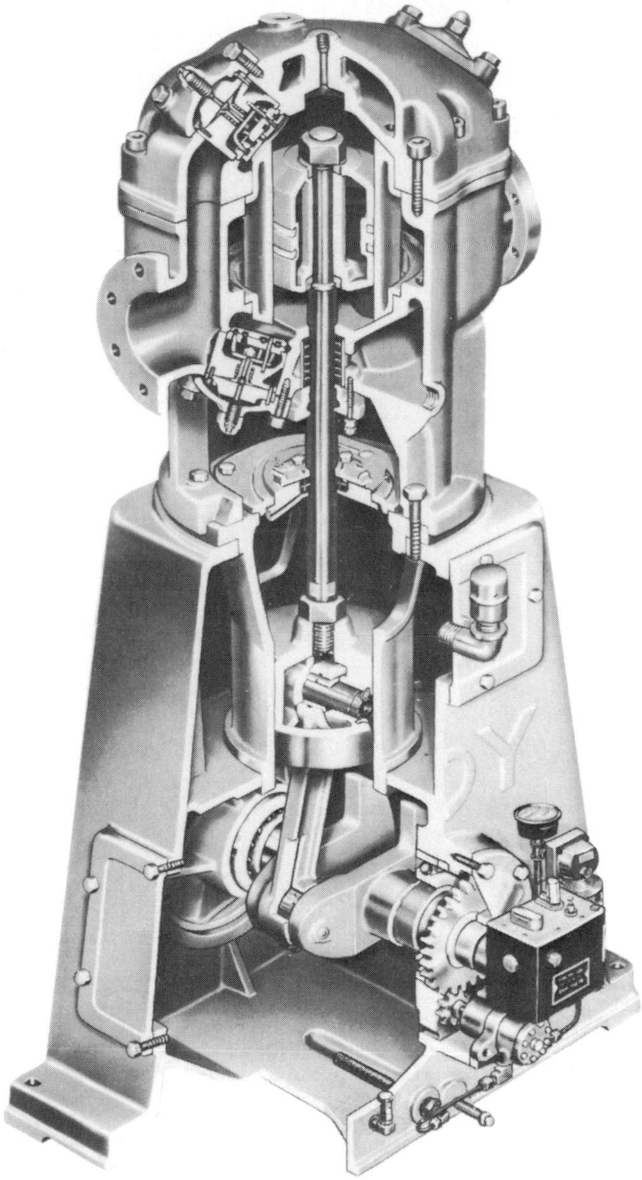

FIGURE 1.3 A vertical compressor. The crank (*bottom*) drives the connecting rod, which moves the crosshead within a guide. The compressor is designed with a crosshead and piston rod so that the piston may be double acting; air is compressed as the piston moves upward and as it moves downward. (*Source:* Joy Manufacturing Company.)

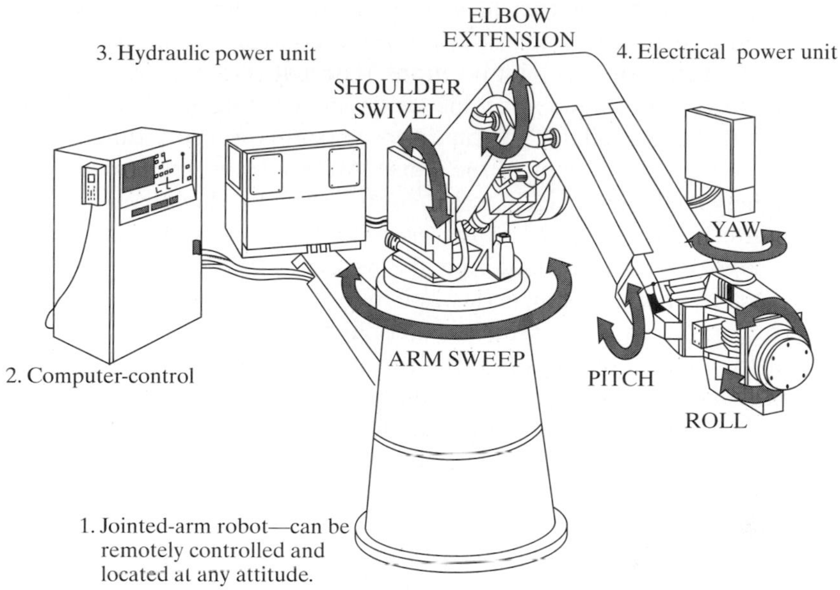

FIGURE 1.4 Industrial robot. (*Source:* Cincinnati Milacron.)

Open-Loop Kinematic Chains

A linkage failing to meet the closed-loop criterion is an open-loop kinematic chain. In this case, one (or more) of the links is connected to only one other link. The industrial robot shown in Figure 1.4 is an open-loop kinematic chain.

Manipulators

Manipulators designed to simulate human arm and hand motion are an example of open kinematic chains. A typical manipulator consists of a supporting base with rigid links connected in series, the final link containing a tool or "hand." Ordinarily, the rigid links are joined by revolute joints or prismatic pairs, although the hand may include a screw pair. Early systems of this type included master–slave-type manipulators for handling radioactive materials. The slave manipulator duplicates the hand–arm motion of a human operator controlling the master manipulator.

Robots

Programmable manipulators, called robots, can follow a sequence of steps directed by a computer program. Unlike machines dedicated to a single task, robots can be retooled and reprogrammed for a variety of tasks. Typical robot tasks include spray painting, assembling parts and welding. The open-chain configuration of robots results in a problem with positional accuracy. This problem is sometimes overcome by using jigs and compliant tooling systems.

It is also possible to achieve accurate positioning by incorporating a sensing system and a feedback system into the robot control system. Internal-state sensors can detect variables such as joint positions. External-state sensors may measure proximity, touch, force, and torque. Machine vision and hearing are external-state sensory capabilities that are available in some robot systems. Sensory-function feedback systems permit adaptive behavior of the robot. A force transducer incorporated into a robot hand may feed back a signal to the control system, which then alters the hand's grasping pattern. Tasks requiring high precision are more often accomplished by numerically controlled machinery designed for specific operations.

Figure 1.4 shows an industrial robot with six revolute joints. This robot has a jointed-arm form, a common robot configuration. Other configurations are shown in Figure 1.5. Part a of the figure is a schematic representing a robot with four revolute joints and one prismatic pair. Part b represents a robot with two revolute joints, a prismatic joint, and a cylinder pair. Note that the cylinder can be replaced by a prismatic joint and a revolute joint. The robot schematic of part c shows three prism joints and one revolute joint. The end effectors may consist of additional links and joints. The **work envelope**, or workspace, is defined by all of the points that the end effector can reach. It can be seen that the type of joints in each of these robot configurations affects the shape of the work envelope. Robots are discussed extensively in Chapter 12.

Linkage

Although some references define *linkages* as kinematic chains joined by only lower pairs, the term is commonly used to identify any assemblage of rigid bodies connected

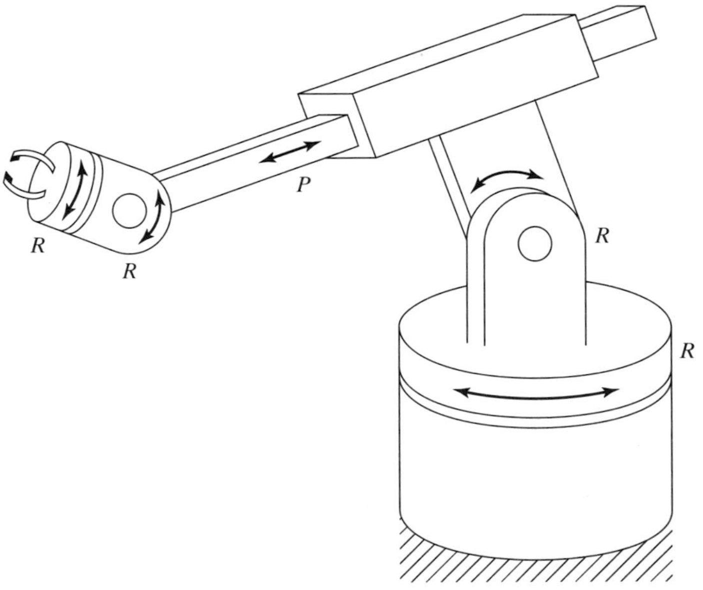

FIGURE 1.5 Schematic diagrams representing various robot configurations. (a) Spherical configuration.

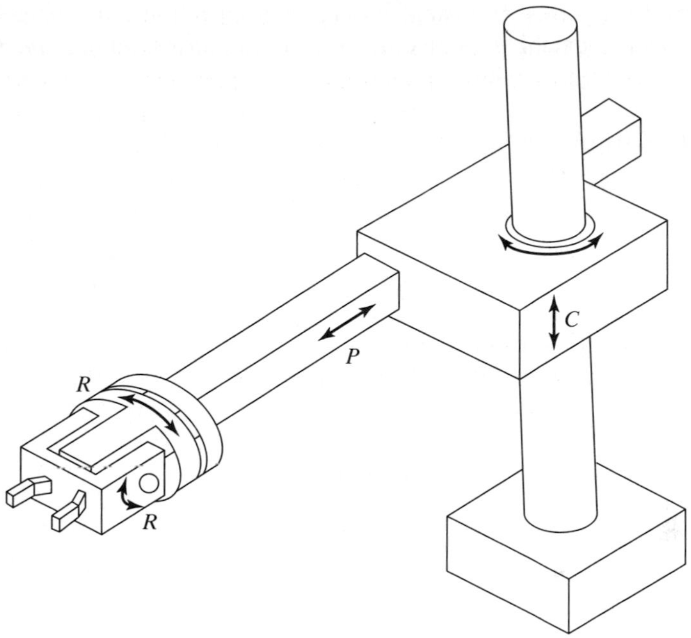

FIGURE 1.5 (b) Cylindrical configuration.

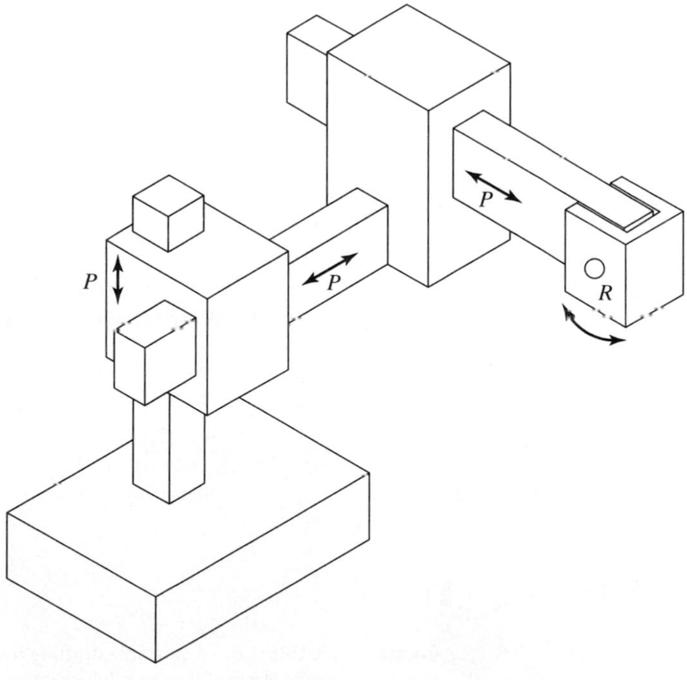

FIGURE 1.5 (c) Rectangular configuration.

by kinematic joints. The same linkage configuration may serve as a component of a mechanism, a machine, or an engine. Thus, the terms *linkage*, *mechanism*, *machine*, and *engine* are often used interchangeably.

Planar Motion and Planar Linkages

If all points in a linkage move in parallel planes, the system undergoes *planar motion* and the linkage may be described as a *planar linkage*. The portable drafting instrument shown in Figure 1.6 is a planar linkage. A skeleton diagram of a planar linkage is formed by projecting all of the link centerlines on one of the planes of motion. The plane of motion of parallelogram linkage $ABCD$, the plane of motion of parallelogram linkage $EFGH$, and the plane of motion of the straightedges are all parallel. In this linkage, planar motion is assured because the axes of revolute joints A, B, C, D, E, F, G, and H are all parallel (i.e., all perpendicular to the plane of the drawing board).

Spatial Motion and Spatial Linkages

The more general case in which motion cannot be described as taking place in parallel planes is called *spatial* motion, and the linkage may be described as a *spatial* or *three-dimensional* (3*D*) *linkage*. The industrial robot of Figure 1.4 is a spatial linkage. To achieve the desired range of motion, the axes of the revolute pairs in the manipulator are arranged to be not all parallel.

Inversion

The absolute motion of a linkage depends on which link is fixed—that is, which link is selected as the frame. If two otherwise equivalent linkages have different fixed links, then each is an *inversion* of the other.

Cycle and Period

A *cycle* is the complete sequence of positions of the links in a mechanism (from some initial position back to that initial position). In a four-stroke-cycle engine, one thermodynamic cycle corresponds to two revolutions or cycles of the crankshaft, but to one revolution of the camshaft and, thus, one cycle of motion of the cam followers and valves. The time required to complete a cycle of motion is called the *period*.

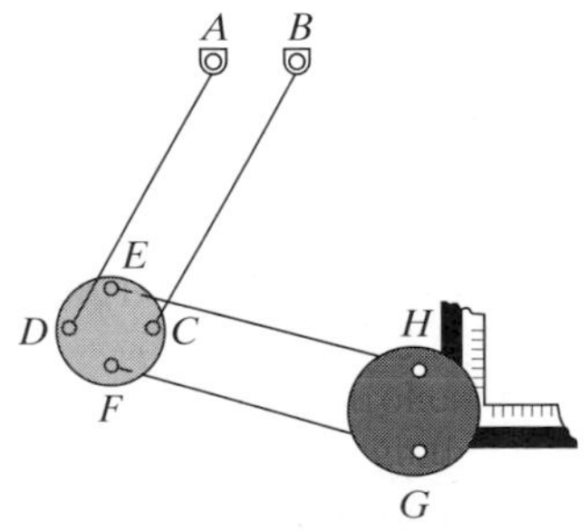

FIGURE 1.6 A portable drafting instrument. The linkage consists of two parallelograms, which permit translation of the straightedge in any direction without rotation.

1.5 DEGREES OF FREEDOM (MOBILITY)

The number of degrees of freedom of a linkage is the number of independent parameters we must specify to determine the position of every link relative to the frame or fixed link. The number of degrees of freedom of a linkage may also be called the *mobility* of the linkage. If the instantaneous configuration of a system may be completely defined by specifying one independent variable, that system has one degree of freedom. Most practical mechanisms have one degree of freedom.

An unconstrained rigid body has six degrees of freedom: translation in three coordinate directions and rotation about three coordinate axes. A body that is restricted to motion in a plane has three degrees of freedom: translation in two coordinate directions and rotation within the plane.

Constraints Due to Joints

Each joint reduces the mobility of a system. A fixed, one-degree-of-freedom joint (e.g., a revolute joint) reduces a link to one degree of freedom. In general, each one-degree-of-freedom joint reduces a system's mobility by providing five constraints; each two-degree-of-freedom joint provides four constraints, and so on. That is, in general, each joint reduces system mobility by $(6 - f_i)$ where f_i is the number of degrees of freedom (connectivity) of the joint. The actual number of degrees of freedom of a mechanism depends on the orientation of the joint. For example, if two or more revolute joints in a mechanism have parallel axes, then the effective number of constraints is reduced.

For a spatial mechanism with n_L links (including one fixed link with zero degrees of freedom), the number of degrees of freedom of the linkage is given by

$$\mathrm{DF}_{\text{spatial}} \geq 6(n_L - 1) - n_c \tag{1.1}$$

where n_c is the total number of constraints. For n_J joints with individual connectivity f_i, we note that

$$n_c \leq 6n_J - \sum_{i=1}^{n_J} f_i, \tag{1.2}$$

from which it follows that

$$\mathrm{DF}_{\text{spatial}} \geq 6(n_L - n_J - 1) + \sum_{i=1}^{n_J} f_i. \tag{1.3}$$

Linkages are often named according to their joint configurations, using the symbols given in Figure 1.2 (R for revolute, S for sphere, etc). For example, Figure 1.7a shows a closed-loop RSSR mechanism, and the robot in Figure 1.4 is called an RRRRRR, or 6-R, open linkage.

Examining the industrial robot of Figure 1.4, we see that there are seven links and six revolute joints, each joint having one degree of freedom and introducing five constraints. Using Eq. (1.1), we obtain

$\mathrm{DF}_{\text{spatial}} \geq 6(7 - 1) - 5 - 5 - 5 - 5 - 5 - 5 = 6$ (the equals sign applies in this case).

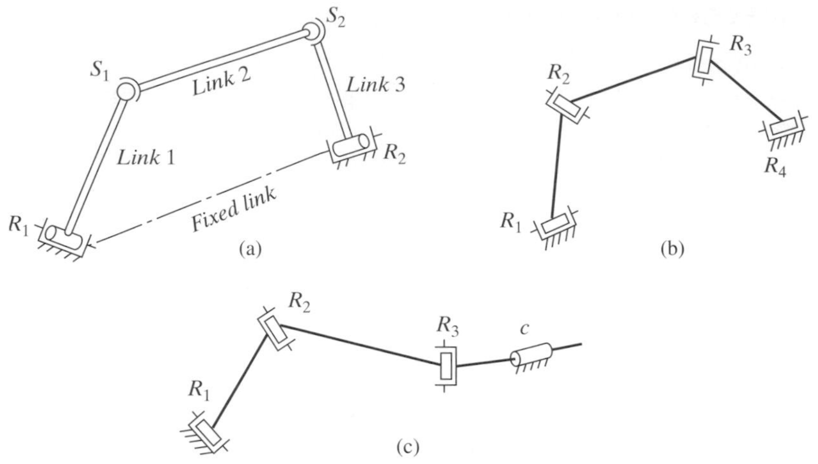

FIGURE 1.7 (a) An RSSR mechanism (two degrees of freedom). (b) An RRRR linkage. (c) An RRRC linkage. In the general case, no relative motion is possible in closed-loop RRRR and RRRC linkages.

FIGURE 1.7 (d) A manipulator arm on a prototype unmanned rover vehicle. (*Source*: NASA).

Alternatively, Eq. (1.3) may be used to obtain the same result:

$$DF_{spatial} \geq 6(7 - 6 - 1) + 1 + 1 + 1 + 1 + 1 + 1 = 6.$$

Next, consider the closed-loop kinematic chain of Figure 1.7a, a general closed-loop RSSR mechanism. The mechanism has four links, two revolute joints, and two spherical joints, as shown in Figure 1.7a. Each revolute joint has one degree of freedom and introduces five constraints, while each spherical joint has three degrees of freedom and introduces three constraints. Thus, from Eq. (1.1), the RSSR mechanism has $6(4 - 1) - 5 - 3 - 3 - 5 = 2$ degrees of freedom. This particular linkage acts, for all practical purposes, as a one-degree-of-freedom linkage if we ignore the degree of freedom that represents the rotation of link 2 about its own axis. If the angular position of link 1 is given, the entire linkage configuration may be determined. Note, however, that this statement assumes that applied forces or inertial effects are present to ensure a prescribed pattern of motion as the linkage passes through limiting positions.

Let one of the spherical joints in the preceding RSSR mechanism be replaced by a universal joint with two degrees of freedom (four constraints). We then form an RSUR mechanism, and the number of degrees of freedom is given by

$$DF \geq 6(4 - 1) - 5 - 3 - 4 - 5 = 1.$$

In a general RRRR linkage, as shown in Figure 1.7b, each joint provides five constraints, and the number of degrees of freedom is given by $DF \geq 6(4 - 1) - 5 - 5 - 5 - 5 = -2$. In a general RRRC linkage, the cylinder joint has two degrees of freedom, providing four constraints. The number of degrees of freedom for the RRRC linkage shown in Figure 1.7c is given by $DF \geq 6(4 - 1) - 5 - 5 - 5 - 4 = -1$. The manipulator arm on the prototype vehicle (Figure 1.7d) is a multi-degree-of-freedom open-loop linkage.

There is no relative motion in the general case for RRRR and RRRC linkages, because they are equivalent to statically indeterminate structures. However, there are important special cases. If all four revolute joints in an RRRR linkage are parallel, then the linkage becomes planar. If the axes of all four revolute joints in an RRRR linkage meet at one point, then the linkage becomes spherical. Both of these special cases are mechanisms with one degree of freedom.

Planar Linkages

Planar linkages, of course, represent a special case. Consider, for example, a mechanism made up of rigid links joined by three revolute joints and a cylinder joint. If the links and joints are oriented so that the links move in parallel planes, this RRRC linkage becomes a *slider-crank linkage*, the planar linkage that represents a major component of piston engines, pumps, compressors, and other common machines. Figure 1.3 shows a vertical compressor, the major components of which may be represented by an RRRC linkage. All points on the crank, connecting rod, and crosshead of the compressor move in parallel planes, and the axes of the revolute joints are also parallel. Thus, we have a slider-crank linkage.

A planar RRRR linkage may be called a four-bar linkage, which, together with other planar link systems, is shown in Figure 1.8.

The joints or pairs that apply to planar linkages are as follows:

	Lower or higher pair	Connectivity (Degrees of freedom of pair in plane motion)
Revolute or pin joint	Lower pair	1
Prism or sliding pair	Lower pair	1
Cam pair	Higher pair	2
Gear pair	Higher pair	2

Actual joints may sometimes be different. For example, the slider-crank linkage may be made up of three revolute joints (crankshaft bearings, crankpin, and wrist pin) and a cylinder pair. The spline-type constraint of a prism pair is unnecessary, since the revolute joints prevent rotation of the piston. If the actual number of degrees of freedom is greater than the minimum determined by using the equation for spatial linkages, the linkage is overconstrained. Overconstraint tends to strengthen a linkage; however, overconstraint can be a disadvantage if manufacturing tolerances are poor.

Determination of Degrees of Freedom for a Planar Linkage

Each unconstrained rigid link has three degrees of freedom in plane motion. A fixed link has zero degrees of freedom. A pin joint connecting two links produces two constraints, since the motion of both links must be equal at the joint (in two coordinate directions). Thus, the number of degrees of freedom for a planar linkage made up of n_L links and n'_J one-degree-of-freedom pairs is given by

$$DF_{\text{planar}} = 3(n_L - 1) - 2n'_J \tag{1.4}$$

or, for n_J joints with individual connectivity f_i,

$$DF_{\text{planar}} = 3(n_L - n_J - 1) + \sum_{i=1}^{n_J} f_i. \tag{1.5}$$

We see that, for the four-bar linkage in Figure 1.8a, $n_L = 4$, $n'_J = 4$, and DF $= 1$. For the five-bar linkage in Figure 1.8b, $n_L = 5$, $n'_J = 5$, and DF $= 2$. The linkage of Figure 1.8d has a double pin at B. Thus, $n_L = 6$, $n'_J = 7$, and DF $= 1$.

Figure 1.8e shows a slider-crank mechanism illustrating a piston engine or a piston pump, where O_1 denotes the crankshaft, link 1 represents the crank, link 2 designates the connecting rod, point C denotes the wrist pin, and link 3 represents the slider or piston that is constrained by the cylinder. The figure illustrates the common special case, an in-line slider crank, where the extended path of wrist pin C goes through crankshaft axis O_1. There are four links, including the slider and frame, and four lower pairs, including the sliding pair. An alternative analysis uses an equivalent linkage. Figure 1.8f shows a four-bar linkage in which point C moves through an arc of

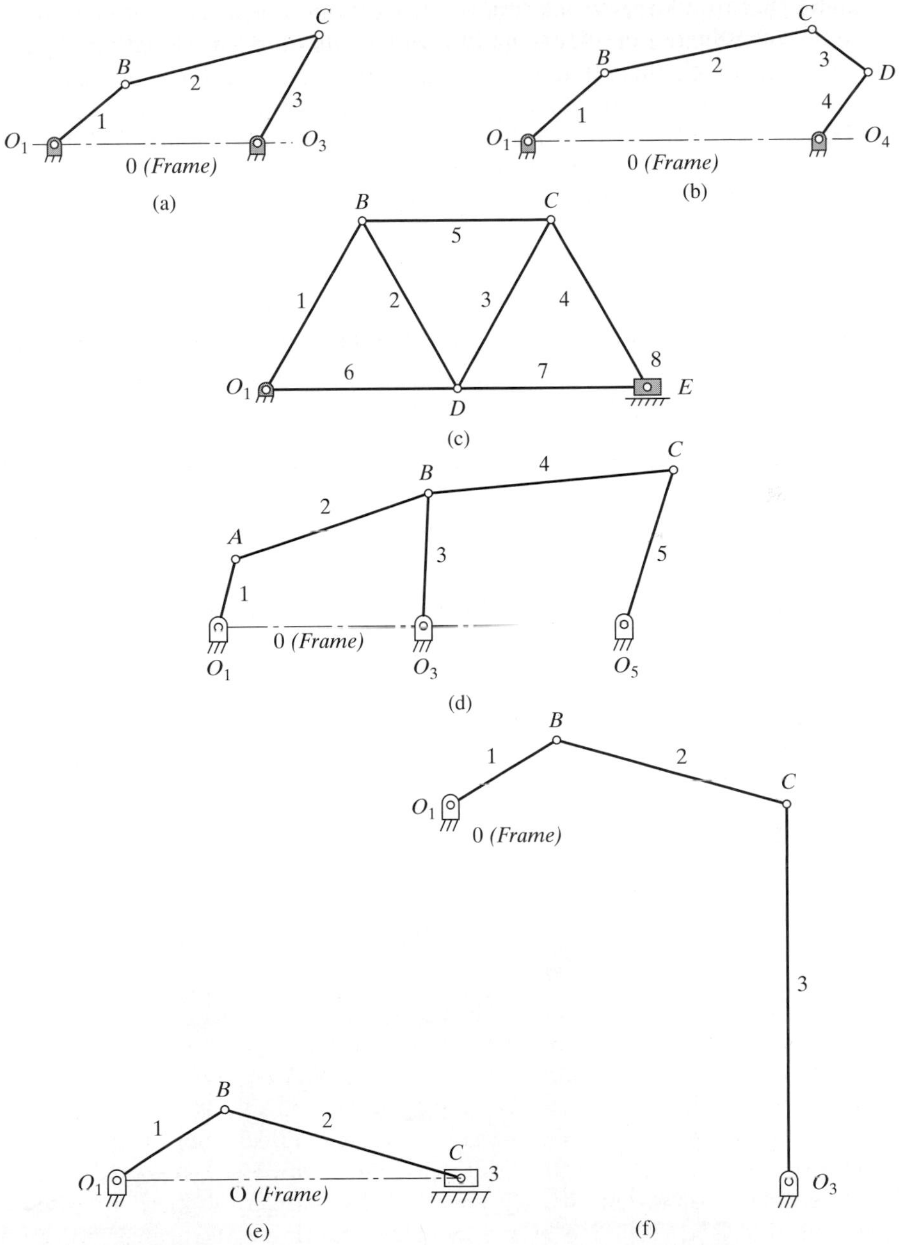

FIGURE 1.8 Skeleton diagrams of planar linkages. (a) Four-bar linkage. (b) Five-bar linkage. (c) Structure. (d) A six-bar linkage with one degree of freedom. (e) A slider-crank mechanism. (f) A four-bar linkage with motion approximating that of a slider-crank mechanism.

radius O_3C. If link 3 is made very long, the motion of the four-bar linkage will approximate that of a slider-crank linkage. If we could construct a linkage by replacing the slider of the slider-crank mechanism with a link of infinite length and perpendicular to the path of C, then that linkage would be equivalent to the slider-crank linkage. Applying Eq. (1.4) to our equivalent linkage, with $n_L = 4$ and $n'_J = 4$ we find that $\text{DF} = 3(4 - 1) - 2(4) = 1$. The equivalent four-bar linkage (and, thus, the slider-crank linkage) has one degree of freedom.

Analyzing the structure in Figure 1.8c in a similar manner and noting the double pins at O_1, B, C, and E and the triple pin at D, we find that $n'_J = 12$, $n_L = 9$, and $\text{DF} = 0$.

If a planar linkage is made up of n'_J one-degree-of-freedom pairs and n''_J two-degree-of-freedom pairs, the number of degrees of freedom of the linkage is given by

$$\text{DF}_{\text{planar}} = 3(n_L - 1) - 2n'_J - n''_J. \tag{1.6}$$

A single gear mesh or the contact point of a cam and follower represents a two-degree-of-freedom higher pair (if the two bodies do not separate). Consider the spur gear differential shown schematically in Figure 1.9. This differential has six links: the frame; gears S_1 and S_2, called *sun gears*; gears P_1 and P_2, called *planet gears*; and link C, the *planet carrier*. There are five independent revolute joints and three gear pairs (S_1P_1, P_1P_2, andP_2S_2). It can be seen that the bearing axes are all parallel and that the spur gear differential is a planar linkage. Using Eq. (1.6), we find that

$$\text{DF}_{\text{planar}} = 3(6 - 1) - 2 \times 5 - 3 = 2 \text{ degrees of freedom.}$$

Thus there are two independent variables. For example, if we specified the motion of both sun gear shafts, that of the planet carrier shaft could be determined.

To achieve balance and reduce gear tooth loading, practical spur gear differentials ordinarily include two to four equally spaced pairs of planet gears. The additional

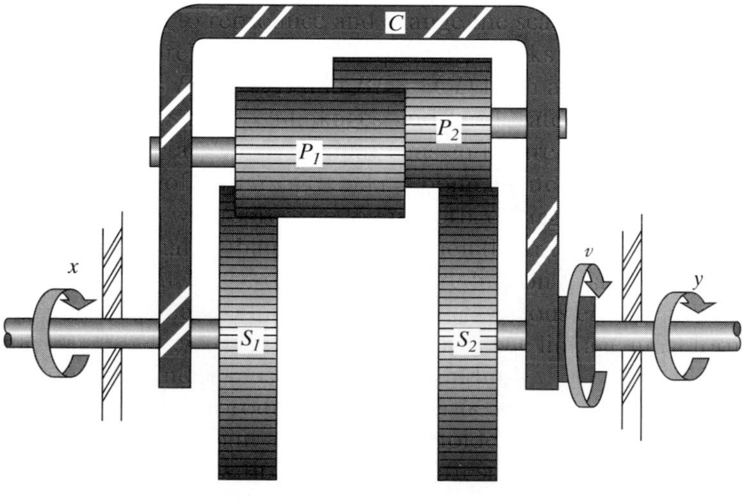

FIGURE 1.9 Schematic of spur gear differential.

pairs of planets do not change the number of degrees-of-freedom. If the gear sizes were arbitrarily chosen, the number of degrees of freedom would be reduced.

One-Degree-of-Freedom Configurations

Planar mechanisms with one degree of freedom are of considerable practical importance. One-degree-of-freedom planar mechanisms made up of lower pairs satisfy Grübler's criterion:

$$2n'_J - 3n_L + 4 = 0. \tag{1.7}$$

Noting that the number of links n_L and one-degree-of-freedom pairs n'_J must be positive integers, we see that n_L must be an even number. For $n_L = 2$, we obtain $n'_J = 1$, the trivial solution that could represent two bars joined by a pin joint. Next, trying four links, we see that the number of joints must be $n'_J = 4$. This solution could represent the four-bar linkage or slider-crank linkage of Figures 1.8a and e. Inversions of the slider-crank linkage are also possible. If there are six links, then seven one-degree-of-freedom lower pairs are required to produce one degree of freedom. For pin joints only, two distinct configurations are possible, as shown in Figures 1.10a and b. Any one of the links may be designated as the frame in each solution. It can be seen that the six-bar linkage of Figure 1.8d may be considered a special case of either of the linkages of Figure 1.10. Determining eight-bar linkage configurations is left to the reader as an exercise.

Spatial linkages are analyzed further in a later section. However, since planar linkages are used most frequently, the word *planar* will be used in the pages that follow only when it is necessary to compare planar and spatial linkages.

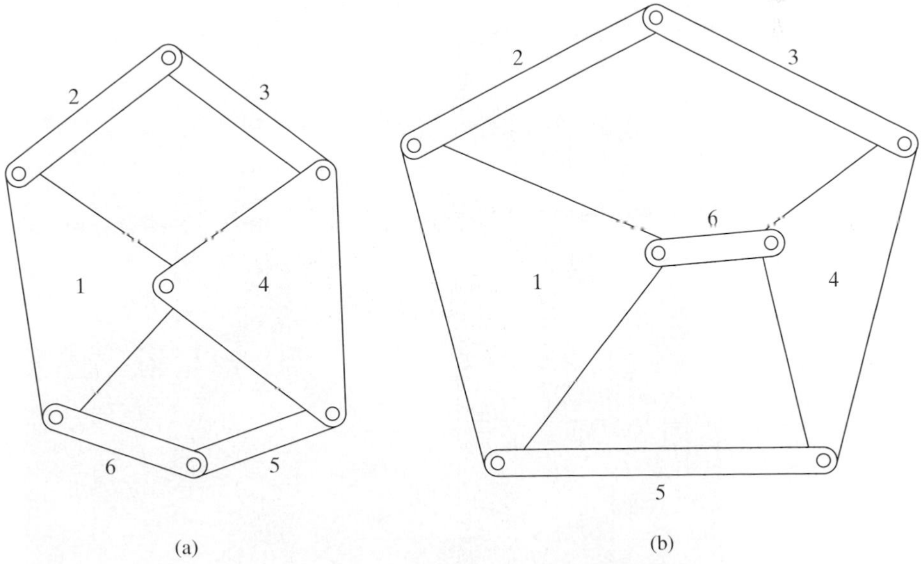

FIGURE 1.10 One-degree-of-freedom six-bar planar linkages. (a) Watt linkage. (b) Stephenson linkage.

Adjustable-Parameter Linkages

We can control input–output relationships by adjusting the length of one or more links in a mechanism. Ingenious designs even enable "on-the-fly" adjustments; that is, we can change a mechanism's characteristics of motion even when the mechanism is operating.

Suppose, for example, we need to vary the output of a pump without changing the rotation speed of the crank. Suppose also that we would like the output to respond to some other variable while the pump is running. A variable-stroke pump with a stroke transformer is one way to satisfy this requirement. Figure 1.11a shows a variable-position pump control cylinder that positions the curved-track stroke transformer. One end of the coupler link slides in the track of the stroke transformer. The wrist pin at the top of the coupler connects to the pump plunger. If the center of curvature of the stroke transformer is close to the wrist pin, then the plunger stroke will be small and the output (fluid

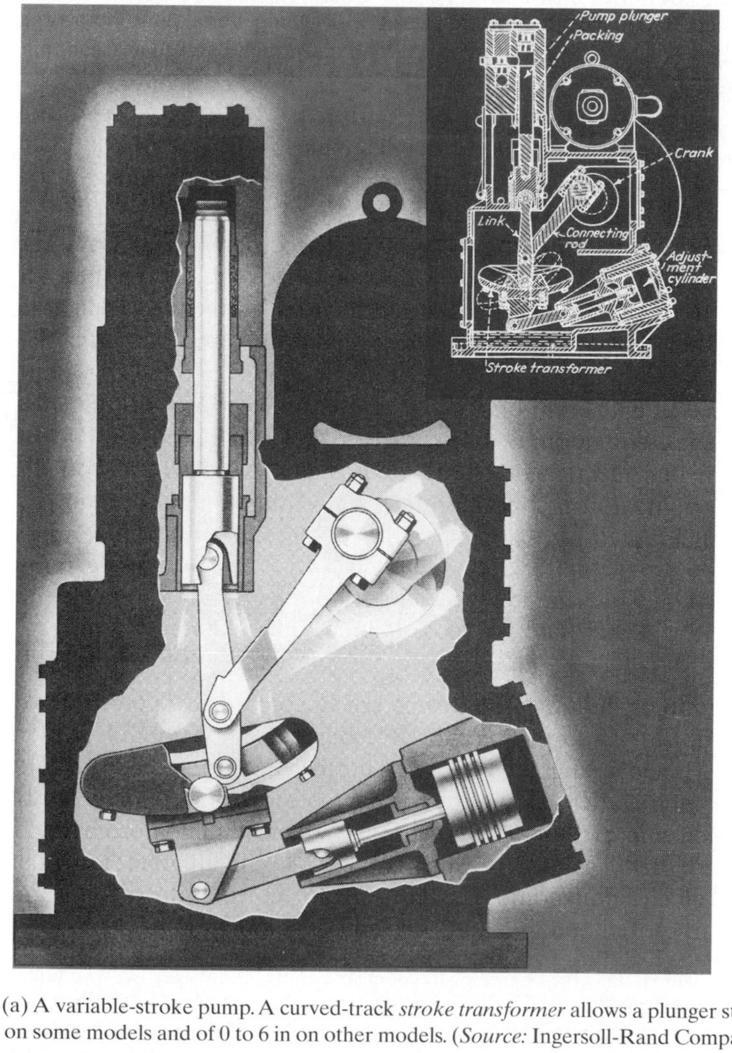

FIGURE 1.11 (a) A variable-stroke pump. A curved-track *stroke transformer* allows a plunger stroke variation of 0 to 2 in on some models and of 0 to 6 in on other models. (*Source:* Ingersoll-Rand Company.)

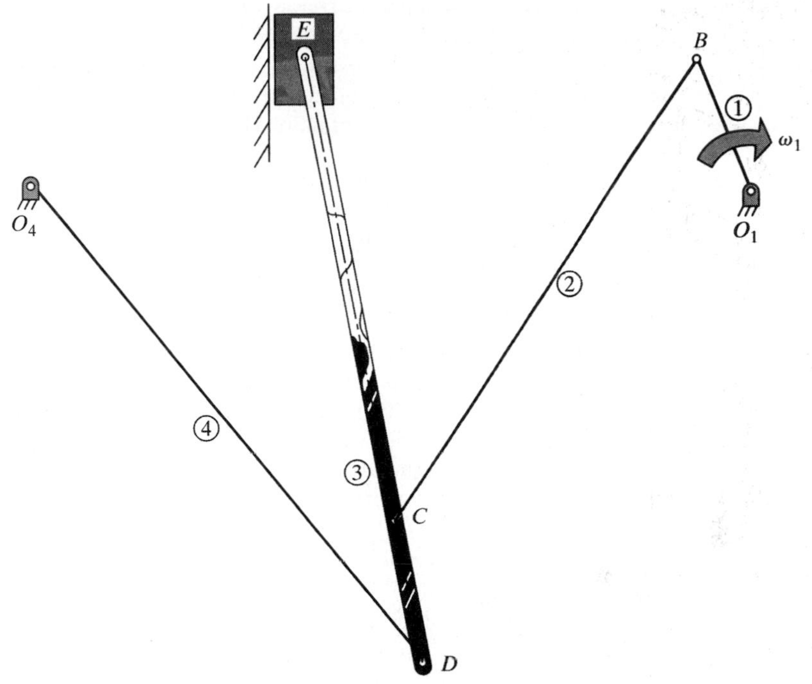

FIGURE 1.11 (b) In this linkage diagram, link 4 replaces the curved track of the stroke transformer. As O_4 is brought closer to E, the movement of the plunger decreases. When O_4 coincides with E, the plunger will become stationary.

flow rate) will be low. If the stroke transformer is adjusted so that its center of curvature falls on the wrist pin, then the plunger will not move and the output will be zero.

When stating the number of degrees of freedom of adjustable-parameter linkages, be sure to state whether the control feature is assumed to be in a fixed position. Figure 1.11b uses an equivalent linkage to represent the variable-stroke pump when the transformer is set for an intermediate stroke.

Although linkages of this type are often hydraulically controlled. One should not neglect other possibilities in designing linkages. Consider, for example, pneumatic or electrical control or combinations of control schemes. If your design requires precise, but infrequent adjustment, consider a manually turned screw.

SAMPLE PROBLEM 1.1

Degrees of Freedom of a Variable-Stroke Pump

a. Determine the number of degrees of freedom for the variable-stroke pump shown in Figure 1.11. Let the adjustment cylinder be fixed in a position that results in an intermediate stroke length.

b. Suppose the adjustment cylinder position is not fixed. Find the number of degrees of freedom.

Solution. **(a)** It can be seen from the figure that motion takes place in a set of parallel planes. Therefore, we have a planar linkage. The number of degrees of freedom can be determined by examining the actual linkage or a schematic diagram of an equivalent linkage. Selecting the latter, we replace the curved stroke transformer by rigid link 4 with a fixed revolute joint (O_4) located at the center of curvature of the stroke transformer. (See part b of the figure). We then have six links, counting the slider and the frame. There are six revolute pairs and a sliding pair, making a total of seven (one-degree-of-freedom) lower pairs. Thus,

$$DF_{\text{planar}} = 3(n_L - 1) - 2n_{J'} = 3(6 - 1) - 2(7) = 1.$$

(b) In this case, we refer to the actual pump configuration. A careful examination shows nine links, including three sliders and the frame. There are eight revolute pairs and three sliding pairs, a total of 11 one-degree-of-freedom pairs. The number of degrees of freedom is given by

$$DF_{\text{planar}} = 3(n_L - 1) - 2n_{J'} = 3(9 - 1) - 2(11) = 2.$$

The implication is that we must specify two variables to define the instantaneous position of the entire linkage. Ordinarily, these variables would be the position of the piston in the pump control cylinder and the instantaneous angular position of link 1, the drive crank.

As noted earlier, the slider-crank linkage used in a piston engine or pump can be an RRRC linkage. This linkage includes three revolute joints with parallel axes, the crankshaft bearings (treated as a single pair), the crankpin, and the wrist pin (joining the connecting rod and piston). The piston and cylinder correspond to a cylinder pair, but the other joints prevent rotation of the piston. A common alternative is an RRCC linkage, in which the piston is free to move a short distance along the wrist pin axis, accommodating misalignment. The RRSC linkage, sometimes used in small pumps, can also operate as a slider-crank. A ball joint (spherical pair) replaces the wrist pin. Note that there is a second degree of freedom: rotation of the piston about the cylinder axis. This motion is trivial and does not affect the operation of the pump as a planar linkage.

1.6 CLASSIFICATION OF CLOSED PLANAR FOUR-BAR LINKAGES: THE GRASHOF CRITERION

Closed planar linkages consisting of four pin-connected rigid links are usually identified simply as **four-bar linkages**. If one of the links can perform a full rotation relative to another link, the linkage is called a **Grashof mechanism**.

Let the length of each link be defined as the distance between the axes of its revolute joints (the centers of its pin joints). Links are characterized by their lengths, where $L_{\max}$ is the longest link, $L_{\min}$ is the shortest link, and L_a and L_b are links of intermediate length. We may immediately eliminate combinations for which

$$L_{\max} \geq L_{\min} + L_a + L_b,$$

since it is obvious that these links could not be assembled to form a closed four-bar linkage.

Suppose we wish to design a **crank-rocker mechanism**—a linkage with a drive crank that rotates continuously, causing a driven crank (rocker) to oscillate through a limited range. Referring to Figure 1.12, we note that limiting positions of link 3, the rocker, occur when the crank (link 1) and the coupler (link 2, the link opposite the

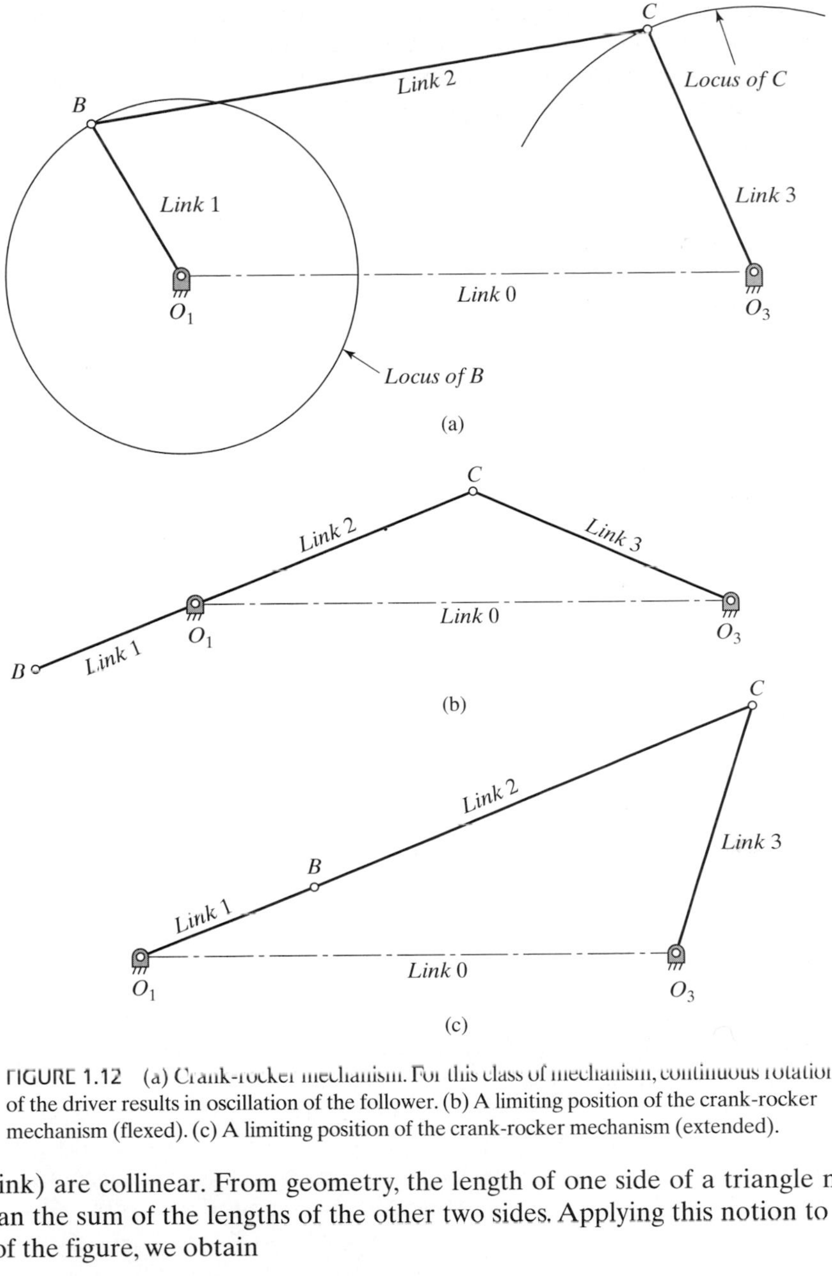

FIGURE 1.12 (a) Crank-rocker mechanism. For this class of mechanism, continuous rotation of the driver results in oscillation of the follower. (b) A limiting position of the crank-rocker mechanism (flexed). (c) A limiting position of the crank-rocker mechanism (extended).

fixed link) are collinear. From geometry, the length of one side of a triangle must be less than the sum of the lengths of the other two sides. Applying this notion to parts b and c of the figure, we obtain

$$L_0 < L_2 - L_1 + L_3,$$
$$L_3 < L_0 + L_2 - L_1,$$

and

$$L_1 + L_2 < L_0 + L_3. \tag{1.8}$$

Those familiar with basic mathematics will recall the special rules that govern inequalities. Adding the first two inequalities and simplifying, we obtain $L_1 < L_2$; that is, the crank must be shorter than the coupler. Using other combinations of the foregoing inequalities, we see that $L_1 = L_{min}$; in other words, the crank is the shortest link in a crank-rocker mechanism. The fixed link, the coupler, or the driven crank may be longest. In every case, the inequalities require that

$$L_{max} + L_{min} < L_a + L_b, \tag{1.9}$$

where L_{max} and L_{min} are the longest and shortest links, respectively, L_a and L_b are each links of intermediate length.

SAMPLE PROBLEM 1.2

Crank-rocker mechanism

Design a mechanism that converts continuous rotation into oscillating motion.

Design decisions. We will try a linkage design with the following link lengths:

fixed link $L_0 = 40$ mm; drive crank $L_1 = 10$ mm;
coupler $L_2 = 30$ mm; driven crank $L_3 = 32$ mm.

Solution. The drive crank is shortest, and $40 + 10 < 30 + 32$, satisfying the crank-rocker criterion. We then test the proposed design by animating the linkage with motion simulation software. (See Figure 1.13.) A motor is placed at the fixed end of the drive crank. The approximate range of link 3 (the rocker, in dark shading) can be seen in the figure.

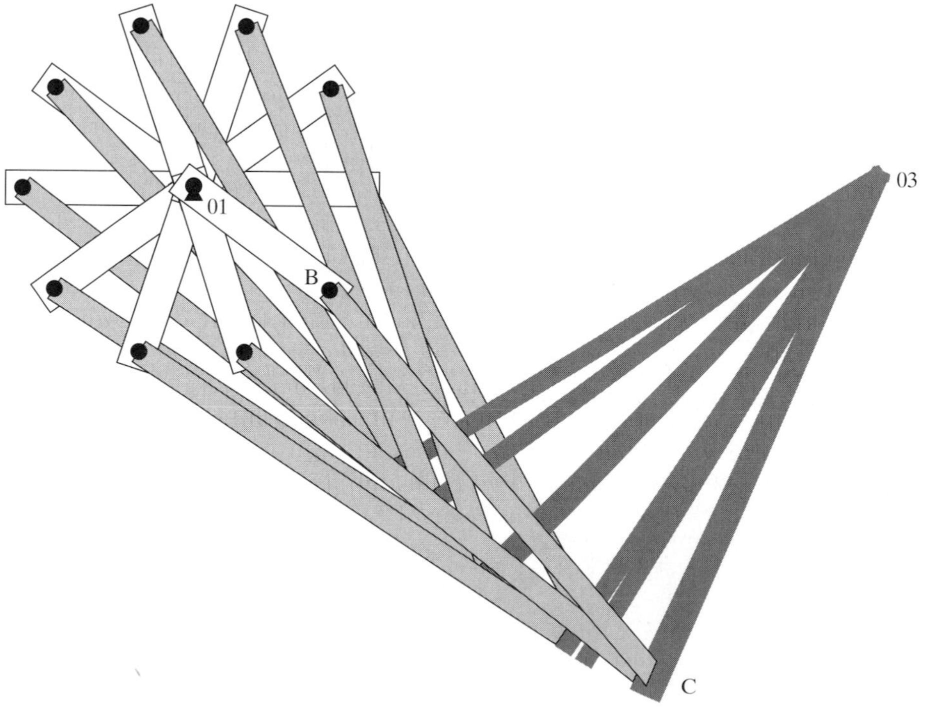

FIGURE 1.13 Crank-rocker mechanism sample problem.

We now consider inversions of the crank-rocker mechanism. Inequality 1.9 is satisfied in each case and the shortest link can rotate continuously relative to the other links. If the fixed link is shortest, the other links can rotate about it. This configuration is called a **drag link mechanism** (or double-crank mechanism).
We can attempt a drag link design with the following link lengths:

fixed link $L_0 = 20\text{ mm}$; drive crank $L_1 = 30\text{ mm}$;

coupler $L_2 = 30\text{ mm}$; driven crank $L_3 = 32\text{ mm}$.

We see that the fixed link is shortest, and Inequality 1.9 is satisfied; that is,

$$32 + 20 < 30 + 30.$$

The proposed design is tested by animating the linkage with motion simulation software. (See Figure 1.14.) A motor running at constant speed is placed at the fixed end of link 1, the drive crank (not shaded). Link 3, the driven crank (dark shading) rotates continuously, but at variable speed.

If the coupler is shortest, this inversion of the crank-rocker mechanism is called a **double-rocker mechanism**. The coupler of a double rocker can rotate continuously while the adjacent links oscillate through a limited range.

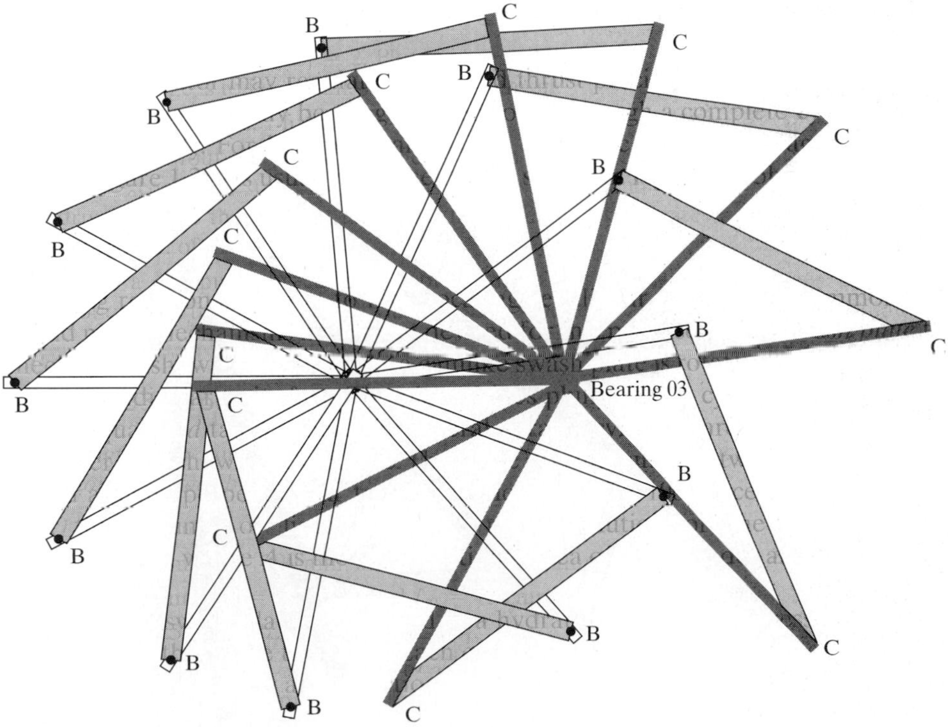

FIGURE 1.14 Animation of a drag link mechanism.

We attempt a double-rocker design, selecting the following link lengths:

fixed link $L_0 = 20$ mm; drive crank $L_1 = 27$ mm;
coupler $L_2 = 7$ mm; driven crank $L_3 = 32$ mm.

We see that the criteria for a double-rocker mechanism are satisfied; the coupler is shortest, and substitution in Inequality 1.9 yields

$$32 + 7 < 27 + 20.$$

Figure 1.15a shows a test of the design using motion simulation software. In this case, a motor mounted on link 1 at point B drives the coupler. Limiting positions of link 3 are shown in parts b and c of the figure.

Parts of the preceding linkages appear to interfere with one another in the animation figures. Computer simulations are usually instructed to ignore collisions when analyzing planar linkages. In designing the actual linkage, we must arrange the bearings so that collisions do not occur. Ingenious designs may be required when complex linkages are based on double-rocker and drag link mechanisms.

A **change-point** or **crossover-position mechanism** results when

$$L_{max} + L_{min} = L_a + L_b. \tag{1.10}$$

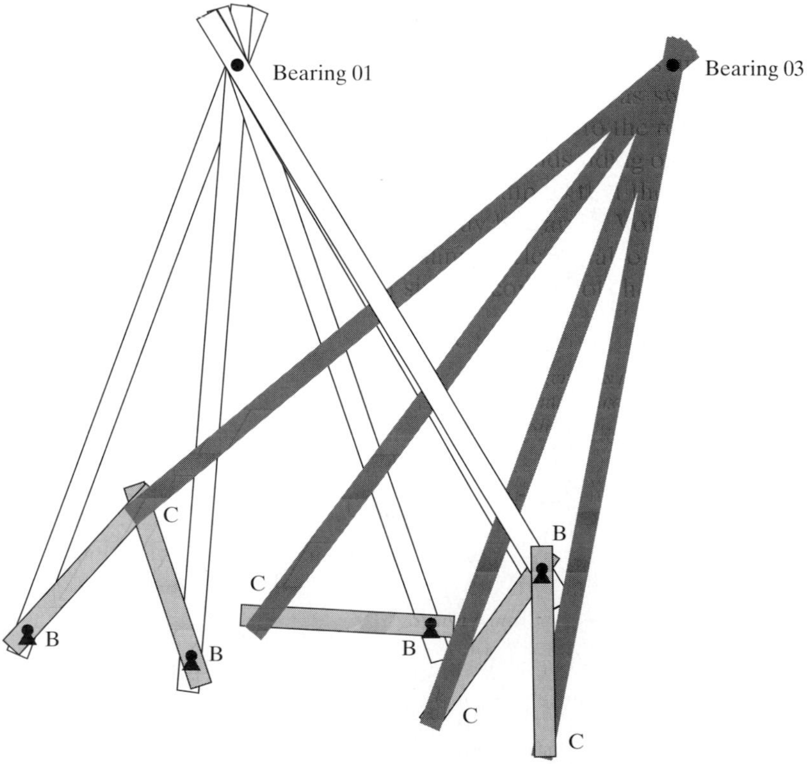

FIGURE 1.15 (a) A double-rocker mechanism with the coupler driven by a motor mounted on link 1.

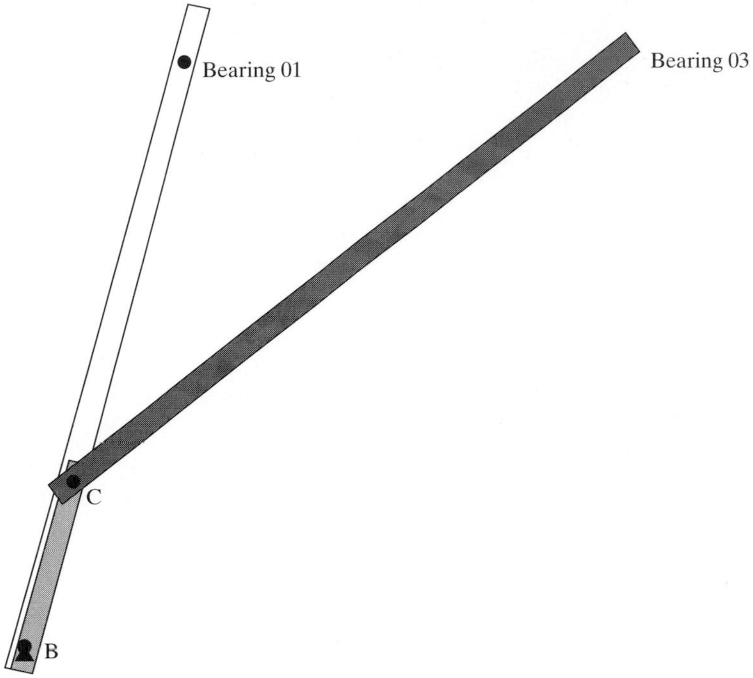

FIGURE 1.15 (b) One limiting position of link 3.

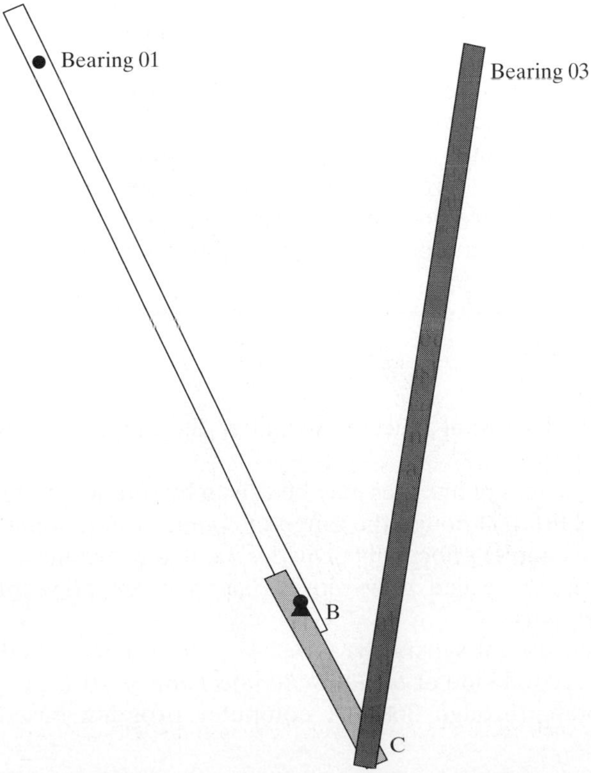

FIGURE 1.15 (c) The other limiting position of link 3.

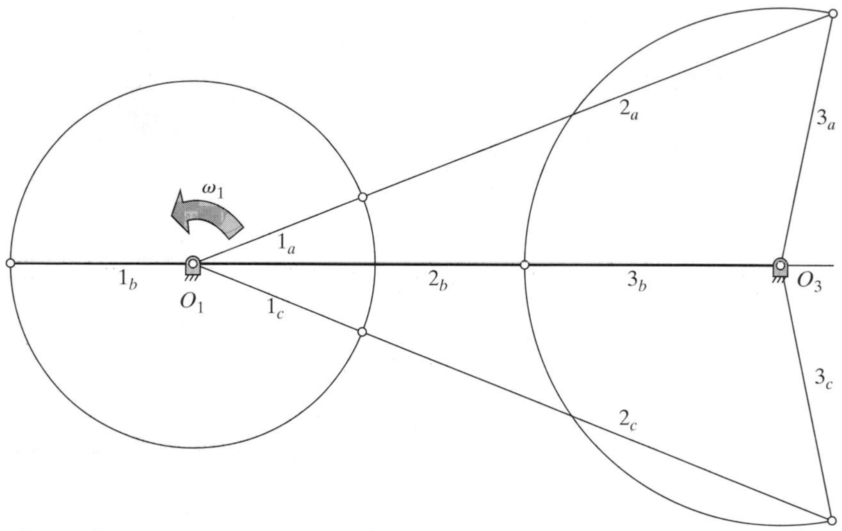

FIGURE 1.16 A crossover-position or change-point linkage.

Figure 1.16 shows a change-point mechanism in which $L_0 = L_{max}$, $L_1 = L_{min}$, and Eq. (1.10) is satisfied. Relative motion of a change-point mechanism may depend on inertia, spring forces, or other forces when the links (in the skeleton diagram) become collinear. Of course, the links in an actual machine operate in parallel planes, not in a single plane. In this example of a change-point mechanism, all of the links become collinear in position *b*. If links 1 and 3 are rotating counterclockwise at this instant, link 3 may continue rotating counterclockwise through the change point because of inertia effects. Alternatively, other forces may cause link 3 to reverse direction, resulting in a "bow-tie" configuration. A parallelogram linkage that has opposite links of equal length is another example of a change-point mechanism.

Crank-rocker, drag link, double-rocker, and change-point mechanisms satisfy the following relationship:

$$L_{max} + L_{min} \leq L_a + L_b \tag{1.11}$$

These mechanisms are called **Grashof mechanisms**, after the investigator who published this criterion in 1883.

Any of the preceding classes of linkages may be driven by rotation of the coupler (the link opposite the fixed link), although the range of coupler rotation may be very limited in some classes. The coupler effectively provides a hinge with a moving center. Coupler-driven linkages may be called *polycentric*. Examples are polycentric door hinges and prosthetic knee joints.

Four-bar linkages that do not satisfy the Grashof criterion are called double-rocker mechanisms of the second kind or *triple-rocker mechanisms*. If $L_{max} + L_{min} > L_a + L_b$, no link can rotate through 360°. A computer program based on the

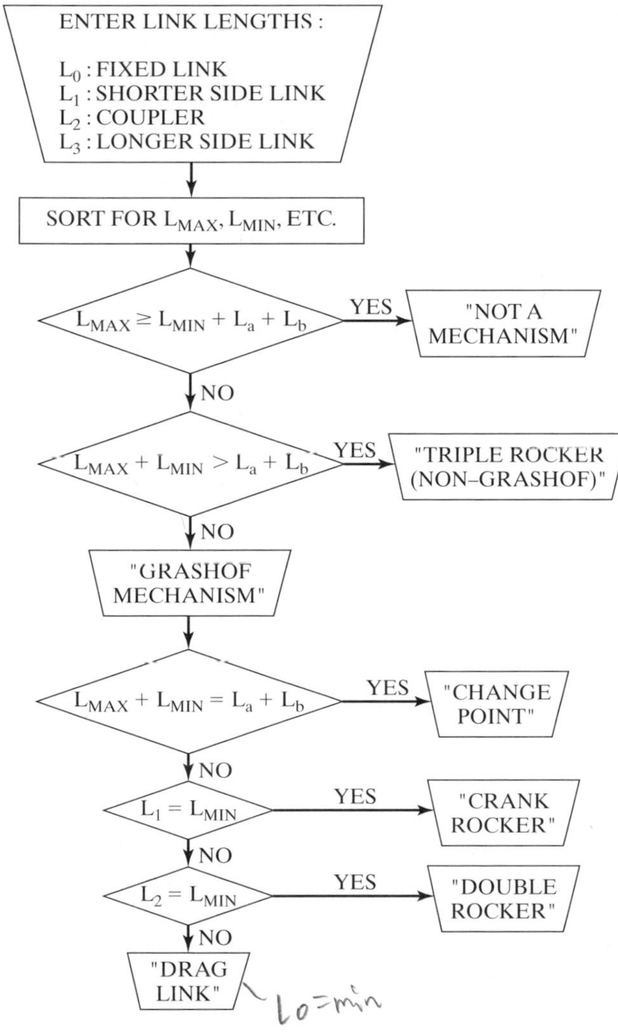

FIGURE 1.17 Flowchart for classifying four-bar linkages according to the characteristics of their motion.

flowchart of Figure 1.17 may be used to classify four-bar linkages by the characteristics of their motion.

SAMPLE PROBLEM 1.3

The Grashof Criterion

This problem concerns the classification of four-bar linkages. Link lengths: L_0, fixed link; L_1, driver crank; L_2, coupler; L_3, follower crank; $L_1 = 100$ mm, $L_2 = 200$ mm, $L_3 = 300$ mm. Find the ranges of values for L_0 if the linkage can be classified as follows:

a. Grashof mechanism

b. Crank-rocker mechanism

c. Drag link mechanism
d. Double-rocker mechanism
e. Change-point mechanism
f. Triple-rocker mechanism

Solution. Using the Grashof inequality, and noting that the crank-rocker, drag link, double-rocker, and change-point mechanisms, are all Grashof mechanisms we begin with the test for the crank-rocker mechanism (b). The condition that the driver crank L_1 be shortest is satisfied by $L_0 > 100$ mm. The inequality $L_{max} + L_{min} < L_a + L_b$ yields $L_0 + 100 < 200 + 300$, from which it follows that $L_0 < 400$ if link 0 is longest. If link 0 is of intermediate length, then $300 + 100 < L_0 + 200$, from which we obtain $200 < L_0$. Thus, 200 mm $< L_0 <$ 400 mm for the crank rocker (answer b).

The drag link criterion (c) requires that link 0 be shortest; that is, $L_0 < 100$. Also, $300 + L_0 < 100 + 200$ is required by the Grashof inequality, from which we get $L_0 < 0$. Thus, no drag link can be formed (answer c).

The double-rocker test (d) requires that the coupler, L_2, be shortest. Thus a double-rocker cannot be formed (answer d).

A change-point mechanism (e) exists if

$$L_{max} + L_{min} = L_a + L_b$$

If L_0 is largest, then

$$L_0 + 100 = 200 + 300,$$

or $L_0 = 400$.

If L_0 is smallest, then $300 + L_0 = 100 + 200$, from which it follows that $L_0 = 0$. (The linkage is not a four-bar mechanism.) If L_0 is of intermediate length, then $300 + 100 = L_0 + 200$, so that $L_0 = 200$. Thus, we have a change-point mechanism if $L_0 = 200$ or 400 mm (answer e).

As noted, any mechanism that meets the criteria b, c, d, and e is a Grashof mechanism. Combining the preceding results, we have 200 mm $\leq L_0 \leq$ 400 mm (answer a).

A triple-rocker mechanism (f) exists when

$$L_{max} + L_{min} > L_a + L_b.$$

If the fixed link L_0 is largest, we have $L_0 + 100 > 200 + 300$, or $L_0 > 400$. If the fixed link L_0 is shortest, we have $300 + L_0 > 200 + 100$, or $L_0 > 0$. If L_0 is of intermediate length, then $300 + 100 > L_0 + 200$, or $200 > L_0$. Noting that no link length may exceed the sum of the lengths of the other three, we have $L_0 < 100 + 200 + 300$, or $L_0 < 600$. Combining these results, we obtain either $0 < L_0 <$ 200 mm or 400 mm $< L_0 <$ 600 mm (answer f).

Cutting devices sometimes incorporate crank-rocker mechanisms. An electric motor drives the shortest link; the longer crank (the rocker) drives an oscillating cutter.

Dead points must be considered when one is designing a rocker-driven crank-rocker mechanism. If the rocker drives, the limiting positions of Figures 1.12b and c are dead-points. Inertia may carry the shorter crank through a dead-point configuration. If the mechanism is stopped at a dead point, we cannot restart it by turning the rocker.

1.7 TRANSMISSION ANGLE

The inequalities that classify four-bar linkages give the extreme theoretical limits of each class of mechanism. Additional limitations apply to the design of practical mechanisms. One important consideration is the *transmission angle,* the angle between the coupler centerline and the driven crank centerline.

Referring to Figure 1.18, suppose the crank (link 1) drives the linkage. The coupler (link 2) transmits a force along its centerline to the driven crank (link 3). If we want to maximize output torque and minimize friction torque, we try to keep transmission angle ϕ near 90°. A transmission angle no less than 40° or 45° and no greater than 135° or 140° is usually satisfactory. Depending on the type of bearing and lubrication, values outside this range may result in binding of the linkage.

Figure 1.18 shows a linkage that satisfies the crank-rocker criteria. However, if link 1 drives, the transmission angle ϕ reaches extreme values, which may prevent the rocker from operating satisfactorily. As link 1 tends to rotate through the position shown in the figure, the direction of force transmitted along link 2 to link 3, results in very little torque on link 3 but a high bearing force at O_3. Wear would probably be excessive. If friction torque exceeded driving torque, the mechanism would jam and could cause the driven crank to buckle. Dimensional tolerances, including looseness at pins and bearings, often tend to worsen the situation. In most cases, then, it is advisable to provide a reasonable "margin of safety" in satisfying the inequalities that determine the motion of a linkage.

Consider the four-bar linkage whose links form a quadrilateral, as in Figure 1.19. For crank angle θ_1, the length L_d of the diagonal of the quadrilateral can be determined by using the law of cosines. For the triangle formed by links 0 and 1 and the diagonal,

$$L_d^2 = L_0^2 + L_1^2 - 2L_0L_1 \cos \theta_1. \tag{1.12}$$

Using the law of cosines for the triangle formed by the diagonal and links 2 and 3, we have

$$L_d^2 = L_2^2 + L_3^2 - 2L_2L_3 \cos \phi. \tag{1.13}$$

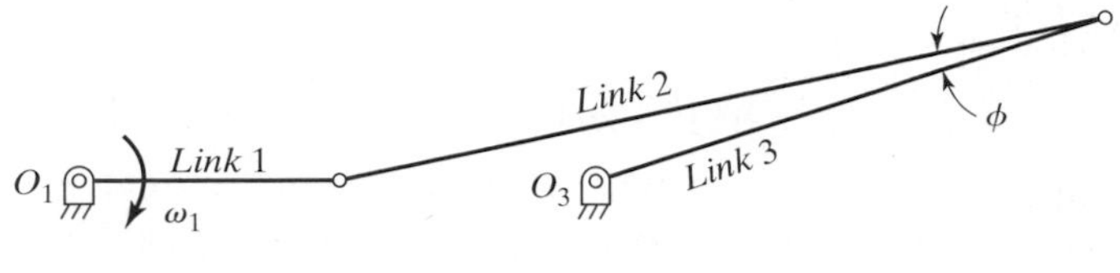

FIGURE 1.18 A mechanism that may fail to operate because of an unsatisfactory transmission angle.

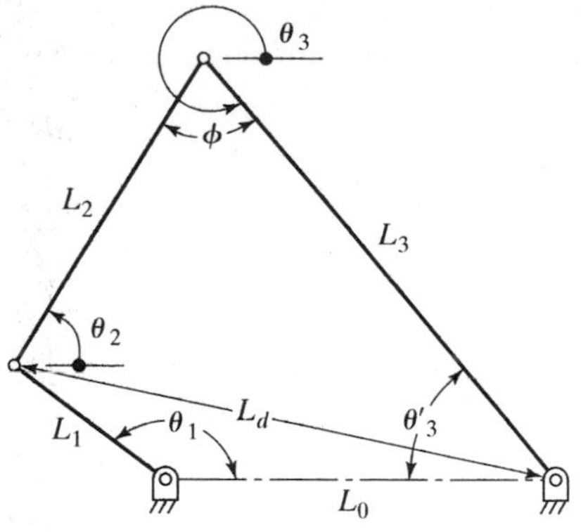

FIGURE 1.19 Determination of transmission angle.

Rearranging the cosine law equation:

$$\cos\phi = \frac{L_2^2 + L_3^2 - L_d^2}{2L_2L_3}. \quad (1.14)$$

Thus, we may obtain transmission angle ϕ at any instant.

We are most interested in extreme values of transmission angle ϕ. For the crank-rocker mechanism, maximum and minimum transmission angles occur when the driver crank and fixed link are collinear. Transmission angle ϕ_{max} corresponds to $L_{d(max)} = L_1 + L_0$, and ϕ_{min} corresponds to $L_{d(min)} = L_0 - L_1$.

SAMPLE PROBLEM 1.4

Transmission Angle

Given the driver crank length $L_1 = 100$ mm, coupler length $L_2 = 200$ mm, and follower length $L_3 = 300$ mm, and considering the transmission angle, find the range of values for the fixed link L_0 if the linkage is to be a crank rocker. In a previous example, we determined that the mechanism theoretically acts as a crank rocker for 200 mm $< L_0 <$ 400 if we put no limit on the transmission angle. Let us make the design decision to limit the transmission angle to $45° \leq \phi \leq 135°$.

Solution. Setting $\phi_{min} = 45°$ and using the law of cosines, we have

$$(L_0 - L_1)^2 = L_2^2 + L_3^2 - 2L_2L_3\cos\phi_{min},$$

or

$$(L_0 - 100)^2 = 200^2 + 300^2 - 2 \times 200 \times 300\cos 45°,$$

so that

$$L_0 = 312.48 \text{ mm}.$$

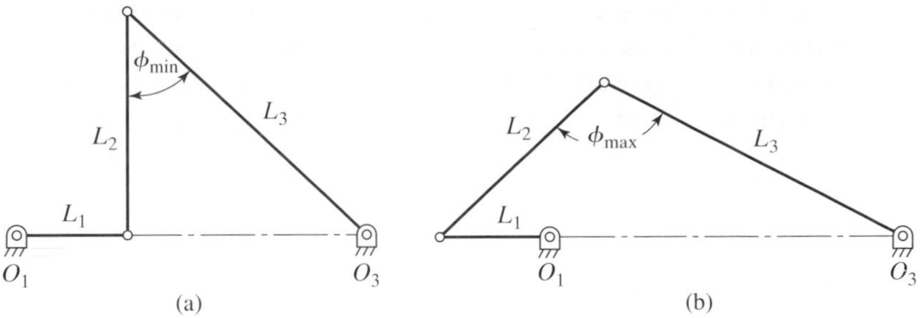

FIGURE 1.20 (a) Minimum value of transmission angle. (b) Maximum value of transmission angle.

Using this value, we obtain

$$\cos\phi_{max} = \frac{L_2^2 + L_3^2 - (L_0 + L_1)^2}{2L_2L_3} = \frac{200^2 + 300^2 - (312.48 + 100)^2}{2 \times 200 \times 300},$$

or $\phi_{max} = 109.54°$, which is within the accepted range.

The results (which could have been determined graphically) are sketched in Figures 1.20a and b. Note that if we had set $\phi_{max} = 135°$ to obtain $L_0 = 363.52$, the value of ϕ_{min} would have been 59.69°. Thus, $312.48 \le L_0 \le 363.52$ is the acceptable range.

Both the follower crank (rocker) and the coupler in a crank-rocker mechanism have a limited range of motion (considerably less than 180° if we require reasonable values of transmission angle).

When links 1 and 2 are collinear, rocker link 3 is at a limiting position. Consider the crank-rocker mechanism designed in the previous example, with $L_0 = 312.48$ mm. Refer to Figure 1.19, except note that links 1 and 2 are collinear and extended. The maximum value of θ'_3 is found as follows:

$$\cos\theta'_3 = \frac{L_0^2 + L_3^2 - (L_1 + L_2)^2}{2L_0L_3} = \frac{312.48^2 + 300^2 - (100 + 200)^2}{2 \times 312.48 \times 300},$$

from which $\theta'_{3(max)} = 58.61°$.

For the other limiting position, with links 1 and 2 collinear and flexed.

$$\cos\theta'_3 = \frac{L_0^2 + L_3^2 - (L_2 - L_1)^2}{2L_0L_3},$$

from which we obtain $\theta'_{3(min)} = 18.65°$, for a range of only 39.96°. Of course, the range of the follower crank can be changed by changing the ratios of the link lengths.

Typical linkage specifications include the range of motion of the output link (the follower crank). We may wish to investigate various options which will lead to a linkage that satisfies our needs. If the transmission angle and range of motion are plotted, we are more likely to approach an optimum design than by using hit-or-miss methods. The sample problem that follows illustrates a method of improving linkage design.

SAMPLE PROBLEM 1.5

Design of crank-rocker linkages.

Design a linkage with a 30° range of output crank motion.

Design decisions. A crank rocker linkage will be used. The fixed link length will be six times the crank length. The transmission angle will be limited to the range $40° \leq \phi \leq 140°$.

Solution. We will determine the proportions of the required design, specifying link lengths R in terms of crank length L_1. Thus, $R_0 = L_0/L_1 = 6$, etc.

Using the Grashof criterion for a crank rocker, we find that

$$R_2 < 5 + R_3,$$

$$R_3 < 5 + R_2, \quad \text{and}$$

$$7 < R_2 + R_3.$$

The Grashof limits are plotted against the linkage proportions in Figure 1.21a. Values of the minimum transmission angle are shown in degrees on the same figure.

Part b of the figure shows both the maximum and the minimum transmission angle plotted against the linkage proportions. The acceptable transmission angle envelope is marked in Figures 1.21b and 1.21c. The range of motion of the follower crank is shown in degrees in Figure 1.21c. Acceptable linkage proportions are given by the part of the 30° range-of-motion curve that falls within the 40° to 140° transmission angle envelope.

Detailed calculations. *Note:* Mathcad™ mathematics software is utilized in this and some other examples. When defining variables and constants, lightface roman type is used and the symbol := is used in place of an equals sign.

Fixed link: $\mathrm{R}_0 := 6$ $\quad \mathrm{N} := 140$

$\mathrm{i} := 0..\mathrm{N}$ $\quad \mathrm{j} := 0..\mathrm{N}$

Coupler: $\mathrm{R}_{2_i} := 1 + \dfrac{\mathrm{i}}{20}$ $\quad$ Driven crank: $\mathrm{R}_{3_j} := 1 + \dfrac{\mathrm{j}}{20}$

Diagonal at minimum transmission angle: $\mathrm{R}_\mathrm{d} := \mathrm{R}_0 - 1$

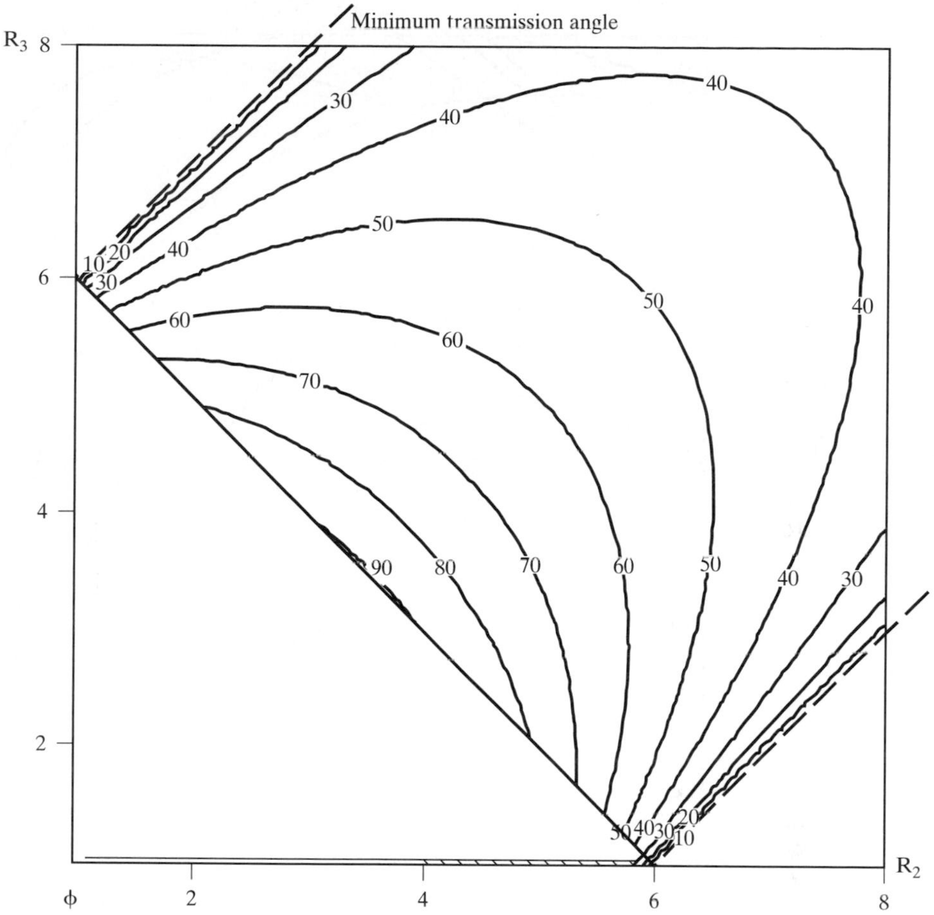

FIGURE 1.21 (a) Design of crank-rocker linkages, sample problem 1.5. Minimum transmission angle and limits based on the Grashof criterion plotted against linkage proportions. (*Note*: Grashof criterion limitations for a crank-rocker mechanism shown with long dashes.)

Diagonal at maximum transmission angle $R_D := R_0 + 1$

Cosine of minimum transmission angle:

$$c(R_2, R_3) := \frac{R_2^2 + R_3^2 - R_d^2}{2 \cdot R_2 \cdot R_3}$$

Cosine of maximum transmission angle:

$$C(R_2, R_3) := \frac{R_2^2 + R_3^2 - R_D^2}{2 \cdot R_2 \cdot R_3}$$

Some values of R_2 and R_3 are not valid (e.g., c (2,2) = −2.125 represents a mechanism that cannot be assembled).

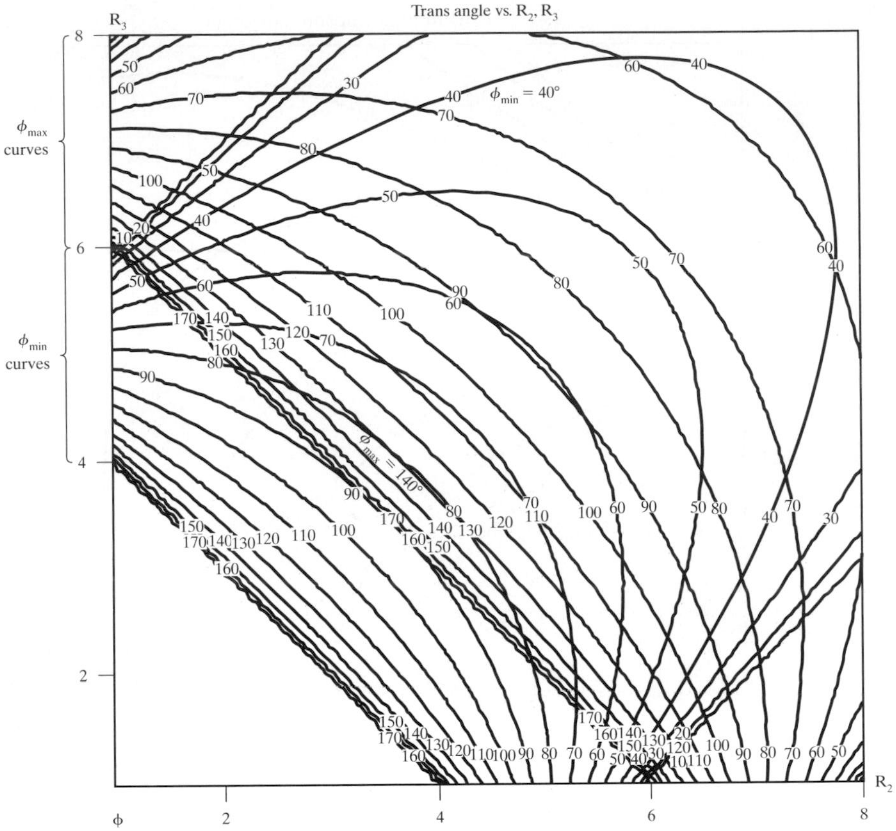

FIGURE 1.21 (b) Maximum and minimum transmission angle plotted against linkage proportions.

Minimum transmission angle:

$$\phi_{i,j} := \frac{\operatorname{acos}\left(c(R_{2_i}, R_{3_j})\right)}{\deg}$$

Maximum transmission angle:

$$\Phi_{i,j} := \frac{\operatorname{acos}\left(C(R_{2_i}, R_{3_j})\right)}{\deg}$$

Limiting positions, interior angle at follower link:

Flexed: $$\theta_f(R_2, R_3) := \operatorname{acos}\left[\frac{R_0^2 + R_3^2 - (R_2 - 1)^2}{2 \cdot R_0 \cdot R_3}\right]$$

Extended: $$\theta_e(R_2, R_3) := \operatorname{acos}\left[\frac{R_0^2 + R_3^2 - (R_2 + 1)^2}{2 \cdot R_0 \cdot R_3}\right]$$

Range: $$\Theta_{i,j} := \frac{\theta_e(R_{2_i}, R_{3_j}) - \theta_f(R_{2_i}, R_{3_j})}{\deg}$$

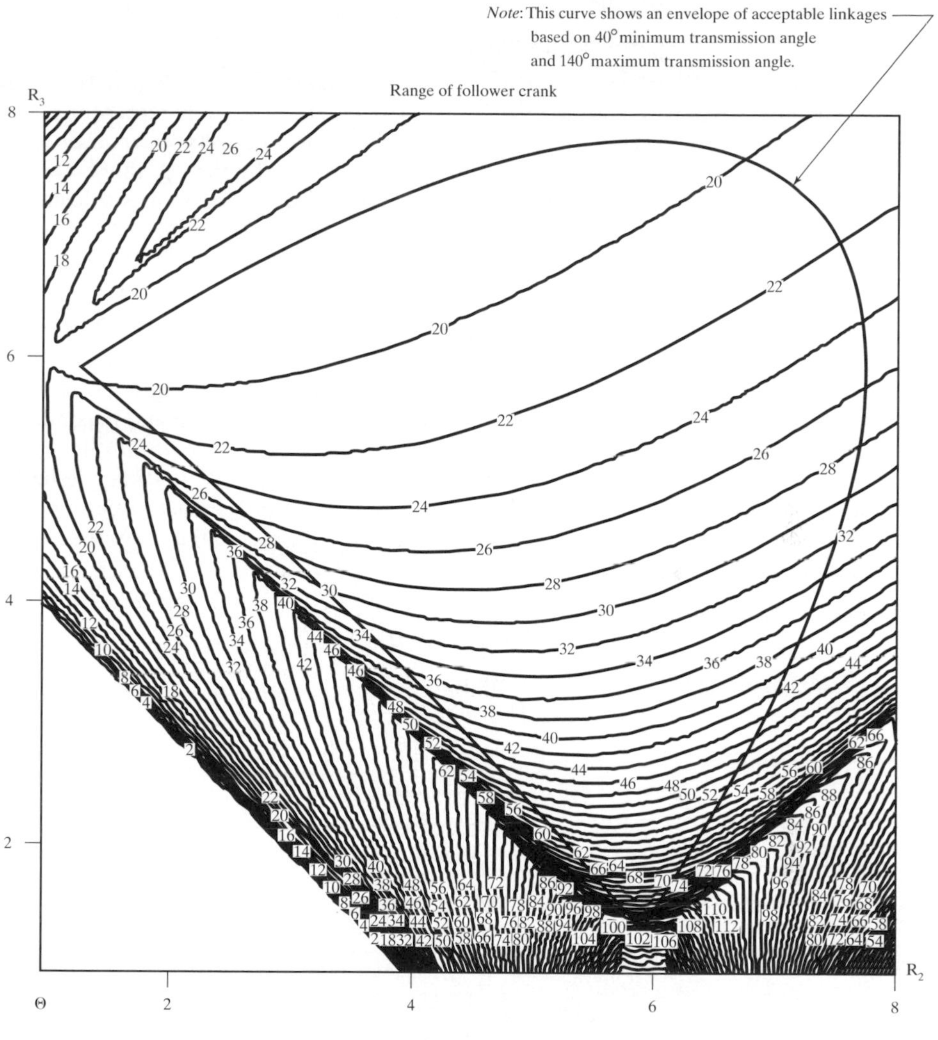

FIGURE 1.21 (c) Range of follower crank motion plotted against linkage proportions.

Instead of selecting just any set of acceptable proportions for a mechanism, we may want to optimize the design. For higher output torque and lower friction, we should attempt to keep the transmission angle close to 90°.

SAMPLE PROBLEM 1.6

Attempting to optimize a crank-rocker design

Optimize the design of a crank-rocker linkage on the basis of the transmission angle. Specifications call for a crank motion with a 30° range of output.

Design decisions. We will use a 50-mm drive crank length and a 300-mm fixed link, (the same ratio as in the previous sample problem).

Solution. Let us look more closely at the transmission angle plots. The narrow region between a minimum transmission angle of 70° and a maximum of 110° may represent a better design. If we plot the range curves on the same sheet, or if we hold both sets of curves up to a light, we see that the 70° to 110° transmission angle region is crossed by the 30° range-of-motion curve at $R_2 = L_2/L_1 \approx 4.7$. The value of R_3 may be read from the plot, or we may calculate the root of

$$\Theta(R_3) - [\theta_e(R_3) - \theta_f(R_3)] = 0,$$

where $\Theta(R_3) = 30° =$ the range of the output crank, $\theta_e(R_3)$, and $\theta_f(R_3) =$ extended and flexed limiting positions, respectively.

The result is $R_3 = L_3/L_1 = 3.864$ for a driven crank length $L_3 = 193.2$ mm. The coupler length is $L_2 = 235$ mm. These dimensions correspond to minimum and maximum transmission angles of 70.7 and 109.3°, respectively.

Recall that a transmission angle near 0° or 180° may cause a linkage to bind. A transmission angle near 90° will usually result in good output torque characteristics and no tendency to bind. We have produced the required output crank motion, and the transmission angle is always close to 90°.

Detailed calculations (using Mathcad™).

Crank-rocker optimization:

Crank length (mm) $L_1 := 50$ Fixed link $L_0 := 300$

$$R_0 = \frac{L_0}{L_1} \qquad R_0 = 6$$

Coupler: $R_2 := 4.7$ $\qquad L_2 := R_2 \cdot L_1 \quad L_2 = 235$

Limiting positions: interior angle at follower link

Flexed: $\qquad \theta_f(R_3) := \operatorname{acos}\left[\dfrac{R_0^2 + R_3^2 - (R_2 - 1)^2}{2 \cdot R_0 \cdot R_3}\right]$

Extended: $\qquad \theta_e(R_3) := \operatorname{acos}\left[\dfrac{R_0^2 + R_3^2 - (R_2 + 1)^2}{2 \cdot R_0 \cdot R_3}\right]$

Range (degrees): $\qquad \Theta(R_3) := 30$

Estimate: $\qquad R_3 := 4 \; R_3 := \operatorname{root}\left(\Theta(R_3) - \dfrac{\theta_e(R_3) - \theta_f(R_3)}{\text{deg}}, R_3\right)$

$$R_3 = 3.864 \quad L_3 = R_3 \cdot L_1 \quad L_3 = 193.185$$

Diagonal at minimum transmission angle: $\qquad R_d := R_0 - 1$

Diagonal at maximum transmission angle: $\qquad R_D := R_0 + 1$

Cosine of minimum transmission angle:

$$c(R_2, R_3) := \frac{R_2^2 + R_3^2 - R_d^2}{2 \cdot R_2 \cdot R_3}$$

Minimum transmission angle (degrees):

$$\phi(R_2, R_3) := \frac{\text{acos}\,(c(R_2, R_3))}{\text{deg}}$$

$$\phi(R_2, R_3) = 70.676$$

Cosine of maximum transmission angle:

$$C(R_2, R_3) := \frac{R_2^2 + R_3^2 - R_D^2}{2 \cdot R_2 \cdot R_3}$$

Maximum transmission angle:

$$\Phi(R_2, R_3) := \frac{\text{acos}\,(C(R_2, R_3))}{\text{deg}}$$

$$\Phi(R_2, R_3) = 109.263$$

We may think of the slider-crank linkage as a four-bar linkage if the slider is replaced by an infinitely long link perpendicular to the sliding path. Then, the transmission angle is defined as the angle between the connecting rod and a perpendicular to the slider path. If the crank can rotate through 360°, the extreme values of transmission angle occur when the crank is perpendicular to the slider path. (The proof of this statement is left as an exercise.)

Spatial linkages may also have transmission angle problems. Consider the RSSR linkage shown in Figure 1.7a. If link 1 drives, then the angle between coupler link 2 and driven crank 3 is of interest. We could compute transmission angle

$$\phi = \arccos\left[\frac{\boldsymbol{r}_2 \cdot \boldsymbol{r}_3}{|\boldsymbol{r}_2||\boldsymbol{r}_3|}\right],$$

where ϕ is the transmission angle.

The numerator of the fraction is the dot product of the vectors representing the coupler and the driven crank, and the denominator is the product of the link lengths. (The dot product and other vector algebra concepts are reviewed briefly in Chapter 2.)

Values of the transmission angle that do not fall in the range

$$40^\circ \leq \phi \leq 140^\circ$$

may indicate that the linkage will jam. The design of linkages with spherical and universal joints is difficult, because the construction of these joints limits their range of motion.

1.8 LIMITING POSITIONS OF SLIDER-CRANK LINKAGES

Limiting positions are of interest for several reasons. The limiting positions of a slider-crank mechanism define the stroke of the piston (slider). The piston has zero velocity at the instant it reaches one of the limiting positions. However, the acceleration of the piston, and consequently the inertial force, is high at that instant. When at a limiting position, a slider-crank mechanism cannot be driven by applying a force to the piston. If a single piston serves as a driver, the linkage may be driven through the limiting position by inertia of the crank. Likewise, the limiting positions define the range of the oscillating crank of a crank-rocker mechanism. The oscillating crank has zero angular velocity and a high value of angular acceleration at the limiting positions.

In-Line Slider-Crank Mechanisms

A slider-crank mechanism with the usual proportions (such that the connecting rod is longer than the crank) has two limiting positions, both occuring when the crank and the connecting rod are collinear (in the skeleton diagram); see Figure 1.22a. When reciprocating steam engines were in common use, these positions were called *dead-center positions, crank (bottom) dead center*, referring to the position with the piston nearest the crankshaft (Figure 1.22a, *left*), and *head (top) dead center*, referring to the position with

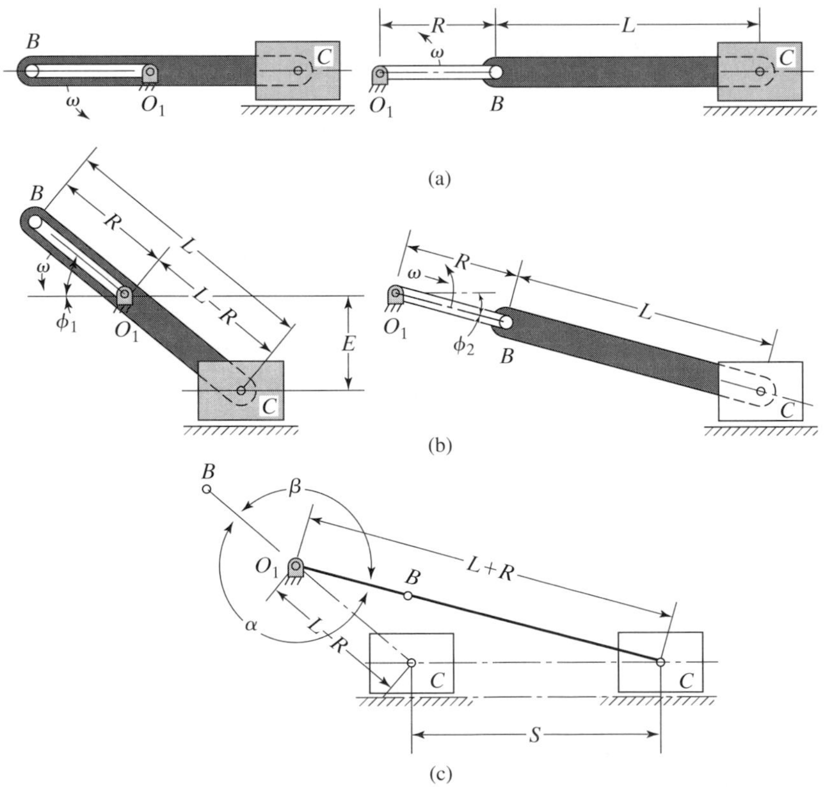

FIGURE 1.22 (a) The two limiting positions of an *in-line* slider-crank mechanism. (b) The two limiting positions of an offset slider-crank mechanism. (c) The limiting positions of an offset slider-crank mechanism superimposed to find the ratio of the time taken for the forward stroke to the time taken for the return stroke.

the piston farthest from the crankshaft (Figure 1.22a, *right*). The piston's direction reverses at these two points.

When the extended path of the wrist pin C goes through the center of the crankshaft O_1 (as in Figure 1.22a), the linkage is called an *in-line slider-crank mechanism.* The stroke, referred to as *piston travel*, equals $2R$, twice the crank length. The crank turns through 180° as the piston moves from left to right and through another and 180° as it returns to the left. If the crank turns at a constant angular velocity ω, the piston takes the same time to move from left to right as it takes to return to the left.

Offset Slider-Crank Mechanisms

The wrist-pin path of the *offset slider-crank mechanism* (see Figure 1.22b) does not extend through the center of the crankshaft. The limiting positions shown represent positions of zero piston velocity, but the angles through which the crank turns between the limiting positions are not equal. If the crank turns counterclockwise, it turns through an angle greater than 180° as the piston moves from left to right and through less than 180° as the piston moves back to the left. If the crank turns counterclockwise at constant angular velocity, the piston takes longer in its stroke to the right than it takes to return to the left. From its limiting position in Figure 1.22a *left* to its limiting position in Figure 1.22b *right*, the crank turns through the angle

$$\alpha = 180^\circ + \phi_1 - \phi_2,$$

as shown in Figure 1.22c. During the return stroke, the crank turns through the angle

$$\beta = 180^\circ - \phi_1 + \phi_2,$$

where

$$\phi_1 = \sin^{-1}\frac{E}{L - R}$$

and

$$\phi_2 = \sin^{-1}\frac{E}{L + R}$$

for crank length R, connecting rod length L, and offset distance E less than $L - R$.

When the crank turns at a constant angular velocity ω, the ratio of the forward to return stroke times is given by α/β. The length of the stroke is

$$S = \sqrt{(L + R)^2 - E^2} - \sqrt{(L - R)^2 - E^2}.$$

The limiting positions of the linkage may be superimposed to form a triangle, as in Figure 1.22c. Using the geometrical fact that the sum of the lengths of any two sides of a triangle exceeds the length of the remaining side, we obtain

$$L - R + S > L + R,$$

from which we see that the stroke length will always exceed $2R$ when the wrist-pin path is offset from the crankshaft. The preceding relationships are valid when both of the following conditions are met: The offset E is less than $L - R$, and R is less than L. Of course, angles α and β and stroke S can be found simply by superimposing the limiting positions of the linkage, as in Figure 1.22c.

1.9 QUICK-RETURN MECHANISMS

Quick-return mechanisms include an oscillating link or reciprocating slider that moves forward slowly and returns quickly (with an input of constant speed). The forward and return directions are arbitrarily assigned as before, to correspond with machine tool usage, in which a forward (working) stroke would have high force capability at low speed and the return stroke could be rapid with no load.

The designation *quick return* has as much to do with the function of a mechanism as with its mode of operation. If there is an intentional difference between the time required for the forward and return strokes, the linkage may be called a *quick-return mechanism*. Most crank-rocker mechanisms exhibit unequal forward and return times for the rocker. If we take advantage of the unequal strokes in designing a piece of machinery, we call the linkage a quick-return mechanism.

The forward and return strokes for the in-line slider-crank mechanism take an equal amount of time, but the offset slider crank acts as a quick-return mechanism.

Other linkage combinations offer considerably more flexibility for quick-return design than does the offset slider crank. The *drag link*, for example, may form part of a mechanism designed for large forward-to-return-time ratios. Figure 1.23 shows four-bar linkage O_1BCO_3, which appears to satisfy the criteria for a drag link mechanism. Slider D represents a machine element that is to have different average velocities for its forward and return strokes, while driving crank 1 turns at constant angular velocity. The two extreme positions of the slider occur when follower link 3 lies along the line of

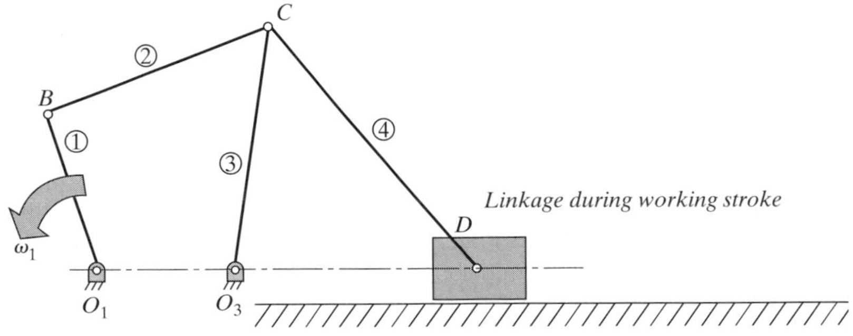

FIGURE 1.23 A drag link mechanism is combined with a slider to form a quick-return mechanism.

centers O_1O_3. Since link 4 is also collinear with the line of centers at both of the extreme positions, we see that the slider *stroke* is twice the length of link 3:

$$S = 2L_3. \tag{1.15}$$

The time for the slider to travel between limiting positions is proportional to the angle between corresponding positions of the driving crank, as long as the angular velocity of the driving crank is constant.

SAMPLE PROBLEM 1.7

Quick-return mechanism based on a drag link

Design a quick-return mechanism with a three-to-one forward-to-return-time ratio.

- **a.** Determine linkage proportions.
- **b.** Specify link lengths for a 180-mm stroke.
- **c.** Are there any special concerns with this design?

Design decisions. The design will be based on a drag link combined with a slider. (See Figure 1.23). We try linkage proportions $R_0 = L_0/L_1 = 0.8$ and $R_3 = L_3/L_1 = 1.4$, with $R_2 = L_2/L_1$ unspecified, where the link lengths are identified as follows: L_0 = fixed link, L_1 = drive crank, L_2 = coupler, and L_3 = driven crank.

Solution.

a. The Grashof criterion for a drag link requires that the fixed link be shortest and that $L_{max} + L_0 < L_a + L_b$.
If the coupler is longest, this equation becomes $L_2 + L_0 < L_1 + L_3$.
Dividing by the drive crank length produces $R_2 + R_0 < 1 + R_3$.
Substituting the foregoing values, we obtain $R_2 < 1.6$.
A similar calculation based on a coupler of intermediate length yields $1.2 < R_2$.

Thus, $1.2 < R_2 < 1.6$, based on the Grashof criterion.

A 3:1 forward-to-return-time ratio requires a return stroke angle $\beta = \pi/2$ rad $= 90°$. During the return stroke, the crank (O_1B) goes from the position shown in Figure 1.24a to the position shown in Figure 1.24b. We seek the value of R_2 that satisfies the equation

$$\pi - \beta_a + \beta_b - \beta = 0 \qquad \text{(all in radians)},$$

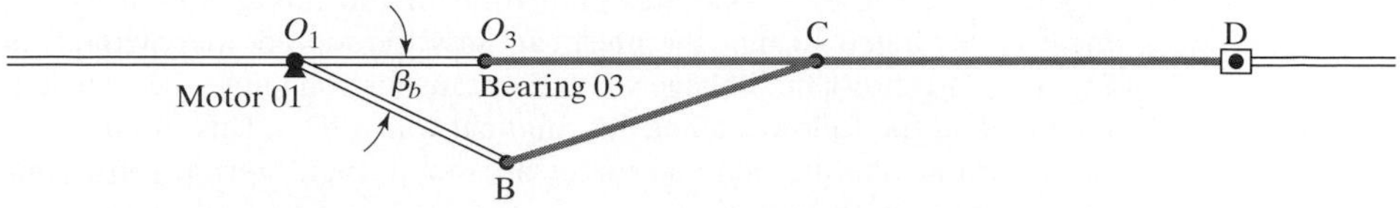

FIGURE 1.24 Quick-return mechanism based on drag link, sample problem 1.7. (a) Limiting position with slider to extreme right.

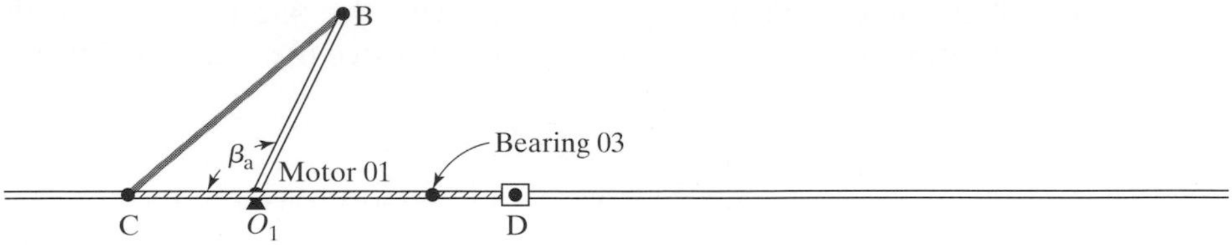

FIGURE 1.24 (b) Limiting position with slider to left.

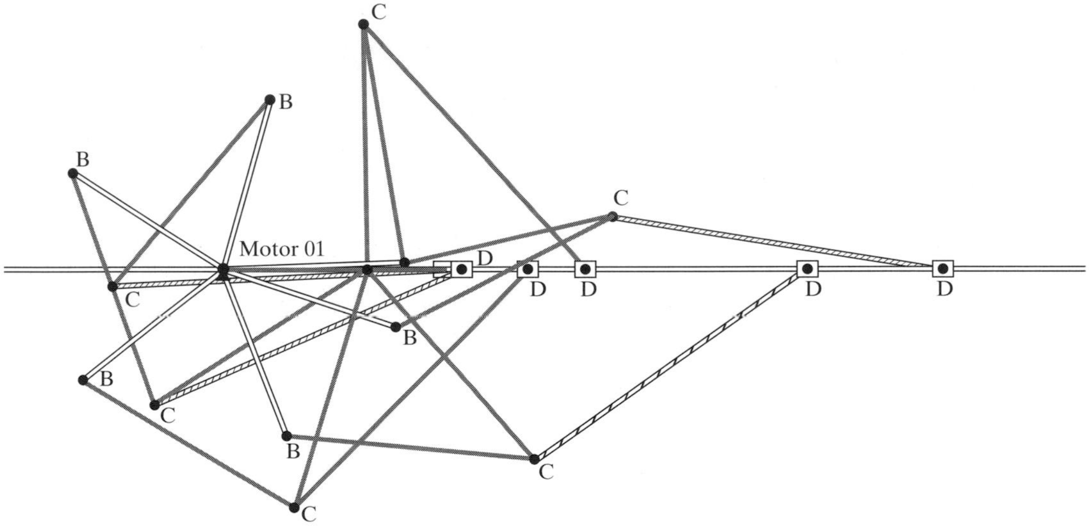

FIGURE 1.24 (c) Animation of mechanism.

where β_a and β_b are internal angles of the triangle formed by the linkage for extreme positions of point C. We use the cosine law to find β_a and β_b. A numerical solution is started with an estimate of R_2 in the middle of the Grashof criterion range. The solution converges on $R_2 = 1.374$.

b. The stroke is twice the driven crank length. Thus, $L_3 = 90$ mm. Then $L_1 = L_3/R_3$, and we find the remaining link lengths. The link joining the driven crank and the slider must be somewhat longer than the driven crank, say, $L_4 = 120$ mm.

c. Figure 1.24c shows a computer animation of the linkage. Bearings and supports must be arranged so that the links can pass by one another without interference. Figure 1.24d shows the linkage when the transmission angle—the angle between the coupler and the follower crank—is minimal (only 8.2°). This situation occurs during the return stroke. If frictional forces are small, then inertial forces may carry the linkage through this position.

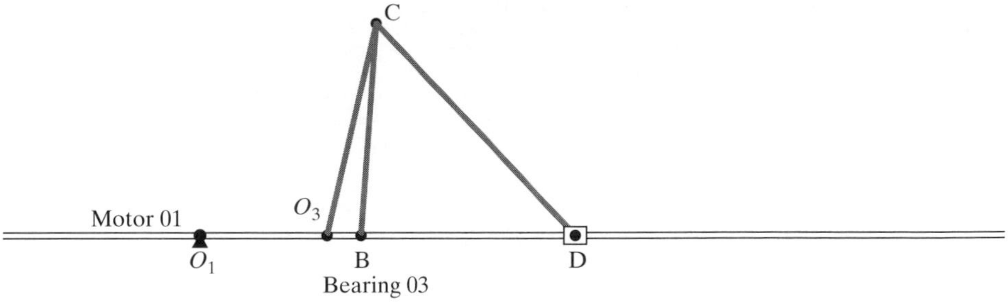

FIGURE 1.24 (d) Minimum transmission angle.

Detailed calculations (using Mathcad™). Select fixed link and driven crank length ratios: $R_0 := .8 \qquad R_3 := 1.4$
Grashof criterion for drag link (fixed link is shortest):

$$R_{2\,min} := R_3 + R_0 - 1 \qquad R_{2\,min} = 1.2 \qquad R_{2\,max} := 1 + R_3 - R_0 \quad R_{2\,max} = 1.6$$

Working-stroke-to-return-time ratio: $T_{wr} := 3$

Range of link 1 during return stroke (rad):

$$\beta := \frac{2 \cdot \pi}{T_{wr} + 1} + \qquad \beta = 1.571 \qquad \frac{\beta}{deg} = 90\,deg$$

Applying the cosine law to find the coupler length ratio: R_2

$$\beta_a(R_2) := \text{acos}\left[\frac{1 + (R_3 - R_0)^2 - {R_2}^2}{2 \cdot (R_3 - R_0)}\right] \qquad \beta_b(R_2) := \text{acos}\left[\frac{1 + (R_3 + R_0)^2 - {R_2}^2}{2 \cdot (R_3 + R_0)}\right]$$

Estimate $R_2 := \dfrac{R_{2\,min} + R_{2\,max}}{2}$

$$R_2 := \text{root}(\pi - \beta_a(R_2) + \beta_b(R_2) - \beta, R_2) \qquad R_2 = 1.374$$

$$\beta_a(R_2) = 2.026 \qquad \frac{\pi - \beta_a(R_2)}{deg} = 63.92\,deg$$

$$\beta_b(R_2) = 0.455 \qquad \frac{\beta_b(R_2)}{deg} = 26.067$$

$$\beta := \pi - \beta_a(R_2) + \beta_b(R_2)\ \beta = 1.571 \qquad \frac{\beta}{deg} = 89.987 \text{ (check)}$$

Required stroke (mm): $S := 180$

Driven crank length (mm): $L_3 := \frac{S}{2}$ $\quad L_3 = 90$

Drive crank length (mm): $L_1 := \frac{L_3}{R_3}$ $\quad L_1 = 64.286$

Fixed link length (mm): $L_0 := R_0 \cdot L_1$ $\quad L_0 = 51.429$

Coupler length (mm): $L_2 := R_2 \cdot L_1$ $\quad L_2 = 88.321$

Minimum transmission angle $\phi_{min} := \text{acos}\left[\frac{L_2^2 + L_3^2 - (L_1 - L_0)^2}{2 \cdot L_2 \cdot L_3}\right]$ $\quad \frac{\phi_{min}}{\text{deg}} = 8.199 \text{ deg}$

Maximum transmission angle: $\phi_{max} := \text{acos}\left[\frac{L_2^2 + L_3^2 - (L_1 + L_0)^2}{2 \cdot L_2 \cdot L_3}\right]$ $\quad \frac{\phi_{max}}{\text{deg}} = 80.913 \text{ deg}$

The preceding problem can also be solved by a trial-and-error graphical method, as illustrated in Figure 1.25. The coupler length is varied in the three trials. The third trial (length B_3C) results in a forward-to-return-time ratio of about three to one.

Sliding contact linkages also form a basis for quick-return mechanisms. Figure 1.26 shows a quick-return mechanism that can be used to drive the cutting tool in a mechanical shaper. Crank 1 is the driver, turning at essentially constant angular velocity, and slider D represents the toolholder. Limiting positions occur when links 1 and 2 are perpendicular. The ratio of the times of the working stroke to the return stroke is equal to the ratio of the angles between corresponding positions of link 1.

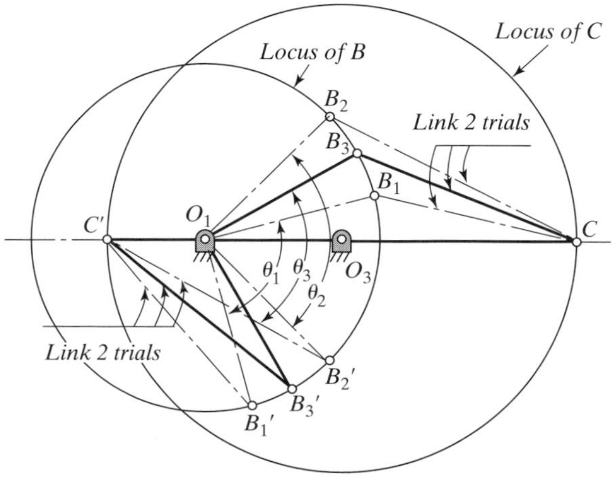

FIGURE 1.25 A sketch of a drag link quick-return mechanism to determine the length of link 2 needed for a three-to-one time ratio. The mechanism is drawn with link 3 in its critical positions.

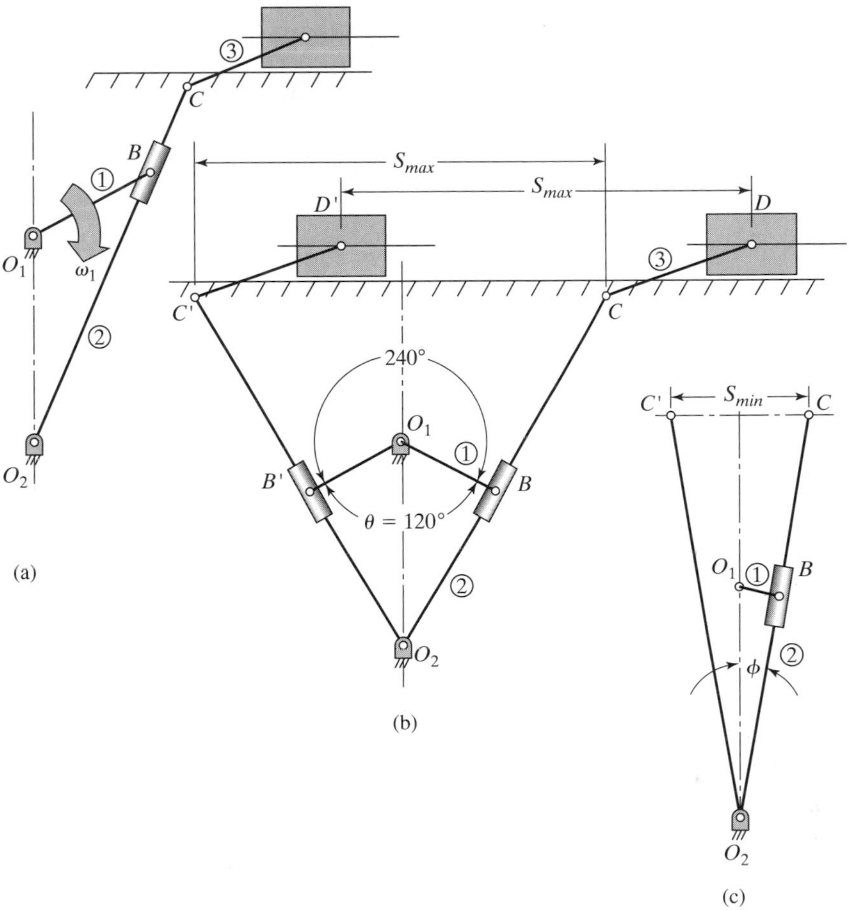

FIGURE 1.26 (a) Quick-return mechanism utilizing a sliding contact linkage. (b) The linkage is shown in its limiting positions (link 1 perpendicular to link 2). The stroke S of the slider is adjusted by changing the length of link 1. (c) Link 1 is shown adjusted to provide a minimum slider stroke.

SAMPLE PROBLEM 1.8

Variable-Stroke Quick-Return Mechanism

Design a mechanism with a stroke that may be varied from 3 to 8 in, having a working-stroke-to-return-stroke time ratio of two to one at maximum stroke length.

Solution. The two-to-one ratio is obtained if the angle θ between limiting positions is given by

$$\frac{360 - \theta}{\theta} = 2, \text{ or } \theta = 120°,$$

as in Figure 1.26b. Often, the key to determining link lengths is to *assign* a reasonable value to *one* or more of the unknown links. The geometric relationships in the linkage are next observed when the linkage is drawn in its limiting positions. The lengths of the remaining unknown links are then obtained. A satisfactory design may be obtained in the first trial. If not, this trial is used as a basis for improving the design.

If distance O_1O_2 is taken to be 4 in, then the maximum length of the drive crank is

$$L_{1(\max)} = 4\sin\left(90° - \frac{\theta}{2}\right) = 2 \text{ in.}$$

Since link 3 lies at the same angle at both limiting positions (the path of D is perpendicular to O_2O_1), the maximum stroke length is

$$S_{\max} = D'D = C'C = 8 \text{ in},$$

from which we obtain the length of link 2:

$$O_2C = \frac{S_{\max}/2}{\sin(90° - \theta/2)} = 8 \text{ in.}$$

The length of link 3 is arbitrarily taken to be 3 in, and the distance from O_1 to the path of D is assumed to be 3.5 in. For the minimum stroke $S_{\min} = 3$ in, the crank must be adjusted to a length of

$$L_{1(\min)} = O_1O_2\sin\phi = O_1O_2\frac{S_{\min}/2}{O_2C} = 0.75 \text{ in},$$

as shown in Figure 1.26c The actual mechanism may differ considerably from the schematic, as long as the motion characteristics are unchanged. Link 1 may be part of a large gear driven by a pinion, in which case the crankpin (B on link 1) will be moved in or out along an adjusting screw. Link 2 may be slotted, so that the crankpin rides within it.

1.10 LINKAGE INTERFERENCE

For convenience in illustrating the motion of plane mechanisms, the mechanisms are shown as if they move within a single plane. Consider a crank-rocker linkage as sketched in Figure 1.12. To avoid interference, the drive crank (link 1) and the coupler (link 2) must operate in two parallel planes. The plane of the drive crank should lie between the plane of the coupler and the plane of the fixed link.

The interference problem encountered in drag link mechanism design is more severe since drive crank, coupler, and follower rotate through 360°. To avoid link interference, the plane of the coupler should be between the planes of the cranks. The fixed bearings of the cranks must be placed on opposite sides of the linkage, with clear space for the linkage to pass between the bearings. Figure 1.27 illustrates one possible configuration schematically. It can be seen that if the plane of the coupler (link 2) is not clear, the linkage would not be able to operate through a complete rotation.

Sometimes, a four-bar linkage forms part of a more complicated linkage. Motion may be transferred from the coupler of a crank-rocker mechanism without much difficulty. Due to the problem of interference associated with the drag link

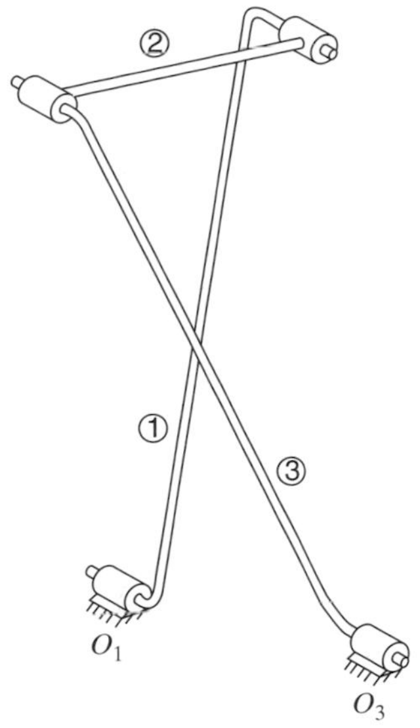

FIGURE 1.27 Configuration for drag link mechanism to avoid interference.

mechanism, and because of the requirement that the coupler plane lie between the cranks, a transfer of motion from the coupler of a drag link mechanism may require complicated arrangements.

The lamination-type impulse drive (described in the section that follows) illustrates motion transfer from one four-bar linkage to another. Considering the equivalent linkage, the rocker of a crank-rocker mechanism acts as the driving link of a second four-bar linkage (which oscillates due to the limited range of the input motion). The drive is made up of several such combinations of mechanisms. Because of space limitations, eccentrics or cams are used instead of conventional cranks.

1.11 MECHANISMS FOR SPECIFIC APPLICATIONS

Before we begin detailed analysis and synthesis of mechanisms, it is worthwhile to consider the basic motion characteristics of some of the commonly available linkages. In the design of a machine, it may be practical to combine simple linkages and other components to obtain the required output-to-input motion relationship. The designer may wish to become familiar with many of the linkage configurations that are in the public domain and should become aware of the proprietary packaged drive trains and other machine components that are available. Then, skill and ingenuity can combine these components for optimum results without a need to "reinvent the wheel." Some familiarity with various classes of available mechanisms will be obtained by leafing through this and other design-oriented books and by using manufacturers' catalogs and engineering periodicals.

Of course, the probable cost advantages of using commercially available components should not prevent the designer from exploring entirely new solutions, even though they may represent significant departures from traditional designs.

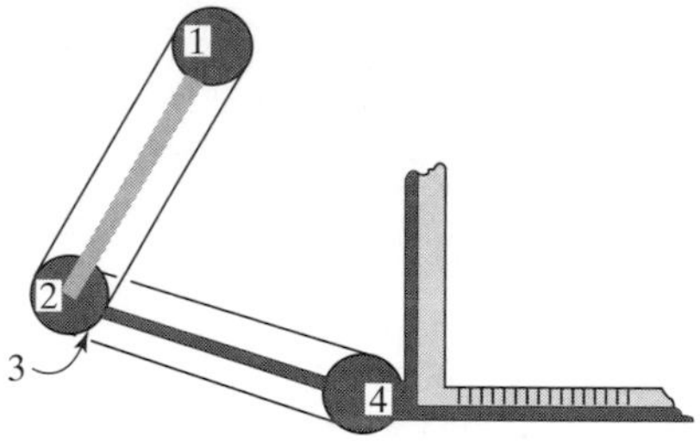

FIGURE 1.28 The belts arranged as shown permit translation of the straightedge, but prevent it from rotating.

Drafting Instruments

The *drafting instrument* using rigid links and pins shown in Figure 1.6 is proportioned so that distances $AB = CD$ and $AD = BC$ form a parallelogram. If the line between fixed centers A and B is horizontal, then DC is also horizontal at all times. Since a straightedge attached at DC would not allow sufficient freedom of movement, another parallelogram linkage is added. A parallelogram linkage can also be used to confine independently suspended automobile wheels to a vertical plane, reducing "tucking under" during turns.

Another drafting system, shown in Figure 1.28, uses tight steel bands (belts) on two pairs of disks with equal diameters. Disk 1 is not permitted to rotate, and as the arm between disks 1 and 2 is moved, disks 2 and 3 translate without rotating. The bands between disks 3 and 4 prevent rotation of disk 4 and the attached straightedges. Applications of this type were more common before the general availability of computer-aided drafting systems.

Pantograph Linkages

The parallelogram also forms the basis for *pantograph linkages*. At one time, pantograph linkages were used to reproduce and change the scale of drawings and patterns. The pantographs of Figure 1.29 are made up of rigid links AC, CD, DE, and EB with pin connections. Lengths $BC = DE$ and $BE = CD$ form a parallelogram. Link BE is parallel to CD at all times for both linkages, and F is located on a line between A and D, making triangles ABF and ACD similar. Thus, in Figure 1.29a the ratio DA/DF is a constant for all positions of the linkage, and if a point located at F is used to trace a pattern, a drawing tool at A will reproduce the pattern, enlarged by the factor DA/DF. If the actual part is to be smaller than the pattern, then the tracing point can be located at A and the drawing tool at F. The result will be a reduction in size of the ratio DF/DA. The pantograph may be made adjustable to produce various enlargement or reduction ratios, provided that the key features are maintained: The linkage must form a parallelogram, and points A, F, and D must lie on a straight line.

If the pattern is to be reproduced full size or nearly full size, point F will serve as the pivot, with D the tracer point and A the toolholder, as in Figure 1.29b. The pattern will be faithfully reproduced with a part-to-pattern size ratio AF/DF, but the orientation will be changed in this case.

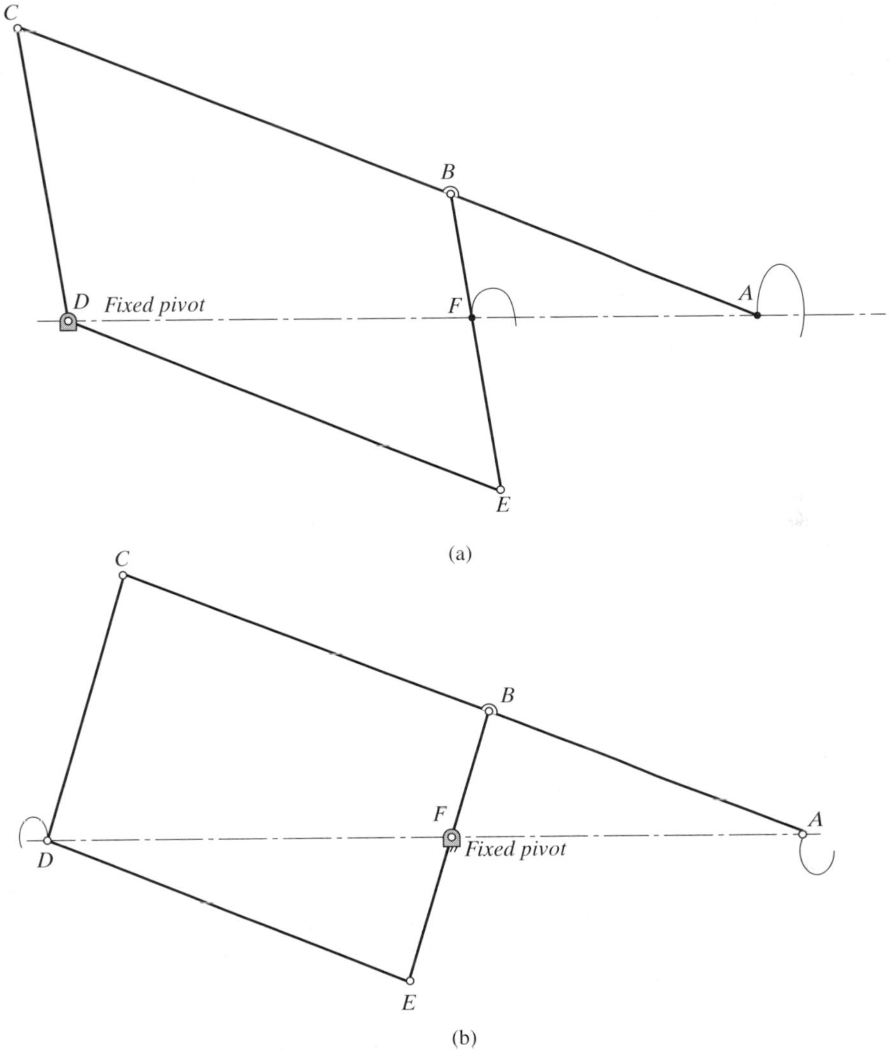

FIGURE 1.29 (a) A pantograph with fixed point D. The pattern can be traced-enlarged by the ratio DA/DF if the tracing point is located at F. Interchanging the tracing point and drawing tool produces a reduced tracing. (b) The pantograph with point F used as the fixed point will produce a tracing approximately the same size as the pattern.

The operation may be automated by using a sensing device to drive the tracing point over the pattern. A number of other linkages are used for similar purposes, including *engine indicators*, which reproduce a pressure signal. Engine or compressor pressure is measured by a small piston operating against a spring in the indicator. The indicator linkage, which resembles a pantograph, magnifies and records the motion of the indicator piston, producing approximately straight-line motion.

SAMPLE PROBLEM 1.9

Pantograph Design

Proportion a linkage to guide an oxyacetylene torch in rough-cutting parts from steel plate. Part dimensions are to be approximately 6 in by 6 in; patterns will be 1.5 times full size.

Solution. A pantograph of the type sketched in Figure 1.29b will be used, so that the pattern will not be too near the cutting torch. The pattern dimensions will be approximately 9 in by 9 in, and the linkage must be designed so that tracing point D moves freely over at least the 9-in-by-9-in area. It can be seen that, by dimensioning the links so that $CB = DE = 12$ in and $CD = BE = 10$ in, the tracing point will cover the required area without nearing its limiting (extreme) positions. So that the size reduction factor of 1/1.5 is obtained, $AB/CB = 1/1.5$, or $AB = 8$ in. Points A, F, and D must form a straight line, from which it follows that $BF/CD = AB/AC$, or $BF = 4$ in, locating the fixed pivot.

For a practical design, it may be necessary to allow the tracing point position A to be adjusted to various positions along the link so that several ratios of pattern to part size can be accommodated. For each position of A, a new point F, the fixed point on link BE, would have to be established to maintain the straight-line relationship between D, F, and A.

Slider-Crank Mechanism

The *slider-crank mechanism* is probably the most common of all mechanisms because of its simplicity and versatility. We are familiar with it in the reciprocating pump and compressor, in which the input rotation is changed to reciprocating motion of the piston. Figure 1.30 shows an air-conditioning and heat pump compressor. In the piston engine, the situation is reversed and the piston is the driver. Of course, if there are several cylinders, the various pistons alternate as driver, and if the engine is a single-cylinder engine, the energy stored in the flywheel and other components actually drives the piston between power strokes. A single slider-crank mechanism and the associated cam and valve train typical of a multicylinder internal-combustion engine are shown in Figure 1.31. Figure 1.32 shows the piston and connecting rod of a small one-cylinder gasoline-powered engine.

Rotating Combustion Engine

The rotating combustion (Wankel) engine in Figure 1.33 is another solution to the same problem with little kinematic resemblance to the conventional piston engine. The three-sided rotor moves eccentrically within a two-lobed engine block. These two parts (the rotor and the shaped block) are equivalent to the pistons, cylinders, combustion chambers, and valve train of an ordinary reciprocating engine. An internal gear, part of the three-lobed rotor, actually acts as a planet gear as it meshes with a smaller, fixed sun gear. Many other configurations of the rotating combustion engine with various numbers of rotor sides and engine block lobes were examined before this design was chosen.

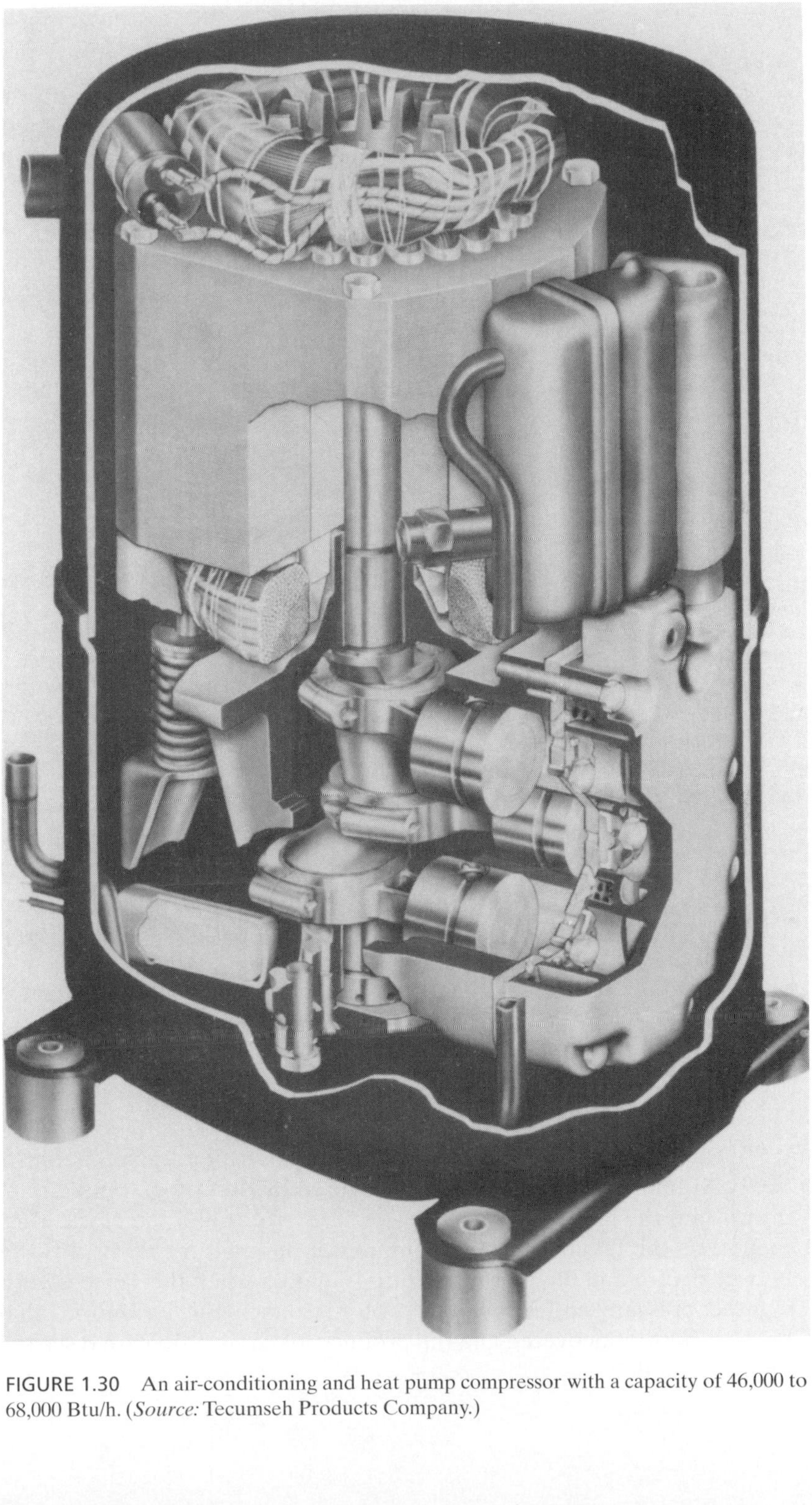

FIGURE 1.30 An air-conditioning and heat pump compressor with a capacity of 46,000 to 68,000 Btu/h. (*Source:* Tecumseh Products Company.)

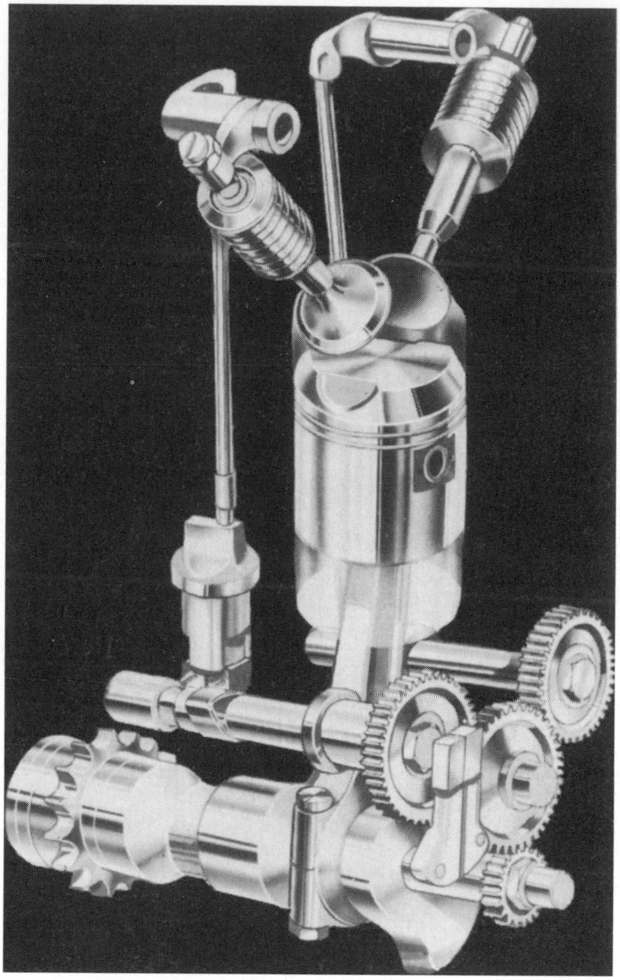

FIGURE 1.31 One cylinder of a reciprocating engine is illustrated, showing the basic slider-crank mechanism, camshaft, and valve train. (*Source:* Curtiss-Wright Corporation.)

The combustion cycle of a rotating combustion engine is illustrated in Figure 1.33b. At intake, an intake port is uncovered by the rotor. A mixture of air and fuel is drawn into the increasing space between the rotor and the block. The eccentric rotor then seals the intake port and compresses the mixture in the now-decreasing space between rotor and block. The mixture is ignited when the space is very small, increasing the pressure and driving the rotor around (the expansion phase). Finally, an exhaust port is uncovered and the products of combustion are discharged. The cycle is then repeated.

FIGURE 1.32 The piston and connecting rod for a one-cylinder engine.

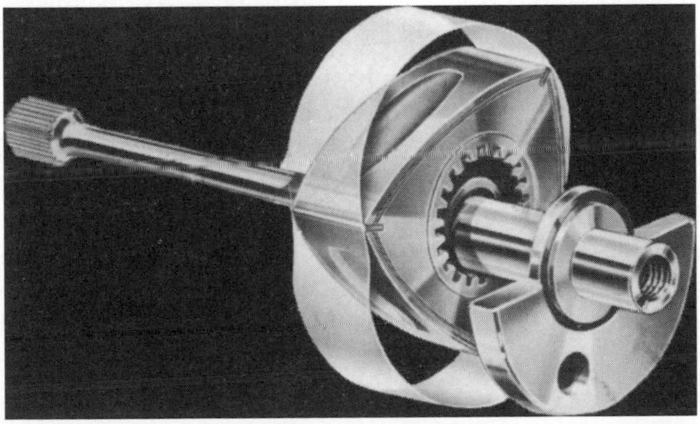

FIGURE 1.33 (a) The rotating combustion engine, showing the three-sided eccentric rotor and internal gear. A major advantage of this engine is its basic simplicity. The rotor and eccentric shaft are the only rotating parts; cams and valves are not required.

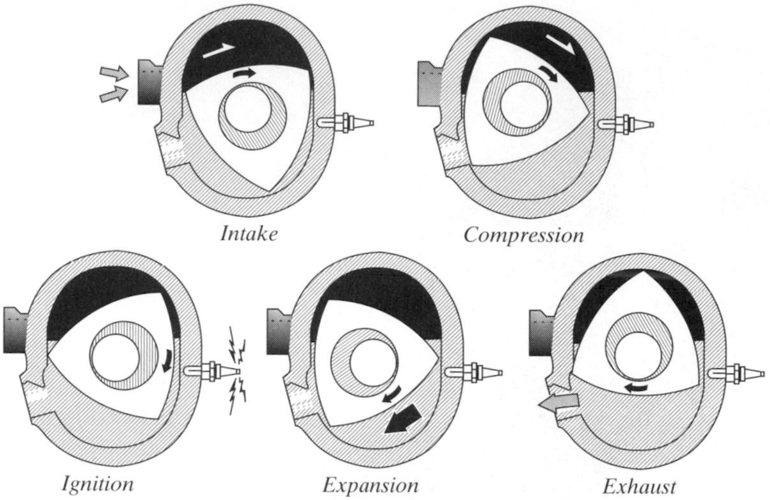

FIGURE 1.33 (b) A complete combustion cycle of the rotating combustion engine. The rotor speed is one-third the eccentric crankshaft speed, maintaining one power impulse for each crankshaft revolution. (*Source:* Curtiss-Wright Corporation.)

In the preceding discussion, we traced only one charge of air and fuel through a complete cycle. The three-sided eccentric rotor and two-lobed engine block, however, correspond to *three* sets of pistons and cylinders. At the time of ignition of the first charge of air and fuel, the intake process is occurring in another chamber. When the first chamber is in the exhaust position, a third chamber is in the intake position. The figure shows the combustion cycle for only one chamber, but at any time, a different phase of the process is occurring in each of the other chambers.

The fixed sun gear and the larger ring gear are shown as circles in Figure 1.33b. Rotation of the crankshaft and eccentric rotor carrier is seen by observing the point of contact between the fixed sun gear and the planetlike internal gear. In observing one thermodynamic cycle of this engine, represented by one rotation of the rotor, we see that the crankshaft (represented by the eccentric carrying the rotor) is given more than one rotation. Actually, the crankshaft is given *three* rotations, where the ratio of internal gear teeth on the rotor to teeth on the fixed gear meshing with it is 1.5 to 1. This result may be determined (with difficulty) by making successive sketches or may be calculated by using principles to be discussed in Chapter 8. The solution is left as an exercise in that chapter.

Fluid Links

Mechanical systems frequently include *fluid links* utilizing hydraulic or pneumatic cylinders or fluid drive transmissions. The backhoe shown in Figure 1.34 uses hydraulic cylinders arranged to give it a wide range of operating positions. Hydraulic feeds are also used for machine tools. By means of a variable delivery pump or a relief valve for

FIGURE 1.34 The motion of this backhoe is determined by several independently controlled mechanisms. (*Source:* Caterpillar.)

control, the operator may regulate speed and thrust precisely. In the case of machine tools, the fluid system may be programmed to go through a complete cycle of operations automatically. For kinematic analysis, a hydraulic cylinder linkage of the type shown in Figure 1.35a is usually represented as shown in Figures 1.35b and c.

Swash Plate

Converting rotational motion to reciprocating rectilinear motion is a common problem, and many mechanisms have been devised for this purpose. In the *swash plate* type of mechanism, shown in Figure 1.36, a camlike swash plate is rotated about an axis that is not perpendicular to its face. The plate drives plungers in a cylinder block. Plunger stroke is equal to $d \tan \phi$, where the several parallel cylinders are arranged in a circle of diameter d, as shown in the figure. The angle ϕ is measured between the swash plate face and a plane perpendicular to the cylinder axes. For 100-percent volumetric efficiency, the volume of liquid pumped per revolution of the swash plate is $Q = ANd \tan \phi$, where A is the cross-sectional area of one cylinder and N is the number of cylinders.

When the swash plate is operated as a hydraulic motor, fluid pressure is applied to the plungers that drive the plate. Each cylinder is alternately connected to the fluid supply and the exhaust by a distribution system operated by the swash plate shaft.

As noted earlier, an *inversion* of a mechanism exhibits the same relative motion as the mechanism, but the links do not have the same absolute motion. The link that is fixed

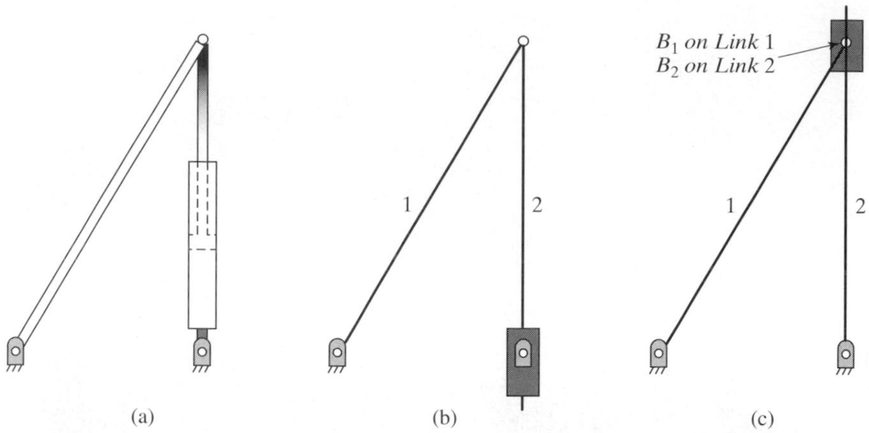

FIGURE 1.35 (a) A linkage that includes a hydraulic cylinder. (b) A kinematic representation of the linkage shown in part a; link 2 slides within a sleeve pinned to the frame. (c) Alternative representation of the linkage shown in part a; link 2 slides within a sleeve pinned to link 1 at point B_1. Point B_2 is taken to be the identical point on link 2 at this instant.

in the original mechanism is not fixed in the inversion. The cylinder block is fixed in the swash plate mechanism of Figure 1.36. If, instead, the cylinder block is rotated and the swash plate fixed, the motion of the plungers relative to the cylinders will not change, and hydraulic fluid will be pumped at the same rate. Figure 1.37 shows an inversion of the basic swash plate mechanism. In this case, the link that acts as swash plate actually rotates, but its rotation is in a plane and is of no significance to the relative motion. This arrangement is kinematically equivalent to the plunger ends riding on a fixed disk.

Volume control is effected by designing the pump so that the angle between the cylinder axes and the plane of the swash plate may be varied. Volume control may be actuated manually, or automatically by a mechanical, electrical, or fluid control device. When the mechanism is used as a motor, a similar control of the offset angle may be

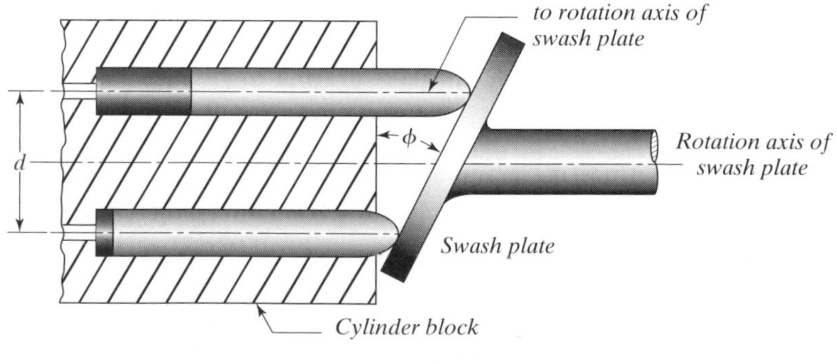

FIGURE 1.36 The swash plate mechanism is but one of a large number of mechanisms designed to convert rotational motion to rectilinear motion.

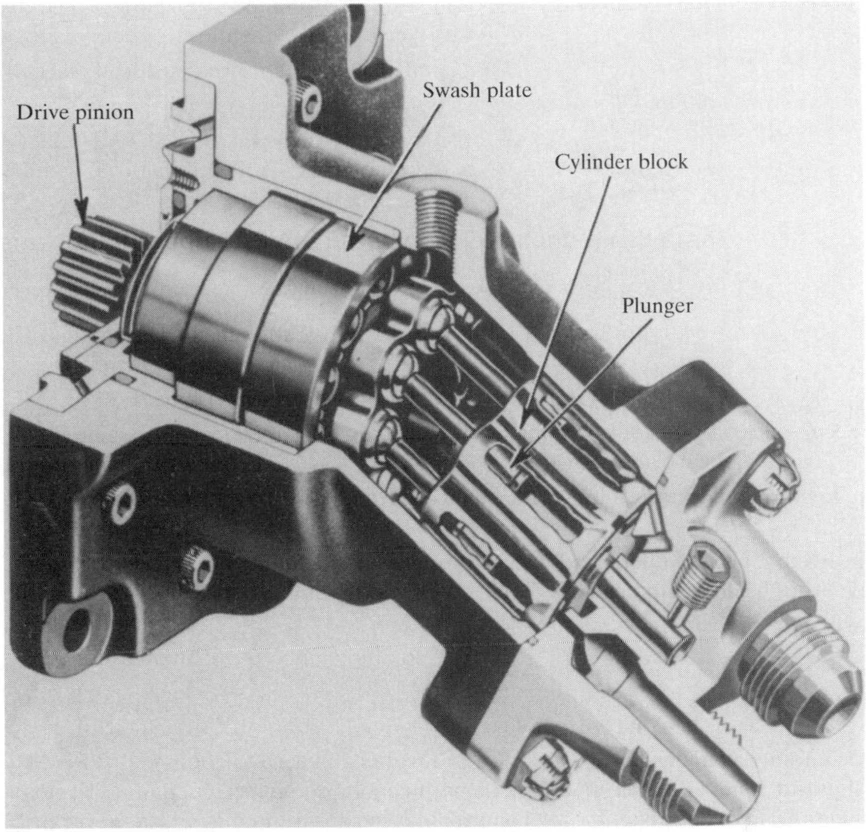

FIGURE 1.37 This *inversion* of a swash plate mechanism has been designed as a *fixed-displacement, piston-type hydraulic pump*. The cylinder block, the driveshaft, and the nine pistons all rotate as a unit. The pump is available with the cylinder block axis offset relative to the driveshaft by 15° to 30°. This offset determines the stroke of the pistons and therefore the flow rate. (*Source:* Sperry Rand Corporation.)

used to change displacement for speed control. An *adjustable-speed transmission* may be assembled from two variable swash plates, one used as a variable-offset *pump* and the other used as a variable-offset *hydraulic motor* (Figure 1.38). Speed is continuously variable over a wide range, with fine-control and high-torque capabilities. The fluid link between the two components allows considerable flexibility in positioning input and output.

Gear Trains

Gear trains are particularly suitable for use at high speeds and in drives with high power ratings. Since gears offer precise speed ratios, they are also used in machine tools and other applications in which precision is required. Differential gears are used to distribute power in automobiles, but may also be used to add or subtract inputs for

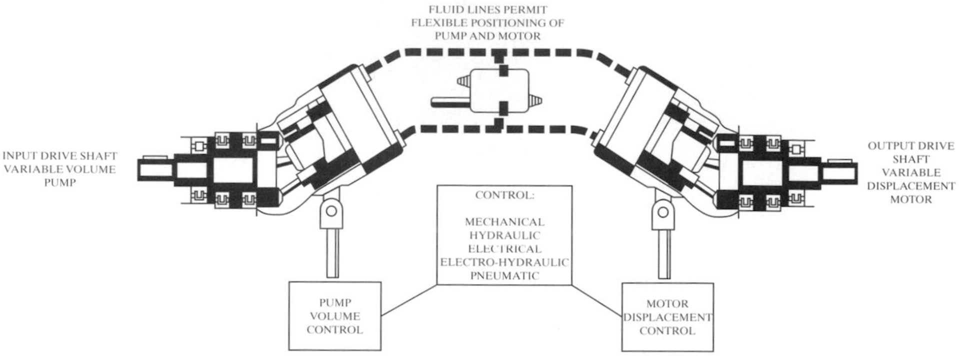

FIGURE 1.38 Two *variable-offset* swash plate mechanisms—one used as a pump and the other used as a motor—are combined to create an adjustable-speed transmission. In the pump, the piston stroke can be varied by changing the angle of offset. Fluid is pumped to the hydraulic motor, operating the pistons that drive the output shaft. (*Source:* Sperry Rand Corporation.)

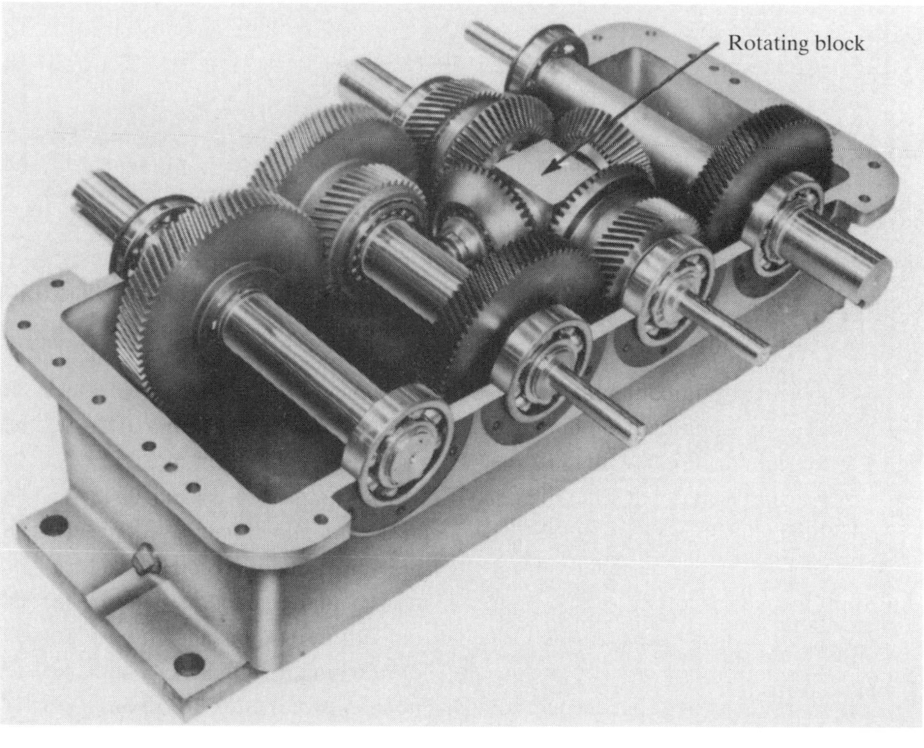

FIGURE 1.39 A differential transmission. (*Source:* Fairchild Industrial Products Division.)

control of certain processes. If two machines are to perform a production-line function in a certain sequence, one machine may drive the other through a differential so that phase adjustment is possible between the operations. Figure 1.39 shows a differential transmission. The differential itself is made up of four bevel gears; the other gears in the transmission are helical gears.

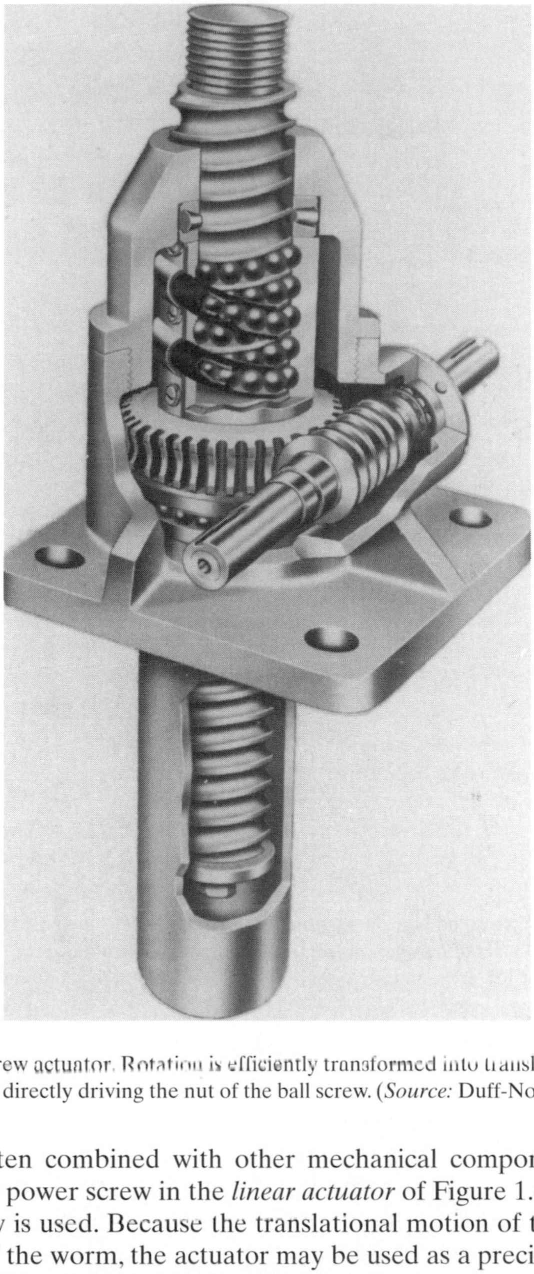

FIGURE 1.40 A ball-screw actuator. Rotation is efficiently transformed into translation through the use of a worm and worm wheel directly driving the nut of the ball screw. (*Source:* Duff-Norton Company.)

Gearing is often combined with other mechanical components. A worm and worm wheel drive a power screw in the *linear actuator* of Figure 1.40. For reduction of friction, a ball screw is used. Because the translational motion of the screw is proportional to rotation of the worm, the actuator may be used as a precision jack or a locating device. Gears will be discussed in detail in later chapters.

Lamination-Type Impulse Drive

Figure 1.41 shows a *lamination-type impulse drive* made up of several linkages. Power is transmitted from an eccentric through an adjustable linkage directly to a one-way clutch on the output shaft. There are several linkage and clutch assemblies, and each assembly operates on its own eccentric. The eccentrics operate in sequence throughout

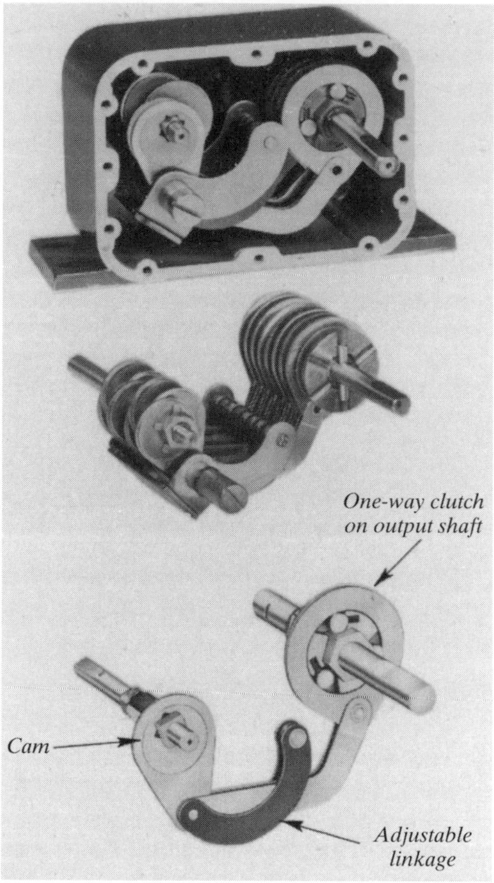

FIGURE 1.41 (a) *Top*: Assembled view of a lamination-type impulse drive with cover plate removed. *Center*: The heart of the unit, which is a set of laminations phased to provide continuous driving. *Bottom*: A single lamination is shown with the important features identified. (*Source:* Zero-Max Ind., Inc., a unit of Barry Wright.)

the entire input cycle to ensure continuous output motion, each linkage driving during its fraction of the input cycle.

Figure 1.41b illustrates the function of the control link. The location of O_3, the control link axis, is adjustable. If it is moved toward O'_3, then link 5 oscillates through a smaller angle for each input rotation. When the control link axis is adjusted to fall on O'_3, the output shaft is stationary.

Reversing input speed direction does not change the output direction. However, output rotation may be reversed if the transmission is equipped with a reversible one-way clutch. When the clutch mechanism is reversed, the magnitude of the output-to-input speed ratio also changes.

The average speed of link 5 of a given linkage assembly during the time that it is driving the output shaft clockwise is not the same as when it drives counterclockwise. If the clutch mechanism and the direction of input rotation are reversed simultaneously

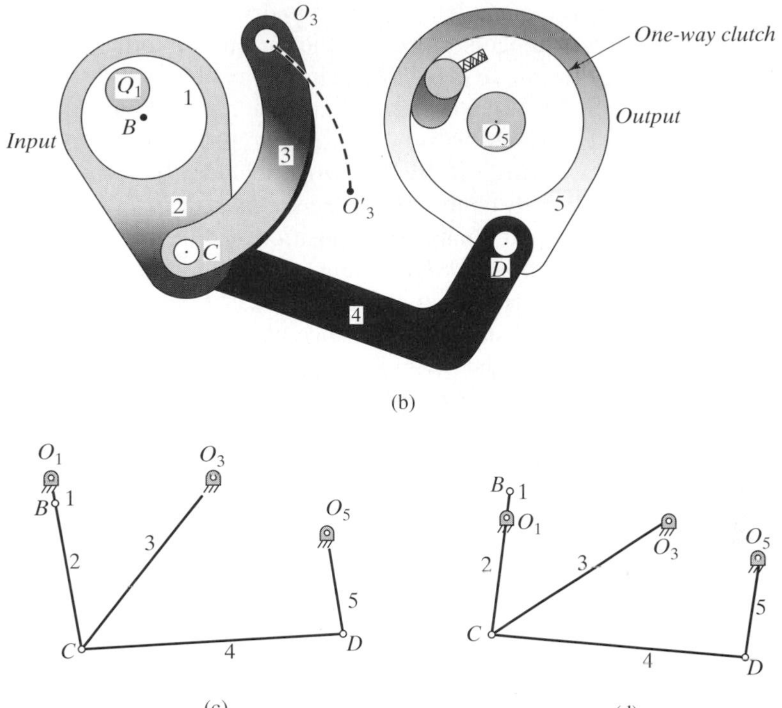

FIGURE 1.41 (b) A single lamination of the lamination type of impulse drive, shown in detail. The location of control axis O_3 governs the output rotation. When the control link axis is adjusted to position O'_3, the output shaft becomes stationary. (c) The equivalent linkage. The linkage is shown in one of its limiting positions. (d) The other limiting position.

the speed ratio does not change. This transmission is designed for input speeds up to 2000 rev/min, and the output-to-input speed ratio may be adjusted from zero to 1/4.

The impulse drive just considered allows for stepless variation of the speed ratio, but pulsations (fluctuations in the output torque or speed) do occur. If the inertial load is relatively high, the one-way clutches permit the load to overrun the driving links, smoothing out pulsations. Linkage flexibility also aids in absorbing transmission pulsations, so that their full effect is not transmitted to the driven machinery.

Figures 1.41c and d show the equivalent mechanism representing the lamination-type of impulse drive in two extreme positions. Links 1, 2, and 3 and the frame (O_1O_3) constitute a four-bar linkage driven by crank 1. Oscillating link 3 also forms part of a second four-bar linkage, along with links 4 and 5 and the frame (O_3O_5).

Oscillating Lawn Sprinkler with Speed Reducer and Variable Stroke Linkage

In order to be competitive in the marketplace, consumer products must be designed for mass-production at low cost. The manufacturer of the lawn sprinkler mechanism shown if Figure 1.42 reduced costs in several ways, including the use of plastic parts, many of which served more than one function.

Water entering the sprinkler is diverted to drive the water turbine shown in Figure 1.42a. A speed reducer is needed because the turbine operates at high speed and does not produce enough torque to drive the sprinkler mechanism. A worm, an integral part of the turbine wheel, drives a worm gear that is an integral part of a second worm. The second worm drives a second worm gear that is directly connected to crank O_1B of a four-bar linkage (see Figure 1.42b). The coupler, link BC, drives oscillating follower crank CO_3, which is directly connected to the sprinkler bar.

A manual adjustment changes the position of link DC, allowing for four sprinkler settings. One position of link DC results in a wide range of motion for the follower crank so that water is distributed over a large area of lawn on both sides of the sprinkler. A second position increases the distance between points C and O_3 (the length of the follower crank), decreasing sprinkler coverage. The other settings limit coverage to only one side of the sprinkler. The designer had to limit the possible positions of adjustment link CD, the link that determines the effective length of oscillating link CO_3. The proportions of linkage O_1BCO_3 must always satisfy the crank-rocker mechanism criteria. In addition, the transmission angle must always lie in an acceptable range.

There are other methods of adjusting output characteristics (speed, stroke length, stroke time ratios, etc.). Some linkages are designed so that mechanical adjustments within the driving linkage itself can be made while the system is operating, often automatically in response to some demand on the system. The *variable-stroke pump* (Figure 1.11) is a mechanism of this type.

Power Screws

There are many ways to convert rotational motion into rectilinear motion. Cams, linkages, rack-and-pinion combinations, and a number of other devices are used. Power screws, one of the most common and precise methods, are frequently employed as machine tool drives in conjunction with gear trains. If a screw with a single thread

(a) (b)

FIGURE 1.42 Oscillating lawn sprinkler linkage with adjustable stoke. (a) Water turbine and speed reducer. (b) Crank rocker mechanism with adjustable output crank length.

engages a nut that is not permitted to rotate, the nut will move relative to the screw a distance equal to the pitch for each screw rotation. (The pitch is the axial distance between adjacent corresponding thread elements.) With a double-thread screw, the nut motion is two pitches, and, in general, the motion of the nut per screw rotation will be the lead (the pitch times the number of threads). When a right-hand screw turns clockwise, the relative motion of the nut is toward the observer; for a left-hand screw turning clockwise, the relative motion of the nut is away from the observer. The nut may be split through an axial plane if it is to be engaged and disengaged from the screw as in a lathe. A split nut also permits adjustment to compensate for wear and eliminate backlash.

Differential Screws

When high-thrust, low-speed linear motion is required, a *differential screw* may be used. Figure 1.43 shows a power screw with leads L_1 for the left half and L_2 for the right half, both right-hand threads. The motion of the slider equals the axial motion of the screw plus the axial motion of the slider with respect to the screw; that is,

$$v = \frac{n}{60}(L_2 - L_1),$$

where v = slider velocity,

n = number of clockwise revolutions per minute of the screw,

L_1 = screw lead at the frame, and

L_2 = screw lead in the slider.

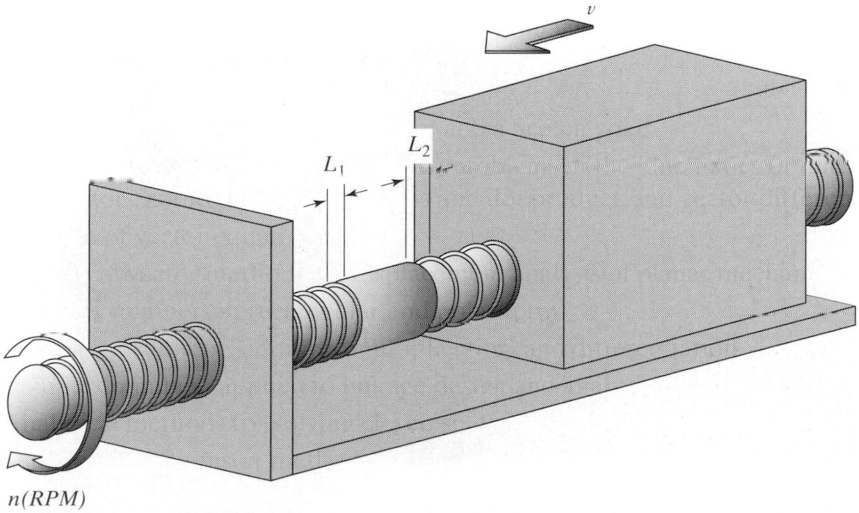

FIGURE 1.43 Differential screw.

For example, a single-thread screw may be cut with 11 threads per inch at the left end and 10 threads per inch at the right end. At 10 rev/min, the slider velocity is

$$v = \frac{10}{60}\left(\frac{1}{10} - \frac{1}{11}\right) = \frac{1}{660} \text{ in/s.}$$

It is more common for power jacks, linear actuators, and other machinery controls to employ a worm drive for low-speed operation. In some cases, the outside of the nut has enveloping worm-wheel teeth cut into it, and the nut is restrained from axial motion by thrust bearings while the screw moves axially. (See Figure 1.44.) Hundreds of jacks of this type were used in a single installation, a linear electron accelerator with a 4-in-diameter by 2-mi-long waveguide that must be kept straight to within 1 mm.

Ball Screws

Ball screws are used when friction must be reduced. The thrust load is carried by balls circulating in helical races, reducing typical friction losses to about 10 percent of the power transmitted. Ball screws must include a ball return to provide a continuous supply of balls between the screw and the nut. Preloading of the nut to eliminate backlash

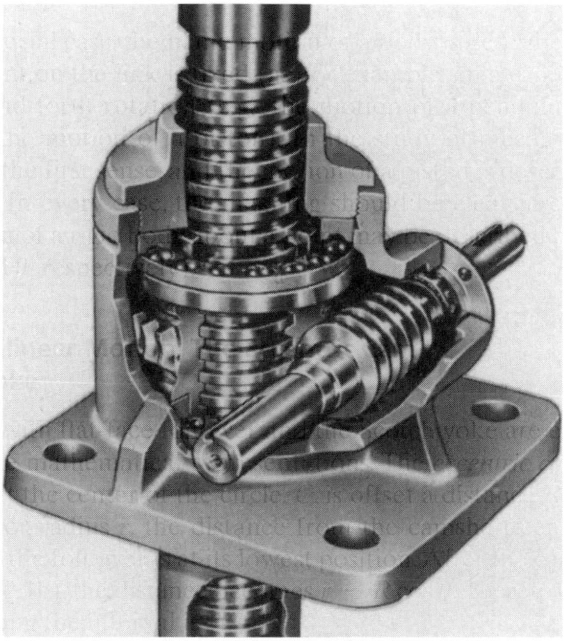

FIGURE 1.44 Machine screw actuator—a worm-gear-driven screw, which may act as an actuator, a precision jack, or a leveling device. Compare this actuator with the ball-screw actuator shown in Figure 1.40. (*Source:* Duff-Norton Company.)

is possible if the ball race in the nut is divided into two sections. A cutaway view of a ball-screw actuator is shown in Figure 1.40.

Special-Use Clutches

Special-use clutches that are self-actuating include centrifugal, torque-limiting, and one-way or overrunning types. Centrifugal clutches are actuated by a mass that locks the clutch parts together at a predetermined speed. Torque-limiting clutches, as the name implies, are released at a predetermined torque. The ball-detent type has a set of steel balls that are held in detents by means of a spring force that determines the limiting torque. Any friction clutch may act as a torque limiter if the contact force is maintained by springs so that slipping occurs at torques above the limiting value.

Sprag-Type Reverse-Locking Clutches

Certain applications require that an input shaft drive the load in either direction, but that the output shaft be prevented from driving the input shaft. This function is performed by the reverse-locking clutch (see Figure 1.45) through specially formed locking members called *sprags*.

Referring to the sectional view in Figure 1.45, assume that the input shaft (which drives the control member) turns counterclockwise. The control member contacts sprag A near the top, pivots it slightly counterclockwise, and thereby frees it from the outer race. The inner race is then driven by sprag A. (Sprag B performs no function during counterclockwise rotation.) Suppose, now, that the output tends to drive counterclockwise with *no* power applied to the input side. Then, the inner race slightly rotates sprag A, forcing it clockwise and jamming it against the fixed outer race, thus locking the system. The identical function is performed by sprag B for clockwise rotation of the clutch.

One-Way Clutches

One-way clutches drive in one direction only, but permit freewheeling if the driven side overspeeds the driver. The clutch operation depends on balls or sprags that roll or slide when relative motion is in one direction, but jam if the direction of relative motion tends to reverse.

Ratchet-and-pawl drives perform a similar function, except that the pawl may engage the ratchet between teeth only. Either a one-way clutch or a ratchet–pawl drive may be used to change oscillation into intermittent one-way rotation. Some machinery-feed mechanisms operate in this manner.

Figure 1.46 illustrates, in principle, the table-feed mechanisms of a mechanical shaper. The lengths of the links are such that link 3 oscillates as link 1; the driver rotates (i.e., the linkage is a crank-rocker mechanism). A spring-held pawl drives the ratchet only during the clockwise motion of link 3 (approximately, but not exactly, half of each cycle). The workpiece table is intermittently fed to the left by the power screw driven by the ratchet. The cutting tool (not shown) moves perpendicular to the direction of motion of the table, but only during the part of the cycle when the table is stationary, to ensure straight cuts.

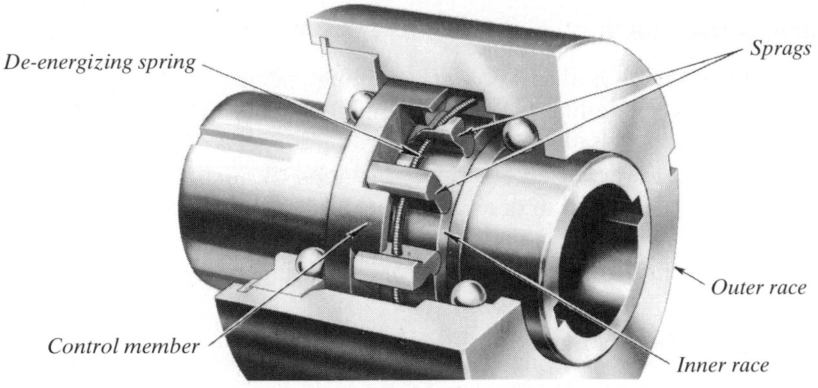

(a)

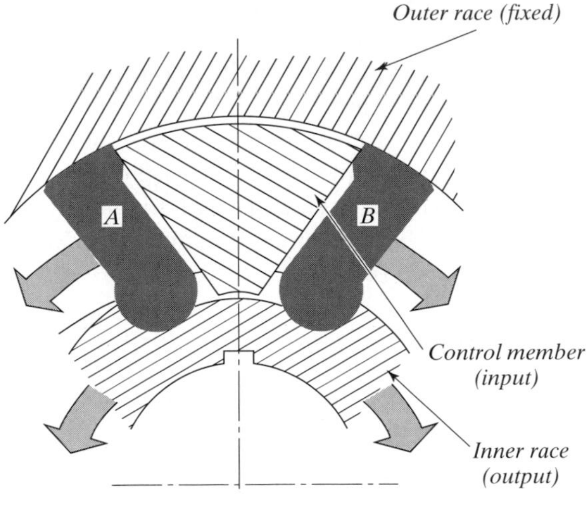

(b)

FIGURE 1.45 (a) Sprag-type reverse-locking clutch. The input (left shaft) drives the load (right shaft) in either direction. When the output shaft tends to drive, the sprags lock it to the outer race. Other sprag configurations are available that permit operation with free-wheeling. (*Source:* Dana Corporation, Formsprag Division.) (b) Sectional view of the sprag-type clutch. As the input begins to rotate counterclockwise, it contacts sprag A. The sprag pivots slightly counterclockwise in its detent, separating from the outer race. The input pushes against the sprag, forcing the inner race (output shaft) to rotate. If the output begins to rotate faster than the input, the sprag is thereby given a slight clockwise motion, jamming the sprag against the fixed outer race and, in turn, locking the output shaft.

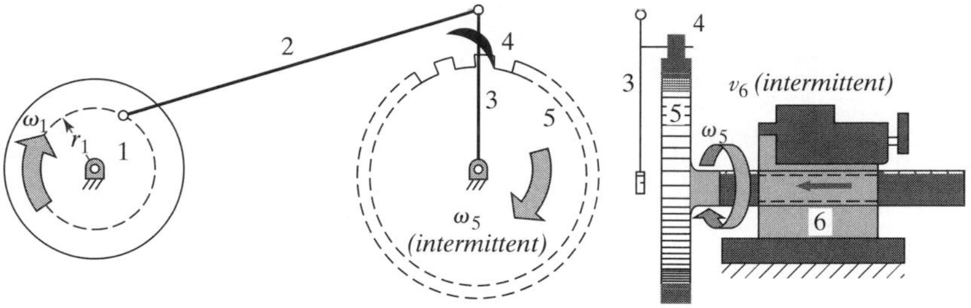

FIGURE 1.46 Ratchet–pawl mechanism (applied to an intermittent drive). As link 1 rotates at constant angular velocity, link 3 oscillates. A pawl (link 4) on link 3 drives the ratchet (5) during the clockwise motion of link 3. The right-hand screw drives the work-table intermittently to the left.

By an increase of the driving crank radius r_1, the angle through which link 3 oscillates is increased. Feed is increased in discrete steps; that is, the rotation of the ratchet per cycle will be an integer multiple of the pitch angle, which is $360°/N$ for N ratchet teeth. Feed is reversed by turning the pawl so that the counterclockwise motion of link 3 rotates the ratchet. Although this action results in a change in instantaneous velocities, the feed per cycle is unchanged.

SAMPLE PROBLEM 1.10

Intermittent Feed Mechanism Design

Design an intermittent feed mechanism to provide rates of feed from 0.010 to 0.024 inch per cycle in increments of 0.002 inch per cycle.

Solution. (There are several solutions to this design problem, each involving many hours of work. We will take the first steps toward a practical design.)

1. A ratchet–pawl mechanism driving a power screw will be selected for our design. The required steps between minimum and maximum feed correspond to ratchet rotations of one pitch angle. For screw lead L inches, the feed per pitch angle is L/N inches. If we use a single-thread power screw with five threads per inch and a 100-tooth ratchet, the required 0.002-in/cycle feed increments are obtained.
2. A linkage with rotating driver crank and oscillating driven crank (similar to that in Figure 1.46) will be used to drive the pawl. The dimensions shown in Figure 1.47 will be provisionally selected, where link 1, the driving crank, is of variable length. Since the feed is 0.200 in per rotation of the screw, the screw must turn through 1/20 rotation (18°) for the 0.010-in/cycle feed. The 0.024-in/cycle feed is obtained by a 43.2° rotation of the screw.
3. For the required range of feeds, the oscillation of link 3 must be at least 43.2° when link 1 is adjusted to maximum length and about 18° when link 1 is adjusted to minimum length. As a trial solution, we might design link 1 so that its length can be adjusted between 0.4 and 1.2 in. The mechanism is shown with link 1 adjusted to 1.2 in, at which setting link 3

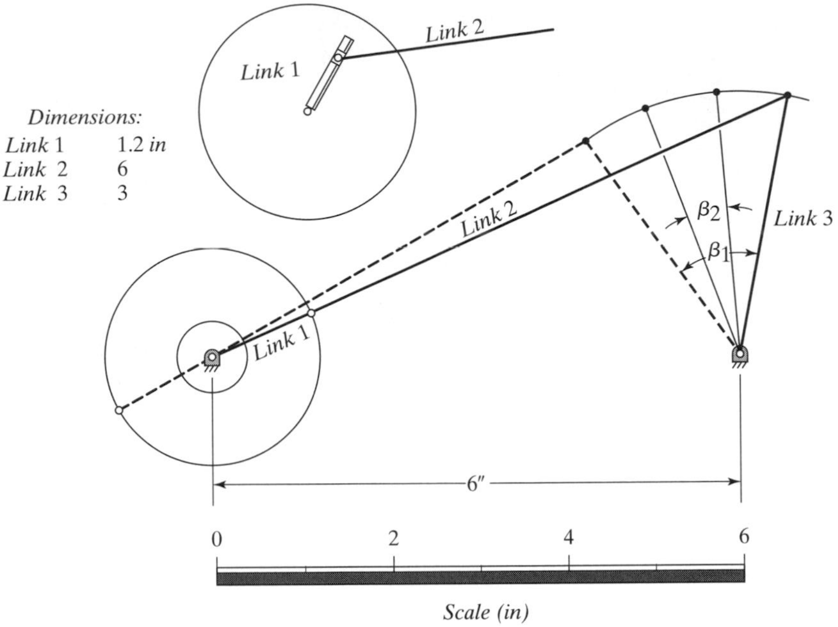

FIGURE 1.47 Sample problem 1.10 for a feed mechanism. Link 3 oscillates through angle β_1 with link 1 adjusted to 1.2 in and through angle β_2 with link adjusted to 0.4 in.

oscillates through angle β_1. Angle β_2, the oscillation corresponding to a 0.4-in length of link 1, is also shown. The trial design has a wider range of feeds than required and is therefore acceptable from that standpoint.

If we were to actually manufacture the mechanism, the next steps in the design process would be to find velocities and accelerations in the mechanism and to specify the members' cross sections. An investigation of tolerances and of stresses and deflections would then be required.

Universal Joints

When the angular relationship between the axes of two drive train elements is variable, the elements may be joined by a flexible coupling, a flexible shaft, or a universal joint. Most flexible couplings are intended only for small amounts of misalignment, and flexible shafts have very limited torque capacity. Where high torque and large misalignments occur, a *universal joint*, shown in Figure 1.48, is the typical solution. The Hooke-type universal joint has a variable output speed ω_2 for misalignment ϕ unequal to zero when the input ω_1 is constant. For the position shown in Figure 1.49, the velocity of point A is $v_A = \omega_1 r$ where ω is in radians per second.

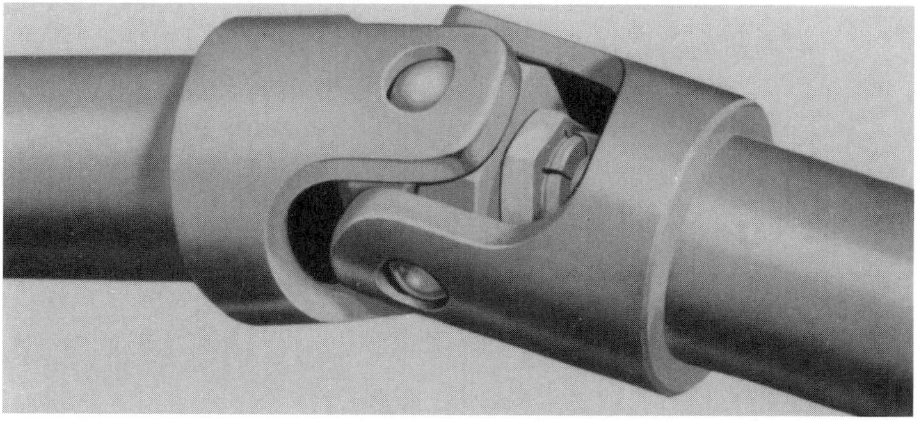

FIGURE 1.48 A preloaded universal joint designed for use in a steering-column-tilting mechanism and similar applications in which backlash is undesirable. The recommended maximum operating angle for this type of universal joint is 18° (*Source:* Bendix Corporation.)

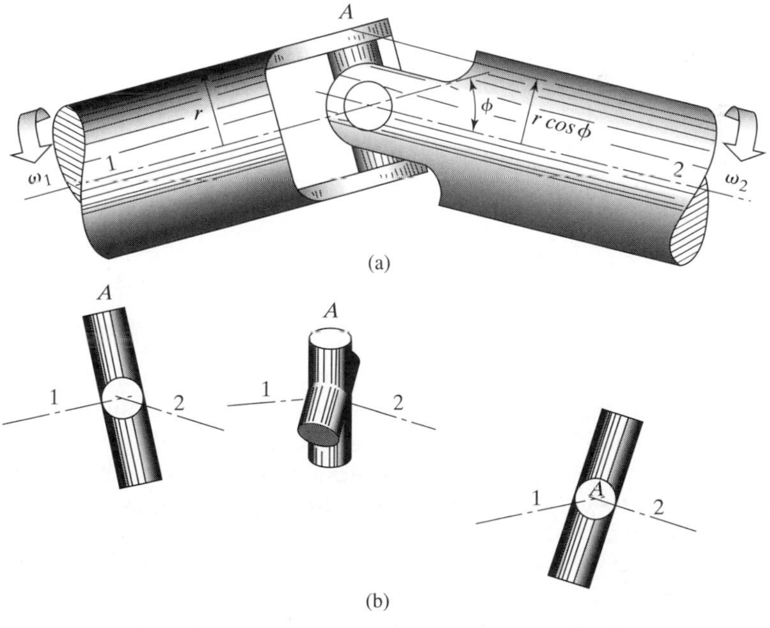

FIGURE 1.49 (a) The Hooke-type universal joint. The misalignment is indicated by the angle ϕ. Velocity ratio ω_2/ω_1 varies instantaneously as the joint rotates. (b) The cross-link of the universal joint is shown as it rotates through 90°.

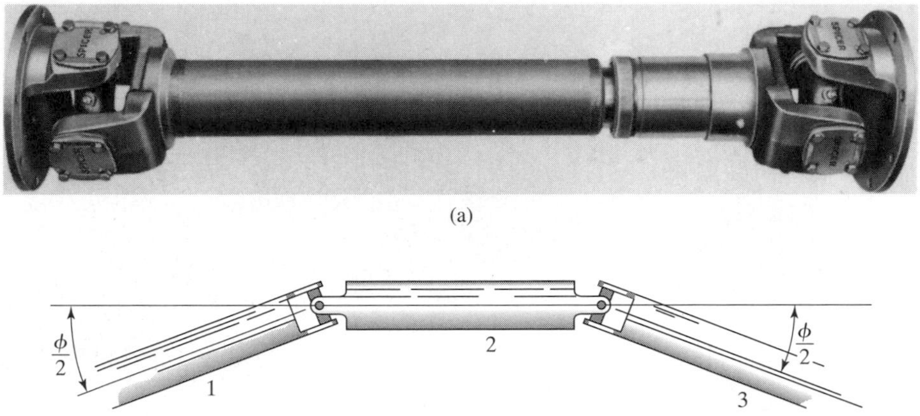

(a)

(b)

FIGURE 1.50 (a) A universal joint for high-torque applications. (*Source:* Dana Corporation.) (b) When two universal joints are used, input and output speeds are equal if each of the universal joints takes half of the misalignment, as shown.

The angular velocity of shaft 2 is maximum at this time and is equal to

$$\omega_2 = \frac{v_A}{r\cos\phi} = \frac{\omega_1}{\cos\phi}.$$

In Figure 1.49b, the cross-link of the universal joint is shown as it rotates through 90°, the last position representing the minimum velocity of shaft 2: $\omega_2 = \omega_1 \cos\phi$.

At high shaft velocities, speed variations may be objectionable, since acceleration and deceleration of the load can cause serious vibration and fatigue. Figure 1.50 shows two Hooke-type universal joints used to join shafts with a total misalignment of ϕ. If the shafts are in the same plane and each joint has a misalignment of $\phi/2$, as in the figure, the input shaft, 1, and the output shaft, 3, travel at the same speed. The intermediate shaft, 2, turns at variable velocity, but if it has a low mass moment of inertia, serious vibration will not result.

An alternative method of avoiding acceleration and deceleration is through the use of a constant-velocity universal joint. A constant-velocity ball joint, seen disassembled in Figure 1.51, is shown in the plane of the misaligned shafts in part b of the figure. Each half of the joint has ball grooves, with pairs of ball grooves intersecting in a plane that bisects the obtuse angle formed by the shafts. Thus, if all ball-groove center radii equal r, the velocity of the center of ball A is given by

$$v_A = \omega_1 r\cos\frac{\phi}{2}$$

and

$$\omega_2 = \frac{v_A}{r\cos\frac{\phi}{2}} = \omega_1.$$

This constant-velocity relationship holds at all times, even as misalignment ϕ changes.

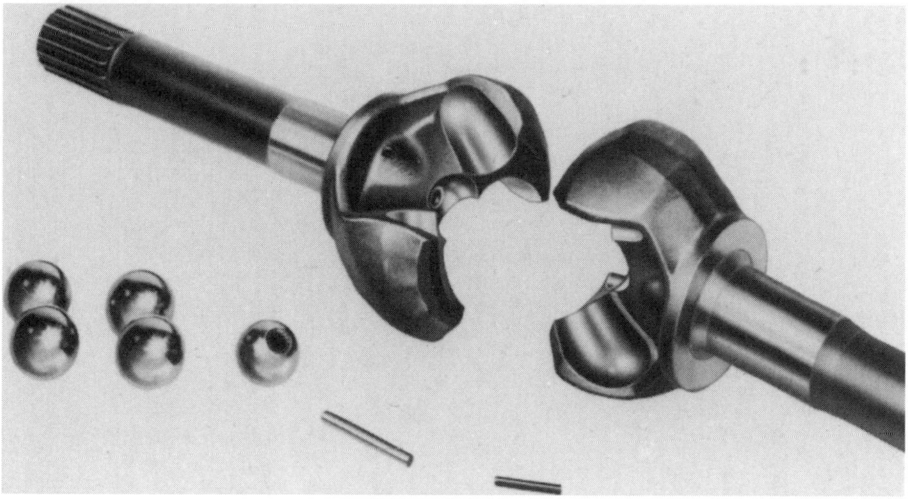

(a)

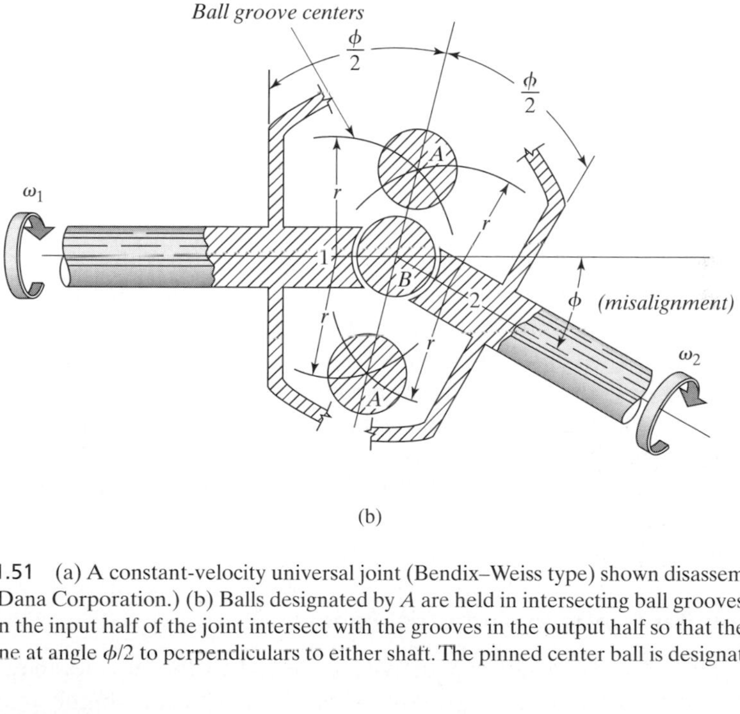

(b)

FIGURE 1.51 (a) A constant-velocity universal joint (Bendix–Weiss type) shown disassembled. (*Source:* Dana Corporation.) (b) Balls designated by A are held in intersecting ball grooves. The grooves in the input half of the joint intersect with the grooves in the output half so that the balls travelin a plane at angle $\phi/2$ to perpendiculars to either shaft. The pinned center ball is designated by B.

Automotive Steering Linkage

The Ackerman-type steering linkage, sketched in Figure 1.52, incorporates a parallelogram linkage made up of the Pitman arm O_1B, the relay rod BE, the idler arm EO_2, and the frame. Tie rods CF and DG are connected to the relay rod and to the steering arms FO_3 and GO_4. The steering arms turn the front wheels about pivoted knuckles O_3 and O_4 when the Pitman arm is rotated by a gear at O_1.

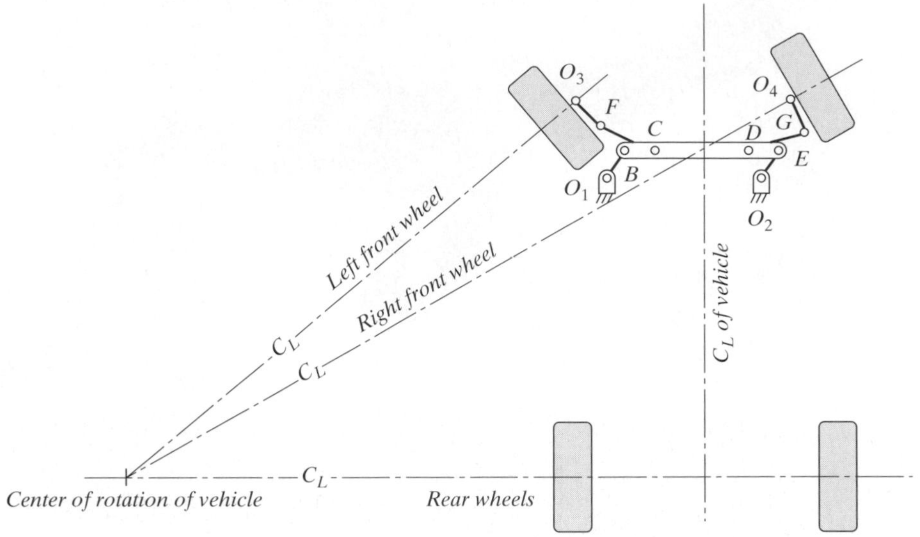

FIGURE 1.52 Automotive steering linkage oriented for a turn. Link nomenclature: O_1B is the Pitman arm (driven by a gear at O_1); BE is the relay rod; EO_2 is the idler arm; CF and DG are tie rods; and FO_3 and GO_4 are steering arms.

To avoid unnecessary tire wear when the vehicle turns, the centerlines of the four wheels should meet as closely as possible at a single point—the center of rotation of the vehicle.

Thus, the Ackerman system is designed so that the wheels do not turn equal amounts. The wheel on the inside of the turn must be rotated through a greater angle about its steering knuckle than the wheel at the outside of the turn in order that the condition on the center of rotation be met.

This problem of the wheels not turning equal amounts accounts, in part, for the complicated linkage design. Another problem is that the steering linkage is not strictly a planar linkage, in that the wheels must follow road contours. Ball studs (ball-and-socket joints) are used at points C, D, F, and G to permit multiaxis rotation. In an alternative design, the steering linkage may be mounted forward of the centerline of the front wheels.

Computer-Controlled Industrial Robots

Demands for increased productivity have led to the development of computer-controlled robots. Figure 1.4 shows an industrial robot with a highly maneuverable six-axis jointed arm. The robot is controlled by a flexible minicomputer program and may be interfaced to peripheral equipment or to an external computer. Although robots of this type have a fixed base, they may be employed in manufacturing operations involving a continuously moving production line. In one such application, the robot arm tracks moving automobile bodies and automatically spot welds them without stopping the production line. The arm works on the front, middle, and rear of the auto body as the body moves through the robot's baseline station. Abort and utility sequences are included in the robot's computer control. The abort sequence directs the robot to exit from the moving part along a pretaught safe path relative to the part. The utility sequence is initiated by an external signal from malfunctioning peripheral equipment so that the robot can take corrective action. For example, the tip of a welding gun may stick to the part, requiring a twisting motion to break it free.

Figure 1.53 shows an application of robotics to drilling and perimeter routing of aircraft panels. A combination of positive-location part fixtures and compliant tooling systems was used to overcome the problem of positional inaccuracy due to joint tolerances and elastic deflection of the multi-degree-of-freedom robot manipulator.

FIGURE 1.53 An application of robotics. (*Source*: General Dynamics.)

1.12 COMPUTER-AIDED LINKAGE DESIGN

The design of linkages for specific applications has always relied heavily on human judgment and ingenuity. This design process may be illustrated by the flowchart given by Sheth and Uicker (1972). (See Figure 1.54.) "Human interaction" includes creativity and possibly lengthy periods of mathematical analysis and computation. While it is unlikely that human creativity can ever be completely replaced, computer-aided design (CAD) can be employed to relieve the designer of many of the routine processes that would otherwise be necessary. Since three-dimensional information can be stored and retrieved in various views by using CAD programs, the construction of tentative physical models can often be eliminated in the design-and-development stage.

Currently, many organizations are integrating engineering design and drafting processes with manufacturing, administration, and other functions. This approach, called *concurrent engineering*, relies heavily on computers and is intended to reduce the interval between the formulation of a design concept and the appearance of the final product.

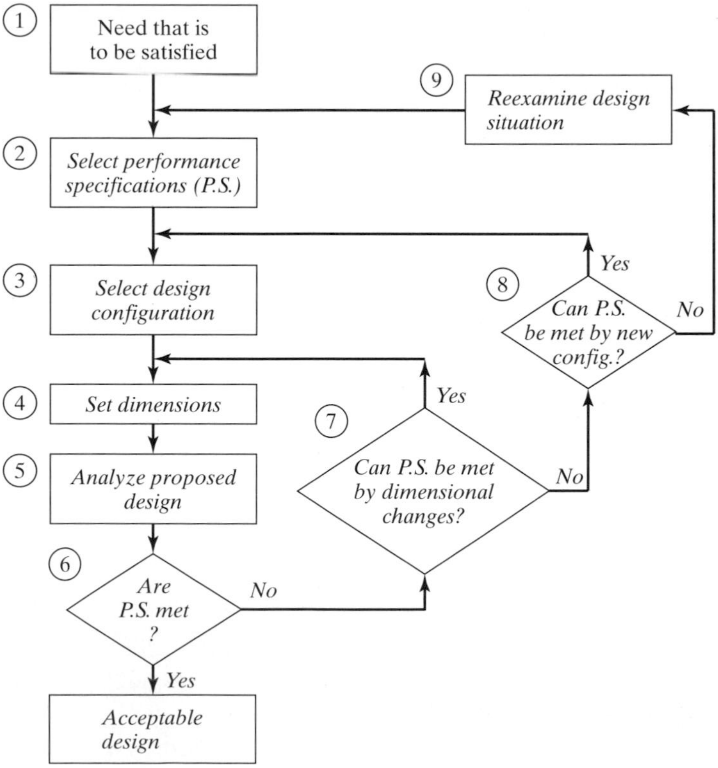

FIGURE 1.54 A portion of the design process.

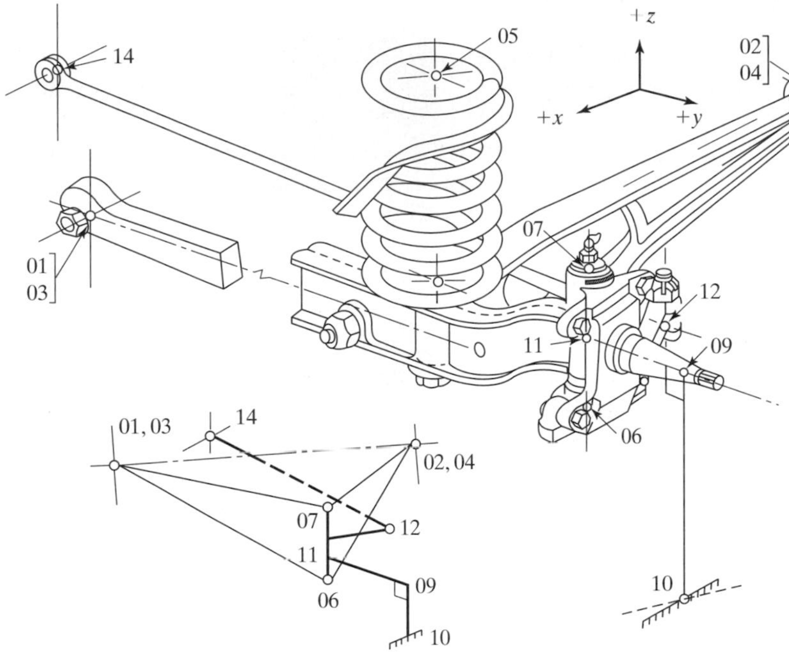

FIGURE 1.55 Suspension system. (*Source:* Chevrolet Motor Division, General Motors Corporation.)

The suspension system shown in Figure 1.55 is an example of the type of problem that may be treated by CAD methods. Most CAD systems can handle open-loop systems, such as robots, as well as closed-loop systems, like four-bar and slider-crank linkages. Features of interest available in one or more CAD programs include kinematic analysis and synthesis of planar and spatial linkages, static and dynamic analysis, and an interactive graphics capability. Additional features of some software include a check for redundant constraints, the capability to handle nonlinear equations, and a zero- and multi-degree-of-freedom capability, as well as a capability to model one-degree-of-freedom systems.

Systems in which the motion is completely specified as a function of time are called *kinematically determinate* systems for purposes of analysis. Examples of kinematically determinate models are a robot with all joint motions specified and a slider-crank mechanism with the position of the crank specified.

In modeling a mechanism design, links are usually considered to be rigid. In some systems, however, link flexibility influences performance. In satellite design, for example, stringent weight requirements may result in a system that undergoes significant structural distortion. Figure 1.56 shows a satellite deploying a flexible antenna extended with a screw jack. In this system, feedback control is employed on the momentum wheels for attitude control. Failure to consider the interaction between the

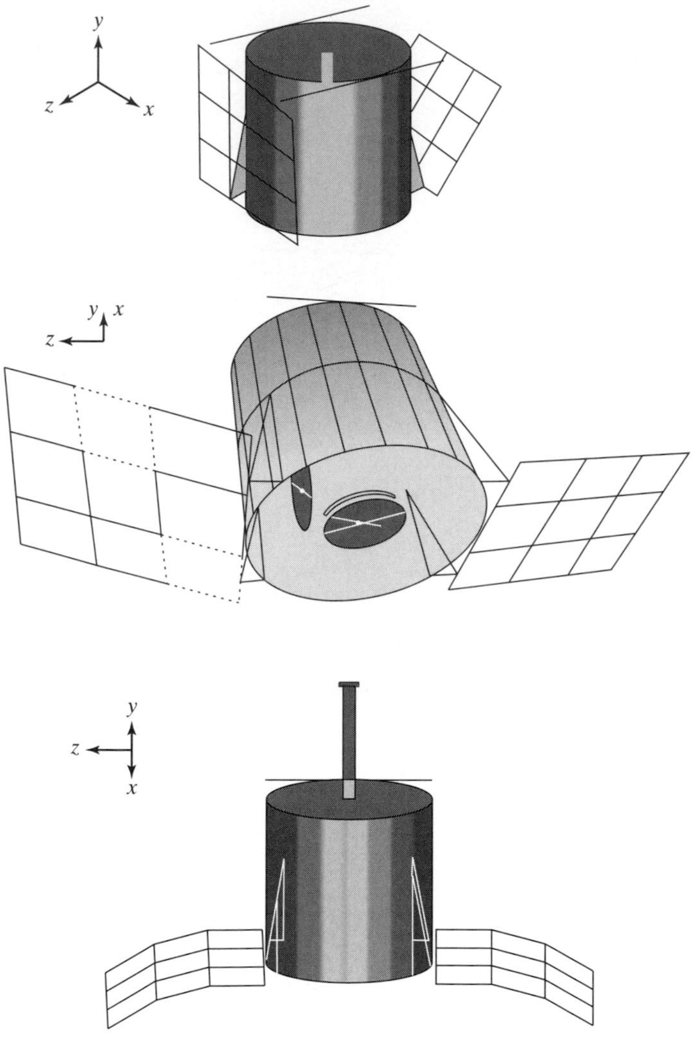

FIGURE 1.56 A satellite system deploying a flexible antenna (modeled by ADAMS™ software). (*Source:* Mechanical Dynamics, Inc.)

flexible antenna and the feedback control system could have resulted in a satellite system that was unstable to the point where it would self-destruct.

Figure 1.57 shows a flowchart for processing kinematics and dynamics information in the IMP (integrated mechanisms program), one of the software packages used for analyzing motion and forces in mechanical systems. Figure 1.58 shows a simulation of a spring-reset plow that was designed to relieve shock loading when an embedded rock is struck. The operation was analyzed and refined with the DRAM program. Another CAD example, a front-end loader, is shown in Figure 1.59.

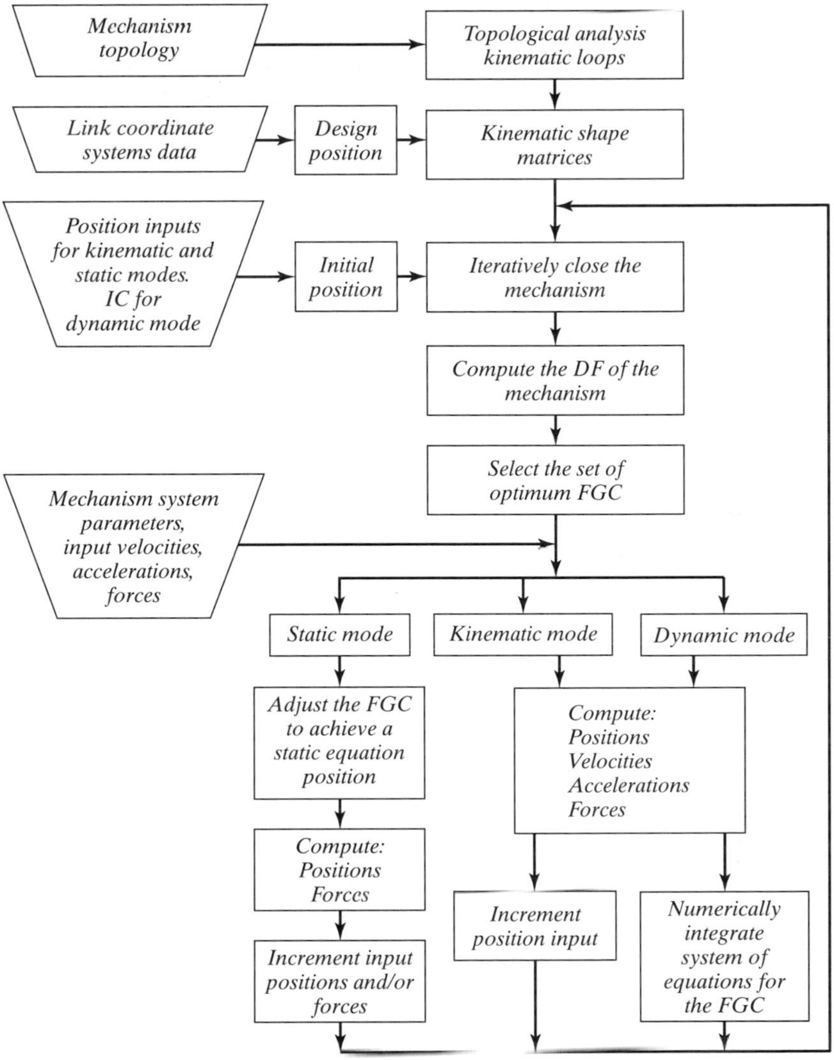

FIGURE 1.57 Flowchart for IMP processing; IC = initial conditions; DF = degrees-of-freedom; FGC = free generalized coordinates. (*Source:* Structural Dynamics Research Corporation.)

Research in Engineering Design Theory and Methodology

Like most engineering design, the design of kinematic systems is a blend of art and science. Some investigators are studying design theory and methodology, attempting to obtain a deeper and more fundamental understanding of the design process. Finger and Dixon (1989) reviewed research in descriptive, prescriptive, and computer-based models of design processes. They summarized studies of how humans create mechanical designs, processes, strategies, and problem-solving methods and reviewed (1) computer-based

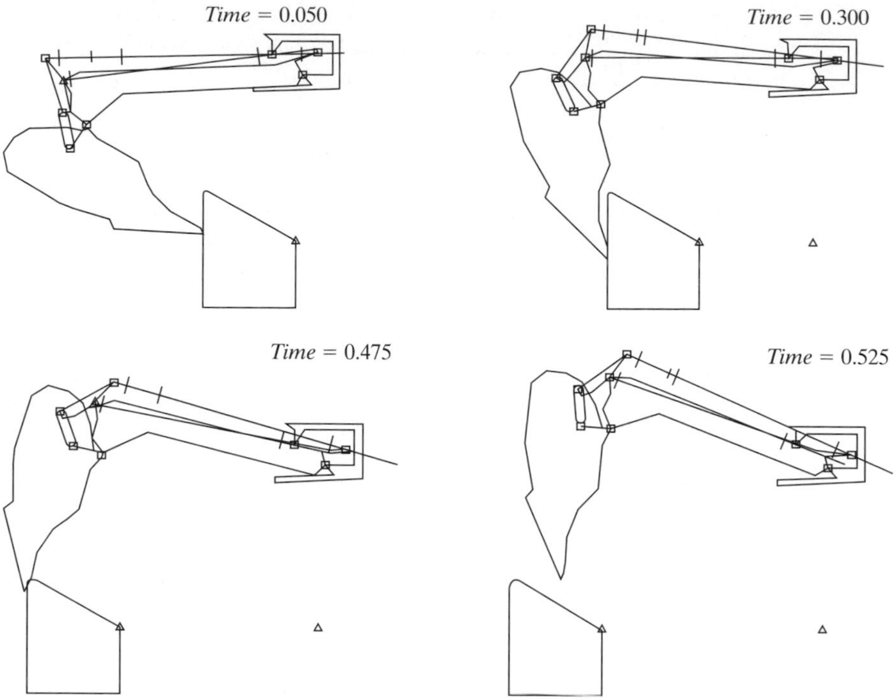

FIGURE 1.58 Analysis of a spring-reset plow, designed with the help of the DRAM program. (*Source:* Mechanical Dynamics, Inc.)

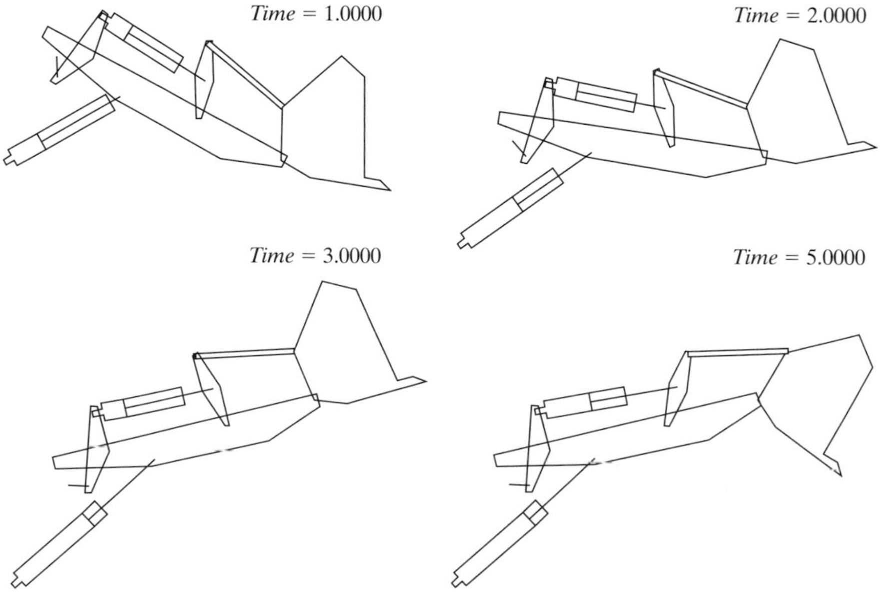

FIGURE 1.59 Computer-aided design applied to construction machinery. (*Source:* Mechanical Dynamics, Inc.)

models that emulate human problem solving, (2) morphological analysis, a methodology to generate and select alternatives, and (3) configuration design, in which a physical concept is transformed into a configuration with a set of attributes.

1.13 COMPUTER-IMPLEMENTED NUMERICAL METHODS

Some engineering problems do not have a simple closed-form solution. Or a closed-form solution may exist for a particular problem, but is not immediately obvious. We are then likely to try a numerical method of solution. Numerical methods differ from trial and error in that each successive approximation in a numerical method is guided by the previous result. Many numerical methods in current use were developed long before computers became available. With computers, however, we may make many iterations to obtain a high degree of accuracy while avoiding hours of tedious calculation. You may choose to skip this section if you use software capable of numerical solutions.

The Newton–Raphson Method

Newton's method, also called the Newton–Raphson method, can be introduced by considering the root of a nonlinear equation in a single variable:

$$F(x) = 0.$$

Figure 1.60 shows a curve that could resemble a plot of $F(x)$ vs. x. [We do not actually plot $F(x)$ vs. x]. Let the first approximation of the root be $x = X$. Unless we were lucky enough to make a perfect guess, $F(X) \neq 0$. It can be seen from the figure that the first approximation can be improved by considering the error $F(X)$ and the slope of the curve $G(X)$. The second approximation of the root is given by

$$X_{\text{new}} = X - F(X)/G(X), \tag{1.16}$$

where $G(X) = dF/dx$ evaluated at $x = X$. The next approximation can be obtained by using the Eq. (1.16), except that the value of X is replaced by the value of X_{new} obtained in the previous step. The process is repeated until $F(X_{\text{new}}) = 0 \pm$ a predetermined tolerance.

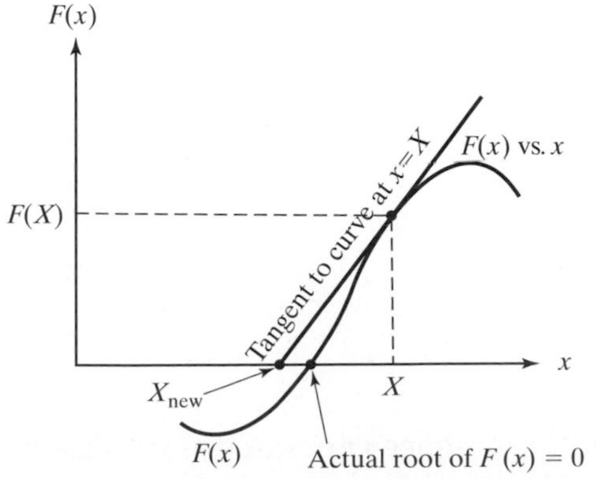

FIGURE 1.60 The Newton–Raphson method.

Success in finding a root depends on the behavior of $F(x)$ in the region of interest. It can be seen that the Newton–Raphson method fails if $G(X) = 0$ at any step. If $F(x) = 0$ has multiple roots, a poor first approximation may result in convergence at a root other than the desired one. Figure 1.61 shows a flowchart outlining the

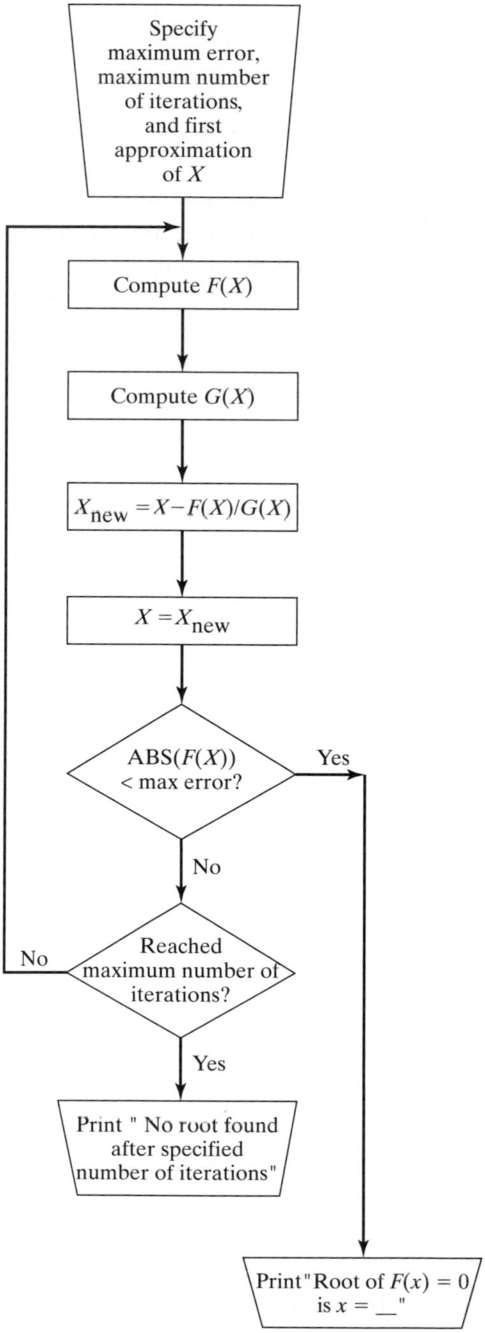

FIGURE 1.61 Flowchart of Newton–Raphson method.

Newton–Raphson method. When a computer is used for iterative processes, a limit may be placed on the number of iterations.

SAMPLE PROBLEM 1.11

An Iterative Solution

Consider an offset slider-crank linkage for which the ratio of connecting rod to crank length is to be 3. Find the eccentricity for which the ratio of stroke to crank length will be 2.7.

Solution. If we superpose sketches of the linkage in its two limiting positions, the resulting triangle can be measured or solved by the law of cosines, leading directly to the answer. Instead, to illustrate a numerical method, it will be assumed that we fail to notice a closed-form solution and resort to the Newton–Raphson method. The problem is described by

$$S = [(L + R)^2 - E^2]^{1/2} - [(L - R)^2 - E^2]^{1/2}.$$

Dividing by R, inserting the given values, and rearranging terms yields

$$F(x) = [16 - x^2]^{1/2} - [4 - x^2]^{1/2} - 2.7 = 0,$$

where $x = E/R$. Differentiating, we obtain

$$G(x) = dF(x)/dx = -x/(16 - x^2)^{1/2} + x/(4 - x^2)^{1/2}.$$

A computer program was written, based on the flowchar of the Newton–Raphson method. Using a tolerance of 10^{-6} and an initial approximation of $x = 1.9$, the program produced the following successive values: $x = 1.821906, 1.800705, 1.799787$, and 1.799786, the final value being the root of $f(x) = 0(\pm 10^{-6})$. The root is then checked by substituting it into the initial equation.

In using numerical methods, the physical problem should be considered before making a first approximation of a root. In the preceding problem, an initial guess of $x > 2$ will result in one term that is the square root of a negative number. An initial guess of $x = 2$ produces an infinite value of $G(x)$ and a message that "no root is found after the specified maximum number of iterations." In this problem, 20 iterations were allowed before the program "gave up" trying to find a root. On observing a sketch of the linkage, one readily sees that values of $x \geq 2$ are not valid.

Other Numerical Methods

Numerical methods are commonly used for solving complicated nonlinear equations. One disadvantage of the Newton–Raphson method is the necessity of obtaining the derivative $dF(x)/dx$.

The **secant method** uses a difference quotient instead of a derivative. However, we are required to supply *two* initial approximations of the root. We begin by making two approximations, X_1 and X_2, for the root of $F(x) = 0$. Then, a new (hopefully improved) approximation of the root is given by

$$X_{\text{new}} = X_2 - F(X_2)/G(X_2),$$

where the difference quotient is given by

$$G(X_2) = [F(X_2) - F(X_1)]/[X_2 - X_1].$$

The procedure continues with X_1 assuming the old value of X_2 and X_2 assuming the value of X_{new}. The iteration continues until the root X_{new} is found, satisfying the equation $F(x) = 0$ within an acceptable tolerance. Otherwise the calculation is stopped after, say, 20 or 30 iterations, with the observation that the process does not converge.

The secant method is available in some mathematical software packages. One such package, Mathcad™, requires the user to input only one estimate of the root after $F(x)$ is defined. The second estimate is taken to be the tolerance if the first estimate is zero. Otherwise the second estimate is given by $X_2 = X_1 + \text{tolerance}$.

Other sections of this text include kinematics problems involving more than one variable. Some of these are sample problems and are solved by numerical methods, with the aid of mathematical software packages.

1.14 MECHANISM DESIGN CONSIDERATIONS

The material in this chapter is intended to form a basis for the analytical work to follow and to introduce some of the analytical tools and approaches used in designing mechanisms. In addition, the more common terminology is brought to the reader's attention to form a common ground for communication.

The design and manufacture of a product by one person working alone is seldom possible and rarely practical. Consider, for example, the complexity of the fully automated machine tool or the case of a relatively simple mechanism to be mass-produced. In either case, many people are involved, due to the interaction of one linkage with the machine as a whole and the relationship between design and manufacture. The designer must transmit ideas to others through mathematical equations and graphical representations, as well as through clear written and oral descriptions.

Past and even present practice relies heavily on ingenious designers taking advantage of their own inventiveness and years of practical experience. But the trend is toward more kinematic analysis and synthesis, including computer-aided optimization. One automobile manufacturer investigated about 8000 linkage combinations in a computer-aided study of four-bar window regulator mechanisms. From those satisfying all of the design requirements (fewer than 500 did), the computer proceeded to select the one "best" linkage, based on a set of predetermined criteria. Computer-aided design cannot replace inventiveness and human judgment, but it can extend the capabilities of an engineer and reduce the tedium of repetitive tasks.

Practical considerations often make it necessary to "freeze" a product design at some stage, thus preventing significant changes. However, a designer should investigate many possible design configurations early in the design process. Suppose, for example, a quick-return mechanism is required for a particular application. Earlier, we noted that offset slider-crank linkages, drag-link–slider-crank combinations, and sliding-contact–slider-crank combinations may be utilized as quick-return mechanisms. There are many other possibilities. For example, we could examine cam-controlled and numerically controlled mechanisms and other combinations that include two or more sliders, in both planar and spatial configurations.

SAMPLE PROBLEM 1.12

Linkage Design

Design a quick-return mechanism with maximum stroke of 170 mm, 40 working strokes per minute, and a forward-to-return ratio of four to one. Use a configuration not previously illustrated in this chapter.

Solution. Many possible configurations can be used as quick-return mechanisms. A one-degree-of-freedom linkage is desired. Applying Grübler's criterion to mechanisms made up of lower pairs, we obtain

$$2n'_J - 3n_L + 4 = 0$$

or

$$n'_J = 3n_L/2 - 2.$$

The following combinations produce an integer number of lower pairs:

n_L	n'_J	n_L	n'_J
2	1	8	10
4	4	10	13
6	7	12	16

One possible combination that includes four revolute pairs, three sliding pairs, and six links is shown in Figure 1.62. The designer may investigate various link lengths and locations for fixed pivot O_1 relative to the slider paths in order to produce a given stroke length and

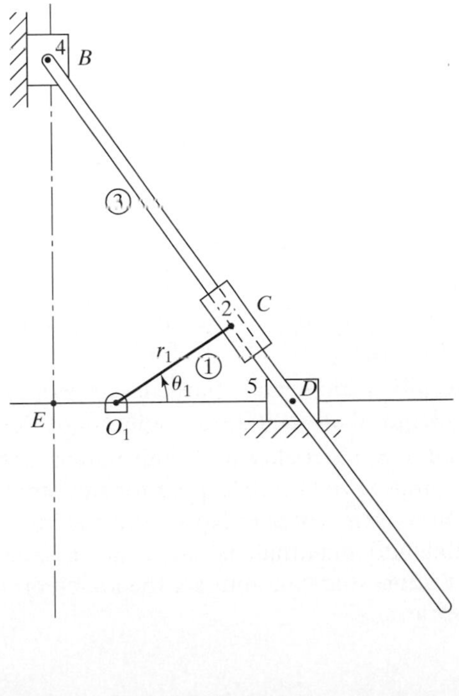

FIGURE 1.62 Three-slider quick-return mechanism.

forward-to-return ratio. Slider link 4 or 5 may be used as a toolholder. For some link-length ratios, one slider will undergo two oscillations per rotation of link 1, while the other slider undergoes one oscillation.

If the intended application of the quick-return mechanism calls for a variable stroke or a variable forward-to-return ratio, the design must include a means to adjust effective link lengths. A screw adjustment of distance O_1C is one possibility. Movement of point O_1 perpendicular to the plane of motion of link 3 provides an alternative means of adjustment. This second option would require a redesign, possibly including spherical pairs. A clever design would allow motion characteristics to be adjusted while the mechanism was operating. Designs of this type require careful placement of links and bearings to avoid interference between moving parts. Design software such as I-DEAS™ with solid modeling and assembly modeling capabilities may be used to aid the task of interference checking.

SUMMARY

The number of degrees of freedom of a mechanism depends on the types of joints or pairs and the number and arrangement of the links. Most practical closed-loop kinematic chains have one degree of freedom, while robots usually have six or more degrees of freedom. The Grashof criterion tells us the theoretical motion characteristics of a four-bar linkage. The actual motion of the linkage may depend on the transmission angle as well. Practical design and analysis of mechanisms is heavily dependent on computers. Animation software permits us to model a proposed design and "put it through its paces." We can then make changes in an attempt to optimize the design.

A Few Review Items

- Identify several planar one-degree-of-freedom linkages.
- How many degrees of freedom does a cylinder pair have? Does the answer depend on whether we are considering a planar or a spatial linkage?
- What is the length of the stroke of a piston engine (in terms of the crank length)?
- Identify the Grashof linkages.
- Write the inequality applicable to a crank-rocker mechanism with a 100-mm drive crank.
- Identify a generally acceptable range for a transmission angle. Explain the possible consequences if the transmission angle falls outside of this range.
- You are asked to design a machine with a "powerful" (large force) working stroke and a quick return. Identify two linkages that could be considered for this design.
- Construct a computer animation of a crank-rocker mechanism. Rotate the drive crank so that the rocker is in one limiting position. Repeat for the other limiting position. What is the range of motion of the rocker? Now show the linkage when the transmission angle is minimal and maximal. Is the transmission angle a potential problem? If so, identify factors that can mitigate the problem.

PROBLEMS

Figures that accompany the problems are indicated with the prefix P. *Otherwise, references apply to figures in the text proper.*

1.1 (a) Find the number of degrees of freedom for the spatial linkage of Figure P1.1 . This open-loop kinematic chain includes cylindrical, prismatic, and spherical pairs.

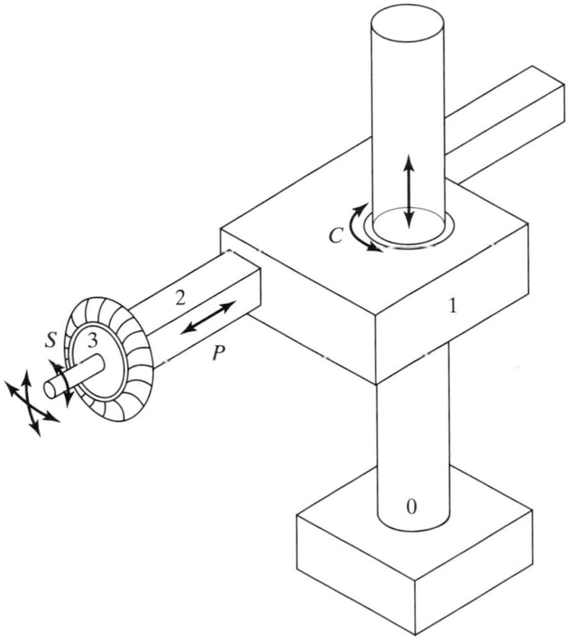

FIGURE P1.1

(b) Figure P1.2 is a schematic representation of a piece of construction machinery. It has two hydraulic cylinders (links 1 and 2, and links 6 and 7), which may be considered

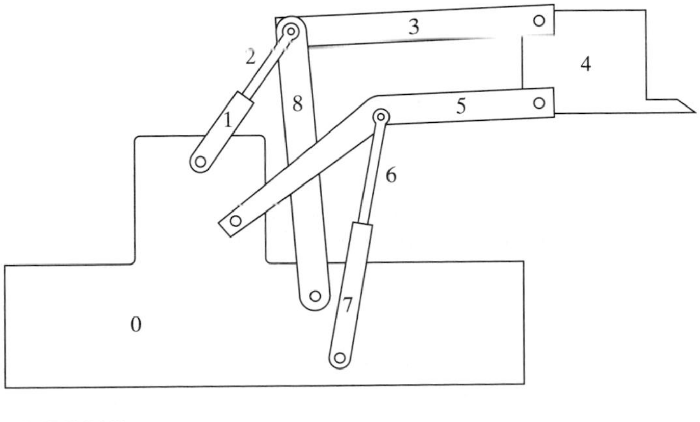

FIGURE P1.2

Problems 1.42 through 1.47 Refer to a Slider-Crank Mechanism

The transmission angle is limited to the range between 45° and 135° (i.e., the angle between the connecting rod and the slider path must fall between −45° and +45°).

1.42 The crank length is 100 mm; find the minimum connecting rod length for zero offset.

1.43 The crank length is 500 mm, and the offset is 100 mm; find the minimum connecting rod length.

1.44 The crank length is 400 mm, and the offset is 200 mm; find the minimum connecting rod length.

1.45 The crank length is 300 mm, and the offset is 50 mm; find the minimum connecting rod length.

1.46 The ratio of the crank length to the connecting rod length is $R/L = 0.5$; find the maximum possible offset.

1.47 Repeat Problem 1.46 for $R/L = 0.25$.

1.48 Sketch the flowchart for a program that designs a crank-rocker mechanism with maximum follower crank range if the transmission angle is limited.

1.49 Write and test a calculator or computer program as designed in Problem 1.48.

1.50 Prove that extreme values of the transmission angle occur when the crank is collinear with the fixed link in a four-bar linkage.

1.51 Prove that the extreme values of the transmission angle occur when the crank is perpendicular to the slider path in a slider-crank linkage.

1.52 In Figures 1.19 and 1.20, let $L_0 = 18$, $L_1 = 7$, $L_2 = 9$, and $L_3 = 17$.

- **(a)** Classify the linkage according to its theoretical motion.
- **(b)** Find extreme values of the transmission angle.
- **(c)** Will the linkage operate as determined in part a?

1.53 Sketch a double-rocker mechanism, showing bearing locations to avoid interference.

1.54 Will a drafting machine of the type shown in Figure 1.28 operate properly if the pulleys are arbitrarily selected to be of different diameters? Show the motion of the straightedges if pulley diameters and $d_1 = 50$ mm, $d_2 = d_3 = d_4 = 100$ mm.

1.55 Design and dimension a pantograph that may be used to double the size of the pattern.

1.56 Design and dimension a pantograph that may be used to increase pattern dimensions by 10 percent. Let the fixed pivot lie between the tracing point and the marking point or toolholder.

1.57 Design and dimension a pantograph that will decrease pattern dimensions by 40 percent.

1.58 Referring to the swash plate pump shown in Figure 1.36, (a) determine the dimensions of a pump with a capacity of 120 ft^3/h at 600 rev/min (assume 100 percent volumetric efficiency), and (b) find the average velocity of the plunger. (There are many possible solutions to this problem.)

1.59 Repeat Problem 1.58 for a capacity of 0.01 m^3/s at 300 rad/s.

1.60 Design a differential power screw for a linear velocity of approximately 0.1 mm/s when the angular velocity is 65 rad/s.

1.61 Design a differential power screw for a linear velocity of approximately 0.0005 in/s when the screw rotates at 60 rev/min.

1.62 Design a ratchet–pawl mechanism to provide feed rates of 0.005 to 0.012 in/cycle in increments of 0.001 in/cycle.

1.63 Repeat Problem 1.62 for feed rates of 1 to 3 mm/cycle in increments of 100 microns per cycle (μm/c).

1.64 A Hooke-type universal joint has a 20° misalignment. Find the output shaft speed range if the input shaft rotates at a constant 1,000 rev/min.

1.65 Repeat Problem 1.64 for a 15° misalignment.

1.66 Find a permissible misalignment of a Hooke-type universal joint if the variation in speed is limited to ±2%.

1.67 Repeat Problem 1.66 for ±3% variation.

Problems 1.68 through 1.75 Are Based on a Constant-Speed Crank

It is suggested that a drag link mechanism be incorporated in the design.

1.68 Design a mechanism with a 10-in stroke and a forward-to-return stroke time ratio of (approximately) two to one.

1.69 Repeat Problem 1.68 for a time ratio of (approximately) 2.5 to 1.

1.70 Design a mechanism with a 150-mm stroke and a forward-to-return stroke time ratio of approximately two to one.

1.71 Design a mechanism with a 100-mm stroke and a forward-to-return time ratio of approximately two to one.

1.72 Design a mechanism with a stroke length that may be varied between 5 and 10 in. The forward-to-return stroke time ratio will be 1.5 to 1 at maximum stroke. Utilize a sliding contact linkage.

1.73 Repeat Problem 1.72 for a time ratio of 2.5 to 1.

1.74 Repeat Problem 1.72 for a stroke of 100 to 200 mm.

1.75 Repeat Problem 1.72 for a stroke of 180 to 280 mm.

1.76 Describe the motion of each link in Figure P1.3. Show the linkage in its limiting positions (corresponding to extreme positions of the slider). Determine the angle through which link 1 turns as the slider moves from the extreme left to the extreme right. Compare your result with the corresponding angle as the slider moves to the left. Find the stroke of the slider (the distance between limiting positions).

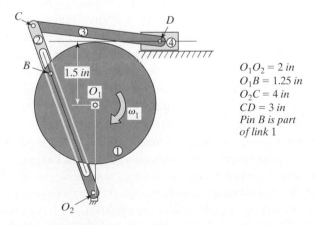

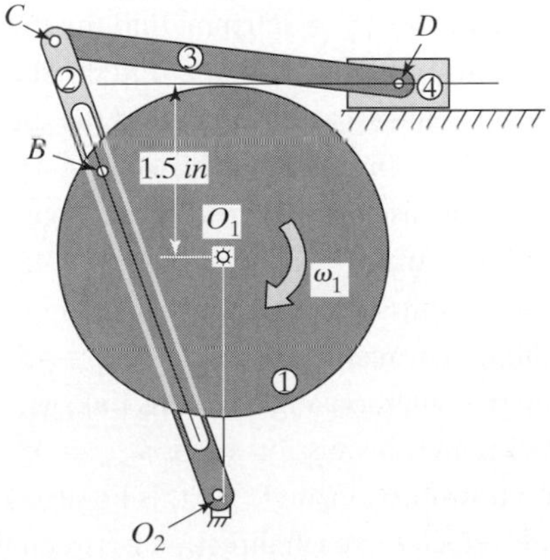

$O_1O_2 = 2$ in
$O_1B = 1.25$ in
$O_2C = 4$ in
$CD = 3$ in
Pin B is part of link 1

FIGURE P1.3

1.77 Repeat Problem 1.76 for Figure P1.4.

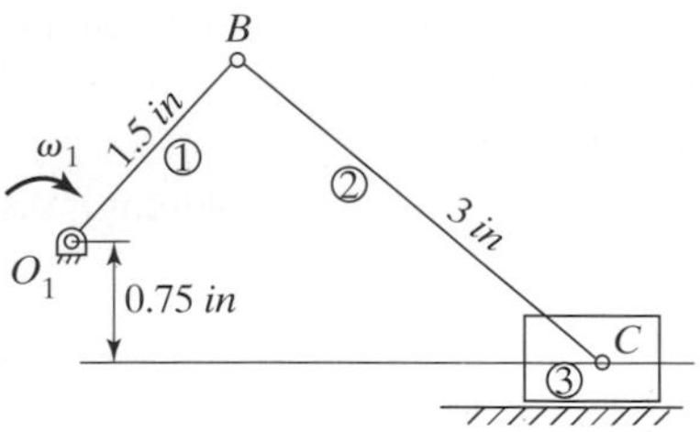

FIGURE P1.4

1.78 Repeat Problem 1.77 for $O_1B = 60$ mm, $BC = 120$ mm, and offset (O_1 to slider path) =20 mm.

Problems 1.79 through 1.83 Refer to Figure 1.22.

1.79 $R = 2$, $L = 5$, and $E = 0.4$.

(a) Find the forward-to-return stroke ratio.

(b) Find the stroke length S.

1.80 Repeat Problem 1.79 for offset $E = 0.8$.

1.81 Repeat Problem 1.79 for offset $E = 0.65$.

1.82 $R = 1$, $L = 3$, and $E = 1$. Plot the path of the midpoint of the connecting rod.

1.83 $R = 150$ mm, $L = 450$ mm, and $E = 150$ mm. Plot the path of the midpoint of the connecting rod.

1.84 Find the number of degrees of freedom of the lamination drive in Figure 1.41 when the control linkage position is given.

Problems 1.85 through 1.92 refer to a linkage that is to act as a crank-rocker mechanism

Link lengths are fixed link L_0, drive crank L_1, coupler L_2, and driven crank L_3.

1.85 $L_0 = 200$ mm, $L_1 = 50$ mm, and $L_3 = 150$ mm; find the theoretical range of L_2.

1.86 $L_0 = 210$ mm, $L_1 = 50$ mm, and $L_3 = 150$ mm; find the theoretical range of L_2.

1.87 $L_0 = 200$ mm, $L_1 = 45$ mm, and $L_3 = 150$ mm; find the theoretical range of L_2.

1.88 $L_0 = 200$ mm, $L_1 = 50$ mm, and $L_3 = 160$ mm; find the theoretical range of L_2.

1.89 **(a)** Find L_2 so that minimum transmission angle $\phi_{min} = 45°$ for the data of Problem 1.85.

(b) Find the maximum transmission angle for this linkage.

1.90 **(a)** Find L_2 so that minimum transmission angle $\phi_{min} = 45°$ for the data of Problem 1.86.

(b) Find the maximum transmission angle for this linkage.

1.91 **(a)** Find L_2 so that minimum transmission angle $\phi_{min} = 45°$ for the data of Problem 1.87.

(b) Find the maximum transmission angle for this linkage.

1.92 **(a)** Find L_2 so that minimum transmission angle $\phi_{min} = 45°$ for the data of Problem 1.88.

(b) Find the maximum transmission angle for this linkage.

1.93 Find the average piston velocity in each direction between limiting positions for a slider-crank linkage with a 150-mm crank length, a 350-mm connecting rod length, and a 100-mm offset. The crank rotates at 240 rad/s clockwise (cw).

1.94 In Figure 1.7a, let spherical pair S_1 be replaced by a cylindrical pair. Find the number of degrees of freedom of the mechanism.

1.95 In Figure 1.7a, let revolute joint R_1 be replaced by a spherical (ball) joint. Find the number of degrees of freedom of the mechanism.

1.96 How many degrees of freedom will a four-bar spatial CCCC linkage have?

1.97 How many degrees of freedom will a four-bar spatial SSSS linkage have?

1.98 Identify five or more spatial four-bar linkages having one degree of freedom. Select linkages that include revolute, prism, helix, cylinder, and sphere joints.

1.99 Show as many one-degree-of-freedom, planar, pin-connected linkage configurations as you can. Use up to eight links.

1.100 Consider an offset slider-crank linkage for which the ratio of connecting rod to crank length is to be 3.

(a) Write a computer program utilizing the Newton–Raphson method or another numerical method to determine the offset for a given value of stroke length.

(b) Find the offset for which the ratio of stroke to crank length will be 2.5.

1.101 Consider an offset slider-crank linkage for which the ratio of connecting rod to crank length is to be 3.

(a) Write a computer program utilizing the Newton–Raphson method or another numerical method to determine the offset for a given value of stroke length.

(b) Find the eccentricity for which the ratio of stroke to crank length will be 2.2.

1.102 Consider an offset slider-crank linkage for which the ratio of connecting rod to crank length is to be 3. Use the Newton–Raphson method or another numerical method to determine the offset for which the ratio of stroke to crank length will be 3.

1.103 Consider the design of four-bar planar crank-rocker linkages. Investigate which linkage proportions are acceptable and which are not. Let the fixed link length L_0 be three times the crank length L_1. Plot the minimum transmission angle vs. R_2 and R_3, where $R_2 = L_2/L_1$, $R_3 = L_3/L_1$, L_2 = coupler length, and L_3 = follower crank length. Plot the maximum transmission angle vs. R_2 and R_3. Plot the range of motion of the follower crank vs. R_2 and R_3.

1.104 Consider the design of four-bar planar crank-rocker linkages. Investigate which linkage proportions are acceptable and which are not. Assume that transmission angles between 40° and 140° are acceptable for the proposed design. Let the fixed link length L_0 be four times the crank length L_1. Plot the minimum transmission angle vs. R_2 and R_3, where $R_2 = L_2/L_1$, $R_3 = L_3/L_1$, L_2 = coupler length, and L_3 = follower crank length. Show the maximum transmission angle on the same plot. Identify the envelope of acceptable linkage proportions based on the transmission angle. Plot the range of motion of the follower crank vs. R_2 and R_3.

1.105 Consider the design of four-bar planar crank-rocker linkages. Investigate which linkage proportions are acceptable and which are not. Assume that transmission angles between 40° and 140° are acceptable for the proposed design. Let the fixed link length L_0 be five times the crank length L_1. Plot the maximum transmission angle vs. R_2 and R_3, where $R_2 = L_2L_1$, $R_3 = L_3/L_1$, L_2 = coupler length, and L_3 = follower crank length. Show the minimum transmission angle on the same plot. Identify the envelope of acceptable linkage proportions, based on the transmission angle. Plot the range of motion of the follower crank vs. R_2 and R_3.

1.106 Design a quick-return mechanism with a five-to-two forward-to-return-time ratio. Determine the linkage proportions of the mechanism. Specify link lengths for a 100-mm stroke. Find the minimum and maximum transmission angles. Design decisions. Base the design on a drag link combined with a slider. Try linkage proportions $R_0 = L_0/L_1 = 0.7$

and $R_3 = L_3/L_1 = 1.35$ with $R_2 = L_2/L_1$ unspecified, where the link lengths are identified as follows: L_0 = fixed link, L_1 = drive crank, L_2 = coupler, and L_3 = driven crank.

1.107 Design a quick-return mechanism with an eight-to-three forward-to-return time-ratio. Determine the linkage proportions of the mechanism. Specify link lengths for a 400-mm stroke. Find the minimum and maximum transmission angles. Design decisions. Base the design on a drag link combined with a slider. Try linkage proportions $R_0 = L_0/L_1 = 0.85$ and $R_3 = L_3/L_1 = 1.15$ with $R_2 = L_2/L_1$ unspecified, where the link lengths are identified as follows: L_0 = fixed link, L_1 = drive crank, L_2 = coupler, and L_3 = driven crank.

1.108 Design a quick-return mechanism with a 16:5 forward-to-return-time ratio. Determine the linkage proportions of the mechanism. Specify link lengths for a 400-mm stroke. Find the minimum and maximum transmission angles. Design decisions. Base the design on a drag link combined with a slider. Try linkage proportions $R_0 = L_0/L_1 = 0.85$ and $R_3 = L_3/L_1 = 1.25$ with $R_2 = L_2/L_1$ unspecified, where the link lengths are identified as follows: L_0 = fixed link, L_1 = drive crank, L_2 = coupler, and L_3 = driven crank.

PROJECTS

Project topics will often be suggested by the instructor's research interests and by current publications. The following topics may also be used as projects.

1.1 Aircraft landing gears sometimes fail to lower into position.

(a) Consider a system that will remedy this problem.

(b) Investigate the feasibility of a redundant landing-gear system that can be deployed if the primary system fails to operate.

1.2 Investigate the possibility of an innovative system to transport skiers to the top of a slope. Try to avoid conventional chairlifts, T-bars, rope tows, gondolas, etc.

1.3 Operators of power tools are sometimes injured because of inadequate guarding or the removal of guards that prevent efficient use of the tool. Design a system to feed wood into a circular saw in such a way that the operator's hands cannot contact the blade.

1.4 Backcountry skiing and Telemark skiing combine aspects of both alpine and Nordic skiing. Backcountry skiers utilize free-heel bindings on skis with metal edges. Some users of this equipment ski on steep slopes.

(a) Consider the design of an innovative release binding for free-heel skiing.

(b) Design a ski brake to stop a released ski.

1.5 Aircraft accidents have been attributed to an improperly latched cargo-hold door. In one or more instances, this problem resulted in loss of pressure in the cargo hold. The then-greater cabin pressure caused the floor between the cabin and the cargo hold to compress. This, in turn, pinched cables needed to steer and control the aircraft. Investigate a new cargo-hold door and latch design.

1.6 Environmental concerns make recycling a necessity. Investigate the design of a system to sort glass and plastic bottles, as well as steel and aluminum cans.

Suggestion: Projects may be assigned to an individual or a group, depending on the instructor's goals. Most mechanical devices in current use are the result of many person-years of effort by experienced engineers. However, student creativity may be stimulated by the demands of the task. It is expected that the mechanical engineering aspects of the design will be given priority. Detailed analysis may be limited to one aspect of the design project if the scope of the project is too large for the time available. The degree of success can be measured by the quality of innovative thinking, analysis, and interpretation, rather than by comparing a proposed design with a commercially available product. As the project progresses, it is expected that it will be necessary to consult several sections in this text as well as other sources of information.

BIBLIOGRAPHY AND REFERENCES

Finger, S., and J. R. Dixon, "A Review of Research in Mechanical Engineering Design. Part I: Descriptive, Prescriptive, and Computer-Based Models of Design Processes," *Research in Engineering Design*, Vol. 1, 1989, pp. 51–67.

Frank, W. H., L. Kloepping, M. St. Norddahl, K. Olaffson, and D. Shook, *TK Solver Plus*, Universal Technical Systems, Rockford, IL, 1988.

Hornbeck, R. W., *Numerical Methods*, Prentice-Hall, Upper Saddle River, NJ, 1975.

Kaufman, R. E., "Mechanism Design by Computer," *Machine Design*, Oct. 26, 1978, pp. 94–100.

Knowledge Revolution, *Working Model™ 2D User's Manual*, Knowledge Revolution, San Mateo, CA, 1996.

Knowledge Revolution, *Working Model™ 3D User's Manual*, Knowledge Revolution, San Mateo, CA, 1997.

Lawry, M. H., *I-DEAS Master Series Student Guide*, Structural Dynamics Research Corp., Milford, OH, 1998.

MathSoft, *Mathcad 2000™ User's Guide*, MathSoft, Inc., Cambridge MA, 1999.

Penton, *Machine Design*, A periodical that includes recent developments related to the design of machines and mechanisms, and useful information about machine components (particularly in advertisements), Penton Publishing, Cleveland, OH.

Pipes, L. A., *Applied Mathematics for Engineers and Physicists*, McGraw-Hill, New York, 1958.

Rubel, A. J., and R. E. Kaufman, "KINSYN III: A New Human Engineered System for Interactive Computer-Aided Design of Planar Linkages," *Transactions of the American Society of Mechanical Engineers, Journal of Engineering for Industry*, vol. 99, no. 2, May 1977.

Scheid, F., *Numerical Analysis*, McGraw-Hill, New York, 1968.

Sheth, P. N., and J. J. Uicker, "IMP (Integrated Mechanisms Program), a Computer-Aided Design Analysis System for Mechanisms and Linkages," *Transactions of the American Society of Mechanical Engineers, Journal of Engineering for Industry*, May 1972.

Stark, P. A., *Introduction to Numerical Methods*, Macmillan, New York, 1970.

Stoer, J., and R. Bulirsch, *Introduction to Numerical Analysis*, Springer, New York, 1980.

Waldron, K. J., "A Method of Studying Joint Geometry," in *Mechanisms and Machine Theory*, vol. 7, Pergamon Press, Elmsford, NY, 1972, pp. 347–353.

Waldron, K. J., *RECSYN: NSF User's Manual*, National Science Foundation Grants ENG 77-22745, SER 78-00500, CME 8002122, 1977.

INTERNET RESOURCES

The Internet can be helpful to those interested in applications that are not covered in this text and as a resource for projects in kinematics and dynamics of machinery. The list that follows includes a small sample of relevant websites and typical products or services. When searching the Internet, try to limit the search to your area of interest by adding modifiers to the key words. For example, a search for *robots* yielded over two million sites, but *pneumatic robot grippers for automotive manufacturing* returned 180 sites.

AUTOMOTIVE DRIVE COMPONENTS

www.ingersoll.com	Ingersoll CM Systems (machinery and systems for crankshaft and camshaft manufacturing)
www.spicerdriveshaft.com	Spicer (vehicle driveshafts–constant velocity and Cardan joints)

CAMS AND PART HANDLERS

www.camcoindex.com	Commercial Cam (cams, index drives, conveyors)

CONSTRUCTION AND EARTH-MOVING EQUIPMENT, CRANES AND LIFT TABLES

www.airtechnical.com	Air Technical Industries (floor cranes, lift trucks, lift tables)
www.bobcat.com	Ingersoll-Rand (excavators, skid-steer loaders, track loaders)
www.caterpillar.com	Caterpillar Products (earth-moving equipment, engines and power systems)
www.coastalcranes.com	Coastal Hydraulic Cranes (telescoping and fixed-boom cranes)

GEARS

www.geartechnology.com	Gear industry buyers guide (gear machines, gear materials, gear drives, software).
www.hdsystems.com	HD Systems (harmonic planetary gearheads)

PROFESSIONAL SOCIETIES AND JOURNALS

www.asme.org	American Society of Mechanical Engineers (*Journal of Mechanical Design* includes kinematics and dynamics of mechanisms)
www.elsevier.com	Elsevier Science (*Mechanism and Machine Theory*, a journal devoted to mechanisms and dynamics of machines)
www.sme.org	Society of Manufacturing Engineers

ROBOTS, GRIPPERS, SENSORS, GUIDED VEHICLE SYSTEMS AND ACCESSORIES

www.abb.com	ABB Group (automated welding equipment)
www.agvp.com	AGV Products (automated guided vehicle systems)
www.keyence.com	Keyence Corp. (machine vision)
www.parker.com	Parker Automation (accessories)
www.robotics.org	Robotic Industries Association (industrial robots, robot safety).
www.robotics-technology.com	European site for robotics information.
www.sankyo.com	Sankyo Robotics (Cartesian coordinate robots, track-mounted robots, and selective compliance assembly robot arms)

SOFTWARE FOR CALCULATION, DESIGN, MANUFACTURING, MOTION SIMULATION AND TESTING

www.eds.com	Structural Dynamics Research Corp. (I-deas software for machine simulation, NC machining)
www.lmsintl.com	LMS International (kinematic and dynamic simulation, data acquisition, virtual testing)
www.mathcad.com	Mathsoft (Mathcad™ calculation software and sample files)
www.workingmodel.com	MSC (Working Model™ simulation software)

CHAPTER 2

Motion in Machinery: Positional Analysis of Planar and Spatial Mechanisms

The piston in a pump or engine is constrained to rectilinear motion. The motion of the crank and connecting rod in the same engine or pump is planar. Robots and automotive steering linkages include links with spatial motion.

Complex-number methods are important analytical tools for analyzing and designing planar linkages. But complex-number methods cannot be used to analyze spatial linkages. Analytical vector methods are used to analyze and design both planar and spatial linkages. Graphical methods are a useful tool for independent verification of analytical work.

Concepts and Methods You Will Learn and Apply While Studying This Chapter

- Motion produced by an eccentric cam and a Scotch yoke
- Vector manipulation required to solve problems in the kinematics of machinery: unit vectors, addition of vectors, cross and dot product, and vector differentiation
- Solution of vector equations
- Analytical vector methods for displacement analysis of planar mechanisms
- Complex numbers in rectangular and polar form
- Complex arithmetic: addition, multiplication, and differentiation
- Complex numbers applied to linkage design and analysis
- Analytical methods for solving closed spatial linkages
- The Newton–Raphson method for two or more variables applied to single- and multiloop linkages (an advanced topic)
- Design of linkages with the aid of animation software: building a simulation of a mechanism, running the simulation, and interpreting results of a design study

2.1 MOTION

When the motion of the elements of a linkage is *not* restricted to a single plane or to a set of parallel planes, the linkage undergoes spatial motion. Three independent coordinates are required to specify the location of a point in spatial motion. A rigid body in spatial motion has six degrees of freedom subject to restrictions imposed by joints. Spatial or three-dimensional linkages often include joints with two or three degrees of freedom, such as cylinder pairs and sphere pairs. A spatial linkage would be required if we were to attempt to duplicate the motion of the human arm.

If the motion of all points in a linkage system is restricted to a plane or to a set of parallel planes, then the motion is *planar*. A point in planar motion is located by two independent coordinates. An unconstrained rigid link has three degrees of freedom in planar motion. We may, for example, identify the x and y coordinates of a point on the link and the angular position of the link centerline. Planar motion may be characterized by two-dimensional vectors. Plane mechanisms, a special case of the more general spatial mechanisms, are of particular interest, because they include major components of the internal-combustion engine, spur gear trains, and most cams, as well as a variety of other mechanisms.

The plane motion of a rigid link may be pure translation (also called rectilinear motion), in which case all points on the link move in the same direction at the same speed. For example, the cam follower in Figure 2.1 undergoes rectilinear motion (i.e., translation along a straight line). Translation of a rigid body in general implies motion in space such that any line connecting two points on the body remains parallel to its original position.

In another special case, the plane motion of a body is described by pure rotation, in which case a point on the link is fixed (as, for example, on a cam). Oscillation refers either to a back-and-forth rotation (e.g., the motion of a pendulum) or a back-and-forth translation (the motion of a piston). In the study of mechanisms, oscillation is commonly used in the first sense, and the motion of a piston is described instead as *reciprocating motion*. In every case, the meaning should be clear in context. Rectilinear motion and rotation of a rigid body about a point may be described by one independent variable (e.g., x and θ, respectively).

Examples of Rectilinear Motion: The Eccentric Cam and the Scotch Yoke

The eccentric cam with flat-face follower and the Scotch yoke are examples of mechanisms having simple mathematical representations. The *eccentric cam* of Figure 2.1a is circular in form, but the center of the circle, C, is offset a distance R from the center of the camshaft O_1. For radius r, the distance from the camshaft center to the follower face is $r - R$ when the follower is at its lowest position. After the cam turns through an angle θ (see Figure 2.1b), the distance becomes $r - R\cos\theta$. Therefore, the displacement of the follower during the interval is

$$x = r - R\cos\theta - (r - R) = R(1 - \cos\theta)$$

for any value of θ, where x is measured from the lowest position.

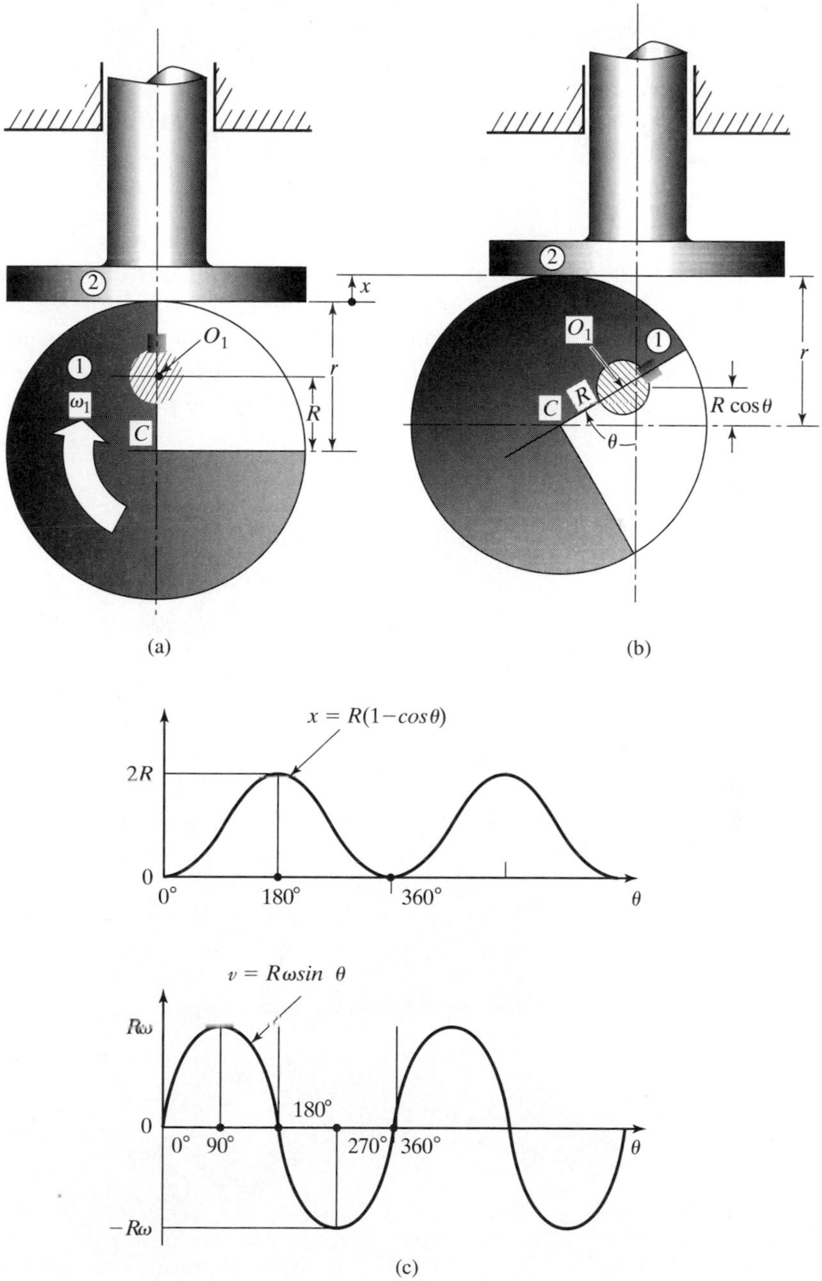

FIGURE 2.1 (a) The cam, link 1, is formed by a circular disk of radius r and eccentricity R. (b) The follower, link 2, undergoes displacement x as the cam rotates through an angle θ. (c) The displacement and velocity of the follower are plotted against cam rotation.

If this expression is differentiated with respect to time, we obtain the follower velocity

$$v = R\omega \sin\theta,$$

where cam angular velocity $\omega = d\theta/dt$ (rad/s). If the cam rotates at constant speed, the follower acceleration is

$$a = R\omega^2 \cos\theta.$$

Displacement and velocity are plotted in Figure 2.1c for constant ω. Note that the velocity is proportional to the eccentricity R, but independent of the cam circle radius r.

In the two-cylinder piston pump of Figure 2.2, the cam raises and lowers a shaft with a piston at each end by acting alternately on two separate follower faces. The cam pushes against the upper follower face, lifting the piston assembly during 180° of

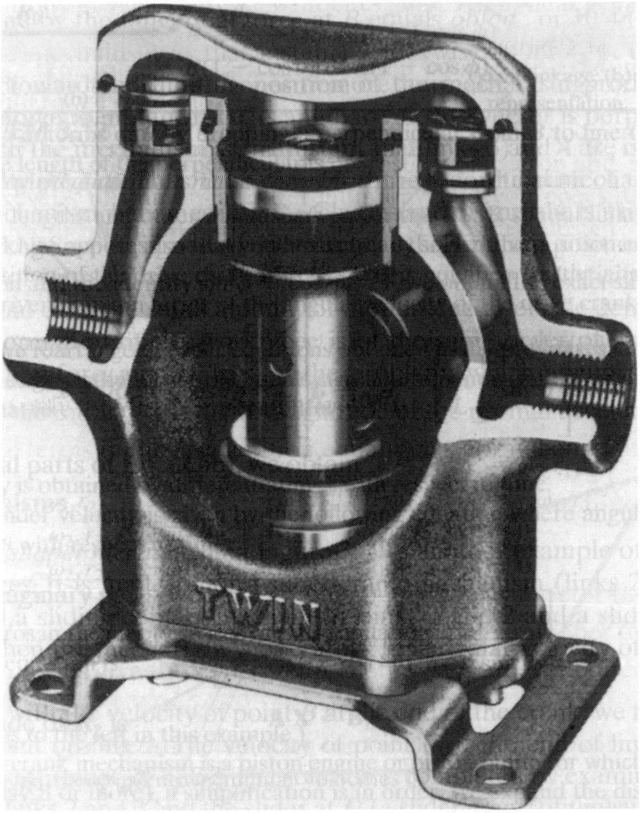

FIGURE 2.2 A two-cylinder cam-type piston pump. An eccentric cam (in the form of a sealed-roller-type bearing) transmits power from crankshaft to follower (the piston drive). The location and velocity of the piston are the same as given for the eccentric cam, except that the roller bearing eliminates sliding. (*Source*: Hypro, a division of Lear Siegler, Inc.)

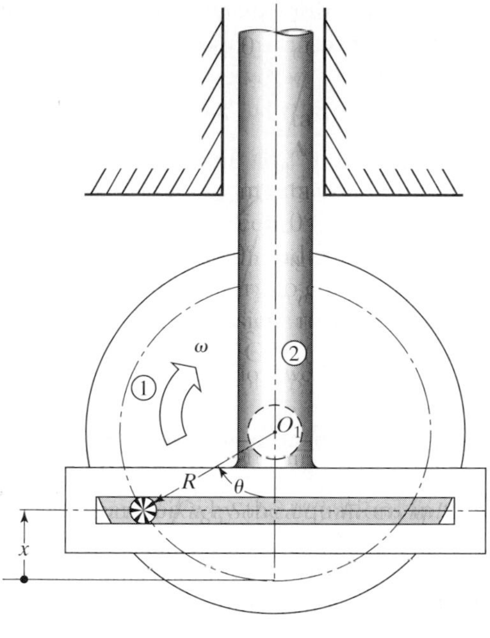

FIGURE 2.3 A Scotch yoke is kinematically equivalent to an eccentric cam.

the cam's rotation. Through the remaining 180° of the rotation, the cam acts against the lower follower face, forcing the piston assembly downward.

The *Scotch yoke* (see Figure 2.3) is kinematically equivalent to the eccentric cam just considered. In this case, link 1, the driver, has a pin on which the slotted follower, link 2, rides. The pin is a distance R from O_1, the axis of the driver. Measuring from the lowest position of the follower, we see that displacement, velocity, and acceleration are given by the equations that were used to describe the eccentric cam.

2.2 VECTORS

Vectors are an important part of the language of mechanism and the other branches of mechanics. They provide us with a graphical and analytical means of representing motion and force. A quantity described by its magnitude, direction, and sense can be considered a vector and can be represented by an arrow. Now, suppose a vector represents a point on a piston that is constrained to move vertically. Then the vector direction is vertical, and with further information, we may determine the vector sense (upward or downward) and vector magnitude.

Graphical and analytical vector methods may be applied to linear displacements, velocities, accelerations, and forces, and to torques and angular velocities and accelerations. Although finite angular displacements possess magnitude and direction, they are not generally considered vectors, because they do not follow the rules of vector addition.

Vectors are usually identified by ***boldface*** type to distinguish them from *scalar* quantities. A line above or below the letter symbol may be used as an alternative identification for a vector. A different designation is suggested for computer use–for example, "L" for scalar link length and "r" for the corresponding vector.

Right-Hand Coordinates and a Sign Convention for Angles

A right-hand coordinate system is used for vectors. The thumb, index finger, and middle finger of the right hand represent, respectively, the mutually perpendicular x-, y-, and z-axes. A vector in an xy-plane can be described by its magnitude (length) and its direction, an angle measured *counterclockwise* from the x-axis. If the vector represents a link rotating in an xy-plane, an increasing angle corresponds to a positive (counterclockwise) angular velocity. Then, the angular-velocity vector is in the z-direction.

Angles in Radians

Most software and programming languages expect angles to be measured in radians. Some software packages will accept either degrees or radians, but be sure that you and the software understand each other.

The Commutative, Associative, and Distributive Laws for Adding Vectors and Multiplying a Vector by a Scalar

In the following laws governing the addition of vectors and the multiplication of a vector by a scalar, $\boldsymbol{A}$, $\boldsymbol{B}$, and $\boldsymbol{C}$ are vectors and m is a scalar quantity:

$$
\begin{aligned}
&\boldsymbol{A} + \boldsymbol{B} = \boldsymbol{B} + \boldsymbol{A} \text{ (commutative law for addition);}\\
&\boldsymbol{A} + (\boldsymbol{B} + \boldsymbol{C}) = (\boldsymbol{A} + \boldsymbol{B}) + \boldsymbol{C} \text{ (associative law for addition);}\\
&m\boldsymbol{A} = \boldsymbol{A}m \text{ (commutative law for multiplication by a scalar);}\\
&m(\boldsymbol{A} + \boldsymbol{B}) = m\boldsymbol{A} + m\boldsymbol{B} \text{ (distributive law for multiplication by a scalar).} \quad (2.1)
\end{aligned}
$$

As noted earlier, finite angular displacements are not generally treated as vectors (or scalars). To see why, consider the motion of an aircraft, with yaw denoted by the angle θ and pitch by the angle ϕ, where both θ and ϕ are moving coordinates referred to the axes of the aircraft. Let the aircraft make a 90° right turn ($\theta = 90°$ cw) and then pitch downward by 90° ($\phi = 90°$) Using any rigid body to represent the aircraft, we see that if the order of these two maneuvers is reversed, then the result is different.

We may run into a similar situation when analyzing the motion of a robot arm. If the order of the commands is changed, the final position may be different. Thus,

$$\theta + \phi \neq \phi + \theta,$$

showing that finite rotations about nonparallel axes do not follow the commutative law for addition.

Unit Vectors

In general, a vector of unit magnitude can be called a unit vector. Thus, $\boldsymbol{A}^u = \boldsymbol{A}/A$ is a unit vector, where $A = |\boldsymbol{A}|$ is the magnitude of vector $\boldsymbol{A}$. Unit vectors $\boldsymbol{i}, \boldsymbol{j}$, and $\boldsymbol{k}$ (or $\boldsymbol{I}$, $\boldsymbol{J}$, and $\boldsymbol{K}$) parallel to the x, y, and z (or X, Y, and Z) coordinate axes, respectively, are

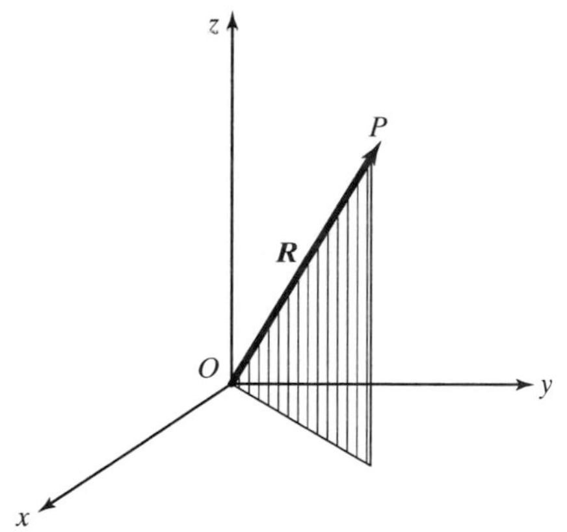

FIGURE 2.4 Vector $\boldsymbol{R}$ locates the position of the point P in the xyz coordinate system.

particularly useful. These unit vectors are also called *rectangular unit vectors*. A right-hand system of mutually perpendicular coordinates is shown in Figure 2.4. A vector may be described in terms of its components along each coordinate axis.

Vector Components

Let the location of a point in space be described by a vector extending from the origin of a coordinate system to the point. For example, point P in Figure 2.4 is located by the vector $\boldsymbol{R}$. The x, y, and z coordinate axes in the figure are mutually perpendicular. Any motion of P will result in a change in the vector $\boldsymbol{R}$, either in its magnitude, its direction, or both.

A plane through P perpendicular to the x-axis intersects that axis at a distance R_x from the origin O. (See Figure 2.5.) The distance R_x is called the *projection* of the vector $\boldsymbol{R}$ on the x-axis, or the x component of $\boldsymbol{R}$. The projections of $\boldsymbol{R}$ on the y-axis and z-axis are labeled R_y and R_z, respectively. Vectors $\boldsymbol{i}, \boldsymbol{j}$, and $\boldsymbol{k}$ are *unit vectors* in the x, y,

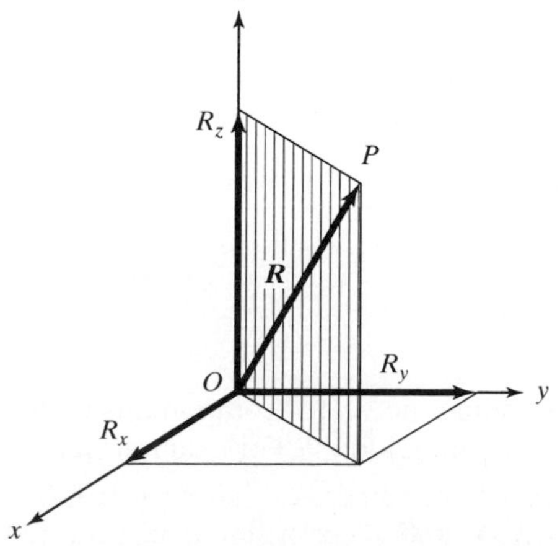

FIGURE 2.5 The vector $\boldsymbol{R}$ can be resolved into components along the x-, y-, and z-axes.

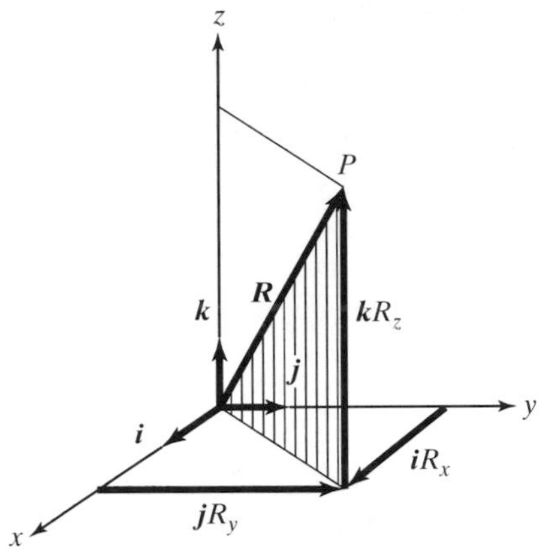

FIGURE 2.6 Each component vector can be considered the product of a unit vector and the scalar magnitude of the component.

and z directions, respectively, as shown in Figure 2.6. That is, each has unit length and is used only to assign a direction. The scalar length R_x multiplied by the unit vector $\boldsymbol{i}$ gives us the vector $\boldsymbol{i}R_x$ of length R_x and parallel to the x-axis. The original vector is given by the vector sum of its components:

$$\boldsymbol{R} = \boldsymbol{i}R_x + \boldsymbol{j}R_y + \boldsymbol{k}R_z. \tag{2.2}$$

Vector Equations

If two vectors are equal, then each component of the first is equal to the corresponding component of the second. Thus, let

$$\mathbf{A} = \boldsymbol{i}A_x + \boldsymbol{j}A_y + \boldsymbol{k}A_z$$

and

$$\boldsymbol{B} = \boldsymbol{i}B_x + \boldsymbol{j}B_y + \boldsymbol{k}B_z.$$

Then

$$\mathbf{A} = \boldsymbol{B} \text{ implies that}$$
$$A_x = B_x, \quad A_y = B_y, \quad \text{and} \quad A_z = B_z.$$

Vector Addition and Subtraction

The addition of vectors simply involves adding the vector components in the x-, y-, and z-directions (i.e., the $\boldsymbol{i}$, $\boldsymbol{j}$, and $\boldsymbol{k}$ components) individually. For example, let

$$\boldsymbol{C} = \mathbf{A} + \boldsymbol{B}$$

where vectors $\boldsymbol{A}$, $\boldsymbol{B}$, and $\boldsymbol{C}$ are represented in terms of their scalar components and rectangular unit vectors. Then

$$\boldsymbol{C} = \boldsymbol{i}(A_x + B_x) + \boldsymbol{j}(A_y + B_y) + \boldsymbol{k}(A_z + B_z).$$

Vector subtraction is similar. If

$$\boldsymbol{D} = \boldsymbol{A} - \boldsymbol{B},$$

then

$$\boldsymbol{D} = \boldsymbol{i}(A_x - B_x) + \boldsymbol{j}(A_y - B_y) + \boldsymbol{k}(A_z - B_z).$$

Graphical Addition and Subtraction of Planar Vectors

Vectors may be added graphically by joining them head to tail. Although graphical procedures may be used to treat both planar and spatial linkages, graphical solutions are most commonly used with planar problems. Consider the vector equation

$$\boldsymbol{A} + \boldsymbol{B} + \boldsymbol{D} + \boldsymbol{E} = \boldsymbol{F}$$

where all of the vectors lie in the same plane, as shown in Figure 2.7a. Beginning at an arbitrary point o, we draw vector $\boldsymbol{A}$ and then successively add vectors $\boldsymbol{B}$, $\boldsymbol{D}$, and $\boldsymbol{E}$, with the tail of each added vector beginning at the head of the vector last drawn.

The vector sum is given by the vector $\boldsymbol{F}$, with tail at o and head drawn to the head of the last vector of the series to be added. The reader can verify that the addition of vectors is commutative; that is, the vectors may be added in any order to obtain the same result: $\boldsymbol{A} + \boldsymbol{B} = \boldsymbol{B} + \boldsymbol{A}$, and so on.

Vector subtraction is sometimes required when we consider relative motion. Thus, if vector $\boldsymbol{G}$ is given by vector $\boldsymbol{A}$ minus vector $\boldsymbol{B}$, we write

$$\boldsymbol{G} = \boldsymbol{A} - \boldsymbol{B} \quad \text{or} \quad \boldsymbol{G} = \boldsymbol{A} + (-\boldsymbol{B}),$$

where vector $(-\boldsymbol{B})$ is identical to vector $\boldsymbol{B}$, except that the sense is reversed, as shown in Figure 2.7b. The second form of the expression for the difference between two vectors is preferred, particularly when many vectors are to be combined.

Determinants

Determinants can be used in vector operations and to solve equations. Determinants are square arrays of elements enclosed in vertical bars, like this second-order determinant:

$$\det \mathbf{A} = \begin{vmatrix} \mathrm{a} & \mathrm{c} \\ \mathrm{b} & \mathrm{d} \end{vmatrix}.$$

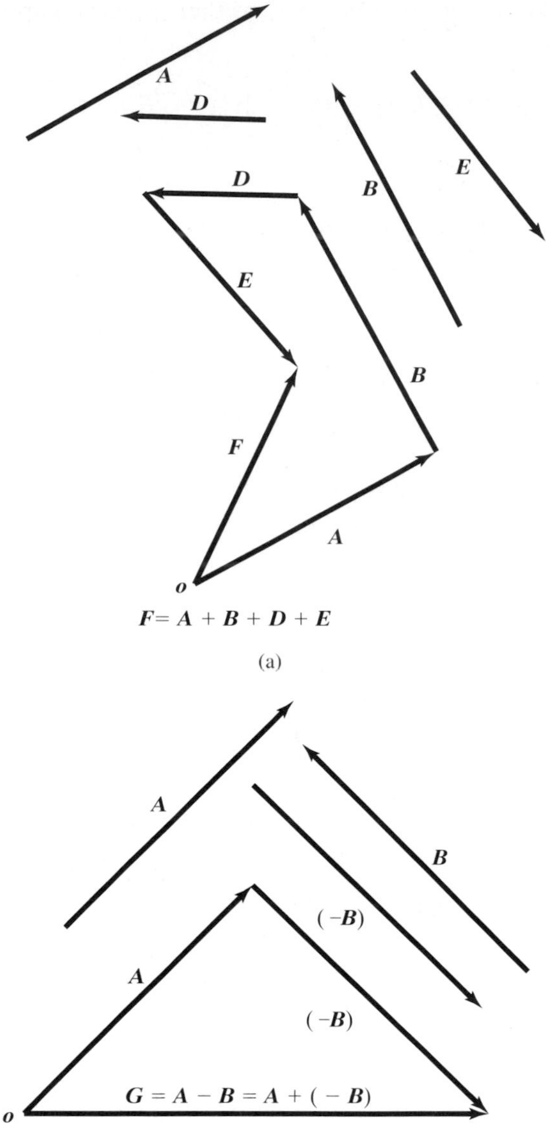

FIGURE 2.7 (a) The graphic addition of several vectors. (b) Vector subtraction. Simply *reverse* the sense of the vector being subtracted, and add it like any other vector.

We evaluate a second order determinant by multiplying terms along the diagonals. The product of the terms on the diagonal running from the lower left to the upper right of the determinant is subtracted from the product of the terms on the diagonal running from the upper left to the lower right of the determinant. This is how it works:

$$\det \mathbf{A} = \begin{vmatrix} a & c \\ b & d \end{vmatrix} = ad - bc.$$

A third-order determinant looks like this:

$$\det \mathbf{B} = \begin{vmatrix} a & d & g \\ b & e & h \\ c & f & i \end{vmatrix}.$$

We can evaluate a third order determinant by repeating the first two columns and multiplying terms along each diagonal. The products of the terms on the diagonals running from the upper left to the lower right are added, and the products of the terms on the diagonals running from the upper right to the lower left are subtracted. This is how a we evaluate a third-order determinant:

$$\det \mathbf{B} = \left|\begin{matrix} a & d & g \\ b & c & h \\ c & f & i \end{matrix}\right|\begin{matrix} a & d \\ b & e \\ c & f \end{matrix} = aei + dhc + gbf - gec - ahf - dbi.$$

A word of caution: The diagonal method will not work for determinants of fourth or higher order.

The vector cross product is expressed as a determinant in the section that follows. Before the general availability of computers, determinants were also a popular method for solving sets of simultaneous equations. (See, for example, the velocity analysis of a four-bar linkage in the next chapter.) Now we are now more likely to use software that is capable of performing vector and matrix operations directly, particularly for problems that result in sets of three or more equations. Another alternative is to use animation software that utilizes powerful built-in numerical methods.

The Vector Cross Product

In addition to the product of a scalar and a vector, two types of products involving vectors alone are defined: the vector or cross product and the scalar or dot product. The *vector cross product*, or simply the *cross product* of two vectors is a vector perpendicular to the plane in which the two vectors lie. The cross product of vectors $\boldsymbol{A}$ and $\boldsymbol{B}$ separated by angle θ is $\boldsymbol{A} \times \boldsymbol{B} = \boldsymbol{C}$, where the magnitude of $\boldsymbol{C}$ is $C = AB \sin \theta$ and the direction of $\boldsymbol{C}$ is given by the right-hand rule, as follows: The thumb and index finger of the right hand are extended in the direction of vectors $\boldsymbol{A}$ and $\boldsymbol{B}$, respectively. If the middle finger is then bent perpendicular to $\boldsymbol{A}$ and $\boldsymbol{B}$, it points in the direction of vector $\boldsymbol{C}$. $\boldsymbol{A} \times \boldsymbol{B}$ is read "$\boldsymbol{A}$ cross $\boldsymbol{B}$." The cross product of two parallel vectors is zero (the null vector) since angle $\theta = 0$. Thus, $\boldsymbol{A} \times \boldsymbol{A} = 0$. Some references identify the null vector as $\mathbf{0}$, using bold type to remind the reader that the null vector can be written as $0\boldsymbol{i} + 0\boldsymbol{j} + 0\boldsymbol{k}$.

Observe that the cross product does not follow the commutative law. Interchanging the order of the vectors changes the sign of the vector product (i.e., $\boldsymbol{B} \times \boldsymbol{A} = -\boldsymbol{A} \times \boldsymbol{B}$). It can be seen that the vector products of unit vectors $\boldsymbol{i}, \boldsymbol{j}$, and $\boldsymbol{k}$ are as follows:

$$\begin{aligned} &\boldsymbol{i} \times \boldsymbol{j} = \boldsymbol{k} \quad \boldsymbol{j} \times \boldsymbol{i} = -\boldsymbol{k} \quad \boldsymbol{j} \times \boldsymbol{k} = \boldsymbol{i} \quad \boldsymbol{k} \times \boldsymbol{j} = -\boldsymbol{i} \\ &\boldsymbol{k} \times \boldsymbol{i} = \boldsymbol{j} \quad \boldsymbol{i} \times \boldsymbol{k} = -\boldsymbol{j} \quad \boldsymbol{i} \times \boldsymbol{i} = \boldsymbol{j} \times \boldsymbol{j} = \boldsymbol{k} \times \boldsymbol{k} = 0 \end{aligned} \tag{2.3}$$

If vectors $\boldsymbol{A}$ and $\boldsymbol{B}$ are expressed in terms of their scalar components and unit vectors as

$$\boldsymbol{A} = \boldsymbol{i}A_x + \boldsymbol{j}A_y + \boldsymbol{k}A_z$$

and

$$\boldsymbol{B} = \boldsymbol{i}B_x + \boldsymbol{j}B_y + \boldsymbol{k}B_z,$$

respectively, then

$$\boldsymbol{A} \times \boldsymbol{B} = \boldsymbol{i}(A_yB_z - A_zB_y) + \boldsymbol{j}(A_zB_x - A_xB_z) + \boldsymbol{k}(A_xB_y - A_yB_x),$$

which may be written more concisely in determinant form as

$$\boldsymbol{A} \times \boldsymbol{B} = \begin{vmatrix} \boldsymbol{i} & \boldsymbol{j} & \boldsymbol{k} \\ A_x & A_y & A_z \\ B_x & B_y & B_z \end{vmatrix}. \tag{2.4}$$

The vector cross product will be used extensively in later chapters. One application of the cross product is determining the velocity of a point on a link rotating about a fixed center at angular velocity

$$\boldsymbol{\omega} = \boldsymbol{i}\omega_x + \boldsymbol{j}\omega_y + \boldsymbol{k}\omega_z.$$

If the vector from the fixed center to the point in question is given by

$$\boldsymbol{r} = \boldsymbol{i}r_x + \boldsymbol{j}r_y + \boldsymbol{k}r_z,$$

then the velocity of the point is given by

$$\boldsymbol{v} = \boldsymbol{\omega} \times \boldsymbol{r} = \begin{vmatrix} \boldsymbol{i} & \boldsymbol{j} & \boldsymbol{k} \\ \omega_x & \omega_y & \omega_z \\ r_x & r_y & r_z \end{vmatrix}. \tag{2.5}$$

SAMPLE PROBLEM 2.1

Rotating Link

Suppose a link that is held in a ball joint (spherical pair) at one end rotates with an instantaneous angular velocity

$$\boldsymbol{\omega} = \boldsymbol{i}2 + \boldsymbol{j}(-1) + \boldsymbol{k}4(\text{rad/s})$$

Find the instantaneous velocity of P, a point on the link defined by the radius vector

$$\boldsymbol{r} = \boldsymbol{i}(-1) + \boldsymbol{j}10 + \boldsymbol{k}2\,(\text{mm}),$$

measured from the ball joint. (See Figure 2.8a.)

Solution. The velocity of the point is given by

$$\boldsymbol{v} = \boldsymbol{\omega} \times \boldsymbol{r} = \begin{vmatrix} \boldsymbol{i} & \boldsymbol{j} & \boldsymbol{k} \\ A_x & A_y & A_z \\ B_x & B_y & B_z \end{vmatrix} = \begin{vmatrix} \boldsymbol{i} & \boldsymbol{j} & \boldsymbol{k} \\ 2 & -1 & 4 \\ -1 & 10 & 2 \end{vmatrix}$$
$$= \boldsymbol{i}(-2 - 40) + \boldsymbol{j}(-4 - 4) + \boldsymbol{k}(20 - 1)$$
$$= \boldsymbol{i}(-42) + \boldsymbol{j}(-8) + \boldsymbol{k}(19)\,(\text{mm/s})$$

The velocity is perpendicular to the plane of $\boldsymbol{\omega}$ and $\boldsymbol{r}$; its magnitude is given by the sum of the squares of the velocity components:

$$v = \sqrt{v_x^2 + v_y^2 + v_z^2} = \sqrt{(-42)^2 + (-8)^2 + 19^2} = 46.79\text{ mm/s}.$$

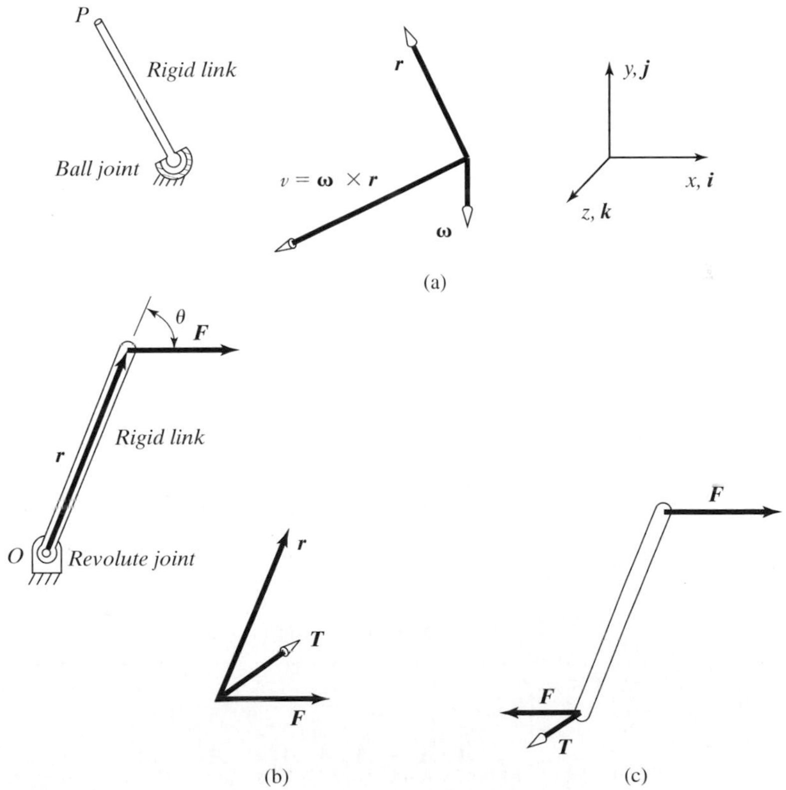

FIGURE 2.8 (a) Velocity of a point on a link, in terms of the cross product. (b) The torque of force $\boldsymbol{F}$ about point O is $\boldsymbol{T} = \boldsymbol{r} \times \boldsymbol{F}$. (c) A link in equilibrium.

Another application of the vector cross product involves torque. For example, in Figure 2.8b, let force $\boldsymbol{F}$ be applied to a rigid link. The resultant torque of $\boldsymbol{F}$ about point O may be represented by the vector product

$$\boldsymbol{T} = \boldsymbol{r} \times \boldsymbol{F} = \begin{vmatrix} \boldsymbol{i} & \boldsymbol{j} & \boldsymbol{k} \\ r_x & r_y & r_z \\ F_x & F_y & F_z \end{vmatrix} \tag{2.6}$$

The magnitude of $\boldsymbol{T}$ is $rF \sin \theta$, and the direction of $\boldsymbol{T}$ is perpendicular to the plane of $\boldsymbol{r}$ and $\boldsymbol{F}$, as shown in the figure. Note that applying the right-hand rule to $\boldsymbol{T} = \boldsymbol{r} \times \boldsymbol{F}$ gives the direction of the resultant torque. If both $\boldsymbol{r}$ and $\boldsymbol{F}$ lie in the xy-plane, then $\boldsymbol{T}$ is represented by a vector in the $\pm z$ direction. The link could be in static equilibrium if a force and torque were applied, as in Figure 2.8c.

Later, you will use the torque cross product for static and dynamic analysis of linkages. The required driving torque can be determined from forces on the driving link.

The Dot or Scalar Product

The dot product of two vectors is a scalar equal to the product of the magnitudes of the vectors and the cosine of the angle between them. Thus, the dot product of vectors $\boldsymbol{A}$ and $\boldsymbol{B}$, separated by angle θ, is given by

$$\boldsymbol{A} \cdot \boldsymbol{B} = AB \cos \theta. \tag{2.7}$$

Hence,

$$\boldsymbol{i} \cdot \boldsymbol{i} = \boldsymbol{j} \cdot \boldsymbol{j} = \boldsymbol{k} \cdot \boldsymbol{k} = 1$$

and

$$\boldsymbol{i} \cdot \boldsymbol{j} = \boldsymbol{j} \cdot \boldsymbol{k} = \boldsymbol{k} \cdot \boldsymbol{i} = 0. \tag{2.8}$$

Let $\boldsymbol{A}$ and $\boldsymbol{B}$ be expressed in terms of their scalar components and unit vectors as

$$\boldsymbol{A} = \boldsymbol{i}A_x + \boldsymbol{j}A_y + \boldsymbol{k}A_z$$

and

$$\boldsymbol{B} = \boldsymbol{i}B_x + \boldsymbol{j}B_y + \boldsymbol{k}B_z.$$

Then

$$\boldsymbol{A} \cdot \boldsymbol{A} = A_x^2 + A_y^2 + A_z^2 = A^2 \tag{2.9}$$

and

$$\boldsymbol{A} \cdot \boldsymbol{B} = A_xB_x + A_yB_y + A_zB_z. \tag{2.10}$$

The dot product of two perpendicular vectors is zero. Thus, since the cross product of two vectors is perpendicular to the plane of the two vectors, we have

$$\boldsymbol{A}\cdot(\boldsymbol{A}\times\boldsymbol{B}) = \boldsymbol{B}\cdot(\boldsymbol{A}\times\boldsymbol{B}) = 0$$

This relationship may be used as a partial check on the calculation of the velocity of a point on a rigid link.

SAMPLE PROBLEM 2.2

The Dot Product

Use the dot product to check the results of a previous example in which

$$\boldsymbol{\omega} = \boldsymbol{i}2 + \boldsymbol{j}(-1) + \boldsymbol{k}4,$$
$$\boldsymbol{r} = \boldsymbol{i}(-1) + \boldsymbol{j}10 + \boldsymbol{k}2,$$

and

$$\boldsymbol{v} = \boldsymbol{\omega}\times\boldsymbol{r} = \boldsymbol{i}(-42) + \boldsymbol{j}(-8) + \boldsymbol{k}19.$$

Solution.

$$\boldsymbol{\omega}\cdot\boldsymbol{v} = (2)(-42) + (-1)(-8) + (4)(19) = 0;$$
$$\boldsymbol{r}\cdot\boldsymbol{v} = (-1)(-42) + (10)(-8) + 2(19) = 0.$$

Thus, $\boldsymbol{v}$ is shown to be perpendicular to the plane of $\boldsymbol{\omega}$ and $\boldsymbol{r}$.

As another example of an application of the dot product, consider a spatial linkage where we require that the rotation of a certain link about its own axis be zero. Designating the link by the vector

$$\boldsymbol{r} = \boldsymbol{i}r_x + \boldsymbol{j}r_y + \boldsymbol{k}r_z$$

and its angular velocity by

$$\boldsymbol{\omega} = \boldsymbol{i}\omega_x + \boldsymbol{j}\omega_y + \boldsymbol{k}\omega_z,$$

we see that the requirement will be satisfied if $\boldsymbol{\omega}\cdot\boldsymbol{r} = 0$, or

$$(\boldsymbol{i}\omega_x + \boldsymbol{j}\omega_y + \boldsymbol{k}\omega_z)\cdot(\boldsymbol{i}r_x + \boldsymbol{j}r_y + \boldsymbol{k}r_z) = \omega_x r_x + \omega_y r_y + \omega_z r_z = 0.$$

Some additional laws relating to the dot product and vector cross product are as follows:

$\boldsymbol{A} \cdot \boldsymbol{B} = \boldsymbol{B} \cdot \boldsymbol{A}$ (the commutative law holds for the dot product, but not the cross product);

$$\boldsymbol{A} \cdot (\boldsymbol{B} + \boldsymbol{C}) = \boldsymbol{A} \cdot \boldsymbol{B} + \boldsymbol{A} \cdot \boldsymbol{C}$$

and

$$\boldsymbol{A} \times (\boldsymbol{B} + \boldsymbol{C}) = \boldsymbol{A} \times \boldsymbol{B} + \boldsymbol{A} \times \boldsymbol{C} \quad \text{(distributive law);} \tag{2.11}$$

$$\boldsymbol{A} \cdot (\boldsymbol{B} \times \boldsymbol{C}) = \boldsymbol{B} \cdot (\boldsymbol{C} \times \boldsymbol{A}) = \boldsymbol{C} \cdot (\boldsymbol{A} \times \boldsymbol{B}) = \begin{vmatrix} A_x & A_y & A_z \\ B_x & B_y & B_z \\ C_x & C_y & C_z \end{vmatrix};$$

and

$$\boldsymbol{A} \times (\boldsymbol{B} \times \boldsymbol{C}) = (\boldsymbol{A} \cdot \boldsymbol{C})\boldsymbol{B} - (\boldsymbol{A} \cdot \boldsymbol{B})\boldsymbol{C}$$

and

$$(\boldsymbol{A} \times \boldsymbol{B}) \times \boldsymbol{C} = (\boldsymbol{A} \cdot \boldsymbol{C})\boldsymbol{B} - (\boldsymbol{B} \cdot \boldsymbol{C})\boldsymbol{A}$$

In solving kinematics and dynamics problems, it is often convenient to use mathematics software with built-in vector functions. Some types of software (e.g., Mathcad™) express vectors in column form:

$$\boldsymbol{A} = \begin{bmatrix} A_x \\ A_y \\ A_z \end{bmatrix}$$

instead of the form $\boldsymbol{i}A_x + \boldsymbol{j}A_y + \boldsymbol{k}A_z$.

We then identify the quantity as a matrix with three rows and one column. The software is capable of adding, subtracting vectors, as well as computing their magnitudes and the dot and cross product. Solutions of equations in matrix form will be considered in a later section. If you have programming skills, you might try writing a program to perform vector manipulation.

SAMPLE PROBLEM 2.3

Vector Operations

Consider the following vectors (in column format):

$$\boldsymbol{A} = \begin{bmatrix} 4 \\ 3 \\ 1 \end{bmatrix}; \boldsymbol{B} = \begin{bmatrix} 2 \\ -1 \\ -5 \end{bmatrix}; \boldsymbol{C} = \begin{bmatrix} 16 \\ -8 \\ -10 \end{bmatrix}; \boldsymbol{D} = \begin{bmatrix} -9 \\ -5 \\ 7 \end{bmatrix};$$

$$\boldsymbol{E} = \begin{bmatrix} 16 \\ -8 \\ -10 \end{bmatrix}.$$

Find $\boldsymbol{A} + \boldsymbol{B} + \boldsymbol{C} + \boldsymbol{D} + \boldsymbol{E}$, $\boldsymbol{A} - \boldsymbol{B} + \boldsymbol{C} - \boldsymbol{D} + \boldsymbol{E}$, $\boldsymbol{A}\cdot\boldsymbol{B}$, $\boldsymbol{A}\cdot\boldsymbol{C}$,

$$|\boldsymbol{A}|,\ |\boldsymbol{B}|,\ (|\boldsymbol{A}|)^2,\quad \boldsymbol{A}\cdot\boldsymbol{A},\ \boldsymbol{A}\times\boldsymbol{A},\quad \boldsymbol{C}\times\boldsymbol{E},\ \boldsymbol{A}\times\boldsymbol{B},$$
$$|\boldsymbol{A}\times\boldsymbol{B}|,\ |\boldsymbol{A}|\cdot|\boldsymbol{B}|,\ \boldsymbol{A}\times\boldsymbol{C},\ \boldsymbol{C}\times\boldsymbol{A},\quad \boldsymbol{A}\cdot(\boldsymbol{A}\times\boldsymbol{C}),$$

and

$$\boldsymbol{C}\cdot(\boldsymbol{A}\times\boldsymbol{C}).$$

Solution. Software with vector capabilities was used to obtain the following results:

$$\boldsymbol{A} + \boldsymbol{B} + \boldsymbol{C} + \boldsymbol{D} + \boldsymbol{E} = \begin{bmatrix} 29 \\ -19 \\ -17 \end{bmatrix};\quad \boldsymbol{A} - \boldsymbol{B} + \boldsymbol{C} - \boldsymbol{D} + \boldsymbol{E} = \begin{bmatrix} 43 \\ -7 \\ -21 \end{bmatrix};$$

$$\boldsymbol{A}\cdot\boldsymbol{B} = 0;\quad \boldsymbol{A}\cdot\boldsymbol{C} = 30;\quad |\boldsymbol{A}| = 5.099;\quad |\boldsymbol{B}| = 5.477;\quad (|\boldsymbol{A}|)^2 = 26;$$

$$\boldsymbol{A}\cdot\boldsymbol{A} = 26;\quad \boldsymbol{A}\times\boldsymbol{A} = \begin{bmatrix} 0 \\ 0 \\ 0 \end{bmatrix};\quad \boldsymbol{C}\times\boldsymbol{E} = \begin{bmatrix} 0 \\ 0 \\ 0 \end{bmatrix};\quad \boldsymbol{A}\times\boldsymbol{B} = \begin{bmatrix} -14 \\ 22 \\ 10 \end{bmatrix};$$

$$|\boldsymbol{A}\times\boldsymbol{B}| = 27.928;\quad |\boldsymbol{A}|\cdot|\boldsymbol{B}| = 27.928;\quad \boldsymbol{A}\times\boldsymbol{C} = \begin{bmatrix} -22 \\ 56 \\ -80 \end{bmatrix};$$

$$\boldsymbol{C}\times\boldsymbol{A} = \begin{bmatrix} 22 \\ -56 \\ 80 \end{bmatrix};\quad \boldsymbol{A}\cdot(\boldsymbol{A}\times\boldsymbol{C}) = 0;\quad \boldsymbol{C}\cdot(\boldsymbol{A}\times\boldsymbol{C}) = 0.$$

The result $\boldsymbol{A}\cdot\boldsymbol{B} = 0$ may be unexpected. Since

$$\boldsymbol{A}\cdot\boldsymbol{B} = \boldsymbol{AB}\cos\theta,$$

where θ is the angle separating the vectors and neither $\boldsymbol{A}$ nor $\boldsymbol{B}$ is zero, then $\cos\theta = 0$. This indicates that vectors $\boldsymbol{A}$ and $\boldsymbol{B}$ are perpendicular to one another. The magnitude of the cross product is given by

$$|\boldsymbol{A}\times\boldsymbol{B}| = \boldsymbol{AB}\sin\theta$$

For this special case, with $\boldsymbol{A}$ and $\boldsymbol{B}$ perpendicular, the cross product equals the product of the magnitudes. The calculations show this.

We note that $\boldsymbol{A}\cdot\boldsymbol{C} = \boldsymbol{C}\cdot\boldsymbol{A}$ and that $\boldsymbol{A}\times\boldsymbol{C} = -\boldsymbol{C}\times\boldsymbol{A}$. These relationships are true in general. Of course, $\boldsymbol{A}\cdot\boldsymbol{A}$, the dot product of a vector with itself, is the square of the magnitude of $\boldsymbol{A}$, and $\boldsymbol{A}\times\boldsymbol{A}$, the cross product of a vector with itself, is always zero. Since vectors $\boldsymbol{C}$ and $\boldsymbol{E}$ are identical, $\boldsymbol{C} \times \boldsymbol{E}$ is zero as well. The cross product $\boldsymbol{A} \times \boldsymbol{C}$ is perpendicular to both vector $\boldsymbol{A}$ and vector $\boldsymbol{C}$. Thus, the scalar triple products $\boldsymbol{A} \cdot (\boldsymbol{A} \times \boldsymbol{C})$ and $\boldsymbol{C} \cdot (\boldsymbol{A} \times \boldsymbol{C})$ are zero.

Differentiation of a Vector

If vector $\boldsymbol{A}$ varies in magnitude and direction with time, then the derivative of $\boldsymbol{A}$ with respect to time is given by

$$\frac{d\boldsymbol{A}}{dt} = \lim_{\Delta t \to 0} \frac{\boldsymbol{A}(t + \Delta t) - \boldsymbol{A}(t)}{\Delta t}, \tag{2.12}$$

where Δt represents an increment in time. Note that $d\boldsymbol{A}/dt$ is a vector and that the numerator of the fraction on the right involves changes in both the magnitude and direction of $\boldsymbol{A}$.

If $\boldsymbol{A}$, $\boldsymbol{B}$, and $\boldsymbol{C}$ are vector functions of time and m is a scalar function of time, then the following rules hold:

$$\begin{aligned}
\frac{d}{dt}(\boldsymbol{A} + \boldsymbol{B}) &= \frac{d\boldsymbol{A}}{dt} + \frac{d\boldsymbol{B}}{dt} \\
\frac{d}{dt}(\boldsymbol{A} \times \boldsymbol{B}) &= \boldsymbol{A} \times \frac{d\boldsymbol{B}}{dt} + \frac{d\boldsymbol{A}}{dt} \times \boldsymbol{B} \\
\frac{d}{dt}(\boldsymbol{A} \cdot \boldsymbol{B}) &= \boldsymbol{A} \cdot \frac{d\boldsymbol{B}}{dt} + \frac{d\boldsymbol{A}}{dt} \cdot \boldsymbol{B} \\
\frac{d}{dt}(m\boldsymbol{A}) &= m\frac{d\boldsymbol{A}}{dt} + \frac{dm}{dt}\boldsymbol{A}
\end{aligned} \tag{2.13}$$

$$\frac{d}{dt}[\boldsymbol{A} \times (\boldsymbol{B} \times \boldsymbol{C})] = \boldsymbol{A} \times \left(\boldsymbol{B} \times \frac{d\boldsymbol{C}}{dt}\right) + \boldsymbol{A} \times \left(\frac{d\boldsymbol{B}}{dt} \times \boldsymbol{C}\right) + \frac{d\boldsymbol{A}}{dt} \times (\boldsymbol{B} \times \boldsymbol{C})$$

Recall that $\boldsymbol{A} \times \boldsymbol{B} = -\boldsymbol{B} \times \boldsymbol{A}$. Thus, the order of the vectors in the vector cross products should be retained when we apply the chain rule of differentiation.

The time differential of a unit vector fixed within a fixed coordinate system is, of course, zero, since neither the magnitude nor the direction of the vector changes. The time differential of a unit vector in a moving coordinate system is not zero. In applications of vectors to velocity and acceleration, we will consider both fixed and moving coordinate systems.

Solution of Vector Equations

Let a vector equation in the form

$$\boldsymbol{r}_0 + \boldsymbol{r}_1 + \boldsymbol{r}_2 + \boldsymbol{r}_3 = 0 \tag{2.14}$$

represent a spatial linkage.

Each vector may be expressed in terms of its x, y, and z components and the unit vectors $\boldsymbol{i}$, $\boldsymbol{j}$, and $\boldsymbol{k}$, such as

$$\boldsymbol{r}_1 = r_{1x}\boldsymbol{i} + r_{1y}\boldsymbol{j} + r_{1z}\boldsymbol{k}$$

and so on. Since the sum of the components in each coordinate direction must equal zero, we have three independent scalar equations:

$$r_{0x} + r_{1x} + r_{2x} + r_{3x} = 0,$$

and so on. Thus, in general, the vector equation $\boldsymbol{r}_0 + \boldsymbol{r}_1 + \boldsymbol{r}_2 + \boldsymbol{r}_3 = 0$ may be solved for three unknown components. For example, we may solve for r_{3x}, r_{3y}, and r_{3z} if the other components are known. Alternatively, each vector could be expressed in terms of its magnitude and two independent angular coordinates. Equation (2.14) could then be solved for any combination of three magnitudes and directions. Planar linkages can be represented by two independent scalar equations. Only two unknown components are permitted if Eq. (2.14) is applied to planar linkages.

If you prefer to solve the nonlincar position equations of a linkage using an iterative process, you will need an initial or trial guess at the solution. A sketch will help. Later, when you determine velocities, you will use linear equations. Then, matrices will be useful. A number of other methods for solving the nonlincar position equations of a linkage include

- graphical methods,
- motion simulation software,
- the dot-product method, and
- the cross-product method. This is a vector elimination method suggested by Chace (1963). Although it seems difficult at first, you can use it to write an efficient noniterative program for solving four-bar linkages.

Solution of Planar Vector Equations

Consider the planar vector equation

$$\boldsymbol{A} + \boldsymbol{B} + \boldsymbol{C} = 0, \tag{2.15}$$

or, in terms of unit vectors ($\boldsymbol{A}^u$, etc.) and magnitudes (A, etc.),

$$\boldsymbol{A}^u A + \boldsymbol{B}^u B + \boldsymbol{C}^u C = 0. \tag{2.16}$$

If the magnitude and direction of the same vector are unknown, then the solution is easily obtained. For example, C is unknown; we use

$$\boldsymbol{C} = -(A_x + B_x)\boldsymbol{i} - (A_y + B_y)\boldsymbol{j} \tag{2.17}$$

or

$$\boldsymbol{C} = -(\boldsymbol{A}\cdot\boldsymbol{i} + \boldsymbol{B}\cdot\boldsymbol{i})\boldsymbol{i} - (\boldsymbol{A}\cdot\boldsymbol{j} + \boldsymbol{B}\cdot\boldsymbol{j})\boldsymbol{j}. \tag{2.18}$$

If the magnitudes of two different vectors are unknown, the vector cross product method, (a vector elimination method suggested by Chace, 1963) may be used. Suppose, for example, magnitudes $A = |\boldsymbol{A}|$ and $B = |\boldsymbol{B}|$ are unknown in the vector equation $\boldsymbol{A}^u A + \boldsymbol{B}^u B + \boldsymbol{C}^u C = 0$. We take the dot product of each term with $\boldsymbol{B}^u \times \boldsymbol{k}$, noting that $\boldsymbol{B}^u \cdot (\boldsymbol{B}^u \times \boldsymbol{k}) = 0$, since vector $\boldsymbol{B}^u$ is perpendicular to vector $\boldsymbol{B}^u \times \boldsymbol{k}$. Thus, we obtain

$$\boldsymbol{A}^u A \cdot (\boldsymbol{B}^u \times \boldsymbol{k}) + \boldsymbol{C} \cdot (\boldsymbol{B}^u \times \boldsymbol{k}) = 0,$$

from which the magnitude of vector $\boldsymbol{A}$ is given by

$$A = \frac{-\boldsymbol{C} \cdot (\boldsymbol{B}^u \times \boldsymbol{k})}{\boldsymbol{A}^u \cdot (\boldsymbol{B}^u \times \boldsymbol{k})}. \tag{2.19}$$

Similarly, the magnitude of $\boldsymbol{B}$ is given by

$$B = \frac{-\boldsymbol{C} \cdot (\boldsymbol{A}^u \times \boldsymbol{k})}{\boldsymbol{B}^u \cdot (\boldsymbol{A}^u \times \boldsymbol{k})}. \tag{2.20}$$

If the vector directions $\boldsymbol{A}^u$ and $\boldsymbol{B}^u$ are unknown, but all vector magnitudes are known, the solution of the equation $\boldsymbol{A} + \boldsymbol{B} + \boldsymbol{C} = 0$ is more complicated. In this case, the results are

$$\boldsymbol{A} = \mp \left[B^2 - \left(\frac{C^2 + B^2 - A^2}{2C} \right)^2 \right]^{1/2} (\boldsymbol{C}^u \times \boldsymbol{k}) + \left[\frac{C^2 + B^2 - A^2}{2C} - C \right] \boldsymbol{C}^u \tag{2.21}$$

and

$$\boldsymbol{B} = \pm \left[B^2 - \left(\frac{C^2 + B^2 - A^2}{2C} \right)^2 \right]^{1/2} (\boldsymbol{C}^u \times \boldsymbol{k}) - \left[\frac{C^2 + B^2 - A^2}{2C} \right] \boldsymbol{C}^u. \tag{2.22}$$

(The significance of the signs before the radical are illustrated in the section that follows.)

When the magnitude of $\boldsymbol{A}$ and the direction of $\boldsymbol{B}$ are unknown, $\boldsymbol{A}$ and $\boldsymbol{B}$ may be found by the equations

$$\boldsymbol{A} = \left[-\boldsymbol{C} \cdot \boldsymbol{A}^u \mp \sqrt{B^2 - \left[\boldsymbol{C} \cdot (\boldsymbol{A}^u \times \boldsymbol{k}) \right]^2} \right] \boldsymbol{A}^u \tag{2.23}$$

and

$$\boldsymbol{B} = -[\boldsymbol{C} \cdot (\boldsymbol{A}^u \times k)](\boldsymbol{A}^u \times \boldsymbol{k}) \pm \sqrt{\boldsymbol{B}^2 - [\boldsymbol{C} \cdot (\boldsymbol{A}^u \times \boldsymbol{k})]^2} \boldsymbol{A}^u. \tag{2.24}$$

This approach uses vector notation throughout, unlike alternative methods that use vector analysis to derive scalar equations. If the method is to be used for computer-aided analysis and design of mechanisms, it is essential to use software with vector capabilities or to write subroutines for that purpose.

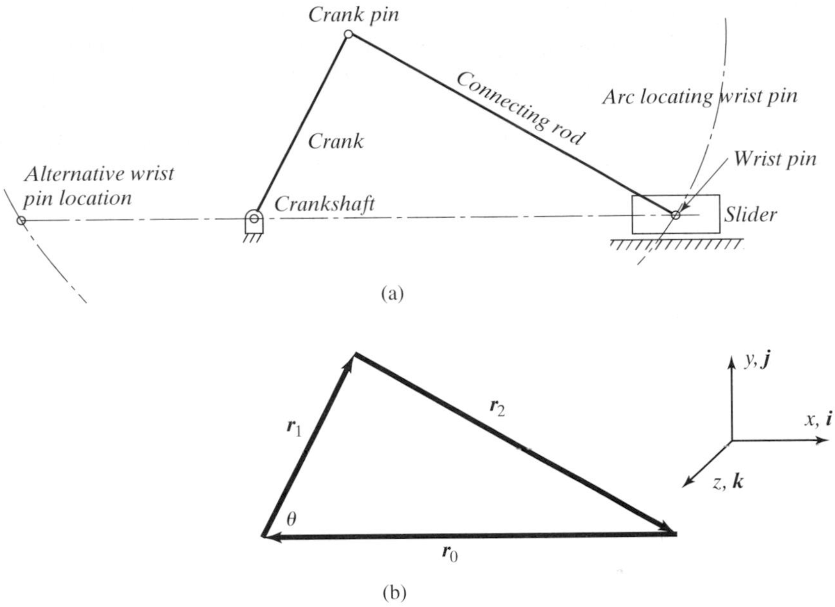

FIGURE 2.9 (a) Slider-crank linkage. (b) Vector representation.

2.3 ANALYTICAL VECTOR METHODS APPLIED TO DISPLACEMENT ANALYSIS OF PLANAR LINKAGES

An in-line slider-crank mechanism has two assembly configurations. The wrist pin can be located by drawing an arc the length of the connecting rod. The two possible wrist pin locations are shown in Figure 2.9a. We usually want an analytical solution. Suppose we have already decided crank and connecting rod lengths. If we are given the position of the crank for the linkage of Figure 2.9b, then the unknowns will be the connecting rod orientation and slider position. These may be found by using the equations of the previous section to solve the vector equation

$$\boldsymbol{r}_0 + \boldsymbol{r}_1 + \boldsymbol{r}_2 = 0. \tag{2.25}$$

SAMPLE PROBLEM 2.4

Slider-Crank Linkage

Suppose an in-line slider-crank linkage (see Figure 2.9a) has a crank described by the vector $\boldsymbol{r}_1 = \boldsymbol{i} + \boldsymbol{j}2$ (at the instant shown) and a connecting rod length $r_2 = 4$. Find the connecting rod orientation $\boldsymbol{r}_2^u$ and the slider position $\boldsymbol{r}_0$. (See Figures 2.9a and b.)

Solution. The slider position is given by the vector

$$\boldsymbol{r}_0 = \left[-\boldsymbol{r}_1 \cdot \boldsymbol{r}_0^u \mp \sqrt{r_2^2 - \left[\boldsymbol{r}_1 \cdot \left(\boldsymbol{r}_0^u \times \boldsymbol{k} \right) \right]^2} \right] \boldsymbol{r}_0^u$$

$$= \left[-(\boldsymbol{i} + \boldsymbol{j}2)\cdot(-\boldsymbol{i}) \mp \sqrt{4^2 - \left[(\boldsymbol{i} + \boldsymbol{j}2)\cdot(-\boldsymbol{i} \times \boldsymbol{k}) \right]^2} \right](-\boldsymbol{i}),$$

where the sign of the root depends on the initial assembly configuration. The positive root applies to the configuration in the figure, yielding $\boldsymbol{r}_0 = -\boldsymbol{i}4.464$. The alternative wrist-pin location, determined by taking the negative root, is $\boldsymbol{r}_0 = +\boldsymbol{i}2.464$.

The vector representing the connecting rod is

$$\begin{aligned} \boldsymbol{r}_2 &= -[(\boldsymbol{i} + \boldsymbol{j}2)\cdot(-\boldsymbol{i} \times \boldsymbol{k})](-\boldsymbol{i} \times \boldsymbol{k}) \\ &\quad \pm \sqrt{4^2 - [(\boldsymbol{i} + \boldsymbol{j}2)\cdot(-\boldsymbol{i} + \boldsymbol{k})]^2}(-\boldsymbol{i}) \\ &= -2\boldsymbol{j} \pm \sqrt{12}(-\boldsymbol{i}). \end{aligned}$$

The negative root applies to the configuration shown, yielding

$$\boldsymbol{r}_2 = \boldsymbol{i}3.464 - \boldsymbol{j}2$$

and

$$\boldsymbol{r}_2^u = \frac{(\boldsymbol{i}3.464 - \boldsymbol{j}2)}{4} = \boldsymbol{i}0.866 - \boldsymbol{j}0.5.$$

For the alternative wrist-pin location, the positive root applies, yielding

$$\boldsymbol{r}_2 = -2\boldsymbol{j} + \sqrt{12}(-\boldsymbol{i}) = -\boldsymbol{i}3.464 - \boldsymbol{j}2$$

and

$$\boldsymbol{r}_2^u = -\boldsymbol{i}0.866 - \boldsymbol{j}0.5.$$

The Four-Bar Linkage

A graphical layout of a four-bar linkage is easily constructed. We require only that the position of one link be given relative to the frame and that the link lengths be known. Then, the linkage may be drawn with the aid of a compass. It can be quite time-consuming, however, to develop the analytical formulas for link positions that are to be used to write a computer program.

Position Analysis Using the Vector Cross Product

Equations 2.21 and 2.22 may be used to find linkage displacements. These equations apply when the directions of vectors $\boldsymbol{A}$ and $\boldsymbol{B}$ are unknown. The four-bar planar linkage of Figure 2.10a is described by the vector equation

$$\boldsymbol{r}_0 + \boldsymbol{r}_1 + \boldsymbol{r}_2 + \boldsymbol{r}_3 = 0. \tag{2.26}$$

Suppose we have already decided the length of each of the links in a tentative design, where link 1 is to be the driver. For a given angular position of link 1, the diagonal vector

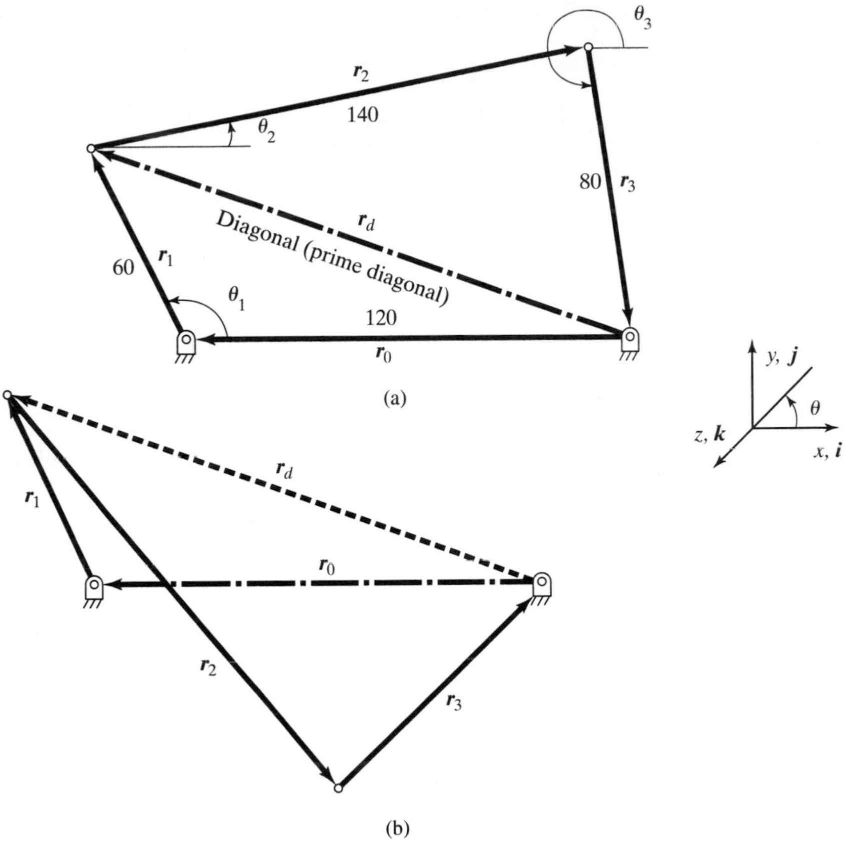

FIGURE 2.10 Four-bar planar linkage. (a) Use the lower set of signs in the position analysis equations for this assembly configuration. This configuration is also called the *open* position. (b) Alternative assembly. Use the upper set of signs. This configuration is also called the *closed* position.

(also called the **prime diagonal**) is given by

$$\boldsymbol{r}_d = \boldsymbol{r}_0 + \boldsymbol{r}_1,$$

and the triangle formed by links 2 and 3 and the diagonal is described by

$$\boldsymbol{r}_d = -(\boldsymbol{r}_2 + \boldsymbol{r}_3). \tag{2.27}$$

Because the lengths of the links are specified and the orientation of link 1 is given, the following substitution

$$\boldsymbol{A} = \boldsymbol{r}_2,$$
$$\boldsymbol{B} = \boldsymbol{r}_3,$$

and

$$\boldsymbol{C} = \boldsymbol{r}_d = \boldsymbol{r}_0 + \boldsymbol{r}_1$$

may be made in Eqs. 2.21 and 2.22; yielding

$$\mathbf{r}_2 = \mp\sqrt{r_3^2 - \left(\frac{r_3^2 - r_2^2 + r_d^2}{2r_d}\right)^2}\left(\mathbf{r}_d^u \times \mathbf{k}\right) + \left(\frac{r_3^2 - r_2^2 + r_d^2}{2r_d} - r_d\right)\mathbf{r}_d^u \tag{2.28}$$

and

$$\mathbf{r}_3 = \pm\sqrt{r_3^2 - \left(\frac{r_3^2 - r_2^2 + r_d^2}{2r_d}\right)^2}\left(\mathbf{r}_d^u \times \mathbf{k}\right) - \left(\frac{r_3^2 - r_2^2 + r_d^2}{2r_d}\right)\mathbf{r}_d^u. \tag{2.29}$$

If the linkage is assembled so that the vector loop $\mathbf{r}_2\mathbf{r}_3\mathbf{r}_d$ is clockwise, then we use the lower set of signs in the preceding equations. (See Figure 2.10a.) If the loop is counterclockwise, we use the upper set of signs (Figure 2.10b).

SAMPLE PROBLEM 2.5

Position Analysis Using the Vector Cross Product

A planar mechanism (see Figure 2.10a) has the following link lengths:

Link 0, fixed: 120 mm;

Link 1, crank: 60 mm;

Link 2, coupler: 140 mm;

Link 3, follower: 80 mm.

Find the orientation of links 2 and 3 at the instant the internal angle between the crank and fixed link is 116.56°.

Solution. If the links are drawn to scale, the joint between links 2 and 3 can be located by using a compass. Alternatively, an analytical vector solution may be obtained as follows. Select the coordinate system so that the fixed link lies in the $-x$ direction (as in Figure 2.10a.) Then the known values are

$$\mathbf{r}_0 = -\mathbf{i}120,$$
$$\mathbf{r}_1 = 60(\mathbf{i}\cos 116.56^\circ + \mathbf{j}\sin 116.56^\circ) = -\mathbf{i}26.83 + \mathbf{j}53.67,$$
$$r_2 = 140,$$

and

$$r_3 = 80.$$

The diagonal is given by the vector

$$\mathbf{r}_d = (\mathbf{r}_0 + \mathbf{r}_1) = -\mathbf{i}146.83 + \mathbf{j}53.67.$$

which has a magnitude $r_d = 156.33$ and unit vector

$$\boldsymbol{r}_d^u = -\boldsymbol{i}0.9392 + \boldsymbol{j}0.3433.$$

The direction of the cross product $\boldsymbol{r}_d^u \times \boldsymbol{k}$ is given by the right-hand rule. The cross-product vector lies in the xy-plane and is perpendicular to $\boldsymbol{r}_d^u$. Expressing the cross product in determinant form, we have

$$\boldsymbol{r}_d^u \times \boldsymbol{k} = \begin{vmatrix} \boldsymbol{i} & \boldsymbol{j} & \boldsymbol{k} \\ -0.9392 & 0.3433 & 0 \\ 0 & 0 & 1 \end{vmatrix} = \boldsymbol{i}0.3433 + \boldsymbol{j}0.9392$$

Using Eqs. 2.28 and 2.29, we define

$$a = \frac{r_3^2 - r_2^2 + r_d^2}{2r_d} = \frac{80^2 - 140^2 + 156.33^2}{2 \times 156.34} = 35.95.$$

When the vector loop $\boldsymbol{r}_2\boldsymbol{r}_3\boldsymbol{r}_d$ is clockwise, the lower set of signs in Eqs. 2.28 and 2.29 applies. The vector representing the coupler is given by

$$\begin{aligned} \boldsymbol{r}_2 &= +\sqrt{r_3^2 - a^2}(\boldsymbol{r}_d^u \times \boldsymbol{k}) + \boldsymbol{r}_d^u(a - r_d) \\ &= \sqrt{80^2 - 35.95^2}(\boldsymbol{i}0.3433 + \boldsymbol{j}0.9392) \\ &\quad + (35.95 - 156.34)(-\boldsymbol{i}0.9392 + \boldsymbol{j}0.3433) \\ &= \boldsymbol{i}137.60 + \boldsymbol{j}25.79 = 140\angle 10.6°. \end{aligned}$$

The vector representing the follower is given by

$$\begin{aligned} \boldsymbol{r}_3 &= -\sqrt{r_3^2 - a^2}(\boldsymbol{r}_d^u \times \boldsymbol{k}) - a\boldsymbol{r}_d^u \\ &= -\sqrt{80^2 - 35.95^2}(\boldsymbol{i}0.3433 + \boldsymbol{j}0.9392) \\ &\quad -39.95(-\boldsymbol{i}0.9392 + \boldsymbol{j}0.3433) \\ &= \boldsymbol{i}9.24 - \boldsymbol{j}79.46 = 80\angle -83.4°. \end{aligned}$$

(Note that the magnitudes of vectors $\boldsymbol{r}_2$ and $\boldsymbol{r}_3$ agree with the given data.)

In some linkage configurations, it is impossible to go from one mode to another without disassembling the links. Applying Grashof-type criteria to check this linkage (see Chapter 1), we find that

$$L_{max} + L_{min} = L_a + L_b$$
$$(140 + 60 = 120 + 80)$$

This is a crossover-position or change-point mechanism. The mechanism may go from one mode to another, depending on inertial forces, spring forces, or other forces. Using Eqs. 2.28 and 2.29 with the upper set of signs for the position where $\boldsymbol{r}_2$, $\boldsymbol{r}_3$, and $\boldsymbol{r}_d$ form a counterclockwise loop, we obtain the vector representing the coupler:

$$\boldsymbol{r}_2 = \boldsymbol{i}88.54 - \boldsymbol{j}108.44 = 140\angle -50.8°.$$

The vector representing the follower is

$$\boldsymbol{r}_3 = \boldsymbol{i}58.30 + \boldsymbol{j}54.68 = 80\angle 43^\circ.$$

The transmission angle $\phi = 0$ at $\theta_1 = 0$. If this linkage is stopped in the $\theta_1 = 0$ position, it cannot be restarted by driving link 1. Special consideration has to be given to problems associated with driving change-point linkages.

Ordinarily, we would want to check the entire range of travel of the linkage. The above example shows how much work is required to find only one linkage position. In next example we will try "working smarter" by using mathematics software to show the performance of the linkage through a full cycle of motion.

SAMPLE PROBLEM 2.6

Checking the performance of a linkage design

In Chapter 1, we attempted to optimize a crank-rocker linkage with a 30° range of output crank motion. The resulting linkage had the following dimensions (mm):

$$L_0 = 300, L_1 = 50, L_2 = 235, \text{and } L_3 = 193.2.$$

Plot the coupler and output crank positions and the transmission angle against the input crank position.

Solution summary. The solution is similar to that for the previous sample problem, except that we let the computer do the tedious work. Vectors in column form represent links, and variables are a function of input angle θ_1 which varies from zero to 2π. Extreme values of the transmission angle, $\phi(0)$ and $\phi(\pi)$, agree with the values in Chapter 1. The results, plotted in degrees and shown in Figure 2.11, also show the range of output crank motion.

Detailed solution. It is customary to show vectors in ***boldface*** type to distinguish them from scalar quantities shown in *lightface* type. The sample problem detailed solutions that utilize mathematics software use a different convention. Both vectors and scalars are shown in lightface type. In the calculations that follow, link lengths are L_0, L_1, etc. and the corresponding vectors are r_0, r_1, etc. The rectangular unit vectors are i, j, and k.

$$L_0 := 300 \quad L_1 := 50 \quad L_2 := 235 \quad L_3 := 193.2 \quad \text{mm}$$

$$\theta_1 := 0, \frac{\pi}{180} \ldots 2 \cdot \pi$$

$$r_0 := \begin{bmatrix} -L_0 \\ 0 \\ 0 \end{bmatrix} \quad r_0 = \begin{bmatrix} -300 \\ 0 \\ 0 \end{bmatrix} \quad r_1(\theta_1) := \begin{bmatrix} L_1 \cdot \cos(\theta_1) \\ L_1 \cdot \sin(\theta_1) \\ 0 \end{bmatrix}$$

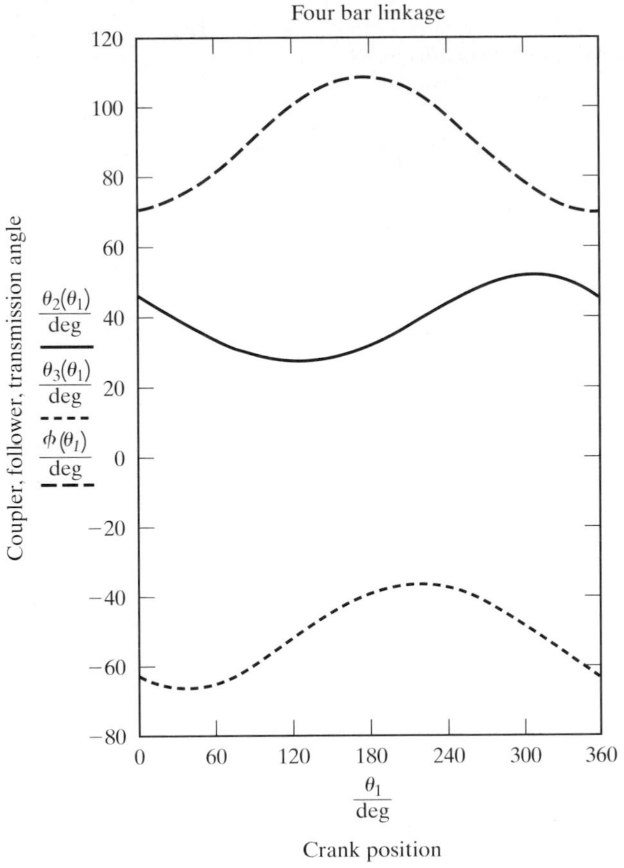

FIGURE 2.11 Checking the performance of a linkage design.

Diagonal vector:

$$r_d(\theta_1) := r_0 + r_1(\theta_1)$$
$$r_{du}(\theta_1) := \frac{r_d(\theta_1)}{|r_d(\theta_1)|}$$

Rectangular unit vectors $\quad i := \begin{bmatrix} 1 \\ 0 \\ 0 \end{bmatrix} \quad j := \begin{bmatrix} 0 \\ 1 \\ 0 \end{bmatrix} \quad k := \begin{bmatrix} 0 \\ 0 \\ 1 \end{bmatrix}$

Define $a(\theta_1) := \dfrac{L_3^2 - L_2^2 + (|r_d(\theta_1)|)^2}{2|r_d(\theta_1)|}$

$q = 1$ for the assembly configuration with vector loop $r_2r_3r_d$ clockwise and -1 for the same vector loop traversed counterclockwise.
$q := 1$
Coupler vector:

$$r_2(\theta_1) := q \cdot \sqrt{L_3^2 - a(\theta_1)^2} \cdot (r_{du}(\theta_1) \times k) + r_{du}(\theta_1) \cdot (a(\theta_1) - |r_d(\theta_1)|)$$
$$\theta_2(\theta_1) := \text{angle}(r_2(\theta_1)_0, r_2(\theta_1)_1)$$

Follower crank vector:

$$r_3(\theta_1) := q \cdot \sqrt{L_3^2 - a(\theta_1)^2} \cdot (r_{du}(\theta_1) \times k) - r_{du}(\theta_1) \cdot a(\theta_1)$$
$$\theta_3(\theta_1) := \text{angle}(r_3(\theta_1)_0, r_3(\theta_1)_1) - 2 \cdot \pi$$

Transmission angle:

$$\phi(\theta_1) := \text{acos}\left[\frac{L_2^2 + L_3^2 - (|r_d(\theta_1)|)^2}{2 \cdot L_2 \cdot L_3}\right]$$

Check transmission angle limits:

$$\frac{\phi(0)}{\text{deg}} = 70.674 \qquad \frac{\phi(\pi)}{\text{deg}} = 109.258 \text{ deg}$$

Position Analysis Using the Dot Product

As an alternative to using the vector cross product, we may analyze linkage displacement by using the dot product. In Figure 2.12, a vector representation of a planar four-bar linkage, the diagonal vector is given by:

$$\boldsymbol{r}_d = \boldsymbol{r}_0 + \boldsymbol{r}_1. \tag{2.30}$$

Suppose we are to determine the linkage position for a given angle θ_1 if all of the link lengths are known. Taking the dot product of each side of this equation with itself, we have

$$\boldsymbol{r}_d \cdot \boldsymbol{r}_d = (\boldsymbol{r}_0 + \boldsymbol{r}_1) \cdot (\boldsymbol{r}_0 + \boldsymbol{r}_1), \tag{2.31}$$

or

$$r_d^2 = r_0^2 + 2r_0 r_1 \cos\phi_a + r_1^2, \tag{2.32}$$

where the angle between vectors $\boldsymbol{r}_0$ and $\boldsymbol{r}_1$ is $\phi_a = 180° - \theta_1$. Thus, Eq. (2.32) is equivalent to the law of cosines:

$$r_d^2 = r_0^2 - 2r_0 r_1 \cos\theta_1 + r_1^2.$$

The direction of the diagonal is given in terms of the x and y components of $\boldsymbol{r}_0$ and $\boldsymbol{r}_1$:

$$\tan\theta_d = \frac{r_{0y} + r_{1y}}{r_{0x} + r_{1x}} \tag{2.33}$$

where $r_{0y} = 0$ if the x-axis is selected to be parallel to the fixed link. For some four-bar linkage proportions, θ_d can be in any of the four quadrants. If Eq. (2.33) is programmed for machine calculations, it is important that θ_d be located in its proper quadrant. This may be possible if a two-argument arctangent function is available wherein both numerator and denominator become inputs and the quadrant of the angle is determined by the signs of the numerator and denominator. As an alternative, we note that

$$\sin\theta_d = \frac{r_{0y} + r_{1y}}{r_d} \tag{2.34}$$

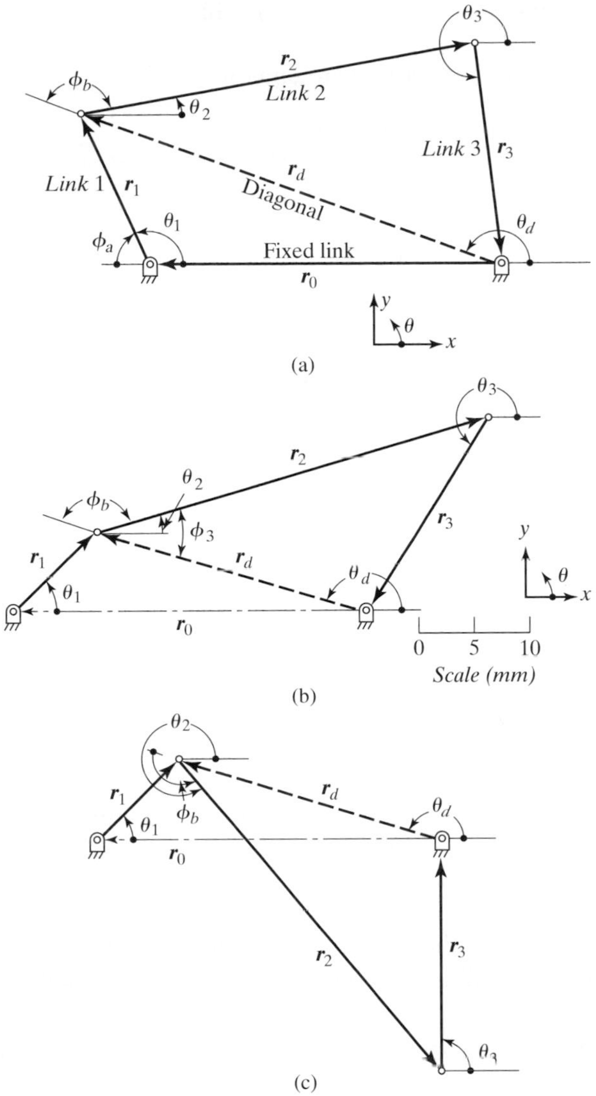

FIGURE 2.12 (a) Four-bar planar linkage; position analysis by dot-product method. (b) Position analysis sample problem. (c) Alternative assembly mode.

and

$$\cos\theta_d = \frac{r_{0x} + r_{1x}}{r_d}. \tag{2.35}$$

It can be shown that

$$\tan\left(\frac{\theta_d}{2}\right) = \frac{1 - \cos\theta_d}{\sin\theta_d}. \tag{2.36}$$

Using the last three expressions instead of Eq. (2.33), we are able to determine the correct quadrant of θ_d.

We continue the analysis with the loop closure equation

$$\boldsymbol{r}_d + \boldsymbol{r}_2 + \boldsymbol{r}_3 = 0, \tag{2.37}$$

or

$$-\boldsymbol{r}_3 = \boldsymbol{r}_d + \boldsymbol{r}_2.$$

Taking the dot product of each side of this equation with itself, we have

$$\boldsymbol{r}_3 \cdot \boldsymbol{r}_3 = (\boldsymbol{r}_d + \boldsymbol{r}_2) \cdot (\boldsymbol{r}_d + \boldsymbol{r}_2), \tag{2.38}$$

or

$$r_3^2 = r_d^2 + 2r_d r_2 \cos\phi_b + r_2^2, \tag{2.39}$$

so that

$$\cos\phi_b = \frac{r_3^2 - r_2^2 - r_d^2}{2r_d r_2}, \quad \text{for} \quad 0 \leq \phi_b \leq 180°. \tag{2.40}$$

Then,

$$\theta_2 = \theta_d \mp \phi_b. \tag{2.41}$$

The sign before ϕ_b depends on the mode of assembly of the linkage. If the vector loop $\boldsymbol{r}_2\boldsymbol{r}_3\boldsymbol{r}_d$ is clockwise, as in Figure 2.12a, the negative sign applies; if counterclockwise, the positive sign applies. In most cases, once the linkage is assembled, the mode will not change.

The position of link 3 may now be found by using the dot product in a similar manner or by using the law of sines. So that errors of angle quadrant are avoided, however, the x and y components of $\boldsymbol{r}_3$ will be determined from the loop equation for the entire linkage, written as

$$\boldsymbol{r}_3 = -(\boldsymbol{r}_0 + \boldsymbol{r}_1 + \boldsymbol{r}_2) \tag{2.42}$$

Then,

$$\sin\theta_3 = \frac{r_{3y}}{r_3}, \tag{2.43a}$$

$$\cos\theta_3 = \frac{r_{3x}}{r_3}, \tag{2.43b}$$

and θ_3 is found from the equation

$$\tan\left(\frac{\theta_3}{2}\right) = \frac{1 - \cos\theta_3}{\sin\theta_3}. \tag{2.44}$$

SAMPLE PROBLEM 2.7

Position Analysis Using the Dot Product

A four-bar linkage has the following link lengths:

Fixed link: $r_0 = 30$ mm;
Drive Crank: $r_1 = 10$ mm;
Coupler: $r_2 = 35$ mm;
Follower: $r_3 = 20$ mm.

Find the position of links 2 and 3 when $\theta_1 = 45°$, as shown in Figure 2.12b.

Solution. We will utilize the equations developed from the position analysis using the dot product, which could easily be programmed if required. From Eq. (2.32), we have

$$r_d^2 = 30^2 + 2 \times 30 \times 10 \cos(180° - 45°) + 10^2.$$

or

$$r_d = 23.99 \text{ mm}.$$

From Eqs. (2.34) through (2.36),

$$\sin\theta_d = \frac{0 + 10\sin 45°}{23.99} = 0.2947,$$

$$\cos\theta_d = \frac{-30 + 10\cos 45°}{23.99} = -0.9588,$$

and

$$\tan\left(\frac{\theta_d}{2}\right) = \frac{(1 + 0.9558)}{0.2947},$$

so that

$$\theta_d = 162.86°.$$

Note that Eq. (2.33) yields

$$\theta_d = \arctan\frac{0 + 10\sin 45°}{-30 + 10\cos 45°} = \arctan(-0.3084),$$

which will usually be evaluated as $-17.14°$. The error would be obvious if we used a sketch, but it might go undetected in machine calculations. From Eq. (2.40), we obtain

$$\cos\phi_b = \frac{20^2 - 35^2 - 23.99^2}{2 \times 23.99 \times 35},$$

so that

$$\phi_b = 146.51°.$$

For the assembly mode shown in Figure 2.12b (the clockwise vector loop),

$$\theta_2 = \theta_d - \phi_b = 162.86 - 146.51 = 16.35°,$$

and for the alternative mode in Figure 2.12c,

$$\theta_2 = \theta_d + \phi_b = 162.86 + 146.51 = 309.37°.$$

Using the loop closure equation, Eq. (2.42), for the x and y components of r_3, we have, for the assembly mode of Figure 2.12b,

$$r_{3x} = -(-30 + 10\cos 45° + 35\cos 16.35°) = -10.66° \text{ mm}$$

and

$$r_{3y} = -(0 + 10\sin 45° + 35\sin 16.35°) = -16.92° \text{ mm}.$$

Then, the position of link 3 is found by using an $xy : r\theta$ conversion, from which it follows that $r_3 = 20\angle -122.21°$. If this conversion is not available, then we may use Eqs. (2.43a) through (2.44), whereupon we obtain

$$\tan\left(\frac{\theta_3}{2}\right) = \frac{1 + 10.66/20}{-16.92/20},$$

or

$$\theta_3 = -122.2°(237.8°).$$

For the alternative mode, from Figure 2.12c,

$$r_{3x} = -(-30 + 10\cos 45° + 35\cos 309.37°) = 0.73 \text{ mm},$$
$$r_{3y} = -(0 + 10\sin 45° + 35\sin 309.37°) = 19.99 \text{ mm},$$

so that

$$\boldsymbol{r}_3 = 20 \text{ mm at } \theta_3 = 87.91°.$$

For a given value of θ_1, the diagonal vector r_d is the same for both assembly modes. Thus, the triangle formed by the diagonal and links 2 and 3 for one mode is congruent to the corresponding triangle for the other mode (but reflected about r_d).

For the dimensions given in this example, link 1, the drive crank, is shortest, and $L_{max} + L_{min} < L_a + L_b$ $(35 + 10 < 30 + 20)$. Thus, we have a crank-rocker linkage as defined in Chapter 1. If this linkage is assembled in one mode (Figure 2.12b), it cannot assume the other mode without reassembly.

The vector methods of position analysis using the dot product and cross product are not limited to one type of linkage. The four-bar linkage was used only as an illustration; other configurations can be solved by using the same principles and a bit of ingenuity.

2.4 COMPLEX NUMBERS

Complex numbers, each made up of a real and an imaginary part, provide an alternative representation for vectors that lie in a plane. Once the rules for handling complex numbers have been mastered, it is fairly easy to apply complex numbers to the analysis of planar linkages. Note that the imaginary quantity is a mathematical artifice. When used together in the analysis of mechanisms, however, the real and imaginary parts of complex numbers represent components of actual dimensions, velocities, and accelerations.

Rectangular Form

A complex number may be written in the rectangular form

$$z = x + jy,$$

where $j = \sqrt{-1}$ is the imaginary unit. The term x represents a real number called the real part of the complex number z. The term y represents a real number called the imaginary part of z.

The *complex plane* shown in Figure 2.13 permits the graphical representation of vectors as complex numbers. Consider a vector $\boldsymbol{R}$ of magnitude R with components R_x and R_y along the real and imaginary axes, respectively. Using angle $\theta = \arctan(R_y/R_x)$ in the complex plane and magnitude $R = \sqrt{R_x^2 + R_y^2}$, vector $\boldsymbol{R}$ may be identified by its real part $R_x = R\cos\theta$ and its imaginary part $R_y = R\sin\theta$ and may be written in the form

$$\boldsymbol{R} = R_x + jR_y. \tag{2.45}$$

Polar Form

It can be shown that

$$e^{j\theta} = \cos\theta + j\sin\theta \tag{2.46}$$

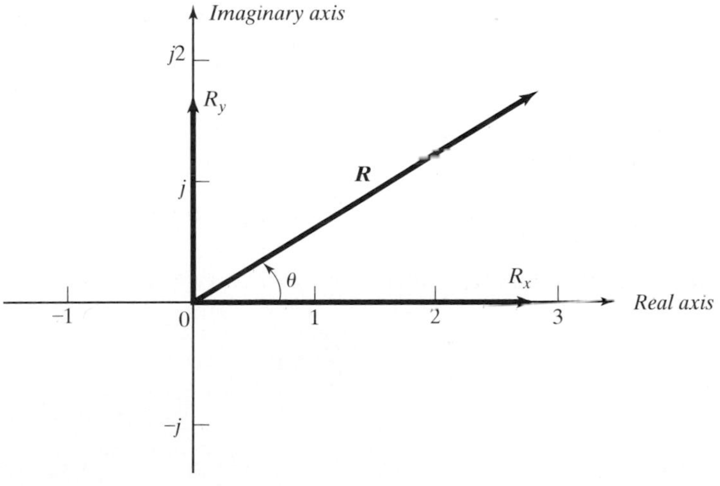

FIGURE 2.13 A vector in the complex plane.

(an identity called the Euler formula). Using this identity and Eq. (2.45), we can express a vector in *polar form* in terms of its magnitude and the complex exponential as

$$\boldsymbol{R} = R(\cos\theta + j\sin\theta) = Re^{j\theta}, \tag{2.47}$$

where θ is in radians.

Complex Arithmetic—Addition

The rectangular form of a complex number is convenient for addition and subtraction. For example, if

$$\boldsymbol{R}_1 = R_{1x} + jR_{1y}$$

and

$$\boldsymbol{R}_2 = R_{2x} + R_{2y},$$

then

$$\boldsymbol{R}_1 + \boldsymbol{R}_2 = (R_{1x} + R_{2x}) + j(R_{1y} + R_{2y}). \tag{2.48}$$

Two complex numbers are equal if and only if the real and imaginary parts of the first are respectively equal to the real and imaginary parts of the second.

Multiplication, Division, and Differentiation

The polar form of a complex number may be more convenient than the rectangular form for multiplication, division, and differentiation. Following the rules of algebra and calculus, if

$$\boldsymbol{R} = Re^{j\theta},$$

then

$$\boldsymbol{R}e^{j\phi} = Re^{j(\theta+\phi)}, \tag{2.49}$$

$$\boldsymbol{R}/e^{j\phi} = Re^{j(\theta-\phi)}, \tag{2.50}$$

and

$$\frac{d\boldsymbol{R}}{dt} = j\omega Re^{j\theta} + e^{j\theta}\frac{dR}{dt}, \qquad \text{where } \omega = \frac{d\theta}{dt}. \tag{2.51}$$

For a rigid link of fixed length R, then

$$\frac{dR}{dt} = 0$$

and

$$\frac{d\boldsymbol{R}}{dt} = j\omega Re^{j\theta} = j\omega\boldsymbol{R}. \tag{2.52}$$

Using the last expression for constant R and noting that j can be expressed in polar form as $j = e^{j\pi/2}$, we obtain

$$\frac{d\boldsymbol{R}}{dt} = \omega R e^{j(\theta+\pi/2)}. \tag{2.53}$$

This result is useful in the velocity analysis of linkages.

When using a calculator, one may employ the rectangular–polar ($xy : R\theta$) conversion feature to put a complex number in a form convenient for addition, differentiation, and so on.

In solving problems involving planar linkages, vector analysis and complex-number methods, as well as other methods, are at our disposal. In some ways, the imaginary unit j resembles a unit vector in the y direction, similar to the unit vector $\boldsymbol{j}$. However, operations with the imaginary unit are different. For example, multiplication of a vector by the unit vector $\boldsymbol{j}$ is defined only by the dot and cross products, neither of which has the same meaning as multiplication by the imaginary unit j.

Multiplication of a complex number by the imaginary unit j represents a counterclockwise rotation of $\pi/2$ rad (90°) in the complex plane. Recalling that $j = e^{j\pi/2}$, we have the following results:

$$\begin{aligned} je^{j\theta} &= e^{j(\theta+\pi/2)}, \\ j^2 &= e^{j\pi} = -1, \\ j^3 &= e^{j3\pi/2} = -j, \\ j^4 &= e^{j2\pi} = +1, \end{aligned} \tag{2.54}$$

and so on.

2.5 COMPLEX-NUMBER METHODS APPLIED TO THE DISPLACEMENT ANALYSIS OF LINKAGES

The displacement, velocity, and acceleration of planar linkages may be analyzed by using complex-number methods. Consider, for example, the sliding contact linkage shown in Figure 2.14a, where link 1 rotates and the slider moves relative to link 2, causing link 2 to oscillate. The linkage can be described at any instant by the vector equation

$$\boldsymbol{R}_2 = \boldsymbol{R}_0 + \boldsymbol{R}_1 \tag{2.55}$$

(see Figure 2.14b), where $\boldsymbol{R}_0$ represents the fixed link O_2O_1, $\boldsymbol{R}_1$ represents the crank O_1B, and $\boldsymbol{R}_2$ represents the portion of link 2 between O_2 and pin B.

If link lengths $\boldsymbol{R}_0$ and $\boldsymbol{R}_1$ are given, Eq. (2.55) may be solved for any given crank angle θ_1. For convenience, we select the real axis in the direction of the fixed link. Then, expressing Eq. (2.55) in complex form, we have

$$\boldsymbol{R}_2 = R_{2x} + jR_{2y} = R_0 + R_{1x} + jR_{1y}, \tag{2.56}$$

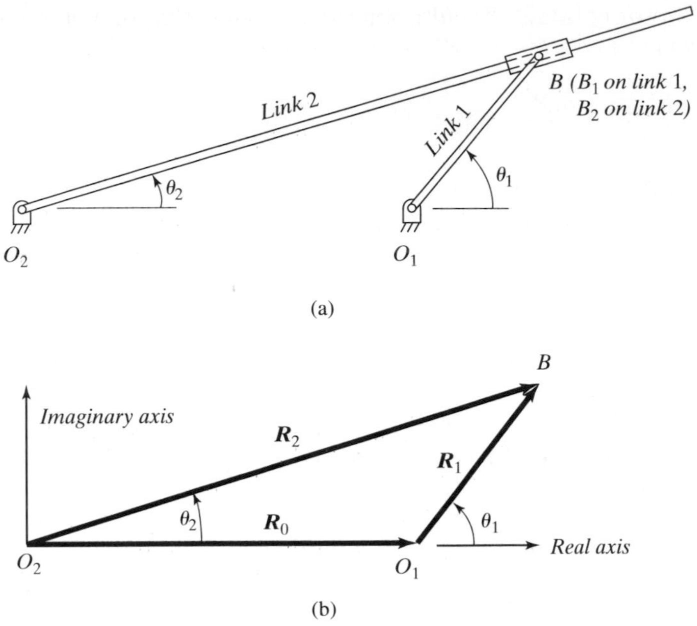

FIGURE 2.14 (a) Sliding contact linkage. (b) Complex plane representation.

where

$$R_{1x} = R_1 \cos \theta_1$$

and

$$R_{1y} = R_1 \sin \theta_1.$$

Equating the real parts of Eq. (2.56), we obtain

$$R_{2x} = R_0 + R_{1x} = R_0 + R_1 \cos \theta_1.$$

Equating the imaginary parts, we have

$$R_{2y} = R_{1y} = R_1 \sin \theta_1,$$

and the magnitude of $\boldsymbol{R}_2$ is

$$\begin{aligned} R_2 &= \sqrt{R_{2x}^2 + R_{2y}^2} \\ &= \sqrt{R_0^2 + 2R_0R_1 \cos \theta_1 + R_1^2 \cos^2\theta_1 + R_1^2 \sin^2 \theta_1} \\ &= \sqrt{R_0^2 + 2R_0R_1 \cos \theta_1 + R_1^2}. \end{aligned} \tag{2.57}$$

The last equation is identical to the law of cosines, which is generally written in terms of the internal angle $180° - \theta_1$. Angle θ_2 may be found from

$$\tan\theta_2 = \frac{R_{2y}}{R_{2x}} = \frac{R_1 \sin\theta_1}{R_0 + R_1 \cos\theta_1}. \tag{2.58}$$

Although the arctangent is actually multivalued, most calculators and computers will give values of θ_2 between 0° and +90° (0 to $\pi/2$ rad) for positive arguments of the function and between 0° and −90° (0 to $-\pi/2$ rad) for negative arguments.

It may be necessary to correct θ_2 for the proper quadrant. It is safer to use a rectangular–polar conversion routine or a two-argument arctangent function such as $\mathrm{ARCTAN}_2(x,y)$ or ANGLE(x,y).

SAMPLE PROBLEM 2.8

Sliding Contact Linkage Solved by the Complex-Number Method

In Figures 2.14a and 2.14b, given link lengths $O_2O_1 = 320$ mm and $O_1B = 170$ mm, locate slider pin B when $\theta_1 = 50°$.

Solution. Pin B is located by Eq. (2.56), from which

$$\begin{aligned}\boldsymbol{R}_2 &= R_{2x} + jR_{2y} = R_0 + R_{1x} + jR_{1y}\\ &= 320 + 170\cos 50° + j170\sin 50° = 429.3 + j130.2.\end{aligned}$$

The magnitude and direction of $\boldsymbol{R}_2$ are given immediately by using a rectangular–polar $(xy{:}r\theta)$ conversion routine available on many calculators. The routine may be programmed to produce the correct angle quadrant:

$$\boldsymbol{R}_2 = 448.6 \text{ mm at } \theta_2 = 16.9°.$$

Or, in complex exponential form, we obtain

$$\boldsymbol{R}_2 = 448.6e^{j0.294}.$$

These values can be checked with Eqs. (2.57) and (2.58), which were avoided by using the $xy{:}r\theta$ conversion. The dimensions of this particular linkage ($R_1 < R_0$) permit values of θ_2 in only the first and fourth quadrants. In a similar linkage, but with $R_1 > R_0$, θ_2 can fall in any quadrant.

Employing the principles illustrated, complex-number methods may be used to analyze many other types of linkages. However, since the complex plane is two dimensional, these methods are not applied to spatial linkages.

Limiting Positions

Limiting positions for some linkages were discussed in Chapter 1. For the sliding contact linkage of Figure 2.14, link 2 oscillates as link 1 rotates continuously. The locus of B_1 (point B on link 1) is a circle of radius R_1 We may find the limiting positions of link

2 graphically by drawing link 2 tangent to that circle. The tangent (representing link 2) will be perpendicular to the radius (representing link 1).

As an alternative, we may use the calculus. Noting that extreme values of θ_2 correspond to extreme values of tan θ_2 for this linkage, we differentiate Eq. (2.58) with respect to θ_1 and set the result equal to zero:

$$\frac{d\tan\theta_2}{d\theta_1} = 0 = R_1\cos\theta_1(R_0 + R_1\cos\theta_1)^{-1}$$
$$-R_1\sin\theta_1(R_0 + R_1\cos\theta_1)^{-2}(-R_1\sin\theta_1).$$

Multiply the above equations by $(R_0 + R_1\cos\theta_1)^2$ and note that $\cos^2\theta_1 + \sin^2\theta_1 = 1$. It follows that the limiting positions correspond to

$$\cos\theta_1 = \frac{-R_1}{R_0}.$$

Thus, the three links form a right triangle with hypotenuse R_0, and, as just determined graphically, links 1 and 2 are perpendicular when link 2 is at a limiting position.

The Geneva Mechanism

The Geneva mechanism (Figure 2.15) provides intermittent motion of the driven link while the driver rotates continuously. It is equivalent to the sliding contact mechanism (Figure 2.14a) during part of its cycle. For the position shown, pin B on the driver is entering the slot on the driven member. Thus, O_1B is equivalent to link 1 in Figure 2.14a, while the slot is equivalent to link 2 for the next quarter rotation of the driven member. Then, the driven member remains stationary until pin B enters the next slot.

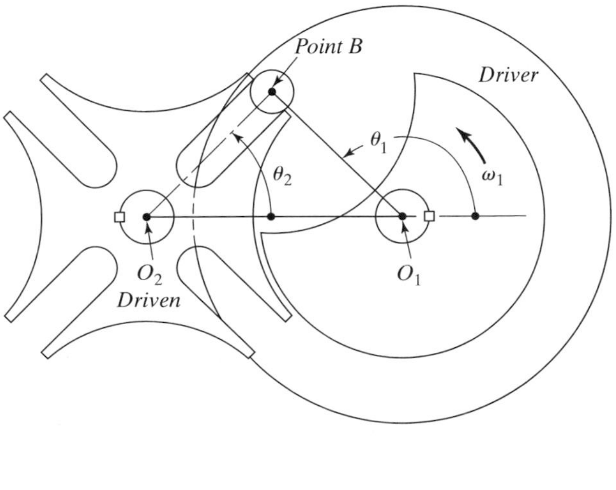

FIGURE 2.15 Geneva mechanism, equivalent to the sliding contact linkage during part of its cycle.

2.6 SPATIAL (THREE-DIMENSIONAL) LINKAGES

The motion of all points in a planar linkage is restricted to a single plane or a set of parallel planes. The motion of points in a spatial linkage is more general. For example, the motion of a given point in a spatial linkage may describe a curve that does not lie in a plane, or the motion of two points in the same spatial linkage may lie in two nonparallel planes.

Kinematic Pairs (Joints)

Spatial linkages employ single-degree-of-freedom joints (e.g., pin joints) and multiple-degree-of-freedom joints (e.g., ball joints). Some common joints were identified in Chapter 1. Many others are possible.

Types of Spatial Mechanisms

Spatial linkages are identified by their joint configuration symbols. For example, a PRCR mechanism consists of a prism (spline), revolute, cylinder, and revolute. A variety of spatial mechanism configurations are shown in Figure 2.16a through i. Limiting positions occur when motion about or along a joint stops and then changes direction. A joint's limiting position defines its range of motion. The absence of a limiting position indicates that continuous motion is possible with respect to that joint. In general, limiting positions in spatial linkages are not as obvious as in simple planar mechanisms.

Analysis of Four-Link Spatial Linkages

Vector methods may be used in the analysis of spatial linkages. For example, if four links (including the frame) form a closed loop as in Figure 2.15i, then the vector equation

$$\boldsymbol{r}_0 + \boldsymbol{r}_1 + \boldsymbol{r}_2 + \boldsymbol{r}_3 = 0 \tag{2.59}$$

may be used to analyze the displacement of the linkage. If each vector is written in terms of its components and the unit vectors $\boldsymbol{i}, \boldsymbol{j}$, and $\boldsymbol{k}$ (i.e., $\boldsymbol{r}_0 = r_{0x}\boldsymbol{i} + r_{0y}\boldsymbol{j} + r_{0z}\boldsymbol{k}$, and so on, then the x, y, and z components of the vectors in the closed loop must each sum to zero. Thus, we have three scalar equations:

$$r_{0x} + r_{1x} + r_{2x} + r_{3x} = 0,$$

$$r_{0y} + r_{1y} + r_{2y} + r_{3y} = 0,$$

and

$$r_{0z} + r_{1z} + r_{2z} + r_{3z} = 0. \tag{2.60}$$

Considering the restraints imposed by the joints and the link lengths, it may be possible to determine the position of the links analytically. One input link position variable is required to solve for displacements in a one-degree-of-freedom spatial linkage. For a

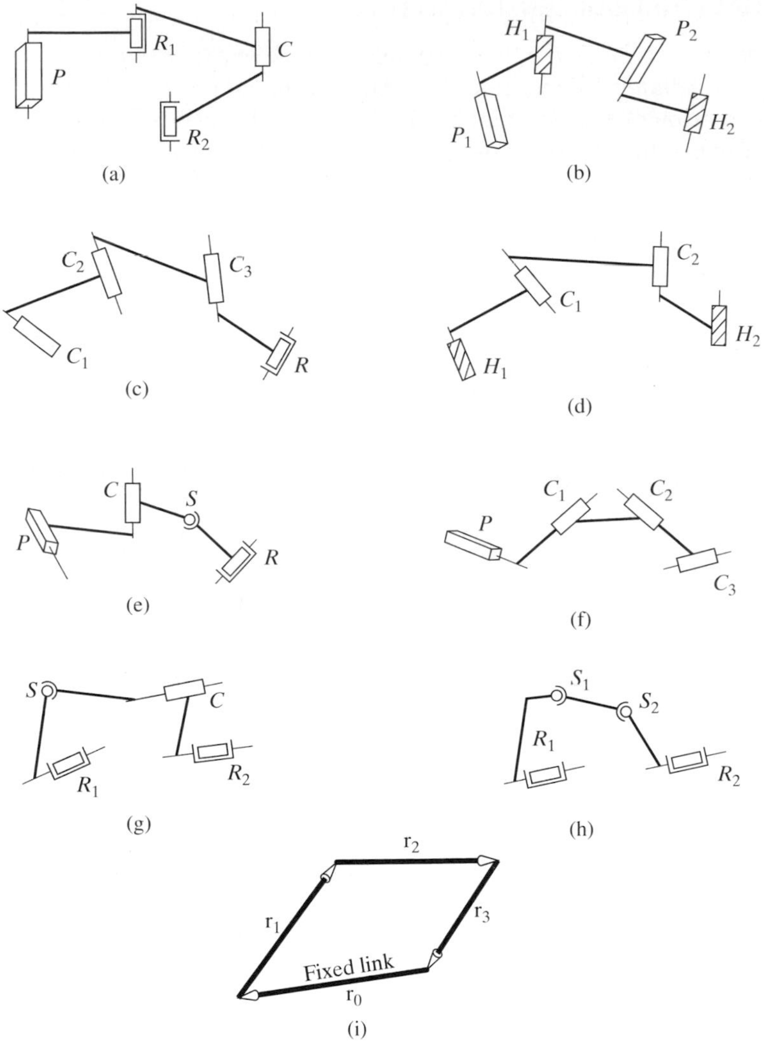

FIGURE 2.16 Four-bar spatial linkages. (a) PRCR mechanism. (b) PHPH mechanism. (c) CCCR mechanism. (d) HCCH mechanism. (e) PCSR mechanism. (f) PCCC mechanism. (g) RSCR mechanism. (h) RSSR mechanism. (i) Typical vector representation of a four-bar closed-loop spatial linkage.

spatial linkage with more than one degree of freedom, more than one input position variable would be required to solve for displacements.

Methods of descriptive geometry may be used as an alternative to an analytical solution for linkage displacements. Consider, for example, an RSSR linkage, as represented in Figure 2.16h. Let the position of the left-hand crank, R_1S_1, be given. Then the locus of possible positions of point S_2 lies on a sphere of radius S_1S_2 and with center at S_1. However, revolute R_2 restricts point S_2 to circular motion. The actual position of S_2

may be determined graphically by constructing the intersection of the sphere and plane loci of S_2 and using the radius of the circle, R_2S_2.

For given input values of velocity and acceleration, it may be possible to determine the velocity and acceleration of every point and the angular velocity and acceleration of every link in a spatial linkage. Vector methods may be used as with planar linkages, but all three coordinate directions must be considered. Straightforward solutions can be obtained for some spatial linkages. Others require ingenuity, as well as considerable time and effort. An example of the latter type, the displacement analysis of a spatial mechanism with seven revolute joints, is given by Duffy and Derby (1979). In analyzing this special case of a 7R (RRRRRRR) mechanism, the authors derived an input–output equation of degree 24, making a major step toward the solution of the general 7R mechanism.

Analysis of a Spatial Linkage Made Up of Two Revolute Pairs and Two Spherical Pairs (an RSSR Linkage)

An RSSR linkage is shown in Figure 2.16h. The number of degrees of freedom for an RSSR linkage is given by

$$\begin{aligned} \mathrm{DF}_{\text{spatial}} &\geq 6(n_L - n_J - 1) + \sum f_i \\ &= 6(4 - 4 - 1) + 1 + 3 + 3 + 1 \\ &= 2. \end{aligned}$$

Inspecting the linkage configuration shown in the figure, we see that one of the degrees of freedom corresponds to rotation of link 2 (link S_1S_2) about its own axis. If this motion is not relevant to the intended application of the mechanism, the RSSR linkage acts essentially as a one-degree-of-freedom linkage.

Linkage Displacements

Referring to Figure 2.16h, we may describe the RSSR linkage by the vector equation

$$\boldsymbol{r}_0 + \boldsymbol{r}_1 + \boldsymbol{r}_2 + \boldsymbol{r}_3 = 0, \qquad \text{(2.59 repeated)}$$

where the vectors form a closed loop as in part i of the figure. Let the position of links 0 and 1 be specified. Then the positions of links 2 and 3 may be identified by three components each, resulting in six unknowns. There are six equations: three from Eq. (2.60) taking into account the x, y, and z directions; two equations based on the lengths of links 2 and 3; and one equation based on the plane of rotation of link 3.

The solution is "easier said than done," because the set of equations is nonlinear. In designing and analyzing spatial linkages, try to select a set of coordinate axes that reduces the number of unknowns.

SAMPLE PROBLEM 2.9

Symbolic Solution for Displacements of an RSSR *Spatial Linkage*

The planes of rotation of the drive crank and driven crank of an RSSR linkage are perpendicular to one another. The fixed bearing of the driven crank lies in the plane of rotation of the drive crank. Find the position of all links in terms of the drive crank position.

Design decisions. We could specify the length of each link and the relative position of the revolute joints. Then a numerical solution is possible for any crank position. However, the numerical solution may be inefficient if we need to plot positions throughout the entire range of motion of the *RSSR* linkage. Therefore, we will attempt a symbolic solution.

First try at a solution. We select coordinates as in Figure 2.17. The drive crank (link 1) rotates in the xy-plane, the coupler (link 2) has general spatial motion, and the follower crank (link 3) rotates in the yz-plane. Revolute joint R_2 is in the plane of motion of the drive crank.

There are three equations based on the vector loop closure equation and two based on link lengths, for a total of five equations:

$$r_{0x} + r_{1x} + r_{2x} = 0;$$
$$r_{0y} + r_{1y} + r_{2y} + r_{3y} = 0;$$
$$r_{2z} + r_{3z} = 0;$$

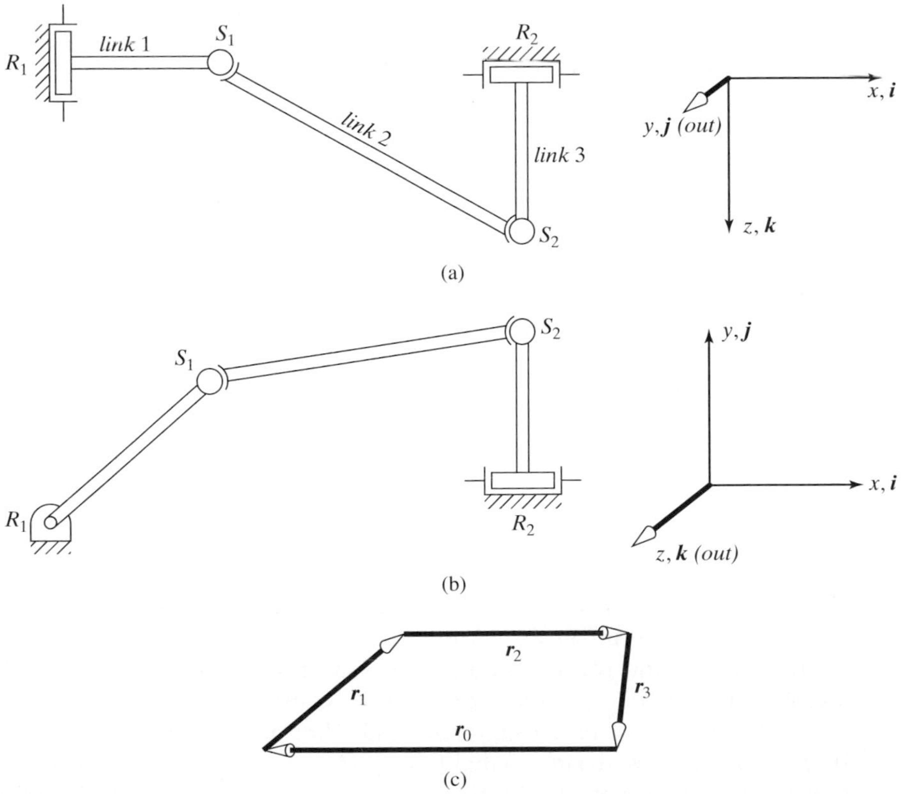

FIGURE 2.17 RSSR spatial linkage (not to scale) (a) Top view (b) Side view (c) Vector representation.

$$r_{2x}^2 + r_{2y}^2 + r_{2z}^2 = r_2^2;$$
$$r_{3y}^2 + r_{3z}^2 = r_3^2. \tag{2.61}$$

There are five unknowns: the x, y, and z components of link vector $\mathbf{r}_2$ and the y and z components of link vector $\mathbf{r}_3$. Known values are the x and y components of $\mathbf{r}_0$, the vector from revolute R_2 to R_1, and the x and y components of link vector $\mathbf{r}_1$, given by

$$r_{1x} = r_1 \cos \theta_1$$

and

$$r_{1y} = r_1 \sin \theta_1.$$

The five equations are fed into a program for solving systems of equations, and the five unknowns are calculated. But the results are disappointing: The symbolic expressions for the unknowns span nine pages, and one solution of r_{3x} is 28 inches long in 10-point type.

Second try at a solution. To shorten the solution, the known vectors are combined into a single vector defined by

$$\mathbf{c} = \mathbf{r}_0 + \mathbf{r}_1,$$

and the set of equations now looks like this:

$$c_x + r_{2x} = 0;$$
$$c_y + r_{2y} + r_{3y} = 0;$$
$$r_{2z} + r_{3z} = 0;$$
$$r_{2x}^2 + r_{2y}^2 + r_{2z}^2 = r_2^2;$$
$$r_{3y}^2 + r_{3z}^2 = r_3^2. \tag{2.62}$$

The symbolic equation solver results include

$$r_{2x} = -c_x, \tag{2.63}$$
$$r_{2y} = (c_x^2 - c_y^2 - r_2^2 + r_3^2)/(2c_y), \tag{2.64}$$
$$r_{3y} = (c_x^2 + c_y^2 - r_2^2 + r_3^2)/(2c_y), \tag{2.65}$$

and rather long solutions for the other two unknowns. However, the last simultaneous equation tells us that

$$r_{3z} = (r_3^2 - r_{3y}^2)^{1/2}. \tag{2.66}$$

Finally, the third simultaneous equation yields the remaining unknown:

$$r_{2z} = -r_{3z}. \tag{2.67}$$

It takes a little longer to analyze the linkage just described if we do not have a symbolic equation solver. We eliminate unknowns and make substitutions in Eqs. (2.62), finally obtaining results equivalent to Eqs. (2.63) through (2.66). Note that we may use the positive or negative square root in the equation for r_{3z}; it depends on how the links are assembled. Our choice affects r_{2z} as well. The motion of some RSSR linkages may resemble that of a planar crank-rocker mechanism. A change in link proportions may cause the linkage to act like a change-point, drag link, or other planar mechanism. If the spatial linkage resembles a planar double or triple rocker, then some crank positions will result in solutions that are not real numbers. Such solutions are not valid; they represent forbidden positions of the crank. (*Caution*: Do not use the Grashof criteria for spatial linkages, as they apply only to planar mechanisms.)

Transmission angle. Recall that we determined the minimum and maximum transmission angles of planar linkages. Transmission angles ranging from about 40° to 140° are generally acceptable. We are concerned if the transmission angle falls outside of that range. Will the output torque be adequate? Will the linkage bind because friction torque on the driven crank exceeds torque due to the force applied by the coupler? The answers depend on how the linkage is used, on the quality and type of bearings and lubrication, and on the effects of inertia.

Transmission metric. Similar concerns apply to machines employing spatial linkages. Consider the vectors representing the coupler and driven crank. Divide the dot product of the two vectors by the product of the absolute values, resulting in what we will call the *transmission metric* T. The transmission metric range,

$$-0.766 \leq T \leq 0.766,$$

is equivalent to the generally acceptable transmission angle range,

$$40^\circ \leq \phi \leq 140^\circ.$$

Alternatively, we can easily convert the transmission metric into an angle.

Suppose the transmission metric falls outside the generally acceptable range? The linkage may still be acceptable: The criterion just warns us of *possible* problems.

SAMPLE PROBLEM 2.10

Design of a spatial linkage

Design a device with a 50-mm output link that oscillates through about 50°. The centerline of the continuously rotating input shaft must be parallel to the plane of motion of the driven link.

Design decisions. There are many possible solutions to this design problem. A pair of bevel gears or a worm and worm gear driving a planar crank-rocker mechanism could be chosen for such an application. An RSSR spatial linkage will be used instead. The driveshaft will lie in the z direction, and the driven crank will oscillate in a yz-plane as in Figure 2.17. However, the linkage in that figure would have very limited motion, so the proportions must be changed. After a number of unsatisfactory tries, we will investigate a linkage with the following proportions:

Drive crank length $r_1 = 25$ mm;
Coupler length $r_2 = 75$ mm;
Driven crank length $r_3 = 50$ mm;

revolute joints R_1 and R_2 are located at (0,0,0) and (10, 70, 0), respectively.
Thus, the components of the fixed link vector are

$$r_{0x} = -10 \text{ mm} \quad \text{and} \quad r_{0y} = -70 \text{ mm}.$$

Solution summary. The equations developed in the previous sample problem using a symbolic solution are employed. The drive crank angular position is identified as θ (without a subscript). The positive root is selected for the z component of the driven crank vector. The negative root represents a different assembly configuration. The coupler and follower crank link vectors are defined in terms of their components. Their vector magnitudes are checked for an arbitrary drive crank position. The good news is that they agree with the specified lengths.

Driven crank position θ_{3x} (in degrees) is plotted against the drive crank position (also in degrees). See Figure 2.18. The approximate range-of-motion requirement appears to be met. The transmission metric is multiplied by 100 for convenience in plotting. The not-so-good news is that the metric extends somewhat beyond generally accepted limits. If the linkage is heavily loaded in this position, we should consider a redesign to improve the transmission metric.

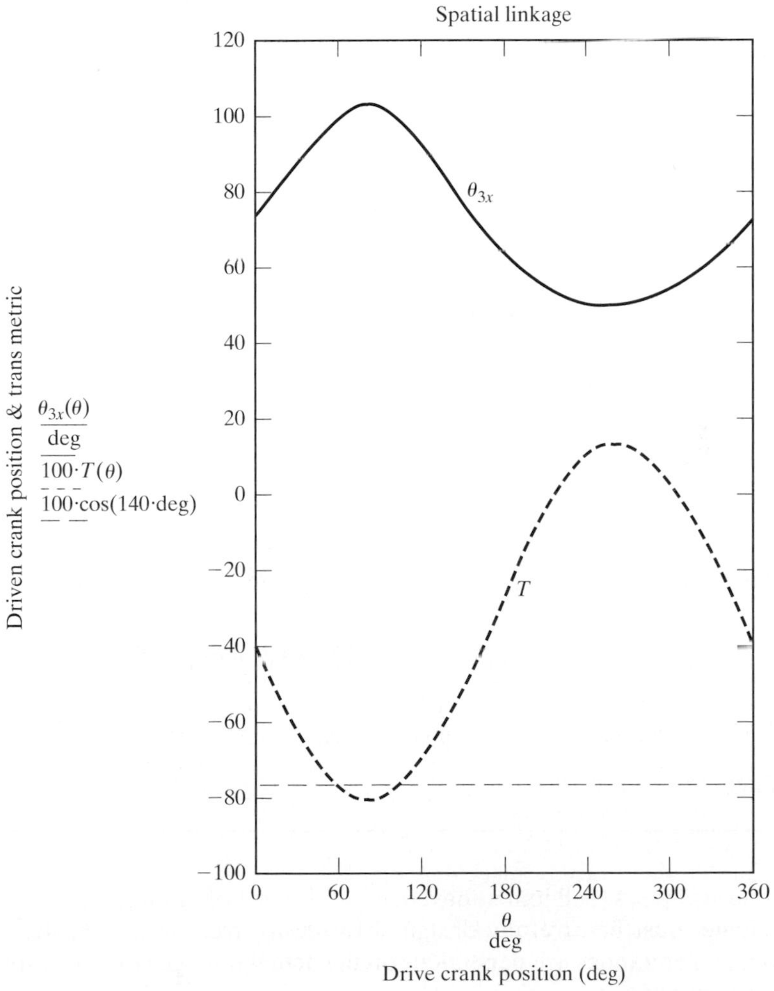

FIGURE 2.18 Driven crank position and transmission metric of spatial linkage.

Solution details **(The software used in this solution does not identify vectors with boldface type).**

Analysis of an RSSR spatial linkage

The drive crank and driven crank rotate in perpendicular planes.

Vector loop: $r_0 + r_1 + r_2 + r_3 = 0,$

Dimensions of links:

Fixed link: $r_{0x} := -10 \qquad r_{0y} := -70$

Drive crank: $r_1 := 25$

Coupler: $r_2 := 75$

Driven crank: $r_3 := 50$

Position Analysis

Drive crank position (subscript omitted): $\theta := 0, \frac{\pi}{18} \ldots 2\pi$

$$r_{1x}(\theta) := r_1 \cdot \cos(\theta) \qquad r_{1y}(\theta) := r_1 \cdot \sin(\theta)$$

Define c = sum of fixed link and drive crank vectors

$$c_x(\theta) := r_{0x} + r_{1x}(\theta) \qquad c_y(\theta) := r_{0y} + r_{1y}(\theta)$$

$$r_{2x}(\theta) := -c_x(\theta) \qquad r_{2y}(\theta) := \frac{c_x(\theta)^2 - c_y(\theta)^2 - r_2^2 + r_3^2}{2c_y(\theta)}$$

$$r_{3y}(\theta) := \frac{-(c_x(\theta)^2 + c_y(\theta)^2 - r_2^2 + r_3^2)}{2c_y(\theta)} \qquad r_{3z}(\theta) := (r_3^2 - r_{3y}(\theta)^2)^{\frac{1}{2}*}$$

$$r_{2z}(\theta) := -r_{3z}(\theta)$$

*** We will select the assembly configuration given by the positive root.**

Driven crank position: $\theta_{3x}(\theta) := \text{angle}(r_{3y}(\theta), r_{3z}(\theta)) \qquad \theta_{3x}(2) = 1.671$

Link vectors: $rv_2(\theta) := \begin{bmatrix} r_{2x}(\theta) \\ r_{2y}(\theta) \\ r_{2z}(\theta) \end{bmatrix} \qquad rv_3(\theta) := \begin{bmatrix} 0 \\ r_{3y}(\theta) \\ r_{3z}(\theta) \end{bmatrix}$

Check results for coupler and driven crank length: $|rv_2(2)| = 75 \qquad |rv_3(2)| = 50$

Transmission metric: $T(\theta) := \frac{rv_2(\theta) \cdot rv_3(\theta)}{|rv_2(\theta)| \cdot |rv_3(\theta)|} \qquad T(80 \cdot deg) = -0.804 \quad T(\pi) = -0.267$

Compare $\cos(140 \cdot deg) = -0.766$

Spherical pairs (ball joints) have a limited range of motion. The actual links in a spatial linkage must be carefully designed to ensure free motion at the joints. The linkage position equations are dependent on the actual linkage configuration and the selection of coordinate axes.

SAMPLE PROBLEM 2.11

Position equations for a different linkage configuration.

Suppose you are considering an RSUR linkage (revolute–spherical–universal–revolute joints). Link 0, the fixed link, lies in the negative x-direction; link 1 rotates in an xy-plane, and link 3 rotates in an xz-plane. Write the position equations in terms of vector coordinates.

Solution. Define a vector

$$\boldsymbol{c} = \boldsymbol{r}_0 + \boldsymbol{r}_1,$$

where $\boldsymbol{c}$ is known for any given value of θ. Then

$$\boldsymbol{c} + \boldsymbol{r}_2 + \boldsymbol{r}_3 = 0,$$

and, considering the plane of rotation of link 3, there are five unknowns: r_{2x}, r_{2y}, r_{2z}, r_{3x}, and r_{3z}. Equating the vector components in each coordinate direction and noting the link lengths, we have the following five equations:

$$\begin{aligned} c_x + r_{2x} + r_{3x} &= 0; \\ c_y + r_{2y} &= 0; \\ r_{2z} + r_{3z} &= 0; \\ r_{2x}^2 + r_{2y}^2 + r_{2z}^2 &= r_2^2; \\ r_{3x}^2 + r_{3z}^2 &= r_3^2. \end{aligned}$$

The five equations can be reduced to a single equation for r_{2x} in terms of known values.

Analysis of a Spatial Linkage Made Up of a Revolute Pair, Two Spherical Pairs, and a Cylinder Pair

Figure 2.19 shows an RSSC spatial linkage in which link RS_1 acts as a crank. The path of the sliding link intersects the plane of the crank at point A, making an angle γ with that plane. Joints C, S_2, and R of this linkage will be in the xy-plane, and link 1 will move within the yz-plane. The link lengths are r_1 (crank RS_1), r_2 (coupler S_1S_2), r_3 (instantaneous distance S_2A on the sliding link), and r_0 (fixed distance AR).

Analytical Solution for Displacements

The links could be identified as vectors and the vector equation

$$\boldsymbol{r}_0 + \boldsymbol{r}_1 + \boldsymbol{r}_2 + \boldsymbol{r}_3 = 0$$

could be used along with constraint equations to solve for displacements. For this configuration, however, it is convenient to express the length of the coupler link, $r_2 = S_1S_2$ in terms of its components in three mutually perpendicular directions:

$$\begin{aligned} r_2^2 &= r_{2x}^2 + r_{2y}^2 + r_{2z}^2 \\ &= (r_3 \sin\gamma)^2 + (r_0 - r_1\cos\theta + r_3\cos\gamma)^2 + (r_1\sin\theta)^2. \end{aligned} \tag{2.68}$$

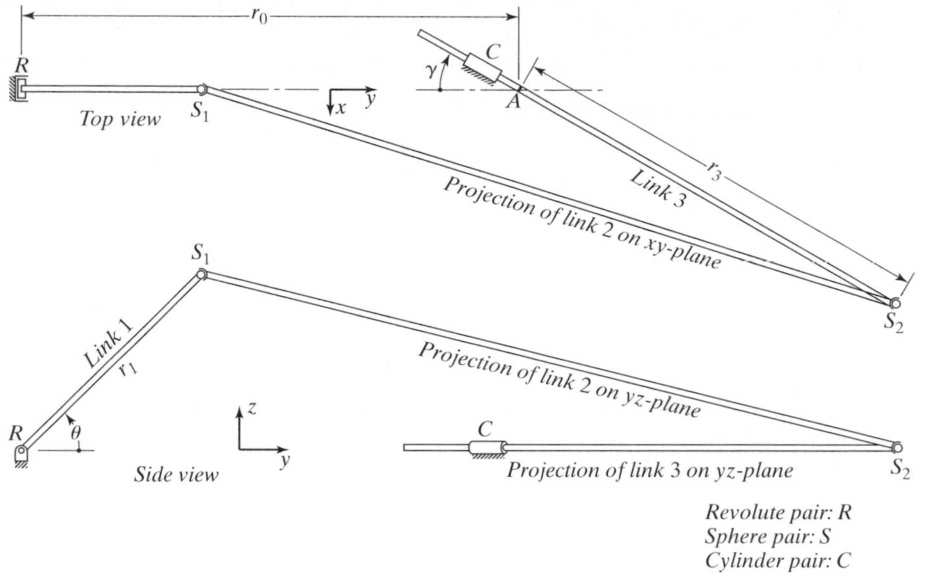

FIGURE 2.19 RSSC spatial linkage.

If crank angle θ is given along with link lengths r_0, r_1, and r_2 and path angle γ, the result is a quadratic equation in r_3:

$$r_3^2 + 2\cos\gamma(r_0 - r_1\cos\theta)r_3 + r_0^2 + r_1^2 - r_2^2 - 2r_0r_1\cos\theta = 0. \tag{2.69}$$

The solution gives the sliding-link position, locating spherical pair (ball joint) S_2. The two roots of the quadratic equation are given by

$$r_3 = \frac{-b \pm \sqrt{b^2 - 4ac}}{2a},$$

where $a = 1$,

$$b = 2\cos\gamma(r_0 - r_1\cos\theta),$$

and

$$c = r_0^2 + r_1^2 - r_2^2 - 2r_0r_1\cos\theta.$$

SAMPLE PROBLEM 2.12

RSSC *Spatial Linkage*

Let dimensions r_0, r_1, and r_2 be given for the RSSC linkage just described. Sketch a flowchart that you can use to find the displacement of the sliding link for every 15° of crank angle θ at various path angles γ.

Solution. See flowchart (Figure 2.20). The mechanism is equivalent to a planar slider-crank linkage when the path angle is zero. For nonzero values of the path angle, we have a spatial linkage. For values of $\gamma = 90°$ or near that value, if the crank RS_1 drives, the linkage is likely to jam.

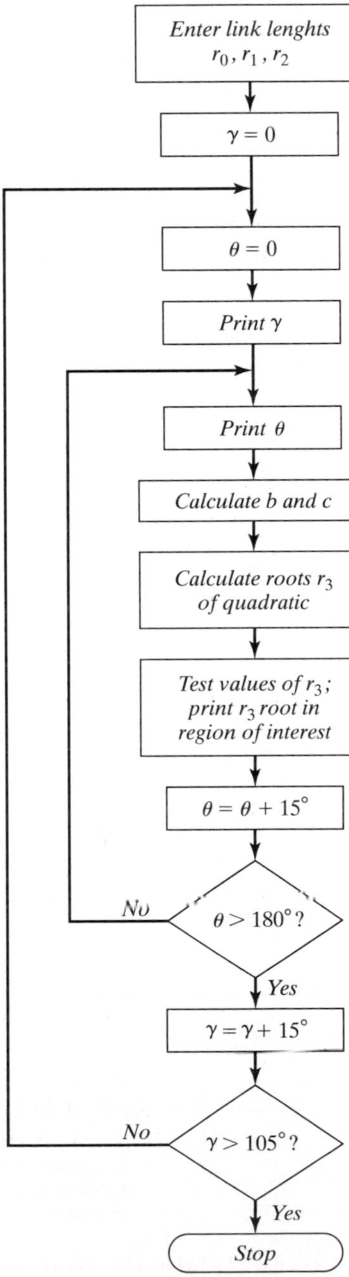

FIGURE 2.20 Flowchart: Displacement of an RSSC spatial linkage.

More complicated spatial linkages are discussed in the technical literature. Lee and Liang (1988a) describe a vector theory for the analysis of spatial linkages that includes displacement, velocity, and acceleration equations for open-chain and closed-loop mechanisms. The same authors (1988b) also analyze displacements of a general spatial seven-link 7R mechanism. Their analysis involves a 16th-degree polynomial input–output equation for displacement in the form of an eight-by-eight determinant. Fanghella (1988) describes the kinematics of spatial linkages by group algebra.

Alternative Analysis of a Spatial Linkage Using Graphical Methods

Spatial linkages may also be solved by methods of descriptive geometry. However, since such graphical methods require so much labor, they are recommended only as a check of computer analysis.

SAMPLE PROBLEM 2.13

Graphical position analysis of a spatial linkage

Find the position of the sliding link of the RSSC spatial mechanism described earlier for path angle $\gamma = 30°$ and crank angle $\theta = 45°$. Let crank length $r_1 = 100$ mm, coupler length $r_2 = 300$ mm, and let point A lie a distance $r_0 = 200$ mm away from the revolute point R, as in Figure 2.21.

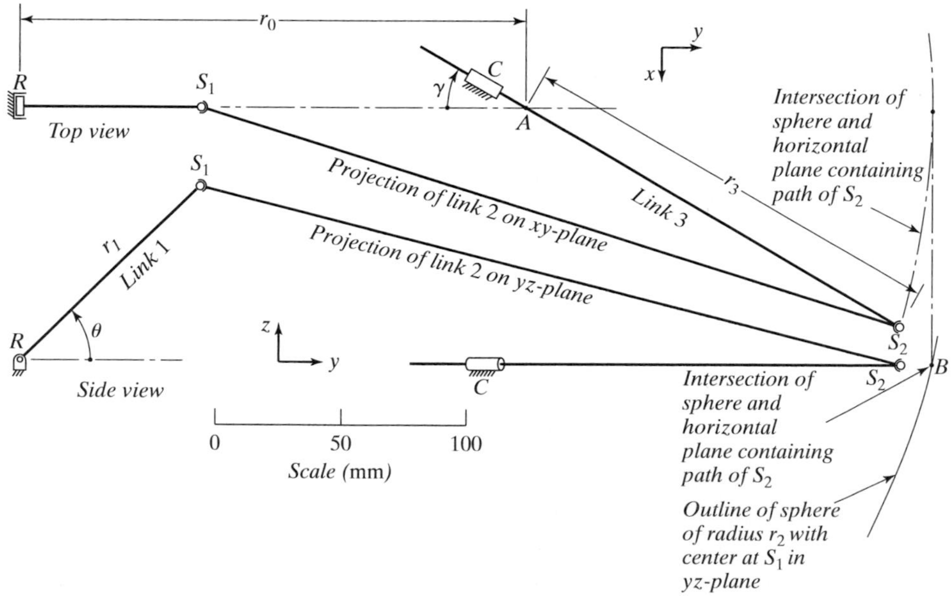

FIGURE 2.21 Alternative analysis of RSSC spatial linkage by graphical methods.

Solution. A side view (yz-plane) and top view (xy-plane) are used. Link 1, the crank, which lies in the yz-plane, is drawn to scale in that plane and projected to the xy-plane. The path of S_2 is located in the xy-plane. Link 2 will not, in general, lie in either the xy- or yz-plane. If link 2 were constrained only at S_1, the locus of all possible points S_2 would lie on a sphere of radius r_2 with center at S_1. A circle with radius 300 mm and center at S_1 is drawn in the yz-plane to represent the outline of the sphere. The intersection of the sphere and the horizontal plane containing the sliding link is a circle whose projection on the yz-plane is a line segment ending at point B, as marked in the figure. Point B is projected upward to the xy-plane, and a circular arc with its center at S_1 is drawn tangent to the projection line. The intersection of the circle and the path of S_2 in the xy-plane locates S_2 and determines the value of r_3.

2.7 COMPUTER-IMPLEMENTED NUMERICAL METHODS OF POSITION ANALYSIS

Linkage displacement relationships tend to be nonlinear. The angular position of the driven crank of a four-bar linkage, for example, is not proportional to the input position. Simple closed-form solutions are available for some mechanisms, while other mechanisms—particularly multiloop linkages—are best solved using iterative numerical methods.

The Newton–Raphson Method for Two or More Variables

In Chapter 1, we utilized the Newton–Raphson method to solve a problem in a single variable. Using the same concept, we may solve linkage problems involving two or more variables. Unfortunately, the Newton–Raphson method for n variables involves an n-by-n matrix of partial derivatives. Those preferring less mathematical complexity may seek closed-form solutions or use preprogrammed numerical routines such as those found in Mathcad™ or other mathematical software.

Suppose a problem is represented by the set of simultaneous equations

$$\begin{aligned} F_1(x_1, x_2, \ldots, x_n) &= 0, \\ F_2(x_1, x_2, \ldots, x_n) &= 0, \\ &\vdots \\ F_n(x_1, x_2, \ldots, x_n) &= 0, \end{aligned} \tag{2.70}$$

or, in vector form,

$$\boldsymbol{F}(x_1, x_2, \ldots x_n) = [F_1, F_2, \ldots F_n] = \mathbf{0}. \tag{2.71}$$

To find the unknown variables, the (state) vector

$$\boldsymbol{x} = [x_1, x_2, \ldots, x_n]$$

we begin by making a first approximation of each of the variables:

$$\boldsymbol{X} = [X_1, X_2, \ldots, X_n].$$

Then we compute the vector $\boldsymbol{F}$ at $\boldsymbol{x} = \boldsymbol{X}$. Unless we are very fortunate, the first approximations will not be correct (i.e., $\boldsymbol{F}(\boldsymbol{X})$ is not equal to zero). As with the Newton–Raphson method applied to one variable, we make a linear adjustment to arrive at what we hope to be a better approximation of $\boldsymbol{x}$.

The second approximation is computed from

$$\boldsymbol{X}_{\text{new}} = \boldsymbol{X} - \boldsymbol{G}^{-1}\boldsymbol{F}(\boldsymbol{X}), \tag{2.72}$$

where

$$\boldsymbol{G} = \begin{bmatrix} \partial F_1/\partial x_1 & \partial F_1/\partial x_2 & \ldots & \partial F_1/\partial x_n \\ \partial F_2/\partial x_1 & \partial F_2/\partial x_2 & \ldots & \partial F_2/\partial x_n \\ \cdots\cdots & \cdots\cdots & \cdots\cdots & \cdots\cdots \\ \cdots\cdots & \cdots\cdots & \cdots\cdots & \cdots\cdots \\ \partial F_n/\partial x_1 & \ldots\ldots & \ldots & \partial F_n/\partial x_n \end{bmatrix}. \tag{2.73}$$

The process is repeated with $\boldsymbol{X}$ replaced by $\boldsymbol{X}_{\text{new}}$ for as many iterations as necessary (i.e., until each component of $\boldsymbol{F}(\boldsymbol{X}) = 0 \pm$ the tolerance). Otherwise the process is stopped after a set number of iterations (say, 20) with a message saying "the process does not converge in 20 iterations."

One's success may depend on the initial guesses of the values. If there is more than one set of roots, a poor first approximation of $\boldsymbol{X}$ may lead to a solution other than the desired one. The determinant of the matrix $\boldsymbol{G}, \boldsymbol{J} = |\boldsymbol{G}|$, is called the Jacobian of the system of simultaneous equations. The Jacobian must not vanish during any of the iterations.

(*Note:* The foregoing discussion is based on an extension of the single-variable Newton–Raphson method. Those desiring a more rigorous approach may refer to Taylor (1955) or Stark (1970).)

SAMPLE PROBLEM 2.14

A Numerical Method Applied to the Four-Bar Linkage

Consider the four-bar linkage of Figure 2.10, where link lengths r_0, r_1, r_2, and r_3 are given and angular positions $\theta_0 = \pi$ rad and $\theta_1 = \pi/3$ rad. Find θ_2 and θ_3.

Solution. In this case, both a graphical solution and a closed-form mathematical solution are possible. However, let us solve the problem by the Newton–Raphson method in order to illustrate the numerical procedures involved. Referring to the figure, we see that the x and y components of diagonal vector $\boldsymbol{r}_d$ are determined, respectively, by

$$r_{dx} = r_0 \cos\theta_0 + r_1 \cos\theta_1 \tag{2.74}$$

and

$$r_{dy} = r_0 \sin \theta_0 + r_1 \sin \theta_1. \tag{2.75}$$

The problem of locating links 2 and 3 can be expressed as two simultaneous equations describing the horizontal and vertical projections of the triangle formed by vectors $\boldsymbol{r}_d$, $\boldsymbol{r}_2$, and $\boldsymbol{r}_3$:

$$r_2 \cos \theta_2 + r_3 \cos \theta_3 = -r_{dx} \tag{2.76}$$

and

$$r_2 \sin \theta_2 + r_3 \sin \theta_3 = -r_{dy}. \tag{2.77}$$

We will utilize Eqs. (2.70) through (2.73), where x_1 becomes θ_2 and x_2 becomes θ_3. Thus, we have

$$F_1 = r_2 \cos \theta_2 + r_3 \cos \theta_3 + r_{dx} = 0$$

and

$$F_2 = r_2 \sin \theta_2 + r_3 \sin \theta_3 + r_{dy} = 0,$$

or, in vector form,

$$\boldsymbol{F} = \begin{bmatrix} r_2 \cos \theta_2 + r_3 \cos \theta_3 + r_{dx} \\ r_2 \sin \theta_2 + r_3 \sin \theta_3 + r_{dy} \end{bmatrix}.$$

In order to form the matrix G, we find $\partial F / \partial \theta_2 = -r_2 \sin \theta_2$, etc., from which it follows that

$$\boldsymbol{G} = \begin{bmatrix} -r_2 \sin \theta_2 & -r_3 \sin \theta_3 \\ r_2 \cos \theta_2 & r_3 \cos \theta_3 \end{bmatrix}.$$

Suppose we are interested in the linkage configuration for which r_d, r_2, and r_3 form a clockwise loop. Then a guess (first approximation) of

$$\boldsymbol{x} = \boldsymbol{X} = \begin{bmatrix} \theta_2 \\ \theta_3 \end{bmatrix} = \begin{bmatrix} 1 \\ 3 \end{bmatrix}$$

seems reasonable. The result is

$$\boldsymbol{F} = \begin{bmatrix} F_1 \\ F_2 \end{bmatrix}_{(x=X)} = \begin{bmatrix} -93.56 \\ 181.06 \end{bmatrix}$$

(instead of the desired value, $\boldsymbol{F} = \boldsymbol{0}$), indicating that the first approximation was not very accurate.

A second approximation of the roots is found by computing

$$\boldsymbol{X}_{\text{new}} = \boldsymbol{X} - \boldsymbol{G}^{-1}\boldsymbol{F} = \begin{bmatrix} 0.072 \\ 4.400 \end{bmatrix}$$

Successively replacing $\boldsymbol{X}$ by $\boldsymbol{X}_{\text{new}}$, the third through sixth approximations are, respectively:

$$\begin{bmatrix} 0.116 \\ 4.077 \end{bmatrix} \quad \begin{bmatrix} 0.077 \\ 4.043 \end{bmatrix} \quad \begin{bmatrix} 0.077 \\ 4.044 \end{bmatrix} \quad \text{and} \quad \begin{bmatrix} 0.077 \\ 4.044 \end{bmatrix}$$

where, for the final value, $\boldsymbol{F} = \boldsymbol{0} \pm$ a small tolerance. Note that there is no change (to three decimal places) between the fifth and sixth approximations. Thus $\theta_2 = 0.077$ rad and $\theta_3 = 4.044$ rad for the desired linkage configuration.

Multiloop Linkages

The link positions of a planar **single-loop linkage** can be described by a single vector equation or two scalar equations describing a skeleton diagram that forms a single closed polygon. The slider-crank linkage and four-bar linkage are examples of single-loop linkages. Planar **multiloop linkages** require one vector equation or two scalar equations for each internal loop. Quick-return mechanisms including the drag-link–slider-crank linkage and the sliding-contact–slider-crank linkage described in Chapter 1 are two-loop linkages.

The degree of difficulty of an analytical solution for displacements of a multiloop linkage depends on the linkage configuration and the given data. Consider the drag-link–slider-crank linkage shown in Figure 2.22. The skeleton diagram forms two independent loops, $O_1BCO_3O_1$ and O_3DEO_3. If the angular position of link 1 is given, the orientation of links 2 and 3 in four-bar linkage O_1BCO_3 can be determined without considering links 4 and 5. Then, using the orientation of link 3, we can easily solve the slider-crank linkage. Closed-form solutions for both the four-bar linkage and slider-crank linkage were given earlier in the chapter.

If the angular position of link 1 is given for the double-slider two-loop linkage of Figure 2.23, the solution cannot be uncoupled. The two kinematic loops are represented by four simultaneous equations. Numerical solutions are suggested for the determination of displacements in linkages of this type.

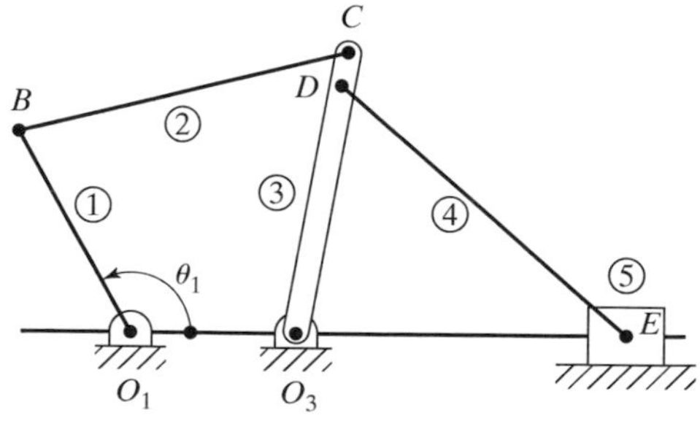

FIGURE 2.22 Drag-link–slider-crank linkage.

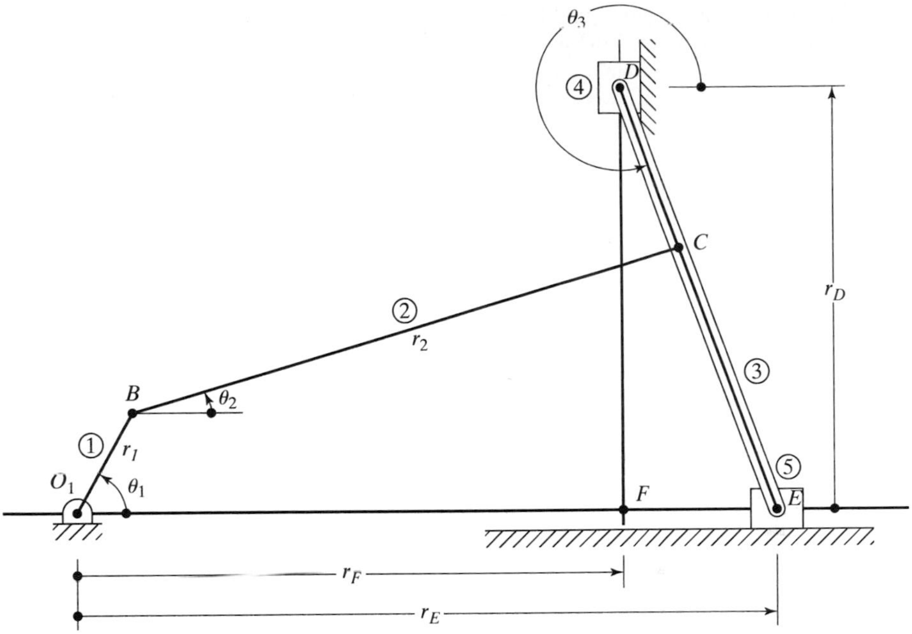

FIGURE 2.23 A double-slider two-loop linkage.

SAMPLE PROBLEM 2.15

Displacement Analysis of a Multiloop Linkage

Consider the double-slider two-loop linkage of Figure 2.23, where $r_1 = 1$, $r_2 = 5.2$, $r_F = 5$, $r_{DE} = 4$, $r_{CD} = 1.5$, and $\theta_1 = \pi/3$ rad. (Distance DE is identified as r_{DE}, etc.) Find θ_2, θ_3, r_D, and r_E.

Solution. The equations describing the horizontal and vertical components of loop O_1BCEO_1 are, respectively,

$$r_1 \cos \theta_1 + r_2 \cos \theta_2 + r_{CE} \cos \theta_3 - r_F = 0$$

and

$$r_1 \sin \theta_1 + r_2 \sin \theta_2 + r_{CE} \sin \theta_3 = 0,$$

and the equations describing the horizontal and vertical components of loop $FDEF$ are, respectively,

$$r_{DE} \cos \theta_3 - r_E + r_F = 0$$

and

$$r_D + r_{DE} \sin \theta_3 = 0.$$

After substituting the given values, we can solve the preceding four equations by the Newton–Raphson method, as in the previous section. However, in this case, the method requires a four-by-four Jacobian matrix. Let us choose another alternative, the Levenberg–Marquart method, which is a quasi-Newtonian method (a variation of the gradient method). The Levenberg–Marquart method is in the public domain (see More, et al., 1980) and is available on mathematics software.

We begin by approximating the unknowns. A trial-and-error graphical solution may be used to generate these approximations for one position. In this case, the approximations are

$$\theta_2 = 0.3, \quad \theta_3 = 5, \quad r_D = 3.8, \quad \text{and} \quad r_E = 6.3$$

(where angles are given in radians). After a few iterations, the program yields

$$\theta_2 = 0.293, \quad \theta_3 = 5.037, \quad r_D = 3.792, \quad \text{and} \quad r_E = 6.274.$$

The graphical solution was more accurate than necessary; a less accurate set of approximations would have yielded the same results. If the linkage must be solved for a number of successive positions, it may not be necessary to make additional graphical approximations. Instead, the results of one solution are likely to be satisfactory as a first approximation of the unknowns after θ_1 is incremented.

SUMMARY

Vectors and complex numbers are a great help in analyzing and designing mechanisms. If vectors $\boldsymbol{A}$ and $\boldsymbol{B}$ are equal, then the x components of $\boldsymbol{A}$ equal the x components of $\boldsymbol{B}$ and so forth. Thus, one vector equation yields three scalar equations for solving spatial linkages and two scalar equations for planar linkages. When dealing with planar linkages, it is convenient to use complex numbers. If two complex numbers are equal, the real part of the first equals the real part of the second, and the imaginary part of the first equals the imaginary part of the second.

Position analysis of four-bar linkages is a difficult problem; the vector cross-product method is recommended. In general, four-bar linkages have two assembly configurations. Change-point mechanisms may shift from one configuration to another.

Motion simulation software saves much of the drudgery of calculating linkage positions. Mathematics software is also used for linkage design and analysis. Although graphical methods are inefficient for detailed design studies of mechanisms, graphical spot-checking is a good way to find errors in computer calculations.

A Few Review Items

- What is a unit vector?
- ...... a rectangular unit vector?
- Describe the dot product.
- ..... the cross product.
- Does the order of the vectors matter?

- State the Euler formula for complex numbers.
- ...... Of what use is it?
- Why do we use vectors to describe spatial linkages?
- Identify a mechanism for producing intermittent rotation. The input is continuous rotation.
- Sketch a multiloop linkage.
- Can the equations for link position be uncoupled?

PROBLEMS

2.1 List three to five machine components whose motion is described by (a) pure translation, (b) pure rotation, and (c) combined rotation and translation.

2.2 The stroke of a Scotch yoke is 60 mm. The driver rotates at 1740 rev/min (constant). Find the following:

(a) Maximum velocity

(b) Maximum acceleration

(c) Maximum jerk (rate of change of acceleration with respect to time)

2.3 Repeat Problem 2.2, except that the driver is to rotate at 3000 rev/min.

2.4 Repeat Problem 2.2, except that the stroke is to be 45 mm.

2.5 Design a two-cylinder, cam-type piston pump to deliver a flow rate of 0.5 m^3/s at 90 rad/s. Let the stroke equal the piston diameter. Assume 80-percent volumetric efficiency.

2.6 Repeat Problem 2.5, except that the flow rate is to be 0.25 m^3/s at 880 rev/min.

2.7 Repeat Problem 2.5, except that the flow rate is to be 120 gal/min at 1760 rev/min.

2.8 Repeat Problem 2.5, except that the flow rate is to be 12 ft^3/min at 700 rev/min.

2.9 A circular cam with 20-mm eccentricity drives a flat-face follower. Plot the displacement, velocity, and acceleration of the follower versus time for $\omega = 200$ rad/s.

2.10 Repeat Problem 2.9 for 28-mm eccentricity and a cam speed of 1500 rev/min.

Problems 2.11 through 2.14 Refer to a Circular Cam with a Flat-Face Follower. The eccentricity of the cam is *R*, and the angular velocity of the follower is ω.

2.11 Find the maximum angular velocity if $R = 5$ mm and the acceleration of the follower cannot exceed g (the acceleration due to gravity).

2.12 Find the maximum eccentricity in millimeters if the acceleration of the follower cannot exceed g (the acceleration due to gravity). The cam rotates at 4000 rev/min.

2.13 Find follower velocity and acceleration amplitude for $R = 0.25$ in at a cam speed of 1200 rev/min.

2.14 Find follower velocity and acceleration amplitude for $\omega = 300$ and $R = 0.5$ in.

Problems 2.15 through 2.27 Refer to Vector Angles Measured Counterclockwise from the Horizontal Axis. *In these problems,*

$\mathbf{A} = 1$ at 30°, $\mathbf{B} = 2$ at 60°, $\mathbf{C} = 1.5$ at 180°, and $\mathbf{D} = 3$ at 225°. All vectors lie in a plane.

2.15 Find $\boldsymbol{A} + \boldsymbol{B}$.

(a) Add the vectors by using trigonometric functions.

(b) Solve graphically.

2.16 Find $\boldsymbol{A} + \boldsymbol{B} + \boldsymbol{D}$. Solve graphically and analytically.

2.17 Find $\boldsymbol{A} + \boldsymbol{B} + \boldsymbol{C} + \boldsymbol{D}$. Solve graphically.

2.18 Find $\boldsymbol{A} + \boldsymbol{B} + \boldsymbol{C} - \boldsymbol{D}$; that is, find $\boldsymbol{A} + \boldsymbol{B} + \boldsymbol{C} + (-\boldsymbol{D})$. Solve graphically and analytically.

2.19 Find the vector product $\boldsymbol{A} \times \boldsymbol{B}$.

2.20 Find the vector product $\boldsymbol{C} \times \boldsymbol{D}$.

2.21 Find the vector product $\boldsymbol{C} \times (\boldsymbol{D} \times \boldsymbol{E})$, where $\boldsymbol{E} = 2.5$, pointing inward in a direction perpendicular to the plane of $\boldsymbol{C}$ and $\boldsymbol{D}$.

2.22 Repeat Problem 2.21, but find $\boldsymbol{D} \times (\boldsymbol{C} \times \boldsymbol{E})$.

2.23 Find $\boldsymbol{A} \cdot \boldsymbol{B}$.

2.24 Find $\boldsymbol{A} \cdot (\boldsymbol{B} + \boldsymbol{C})$.

2.25 Find $\boldsymbol{A} \cdot (\boldsymbol{B} \times \boldsymbol{C})$.

2.26 Find $\boldsymbol{B} \cdot (\boldsymbol{C} \times \boldsymbol{A})$.

2.27 Find $\boldsymbol{C} \cdot (\boldsymbol{A} \times \boldsymbol{B})$.

In Problems 2.28 through 2.30, *Let* $\boldsymbol{\omega} = (5 + 3t)\boldsymbol{k}$ *and* $\boldsymbol{r} = \boldsymbol{i}r_x(t) + \boldsymbol{j}r_y(t)$, *where* **i**, **j**, *and* **k** *are unit vectors in a fixed coordinate system.*

2.28 Find $\boldsymbol{\omega} \times \boldsymbol{r}$.

2.29 Find $\dfrac{d}{dt}(\boldsymbol{\omega} \times \boldsymbol{r})$.

2.30 Find $\dfrac{d}{dt}(\boldsymbol{\omega} \times \boldsymbol{r})$ if $\boldsymbol{i}$ and $\boldsymbol{j}$ are vectors in a moving coordinate system.

2.31 The Immelman turn was a World War I aircraft maneuver used to gain altitude while turning to fly in the opposite direction. The turn consists of a half loop followed by a half roll to resume normal a level position. Describe the maneuver in terms of pitch and roll coordinates referred to the aircraft. Consider reversing the order of the rotations. Does the commutative law of vector addition apply?

In Problems 2.32 through 2.34,

$$\boldsymbol{r}_1 + \boldsymbol{r}_2 + \boldsymbol{r}_3 = 0.$$

2.32 $\boldsymbol{r}_1 = 10\boldsymbol{i} + 25\boldsymbol{j}$, and $\boldsymbol{r}_2 = 40\boldsymbol{i} - 20\boldsymbol{j} + 10\boldsymbol{k}$; find $\boldsymbol{r}_3$.

2.33 Vectors $\boldsymbol{r}_1$, $\boldsymbol{r}_2$, and $\boldsymbol{r}_3$ represent a planar linkage, with

$$\frac{r_{1y}}{r_{1x}} = 2, \quad \frac{r_{2y}}{r_{2x}} = 0.5, \text{ and } \quad r_3 = i50 + j75.$$

Find $\boldsymbol{r}_1$ and $\boldsymbol{r}_2$ by using the vector cross product method suggested by Chace (see section 2.2 and 2.3).

2.34 Repeat Problem 2.33, except that $r_3 = i5 - j8$.

2.35 A force $F = 22$ N acts on a bar of length $r = 180$ mm (see Figure 2.8b), where $\theta = 68°$. Find the torque about point O.

2.36 Repeat Problem 2.35 for $\theta = 2$ rad.

2.37 An in-line slider-crank linkage has a crank length r_1 and connecting rod length $r_2 = 1.5r_1$. (See Figure 2.9.) Find the connecting rod and slider position when $\theta = 40°$ by using analytical vector methods.

2.38 Repeat Problem 2.37 for $\theta = 140°$.

2.39 Plot the slider position versus θ for an in-line slider-crank linkage for which the ratio of connecting rod to crank length is 1.5. Let $\theta = 0, \pi/9, 2\pi/9, \pi/3$ rad, and so on.

2.40 The link lengths of a planar four-bar mechanism are $r_0 = 120$, $r_1 = 60$, $r_2 = 140$, and $r_3 = 80$. Find the orientation of links 2 and 3 when the internal angle between the crank and the fixed link is 30° and the linkage is in the open phase (i.e., the coupler does not cross the fixed link). Use the vector cross-product method.

2.41 Repeat Problem 2.40 for a 60° internal angle.

2.42 Repeat Problem 2.40 by using the dot-product method.

2.43 Repeat Problem 2.41 by using the dot product method.

2.44 Repeat Problem 2.40 for the crossed phase.

2.45 Repeat Problem 2.40 for the crossed phase, using the dot-product method.

In Problems 2.46 and 2.47,

$$R_1 = 200 \text{ mm at } \theta_1 = \frac{\pi}{3} \text{ rad}$$

and

$$R_2 = 150 \text{ mm at } \theta_2 = \frac{5\pi}{3} \text{ rad.}$$

2.46 **(a)** Express R_1 and R_2 in complex rectangular form.
(b) Find R_0, where $R_0 + R_1 = R_2$.
(c) Express R_0 in polar form.

2.47 Find dR_1/dt if R_1 is constant in magnitude and $d\theta_1/dt = 120$ rad/s.

2.48 Repeat Sample Problem 2.8 for $\theta_1 = 80°$.

2.49 Repeat Problem 2.48 for $\theta_1 = 110°$.

2.50 Repeat Problem 2.37, using complex-number methods.

2.51 Repeat Problem 2.37 with $\theta = 140°$, using complex-number methods.

2.52 For the *RSSC* linkage described in Sample Problem 2.13. Use an analytical method. Let path angle $\gamma = 40°$. Find the displacement of the sliding link for crank angle $\theta = 60°$.

2.53 Repeat Problem 2.52 for $\theta = 0°, 15°, 30°$, and so on. Use a computer.

2.54 Solve Problem 2.52, using methods of descriptive geometry.

2.55 Use vectors A, B, C, D, and E given in Sample Problem 2.3, and determine $A + B + C + D$, $A - C - D - E$,
$B \cdot C$, $C \cdot B$, $|B|$, $|C|$, $(|B|)^2$, $B \cdot B$, $B \times B$, $C \times D$, $D \times C$, and $C \cdot (C \times D)$.

2.56 Use vectors $\boldsymbol{A}$, $\boldsymbol{B}$, $\boldsymbol{C}$, $\boldsymbol{D}$, and $\boldsymbol{E}$ given in Sample Problem 2.3, and determine $\boldsymbol{A}+\boldsymbol{B}+\boldsymbol{C}+\boldsymbol{E}$, $\boldsymbol{A}+\boldsymbol{B}-\boldsymbol{C}-\boldsymbol{D}-\boldsymbol{E}$, $\boldsymbol{D}\cdot\boldsymbol{E}$, $\boldsymbol{E}\cdot\boldsymbol{D}$, $|\boldsymbol{D}|$, $|\boldsymbol{E}|$, $(|\boldsymbol{D}|)^2$, $\boldsymbol{D}\cdot\boldsymbol{D}$, $\boldsymbol{D}\times\boldsymbol{D}$, $\boldsymbol{E}\times\boldsymbol{D}$, $\boldsymbol{D}\times\boldsymbol{E}$, and $\boldsymbol{E}\cdot(\boldsymbol{E}\times\boldsymbol{D})$.

2.57 Use vectors $\boldsymbol{A}$, $\boldsymbol{B}$, $\boldsymbol{C}$, $\boldsymbol{D}$, and $\boldsymbol{E}$ given in Sample Problem 2.3, and determine $\boldsymbol{D}+\boldsymbol{B}+\boldsymbol{C}+\boldsymbol{E}$, $\boldsymbol{B}+\boldsymbol{B}-\boldsymbol{C}-\boldsymbol{D}-\boldsymbol{E}-\boldsymbol{A}$, $\boldsymbol{B}\cdot\boldsymbol{D}$, $\boldsymbol{D}\cdot\boldsymbol{B}$, $|\boldsymbol{D}|$, $|\boldsymbol{B}|$, $(|\boldsymbol{E}|)^2$, $\boldsymbol{E}\cdot\boldsymbol{E}$, $\boldsymbol{E}\times\boldsymbol{E}$, $\boldsymbol{D}\cdot\boldsymbol{D}$, $\boldsymbol{E}\cdot(\boldsymbol{E}\times\boldsymbol{B})$, and $\boldsymbol{B}\times\boldsymbol{D}$.

2.58 Consider an RSSR linkage similar to that in Figure 1.7a, where the link lengths are $r_0 = 32$, $r_1 = 10$, $r_2 = 28$, and $r_3 = 20$. Link 0 lies on the x-axis, link 1 rotates in the xy-plane, and link 3 rotates in the xz-plane. Plot the vector components representing the positions of links 2 and 3 against angular position θ of link 1.

2.59 Consider an RSSR linkage similar to that in Figure 1.7a, where the link lengths are $r_0 = 62$, $r_1 = 20$, $r_2 = 55$, and $r_3 = 45$. Link 0 lies on the x-axis, link 1 rotates in the xy-plane, and link 3 rotates in the xz-plane. Plot the vector components representing the positions of links 2 and 3 against angular position θ of link 1.

2.60 Consider a four-bar linkage for an assembly configuration with r_d, r_2, and r_3 forming a counterclockwise loop. The link lengths are $r_0 = 120$, $r_1 = 60$, $r_2 = 140$, and $r_3 = 80$. Let $\theta_0 = \pi$ rad. Find θ_2 and θ_3 at the instant that $\theta_1 = \pi/3$ rad. Use the Newton–Raphson method. A first approximation may be obtained by sketching the linkage.

2.61 Consider a four-bar linkage for the assembly configuration with r_d, r_2, and r_3 forming a clockwise loop. The link lengths are to be $r_0 = 3$, $r_1 = 1$, $r_2 = 3.6$, and $r_3 = 2.1$. Let $\theta_0 = \pi$ rad. Find θ_2 and θ_3 at the instant that $\theta_1 = \pi/4$ rad. Use the Newton–Raphson method. A first approximation may be obtained by sketching the linkage.

2.62 Consider the double-slider two-loop linkage illustrated in Figure 2.23, where $r_1 = 1$, $r_2 = 5.2$, $r_F = 5$, $r_{DE} = 4$, $r_{CD} = 1.5$, and $\theta_1 = \pi/6$ rad. (Distance DE is identified as r_{DE}, etc.) Find θ_2, θ_3, r_D, and r_E. Use a numerical method.

2.63 Consider the double-slider two-loop linkage illustrated in Figure 2.23, where $r_1 = 1$, $r_2 = 5.2$, $r_F = 5$, $r_{DE} = 4$, $r_{CD} = 1.5$, and $\theta_1 = \pi/2$ rad. (Distance DE is identified as r_{DE}, etc.) Find θ_2, θ_3, r_D, and r_E. Use a numerical method.

2.64 Consider the double-slider two-loop linkage illustrated in Figure 2.23, where $r_1 = 1$, $r_2 = 5.2$, $r_F = 5$, $r_{DE} = 4$, $r_{CD} = 1.5$, and $\theta_1 = 2\pi/3$ rad. (Distance DE is identified as r_{DE}, etc.) Find θ_2, θ_3, r_D, and r_E. Use a numerical method.

2.65 Consider the double-slider two-loop linkage illustrated in Figure 2.23, where $r_1 = 20$, $r_2 = 115$, $r_F = 100$, $r_{DE} = 85$, $r_{CD} = 40$, and $\theta_1 = \pi/4$ rad. (Distance DE is identified as r_{DE}, etc.) Find θ_2, θ_3, r_D, and r_E. Use a numerical method.

2.66 Consider the double-slider two-loop linkage illustrated in Figure 2.23, where $r_1 = 20$, $r_2 = 115$, $r_F = 100$, $r_{DE} = 85$, $r_{CD} = 35$, and $\theta_1 = 2\pi/3$ rad. (Distance DE is identified as r_{DE}, etc.) Find θ_2, θ_3, r_D, and r_E. Use a numerical method.

2.67 A four-bar linkage has the following dimensions (mm):

$$L_0 = 180,\ L_1 = 30,\ L_2 = 174,\ \text{and}\ L_3 = 81.$$

The linkage is assembled so that the vector loop $\boldsymbol{r}_2\,\boldsymbol{r}_3\,\boldsymbol{r}_d$ is clockwise. Check extreme values of the transmission angle. Plot the coupler and output crank positions and the transmission angle against the input crank position.

2.68 A four-bar linkage has the following dimensions (mm):

$L_0 = 120$, $L_1 = 20$, $L_2 = 114$, and $L_3 = 40$. The linkage is assembled so that the vector loop $\boldsymbol{r}_2\,\boldsymbol{r}_3\,\boldsymbol{r}_d$ is counterclockwise. Check extreme values of the transmission angle. Plot the coupler and output crank positions and the transmission angle against the input crank position.

2.69 A four-bar linkage has the following dimensions (mm):
$L_0 = 210$, $L_1 = 35$, $L_2 = 196$, and $L_3 = 84$. The linkage is assembled so that the vector loop $\mathbf{r}_2\mathbf{r}_3\mathbf{r}_d$ is counterclockwise. Check extreme values of the transmission angle. Plot the coupler and output crank positions and the transmission angle against the input crank position.

2.70 Design a mechanism with a 60-mm output link that rotates through an angle of about 55°. The centerline of the continuously rotating input shaft must be 10 mm away from, and parallel to, the plane of motion of the driven link. As a design decision, try an RSSR spatial linkage similar to Figure 2.17, except that $r_1 = 33$, $r_2 = 88$, $r_3 = 60$, $r_{0x} = 10$, and $r_{0y} = -95$ (all dimensions in mm).
Plot and tabulate the output link (driven crank) position and the transmission metric against the drive crank position. Compare your results with the desired output link motion and generally accepted limits of the transmission metric.

2.71 Design a mechanism with a 115-mm output link that rotates through an angle of about 50°. The centerline of the continuously rotating input shaft must be 20 mm away from, and parallel to, the plane of motion of the driven link.
As a design decision, try an RSSR spatial linkage similar to Figure 2.17, except that the assembly configuration will be described by the negative root of r_{3z}. Let $r_1 = 55$, $r_2 = 190$, $r_3 = 115$, $r_{0x} = 20$, and $r_{0y} = -180$ (all dimensions in mm).
Plot and tabulate the output link (driven crank) position and the transmission metric against the drive crank position. Compare the results with desired output link motion and generally accepted limits of the transmission metric.

PROJECTS

See Projects 1.1 to 1.6 and suggestions in Chapter 1. Select a project at this time, or continue with the previously selected project. Establish a set of performance requirements for the project. Examine the linkages involved in the chosen project. Describe and plot motion characteristics of the linkages. Make use of computer software wherever practical. Evaluate the linkages in terms of performance requirements.

BIBLIOGRAPHY AND REFERENCES

Baker, J. E., "Limit Positions of Spatial Linkages via Connectivity Sum Reduction," *Transactions of the American Society of Mechanical Engineers, Journal of Mechanical Design*, vol. 101, July 1979, pp. 504–508.

Chace, M. A., "Vector Analysis of Linkages," *Transactions of the American Society of Mechanical Engineers, Journal of Engineering for Industry, Series B 55*, no. 3, August 1963, pp. 289–297.

Denavit, J., R. S. Hartenberg, R. Razi, and J. J. Uicker, "Velocity, Acceleration and Static-Force Analyses of Spatial Linkages," *Transactions of the American Society of Mechanical Engineers, Journal of Applied Mechanics*, December 1965, pp. 903–910.

Duffy, J., and S. Derby, "Displacement Analysis of a Spatial 7R Mechanism—a Generalized Lobster's Arm," *Transactions of the American Society of Mechanical Engineers, Journal of Mechanical Design*, vol. 101, April 1979, pp. 224–228.

Fanghella, P., "Kinematics of Spatial Linkages by Group Algebra: A Structure-Based Approach," *Mechanism and Machine Theory*, vol. 23, no. 3, 1988, pp. 171–184.

Knowledge Revolution, *Working Model™ 2D User's Manual*, Knowledge Revolution, San Mateo CA, 1996.

Knowledge Revolution, *Working Model™ 3D User's Manual*, Knowledge Revolution, San Mateo CA, 1998.

Lee, H.-Y., and C.-G. Liang, "A New Vector Theory for the Analysis of Spatial Mechanisms," *Mechanism and Machine Theory*, vol. 23, no. 3, 1988a, pp. 209–218.

Lee, H.-Y., and C.-G. Liang, "Displacement Analysis of the General Seven-Link 7-R Mechanism," *Mechanism and Machine Theory*, vol. 23, no. 3, 1988b, pp. 219–226.

MathSoft, *Mathcad 2000™ User's Guide*, MathSoft, Inc., Cambridge MA, 1999.

Mischke, C. R., *Elements of Mechanical Analysis*, Addison-Wesley, Reading, MA, 1963.

More, J., B. Garbow, and K. Hillstrom, *Users Guide to MINPACK I*, ANL-80-74, Argonne National Laboratory, Argonne, IL, 1980.

Stark, P. A., *Introduction to Numerical Methods*, Macmillan, New York, 1970, pp. 115–147.

Taylor, A. E., *Advanced Calculus*, Ginn, Boston, 1955.

Note also the Internet references listed under the heading of **Software for calculation, design, manufacturing, motion simulation and testing** found at the end of Chapter 1. You will find many more sites related to the topics of this chapter by searching the Internet yourself.

C H A P T E R 3

Velocity Analysis of Mechanisms

Velocity is a vector quantity, having both magnitude and direction. We need to know the velocity of points on a mechanism and the angular velocity of links. Both average and instantaneous velocity are important design criteria. Velocity analysis precedes acceleration and dynamic analyses, necessary steps in the design of high-speed machinery.

Concepts You Will Learn and Apply While Studying This Chapter

- Instantaneous and average velocity of a point on a mechanism
- Angular velocity of a link
- Relative velocity
- Analytical vector methods for design and analysis of planar and spatial mechanisms
- Matrix methods applied to spatial mechanisms
- Complex-number methods for design and analysis of planar mechanisms
- Graphical methods for design and analysis of mechanisms
- Velocity analysis of slider-crank linkages, planar and spatial four-bar linkages, sliding contact linkages, and combinations of these.
- Practical applications of the basic linkages

3.1 BASIC CONCEPTS

Velocity is a vector representing the change in position of a moving point, divided by the time interval during which the point changes its position. If the time interval is finite, the result is the **average velocity**

$$\boldsymbol{v}_{\text{average}} = \frac{\Delta \boldsymbol{s}}{\Delta t},$$

where $\Delta \mathbf{s}$ is the change in position and Δt = change in time. If the time interval is infinitesimal, we have the **instantaneous velocity**

$$\boldsymbol{v} = \text{limit}_{\Delta t \to 0} \frac{\Delta \boldsymbol{s}}{\Delta t} = \frac{d\boldsymbol{s}}{dt}.$$

Since we are concerned largely with instantaneous values, the term *velocity* will refer to instantaneous velocity unless otherwise noted.

Average speed is a scalar quantity equal to the total distance traveled divided by the time interval. Consider an automotive piston as it travels between limiting positions during one-half crankshaft rotation. Average speed and average velocity of the piston are both given by the stroke divided by the time for one-half crankshaft rotation. Velocity will also indicate the direction of travel. For a full crankshaft rotation, average speed is twice the stroke divided by the time for a full rotation. But the piston has returned to its original position, and the average velocity is zero.

A vector representing the change in angular position of a body divided by the time interval during which the body changes its angular position is called the **angular velocity**

$$\boldsymbol{\omega} = \text{limit}_{\Delta t \to 0} \frac{\Delta \theta}{\Delta t} = \frac{d\theta}{dt}.$$

Angular velocity is sometimes treated as a scalar in dealing with planar linkages. Analytical and graphical vector methods, including representing of vectors in complex form, are useful in velocity studies related to linkage design.

Velocity of a Point

Let the location of a point be described by a vector $\boldsymbol{R}$. In Figure 3.1, consider point P, which moves along curve C through a displacement $d\boldsymbol{R}$ during a time interval dt. The new position vector is then $\boldsymbol{R} + d\boldsymbol{R}$, representing a change in the direction of $\boldsymbol{R}$, a change in the magnitude of $\boldsymbol{R}$, or both. If we allow the time interval dt to become infinitesimal, the corresponding infinitesimal displacement of $d\boldsymbol{R}$ lies on the curve C. Then, the instantaneous velocity of point P is given by

$$\boldsymbol{v} = \frac{d\boldsymbol{R}}{dt} \tag{3.1}$$

where the direction of $\boldsymbol{v}$ is given by a tangent to curve C at P. A dot above a vector or scalar quantity is sometimes used to indicate differentiation with respect to time; thus, $d\boldsymbol{R}/dt$ becomes $\dot{\boldsymbol{R}}$.

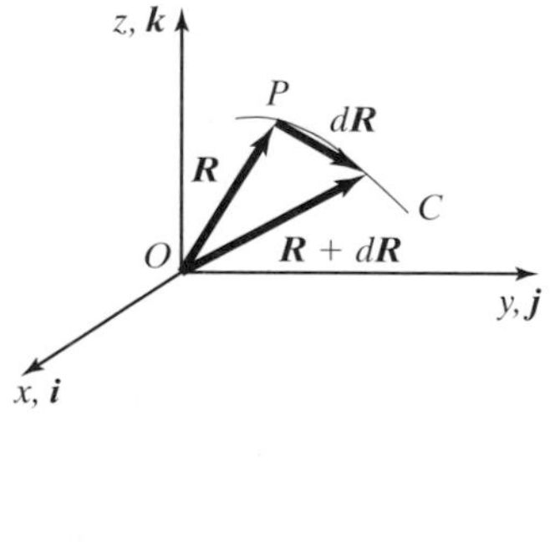

FIGURE 3.1 Velocity of a point.

In general, the vector $d\boldsymbol{R}$ represents a change in the x, y, and z components of $\boldsymbol{R}$. The velocity of point P may be expressed in terms of the components R_x, R_y, and R_z and the unit vectors $\boldsymbol{i}, \boldsymbol{j}$, and $\boldsymbol{k}$, parallel to the coordinate axes, as noted in Chapter 2. Then,

$$\boldsymbol{v} = \dot{\boldsymbol{R}} = \boldsymbol{i}\dot{R}_x + \boldsymbol{j}\dot{R}_y + \boldsymbol{k}\dot{R}_z \tag{3.2}$$

if the x, y, z coordinate system is stationary.

Angular Velocity

Angular velocity may be treated as a vector quantity. Consider a link whose angular position changes at a rate of

$$\boldsymbol{\omega} = \omega_x\boldsymbol{i} + \omega_y\boldsymbol{j} + \omega_z\boldsymbol{k} \text{ radians per second}$$

(where $\omega = 2\pi n/60$, for a rotation speed of n revolutions per minute). The direction of vector $\boldsymbol{\omega}$ is perpendicular to the plane of rotation, and its sense is found by curving the fingers of the right hand in the direction of rotation. The thumb then points in the direction of vector $\boldsymbol{\omega}$. Alternatively, consider a right-hand screw rotating clockwise. The direction of the vector $\boldsymbol{\omega}$ is the direction of advance along the screw axis. For a body that rotates in the xy-plane, $\boldsymbol{\omega}$ will be in the $\pm z$ direction. In Figure 3.2 for example, $\boldsymbol{\omega} = \omega\boldsymbol{k}$.

Motion of a Rigid Body about a Fixed Axis (Without Translation)

Consider a rigid body rotating about an axis that is fixed in a stationary coordinate system. (See Figure 3.3.) Then, angular velocity $\boldsymbol{\omega}$ has a fixed direction (along that axis). If a point P that is fixed in the body is identified by vector $\boldsymbol{R}$, then P moves in a curved

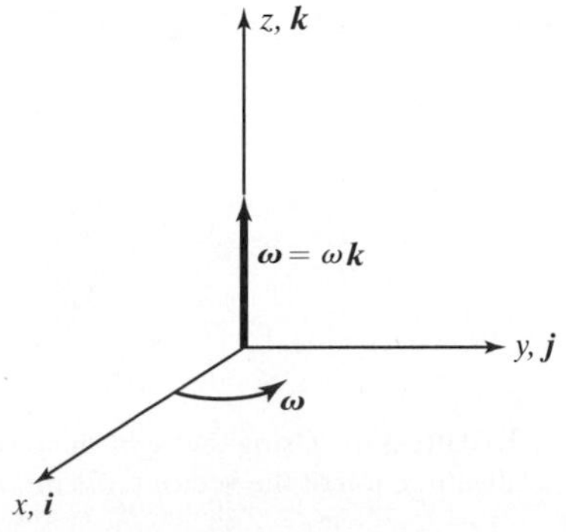

FIGURE 3.2 Rotation in the xy-plane.

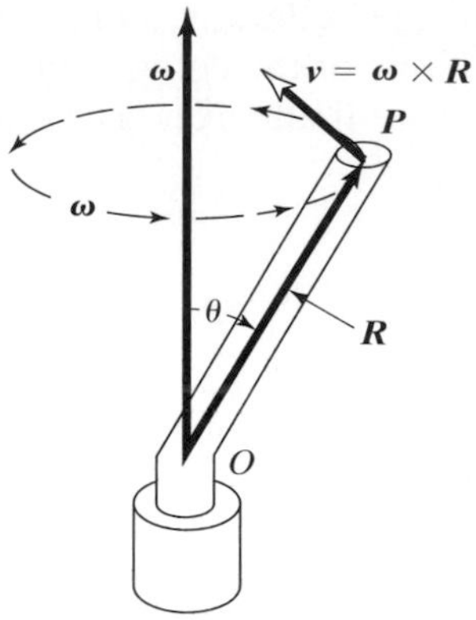

FIGURE 3.3 Motion of a rigid body motion about a fixed axis.

path of radius $R \sin \theta$ about that axis, where θ is the angle between the axis of rotation and $\boldsymbol{R}$. The speed (i.e., magnitude of the velocity) of P is given by

$$v = \omega R \sin \theta. \tag{3.3}$$

Compare this result with the vector cross product identified in Chapter 2. Thus, we may write

$$\dot{\boldsymbol{R}} = \boldsymbol{v} = \boldsymbol{\omega} \times \boldsymbol{R}, \tag{3.4}$$

since the direction of the velocity is perpendicular to the plane of $\boldsymbol{\omega}$ and $\boldsymbol{R}$ and is given by the right-hand rule. The thumb of the right hand is pointed in the $\boldsymbol{\omega}$ direction, and the index finger in the $\boldsymbol{R}$ direction. (See Figure 3.4.) Velocity is then in the direction of the third finger.

If $\boldsymbol{\omega}$ varies (with time) in magnitude or direction, the velocity is still given by the preceding equations, provided that $d\boldsymbol{\omega}/dt$ is finite. Thus, the instantaneous velocity of a point P in a rigid body that rotates about a fixed point (e.g., a ball joint) with instantaneous angular velocity $\boldsymbol{\omega}$ is given by

$$\boldsymbol{v} = \boldsymbol{\omega} \times \boldsymbol{R},$$

where $\boldsymbol{R}$ is measured from any point on the instantaneous axis of rotation. Thus, for a link that moves in a plane about a stationary revolute joint, $\boldsymbol{\omega}$ and $\boldsymbol{R}$ are perpendicular,

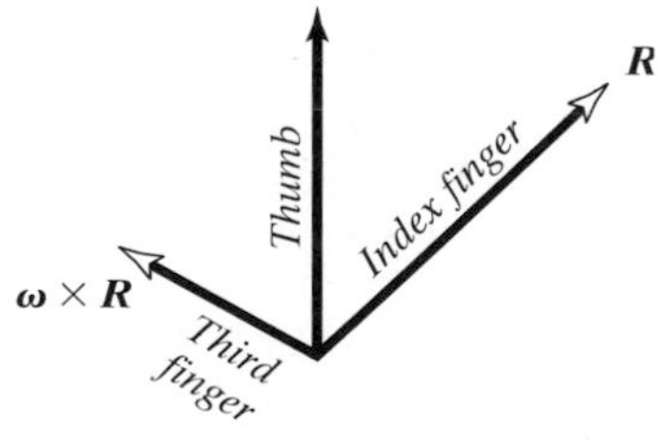

FIGURE 3.4 Using the right-hand rule to find the direction of the vector cross product.

and the speed of a point in the link is given by the product ωR. If we consider sliding along a rotating link, or if we consider links that are not fixed at a point, additional terms enter into our analysis.

SAMPLE PROBLEM 3.1

Surface Speed

Surface speeds of from 800 to 2,000 ft/min (4,064 to 10,150 mm/s) are recommended for milling aluminum. Find the *corresponding speeds* (i.e., the *angular velocities*), *in revolutions per minute*, for a 4-in- (101.5-mm-) diameter milling cutter.

Solution. Let the radius of the cutter be $R = 2$ in. The lowest surface speed is

$$v = 800 \text{ ft/min} \times 12 \text{ in/ft} \times 1 \text{ min/60 s} = 160 \text{ in/s}.$$

The angular velocity vector and the radius vector are perpendicular to each other. Thus, $\boldsymbol{v} = \boldsymbol{\omega} \times \boldsymbol{R} = \omega R$ tangent to the surface, and

$$\omega = \frac{v}{R} = \frac{160 \text{ in/s}}{2 \text{ in}} = 80 \text{ rad/s}.$$

Next, divide 80 rad/s by 0.1047 (rad/s) / (1 rev/min) to obtain 764 rev/min, which is the minimum value of the angular speed. Similarly, 2,000 ft/min gives us a maximum value of 1,910 rev/min.

SAMPLE PROBLEM 3.2

An In-Line Slider-Crank Mechanism

a. Determine the velocity of the piston in a pump modeled as an in-line slider-crank linkage.

b. Let $R = 2$ in, $L = 3.76$ in, $\theta = 70°$, and $\omega = 10$ rad/s. Find the slider velocity analytically.

c. An in-line slider-crank mechanism has a crank length of 200 mm and a connecting-rod length of 560 mm. The crank rotates at a constant angular velocity of 50 rad/s counterclockwise. Find the average slider velocity during one stroke.

Solution. **(a)** An analytical examination of the in-line slider-crank mechanism shows that the mechanism has some resemblance to the Scotch yoke considered in Chapter 2. In fact, the Scotch yoke can be considered a special case of the slider crank: a slider-crank linkage with an infinite connecting rod. It is the connecting rod and the angle ϕ that it forms with the slider path which complicates our analytical solution. Figure 3.5 shows the in-line slider crank first in its extreme extended position (top dead center) and then in a general position with angular displacement θ of the crank.

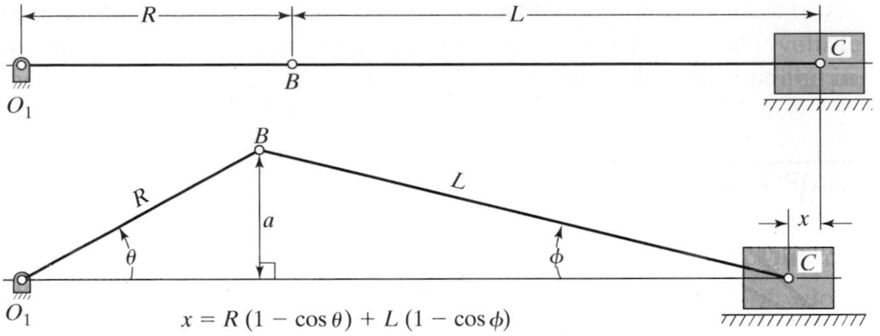

FIGURE 3.5 In-line slider-crank mechanism, shown first in its extended position (*top*) and then an instant later, when the crank has moved through an angle θ (*bottom*).

Measuring piston displacement x from the original position, we have

$$x = R + L - (R\cos\theta + L\cos\phi)$$
$$= R(1 - \cos\theta) + L(1 - \cos\phi).$$

We may express ϕ in terms of θ by dropping a perpendicular from B to line O_1C, forming two right triangles. The length of the perpendicular is

$$a = R\sin\theta = L\sin\phi.$$

Using this equation and the identity $\sin^2\phi + \cos^2\phi = 1$, we obtain the exact slider displacement in terms of θ only:

$$x = R(1 - \cos\theta) + L\left[1 - \sqrt{1 - \left[\frac{R}{L}\right]^2 \sin^2\theta}\right].$$

The slider velocity is obtained by differentiating x with respect to time.

The exact slider velocity is given by the following equation where angular speed ω is the rate of change of θ with respect to time:

$$v = R\omega\sin\theta\left[1 + \left(\frac{R}{L}\right)\frac{\cos\theta}{\sqrt{1 - (R/L)^2\sin^2\theta}}\right].$$

(Positive velocity is to the left in this example.)

If the slider-crank mechanism is a piston engine or piston pump for which the ratio of L to R is fairly large (say, 3 or more), a simplification is in order. We expand the displacement equation by the binomial theorem, retaining only the terms

$$x = R\left[1 - \cos\theta + \left(\frac{1}{2}\right)\left(\frac{R}{L}\right)\sin^2\theta\right].$$

Using this equation or simplifying the velocity equation directly, we obtain the approximate slider velocity

$$v = R\omega \sin\theta\left[1 + \left(\frac{R}{L}\right)\cos\theta\right].$$

From the trigonometric identity $\sin\theta \cdot \cos\theta = \frac{1}{2}\sin(2\theta)$, we may write the approximate velocity equation in the form

$$v = R\omega\left[\sin\theta + \frac{1}{2}(R/L)\sin(2\theta)\right].$$

(b) Using the exact equation, we have

$$v = 2(10)\sin 70^\circ\left[1 + \left(\frac{2}{3.76}\right)\frac{\cos 70^\circ}{\sqrt{1 - (2/3.76)^2(\sin 70^\circ)^2}}\right]$$

$$= 22.8 \text{ in/s to the left.}$$

From the approximate velocity equation, we obtain $v = 22.2$ in/s, which is a fairly good approximation, given that the ratio L/R is *not* within the recommended range for the approximate equation.

(c) We might be tempted to integrate an expression for velocity or to average values over an entire plot. The exact solution, however, is simply the stroke, $2R$, divided by the time taken to complete half of one cycle. For any in-line slider-crank mechanism with a constant crank speed, the slider speed becomes

$$v_{\text{av}} = \frac{2R}{\pi/\omega}.$$

Thus, the average slider speed is given by

$$v_{\text{av}} = \frac{2 \times 200 \times 50}{\pi} = 6366 \text{ mm/s.}$$

In the past, approximate solutions to kinematics problems were used to save time calculating. Today, with the general availability of computers, you might wonder why we used an approximate equation at all in the preceding example. The answer is that the approximate velocity equation is easy to differentiate; we can then find acceleration and inertial forces in a useful form for designing vibration isolation.

Parameter Studies

Parameter studies are an aid in selecting optimum linkage dimensions and speeds. We might examine velocities in a particular class of linkages (the slider crank, for example), without specifying actual dimensions or speeds. To be as general as possible, a family of curves of velocity versus crank angle can be plotted, each curve for a different ratio of connecting-rod length L to crank length R. (See Figure 3.6a.) To normalize

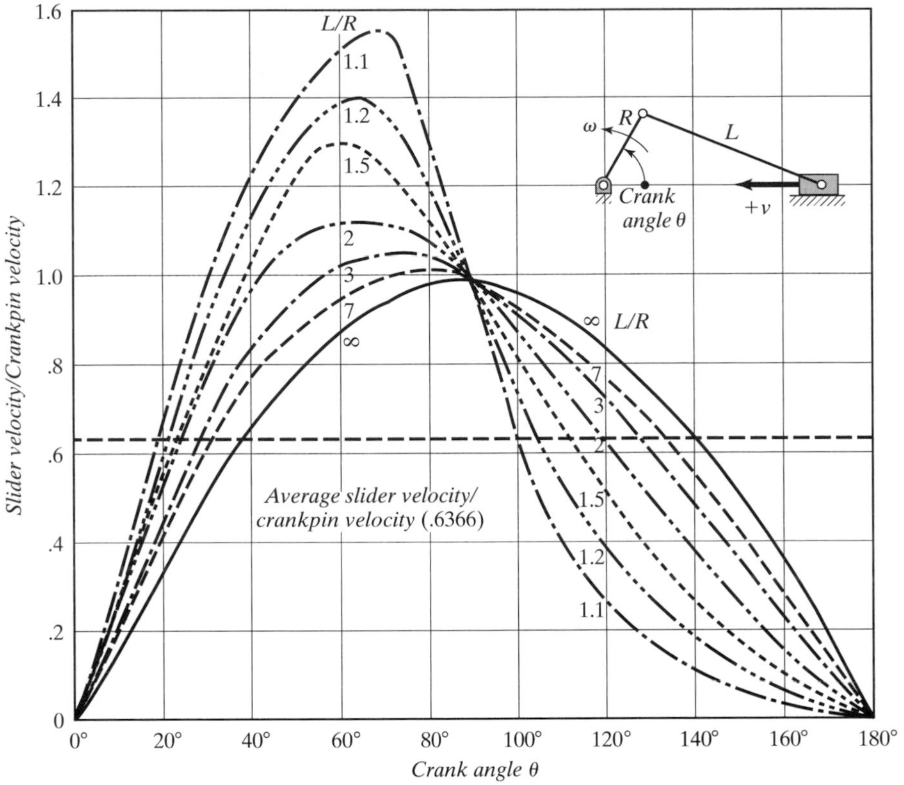

FIGURE 3.6 (a) The family of curves shown here represents those obtained from an in-line slider-crank mechanism *parameter study*. The ratio of slider velocity to crankpin velocity is plotted against the crank angle for various L/R ratios. The slider velocity at any position of the mechanism is found by multiplying the ordinate (slider velocity/crankpin velocity) by the actual value of $R\omega$.

the results, the product of crank angular velocity and crank length may be assigned a value of unity. Later, all velocities will be multiplied by the actual value of ωR to obtain slider velocity.

Figure 3.6a is a family of curves of slider velocity versus crank angle for an in-line slider-crank mechanism. The $L/R = \infty$ curve is a sine wave, representing the actual velocity of a Scotch yoke mechanism or the limiting velocity relationship for a connecting-rod length many times greater than the crank length. Note how closely the curve $L/R = 7$ resembles the sine curve. To obtain curves for which the ratio of connecting rod length to crank length is near unity, the exact analytical solution is preferred.

Figure 3.6b shows the normalized piston velocity $v/(\omega R)$ versus the crank angle for an in-line slider-crank mechanism over a full cycle of motion. In this plot, L/R ratios range from 1.2 to 2, and piston velocity toward the crank is shown below the axis. It can be seen that all the plots for the in-line slider-crank mechanism are antisymmetric about a crank angle of π radians.

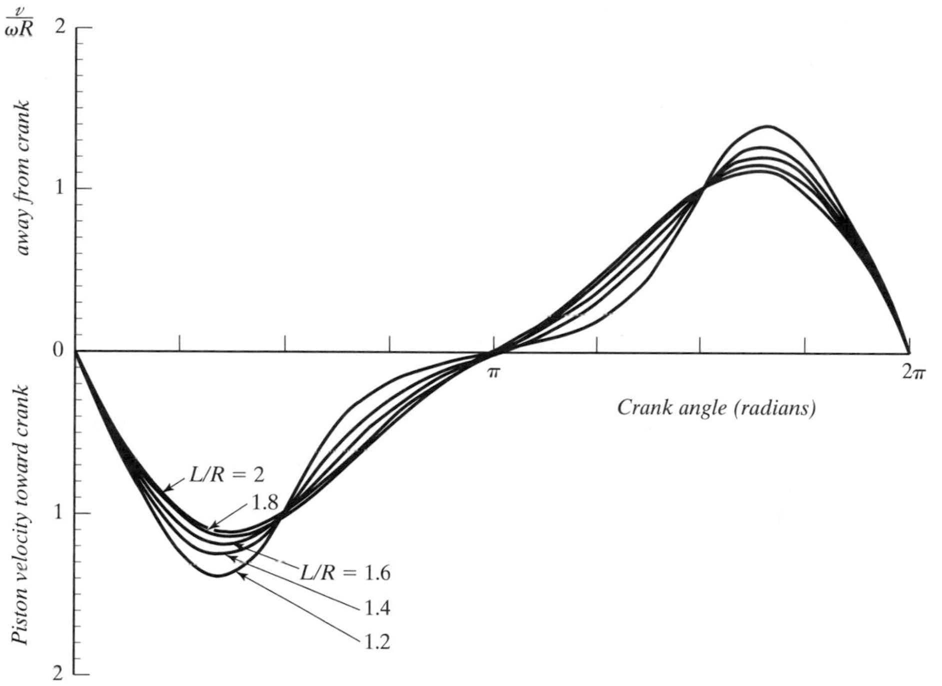

FIGURE 3.6 (b) Normalized piston velocity vs. crank position for various ratios L/R (in-line slider-crank mechanism).

SAMPLE PROBLEM 3.3

Specification of a Linkage to Satisfy Velocity Conditions

Let us specify the dimensions of a linkage to meet a simple set of requirements. Suppose a certain process requires rectilinear motion with a velocity between 75 and 100 in/s in one direction during at least 15 percent of each cycle. The linkage is to be driven by a shaft that turns at 600 rev/min.

Solution. The requirements are not very rigid, and therefore, several mechanisms would be satisfactory, but we will consider the in-line crank mechanism that has already been examined in detail. Checking the curve representing $L/R = 3$ in Figure 3.6a, we see that this connecting-rod-to-crank-length ratio may be satisfactory. For that curve, $v/(R\omega)$ ranges from about 0.8 to 1.05 during the interval from $\theta = 40°$ to $\theta = 110°$ (which is greater than 15 percent of one cycle). If we let $v/(R\omega) = 0.8$ correspond to $v = 75$ in/s when

$$\omega = \frac{2\pi}{60} \times 600 \text{ rev/min} = 62.8 \text{ rad/s},$$

then

$$R = \frac{v}{0.8\omega} = \frac{75}{0.8 \times 62.8} = 1.5 \text{ in.}$$

At maximum velocity, $v/(R\omega) = 1.05$, or $v \approx 99$ in/s, which is within the required range. The tentative solution, then, is an in-line slider-crank mechanism with crank length $R = 1.5$ in and connecting-rod length $L = 3R = 4.5$ in, and the conditions are satisfied between crank angles of $\theta = 40°$ and $110°$ (approximately). Since the curves are only approximate, the next step is to determine the velocity accurately during the chosen interval.

In almost every practical design situation, the first step involves sketching as many linkages as possible that might be suitable. The only limits are the designer's creativity and experience. Then, the motion characteristics of the linkages are analyzed, first to ascertain whether the displacement pattern meets all requirements and then to check the velocity and acceleration of the mechanism.

3.2 MOVING COORDINATE SYSTEMS AND RELATIVE VELOCITY

It is sometimes convenient to establish a coordinate system that translates or rotates along with a moving link. In most cases, we then refer velocities and accelerations back to a fixed coordinate system.

Consider the two coordinate systems of Figure 3.7. Coordinate axes X, Y, and Z and the corresponding unit vectors $\boldsymbol{I}$, $\boldsymbol{J}$, and $\boldsymbol{K}$ are fixed. (For most work with mechanisms, this would mean that the XYZ coordinate system is an inertial reference frame; that is, it does not move with respect to the earth.) The origin o of coordinate system xyz is defined by the position vector $\boldsymbol{R}_o$. Unit vectors $\boldsymbol{i}$, $\boldsymbol{j}$, and $\boldsymbol{k}$ for this set of moving axes lie along, and move with, the x-, y-, and z-axes, respectively. The xyz–$\boldsymbol{ijk}$ system may translate and/or rotate in any direction. A point P in a linkage is described by the vector $\boldsymbol{r}$ (the position vector $o\boldsymbol{P}$) in the moving coordinate system xyz. The total *position vector* of P is

$$\boldsymbol{R} = \boldsymbol{R}_o + \boldsymbol{r}, \tag{3.5}$$

measured from the origin of the fixed coordinates.

Expressing vectors $\boldsymbol{R}_o$ and $\boldsymbol{r}$ in terms of their components and corresponding unit vectors, the radius vector to point P is given by

$$\boldsymbol{R} = R_{0X}\boldsymbol{I} + R_{0Y}\boldsymbol{J} + R_{0Z}\boldsymbol{K} + r_x\boldsymbol{i} + r_y\boldsymbol{j} + r_z\boldsymbol{k}. \tag{3.6}$$

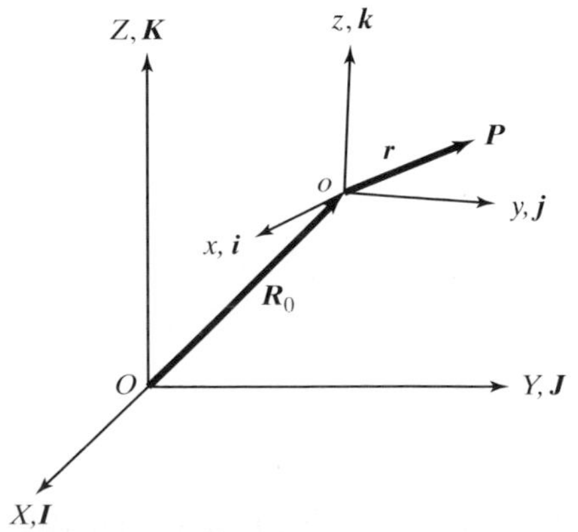

FIGURE 3.7 A moving coordinate system. System xyz moves within fixed system XYZ. Point P moves within xyz. The absolute position of point P is given by the vector sum $\boldsymbol{R}_o + \boldsymbol{r}$, where $\boldsymbol{r}$ is the position of P with respect to the moving system and $\boldsymbol{R}_o$ locates the origin of the moving system.

Note that in Eq. 3.6, $\boldsymbol{R}_o$ is written in terms of the fixed coordinate system, while $\boldsymbol{r}$ is written in terms of the moving coordinate system. The *velocity* of point P is given by the rate of change of $\boldsymbol{R}$ with respect to time:

$$\dot{\boldsymbol{R}} = \dot{R}_{0X}\boldsymbol{I} + \dot{R}_{0Y}\boldsymbol{J} + \dot{R}_{0Z}\boldsymbol{K} + \dot{r}_x\boldsymbol{i} + \dot{r}_y\boldsymbol{j} + \dot{r}_z\boldsymbol{k} + r_x\dot{\boldsymbol{i}} + r_y\dot{\boldsymbol{j}} + r_z\dot{\boldsymbol{k}}. \tag{3.7}$$

The first vector on the right side of the equation is the rate of change of $\boldsymbol{R}_o$ in the X direction—the X component of the velocity of o. Since the X, Y, and Z coordinate frame is fixed, unit vectors $\boldsymbol{I}, \boldsymbol{J}$, and $\boldsymbol{K}$ do not change, and the velocity of o is given completely by the first three vectors on the right of the equation. The sum of these vectors will be identified by the symbol $\dot{\boldsymbol{R}}_o$. The next three vectors, $\dot{r}_x\boldsymbol{i}$ and so on, represent the rate of change in the $\boldsymbol{r}$ vector *with respect to the moving coordinates*, or the velocity of P relative to the moving coordinate system xyz. The sum of these vectors will be identified by the symbol $\dot{\boldsymbol{r}}_r$. (The velocity of P relative to the xyz system will be denoted by the use of the subscript r.)

The last three vectors of Eq. 3.7 represent the effect of the rotating coordinate system (xyz) in any expression for the absolute velocity of P. Unit vectors $\boldsymbol{i}, \boldsymbol{j}$, and $\boldsymbol{k}$ are fixed relative to the moving xyz system (i.e., $\boldsymbol{i}, \boldsymbol{j}, \boldsymbol{k}$ move *with* the xyz system). Relative to the fixed XYZ system, however, unit vectors $\boldsymbol{i}, \boldsymbol{j}$, and $\boldsymbol{k}$ rotate; thus, their positions relative to fixed system XYZ are functions of time.

The first derivative of a vector of constant magnitude is the cross product of the angular velocity of the vector (i.e., the angular velocity of the moving coordinate system) and the vector itself. Thus, for the last three vectors in Eq. 3.7,

$$r_x\dot{\boldsymbol{i}} = r_x(\boldsymbol{\omega} \times \boldsymbol{i}),$$

$$r_x\dot{\boldsymbol{j}} = r_x(\boldsymbol{\omega} \times \boldsymbol{j}),$$

and

$$r_x\dot{\boldsymbol{k}} = r_x(\boldsymbol{\omega} \times \boldsymbol{k}), \tag{3.8}$$

so that

$$\begin{aligned} r_x\dot{\boldsymbol{i}} + r_y\dot{\boldsymbol{j}} + r_z\dot{\boldsymbol{k}} &= r_x(\boldsymbol{\omega} \times \boldsymbol{i}) + r_y(\boldsymbol{\omega} \times \boldsymbol{j}) + r_z(\boldsymbol{\omega} \times \boldsymbol{k}), \\ &= \boldsymbol{\omega} \times (r_x\boldsymbol{i} + r_y\boldsymbol{j} + r_z\boldsymbol{k}) \\ &= \boldsymbol{\omega} \times \boldsymbol{r}. \end{aligned} \tag{3.9}$$

Therefore, the last three vectors of Eq. 3.7 can be replaced by the vector product $\boldsymbol{\omega} \times \boldsymbol{r}$, where $\boldsymbol{\omega}$ is the angular velocity of the xyz coordinate system and $\boldsymbol{r}$ is the position vector of P in the xyz system. Recall, from the previous section, that the cross product

$$\boldsymbol{\omega} \times \boldsymbol{r}$$

represents the velocity of a point on a rigid body rotating about a fixed axis.

The velocity of $\boldsymbol{R}$ may be expressed more concisely as

$$\dot{\boldsymbol{R}} = \dot{\boldsymbol{R}}_o + \dot{\boldsymbol{r}}_r + \boldsymbol{\omega} \times \boldsymbol{r}, \tag{3.10}$$

where $\dot{\boldsymbol{R}}$ = absolute velocity of point relative to P (XYZ),
$\dot{\boldsymbol{R}}_o$ = velocity of the origin o of the xyz system,
$\dot{\boldsymbol{r}}_r$ = velocity of point P relative to the xyz system,

and

$\boldsymbol{\omega} \times \boldsymbol{r}$ = cross product of the angular velocity of the moving system xyz in the XYZ system and the position vector $\boldsymbol{r}$.

Relative Velocity from Another Viewpoint

In the preceding section, we referred to absolute velocity—that is, a velocity measured in a fixed coordinate system (an inertial reference frame). In addition, a relative velocity was identified—the velocity of a point with respect to a moving coordinate system. In the study of mechanisms, it is sometimes useful to describe the velocity of a point by referring to another moving point. In this regard, consider nonstationary points B and C. The term $\boldsymbol{v}_{CB}$ is defined as *the absolute vector velocity of C minus the absolute vector velocity of B*; that is,

$$\boldsymbol{v}_{CB} = \boldsymbol{v}_C - \boldsymbol{v}_B. \tag{3.11}$$

Frequently, $\boldsymbol{v}_{CB}$ is referred to as the velocity of C *relative to B* or the velocity of C *with respect to B*. Other terms include *velocity difference* and the velocity of C *about B*. Some works use a different notation, such as $\boldsymbol{v}_{C/B}$, instead of $\boldsymbol{v}_{CB}$. When the terms "relative to," "with respect to," and "about" are used, it is understood that motion is viewed from an inertial, or nonrotating, reference frame. An observer in a rotating reference frame would not, in general, detect the correct relative velocity as just defined. In the study of mechanisms, the earth is most often selected as a "stationary reference frame." However, problems of spaceflight and even problems of long-range ballistics on the earth require that the earth's motion be considered.

Equation 3.11 may be written in the equivalent form

$$\boldsymbol{v}_C = \boldsymbol{v}_B + \boldsymbol{v}_{CB}, \tag{3.12}$$

where the plus sign indicates the *vector* sum. For a simple example of relative velocity, let an aircraft carrier B move northward with a velocity $\boldsymbol{v}_B$ = 15 knots (7.72 m/s). Suppose an aircraft C on the flight deck has a velocity of $\boldsymbol{v}_{CB}$ = 25 knots (12.86 m/s) relative to the carrier. The direction of the path of the aircraft across the flight deck differs from the direction of the velocity of the aircraft carrier by 20°, as shown in Figure 3.8. The vector $\boldsymbol{v}_B$, which represents the velocity of the carrier, is drawn to a convenient scale, starting at an arbitrary point o. Then the vector $\boldsymbol{v}_{CB}$, which represents the velocity of the aircraft *relative to the carrier*, is drawn, beginning at the head of vector $\boldsymbol{v}_B$. The vector sum

$$\boldsymbol{v}_B + \boldsymbol{v}_{CB} = \boldsymbol{v}_C$$

is the vector beginning at o, with its head at the head of $\boldsymbol{v}_{CB}$.

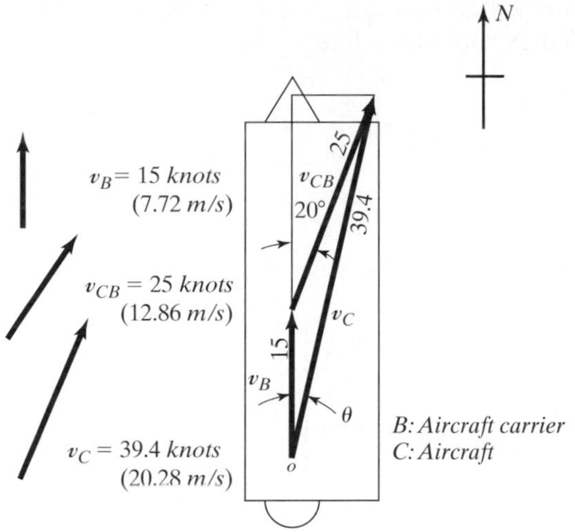

FIGURE 3.8 Relative velocity.

The concept of relative motion also applies to machine operations, as in the case of the motion of a robot manipulator on a fixed base, where the manipulator is to interact with an assembly on a moving production line, or a lathe tool that moves axially to cut a helical thread in a rotating workpiece.

3.3 MATRIX AND DETERMINANT CONCEPTS USEFUL IN THE STUDY OF KINEMATICS AND DYNAMICS OF MACHINERY

Many kinematics and dynamics problems can be reduced to a set of linear equations. Matrix notation may make these problems easier to solve. If there are more than two equations in the set, we can "work smart" by using software with matrix capability. This section provides only a brief introduction to the matrix methods we need for kinematics and dynamics. If you are convinced that matrices provide a convenient and powerful approach to engineering problems, look for books with *matrices*, *linear algebra*, or *linear analysis* in the title.

A Few Definitions

A matrix is an array of elements in rows and columns. The form of a matrix with m rows and n columns is

$$\mathbf{A} = \begin{bmatrix} a_{11} & a_{12} & \cdots & a_{1n} \\ a_{21} & a_{22} & \cdots & a_{2n} \\ \cdots & \cdots & \cdots & \cdots \\ a_{m1} & a_{m2} & \cdots & a_{mn} \end{bmatrix}.$$

A matrix with one column is called a *vector* or a column matrix. In ordinary three-dimensional space such a matrix could look like

$$\mathbf{r_2} = \begin{bmatrix} r_{2x} \\ r_{2y} \\ r_{2z} \end{bmatrix}.$$

Or, a set of angular velocities could form the vector:

$$\mathbf{X} = \begin{bmatrix} \omega_{2x} \\ \omega_{2y} \\ \omega_{2z} \\ \omega_3 \end{bmatrix}.$$

Note that vectors are not limited to describing links and velocities in a plane or in three dimensional space. You may prefer to call the preceding vector a state vector.

An identity matrix (usually labeled **I**) has ones as the diagonal elements and zeros elsewhere. The identity matrix of order four is thus

$$\mathbf{I} = \begin{bmatrix} 1 & 0 & 0 & 0 \\ 0 & 1 & 0 & 0 \\ 0 & 0 & 1 & 0 \\ 0 & 0 & 0 & 1 \end{bmatrix}.$$

The inverse of matrix **A** is labeled $\mathbf{A}^{-1}$. Multiplying a matrix by its inverse results in the identity matrix; that is,

$$\mathbf{A}^{-1}\mathbf{A} = \mathbf{I}.$$

Matrix Multiplication

Matrix **A** may be multiplied by matrix **X** (in the order **AX**) if the number of columns in **A** equals the number of rows in **X**. Suppose **A** has r_A rows and c_A columns, and **X** has r_X rows and c_X columns. Then the product $\mathbf{AX} = \mathbf{B}$ is defined if c_A equals r_X. Note that **B** has r_A rows and c_X columns.

Now consider an n-by-n square matrix

$$\mathbf{A} = \begin{bmatrix} a_{11} & a_{12} & \cdots & a_{1n} \\ a_{21} & a_{22} & \cdots & a_{2n} \\ \cdots & \cdots & \cdots & \cdots \\ a_{n1} & a_{n2} & \cdots & a_{nn} \end{bmatrix}$$

and an n-row column matrix

$$\mathbf{X} = \begin{bmatrix} x_1 \\ x_2 \\ \cdots \\ x_n \end{bmatrix}.$$

We can form the product $\mathbf{AX} = \mathbf{B}$, where $\mathbf{B}$ is a column matrix with n elements. The ith element of $\mathbf{B}$ is

$$b_i = \sum_{j=1}^{n} a_{ij} \cdot x_j.$$

For example, if $\mathbf{A}$ is a four-by-four matrix and $\mathbf{X}$ is a column matrix with four terms, then the second term in $\mathbf{B}$ is

$$b_2 = a_{21}x_1 + a_{22}x_2 + a_{23}x_3 + a_{24}x_4.$$

Using Matrices to Solve a Set of Linear Equations

Problems in kinematics and dynamics sometimes result in a set of linear equations. The matrix method works like this:

a. Arrange the equations in the form

$$\begin{aligned} a_{11}x_1 + a_{12}x_2 + a_{13}x_3 + a_{14}x_4 &= b_1, \\ a_{21}x_1 + a_{22}x_2 + a_{23}x_3 + a_{24}x_4 &= b_2, \\ a_{31}x_1 + a_{32}x_2 + a_{33}x_3 + a_{34}x_4 &= b_3, \\ a_{41}x_1 + a_{42}x_2 + a_{43}x_3 + a_{44}x_4 &= b_4, \end{aligned}$$

where $x_1 \ldots x_4$ are unknown, but we know the values of $a_{11} \ldots . a_{44}$ and $b_1 \ldots b_4$. Of course, although four simultaneous equations are shown, we may have any number of equations.

b. Express the simultaneous equations as the matrix equation

$$\mathbf{AX} = \mathbf{B}$$

where the known matrices $\mathbf{A}$ and $\mathbf{B}$ and the unknown matrix $\mathbf{X}$ are defined as follows:

$$\mathbf{A} = \begin{bmatrix} a_{11} & a_{12} & a_{13} & a_{14} \\ a_{21} & a_{22} & a_{23} & a_{24} \\ a_{31} & a_{32} & a_{33} & a_{34} \\ a_{41} & a_{42} & a_{43} & a_{44} \end{bmatrix} \quad \mathbf{X} = \begin{bmatrix} x_1 \\ x_2 \\ x_3 \\ x_4 \end{bmatrix} \quad \mathbf{B} = \begin{bmatrix} b_1 \\ b_2 \\ b_3 \\ b_4 \end{bmatrix}.$$

Note that this product conforms to the foregoing rules for matrix multiplication.

c. Multiply the matrix equation $\mathbf{AX} = \mathbf{B}$ by the inverse matrix $\mathbf{A}^{-1}$ to get

$$\mathbf{A}^{-1}\mathbf{AX} = \mathbf{A}^{-1}\mathbf{B}.$$

Recall that $\mathbf{A}^{-1}\mathbf{A} = \mathbf{I}$, the identity matrix. Multiplying by the identity matrix does not change $\mathbf{X}$, and the result is

$$\mathbf{X} = \mathbf{A}^{-1}\mathbf{B}.$$

d. Calculate the set of unknowns, given in column matrix $\mathbf{X}$, by means of the preceding equation.

Working Efficiently with Matrices

- Arrange the equations so that the coefficients of the unknowns line up; insert zeros if necessary.
- Form matrices $\mathbf{A}$ and $\mathbf{B}$.
- Identify the terms in the unknown matrix $\mathbf{X}$ (as a comment).
- If available, use software that can compute $\mathbf{X} = \mathbf{A}^{-1}\mathbf{B}$ directly. Calculating the inverse of a large matrix is a long and boring task without such help.

You might wonder why we did not use matrices for position analysis of spatial linkages. The matrix methods we have presented are useful for describing *linear* relationships. The spatial linkage position equations contained unknown terms like r_{2x} and r_{2x}^2, so that the set of equations is not linear. Sometimes nonlinear problems are attacked with a combination of matrix methods and iterative numerical procedures. See, for example, the section on the Newton–Raphson method for two or more variables in Chapter 2.

Determinants: Cramer's Rule, an Alternative Method for Solving Simultaneous Equations

Besides being used to express the vector cross product, determinants can be employed to solve a set of simultaneous equations. However, employing determinant methods for large sets of equations is an inefficient use of time when software with matrix capability is available.

Consider a set of nonhomogeneous linear equations arranged as follows:

$$\begin{aligned} a_{11}x_1 + a_{12}x_2 + \ldots + a_{1n}x_n &= b_1 \\ a_{21}x_1 + a_{22}x_2 + \ldots + a_{2n}x_2 &= b_2 \\ &\vdots \\ a_{n1}x_1 + \qquad \ldots + a_{nn}x_n &= b_n \end{aligned}$$

These equations can be written in matrix form as

$$\mathbf{AX} = \mathbf{B},$$

where the terms are defined as in the preceding paragraph.

The unknowns are given by

$$
\begin{aligned}
x_1 &= D_1/D, \\
x_2 &= D_2/D, \\
&\vdots \\
x_n &= D_n/D,
\end{aligned}
$$

where D = det **A**, the determinant of the **A** matrix,
D_1 = the determinant of the matrix formed by the **A** matrix with the first column replaced by the **B** matrix (i.e., the elements of the **B** vector),
D_2 = the determinant of the matrix formed by the **A** matrix with the second column replaced by the **B** matrix,
D_n = the determinant of the matrix formed by the **A** matrix with the nth column replaced by the **B** matrix.

This method, called Cramer's rule, will be illustrated in a sample problem and used later to find angular velocities in a four-bar linkage.

SAMPLE PROBLEM 3.4

Using determinants to solve a set of linear equations

Solve the following set of simultaneous equations:

$$
\begin{aligned}
3u + 4v &= 25.5; \\
u + 5v + w &= 41.5; \\
10u + 2v + 2w &= 39.
\end{aligned}
$$

Solution. The equations are linear in u, v, and w. We will write them in the form

$$\mathbf{AX} = \mathbf{B},$$

where

$$\mathbf{A} := \begin{bmatrix} 3 & 4 & 0 \\ 1 & 5 & 1 \\ 10 & 2 & 2 \end{bmatrix} \quad \mathbf{B} := \begin{bmatrix} 25.5 \\ 41.5 \\ 39 \end{bmatrix} \quad \text{and} \quad \mathbf{X} := \begin{bmatrix} u \\ v \\ w \end{bmatrix}.$$

We then calculate the determinants to find the unknowns u, v, and w:

$$D := |\mathbf{A}| \quad D = 56$$

$$D_1 := \left| \begin{bmatrix} 25.5 & 4 & 0 \\ 41.5 & 5 & 1 \\ 39 & 2 & 2 \end{bmatrix} \right| \quad D_1 = 28 \quad u := \frac{D_1}{D} \quad u = 0.5$$

$$D_2 := \left|\begin{bmatrix} 3 & 25.5 & 0 \\ 1 & 41.5 & 1 \\ 10 & 39 & 2 \end{bmatrix}\right| \quad D_2 = 336 \quad v := \frac{D_2}{D} \quad v = 6$$

$$D_3 := \left|\begin{bmatrix} 3 & 4 & 25.5 \\ 1 & 5 & 41.5 \\ 10 & 2 & 39 \end{bmatrix}\right| \quad D_3 = 616 \quad w := \frac{D_3}{D} \quad w = 11$$

SAMPLE PROBLEM 3.5

Using matrix methods

Is there a quicker way to solve the previous set of equations?

Solution. Yes, use the form $\mathbf{X} := \mathbf{A}^{-1}\mathbf{B}$.

The answer is immediate: $\mathbf{X} = \begin{bmatrix} 0.5 \\ 6 \\ 11 \end{bmatrix}$.

The elements of the **X** vector may be numbered beginning with zero. That is,

$$u := \mathrm{X}_0 \quad u = 0.5 \quad v := \mathrm{X}_1 \quad v = 6 \quad w := \mathrm{X}_2 \quad w = 11$$

3.4 APPLICATION OF ANALYTICAL VECTOR AND MATRIX METHODS TO LINKAGES

Analytical vector methods may be used to find velocities in planar and spatial linkages. Basically, it is necessary to determine the link orientations by first solving the position equation and then differentiating the position equation with respect to time. If the linkage may be described by the vector polygon

$$\boldsymbol{r}_0 + \boldsymbol{r}_1 + \boldsymbol{r}_2 + \boldsymbol{r}_3 = 0,$$

then differentiation with respect to time yields the velocity equation

$$\dot{\boldsymbol{r}}_0 + \dot{\boldsymbol{r}}_1 + \dot{\boldsymbol{r}}_2 + \dot{\boldsymbol{r}}_3 = 0. \tag{3.13}$$

The form of the solution depends on the given data and the type of linkage. The solution for a linkage with sliding pairs is somewhat different from the solution for a linkage with revolute joints only.

Four-Bar Linkage

Consider a four-bar planar linkage represented by vectors, as shown in Figure 3.9. All the links are rigid and there is no sliding contact. For the frame, $\dot{\boldsymbol{r}}_0 = 0$.

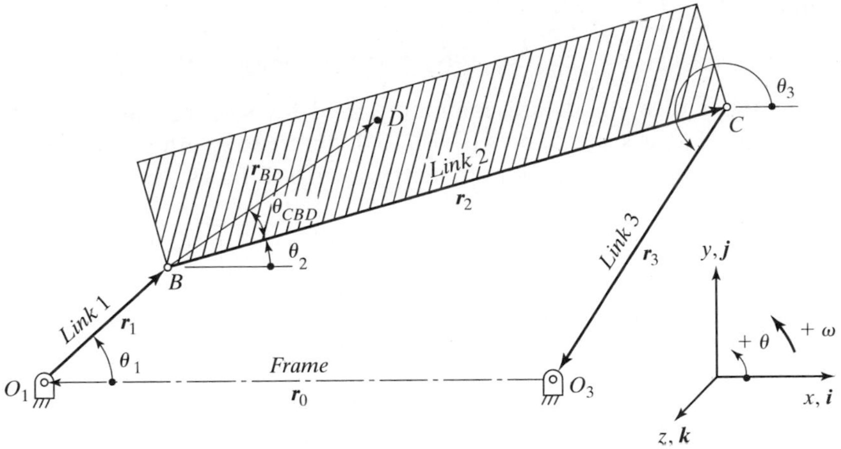

FIGURE 3.9 Analytical study of a four-bar linkage.

For the four-bar linkage, the remaining velocity terms are given by the cross products, and the last equation becomes

$$\boldsymbol{\omega}_1 \times \boldsymbol{r}_1 + \boldsymbol{\omega}_2 \times \boldsymbol{r}_2 + \boldsymbol{\omega}_3 \times \boldsymbol{r}_3 = 0. \tag{3.14}$$

For this planar linkage, the coordinate axes are selected so that vectors $\boldsymbol{r}_1$, $\boldsymbol{r}_2$, and $\boldsymbol{r}_3$ have components in the x and y directions and angular velocities $\boldsymbol{\omega}$ in the z direction. Thus, examining the first term in Eq. (3.14) as typical, we have

$$\boldsymbol{\omega}_1 = \boldsymbol{k}\omega_1$$

and

$$\boldsymbol{r}_1 = \boldsymbol{i}r_{1x} + \boldsymbol{j}r_{1y}.$$

Using determinant form for the cross product, as described in Chapter 2, we obtain

$$\boldsymbol{\omega}_1 \times \boldsymbol{r}_1 = \begin{vmatrix} \boldsymbol{i} & \boldsymbol{j} & \boldsymbol{k} \\ 0 & 0 & \omega_1 \\ r_{1x} & r_{1y} & 0 \end{vmatrix} = -\boldsymbol{i}\omega_1 r_{1y} + \boldsymbol{j}\omega_1 r_{1x}. \tag{3.15}$$

Since the remaining terms in Eq. 3.14 are similar in form, that equation may be written as

$$-\boldsymbol{i}(\omega_1 r_{1y} + \omega_2 r_{2y} + \omega_3 r_{3y}) + \boldsymbol{j}(\omega_1 r_{1x} + \omega_2 r_{2x} + \omega_3 r_{3x}) = 0. \tag{3.16}$$

The $\boldsymbol{i}$ component and the $\boldsymbol{j}$ component of Eq. (3.16) must each separately equal zero, yielding the two simultaneous equations

$$\omega_1 r_{1y} + \omega_2 r_{2y} + \omega_3 r_{3y} = 0$$

and

$$\omega_1 r_{1x} + \omega_2 r_{2x} + \omega_3 r_{3x} = 0. \tag{3.17}$$

A number of methods are commonly used for solving simultaneous equations. For example, if ω_1 is known and we wish to determine the other angular velocities, we may write the preceding equations in matrix form as follows:

$$\begin{bmatrix} r_{2y} & r_{3y} \\ r_{2x} & r_{3x} \end{bmatrix} \begin{bmatrix} \omega_2 \\ \omega_3 \end{bmatrix} = -\omega_1 \begin{bmatrix} r_{1y} \\ r_{1x} \end{bmatrix}. \tag{3.18}$$

The angular velocity of link 2, the coupler, is given by the determinant expression

$$\omega_2 = \frac{-\omega_1}{D} \begin{vmatrix} r_{1y} & r_{3y} \\ r_{1x} & r_{3x} \end{vmatrix},$$

where

$$D = \begin{vmatrix} r_{2y} & r_{3y} \\ r_{2x} & r_{3x} \end{vmatrix},$$

and the angular velocity of link 3 is given by

$$\omega_3 = \frac{-\omega_1}{D} \begin{vmatrix} r_{2y} & r_{1y} \\ r_{2x} & r_{1x} \end{vmatrix}.$$

Expanding these expressions, we have

$$\omega_2 = \frac{-\omega_1 (r_{1y} r_{3x} - r_{3y} r_{1x})}{r_{2y} r_{3x} - r_{3y} r_{2x}} \tag{3.19}$$

and

$$\omega_3 = \frac{-\omega_1 (r_{2y} r_{1x} - r_{1y} r_{2x})}{r_{2y} r_{3x} - r_{3y} r_{2x}}. \tag{3.20}$$

Since link 1 in Figure 3.9 rotates about fixed point O_1, the velocity of any point on link 1 is given by $\boldsymbol{\omega}_1 \times \boldsymbol{r}$, where $\boldsymbol{r}$ is the vector measured from O_1 to the point in question. Link 2, the coupler, has no fixed point. The velocity of an arbitrary point D on link 2 may be found by using Eq. 3.10, where link 2 is fixed in a rotating coordinate system with origin at B. Then,

$$\dot{\boldsymbol{R}}_0 = \boldsymbol{\omega}_1 \times \boldsymbol{r}_1,$$
$$\dot{\boldsymbol{r}}_r = 0,$$
$$\boldsymbol{\omega} \times \boldsymbol{r} = \boldsymbol{\omega}_2 \times \boldsymbol{r}_{BD},$$

and the velocity of point D is given by

$$\boldsymbol{v}_D = \boldsymbol{\omega}_1 \times \boldsymbol{r}_1 + \boldsymbol{\omega}_2 \times \boldsymbol{r}_{BD}, \tag{3.21}$$

where the first term on the right represents the velocity of point B and the last term the velocity of D with respect to B. The same result could be obtained by noting that the position of point D could be described by the equation

$$\boldsymbol{r}_D = \boldsymbol{r}_1 + \boldsymbol{r}_{BD},$$

where vectors $\boldsymbol{r}_1$ and $\boldsymbol{r}_{BD}$ have fixed magnitude. Note that point D does not move relative to link 2.

Point C in Figure 3.9 represents the revolute (pin) joint between links 2 and 3. The velocity of C can be found by the equation

$$\boldsymbol{v}_C = \boldsymbol{\omega}_1 \times \boldsymbol{r}_1 + \boldsymbol{\omega}_2 \times \boldsymbol{r}_2, \tag{3.22}$$

where the first term on the right represents the velocity of B and the last term the velocity of C with respect to B. (Compare $\boldsymbol{v}_D$ in Eq. 3.21.) An alternative expression is

$$\boldsymbol{v}_C = \boldsymbol{\omega}_3 \times (-\boldsymbol{r}_3), \tag{3.23}$$

since link 3 rotates about fixed center O_3 and $-\boldsymbol{r}_3$ represents the radius vector O_3C. If we subtract Eq. 3.23 from Eq. 3.22, the result is Eq. 3.14.

Equations 3.22 and 3.23 form the basis of the graphical relative velocity and velocity polygon methods for plane linkages of this type. For example, if the velocity of point B (Figure 3.9) is known, we use the fact that the cross products are perpendicular to the link vectors to find the velocity of point C.

Although the equations we have developed in this section are general for the four-bar planar linkage, the solutions are actually *instantaneous velocities* and *instantaneous angular velocities*. Even if $\boldsymbol{\omega}_1$ is constant in magnitude, only the magnitude of v_B will be constant. Due to the changing position, $\boldsymbol{\omega}_2$ and $\boldsymbol{\omega}_3$ will, in general, vary, as will the velocities of points on links 2 and 3.

SAMPLE PROBLEM 3.6

Linkage Velocities by Analytical Vector Methods

Referring to Figure 3.9, let $\omega_1 = 100$ rad/s ccw, $\theta_1 = 45°$, $\theta_{CBD} = 20°$ (constant), $r_0 = 30$ mm, $r_1 = 10$ mm, $r_2 = 35$ mm, $r_3 = 20$ mm, and $r_{BD} = 15$ mm. Find $\boldsymbol{\omega}_2$, $\boldsymbol{\omega}_3$, $\boldsymbol{v}_C$, and $\boldsymbol{v}_D$.

Solution. In chapter 2, for the assembly configuration shown, we determined from the position analysis that $\theta_2 = 16.35°$ and $\theta_3 = 237.81°$. The components of the link vectors are

$$r_x = r\cos\theta \text{ and } r_y = r\sin\theta.$$

Programming a polar–rectangular and rectangular–polar conversion subroutine or using a preprogrammed one saves time. For this linkage at the instant considered,

$$r_{1x} = 7.0711, \quad r_{1y} = 7.0711,$$
$$r_{2x} = 33.590, \quad r_{2y} = 9.853,$$

and

$$r_{3x} = -10.660, \quad r_{3y} = -16.922.$$

Using these values in Eqs. 3.19 and 3.20, we find that

$$\omega_2 = \frac{-100(-7.0711 \times 10.660 + 16.922 \times 7.0711)}{-9.853 \times 10.660 + 16.992 \times 33.590}$$
$$= -9.567 \text{ rad/s} \quad (9.567 \text{ rad/s cw})$$

and

$$\omega_3 = \frac{-100(9.853 \times 7.0711 - 7.0711 \times 33.590)}{-9.853 \times 10.660 + 16.992 \times 33.590}$$
$$= 36.208 \text{ rad/s ccw.}$$

The velocity of point C is given by the vector sum of the velocity of B and the velocity of C with respect to B (Eq. 3.22). Noting that $\boldsymbol{\omega}_1 = 100\boldsymbol{k}$ and $\boldsymbol{\omega}_2 = -9.567\boldsymbol{k}$, we have

$$\boldsymbol{v}_C = \begin{vmatrix} \boldsymbol{i} & \boldsymbol{j} & \boldsymbol{k} \\ 0 & 0 & 100 \\ 7.0711 & 7.0711 & 0 \end{vmatrix} + \begin{vmatrix} \boldsymbol{i} & \boldsymbol{j} & \boldsymbol{k} \\ 0 & 0 & -9.567 \\ 33.590 & 9.853 & 0 \end{vmatrix}$$
$$= -\boldsymbol{i}612.83 + \boldsymbol{j}385.80 = 724.16 \text{ mm/s} \angle 147.8°.$$

Note that the velocity vector is perpendicular to the link vector.

Using, instead, the angular velocity and length of link 3 (Eq. 3.23), we obtain

$$\boldsymbol{v}_C = \begin{vmatrix} \boldsymbol{i} & \boldsymbol{j} & \boldsymbol{k} \\ 0 & 0 & 63.208 \\ 10.660 & 16.922 & 0 \end{vmatrix},$$

which differs from the previous solution only because of round-off error. This alternative method provides a partial check of our arithmetic operations or of our coding of the program.

In finding the velocity of point D, which lies in link 2, we note that the components of the vector location are defined by the angle $\theta_2 + \theta_{CBD} = 16.35 + 20°$, from which it follows that

$$r_{BDx} = 12.081 \quad \text{and} \quad r_{BDy} = 8.891.$$

Adding the vector velocity of B and the vector velocity of D with respect to B, (Eq. 3.21), we obtain

$$\boldsymbol{v}_D = \begin{vmatrix} \boldsymbol{i} & \boldsymbol{j} & \boldsymbol{k} \\ 0 & 0 & 100 \\ 7.0711 & 7.0711 & 0 \end{vmatrix} + \begin{vmatrix} \boldsymbol{i} & \boldsymbol{j} & \boldsymbol{k} \\ 0 & 0 & -9.567 \\ 12.081 & 8.891 & 0 \end{vmatrix}$$
$$= -\boldsymbol{i}622.04 + \boldsymbol{j}591.53 = 858.4 \text{ mm/s} \angle 136.4°.$$

Inverse Matrix Solution Using Mathematics Software

Equation 3.18 may be written in the form

$$\mathbf{AX} = \mathbf{B}. \quad (3.24)$$

Let the numerical values of the elements of matrix **A** and state vector **B** be specified. Then state vector **X**, representing the angular velocities of links 2 and 3, is the only unknown. Hence,

$$\mathbf{X} = \mathbf{A}^{-1}\mathbf{B}. \quad (3.25)$$

A direct solution of Eq. 3.25 is easily obtained with the use of mathematics software.

3.5 USING A SPREADSHEET TO SOLVE PROBLEMS IN KINEMATICS

An electronic spreadsheet program is a convenient tool for analyzing a linkage in a series of positions. Spreadsheets allow for rapid evaluation of potential design changes, and include built-in plotting routines. One disadvantage is that spreadsheet formulas are written in terms of cell references, while we are accustomed to writing equations in terms of physical parameters.

Spreadsheets allow copying of formulas in a given range of cells to an additional range of cells. If the cell reference is not to change when a formula is copied to another cell, then an **absolute cell reference** is used. A **relative cell reference** is the default form. The column letter or row number of a relative cell reference changes according to the cell into which a formula is copied. Sample Problem 3.7 is intended to illustrate an application of spreadsheets to kinematics. Complete instructions for manipulating spreadsheets are found in the manuals that accompany the software.

SAMPLE PROBLEM 3.7

Utilizing a Spreadsheet to Plot Velocities in a Four-Bar Linkage

Let crank angle θ_1 vary from 0° to 360° in 5° increments in the four-bar linkage described in Sample Problem 3.6. Plot the angular velocities of the coupler and follower crank and the velocity of point D vs. θ_1.

Solution. We begin by listing the given data in cells that are identified by their column letter and row number.

The initial value of crank angle θ_1 is zero, and it is incremented 5° in each succeeding row. Equations are entered in cell format according to the spreadsheet manual. Angle functions usually require arguments in radians. The job is not finished until we title the graphs, label the axes, and label the curves or include a key.

Figure 3.10 shows the coupler and follower angular velocity plotted against the crank position. Figure 3.11 shows the velocity of point D on the coupler of the linkage. The x and y components are shown, as well as the resultant velocity.

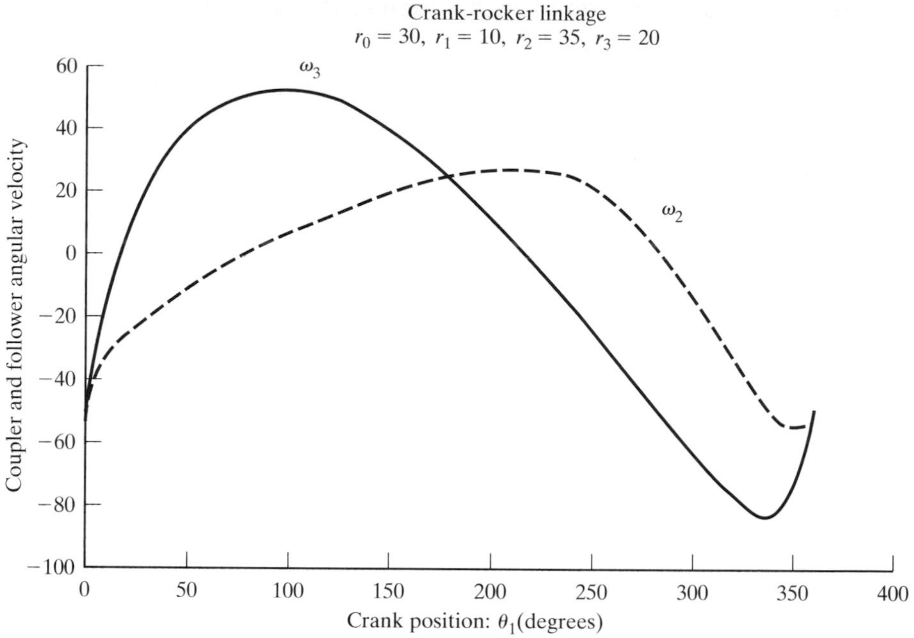

FIGURE 3.10 Angular velocities.

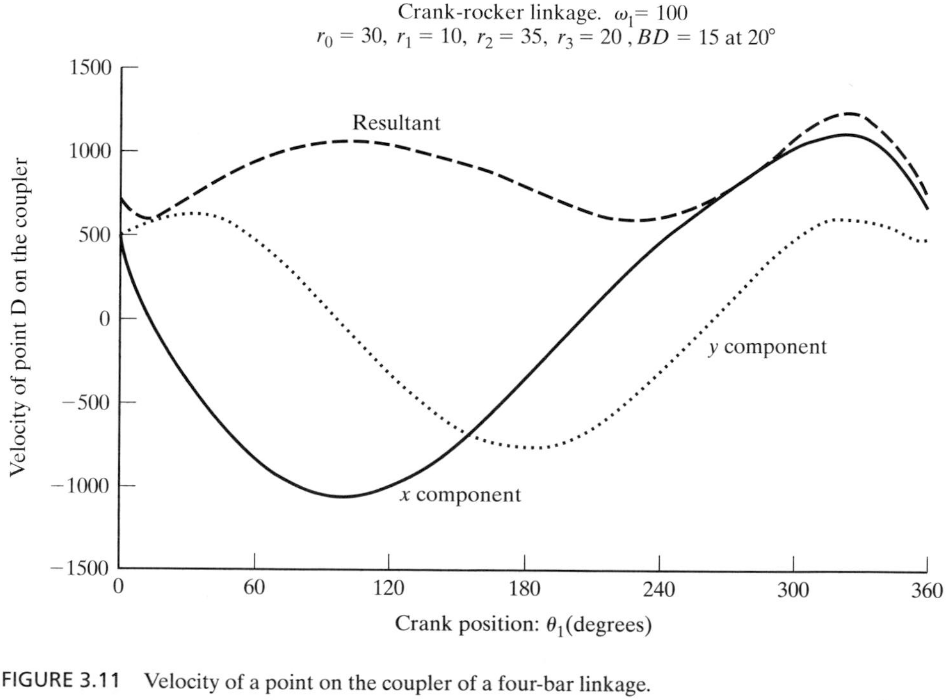

FIGURE 3.11 Velocity of a point on the coupler of a four-bar linkage.

3.6 MATHEMATICS SOFTWARE APPLIED TO VECTOR SOLUTIONS OF KINEMATICS PROBLEMS

If we choose to solve kinematics problems in vector form, we may write vector-manipulation routines or use commercially available mathematics software. Consider the offset slider-crank linkage sketched in Figure 3.12. This linkage can be described by the vector equation

$$\boldsymbol{e} + \boldsymbol{r}_1 + \boldsymbol{r}_2 + \boldsymbol{r}_0 = 0, \tag{3.26}$$

where the magnitude of $\boldsymbol{e}$ is the offset—that is, the distance from the centerline of the slider path to crank bearing O_1.

Position Analysis

If the offset crank length, connecting-rod length, and crank position are given, then it is convenient to rewrite Eq. (3.26) as

$$\boldsymbol{r}_3 + \boldsymbol{r}_2 + \boldsymbol{r}_0 = 0, \tag{3.27}$$

where $\boldsymbol{r}_3 = \boldsymbol{e} + \boldsymbol{r}_1$. Noting that the magnitude of $\boldsymbol{r}_0$ is unknown and the direction of $\boldsymbol{r}_2$ is unknown, we see that the solution is given by Eqs. (2.23) and (2.24), where $\boldsymbol{r}_0 = \boldsymbol{A}$, $\boldsymbol{r}_2 = \boldsymbol{B}$, and $\boldsymbol{r}_3 = \boldsymbol{e} + \boldsymbol{r}_1 = \boldsymbol{C}$.

Velocity Analysis

We note that the magnitudes of $\boldsymbol{r}_1$ and $\boldsymbol{r}_2$ are constant, that $\boldsymbol{r}_0$ has a fixed direction, and that $\boldsymbol{e}$ is constant in both magnitude and direction. Differentiating Eq. (3.26), we obtain

$$\boldsymbol{\omega}_1 \times \boldsymbol{r}_1 + \boldsymbol{\omega}_2 \times \boldsymbol{r}_2 - \boldsymbol{v}_c = 0, \tag{3.28}$$

where $dr_0/dt = -v_c$, the slider velocity, which is in the x direction. The angular velocity vectors are in the z direction. To eliminate v_c in Eq. (3.28), we take the dot product of each term with $\boldsymbol{j}$, from which we get

$$\omega_1(\boldsymbol{k} \times \boldsymbol{r}_1)\cdot\boldsymbol{j} + \omega_2(\boldsymbol{k} \times \boldsymbol{r}_2)\cdot\boldsymbol{j} = 0. \tag{3.29}$$

Using vector manipulation rules from Chapter 2, we have

$$(\boldsymbol{k} \times \boldsymbol{r})\cdot\boldsymbol{j} = \boldsymbol{r}\cdot(\boldsymbol{j} \times \boldsymbol{k})$$

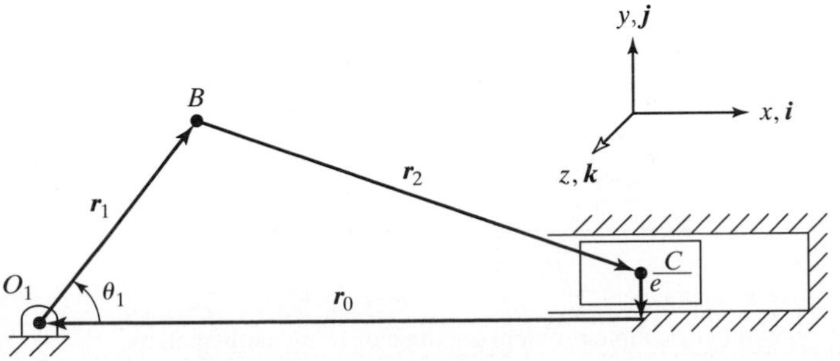

FIGURE 3.12 Offset slider crank linkage.

and

$$j \times k = i.$$

Equation (3.29) is equivalent to

$$\omega_1 r_1 \cdot i + \omega_2 r_2 \cdot i = 0,$$

from which it follows that

$$\omega_2 = -\omega_1 r_1 \cdot i / (r_2 \cdot i). \tag{3.30}$$

Equation (3.28) may now be solved for v_c.

SAMPLE PROBLEM 3.8

Velocity Analysis of an Offset Slider-Crank Linkage Using Vector Methods

a. Referring to Figure 3.12, let $R_2/R_1 = 1.8$ and $e/R_1 = -0.5j$, where R_1 = crank length and R_2 = connecting-rod length. The angular velocity of the crank is constant. Plot the slider position, slider velocity, and angular velocity of the connecting rod vs. crank angle θ_1.

b. Examine the effect of varying the offset. Let $R_2/R_1 = 2$.

Solution Summary. **(a)** Commercially available mathematics software was used to obtain the solution shown in Figure 3.13. In this solution, we set $R_1 = 1$ and $\omega_1 = 1$. Then the results apply to other slider-crank linkages with the same proportions if r_0 is multiplied by the actual R_1, ω_2 by

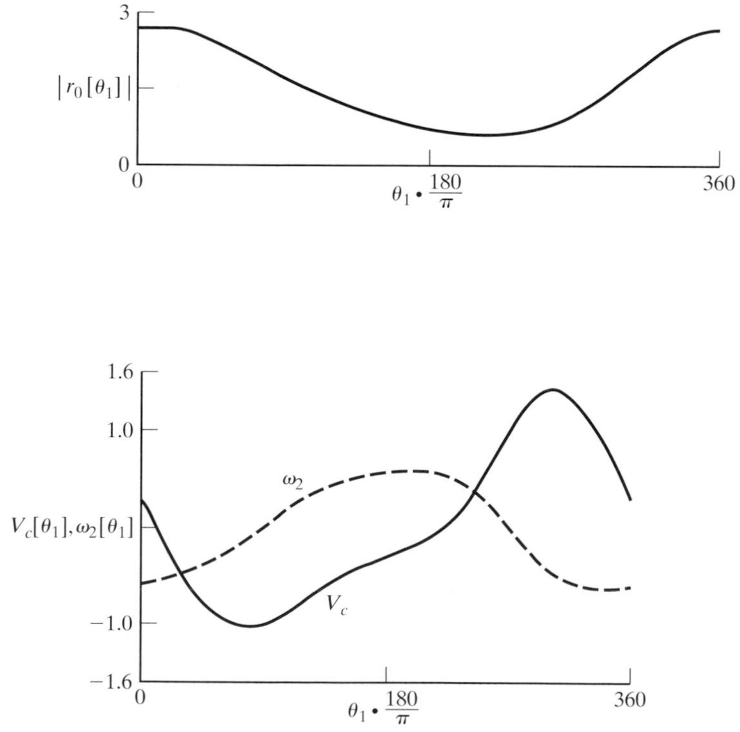

FIGURE 3.13 Solution to sample problem obtained by using mathematics software.

the actual ω_1, and v_c by the actual $\omega_1 R_1$. Crank angle θ_1 is to range from 0 to 2π rad in steps of $\pi/9$. Known vectors $\boldsymbol{e}, \boldsymbol{r}_0^u, \boldsymbol{r}_1$, and $\boldsymbol{r}_3$ are expressed in column form. Rectangular unit vectors $\boldsymbol{i}, \boldsymbol{j}$, and $\boldsymbol{k}$ are also defined in column form. Equations (2.23) and (2.24) are then used to determine $\boldsymbol{r}_0$ and $\boldsymbol{r}_2$. The magnitude of $\boldsymbol{r}_0$ is tabulated and plotted against θ_1 (which is converted to degrees). Connecting rod angular velocity $\boldsymbol{\omega}_2$ and slider velocity $\boldsymbol{v}_c$ are then calculated by Eqs. (3.30) and (3.28) and plotted. Even though $\boldsymbol{v}_c$ lies only in the $\pm x$ direction, it has been stored as a vector. In order to tabulate and plot its magnitude and sense, we compute

$$V_c = \boldsymbol{v}_c \cdot \boldsymbol{i}.$$

(b) Figure 3.14 shows the effect of varying offset $\boldsymbol{e}$. In this plot, slider velocity V_c is plotted against crank angle θ_1 for a slider-crank linkage for which $R_2/R_1 = 2$.

Detailed calculations. Slider-crank linkage. $c_0 = 1$ (clockwise configuration positive)

$$\theta_1 = 0. \frac{\pi}{9} .. 2 \cdot \pi \quad R_1 = 1 \quad R_2 = 1.8 \quad \boldsymbol{e} = \begin{bmatrix} 0 \\ -0.5 \\ 0 \end{bmatrix} \quad \boldsymbol{r}_{0u} = \begin{bmatrix} -1 \\ 0 \\ 0 \end{bmatrix}$$

$$\boldsymbol{r}_1[\theta_1] = \begin{bmatrix} R_1 & \cdot & \cos[\theta_1] \\ R_1 & \cdot & \sin[\theta_1] \\ & 0 & \end{bmatrix} \quad \boldsymbol{i} = \begin{bmatrix} 1 \\ 0 \\ 0 \end{bmatrix} \quad \boldsymbol{j} = \begin{bmatrix} 0 \\ 1 \\ 0 \end{bmatrix} \quad \boldsymbol{k} = \begin{bmatrix} 0 \\ 0 \\ 1 \end{bmatrix}$$

$$\boldsymbol{r}_3[\theta_1] - \boldsymbol{r}_1[\theta_1] + \boldsymbol{e}$$

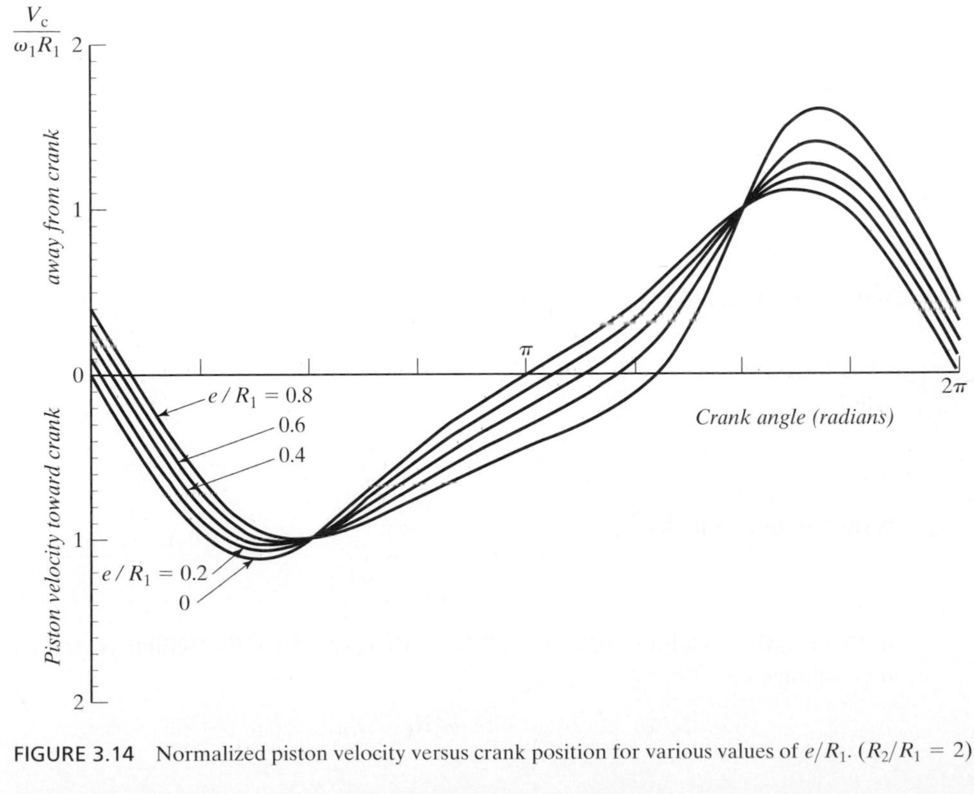

FIGURE 3.14 Normalized piston velocity versus crank position for various values of e/R_1. ($R_2/R_1 = 2$).

$$r_0[\theta_1] = \left[-r_3\left[\theta_1\right] \cdot r_{0u} + c_0 \cdot \left[R_2^2 - \left[r_3\left[\theta_1\right] \cdot \left[r_{0u} \times k \right] \right]^2 \right]^{1/2} \right] \cdot r_{0u}$$

$$a[\theta_1] = -[r_3[\theta_1] \cdot [r_{0u} \times k]] \cdot [r_{0u} \times k]$$

$$r_2[\theta_1] = a[\theta_1] - c_0 \cdot \left[R_2^2 - \left[r_3[\theta_1] \cdot \left[r_{0u} \times k \right] \right]^2 \right]^{1/2} \cdot r_{0u}$$

$$\omega_1 = 1 \qquad \omega_2[\theta_1] = -\omega_1 \cdot \frac{r_1[\theta_1] \cdot i}{r_2[\theta_1] \cdot i}$$

$$v_c[\theta_1] = \omega_1 \cdot k \times r_1[\theta_1] + \omega_2[\theta_1] \cdot k \times r_2[\theta_1] \quad V_c[\theta_1] = v_c[\theta_1] \cdot i$$

3.7 COMPLEX-NUMBER METHODS APPLIED TO VELOCITY ANALYSIS

Complex numbers are a convenient form for representing vectors. They may be used to develop analytical solutions to linkage velocity problems. With complex-number methods, we are limited, of course, to planar linkages. First, the loop closure (displacement) equation is solved for unknown directions and magnitudes. Then, the displacement equation is differentiated with respect to time to obtain the velocity equation.

Consider the sliding contact linkage shown in Figures 3.15a and b, where the slider moves along link 2. The linkage is described by the equation

$$\boldsymbol{R}_2 = \boldsymbol{R}_0 + \boldsymbol{R}_1, \tag{3.31}$$

where $\boldsymbol{R}_0$, representing the frame, is fixed in magnitude and direction. Vector $\boldsymbol{R}_1$, representing the rotating crank, has constant magnitude, but the magnitude of $\boldsymbol{R}_2$ changes with time. If the real axis is selected to be parallel to the fixed link, we may write

$$R_0 + R_1 e^{j\theta_1} = R_2 e^{j\theta_2}. \tag{3.32}$$

Differentiating with respect to time, we have the complex velocity equation

$$j\omega_1 R_1 e^{J\theta_1} = j\omega_2 R_2 e^{j\theta_2} + v_{B_1B_2} e^{j\theta_2}, \tag{3.33}$$

where

$$\omega = \frac{d\theta}{dt}$$

is the angular velocity,

$$v_{B_1B_2} = \frac{dR_2}{dt}$$

is the relative (sliding) velocity of B_1 with respect to B_2 (which is positive if R_2 is increasing),

$$\omega_1 R_1 = v_{B_1}$$

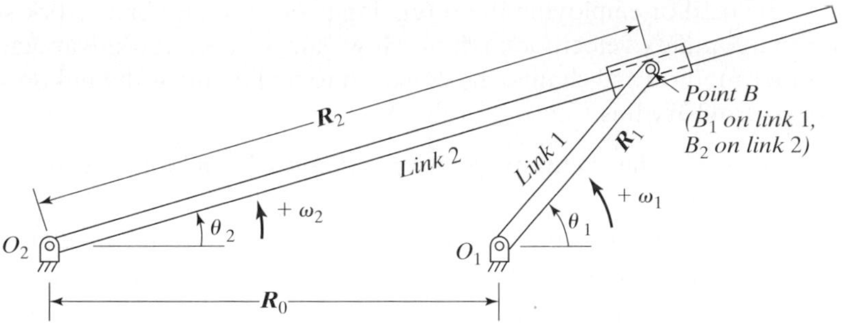

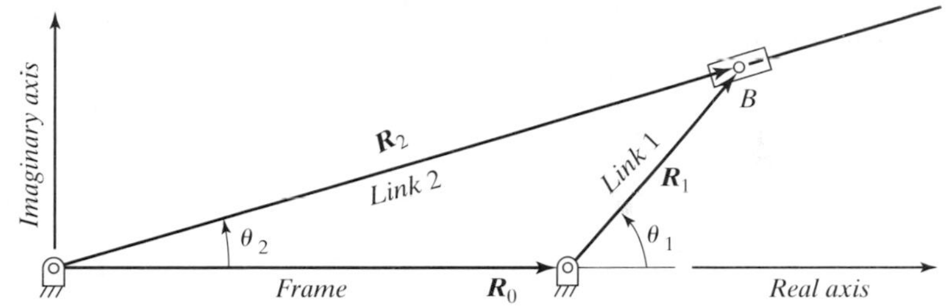

FIGURE 3.15 (a) Schematic of sliding contact linkage. (b) Vector representation.

and

$$\omega_2 R_2 = v_{B_2}.$$

Thus, we have the vector equation

$$\boldsymbol{v}_{B_1} = \boldsymbol{v}_{B_2} + \boldsymbol{v}_{B_1B_2}, \tag{3.34}$$

with the complex exponentials indicating the directions. Note that the order of the subscripts is critical ($\boldsymbol{v}_{B_2B_1} = -\boldsymbol{v}_{B_1B_2}$).

Referring to Eq. (3.33), suppose link lengths R_0 and R_1 are given along with the orientation and angular velocity of link 1. Then R_2 and θ_2 may be found by solving the displacement equation. (See Chapter 2.) Two unknowns remain—the angular velocity of link 2 and the sliding velocity—both of them part of complex expressions.

We could now find the unknowns as follows:

- Use the Euler formula (given in Chapter 2) to convert the equation to rectangular form.
- Separate the real and imaginary parts. Note that we now have two simultaneous equations; that is, both unknowns appear in both equations.
- Use Software like Mathcad™ to solve the simultaneous equations for the sliding velocity and the angular velocity of link 2.

Instead of employing the foregoing procedure, we can "work smart" by examining the complex velocity equation. How can we take full advantage of the complex-number method and change the equation to make one of the unknowns appear only in a real term? Try this:

- Multiply the complex velocity equation by a quantity that will separate the unknowns.
- Use the Euler formula to convert the equation to rectangular form.
- Separate the real and imaginary parts. Note that the unknowns appear in separate equations. Thus, we have the solution without solving simultaneous equations.

Here are the details: We begin by multiplying both sides of Eq. (3.33) by $e^{-j\theta_2}$. This step is equivalent to rotating the coordinate system through an angle θ_2.

The result is

$$j\omega_1 R_1 e^{j(\theta_1 - \theta_2)} = j\omega_2 R_2 + v_{B_1B_2}. \tag{3.35}$$

Expressing the exponential in rectangular form (via the Euler formula), we obtain

$$j\omega_1 R_1 \cos(\theta_1 - \theta_2) - \omega_1 R_1 \sin(\theta_1 - \theta_2) = j\omega_2 R_2 + v_{B_1B_2}. \tag{3.36}$$

The imaginary parts of Eq. (3.36) yield

$$\omega_1 R_1 \cos(\theta_1 - \theta_2) = \omega_2 R_2,$$

from which we obtain the angular velocity of link 2:

$$\omega_2 = \omega_1 R_1 \cos(\theta_1 - \theta_2)/R_2. \tag{3.37}$$

Equating the real parts, we obtain the sliding velocity—the relative velocity of B_1 with respect to B_2:

$$v_{B_1B_2} = -\omega_1 R_1 \sin(\theta_1 - \theta_2). \tag{3.38}$$

For $R_1 < R_0$, link 2 oscillates while link 1 rotates. We see that $\omega_2 = 0$ when $\cos(\theta_1 - \theta_2) = 0$, or $\theta_1 - \theta_2 = \pi/2, 3\pi/2, 5\pi/2$, and so on. Alternatively, these limiting positions (link 1 perpendicular to link 2) may be obtained by sketching the linkage. The magnitude of sliding velocity $v_{B_1B_2}$ is maximum at the limiting positions.

SAMPLE PROBLEM 3.9

Analysis of a Sliding Contact Linkage by means of Complex Numbers

We will examine velocities in a sliding contact linkage, utilizing the results of complex number analysis. Referring to Figure 3.15, let $\omega_1 = 20$ rad/s counterclockwise (constant), $R_0 = 40$ mm, and $R_1 = 20$ mm. Find R_2, θ_2, ω_2, and $v_{B_1B_2}$ for the instant when $\theta_1 = 75°$.

Solution. Using the displacement equations from Chapter 2, we proceed as follows. The slider position is

$$\begin{aligned} R_2 &= \sqrt{R_0^2 + R_1^2 + 2R_0R_1 \cos\theta_1} \\ &= \sqrt{40^2 + 20^2 + 2 \times 40 \times 20 \times \cos 75^\circ} = 49.13 \text{ mm}, \end{aligned}$$

and the position of link 2 is

$$\theta_2 = \arcsin\left(\frac{R_1 \sin\theta_1}{R_2}\right) = \arcsin\left(\frac{20 \sin 75^\circ}{49.13}\right) = 23.15^\circ.$$

For this linkage, the proportions are such that θ_2 is limited to the first and fourth quadrants, and the preceding result is correct. A safer procedure utilizes the function $\arctan_2$. Alternatively, compute

$$\cos\theta_2 = \frac{R_1 \cos\theta_1 + R_0}{R_2},$$

$$\sin\theta_2 = \frac{R_1 \sin\theta_1}{R_2},$$

and

$$\tan\left(\frac{\theta_2}{2}\right) = \frac{1 - \cos\theta_2}{\sin\theta_2},$$

which yields the same value of θ_2. The angular velocity is

$$\begin{aligned} \omega_2 &= \frac{\omega_1 R_1 \cos(\theta_1 - \theta_2)}{R_2} = \frac{20 \times 20 \cos(75^\circ - 23.15^\circ)}{49.13} \\ &= +5.03 \text{ rad/s (ccw)}, \end{aligned}$$

and the sliding velocity (the motion of the slider relative to link 2) is

$$\begin{aligned} v_{B_1B_2} &= \frac{dR_2}{dt} = -\omega_1 R_1 \sin(\theta_1 - \theta_2) \\ &= -20 \times 20 \sin(75^\circ - 23.15^\circ) = -314.6 \text{ mm/s} \end{aligned}$$

(i.e., 314.6 mm/s along link 2 toward O_2).

3.8 SPATIAL LINKAGES: VECTOR AND MATRIX METHODS

Consider a spatial linkage with $n \geq 3$ links that form a closed loop. If links are represented by vectors $\boldsymbol{r}_0$, $\boldsymbol{r}_1$, and so on, the three-dimensional loop closure equation

$$\boldsymbol{r}_0 + \boldsymbol{r}_1 + \boldsymbol{r}_2 + \cdots + \boldsymbol{r}_n = 0 \tag{3.39}$$

may be used to describe the linkage. Differentiating with respect to time yields the velocity equation

$$\dot{\boldsymbol{r}}_0 + \dot{\boldsymbol{r}}_1 + \dot{\boldsymbol{r}}_2 + \cdots + \dot{\boldsymbol{r}}_n = 0. \tag{3.40}$$

For a vector of fixed magnitude, an $\dot{\boldsymbol{r}}$ term may be replaced by $\boldsymbol{\omega} \times \boldsymbol{r}$. When sliding occurs, the relative velocity of coincident points must be considered.

The velocity vector equation for a spatial linkage is, in general, three dimensional rather than two dimensional, as for a planar linkage. Carrying out a comprehensive analysis of a full cycle of motion of a spatial linkage is a formidable task. Software with matrix capabilities or three-dimensional motion simulation software can eliminate hours of repetitive calculations.

Analysis of an RSSC Spatial Linkage

For a given value of θ, the positions of all links in the RSSC linkage of Figure 3.16 may be found. Both graphical and analytical solutions are described in Chapter 2. If the angular velocity of link 1 is known, the velocity of spherical pair S_2 is given by

$$\boldsymbol{v}_{S_1} = \boldsymbol{v}_{S_1} + \boldsymbol{v}_{S_2S_1} = \boldsymbol{\omega}_1 \times \boldsymbol{r}_1 + \boldsymbol{\omega}_2 \times \boldsymbol{r}_2. \tag{3.41}$$

For the configuration shown, $\boldsymbol{v}_{S_2}$ lies in the xy-plane and

$$\frac{v_{S_2x}}{v_{S_2y}} = \tan\gamma,$$

a constant. The spherical joints permit link 2 to rotate about its own axis. This motion may be set equal to zero without affecting other kinematic aspects of the problem. Thus, we may write $\boldsymbol{\omega}_2 \cdot \boldsymbol{r}_2 = 0$. These relationships result in a set of simultaneous equations from which the unknown velocities are determined.

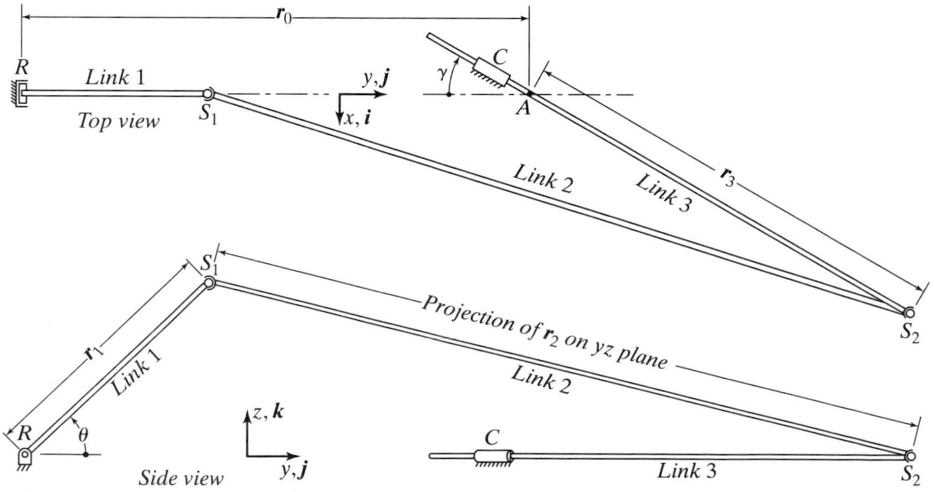

FIGURE 3.16 *RSSC* spatial linkage.

SAMPLE PROBLEM 3.10

RSSC Spatial Linkage

Here, a spatial linkage is analyzed with the use of vector methods. Find link orientations and velocities for the RSSC spatial linkage of Figure 3.16 for $\gamma = 30°, \theta = 45°$, and crank angular velocity $\omega_1 = 50$ rad/s counterclockwise; let $\boldsymbol{r}_0 = 200$ mm, $\boldsymbol{r}_1 = 100$ mm, and $\boldsymbol{r}_2 = 300$ mm. Use vector notation.

Solution. The links may be described in terms of unit vectors as follows:

$$\begin{aligned} \boldsymbol{r}_0 &= 0\boldsymbol{i} & &-200\boldsymbol{j} & &+ 0\boldsymbol{k}; \\ \boldsymbol{r}_1 &= 0\boldsymbol{i} & &+100(\cos\theta)\boldsymbol{j} & &+100(\sin\theta)\boldsymbol{k}; \\ \boldsymbol{r}_2 &= r_{2x}\boldsymbol{i} & &+ r_{2y}\boldsymbol{j} & &+ r_{2z}\boldsymbol{k}; \\ \boldsymbol{r}_3 &= r_{3x}\boldsymbol{i} & &+ r_{3y}\boldsymbol{j} & &+ 0\boldsymbol{k}. \end{aligned}$$

The equation $\boldsymbol{r}_0 + \boldsymbol{r}_1 + \boldsymbol{r}_2 + \boldsymbol{r}_3 = 0$ applies to each vector direction. Thus, summing the $\boldsymbol{i}$ components, we have $r_{3x} = -r_{2x}$. Since $r_{3x} = r_{3y}\tan\gamma$, we also have $r_{2x} = -r_{3y}\tan\gamma$. From the $\boldsymbol{j}$ components, we obtain $r_{2y} = 200 - 100\cos\theta - r_{3y}$, and from the $\boldsymbol{k}$ components, $r_{2z} = -100\sin\theta$. The length of link 2 is 300 mm, so

$$r_{2x}^2 + r_{2y}^2 + 2_{2z}^2 = 300^2.$$

Substituting the given values of θ and γ, we have a quadratic equation in r_{3y}. The root of the equation that corresponds to the assembly configuration in the figure is

$$r_{3y} = -149.23.$$

Using the aforementioned equations, we may then write

$$\begin{aligned} \boldsymbol{r}_0 &= 0\boldsymbol{i} - 200\boldsymbol{j} + 0\boldsymbol{k}; \\ \boldsymbol{r}_1 &= 0\boldsymbol{i} + 70.71\boldsymbol{j} + 70.71\boldsymbol{k}; \\ \boldsymbol{r}_2 &= 86.26\boldsymbol{i} + 278.52\boldsymbol{j} - 70.71\boldsymbol{k}; \\ \boldsymbol{r}_3 &= -86.16\boldsymbol{i} - 149.23\boldsymbol{j} + 0\boldsymbol{k}. \end{aligned}$$

The velocity of link 3, the sliding link, is given by

$$\boldsymbol{v}_{S_2} = \boldsymbol{v}_{S_1} + \boldsymbol{v}_{S_2S_1}$$

where

$$\begin{aligned} \boldsymbol{v}_{S_1} &= \boldsymbol{\omega}_1 \times \boldsymbol{r}_1, \\ \boldsymbol{v}_{S_2S_1} &= \boldsymbol{\omega}_2 \times \boldsymbol{r}_2, \\ \boldsymbol{v}_{S_2} &= v_{S_2x}\boldsymbol{i} + v_{S_2y}\boldsymbol{j} + 0\boldsymbol{k}, \\ \boldsymbol{\omega}_1 &= 50\boldsymbol{i} + 0\boldsymbol{j} + 0\boldsymbol{k}, \end{aligned}$$

and

$$\boldsymbol{\omega}_2 = \omega_{2x}\boldsymbol{i} + \omega_{2y}\boldsymbol{j} + \omega_{2z}\boldsymbol{k}.$$

We ignore the rotation of link 2 about its own axis; this motion does not affect that of the rest of the linkage. Thus, we may set the component of $\boldsymbol{\omega}_2$ in the $\boldsymbol{r}_2$ direction equal to zero, which is equivalent to stating that $\boldsymbol{\omega}_2 \cdot \boldsymbol{r}_2 = 0$, from which it follows that

$$\boldsymbol{\omega}_2 \cdot \boldsymbol{r}_2 = 86.16\omega_{2x} + 278.52\omega_{2y} - 70.71\omega_{2z} = 0.$$

The motion of link 3 is limited by the cylindrical joint, so that

$$v_{s_2x} = v_{s_2y} \tan \gamma.$$

This last equation may be used to reduce the problem to a system of four unknowns and four equations.

The velocities are given by

$$\begin{aligned} \boldsymbol{v}_{s_1} = \boldsymbol{\omega}_1 \times \boldsymbol{r}_1 &= \begin{vmatrix} \boldsymbol{i} & \boldsymbol{j} & \boldsymbol{k} \\ \omega_{1x} & \omega_{1y} & \omega_{1z} \\ r_{1x} & r_{1y} & r_{1z} \end{vmatrix} \\ &= \begin{vmatrix} \boldsymbol{i} & \boldsymbol{j} & \boldsymbol{k} \\ 50 & 0 & 0 \\ 0 & 70.71 & 70.71 \end{vmatrix} \\ &= 0\boldsymbol{i} - 3535.5\boldsymbol{j} + 3535.5\boldsymbol{k}, \end{aligned}$$

$$\begin{aligned} \boldsymbol{v}_{s_2s_1} = \boldsymbol{\omega}_2 \times \boldsymbol{r}_2 &= \begin{vmatrix} \boldsymbol{i} & \boldsymbol{j} & \boldsymbol{k} \\ \omega_{2x} & \omega_{2y} & \omega_{2z} \\ r_{2x} & r_{2y} & r_{2z} \end{vmatrix} \\ &= \begin{vmatrix} \boldsymbol{i} & \boldsymbol{j} & \boldsymbol{k} \\ \omega_{2x} & \omega_{2y} & \omega_{2z} \\ 86.16 & 278.52 & -70.71 \end{vmatrix} \\ &= (-70.71\omega_{2y} - 278.52\omega_{2z})\boldsymbol{i} \\ &\quad +(86.16\omega_{2z} + 70.71\omega_{2x})\boldsymbol{j} \\ &\quad +(278.52\omega_{2x} - 86.16\omega_{2y})\boldsymbol{k}, \end{aligned}$$

and

$$\boldsymbol{v}_{s_2} = 0.5774 v_{s_2y}\boldsymbol{i} + v_{s_2y}\boldsymbol{j} + 0\boldsymbol{k}.$$

The velocity equation

$$\boldsymbol{v}_{s_2} = \boldsymbol{v}_{s_1} + \boldsymbol{v}_{s_2s_1}$$

applies in each vector direction.

From the $\boldsymbol{i}$ components, we have

$$0.5774 v_{s_2y} = 0 - 70.71\omega_{2y} - 278.52\omega_{2z},$$

from the $\boldsymbol{j}$ components,

$$v_{s_2y} = -3535.5 + 86.16\omega_{2z} + 70.71\omega_{2x},$$

and from the $\boldsymbol{k}$ components,

$$0 = 3535.5 + 278.52\omega_{2x} - 86.16\omega_{2y}.$$

Combining these equations with the result of $\boldsymbol{\omega}_2 \times \boldsymbol{r}_2 = 0$, we get

$$\begin{bmatrix} 0 & 70.71 & 278.52 & 0.5774 \\ 70.71 & 0 & 86.16 & -1 \\ 278.52 & -86.16 & 0 & 0 \\ 86.16 & 278.52 & -70.71 & 0 \end{bmatrix} \begin{bmatrix} \omega_{2x} \\ \omega_{2y} \\ \omega_{2z} \\ v_{S_2y} \end{bmatrix} = \begin{bmatrix} 0 \\ 3535.5 \\ -3535.5 \\ 0 \end{bmatrix}.$$

It is most convenient to use a calculator or computer program to solve this set of linear simultaneous equations. The results are

$$\omega_{2x} = -11.190,$$
$$\omega_{2y} = 5.091,$$
$$\omega_{2z} = 6.505,$$

and

$$v_{S_2y} = -3761.$$

Noting the direction of v_{S_2}, we have $v_{S_2x} = -2{,}172$. The angular velocity of link 2 is given by

$$\boldsymbol{\omega}_2 = -11.190\boldsymbol{i} + 5.091\boldsymbol{j} + 6.505\boldsymbol{k}.$$

The velocity of the sliding link is given by

$$\boldsymbol{v}_{S_2} = -2172\boldsymbol{i} - 3761\boldsymbol{j}.$$

The relative velocity is

$$\boldsymbol{v}_{S_2S_1} = \boldsymbol{\omega}_2 \times \boldsymbol{r}_2 = -2{,}172\boldsymbol{i} - 231\boldsymbol{j} - 3{,}535\boldsymbol{k}.$$

We may now check to ensure that the vector velocity equation $\boldsymbol{v}_{S_2} = \boldsymbol{v}_{S_1} + \boldsymbol{v}_{S_2S_1}$ is satisfied.

For the special case in which path angle $\gamma = 0$, we have, of course, an in-line planar slider-crank linkage. The displacement equation reduces to

$$r_2^2 = (r_0 - r_1\cos\theta + r_3)^2 + (r_1\sin\theta)^2,$$

where point A is undefined but $r_1 + r_3$ is the distance between revolute R and ball joint S_2. For $\gamma = 0$, the two ball joints could be replaced by revolute joints, since the motion is restricted to a plane. For this special case, the velocity equation reduces to

$$\frac{dr_3}{dt} = -r_1\sin\theta\frac{d\theta}{dt}\left[1 + \frac{r_1\cos\theta}{r_0 + r_3 - r_1\cos\theta}\right].$$

It can be shown that the preceding equation is identical (except for the sign convention) to the planar slider-crank velocity equation derived earlier for crank length R and connecting rod length L:

$$v = R\omega\sin\theta\left[1 + \left(\frac{R}{L}\right)\frac{\cos\theta}{\sqrt{1 - (R/L)^2\sin^2\theta}}\right].$$

Matrix Methods Applied to Velocity Analysis of an RSSR Spatial Linkage

An RSSR spatial linkage is described by the vector position equation

$$r_0 + r_1 + r_2 + r_3 = 0,$$

where the links are identified as in Figure 3.17. In solving spatial linkages, position analysis is a difficult task. Methods are described in Chapter 2, but specific solutions given in that chapter apply only to certain linkage configurations.

Noting that all links have a fixed length and that link 0 is stationary, we differentiate the position equation to get the velocity equation

$$\boldsymbol{\omega}_1 \times \mathbf{r}_1 + \boldsymbol{\omega}_2 \times \mathbf{r}_2 + \boldsymbol{\omega}_3 \times \mathbf{r}_3 = 0. \quad (3.42)$$

Now, suppose the link lengths and configuration are specified, and we have already solved the position equation. Let drive crank position and the angular velocity be given. Then there are four unknowns in the velocity equation: the three components of the angular velocity of link 2 (the coupler) and the angular velocity of link 3 (the follower crank). But that vector equation is only worth the three scalar equations we get by expanding it and separately equating its **i**, **j**, and **k** components.

We need a fourth equation. Recall that an RSSR spatial linkage has two degrees of freedom, but we do not usually care about rotation of the coupler about its own axis.

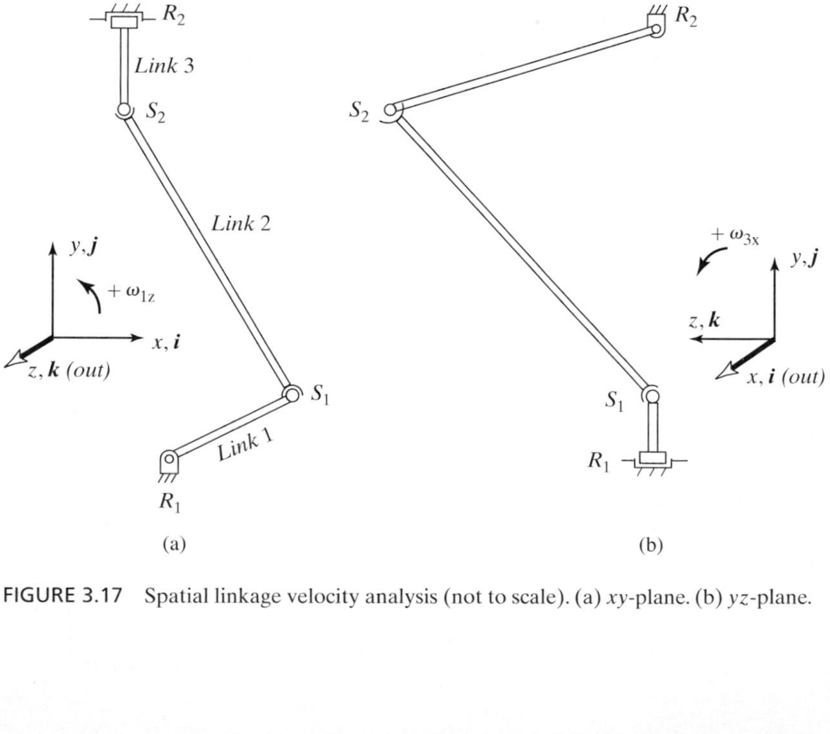

FIGURE 3.17 Spatial linkage velocity analysis (not to scale). (a) xy-plane. (b) yz-plane.

If we say that link 2 does not rotate about its own axis, then the dot product of the angular velocity and the link vector is zero:

$$\boldsymbol{\omega}_2 \cdot \boldsymbol{r}_2 = 0. \tag{3.43}$$

The right side of this equation could be any number. The coupler has a ball joint at both ends; its rotation about its own axis is irrelevant if we care only about the motion of the output crank vs. the drive crank.

SAMPLE PROBLEM 3.11

Velocity analysis : expanding the vector equations

The drive crank and the driven crank of an RSSR linkage rotate in perpendicular planes, and the fixed bearing of the driven crank is in the plane of the driver. Write a set of scalar equations from which you can find the follower crank velocity in terms of the driver velocity.

Decisions. We will set up the coordinate axes as in Figure 3.17. Then, the drive crank vector has components in the x and y directions and rotates about the z-axis. The driven crank has components in the y and z directions, and its angular velocity vector is in the $\pm x$ direction.

Solution. We first compute the cross products of the angular velocity and link vectors:

$$\boldsymbol{\omega}_1 \times \mathbf{r}_1 = \begin{vmatrix} \mathbf{i} & \mathbf{j} & \mathbf{k} \\ 0 & 0 & \omega_{1z} \\ r_{1x} & r_{1y} & 0 \end{vmatrix} = \mathbf{j}\,\omega_{1z}\, r_{1x} - \mathbf{i}\omega_{1z}\, r_{1y};$$

$$\boldsymbol{\omega}_2 \times \mathbf{r}_2 = \begin{vmatrix} \mathbf{i} & \mathbf{j} & \mathbf{k} \\ \omega_{2x} & \omega_{2y} & \omega_{2z} \\ r_{2x} & r_{2y} & r_{2z} \end{vmatrix} = \begin{matrix} \mathbf{i}(\omega_{2y} r_{2z} - \omega_{2z} r_{2y}) + \\ \mathbf{j}(\omega_{2z} r_{2x} - \omega_{2x} r_{2z}) + \\ \mathbf{k}(\omega_{2x} r_{2y} - \omega_{2y} r_{2x}); \end{matrix}$$

$$\boldsymbol{\omega}_3 \times \mathbf{r}_3 = \begin{vmatrix} \mathbf{i} & \mathbf{j} & \mathbf{k} \\ \omega_{3x} & 0 & 0 \\ 0 & r_{3y} & r_{3z} \end{vmatrix} = -\mathbf{j}\omega_{3x}\, r_{3z} + \mathbf{k}\omega_{3x}\, r_{3y}.$$

Adding the cross products, we have

$$\begin{aligned} \boldsymbol{\omega}_1 \times \mathbf{r}_1 + \boldsymbol{\omega}_2 \times \mathbf{r}_2 + \boldsymbol{\omega}_3 \times \mathbf{r}_3 = {} & \mathbf{i}(-\omega_{1z} r_{1y} + \omega_{2y} r_{2z} - \omega_{2z} r_{2y}) \\ & + \mathbf{j}(\omega_{1z} r_{1x} + \omega_{2z} r_{2x} - \omega_{2x} r_{2z} - \omega_{3x} r_{3z}) \\ & + \mathbf{k}(\omega_{2x} r_{2y} - \omega_{2y} r_{2x} - \omega_{3x} r_{3y}) = 0. \end{aligned}$$

The **i**, **j**, and **k** components of this equation each equal zero.
The terms preceded by **i** give us one scalar equation:

$$-\omega_{1z} r_{1y} + \omega_{2y} r_{2z} - \omega_{2z} r_{2y} = 0,$$

the **j** terms another,

$$\omega_{1z} r_{1x} + \omega_{2z} r_{2x} - \omega_{2x} r_{2z} - \omega_{3x} r_{3z} = 0,$$

and the **k** terms a third,

$$\omega_{2x}r_{2y} - \omega_{2y}r_{2x} + \omega_{3x}r_{3y} = 0.$$

We need four equations, however, because there are four unknowns. The fourth equation is

$$\boldsymbol{\omega}_2 \cdot \mathbf{r}_2 = \omega_{2x}r_{2x} + \omega_{2y}r_{2y} + \omega_{2z}r_{2z} = 0.$$

We are now prepared to find the four unknowns: the x, y, and z components of the coupler angular velocity and ω_{3x}, the angular velocity of the driven crank.

The good news is that we now have the correct number of equations and unknowns, and the equations are linear. We can thus use matrix methods to solve the set of equations. The bad news is that analyzing a full cycle of motion of the *RSSR* spatial linkage requires hundreds of calculations, and we would not attempt the task without a computer. Also, different linkage types may require different sets of position and velocity equations.

SAMPLE PROBLEM 3.12

Spatial linkage velocity equations in matrix form

Suppose that link lengths and other data are given for the RSSR spatial linkage considered in Sample Problem 3.11. Suppose also that you have already solved the position equations in terms of the angular position of the drive crank. Write a matrix equation for the angular velocities of the coupler and driven crank in terms of the angular velocity of the drive crank.

Solution. First, we rearrange the four equations obtained in the previous sample problem. The unknown quantities go to the left of the equals sign and the knowns to the right. The equations are arranged so that like ω terms line up as follows:

$$\begin{array}{ccccl}
 & \omega_{2y}r_{2z} & -\omega_{2z}r_{2y} & & = \omega_{1z}r_{1y}; \\
-\omega_{2x}r_{2z} & & +\omega_{2z}r_{2x} & -\omega_{3x}r_{3z} & = -\omega_{1z}r_{1x}; \\
\omega_{2x}r_{2y} & -\omega_{2y}r_{2x} & & +\omega_{3x}r_{3y} & = 0; \\
\omega_{2x}r_{2x} & +\omega_{2y}r_{2y} & +\omega_{2z}r_{2z} & & = 0.
\end{array}$$

(It helps to put zeros in the empty spaces of these equations.)

The matrix equation is

$$\mathbf{A}\,\mathbf{X} = \mathbf{B},$$

where the column matrix of unknown angular velocities is

$$\mathbf{X} = \begin{bmatrix} \omega_{2x} \\ \omega_{2y} \\ \omega_{2z} \\ \omega_{3x} \end{bmatrix}.$$

The **A** matrix is constructed so that it represents the coefficients of the left side of the four aforesaid equations:

$$\mathbf{A} = \begin{bmatrix} 0 & r_{2z} & -r_{2y} & 0 \\ -r_{2z} & 0 & r_{2x} & -r_{3z} \\ r_{2y} & -r_{2x} & 0 & r_{3y} \\ r_{2x} & r_{2y} & r_{2z} & 0 \end{bmatrix}.$$

The known quantities on the right of the four equations form the column matrix:

$$\mathbf{B} = \begin{bmatrix} \omega_{1z} \cdot r_{1y} \\ -\omega_{1z} \cdot r_{1x} \\ 0 \\ 0 \end{bmatrix}.$$

You can see why the equations in the forgoing sample problem are lined up on the ω_{2x} terms, etc. If you are not comfortable with this configuration, be sure to review sections of this chapter on matrices, or review applicable sections in one of your mathematics books. Your time will be well spent. "Working smart" involves using matrices to solve problems in the kinematics and dynamics of machinery; you may have opportunities to use matrices throughout your engineering career.

Checking for errors. Unless you are very lucky or talented, the results of complicated calculations are likely to be wrong the first time around. Errors in entering data and equations are the source of most "computer mistakes." Occasionally, software and hardware introduce errors. Insert simple tests into your programs. Is the magnitude of a link vector equal to the actual link length? Is the velocity zero when the slope of the position curve is horizontal? Compare velocity with change in position divided by change time: (over a short interval).

The chain rule. Do you remember the chain rule for differentiation? Suppose driven crank position θ_3 is known as a function of drive crank position θ_1. Using a special case of the chain rule, the follower crank velocity is found by calculating

$$\omega_3 = d\theta_3/dt = (d\theta_3/d\theta_1)(d\theta_1/dt) = \omega_1(d\theta_3/d\theta_1).$$

Numerical differentiation. Take an arbitrary drive crank position. If the driven crank velocity calculated by the matrix method and by numerical differentiation do not produce comparable results, something must be wrong. For a small time interval Δt, the angular velocity of the driven crank is approximated by

$$\omega_3 \approx \Delta\theta_3/\Delta t \approx (\Delta\theta_3/\Delta\theta_1) \cdot (\Delta\theta_1/\Delta t) \approx \omega_1 \cdot (\Delta\theta_3/\Delta\theta_1),$$

where $\Delta\theta_3/\Delta\theta_1$ is calculated from a θ_3 vs. θ_1 plot or table, using a small interval $\Delta\theta_1$. Compare this approximation of ω_3 with the calculated value (matrix method) in the middle of the interval.

Numerical differentiation is illustrated in the next example. Be sure to check units. If measuring the slope of a curve, check the horizontal and vertical scales.

SAMPLE PROBLEM 3.13

Calculating spatial linkage velocity

A mechanism is needed with a 115-mm output link that oscillates through a range of about 50°. The input shaft rotation speed is $150/\pi$ rpm (5 rad/s). The input shaft is parallel to the plane of the output link at a distance of 20 mm. Design the mechanism and find the angular velocity of the output link. Check the transmission metric and check the angular velocity by numerical differentiation.

Design decisions. We will select an RSSR spatial linkage similar to that in Figure 3.17. After a number of tries, the following dimensions are chosen:

$$\text{drive crank length } r_1 = 55 \text{ mm;}$$
$$\text{coupler length } r_2 = 190 \text{ mm;}$$
$$\text{driven crank } r_3 = 115 \text{ mm (required);}$$

revolute joints: R_1 located at $(0, 0, 0)$ and R_2 at $(-20, 180, 0)$, from which fixed link components are $r_{0x} = 20$ and $r_{0y} = -180$

Solution summary. The first part of the solution is based on the analysis of an RSSR linkage in Chapter 2. The range of motion of the output link approximates the desired value, and transmission metric T is acceptable, as shown on the graph (Figure 3.17c). Drive crank position θ_{1z} is identified simply as θ. The four unknown angular velocities are computed for each value of θ, using the equation

$$\mathbf{X} = \mathbf{A}^{-1}\mathbf{B}$$

where matrices **X**, **A**, and **B** are defined in Sample Problem 3.12.

Although we must consider the motion of coupler link 2 to solve the problem, we are interested only in the results for driven crank link 3. The angular velocity of the driven crank is the last element of the **X** matrix. If the elements are numbered 0, 1, 2, and 3, then

$$\omega_{3x} = X_3.$$

The graph shows ω_{3x} divided by the drive crank angular velocity ω_{1z}. We see that a zero slope of the curve of the driven crank position corresponds to zero angular velocity.

Let us obtain the average angular velocity of the driven crank for the interval between 90° and 110° drive crank positions. This velocity is given by the change in θ_{3x} divided by the change in time. The value is -0.311 rad/s, a rough approximation of the matrix-generated value at 100°, $\omega_{3x} = -0.32$ rad/s. The accuracy should improve if the interval is decreased. Results are also checked using a derivative algorithm to obtain

$$\omega_{3x} = \omega_{1z} \cdot d\theta_{3x}/d\theta_{1z}.$$

The graph shows no significant difference between the value computed by using the derivative algorithm and that obtained with the matrix solution. Note that the horizontal axis of the graph is the drive crank angle in degrees. The driven crank angle is in radians, and the other two curves are dimensionless. (*Caution*: The mixed units are for presentation only; be sure to use consistent units in your calculations.)

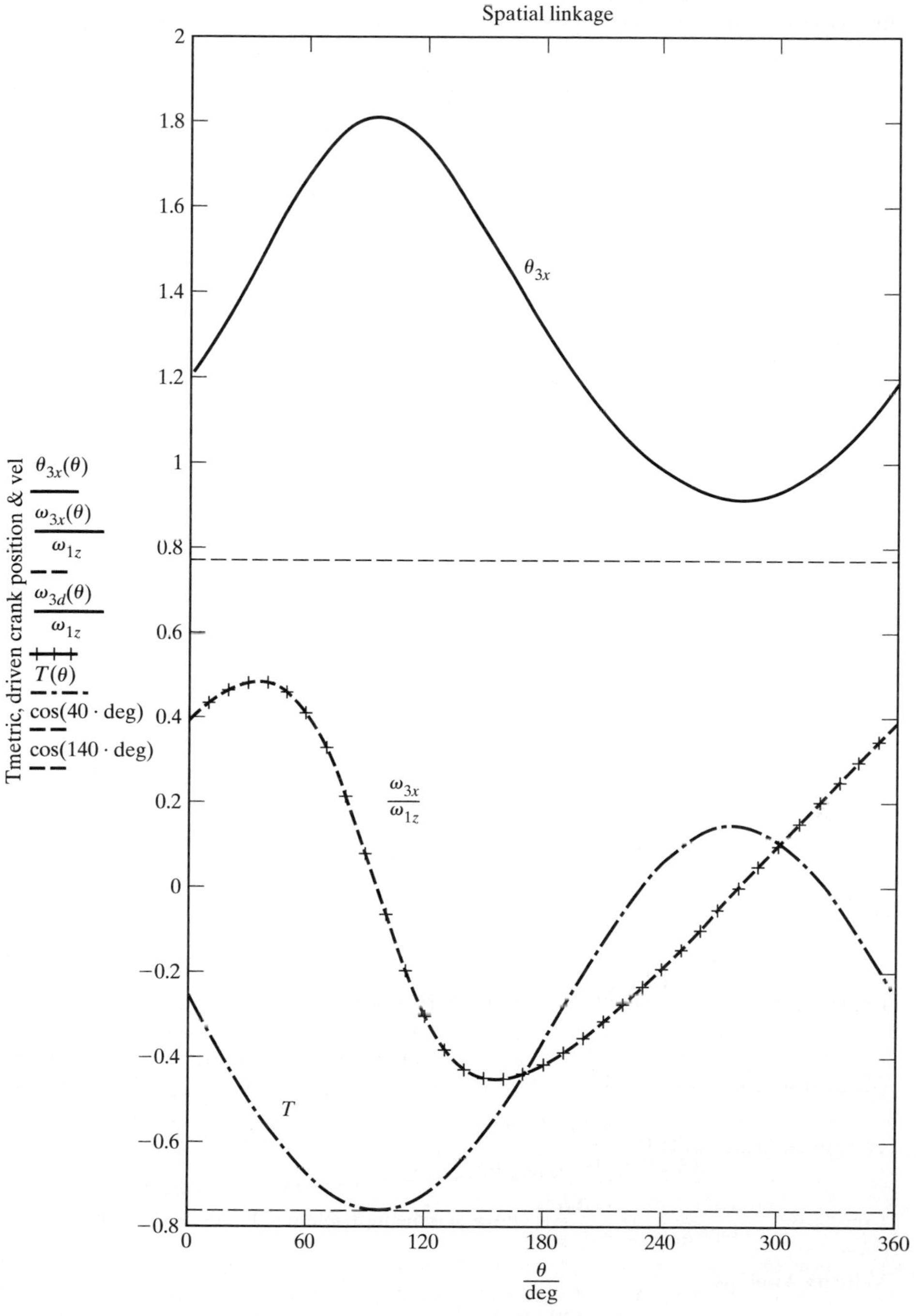

FIGURE 3.17 (c) Spatial linkage motion.

Solution details **(The software used to solve this problem does not identify vectors with bold-face type).**

The drive crank and driven crank rotate in perpendicular planes.

Vector loop: $r_0 + r_1 + r_2 + r_3 = 0$, where $r_0 = R_2 - R_1$

Dimensions of links:

Fixed link $r_{0x} := 20 \quad r_{0y} := -180$

Drive crank $r_1 := 55$

Coupler $r_2 := 190$

Driven crank $r_3 := 115$

Position Analysis

Drive crank position (subscript omitted) $\theta := 0, \frac{\pi}{18} .. 2\pi$

$$r_{1x}(\theta) := r_1\cos(\theta) \quad r_{1y}(\theta) := r_1 \cdot \sin(\theta)$$

Define c = sum of fixed link and drive crank vectors. Then

$$c_x(\theta) := r_{0x} + r_{1x}(\theta) \quad c_y(\theta) := r_{0y} + r_{1y}(\theta)$$

$$r_{2x}(\theta) := -c_x(\theta) \qquad r_{2y}(\theta) := \frac{c_x(\theta)^2 - c_y(\theta)^2 - r_2^2 + r_3^2}{2c_y(\theta)}$$

$$r_{3y}(\theta) := \frac{-(c_x(\theta)^2 + c_y(\theta)^2 - r_2^2 + r_3^2)}{2c_y(\theta)} \qquad r_{3z}(\theta) := (r_3^2 - r_{3y}(\theta)^2)^{\frac{1}{2}*}$$

$$r_{2z}(\theta) := -r_{3z}(\theta)$$

* We will select the assembly configuration given by the positive root.

Driven crank position $\theta_{3x}(\theta) := \text{angle}(r_{3y}(\theta), r_{3z}(\theta)) \qquad \theta_{3x}(2) = 1.772$

Link vectors $rr_2(\theta) := \begin{bmatrix} r_{2x}(\theta) \\ r_{2y}(\theta) \\ r_{2z}(\theta) \end{bmatrix} \qquad rr_3(\theta) := \begin{bmatrix} 0 \\ r_{3y}(\theta) \\ r_{3z}(\theta)s \end{bmatrix}$

Check results for coupler and driven crank length: $|rr_2(2)| = 190 \qquad |rr_3(2)| = 115$

Transmission metric $T(\theta) := \dfrac{rr_2(\theta) \cdot rr_3(\theta)}{|rr_2(\theta)| \cdot |rr_3(\theta)|}$

Compare $\begin{matrix} \cos(40 \cdot \text{deg}) = 0.766 \\ \cos(140 \cdot \text{deg}) = -0.766 \end{matrix}$ There may be a problem if T falls outside this range.

Velocity Analysis

Drive crank speed (rpm) $n := \dfrac{150}{\pi}$

Drive crank angular velocity rad/s $\omega_{1z} := \dfrac{\pi \cdot n}{30} \qquad \omega_{1z} = 5$

Matrix equation $\mathbf{AX} = \mathbf{B}$, where

$$X = \begin{bmatrix} \omega_{2x} \\ \omega_{2y} \\ \omega_{2z} \\ \omega_{3x} \end{bmatrix} \quad A(\theta) := \begin{bmatrix} 0 & r_{2z}(\theta) & -r_{2y}(\theta) & 0 \\ -r_{2z}(\theta) & 0 & r_{2x}(\theta) & -r_{3z}(\theta) \\ r_{2y}(\theta) & -r_{2x}(\theta) & 0 & r_{3y}(\theta) \\ r_{2x}(\theta) & r_{2y}(\theta) & r_{2z}(\theta) & 0 \end{bmatrix} \quad B(\theta) = \begin{bmatrix} \omega_{1z} \cdot r_{1y}(\theta) \\ -\omega_{1z} \cdot r_{1x}(\theta) \\ 0 \\ 0 \end{bmatrix}$$

Solve for angular velocities $X(\theta) := A(\theta)^{-1} \cdot B(\theta)$

Select element 3 of X matrix to find angular velocity of driven crank: $\omega_{3x}(\theta) := X(\theta)_3$

(Elements are numbered 0, 1, 2, 3.)

Find the angular velocity of the driven crank by differentiating the angular position. Use the chain rule: $\omega_{3d}(\theta) := \omega_{1z} \cdot \left(\frac{d}{d\theta} \theta_{3x}(\theta) \right)$.

Approximate the angular velocity of the driven crank (rad/s) when the drive crank is at 100°. Divide the change in position (rad) by the time interval (s):

$$\omega_{3\times(100)} := \frac{(\theta_{3x}(110 \cdot \text{deg}) - \theta_{3x}(90 \cdot \text{deg})) \cdot \omega_{1z}}{110 \cdot \text{deg} - 90 \cdot \text{deg}}$$

Rough approximation: $\omega_{3x(100)} = -0.311$

Value calculated from matrix solution: $\omega_{3x}(100 \cdot \text{deg}) = -0.32 \quad \frac{\text{rad}}{\text{s}}$

Software packages that treat mechanical systems are useful in solving complicated spatial linkages. Figure 3.18 is a composite drawing of a satellite deploying panels mounted on flexible arms. It was necessary to find the speed of deployment of the arms, which were driven by highly nonlinear rotary springs. The solution to this problem involves the response of the flexible system to forces that, in turn, depend on the

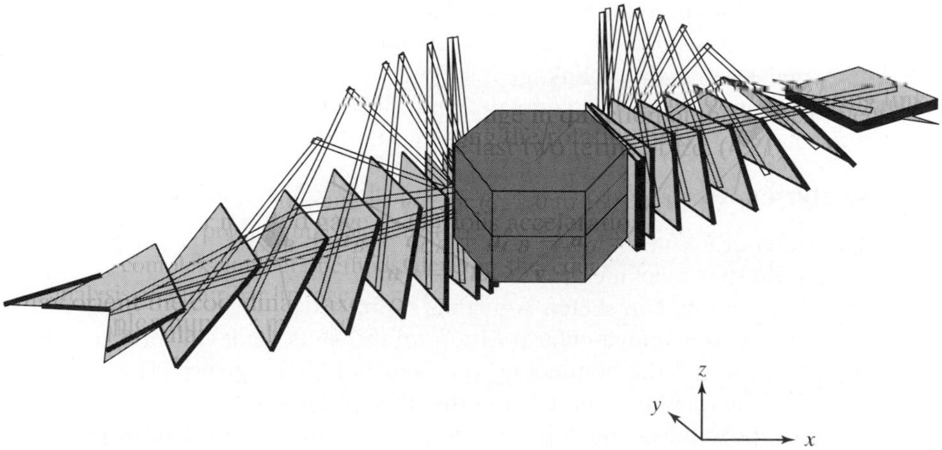

FIGURE 3.18 A satellite deploying flexible arms (modeled by ADAMS™software). (*Source:* Mechanical Dynamics, Inc.)

instantaneous position of the arms. The superimposed display of the deployment history shown in the figure was obtained with ADAMS™ software after entering the results of tests of the springs and other system data.

3.9 GRAPHICAL ANALYSIS OF LINKAGE MOTION UTILIZING RELATIVE VELOCITY

It is desirable to have an independent method of solving equations of motion that may be used to test and debug computer programs and programmable calculator procedures. Graphical solutions are ideal for this purpose. In addition, they provide insight into kinematic problems in a way that analytical solutions cannot.

Earlier in the chapter, we defined relative velocity as a difference between velocities. For example, for points B and C on the same rigid link, the relative velocity $\boldsymbol{v}_{CB}$ must be perpendicular to line BC between the points. This is demonstrated by the cross product relationship

$$\dot{\boldsymbol{r}} = \boldsymbol{\omega} \times \boldsymbol{r},$$

where $\boldsymbol{\omega}$ is the angular velocity of the link and vector $\boldsymbol{r}$ (of constant magnitude) represents line BC. If the relative velocity were not perpendicular to line BC, there would be a component of $\boldsymbol{v}_{CB}$ along line BC, representing a change in length. Obviously, real links deflect due to load, but these small strains are ordinarily negligible compared to the rigid-body motion.

When sliding occurs, we consider the motion of instantaneously coincident points. Then, for coincident points B_1 on link 1 and B_2 on link 2, relative velocity $\boldsymbol{v}_{B_2B_1}$ is tangent to the relative path of the motion. That is, the relative velocity of two points is tangent to the path that one point traces on the link on which the second point is defined. Both of these relationships are utilized to find velocities in linkages by graphical means.

Before a velocity analysis is performed, a position analysis must be made to determine the direction of all links. As observed in the previous sections, an analytical position analysis may be the most difficult part of the entire analysis. Graphical position analysis of a planar linkage is simple, since it is necessary only to draw the linkage to scale, generally using a compass, scale, protractor, and straightedge.

Analyzing Motion of the In-Line Slider-Crank Mechanism

The slider-crank mechanism is a basic part of reciprocating engines, pumps, compressors, and other machines. (See Figure 3.19.) Figure 3.20a is a representation of an in-line slider crank. The sketch is further simplified in Figure 3.20b by showing only the centerlines, sizes, and angular positions of the links. Link O represents the frame, link 1 the crank, link 2 the connecting rod, and link 3 the piston. The crankshaft center is point O_1, the crankpin point B and the wrist pin point C.

Suppose that point B has a velocity of 20 in/s as link 1 turns counterclockwise. A velocity scale is selected that will result in vectors large enough for accurate results. Velocities $\boldsymbol{v}_B$ and $\boldsymbol{v}_{CB}$ are drawn perpendicular to lines O_1B and BC, respectively, as

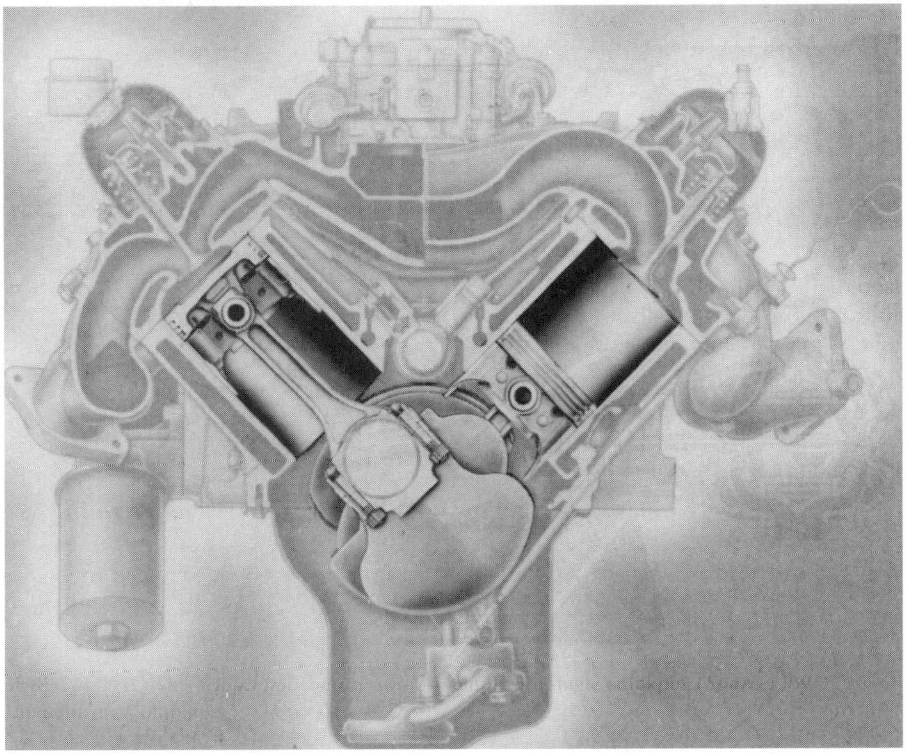

FIGURE 3.19 This sectional view of a V-8 engine shows two pistons and connecting rods (slider-crank mechanisms) at their extreme positions. The crankshaft represents the crank of the mechanisms. (*Source:* General Motors Corporation.)

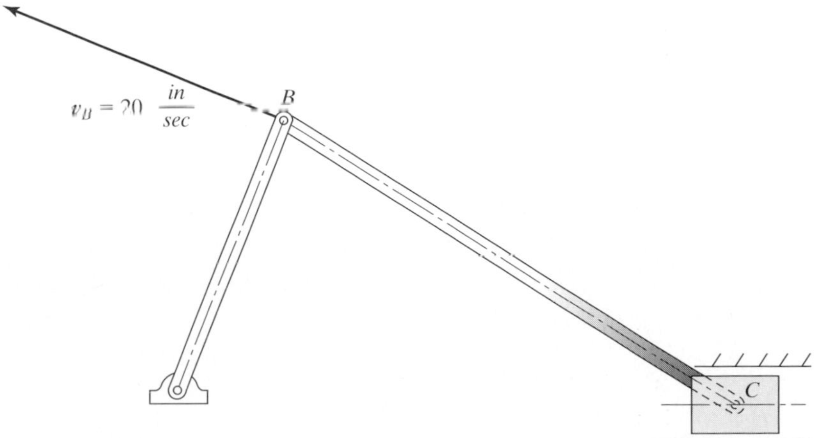

FIGURE 3.20 (a) Simplified sketch of an in-line slider-crank mechanism.

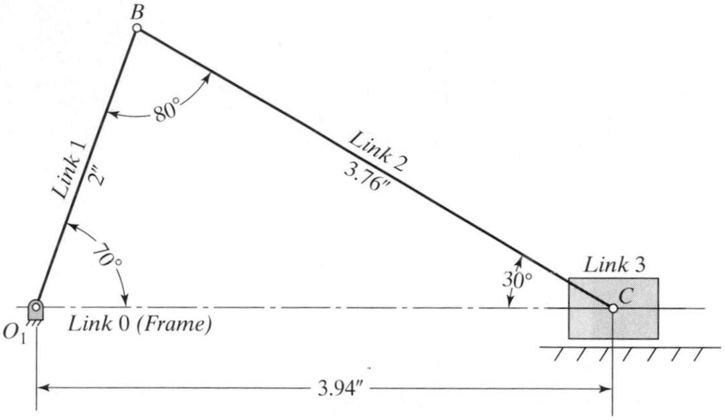

FIGURE 3.20 (b) Skeleton diagram of the mechanism.

shown in Figure 3.20c, an exploded view of the mechanism. The direction of the velocity of C is horizontal, because the piston is constrained to move within the cylinder.

The single arrowhead of $\mathbf{v}_B$ in Figure 3.20c indicates that $\mathbf{v}_C$ has been drawn to scale to represent a known magnitude and direction. Vectors $\mathbf{v}_{CB}$ and $\mathbf{v}_C$ are given *double arrowheads*, indicating that, while their directions are known, their magnitudes are not. When we draw a vector of unknown magnitude, we will call it a *trial vector*. The term *magnitude* will be interpreted to mean both vector length and vector sense. Thus, trial vector $\mathbf{v}_c$ may be to the left, as shown, or to the right, while $\mathbf{v}_{CB}$ may be oriented as shown, or it may be in exactly the opposite direction.

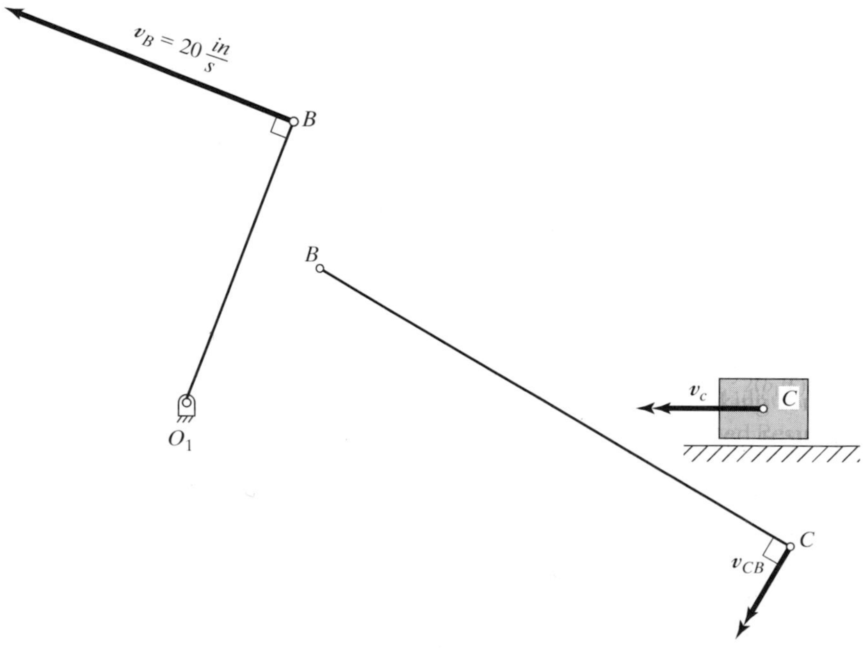

FIGURE 3.20 (c) Velocity vector $\mathbf{v}_B$ is perpendicular to link O_1B, $\mathbf{v}_{CB}$ is perpendicular to link BC, and $\mathbf{v}_C$ is constrained to a straight line as shown.

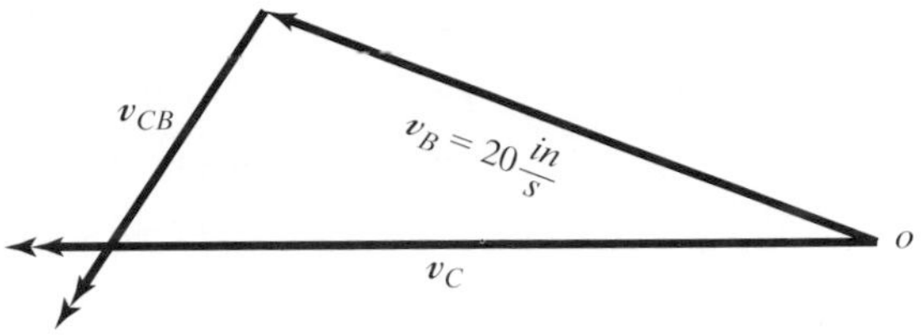

FIGURE 3.20 (d) The vectors can be added together to form a *velocity diagram*. Absolute velocity vectors (velocities relative to the fixed frame) are drawn from a common reference point o. The relative velocity vector is drawn starting at the head of the known absolute velocity vector and with the correct direction, ($\perp BC$). The points of intersection determine the magnitudes of the unknown vectors.

The solution to the problem of finding $\mathbf{v}_C$, the velocity of the piston, is again based on the vector equation $\mathbf{v}_C = \mathbf{v}_B + \mathbf{v}_{CB}$. Beginning at an arbitrary point o in Figure 3.20d, the vector $\mathbf{v}_B$ is drawn to scale. Then, trial vector $\mathbf{v}_{CB}$ is added to $\mathbf{v}_C$, starting at the head of $\mathbf{v}_B$. Next, trial vector $\mathbf{v}_C$ is drawn beginning at the point o. Since $\mathbf{v}_C = \mathbf{v}_B + \mathbf{v}_{CB}$, we have the equivalent of two simultaneous equations, one representing the line $\mathbf{v}_C$ and the other the line $\mathbf{v}_{CB}$ added to $\mathbf{v}_B$. The solution is represented by the intersection of the two lines. In Figure 3.20e, the double arrowheads have been replaced by single arrowheads, since the magnitudes of the relative velocity $\mathbf{v}_{CB}$ and the piston velocity $\mathbf{v}_C$ are determined by the construction. The vector lengths are measured, and, with the use of the velocity scale, the velocities represented are written directly on the figure. We note that the piston velocity is 22.8 in/s to the left.

Figure 3.20d has been redrawn in Figure 3.20e only to illustrate the steps in obtaining a solution; in practice, the construction would simply be "cleaned up" and darkened for clarity. It can be seen that the solution does not depend on our ability to guess the correct sense of $\mathbf{v}_{CB}$ and $\mathbf{v}_C$. If, for example, $\mathbf{v}_C$ were assumed to be to the right, there would be no intersection, and we would then try drawing that vector in the opposite direction and obtain the correct solution.

Sometimes the required accuracy is greater than can be obtained by a simple graphical solution. Or we may wish to make a velocity analysis based on freehand sketches without using drafting tools. In the previous sample problem, let crank angle BO_1C be 70°.

The law of sines states that the ratio of the length of a side to the sine of the opposite angle is the same for all three sides in a triangle. Thus,

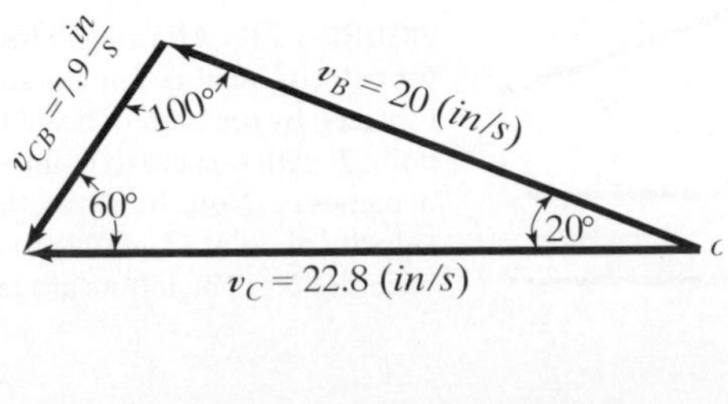

FIGURE 3.20 (e) The completed velocity diagram.

for the triangle formed by the linkage,

$$\frac{BC}{\sin(BO_1C)} = \frac{O_1B}{\sin(O_1CB)},$$

or

$$\frac{3.76}{\sin 70} = \frac{2}{\sin(O_1CB)},$$

from which angle $O_1CB = 30°$.

In the velocity diagram, $\boldsymbol{v}_C$ is horizontal. Vector $\boldsymbol{v}_B$, which is perpendicular to crank O_1B, makes a 20° angle with $\boldsymbol{v}_C$. Vector $\boldsymbol{v}_{CB}$, perpendicular to link BC, makes a 60° angle with $\boldsymbol{v}_C$. Since the sum of the internal angles of any triangle is 180°, we obtain the remaining angle in the velocity polygon: 100°. Using the law of sines once more, we have

$$\frac{v_B}{\sin 60°} = \frac{v_{CB}}{\sin 20°} = \frac{v_C}{\sin 100°}.$$

Crankpin velocity $\boldsymbol{v}_B$ was given as 20 in/s. Substituting this value in to the preceding equation, we obtain relative velocity $\boldsymbol{v}_{CB} = 7.9$ in/s and piston velocity $\boldsymbol{v}_C = 22.8$ in/s to the left.

3.10 THE VELOCITY POLYGON

If the relative-velocity vector $\boldsymbol{v}_{CB}$ is replaced by the vector $\boldsymbol{bc}$ with $\boldsymbol{c}$ at the head of $\boldsymbol{v}_{CB}$, we have the basis for an alternative form of notation for the method of relative velocity: the *velocity polygon*. (See Figure 3.21.) Absolute velocity $\boldsymbol{v}_B$ might as well have been called $\boldsymbol{v}_{BO}$, the velocity of point B with respect to the frame O. Thus, $\boldsymbol{v}_B$ is replaced by $\boldsymbol{ob}$ and, likewise, $\boldsymbol{v}_C$ (or $\boldsymbol{v}_{CO}$) by $\boldsymbol{oc}$. The velocity equation $\boldsymbol{v}_C = \boldsymbol{v}_B + \boldsymbol{v}_{CB}$ now becomes

$$\boldsymbol{oc} = \boldsymbol{ob} + \boldsymbol{bc}. \tag{3.44}$$

Note that we identify $\boldsymbol{ob}$ with the actual velocity in millimeters per second or inches per second, rather than letting $\boldsymbol{ob}$ mean a length in inches on a sketch.

The velocity polygon constructed in Figure 3.22 is based on the linkage of Figures 3.20a and b. In addition to $\boldsymbol{oc}$, the piston velocity, we have determined $\boldsymbol{bc}$, the velocity

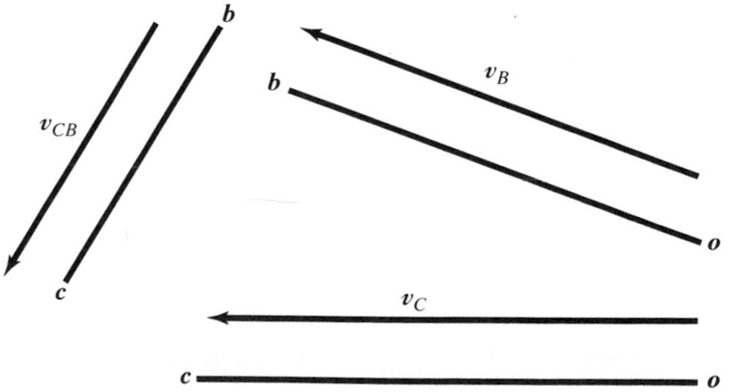

FIGURE 3.21 Alternative form of notation for velocity vectors. The notation $\boldsymbol{v}_B$ is replaced by $\boldsymbol{ob}$; both indicate the velocity of point B with respect to point O. Thus, $\boldsymbol{v}_C$ becomes $\boldsymbol{oc}$. Note, however, that $\boldsymbol{v}_{CB}$, the velocity of point C relative to point B, becomes $\boldsymbol{bc}$. (The letters are reversed.)

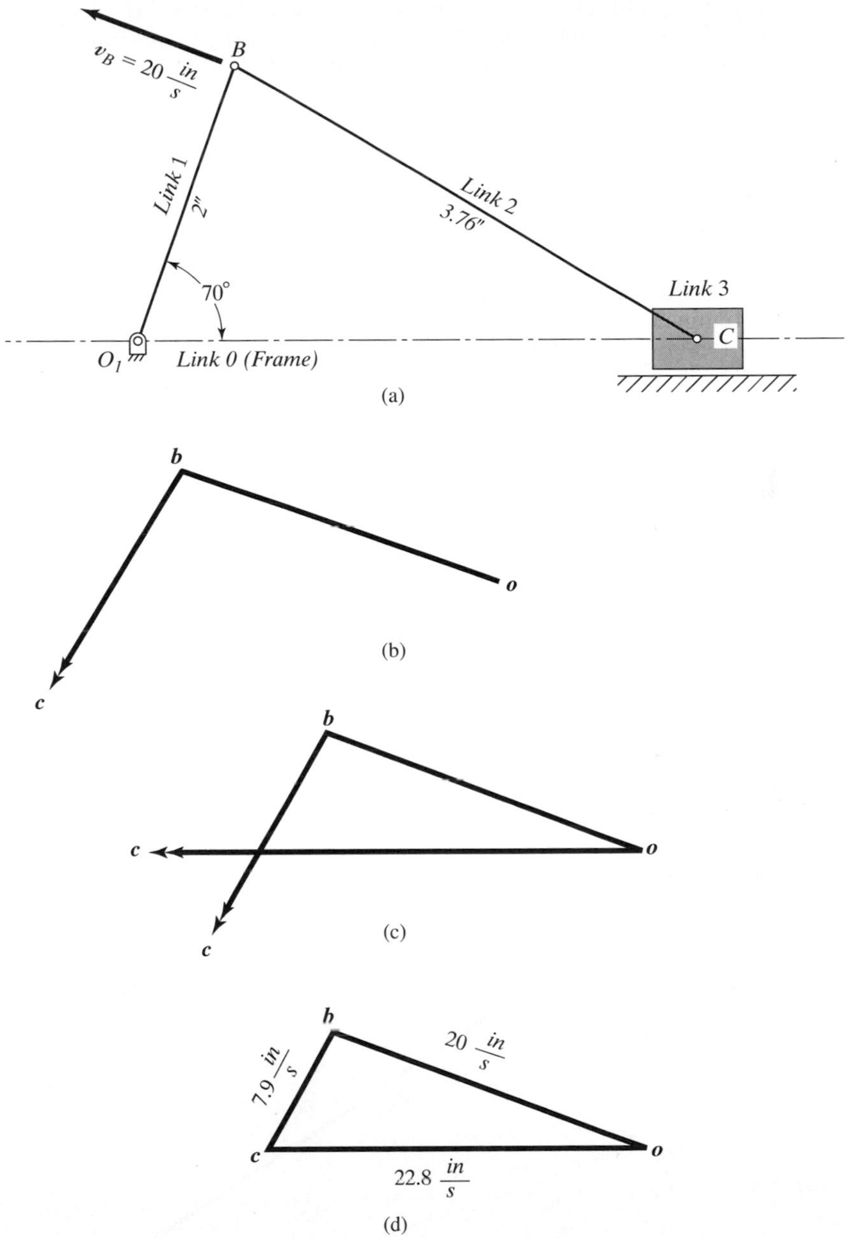

FIGURE 3.22 (a) Slider-crank linkage redrawn from Figure 3.20. (b) Velocity of point B (given) is drawn to scale as vector ***ob*** perpendicular to crank O_1B. From the head of ***ob***, vector ***bc***, the velocity of point C relative to point B, is drawn perpendicular to link BC. We do not know the magnitude of vector ***bc***. (c) Point C is constrained to move horizontally. Thus, vector ***oc*** is drawn horizontally to intersect vector ***bc***. The intersection determines the lengths (magnitudes) of the unknown vectors. (d) The completed velocity polygon.

of the wrist pin C relative to the crankpin B. If ***bc*** is not zero, the motion of the connecting rod BC includes rotation. The value of ω_2, the angular velocity of rod BC, is determined just as it is with links having a fixed center of rotation and is equal to the relative velocity divided by the distance BC. Thus,

$$\omega_2 = \frac{bc}{BC}, \tag{3.45}$$

where bc is the magnitude of the velocity of C relative to B and BC is the distance between pins B and C on the actual linkage, not a distance on the sketch. Substituting in to Eq. (3.45), we find that

$$\omega_2 = \frac{7.9 \text{ in/s}}{3.76 \text{ in}} = 2.1 \text{ rad/s}.$$

The *direction of the angular velocity* of rod BC is found by locating of C relative to B at point C on the linkage sketch. The direction of the relative velocity given by the order of the letters ***bc*** is downward and to the left, as shown by the velocity diagram in Figure 3.22d. Therefore, ω_2 is clockwise, as shown in Figure 3.23a. Note that while the piston velocity ***oc*** must be horizontal, relative velocity ***bc*** must be perpendicular to BC.

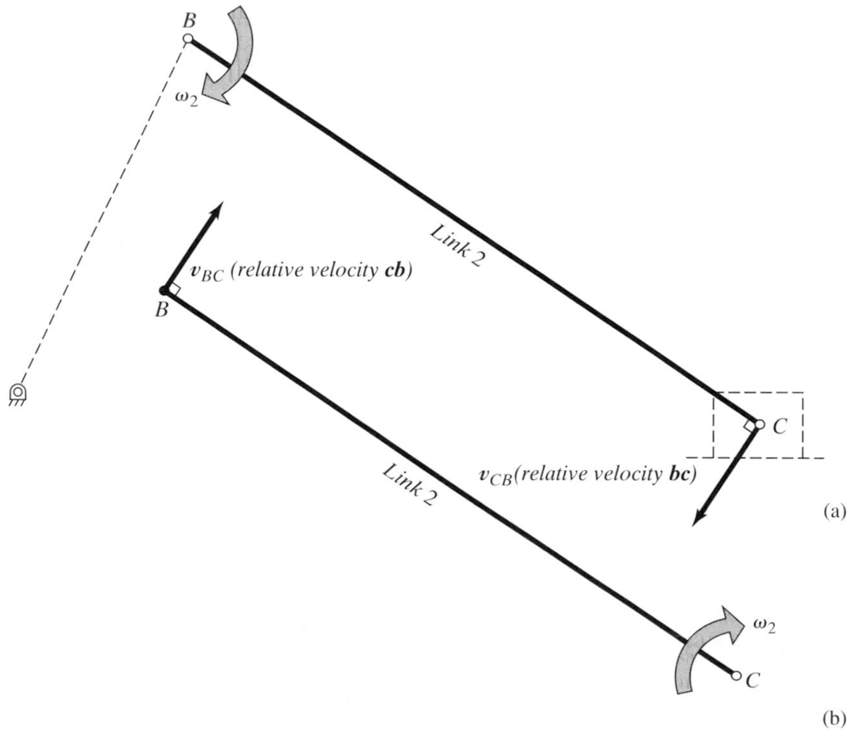

FIGURE 3.23 (a) Determining the angular velocity of link BC from ***bc***, the velocity of point C relative to point B. (b) An alternative method.

Alternatively, the velocity of B relative to C may be located at point B to find the direction of ω_2. Since $\boldsymbol{cb} = -\boldsymbol{bc}$ (Figure 3.23b), the magnitude of ω_2 has the same value as that found earlier.

Noting the order of the letters ***cb*** and inspecting Figure 3.22d, we see that ***cb*** is upward and to the right; therefore, ω_2 is clockwise. The result is the same whether we consider ***bc*** at C or ***cb*** at B. In fact, the angular velocity of a link can be determined from the relative velocity of any two points on the link.

Layout Techniques

A few words about layout may be helpful at this point. In many linkage problems, the velocity polygon can be drawn with sufficient accuracy that measurements may be taken directly from it. But the care used in drawing the velocity polygon bears heavily on the results.

The mechanism that is to be analyzed should be sketched in skeleton form. Only link centerlines, pins, fixed centers, and sliders are shown. If a sketch must be copied, the use of tracing paper or dividers is preferred. The scale of the drawing is indicated, and the length of each link (from pin to pin) is shown directly on the link.

When one is working at a desk and using letter-size paper, lines can be drawn parallel and perpendicular to one another by using two triangles, as shown in Figure 3.24. The paper should be taped in place to keep it from slipping, since accuracy in both the directions and lengths of vectors is critical. A ruler with decimal graduations is preferable to one graduated in sixteenths and thirty-seconds. The velocity polygon (and, later, the acceleration and force polygons) should be on the same sheet as the sketch of the

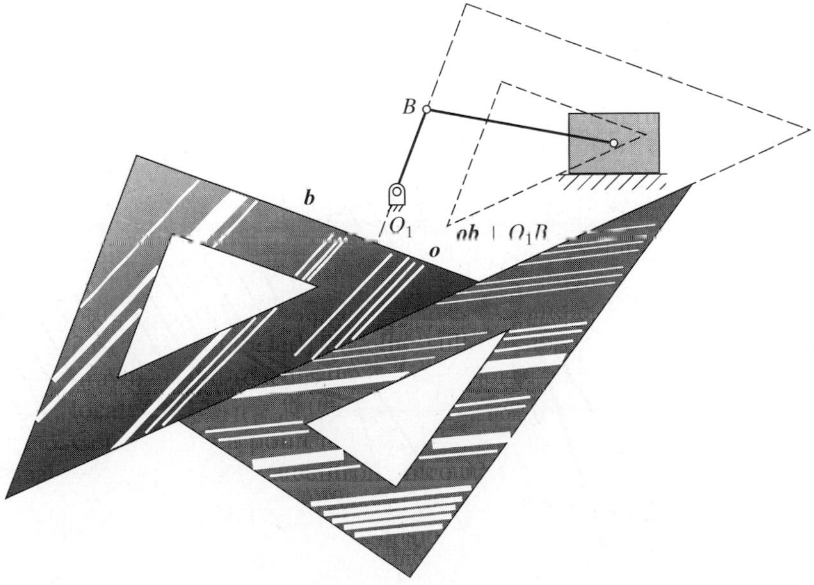

FIGURE 3.24 Vectors are drawn perpendicular to the respective links of the linkage diagram by using two triangles.

mechanism, in order to avoid vector orientation errors. A graphical solution for one link position may be used to check a computer solution.

Velocity Image

The utility of the velocity polygon notation is illustrated by problems in which several points lie on the same link. Consider a rigid link in plane motion, such as BCD of Figure 3.25. On any rigid link, each relative velocity is perpendicular to the line between the points considered. Thus,

$$\mathbf{bc} \perp BC,$$

$$\mathbf{bd} \perp BD,$$

and

$$\mathbf{cd} \perp CD$$

satisfy the conditions for similar triangles.

Triangle ***bcd*** of the velocity polygon is similar to triangle BCD, the rigid link, and we call ***bcd*** the *velocity image* of rigid link BCD. As a result,

$$\frac{bc}{BC} = \frac{bd}{BD} = \frac{cd}{CD}, \tag{3.46}$$

since corresponding sides of similar triangles are proportional. In order to draw ***bcd*** to the correct scale, however, we must know one of the relative velocities—for example, relative velocity ***bc***. For any configuration of points on a rigid link, the velocity polygon contains the exact image, except for its size and orientation.

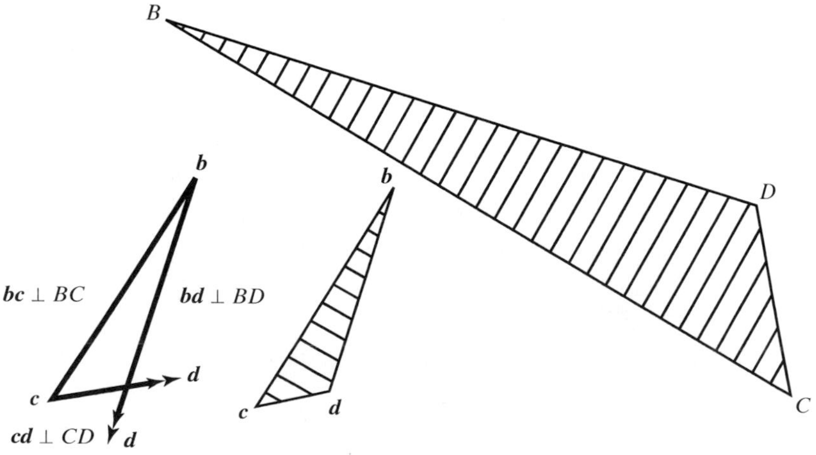

FIGURE 3.25 The velocity image of a link. The vectors are drawn perpendicular to the lines connecting the points on the link.

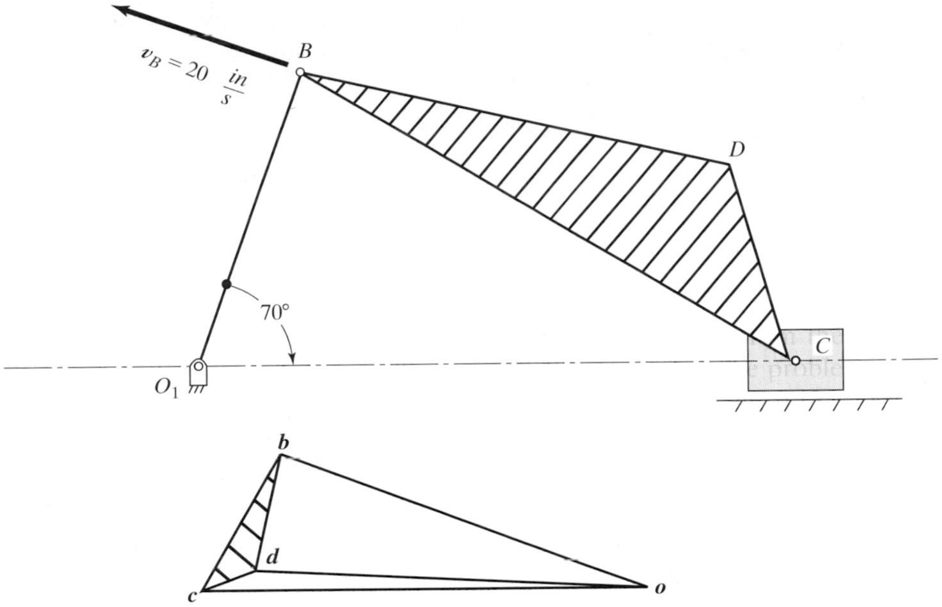

FIGURE 3.26 The velocity image of the connecting rod of a slider-crank mechanism.

Figure 3.26 shows a mechanism containing link BCD. Let distances O_1B, BC, and O_1C and the velocity of B be the same as in the example illustrated in Figure 3.22. Then we can take the velocity polygon ***obc*** directly from that figure. Drawing ***bd*** perpendicular to BD and ***cd*** perpendicular to CD, as in Figure 3.25, we obtain ***bcd***, the image of BCD directly on the velocity polygon. The absolute velocity of D is found by measuring ***od***, a vector of about 20 in/s in magnitude, to the left and slightly upward. (The reader is again reminded that the velocity image principle applies only to points that lie on the same rigid link.)

The path of point D is neither circular (like that of B) nor a straight line (like that of C), as can be shown by drawing the mechanism at several different crank positions. A point such as D on a mechanism may provide just the right motion required to perform a given task. In the design of machinery, it is often necessary to investigate a large number of mechanisms before the desired input–output relationship is obtained.

Let us now consider the velocity image of three points, B, C, and E, lying on a straight line, all on the same rigid link, as in Figure 3.27. Let the link have planar motion, which includes, in general, both rotation at an angular velocity ω and translation. The rotation gives us the following relative velocities:

$$\boldsymbol{bc} = BC\omega, \perp BC,$$

$$\boldsymbol{be} = BE\omega, \perp BE,$$

and

$$\boldsymbol{ec} = EC\omega, \perp EC. \tag{3.47}$$

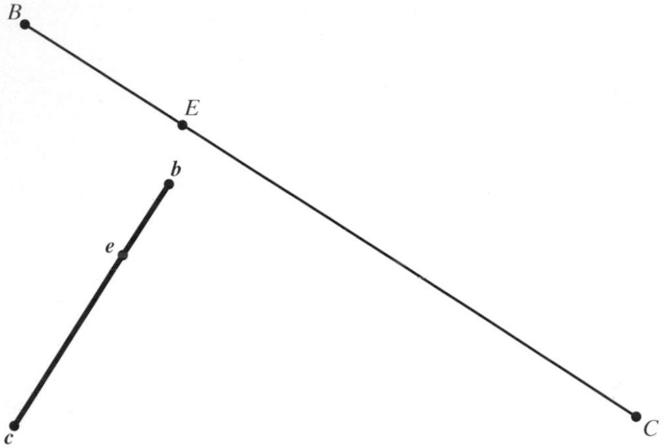

FIGURE 3.27 The velocity image of three points lying on a line in a rigid link is itself a line.

Dividing the second of Eqs. (3.47) by the first and the third by the first, we obtain

$$\frac{be}{bc} = \frac{BE}{BC} \tag{3.48}$$

and

$$\frac{ec}{bc} = \frac{EC}{BC}. \tag{3.49}$$

In practice, the velocity image of points B, E, and C on one rigid link is obtained by using either Eq. (3.48) or Eq. (3.49) and the fact that the order of ***b***, ***e***, and ***c*** in the velocity polygon is the same as that of B, E, and C on the link. Now, recall the preceding discussion in which it was mentioned that the velocity image and the link were similar triangles. If the points under consideration lie on a straight line, then we have the special case of triangles with angles 0°, 0°, and 180°.

Let us examine the velocity of a point E lying on connecting rod BC, as in Figure 3.28a. The linkage is identical to that of Figure 3.22 (except for the addition of point E), and B will again be given a velocity of 20 in/s. The velocity polygon in Figure 3.28b may be taken directly from Figure 3.22, leaving only point ***e*** to be found. From Eq. 3.48, note that

$$\frac{be}{bc} = \frac{BE}{BC}, \quad \text{or} \quad \frac{be}{7.9 \text{ in/s}} = \frac{1 \text{ in}}{3.76 \text{ in}},$$

from which $be = 2.1$ in/s.

Point ***e*** is located a distance from ***b*** corresponding to 2.1 in/s. Since E falls between B and C, ***e*** falls between ***b*** and ***c***. Scaling the vector ***oe***, we find the velocity of E to be approximately 20.6 in/s upward and to the left. (See Figure 3.28c.)

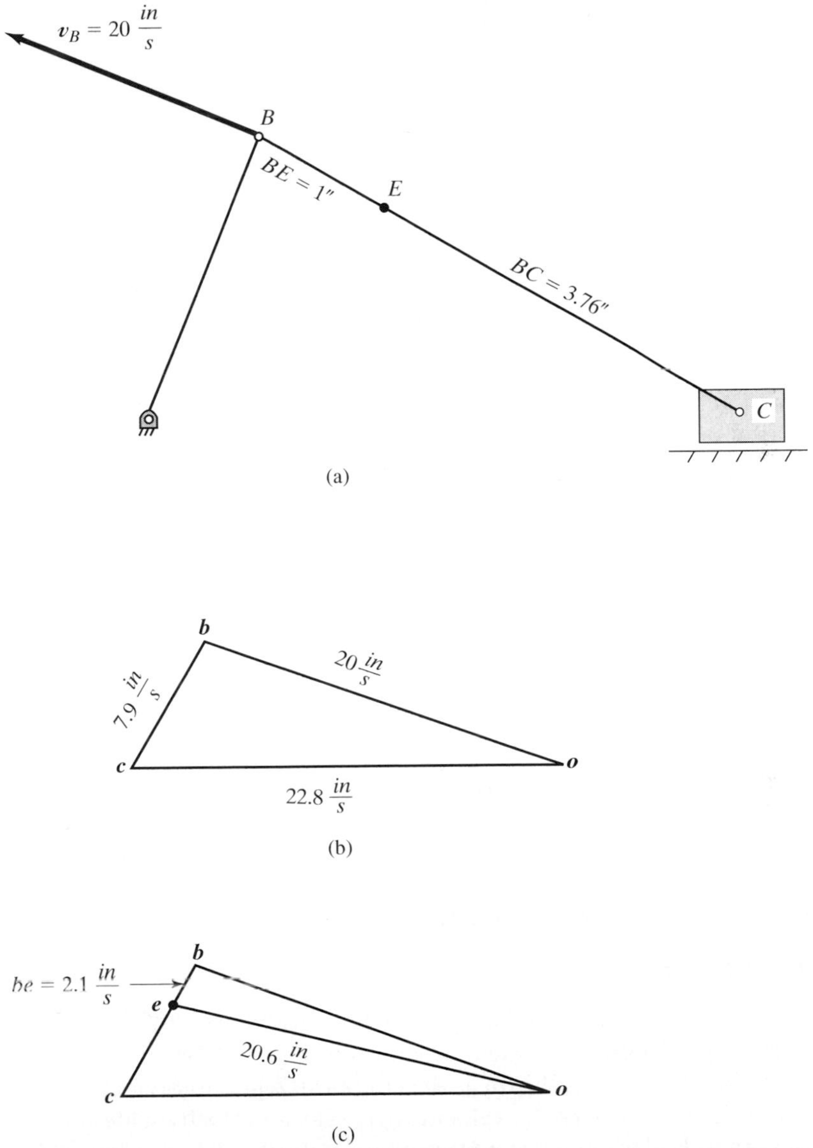

FIGURE 3.28 (a) The problem is to find the velocity of point E (which could be the center of gravity) at the instant when the linkage is in the position shown. (b) Velocity polygon ***obc*** (c) With the velocity image principle, point e is located on vector ***bc***, the velocity image of link BC. The velocity of point E is found by drawing line ***oe*** on the velocity polygon and measuring the length of the line.

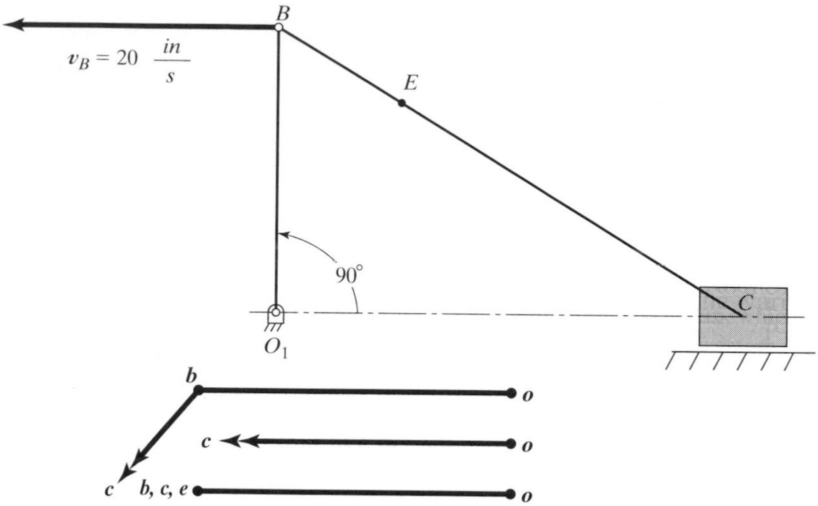

FIGURE 3.29 The connecting rod is shown at the instant when its angular velocity is zero. *At this instant, there is no relative velocity* ($bc = 0$). The velocity image of the connecting rod therefore shrinks to a single point. While *ob* and *oc* are shown parallel, they are actually collinear.

While the velocity image relationships hold in every case, the velocity image of a link undergoing translation shrinks to a single point. This is true both in the case of a slider that always translates and in the case of a connecting rod at the instant when its angular velocity is zero. The latter case is illustrated by the in-line slider-crank mechanism of Figure 3.29. When the crank angle is 90°, ***ob***, the velocity of point *B*, is horizontal. The slider velocity ***oc*** is horizontal also, and thus, ***ob*** and ***oc*** are collinear for an instant.

3.11 GRAPHICAL ANALYSIS OF BASIC LINKAGES

The Four-Bar Linkage

Graphical velocity analysis of a four-bar linkage differs little from the analysis of a slider-crank mechanism. For the mechanism of Figure 3.30a, the velocity of *B* is given as 300 mm/s at the instant shown as the crank rotates counterclockwise.

In order to find all of the velocities, we select a velocity scale, and the vector ***ob*** is drawn perpendicular to O_1B to represent the velocity of *B*. Relative velocity vector ***bc*** of unknown length is drawn perpendicular to *BC*, starting at ***b***. The velocity polygon is completed by drawing ***oc*** beginning at ***o*** and perpendicular to O_3C. The last two steps locate ***c*** on the velocity polygon. Figure 3.30c shows the velocity polygon with all construction lines removed and the values of the vectors shown directly on the polygon. We see that

$$\boldsymbol{bc} = \mathbf{v}_{CB} = 101 \text{ mm/s} \perp BC \text{ (downward and to the right)}$$

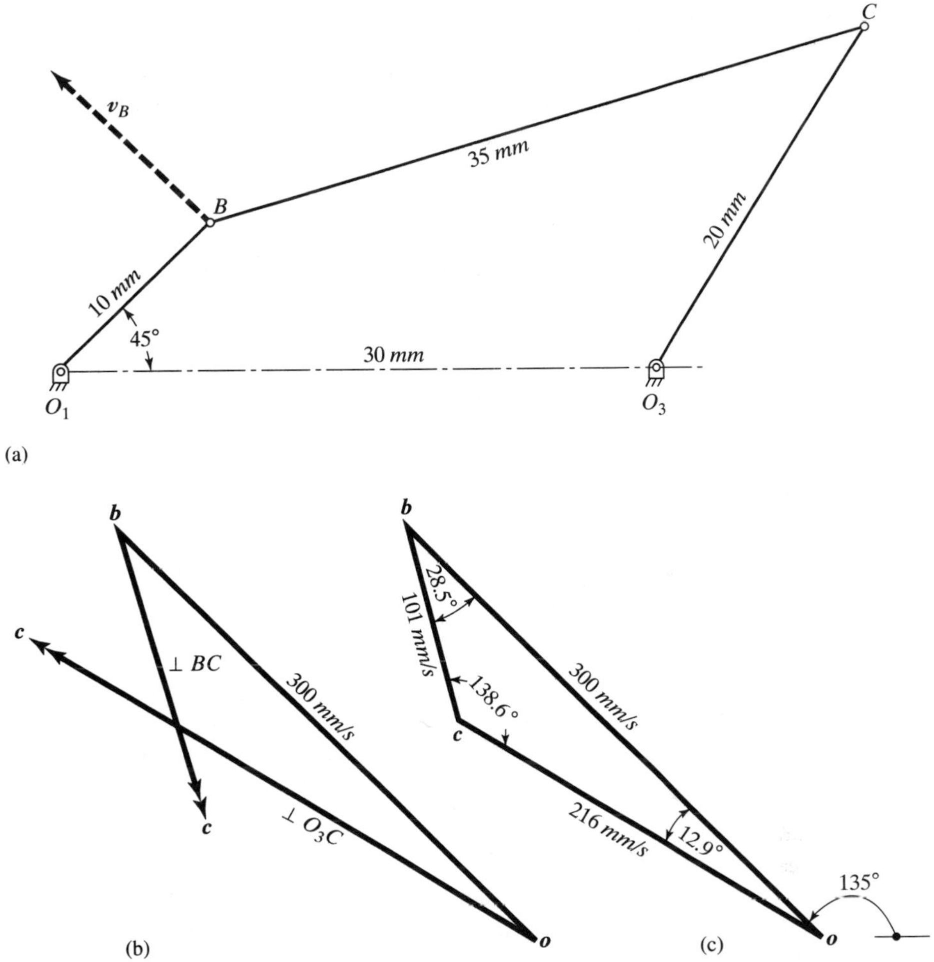

FIGURE 3.30 (a) A four-bar linkage. The velocity of point B is given. (b) While we know the magnitude of only one of the vectors, we know the directions (perpendicular to the links) of all the vectors. (c) The completed velocity polygon. The points of intersection determine the unknown vector quantities.

and

$$\mathbf{oc} = \mathbf{v}_C = 216\,\text{mm/s} \perp O_3C \text{ (upward and to the left).}$$

From these results, the angular velocities of coupler 2 and follower 3 can be determined as

$$\omega_2 = \frac{bc}{BC} = \frac{101}{35} = 2.89\,\text{rad/s cw}$$

and

$$\omega_3 = \frac{oc}{O_3C} = \frac{216}{20} = 10.8 \text{ rad/s ccw},$$

respectively.

The preceding values can also be found by drawing line BO_3 on the mechanism to form two triangles and solving the triangles by the law of cosines and the law of sines. We obtain

$$z^2 = x^2 + y^2 - 2xy \cos Z$$

and

$$\frac{\sin X}{x} = \frac{\sin Y}{y} = \frac{\sin Z}{z}, \tag{3.50}$$

where X, Y, and Z represent the internal angles opposite sides x, y, and z, respectively. In addition, we use the relation $X + Y + Z = 180°$ for any triangle and note that velocity directions differ from link directions by 90° to draw velocity polygon ***obc***.

Up to this point, the velocity polygon was used to analyze the motion of mechanisms without sliding contact (four-bar linkages) and mechanisms in which there is sliding along a fixed path (the piston engine and other slider-crank mechanisms). We will now solve mechanisms in which one link slides along a rotating link.

Analyzing Sliding Contact Linkages

Sliding contact exists between slider and frame in the slider-crank mechanism. In cams, gears, and certain other mechanisms, moving links slide on one another. If a point on one link slides in a curved path on a second link, the relative velocity of the common points is tangent to the path described on the second link. In the example that follows, the relative path is straight.

The mechanism of Figure 3.31a has a slider pinned to link 1. The slider is constrained to slide along link 2. This mechanism is basic to the mechanically driven shaper and is utilized in combination with other linkages, like the backhoe shown in Figure 3.32. The key to solving problems of this type is the designation of a double point B. (See Figure 3.31a.) B_1 is a point on the slider and on link 1, and B_2 is a common point on link 2. While, at this instant, B_1 and B_2 are the same point, B_1 moves relative to B_2 by sliding along link 2. Thus, the direction of relative velocity $\boldsymbol{b_2b_1}$ (the velocity of B_1 with respect to B_2) is along link 2. Relative velocity $\boldsymbol{b_1b_2}$ (equal and opposite to $\boldsymbol{b_2b_1}$) is therefore also along link 2.

Now, contrast the relative velocity of two coincident points on different links with the relative velocity of two points on the same link. In cases where the two points lie on the same rigid link (considered in earlier sections), the relative velocity is perpendicular to the line between the two points.

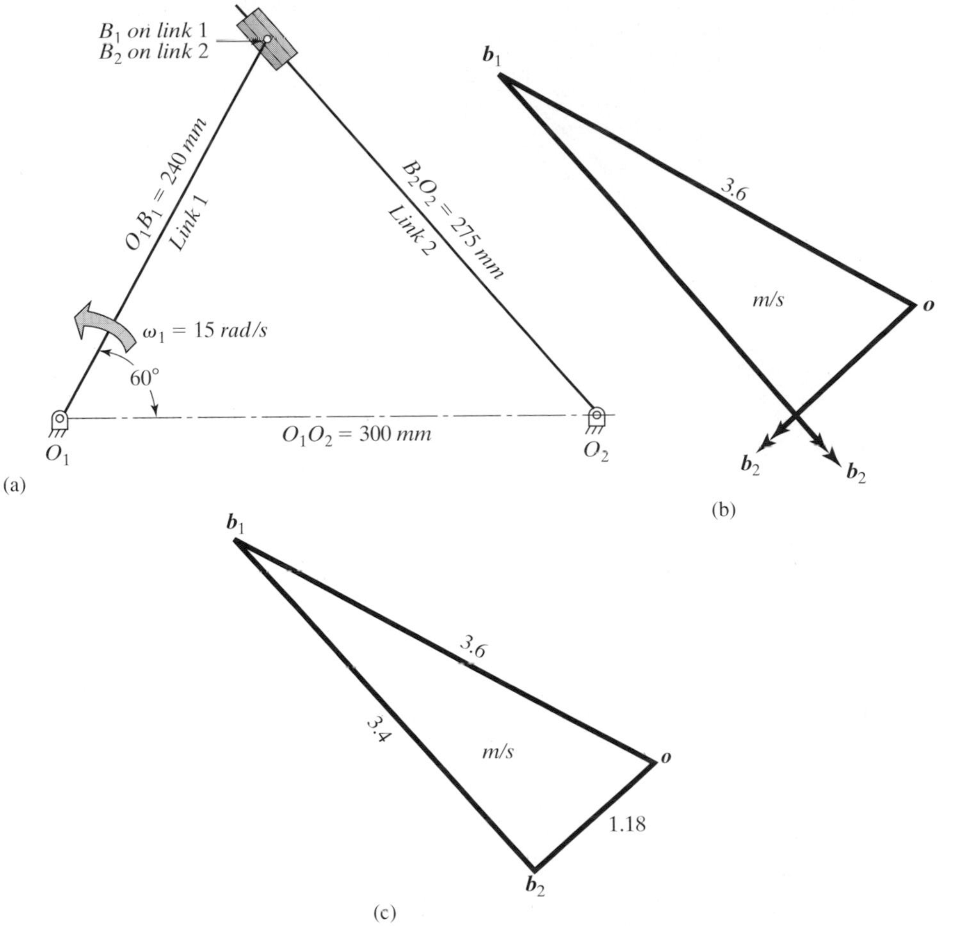

FIGURE 3.31 (a) A sliding contact linkage. The slider is pinned to link 1 and slides along link 2. (b) The velocity polygon for a sliding contact linkage. Note that the relative velocity vector $\boldsymbol{b_2b_1}$ is drawn parallel to link 2, since the motion of B_1 relative to B_2 must be along link 2 at any given instant. (c) The completed polygon.

Suppose link 1 of Figure 3.31a rotates counterclockwise at 15 rad/s, making the velocity of B_1 equal to 3.6 m/s, perpendicular to O_1B_1, upward and to the left. To solve the mechanism, we use the relationship

$$\boldsymbol{v}_{B2} = \boldsymbol{v}_{B1} + \boldsymbol{v}_{B2B1}, \quad \text{or} \quad \boldsymbol{ob}_2 = \boldsymbol{ob}_1 + \boldsymbol{b}_1\boldsymbol{b}_2. \tag{3.51}$$

Vector $\boldsymbol{ob}_1$ (representing the velocity of B_1) is drawn to scale in Figure 3.31b, beginning at an arbitrary pole point $\boldsymbol{o}$. Sliding velocity vector $\boldsymbol{b}_1\boldsymbol{b}_2$ is added to $\boldsymbol{ob}_1$ beginning at $\boldsymbol{b}_1$. The direction of vector $\boldsymbol{b}_1\boldsymbol{b}_2$ is parallel to link 2, but the length and sense of this vector are unknown. Thus, we draw trial vector $\boldsymbol{b}_1\boldsymbol{b}_2$ with a double arrowhead at $\boldsymbol{b}_2$ (not caring about the sense of the vector, because, if our original guess of the sense is wrong, we later reverse the vector to obtain the velocity polygon). Trial vector

FIGURE 3.32 Backhoe. (*Source:* Caterpillar Products.)

$\boldsymbol{ob_2}$ (also of unknown sense and magnitude) is drawn starting at pole point $\boldsymbol{o}$ and perpendicular to O_2B_2. The length (and, if necessary, the sense) of both vectors $\boldsymbol{b_1b_2}$ and $\boldsymbol{ob_2}$ is corrected, completing the velocity polygon $\boldsymbol{ob_1b_2}$ in Figure 3.31c.

Scaling vector $\boldsymbol{b_1b_2}$, we find the sliding velocity to be 3.4 m/s; thus, link 2 moves at a speed of 3.4 m/s downward and to the right relative to the slider. The order of the subscripts is important. Vector $\boldsymbol{b_2b_1}$ refers to the velocity of point B on link 1 relative to the coincident point on link 2. Thus, the slider moves upward and to the left at 3.4 m/s relative to link 2. The velocity of B_2 on link 2 scales to 1.18 m/s downward and to the left, from which we obtain the relationship

$$\omega_2 = \frac{ob_2}{O_2B_2} = \frac{1.18\text{ m/s}}{0.275\text{m}} = 4.29\text{ rad/s}$$

counterclockwise. The method for determining the velocities of the sliding contact linkage is essentially the same, even if links 1 and 2 are curved. Velocity vector $\boldsymbol{ob_1}$ is perpendicular to O_1B_1, and $\boldsymbol{ob_2}$ is perpendicular to O_2B_2. Sliding velocity $\boldsymbol{b_1b_2}$ is in the direction of the relative path; that is, $\boldsymbol{b_1b_2}$ is *tangent* to the instantaneous path of B_1 on link 2 (at B_1). If greater accuracy is desired, the velocity polygon may be solved analytically by using the law of sines, as was done in an earlier example.

Sliding velocity is of particular interest because of friction and wear considerations. (Some references state the coefficient of friction in terms of sliding velocity.) In addition, we must find the sliding velocity in order to compute the Coriolis acceleration. This phase of the problem is treated in the next chapter.

Comparison of Results with an Analytical Solution

A sliding contact linkage was analyzed previously (Section 3.7) by using complex-number methods. With the notation of that section (see Figure 3.15), we have $\theta_1 = 240°$, $R_0 = 300$ mm, $R_1 = 240$ mm, and $\omega_1 = 15$ rad/s, from which we obtain

$$R_2 = \sqrt{R_0^2 + R_1^2 - 2R_0R_1 \cos\theta_1} = 275.0 \text{ mm},$$

$$\theta_2 = \arcsin\left(\frac{R_1 \sin\theta_1}{R_2}\right) = -49.1°,$$

$$\omega_2 = \frac{\omega_1 R_1 \cos(\theta_1 - \theta_2)}{R_2} = 4.28 \text{ rad/s (ccw)},$$

$$v_{B_2} = \omega_2 R_2 = 1{,}178 \text{ mm/s} \quad (1.178 \text{ m/s}),$$

and

$$v_{B_1B_2} = -\omega_1 R_1 \sin(\theta_1 - \theta_2) = 3{,}402 \text{ mm/s}$$

(3.402 m/s along link 2, away from point O_2). All of these values correspond closely to the graphical (velocity polygon) solution.

The reader should be alert for mechanisms that are kinematically equivalent to the sliding contact linkage of Figure 3.31. Two examples are a variable-displacement pump in which the plungers move within a rotating cylinder block and the Geneva mechanism, in which a pin on a rotating wheel (the driver) enters radial slots in the driven member, giving it intermittent rotation as the driver rotates at constant velocity.

Cams and Cam Followers

Almost any motion–time relationship may be generated by using one or more cams. Usually, the cam rotates at constant angular velocity, giving the follower reciprocating or oscillating motion having some predetermined sequence. The design of practical high-speed cams is discussed in Chapter 5. You can use the velocity polygon method to analyze cam follower motion as a partial check of your design.

When sliding occurs between cam and follower, the key to solving for velocities is again a double point where the two make contact, and the solution proceeds as with other sliding contact mechanisms. The velocity of the point of contact of the follower is equal to the vector sum of the velocity of the point of contact on the cam plus the sliding velocity. If B is the point of tangency, the latter statement may be expressed symbolically as

$$\mathbf{ob_2} = \mathbf{ob_1} + \mathbf{b_1b_2},$$

where subscripts 1 and 2 refer, respectively, to the cam and the follower.

In the case of a cam with an oscillating follower (Figure 3.33), $\boldsymbol{ob_2}$ is perpendicular to a line between the center of rotation of the follower and the contact point and $\boldsymbol{ob_1}$ is perpendicular to a line between the center of rotation of the cam and the contact point. Sliding velocity $\boldsymbol{b_1b_2}$ is parallel to the common tangent to the cam and the follower.

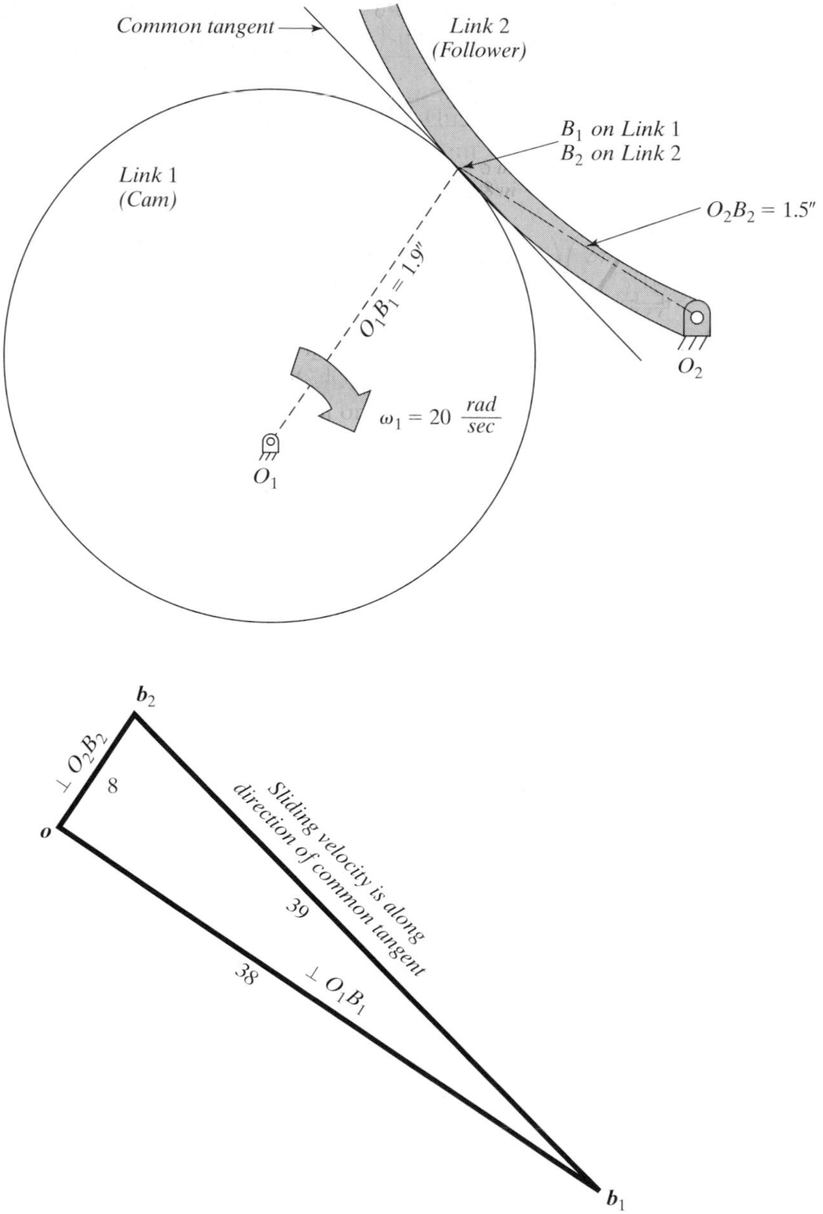

FIGURE 3.33 A cam with an oscillating follower. To obtain the velocity polygon, $\boldsymbol{ob_2}$ is drawn perpendicular to O_2B_2, and $\boldsymbol{ob_1}$ is drawn perpendicular to O_1B_1. Sliding velocity $\boldsymbol{b_1b_2}$ is drawn parallel to the common tangent at B_1B_2.

Figure 3.33 shows a cam formed by an eccentric circle. Let the angular velocity of the cam be 20 rad/s, and at the instant shown, let the distance from the center of rotation of the cam to the point where it makes contact with the follower be 1.9 in. Thus, velocity $\boldsymbol{ob}_1 = 38$ in/s is drawn to scale (to the right and downward, beginning at an arbitrary pole point $\boldsymbol{o}$). Trial vector $\boldsymbol{b}_1\boldsymbol{b}_2$ is added to $\boldsymbol{ob}_1$, and trial vector $\boldsymbol{ob}_2$ is drawn beginning at point $\boldsymbol{o}$. The two trial vectors are made to intersect, and the intersection is labeled $\boldsymbol{b}_2$.

The sliding velocity, scaled from the velocity polygon, is 39 in/s, with B_2 on the follower sliding upward and to the left with respect to B_1 on the cam (or B_1 on the cam sliding downward and to the right with respect to B_2 on the follower). The velocity of B_2, found by scaling vector $\boldsymbol{ob}_2$, is 8 in/s upward and to the right, so that follower angular velocity

$$\omega_2 = \frac{\boldsymbol{ob}_2}{O_2B_2} = \frac{8 \text{ in/s}}{1.5 \text{ in}} = 5.3 \text{ rad/s}$$

clockwise. At this instant, then, the ratio of the follower angular velocity to the cam angular velocity is

$$\frac{\omega_2}{\omega_1} = \frac{5.3 \text{ rad/s}}{20 \text{ rad/s}} = +0.27,$$

where a positive sign is used when both turn in the same direction.

An Equivalent Linkage for a Cam Mechanism

If, at the point of contact, the cam and follower both have finite radii, then a four-bar linkage may be used to analyze the motion. Referring to Figure 3.33, for example, we see that the equivalent links would form a crank-rocker mechanism as sketched in Figure 3.34. The driver crank would consist of a link from O_1 to the center of curvature of the cam at the point of contact (C, the center of the circular cam in this case). The driven crank (the rocker) would consist of a link from O_2 to D, the center of curvature of the follower at the point of contact. Coupler CD extends from one center of curvature to the other.

We have constructed an equivalent linkage wherein the driver crank of the four-bar linkage has the same motion as the cam and the driven crank of the four-bar linkage has the same motion as the rotating cam follower. The coupler has no counterpart on the cam–follower system. If the cam is circular and the follower has the form of a circular arc, then the equivalent linkage dimensions are constant. The radius of curvature varies in most practical cams, however, so the equivalent linkage is of limited value because its dimensions change as the radius of the cam follower change at the point of contact.

Friction Drives

Motion may be transmitted between two shafts by disks that roll on one another. Consider the friction drive of Figure 3.35, where P is a point common to both disks.

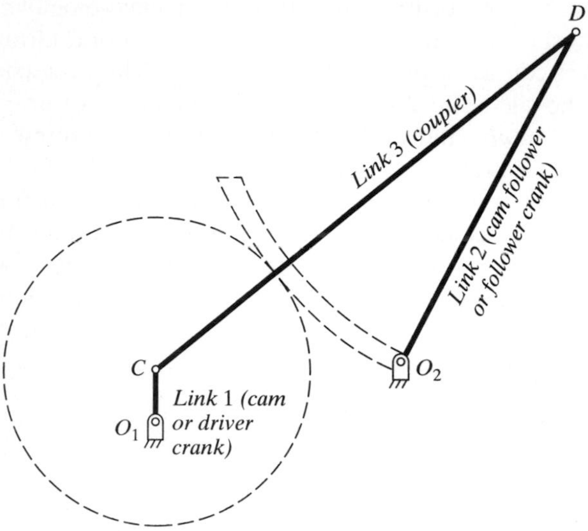

FIGURE 3.34 A four-bar linkage, equivalent to a cam and follower.

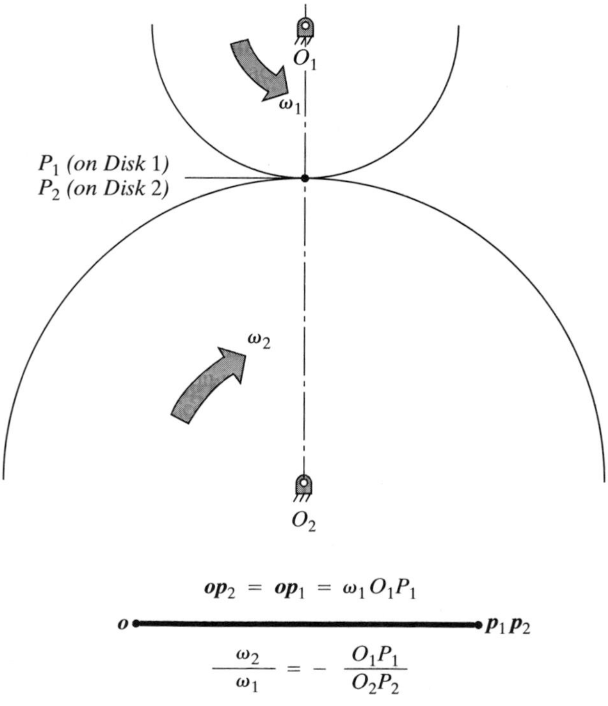

FIGURE 3.35 A friction drive. There is no relative velocity ($p_1p_2 = 0$) if the disks roll without slipping.

If the two disks roll together without slipping, the relative velocity p_1p_2 *is zero, and Eq. (3.51) becomes*

$$\boldsymbol{op}_2 = \boldsymbol{op}_1 = \omega_1 O_1 P_1. \tag{3.52}$$

Since $\omega_2 = op_2/O_2P_2$, *but in a direction opposite that of* ω_1, *we have the angular velocity ratio*

$$\frac{\omega_2}{\omega_1} = -\frac{O_1P_1}{O_2P_2}. \tag{3.53}$$

A drive of this type is satisfactory for low-power applications. However, when large torques are involved, the designer might turn to a gear drive, sacrificing the simplicity of a friction drive to ensure that power will be transmitted under all conditions.

Straight-Toothed Spur Gears

Spur gear velocities may be found by examining a pair of teeth *at their point of contact.* The velocity of that point on the driven gear is the vector sum of the velocity of the same point on the driver and the relative velocity. The equation used earlier for cams and mechanisms that include sliding also applies to gears: $\boldsymbol{ob}_2 = \boldsymbol{ob}_1 + \boldsymbol{b}_1\boldsymbol{b}_2$. In this case, the common point, or point of contact, is B, with subscripts 1 and 2 referring to the driver and driven gear, respectively.

Figure 3.36a shows a pair of involute spur gears with contact occurring at point B. Velocity vector $\boldsymbol{ob}_1 = \omega_1 O_1 B_1$ is drawn perpendicular to O_1B_1 in Figure 3.36b. Then, trial vector $\boldsymbol{ob}_2$ is drawn, beginning at $\boldsymbol{o}$ and perpendicular to O_2B_2. Trial vector $\boldsymbol{b}_1\boldsymbol{b}_2$ (the sliding velocity) is drawn from $\boldsymbol{b}_1$ (parallel to the common tangent to the teeth, where the two vectors make contact) until it intersects trial vector $\boldsymbol{ob}_2$. The true location of $\boldsymbol{b}_2$ is thus found, and the angular velocity of gear 2 is given by $\omega_2 = ob_2/O_2B_2$. Figure 3.36c shows the velocity polygon for contact on the line of centers (the pitch point).

The foregoing construction would not be used to find angular velocities, since the ratio of the angular velocities of a pair of gears is given simply by the inverse ratio of the numbers of teeth, N_1 and N_2, on the gears:

$$\frac{|\omega_2|}{|\omega_1|} = \frac{N_1}{N_2}. \tag{3.54}$$

This relationship, which holds for any pair of gears except those in planetary trains, will be derived in Chapters 6 and 7.

Helical Gears

While the tooth elements of straight spur gears are parallel to the gear shaft, helical gear tooth elements are not. When a pair of helical gears are mounted on parallel shafts, the sliding velocity vector for any contact point will lie in a plane perpendicular to the shafts (as with straight spur gears). This is not the case with crossed helical gears.

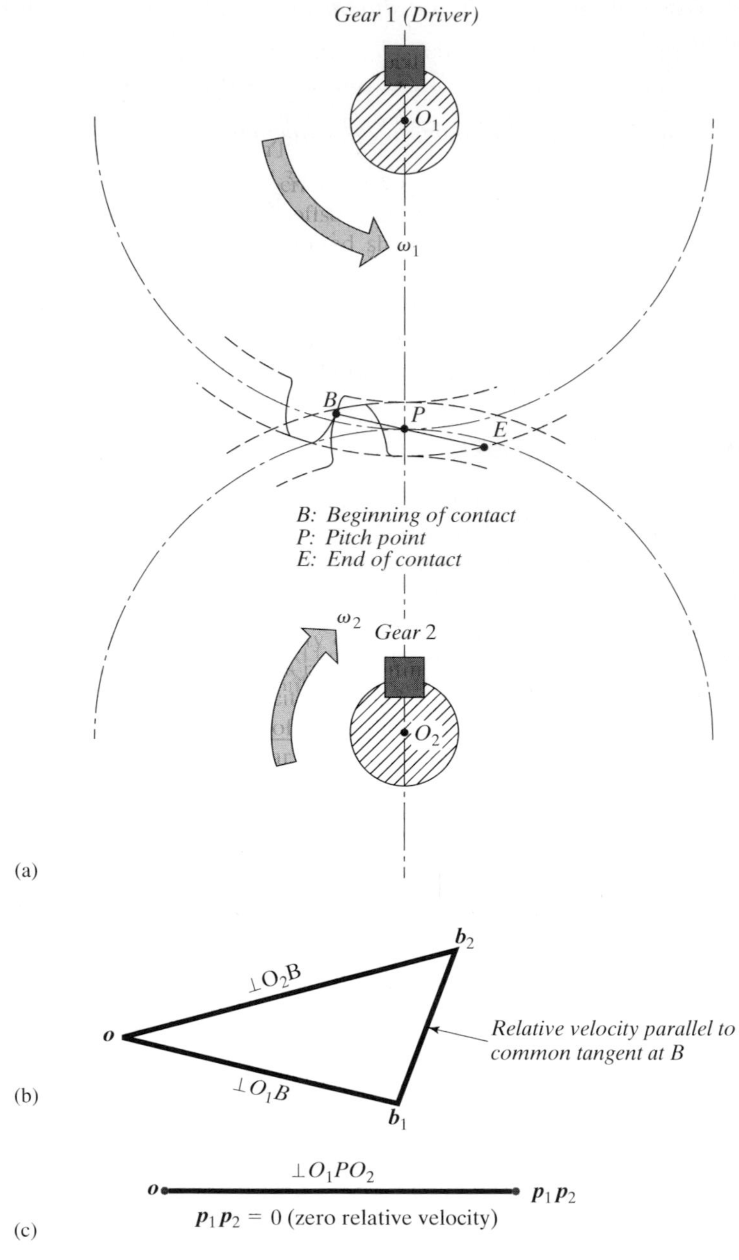

FIGURE 3.36 (a) A pair of spur gears. For clarity, only two teeth are shown. Point B is the point of initial contact (B_1 on gear 1 and B_2 on gear 2). (b) The velocity polygon. Vectors $\boldsymbol{ob}_1$ and $\boldsymbol{ob}_2$ are drawn perpendicular, respectively, to O_1B_1 and O_2B_2. Relative velocity vector $\boldsymbol{b}_1\boldsymbol{b}_2$ is drawn from $\boldsymbol{b}_1$ parallel to the common tangent to the teeth at the point of contact, B. (c) The velocity polygon when contact occurs at the pitch point P.

Crossed Helical Gears

In the case of crossed helical gears (helical gears on nonparallel shafts), the sliding velocity at a general point of contact has a component across the face of the gears. (See Figure 3.37a.) Let a pair of crossed helical gears make contact at a point B, the pitch point (B_1 on gear 1, B_2 on gear 2), which lies on a perpendicular common to the two shafts. The velocity of B_1, represented by $\boldsymbol{ob_1}$, is $\omega_1 d_1/2$, perpendicular to shaft 1. The velocity of B_2, unknown at this time, is represented by $\boldsymbol{ob_2}$, perpendicular to shaft 2, using a double arrowhead. Velocity $\boldsymbol{ob_2} = \boldsymbol{ob_1} + \boldsymbol{b_1b_2}$, where the relative velocity $\boldsymbol{b_1b_2}$ must be the sliding velocity *parallel to the gear tooth* in a plane through B that is parallel to both shafts. (See Figure 3.37b.) The point $\boldsymbol{b_2}$ is therefore located by drawing $\boldsymbol{b_1b_2}$ *parallel to the gear tooth* until it meets the line $\boldsymbol{ob_2}$. (See Figure 3.37c.) The pitch-line

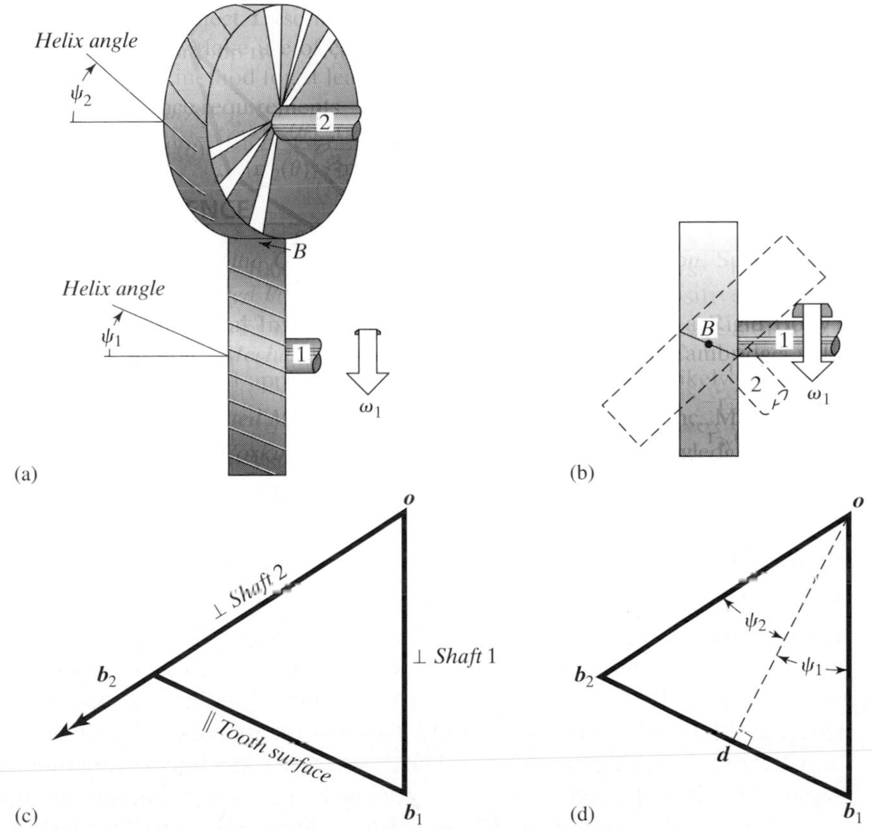

FIGURE 3.37 (a) Crossed helical gears. Pitch diameters of gears 1 and 2 are d_1 and d_2, respectively. (b) Sectional view of the crossed helical gears. Gear 2 is shown in dashed lines. (c) Velocity polygon showing pitch line velocities $\boldsymbol{ob_1}$ and $\boldsymbol{ob_2}$ and the sliding velocity $\boldsymbol{b_1b_2}$. (d) Line $\boldsymbol{od}$ is perpendicular to sliding velocity $\boldsymbol{b_1b_2}$.

velocity of gear 2, $\boldsymbol{ob}_2$, may then be scaled to find angular velocity

$$\omega_2 = \frac{\boldsymbol{ob}_2}{d_2/2}.$$

We see from the direction of $\boldsymbol{ob}_2$ that gear 2 turns clockwise when viewed from the right.

Greater accuracy may be obtained analytically. In Figure 3.37d, a perpendicular $\boldsymbol{od}$ from $\boldsymbol{o}$ to line $\boldsymbol{b}_1\boldsymbol{b}_2$ forms angles ψ_1 and ψ_2. It can be seen that $od = ob_1 \cos \psi_1 = ob_2 \cos \psi_2$. Substituting $ob_1 = \omega_1 d_1/2$ and $ob_2 = \omega_2 d_2/2$, the velocity ratio

$$\frac{n_2}{n_1} = \frac{\omega_2}{\omega_1} = \frac{d_1 \cos \psi_1}{d_2 \cos \psi_2} \tag{3.55}$$

is obtained. Alternatively, the tooth numbers, if known, may be used to find the same ratio:

$$\left|\frac{n_2}{n_1}\right| = \frac{N_1}{N_2}. \tag{3.56}$$

A sectional view through the point of contact will aid in establishing the direction of rotation.

When we consider contact at a point other than the pitch point, additional sliding velocity components must be considered. The ratio of the angular velocities, however, is constant and may be found by one of the methods just described.

Gears and cams are among the most commonly used and most versatile mechanisms. The preceding material on velocities only touches on the problem of the analysis and design of gears and cams, which is covered in considerably more detail in other sections.

3.12 ANALYZING COMBINATIONS OF BASIC LINKAGES

Toggle Linkage

Many practical multilink mechanisms are made up of basic linkage combinations such as the slider-crank and the four-bar mechanism. The *toggle linkage* shown in Figure 3.38a is an example of a mechanism of this type; the toggle principle is applied in ore crushers and in essentially static linkages that act as clamps. The linkage analysis is made by considering the basic linkages separately. To solve for the velocities of the mechanism, we may begin by ignoring the slider and its connecting rod while solving the four-bar linkage separately. Then, the velocity polygon may be completed by finding the slider velocity.

In this example, link 1 will be 100 mm long and ω_1 will be 100 rad/s clockwise. Examining links 1, 2, and 3 alone (Figure 3.38b), we draw the following vectors to some convenient scale:

Vector $\boldsymbol{ob} = \omega_1 O_1B = 10\ \text{m/s} \perp O_1B$ (downward and to the left)
Trial vector $\boldsymbol{oc} \perp O_3C$
Trial vector $\boldsymbol{bc} \perp BC$ (Figure 3.38c)

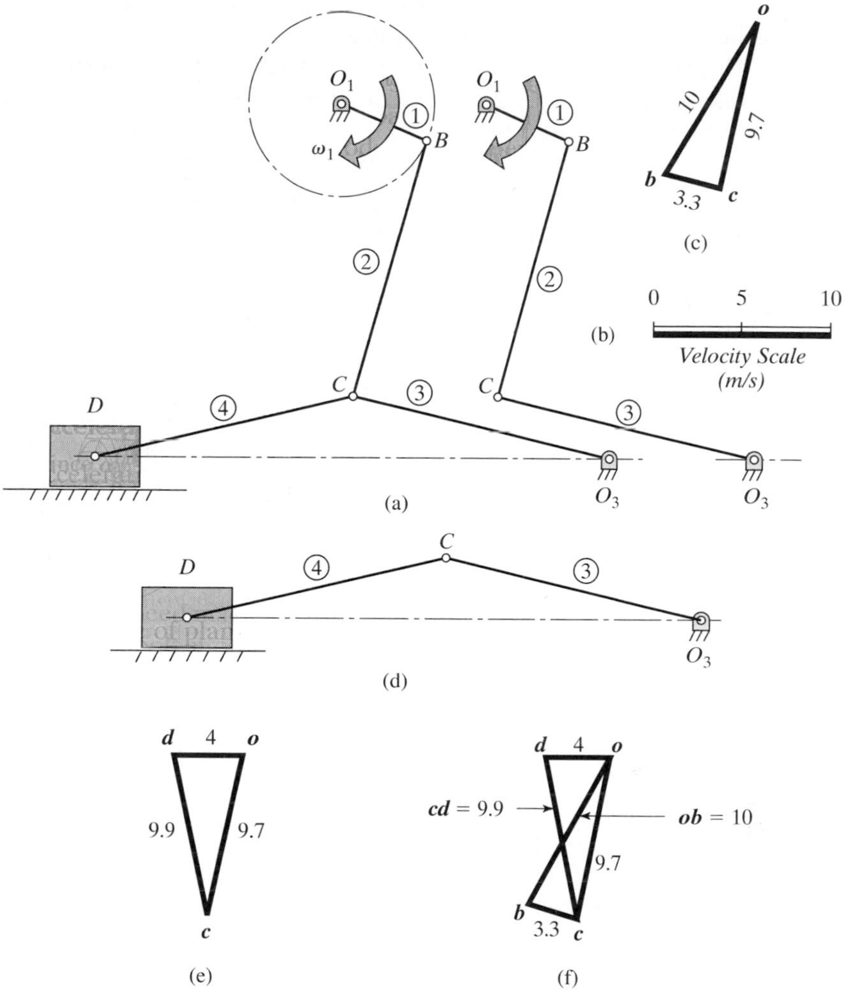

FIGURE 3.38 (a) This toggle mechanism is a combination of two basic linkages, the four-bar mechanism and the slider-crank mechanism. (b) The velocities for the entire toggle mechanism are found by solving for the velocities of the component mechanisms. Thus, the four-bar linkage component is considered first (link 4 is not shown). (c) After a suitable velocity scale is selected, the velocity polygon for the four-bar mechanism is drawn. (d) The remainder of the toggle mechanism, the slider crank, is considered next. (e) The velocity polygon for the slider crank is drawn. Note that link O_3C is common to both mechanisms. The velocity of point C, ***oc***, was already determined in the four-bar polygon. (f) The velocity polygon for the entire mechanism.

Velocity point ***c*** is located at the intersection of trial vectors ***oc*** and ***bc***, completing the velocity polygon for links 1, 2, and 3.

In order to illustrate the mental process involved in solving multilink mechanisms, we will now consider links 3 and 4 and the slider separately (Figure 3.38d). Beginning at a new pole point (***o*** in Figure 3.38e), vector ***oc*** is redrawn, its direction and magnitude taken from the four-bar polygon. Then, trial vectors are drawn, ***cd*** $\perp CD$ and ***od*** in the direction of the path of the slider, from which velocity point ***d*** is located. Slider velocity at this instant is given by ***od***, 4 m/s to the left.

The preceding solution, in two parts, is given for demonstration purposes only. After velocity polygon ***obc*** is completed, it is faster and slightly more accurate to continue by finding point ***d*** on the same polygon. Figure 3.38f shows velocity polygon ***obcd*** found in this manner, a compact velocity representation for the entire mechanism. An additional advantage of using the complete polygon is that it is only necessary to join points ***b*** and ***d*** to find the velocity of *D* with respect to *B* if this velocity is important in a particular machine.

An important feature of the toggle mechanism is its ability to produce high values of force at the slider with relatively low torque input. While the study of mechanisms is concerned primarily with motion, forces are of great importance to the designer and are intimately related to motion analysis. If a rigid mechanism has a single input and a single output with negligible losses, the rate of energy input equals the rate of energy output. Force ratios are the inverse of velocity ratios when inertia effects are negligible. Specifically, in the toggle linkage at the instant shown, the horizontal force at *D* divided by the tangential force at *B* equals *ob*/*od* , or 10/4 (the mechanical advantage of the mechanism at that instant). Clockwise rotation of link 1 from the position shown toward the limiting position of the mechanism produces very high ratios of output force to input torque.

If we sketch the mechanism at the instant that links 3 and 4 are collinear, we see that slider velocity ***od*** = 0. Thus, the ratio *ob*/*od*, the theoretical mechanical advantage of the toggle mechanism, becomes infinite. Clamps and ore crushers using the principle of the toggle linkage are designed to operate near this limiting position. Actual forces at the slider are, of course, finite, due to bearing clearances and elasticity of the linkage. In this exceptional case, the small amount of motion due to elastic deformation of the linkage and deformation of the workpiece is of the same order of magnitude as the slider motion. Therefore, any analysis of the problem (which assumes perfectly rigid links) must serve only as a first approximation.

Shaper Mechanisms

The *mechanical-shaper mechanism* in Figure 3.39 is another example of a combination of simple linkages. It is made up of a slider-crank mechanism (links 2 and 3 and the slider at *D*) and a sliding contact mechanism (links 1 and 2 and a slider at *B* moving along link 2). When the mechanism is operating, the angular velocity of link 1 is essentially constant.

Beginning with the velocity of point *B* at the end of the crank, we find the velocity of a common point on link 2. The velocity of point *C* at the end of link 2 is found by forming the velocity image of link 2. The solution is completed by examining the mechanism formed by links 2 and 3 and the slider at *D* (a slider crank of unusual proportions).

A detailed solution follows for the instant shown, with the angular velocity of link 1 equal to 10 rad/s clockwise.

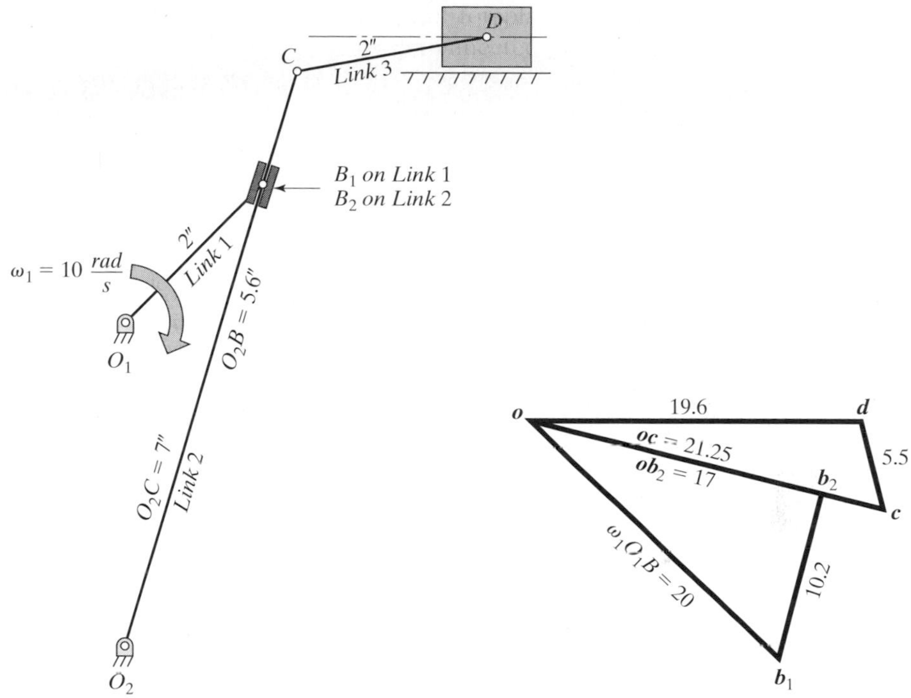

FIGURE 3.39 A mechanical-shaper mechanism is another combination of two simple mechanisms. The velocity diagram is constructed as usual. To find the magnitude of *oc*, however, we must use velocity ob_2 and the proportionality of link 2 (O_2C to O_2B_2).

STEP 1. Select a reasonable velocity scale and draw velocity vector $\boldsymbol{ob_1} = \omega_1 O_1 B_1 = 20$ in/s perpendicular to link 1 (downward and to the right), as shown in Figure 3.39.

STEP 2. Draw trial vector $\boldsymbol{ob_2}$ perpendicular to link 2 and trial vector $\boldsymbol{b_1 b_2}$ parallel to link 2, locating $\boldsymbol{b_2}$.

STEP 3. Use the proportionality equation $\boldsymbol{oc}/\boldsymbol{ob_2} = O_2C/O_2B_2$ to locate point $\boldsymbol{c}$ on the velocity polygon.

STEP 4. Draw trial vector ***od*** from ***o*** parallel to the path of the slider D and trial vector ***cd*** from ***c*** perpendicular to link CD, locating velocity point ***d***. The velocity of D (i.e., the velocity of the shaper tool) is 19.6 in/s to the right at the instant shown, as given by the scaled vector ***od***. When angular velocity ω_1 is constant and clockwise, the average velocity of point D is greater when D moves to the left than when it moves to the right. This feature of the shaper ensures a slow, powerful cutting stroke and a quick return. The stroke length (the distance between extreme positions of D) is varied by adjusting the length of link 1.

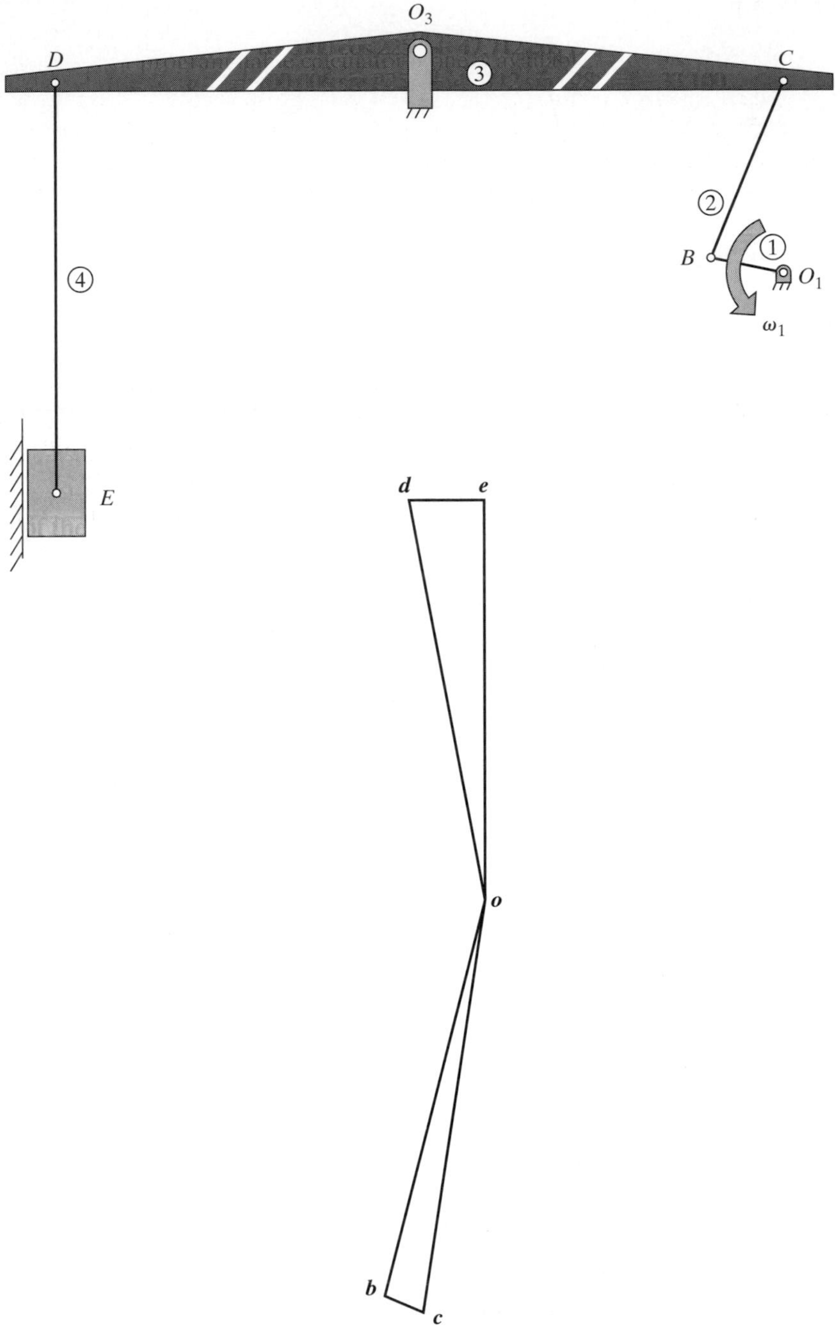

FIGURE 3.40 The beam pump. This mechanism may be analyzed by first solving the four-bar linkage (links 1, 2, 3, and the frame) and then drawing the velocity image of link 3. The slider-crank part of the mechanism (links 3 and 4 and the slider and frame) is then solved.

Beam Pump

The combination four-bar and slider-crank mechanism of Figure 3.40 forms a *beam pump*. The connecting rod (4) moves practically along its own axis and may be made very long. If we know the angular velocity of crank 1, it is a simple matter to construct velocity polygon ***obc***. Then velocity image ***cod*** is drawn similar to CO_3D on link 3. The velocity polygon is completed by locating point ***e***, corresponding to the slider. The slider velocity is then given by vector ***oe***.

Multiple Slider-Crank Mechanisms

Single-cylinder internal-combustion engines provide adequate power for lawn mowers, portable tools, and similar applications. The power output, however, is limited by the size of the cylinder. If we were to design a single-cylinder engine with a capacity of several hundred horsepower, the piston, connecting rod, and crank might be unreasonably large. At full speed, inertial forces could be a serious problem. Furthermore, the single power stroke per revolution in the case of a two-stroke-cycle engine (or one power stroke for each two revolutions in the case of a four-stroke-cycle engine) might cause unacceptable fluctuations in speed, even when a flywheel is used. A pump or an air compressor poses the same problems, particularly when high capacity or uniform output is called for. One solution is a design with several separate cylinders, which might be in-line (with parallel axes) or in a V arrangement (two separate banks of cylinders at an angle to one another).

The multicylinder high-pressure pump shown in front and end-section views in Figure 3.41 is an example of a mechanism combination, and Figure 3.42 is a sketch representing two cylinders of a V-block engine or pump. Ordinarily, all of the slider-crank linkages in a multicylinder engine are identical, except for their instantaneous link orientation. For example, we might examine piston velocity as a function of crank position for a single piston of an eight-cylinder engine. If the crankshaft speed is maintained, the results would apply equally to the other pistons, the only difference being the individual cylinder orientation and the phasing of the motion.

In Figures 3.41 and 3.42, each connecting rod is attached to a *separate* crankpin. There are, however, several variations of the crank–connecting-rod arrangement. Figure 3.43a shows an alternative configuration with both connecting rods (BC and BD) attached to a *single* crankpin, B. A practical example of this alternative arrangement is shown in the cutaway view of the two-stage, V-type compressor in Figure 3.43b. A similar arrangement is seen in the cutaway view of the four-cylinder semiradial compressor shown in Figure 3.43c, where the four connecting rods are again attached to a common crankpin.

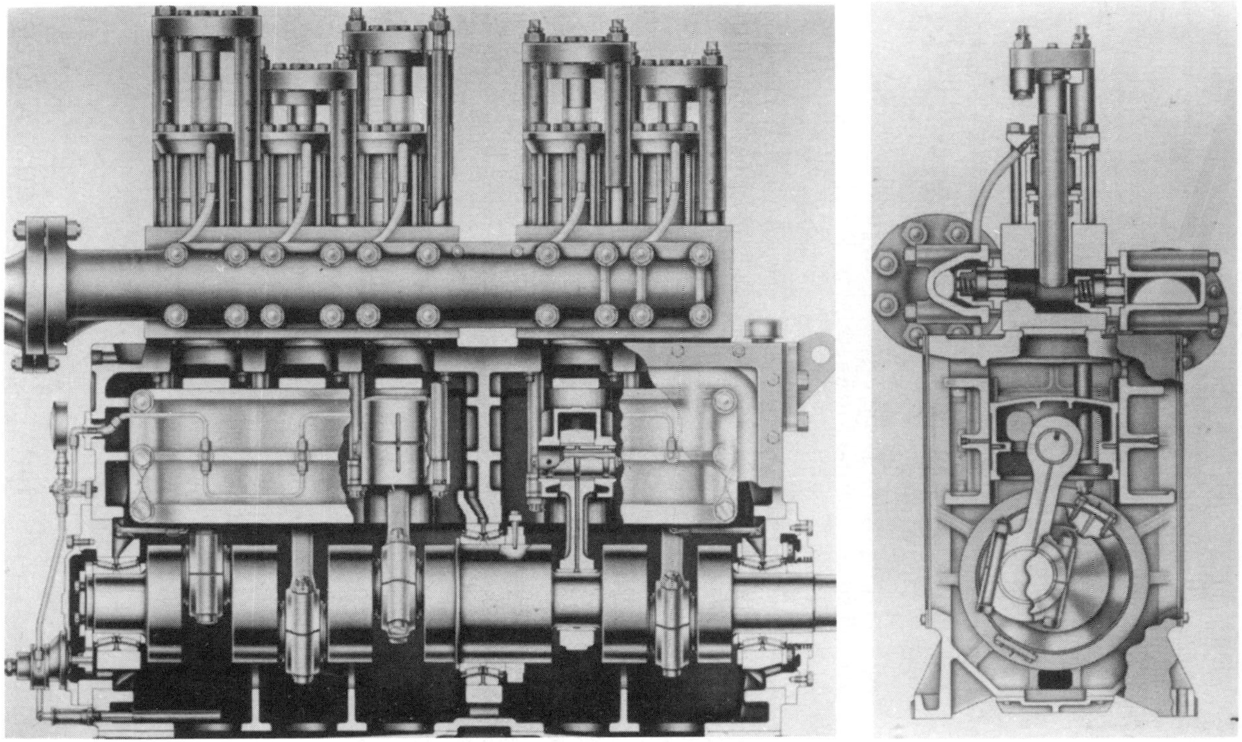

FIGURE 3.41 A high-pressure pump. The in-line slider-crank mechanism is basic to several types of machines. In this case, five slider-crank mechanisms are used to produce a continuous flow at up to several thousand pounds per square inch. At any instant, each linkage is at a different point in its stroke. As seen in the end view (*right*), the crank drives the connecting rod, which moves the crosshead. "Trombone" side rods connect the crossheads to the plungers, which enter the top of the cylinder. (*Source*: Cooper Energy Services.)

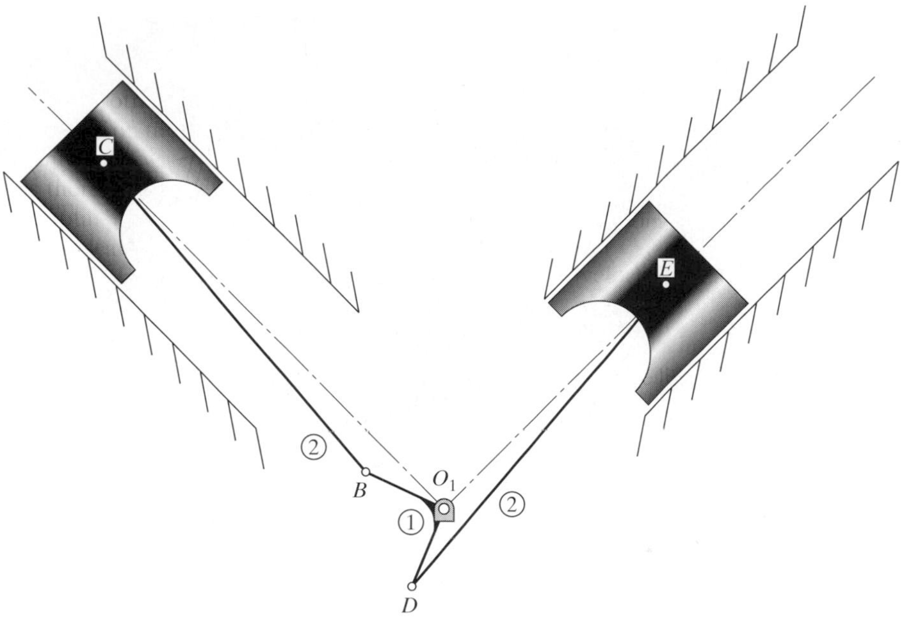

FIGURE 3.42 A typical multicylinder engine or compressor configuration. This V design shows two slider-crank linkages on a common crankshaft, but with separate crankpins at different axial locations along the shaft.

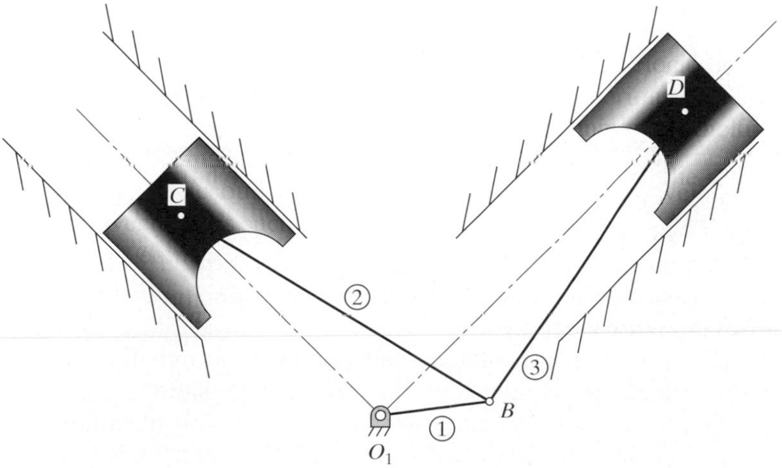

FIGURE 3.43 (a) Note that in this variation of the crank–connecting-rod arrangement, both connecting rods are attached to a single crankpin (at different axial locations).

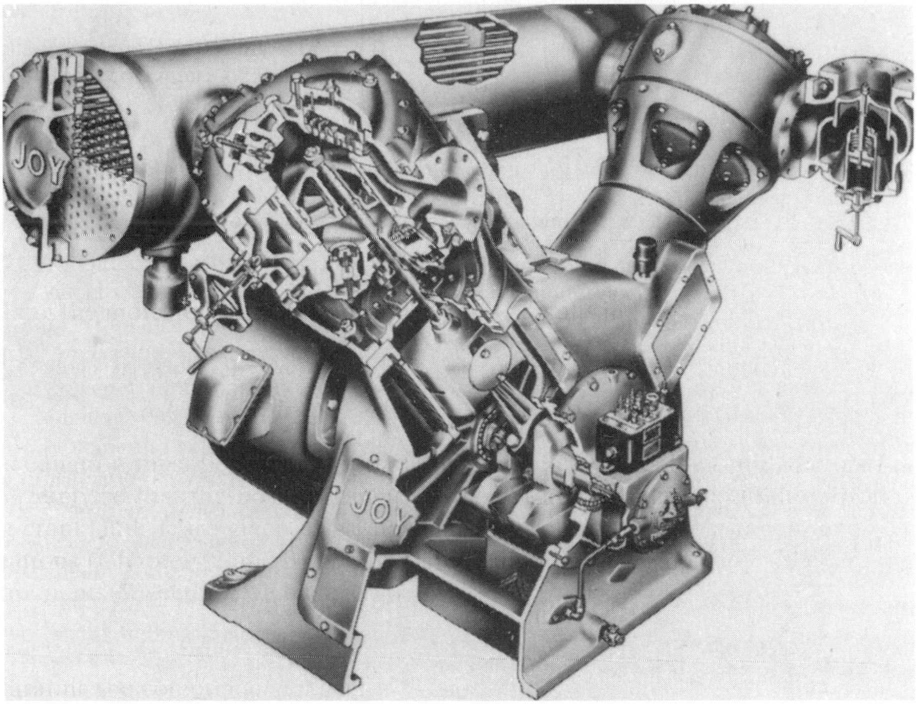

FIGURE 3.43 (b) This V type of compressor is a commercial application of the single crankpin arrangement. Both linkages are similar kinematically, except as regards their instantaneous position. This unit compresses air in two stages. In the first stage, the large-diameter piston compresses the air, which is then cooled and brought to a still higher pressure by the small (upper) piston in the second stage. (*Source:* Joy Manufacturing Company.)

Articulated Connecting Rods

When several cylinders are to be arranged radially in an engine or compressor, still another arrangement, the *articulated connecting rod*, may be used. The articulated connecting-rod linkage, sketched in Figure 3.44a, consists of a single crank pinned to a *master connecting rod*. The connecting rods of the remaining cylinders are in turn pinned (at different points) to the master connecting rod. A practical example is shown in Figure 3.44b.

The use of an articulated connecting rod permits the design of a multicylinder engine with all the cylinder centerlines in a single plane.

Graphical methods were used to analyze several multiloop mechanisms illustrated in this section. Analytical vector methods and complex-number methods may be used as well, particularly if many positions are to be considered. The key to solving these linkages is similar in each case. We begin by considering a single basic linkage, which is a component of the more complex system, and then proceed through a common point to solve the next component of the system.

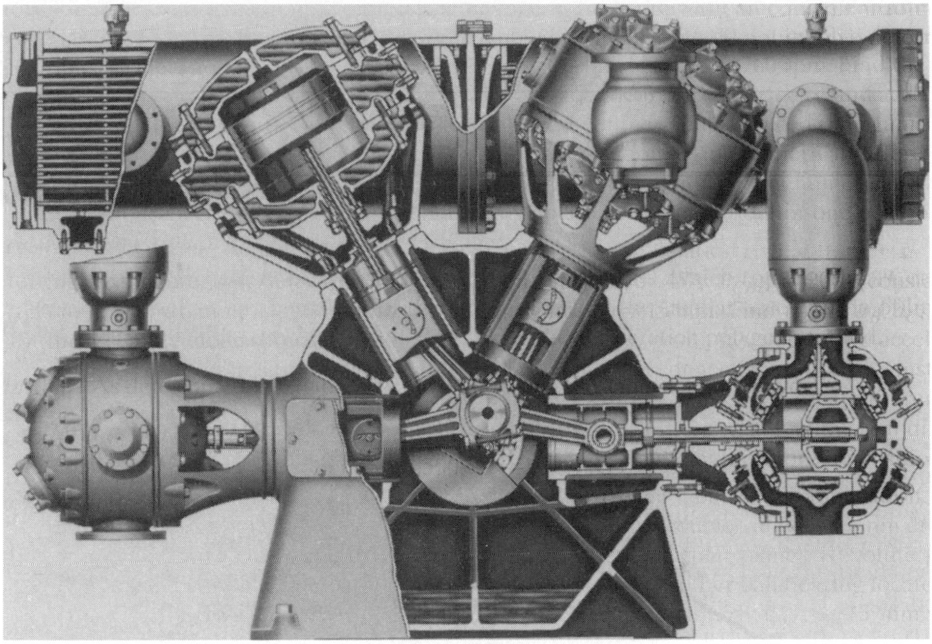

FIGURE 3.43 (c) This *semiradial compressor* also employs a single crankpin. (*Source:* Joy Manufacturing Company.)

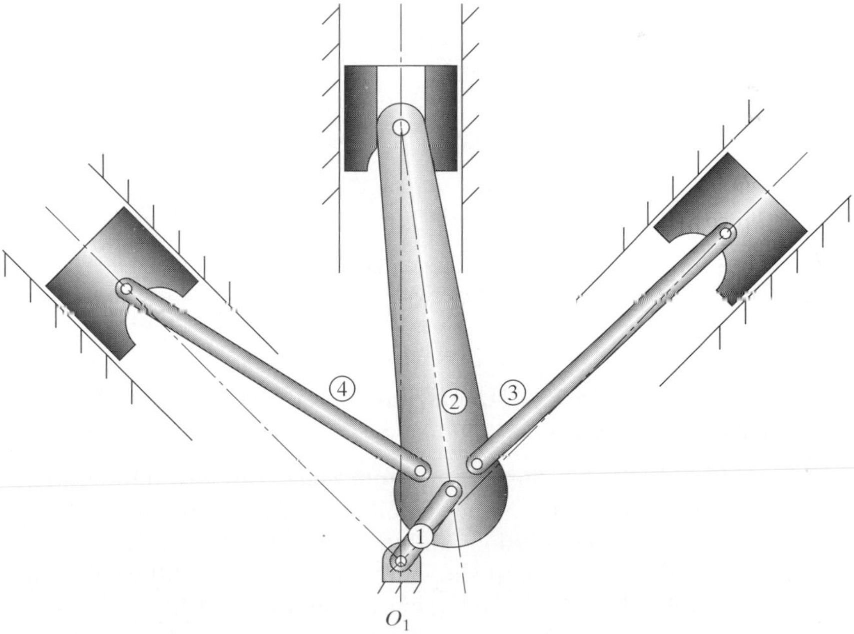

FIGURE 3.44 (a) The *articulated* connecting rod. In this crank–connecting-rod arrangement, the crank is pinned to a master connecting rod. The rest of the connecting rods are, in turn, pinned to the master rod.

FIGURE 3.44 (b) Commercial application of an articulated connecting rod. The centerlines of the cylinder compressor are arranged radially in a single plane perpendicular to the crankshaft axis for better balance. (*Source:* Worthington Group, McGraw-Edison Company.)

3.13 ANALYZING LINKAGES THROUGH TRIAL SOLUTIONS AND INVERSE METHODS

This section treats mechanisms that cannot be easily solved by straightforward graphical methods. If you are using the velocity polygon method to verify an analytical solution, you may find trial solutions and inverse methods unnecessary.

A variable-stroke pump, shown in Figure 3.45a, is difficult to analyze by straightforward graphical methods. Referring to Figure 3.45b, link 4 is an equivalent link; in the actual pump, point D represents the pin in the guide block riding in a curved track (of radius O_4D). The stroke length (the length of the path of E) may be varied from zero to a maximum value by tilting the curved track, which is equivalent to changing

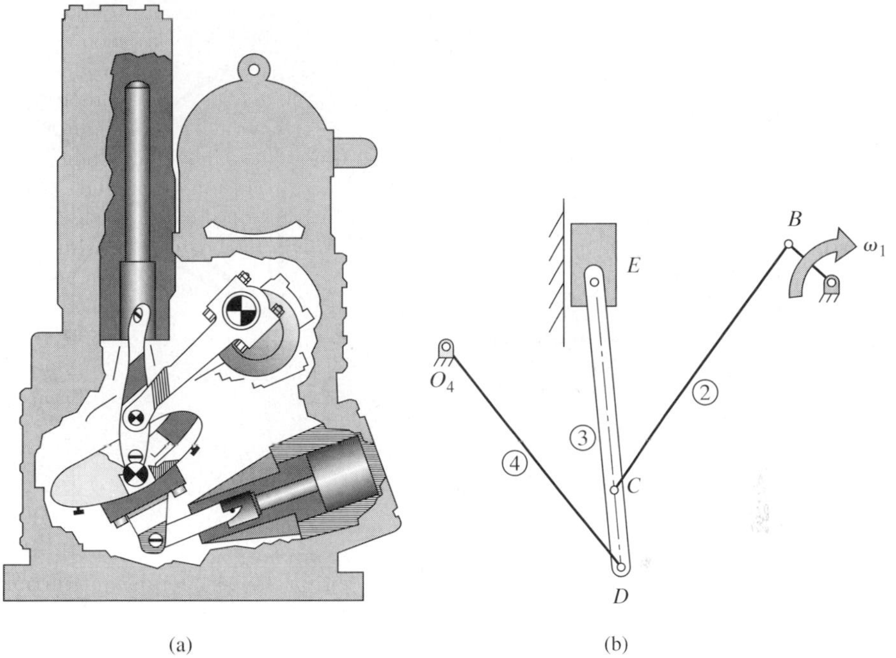

FIGURE 3.45 (a) A variable-stroke pump. This pump is identical to the pump shown in Chapter 1. Here, the adjustment cylinder is set so that the stroke transformer provides a maximum plunger stroke. (*Source:* Ingersoll-Rand Company.) (b) The equivalent linkage for the variable-stroke pump.

the location of point O_4. The control mechanism may be manually operated or may incorporate an air-operated plunger to rotate the track automatically in response to a remote signal.

When the position of the curved track is set, the mechanism has one degree of freedom. If the motion of one link is specified, it should be possible to describe the motion of the entire linkage.

Trial Solution Method

Suppose it is necessary to find the velocity of the piston of the mechanism in Figures 3.45a and b when the velocity of crankpin B is 100 in/s with crank O_1B rotating clockwise. A *trial solution* procedure follows.

STEP 1. Select a convenient velocity scale, and draw $\boldsymbol{ob}$ = 100 in/s perpendicular to O_1B, as in Figure 3.45c.

STEP 2. It is usually best to indicate all known vectors and vector directions on the sketch. Accordingly, draw the following trial vectors:

od perpendicular to O_4D
oe in the direction of sliding (vertical)
bc perpendicular to BC (***bc*** is added to ***ob***)

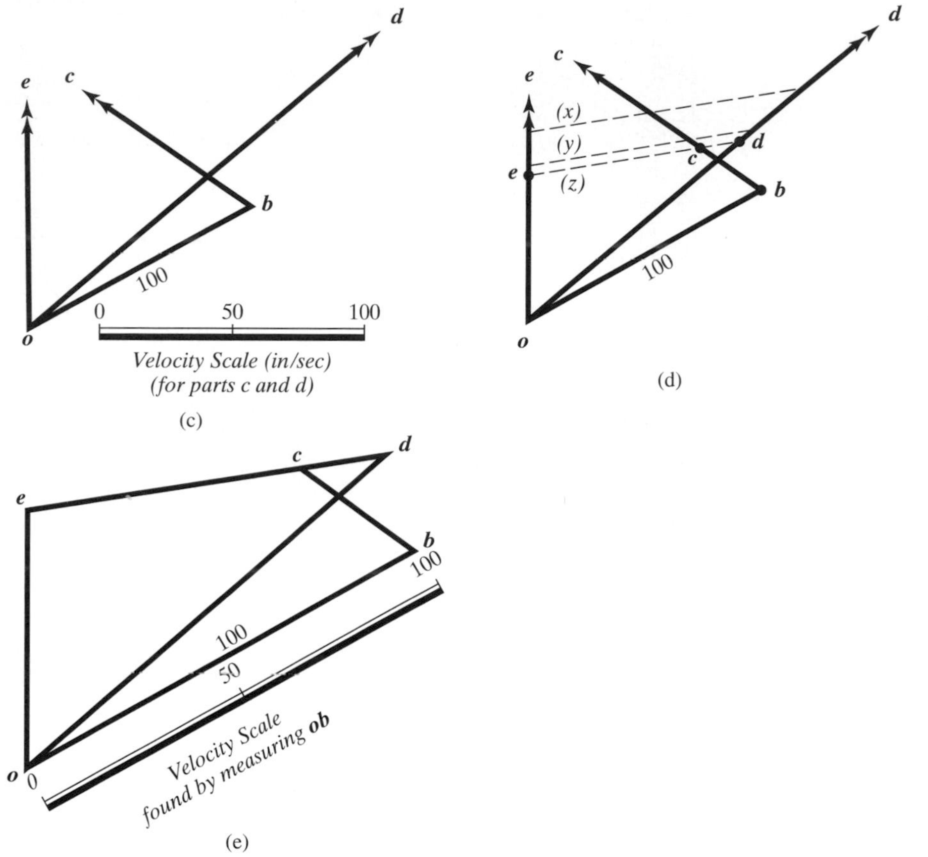

FIGURE 3.45 (c) After a velocity scale is selected, the velocity polygon is begun. However, the location of the velocity image *dce* is not immediately obvious. (d) Trial solution method. (e) Inverse method.

STEP 3. In a straightforward problem, we would continue the polygon by finding another velocity point, say, point ***c***. The following vector equations might be used:

$$\boldsymbol{oc} = \boldsymbol{ob} + \boldsymbol{bc} \quad \text{and} \quad \boldsymbol{oc} = \boldsymbol{oe} + \boldsymbol{ec}.$$

But the magnitudes of ***bc***, ***oe***, and ***ec*** are unknown, and the magnitude and direction of ***oc*** are unknown; there are too many unknowns to utilize the equations. A similar problem exists with the vector equation $\boldsymbol{oc} = \boldsymbol{od} + \boldsymbol{dc}$, since the magnitudes of ***od*** and ***dc*** are also unknown.

STEP 4. The velocity image relationship is the missing tool; with it, a solution is possible. On the velocity polygon, ***dce*** is the (straight-line) image of *DCE*, three points on rigid link 3. The relationship is used by noting that

d, ***c***, and ***e*** lie on a line perpendicular to line DCE with ***c*** between ***d*** and ***e***, proportioning the line by the equation $dc/de = DC/DE$. Now that we have the direction and relative proportions of line ***dce***, we may satisfy the velocity image condition by trial and error. Line x in Figure 3.45d is the first approximation. This line intercepts trial vectors ***od***, ***bc***, and ***oe*** in such a way that the ratio dc/de would be too large. We can see that the ratio is decreased as the trial line is moved downward. Line y, the second approximation, is somewhat better, and line z, the third approximation, closely satisfies the required relationship, $dc/de = DC/DE$. Line z thus completes the polygon, and velocity points ***d***, ***c***, and ***e*** are located where trial vectors ***od***, ***bc***, and ***oe*** intercept the line. The piston velocity is given by vector ***oe***.

Trial solutions are required more frequently in engineering practice than is indicated by typical academic assignments. In the academic situation, a shortage of time favors the use of problems in which the answer is obtained directly. The reader should, however, be prepared for both types of problem.

Inverse Method

The piston velocity of the variable-speed pump may also be found by a method that avoids the inconvenience and potential error involved in making several approximations. The reader may have observed in Figure 3.45c that, had the piston velocity been given instead of the crankpin velocity, we could draw the velocity polygon directly. The solution would proceed from the slider-crank mechanism, to the velocities of E and D, to the velocity of C by proportion, and thence to the velocity of B. Let us, then, solve the problem by an *inverse method* (i.e., we assume the answer at the beginning). The following steps constitute an entirely new solution to the problem without making use of the trial solution:

STEP 1. Represent piston velocity ***oe*** by a vertical vector of arbitrary length (Figure 3.45e). Of course, we cannot give ***oe*** an actual velocity or select a scale since we would undoubtedly guess wrong unless we had already solved the problem by another method.

STEP 2. Draw trial vectors ***od*** perpendicular to O_4D and ***de*** perpendicular to DE, to locate velocity point ***d***. Velocity point ***c*** is located between ***d*** and ***e*** on line ***de*** by using the proportion $dc/de = DC/DE$. Trial vectors ***cb*** perpendicular to CB and ***ob*** perpendicular to O_1B locate velocity point ***b***, completing the polygon.

STEP 3. Finally, the scale of the velocity polygon must be determined from the given data. In this problem, the velocity of the crankpin ***ob*** = 100 in/s. The length representing ***ob*** on the velocity polygon becomes 100 in/s, and all other vectors are scaled accordingly and labeled with their correct velocities. Measurements taken directly from the velocity polygon give $oe/ob = 0.58$, from which piston velocity ***oe*** $= 0.58 \times 100 = 58$ in/s upward at the instant depicted.

3.14 CENTROS

Traditionally, the centro method has been of considerable interest. The method consists of locating a point, the **centro**, that has the same (vector) velocity in two links. The centro is then used to relate unknown velocities to known velocities.

The centro method does not lead directly to acceleration analysis. Thus, while the centro method can provide some useful insight into mechanism velocities, it is generally not the best method of analysis. However, the centro method can add to our understanding of mechanisms.

The pin connecting two links in a mechanism is obviously the centro of the two, since it has the same velocity in both. (See Figure 3.46a.) If two bodies (two disks, for instance) roll on one another without sliding, then the instantaneous point of contact is the centro. The centros just mentioned are *observed centros* and are to be located and labeled before any others are found by construction. The centro label will consist of the two link numbers, with the smaller number written first. The point having the same velocity in both link 0 and link 1, for example, will be labeled centro 01. The label 10 would apply to the same centro, but we will avoid duplication by always labeling this centro 01.

Since each pair of links has a centro, three links, which can form three different pairs, will have three centros. In general, given sufficient information, we should be able to find $n(n - 1)/2$ centros for a linkage with n links. In the three-link case, links 0, 1, and 2 form three centros, which we will label 01, 02, and 12. Figure 3.46b shows three links, including 0, the frame. Thus, we have centros 01 (where link 1 is pinned to the frame), 02 (where link 2 is pinned to the frame), and 12 (which we will attempt to locate).

An arbitrary point B that does not lie on (line) $\overline{01 \quad 02}$ is selected as a possible location for centro 12. The bar on $\overline{01 \quad 02}$ will be used to represent a line in the mathematical sense (extending infinitely on both sides of the line segment between 01 and 02).

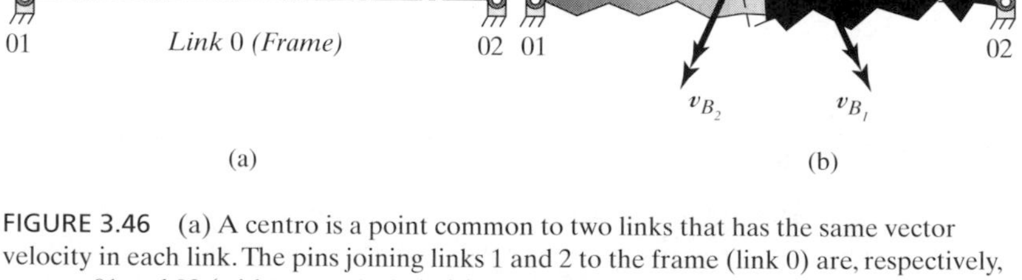

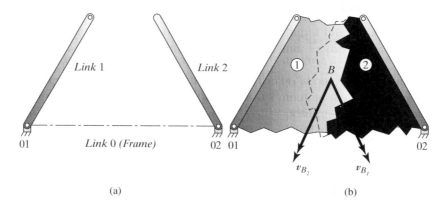

FIGURE 3.46 (a) A centro is a point common to two links that has the same vector velocity in each link. The pins joining links 1 and 2 to the frame (link 0) are, respectively, centros 01 and 02 (with zero velocity). (b) Three links have three centros, all of which lie on a straight line. To prove this theorem, let us assume that B, which is common to extended links 1 and 2, is the third centro. Vectors $\mathbf{v}_{B_1}$ and $\mathbf{v}_{B_2}$ violate the requirement for a centro that the velocity vector must be *identical* for both links. To meet this basic requirement, centro 12 cannot occur anywhere except on $\overline{01 \quad 02}$.

Let us imagine an extension of links 1 and 2 so that B may lie in both. Then the velocity of B in link 1, $\mathbf{v}_{B_1}$, is perpendicular to $\overline{01\,B}$. Likewise, the velocity of B in link 2, $\mathbf{v}_{B_2}$, is perpendicular to $\overline{02\,B}$. (The magnitudes of these velocities are unknown.) Since $\overline{01\,B}$ and $\overline{02\,B}$ are neither parallel nor collinear, $\mathbf{v}_{B_1}$ and $\mathbf{v}_{B_2}$ cannot be the same; the directions of $\mathbf{v}_{B_1}$ and $\mathbf{v}_{B_2}$ are different. Certainly, then, B is not the centro 12.

Kennedy's Theorem

We can see from the preceding discussion that centro 12 cannot be any point that does *not* lie on $\overline{01 \quad 02}$. A similar examination of the direction of the velocity of points lying on $\overline{01 \quad 02}$ shows that centro 12 may lie somewhere on $\overline{01 \quad 02}$. Considering the infinite extent of $\overline{01 \quad 02}$, some point on it must have the same velocity (both magnitude and direction) in both link 1 and link 2. Thus, we have Kennedy's theorem (the three-link theorem) in a nutshell: *Three links have three centros that lie on a line.*

Kennedy's theorem applies except in trivial cases (e.g., a linkage without any relative motion), but it does not assure us of finding all the centros that exist in a theoretical sense. As for the original problem of actually locating centro 12, we have failed because there is a need for additional data.

Centros of a Four-Bar Linkage

The mechanism of Figure 3.47a has four links, including the frame. There are six possible pairs formed by four numbers and, hence, six centros: the observed centros 01, 12, 23, and 03 and the two remaining centros, 02 and 13, which must be located by construction. A procedure for finding the unobserved centros is as follows:

STEP 1. The three-link theorem will be used to draw a line on which centro 02 must lie. For the necessary three links, we must include links 0 and 2, since we are looking for centro 02; then, either of the remaining links will do. Using links 0, 2, and 1, we have centros 01 and 12 (both already labeled) and 02 (the unknown), all three on a line. Line $\overline{01 \quad 12}$ (of arbitrary length) is drawn and labeled 02 for our unknown centro.

STEP 2. The three-link theorem is used again to draw another line to locate centro 02. Links 0 and 2 must again be included, this time along with the other remaining link, link 3. Links 0, 2, and 3 have three centros, 03 and 23 (already labeled) and the still unknown centro 02. Line $\overline{03 \quad 23}$ is drawn until it reaches the extension of $\overline{01 \quad 12}$, at which point centro 02 is located.

STEP 3. Centro 02 is a point in link 0, the frame, and in link 2, the coupler. By the definition of a centro, centro 02 has the same velocity in both links. While it is not actually a part of either, it is considered to be in the links for velocity analysis. Link 0 is fixed, and thus, the point common to link 0 and link 2, centro 02, is a point in link 2, which is (instantaneously) stationary.

STEP 4. Since link 2 rotates about centro 02, as in Figure 3.47a, the velocities of points on link 2 are proportional to their distances from 02. We may then

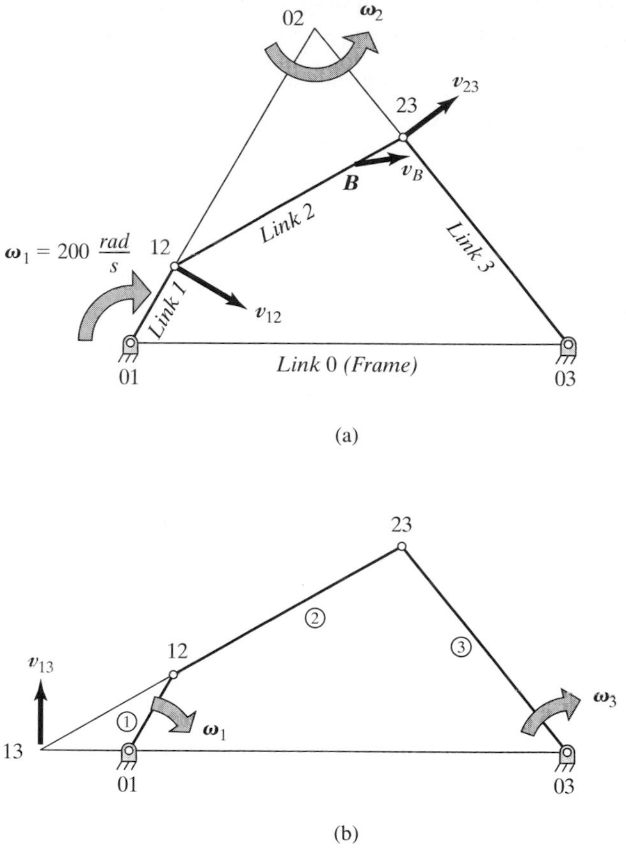

FIGURE 3.47 (a) Locating unknown centro 02 using the three-link (Kennedy's) theorem. Centro 02 is determined by the intersection of the extensions of $\overline{01\ \ 12}$ and $\overline{03\ \ 23}$. (b) Unknown centro 13 is determined by the intersection of $\overline{01\ \ 03}$ and $\overline{12\ \ 23}$.

state the ratio of the magnitudes of the velocities of points 23 and 12 as

$$\frac{v_{23}}{v_{12}} = \frac{02\text{–}23}{02\text{–}12}, \tag{3.57}$$

where 02–23 and 02–12 are distances scaled from the linkage drawing. The magnitude of velocity $\mathbf{v}_{12}$ is given by the length of link 1 times ω_1. Using this, we find the magnitude of velocity $\mathbf{v}_{23}$ by Eq. (3.57). If ω_1 is clockwise, then $\mathbf{v}_{12}$ is to the right and downward. Observing the location of 02, the instantaneous center of link 2, we see that the link must rotate counterclockwise at the instant in equation (since $\mathbf{v}_{12}$ is to the right and downward). Velocity $\mathbf{v}_{23}$, then, is to the right and upward (perpendicular to $\overline{03\ \ 23}$.)

STEP 5. The magnitude of the angular velocity of link 2 is given by

$$\omega_2 = \frac{v_{12}}{02\ -\ 12},$$

where 02–12 is the *actual distance* from point 12 to centro 02 of link 2 for the full-size mechanism. The velocity of an arbitrary point B on link 2 of Figure 3.47a is given by

$$\frac{v_B}{v_{12}} = \frac{02-B}{02-12}, \tag{3.58}$$

where lengths 02—B and 02–12 are scaled from the diagram.

Velocity $\mathbf{v}_B$ is perpendicular to 02–B and is to the right and upward, with its sense determined in the same manner as that of $\mathbf{v}_{23}$.

STEP 6. The angular velocity of link 3 follows immediately from the foregoing calculations, but we will use centro 13 to complete the problem in order to complete our illustration of the centro method. Centro 13 is located in a manner similar to the procedure for locating centro 02. In this case, lines $\overline{03\quad 23}$ and $\overline{12\quad 23}$ both contain centro 13. Now, 13 is a centro in the general sense, having the same nonzero velocity in links 1 and 3. Using that property, we have

$$v_{13} = \omega_1\ 01-13 \quad \text{and} \quad v_{13} = \omega_3\ 03-13.$$

Equating these two expressions, we obtain

$$\omega_3(03-13) = \omega_1(01-13) \quad \text{or} \quad \frac{\omega_3}{\omega_1} = \frac{01-13}{03-13}, \tag{3.59}$$

which might be expressed in words as follows: *For two links with fixed pivots, the angular velocities of the two links are inversely proportional to distances from the respective fixed pivots to the common centro.* In Figure 3.47b, we see that v_{13} is upward when ω_1 is clockwise. But then ω_3 is seen to be clockwise also. In general, when the common centro falls between the fixed centers of a pair of links, one link turns clockwise and the other counterclockwise; otherwise, both turn clockwise or both turn counterclockwise. The reader will observe that the pitch point for a pair of gears and the tangent point for a pair of friction disks represent the common centro. In these examples, the previous expression boils down to this: *Angular velocities are inversely proportional to radii.*

Referring to the skeleton diagram of a planar four-bar linkage, if links 0 and 1 are collinear, then centros 02 and 03 are coincident. Note that links 2 and 3 are joined at centro 23. As a result, the angular velocities of links 2 and 3 are equal at the instant that links 0 and 1 are collinear. This condition serves as a partial check of the angular velocity plots shown earlier in the chapter.

Analyzing a Slider-Crank Mechanism

The slider-crank mechanism in Figure 3.48 is solved by first examining the equivalent linkage shown in the figure. One readily sees that the slider may be replaced by a link of infinite length perpendicular to the slider path. The slider moves in a horizontal path.

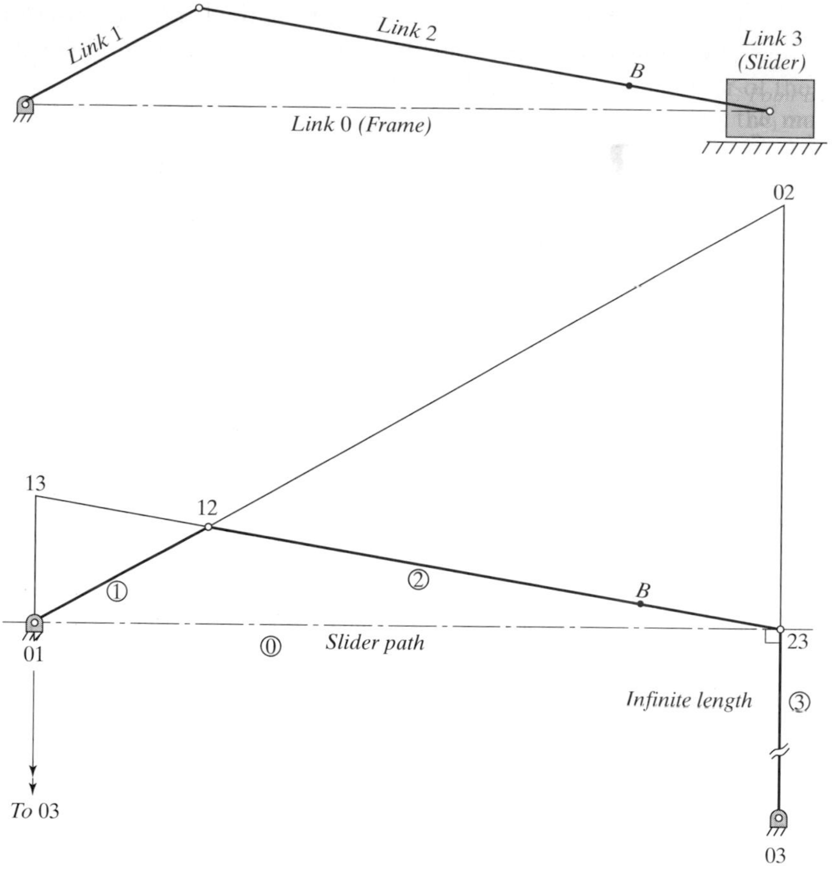

FIGURE 3.48 For the use of centros to find velocities in a slider-crank mechanism, an equivalent linkage must be used, in which the slider is replaced by a link of infinite length. The solution then proceeds as for a four-bar linkage.

Its motion can be duplicated by an equivalent link, link 3, which is vertical. Centro 03, the "fixed center" of the equivalent link, is shown an infinite distance below the slider path.

SUMMARY

Animation software, analytical vector methods, and complex-number methods are important tools in the velocity analysis of planar linkages. The velocity polygon is a useful graphical check of analytical work. Analytical vector methods and matrix methods are used to solve spatial linkages. Complex-number methods are not applicable to spatial linkages, and graphical methods are impractical for the detailed analysis of spatial linkages. Designing mechanisms often requires analysis through a full range of motion, a task that calls for animation software or mathematics software.

Computer-generated plots and tables are very convincing. But our work is not finished until we check and interpret those plots and tables. Try to include simple validity checks within computer programs. Add comments to help identify valid solutions. Include checks to identify design parameters outside of generally accepted ranges and designs that do not meet motion requirements. And, finally, be sure to indicate units and test for consistency.

A Few Review Items

- What is the direction of the angular velocity vectors in a planar mechanism? Must we express angular velocity in radians per second?
- Write the basic vector velocity equation for a planar four-bar linkage.
- .... a slider-crank linkage.
- .... a sliding contact linkage.
- Explain the differences in the three equations you have written down.
- Write the basic velocity equation, in complex polar form, for a planar four-bar linkage.
- Repeat the preceding for a slider-crank linkage.
- .... a sliding contact linkage.
- Why not use this form for spatial linkages?
- Write the basic vector velocity equation for an *RSSR* spatial linkage.
- Identify the unknown angular-velocity components if the drive crank angular velocity is given.
- Do we need an additional equation to solve for these unknowns?
- Give the form of the matrix equation needed to solve for the unknown angular velocity components.
- Write the velocity polygon equation for a slider-crank linkage. The crankpin is identified by B and the wrist pin by C.
- Describe the velocity image principle for three points that do not lie on a line.
- ..... for three points that lie on a line.
- Can you combine these two cases in a single principle?

PROBLEMS

Some of the problems in this chapter require calculating and plotting results for many linkage positions. It is suggested that animation software, mathematics software, or a spreadsheet be used. If the problems are solved only with the aid of a calculator, one or two linkage positions may be selected to avoid repetitious calculations.

3.1 Find the velocity of a point P in a rigid body with angular velocity $\boldsymbol{\omega} = \boldsymbol{i}10 + \boldsymbol{j}15 + \boldsymbol{k}20$. The body rotates about fixed point O_1, and the radius vector $\boldsymbol{O_1P}$ is given by $\boldsymbol{r} = -\boldsymbol{i}100 + \boldsymbol{j}50 + \boldsymbol{k}60$.

3.2 Repeat Problem 3.1, except that $\boldsymbol{\omega} = \boldsymbol{i}200 + \boldsymbol{j}0 - \boldsymbol{k}150$.

3.3 Find the velocity of a point on the circumference of a 30-in-diameter flywheel rotating at 800 rev/min.

3.4 Low-carbon steel is turned on a lathe at typical surface speeds of 100 to 400 ft/min. Find the corresponding lathe spindle speeds in revolutions per minute for **(a)** a 2-in-diameter bar and **(b)** a 3-in-diameter bar.

3.5 Write an equation to determine lathe spindle speed n (in revolutions per minute) when bar diameter d (in inches) and surface speed s (in feet per minute) are given. Is the same equation valid if n represents the speed of a milling cutter of diameter d?

3.6 A point P is described in terms of a fixed coordinate system XYZ with unit vectors $\boldsymbol{I}, \boldsymbol{J}$, and $\boldsymbol{K}$ and a moving coordinate system xyz with unit vectors $\boldsymbol{i}, \boldsymbol{j}$, and $\boldsymbol{k}$. At a given instant, the location of the origin of the moving system is $12\boldsymbol{I} + 5\boldsymbol{J}$, and the velocity of the origin of the moving system is $80\boldsymbol{I} - 90\boldsymbol{J}$. The velocity of P relative to the moving system is $50\boldsymbol{i} + 45\boldsymbol{j}$; the radius of point P is $3\boldsymbol{i} + 4\boldsymbol{j}$ in the moving system, which rotates at angular velocity $\boldsymbol{\omega} = 100\boldsymbol{k}$. Find the velocity of point P in the fixed system if the x-axis is rotated 30° counterclockwise from the X-axis.

3.7 Repeat Problem 3.6, except that $\boldsymbol{\omega} = -95\boldsymbol{k}$.

3.8 A 500-mm-diameter wheel rolls in a straight path, rotating at 800 rev/min. Find relative velocity $\boldsymbol{v}_{CB}$, where C is the center of the wheel and B is at the top.

3.9 Repeat Problem 3.8 for a 16-in-diameter wheel.

3.10 Velocity $\boldsymbol{v}_B = 15$ in/s at 45°. Velocity $\boldsymbol{v}_{CB} = 20$ in/s at 0°. Find $\boldsymbol{v}_C$. (Use $v_C = v_B + v_{CB}$.)

3.11 Velocity $\boldsymbol{v}_C = 30$ in/s at 0°. Relative velocity $\boldsymbol{v}_{CB} = 10$ in/s at 135°. Find $\boldsymbol{v}_B$. [Use $\boldsymbol{v}_B = \boldsymbol{v}_C + (-\boldsymbol{v}_{CB})$.]

3.12 Velocity $\boldsymbol{v}_B = 20$ in/s at 45°. Relative velocity $\boldsymbol{v}_{CB}$ is an unknown vector at 315°. Velocity v_C is an unknown vector at 0°. Find $\boldsymbol{v}_{CB}$ and $\boldsymbol{v}_C$. (Use $\boldsymbol{v}_C = \boldsymbol{v}_B + \boldsymbol{v}_{CB}$.)

3.13 Referring to Figure 3.9, let $r_0 = 30$ mm, $r_1 = 10$ mm, $r_2 = 35$ mm, $r_3 = 20$ mm, and $\theta_1 = 45°$. Calculate θ_2, θ_3, $\boldsymbol{v}_C$, ω_2, and ω_3 for $\omega_1 = 30$ rad/s ccw. Use analytical vector methods.

3.14 Repeat Problem 3.13 for $\theta_1 = 60°$.

3.15 Plot ω_3/ω_1 versus θ_1 for the linkage of Figure 3.9 if $r_0/r_1 = 3$, $r_2/r_1 = 3.5$, and $r_3/r_1 = 2$. Use a computer or a programmable calculator.

3.16 Referring to Figure 3.9, let $r_0 = 30$ mm, $r_1 = 10$ mm, $r_2 = 35$ mm, $r_3 = 20$ mm, $r_{BD} = 15$ mm, $\theta_{CBD} = 20°$, and $\theta_1 = 30°$. Determine $v_D/(\omega_1 r_1)$. Use analytical vector methods.

3.17 Repeat Example Problem 3.7, but use analytical vector methods.

3.18 Refer to Figure 3.15. Find ω_2/ω_1 by using analytical vector methods. Write a computer or calculator program to determine ω_2/ω_1 for a series of values of θ_1, where ω_1 is constant and counterclockwise.

3.19 Refer to Figure 3.15. Let $R_1 = 30$ mm, $O_1O_2 = 50$ mm, $\omega_1 = 10$ rad/s ccw, and $\theta_1 = 15°$. Find **(a)** O_2B, **(b)** θ_2, and **(c)** ω_2. Use complex-number methods.

3.20 Repeat Problem 3.19 for $\theta_1 = 30°$.

3.21 Repeat Problem 3.19 for $\theta_1 = 45°$.

3.22 Repeat Problem 3.13, but use complex-number methods.

3.23 Refer to Figure 3.15. Calculate ω_2/ω_1 for $\theta_1 = 0$ to 180° in 15° steps if $R_0/R_1 = 3$. Use a computer or a programmable calculator.

3.24 Refer to Figure P3.1, which shows an $RSSR$ spatial linkage. With R_1 as the origin, the x-, y-, and z-coordinates of the joints are given, respectively, as follows:

R_1: 0, 0, 0 $\quad$ S_1: − 25, 0, 35

R_2: − 20, 95, 0 $\quad$ S_2: − 20, 80, 40

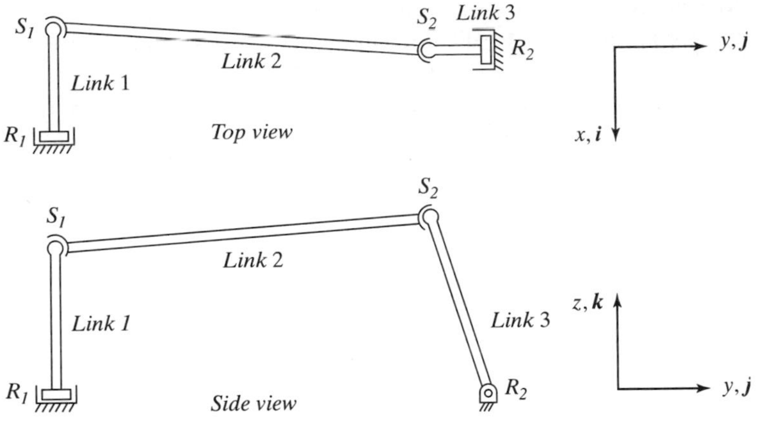

FIGURE P3.1 *RSSR* spatial linkage.

(all in millimeters). Link 1 rotates at $\omega_1 = 25$ rad/s (constant) in the xz-plane. Sphere joint S_1 is moving away from the observer at this instant. Link 3 rotates in the yz-plane. Find velocities v_{S_1} and v_{S_2} and angular velocity ω_3. Set the angular velocity of link 2 about its own axis to 0.

3.25 Repeat Problem 3.24, except that $\omega_3 = 10$ rad/s cw (constant). Find $\mathbf{v}_{S_1}$, $\mathbf{v}_{S_2}$, and ω_1.

3.26 **(a)** Find the crank position corresponding to the maximum piston velocity for an in-line slider-crank mechanism. Crank speed ω is constant, and the ratio of the connecting rod to the crank length is $L/R = 2$.

(b) Find the maximum piston velocity in terms of R and ω.

3.27 Repeat Problem 3.26 for $L/R = 1.5$.

In Problems 3.28 through 3.61,

(a) *write the appropriate vector equation,*

(b) *solve the equation graphically unless directed otherwise, using velocity polygon notation,*

(c) *dimension all vectors of the velocity polygon,* and

(d) *express angular velocities in radians per second, and indicate their directions.*

3.28 In Figure P3.2, $\theta = 45°$ and $\omega_1 = 10$ rad/s. Draw and dimension the velocity polygon. Find ω_2.

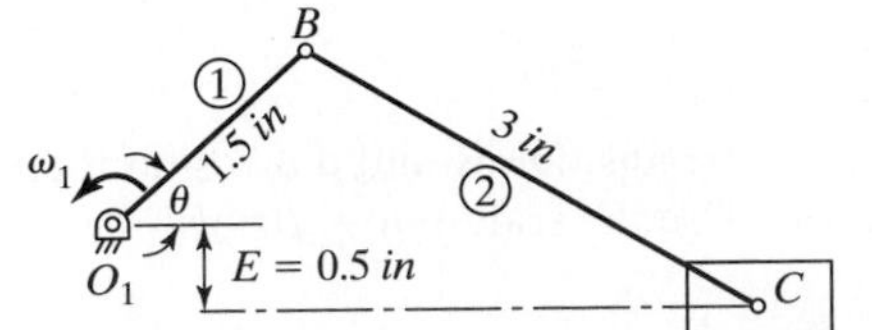

FIGURE P3.2

3.29 Repeat Problem 3.28 for $\theta = 120°$.

3.30 Repeat Problem 3.28 for the mechanism in the limiting position (with C to the extreme right).

3.31 In Figure P3.3, $\omega_1 = 100$ rad/s. Draw and dimension the velocity polygon. Find v_D and ω_2. Use the scale 1 in = 100 in/s.

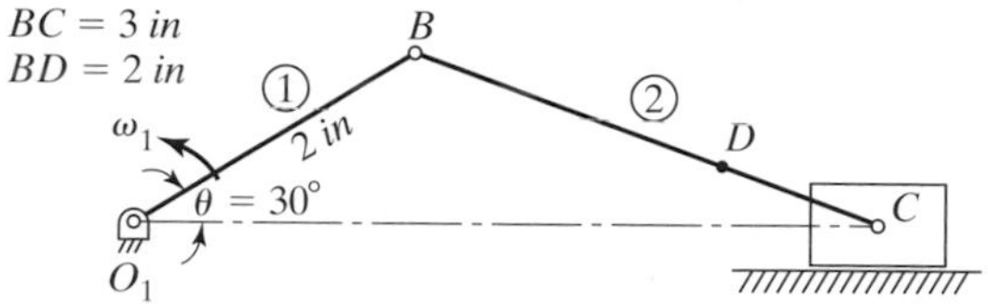

FIGURE P3.3

3.32 In Figure P3.4, $\omega_1 = 50$ rad/s. Draw and dimension the velocity polygon. Find v_D, ω_2, and ω_3. Use the scale 1 in = 50 in/s.

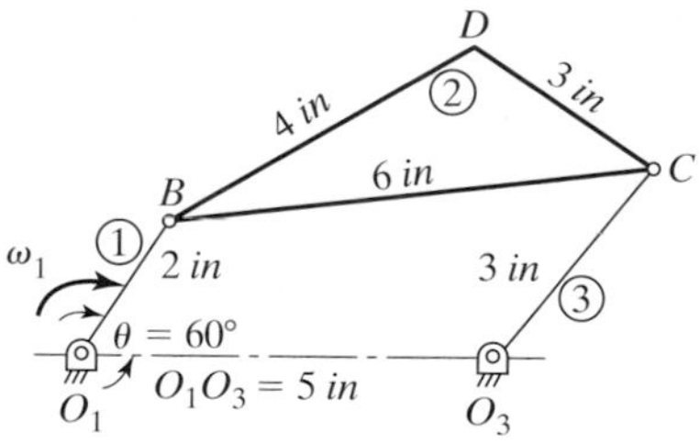

FIGURE P3.4

3.33 In Figure P3.5, $\omega_1 = 20$ rad/s.

(a) Draw and dimension the velocity polygon for the limiting position shown. Find relative velocity v_{CB}. Use the scale 1 in = 10 in/s.

(b) Repeat the problem for the other limiting position.

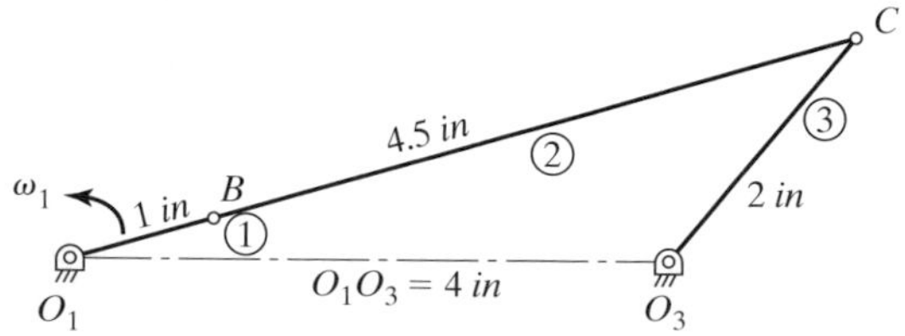

FIGURE P3.5

3.34 In Figure P3.6, $\theta = 105°$ and $\omega_2 = 20$ rad/s. Draw and dimension the velocity polygon. Identify the sliding velocity. Find ω_1. Use the scale 1 in = 20 in/s.

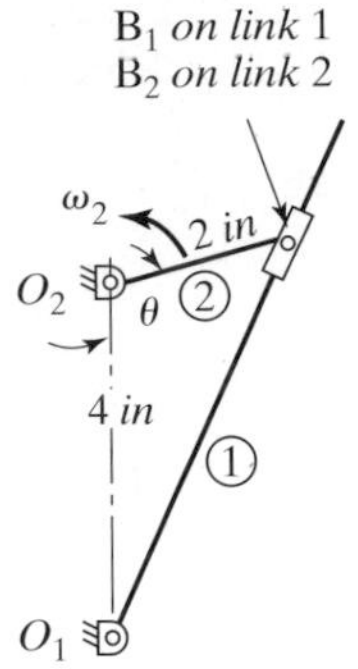

FIGURE P3.6

3.35 Repeat Problem 3.34 for $\theta = 30°$.

3.36 In Figure P3.3, $\omega_1 = 100$ rad/s.

(a) Find $\boldsymbol{v}_C$ *analytically* for the position shown.

(b) Find $\boldsymbol{v}_C$ analytically at both limiting positions.

3.37 In Figure P3.7, let the angular velocity of the crank be ω.

(a) Draw the velocity polygon for the position shown. Identify relative velocity v_{CB}.

(b) Repeat for the other limiting position.

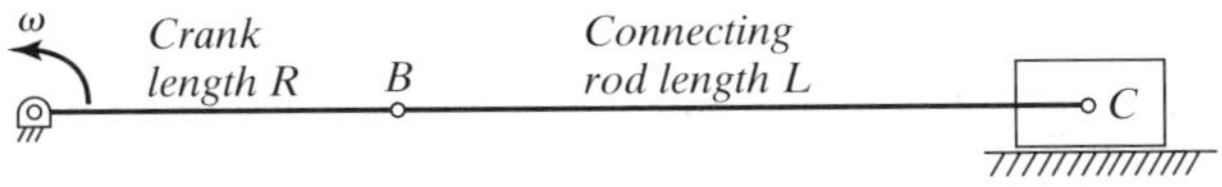

FIGURE P3.7

3.38 In Figure P3.8, $\omega_1 = 100$ rad/s. Draw and dimension the velocity polygon. Use the scale 1 in = 100 in/s.

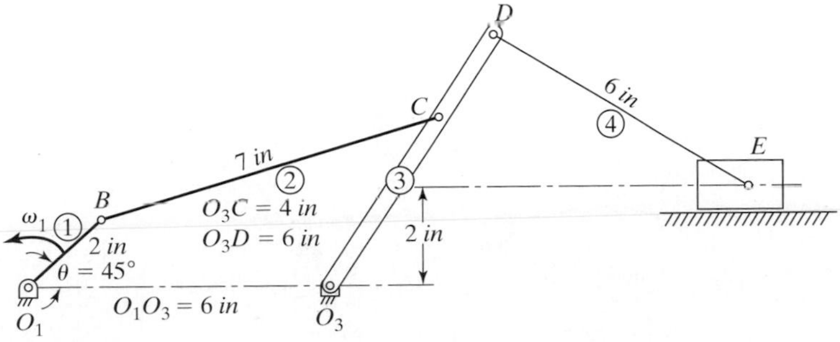

FIGURE P3.8

3.39 In Figure P3.9, $\theta = 135°$ and $\omega_1 = 10$ rad/s. Draw and dimension the velocity polygon. Identify the follower velocity and the sliding velocity. Use the scale 1 in = 5 in/s.

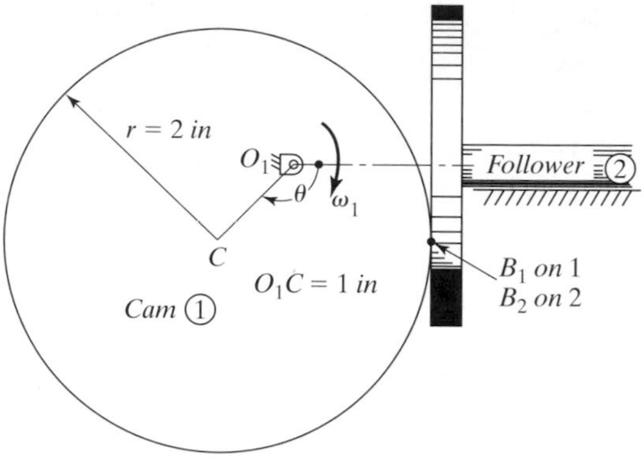

FIGURE P3.9

3.40 Repeat Problem 3.39 for $\theta = 30°$.

3.41 In Figure P3.10, $\omega_1 = 35$ rad/s. Draw the velocity polygon. Use the scale 1 in = 10 in/s. Find ω_2 and ω_3. Find the velocity of the midpoint of each link.

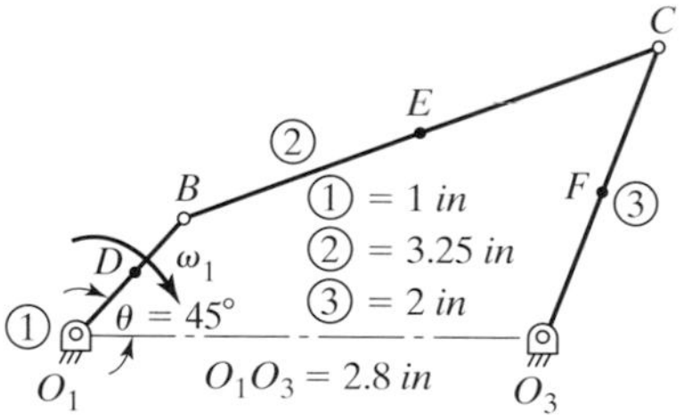

FIGURE P3.10

3.42 In Figure P3.11, $\omega_1 = 20$ rad/s.

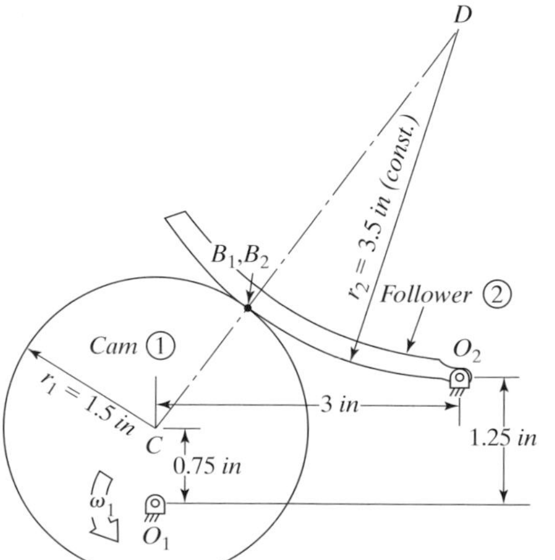

FIGURE P3.11

(a) Draw and dimension the velocity polygon. Find ω_2 and identify the sliding velocity.

(b) Note that CD is a fixed distance. Thus, we can use the equivalent linkage O_1CDO_2. Draw the velocity polygon, and again, find the angular velocity of the follower (represented by DO_2).

3.43 Consider a pair of involute spur gears with a 20° pressure angle. Let the driver speed be 300 rev/min clockwise and the driven gear speed 1,000 rev/min counterclockwise. Find the sliding velocity when contact occurs 1.2 in from the pitch point.

3.44 In Figure P3.12, $\theta = 105°$ and $\omega_2 = 20$ rad/s. Draw and dimension the velocity polygon, using the scale 1 in = 20 in/s.

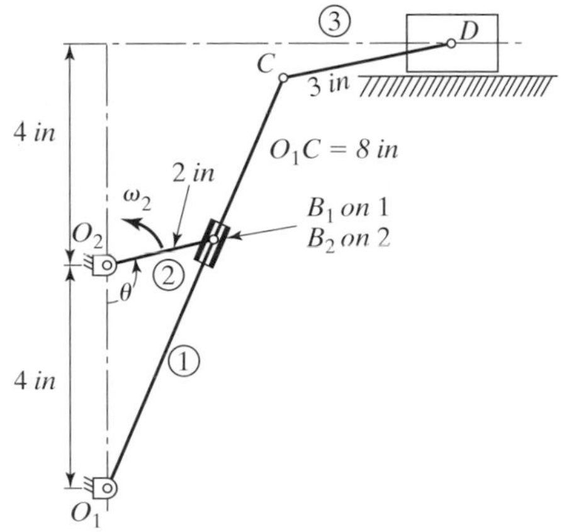

FIGURE P3.12

3.45 Repeat Problem 3.44 for $\theta = 30°$.

3.46 In Figure P3.9, $\omega_1 = 10$ rad/s. Find the follower velocity analytically when **(a)** $\theta = 135°$ and **(b)** $\theta = 30°$.

3.47 In Figure P3.13, ω_1(the angular velocity of O_1B_1) = 30 rad/s. Draw and dimension the velocity polygon, using the scale 1 in = 20 in/s. Identify the sliding velocity and find ω_2, the angular velocity of O_2B_2.

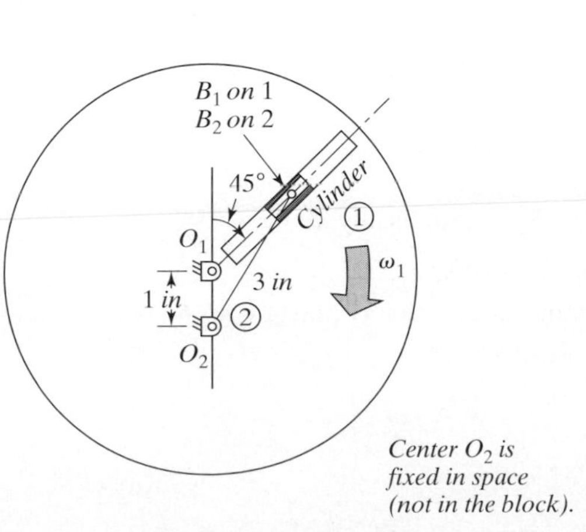

FIGURE P3.13

3.48 For Figure P3.14, draw and dimension the velocity polygon, using the scale 1 in = 100 in/s. Find ω_2 and v_D. Points B, C, and D lie on the same rigid link.

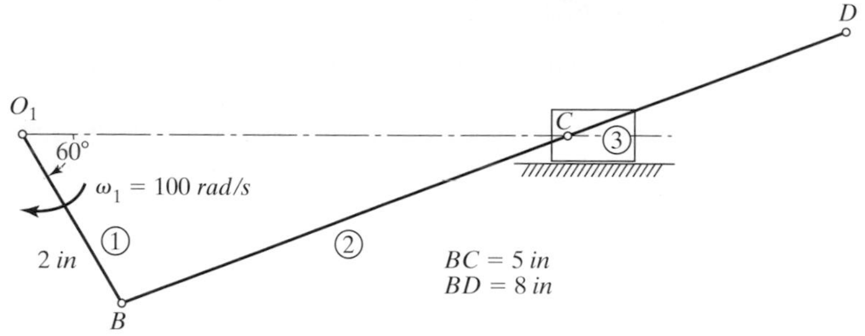

FIGURE P3.14

3.49 Locate all of the centros in Figure P3.14. Using centro 13, write an expression for the slider velocity in terms of ω_1. Calculate $\boldsymbol{v}_C$.

3.50 In Figure P3.14, use centro 02 in order to write an expression for **(a)** ω_2 in terms of v_B; **(b)** ω_2/ω_1; **(c)** v_C in terms of ω_2; and **(d)** v_D in terms of ω_2. **(e)** Calculate ω_2, v_C, and v_D.

3.51 In Figure P3.15, $\omega_5 = 15$ rad/s. Use a vector 3 in long to represent the velocity of B. Complete the velocity polygon and determine the velocity scale. Dimension the polygon and find the angular velocity of each link.

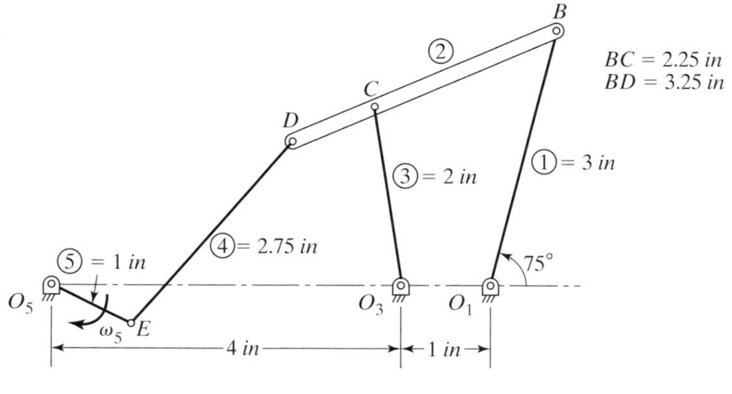

FIGURE P3.15

3.52 Refer to Figure P3.16. Solve graphically for $\theta_1 = 15°$.

(a) Draw the linkage to a 1:1 scale.

(b) Let 1 mm = 5 mm/s; draw velocity polygon $\boldsymbol{ob_1b_2}$. Add point c, where $O_2C = 100$ m.

(c) Find v_{B2}.

(d) Find v_c.

(e) Find ω_2.

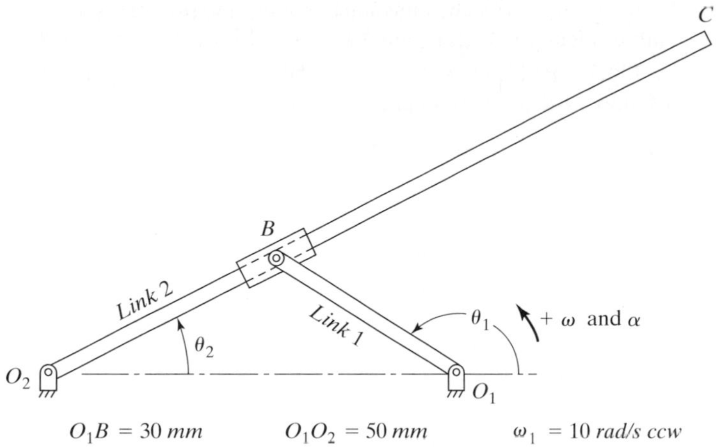

FIGURE P3.16

3.53 Repeat Problem 3.52 for $\theta_1 = 30°$.

3.54 Repeat Problem 3.52 for $\theta_1 = 45°$.

3.55 Refer to Figure P3.16. Find θ_1 when $\omega_2 = 0$.

3.56 Refer to Figure P3.17.

(a) Draw velocity polygon ***obc***.

(b) Find ω_2.

(c) Locate d on the velocity polygon. Find v_D.

(d) Identify angles θ and ϕ in your solution. Find velocity v_c and relative velocity bc in terms of ω_1, O_1B, θ, and ϕ.

O_1B = 120 mm
BC = 200 mm
θ = 15°
ω_1 = 500 rad/s ccw
BD = 40 mm

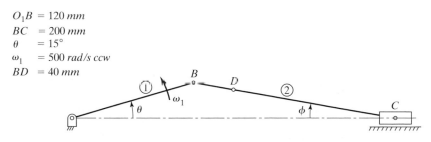

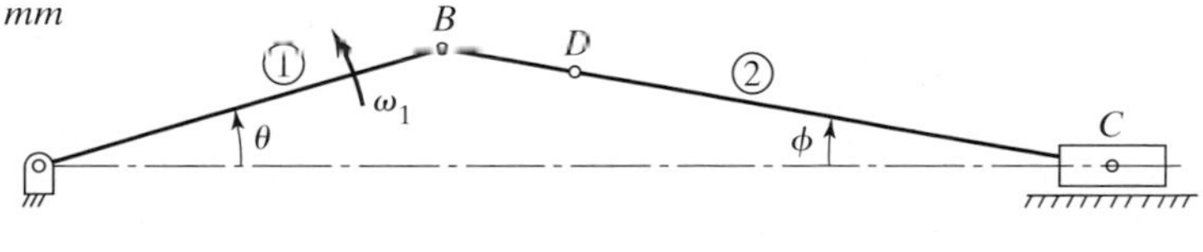

FIGURE P3.17

3.57 A four-bar planar linkage has the following dimensions:

$$r_0\,(\text{frame}) = 4;$$
$$r_1\,(\text{driver}) = 1.25;$$
$$r_2\,(\text{coupler}) = 5.25;$$
$$r_3\,(\text{driven link}) = 2.5.$$

If $\omega_1 = 5.75$ rad/s cw (constant), plot ω_2 and ω_3 versus θ. Use a computer or programmable calculator. Check values at $\theta = 120°$ by using a velocity polygon.

3.58 In Figure P3.18, $\theta_2 = 135°$, $\omega_2 = 500$ rad/s ccw (constant), $O_1O_2 = 300$ mm, $O_1C = 400$ mm, $O_2B = 150$ mm, and $CD = 160$ mm.

(a) Draw velocity polygon $\boldsymbol{ob_1b_2}$.

(b) Find $\boldsymbol{oc}$.

(c) Find ω_1.

(d) Add $\boldsymbol{c}$ to the polygon.

(e) Add $\boldsymbol{d}$ to the polygon.

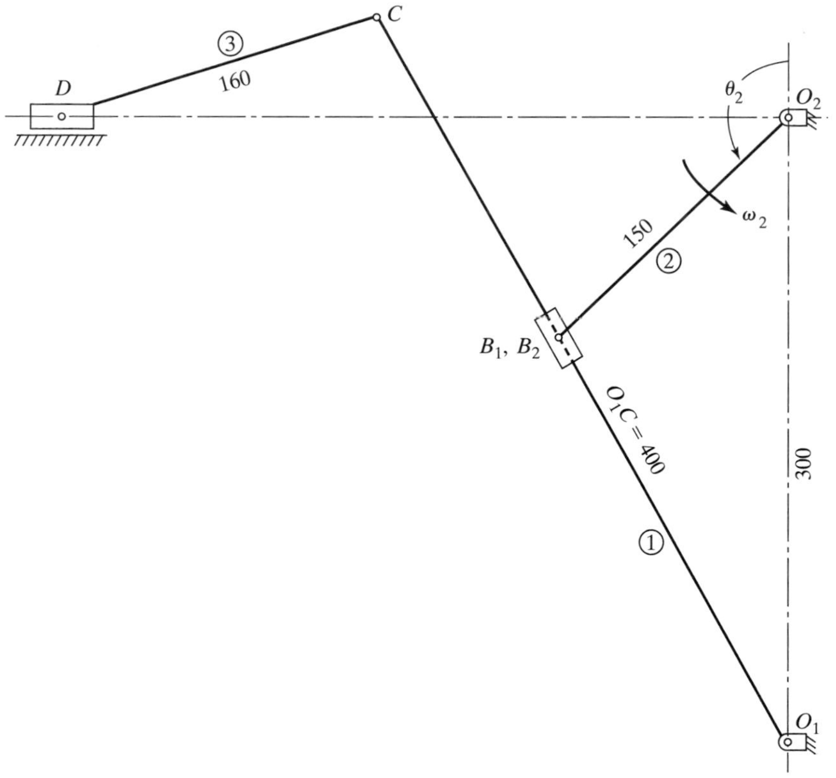

FIGURE P3.18

3.59 In Problem 3.58, let $\theta_2 = 120°$.

(a) Find v_{B_1}, v_D, and ω_1 at $\theta_2 = 120°$.

(b) For the interval $120° < \theta_2 < 135°$, find the average angular acceleration of link 1 and the average acceleration of point D.

3.60 Refer to Figure 3.15. Let $R_1 = 30$ and $O_1O_2 = 50$. Find θ_1 when $\omega_2 = 0$.

3.61 **(a)** Derive equations for position angles and angular velocities of a four-bar planar linkage. Use complex-number methods. Refer to Figure 3.9.

(b) Given $\theta_1 = 60°$, $r_0 = 4$, $r_1 = 2$, $r_2 = 4.5$, $r_3 = 3$, and $\omega_1 = 50$ rad/s cw, find ω_2 and ω_3, using the equations derived above.

(c) Check your results, using graphical methods.

3.62 Solve Problem 3.28 by analytical vector methods.

3.63 Solve Problem 3.29 by analytical vector methods.

3.64 Solve Problem 3.28 by complex-number methods.

3.65 Solve Problem 3.29 by complex-number methods.

3.66 A four-bar linkage has dimensions $r_0 = 312.48$, $r_1 = 100$, $r_2 = 200$, and $r_3 = 300$. The assembly is such that the vector loop $r_2 r_3 r_d$ is clockwise. Let the angular velocity of the crank be unity, and let the angular acceleration of the crank be zero.

(a) Tabulate the link positions, transmission angle, and coupler and follower crank angular velocities for values of θ_1 from 0 to 360°.

(b) Plot the coupler and follower crank angular velocities for values of θ_1 from 0 to 360°.

3.67 A four-bar linkage has dimensions $r_0 = 312.48$, $r_1 = 100$, $r_2 = 200$, and $r_3 = 300$. The assembly is such that the vector loop $r_2 r_3 r_d$ is counterclockwise. Let the angular velocity of the crank be unity, and let the angular acceleration of the crank be zero.

(a) Tabulate the link positions, transmission angle, and coupler and follower crank angular velocities for values of θ_1 from 0 to 360°.

(b) Plot the coupler and follower crank angular velocities for values of θ_1 from 0 to 360°.

3.68 A four-bar linkage has dimensions: $r_0 = 37$, $r_1 = 10$, $r_2 = 25$, and $r_3 = 40$. Point D lies on the coupler at a distance of 15 from the crankpin, at an angle of 20°. The assembly is such that the vector loop $r_2 r_3 r_d$ is clockwise. Let the angular velocity of the crank be 100 rad/s, and let the angular acceleration of the crank be zero.

(a) Tabulate the link positions, transmission angle, coupler and follower crank angular velocities, and velocity of point D for values of θ_1 from 0 to 360°.

(b) Plot the coupler and follower crank angular velocities for values of θ_1 from 0 to 360°.

(c) Plot the velocity of D and its x and y components.

3.69 A four-bar linkage has dimensions $r_0 = 38$, $r_1 = 13$, $r_2 = 27$, and $r_3 = 41$. Point D lies on the coupler at a distance of 12 from the crankpin, at an angle of 20°. The assembly is such that the vector loop $r_2 r_3 r_d$ is clockwise. Let the angular velocity of the crank be 50 rad/s, and let the angular acceleration of the crank be zero.

(a) Tabulate the link positions, transmission angle, coupler and follower crank angular velocities, and velocity of point D for values of θ_1 from 0 to 360°.

(b) Plot the coupler and follower crank angular velocities for values of θ_1 from 0 to 360°.

(c) Plot the velocity of D and its x and y components.

3.70 Consider an offset slider-crank linkage for which the connecting-rod-to-crank-length ratio is $R_2/R_1 = 2.5$ and the offset ratio $e/R_1 = -j0.4$. Tabulate and plot normalized slider position r_0/R_1, normalized slider velocity $v_c/(\omega_1 R_1)$, and angular velocity ratio ω_2/ω_1, all against the crank position. Assume that the angular velocity of the crank is constant. (*Suggestion*: Write a vector manipulation routine or use commercially available mathematics software.)

3.71 Consider an offset slider-crank linkage for which the connecting-rod-to-crank-length ratio is $R_2/R_1 = 2$ and the offset ratio $\boldsymbol{e}/R_1 = -\boldsymbol{j}0.5$. Tabulate and plot normalized slider position r_0/R_1, normalized slider velocity $v_c/(\omega_1 R_1)$, and angular-velocity ratio ω_2/ω_1, all against the crank position. Assume that the angular velocity of the crank is constant. (*Suggestion*: Write a vector manipulation routine or use commercially available mathematics software.)

3.72 Consider an offset slider-crank linkage for which the connecting-rod-to-crank-length ratio is $R_2/R_1 = 3$ and the offset ratio $\boldsymbol{e}/R_1 = -\boldsymbol{j}0.7$. Tabulate and plot the normalized slider position r_0/R_1, normalized slider velocity $v_c/(\omega_1 R_1)$, and angular-velocity ratio ω_2/ω_1, all against the crank position. Assume that the angular velocity of the crank is constant. (*Suggestion*: Write a vector manipulation routine or use commercially available mathematics software.)

3.73 Consider an RSSR linkage similar to that in Figure 1.6a, where the link lengths are $r_0 = 4$, $r_1 = 1$, $r_2 = 3.5$, and $r_3 = 2.5$. Link 0 lies on the x-axis. Link 1 rotates in the xy-plane with an angular velocity of 1 rad/s (constant), and link 3 rotates in the xz-plane. Plot the vector components representing the angular velocity of link 2 and the angular position and angular velocity of link 3 against angular position θ of link 1. Tabulate the resultant angular velocity of link 2, the angular position of link 3, and the angular velocity of link 3 against the angular position of link 1.

3.74 Consider an RSSR linkage similar to that in Figure 1.6a, where link the lengths are $r_0 = 3.2$, $r_1 = 1$, $r_2 = 2.8$, and $r_3 = 2$. Link 0 lies on the x-axis. Link 1 rotates in the xy-plane with an angular velocity of 1 rad/s (constant), and link 3 rotates in the xz-plane. Plot the vector components representing the angular velocity of link 2 and the angular position and angular velocity of link 3 against angular position θ of link 1. Tabulate the resultant angular velocity of link 2, the angular position of link 3, and the angular velocity of link 3 against the angular position of link 1.

3.75 We would like to design a mechanism with a 2-in output link that oscillates through a range of about 105°. The input shaft rotation speed is 20 rad/s. The input shaft is parallel to the plane of the output link, at a distance of 0.2 in. Design the mechanism and find the angular velocity of the output link. Check the transmission metric and check the angular velocity by numerical differentiation. Design decisions. Try an *RSSR* spatial linkage with the following dimensions:

drive crank length $r_1 = 1.5$ in;

coupler length $r_2 = 3.4$ in;

driven crank $r_3 = 2$ in (required);

revolute joints: R_1 located at $(0, 0, 0)$ and R_2 at $(-0.2, -3, 0)$.

(*Note*: The desired range of motion and the decisions you make may result in a transmission metric that is outside of generally accepted limits.)

3.76 A mechanism is needed with a 110-mm output link that oscillates through a range of about 48°. The input shaft rotation speed is 10 rad/s. The input shaft is parallel to the plane of the output link, at a distance of 15 mm. Design the mechanism and find the angular velocity of the output link. Check the transmission metric and check the angular velocity by numerical differentiation.

Design decisions. Select an RSSR spatial linkage with the following dimensions:

drive crank length $r_1 = 50$ mm;

coupler length $r_2 = 180$ mm;

driven crank $r_3 = 110$ mm (required);

fixed link components are $r_{0x} = -15$ and $r_{0y} = -175$.

3.77 A mechanism is needed with a 55-mm output link that oscillates through a range of about 45°. The input shaft rotation speed is 300 rpm. The input shaft is parallel to the plane of the output link, at a distance of 30 mm. Design the mechanism and find the angular velocity of the output link. Check the transmission metric and check the angular velocity by numerical differentiation.

Design decisions. Select an RSSR spatial linkage with the following dimensions:

drive crank length $r_1 = 20$ mm;

coupler length $r_2 = 155$ mm;

driven crank $r_3 = 55$ mm (required);

fixed link components are $r_{0x} = 30$ and $r_{0y} = 140$.

PROJECTS

See Projects 1.1 through 1.6 and the suggestions in Chapter 1. Examine linkages involved in the chosen project. Describe and plot the velocity and angular velocity characteristics of the linkages. Make use of computer software wherever practical. Check your results by a graphical method for at least one linkage position. Evaluate the linkage in terms of the performance requirements.

BIBLIOGRAPHY AND REFERENCES

Angeles, J., *Spatial Kinematic Chains: Analysis–Synthesis–Optimization*, Springer, New York, 1982.

Hirschhorn, J., "A Graphical Investigation of the Velocity Pattern of a Rigid Body in Three-Dimensional Motion," *Mechanism and Machine Theory*, Penton, Cambridge, vol. 23, no. 3, 1988, pp. 185–189.

JML Research, Inc., *Integrated Mechanisms Program*, JML Research, Inc., Madison, WI, 1988.

Knowledge Revolution, *Working Model™2D User's Manual*, Knowledge Revolution, San Mateo, CA, 1996.

Knowledge Revolution, *Working Model™3D User's Manual*, Knowledge Revolution, San Mateo, CA, 1998.

Lee, H.-Y., and C.-G. Liang, "A New Vector Theory for the Analysis of Spatial Mechanisms," *Mechanism and Machine Theory*, vol. 23, no. 3, 1988, pp. 209–218.

MathSoft, *Mathcad 2000™User's Guide*, MathSoft, Inc., Cambridge, MA, 1999.

Mechanical Dynamics, Inc., *ADAMS Applications Manual*, Mechanical Dynamics, Inc., Ann Arbor, MI, 1987.

Mechanical Dynamics, Inc., *ADAMS User's Manual*, Mechanical Dynamics, Inc., Ann Arbor, MI, 1987.

Mischke, C. R., *Elements of Mechanical Analysis*, Addison-Wesley, Reading, MA, 1963.

CHAPTER 4

Acceleration Analysis of Mechanisms

The acceleration of a point is a vector representing the change in velocity per unit time. Velocity is a vector, so changes in its magnitude and direction both contribute to acceleration. In general, angular velocity and angular acceleration are also vectors. However, they may be treated as scalars in planar mechanisms.

Concepts You Will Learn and Apply When Studying This Chapter

- Acceleration of a point on a rotating link
- Acceleration in a moving coordinate system
- Normal, tangential, and Coriolis acceleration
- Analytical vector methods for finding accelerations in linkages
- Complex-number methods for finding accelerations in linkages
- The acceleration polygon, a graphical vector method
- Acceleration image
- Equivalent linkages
- Linkage combinations
- Matrix methods for determining accelerations in spatial linkages
- Practical applications
- Computational techniques for "working smart"
- Interpretation and assessment of results

4.1 BASIC CONCEPTS

Acceleration in linkages is of particular importance because inertial forces are proportional to rectilinear acceleration and inertial torques are proportional to angular accelerations. Graphical and analytical vector techniques, including representing vectors in complex form, are useful in determining linkage accelerations.

Acceleration of a Point

Consider a point moving along a curve in three-dimensional space and located by a vector $\boldsymbol{R}$. The acceleration $\boldsymbol{a}$ of the point is given by the rate of change in velocity with respect to time:

$$\boldsymbol{a} = \frac{d\boldsymbol{v}}{dt} = \frac{d^2\boldsymbol{R}}{dt^2}. \tag{4.1}$$

Acceleration may be expressed in terms of its x, y, and z components and their respective unit vectors $\boldsymbol{i}$, $\boldsymbol{j}$, and $\boldsymbol{k}$ in a fixed coordinate system:

$$\boldsymbol{a} = \boldsymbol{i}\ddot{R}_x + \boldsymbol{j}\ddot{R}_y + \boldsymbol{k}\ddot{R}_z, \tag{4.2}$$

where two dots above the variable represent the second derivative with respect to time.

Angular Acceleration

Angular acceleration $\boldsymbol{\alpha}$ represents the rate of change in angular velocity $\boldsymbol{\omega}$ with respect to time. In general,

$$\boldsymbol{\alpha} = \frac{d\boldsymbol{\omega}}{dt} = \alpha_x\boldsymbol{i} + \alpha_y\boldsymbol{j} + \alpha_z\boldsymbol{k}. \tag{4.3}$$

For the special case of planar motion, the vector direction of $\boldsymbol{\alpha}$ is perpendicular to the plane of rotation. For motion in the xy-plane, vector $\boldsymbol{\alpha}$ is in the $\pm z$ direction; that is, $\boldsymbol{\alpha} = \alpha\boldsymbol{k}$.

SAMPLE PROBLEM 4.1

Average Angular Acceleration

An automobile accelerates from 0 to 60 mi/h (0 to 96.56 km/h) in 15 s. *Find the average angular acceleration of the rear axle.* The tires have a 13-in. (330.2-mm) outer radius.

Solution. This problem is equivalent to a dynamometer test, where the 60-mi/h speed is the linear velocity of a point on the tread of the tire. Let us convert this speed to more manageable dimensions:

$$60 \text{ mi/h} \times \frac{5{,}280 \text{ ft}}{1 \text{ mi}} \times \frac{1 \text{ h}}{3{,}600 \text{ s}} \times \frac{12 \text{ in}}{1 \text{ ft}} = 1{,}056 \text{ in/s}.$$

Using the velocity equation $\boldsymbol{v} = \boldsymbol{\omega} \times \boldsymbol{R}$, we obtain the angular velocity magnitude

$$\omega = \frac{v}{R} = \frac{1{,}056 \text{ in/s}}{13 \text{ in}} = 81.2 \text{ rad/s}.$$

Average angular acceleration is defined as the rate of change of the angular velocity with respect to time. Therefore,

$$\alpha_{av} = \frac{\omega(\text{final}) - \omega(\text{initial})}{\text{time interval}}$$
$$= \frac{81.2 - 0 \text{ rad/s}}{15 \text{ s}} = 5.41 \text{ rad/s}^2.$$

Motion of a Rigid Body about a Fixed Point

This important special case occurs frequently in linkage analysis. As noted in Chapter 3, the velocity of a point in a rigid body rotating about a fixed point is given by

$$\dot{\boldsymbol{R}} = \boldsymbol{v} = \boldsymbol{\omega} \times \boldsymbol{R}, \tag{4.4}$$

where $\boldsymbol{R}$ is the vector from the fixed point to the point in question. Differentiating $\boldsymbol{v}$ with respect to time, we obtain the acceleration of the moving point:

$$\dot{\boldsymbol{v}} = \dot{\boldsymbol{\omega}} \times \boldsymbol{R} + \boldsymbol{\omega} \times \dot{\boldsymbol{R}}. \tag{4.5}$$

Using Eqs. (4.3) and (4.4), we may write Eq. (4.5) as follows:

$$\boldsymbol{a} = \boldsymbol{\alpha} \times \boldsymbol{R} + \boldsymbol{\omega} \times (\boldsymbol{\omega} \times \boldsymbol{R}). \tag{4.6}$$

Planar motion. Suppose a rigid link is connected to a frame by a revolute pair with an axis perpendicular to the link (one or more journal bearings or ball bearings). Then that link moves in a plane, and its angular-velocity vector is perpendicular to its plane of motion. The first vector on the right in Eq. (4.6), $\boldsymbol{\alpha} \times \mathbf{R}$, is the tangential acceleration for planar motion. This vector is tangent to the path of the point on the body (perpendicular to radius vector $\mathbf{R}$), as shown in Figure 4.1. The magnitude of the tangential acceleration is

$$a^t = \alpha\mathrm{R}.$$

The second vector on the right in Eq. (4.6), $\boldsymbol{\omega} \times (\boldsymbol{\omega} \times \mathbf{R})$, is the normal acceleration. This vector is normal to the path of the point for planar motion. Its direction is parallel to, but opposite, the radius vector. The magnitude of the normal acceleration is

$$a^n = \omega^2 R = \frac{V^2}{R}.$$

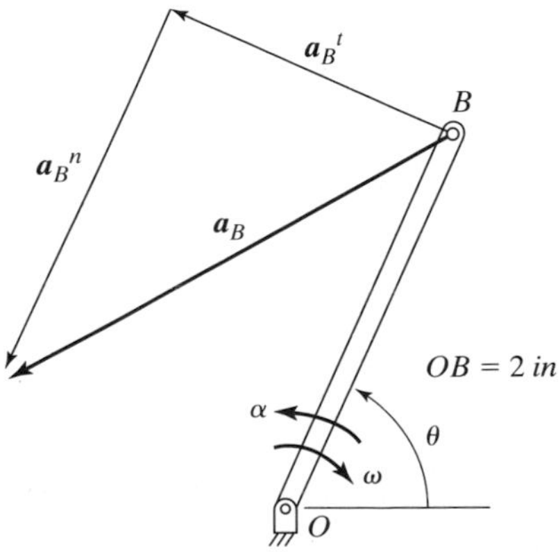

FIGURE 4.1 Acceleration of a point.

Spatial motion. The preceding equations for planar motion do not apply to spatial motion. For example, if a rigid body is connected to a frame by a spherical pair (ball joint) then the angular-velocity vector and the radius vector are not necessary perpendicular. We would then use the vector cross-product form (Eq. 4.6).

SAMPLE PROBLEM 4.2

Acceleration of a point in planar motion

Point B on a rigid body is 2 in (50.8 mm) from center of rotation, O, as shown in Figure 4.1. Point O represents a revolute pair; the body has planar motion. At the instant shown, the angular acceleration is 750,000 rad/s^2 counterclockwise, and the angular velocity is 1,000 rad/s clockwise. Find the acceleration of point B at this instant.

Solution. The normal acceleration of point B is

$$\begin{aligned} \boldsymbol{a}_B^n &= \boldsymbol{\omega} \times (\boldsymbol{\omega} \times \boldsymbol{R}) = (-1000 \text{ rad/s})^2(2 \text{ in}) \\ &= 2{,}000{,}000 \text{ in/s}^2 \angle \theta + \pi \, (\text{along } BO \text{ toward } O) \end{aligned}$$

The tangential acceleration is given by

$$\begin{aligned} \boldsymbol{a}_B^t &= \dot{\boldsymbol{\omega}} \times \boldsymbol{R} = (750{,}000 \text{ rad/s}^2)\,(2 \text{ in}) \\ &= 1{,}500{,}000 \text{ in/s}^2 \angle \theta + \pi/2, \end{aligned}$$

perpendicular to OB to the left, since α is counterclockwise.

Adding the vectors, we obtain the total acceleration of point B:

$$\begin{aligned} \boldsymbol{a}_B &= \boldsymbol{a}_B^t + \boldsymbol{a}_B^n = \sqrt{(a_B^t)^2 + (a_B^n)^2} \\ &= \sqrt{(1{,}500{,}000)^2 + (2{,}000{,}000)^2} \\ &= 2{,}500{,}000 \text{ in/s}^2, \end{aligned}$$

to the left and downward.

Moving Coordinate Systems

A more general case of linkage motion may be described by first considering a point within a moving coordinate system (Figure 4.2). A coordinate system xyz with respective unit vectors $\boldsymbol{i}, \boldsymbol{j}$, and $\boldsymbol{k}$ moves within a fixed system XYZ with respective unit vectors $\boldsymbol{I}, \boldsymbol{J}$, and $\boldsymbol{K}$. The velocity of a point P may be described by

$$\dot{\boldsymbol{R}} = \dot{\boldsymbol{R}}_0 + \dot{\boldsymbol{r}}_r + \boldsymbol{\omega} \times \boldsymbol{r} \tag{4.7}$$

(repeated from Chapter 3), where

$\dot{\boldsymbol{R}}$ = absolute velocity of point P relative to XYZ,
$\dot{\boldsymbol{R}}_0$ = velocity of the origin o of the xyz system,
$\dot{\boldsymbol{r}}_r$ = velocity of point P relative to the xyz system, and
$\boldsymbol{\omega} \times \boldsymbol{r}$ = cross product of the angular velocity of the rotating system xyz in the XYZ system and the position vector $\boldsymbol{r}$

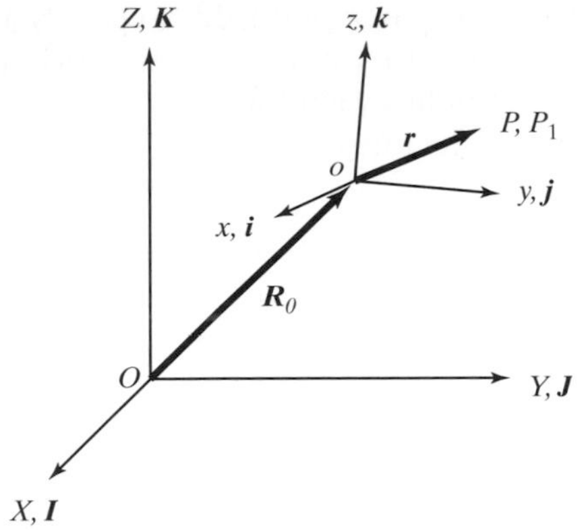

FIGURE 4.2 A moving coordinate system: System xyz moves within fixed system XYZ.

Differentiating the first term on the right with respect to time, we have

$$\frac{d}{dt}\dot{\boldsymbol{R}}_0 = \ddot{\boldsymbol{R}}_0. \tag{4.8}$$

The next term could be written

$$\dot{\boldsymbol{r}}_r = \dot{r}_{rx}\boldsymbol{i} + \dot{r}_{ry}\boldsymbol{j} + \dot{r}_{rz}\boldsymbol{k}. \tag{4.9}$$

Differentiating, we get

$$\frac{d\dot{\boldsymbol{r}}_r}{dt} = \ddot{r}_{rx}\boldsymbol{i} + \ddot{r}_{ry}\boldsymbol{j} + \ddot{r}_{rz}\boldsymbol{k} + \dot{r}_{rx}\dot{\boldsymbol{i}} + \dot{r}_{ry}\dot{\boldsymbol{j}} + \dot{r}_{rz}\dot{\boldsymbol{k}}. \tag{4.10}$$

The differential of a unit vector with respect to time is the cross product of the angular velocity of the moving coordinate system and the unit vector; for example, $\dot{\boldsymbol{i}} = \boldsymbol{\omega} \times \boldsymbol{i}$. Then $\dot{r}_{rx}\dot{\boldsymbol{i}} = \dot{r}_{rx}(\boldsymbol{\omega} \times \boldsymbol{i}) = \boldsymbol{\omega} \times \dot{r}_{rx}\boldsymbol{i}$, and so on. Consequently,

$$\frac{d\dot{\boldsymbol{r}}_r}{dt} = \ddot{r}_{rx}\boldsymbol{i} + \ddot{r}_{ry}\boldsymbol{j} + \ddot{r}_{rz}\boldsymbol{k} + \omega \times \dot{r}_{rx}\boldsymbol{i} + \omega \times \dot{r}_{ry}\boldsymbol{j} + \omega \times \dot{r}_{rz}\boldsymbol{k}, \tag{4.11}$$

which, from Eq. (4.9), may be written

$$\frac{d\dot{\boldsymbol{r}}_r}{dt} = \ddot{\boldsymbol{r}}_r + \boldsymbol{\omega} \times \dot{\boldsymbol{r}}_r. \tag{4.12}$$

The last term on the right side of Eq. (4.7) may be written

$$\boldsymbol{\omega} \times \boldsymbol{r} = \boldsymbol{\omega} \times (r_x\boldsymbol{i} + r_y\boldsymbol{j} + r_z\boldsymbol{k}). \tag{4.13}$$

Differentiating, we have

$$\frac{d}{dt}(\boldsymbol{\omega} \times \boldsymbol{r}) = \dot{\boldsymbol{\omega}} \times (r_x\boldsymbol{i} + r_y\boldsymbol{j} + r_z\boldsymbol{k}) + \boldsymbol{\omega} \times (\dot{r}_{rx}\boldsymbol{i} + \dot{r}_{ry}\boldsymbol{j} + \dot{r}_{rz}\boldsymbol{k})$$
$$+ \boldsymbol{\omega} \times (r_x\dot{\boldsymbol{i}} + r_y\dot{\boldsymbol{j}} + r_z\dot{\boldsymbol{k}}).$$

Noting that $\dot{\boldsymbol{i}} = \boldsymbol{\omega} \times \boldsymbol{i}$ and so on, as in the previous case, and combining terms, we obtain

$$\frac{d}{dt}(\boldsymbol{\omega} \times \boldsymbol{r}) = \dot{\boldsymbol{\omega}} \times \boldsymbol{r} + \boldsymbol{\omega} \times \dot{\boldsymbol{r}}_r + \boldsymbol{\omega} \times (\boldsymbol{\omega} \times \boldsymbol{r}). \tag{4.14}$$

Then, from Eqs. (4.8), (4.12), and (4.14), the *total acceleration of point P* in fixed coordinate system XYZ is

$$\ddot{\boldsymbol{R}} = \ddot{\boldsymbol{R}}_0 + \dot{\boldsymbol{\omega}} \times \boldsymbol{r} + \boldsymbol{\omega} \times (\boldsymbol{\omega} \times \boldsymbol{r}) + \ddot{\boldsymbol{r}}_r + 2\boldsymbol{\omega} \times \dot{\boldsymbol{r}}_r. \tag{4.15}$$

The first term on the right of Eq. (4.15) is the total acceleration of the origin o of the moving coordinates; the next two terms give the acceleration of P_1 relative to o, where P_1 is a point instantaneously coincident with P and having no motion relative to the moving coordinates xyz. The last two terms in Eq. (4.15) represent the motion of P relative to P_1. It is important to remember that ω and $\dot{\omega}$ refer, respectively, to the angular velocity and the angular acceleration of the moving coordinate system.

Problems involving spatial linkages (mechanisms involving motion that does not lie in a plane or in a set of parallel planes) require that Eq. (4.15) be applied in its general form. In plane mechanisms, the vector products take the following forms: $\dot{\boldsymbol{\omega}} \times \boldsymbol{r}$ becomes αr, the tangential acceleration; $\boldsymbol{\omega} \times (\boldsymbol{\omega} \times \boldsymbol{r})$ becomes $\omega^2 r$, the normal acceleration; and $2\boldsymbol{\omega} \times \dot{\boldsymbol{r}}_r$ becomes $2\omega v$, the Coriolis acceleration. The Coriolis acceleration term appears when sliding occurs along a rotating link. From Figure 4.2, the term $\boldsymbol{v} = \dot{\boldsymbol{r}}_r$ is the velocity of point P relative to a point P_1 that is instantaneously coincident with P, but that has no motion relative to the moving coordinates. Normal acceleration, tangential acceleration, and Coriolis acceleration will appear later, as we use graphical and analytical methods to investigate the motion of linkages.

Relative Acceleration

Relative acceleration is a useful concept for graphical solutions to planar linkages. In Eq. (4.15), the acceleration of point P is described in terms of the acceleration of the origin o of a set of moving coordinates and four terms representing the difference between the acceleration of P and the acceleration of the origin o of a set of moving coordinates. Consistent with the terminology used for velocities, the acceleration difference is called the acceleration of P relative to o or the acceleration of P *with respect to o*.

The special case involving two points on the same rigid link is frequently encountered. Consider link BC of Figure 4.3; this link is not fixed at any point. If the acceleration of point B is known, we may find the acceleration of any point C on the link by adding *the acceleration of point C with respect to B to the acceleration of B.* Symbolically, then, the acceleration of point C is given by the expression

$$\boldsymbol{a}_C = \boldsymbol{a}_B + \boldsymbol{a}_{CB}, \tag{4.16}$$

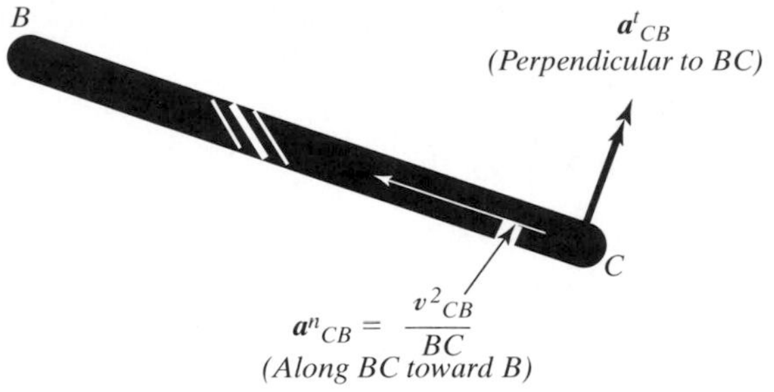

FIGURE 4.3 The acceleration of point C relative to point B, $\boldsymbol{a}_{CB}$, is shown broken into its normal and tangential components. The vector representing the normal component, $\boldsymbol{a}^n_{CB}$, lies along BC and is directed toward B; the vector representing the tangential component, $\boldsymbol{a}^t_{CB}$, is perpendicular to BC.

where $\boldsymbol{a}_{CB}$, the acceleration of point C with respect to point B, may be broken into its normal and tangential components as follows:

$$\boldsymbol{a}_{CB} = \boldsymbol{a}^n_{CB} + \boldsymbol{a}^t_{CB}. \tag{4.17}$$

The normal acceleration of C with respect to B is given by

$$\boldsymbol{a}^n_{CB} = \boldsymbol{\omega}_{BC} \times (\boldsymbol{\omega}_{BC} \times \boldsymbol{BC}), \tag{4.18}$$

and the tangential acceleration is given by

$$\boldsymbol{a}^t_{CB} = \boldsymbol{\alpha}_{BC} \times \boldsymbol{BC}. \tag{4.19}$$

Using Eqs. (4.18) and (4.19), we have

$$\boldsymbol{a}_C = \boldsymbol{a}_B + \boldsymbol{\omega}_{BC} \times (\boldsymbol{\omega}_{BC} \times \boldsymbol{BC}) + \boldsymbol{\alpha}_{BC} \times \boldsymbol{BC}. \tag{4.20}$$

If link BC moves in a plane, the magnitude of the angular velocity of link BC is given by

$$\omega_{BC} = \frac{v_{CB}}{CB}.$$

The magnitude of the normal acceleration is

$$a^n_{CB} = \omega^2_{BC} BC = \frac{v^2_{CB}}{BC},$$

and that of the tangential acceleration is

$$a^t_{CB} = \alpha_{BC} BC.$$

4.2 ANALYSIS OF A FOUR-BAR LINKAGE BY ANALYTICAL VECTOR METHODS

The vector equations developed in the preceding section may be applied to the analysis of linkages. Consider the four-bar planar linkage of Figure 4.4. The loop equation

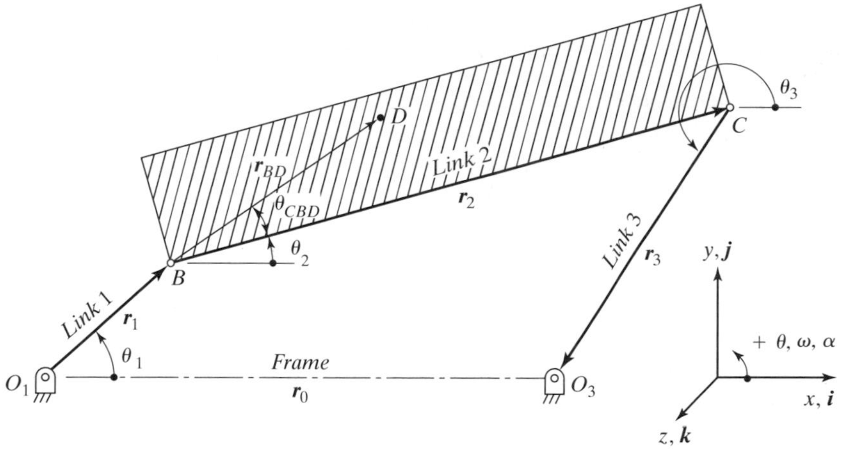

FIGURE 4.4 Analytical study of a four-bar linkage.

for the linkage,

$$\boldsymbol{r}_0 + \boldsymbol{r}_1 + \boldsymbol{r}_2 + \boldsymbol{r}_3 = 0,$$

was solved in Chapter 2 to determine relative link positions. Differentiating the loop equation and making the substitutions indicated in Chapter 3, we obtain the velocity equation [Eq. (3.14), repeated]. This equation was solved in Chapter 3.

$$\boldsymbol{\omega}_1 \times \boldsymbol{r}_1 + \boldsymbol{\omega}_2 \times \boldsymbol{r}_2 + \boldsymbol{\omega}_3 \times \boldsymbol{r}_3 = 0$$

Differentiating the velocity equation, while noting that the links are fixed in length, we obtain the acceleration equation

$$\begin{aligned} &\boldsymbol{\alpha}_1 \times \boldsymbol{r}_1 + \boldsymbol{\omega}_1 \times (\boldsymbol{\omega}_1 \times \boldsymbol{r}_1) + \boldsymbol{\alpha}_2 \times \boldsymbol{r}_2 + \boldsymbol{\omega}_2 \times (\boldsymbol{\omega}_2 \times \boldsymbol{r}_2) \\ &\quad + \boldsymbol{\alpha}_3 \times \boldsymbol{r}_3 + \boldsymbol{\omega}_3 \times (\boldsymbol{\omega}_3 \times \boldsymbol{r}_3) = 0. \end{aligned} \tag{4.21}$$

The $\boldsymbol{\alpha} \times \boldsymbol{r}$ terms in this equation account for the change in angular velocity of each link (causing a change in magnitude of the velocity vector).
The $\boldsymbol{\omega} \times (\boldsymbol{\omega} \times \boldsymbol{r})$ terms account for the change in direction of the velocity vector.

The sense of vector $\boldsymbol{r}_3$ is such that the last two terms of Eq. (4.21) represent $-\boldsymbol{a}_C$. Thus, the equation is equivalent to

$$\boldsymbol{a}_B^t + \boldsymbol{a}_B^n + \boldsymbol{a}_{CB}^t + \boldsymbol{a}_{CB}^n = \boldsymbol{a}_C^t + \boldsymbol{a}_C^n. \tag{4.22}$$

We may orient the coordinate axes so that the linkage lies in the xy-plane, with angular velocity and angular accelerations given in the form $\boldsymbol{\omega} = \omega \boldsymbol{k}$ and $\boldsymbol{\alpha} = \alpha \boldsymbol{k}$. Then, typical terms in Eq. (4.21) have the form

$$\boldsymbol{\alpha} \times \boldsymbol{r} = \begin{vmatrix} \boldsymbol{i} & \boldsymbol{j} & \boldsymbol{k} \\ 0 & 0 & \alpha \\ r_x & r_y & 0 \end{vmatrix} = \alpha(\boldsymbol{j} r_x - \boldsymbol{i} r_y),$$

$$\boldsymbol{\omega} \times \boldsymbol{r} = \begin{vmatrix} \boldsymbol{i} & \boldsymbol{j} & \boldsymbol{k} \\ 0 & 0 & \omega \\ r_x & r_y & 0 \end{vmatrix} = \omega(\boldsymbol{j}r_x - \boldsymbol{i}r_y),$$

and

$$\boldsymbol{\omega} \times (\boldsymbol{\omega} \times \boldsymbol{r}) = \begin{vmatrix} \boldsymbol{i} & \boldsymbol{j} & \boldsymbol{k} \\ 0 & 0 & \omega \\ -\omega r_y & -\omega r_x & 0 \end{vmatrix} = -\omega^2(\boldsymbol{i}r_x + \boldsymbol{j}r_y).$$

Making the indicated substitutions in Eq. (4.21) results in

$$\begin{aligned} \alpha_1(-\boldsymbol{i}r_{1y} + \boldsymbol{j}r_{1x}) - \omega_1{}^2(\boldsymbol{i}r_{1x} + \boldsymbol{j}r_{1y}) + \alpha_2(-\boldsymbol{i}r_{2y} + \boldsymbol{j}r_{2x}) - \omega_2^2(\boldsymbol{i}r_{2x} + \boldsymbol{j}r_{2y}) \\ + \alpha_3(-\boldsymbol{i}r_{3y} + \boldsymbol{j}r_{3x}) - \omega_3^2(\boldsymbol{i}r_{3x} + \boldsymbol{j}r_{3y}) = 0. \end{aligned} \tag{4.23}$$

This equation must be satisfied separately for the coefficients of unit vectors $\boldsymbol{i}$ and $\boldsymbol{j}$. Now, suppose that the position, angular velocity, and angular acceleration of the driving crank, link 1, are given. If we have already solved the displacement and velocity equations (Chapters 2 and 3), the remaining unknowns are α_2 and α_3. Separating the components of vector $\boldsymbol{i}$, we have

$$\alpha_2 r_{2y} + \alpha_3 r_{3y} = -\alpha_1 r_{1y} - \omega_1{}^2 r_{1x} - \omega_2^2 r_{2x} - \omega_3^2 r_{3x}. \tag{4.24}$$

Separating the components of vector $\boldsymbol{j}$ yields

$$\alpha_2 r_{2x} + \alpha_3 r_{3x} = -\alpha_1 r_{1x} + \omega_1{}^2 r_{1y} + \omega_2^2 r_{2y} + \omega_3^2 r_{3y}. \tag{4.25}$$

These two simultaneous equations may be solved by elimination or another convenient method. For example, we may use the matrix form $\boldsymbol{AX} = \boldsymbol{B}$; that is,

$$\begin{bmatrix} r_{2y} & r_{3y} \\ r_{2x} & r_{3x} \end{bmatrix} \begin{bmatrix} \alpha_2 \\ \alpha_3 \end{bmatrix} = \begin{bmatrix} a \\ b \end{bmatrix}, \tag{4.26}$$

where a represents the right side of Eq. (4.24) and b the right side of Eq. (4.25). Then the solution is given by

$$\mathbf{X} = \begin{bmatrix} \alpha_2 \\ \alpha_3 \end{bmatrix} = \mathbf{A}^{-1}\mathbf{B}.$$

Alternatively, using determinants, we have, for the angular acceleration of the coupler,

$$\alpha_2 = \frac{1}{D}\begin{vmatrix} a & r_{3y} \\ b & r_{3x} \end{vmatrix} = \frac{ar_{3x} - br_{3y}}{D},$$

and for the angular acceleration of the follower crank,

$$\alpha_3 = \frac{1}{D}\begin{vmatrix} r_{2y} & a \\ r_{2x} & b \end{vmatrix} = \frac{br_{2y} - ar_{2x}}{D},$$

where

$$D = \begin{vmatrix} r_{2y} & r_{3y} \\ r_{2x} & r_{3x} \end{vmatrix} = r_{2y}r_{3x} - r_{2x}r_{3y}. \tag{4.27}$$

SAMPLE PROBLEM 4.3

Accelerations in a Four-Bar Linkage

Referring to Figure 4.4, let $\omega_1 = 100$ rad/s (constant), $\theta_1 = 45°$, $\theta_{CBD} = 20°$ (constant), $r_0 = 30$ mm, $r_1 = 10$ mm, $r_2 = 35$ mm, $r_3 = 20$ mm, and $r_{BD} = 15$ mm. Find α_2, α_3, a_B, a_C, and a_D for the assembly mode shown.

Solution. As determined in Chapters 2 and 3 and using the same data, we have

$$r_{1x} = 7.0711, \quad r_{1y} = 7.0711,$$

$$r_{2x} = 33.590, \quad r_{2y} = 9.853,$$
$$r_{3x} = -10.660, \quad r_{3y} = -16.929,$$

and

$$\omega_2 = -9.567, \text{ and } \omega_3 = 36.208.$$

Using these values, we obtain

$$\begin{aligned}
a &= -\alpha_1 r_{1y} - \omega_1{}^2 r_{1x} - \omega_2^2 r_{2x} - \omega_3^2 r_{3x} \\
&= 0 - 100^2(7.0711) - 9.567^2(33.590) - 36.208^2(-10.660) \\
&= -59{,}810 \text{ mm/s}^2, \\
b &= -\alpha_1 r_{1x} + \omega_1{}^2 r_{1y} + \omega_2^2 r_{2y} + \omega_3^2 r_{3y} \\
&= 0 + 100^2(7.0711) + 9.567^2(9.853) + 36.208^2(-16.929) \\
&= 49{,}381 \text{ mm/s}^2, \\
D &= r_{2y}r_{3x} - r_{2x}r_{3y} = 9.853(-10.660) - 33.590(-16.929) \\
&= 463.80, \\
\alpha_2 &= \frac{ar_{3x} - br_{3y}}{D} = \frac{(-59{,}810)(-10.660) - (49{,}381)(-16.929)}{463.80} \\
&= 3{,}180 \text{ rad/s}^2,
\end{aligned}$$

and

$$\begin{aligned}
\alpha_3 &= \frac{br_{2y} - ar_{2x}}{D} = \frac{(49{,}381)(9.853) - (-58{,}600)(33.590)}{463.80} \\
&= 5{,}386 \text{ rad/s}^2.
\end{aligned}$$

Once the angular accelerations are determined, we may find the acceleration of any point on the linkage. The acceleration of point B is given by

$$\begin{aligned}
\boldsymbol{a}_B &= \boldsymbol{\alpha}_1 \times \boldsymbol{r}_1 + \boldsymbol{\omega}_1 \times (\boldsymbol{\omega}_1 \times \boldsymbol{r}_1) = -\omega_1{}^2(\boldsymbol{i}r_{1x} + \boldsymbol{j}r_{1y}) \\
&= -100^2(\boldsymbol{i}7.0711 + \boldsymbol{j}7.0711) = \boldsymbol{i}70{,}711 - \boldsymbol{j}70{,}711 \\
&= 100{,}000 \text{ mm/s}^2 \angle -135°.
\end{aligned}$$

The acceleration of point C, as determined directly from its position on link 3, is

$$\begin{aligned} \boldsymbol{a}_C &= \boldsymbol{\alpha}_3 \times (-\boldsymbol{r}_3) + \boldsymbol{\omega}_3 \times (\boldsymbol{\omega}_3 \times (-\boldsymbol{r}_3)) \\ &= \alpha_3(-\boldsymbol{j}r_{3x} + \boldsymbol{i}r_{3y}) - \omega_3^2(-\boldsymbol{i}r_{3x} - \boldsymbol{j}r_{3y}) \\ &= 5{,}386(\boldsymbol{j}10.660 - \boldsymbol{i}16.929) - 36.208^2(\boldsymbol{i}10.660 + \boldsymbol{j}16.929) \\ &= -\boldsymbol{i}105{,}100 + \boldsymbol{j}35{,}200 = 110{,}900 \text{ mm/s}^2\angle 161.5°. \end{aligned}$$

As an alternative, we may calculate both the acceleration of point C in terms of the acceleration of B and the acceleration of C with respect to B. We have

$$\boldsymbol{a}_C = \boldsymbol{a}_B + \boldsymbol{a}_{CB},$$

where

$$\boldsymbol{a}_{CB} = \boldsymbol{\alpha}_2 \times \boldsymbol{r}_2 + \boldsymbol{\omega}_2 \times (\boldsymbol{\omega}_2 \times \boldsymbol{r}_2) = \alpha_2(\boldsymbol{j}r_{2x} - \boldsymbol{i}r_{2y}) - \omega_2^2(\boldsymbol{i}r_{2x} + \boldsymbol{j}r_{2y}).$$

Substituting the values obtained earlier, we get

$$\begin{aligned} \boldsymbol{a}_C &= -\boldsymbol{i}70{,}711 - \boldsymbol{j}70{,}711 + 3{,}180(\boldsymbol{j}33.590 - \boldsymbol{i}9.835) \\ &\quad - 9.567^2(\boldsymbol{i}33.590 + \boldsymbol{j}9.835), \end{aligned}$$

which differs from the previous result due to rounding errors alone.

The acceleration of point D in the coupler is given by

$$\boldsymbol{a}_D = \boldsymbol{\alpha}_1 \times \boldsymbol{r}_1 + \boldsymbol{\omega}_1 \times (\boldsymbol{\omega}_1 \times \boldsymbol{r}_1) + \boldsymbol{\alpha}_2 \times \boldsymbol{r}_{BD} + \boldsymbol{\omega}_2 \times (\boldsymbol{\omega}_2 \times \boldsymbol{r}_{BD}),$$

where the radius vector extending from B to D is

$$\boldsymbol{r}_{BD} = r_{BD}\angle\theta_2 + \theta_{CBD} = 15\,\text{mm}\,\angle 16.35 + 20° = \boldsymbol{i}12.081 + \boldsymbol{j}8.891,$$

so that

$$\begin{aligned} \boldsymbol{a}_D &= -100^2(\boldsymbol{i}7.0711 + \boldsymbol{j}7.0711) + 3{,}180(\boldsymbol{j}12.081 - \boldsymbol{i}8.891) \\ &\quad - 9.567^2(\boldsymbol{i}12.081 + \boldsymbol{j}8.891) \\ &= -\boldsymbol{i}100{,}100 - \boldsymbol{j}33{,}100 = 105{,}400 \text{ mm/s}^2\angle -161.7°. \end{aligned}$$

4.3 ACCELERATION ANALYSIS WITH A SPREADSHEET

If the displacement and velocity formulas for a given type of linkage are already programmed on a spreadsheet, then the acceleration formulas can be added with little difficulty. To analyze a four-bar linkage, we may use the equations in the previous section, converting them to spreadsheet form. The results of the analysis may be plotted with the spreadsheet plotting routines. The graphical results aid in checking for programming errors, since inconsistencies are more easily detected in plotted results than in tabulated results.

SAMPLE PROBLEM 4.4

Utilizing a Spreadsheet to Determine Angular Accelerations in a Four-Bar Linkage

For the four-bar linkage described in Sample Problem 4.3, let the angular velocity of the crank be 100 rad/s (constant and counterclockwise). Tabulate and plot the angular velocities and angular accelerations of links 2 and 3 against the crank angle.

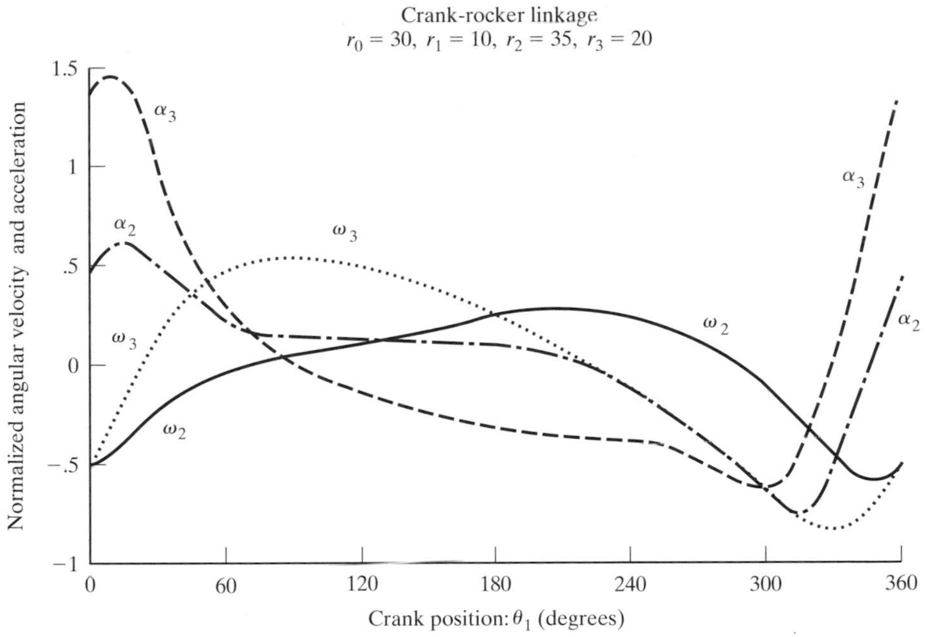

FIGURE 4.5 Velocities and accelerations in a crank-rocker linkage.

Solution. The equations used in Sample Problem 4.3 were converted into spreadsheet formulas. The crank position was changed in 5° increments, and the formulas were copied throughout the spreadsheet.

For plotting purposes, angular velocities are normalized by dividing by the crank angular velocity, and angular accelerations are normalized by dividing by the square of the crank angular velocity. The plotted results are shown in Figure 4.5. Note that the angular velocities of links 2 and 3 are equal at $\theta_1 = 0°$ and also at $\theta_1 = 180°$. We observed that this was the case when we examined the centros of a four-bar linkage. Note also that zero angular acceleration of a given link corresponds to an angular velocity extremum (a maximum or minimum).

4.4 VECTOR MANIPULATION WITH MATHEMATICS SOFTWARE

If we choose to work directly with vectors to solve linkage acceleration problems, we may write programs for vector manipulation or use commercially available mathematics software. In most cases, the vector solution will require less calculation on our part, but will require more computer time than solutions in scalar form.

Consider the offset slider-crank linkage of Figure 4.6a. The linkage may be described by the following position and velocity equations, as given in Chapters 2 and 3:

$$\boldsymbol{e} + \boldsymbol{r}_1 + \boldsymbol{r}_2 + \boldsymbol{r}_0 = 0$$

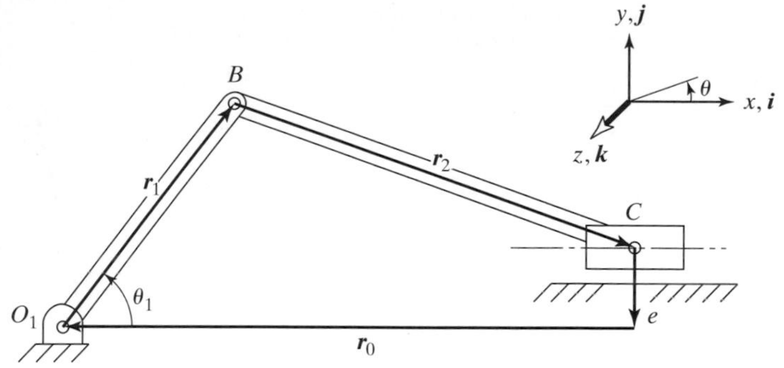

FIGURE 4.6 (a) Offset slider crank linkage.

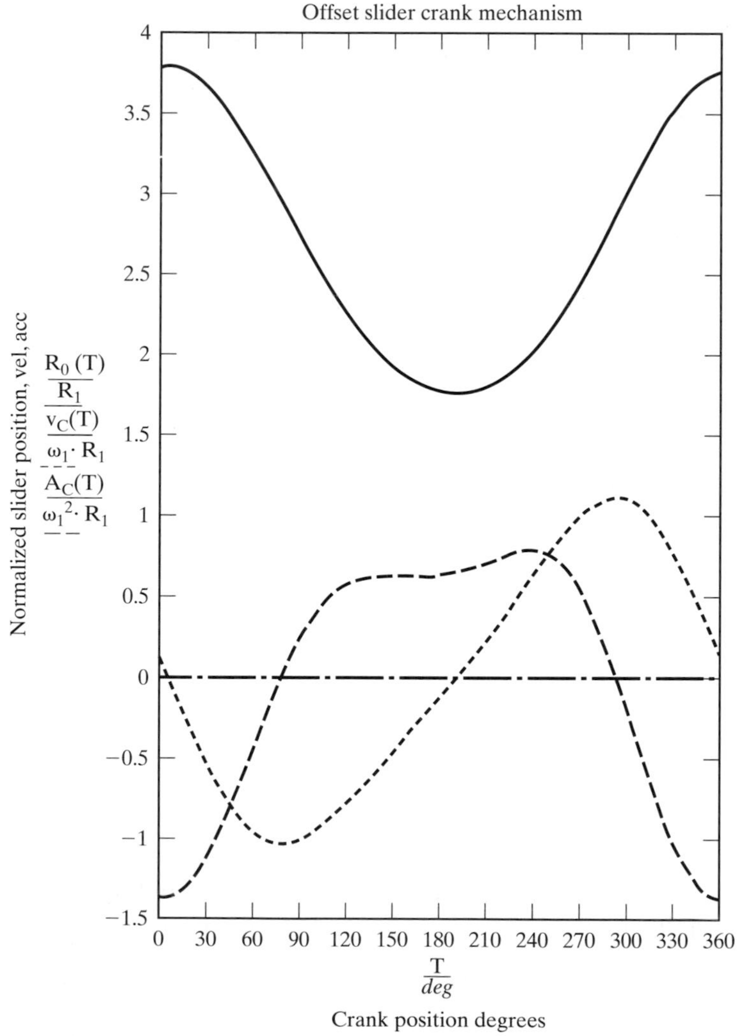

FIGURE 4.6 (b) Slider motion.

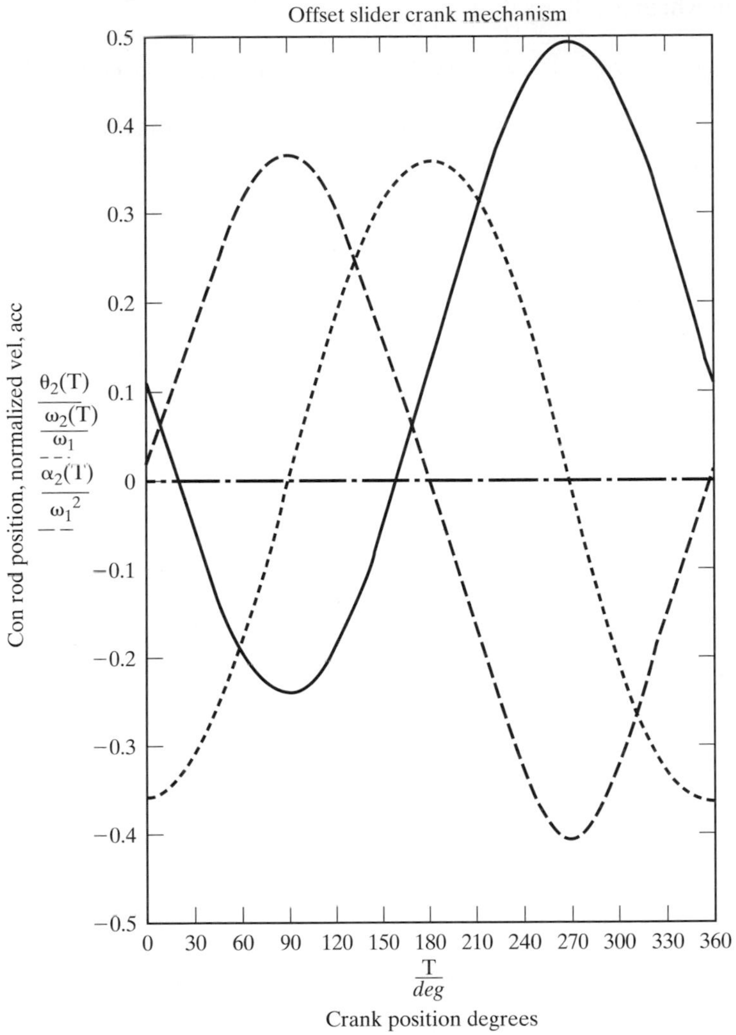

FIGURE 4.6 (c) Connecting-rod motion.

and

$$\boldsymbol{\omega}_1 \times \boldsymbol{r}_1 + \boldsymbol{\omega}_2 \times \boldsymbol{r}_2 - \boldsymbol{v}_c = 0.$$

Differentiating the latter equation with respect to time, we obtain

$$\boldsymbol{\omega}_1 \times (\boldsymbol{\omega}_1 \times \boldsymbol{r}_1) + \boldsymbol{\omega}_2 \times (\boldsymbol{\omega}_2 \times \boldsymbol{r}_2) + \boldsymbol{\alpha}_2 \times \boldsymbol{r}_2 - \boldsymbol{a}_c = 0 \tag{4.28}$$

if the angular velocity of the crank is constant.

Noting that the acceleration of the slider lies along the x-axis, we may eliminate the last term in the Eq. (4.28) by taking the dot product of each term with the unit vector $\boldsymbol{j}$. The result is

$$\boldsymbol{\omega}_1 \times (\boldsymbol{\omega}_1 \times \boldsymbol{r}_1)\cdot\boldsymbol{j} + \boldsymbol{\omega}_2 \times (\boldsymbol{\omega}_2 \times \boldsymbol{r}_2)\cdot\boldsymbol{j} + \boldsymbol{\alpha}_2 \times \boldsymbol{r}_2\cdot\boldsymbol{j} = 0,$$

from which it follows that

$$\alpha_2 = [-\boldsymbol{\omega}_1 \times (\boldsymbol{\omega}_1 \times \boldsymbol{r}_1)\cdot \boldsymbol{j} - \boldsymbol{\omega}_2 \times (\boldsymbol{\omega}_2 \times \boldsymbol{r}_2)\cdot \boldsymbol{j}]/(\boldsymbol{k} \times \boldsymbol{r}_2 \cdot \boldsymbol{j}), \qquad (4.29)$$

where $\boldsymbol{\omega} = \omega \boldsymbol{k}$ and $\boldsymbol{\alpha} = \alpha \boldsymbol{k}$. The slider acceleration may now be obtained by rearranging Eq. (4.28) as follows:

$$\boldsymbol{a}_c = \boldsymbol{\omega}_1 \times (\boldsymbol{\omega}_1 \times \boldsymbol{r}_1) + \boldsymbol{\omega}_2 \times (\boldsymbol{\omega}_2 \times \boldsymbol{r}_2) + \boldsymbol{\alpha}_2 \times \boldsymbol{r}_2. \qquad (4.30)$$

Displaying the results graphically. To display our results, we want scalars. If we used the absolute values of vectors $\boldsymbol{r}_0$, $\boldsymbol{v}_C$, and $\boldsymbol{a}_C$ in plots and tables, the directions would be lost. If the slider moves horizontally, we can use dot products to define position, velocity, and acceleration scalars:

$$R_0 = -\boldsymbol{r}_0 \cdot \boldsymbol{i};$$
$$V_C = \boldsymbol{v}_C \cdot \boldsymbol{i};$$
$$A_C = \boldsymbol{a}_C \cdot \boldsymbol{i}.$$

The velocity and acceleration vectors are already defined as positive to the right (the $\boldsymbol{i}$, or +x, direction). The sign change makes R_0 consistent with that definition. When several curves are displayed on the same graph, we usually need to scale the numbers. One suggestion is that we plot normalized values, using the dimensionless quantities

$$R_0/R_1, \quad V_C/(\omega_1 R_1), \quad A_C/(\omega_1^2 R_1), \quad \omega_2/\omega_1, \quad \text{and} \quad \alpha_2/\omega_1^2\,.$$

Then, all the curves will usually be of the same order of magnitude. Note that the second and third terms are equivalent to dividing the wrist-pin velocity and acceleration by the crankpin velocity and normal acceleration.

Sign convention for angles. Remember the sign convention: Counterclockwise is positive for angular position, velocity, and acceleration. That is, angular velocity or acceleration in the +z ($\boldsymbol{k}$) direction is positive. For spatial linkages, angular velocity and acceleration vector components in the $+x$, $+y$, and $+z$ directions (the $\boldsymbol{i}$, $\boldsymbol{j}$, and $\boldsymbol{k}$ directions) are positive.

Verifying results. Murphy's law, "If anything can go wrong, it will," is not entirely a joke. One antidote is to check results frequently. Computers make checking easy. For example, we can check the value of $\boldsymbol{r}_0$ and the vector position equation at some crank angle. Does $\boldsymbol{r}_0$ agree with its representation in a freehand sketch? Does the position equation represent a closed vector loop, giving us a zero vector? And do the units check?

Numerical differentiation and the chain rule. Numerical differentiation provides additional verification. In Chapter 3, we used the chain rule with numerical differentiation to check the angular velocity of one link in a spatial linkage. Using the chain rule and numerical differentiation for the offset slider crank with constant crank speed, we have

$$V_C = \omega_1\, dR_0/d\theta_1$$
$$A_C = \omega_1^2\, d^2R_0/d\theta_1^2\,,$$
$$\omega_2 = \omega_1\, d\theta_2/d\theta_1,$$

and

$$\alpha_2 = \omega_1^2\, d^2\theta_2/d\theta_1^2\,.$$

Do these values check the results from vector velocity and acceleration methods?

SAMPLE PROBLEM 4.5

Acceleration analysis of an offset slider-crank linkage: direct vector manipulation using mathematics software

An offset slider crank linkage similar to that in Figure 4.6a is described by the vector equation

$$\boldsymbol{r_1} + \boldsymbol{e} + \boldsymbol{r_2} + \boldsymbol{r_0} = \boldsymbol{0}.$$

Crank length = 15, connecting rod length = 42, and offset vector

$$\boldsymbol{e} = \begin{bmatrix} 0 \\ -5 \\ 0 \end{bmatrix} \quad \text{(all mm)}.$$

The crank rotates counterclockwise at a constant speed of 500 rpm. Find link position, velocity, and acceleration vectors. Plot the slider position, velocity, and acceleration and the angular position, velocity, and acceleration of the connecting rod (all against the crank position). Check your results.

Solution summary. We will use lowercase letters for vectors and uppercase for scalars in most cases. The computer knows the difference because vectors are identified as matrices with three rows and one column. For convenience, crank position θ_1 (radians) is replaced by T.

The crank angle goes from zero to 2π, and the vector crank position is easily calculated. We add it to the given offset vector and call the resulting vector $\boldsymbol{r_3}$. Position vectors are calculated from the equations in Chapter 2, with one magnitude (R_0) and one direction (θ_2) unknown. We are lucky here; the four vectors add to zero for an arbitrary value of θ_1 (1 rad). Sometimes, the sum is a small value, say 10^{-15}, which just represents rounding error. The connecting-rod angle is calculated with the two-argument Mathcad function *angle* $(r_{2(0)}, r_{2(1)})$, where the x, y, and z components of vector $\boldsymbol{r_2}$ are numbered 0, 1, 2, and respectively. This function gives arctangent values $0 \le \theta \le 2\pi$ rad. If θ_2 jumps around from 0 to 2π rad, the appearance of the graph can be improved by an *IF statement* in the form

$$\text{if}\,(\theta_2 \le \pi, \theta_2, \theta_2 - 2\pi).$$

If the inequality is true, the value after the first comma holds; if not, the value after the second comma does. That is, if the calculated value is less than or equal to π, it is used as is; if not, we subtract 2π. Programming languages and other software may use a different two-argument arctangent function (ARCTAN_2, for example), which may yield angle values between $-\pi$ and π. If you are unfamiliar with the software, be sure to read the instructions and use the *help* screens.

Velocity and acceleration vectors are calculated using cross products. The magnitudes of the slider velocity and the slider acceleration are given by the dot product of the vector and the unit vector ***i***. Both velocity and acceleration values at an arbitrary crank angle agree with results of numerical differentiation (subscripted n) using the chain rule.

The results are plotted in dimensionless form. It is encouraging to note that zero slope of the velocity curve corresponds to zero acceleration, etc. It is easy to convert back to actual values. For example, a value taken from the acceleration curve is multiplied by $\omega_1^2 R_1$, where $\omega_1 = 500\,\pi/30$ and $R_1 = 15$, to obtain A_c (mm/s^2).

Solution details **(The software used to solve this problem does not identify vectors with boldface type).**

Units: mm, sec, rad.

Vector equation $r_3 + r_2 + r_0 = 0$, where $r_3 = r_1 + e$

Let $T = \theta_1 =$ crank position (radians): $T := 0, \frac{\pi}{72}, 2\pi$

Given:

Crank length $R_1 := 15$ Connecting-rod length $R_2 := 42$

Offset $e := \begin{bmatrix} 0 \\ -5 \\ 0 \end{bmatrix}$ r_0 unit vector $r_{0u} := \begin{bmatrix} -1 \\ 0 \\ 0 \end{bmatrix}$

Crank speed-rpm: $n_1 := 500$

Angular velocity, rad/s: $\omega_1 := \frac{\pi \cdot n_1}{30}$ $\omega_1 = 52.36$

Rectangular unit vectors: $i := \begin{bmatrix} 1 \\ 0 \\ 0 \end{bmatrix}$ $j := \begin{bmatrix} 0 \\ 1 \\ 0 \end{bmatrix}$ $k := \begin{bmatrix} 0 \\ 0 \\ 1 \end{bmatrix}$

Angular acceleration: $\alpha_1 := 0$

Position analysis

The magnitude of r_0 and the direction of r_2 are unknown.

Crank vector: $r_1(T) := \begin{bmatrix} R_1 \cdot \cos(T) \\ R_1 \cdot \sin(T) \\ 0 \end{bmatrix}$ Add offset, define vector: $r_3(T) := r_1(T) + e$

Slider position vector: $r_0(T) := \left[-r_3(T) \cdot r_{0u} + \left[R_2^2 - [r_3(T) \cdot (r_{0u} \times k)]^2 \right]^{\frac{1}{2}} \right] \cdot r_{0u}$

Scalar for plotting: $R_0(T) := -r_0(T) \cdot i$ (positive to right)

For convenience, define $A(T) := r_3(T) \cdot (r_{0u} \times k)$

Connecting-rod vector: $r_2(T) := -A(T) \cdot (r_{0u} \times k) - (R_2^2 - A(T)^2)^{\frac{1}{2}} \cdot r_{0u}$

Angular position: $q(T) := \text{angle}(r_2(T)_0, r_2(T)_1)$

For plotting: $\theta_2(T) := \text{if}(q(T) \le \pi, q(T), q(T) - 2\pi)$

Check vector closure: $e + r_0(1) + r_1(1) + r_2(1) = \begin{bmatrix} 0 \\ 0 \\ 0 \end{bmatrix}$

Velocity analysis

Connecting-rod angular velocity: $\omega_2(T) := \frac{-\omega_1 \cdot (r_1(T) \cdot i)}{r_2(T) \cdot i}$

Slider velocity vector (positive to right): $V_C(T) := \omega_1 \cdot (k \times r_1(T)) + \omega_2(T) \cdot (k \times r_2(T))$
Scalar: $V_C(T) := V_C(T) \cdot i$

Numerical differentiation: $V_{Cn}(T) := \omega_1 \cdot \left(\frac{d}{dT} R_0(T)\right) \quad \omega_{2n}(T) := \omega_1 \cdot \left(\frac{d}{dT} \theta_2(T)\right)$

Check: $V_C(1) = -739.201 \quad V_{Cn}(1) = -739.201$ mm/s
$\omega_2(1) = -10.274 \quad \omega_{2n}(1) = -10.274$ rad/s

Acceleration analysis

Connecting rod angular acceleration: $\alpha_2(T) := \frac{-\omega_1 \cdot k \times (\omega_1 \cdot k \times r_1(T)) \cdot j - \omega_2(T) \cdot k \times (\omega_2(T) \cdot k \times r_2(T)) \cdot j}{k \times r_2(T) \cdot j}$

Slider acceleration vector: $a_C(T) := (\omega_1 \cdot k) \times [(\omega_1 \cdot k) \times r_1(T)] \cdots$
$+(\omega_2(T) \cdot k) \times (\omega_2(T) \cdot k \times r_2(T)) + \alpha_2(T) \cdot k \times r_2(T)$

Scalar: $A_C(T) := a_C(T) \cdot i$

Numerical differentiation:

$$A_{Cn}(T) := \omega_1^2 \frac{d}{dT}\left(\frac{d}{dT} R_0(T)\right) \quad \alpha_{2n}(T) := \omega_1^2 \frac{d}{dT}\left(\frac{d}{dT} \theta_2(T)\right)$$

Check: $A_C(1) = -2.034 \cdot 10^4 \qquad \alpha_2(1) = 818.339$

$$A_{Cn}(1) = -2.034 \cdot 10^4 \frac{mm}{s^2} \quad \alpha_{2n}(1) = 818.339 \frac{rad}{s^2}$$

Engineering Significance. Slider crank linkages, which include piston engines and pumps, are an important class of machine components. In most cases, they are in-line; that is, the piston (slider path) centerline intersects the center of the crankshaft.

The equations in the detailed solution just presented apply to the in-line case if we set offset equal to zero. With a zero offset, the slider position, velocity, and acceleration are symmetric or antisymmetric about crank position $\theta_1 = \pi$ rad (180°). The same applies to the connecting-rod position, angular velocity, and angular acceleration. None of the curves showed symmetry or antisymmetry in the foregoing offset slider-crank example.

Inertial forces and torques are critical in the design of high-speed machinery. These inertial effects often exceed applied forces and torques. We need accelerations and angular accelerations to determine inertial effects. (Methods of analysis and design are discussed in a later chapter.)

4.5 COMPLEX-NUMBER METHODS APPLIED TO ACCELERATION ANALYSIS

As illustrated in Chapters 2 and 3, complex numbers are a convenient form for representing the vectors that model planar linkage elements and velocities. It follows, therefore,

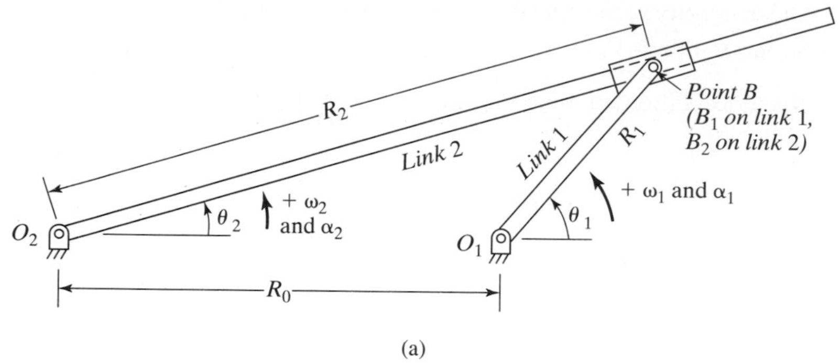

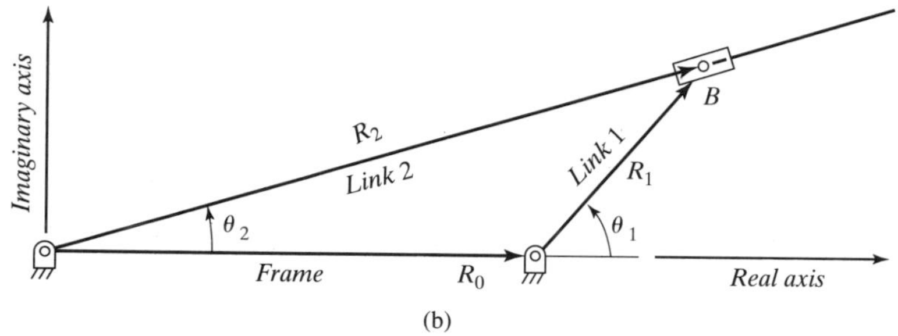

FIGURE 4.7 (a) Schematic for sliding contact linkage. (b)Vector representation.

that complex number methods can be applied to acceleration analysis of planar linkages. Consider the sliding contact linkage described in Section 3.7 and shown in Figures 4.7a and b. The displacement equation is given by

$$R_0 + R_1 e^{j\theta_1} = R_2 e^{j\theta_2}$$

and the velocity equation by

$$j\omega_1 R_1 e^{j\theta_1} = j\omega_2 R_2 e^{j\theta_2} + v_{B_1B_2} e^{j\theta_2},$$

as in Chapter 3.

Differentiating with respect to time, we obtain the acceleration equation:

$$- \omega_1{}^2 R_1 e^{j\theta_1} + j\frac{d\omega_1}{dt} R_1 e^{j\theta_1}$$
$$= -\omega_2^2 R_2 e^{j\theta_2} + j2\omega_2 v_{B_1B_2} e^{j\theta_2} + j\frac{d\omega_2}{dt} R_2 e^{j\theta_2} + \frac{dv_{B_1B_2}}{dt} e^{j\theta_2}. \qquad (4.31)$$

Thus, we have the vector equation

$$\boldsymbol{a}^n_{B_1} + \boldsymbol{a}^t_{B_1} = \boldsymbol{a}^n_{B_2} + \boldsymbol{a}^t_{B_2} + \boldsymbol{a}^c_{B_1B_2} + \boldsymbol{a}^t_{B_1B_2}, \qquad (4.32)$$

where the vector magnitudes are the normal accelerations

$$a^n_{B_1} = \omega_1^2 R_1 \quad \text{and} \quad a^n_{B_2} = \omega_2^2 R_2,$$

the angular acceleration

$$\frac{d\omega}{dt} = \alpha,$$

the tangential accelerations

$$\alpha_1 R_1 = a^t_{B_1} \quad \text{and} \quad \alpha_2 R_2 = a^t_{B_2},$$

the Coriolis acceleration

$$2\omega_2 v_{B_1B_2} = a^c_{B_1B_2},$$

and the relative (tangential) acceleration of B_1 with respect to B_2

$$\frac{dv_{B_1B_2}}{dt} = a^t_{B_1B_2}$$

which is positive if dR_2/dt is increasing (note again that the order of the subscripts is critical).

Solving the Complex Acceleration Equation

In a problem of this type, it is likely that link lengths R_0 and R_1 would be specified, as would the angular velocity and acceleration of link 1. Then R_2 and θ_2 can be found for given values of θ_1 by using the displacement equations as in Chapter 2. Similarly, angular velocity ω_2 and relative velocity $v_{B_1B_2}$ can be found as in Chapter 3. The remaining unknowns in Eq. 4.31 are $d\omega_2/dt$ and $dv_{B_1B_2}/dt$. All of the terms in Eq. 4.31 are, in general, complex. If each term is multiplied by $e^{-j\theta_2}$, the equation then takes the form

$$(-\omega_1{}^2 + j\alpha_1)R_1 e^{j(\theta_1-\theta_2)} = -\omega_2^2 R_2 + j2\omega_2 v_{B_1B_2} + j\alpha_2 R_2 + a^t_{B_1B_2}, \tag{4.33}$$

where $\alpha_2 = d\omega_2/dt$ and $a^t_{B_1B_2} = dv_{B_1B_2}/dt$. The two unknowns can now be separated, since the term containing $a^t_{B_1B_2}$ is real and the term containing α_2 is imaginary. Using the Euler formula

$$e^{j(\theta_1-\theta_2)} = \cos(\theta_1 - \theta_2) + j\sin(\theta_1 - \theta_2) \tag{4.34}$$

and noting that $j^2 = -1$, we equate the real parts of the resulting equation to obtain

$$a^t_{B_1B_2} = -\omega_1^2 R_1 \cos(\theta_1 - \theta_2) + \alpha_1 R_1 \sin(\theta_2 - \theta_1) + \omega_2^2 R_2. \tag{4.35}$$

Equating the imaginary parts yields

$$\alpha_2 = \frac{1}{R_2}\left[\omega_1^2 R_1 \sin(\theta_2 - \theta_1) + \alpha_1 R_1 \cos(\theta_2 - \theta_1) - 2\omega_2 v_{B_1B_2}\right]. \tag{4.36}$$

You may find complex numbers an unpleasant dose of mathematics. Our solution to a linkage problem included the differentiation of complex quantities, multiplication by a complex quantity, applying the Euler formula, and separating real and imaginary parts. In return, we get separate equations for the unknowns. The complex-number method eliminated the need for vector manipulation and solutions of simultaneous equations by matrix or determinant methods.

We had to multiply the sliding contact linkage equations by a certain complex quantity to get the needed results. But look before you leap: The object is to get at least one unknown term in a purely real or purely imaginary expression. A different linkage may call for multiplication by a different complex quantity. Or the variables may already be separated; if so, then skip a step.

If you can manipulate complex quantities, you have a powerful tool for solving problems in many engineering fields. Look for new applications, but remember that you are limited to two-dimensional problems. Complex-number methods work well with planar linkages, for example, but not spatial linkages.

SAMPLE PROBLEM 4.6

Accelerations in a Sliding Contact Linkage

Referring to Figure 4.7a and b, let $\omega_1 = 20$ rad/s (constant and counterclockwise), $R_0 = 40$ mm, and $R_1 = 20$ mm. Find $a^t_{B_1B_2}$ and α_2 at $\theta_1 = 75°$.

Solution. Using the equations of Chapters 2 and 3, we obtain $R_2 = 49.13$ mm, $\theta_2 = 23.15$, $\omega_2 = 5.03$ rad/s ccw, and $v_{B_1B_2} = -314.6$ mm/s.
The relative acceleration is found by noting that ω_1 is constant ($\alpha_1 = 0$):

$$\begin{aligned} a^t_{B_1B_2} &= -\omega_1^2 R_1 \cos(\theta_1 - \theta_2) + \alpha_1 R_1 \sin(\theta_2 - \theta_1) + \omega_2^2 R_2 \\ &= -20^2 \times 20 \cos(75° - 23.15°) + 5.03^2 \times 49.13 = -3{,}698.7 \text{ mm/s}^2 \end{aligned}$$

(i.e., 3698.7 mm/s^2 along link 2 toward O_2). The angular acceleration is

$$\begin{aligned} \alpha_2 &= \frac{\omega_1^2 R_1 \sin(\theta_2 - \theta_1) + \alpha_1 R_1 \cos(\theta_2 - \theta_1) - 2\omega_2 v_{B_1B_2}}{R_2} \\ &= \frac{20^2 \times 20 \sin(23.15° - 75°) - 2 \times 5.03(-314.6)}{49.13} = -63.63 \text{ rad/s}^2 \end{aligned}$$

(i.e., $\alpha_2 = 63.63$ rad/s^2 clockwise).

Engineering Applications. Let the input link of the sliding contact linkage in sample problem 4.6 rotate at constant speed. The output link oscillates between limiting positions, but the average speed in the clockwise direction will not equal the average speed in the counterclockwise direction. The oscillating link can be joined to other links to form a quick-return mechanism. If the linkage is used as a shaper, our objective would be a high-force stroke for cutting metal and a quick return. We need to know cutting speeds, and we need accelerations to find forces and torques, particularly at high speeds.

Some designs call for lightweight materials to reduce inertial forces and torques. Other applications include a flywheel or other high mass-moment-of-inertia parts to store energy and reduce fluctuations in speed. Shapers, punch presses, and engines are a few of the many machines utilizing flywheels.

4.6 THE ACCELERATION POLYGON

Planar linkage acceleration problems may be solved with the aid of motion simulation software (e.g., Working Model), by analytical vector methods, by complex-number methods, or by numerical differentiation of the position and velocity values. The acceleration polygon, a graphical vector method, is another alternative; you may find it useful for spot-checking the results of a different method. If you want to find the accelerations of a linkage through a full range of motion, total reliance on graphical methods is not "working smart."

Analysis of Slider-Crank Mechanisms

The acceleration polygon is analogous to the velocity polygon discussed in Chapter 3. The vector polygon provides us with a convenient method of finding unknown vectors through their relationship to known (easily calculated) vectors. In the current situation, the vectors being considered are accelerations. The acceleration polygon is simply the graphical expression of the acceleration vector equation, Eq. (4.22), where $\boldsymbol{a}_C = \boldsymbol{a}_C^t$. Vector $\boldsymbol{a}_C^n = 0$, since the slider moves in a straight path.

SAMPLE PROBLEM 4.7

Acceleration Polygon for a Slider-Crank Linkage

Figure 4.8 shows a slider-crank linkage that was examined in the preceding chapter. We want to find the acceleration of point C on the slider.

Solution. The velocity polygon in Figure 4.8b is taken from Chapter 3. Since the crank has a constant angular velocity,

$$\boldsymbol{a}_B^t = \boldsymbol{\alpha}_1 \times \boldsymbol{O_1B} = 0.$$

Thus, the total acceleration of B is

$$\boldsymbol{a}_B = \boldsymbol{a}_B^n = \boldsymbol{\omega}_1 \times (\boldsymbol{\omega}_1 \times \boldsymbol{O_1B}),$$

with the magnitude given by

$$a_B^n = \omega_1^2 O_1B = \frac{v_B^2}{O_1B} = \frac{(ob)^2}{O_1B} = \frac{(20 \text{ in/s})^2}{2 \text{ in}}$$
$$= 200 \text{ in/s}^2$$

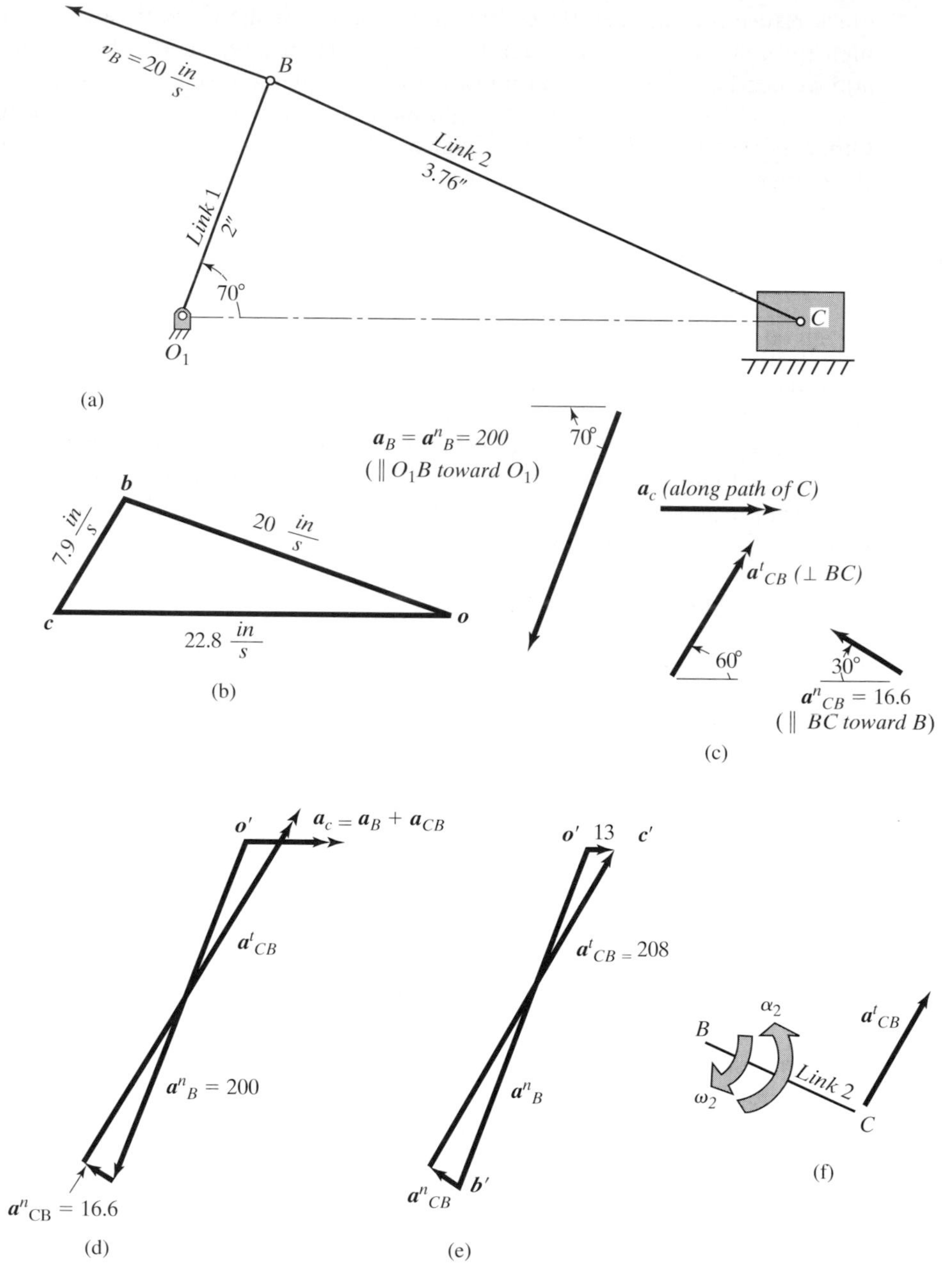

FIGURE 4.8 (a) Slider-crank mechanism drawn to scale. (b) The velocity polygon. (c) The directions of the acceleration components are identified by inspecting the orientation of the linkage. If vector magnitudes can be determined (with the aid of the velocity polygon and link lengths), the vectors are drawn to scale. If the vector magnitudes are unknown, they are drawn with double arrowheads. (d) The acceleration polygon is begun. (e) Accelerations $\boldsymbol{a}_C$ and $\boldsymbol{a}^t_{CB}$ are scaled directly from the acceleration polygon. (f) Knowing tangential acceleration $\boldsymbol{a}^t_{CB}$, and knowing the length of link BC, we can find the angular acceleration of the link.

TABLE 4.1 Tabulation for the Vector Acceleration Polygon of a Slider-Crank Mechanism with Uniform Crank Velocity

Vector	$\boldsymbol{a}_C =$	$\boldsymbol{a}_B^n$	$+\boldsymbol{a}_B^t$	$+\boldsymbol{a}_{CB}^n$	$+\boldsymbol{a}_{CB}^t$
Vector magnitude	?	$\frac{(ob)^2}{O_1B}$	$\alpha_1 O_1 B$	$\frac{(bc)^2}{BC}$	?
Vector direction	‖ path of C	$\| O_1$ B toward O_1	$\perp O_1B$	$\| BC$ toward B	$\perp BC$
Vectors used to construct polygon	? ⟶⟶	200 in/s^2 ↙	0	16.6 in/s^2 ↖	? ↗

in a direction parallel to O_1B toward O_1. (See Figure 4.8c and the values of acceleration listed in Table 4.1.)

Before constructing the actual acceleration polygon, it will prove helpful to note the directions of the acceleration vector, which are apparent from the linkage drawn in Figure 4.8a. After noting the linkage orientation, the restraints on the mechanism, and the given data, we can identify the following acceleration vectors (Figure 4.8c):

$\boldsymbol{a}_C$, along the horizontal path to which the slider is constrained;
$\boldsymbol{a}_B^n$, parallel to O_1B and toward fixed point O_1;
$\boldsymbol{a}_B^t = 0$, since crank O_1B rotates with constant angular velocity;
$\boldsymbol{a}_{CB}^n$, parallel to link BC and directed toward B; and
$\boldsymbol{a}_{CB}^t$, perpendicular to link BC.

The procedure for constructing the acceleration polygon is similar in some ways to that for constructing the velocity polygon. The first step, shown in Figure 4.8d, includes the selection of an acceleration scale that will result in an acceleration polygon of reasonable size. Of course, the accelerations are not all known at this time, but it may be assumed that they are of the same order of magnitude as $\boldsymbol{a}_B$, the acceleration of the crankpin.

In Figure 4.8d, the vector $\boldsymbol{a}_B^n$ has already been drawn. The tangential acceleration of the crankpin is zero in the special case under consideration, eliminating the term $\boldsymbol{a}_B^t$ from Eq. (4.22). (*Note*: It does not necessarily follow that the tangential acceleration of C with respect to B is likewise zero; in fact, it will be shown that $\boldsymbol{a}_{CB}^t$ is quite large in this example.) Since $a_B^t = 0$, we must next evaluate $\boldsymbol{a}_{CB}^n$, the normal acceleration of C with respect to B, which is given by

$$\boldsymbol{a}_{CB}^n = \boldsymbol{\omega}_2 \times (\boldsymbol{\omega}_2 \times \boldsymbol{BC}).$$

The magnitude is

$$a_{CB}^n = \omega_2^2 BC = \frac{v_{CB}^2}{BC} = \frac{(bc)^2}{BC} = \frac{7.9^2}{3.76} = 16.6 \text{ in/s}^2,$$

and the direction is opposite that the vector $\boldsymbol{BC}$. That is, $\boldsymbol{a}_{CB}^n$ will lie along the connecting rod BC, directed toward B. Vector $\boldsymbol{a}_{CB}^n$ is then drawn to scale and added to the head of vector $\boldsymbol{a}_B^n$ (Figure 4.8d), parallel to BC and directed toward point B. The final vector, $\boldsymbol{a}_{CB}^t$, is added at the head of, and perpendicular to, $\boldsymbol{a}_{CB}^n$ to complete the vector sum of Eq. (4.22). A double arrowhead is used in Figure 4.8d to indicate that the length of $\boldsymbol{a}_{CB}^t$ is not yet known. The sum $\boldsymbol{a}_B^n + \boldsymbol{a}_{CB}^n + \boldsymbol{a}_{CB}^t$ represents the total acceleration of C; the true direction of the acceleration of C is horizontal. (The direction of C was obvious at the outset and was drawn as a horizontal vector in Figure 4.8c. We therefore draw $\boldsymbol{a}_C$ horizontally (to the right to close the polygon) from pole point o' in the acceleration polygon of Figure 4.8d, again using a double arrowhead, since the magnitude of a_C is unknown. Both vectors $\boldsymbol{a}_C$ and $\boldsymbol{a}^t_{CB}$ end where they intersect, a point we label c'. Measuring the lengths of each on the acceleration scale, we find that

$$\boldsymbol{a}^t_{CB} = 208 \text{ in/s}^2 \text{ (upward and to the right)}$$

and

$$\boldsymbol{a}_C = 13 \text{ in/s}^2 \text{ (to the right)},$$

as shown in Figure 4.8e. If greater accuracy is required, we may compute the exact angles and use trigonometric functions or employ analytical or computer methods from the start.

Knowing the length BC and the tangential acceleration of C with respect to B, we can find the angular acceleration of the connecting rod, link 2. From the formula for tangential acceleration, $a^t = r\alpha$, we obtain

$$\alpha_2 = \frac{a^t}{r} = \frac{a^t_{CB}}{BC} = \frac{208 \text{ in/s}^2}{3.76 \text{ in}} = 55 \frac{\text{rad}}{\text{s}^2}.$$

The method for finding the direction of α_2 is similar to that for finding the direction of ω. Tangential acceleration vector a^t_{CB} is placed at C on link BC, as in Figure 4.8f. We see immediately that α_2 is counterclockwise (opposite the direction of ω_2 found in the preceding chapter). Thus, at this instant, the angular acceleration α_2 is opposing the angular velocity ω_2, which means that ω_2 is decreasing. (The reader will recall that ω_1, the angular velocity of the crank, is constant in this example.)

Let us now review the preceding steps for finding accelerations of the slider-crank mechanism in Figure 4.8a.

STEP 1. Draw the linkage to scale. Draw the velocity polygon obc representing the solution of the vector equation

$$\boldsymbol{v}_C = \boldsymbol{v}_B + \boldsymbol{v}_{CB}, \quad \text{or} \quad \boldsymbol{oc} = \boldsymbol{ob} + \boldsymbol{bc}.$$

STEP 2. Solve the general acceleration vector equation for the slider-crank mechanism graphically, as demonstrated in Table 4.1.

STEP 3. The acceleration of B is labeled $\boldsymbol{o'b'}$. The prime indicates that the vector is an acceleration and not a velocity. In this case, $\boldsymbol{a}^t_B = 0$, since $\alpha_1 = 0$, from which it follows that $\boldsymbol{a}_B = \boldsymbol{a}^n_B$ only. To $\boldsymbol{a}_B$, we add vectors $\boldsymbol{a}^n_{CB}$ (the magnitude is found with the aid of the velocity polygon) and $\boldsymbol{a}^t_{CB}$ (drawn perpendicular to BC and of unknown magnitude). The intersection of $\boldsymbol{a}_C$ and $\boldsymbol{a}^t_{CB}$ completes the polygon and determines the magnitude of each of those vectors.

STEP 4. The acceleration vectors have been identified by their components (e.g., $\boldsymbol{a}^n_B$ and $\boldsymbol{a}^t_B$) and by an acceleration polygon notation patterned after the velocity polygon notation. For the linkage under consideration,

$$\boldsymbol{a}_B = \boldsymbol{o'b'}$$

and

$$\boldsymbol{a}_{CB} = \boldsymbol{b'c'}.$$

Although not shown in the acceleration polygon, the normal and tangential components of a_{CB} could be replaced by a single vector representing their sum and extending from $\boldsymbol{b}'$ to $\boldsymbol{c}'$ ($\boldsymbol{b}'\boldsymbol{c}'$). Also,

$$\boldsymbol{a}_C = \boldsymbol{o}'\boldsymbol{c}'.$$

Note the reversal of letters in acceleration polygon notation: Acceleration $\boldsymbol{a}_{CB}$ becomes $\boldsymbol{b}'\boldsymbol{c}'$, just as velocity $\boldsymbol{v}_{CB}$ becomes $\boldsymbol{bc}$ in velocity polygon notation. The acceleration polygon will be used to advantage later, when we consider the acceleration image.

Comparison with an Analytical Solution

In Chapter 3, the velocity of the slider of an in-line slider-crank linkage was approximated by

$$v = R\omega \sin\theta\left[1 + \left(\frac{R}{L}\right)\cos\theta\right].$$

Differentiating the approximate equation for the velocity of the slider of a slider-crank mechanism, we obtain the approximate slider acceleration

$$a = R\omega^2\left[\cos\theta + \left(\frac{R}{L}\right)\cos 2\theta\right] \tag{4.37}$$

if angular acceleration of the crank is zero. The preceding two equations give a positive value for velocity and acceleration directed toward the crankshaft and a negative value for velocity and acceleration directed away from the crankshaft. Both equations are valid for the in-line slider crank when the crank speed is constant and the ratio L/R does not approach a value of unity, say, L/R is greater than or equal to 3. For the data given in the foregoing example, the approximate slider acceleration is $a_C = -13.1(13.1\ \text{in/s}^2$ to the right), which corresponds closely to the result obtained using the acceleration polygon.

SAMPLE PROBLEM 4.8

Linkage with Angular Acceleration of the Crank

In this problem, we consider an acceleration analysis of the slider-crank linkage with angular acceleration of the crank. Find the accelerations for the linkage of Figure 4.9a. The data given are the same as those for the preceding problem, except that link 1 does not have a constant angular velocity, but instead accelerates at a rate $\alpha_1 = 40\ \text{rad/s}^2$ counterclockwise.

Solution. The addition of an angular acceleration has no effect on the instantaneous velocity, so the velocity polygon remains unchanged. (See Figure 4.9b.) We will again use Eq. (4.22). In this case, however, the tangential acceleration of point B does not equal zero:

$$\boldsymbol{a}_B^t = \boldsymbol{\alpha}_1 \times \boldsymbol{O_1B} = (40\ \text{rad/s}^2)(2\ \text{in}) = 80\ \text{in/s}^2 \perp O_1B.$$

Part c of the figure shows how angular acceleration affects the acceleration polygon.

Scaling the vectors in Figure 4.9c, we obtain the slider acceleration

$$\boldsymbol{a}_C = 70\ \text{in/s}^2 \quad \text{(to the left)}$$

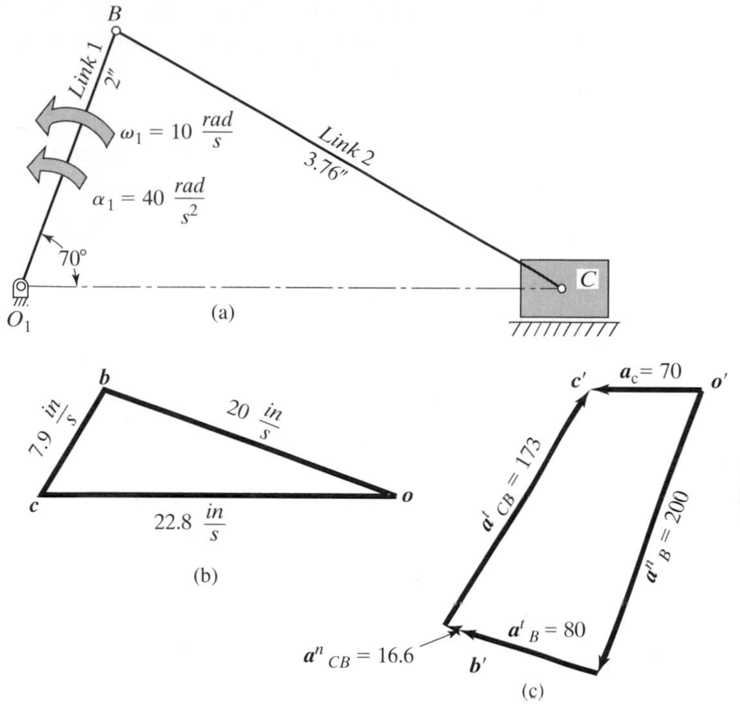

FIGURE 4.9 (a) The mechanism of Figure 4.8 is redrawn. This time, the crank is given an angular acceleration instead of a constant angular velocity. (b) The addition of an angular acceleration does not affect the velocity polygon for the instant considered. (c) The acceleration polygon with angular acceleration of the crank. The expression for $\boldsymbol{a}_C$ is now equal to the vector sum $\boldsymbol{a}^n_B + \boldsymbol{a}^t_B + \boldsymbol{a}^n_{CB} + \boldsymbol{a}^t_{CB}$.

and the tangential component of relative acceleration

$$\boldsymbol{a}^t_{CB} = 173 \text{ in/s}^2 \quad \text{(upward and to the right)}.$$

From the latter acceleration, we can also obtain the angular acceleration of the connecting rod:

$$\alpha_2 = \frac{a^t_{CB}}{CB} = 46 \text{ rad/s}^2 \quad \text{(counterclockwise)}.$$

Acceleration Image

In Sample Problem 4.8, we determined the accelerations of the crankpin B and the slider C. We may also wish to find the acceleration of another point on the crank or connecting rod. The acceleration of the center of gravity, for example, would be used to perform a dynamic analysis of a link, or, in a more complicated linkage, an intermediate point on a link that serves as a connecting point would be investigated. We may resort to the acceleration image method (similar to the velocity image method) in order to find the acceleration of any point.

Consider any three points B, C, and D that lie on the same rigid link, as shown in Figure 4.10a. Let the link have an angular velocity ω_2 and an angular acceleration α_2. Then, the magnitudes of accelerations are

$$a_{CB}^n = \omega_2^2 BC \quad \text{and} \quad a_{CB}^t = \alpha_2 BC.$$

The total acceleration of C with respect to B, a_{CB}, is the vector sum of its normal and tangential components. The magnitude of this vector is given by the expression

$$a_{CB} = b'c' = \sqrt{(a_{CB}^n)^2 + (a_{CB}^t)^2} = \sqrt{(\omega_2^2 BC)^2 + (\alpha_2 BC)^2},$$

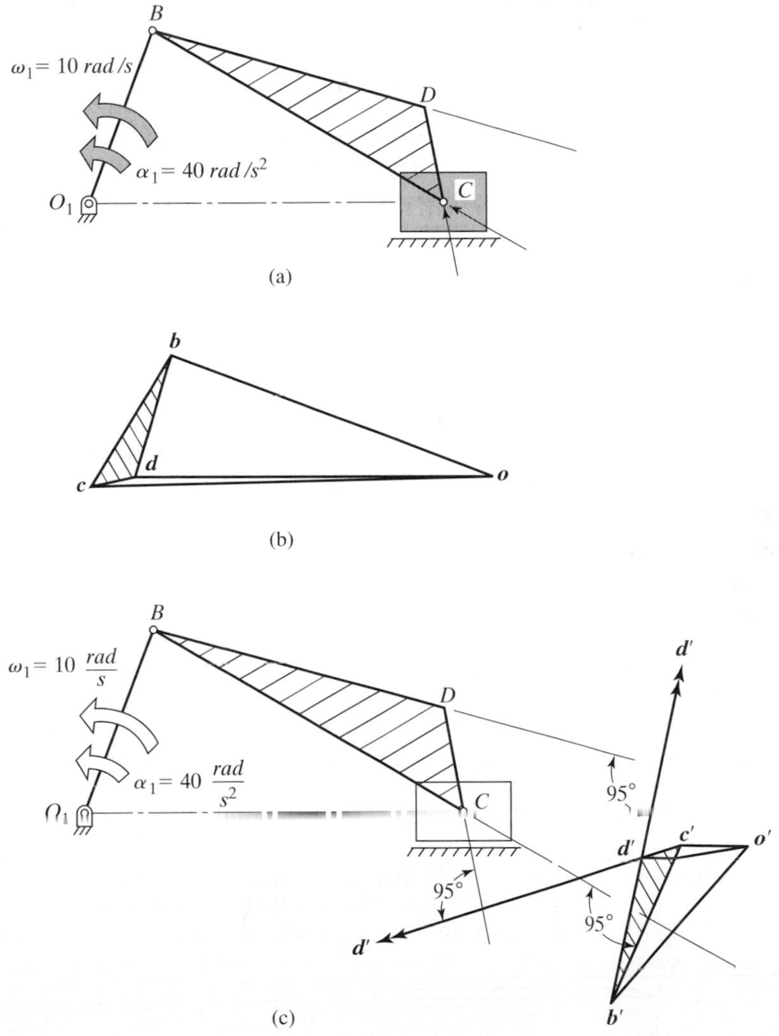

FIGURE 4.10 Acceleration image. The slider-crank mechanism of Figure 4.9a is repeated here. The dimensions of the linkage and the motion of the crank remain unchanged. We are interested in finding the acceleration of an arbitrary point D on the connecting rod. (b) The velocity polygon for the slider-crank mechanism, showing the velocity image of the connecting rod. (c) The acceleration image of BCD is constructed.

from which it follows that

$$b'c' = BC\sqrt{\omega_2^4 + \alpha_2^2}.$$

Similarly, the magnitudes of the other relative accelerations for the connecting rod are

$$b'd' = BD\sqrt{\omega_2^4 + \alpha_2^2}$$

and

$$c'd' = CD\sqrt{\omega_2^4 + \alpha_2^2},$$

from which we obtain the following convenient acceleration image relationships:

$$\frac{b'd'}{b'c'} = \frac{BD}{BC}, \frac{c'd'}{b'c'} = \frac{CD}{BC}, \frac{b'd'}{c'd'} = \frac{BD}{CD}. \tag{4.38}$$

Equation (4.38) may be summarized by stating that triangle $\boldsymbol{b'c'd'}$ (the acceleration image of BCD) is similar to triangle BCD for any points B, C, and D on the same rigid link. The angle relationship between a line connecting two points on a rigid link and the relative acceleration of those points depends on the angular acceleration α and the angular velocity ω and is the same for any pair of points on the same rigid link. In the sample problems that follow, we will utilize this relationship without having to calculate α and ω.

SAMPLE PROBLEM 4.9

Acceleration Image

Three points, B, C, and D, lie on the rigid link shown in Figure 4.10a, but do not lie on a straight line. Using the acceleration image method and the data given in the illustration, find the acceleration of point D of the mechanism.

Solution. This problem and the problem of Figure 4.9a are identical, except for the addition of an arbitrary point D. The velocity polygon, including the velocity image, is constructed (Figure 4.10b) as described in Chapter 3. The acceleration polygon $\boldsymbol{o'b'c'}$ (Figure 4.10c) is taken directly from Figure 4.9c, but the normal and tangential components of $\boldsymbol{a}_B$ and $\boldsymbol{a}_{CB}$ have been omitted here to clarify the construction.

We observe in Figure 4.10c that the relative acceleration vector $\boldsymbol{b'c'}$ (forming one leg of the required acceleration image) lies in the direction of line BC, rotated approximately 95° counterclockwise. Since the acceleration image $\boldsymbol{b'c'd'}$ and link BCD are similar triangles, each leg of the acceleration image will make a 95° angle with its respective side in the linkage drawing. Beginning at $\boldsymbol{b'}$, we construct the acceleration image by first drawing trial vector $\boldsymbol{b'd'}$, determining its direction by rotating line BD 95° counterclockwise. Then, we draw trial vector $\boldsymbol{c'd'}$ from point $\boldsymbol{c'}$; its direction is also found by rotating line CD 95° counterclockwise. Point $\boldsymbol{d'}$ is thus determined by the intersection of trial vectors $\boldsymbol{b'd'}$ and $\boldsymbol{c'd'}$, as in Figure 4.10c, completing triangle $\boldsymbol{b'c'd'}$, the acceleration image of link BCD on the acceleration polygon. The acceleration of D is thus given by vector $\boldsymbol{o'd'}$. Using the acceleration scale, we find that $\boldsymbol{a}_D = \boldsymbol{o'd'} = 113 \text{ in/s}^2$, slightly downward to the left.

SAMPLE PROBLEM 4.10

Acceleration Image of Three Points on a Line

In this problem, we are required to find the acceleration of point E, which lies on the line BC in Figure 4.11a.

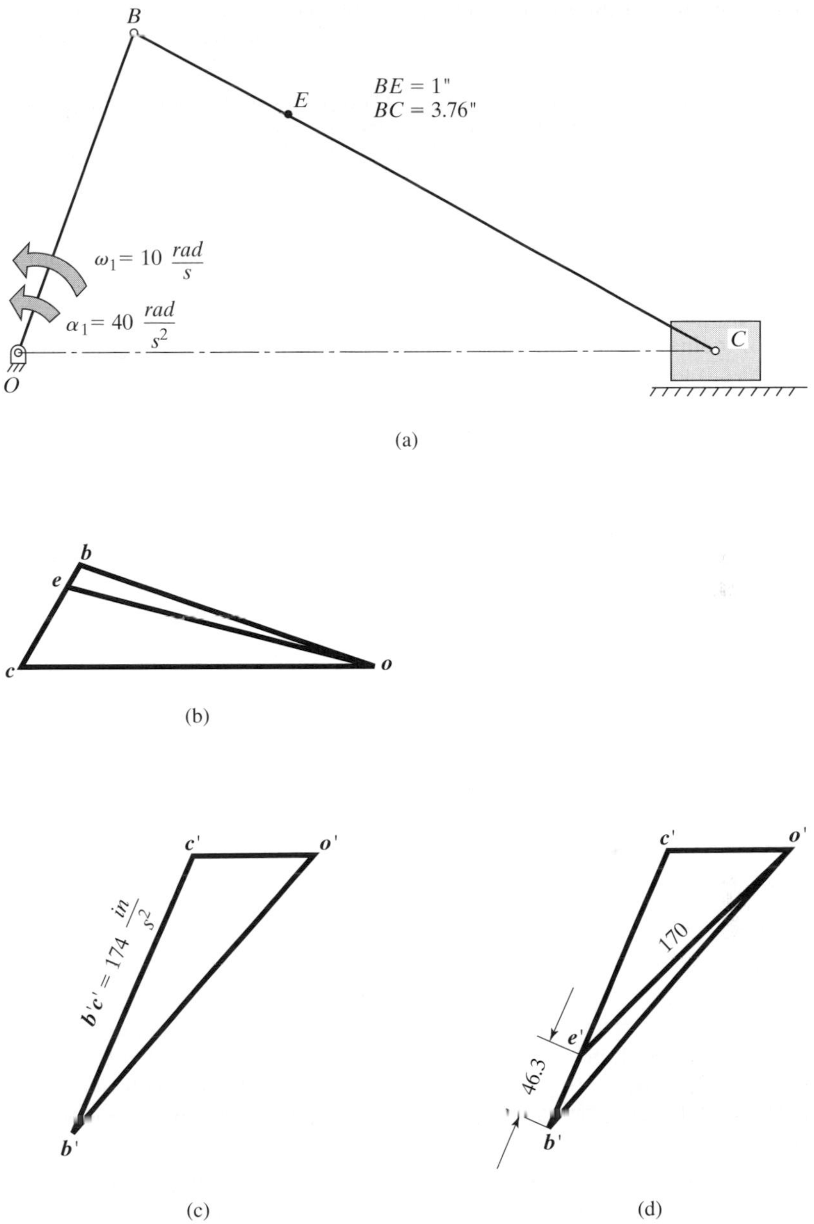

FIGURE 4.11 (a) Figure 4.9 is repeated again. We want to find the acceleration of a point E lying on line BC. (b) The velocity polygon for the linkage at the instant shown. (c) The acceleration polygon for the linkage at the instant shown. (d) Once the acceleration polygon is constructed, the position of e' (which must lie along $b'c'$) is determined by the proportionality equation $b'e'/b'c' = BE/BC$. Drawing vector $\boldsymbol{o'e'}$ then gives us the magnitude and direction of the acceleration of E.

Solution. Again, this problem and the problem of Figure 4.9a are identical except for the addition of a point E along line (link) BC. We are again spared the necessity of constructing the velocity and acceleration polygons for the mechanism. See Figures 4.11b and c. Acceleration polygon $\boldsymbol{o'b'c'}$ is again taken directly from Figure 4.9c. This problem is simpler than the preceding problem because no additional construction is necessary after the acceleration polygon is constructed. Since point E lies on line BC, we know that e' must lie somewhere on vector $\boldsymbol{b'c'}$. A proportion similar to one of Eq. 4.38 gives us the desired acceleration image relationship:

$$\frac{b'e'}{b'c'} = \frac{BE}{BC}$$

from which we find that the acceleration of E relative to B is

$$b'e' = b'c'\left(\frac{BE}{BC}\right) = \left(174\ \text{in/s}^2\right)\left(\frac{1\ \text{in}}{3.76\ \text{in}}\right) = 46.3\ \text{in/s}^2.$$

This locates point e' on line $\boldsymbol{b'c'}$. (Note that e' lies between $\boldsymbol{b'}$ and $\boldsymbol{c'}$, just as E lies between B and C.) Vector $\boldsymbol{o'e'}$ is then drawn to obtain the acceleration of E, as shown in Figure 4.11d. Measuring $\boldsymbol{o'e'}$ against the acceleration scale, we obtain

$$\boldsymbol{a}_E = \boldsymbol{o'e'} = 170\ \text{in/s}^2 \quad \text{(to the left and downward).}$$

The image principle illustrates the analogy between velocity polygons and acceleration polygons. The acceleration image principle (as well as the velocity image examined in Chapter 3) applies to a set of points on any rigid link, whether the link acts as a crank rotating about a fixed point or as a connecting rod. The only restriction is that the points considered must all lie on the same rigid link.

Graphical Analysis of the Four-Bar Linkage

The acceleration analysis of a four-bar linkage requires no new concepts. Referring to Figure 4.12a, for example, we may again relate accelerations by the vector equation

$$\boldsymbol{a}_C = \boldsymbol{a}_B + \boldsymbol{a}_{CB},$$

just as for the slider-crank mechanism, but with one additional complication: Each of the acceleration vectors will have, in general, both a normal and a tangential component, and the equation will take the form

$$\boldsymbol{a}_C = \boldsymbol{a}_C^n + \boldsymbol{a}_C^t = \boldsymbol{a}_B^n + \boldsymbol{a}_B^t + \boldsymbol{a}_{CB}^n + \boldsymbol{a}_{CB}^t \qquad \text{(4.22 repeated)}$$

The first step of the acceleration analysis of a four-bar linkage is to construct the skeleton drawing and the velocity polygon. The dimensions of the linkage, together with the velocities taken from the velocity polygon allow us to calculate the normal components of acceleration of the links. As we have seen, those components are usually the starting point for the acceleration polygon.

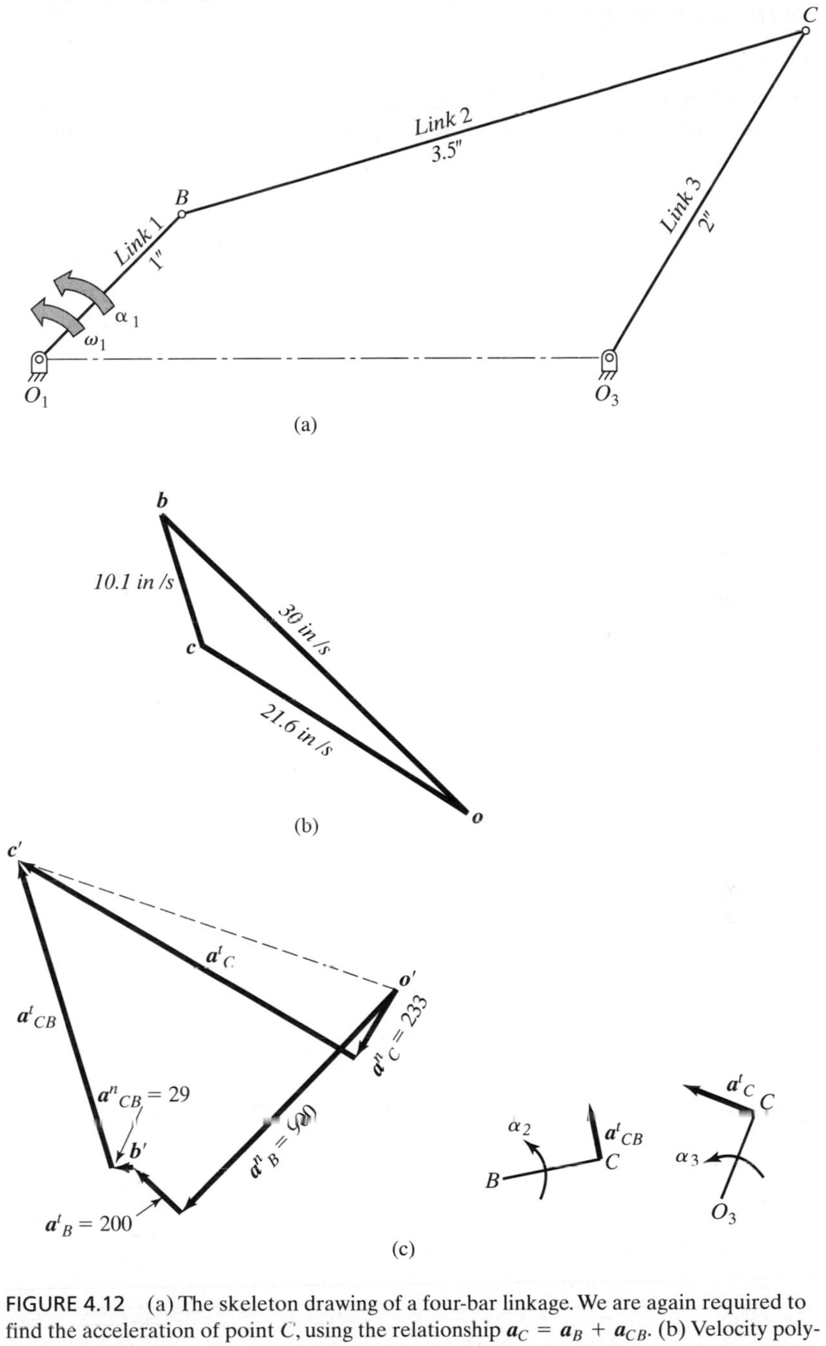

FIGURE 4.12 (a) The skeleton drawing of a four-bar linkage. We are again required to find the acceleration of point C, using the relationship $\boldsymbol{a}_C = \boldsymbol{a}_B + \boldsymbol{a}_{CB}$. (b) Velocity polygon for the four-bar mechanism. (c) Acceleration polygon for the four-bar mechanism. The presence of the many vectors makes a method of tabulation of the various vectors desirable to ensure correct vector addition and orientation.

SAMPLE PROBLEM 4.11

Four-Bar Linkage

Figure 4.12a shows a four-bar linkage. The lengths of all the links are indicated on the skeleton drawing. The crank has an angular velocity $\omega_1 = 30\,\text{rad/s}$ and an angular acceleration $\alpha_1 = 200\,\text{rad/s}^2$. We seek an acceleration analysis of the linkage.

Solution. The velocity polygon is constructed in Figure 4.12b after a suitable scale is selected. The velocities are then indicated directly on the polygon. The skeleton drawing and the velocity polygon give us the information needed to determine the normal components of acceleration. Given the angular acceleration of link 1, we can calculate the tangential component of acceleration for point B, after which we can put together an acceleration vector table (Table 4.2) and begin the construction of the acceleration polygon. Velocities $\boldsymbol{ob}$, $\boldsymbol{oc}$, and $\boldsymbol{bc}$ are taken from the velocity polygon in Figure 4.12b.

We now construct the acceleration polygon (see Figure 4.12c), adding the vectors in the order indicated in Table 4.2. Beginning at the pole point $\boldsymbol{o}'$, we draw $\boldsymbol{a}_C^n$ to the scale selected. To $\boldsymbol{a}_C^n$, we add trial vector $\boldsymbol{a}_C^t$. The head of $\boldsymbol{a}_C^t$ may be labeled $\boldsymbol{c}'$, but we do not as yet know the true magnitude of that vector. Again, beginning at $\boldsymbol{o}'$, we add the vectors on the right side of Eq. (4.22) in the order indicated. The sum $\boldsymbol{a}_B^n + \boldsymbol{a}_B^t = \boldsymbol{a}_B$ (or $\boldsymbol{o}'\boldsymbol{b}'$); thus, the head of $\boldsymbol{a}_B^t$ is labeled $\boldsymbol{b}'$. Adding the last two of the four vectors on the right side of the equation, which includes trial vector $\boldsymbol{a}_{CB}^t$, we again obtain $\boldsymbol{a}_C$ (or $\boldsymbol{o}'\boldsymbol{c}'$). Point $\boldsymbol{c}'$ is located at the intersection of the trial vectors $\boldsymbol{a}_C^t$ and $\boldsymbol{a}_{CB}^t$, completing the polygon and determining in turn the magnitude of each tangential component. Thus, $\boldsymbol{a}_C^t = 1{,}110\,\text{in/s}^2$ and $\boldsymbol{a}_{CB}^t = 930\,\text{in/s}^2$. Using the acceleration scale, we can obtain the acceleration of point C, viz.,

$$\boldsymbol{a}_C = \boldsymbol{a}_C^n + \boldsymbol{a}_C^t = \boldsymbol{o}'\boldsymbol{c}' = 1{,}134\ \text{in/s}^2,$$

to the left and upward.

Using the tangential acceleration of point C, we can find the angular acceleration of link 3 (O_3C). We obtain

$$\alpha_3 = \frac{a_C^t}{O_3C} = \frac{1{,}110\ \text{in/s}^2}{2\ \text{in}} = 555\ \text{rad/s}^2\ \text{(counterclockwise)}.$$

Similarly, the angular acceleration of link 2 ($\boldsymbol{BC}$) is given by

$$\alpha_2 = \frac{a_{CB}^t}{CB} = \frac{930\ \text{in/s}^2}{3.5\ \text{in}} = 266\ \text{rad/s}^2\ \text{(counterclockwise)}.$$

TABLE 4.2 Vector Tabulation for the Acceleration Analysis of a Four-Bar Mechanism, Figure 4.12

Vector	$\boldsymbol{a}_C^n$	$+\boldsymbol{a}_C^t$	$= \boldsymbol{a}_B^n$	$+\boldsymbol{a}_B^t$	$+\boldsymbol{a}_{CB}^n$	$+\boldsymbol{a}_{CB}^t$
Vector magnitude	$\frac{(oc)^2}{O_3C}$	?	$\frac{(ob)^2}{O_1B}$	$\alpha_1 O_1B$	$\frac{(bc)^2}{BC}$	?
Vector direction	$\parallel O_3C$ toward O_3	$\perp O_3C$	$\parallel O_1B$ toward O_1	$\perp O_1B$	$\parallel BC$ toward B	$\perp BC$
Vectors used to construct polygon	233 in/s² ↙	? ↖	900 in/s² ↙	200 in/s² ↖	29 in/s² ↙	? ↖

An Analytical Solution Based on the Acceleration Polygon

It is possible to increase the precision of the analysis while still using the acceleration polygon concept. Sample Problem 4.12 shows an acceleration-polygon-based analytical solution for one position of a four-bar linkage. If you use the polygon method for checking the analytical vector method, the complex-number method, or results from animation software, you may decide that the increased precision is not worth the computational difficulty of the method illustrated next.

SAMPLE PROBLEM 4.12

Acceleration-Polygon-Based Analytical Solution

Refer to Figure 4.4 and Sample Problem 4.3. Solve for accelerations analytically, but use an acceleration polygon approach.

Solution. Since $\boldsymbol{\alpha}_1 = 0$, the acceleration of point B is given by

$$\boldsymbol{o'b'} = \boldsymbol{a}_B = \omega_1{}^2 r_1 = 100^2 \times 10 = -100{,}000\angle\theta_1.$$

Also, after determining velocities, we find

$$\boldsymbol{a}^n_{CB} = \frac{bc^2}{BC} = -3193\angle\theta_2$$

$$\boldsymbol{a}^n_C = \frac{oc^2}{O_3C} = 26{,}260\angle\theta_3.$$

An arbitrary pole point $\boldsymbol{o}'$ is selected, with coordinates $x = 0$, $y = 0$. Vector $\boldsymbol{a}^n_C$ is drawn. The coordinates of its head are identified as

$$x_1 = -a^n_C \cos\theta_3 = 26{,}260 \cos(-122.21°) = -13{,}997$$

and

$$y_1 = -a^n_C \sin\theta_3 = 26{,}260 \sin(-122.21°) = -22{,}219$$

as shown in Figure 4.13. Vector $\boldsymbol{o'b'}$ is drawn, and $\boldsymbol{a}^n_{CB}$ is added to it. The coordinates of the head of $\boldsymbol{a}^n_{CB}$ are identified as

$$\begin{aligned} x_2 &= -o'b' \cos\theta_1 - a^n_{CB} \cos\theta_2 \\ &= -100{,}000 \cos 45° - 3193 \cos 16.35° = -73{,}775 \end{aligned}$$

and

$$\begin{aligned} y_2 &= -o'b' \sin\theta_1 - a^n_{CB} \sin\theta_2 \\ &= -100{,}000 \sin 45° - 3193 \sin 16.35° = -71{,}610. \end{aligned}$$

At x_1, y_1, we add vector $\boldsymbol{a}^t_C$ with unknown magnitude but with slope

$$m = \tan(\theta_3 - 90°) = \tan(237.79 - 90°) = -0.62997$$

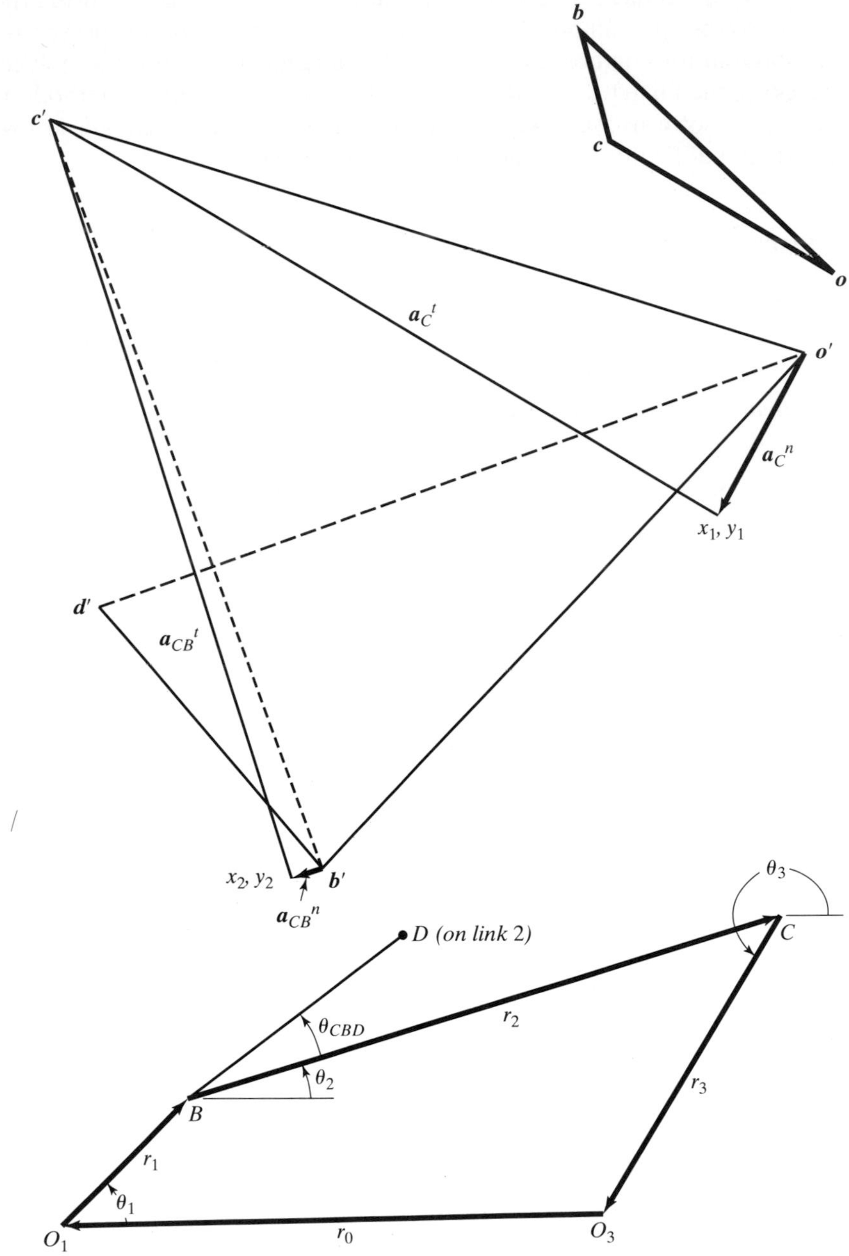

FIGURE 4.13 Analytical solution of a four-bar linkage based on the acceleration polygon.

a_C^t lies on a line described by

$$y = y_1 + m(x - x_1) = -22{,}219 - 0.62997(x + 13{,}997).$$

At x_2, y_2, we add vector $\boldsymbol{a}_{CB}^t$ with unknown magnitude, but with slope

$$n = \tan(\theta_2 + 90°) = \tan(16.5 + 90°) = -3.4087.$$

$\boldsymbol{a}_{CB}^t$ lies on a line described by

$$y = y_2 + n(x - x_2) = -71{,}610 - 3.4087(x + 73{,}775).$$

The intersection of vectors $\boldsymbol{a}_C^t$ and $\boldsymbol{a}_{CB}^t$ locates point c' on the acceleration polygon. Equating the right sides of the two equations of the lines describing $\boldsymbol{a}_C^t$ and $\boldsymbol{a}_{CB}^t$, we find the x-coordinate of $\boldsymbol{a}_C$:

$$x = a_{C_x} = -105{,}100.$$

Substituting the value of x into one of the preceding equations, we have

$$y = a_{C_y} = 35{,}174,$$

from which we find that

$$\boldsymbol{a}_C = \boldsymbol{o'c'} = 110{,}800 \text{ mm/s}^2\angle 161.5°.$$

The relative acceleration is given by

$$\boldsymbol{b'c'} = \boldsymbol{a}_{CB} = \boldsymbol{a}_C - \boldsymbol{a}_B,$$

from which we obtain

$$a_{CB_x} = -105{,}102 - 100{,}000 \cos 225°,$$
$$a_{CB_y} = 35{,}174 - 100{,}000 \sin 225°,$$

and

$$\boldsymbol{a}_{CB} = 111{,}330 \text{ mm/s}^2\angle 108°.$$

Using the image principle, we find the relative acceleration of two points on the coupler:

$$a_{BD} = b'd' = \frac{b'c'BD}{BC} = \frac{(111{,}330)(15)}{35}$$
$$= 47{,}712 \text{ mm/s}^2\angle 108° + 20°.$$

The acceleration of point D on the coupler is given by

$$\boldsymbol{a}_D = \boldsymbol{o'd'} = \boldsymbol{o'b'} + \boldsymbol{b'd'}$$

from which it follows that

$$a_{D_x} = 100{,}000 \cos 225^\circ + 47{,}712 \cos 128^\circ = -100{,}100,$$
$$a_{D_y} = 100{,}000 \sin 225^\circ + 47{,}712 \sin 128^\circ = -33{,}100,$$

and

$$a_D = 105{,}400 \text{ mm/s}^2 \angle -162^\circ.$$

4.7 EQUIVALENT LINKAGES

Equivalent linkages (which duplicate the motion of an actual mechanism) are sometimes useful in velocity and acceleration analysis. Figure 4.14 illustrates a curved-wing air pump, a mechanism that is obviously not a four-bar linkage. The pump has four evenly spaced wings, but for purposes of analysis, we need only consider the motion of one of these wings. The key to arriving at an equivalent linkage is to examine the forces

FIGURE 4.14 A curved-wing air pump. Air is carried from inlet to outlet by the curved wings, which are held against the housing by acceleration forces. (*Source:* ITT Pneumotive.)

acting on the mechanism and the restraints that restrict the mechanism to its specific path. These forces and restraints are replaced by links that are arranged so that the linkage duplicates the motion of the actual mechanism.

Figure 4.15a shows only one of the four wings (dashed lines) and its equivalent linkage (solid lines). Link 1 represents the driver crank, link 2 the wing. Point O_3 is the geometric center of the housing, and link 3 represents the distance from the center of the housing to the point of contact between wing and housing. (Actually, the wing is restrained by inertial forces to follow the curvature of the housing. At the instant being considered, this restraint can be considered a rigid link that forces point C to rotate about a circle of radius O_3C.) Link 3 does not exist on the actual pump, but it is essential to the equivalent linkage that we have devised, which is the familiar four-bar linkage. In sample Problem 4.13, we will assign dimensions to the pump, assume a reasonable rotating speed, and analyze the acceleration of this mechanism.

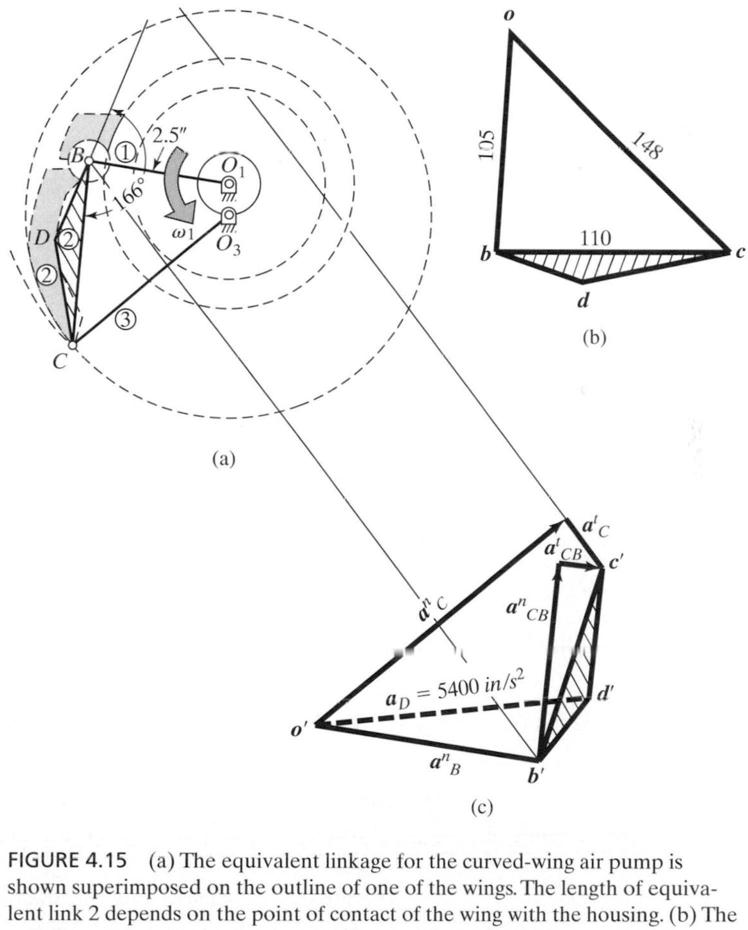

FIGURE 4.15 (a) The equivalent linkage for the curved-wing air pump is shown superimposed on the outline of one of the wings. The length of equivalent link 2 depends on the point of contact of the wing with the housing. (b) The velocity polygon for the four-bar (equivalent) linkage. (c) The acceleration polygon for the four-bar (equivalent) linkage.

SAMPLE PROBLEM 4.13

Equivalent Linkages

The pump shown in Figure 4.14 rotates at a constant 400 rev/min. The wing pins rotate about a circle of radius 2.5 in. The pump is drawn to scale. Find the angular acceleration of the wing and the acceleration of its center of gravity.

Solution. The equivalent linkage is drawn as shown in Figure 4.15a. The lengths of links 2 and 3 are obtained from skeleton drawing.

The velocity polygon in Figure 4.15b is based on the given speed of 400 rev/min and a given length of 2.5 in for link 1. The velocity polygon represents the solution of the vector equation

$$\boldsymbol{v}_C = \boldsymbol{v}_B + \boldsymbol{v}_{CB}, \quad \text{or} \quad \boldsymbol{oc} = \boldsymbol{ob} + \boldsymbol{bc}.$$

Point D of our equivalent linkage represents the center of gravity of the curved wing. Equivalent link 2 (BCD) forms the velocity image $\boldsymbol{bcd}$ on the velocity polygon. (This example is intended only to illustrate principles of acceleration analysis; the dimensions used may not correspond to those of an actual pump.) The velocities of B and C and the relative velocity $\boldsymbol{bc}$ shown on the velocity polygon are obtained in the usual manner and are used to construct the basic acceleration polygon $\boldsymbol{o'b'c'}$, in the order given in Table 4.3.

The magnitudes of the normal accelerations, as calculated in the table, indicate the need for an acceleration scale on the order of 1 in = 2,000 in/s^2. Since link 1 rotates at constant angular velocity (there is no angular acceleration), $\boldsymbol{a}_B^n$ represents the total acceleration of point B, $\boldsymbol{o'b'}$. Beginning with this acceleration, we draw $\boldsymbol{o'b'}$ parallel to O_1B. (See Figure 4.15c.) To $o'b'$, we add known relative acceleration $\boldsymbol{a}_{CB}^n$, parallel to BC, and trial vector $\boldsymbol{a}_{CB}^t$, perpendicular to BC, as shown in Figure 4.15c. This completes the addition of the vectors on the right side of the equation.

Starting again at pole point $\boldsymbol{o'}$, we draw the known acceleration $\boldsymbol{a}_C^n$ parallel to O_3C. To a_C^n, we add trial vector $\boldsymbol{a}_C^t$, perpendicular to O_3C. The basic acceleration equation is satisfied when we locate $\boldsymbol{c'}$ at the intersection of the trial vectors $\boldsymbol{a}_C^t$ and $\boldsymbol{a}_{CB}^t$. The angular acceleration of the wing is then given by

$$\alpha_2 = \frac{a_{CB}^t}{CB} = 265 \text{ rad/s}^2 \quad \text{(counterclockwise)},$$

where we have used values for $\boldsymbol{a}_{CB}$ and BC scaled from the illustration.

TABLE 4.3 Tabulation for the Vector Acceleration Analysis of the Air Pump Equivalent Linkage

Vector	$\boldsymbol{a}_C^n$	$+\boldsymbol{a}_C^t$	$=\boldsymbol{a}_B^n$	$+\boldsymbol{a}_B^t$	$+\boldsymbol{a}_{CB}^n$	$+\boldsymbol{a}_{CB}^t$
Vector magnitude	$\dfrac{(oc)^2}{O_3C}$	?	$\dfrac{(ob)^2}{O_1B}$	$\alpha_1 O_1B$	$\dfrac{(bc)^2}{BC}$	?
Vector direction	$\parallel O_3C$ toward O_3	$\perp O_3C$	$\parallel O_1B$ toward O_1	$\perp O_1B$ when $\alpha_1 \neq 0$	$\parallel BC$ toward B	$\perp BC$
Vectors used to construct polygon	6080 in/s^2 ↗	? ↘	4400 in/s^2 ↘	0 for $\alpha_1 = 0$	3560 in/s^2 ↑	? →

To find the acceleration of the center of gravity, D, we construct the acceleration image of link 2. Points B, C, and D all lie on the same rigid link, permitting us to use the *acceleration image principle* to find $\boldsymbol{a}_D$. Points $\boldsymbol{b}'$ and $\boldsymbol{c}'$ on the acceleration polygon are joined to form the image of BC. Using a protractor, we see that the orientation of $\boldsymbol{b}'\boldsymbol{c}'$ on the acceleration polygon is given by rotating BC on the skeleton linkage 166° counterclockwise. Similarly, BD and CD are rotated 166° counterclockwise to obtain the directions of $\boldsymbol{b}'\boldsymbol{d}'$ and $\boldsymbol{c}'\boldsymbol{d}'$, respectively. The acceleration image is thereby completed, locating $\boldsymbol{d}'$ at the intersection of $\boldsymbol{b}'\boldsymbol{d}'$ and $\boldsymbol{c}'\boldsymbol{d}'$. Vector $\boldsymbol{o}'\boldsymbol{d}'$ represents the total acceleration of the center of gravity of the wing, point D. Measuring $\boldsymbol{o}'\boldsymbol{d}'$ on the acceleration scale, we obtain

$$\boldsymbol{a}_D = 5{,}400 \text{ in/s}^2,$$

to the right and slightly upward.

The construction used in the preceding problem (rotating the links to determine the position of the acceleration image) is equivalent to simply transferring angles BCD and CBD to the acceleration polygon (as angles $\boldsymbol{b}'\boldsymbol{c}'\boldsymbol{d}'$ and $\boldsymbol{c}'\boldsymbol{b}'\boldsymbol{d}'$, respectively). The acceleration image triangle $\boldsymbol{b}'\boldsymbol{c}'\boldsymbol{d}'$ is similar to triangle BCD. Since we read BCD going around the link clockwise, $\boldsymbol{b}'\boldsymbol{c}'\boldsymbol{d}'$ must appear in that same order, reading clockwise around the acceleration image.

The equivalent linkage that we have used to analyze the air pump is valid for any instant, so long as point C on the wing contacts the housing. The equivalent linkage may not be used, however, for the portion of the cycle during which point C leaves the housing. (See the wing at the top of the mechanism.)

4.8 GRAPHICAL ANALYSIS OF SLIDING CONTACT LINKAGES

Coriolis Acceleration

In the preceding graphical studies, we considered the acceleration of a slider moving along a fixed path (e.g., a piston in a cylinder), and we also considered the acceleration of a point on a rotating link (e.g., either of the moving pin joints in a four-bar linkage). We will now look into the case of a link that slides along a rotating member. When these two conditions are met (i.e., a rotating path and a point that has a velocity relative to that path), there exists an additional acceleration component, the Coriolis acceleration. Thus, the total acceleration of a point on the slider in Figure 4.16a consists of the following:

1. Normal and tangential accelerations of a coincident point on the rotating link
2. Relative acceleration of the slider along the rotating link
3. *Coriolis acceleration*

The analytical vector method gave us Coriolis acceleration in the form

$$\boldsymbol{a}^{\mathrm{C}} = 2\boldsymbol{\omega} \times \boldsymbol{v},$$

and the complex number method yielded the form

$\boldsymbol{a}^{\mathrm{C}} = j2\omega\, v\, e^{j\theta}$
where ω = angular velocity of the rotating path and
v = relative velocity.

Terms must be identified by subscripts as illustrated in the sample problems that follow. For planar linkages, the direction of the Coriolis acceleration is found by rotating

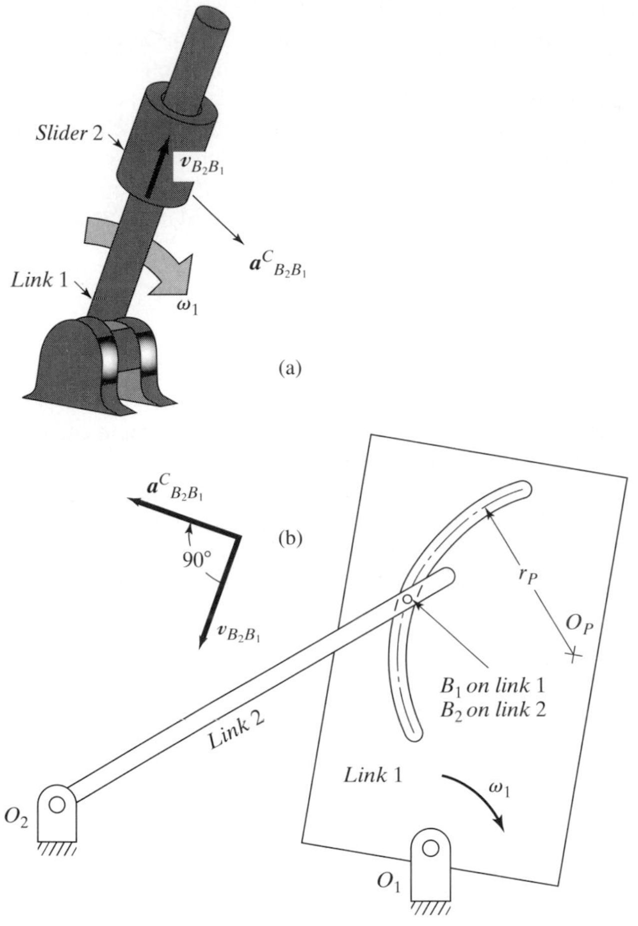

FIGURE 4.16 Coriolis acceleration due to sliding contact along a rotating path. (a) Straight path. (b) Curved path (note direction of relative velocity).

the relative velocity vector 90° in the direction of the rotating path. (Check this result with the analytical vector and complex-number forms used in previous sections.)

SAMPLE PROBLEM 4.14

Coriolis Acceleration

Link 1 in Figure 4.16b has an angular velocity $\omega_1 = 20$ rad/s clockwise. The velocity of point B_2 on link 2 with respect to a coincident point on link 1 is $\boldsymbol{v}_{B_2B_1} = 30$ in/s in the direction shown. Find the Coriolis acceleration.

Solution. The Coriolis component of the slider acceleration is given by

$$\boldsymbol{a}^c_{B_2B_1} = 2\boldsymbol{\omega}_1 \times \boldsymbol{v}_{B_2B_1} = 2(20 \text{ rad/s})(30 \text{ in/s}) = 1{,}200 \text{ in/s}^2.$$

The direction of $\boldsymbol{a}^c_{B_2B_1}$ is found by rotating relative velocity vector $\boldsymbol{v}_{B_2B_1}$ in the direction of $\boldsymbol{\omega}_1$ (clockwise) by 90°. In this example, $\boldsymbol{a}^c_{B_2B_1}$ is to the left and upward as shown. Note that the subscripts of $\boldsymbol{a}^c$ and $\boldsymbol{v}$ must correspond. We use angular velocity ω_1 and relative velocity $v_{B_2B_1}$, since link 1 guides the slider (i.e., we know the relative path of link 2 along link 1).

Note, however, that we may replace the linkage in the figure with an equivalent four-bar linkage ($O_1O_pBO_2$). Acceleration analysis of the equivalent linkage would not involve determining Coriolis acceleration.

SAMPLE PROBLEM 4.15

Coriolis Acceleration in a Sliding Contact Linkage

The slider in the mechanism of Figure 4.17a travels along link 1, which rotates. Thus, Coriolis acceleration is involved in an analysis of this linkage. To find the angular acceleration of link 1, we will find the tangential acceleration of B_1 by using the acceleration polygon method.

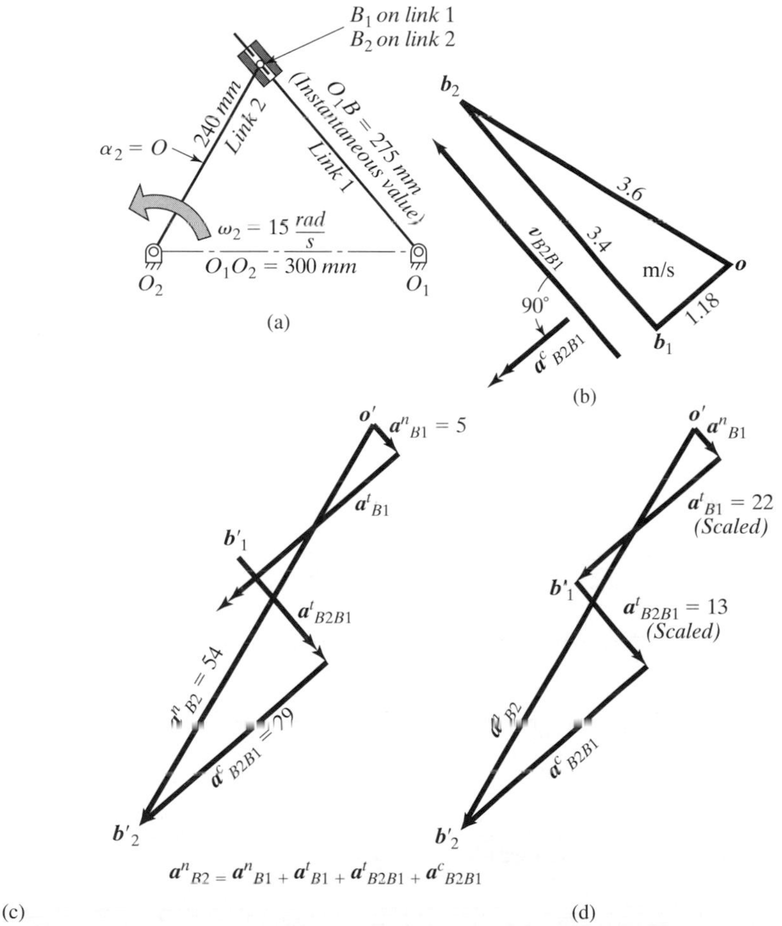

FIGURE 4.17 (a) A sliding contact linkage. (b) The velocity polygon for the sliding contact linkage. When the direction of the relative velocity is determined, the direction of the Coriolis acceleration is found by rotating the relative velocity vector 90° in the direction of the angular velocity of the link on which the slider rides. (c) The acceleration polygon. (d) The "cleaned-up" acceleration polygon.

Solution. Velocities were computed for the linkage in Chapter 3, but the link numbers have been changed here. In this problem, the continuously rotating crank is identified as link 2. Point B_2 represents the pin (revolute pair) joining link 2 to the slider. We have, from the velocity polygon (Figure 4.17b),

$$\boldsymbol{v}_{B_1} = \boldsymbol{ob}_1 = 1.18 \text{ m/s} \quad \text{(downward to the left)},$$
$$\boldsymbol{v}_{B_2} = \boldsymbol{ob}_2 = 3.6 \text{ m/s} \quad \text{(upward to the left)},$$

and

$$\boldsymbol{v}_{B_2B_1} = \boldsymbol{b}_1\boldsymbol{b}_2 = 3.4 \text{ m/s} \quad \text{(upward to the left)}.$$

In solving for accelerations, we will express the acceleration of a point on the slider as the vector sum of the acceleration of a coincident point on link 1 and the relative acceleration of the slider on the link, which includes Coriolis acceleration.

We may apply Eq. (4.15), where point B_1 is taken as the origin of a coordinate system that rotates with link 1. Note that we have selected a coordinate system rotating with link 1 because sliding occurs along that link. Then

$$\ddot{\boldsymbol{R}} = \boldsymbol{a}^n_{B_2} + \boldsymbol{a}^t_{B_2},$$
$$\ddot{\boldsymbol{R}}_0 = \boldsymbol{a}^n_{B_1} + \boldsymbol{a}^t_{B_1},$$
$$\ddot{\boldsymbol{r}}_r = \boldsymbol{a}^t_{B_2B_1},$$

and

$$2\boldsymbol{\omega} \times \dot{\boldsymbol{r}}_r = \boldsymbol{a}^c_{B_2B_1}$$

The remaining terms in Eq. 4.15 are zero, since the relative path is straight (along link 1). Thus, we have

$$\boldsymbol{a}^n_{B_2} + \boldsymbol{a}^t_{B_2} = \boldsymbol{a}^n_{B_1} + \boldsymbol{a}^t_{B_1} + \boldsymbol{a}^t_{B_2B_1} + \boldsymbol{a}^c_{B_2B_1}. \tag{4.39}$$

It can be seen that when we refer to coincident points on two different links (B_1 and B_2 in this problem), the solution differs from that obtained from an analysis of linkages using two different points on the same rigid link (e.g., B and C in Eq. (4.22)).

For this problem, the condition $\alpha_2 = 0$ eliminates the term $\boldsymbol{a}^t_{B_2}$. We will now set up the vector table (Table 4.4) and proceed to solve for the remaining terms in the acceleration equation.

TABLE 4.4 Tabulation for the Vector Acceleration Analysis of a Sliding Contact Linkage, Figure 4.17

Vector	$\boldsymbol{a}^n_{B_2}$	$=\boldsymbol{a}^n_{B_1}$	$+\boldsymbol{a}^t_{B_1}$	$+\boldsymbol{a}^t_{B_2B_1}$	$+\boldsymbol{a}^c_{B_2B_1}$
Vector magnitude	$\frac{(ob_2)^2}{O_2B}$	$\frac{(ob_1)^2}{O_1B}$	?	?	$2(b_1b_2)\omega_1$
Vector direction	$\parallel O_2B$ toward O_2	$\parallel O_1B$ toward O_1	$\perp O_1B$	$\parallel$ link 1 (the relative slider path)	direction of b_1b_2 rotated 90° counterclockwise
Vectors used to construct polygon	54 m/s² ↙	5 m/s²? ↘	? ↙	? ↘	29 m/s² ↙

Since all vector directions are known and we require only two magnitudes, $\boldsymbol{a}^t_{B_1}$ and $\boldsymbol{a}^t_{B_2B_1}$, we are prepared to construct the acceleration polygon. Beginning at an arbitrary point $\boldsymbol{o}'$ in Figure 4.17c, we draw $\boldsymbol{a}^n_{B_2}$ to a convenient scale. Vector $\boldsymbol{a}^n_{B_2}$ represents the total acceleration of B_2 in this problem; the head of $\boldsymbol{a}^n_{B_2}$ is labeled $\boldsymbol{b}'_2$. Then, working with the right side of the Eq. (4.39) and again beginning at $\boldsymbol{o}'$, we draw $\boldsymbol{a}^n_{B_1}$ *and add trial vector* $\boldsymbol{a}^t_{B_1}$.

It would be convenient to continue by adding $\boldsymbol{a}^t_{B_2B_1}$, the next term in the equation, to the head of $\boldsymbol{a}^t_{B_1}$, but we do not know where to begin because the head of $\boldsymbol{a}^t_{B_1}$ has not been located. Instead, we observe that both sides of the equation represent $\boldsymbol{o}'\boldsymbol{b}'_2$, the acceleration of B_2. Then, the last term, $\boldsymbol{a}^c_{B_2B_1}$, may be put in its logical place, with its head at point $\boldsymbol{b}'_2$. Working backwards, we place the next-to-last term, trial vector $\boldsymbol{a}^t_{B_2B_1}$, so that its head is at the tail of $\boldsymbol{a}^c_{B_2B_1}$. The intersection of the trial vectors locates the tail of $\boldsymbol{a}^t_{B_2B_1}$ and the head of $\boldsymbol{a}^t_{B_1}$, and we label that point $\boldsymbol{b}'_1$. Figure 4.17d shows the "cleaned-up" acceleration polygon with scaled values of $\boldsymbol{a}^t_{B_1}$ and $\boldsymbol{a}^t_{B_2B_1}$. Having obtained the tangential acceleration of B_1, we find that the solution to the problem, the angular acceleration of link 1, is given by

$$\alpha_1 = \frac{\boldsymbol{a}^t_{B_1}}{O_1B} = 22/0.275 = 80 \text{ rad/s}^2.$$

Transferring tangential acceleration $\boldsymbol{a}^t_{B_1}$ to point B_1 on link 1, we see that the direction of α_1 is counterclockwise.

Comparison of Results with an Analytical Solution

Sample Problem 4.15 may be solved analytically, using equations derived by complex-number methods in Section 4.5. Changing the equations so that link numbers and data correspond to those of Figures 4.17a–d, with $R_1 = O_1B$ and $R_2 = O_2B$, we find the relative acceleration of the coincident points (using results obtained in Section 3.11):

$$\begin{aligned} \boldsymbol{a}^t_{B_2B_1} &= -\omega_2^2 R_2 \cos(\theta_2 \quad \theta_1) + \alpha_2 R_2 \sin(\theta_1 - \theta_2) + \omega_1^2 R_1 \\ &= -12.62 \text{ m/s}^2. \end{aligned}$$

The slider acceleration relative to link 1 is toward O_1; that is, the outward relative velocity is decreasing. The angular acceleration of link 1 is

$$\begin{aligned} \alpha_1 &= \frac{\omega_2^2 R_2 \sin(\theta_1 - \theta_2) + \alpha_2 R_2 \cos(\theta_1 - \theta_2) - 2\omega_1 v_{B_2B_1}}{R_1} \\ &= 79.53 \text{ rad/s}^2 \quad \text{(counterclockwise)}, \end{aligned}$$

from which it follows that the tangential acceleration of B_1 on link 1, $a^t_{B_1} = \alpha_1 R_1 = 21.9 \text{ m/s}^2$, downward and to the left. The results compare closely with those obtained in the graphical solution.

The significance of the order of the subscripts will be noted once more. In Sample Problem 4.15, the solution takes the form

$$\boldsymbol{a}_{B_2} = \boldsymbol{a}_{B_1} + \boldsymbol{a}_{B_2B_1}.$$

The Coriolis component is therefore $\boldsymbol{a}^c_{B_2B_1} = 2\boldsymbol{\omega}_1 \times \boldsymbol{v}_{B_2B_1}$, where $\boldsymbol{\omega}_1$ is the angular velocity of the link along which the slider slides. The subscripts of $\boldsymbol{a}^n_C$ and v must correspond.

If link 1 were curved, or if the end of link 2 were sliding in a curved slot in link 1, then a normal relative acceleration term would be added in Eq. 4.39, resulting in the equation

$$\boldsymbol{a}^n_{B_2} + \boldsymbol{a}^t_{B_2} = \boldsymbol{a}^n_{B_1} + \boldsymbol{a}^t_{B_1} + \boldsymbol{a}^n_{B_2B_1} + \boldsymbol{a}^t_{B_2B_1} + \boldsymbol{a}^c_{B_2B_1}. \quad (4.40)$$

In Figure 4.16b, the relative path radius is r_p. Acceleration $\boldsymbol{a}^n_{B_2B_1}$ is given by $v^2_{B_2B_1}/r_p$, which is directed from B_1 toward O_p. A similar situation occurs with an oscillating roller follower and a disk cam. In cases such as these, an equivalent linkage may be used as an alternative form of solution.

SAMPLE PROBLEM 4.16

Vane Pump

The straight-wing air pump (Figure 4.18a) has four sliding vanes that are held against the housing by inertial forces. Let us examine a single vane (Figure 4.18b) and find the acceleration of a point B on that vane.

Solution. The housing is circular, and as long as the vane makes contact with the housing, the point of contact will describe a circle. Thus, we may introduce an artificial link 2 with its center at O_2, the center of the housing, and its length equal to O_2B. The equivalent mechanism is shown in Figure 4.18c, in which link 1 represents the vane guide with center of rotation O_1. Let the angular velocity of the vane guide be a constant value $\boldsymbol{\omega}_1$ (given).

FIGURE 4.18 (a) This straight-wing air pump can be considered a sliding contact mechanism. The machine is used as a compressor or a vacuum pump and is operated at 500 to 1,500 revolutions per minute. Acceleration forces hold the four sliding vanes against the housing. (Source: ITT Pneumotive.)

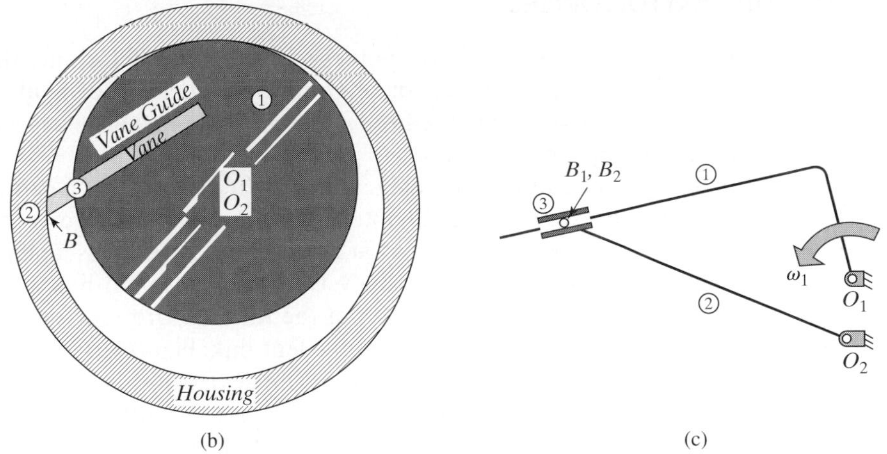

FIGURE 4.18 (b) A single vane of the pump is examined. (c) The equivalent linkage for the mechanism.

TABLE 4.5 Tabulation for the Vector Acceleration Analysis of the Air Pump Equivalent Linkage of Figure 4.17c

Vector	$\boldsymbol{a}^n_{B_2}$	$\boldsymbol{a}^t_{B_2}$	$=\boldsymbol{a}^n_{B_1}$	$+\boldsymbol{a}^c_{B_2B_1}$	$+\boldsymbol{a}^t_{B_2B_1}$
Vector magnitude	$\frac{(ob_2)^2}{O_2B}$	?	$\frac{(ob_1)^2}{O_1B}$	$2(b_1b_2)\omega_1$	?
Vector direction	$\parallel O_2B$ toward O_2	$\perp O_2B$	$\parallel O_1B$ toward O_1	$\perp$ path of slider on link 1 (found by rotating b_1b_2 90° in direction of ω_1)	$\parallel$ path of slider on link 1

Equation (4.39) applies to sliding contact problems of the type

$$\boldsymbol{a}^n_{B_2} + \boldsymbol{a}^t_{B_2} = \boldsymbol{a}^n_{B_1} + \boldsymbol{a}^t_{B_1} + \boldsymbol{a}^t_{B_2B_1} + \boldsymbol{a}^c_{B_2B_1}.$$

The data for this problem differ only slightly from data given in the previous problem. In this case, angular velocity ω_1 is constant for the vane guide (link 1); angular acceleration α_1 is therefore zero, and we can eliminate the term for the tangential acceleration, $\boldsymbol{a}^t_{B_1}$. The angular acceleration α_2 of equivalent link 2 is unknown, and the tangential component of acceleration for B_2, $\boldsymbol{a}^t_{B_2}$, will not, in general, equal zero.

Equation (4.39) can now be rewritten as in Table 4.5. The components making up the acceleration polygon are handled most easily when they are added in the order indicated in the table. After we specify $\boldsymbol{\omega}_1$ and construct a velocity polygon, we construct the acceleration polygon for the straight-wing air pump.

4.9 CAMS AND CAM FOLLOWERS

Cam design is ordinarily a process of synthesis in which we determine the cam shape required to meet a predetermined set of conditions on displacement, velocity, and acceleration. The process is examined for a number of follower-displacement-versus-cam-angle relationships in Chapter 5.

In some cases, an equivalent linkage may be used to analyze a cam and follower by graphical methods. As noted in Chapter 3, if a cam and follower both have finite radii at the point of contact, then a four-bar linkage may be used as an equivalent linkage for velocity analysis. Acceleration analysis of the equivalent linkage then follows. Angular velocity and angular acceleration of the follower are represented by angular velocity and angular acceleration of the equivalent link. Figure 4.19a shows a cam and a roller follower rotating on an oscillating arm. The device may be analyzed by examining an equivalent four-bar linkage, depicted in Figure 4.19b. If the cam surface consists of a series of circular arcs, then a different four-bar linkage is needed to represent each segment of the surface. In that case, there may be jumps (instantaneous changes) in the angular acceleration of the follower arm. A cam that produces follower acceleration jumps is unacceptable for high-speed operation.

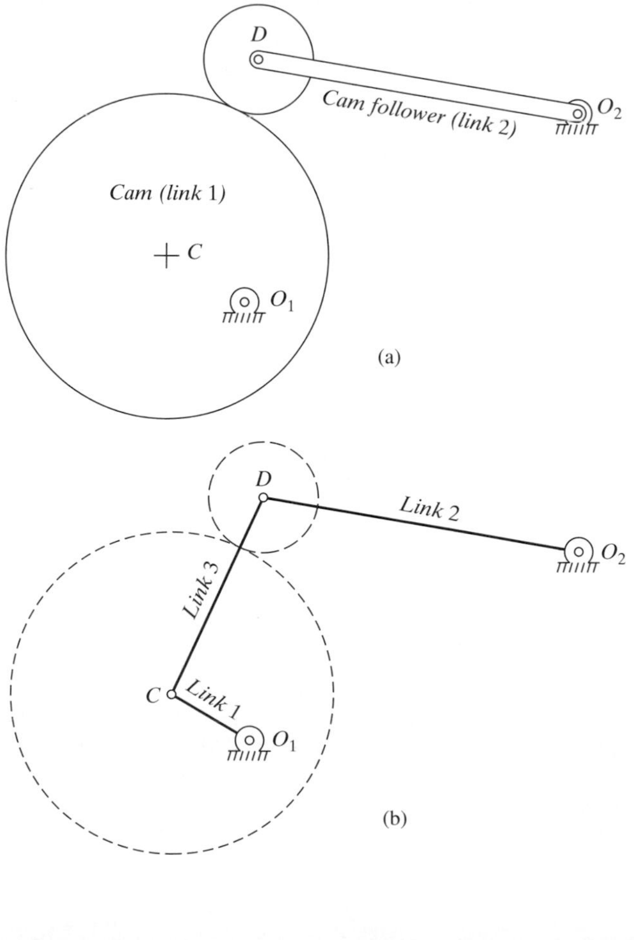

FIGURE 4.19 (a) Cam and roller follower on oscillating arm. (b) Equivalent linkage.

4.10 ANALYZING COMBINATIONS OF BASIC LINKAGES

In most cases, mechanisms made up of more than four links may be broken down into simple basic linkages and solved in a straightforward manner for velocities and accelerations.

If we need to know accelerations for a full range of motion, we can choose motion simulation software, analytical vector methods, or complex-number methods. An acceleration polygon may be used to check analytical results and lend insight into the problem.

Consider a toggle mechanism, a linkage that may be used to apply large forces when it is near the end of its stroke. Figure 4.20a shows a skeleton diagram of such a mechanism.

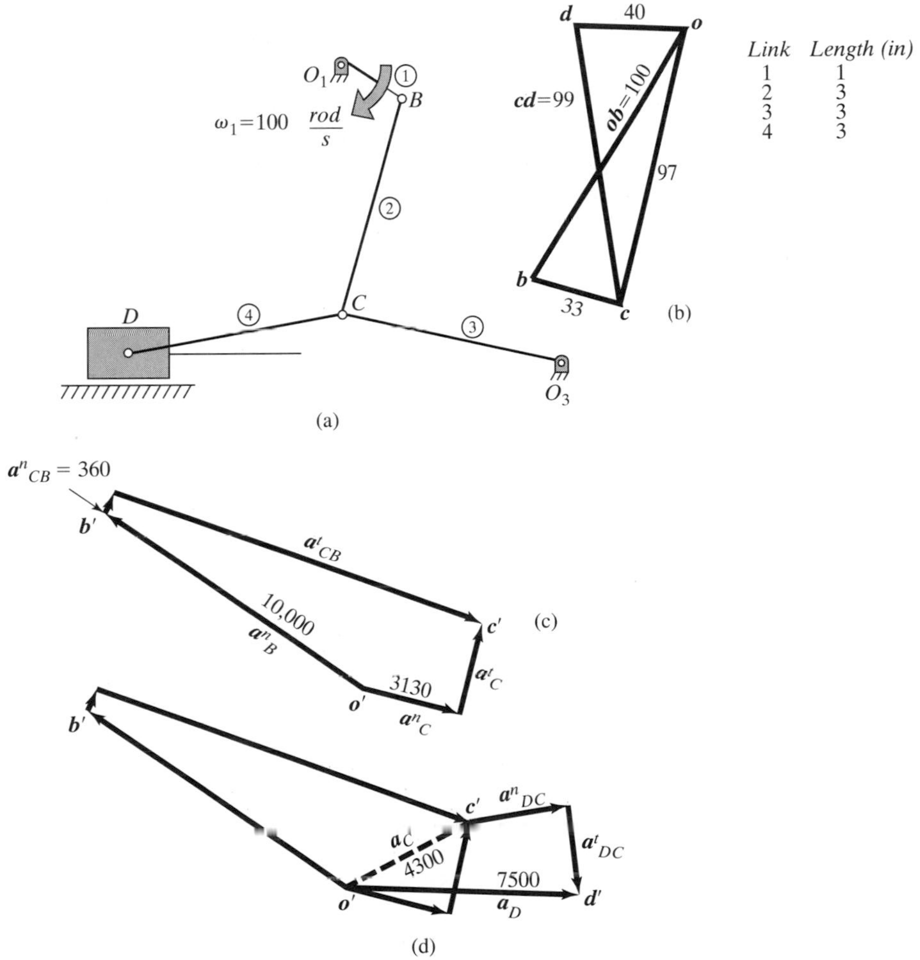

FIGURE 4.20 (a) The toggle mechanism is analyzed by considering it to be made up of two simpler mechanisms: a four-bar linkage (1, 2, 3, and frame): and a slider-crank mechanism (3, 4, and frame). (b) The velocity polygon for the entire mechanism, easily constructed as a single polygon. (We know the length and angular velocity of link 1.) (c) The acceleration polygon for the four-bar linkage portion of the toggle mechanism. (d) The acceleration polygon for the entire toggle mechanism is completed by adding the acceleration vectors for the slider crank to the acceleration polygon for the four-bar mechanism.

TABLE 4.6 Tabulation for the Vector Acceleration Polygon for the Toggle Mechanism, Part I

Vector	$\boldsymbol{a}_C^n$	$+\boldsymbol{a}_C^t$	$= \boldsymbol{a}_B^n$	$+\boldsymbol{a}_{CB}^n$	$+\boldsymbol{a}_{CB}^t$
Vector magnitude	$\frac{(oc)^2}{O_3C}$	?	$\frac{(ob)^2}{O_1B}$	$\frac{(bc)^2}{BC}$	?
Vector direction	$\parallel O_3C$ toward O_3	$\perp O_3C$	$\parallel O_1B$ toward O_1	$\parallel BC$ toward B	$\perp BC$
Vectors used to construct polygon	3130 in/s^2 ↘	? ↗	10,000 in/s^2 ↖	360 in/s^2 ↗	? ↘

The velocity polygon in Figure 4.20b is constructed according to the methods given in Chapter 3. Using values from the velocity polygon, we will proceed to construct the acceleration polygon for part of the toggle mechanism—the four-bar linkage made up of links 1, 2, and 3 and the frame. If link 1 rotates at constant angular velocity, then $\boldsymbol{a}_B^t = 0$, and the acceleration equation for a four-bar mechanism reduces to the equation shown in Table 4.6. Noting the magnitudes of the accelerations as indicated in the table, we select a convenient scale. An acceleration polygon is drawn for the four-bar linkage made up of links 1, 2, and 3 and the frame. (See Figure 4.20c.) The two unknown tangential components, $\boldsymbol{a}_C^t$ and $\boldsymbol{a}_{CB}^t$, may then be scaled from the completed polygon.

To complete the acceleration analysis, we must now consider the rest of the mechanism—the slider crank made up of links 3 and 4 and the frame. We can set up the general equation for the slider acceleration $\boldsymbol{a}_D$ as equal to the acceleration of point C plus the acceleration of D with respect to C. The formula is given in Table 4.7.

The first vector of the slider-crank acceleration polygon, $\boldsymbol{a}_C$, has already been constructed in the four-bar linkage polygon. Adding the remaining slider-crank acceleration vectors to the acceleration polygon in Figure 4.20c, we find slider acceleration $\boldsymbol{a}_D = 7{,}500\ \text{in/s}^2$ to the right (measured on the completed acceleration polygon in Figure 4.20d). The direction of the slider acceleration ($\boldsymbol{o'd'}$ on the acceleration polygon) is opposite that of the slider velocity ($\boldsymbol{od}$ from the velocity polygon); that is, the slider is slowing down.

TABLE 4.7 Tabulation for the Vector Acceleration Polygon for the Toggle Mechanism, Part II

Vector	$\boldsymbol{a}_D =$	$\boldsymbol{a}_C$	$+\boldsymbol{a}_{DC}^n$	$+\boldsymbol{a}_{DC}^t$
Vector magnitude	?	Vector $\boldsymbol{o'c'}$ (already found)	$\frac{(cd)^2}{CD}$	?
Vector direction	Along slider	See figure	$\parallel CD$ toward C	$\perp CD$
Vectors used to construct polygon	? ⟶	4300 in/s^2 ↗	3270 in/s^2 ↗	? ↘

4.11 TRIAL SOLUTION METHOD APPLIED TO LINKAGE ACCELERATION ANALYSIS

A straightforward solution may not be apparent in the analysis of some mechanisms. As with problems in velocity analysis, inverse and trial solutions can be attempted. Since we must deal with both tangential and normal acceleration components, a detailed analysis is often very time-consuming, suggesting the use of computer-aided procedures. The problem that follows illustrates a graphical procedure that will apply to certain types of problems.

Figure 4.21a shows a mechanism with six links and seven revolute joints. Using Grübler's criterion (see Chapter 1), we find that there is one degree of freedom. Thus, if the motion of one link is specified, we should be able to describe the motion of the entire mechanism.

Let the velocity of point B be given, and let link 1 rotate clockwise at the instant shown. Although a direct method would be used if $\boldsymbol{v}_D$ or $\boldsymbol{v}_E$ were given, since $\boldsymbol{v}_B$ is given in this example, we will use the inverse method to complete the velocity polygon. (See Chapter 3.) Using the relationship $\boldsymbol{oe} = \boldsymbol{od} + \boldsymbol{de}$, we ascribe an arbitrary length to vector $\boldsymbol{od}$ (Figure 4.21b). Then, velocity point $\boldsymbol{e}$ is located by noting that $\boldsymbol{oe}$ is horizontal and $\boldsymbol{de}$ is perpendicular to DE. Velocity point $\boldsymbol{c}$ is found from the image principle $dc/de = DC/DE$. The vector equation $\boldsymbol{ob} = \boldsymbol{oc} + \boldsymbol{cb}$, where $\boldsymbol{cb} \perp CB$, allows us to complete the velocity polygon. The scale is obtained by comparing vector $\boldsymbol{ob}$ on the polygon with the given velocity $\boldsymbol{v}_B$. Thus, the velocity scale is $\boldsymbol{v}_B$ (given)/$\boldsymbol{ob}$ (measured), and it follows that $\boldsymbol{v}_E = \boldsymbol{oe}$ (measured) $\times$ (velocity scale), $\boldsymbol{v}_{CB} = \boldsymbol{bc}$ (measured) $\times$ (velocity scale), and so on.

Trial Solution Method

The given data in a problem may be such that neither straightforward nor inverse graphical solutions are practical. This would be the case if the velocity of point B and the angular acceleration of link 1 were specified in Figure 4.21a. We may then proceed by a set of approximations of one of the vector magnitudes, attempting to satisfy the vector equations.

For example, let the velocity of point B in Figure 4.21a be 4 m/s at the instant shown, with ω_1 constant and clockwise. Then the acceleration of point D may be expressed as in Table 4.8. In addition, we see that $\boldsymbol{a}_E(\boldsymbol{o'e'})$ is horizontal and that $d'c'e'$ is the image of DCE, that is, d', c', and e' lie on a line and $d'c'/d'e' = DC/DE$.

Using Table 4.8, we draw $\boldsymbol{a}_B^n = o'b'$ to a convenient scale, add $\boldsymbol{a}_{CB}^n$, and then add a first approximation of $\boldsymbol{a}_{CB}^t$. Vector $\boldsymbol{a}_{DC}^n$ is then added, along with trial vector $\boldsymbol{a}_{CD}^t$, which intersects the vertical trial vector $o'd'$ at d'. Line $d'c'$ is extended to intersect horizontal trial vector $\boldsymbol{o'e'}$ at e'. We then check the acceleration image proportion $d'c'/d'e' = DC/DE$. If it is not satisfied, we make a second approximation of $\boldsymbol{a}_{CB}^t$, attempting to obtain the correct proportion. Several approximations may be required for each of the mechanism positions we wish to examine.

Figure 4.21c shows an acceleration polygon that satisfies the equations with reasonable accuracy. For the data given in this problem, we obtain $\boldsymbol{a}_E = 132 \text{ m/s}^2$ to the left and $\boldsymbol{a}_D = 97 \text{ m/s}^2$ upward.

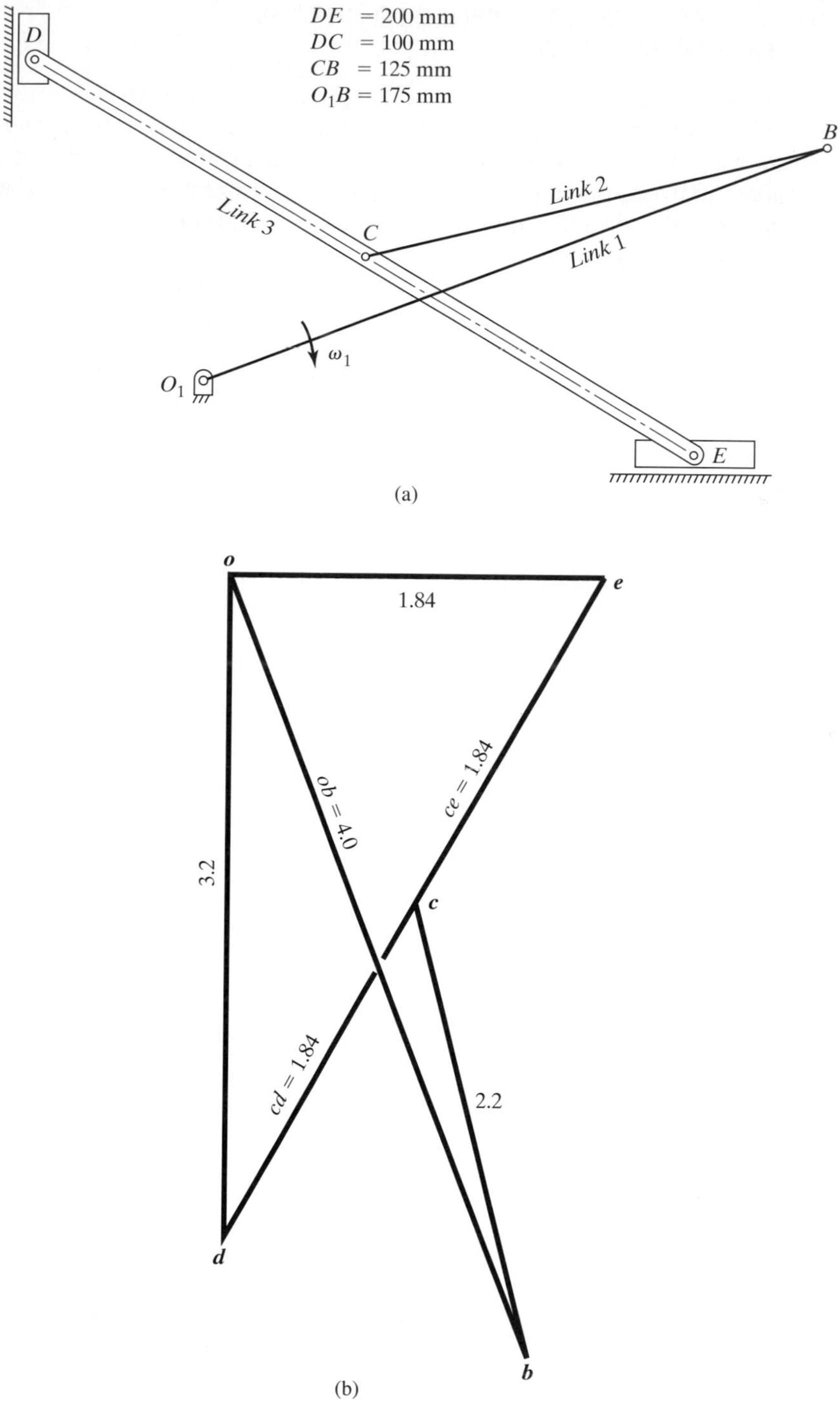

FIGURE 4.21 (a) A six-bar linkage. (b) Velocity polygon; velocities in meters per second.

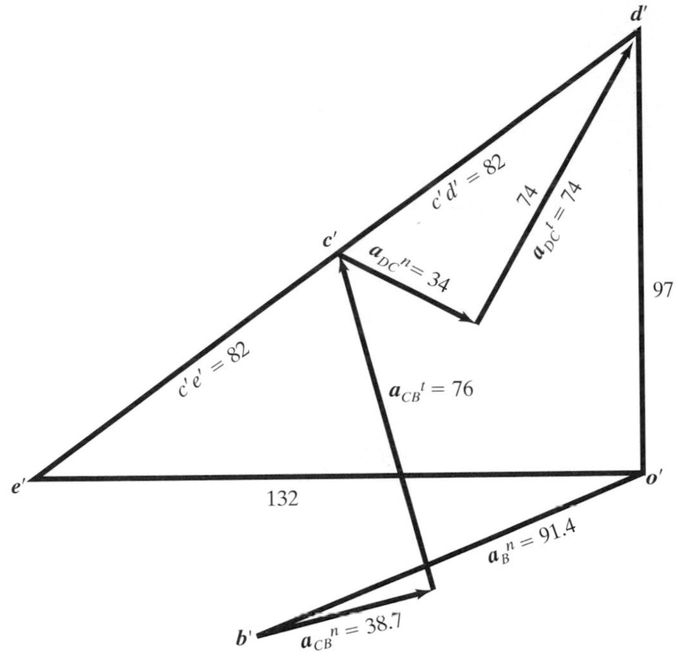

FIGURE 4.21 (c) Acceleration polygon for constant ω_1, accelerations in meters per second squared.

TABLE 4.8 Acceleration by a Trial Solution Method

	$o'd'$	$=$	$o'b'$	$+$	$b'c'$	$+$	$c'd'$
Vector	$\boldsymbol{a}_D$	$=\boldsymbol{a}^n_B$	$+\boldsymbol{a}^t_B$	$+\boldsymbol{a}^n_{CB}$	$+\boldsymbol{a}^t_{CB}$	$+\boldsymbol{a}^n_{DC}$	$+\boldsymbol{a}^t_{DC}$
Vector magnitude	?	$\frac{ob^2}{O_1B}$	$\alpha_1 O_1B_1$	$\frac{bc^2}{BC}$	$\alpha_2 BC$	$\frac{cd^2}{CD}$	$\alpha_3 CD$
Vector direction	Along slider path	$\parallel O_1B$ toward O_1	$\perp O_1B$	$\parallel BC$ toward B	$\perp BC$	$\parallel CD$ toward C	$\perp CD$
Vectors used to construct polygon (m/s^2)	? ↑	91.4 ↙	0	38.7 ↗	? ↖	34 ↘	? ↗

4.12 LIMITING POSITIONS

Limiting positions of the slider-crank mechanism and the crank-rocker mechanism were discussed in previous sections. As the driver crank rotates continuously, the follower link (the slider or driven crank) stops and changes direction at the limiting position. At that position, the follower velocity is zero, but the follower acceleration, in general, is not equal to zero.

Consider, for example, the in-line slider-crank mechanism of Figure 4.22, with the crank rotating at a constant angular velocity ω. In the limiting position in (Figure 4.23a), the crankpin velocity is given by $\boldsymbol{ob} = \omega R$, perpendicular to O_1B (downward).

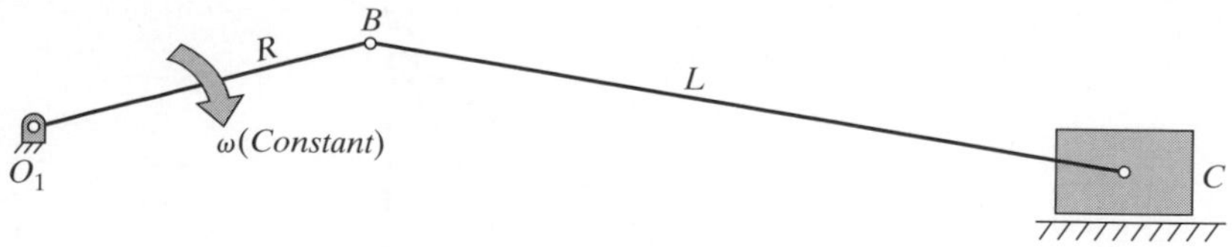

FIGURE 4.22 The in-line slider-crank mechanism. The crank R rotates at constant angular velocity ω.

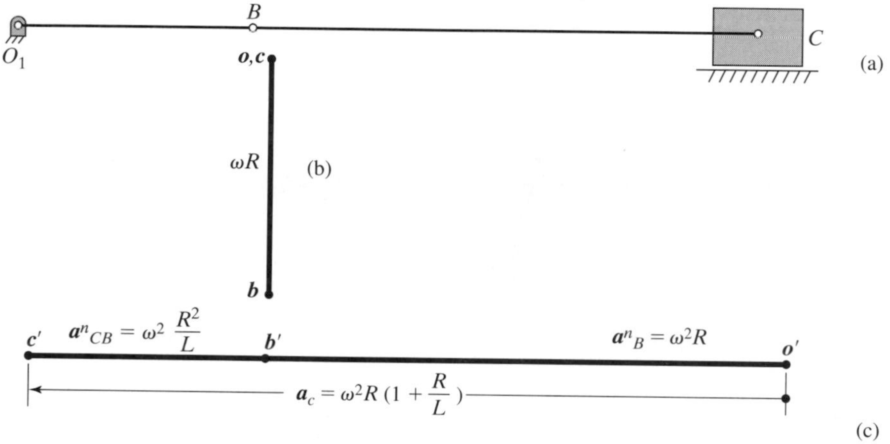

FIGURE 4.23 (a) The mechanism is shown in its extended limiting position. (b) Velocity polygon at the instant the mechanism is in its limiting position. (c) Accelerating polygon for the mechanism in its limiting position.

In forming the velocity polygon for the linkage at its limiting position in Figure 4.23b, we note that relative velocity bc is collinear with velocity $\boldsymbol{ob}$, since the crank (O_1B) and connecting rod (BC) form a straight line. Point C travels only in a horizontal path; therefore, velocity point C must lie at pole point o to correctly position $\boldsymbol{bc}$ perpendicular to BC. The velocity polygon tells us what we already knew: The velocity of the slider is zero. In addition, it gives us the relative velocity:

$$\boldsymbol{bc} = \omega R \text{ (upward).}$$

If the crank has a constant angular velocity, the acceleration equation for the slider may be written as shown in Table 4.9. But a nonzero value of $\boldsymbol{a}^t_{CB}$, is clearly inconsistent with the acceleration equation, since there are no vertical acceleration components. Therefore, $\boldsymbol{a}^t_{CB} = 0$, and our vector equation becomes the simple scalar equation

$$\boldsymbol{a}_C = \boldsymbol{a}^n_B + \boldsymbol{a}^n_{CB} = \omega^2 R\left(1 + \frac{R}{L}\right), \tag{4.41}$$

for the limiting position with the mechanism extended. (See Figure 4.23c.) When the slider is at or near the extreme right position, its acceleration is to the left, as shown in

TABLE 4.9 Tabulation for the Vector Acceleration Analysis of an In-Line Slider Crank in a Limiting Position (Figure 4.23a)

Vector	$\boldsymbol{a}_C =$	$\boldsymbol{a}_B^n$	$+\boldsymbol{a}_{CB}^n$	$+\boldsymbol{a}_{CB}^t$
Vector magnitude	?	$\frac{(ob)^2}{R}$ or $\omega^2 R$	$\frac{(bc)^2}{BC}$	?
Vector direction	Along slider path	$\Vert O_1B$ toward O_1	$\Vert BC$ toward B	$\perp BC$
Vectors used to construct polygon	? $\twoheadleftarrow$	$\omega^2 R$ $\leftarrow$	$\omega^2 \frac{R^2}{L}$ $\leftarrow$	? $\downarrow$

the acceleration polygon. For crank angular acceleration $\alpha \neq 0$, we must include the tangential acceleration $\boldsymbol{a}_B^t = \alpha R$, which is balanced by the equal and opposite vector $\boldsymbol{a}_{CB}^t$. Therefore, Eq. (4.41) holds for the limiting position in Figure 4.23a for constant or variable crank angular velocity.

The other limiting position, with B and C at opposite sides of O_1, results in a slider acceleration

$$\boldsymbol{a}_C = \omega^2 R\left(1 - \frac{R}{L}\right). \tag{4.42}$$

When the slider is at or near its extreme left position, its acceleration is to the right.

The results of the acceleration polygon method are exact. If we substitute $\theta = 0$ and $\theta = 180°$ in the approximate acceleration equation [Eq. 4.37], the results are identical to Eqs. (4.41) and (4.42). Thus, the approximate analytical expressions are exact at $\theta = 0$ and $\theta = 180°$. (The reader is reminded that Eqs. (4.41) and (4.42) were derived for the in-line slider-crank mechanism in its limiting positions; the offset slider-crank mechanism and other linkages will require separate analyses.)

4.13 SPATIAL LINKAGES

Many of the principles used to analyze planar linkages apply to spatial linkages as well. Although both graphical and analytical methods of spatial linkage displacement analysis were illustrated in Chapter 2, analytical methods of velocity and acceleration analysis are generally found most practical. For analytical vector methods, velocity and acceleration terms in the form $\boldsymbol{\alpha} \times \boldsymbol{r}$, $\boldsymbol{\omega} \times (\boldsymbol{\omega} \times \boldsymbol{r})$, and so on, must, of course, be applied in their general, three-dimensional form

$$\boldsymbol{\alpha} \times \boldsymbol{r} = \begin{vmatrix} \boldsymbol{i} & \boldsymbol{j} & \boldsymbol{k} \\ \alpha_x & \alpha_y & \alpha_z \\ r_x & r_y & r_z \end{vmatrix},$$

and so on.

Analysis of an RSSC Spatial Linkage

An RSSC spatial linkage was described previously. (See Chapter 2 and Figure 4.24.) In our description, we avoided vector methods, writing a quadratic equation in terms of link lengths and positions:

$$r_3^2 + 2\cos\gamma(r_0 - r_1\cos\theta)r_3 + r_0^2 + r_1^2 - r_2^2 - 2r_0r_1\cos\theta = 0.$$

In this displacement equation, note that link lengths r_0, r_1, and r_2 and path angle γ are constants. Differentiating with respect to time, we have

$$2r_3\frac{dr_3}{dt} + 2\cos\gamma\, r_1r_3\sin\theta\frac{d\theta}{dt} + 2\cos\gamma\,(r_0 - r_1\cos\theta)\frac{dr_3}{dt} + 2r_0r_1\sin\theta\frac{d\theta}{dt} = 0, \quad (4.43)$$

from which we obtain the velocity of the sliding link, viz.,

$$\frac{dr_3}{dt} = \frac{-\omega_1r_1\sin\theta\,(r_0 + r_3\cos\gamma)}{(r_0 - r_1\cos\theta)\cos\gamma + r_3}, \quad (4.44)$$

where $\omega_1 = d\theta/dt$, the angular velocity of the crank. Differentiating Equation (4.43) or (4.44) with respect to time and simplifying the result, we obtain the acceleration of the sliding link, viz.,

$$\frac{d^2r_3}{dt^2} = -\left[\frac{\begin{array}{c}(dr_3/dt)[(dr_3/dt) + 2\omega_1r_1\cos\gamma\sin\theta] \\ +r_1(r_0 + r_3\cos\gamma)(\omega_1{}^2\cos\theta + \alpha_1\sin\theta)\end{array}}{r_3 + (r_0 - r_1\cos\theta)\cos\gamma}\right], \quad (4.45)$$

where $\alpha_1 = d^2\theta/dt^2$, the angular acceleration of the crank.

SAMPLE PROBLEM 4.17

RSSC *Spatial Linkage*

For the velocity and acceleration analysis of an RSSC spatial linkage, let the RSSC linkage of Figure 4.24 have the proportions

$$\frac{r_0}{r_1} = 2 \quad\text{and}\quad \frac{r_2}{r_1} = 3,$$

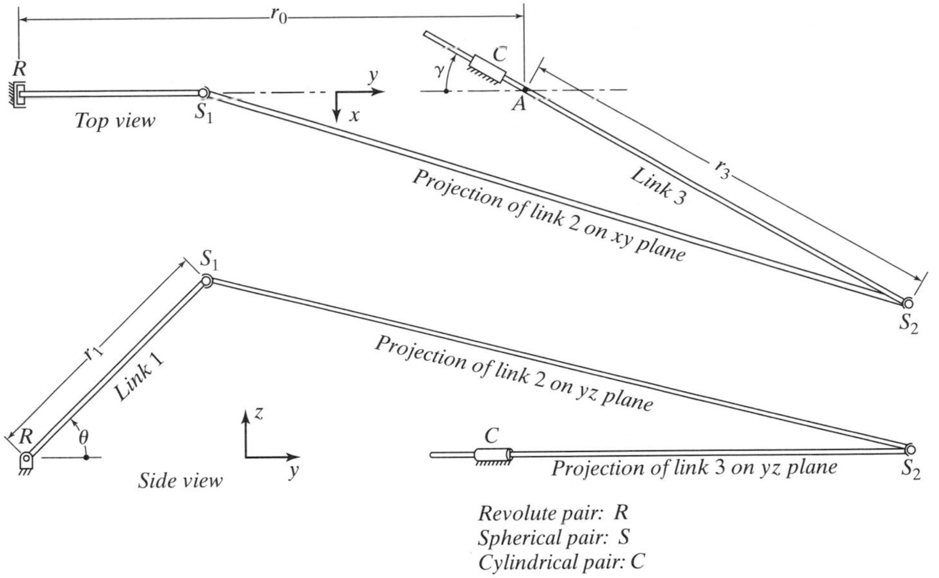

FIGURE 4.24 RSSC spatial linkage.

where crank angular velocity ω_1 is constant and the sliding link (3) and revolute joint R lie in the xy-plane. Plot values of the sliding link velocity and acceleration for various values of crank angle θ, where path angle $\gamma = 30°$.

Solution. The results will be plotted in dimensionless form. Substituting

$$r_0 = \frac{r_0}{r_1} = 2, \quad r_1 = 1, \quad \text{and} \quad r_2 = \frac{r_2}{r_1} = 3$$

and the given values of θ and γ, we may solve the quadratic position equation. The root of interest represents r_3/r_1, the ratio of the slider displacement to the crank length.

Using this root and $\omega_1 = 1$ in Eq. (4.44), the result dr_3/dt represents $(dr_3/dt)/(\omega_1 r_1)$, the dimensionless ratio of the slider velocity to the velocity of joint S_1.

We use this value of dr_3/dt and the preceding substitutions in Eq. (4.45), noting that crank angular acceleration $\alpha_1 = 0$. The result, d^2r_3/dt^2, represents $(d^2r_3/dt^2)/(\omega_1{}^2 r_1)$, the dimensionless ratio of the slider acceleration to the normal acceleration of joint S_1. The results are plotted in Figure 4.25 for path angle $\gamma = 30°$.

For the linkage of this example, when path angle $\gamma = 90°$ and crank angle $\theta = 180°$, velocity $V_3 = 0/0$ (indeterminate), and A_3 is infinite. In that case, the velocity of the sliding link cannot be determined by further examination of the limit (e.g., by l'Hôpital's rule), because the defect exists in the configuration of the mechanism. Since an infinite slider acceleration would require an infinite force, this linkage cannot be operated near crank angle $\theta = 180°$ when path angle $\gamma = 90°$.

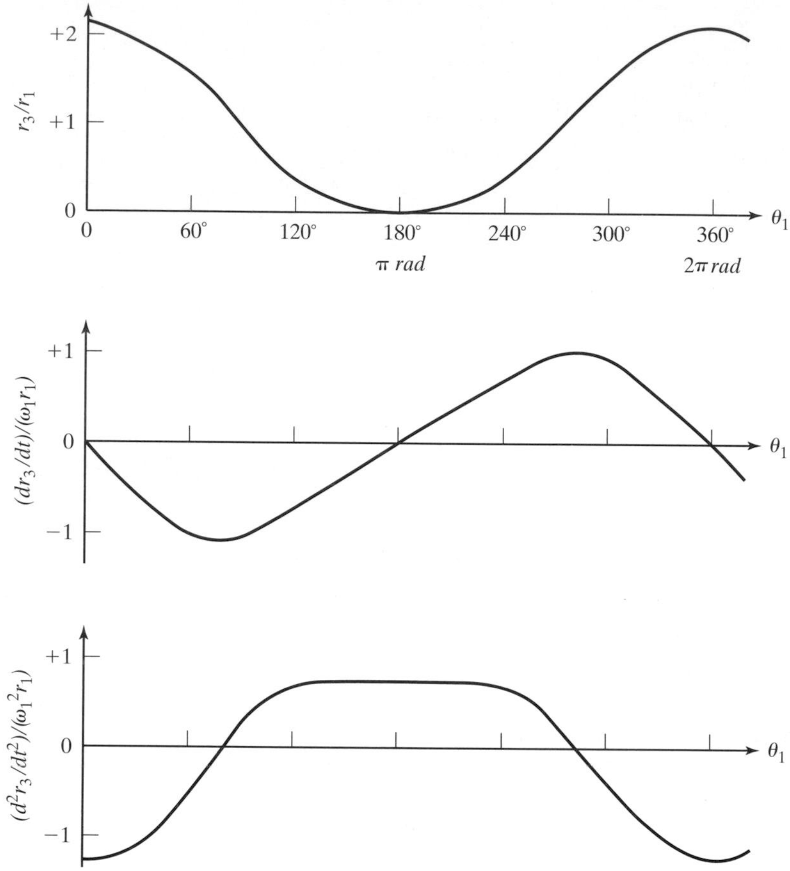

FIGURE 4.25 Displacement, velocity, and acceleration of an *RSSC* linkage.

Vector Methods Applied to an RSSC Linkage

As an alternative to the analytical scalar method, an analytical vector solution could have been used to solve the RSSC linkage. The method uses displacements we have already determined. (See Chapters 2 and 3.)

Accelerations may be determined by using the vector equation

$$\boldsymbol{a}_{s_2}^n + \boldsymbol{a}_{s_2}^t = \boldsymbol{a}_{s_1}^n + \boldsymbol{a}_{s_1}^t + \boldsymbol{a}_{s_2 s_1}^n + \boldsymbol{a}_{s_2 s_1}^t, \tag{4.46}$$

where

$$\left.\begin{aligned} \boldsymbol{a}_{s_1}^n &= \boldsymbol{\omega}_1 \times (\boldsymbol{\omega}_1 \times \boldsymbol{r}_1), \\ \boldsymbol{a}_{s_1}^t &= \boldsymbol{\alpha}_1 \times \boldsymbol{r}_1, \\ \boldsymbol{a}_{s_2 s_1}^n &= \boldsymbol{\omega}_2 \times (\boldsymbol{\omega}_2 \times \boldsymbol{r}_2), \end{aligned}\right\} \tag{4.47}$$

and

$$\boldsymbol{a}_{s_2 s_1}^t = \boldsymbol{\alpha}_2 \times \boldsymbol{r}_2.$$

In the previous example of an RSSC spatial linkage, $\boldsymbol{a}^n_{s_2} = 0$, since S_2 travels in a straight path and the direction of $\boldsymbol{a}^t_{s_2}$ is along that path. Taking the angular acceleration of link 2 about its own axis to be zero, we have the dot product $\boldsymbol{\alpha}_2 \cdot \boldsymbol{r}_2 = 0$. If $\boldsymbol{\omega}_1$ and $\boldsymbol{\alpha}_1$ are given, we may solve for the unknown vectors $\boldsymbol{\alpha}_2$ and $\boldsymbol{\alpha}^t_{s_2}$.

Matrix and Vector Methods Applied to Acceleration Analysis of an RSSR Spatial Linkage

In previous chapters, we determined the vector position equation

$$\mathbf{r}_0 + \mathbf{r}_1 + \mathbf{r}_2 + \mathbf{r}_3 = 0$$

and the vector velocity equation

$$\boldsymbol{\omega}_1 \times \mathbf{r}_1 + \boldsymbol{\omega}_2 \times \mathbf{r}_2 + \boldsymbol{\omega}_3 \times \mathbf{r}_3 = 0$$

for an RSSR spatial linkage.

We differentiate the velocity equation to get the acceleration equation. (Recall the chain rule for differentiation.) Since all of the **r** vectors have constant magnitude, differentiation of a typical velocity term looks like this:

$$d/dt[\boldsymbol{\omega} \times \mathbf{r}] = \boldsymbol{\alpha} \times \mathbf{r} + \boldsymbol{\omega} \times (\boldsymbol{\omega} \times \mathbf{r}),$$

and the full acceleration equation is

$$\begin{aligned}\boldsymbol{\alpha}_1 \times \mathbf{r}_1 + \boldsymbol{\omega}_1 \times (\boldsymbol{\omega}_1 \times \mathbf{r}_1) + \boldsymbol{\alpha}_2 \times \mathbf{r}_2 + \boldsymbol{\omega}_2 \times (\boldsymbol{\omega}_2 \times \mathbf{r}_2)\\ + \boldsymbol{\alpha}_3 \times \mathbf{r}_3 + \boldsymbol{\omega}_3 \times (\boldsymbol{\omega}_3 \times \mathbf{r}_3) = 0.\end{aligned} \tag{4.48}$$

Suppose the link lengths, configuration, and crank speed are specified and we have already solved the position and velocity equations. Let the drive crank angular velocity and acceleration be given. Then there are four unknowns in the acceleration equation: the three components of the angular acceleration of link 2, the coupler, and the angular acceleration of link 3, the driven crank. By expanding the vector acceleration equation, and separating **i**, **j**, and **k** components, we get three scalar equations. We need a fourth equation, however: An RSSR spatial linkage has two degrees of freedom, but we do not usually care about rotation of the coupler about its own axis. If we arbitrarily say that the component of coupler angular velocity along the coupler axis is always zero, then the dot product of the angular acceleration with the link vector is also zero, giving us the fourth equation:

$$\boldsymbol{\alpha}_1 \cdot \mathbf{r}_2 = 0. \tag{4.49}$$

SAMPLE PROBLEM 4.18

Acceleration Analysis: Expanding the Vector Equations

The drive crank of an RSSR linkage rotates at constant speed. The planes of rotation of the drive crank and the driven crank are perpendicular, and the fixed bearing of the driven crank is in the plane of the driver. Write a set of scalar equations from which you can find the angular acceleration of the driven crank.

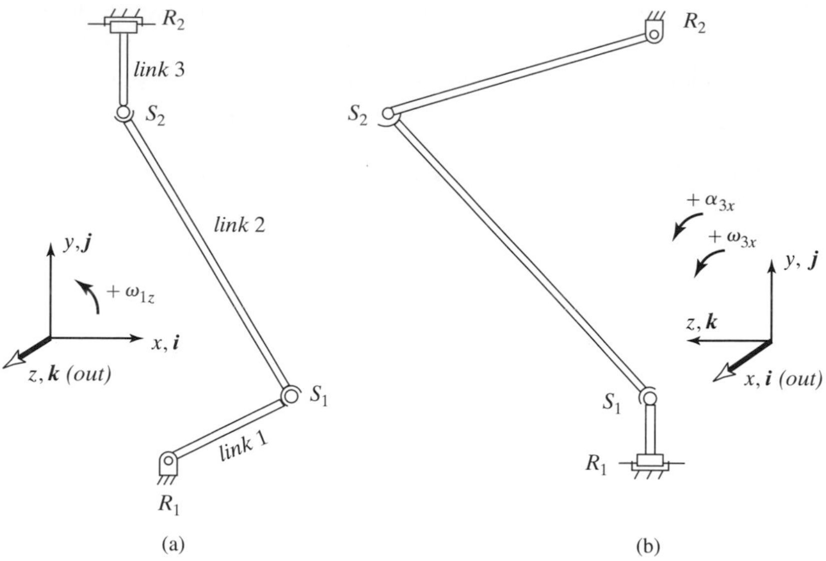

FIGURE 4.26 Acceleration analysis of an RSSR spatial linkage.

Decisions We will set up the coordinate axes as in Figure 4.26. Then, the drive crank vector has components in the x and y directions and rotates about the z-axis. The driven crank has components in the y and z directions, and its angular velocity vector is in the $\pm x$ direction. The unknowns are the angular acceleration of link 3 and the three components of angular acceleration of link 2. Suppose that all other quantities are given or have already been calculated using methods given in Chapters 2 and 3.

Solution. We first compute the cross products of the angular acceleration and link vectors:

$$\boldsymbol{\alpha}_1 \times \mathbf{r}_1 = \begin{vmatrix} \mathbf{i} & \mathbf{j} & \mathbf{k} \\ 0 & 0 & \alpha_{1z} \\ r_{1x} & r_{1y} & 0 \end{vmatrix} = \mathbf{j}\,\alpha_{1z}\,r_{1x} - \mathbf{i}\,\alpha_{1z}\,r_{1y} = 0 \text{ if crank speed is constant;}$$

$$\boldsymbol{\alpha}_2 \times \mathbf{r}_2 = \begin{vmatrix} \mathbf{i} & \mathbf{j} & \mathbf{k} \\ \alpha_{2x} & \alpha_{2y} & \alpha_{2z} \\ r_{2x} & r_{2y} & r_{2z} \end{vmatrix} = \begin{array}{l} \mathbf{i}(\alpha_{2y}\,r_{2z} - \alpha_{2z}\,r_{2y}) \\ +\mathbf{j}(\alpha_{2z}\,r_{2x} - \alpha_{2x}\,r_{2z}) \\ +\mathbf{k}(\alpha_{2x}\,r_{2y} - \alpha_{2y}\,r_{2x}); \end{array}$$

$$\boldsymbol{\alpha}_3 \times \mathbf{r}_3 = \begin{vmatrix} \mathbf{i} & \mathbf{j} & \mathbf{k} \\ \alpha_{3x} & 0 & 0 \\ 0 & r_{3y} & r_{3z} \end{vmatrix} = -\mathbf{j}\,\alpha_{3x}\,r_{3z} + \mathbf{k}\,\alpha_{3x}\,\alpha_{3y}$$

In Chapter 2, we noted that

$$\mathbf{A} \times (\mathbf{B} \times \mathbf{C}) = (\mathbf{A} \cdot \mathbf{C})\mathbf{B} - (\mathbf{A} \cdot \mathbf{B})\mathbf{C},$$

from which it follows that

$$\boldsymbol{\omega} \times (\boldsymbol{\omega} \times \mathbf{r}) = (\boldsymbol{\omega} \cdot \mathbf{r})\boldsymbol{\omega} - (\boldsymbol{\omega} \cdot \boldsymbol{\omega})\mathbf{r}. \tag{4.50}$$

For the drive crank, the angular velocity vector is perpendicular to the radius vector, so

$$\boldsymbol{\omega}_1 \cdot \mathbf{r}_1 = \mathbf{0},$$

Likewise, for the driven crank;

$$\boldsymbol{\omega}_3 \cdot \mathbf{r}_3 = \mathbf{0},$$

and we have arbitrarily set

$$\boldsymbol{\omega}_2 \cdot \mathbf{r}_2 = \mathbf{0},$$

for the coupler.

Thus, for all three moving links, the form of the $\boldsymbol{\omega} \times (\boldsymbol{\omega} \times \mathbf{r})$ part of the equation reduces to

$$\boldsymbol{\omega} \times (\boldsymbol{\omega} \times \mathbf{r}) = (\boldsymbol{\omega} \cdot \mathbf{r})\boldsymbol{\omega} - (\boldsymbol{\omega} \cdot \boldsymbol{\omega})\mathbf{r} = -(\boldsymbol{\omega} \cdot \boldsymbol{\omega})\mathbf{r} = -\omega^2 \mathbf{r},$$

contributing

$$-\omega_{1z}^2 (\mathbf{i}\, r_{1x} + \mathbf{j}\, r_{1y}) - \omega_2^2\, (\mathbf{i}\, r_{2x} + \mathbf{j}\, r_{2y} + \mathbf{k}\, r_{2z}) - \omega_{3x}^2 (\mathbf{j}\, r_{3y} + \mathbf{k}\, r_{3z})$$

where

$$\omega_2^2 = \omega_{2x}^2 + \omega_{2y}^2 + \omega_{2z}^2.$$

Adding the terms in the acceleration equation,

$$\begin{aligned}
\boldsymbol{\alpha}_1 \times \mathbf{r}_1 &+ \boldsymbol{\omega}_1 \times (\boldsymbol{\omega}_1 \times \mathbf{r}_1) + \boldsymbol{\alpha}_2 \times \mathbf{r}_2 + \boldsymbol{\omega}_2 \times (\boldsymbol{\omega}_2 \times \mathbf{r}_2) + \boldsymbol{\alpha}_3 \times \mathbf{r}_3 + \boldsymbol{\omega}_3 \times (\boldsymbol{\omega}_3 \times \mathbf{r}_3) \\
&= \mathbf{i}(\alpha_{2y}\, r_{2z} - \alpha_{2z}\, r_{2y}) + \mathbf{j}(\alpha_{2z}\, r_{2x} - \alpha_{2x}\, r_{2z}) + \mathbf{k}(\alpha_{2x}\, r_{2y} - \alpha_{2y}\, r_{2x}) \\
&\quad -\mathbf{j}\alpha_{3x}\, r_{3z} + \mathbf{k}\, \alpha_{3x}\, r_{3y} - \omega_{1z}^2 (\mathbf{i}\, r_{1x} + \mathbf{j}\, r_{1y}) - \omega_2^2\, (\mathbf{i}\, r_{2x} + \mathbf{j}\, r_{2y} + \mathbf{k}\, r_{2z}) \\
&\quad -\omega_{3x}^2 (\mathbf{j}\, r_{3y} + \mathbf{k}\, r_{3z}) = 0.
\end{aligned} \tag{4.51}$$

Now, we separate the **i**, **j**, and **k** components, setting each equal to zero to get three scalar equations:

$$\alpha_{2y}\, r_{2z} - \alpha_{2z}\, r_{2y} - \omega_{1z}^2\, r_{1x} - \omega_2^2\, r_{2x} = 0 \qquad \text{(from the } \mathbf{i} \text{ terms);}$$

$$\alpha_{2z}\, r_{2x} - \alpha_{2x}\, r_{2z} - \alpha_{3x}\, r_{3z} - \omega_{1z}^2\, r_{1y} - \omega_2^2\, r_{2y} - \omega_{3x}^2\, r_{3y} = 0 \qquad \text{(the } \mathbf{j} \text{ terms);}$$

$$\alpha_{2x}\, r_{2y} - \alpha_{2y}\, r_{2x} + \alpha_{3x}\, r_{3y} - \omega_2^2\, r_{2z} - \omega_{3x}^2\, r_{3z} = 0 \qquad \text{(the } \mathbf{k} \text{ terms).}$$

The fourth equation is

$$\boldsymbol{\alpha}_2 \cdot \mathbf{r}_2 = \alpha_{2x}\, r_{2x} + \alpha_{2y}\, r_{2y} + \alpha_{2z}\, r_{2z} = 0.$$

The analysis of a full cycle of motion for the spatial linkage requires a computer solution. We will "work smart" and put the equations in matrix form.

SAMPLE PROBLEM 4.19

Spatial Linkage Acceleration Equations in Matrix Form

Assume that the angular accelerations are the only unknowns in the preceding four equations. Write the equations in matrix form.

Solution. The four equations are rearranged so that like α's are aligned and the known values are at the right:

$$\begin{aligned}
\alpha_{2y} r_{2z} - \alpha_{2z} r_{2y} \qquad\qquad &= \omega_{1z}^2 r_{1x} + \omega_2^2 r_{2x};\\
-\alpha_{2x} r_{2z} \qquad + \alpha_{2z} r_{2x} - \alpha_{3x} r_{3z} &= \omega_{1z}^2 r_{1y} + \omega_2^2 r_{2y} + \omega_{3x}^2 r_{3y};\\
\alpha_{2x} r_{2y} - \alpha_{2y} r_{2x} \qquad + \alpha_{3x} r_{3y} &= \omega_2^2 r_{2z} + \omega_{3x}^2 r_{3z};\\
\alpha_{2x} r_{2x} + \alpha_{2y} r_{2y} + \alpha_{2z} r_{2z} \qquad &= 0.
\end{aligned}$$

The matrix equation is

$$\mathbf{A\,Y} = \mathbf{C}, \tag{4.52}$$

where the column matrix of unknown angular accelerations is

$$\mathbf{Y} = \begin{bmatrix} \alpha_{2x} \\ \alpha_{2y} \\ \alpha_{2z} \\ \alpha_{3x} \end{bmatrix}.$$

The matrix of coefficients of the unknown quantities is

$$\mathbf{A} = \begin{bmatrix} 0 & r_{2z} & -r_{2y} & 0 \\ -r_{2z} & 0 & r_{2x} & -r_{3z} \\ r_{2y} & -r_{2x} & 0 & r_{3y} \\ r_{2x} & r_{2y} & r_{2z} & 0 \end{bmatrix}.$$

Does this matrix look familiar? Check the velocity analysis of the *RSSR* linkage in Chapter 3. Finally, the column matrix of known values is

$$\mathbf{C} = \begin{bmatrix} \omega_{1z}^2 r_{1x} + \omega_2^2 r_{2x} \\ \omega_{1z}^2 r_{1y} + \omega_2^2 r_{2y} + \omega_{3x}^2 r_{3y} \\ \omega_2^2 r_{2z} + \omega_{3x}^2 r_{3z} \\ 0 \end{bmatrix}.$$

To find the unknown angular accelerations, we compute

$$\mathbf{Y} = \mathbf{A}^{-1}\mathbf{C}. \tag{4.53}$$

Checking for errors. Spatial linkage analysis is complicated, requiring hundreds of keystrokes, with many chances of error. Simple tests may help in detecting errors. Is the acceleration zero when the slope of the velocity curve is horizontal? Be sure to check units. If you are measuring slope on a plot, check the horizontal and vertical scales.

Numerical Differentiation

For constant drive crank speed, the angular acceleration of the driven crank is given by

$$\alpha_3 = d\omega_3/dt = (d\omega_3/d\theta_1)\cdot(d\theta_1/dt) = \omega_1{}^2\cdot(d^2\theta_3/d\theta_1{}^2). \tag{4.54}$$

Take an arbitrary drive crank position. If calculations of the driven crank acceleration by the matrix method and by numerical differentiation do not produce comparable results, something must be wrong.

SAMPLE PROBLEM 4.20

Calculating Spatial Linkage Accelerations

Consider an RSSR spatial linkage with a 60-mm drive crank, 215-mm coupler, and 130-mm driven crank. The drive crank and driven crank rotate in perpendicular planes. Revolute joints R_1 is a ball bearing located at (0, 0, 0); R_2 is another ball bearing located at (15, 210, 0) (mm), from which fixed link components are $r_{0x} = -15$ and $r_{0y} = -210$ mm. The drive crank rotates counterclockwise at a constant speed of 1,000 rpm.

Find the link positions in terms of the drive crank position. Use matrix methods to find angular velocities and accelerations. Plot the driven crank position, angular velocity, and angular acceleration against the drive crank position. Express links, angular velocities, and angular accelerations as vectors. Is the vector link closure equation satisfied? It is convenient to evaluate this and other tests for one arbitrary position, say, $\theta_1 =$ one radian. Have you set the coupler angular velocity about the coupler axis equal to zero? Does the dot product of the two vectors satisfy this condition? Have you set the coupler angular acceleration about the coupler axis equal to zero? Does the dot product of the two vectors satisfy this condition as well? Form the basic vector *acceleration* equation in cross-product form. Is the equation satisfied? Find the angular velocity and angular acceleration of the driven crank by numerical differentiation. Plot the results along with the results from matrix analysis.

Design decisions. We will select an RSSR spatial linkage similar to Figure 4.26. The drive crank will rotate in the xy-plane and the driven crank in the yz-plane.

Solution. The first part of the solution is based on the analysis of an RSSR linkage in Chapters 2 and 3. Drive crank position θ_{1z} is identified simply as θ. The four unknown angular accelerations are computed for each value of θ, using the equation

$$\mathbf{Y} = \mathbf{A}^{-1}\mathbf{C},$$

where matrices **Y**, **A**, and **C** are defined in sample Problem 4.19.

Although we must consider the acceleration of coupler link 2 to solve the problem, we are interested only in the results for driven crank link 3. If the **Y** matrix elements are numbered 0, 1, 2, 3, then

$$\alpha_{3x} = Y_3.$$

The graph in Figure 4.27 shows α_{3x} divided by the square of the drive crank angular velocity, so that the plots have the same order of magnitude. A value taken from the curve is multiplied by $(\omega_{1z}\ \text{rad/s})^2$ to obtain the angular acceleration. We see that a zero slope of the driven crank angular velocity curve corresponds to zero angular acceleration. The results are also checked using a derivative algorithm. The curves based on numerical results cannot be distinguished from the matrix-generated results. Note that the horizontal axis of the graph is the drive crank angle in degrees. The driven crank angle is in radians, and the other curves are dimensionless. The mixed units are for presentation only; be sure to use consistent units in your calculations. Finally, the acceleration equation in terms of vector cross products is satisfied (except for a small rounding error).

Solution details

Acceleration analysis of an RSSR spatial linkage

The drive crank and the driven crank rotate in perpendicular planes.

Units: mm, rad, seconds

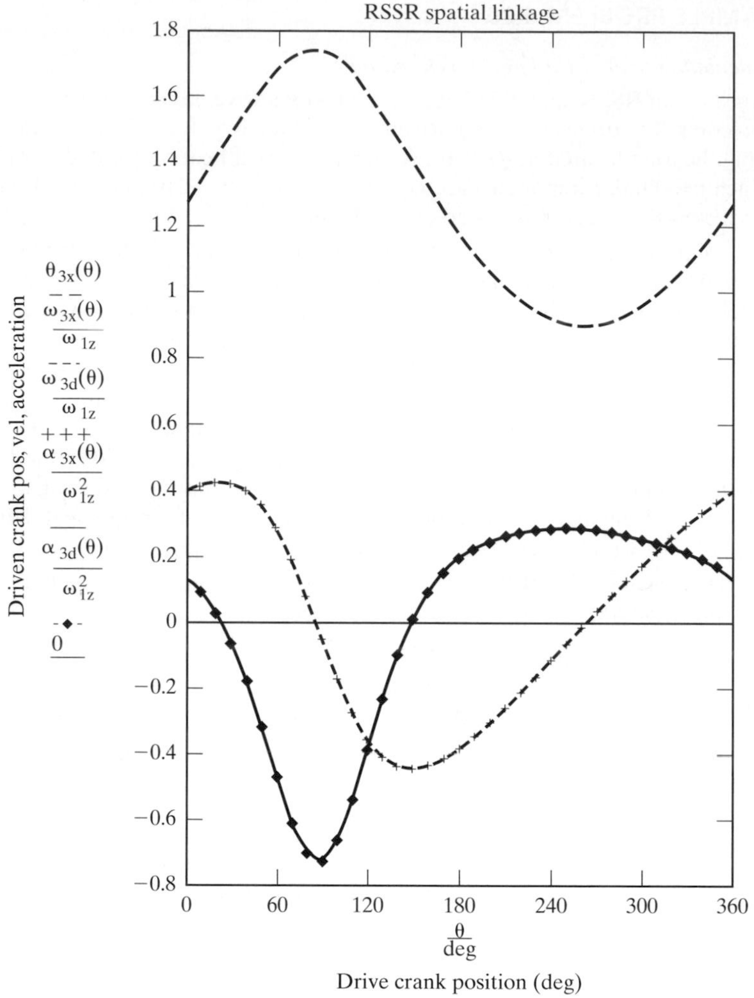

FIGURE 4.27 Spatial linkage driven crank angular position, velocity, and acceleration.

Vector loop: $\mathbf{r}_0 + \mathbf{r}_1 + \mathbf{r}_2 + \mathbf{r}_3 = 0$, where $\mathbf{r}_0 = \mathbf{R}_2 - \mathbf{R}_1$

Dimensions of links: Fixed link $r_{0x} := -15$ $\quad r_{0y} := -210$

Drive crank $r_1 := 60$ Coupler $r_2 := 215$ Driven crank $r_3 := 130$

Position Analysis

Drive crank position (subscript omitted) $\theta := 0, \dfrac{\pi}{18} .. 2\pi$

$$r_{1x}(\theta) := r_1 \cdot \cos(\theta) \qquad r_{1y}(\theta) := r_1 \cdot \sin(\theta)$$

Define c = sum of fixed link and drive crank vectors

$$c_x(\theta) := r_{0x} + r_{1x}(\theta) \qquad c_y(\theta) := r_{oy} + r_{1y}(\theta)$$

$$r_{2x}(\theta) := -c_x(\theta) \qquad r_{2y}(\theta) := \frac{c_x(\theta)^2 - c_y(\theta)^2 - r_2^{\,2} + r_3^{\,2}}{2c_y(\theta)}$$

$$r_{3y}(\theta) := \frac{-(c_x(\theta)^2 + c_y(\theta)^2 - r_2^{\,2} + r_3^{\,2})}{2c_y(\theta)} \qquad r_{3z}(\theta) := \left(r_3^{\,2} - r_{3y}(\theta)^2\right)^{\frac{1}{2}*} \quad r_{2z}(\theta) = -r_{3z}(\theta)$$

* We will select an assembly configuration given by the positive root.

Driven crank position $\theta_{3x}(\theta) := \text{angle}(r_{3y}(\theta), r_{3z}(\theta))$

Link vectors $rr_0 := \begin{bmatrix} r_{0x} \\ r_{0y} \\ 0 \end{bmatrix} \quad rr_1(\theta) := \begin{bmatrix} r_{1x}(\theta) \\ r_{1y}(\theta) \\ 0 \end{bmatrix} \quad rr_2(\theta) := \begin{bmatrix} r_{2x}(\theta) \\ r_{2y}(\theta) \\ r_{2z}(\theta) \end{bmatrix} \quad rr_3(\theta) := \begin{bmatrix} 0 \\ r_{3y}(\theta) \\ r_{3z}(\theta) \end{bmatrix}$

Check closure of vector loop $rr_0 + rr_1(1) + rr_2(1) + rr_3(1) = \begin{bmatrix} 0 \\ 4.263 \cdot 10^{-14} \\ 0 \end{bmatrix}$

Transmission metric $T(\theta) := \dfrac{rr_2(\theta) \cdot rr_3(\theta)}{|rr_2(\theta)| \cdot |rr_3(\theta)|}$

Velocity Analysis

Drive crank speed rpm: $n := 1000 \qquad \omega_{1z} := \dfrac{\pi \cdot n}{30} \qquad \omega_{1z} = 104.72$ rad/s

Matrix equation $\mathbf{AX} = \mathbf{B}$

where $X = \begin{bmatrix} \omega_{2x} \\ \omega_{2y} \\ \omega_{2z} \\ \omega_{3x} \end{bmatrix} \qquad A(\theta) := \begin{bmatrix} 0 & r_{2z}(\theta) & -r_{2y}(\theta) & 0 \\ -r_{2z}(\theta) & 0 & r_{2x}(\theta) & -r_{3z}(\theta) \\ r_{2y}(\theta) & -r_{2x}(\theta) & 0 & r_{3y}(\theta) \\ r_{2x}(\theta) & r_{2y}(\theta) & r_{2z}(\theta) & 0 \end{bmatrix}$

$$B(\theta) := \begin{bmatrix} \omega_{1z} \cdot r_{1y}(\theta) \\ -\omega_{1z} \cdot r_{1x}(\theta) \\ 0 \\ 0 \end{bmatrix}$$

Solve matrix equation for angular velocities $X(\theta) := A(\theta)^{-1} \cdot B(\theta)$ (Elements are numbered 0, 1, 2, 3)
Coupler angular velocity $\omega_{2x}(\theta) := X(\theta)_0 \quad \omega_{2y}(\theta) := X(\theta)_1 \quad \omega_{2z}(\theta) := X(\theta)_2$
Driven crank $\omega_{3x}(\theta) := X(\theta)_3$

Coupler angular velocity magnitude $\omega_2(\theta) := \left(\omega_{2x}(\theta)^2 + \omega_{2y}(\theta)^2 + \omega_{2z}(\theta)^2\right)^{\frac{1}{2}}$
Angular velocity vectors

Drive crank $\omega_1 := \begin{bmatrix} 0 \\ 0 \\ \omega_{1z} \end{bmatrix} \quad$ Coupler $\Omega_2(\theta) := \begin{bmatrix} \omega_{2x}(\theta) \\ \omega_{2y}(\theta) \\ \omega_{2z}(\theta) \end{bmatrix} \quad$ Driven crank $\omega_3(\theta) := \begin{bmatrix} \omega_{3x}(\theta) \\ 0 \\ 0 \end{bmatrix}$

Check at theta $= 1$ $\omega_2(1) = 24.94$ $|\Omega_2(1)| = 24.94$

We set coupler rotation about its own axis $= 0$. Check this. $\Omega_2(1)\cdot rr_2(1) = 2.063\cdot 10^{-13}$

Acceleration Analysis

Drive crank angular acceleration $\alpha_1 := 0$

Matrix equation $\mathbf{AY} = \mathbf{C}$

$$\text{where } Y = \begin{bmatrix} \alpha_{2x} \\ \alpha_{2y} \\ \alpha_{2z} \\ \alpha_{3x} \end{bmatrix} \qquad A(\theta) := \begin{bmatrix} 0 & r_{2z}(\theta) & -r_{2y}(\theta) & 0 \\ -r_{2z}(\theta) & 0 & r_{2x}(\theta) & -r_{3z}(\theta) \\ r_{2y}(\theta) & -r_{2x}(\theta) & 0 & r_{3y}(\theta) \\ r_{2x}(\theta) & r_{2y}(\theta) & r_{2z}(\theta) & 0 \end{bmatrix}$$

$$C(\theta) := \begin{bmatrix} \omega_{1z}^2\cdot r_{1x}(\theta) + \omega_2(\theta)^2\cdot r_{2x}(\theta) \\ \omega_{1z}^2\cdot r_{1y}(\theta) + \omega_2(\theta)^2\cdot r_{2y}(\theta) + \omega_{3x}(\theta)^2\cdot r_{3y}(\theta) \\ \omega_2(\theta)^2\cdot r_{2z}(\theta) + \omega_{3x}(\theta)^2\cdot r_{3z}(\theta) \\ 0 \end{bmatrix}$$

Solve matrix equation for angular accelerations $X(\theta) := A(\theta)^{-1}\cdot B(\theta)$ (Elements are numbered 0, 1, 2, 3)

Coupler $\alpha_{2x}(\theta) := Y(\theta)_0$ $\alpha_{2y}(\theta) := Y(\theta)_1$ $\alpha_{2z}(\theta) := Y(\theta)_2$

Driven crank $\alpha_{3x}(\theta) := Y(\theta)_3$

Angular acceleration of the coupler and driven crank (in vector form)

$$\alpha_2(\theta) := \begin{bmatrix} \alpha_{2x}(\theta) \\ \alpha_{2y}(\theta) \\ \alpha_{2z}(\theta) \end{bmatrix} \qquad \alpha_3(\theta) := \begin{bmatrix} \alpha_{3x}(\theta) \\ 0 \\ 0 \end{bmatrix}$$

Find the angular velocity and angular acceleration of the driven crank by differentiating the angular position. Use the chain rule. $\omega_{3d}(\theta) := \omega_{1z}\cdot\left[\frac{d}{d\theta}\theta_{3x}(\theta)\right]$ $\alpha_{3d}(\theta) := \omega_{1z}\cdot\left[\frac{d}{d\theta}\omega_{3d}(\theta)\right]$

Check: $\alpha_{3x}(1) = -4.711\cdot 10^3$ $\alpha_{3d}(1) = -4.711\cdot 10^3$

We set the angular acceleration of the coupler about the coupler axis $= 0$. Check this. $\alpha_2(1)\cdot rr_2(1) = 2.184\cdot 10^{-11}$

Check the vector acceleration equation

$$a(\theta) := \omega_1 \times (\omega_1 \times rr_1(\theta)) + \Omega_2(\theta) \times (\Omega_2(\theta) \times rr_2(\theta)) + \omega_3(\theta) \times (\omega_3(\theta) \times rr_3(\theta))$$

$$b(\theta) := \alpha_2(\theta) \times rr_2(\theta) + \alpha_3(\theta) \times rr_3(\theta) \qquad a(1) + b(1) = \begin{bmatrix} 0 \\ 0 \\ 4.366\cdot 10^{-11} \end{bmatrix}$$

Mechanical Systems Software Packages

Some design studies involve acceleration analysis, mechanism dynamics, and closed-loop feedback control. Software packages may be used to aid in reducing the massive programming effort that would be required to model the mechanisms involved. Figure 4.28a illustrates a design study involving human–machine interaction, combining

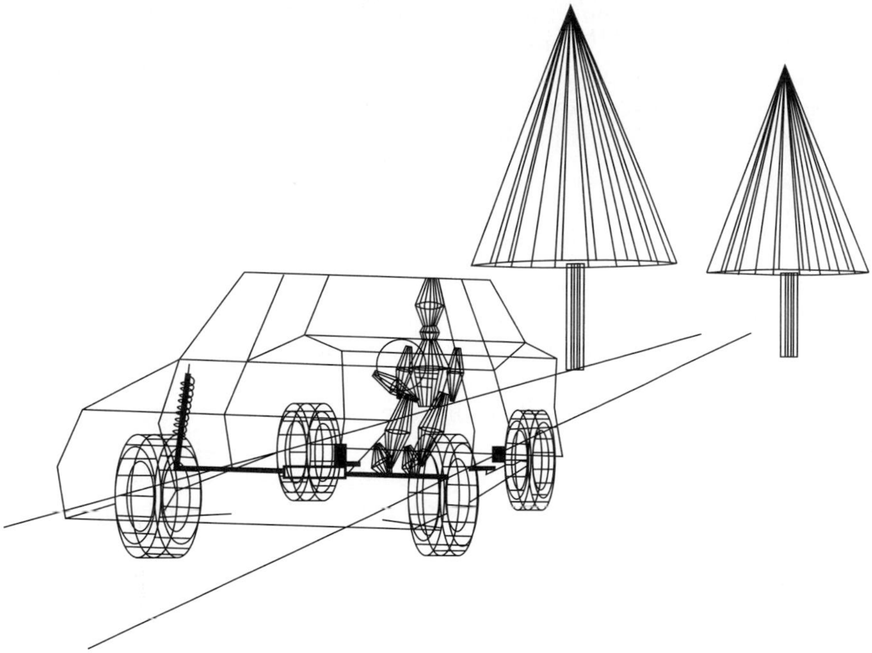

(a)

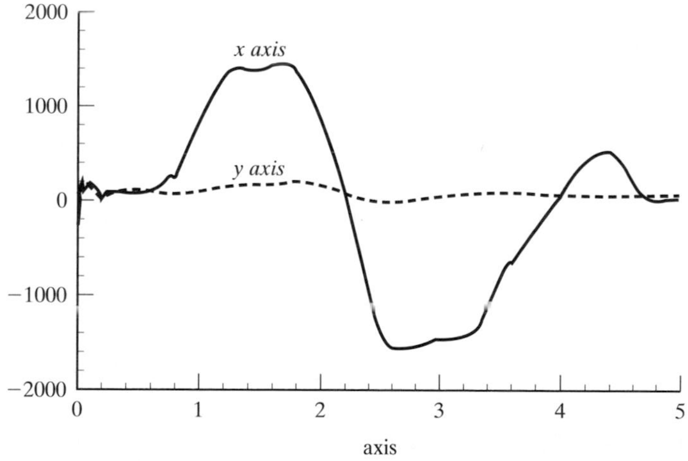

(b)

FIGURE 4.28 (a) Human–machine interaction. (*Source:* Mechanical Dynamics, Inc., modeled by ADAMS^{TM} software.) (b) Acceleration at the driver mass center. (*Source:* Mechanical Dynamics, Inc., modeled by ADAMS^{TM} software.)

vehicular analysis and anthropomorphic characteristics. The study utilized ADAMS/ANDROID™ and ADAMS/TIRE™ to model a driver and vehicle, respectively. Using closed-loop feedback control, the upper musculature of the driver was programmed to move on the basis of an a obstacle sensed on the road. Figure 4.28b shows computed accelerations at the driver center of mass during a maneuver to avoid a tree.

SUMMARY

Velocity is a vector quantity; a change in the magnitude or direction of velocity results in acceleration. Acceleration analysis is the foundation of dynamic analysis and is an essential step in the design of high-speed mechanisms, including piston engines and pumps, quick-return mechanisms, and production machinery.

Animation software shows linkage positions and includes "meters" that plot position, velocity, and acceleration. Complex-number methods are efficient for the acceleration analysis of planar mechanisms. The acceleration polygon (constructed for one linkage position) provides an independent check. Analytical vector methods can be used to solve both planar and spatial linkages. Matrix methods combined with vector methods make it possible to "work smart" when we need to examine a spatial linkage through its full range of motion.

Try to interpret and check results at every step. Does the acceleration plot agree with the velocity and position plots? Do the results of vector and matrix methods check with the results of numerical differentiation? Did you check the units? Is the linkage suitable for its intended purpose? Can you suggest a redesign?

A Few Review Items

- Sketch a planar four-bar linkage, identifying links and revolute joints.
- Write the basic acceleration equation for the four-bar linkage in terms of vector cross products.
- Identify the terms in the acceleration equation as normal and tangential acceleration components.
- What conditions give rise to Coriolis acceleration?
- Sketch an example of Coriolis acceleration.
- Write the Coriolis acceleration component in analytical vector form.
- Write the Coriolis acceleration component in complex-number form.
- Do the two forms give the same vector (both magnitude and direction)?
- Sketch a slider-crank linkage.
- Write the basic acceleration equation for the slider crank linkage in terms of vector cross products.
- Write the basic acceleration equation for the slider-crank linkage in terms of complex numbers in polar form.
- State the acceleration image principle.
- Write the acceleration equation for an RSSR spatial linkage in cross-product form.

- What do we need to know to solve the preceding equation?
- List the steps in the solution.

Some of the problems in this chapter require calculations and the plotting of results for many linkage positions. It is suggested that animation software, mathematics software, or a spreadsheet be used. If the problems are solved only with the aid of a calculator, one or two linkage positions may be selected to avoid repetitious calculations.

PROBLEMS

4.1 Point P lies in a rigid body that rotates at angular velocity

$$\boldsymbol{\omega} = \boldsymbol{i}10 - \boldsymbol{j}20 + \boldsymbol{k}30$$

and angular acceleration

$$\boldsymbol{\alpha} = -\boldsymbol{i}20 - \boldsymbol{j}40 + \boldsymbol{k}80.$$

The body rotates about fixed point O_1, and the radius vector O_1P is given by

$$\boldsymbol{r} = -\boldsymbol{i}50 + \boldsymbol{j}100 - \boldsymbol{k}30.$$

Find the acceleration of P. Unit vectors $\boldsymbol{i}, \boldsymbol{j}$, and $\boldsymbol{k}$ lie in a fixed coordinate system.

4.2 Repeat Problem 4.1 for

$$r = -\boldsymbol{i}40 - \boldsymbol{j}80 + \boldsymbol{k}100.$$

4.3 Find the average acceleration as a piston increases its speed from 120 in/s to 140 in/s during a 0.01-s interval.

4.4 A body moving at 1,000 mi/h is brought to a stop in 0.05 s. Find the average acceleration in inches per second squared.

4.5 Find the average angular acceleration (in radians per second squared) as a flywheel goes **(a)** from zero to 1,000 rev/min in 20 s and **(b)** from 1,000 to 990 rev/min in 0.5 s.

4.6 A vehicle traveling at 50 km/h makes a turn with a 12-meter radius. Find the normal acceleration of the vehicle.

4.7 Repeat Problem 4.6 for a 10-meter radius.

4.8 In Figure 4.1, let $OB = 125$ mm, $\omega = 24$ rad/s cw, and $\alpha = 30$ rad/s^2 ccw. Find the velocity and acceleration of point B.

4.9 Repeat Problem 4.8 for $\alpha = 100$ rad/s^2 cw.

4.10 Referring to Figure 4.2, let the $\boldsymbol{I}$-, $\boldsymbol{J}$-, and $\boldsymbol{K}$-axes be initially coincident with the i-, j-, and k- axes (respectively). Let $R_0 = 40\boldsymbol{i}$, $\dot{R}_0 = 10\boldsymbol{i}$, $\ddot{R}_0 = -20\boldsymbol{i}$, $\boldsymbol{r} = 20\boldsymbol{i} + 60\boldsymbol{j}$, $\dot{\boldsymbol{r}}_r = 5\boldsymbol{i} + 15\boldsymbol{j}$, $\ddot{\boldsymbol{r}}_r = 10\boldsymbol{i} + 30\boldsymbol{j}$, $\boldsymbol{\omega} = 150\ \boldsymbol{k}$, and $\dot{\boldsymbol{\omega}} = 80\boldsymbol{k}$, where millimeter, radian, and second units are used. Find the location, velocity, and acceleration of point P.

4.11 Repeat Problem 4.10 with the same data,except that $\boldsymbol{R}_0 = 0$, $\dot{\boldsymbol{R}}_0 = 10\boldsymbol{k}$, and $\ddot{\boldsymbol{R}}_0 = 20\boldsymbol{k}$.

4.12 Refer to Figure 4.4 and Sample Problem 4.3. Find the acceleration of point D for $\theta_{CBD} = 20°$ and $r_{BD} = 15$ mm. Solve for $\theta_1 = 120°$ by analytical vector methods.

4.13 Refer to Figure 4.4 and Sample Problem 4.3. Find the acceleration of C by using relative acceleration—that is, $\boldsymbol{a}_C = \boldsymbol{a}_B + \boldsymbol{a}_{CB}$, for $\theta_1 = 120°$. Compare your result with the value obtained by using α_3 and ω_3. Solve by analytical vector methods.

4.14 Refer to Figure 4.4 and Sample Problem 4.3. Using the linkage proportions given, determine the ratios ω_3/ω_1 and α_3/ω_1^2 for a range of values of θ_1 if ω_1 is constant. Use a computer or programmable calculator if one is available.

4.15 Repeat Problem 4.14, except that $r_0 = 10$ mm, $r_1 = 35$ mm, $r_2 = 20$ mm, and $r_3 = 30$ mm.

4.16 Plot the slider acceleration versus the crank angle for an in-line slider-crank linkage with connecting-rod-to-crank-length ratio $L/R = 3$. Assume that the crank speed is constant. Use analytical vector methods and a computer or programmable calculator.

4.17 Repeat Problem 4.16 for $L/R = 2$ and offset $E/R = 0.5$.

4.18 In Figure P3.2, $\theta = 45°$ and $\omega_1 = 10$ rad/s (constant). Draw the acceleration polygon, using the scale 1 in = 100 in/s^2. Find α_2.

4.19 Repeat Problem 4.18 for $\theta = 120°$.

4.20 Repeat Problem 4.18 for the mechanism in the limiting position (with point C to the extreme right).

4.21 In Figure P3.3, $\theta = 30°$ and $\omega_1 = 100$ rad/s (constant). Draw the acceleration polygon, using the scale 1 in = 10,000 in/s^2. Find $\boldsymbol{a}_D$ and α_2.

4.22 Repeat Problem 4.21 for $\alpha_1 = 2{,}000$ rad/s^2 counterclockwise.

4.23 In Figure P3.4, $\theta = 60°$, $\omega_1 = 50$ rad/s clockwise, and $\alpha_1 = 500$ rad/s^2 counterclockwise. Draw the acceleration polygon, using the scale 1 in = 2,000 in/s^2. Find $\boldsymbol{a}_D$, α_2, and α_3.

4.24 Repeat Problem 4.23 for $\alpha_1 = 0$.

4.25 In Figure P3.5, $\omega_1 = 20$ rad/s counterclockwise and $\alpha_1 = 100$ rad/s^2 counterclockwise. Draw the acceleration polygon, using the scale 1 in = 200 in/s^2. Find α_2 and α_3.

4.26 Repeat Problem 4.25 for the mechanism in the other limiting position, using the scale 1 in = 100 in/s^2.

4.27 In Figure P3.6, $\theta_2 = 105°$ and $\omega_2 = 20$ rad/s (constant). Draw the acceleration polygon, using the scale 1 in = 200 in/s^2. Find α_1.

4.28 Repeat Problem 4.27 for $\theta_2 = 30°$, using the scale 1 in = 500 in/s^2.

4.29 In Figure P3.3, $\omega_1 = 100$ rad/s (constant). Solve for $\boldsymbol{a}_C$ analytically **(a)** for $\theta = 30°$ and **(b)** for both limiting positions. (Note that the approximate analytical expression is not intended for a linkage with the dimensions of the linkage of Figure P3.3.) Compare your results with a graphical analysis.

4.30 In Figure P3.7, the crank is given an angular velocity ω and an angular acceleration α (both counterclockwise).

(a) Draw the acceleration polygon for the position shown. Write an expression for $\boldsymbol{a}_C$ in terms of ω, R, and L.

(b) Draw the acceleration polygon for the other limiting position.

(c) Examine the effect of α on the value of $\boldsymbol{a}_C$ when the mechanism is in a limiting position.

4.31 In Figure P3.8, $\omega_1 = 100$ rad/s (constant). Draw the acceleration polygon, using the scale 1 in = 10,000 in/s^2.

4.32 Repeat Problem 4.31 for $\alpha_1 = 2000$ rad/s^2.

4.33 In Figure P3.9, $\theta = 135°$ and $\omega_1 = 10$ rad/s (constant). Note that distance B_1C on the cam is constant. Thus, we may take C as a double point: C_1 on the cam and C_2 on the follower. Draw the acceleration polygon, according to the scale 1 in = 50 in/s^2.

4.34 Repeat Problem 4.33 for $\theta = 30°$.

4.35 In Figure P3.10, $\omega_1 = 35$ rad/s (constant). Draw the acceleration polygon, using the scale 1 in = 200 in/s^2. Find the acceleration of the midpoint of each link (D, E, and F).

4.36 In Figure P3.11, $\omega_1 = 20$ rad/s counterclockwise (constant). Note that CD is a fixed distance. Thus, the cam and follower can be replaced by an equivalent four-bar linkage. O_1CDO_2. Draw the velocity and acceleration polygons for the equivalent linkage. Find α_2, the angular acceleration of the follower (represented by O_2D).

4.37 In Figure P3.12, $\theta = 105°$ and $\omega_2 = 20$ rad/s (constant). Draw the acceleration polygon, using the scale 1 in = 200 in/s^2. Find α_3.

4.38 Repeat Problem 4.37 for $\theta = 30°$. Use the scale 1 in = 500 in/s^2.

4.39 In Figure P3.13, $\omega_1 = 30$ rad/s (constant). Draw the velocity and acceleration polygons, using the scales 1 in = 20 in/s and 1 in = 1,000 in/s^2. Find α_2.

4.40 In Figure P3.14, $\omega_1 = 100$ rad/s and $\omega_1 = 2{,}000$ rad/s^2 (both clockwise). Draw the velocity and acceleration polygons, using the scales 1 in = 100 in/s and 1 in = 10,000 in/s^2. Identify $\boldsymbol{a}_D$ and find α_2.

4.41 An in-line slider-crank linkage has a crank length of 4 in and a connecting rod length of 10 in. The maximum slider acceleration occurs at the limiting position, with the slider farthest from the crankshaft. At what crank velocity will the slider acceleration reach 50,000 in/s^2?

4.42 In Figure P3.15, $\omega_5 = 15$ rad/s and $\alpha_1 = 0$. Draw the acceleration polygon, using the scale 1 in = 50 in/s^2. Explain the difficulty encountered in solving this problem where α_5 is specified instead of α_1.

4.43 In Figure P4.1, draw the acceleration polygon for the limiting position shown. Let $\omega_2 = 100$ rad/s (constant). Use the scales 1 in = 100 in/s and 1 in = 10,000 in/s^2. Find α_1 and α_3.

4.44 In Figure P4.2, $\omega_1 = 200$ rad/s cw and $\alpha_1 = 4{,}000$ rad/s^2 cw. Find the acceleration of points B, C, and D and the angular acceleration of link 2.

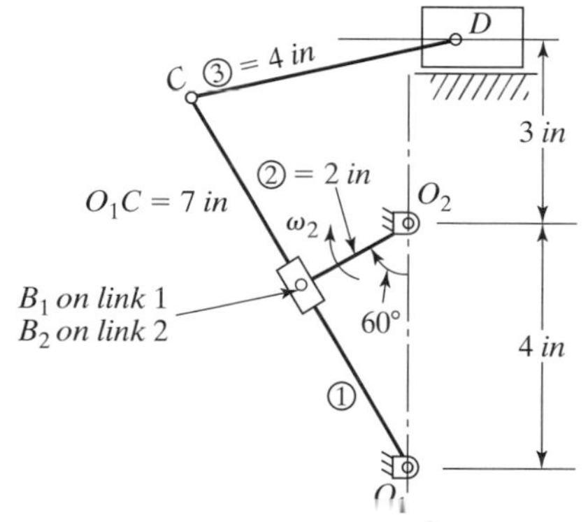

FIGURE P4.1

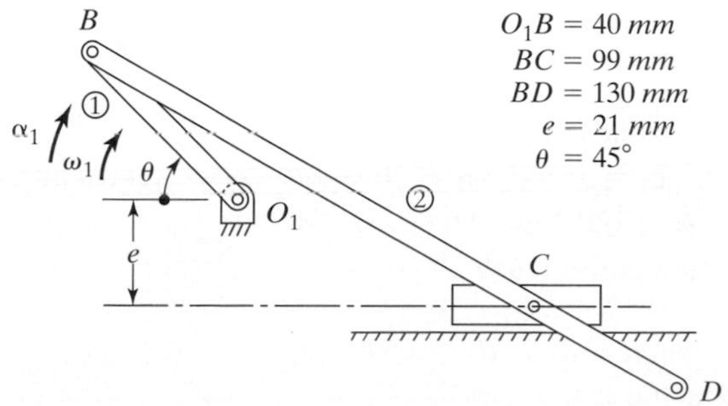

FIGURE P4.2

4.45 Write an expression for the slider acceleration in a slider-crank mechanism if the angular velocity of the crank is not constant.

4.46 **(a)** Find the crank position corresponding to maximum piston acceleration for an in-line slider-crank mechanism. Assume that the crank speed is constant and the ratio $L/R = 2$ of the length to the connecting rod crank length.

(b) Find the maximum piston acceleration in terms of R and ω.

4.47 Repeat Problem 4.46 for $L/R = 1.5$.

4.48 Refer to Figure P3.3, and let $O_1B = 6$ cm, $BC = 15$ cm, $\theta = 210°$, and $\omega_1 = 150$ rad/s (constant). Draw and dimension the velocity and acceleration polygons.

4.49 Determine positions, velocities, and accelerations for the offset sliding contact linkage of Figure P4.3. Assume that r_0, r_1, e, θ_1, ω_1, and α_1 are given. Use complex-number methods.

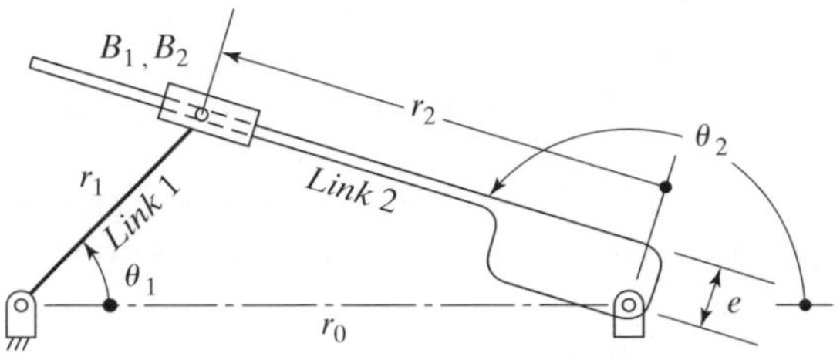

FIGURE P4.3

4.50 Refer to Figure P4.4, the Oldham coupling. Find r_1, r_2, $\dot{r}_1$, $\dot{r}_2$, $\ddot{r}_1$, and $\ddot{r}_2$ in terms of r_0, θ_1, ω_1, α_1, and fixed angle ϕ. Use complex-number methods.

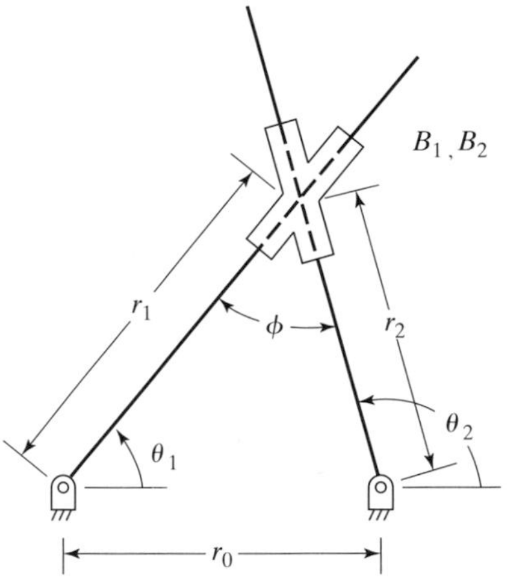

FIGURE P4.4

4.51 Repeat Problem 4.50 by analytical vector methods.

4.52 In Figure P3.18, $\theta_2 = 135°$, $\omega_2 = 500$ rad/s ccw (constant), $O_1O_2 = 300$ mm, $O_1C = 400$ mm, $O_2B = 150$ mm, and $CD = 160$ mm.

(a) Draw acceleration polygon $o'b'_1b'_2$.

(b) Locate c'.

(c) Complete polygon $o'b'_1b'_2c'd'$.

(d) Find α_1 and a_D.

4.53 In Figure P3.2, let $\theta = 45°$, $O_1B = 80$ mm, $BC = 300$ mm, $E = 40$ mm, $\omega_1 = 100$ rad/s cw, and $\alpha_1 = 0$. Draw the velocity and acceleration polygons. Show all accelerations. Find α_2.

4.54 Repeat Problem 4.53 for $\omega_1 = 100$ rad/s ccw.

4.55 Refer to Figure P3.16. Let $\theta_1 = 45°$, $O_1O_2 = 100$ mm, $O_1B_1 = 40$ mm, $O_2C = 150$ mm, $\omega_1 = 10$ rad/s cc, and $\alpha_1 = 0$. Draw the velocity and acceleration polygons. Show all velocities and accelerations.

4.56 Repeat Problem 4.55. Use complex-number methods.

4.57 A slider-crank linkage is similar to Fig. 4.9, except that $O_1B = 60$ mm, $BC = 150$ mm, $\theta_1 = 120°$, and $\omega_1 = 150$ rad/s (constant). Draw the velocity and acceleration polygons. Show all values.

4.58 Repeat Problem 4.57 for $\theta = 180°$.

4.59 Repeat Problem 4.57 by analytical vector methods.

4.60 Refer to Figure P4.5. Let $O_1B_1 = 28$ mm, $O_2B_2 = 58$ mm, $O_2C = 83$ mm, and $O_1O_2 = 80$ mm. Draw the velocity and acceleration polygons. Show all values. Find α_2.

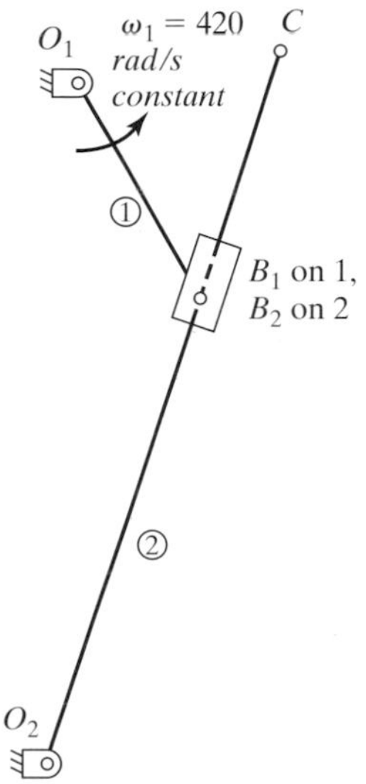

FIGURE P4.5

4.61 Refer to Figure 4.7. Let $R_1 = 30$ mm, $O_1O_2 = 50$ mm, $\omega_1 = 10$ rad/s ccw (constant), and $\theta_1 = 15°$. Find α_2 and $a^t_{B_1B_2}$. Use complex-number methods.

4.62 Repeat Problem 4.61 for $\theta_1 = 30°$.

4.63 Refer to Figure 4.7. Compute α_2/ω_1^2 for $\theta_1 = 0$ to 180° in 15° steps if $R_0/R_1 = 3$. Use a computer or a programmable calculator.

4.64 Using complex-number methods, develop the equations required for an acceleration analysis of a four-bar linkage.

4.65 Consider a four-bar linkage in which drive crank length $r_1 = 10$ mm, coupler length $r_2 = 35$ mm, driven crank length $r_3 = 40$ mm, and fixed link length $r_0 = 30$ mm. Find accelerations when $\theta = 45°$ and $\omega_1 = 30$ rad/s ccw (constant) and the linkage is in the open position.

4.66 Repeat Problem 4.65 for $r_1 = 15$ mm.

4.67 Refer to Figure 4.24, which depicts a spatial linkage. Let $r_0 = 2.5$, $r_1 = 1$, $r_2 = 2.5$, $\omega_1 = 100$ ccw, and $\gamma = 45°$. Find the acceleration of link 3 when $\theta = 60°$.

4.68 In Problem 4.67, find $a_3/(\omega_1^2 r_1)$ for one complete cycle of motion (in 30° steps). Use a computer or a programmable calculator.

4.69 Refer to Figure 4.24. Let $r_0 = 6$, $r_1 = 1$, $r_2 = 9$, and $\gamma = 90°$. Find $a_3/(\omega_1^2 r_1)$ for one complete cycle of motion (in 30° steps). Use a computer or a programmable calculator.

4.70 Consider a four-bar linkage for which $r_0 = 3.5$, $r_1 = 1.2$, $r_2 = 2.2$, and $r_3 = 3.1$. The angular velocity of the crank is 400 rad/s (constant and counterclockwise). Tabulate and plot angular velocities and angular accelerations of links 2 and 3 against the crank angle. (*Suggestion:* Use a spreadsheet program.)

4.71 Consider a four-bar linkage for which $r_0 = 38$, $r_1 = 13$, $r_2 = 27$, and $r_3 = 41$. The angular velocity of the crank is 50 rad/s (constant and counterclockwise). Tabulate and plot angular velocities and angular accelerations of links 2 and 3 against the crank angle. (*Suggestion:* Use a spreadsheet program.)

4.72 Consider a four-bar linkage for which $r_0 = 312.48$, $r_1 = 100$, $r_2 = 200$, and $r_3 = 300$. The angular velocity of the crank is 250 rad/s (constant and counterclockwise). Tabulate and plot angular velocities and angular accelerations of links 2 and 3 against the crank angle. (*Suggestion:* Use a spreadsheet program.)

4.73 Consider a four-bar linkage for which $r_0 = 37$, $r_1 = 10$, $r_2 = 25$, and $r_3 = 40$. The angular velocity of the crank is 100 rad/s (constant and counterclockwise). Tabulate and plot angular velocities and angular accelerations of links 2 and 3 against the crank angle. (*Suggestion:* Use a spreadsheet program.)

4.74 A slider-crank linkage has a connecting-rod-to-crank-length ratio of 2 and an offset-to-crank-length ratio of 0.5. Tabulate and plot the connecting-rod angular velocity and angular acceleration and the slider velocity and acceleration, all against the crank angle. Assume that the crank angular velocity is constant. (*Suggestion:* Let crank length = 1 and crank angular velocity = 1, and normalize the results on these parameters.) Consider the effect of other values of the crank length and angular velocity. Solve the problem by computer, using vector methods.

4.75 A slider-crank linkage has a connecting-rod-to-crank-length ratio of 3 and an offset-to-crank-length ratio of 0.5. Tabulate and plot the connecting-rod angular velocity and angular acceleration, and the slider velocity and acceleration, all against the crank angle. Assume that the crank angular velocity is constant. (*Suggestion:* Let crank length = 1 and crank angular velocity = 1, and normalize the results on these parameters.) Consider the effect of other values of the crank length and angular velocity. Solve the problem by computer, using vector methods.

4.76 A slider-crank linkage has a connecting-rod-to-crank-length ratio of 2.5 and an offset-to-crank-length ratio of 0.6. Tabulate and plot the connecting-rod angular velocity and angular acceleration, and the slider velocity and acceleration, all against the crank angle. Assume that the crank angular velocity is constant. (*Suggestion:* Let crank length = 1 and crank angular velocity = 1, and normalize the results on these parameters.) Consider the effect of other values of the crank length and angular velocity. Solve the problem by computer, using vector methods.

4.77 An offset slider-crank linkage is described by the vector equation $\mathbf{r}_1 + \mathbf{e} + \mathbf{r}_2 + \mathbf{r}_0 = 0$. The crank length = 25, connecting-rod length = 45, and offset vector $\mathbf{e} = -10\mathbf{j}$ (all in mm). The crank rotates counterclockwise at a constant angular velocity of 100 rad/s. Find the link position, velocity, and acceleration vectors. Plot the slider position, velocity, and acceleration and angular position, velocity, and acceleration of the connecting rod (all against the crank position). Check the vector closure for one position. Using numerical differentiation, check the slider velocity and acceleration for one position. Using numerical differentiation, check the connecting-rod angular velocity and acceleration for one position.

4.78 An offset slider-crank linkage is described by the vector equation $\mathbf{r}_1 + \mathbf{e} + \mathbf{r}_2 + \mathbf{r}_0 = 0$. The crank length = 20, connecting-rod length = 32, and offset vector $\mathbf{e} = 6\mathbf{j}$ (all in mm). The crank rotates counterclockwise at a constant speed of 1,760 rpm. Find the link position, velocity, and acceleration vectors. Plot the slider position, velocity, and acceleration and angular position, velocity, and acceleration of the connecting rod (all against the crank position). Check the vector closure for one position. Using numerical differentiation, check the slider velocity and acceleration for one position. Using numerical differentiation, check the connecting-rod angular velocity and acceleration for one position.

4.79 An offset slider-crank linkage is described by the vector equation $\mathbf{r}_1 + \mathbf{e} + \mathbf{r}_2 + \mathbf{r}_0 = 0$. The crank length = 50, connecting-rod length = 110, and offset vector $\mathbf{e} = -20\mathbf{j}$ (all in mm). The crank rotates counterclockwise at a constant speed of 1,000 rpm. Find the link position, velocity, and acceleration vectors. Plot the slider position, velocity, and acceleration and angular position, velocity, and acceleration of the connecting rod (all against the crank position). Check the vector closure for one position. Using numerical differentiation, check the slider velocity and acceleration for one position. Using numerical differentiation, check the connecting-rod angular velocity and acceleration for one position.

4.80 Suppose you are investigating an RSSR spatial linkage for a particular application. The tentative design consists of a 50-mm drive crank, a 180-mm coupler, and a 110-mm driven crank. The drive crank and driven crank rotate in perpendicular planes. Revolute joint R_1 is a ball bearing located at (0, 0, 0), revolute joint R_2 is another ball bearing located at (15, 175, 0) (mm), from which the fixed link components are $r_{ox} = -15$ mm and $r_{oy} = -175$ mm. The drive crank rotates counterclockwise at a constant angular velocity of 10 rad/s. Find the link positions in terms of the drive crank position. Use matrix methods to find the angular velocities and accelerations. Plot the driven crank position, angular velocity, and angular acceleration against the drive crank position. Express links, angular velocities, and angular accelerations as vectors. Is the vector link closure equation satisfied? You may evaluate this and other tests for one arbitrary position, say, $\theta_1 = 1$ radian. Have you set the coupler angular *velocity* about the coupler axis equal to zero? Does the dot product of the two vectors satisfy this condition? Have you set the coupler angular *acceleration* about the coupler axis equal to zero? Does the dot product of the two vectors satisfy this condition as well? Form the basic vector acceleration equation in cross-product form. Is the equation satisfied? Find the angular velocity and angular acceleration of the driven crank by numerical differentiation. If your software has a numerical differentiation capability, plot the results along with the results from matrix analysis. Otherwise, check for one position. Among your design decisions, select an *RSSR* spatial linkage similar to the one in this chapter. The drive crank will rotate in the *xy*-plane and the driven crank in the *yz*-plane.

4.81 Suppose you are considering an RSSR spatial linkage for a particular application. The tentative design consists of a 24-mm drive crank, a 230-mm coupler, and a 140-mm driven crank. The drive crank and driven crank rotate in perpendicular planes. Fixed link components are $r_{ox} = -18$ mm and $r_{oy} = -215$ mm. The drive crank rotates counterclockwise at a constant speed of 880 rpm. Find the link positions in terms of the drive crank position. Use matrix methods to find the angular velocities and accelerations. Plot the driven crank position, angular velocity, and angular acceleration against the drive crank position. Express links, angular velocities, and angular accelerations as vectors. Is the vector link closure equation satisfied? You may evaluate this and other tests for one arbitrary position, say $\theta_1 = 1$ radian. Have you set the coupler angular *velocity* about the coupler axis equal to zero? Does the dot product of the two vectors satisfy this condition? Have you set the coupler angular *acceleration* about the coupler axis equal to zero? Does the dot product of the two vectors satisfy this condition as well? Form the basic vector acceleration equation in cross-product form. Is the equation satisfied? Find the angular velocity and angular acceleration of the driven crank by numerical differentiation. If your software has a numerical differentiation capability, plot the results along with the results from matrix analysis. Otherwise, check for one position. Among your design decisions, select an RSSR spatial linkage similar to the one in this chapter. The drive crank will rotate in the *xy*-plane and the driven crank in the *yz*-plane.

4.82 Suppose you are designing an RSSR spatial linkage. The tentative design consists of a 90-mm drive crank, a 340-mm coupler, and a 200-mm driven crank. The drive crank and driven crank rotate in perpendicular planes. Fixed link components are $r_{ox} = 20$ mm and $r_{oy} = -305$ mm. The drive crank rotates counterclockwise at a constant speed of 200 rpm. Find the link positions in terms of the drive crank position. Use matrix methods to find the angular velocities and accelerations. Plot the driven crank position, angular velocity, and angular acceleration against the drive crank position. Express links, angular velocities, and angular accelerations as vectors. Is the vector link closure equation satisfied? You may evaluate this and other tests for one arbitrary position, say, $\theta_1 = 1$ radian. Have you set the coupler angular *velocity* about the coupler axis equal to zero? Does the dot product of the two vectors satisfy this condition? Have you set the coupler angular *acceleration* about the coupler axis equal to zero? Does the dot product of the two vectors satisfy this condition as well? Form the basic vector acceleration equation in cross-product form. Is the equation satisfied? Find the angular velocity and angular acceleration of the driven crank by numerical differentiation. If your software has a numerical differentiation capability, plot the results along with the results from matrix analysis. Otherwise, check for one position. Among your design decisions, select an *RSSR* spatial linkage similar to the one in this chapter. The drive crank will rotate in the *xy*-plane and the driven crank in the *yz*-plane.

PROJECTS

See Projects 1.1 through 1.6 and the suggestions in Chapter 1.

Describe and plot acceleration and angular acceleration characteristics of linkages in your design. Make use of computer software wherever practical. Check the results by a graphical method for at least one linkage position. Evaluate the linkage in terms of its performance requirements.

BIBLIOGRAPHY AND REFERENCES

Chace, M. A., "Vector Analysis of Linkages," *Transactions of the American Society of Mechanical Engineers, Journal of Engineering for Industry*, vol. 55, ser. B, August 1963, pp. 289–297.

JML Research, Inc., *Integrated Mechanisms Program*, JML Research, Inc., Madison, WI, 1988.

Knowledge Revolution, *Working Model™ 2D User's Manual*, Knowledge Revolution, San Mateo, CA, 1996.

Knowledge Revolution, *Working Model™ 3D User's Manual*, Knowledge Revolution, San Mateo, CA, 1998.

Lawry, M. H., *Master Series Student Guide*, Structural Dynamics Research Corp., Milford, OH, 2d. ed., 1998.

MathSoft, *Mathcad2000™ User's Guide*, MathSoft, Inc., Cambridge, MA, 1999.

Mechanical Dynamics, Inc., *ADAMS Applications Manual*, Mechanical Dynamics Inc., Ann Arbor, MI, 1989.

C H A P T E R 5

Design and Analysis of Cam-and-Follower Systems

In its most common form, a cam mechanism consists of a rotating disk that drives an oscillating follower. The required follower motion determines the shape of the cam. A few cam and follower configurations are shown in Figure 5.1a-d. An engine with dual camshafts is shown in Figure 5.1e. These and hundreds of other cam-and-follower designs are used in transportation, industry, and consumer products. A few of the many applications

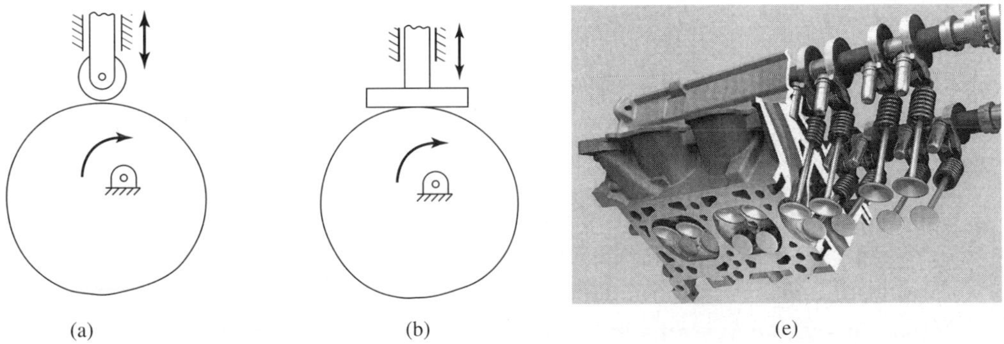

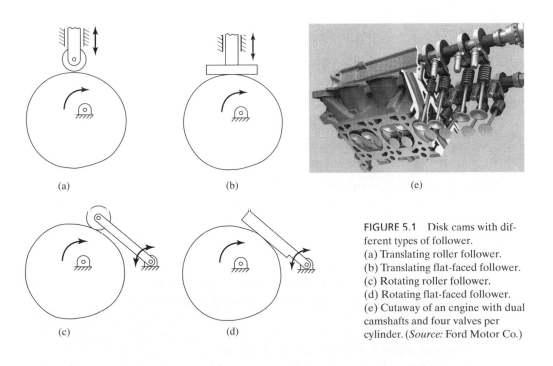

FIGURE 5.1 Disk cams with different types of follower.
(a) Translating roller follower.
(b) Translating flat-faced follower.
(c) Rotating roller follower.
(d) Rotating flat-faced follower.
(e) Cutaway of an engine with dual camshafts and four valves per cylinder. (*Source:* Ford Motor Co.)

of cam-and-follower systems include the precise control of intake and exhaust valves, the operation of home appliances, and the control of manufacturing processes.

Concepts You Will Learn and Apply When Studying This Chapter

- Follower selection criteria for various applications
- Motion characteristics that define a good high-speed cam-and-follower system
- Developing cam profiles to satisfy critical criteria
- Pressure angle
- Practical considerations in the design of cam and follower systems
- Special-purpose cam-and-follower systems
- The theory of envelopes

5.1 INTRODUCTION

Since the cam mechanism has its motion prescribed, it is a good example of kinematic synthesis. That is, rather than requiring the engineer to analyze a mechanism to determine its motion, the cam follower has a predetermined motion, and the engineering process consists of designing the cam so that it gives the desired motion. It is theoretically possible to obtain almost any type of follower motion by properly designing the cam. However, practical design considerations often necessitate modifications of desired follower motions.

Some Important Applications of Cam-and-Follower Systems

A two-cylinder eccentric-cam-type piston pump was shown in Chapter 2. Cams are also used to drive fuel injector pumps, particularly in diesel (compression-ignition) engines. The high pressures required for diesel fuel injection cause high contact stresses on the cam. A roller follower is used to reduce wear. Most cam-and-follower systems are designed to control a process, not to transmit significant power.

Valve-Operating Systems in Internal Combustion Engines

The thermodynamic cycle of a four-stroke-cycle engine (commonly called a four-stroke engine) involves two revolutions of the crankshaft. For a cylinder oriented as in Figure 1.31, the cycle progresses as follows:

- *Intake stroke (induction).* A cam system opens the inlet valve to draw in a charge of air as the piston moves downward in the cylinder. Fuel may be mixed with air in a spark-ignition (S-I) engine at an air–fuel ratio of, say, 15:1.
- *Compression stroke.* The intake and exhaust valves are closed. The piston moves upward toward top dead center (tdc). In diesel engines, fuel is injected before tdc. Ignition occurs before tdc in diesel and S-I engines. Typical compression ratios range from 7.5:1 to 10:1 for S-I engines and from 12:1 to 24:1 in turbocharged diesels.

- *Power stroke (expansion)*. Combustion raises temperature and pressure, forcing the piston down. A cam system opens the exhaust valve at the end of the power stroke. In a single-cylinder engine, energy is stored in a flywheel (and other masses) during the power stroke. Some of that energy is dissipated during the other three strokes. Designers of multicylinder engines attempt to select a sequence of power strokes (i.e., a firing order) that provides smooth operation and minimum vibration.
- *Exhaust stroke*. The exhaust valve is held open while the piston rises and expels exhaust gases.

Large volumes of gases are involved, and timing is critical. Reliability, smooth operation, clean operation, and efficiency all depend on a well-designed cam-and-follower system to control gas flow through the intake and exhaust valves.

Each cylinder in a four-stroke-cycle engine will have one or two intake valves and one or two exhaust valves. The valves are controlled by cams arranged along one or two camshafts. The camshaft speed must be half the crankshaft speed. The camshaft is driven by a chain, by a pair of gears, or by a toothed belt. Figure 5.2a illustrates an engine in cross section, showing a cam, a cam follower, and a valve. The cam follower is held against the cam profile by a spring. Figure 5.2b shows the camshaft of an eight-cylinder engine.

A two-stroke-cycle engine can be designed with a spring-loaded intake valve, and exhaust and transfer ports, eliminating the need for cam-operated valves. A two-stroke-cycle engine has one power stroke every revolution, giving it roughly twice the

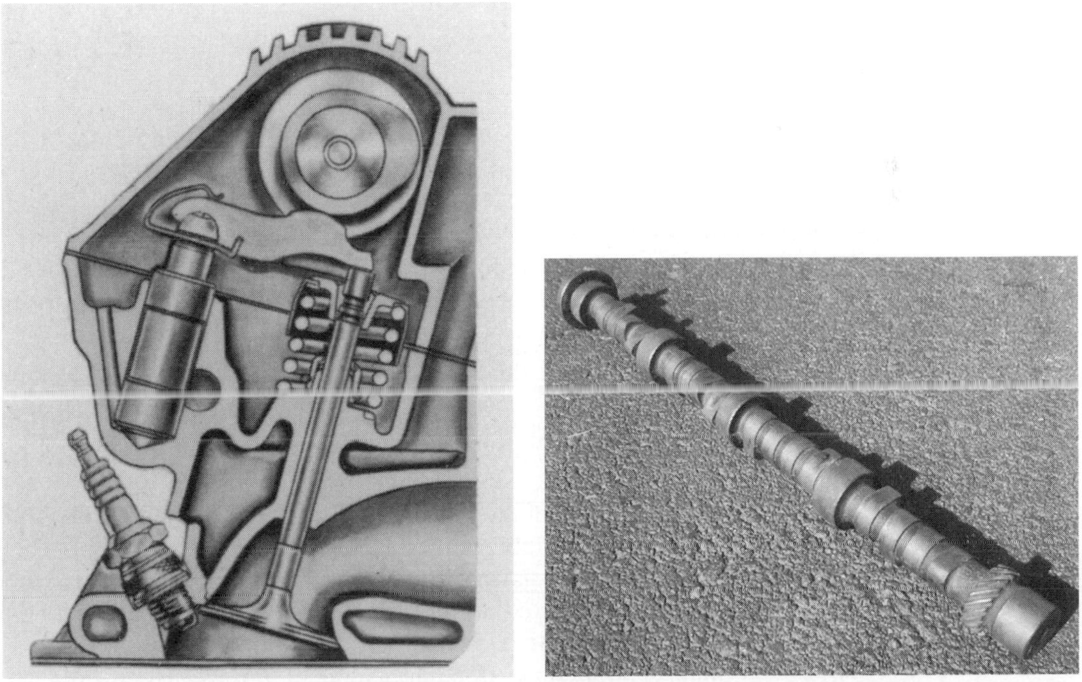

(a) (b)

FIGURE 5.2 (a) Engine cross section showing cam and follower. (*Source:* General Motors Corp.) (b) Camshaft.

power-to-weight ratio of a four-stroke-cycle engine. But the disadvantages of a two-stroke-cycle engine—increased air pollution, higher noise levels, and poorer efficiency—outweigh the advantages for most applications.

Other Cam System Applications

Cams are also used to operate switches and for other light-duty service. One-, two-, three-, and four-lobe cams are commonly stocked. Adjustable cams are also available and consist of two thin steel cams that may be adjusted relative to one another to vary the "open" and "close" interval.

Figure 5.3a shows an early design of an apple peeler with one cam to guide the cutter blade and another cam to eject the apple. In the close-up photo, Figure 5.3b, the follower is about to contact the stationary blade guide cam. The three dimensional cam, Figure 5.3c, is designed so that follower position is a function of two input variables, axial position of the cam relative to the follower and cam rotation angle. Several custom cams designed for special motion requirements are shown in Figure 5.3d.

Cams appear in many other forms. Look in Chapter 8 for the figure showing a limited-slip automotive differential. This differential is designed to apply the greatest torque to the wheel with the best traction. To accomplish that objective, cross pins move up cam surfaces to engage disk clutches. A variable-pitch pulley is also shown in Chapter 8. The pulley incorporates a cam to increase pressure on the side of the pulley to compensate for increases in torque.

Terminology

The following are a few commonly used terms that relate to disk cam design:

- *Stroke (or travel or throw):* The distance or angle of maximum follower travel.
- *Rise:* The interval during which the follower is moving away from the center of the cam. If the follower is below the cam, motion away from the cam center is still called *rise* even though such motion is downward.
- *Dwell:* An interval during which the follower is stationary. A dwell following a rise is called a *top dwell.*
- *Return:* The interval during which the follower moves toward the center of the cam.
- Intervals are measured in terms of the cam rotation angle rather than actual time in seconds, because we must maintain the same cam-and-follower relationships regardless of the cam speed. It is safest to measure angles in radians; when degrees are used, one must be sure that the computer can handle them. And one must check units in determining the follower velocity and acceleration.
- *Base circle:* The smallest circle that can be drawn tangent to the cam profile with center at the camshaft axis.

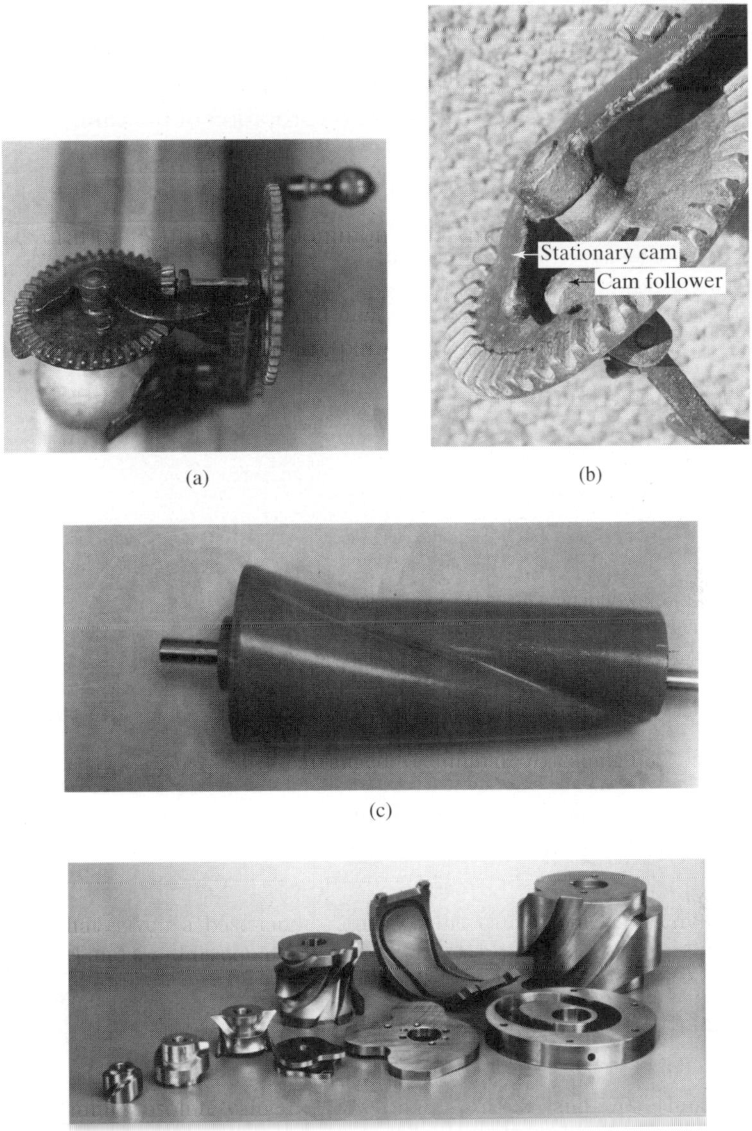

(a) (b) (c) (d)

FIGURE 5.3 (a) An apple peeler (patented in 1898) utilizes two cams. (b) The apple peeler blade guide follower rotates with the crown gear. It is about to contact the stationary cam. (c) A three-dimensional cam. (d) Custom cams designed for special motion requirements. (*Source:* Commercial Cam Co.)

Disk Cam Design and Manufacture

The design and manufacture of a disk cam may involve the following steps:

- Select the type of follower: a translating roller follower, rotating flat follower, etc.
- Determine the motion requirements of the follower: stroke, rise, top dwell, return, dwell, etc.

- Select a disk with reasonable dimensions.
- Select the form of motion during the rise and return: cycloidal, 3–4–5 polynomial, etc. (We will compare forms of motion later.)
- Program the chosen motion for one rotation of the cam.
- Select a grinder or milling cutter with a shape similar to that of the follower (e.g., a grinder with the same diameter as that of a roller follower).
- Use a computer numerically controlled (CNC) process to position the cutter to each desired follower position. Rotate the cam a fraction of a degree, and reposition the cutter until the entire cam is generated. Smooth any projections on the working surface. The result is a prototype cam.
- Select or develop a process to duplicate the prototype cam curve if more cams are required.

The foregoing steps are greatly oversimplified; manufacturing details are beyond the scope of this book.

5.2 GRAPHICAL CAM DESIGN

The design and manufacture of precision cams is not dependent on graphical methods. The key is *knowing where we want the cam follower* for each angular position of the cam. Nevertheless, graphical methods give us insight into practical cam design.

Graphical Design of a Cam with a Radial Translating Roller Follower. It is difficult to visualize generation of a cam profile while rotating the cam-to-be. Instead, we invert the situation, generating the cam as if it were stationary, while the follower translates and assumes various positions about the cam. The relative motion is the same. The process is illustrated by Figure 5.4.

Suppose we need to design a cam for a control system. Our design requires a translating follower with a given stroke. Let the follower position vs. the cam position be specified through an entire cam rotation. Here is one possible graphical solution:

- Select a radial translating roller follower. The axis of a *radial* follower intersects the center of the camshaft. The roller is intended to reduce wear on the cam.
- Draw radial lines at intervals of equal angles, as in the figure. These lines represent positions of the follower relative to the cam.
- Select the diameter of the roller.
- Locate the roller center a reasonable distance from the camshaft center. Call this point 0, the follower location before the rise.
- Locate the roller at position 1, position 2, etc., according to specified values of the follower position vs. the cam angle (governed by the requirements of the application).
- Draw a smooth curve tangent to all of the roller circles. This is a rough approximation of the cam profile.
- Check for unacceptable cam characteristics (e.g., a cusp).
- Check the maximum pressure angle (the angle between the follower axis and a normal to the cam profile, as shown in Figures 5.4 and 5.5). If the pressure angle is

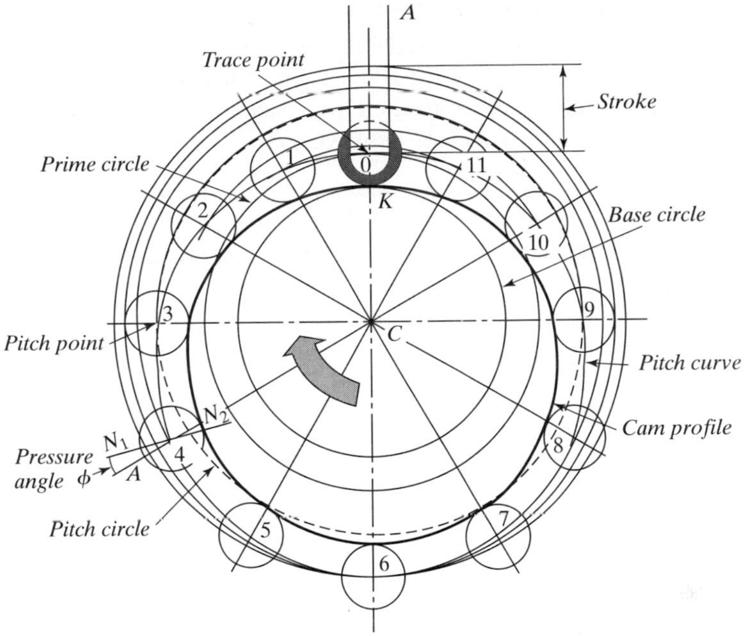

FIGURE 5.4 Cam layout for a roller follower. The important terms illustrated in the figure are as follows: *stroke*—maximum follower rise; *trace point*—reference point for follower displacement, located at roller center; *base circle*—smallest circle having its center at the cam center of rotation and that can be drawn tangent to the cam profile; *pitch curve*—curve drawn through the center of the roller at various positions around the circumference of the cam; *cam profile*—actual shape of the disk cam surface; *pressure angle*—angle between the line of travel of the follower and a normal drawn to the pitch curve; *pitch point*—that point on the pitch curve having the largest pressure angle; *pitch circle*— circle drawn through the pitch point and having its center at the cam center *C*; *prime circle*—smallest circle having its center at the cam center and that can be drawn tangent to the pitch curve.

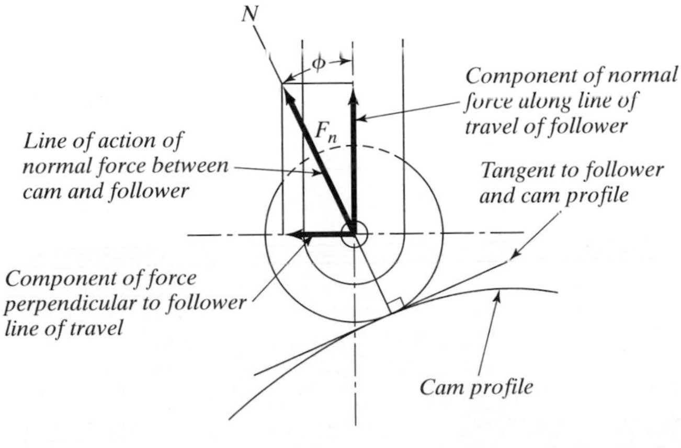

FIGURE 5.5 The relationship between the contact force and the pressure angle for a roller follower constrained to move vertically.

large, then the lateral force on the follower may prevent smooth motion, or the follower may even jam. Some designers try to limit the pressure angle to 30° or less.

- Consider changing the diameter of the roller or the diameter of the base circle if necessary. An increase in the base circle diameter decreases the maximum pressure angle. The trade-off, however, is a larger cam, which may result in other design complications.

You may want to use more radial lines, particularly to improve the accuracy of the pressure angle measurement. But do not get carried away with a quest for graphical accuracy: The purpose of the graphical presentation is to make predesign decisions and to gain an understanding of cam design. Precision cams are not generated from graphical presentations.

Graphical Design of a Cam with an Offset Roller Follower. Cam systems are sometimes designed with offset followers. The intent is to reduce the lateral forces on a cam follower during the rise portion of the cycle. If spring forces govern the return, the transmission angle may not be critical during that interval. Finding the optimum offset requires careful analysis; careless design could worsen the situation. You might want to check the effect of offsetting the follower if the camshaft can reverse direction. Will the offset result in an unacceptable transmission angle if the camshaft is reversed?

Figure 5.6 illustrates the procedure for designing a cam with an offset follower. Note cam rotation direction and direction of the offset. The points 0 to 4 on the follower are the follower displacements for a given type of motion. The total displacement of the follower is to occur in 120° of cam rotation.

A circle, called the *offset circle*, is drawn with its center at C and with a radius equal to the distance between the cam centerline and the follower centerline. One

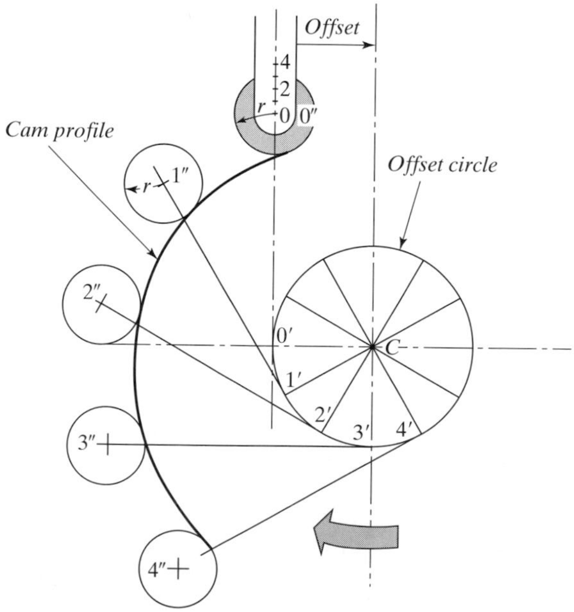

FIGURE 5.6 The offset, translating roller follower, shown may have less side thrust than an in-line or radial roller follower.

hundred twenty degrees of this circle, during which the follower motion is to occur, is divided into four equal parts of 30° each. Then, points $0'$, $1'$, $2'$, $3'$, and $4'$ are located, beginning on the horizontal axis of the offset circle.

Next, lines tangent to the offset circle (such lines are also perpendicular to the radial lines of the circle) are drawn through the primed points. When the measured distance $0'1$ is laid off along the tangent line from $1'$, point $1''$ is located. The distance from $0'$ to 2 is then laid off from $2'$ to locate point $2''$, and so on. The double-primed points, thus obtained, are used as the centers of circles having the same radius as the follower. A smooth curve tangent to these circles approximates the required cam profile.

5.3 THE HEAVISIDE STEP FUNCTION AND RELATED FUNCTIONS: IF-FUNCTION, AND-GATE, OR-GATE, AND INTERVAL FUNCTION

Cams are designed to do what we want when we want it. The follower motion during one cam revolution may include a rise, a dwell, a return, and another dwell. There are four intervals, so the follower position is represented by four equations. If the top dwell (between rise and return) is eliminated, there are three intervals and three equations. If rise and return are symmetric, we can simplify the problem to two equations. We can reduce the problem to one equation by eliminating the dwell. A cam without a dwell has only a few applications (e.g., the cam-type pump).

We can combine two or more equations by using the Heaviside step function and related functions. The IF-function, the AND-gate, the OR-gate, and an interval function may be useful in analyzing cams and plotting results. These functions can help you solve other engineering problems, too. (You may want to check other references to learn more about Boolean functions and logical functions.)

The Heaviside step function. The Heaviside function is a unit step defined as

$$\Phi(x) = 0 \text{ for } x < 0$$

and

$$\Phi(x) = 1 \text{ for } x \geq 0. \tag{5.1}$$

The function

$$\Phi(\theta - a) = 0 \text{ for } \theta < a$$

and

$$\Phi(\theta - a) = 1 \text{ for } \theta \geq a \tag{5.2}$$

turns on (initiates) a unit step at $\theta - a$. We use this function to turn on other functions; for example, $\Phi(\theta - a) \sin(\theta - a)$ turns on a sine wave, beginning at $\theta = a$.

Interval Function. We can construct an interval function from two Heaviside step functions by defining

$$\text{H}(\theta, a, b) = \Phi(\theta - a) - \Phi(\theta - b). \tag{5.3}$$

Then,

$$H(\theta, a, b) = 1 \text{ for } a \leq \theta \leq b$$

and

$$H(\theta, a, b) = 0 \text{ elsewhere.} \tag{5.4}$$

The interval function $H(\theta, a, b)$ can be used to turn on another function at $\theta = a$ and turn it off at $\theta = b$.

The IF-Function. An IF-function, or IF-statement, tests a condition and returns one answer if the condition is satisfied and another if not. For example, we may write

$$\Phi_1(\theta, a) = \text{if}(\theta \geq a, 1, 0). \tag{5.5}$$

If the condition $\theta \geq a$ is true, then the value after the first comma applies; if $\theta \geq a$ is false, the value after the second comma applies. (Note that Eq. (5.5) is in Mathcad™ format. For other computer software or programming languages, consult the manual for the correct IF-function format.) Comparing the results of the preceding IF-function to the Heaviside step function, we see that they are equivalent; that is,

$$\Phi_1(\theta, a) = \Phi(\theta - a).$$

The AND-Gate. An AND-gate is formed by multiplying two conditions together. For example, let us define an interval function

$$H_1(\theta, a, b) = \text{if } [(\theta \geq a) \cdot (\theta \leq b), 1, 0]. \tag{5.6}$$

If the first *and* second condition are true, then the value after the first comma applies; otherwise the value after the second comma applies. The interval function of Eq. (5.6), written in terms of an IF-function and an AND-gate, produces the same results as the interval function written in terms of Heaviside step functions.

The OR-Gate. We form an OR-gate by adding two conditions together. Suppose we want a function turned on for the interval $a \leq \theta \leq b$. Then an alternative definition of the interval function is given by

$$H_2(\theta, a, b) = \text{if } [(\theta < a) + (\theta > b), 0, 1]. \tag{5.7}$$

If *either* the first *or* the second condition is true, the value after the first comma applies; if neither condition is true, the value after the second comma applies.

Because of the importance of logical decisions to engineering and other fields, programming languages and software often have literal AND- and OR-functions. For example, a statement could take a form like

IF [first condition] AND [second condition]

THEN [result A or subroutine A]

ELSE [result B or subroutine B].

We are not limited to only two conditions; sometimes we need to combine three or more conditions and both AND- and OR-functions.

SAMPLE PROBLEM 5.1

Applying the Heaviside Step Function and Construction of an Interval Function

Write statements to describe step and interval functions. Plot the following functions for the range $0 \leq \theta \leq 2\pi$:

- A Heaviside (unit) step function beginning at $\theta = 0.5$.
- An interval function (based on the Heaviside function) that can be used to turn another function on at $\theta = 1$ and off at $\theta = 3$.
- A unit step function (formed from an IF-statement), beginning at $\theta = 2$.
- An interval function (based on an IF-statement and an AND gate) to turn another function on at $\theta = 3$ and off at $\theta = 5$.
- An interval function (based on an IF-statement and an OR gate) to turn on another function at $\theta = 4$ and off at $\theta = 5.5$.

Solution summary. Consult the manual or help screen for the forms of IF-, AND-, and OR-statements in the programming language or software you prefer. The statements that follow are written in Mathcad^{TM} format. Each step and interval function has unit magnitude. Eight is added to the first, six to the second, and so on, so that the plots (Figure 5.7) appear in order from top to bottom.

Solution details.

Let $\theta := 0, \dfrac{\pi}{180} \cdot\cdot 2\pi$

The Heaviside step function $\Phi(\theta - a)$ turns on a unit step at $\theta = a$.

We can define an *interval function* to turn a unit step function on at a and off at b:

$$\mathrm{H}(\theta, \mathrm{a}, \mathrm{b}) := \Phi(\theta - \mathrm{a}) - \Phi(\theta - \mathrm{b})$$

An IF statement can be used to construct the Heaviside step function

$$\Phi_1(\theta, \mathrm{a}) := \mathrm{if}\ (\theta \geq \mathrm{a}, 1, 0)$$

The first value after the comma applies if $\theta \geq$ a; otherwise the second value applies.

An AND-gate is formed by multiplying two conditions together.

The INTERVAL function can be constructed with an IF-statement and an AND-gate:

$$\mathrm{H}_1(\theta, \mathrm{a}, \mathrm{b}) := \mathrm{if}\ [(\theta \geq \mathrm{a}) \cdot (\theta \leq \mathrm{b}), 1, 0]$$

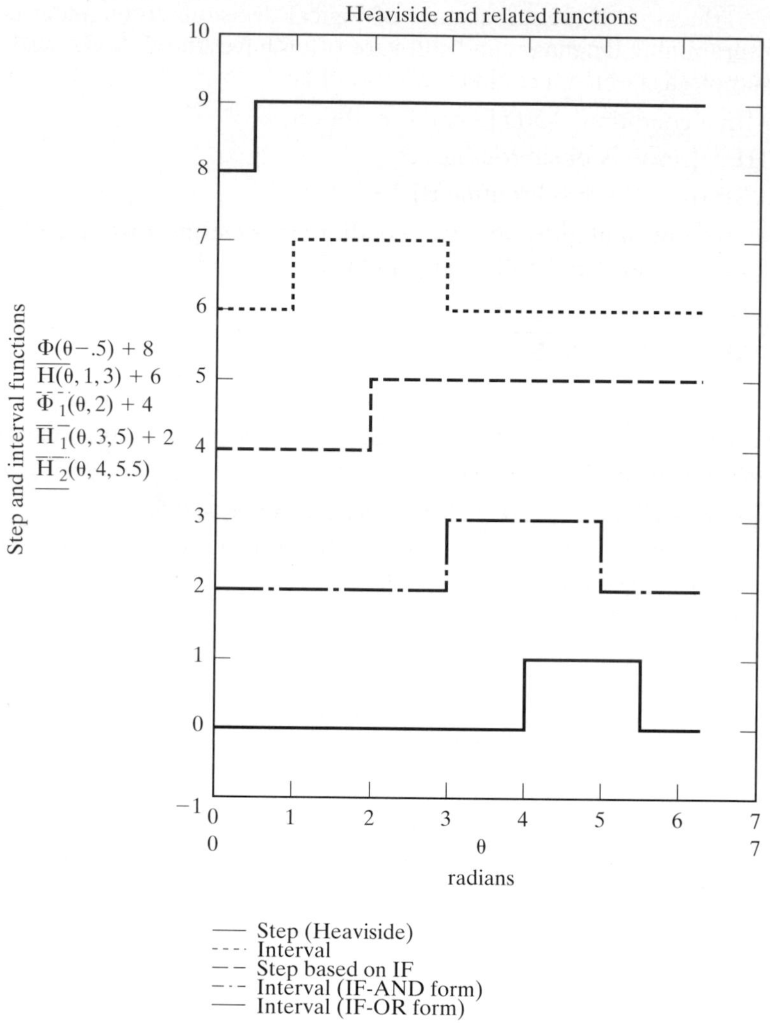

FIGURE 5.7 Heaviside and related functions.

An OR-gate is formed by adding two conditions together.

The INTERVAL function can also be constructed with an IF-statement and an OR-gate:

$$H_2(\theta, a, b) := \text{if}[(\theta < a) + (\theta > b), 0, 1]$$

5.4 CAM DESIGN IN TERMS OF FOLLOWER MOTION

It is easy to design cams for low-speed operation. When accelerations and inertial forces are low, we can use any reasonable cam profile. But when we design high-speed cams, the follower velocity and acceleration become more important. **Jerk**, the differential of the follower acceleration with respect to time, is also considered in designing cams, and some designers even consider the differential of jerk with respect to time.

Possible Forms of Follower Displacement

Consider these follower-motion vs. cam-position relationships:

- Uniform motion (resulting in constant velocity)
- Parabolic motion (resulting in constant acceleration)
- Harmonic motion
- Cycloidal motion
- 3–4–5 polynomial motion
- Higher order polynomial motion

Suppose we define the following terms:

s = follower displacement (mm or in)
h = total rise = maximum follower displacement (mm or in)
α = cam rotation during rise (rad)
θ = cam position (rad)
x = θ/α during the rise

The term *rise* refers to motion away from the cam center; the actual motion in not necessarily upward.

Uniform Motion

Uniform motion results in constant velocity; that is, the follower displacement during the rise has the form

$$s = hx. \tag{5.8}$$

Parabolic Motion

Parabolic motion results in constant acceleration; that is, the follower displacement during the rise has the form

$$s = hx^2. \tag{5.9}$$

Harmonic Motion

In harmonic motion, the follower displacement during the rise interval for parabolic motion has the form

$$s = (h/2)(1 - \cos\pi x) \tag{5.10}$$

Follower Return

We can look at the return as a reflection of the rise; that is, we can replace

$$x = \theta/\alpha$$

in the preceding equations with

$$x = (\theta_f - \theta)/\alpha,$$

where α = cam rotation during return (rad)

and

$$\theta_f = \text{cam position at end of return (rad)}.$$

Problems with Harmonic and Parabolic Motion

Harmonic motion works well if the motion is continuous. The cam-type pump, described previously, is one application of harmonic (sinusoidal) motion. We might be tempted to incorporate harmonic motion into the design of a cam for intermittent motion. Figure 5.8 shows harmonic follower motion with dwells. We see that there are jumps in the acceleration plot, resulting in infinite jerk. The acceleration jumps result in abrupt changes in inertial force, generally considered unsatisfactory for high-speed cam applications. Figure 5.9 shows a similar problem if we use intervals of constant acceleration (parabolic motion).

It is obvious that uniform (constant-velocity) motion has even more severe problems.

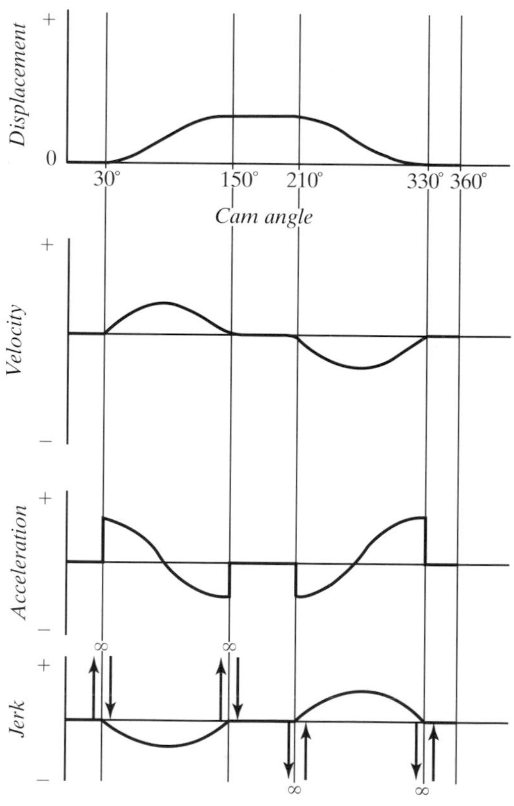

FIGURE 5.8 The follower displacement, velocity, acceleration, and jerk diagrams for simple harmonic motion are shown. The follower dwells from 0 to 30°, rises with simple harmonic motion from 30° to 150°, dwells from 150° to 210°, returns with simple harmonic motion from 210° to 330°, and dwells from 330° to 360°. Note the effect on acceleration and jerk.

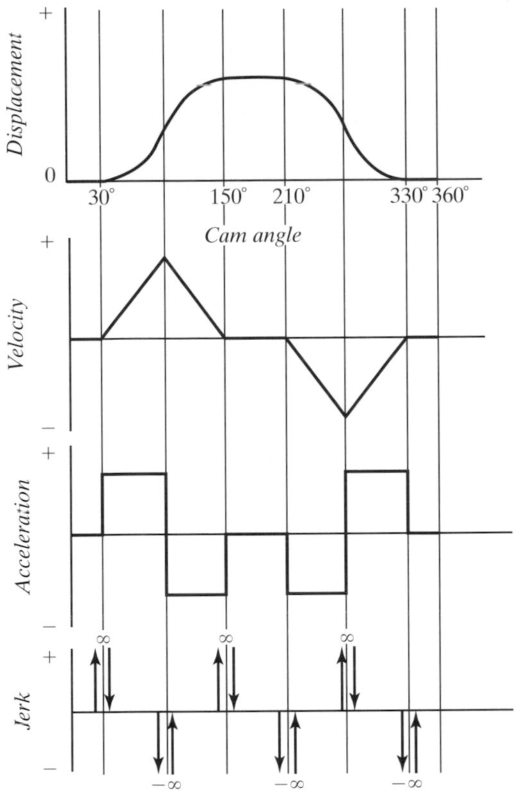

FIGURE 5.9 The follower displacement, velocity, acceleration, and jerk diagrams for parabolic motion are shown. The follower dwells from 0 to 30°, rises with constant positive acceleration from 30° to 90°, continues to rise with constant negative acceleration from 90° to 150°, dwells from 150° to 210°, returns with constant negative acceleration from 210° to 270°, completes the return with constant positive acceleration from 270° to 330°, and then dwells from 330° to 360°.

Follower Motion for High-Speed Cams

"Good" High-Speed Cams

Design criteria for a cam and follower depend on the application. The following is one set of criteria we can use to define a "good" high-speed cam:

- zero velocity at the beginning and end of the follower rise
- zero acceleration at the beginning and end of the rise
- zero velocity at the beginning and end of the return
- zero acceleration at the beginning and end of the return

Suppose we want a total rise h (mm or in) as the cam turns through the interval $\theta = 0$ to $\theta = \alpha$ (rad). Then, the conditions on the follower motion during the rise are

$$s(0) = v(0) = a(0) = 0,$$
$$s(\alpha) = h, \text{ and } v(\alpha) = a(\alpha) = 0, \tag{5.11}$$

where $s(\theta)$, $v(\theta)$, and $a(\theta)$ are, respectively, the follower position, velocity, and acceleration.

5.5 CYCLOIDAL CAMS

Designers found that a ramp function combined with a sinusoid was the basis for a good high-speed cam. This combination is called cycloidal motion. The follower displacement, as the cam position changes from $\theta = 0$ to $\theta = \alpha$ (rad) is given by

$$s = h\left(\frac{\theta}{\alpha}\right) - \left(\frac{h}{2\pi}\right)\sin\left(2\pi\frac{\theta}{\alpha}\right) \quad 0 \le \theta \le \alpha, \tag{5.12}$$

where α is the angle the cam turns through while the follower receives its total lift and θ is the cam angle during which the displacement s occurs.

Differentiating the displacement equation gives the expression for velocity:

$$v = \frac{ds}{dt} = \frac{h\,d\theta}{\alpha\,dt} - \left(\frac{h}{2\pi}\right)\cos\left(2\pi\frac{\theta}{\alpha}\right)\left(\frac{2\pi\,d\theta}{\alpha\,dt}\right).$$

But $d\theta/dt = \omega$; therefore,

$$v = \frac{h\omega}{\alpha} - \left(\frac{h\omega}{\alpha}\right)\cos\left(2\pi\frac{\theta}{\alpha}\right). \tag{5.13}$$

The acceleration is obtained by differentiating the velocity equation:

$$a = \frac{dv}{dt} = h\left(\frac{2\pi\omega^2}{\alpha^2}\right)\sin\left(2\pi\frac{\theta}{\alpha}\right). \tag{5.14}$$

Finally, the jerk is obtained by differentiating the acceleration equation:

$$j = \frac{da}{dt} = \left(\frac{h4\pi^2\omega^3}{\alpha^3}\right)\cos\left(2\pi\frac{\theta}{\alpha}\right). \tag{5.15}$$

Suppose we need a good high-speed cam-and-follower system to provide a rise–dwell–return–dwell (RDRD) motion. Let us select cycloidal motion and examine the follower position, velocity, acceleration, and jerk during the rise. It is convenient to write the follower position equation in the form

$$s_c(x) = h[x - \sin(2\pi x)/(2\pi)]. \tag{5.16}$$

If the total rise h occurs while the cam rotates from $\theta = 0$ to $\theta = \alpha$, then

$$x = \theta/\alpha$$

Differentiating with respect to time, we obtain the velocity, acceleration, and jerk. Be sure to use the chain rule, $ds/dt = (ds/dx)\cdot(dx/dt)$, etc., to get the following motion characteristics:

Velocity $$v_c(x) = \frac{\omega}{\alpha}\frac{d}{dx}s_c(x) = \frac{h\cdot\omega}{\alpha}\cdot(1 - \cos(2\pi x)) \tag{5.17}$$

Acceleration $$a_c(x) = \frac{\omega}{\alpha}\frac{d}{dx}v_c(x) = 2 \cdot \pi \cdot h \cdot \left(\frac{\omega}{\alpha}\right)^2 \cdot \sin(2\pi x) \qquad (5.18)$$

Jerk $$j_c(x) = \frac{\omega}{\alpha}\frac{d}{dx}a_c(x) = 4\pi^2 h \cdot \left(\frac{\omega}{\alpha}\right)^3 \cdot \cos(2\pi x) \qquad (5.19)$$

We now want to check out our cam for the full RDRD cycle. First, we define an interval function based on the Heaviside step function, as discussed in an earlier section:

$$\mathrm{H}(\theta, a, b) = \Phi(\theta - a) - \Phi(\theta - b)$$

Our interval function turns another function on at $\theta = a$ and off at $\theta = b$. Let the intervals be as follows:

rise: $\theta_0 \le \theta \le \theta_1$; top dwell: $\theta_1 \le \theta \le \theta_2$;

return: $\theta_2 \le \theta \le \theta_3$; dwell: $\theta_3 \le \theta \le$ end of cycle.

We usually set $\theta = \theta_0 = 0$ at the beginning of the cycle. Then, $\theta = 2\pi$ at the end of the cycle. The following equations describe the follower motion for the whole RDRD cycle—that is, a full rotation of the cam:

Position

$$S_c(\theta) = \mathrm{H}(\theta, \theta_0, \theta_1) \cdot s_c\left(\frac{\theta}{\theta_1 - \theta_0}\right) + h \cdot \mathrm{H}(\theta, \theta_1, \theta_2) + \mathrm{H}(\theta, \theta_2, \theta_3) \cdot s_c\left(\frac{\theta_3 - \theta}{\theta_3 - \theta_2}\right) \qquad (5.20)$$

Velocity $$V_c(\theta) = \mathrm{H}(\theta, \theta_0, \theta_1) \cdot v_c\left(\frac{\theta}{\theta_1 - \theta_0}\right) - \mathrm{H}(\theta, \theta_2, \theta_3) \cdot v_c\left(\frac{\theta_3 - \theta}{\theta_3 - \theta_2}\right) \qquad (5.21)$$

Acceleration $$A_c(\theta) = \mathrm{H}(\theta, \theta_0, \theta_1) \cdot a_c\left(\frac{\theta}{\theta_1 - \theta_0}\right) + \mathrm{H}(\theta, \theta_2, \theta_3) \cdot a_c\left(\frac{\theta_3 - \theta}{\theta_3 - \theta_2}\right) \qquad (5.22)$$

Jerk $$J_c(\theta) = \mathrm{H}(\theta, \theta_0, \theta_1) \cdot j_c\left(\frac{\theta}{\theta_1 - \theta_0}\right) - \mathrm{H}(\theta, \theta_2, \theta_3) \cdot j_c\left(\frac{\theta_3 - \theta}{\theta_3 - \theta_2}\right) \qquad (5.23)$$

Do the equations make sense? Check the position equation first. The first term on the right multiplies the interval function by the rise equation. We must cut off the rise equation at $\theta = \theta_1$, because the equation does not apply in the intervals that follow. The next term gives the follower a constant displacement h during the top dwell. The third term on the right describes the return. If the rise and return intervals are equal, the return displacement is a mirror image of the rise. We use the same displacement equation, except that the argument is now

$$x = (\theta_3 - \theta)/\alpha,$$

where $\alpha = \theta_1 - \theta_0 = \theta_3 - \theta_2$.

There is no fourth term; the follower position is zero for the final dwell.

Next, we check the velocity equation. The first term on the right is patterned after the displacement equation. Terms representing the dwells are unnecessary, because the velocity is zero. The last term is the return. Remember the chain rule?

If $x = (\theta_3 - \theta)/\alpha$,

then $dx/dt = -\omega/\alpha$,

not $+\omega/\alpha$, as in the rise.

The sign change appears before the interval function H.

The plus sign before the H in the return part of the acceleration equation reflects two sign changes. Three sign changes apply to the return part of the jerk equation.

A good engineer works "smart." Try to copy and "recycle" equations to save time. Use *find* and *replace* functions to edit equations if those functions are available. Let the computer do the work; you do the thinking. These forms of equation can be used to describe the motion of other cams and may help in other engineering problems. You may think of a way to make them more efficient.

SAMPLE PROBLEM 5.2

Design of a Cycloidal Cam

We need a cam with a 12-mm rise. The rise and return intervals are both 110°, and the top dwell is 35°. The cam rotation speed will be 880 rpm. Design the cam. Plot the follower position, velocity, acceleration, and jerk. Check your results.

Design decisions. Cycloidal motion will be selected, because we want smooth, quiet operation. If we can describe the displacement for the full range of motion, the cam can be generated.

Solution summary. Figure 5.10 shows the follower position, velocity, acceleration, and jerk plotted against the cam position in degrees. The units used in the detailed calculations are millimeters, radians, and seconds.

α = interval of the rise (radians),

ω = angular velocity of the cam (rad/s) = $n\pi/30$,

n = cam speed (rpm).

The position equation is written with argument x, where $x = \theta/\alpha$ for the rise. Position $s_c(0) = 0$ (at the beginning of the rise) and $s_c(1) = 12$ (at the end of the rise), as required.

The velocity and acceleration boundary conditions are $v(0) = v(1) = a(0) = a(1) = 0$. (These should be checked as well.)

The calculated value of the jerk at $x = 0$ compares with an approximation obtained from a change in acceleration over a short time interval.

Detailed solution.

Units: mm, sec, rad

Rise h := 12 Cam speed $\omega := 880 \cdot \dfrac{\pi}{30}$ rad/s

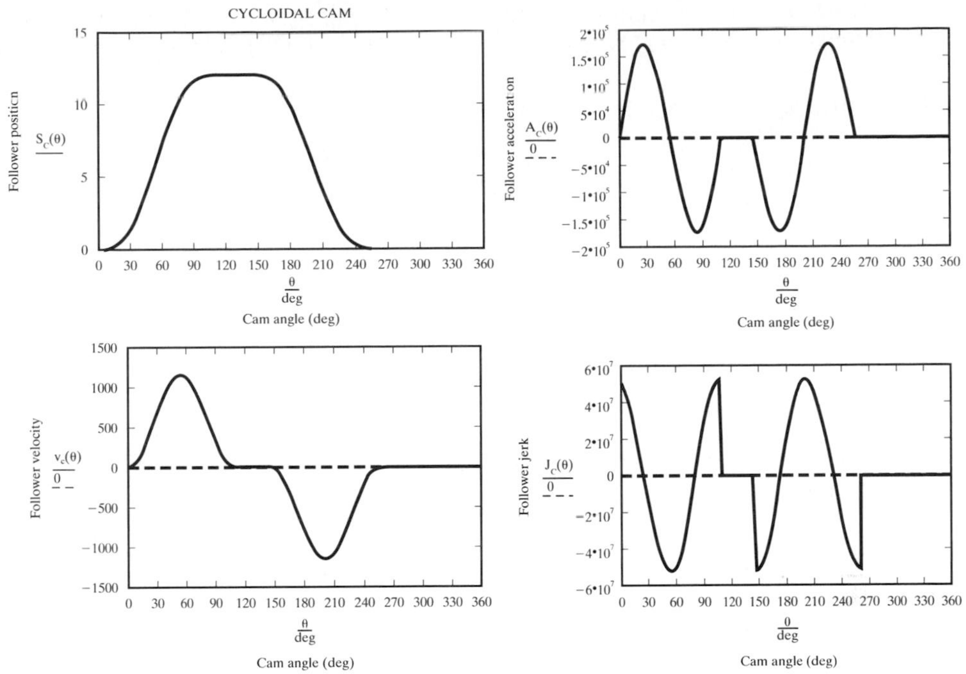

FIGURE 5.10 Follower position, velocity, acceleration, and jerk for a cycloidal cam.

Rise and return interval $\alpha = \dfrac{110\pi}{180}$ rad

Dwell interval $\beta = \dfrac{35\pi}{180}$ Cam position $\theta := 0, \dfrac{\pi}{180} .. 2\pi$

Begin rise $\theta_0 := 0$ End rise, begin dwell $\theta_1 := \theta_0 + \alpha \quad \theta_1 = 1.92$
Begin return $\theta_2 := \theta_1 + \beta$ End return $\theta_3 := \theta_2 + \alpha \quad \theta_3 = 4.451$

Rise time $t_1 := \dfrac{\alpha}{\omega} \quad t_1 = 0.021$

Follower position $s_c(x) := h \cdot \left(x - \dfrac{1}{2\pi} \cdot \sin(2\pi x) \right)$

Check boundary conditions (BC's) $s_c(0) = 0 \qquad s_c(1) = 12$

Velocity $v_c(x) := \dfrac{\omega}{\alpha} \cdot \dfrac{d}{dx} s_c(x) \qquad v_c(x) := \dfrac{h \cdot \omega}{\alpha} \cdot \left(1 - \cos(2\pi x)\right)$

Check BC's $\quad v_c(0) = 0 \qquad v_c(1) = 0$

Compare $v_c(.5) = 1.152 \cdot 10^3 \qquad v_{avg} := \dfrac{s_c(1) - s_c(0)}{t_1} \quad v_{avg} = 576$

Acceleration $a_c(x) := \frac{\omega}{\alpha} \cdot \frac{d}{dx} v_c(x)$ $\quad a_c(x) := 2 \cdot \pi \cdot h \left(\frac{\omega}{\alpha}\right)^2 \cdot \sin(2\pi x)$

Check BC's $a_c(0) = 0$ $\quad a_c(1) = -4.255 \cdot 10^{-11}$

Jerk $j_c(x) := \frac{\omega}{\alpha} \cdot \frac{d}{dx} a_c(x)$ $\quad j_c(x) := 4\pi^2 h \cdot \left(\frac{\omega}{\alpha}\right)^3 \cdot \cos(2\pi x)$

$j_c(0) = 5.239 \cdot 10^7$ $\quad j_c(0.5) = -5.239 \cdot 10^7$ $\quad j_c(1) = 5.239 \cdot 10^7$

Set Jerk $= 0$ Estimate $x := 2$

$X := \text{root}(j_c(x), x)$ $\quad X = 0.25$ $\quad$ Max acceleration: $a_c(X) = 1.737 \cdot 10^5$

Compare with j at 0 $j_0 := \frac{a_c(.001) - a_c(0)}{.001} \cdot \frac{\omega}{\alpha}$ $\quad j_0 = 5.239 \cdot 10^7$ avg for .001 rad

Define an interval function in terms of Heaviside step functions $H(\theta, a, b) := \Phi(\theta - a) - \Phi(\theta - b)$. Then, describe follower motion for a full cam rotation in terms of the interval function as in Eqs. (5.20–5.23).

Polynomial Motion

As was pointed out earlier, operation at high cam speeds can be improved by eliminating discontinuities in the derivatives of the follower motion with respect to time. An example of this principle, as applied to the rise–dwell—return–dwell motion that we have been discussing, is cycloidal motion. In that case, there are no discontinuities in either velocity or acceleration, and therefore, both acceleration and jerk remain finite. Hence, cycloidal motion is suitable for high-speed operation.

Polynomial motion is another option. The displacement equation for general polynomial motion can be written as

$$s = C_0 + C_1(\theta - \theta_i) + C_2(\theta - \theta_i)^2 + C_3(\theta - \theta_i)^3 + C_4(\theta - \theta_i)^4 + \cdots + C_N(\theta - \theta_i)^N$$
$$= \sum_{k=0}^{N} C_k(\theta - \theta_i)^k,$$

where s is the follower displacement, θ is the angular position of the cam, and θ_i is the initial cam angle at the beginning of the polynomial motion. In other words, the difference $(\theta - \theta_i)$ is the cam rotation occurring during displacement s. N is referred to as the degree of the polynomial, and $(N + 1)$ is the number of terms in the polynomial expression. The velocity and acceleration equations are obtained by successive differentiation.

5.6 DESIGN OF GOOD HIGH-SPEED POLYNOMIAL CAMS

Suppose we need a cam to produce RDRD motion. A good high-speed cam must operate smoothly. For RDRD motion of the follower, the rise and return intervals should begin and end with zero velocity and acceleration. We found earlier that a cycloidal cam can meet these criteria. If the follower rises a total distance h while the cam position changes from $\theta = 0$ to $\theta = \alpha$ (rad), then there are six boundary conditions:

At $\theta = 0$, follower position, velocity, and acceleration $= 0$;

at $\theta = \alpha$, follower position $= h$ and velocity and acceleration $= 0$.

We need six arbitrary constants to satisfy the six boundary conditions.

A fifth-order polynomial should do the job. To simplify typing, the polynomial can be written in terms of x. For the rise interval,

$$x = \theta/\alpha,$$

and the follower position equation is

$$s = C_0 + C_1x + C_2x^2 + C_3x^3 + C_4x^4 + C_5x^5. \tag{5.24}$$

SAMPLE PROBLEM 5.3

Design of a Good High-Speed Polynomial Cam

We cannot use Eq. (5.24) as is;. we need the six arbitrary constants.

Design decisions. We will apply the boundary conditions that are likely to result in a good high-speed RDRD cam.

Solution summary. The first step is to differentiate the follower position equation with respect to time to obtain the velocity. Remember the chain rule; we used it when we examined the cycloidal cam. We differentiate the velocity equation to get the acceleration. We differentiate the acceleration equation to get the jerk, for good measure. We do not need jerk now, but it might come in handy later.

Next, we apply the six boundary conditions, where $x = 0$ at the beginning of the rise and $x = 1$ at the end:

$$s(0) = v(0) = a(0) = 0$$
$$s(1) = h;$$
$$v(1) = a(1) = 0$$

Noting that $\omega/\alpha \neq 0$, we get six simultaneous linear equations, represented in matrix form by

$$\mathbf{AX} = \mathbf{B}.$$

We premultiply by the inverse of $\mathbf{A}$ to put the equation in solvable form:

$$\mathbf{A}^{-1}\mathbf{AX} = \mathbf{X} = \mathbf{A}^{-1}\mathbf{B}. \tag{5.25}$$

After filling in the six-by-six matrix **A** and column matrix **B**, we solve for **X**, a column matrix of the six arbitrary constants. The nonzero constants are

$$C_4 = 10h, \quad C_5 = -15h, \quad \text{and} \quad C_6 = 6h. \tag{5.26}$$

Detailed solution. For the fifth order polynomial cam, s is the follower position and $x = \theta/\alpha$ = cam position/interval. We then have the following results:

Position $s(x) := (C_0 + C_1 \cdot x + C_2 \cdot x^2 + C_3 \cdot x^3 + C_4 \cdot x^4 + C_5 \cdot x^5)$

$$v(x) := \frac{\omega}{\alpha} \cdot \frac{d}{dx} s(x)$$

Velocity $v(x) := \dfrac{\omega}{\alpha} \cdot \dfrac{d}{dx} s(x)$

$$v(x) \rightarrow \frac{\omega}{\alpha} \cdot (C_1 + 2 \cdot C_2 \cdot x + 3 \cdot C_3 \cdot x^2 + 4 \cdot C_4 \cdot x^3 + 5 \cdot C_5 \cdot x^4)$$

Acceleration $a(x) := \dfrac{\omega}{\alpha} \cdot \dfrac{d}{dx} \cdot \dfrac{\omega}{\alpha} \cdot (C_1 + 2 \cdot C_2 \cdot x + 3 \cdot C_3 \cdot x^2 + 4 \cdot C_4 \cdot x^3 + 5 \cdot C_5 \cdot x^4)$

$$a(x) \rightarrow \frac{\omega^2}{\alpha^2} \cdot (2 \cdot C_2 + 6 \cdot C_3 \cdot x + 12 \cdot C_4 \cdot x^2 + 20 \cdot C_5 \cdot x^3)$$

Jerk $j(x) := \dfrac{\omega}{\alpha} \cdot \dfrac{d}{dx} \cdot \dfrac{\omega^2}{\alpha^2} \cdot (2 \cdot C_2 + 6 \cdot C_3 \cdot x + 12 \cdot C_4 \cdot x^2 + 20 \cdot C_5 \cdot x^3)$

$$j(x) \rightarrow \frac{\omega^3}{\alpha^3} \cdot (6 \cdot C_3 + 24 \cdot C_4 \cdot x + 60 \cdot C_5 \cdot x^2)$$

Boundary conditions at $x = 0$: $s = 0$; $v = 0$; $a = 0$;

at $x = 1$: $s = h$; $v = 0$; $a = 0$

Position $s(0) \rightarrow C_0$ $\quad s(1) \rightarrow C_0 + C_1 + C_2 + C_3 + C_4 + C_5$

Since ω/α is not equal to zero, we can apply the velocity and acceleration boundary conditions to the first and second derivatives of s with respect to x. We obtain the following equations:

First derivative: $s_x(x) := C_1 + 2 \cdot C_2 \cdot x + 3 \cdot C_3 \cdot x^2 + 4 \cdot C_4 \cdot x^3 + 5 \cdot C_5 \cdot x^4$

$$s_x(0) \rightarrow C_1$$

$$s_x(1) \rightarrow C_1 + 2 \cdot C_2 + 3 \cdot C_3 + 4 \cdot C_4 + 5 \cdot C_5$$

Second derivative: $s_{xx}(x) := 2 \cdot C_2 + 6 \cdot C_3 \cdot x + 12 \cdot C_4 \cdot x^2 + 20 \cdot C_5 \cdot x^3$

$$s_{xx}(0) \rightarrow 2 \cdot C_2$$

$$s_{xx}(1) \rightarrow 2 \cdot C_2 + 6 \cdot C_3 + 12 \cdot C_4 + 20 \cdot C_5$$

Applying the six boundary conditions yields these equations:

$$C_0 := 0$$
$$C_0 + C_1 + C_2 + C_3 + C_4 + C_5 := h$$
$$C_1 := 0$$
$$C_1 + 2 \cdot C_2 + 3 \cdot C_3 + 4 \cdot C_4 + 5 \cdot C_5 := 0$$

$$2 \cdot C_2 := 0$$
$$2 \cdot C_2 + 6 \cdot C_3 + 12 \cdot C_4 + 20 \cdot C_5 := 0$$

In matrix form, $AX = B$, where

$$A := \begin{bmatrix} 1 & 0 & 0 & 0 & 0 & 0 \\ 1 & 1 & 1 & 1 & 1 & 1 \\ 0 & 1 & 0 & 0 & 0 & 0 \\ 0 & 1 & 2 & 3 & 4 & 5 \\ 0 & 0 & 2 & 0 & 0 & 0 \\ 0 & 0 & 2 & 6 & 12 & 20 \end{bmatrix} \quad B := \begin{bmatrix} 0 \\ h \\ 0 \\ 0 \\ 0 \\ 0 \end{bmatrix} \quad X := \begin{bmatrix} C_0 \\ C_1 \\ C_2 \\ C_3 \\ C_4 \\ C_5 \end{bmatrix} = A^{-1}B \rightarrow \begin{bmatrix} 0 \\ 0 \\ 0 \\ 10 \cdot h \\ -15 \cdot h \\ 6 \cdot h \end{bmatrix}.$$

Substituting the constants we have just found into the displacement equation results in

$$s(x) := h(10 \cdot x^3 - 15x^4 + 6x^5).$$

Cams based on this equation are called 3–4–5 cams because of the third-, fourth-, and fifth-order terms. Such cams they have been cut on numerically controlled machines and used in high-speed machinery for many years. Is the solution an example of working smart? Let us look at other options.

Software with symbolic capability was used to solve for the 3–4–5 cam coefficients. But any engineer can easily differentiate a polynomial, and with the h factored out, the matrix equation could have been solved numerically.

We could even do the job by hand if the computer crashed. The three boundary conditions at $x = 0$ lead to $C_0 = C_1 = C_2 = 0$. The problem then reduces to three simultaneous linear equations that can be solved with the use of determinants. (See your mathematics text or Chapter 3 of this book)

Completing and Checking the 3–4–5 Polynomial Cam Design. We need to specify the rise, top dwell, return, and dwell intervals. Then, the position equation can be expressed in terms of interval functions as was done for a cycloidal cam. After we specify the follower shape and base circle, it may be possible to cut a master cam directly from the position equation.

The design is not complete without checking the follower velocity, acceleration, and jerk for an entire cam rotation. The underlying equations are different, but the form for the 3–4–5 cam is similar to that for the cycloidal cam. Do you remember why there are minus signs before the return terms in the velocity and jerk equations (for the cycloidal cam)? Will the 3–4–5 cam follower equations have minus signs?

SAMPLE PROBLEM 5.4

The 3–4–5 Cam for a Full Cycle of Motion

Design a cam to produce a 10-mm rise in 120° of cam rotation, dwell 45°, and return in 120°. The cam speed is 400 rpm. Verify the boundary conditions and plot the follower motion.

Design decision. We will use a 3–4–5 polynomial cam.

Solution summary. Except for the basic polynomial equations, this cam closely resembles a cycloidal cam. If we have already designed a cycloidal cam, we can copy many of the equations from that file with only a few changes.

The results are shown in Figure 5.11.

Detailed solution.

3–4–5 Polynomial cam

Units: mm, sec, rad

Rise $h := 10$ Cam speed $\omega := \frac{400\pi}{30}$ rad/s

Rise and return interval $\alpha := 120\frac{\pi}{180}$ $\quad \alpha = 2.094$ rad

Dwell interval $\beta := 45\frac{\pi}{180}$ $\quad$ Cam position $\theta := 0, \frac{\pi}{180} .. 2\pi$

Begin rise $\theta_0 := 0$ End rise, begin dwell $\theta_1 := \theta_0 + \alpha$ $\quad \theta_1 = 2.094$

Begin return $\theta_2 := \theta_1 + \beta$ $\quad$ End return $\theta_3 := \theta_2 + \alpha$ $\quad \frac{\theta_3}{deg} = 285$

Rise time $t_1 := \frac{\alpha}{\omega}$ $\quad t_1 = 0.05$

Follower position $s(x) := h(10 \cdot x^3 - 15x^4 + 6x^5)$

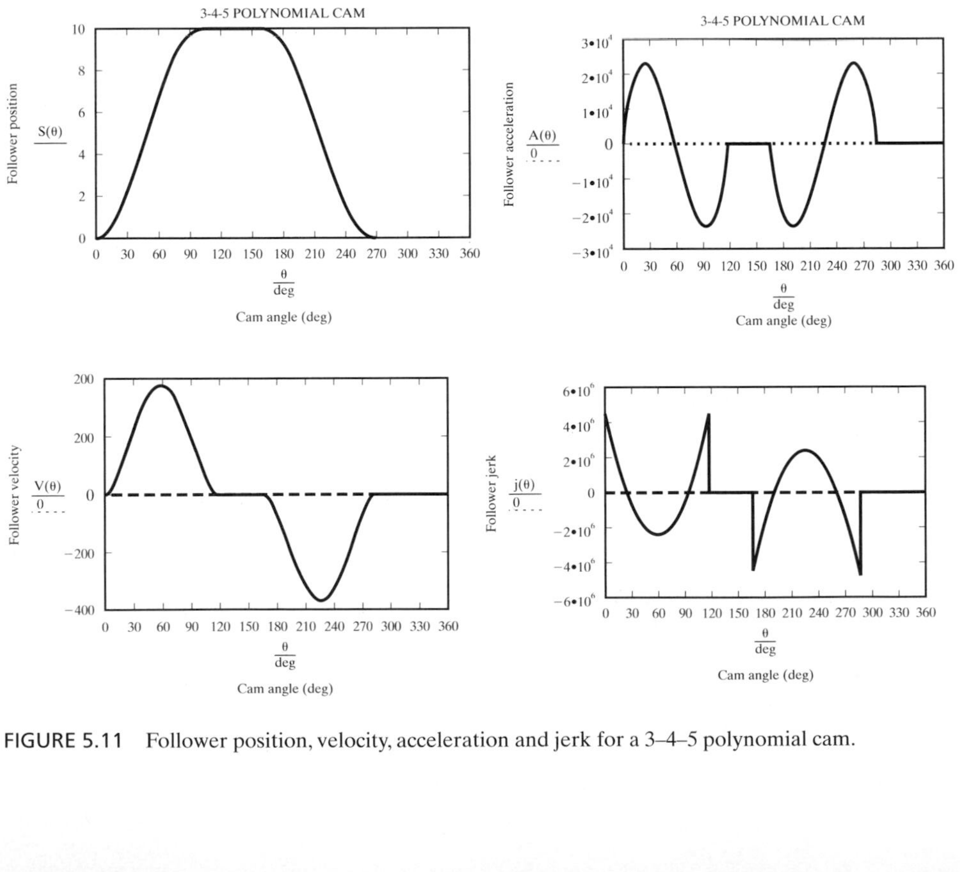

FIGURE 5.11 Follower position, velocity, acceleration and jerk for a 3–4–5 polynomial cam.

Check BC's $s(0) = 0 \quad s(1) = 10$

Velocity $vel := \frac{\omega}{\alpha} \cdot \frac{d}{dx} s(x) \qquad v(x) := h \cdot \frac{\omega}{\alpha} \cdot (30 \cdot x^2 - 60x^3 + 30x^4)$

Check BC's $v(0)=0$ $v(1)=0$

Compare $v(.5) = 375$ $v_{avg} := \dfrac{s(1) - s(0)}{t_1} \qquad v_{avg} = 200$

Acceleration: $acc := \frac{\omega}{\alpha} \cdot \frac{d}{dx} v(x) \qquad a(x) := h \cdot \left(\frac{\omega}{\alpha}\right)^2 \cdot (60 \cdot x - 180x^2 + 120x^3)$

Check BC's $a(0) = 0 \quad a(1) = 0$

Jerk: $jerk := \frac{\omega}{\alpha} \cdot \frac{d}{dx}\left[\frac{\omega^2}{\alpha^2} \cdot h \cdot (60 \cdot x - 180 \cdot x^2 + 120 \cdot x^3)\right] \quad j(x) := h \cdot \left(\frac{\omega}{\alpha}\right)^3 \cdot (60 - 360x + 360x^2)$

$j(0) = 4.8 \cdot 10^6 \quad j(.5) = -2.4 \cdot 10^6 \quad j(1) = 4.8 \cdot 10^6$

Compare approximation $j_0 := \dfrac{a(.001) - a(0)}{.001} \cdot \dfrac{\omega}{\alpha} \quad j_0 = 4.786 \cdot 10^6$

Set jerk $= 0$ to find maximum acceleration Estimate $x := .25$

$X := root(j(x), x)$ $X = 0.211$ Max acceleration: $a(X) = 2.309 \cdot 10^4$

Compare approximation $a_{max} := \dfrac{v(.212) - v(.210)}{.002} \cdot \dfrac{\omega}{\alpha} \qquad a_{max} = 2.309 \cdot 10^4$

Define an interval function in terms of Heaviside step functions $H(\theta, a, b) := \Phi(\theta - a) - \Phi(\theta - b)$. Then describe follower motion for a full rotation of the cam. Use an equation form similar to Eqs. (5.20–5.23).

For a given set of motion requirements, we could select either a 3–4–5 polynomial cam or a cycloidal cam. If we used the same follower system and equal base circles, it would be nearly impossible to tell the cams apart. Some differences are apparent in the follower acceleration and jerk. Maximum acceleration occurs in the cycloidal cam follower at the quarter point of the rise—that is, at $x = \theta/\alpha = \frac{1}{4}$. Other acceleration extrema occur at the three-quarters point of the rise at and the one-quarter and three-quarter points of the return. By setting jerk $= 0$ in Sample Problem 5.4, we found the first acceleration extremum at $x = 0.211$ for the 3–4–5 cam. To compare the two types of cam, we normalize the motion curves, an operation that is equivalent to setting $h = \omega = \alpha = 1$. The result is shown in Figure 5.12. Note that the maximum acceleration is higher for the cycloidal cam, but the maximum jerk is higher for the 3–4–5 cam.

Rise–Return–Dwell (RRD) Cams

Sometimes a top dwell is unnecessary because we want the cam follower to begin the return as soon as it reaches its maximum position. One possible set of boundary conditions for the rise is then

$$s(0) = v(0) = a(0) = 0,$$

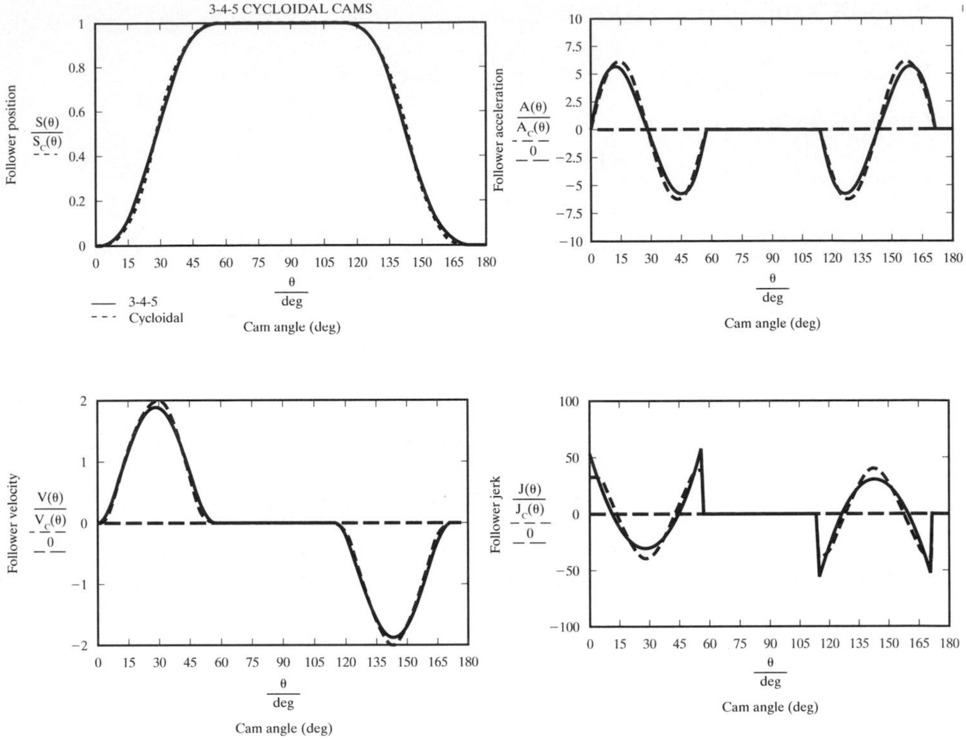

FIGURE 5.12 Follower position, velocity, acceleration and jerk. A 3–4–5 polynomial cam (solid line) is compared with a cycloidal cam (short dashes).

and

$$s(1) = h, \quad v(1) = 0.$$

A fourth-order polynomial in $x = \theta/\alpha$ will satisfy the five boundary conditions. You may want to determine the five arbitrary constants for the fourth-order RRD cam and compare the follower motion with that for a 3–4–5 polynomial cam or a cycloidal cam.

Higher order polynomials for improved motion characteristics

Is it possible to make the cam follower motion smoother by using higher order polynomials? Certainly, we can try. If we want an RRD cam, we can "design smart" by setting $x = 0$ at the top of the rise. Then, for equal rise and return intervals, we have a single rise–return position equation that is symmetrical about $x = 0$. Symmetry suggests an even function—that is, $s(-x) = s(x)$; we need only the even powers of the rise–return position polynomial. The other motion curves have symmetry or antisymmetry about $x = 0$. Next, we decide what the boundary conditions should be. Remember that the number of arbitrary constants must equal the number of nontrivial boundary conditions. Some boundary conditions are redundant—already satisfied because we constructed the position equation out of even powers only.

SAMPLE PROBLEM 5.5

Specifications for an Eighth-Order Rise–Return–Dwell Cam

We need a cam that meets the following follower-motion requirements:

- At the beginning of the rise and the end of the return, $s = v = a = j = 0$.
- At the end of the rise, $s = h$ and $v = j = 0$. (There is no top dwell, and the acceleration is unspecified.)

Suggest the form of a position equation that can satisfy these requirements. Find the arbitrary constants, and plot the follower motion for a 15-mm rise. The cam speed is 1200 rpm, and the rise and return intervals are each 105°. The follower is to dwell for the remainder of the cycle.

Design decision. We will base the cam design on an eighth-order polynomial.

Solution summary. Let us take advantage of the rise–return symmetry by setting $x = 0$ at the end of the rise. Note that odd powers of the position polynomial are antisymmetric, so we will leave them out. Then, the eighth-order polynomial has the form

$$s = C_0 + C_2x^2 + C_4x^4 + C_6x^6 + C_8x^8,$$

where $x = (\theta - \alpha)/\alpha \quad -1 \leq x \leq 0$ for the rise; $0 \leq x \leq 1$ for the return;

$\alpha =$ rise interval $=$ return interval;

$\theta =$ cam position, $0 \leq \theta \leq 2\pi$;

$0 \leq \theta \leq \alpha$ for the rise; $\alpha \leq \theta \leq 2\alpha$ for the return; and

$2\alpha \leq \theta \leq 2\pi$ for the dwell following the return.

The follower position equation is differentiated to obtain the velocity; again, remember the chain rule:

$$v = ds/dt = (ds/dx)(dx/dt) = (\omega/\alpha)(ds/dx).$$

Similar equations apply to the acceleration and the jerk. Two boundary conditions:

$$v(0) = 0 \quad \text{and} \quad j(0) = 0$$

should be identically satisfied, but when you see a statement like this, it is best to check for yourself, because books are not error-free. The remaining five boundary conditions give us the five simultaneous equations we need to find the arbitrary constants. In matrix form, we have

$$\mathbf{AX} = \mathbf{B}, \quad \text{or} \quad \mathbf{X} = \mathbf{A}^{-1}\mathbf{B},$$

where the **X**-matrix is a column vector of arbitrary constants $C_0, C_2 \ldots C_8$, as defined in the detailed solution.

Solving for **X**, we find the five constants and substitute back into the position equation to get

$$s = h(1 - 4x^2 + 6x^4 - 4x^6 + x^8). \tag{5.27}$$

After specifying the rise h, rise and return intervals α, type of follower, radius of the base circle, etc., the manufacture of our eighth-order cam can be based on the equation

$$S = [1 - \Phi(\theta - 2\alpha)] \cdot s[(\theta - \alpha)/\alpha],$$

where the Heaviside step function $\Phi(\theta - 2\alpha)$ turns off the rise–return equation at $\theta = 2\alpha$, the end of the return and start of the dwell.

Similar equations are used to plot the follower velocity, acceleration, and jerk for the specified cam speed. Note that there are no discontinuities in jerk at the transition points from dwell to rise and return to dwell; at those points, jerk = 0. Figure 5.13 shows the follower position, velocity, acceleration, and jerk for the eighth-order polynomial cam.

Detailed solution. For the eighth-order polynomial cam, s is the follower position, θ is the cam position, and α is both the rise interval and the return interval.

We use even powers (an even function) for symmetry about $x = 0$:

Position $\quad s(x) := C_0 + C_2 \cdot x^2 + C_4 \cdot x^4 + C_6 \cdot x^6 + C_8 \cdot x^8$

Velocity $\quad v(x) := \dfrac{\omega}{\alpha} \cdot \dfrac{d}{dx} s(x)$

$$v(x) \rightarrow \frac{\omega}{\alpha} \cdot \left(2 \cdot C_2 \cdot x + 4 \cdot C_4 \cdot x^3 + 6 \cdot C_6 \cdot x^5 + 8 \cdot C_8 \cdot x^7\right)$$

Acceleration $\quad a(x) := \dfrac{\omega}{\alpha} \cdot \dfrac{d}{dx} \dfrac{\omega}{\alpha} \cdot \left(2 \cdot C_2 \cdot x + 4 \cdot C_4 \cdot x^3 + 6 \cdot C_6 \cdot x^5 + 8 \cdot C_8 \cdot x^7\right)$

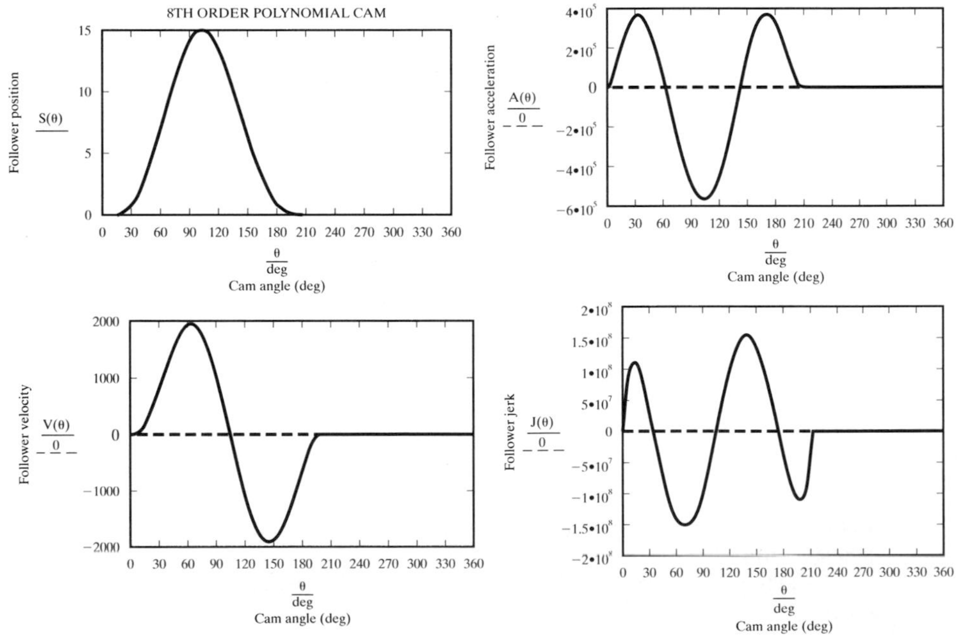

FIGURE 5.13 Follower position, velocity, acceleration, and jerk for an eighth-order polynomial cam.

$$a(x) \rightarrow \frac{\omega^2}{\alpha^2} \cdot \left(2 \cdot C_2 + 12 \cdot C_4 \cdot x^3 + 30 \cdot C_6 \cdot x^4 + 56 \cdot C_8 \cdot x^6\right)$$

Jerk $$j(x) := \frac{\omega}{\alpha} \cdot \frac{d}{dx} \frac{\omega^2}{\alpha^2} \cdot \left(2 \cdot C_2 + 12 \cdot C_4 \cdot x^2 + 30 \cdot C_6 \cdot x^4 + 56 \cdot C_8 \cdot x^6\right)$$

$$j(x) \rightarrow \frac{\omega^3}{\alpha^3} \cdot \left(24 \cdot C_4 \cdot x + 120 \cdot C_6 \cdot x^3 + 336 \cdot C_8 \cdot x^5\right)$$

Since ω/α is not equal to zero, we can apply the velocity, acceleration, and jerk boundary conditions to the derivatives of s with respect to *x* (the terms in parentheses).
Boundary conditions at x = 0: s = h; v = 0*; a unspecified; j = 0*
Boundary conditions at x = 1: s = 0; v = 0; a = 0; j = 0;

Applying the five nontrivial boundary conditions yields the following equations:

$$C_0 := h$$
$$C_0 + C_2 + C_4 + C_6 + C_8 := 0$$
$$2 \cdot C_2 + 4 \cdot C_4 + 6 \cdot C_6 + 8 \cdot C_8 := 0$$
$$2 \cdot C_2 + 12 \cdot C_4 + 30 \cdot C_6 + 56 \cdot C_8 := 0$$
$$24 \cdot C_4 + 120 \cdot C_6 + 336 \cdot C_8 := 0$$

These five equations can be expressed in matrix form as AX = B, where

$$A := \begin{bmatrix} 1 & 0 & 0 & 0 & 0 \\ 1 & 1 & 1 & 1 & 1 \\ 0 & 2 & 4 & 6 & 8 \\ 0 & 2 & 12 & 30 & 56 \\ 0 & 0 & 24 & 120 & 336 \end{bmatrix} \quad B := \begin{bmatrix} h \\ 0 \\ 0 \\ 0 \\ 0 \end{bmatrix}$$

$$X := A^{-1}B \quad X \rightarrow \begin{bmatrix} h \\ -4 \cdot h \\ 6 \cdot h \\ -4 \cdot h \\ h \end{bmatrix} \text{ where X is defined as } \begin{bmatrix} C_0 \\ C_2 \\ C_4 \\ C_6 \\ C_8 \end{bmatrix}$$

$$X := \begin{bmatrix} h \\ -4 \cdot h \\ 6 \cdot h \\ -4 \cdot h \\ h \end{bmatrix}$$

For h := 15, the constants are

$C_0 := X_0 \quad C_0 = 15 \quad C_2 := X_1 \quad C_2 = -60 \quad C_4 := X_2 \quad C_4 = 90$

$C_6 := X_3 \quad C_6 = -60 \quad C_8 := X_4 \quad C_8 = 15$

* Satisfied identically due to form of s.

The constants we have just found are substituted into the displacement equation:

Displacement for rise and return is $s(x) := h \cdot (1 - 4x^2 + 6x^4 - 4x^6 + x^8)$,
where x ranges from −1 to 1 as θ ranges from 0 to 2 α

There is a dwell between $\theta = 2\,\alpha$ and $\theta = 2\pi$

We can turn off the equation with a Heaviside step function. Let

$$\omega := 1200\frac{\pi}{30} \quad \alpha := \frac{105 \cdot \pi}{180} \quad \theta := 0, \frac{\pi}{180} .. 2\pi$$

Displacement $S(\theta) := (1 - \Phi(\theta - 2\alpha)) \cdot s\left(\frac{\theta - \alpha}{\alpha}\right)$

Velocity

$$v(x) := \frac{\omega}{\alpha} \cdot (2 \cdot C_2 \cdot x + 4 \cdot C_4 \cdot x^3 + 6 \cdot C_6 \cdot x^5 + 8 \cdot C_8 \cdot x^7) \quad V(\theta) := (1 - \Phi(\theta - 2\alpha)) \cdot v\left(\frac{\theta - \alpha}{\alpha}\right)$$

Acceleration

$$a(x) := \frac{\omega^2}{\alpha^2} \cdot (2 \cdot C_2 + 12 \cdot C_4 \cdot x^2 + 30 \cdot C_6 \cdot x^4 + 56 \cdot C_8 \cdot x^6) \quad A(\theta) := (1 - \Phi(\theta - 2\alpha)) \cdot a\left(\frac{\theta - \alpha}{\alpha}\right)$$

Jerk

$$j(x) := \frac{\omega^3}{\alpha^3} \cdot (24 \cdot C_4 \cdot x + 120 \cdot C_6 \cdot x^3 + 336 \cdot C_8 \cdot x^5) \quad J(\theta) := (1 - \Phi(\theta - 2\alpha)) \cdot j\left(\frac{\theta - \alpha}{\alpha}\right)$$

5.7 ANALYTICAL CAM DESIGN BASED ON THE THEORY OF ENVELOPES

Suppose we know where we want the follower for each cam position. Suppose also that, for a cam operating at high speeds, we have satisfied the boundary conditions necessary to avoid shock loading. Is this information sufficient to generate a cam (or a prototype cam) using numerically controlled (NC) or computer numerically controlled (CNC) machinery? The answer is *no*: We need to make some design decisions, including the following:

- type of follower we want
- follower offset (if any)
- cam base circle
- other decisions based on forces on the cam-and-follower system, including inertial loading

But suppose the design decisions are wrong? Can we find that out before we manufacture an unsatisfactory cam? Here are some ways we can check our design:

- Sketch the cam with drafting materials as discussed earlier. If there are abrupt changes in curvature, the actual follower motion will not be as prescribed.
- Use a commercially available computer-aided design (CAD) program for cams, possibly integrated with a computer-aided manufacturing (CAM) program.
- Use the theory of envelopes. An introduction to that advanced topic and some simple examples follow.

Theory of Envelopes

The process of analytically generating a profile for a disk cam closely parallels the graphical approach that has been presented. As in the graphical approach, the desired positions of the follower are determined for an inversion of the cam-and-follower system in which the cam is held stationary. The cam that will produce the desired motion is then obtained by fitting a tangent curve to the follower positions. The analytical approach, however, can contemplate a virtually unlimited number of continuous follower positions, as opposed to a finite number of discrete positions in a graphical layout.

The basis for the approach is the theory of envelopes, from calculus. Consider the series of follower positions shown in Figure 5.14a. The follower is depicted as a circular roller follower. However, any shape of follower including a flat face, can be considered. The set of all follower positions describes a family of curves, circles in this case. The boundary of the family of follower curves is referred to as the envelope and is the cam profile. Notice that, in this example, the envelope consists of two curves, indicating that there are two possible cam profiles: an inner profile and an outer profile. The mathematical theory for determining the envelope is described in the paragraphs that follow.

A family of curves in the xy-plane can be expressed mathematically as

$$F(x, y, \lambda) = 0, \tag{5.28}$$

where λ is called the *parameter of the family*. λ distinguishes the member curves from one another, and for a particular value of λ, Eq. (5.28) defines one member of the family of curves. For example, in Figure 5.14a, the parameter λ represents the location of the center of the constant-radius circular follower. In Figure 5.14b, two member curves, corresponding to arbitrary parameter values λ_1 and λ_2, are shown. It can be seen that points lying on the envelope also lie on the curves, and therefore, the x- and y-coordinates of the envelope must satisfy Eq. (5.28).

Consider the following equation involving the partial derivative of function F with respect to parameter λ:

$$\frac{\partial F}{\partial \lambda}(x, y, \lambda) = 0. \tag{5.29}$$

Equation (5.29) represents a second family of curves with parameter λ. It can be shown that each member curve of that equation intersects the corresponding member of Eq. (5.28) at the envelope. Therefore, the simultaneous solution of Eqs. (5.28) and (5.29)

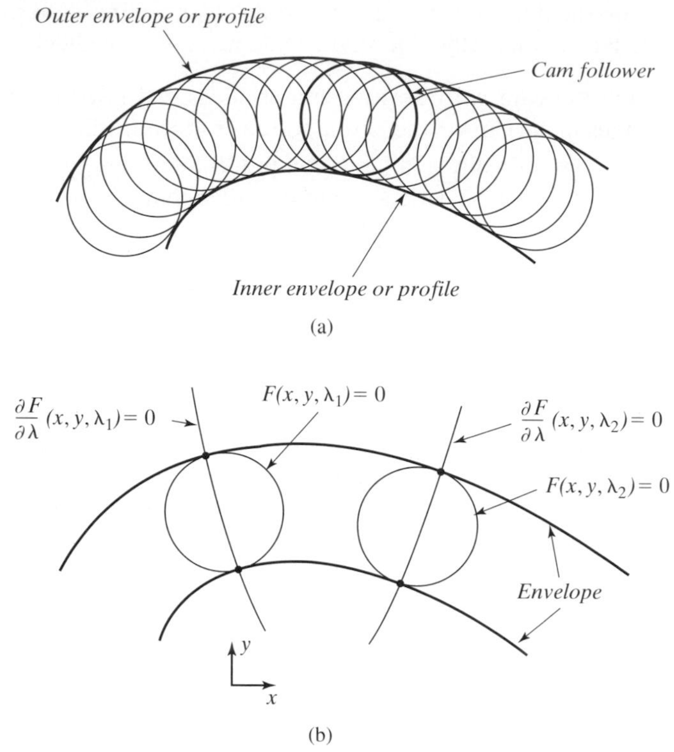

FIGURE 5.14 (a) The family of circles represents the positions of a roller follower as it moves relative to a cam. The boundary, or envelope, of this family of curves is therefore the cam profile. (b) The mathematical basis for describing the cam profile by means of the theory of envelopes is shown.

defines the envelope. This solution is found either by eliminating the parameter λ or by expressing x and y in terms of λ.

SAMPLE PROBLEM 5.6

The Envelope of a Family of Circles

Figure 5.15 contains a family of circles, each having a radius of 1.0 and a center lying on a 45° line in the xy-plane. Determine the envelope of this family of curves.

Solution. The curves can be expressed mathematically as

$$F(x, y, \lambda) = (x - \lambda)^2 + (y - \lambda)^2 - 1 = 0, \tag{5.28a}$$

where a particular value of λ defines a circle of radius 1.0 centered at point $x = \lambda$, $y = \lambda$. Equation (5.29) takes the form

$$\frac{\partial F}{\partial \lambda}(x, y, \lambda) = -2(x - \lambda) - 2(y - \lambda) = 0. \tag{5.29a}$$

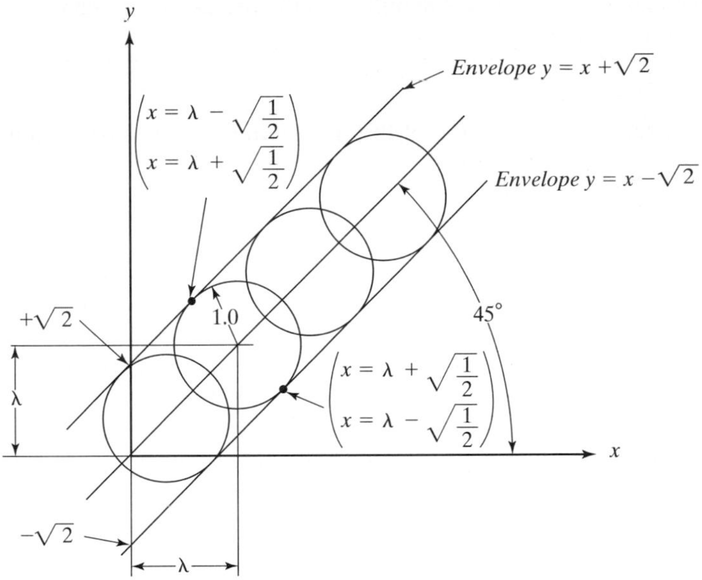

FIGURE 5.15 A family of circles, each having a radius of 1.0 and a center lying on the line $y = x$. As shown, the envelope of this family of curves consists of two straight lines.

From Eq. (5.29a), we obtain

$$\lambda = \frac{x + y}{2}.$$

Substituting into Eq. (5.28a) yields

$$\left[x - \left(\frac{x + y}{2}\right)\right]^2 + \left[y - \left(\frac{x + y}{2}\right)\right]^2 - 1 = 0.$$

and solving for y, we have

$$y = x \pm \sqrt{2}.$$

This equation defines the envelope: a pair of straight lines with 45° slopes and y-intercepts of $+\sqrt{2}$ and $-\sqrt{2}$, as shown in Figure 5.15.

Alternatively, the envelope can be found in parametric form as a function of λ. From Eq. (5.29a), we get

$$y = 2\lambda - x.$$

Substituting into Eq. (5.28a) and solving for x, we have

$$x = \lambda \pm \sqrt{\frac{1}{2}}$$

and then

$$y = \lambda \mp \sqrt{\frac{1}{2}}.$$

Substituting a range of values for λ, we readily see that these equations define the same envelope given previously and shown in Figure 5.15.

The sections that follow present the application of the theory of envelopes to common cam-follower types.

Disk Cam with Translating Flat-Faced Follower

A cam with a translating flat-faced follower is shown in Figure 5.16. The figure depicts an inversion of the cam-and-follower mechanism wherein the cam is fixed and the follower moves relative to it. In normal operation, the cam would rotate and the follower would translate in a guideway along the y-axis. In either case, the relative motion between cam and follower is the same. The cam, having base-circle radius r_b, is assumed to rotate in the clockwise direction under normal operation. Thus, for a cam rotation θ, the follower will rotate counterclockwise relative to the cam through angle θ while experiencing a translational displacement s, as shown in the figure. In this and the sections that follow, it is assumed that the follower displacement is a known function of the cam angle, as would be true in an actual design situation.

The equation of the family of straight lines (the follower face) generating the cam profile envelope is given by

$$y = mx + b, \tag{5.30}$$

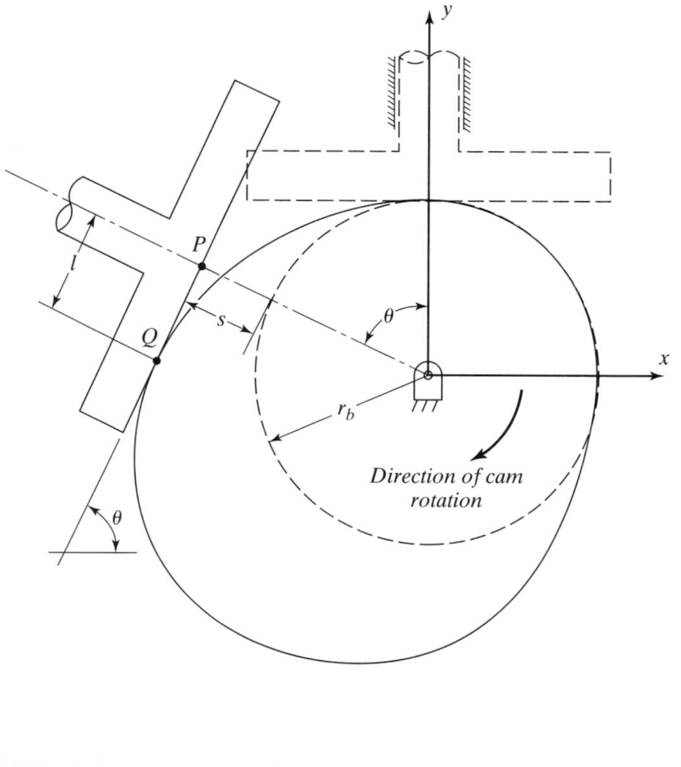

FIGURE 5.16 A disk cam with a translating flat-faced follower. The figure shows the motion of the follower relative to the cam. This motion consists of follower translation s during cam rotation θ. Point P is the point on the follower face coinciding with the follower centerline, and point Q is the instantaneous point of contact between the cam and follower.

where m is the slope and b is the y-intercept of the line. The origin of the xy-coordinate system is positioned at the center of the base circle, which is also the pivot point of the cam. From inspection of Figure 5.16, we see that

$$m = \tan\theta.$$

The coordinates of point P, the intersection of the face of the follower and its axis, are given by

$$x = -(r_b + s)\sin\theta \tag{5.31}$$

and

$$y = (r_b + s)\cos\theta, \tag{5.32}$$

where s, the displacement of the follower, is a prescribed function of cam angle θ. Point P is on the line described by Eq. (5.30). Substituting Eqs. (5.31) and (5.32) into Eq. (5.30) and solving for b, we have

$$b = \frac{(r_b + s)}{\cos\theta},$$

and Eq. (5.30) becomes

$$y = \frac{x\sin\theta + (r_b + s)}{\cos\theta}.$$

Rearranging terms, we find that the family of straight lines (follower positions) generating the cam profile envelope is given by

$$F(x, y, \theta) = y\cos\theta - x\sin\theta - r_b - s = 0, \tag{5.33}$$

where θ is the parameter of the family; that is, each value of θ represents a different follower position and corresponding straight line.

Differentiating Eq. (5.33) yields

$$\frac{\partial F}{\partial\theta} = -y\sin\theta - x\cos\theta - \frac{ds}{d\theta} = 0, \tag{5.34}$$

where the quantity $ds/d\theta$ can be evaluated from the known displacement function. Solving Eqs. (5.33) and (5.34) simultaneously leads to the following expressions for the cam profile coordinates:

$$x = -(r_b + s)\sin\theta - \frac{ds}{d\theta}\cos\theta \tag{5.35}$$

and

$$y = (r_b + s)\cos\theta - \frac{ds}{d\theta}\sin\theta \tag{5.36}$$

Equations (5.35) and (5.36) give the coordinates of the cam-and-follower contact point (point Q in Fig. 5.16) for cam angle θ. The distance l between points P and Q is the perpendicular distance from the follower centerline to the contact point. From Eqs. (5.31), (5.32), (5.35), and (5.36), we obtain

$$l = \sqrt{(x_P - x_Q)^2 + (y_P - y_Q)^2} = \frac{ds}{d\theta}. \tag{5.37}$$

The maximum value of l can be used in determining dimensions for the follower face. Equation. (5.37) can be rewritten as

$$l = \frac{ds}{dt}\frac{dt}{d\theta} = \frac{v}{\omega}, \tag{5.38}$$

which is the translational follower velocity divided by the rotational cam velocity. Thus, for a cam with constant angular velocity, the maximum value of l occurs when the follower velocity is at a maximum. It is noteworthy that l is independent of the base-circle radius r_b.

SAMPLE PROBLEM 5.7

Design of a Disk Cam with a Translating Flat-Faced Follower

Design a disk cam to produce the following motion of a translating flat-faced follower: a rise through distance h with simple harmonic motion during 180° of rotation, followed by a return, also with simple harmonic motion, during the remaining 180° of cam rotation.

Solution. For this special case, simple harmonic follower motion is given by

$$s = \frac{h}{2} - \left(\frac{h}{2}\right)\cos\theta,$$

where $\alpha = \pi$ and the equation holds for both the rise and return motions. The derivative of displacement s with respect to θ is

$$\frac{ds}{d\theta} = \left(\frac{h}{2}\right)\sin\theta.$$

Substituting into Eqs. (5.35) and (5.36) yields the following expressions for the cam profile coordinates:

$$x = -\left(r_b + \frac{h}{2}\right)\sin\theta$$

and

$$y = \left(r_b + \frac{h}{2}\right)\cos\theta - \frac{h}{2}.$$

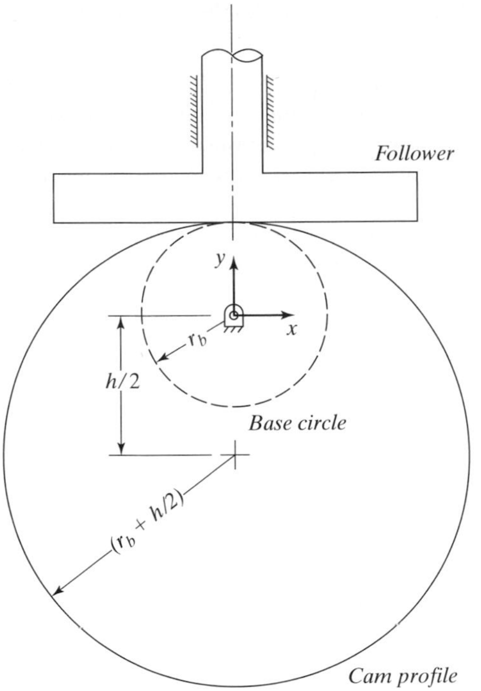

FIGURE 5.17 The cam and follower of Sample Problem 5.7. The cam profile in this case is an offset circle, which will produce simple harmonic rise and return motion of the follower.

Alternatively, these equations can be rearranged to eliminate θ, leading to

$$x^2 + \left(y + \frac{h}{2}\right)^2 = \left(r_b + \frac{h}{2}\right)^2.$$

It can be seen that, given a base-circle radius r_b, the cam profile is a circle with center at ($x = 0$, $y = -h/2$) and a radius equal to $r_b + h/2$. From Eq. (5.37),

$$l = \frac{ds}{d\theta} = \left(\frac{h}{2}\right) \sin\theta,$$

which has a maximum absolute value of $h/2$ when θ equals 90° and 270°. Thus, the width of the follower face should be made greater than h. Figure 5.17 shows the offset circular cam that will produce the prescribed motion.

SAMPLE PROBLEM 5.8

Design of a Disk Cam with a Translating Flat-Faced Follower (to produce a different motion pattern).

Design a disk cam to produce the following motion of a translating flat-faced follower: a dwell during 30° of cam rotation, a 2-in rise with parabolic motion during the next 150° of rotation, a second dwell during the next 60° of rotation, and a 2-in return with simple harmonic motion during the final 120° of cam rotation. The base-circle radius is to be 3 in.

Solution. Figures 5.18a and 5.18b show the results of a typical computer program used to generate the required cam profile. Such a program would include analytical expressions for profile coordinates and follower motions. In this example, equations for simple harmonic motion and parabolic motion have been combined with Eqs. (5.35) and (5.36) to produce the results shown.

Figure 5.18a lists some of the computed profile coordinates and shows a computer-generated drawing of the cam and follower. Figure 5.18b contains computer graphics plots of the follower displacement, velocity, and acceleration corresponding to the prescribed motion for a complete cam rotation. It can be seen that the prescribed pattern of motion is not acceptable for high-speed operation.

CAM ANGLE	x COORDINATE	y COORDINATE
1	−0.05236	2.99954
11	−0.57243	2.94488
21	−1.07511	2.80074
31	−1.56267	2.56116
41	−2.15141	2.13334
51	−2.66160	1.60483
61	−3.07945	0.98491
71	−3.39105	0.28425
81	−3.58230	−0.48454
91	−3.63926	−1.30640
101	−3.54859	−2.16325
111	−3.37396	−2.80082
121	−3.13631	−3.28674
131	−2.79649	−3.75364
141	−2.35896	−4.17560
151	−1.83482	−4.52876
161	−1.24093	−4.79293
171	−0.59879	−4.95290
181	0.08729	−4.99924
191	0.95407	−4.90813
201	1.79187	−4.66789
211	2.57522	−4.28582
221	3.28032	−3.77352
231	3.88576	−3.14657
241	4.35378	−2.45819
251	4.54998	−2.01720
261	4.67030	−1.53317
271	4.70664	−1.00603
281	4.64644	−0.43967
291	4.47496	0.15550
301	4.17852	0.76141
311	3.74805	1.35251
321	3.18239	1.89769
331	2.49063	2.36350
341	1.69289	2.71813
351	0.81947	2.93559

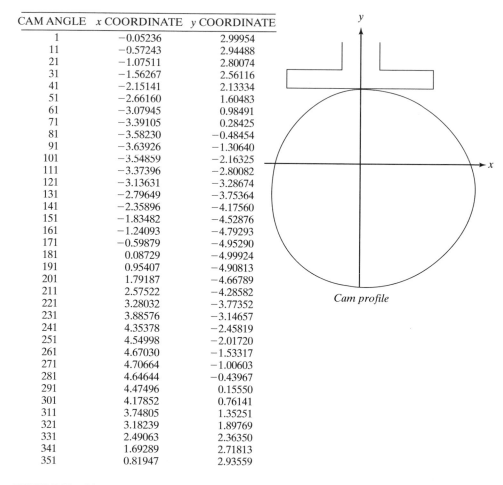

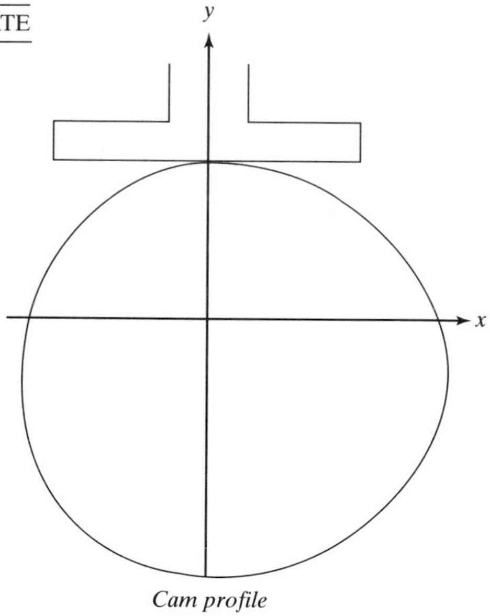

FIGURE 5.18 (a) Cam profile coordinates and layout for Sample Problem 5.8.

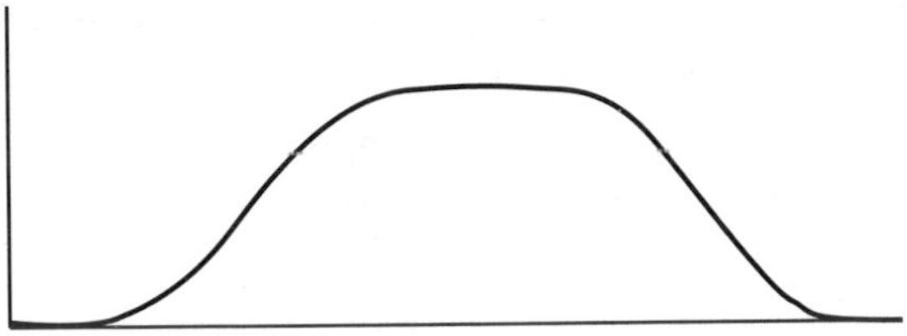

Follower displacement versus cam angle

Base radius: 3.0000
Type of motion and follower type: *Tran flat*
Number of segments: 4

Motion	*Degrees*	*Displacement*
DWLL	30	0.
PRAB	150	2.000
DWLL	60	0.
SMHM	120	−2.000

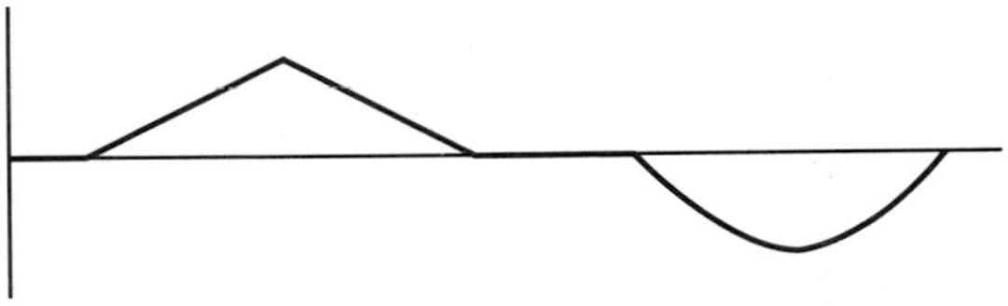

Follower velocity versus cam angle

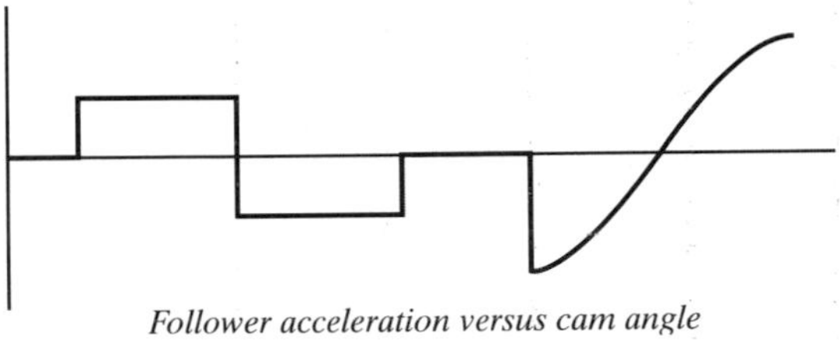

Follower acceleration versus cam angle

FIGURE 5.18 (b) Follower displacement, velocity, and acceleration for Sample Problem 5.8.

Disk Cam with Translating, Offset Roller Follower

The configuration for the disk cam with a translating, offset roller follower type is shown in Figure 5.19. The follower has a roller of radius r_f and an eccentricity, or offset, e; radial follower motion is a special case of this configuration in which offset $e = 0$. As before, the base-circle radius is r_b, and the follower displacement is s, which is a prescribed function of cam angle θ.

The equation for the family of circles described by the follower roller is

$$F(x, y, \theta) = (x - x_c)^2 + (y - y_c)^2 - r_f^2 = 0, \tag{5.39}$$

where x_c and y_c are the x- and y-coordinates, respectively, of the roller center c. For the arbitrary position shown in Figure 5.19 corresponding to cam angle θ,

$$x_c = -(r_b + r_f)\sin(\theta + \beta) - s \sin \theta \tag{5.40}$$

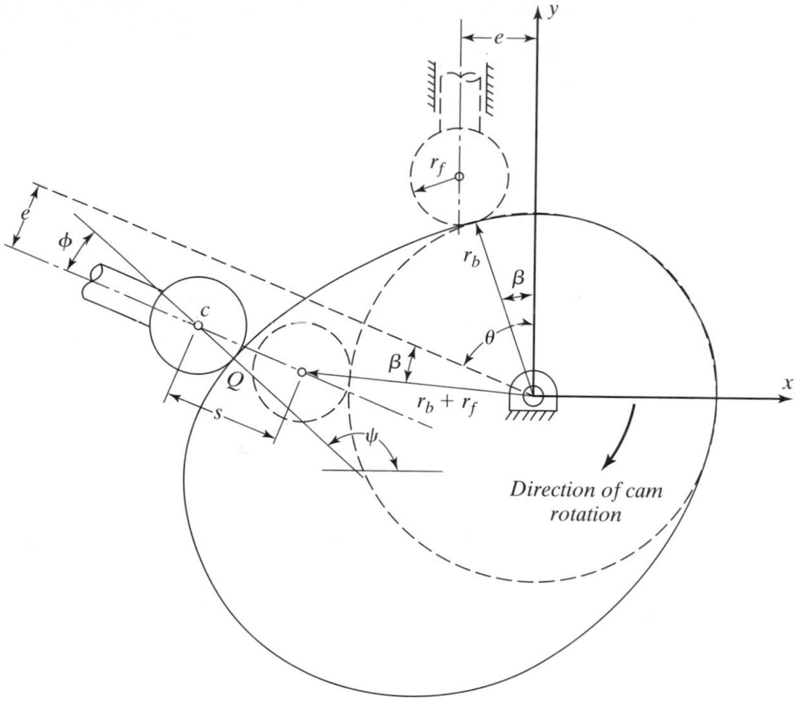

FIGURE 5.19 A disk cam with a translating, offset roller follower. Angle β is a function of the base-circle radius r_b, the roller radius r_f, and the offset e. Angle ϕ is the pressure angle. The follower translates through distance s as the cam rotates through angle θ.

and

$$y_c = (r_b + r_f)\cos(\theta + \beta) + s\cos\theta, \tag{5.41}$$

where

$$\beta = \arcsin\left(\frac{e}{r_b + r_f}\right). \tag{5.42}$$

Substituting Eqs. (5.40) and (5.41) into Eq. (5.39), we have

$$\begin{aligned} F(x, y, \theta) &= [x + (r_b + r_f)\sin(\theta + \beta) + s\sin\theta]^2 \\ &\quad + [y - (r_b + r_f)\cos(\theta + \beta) - s\cos\theta]^2 - r_f^2 = 0. \end{aligned} \tag{5.43}$$

From Eq. (5.39), the partial-derivative equation is

$$\frac{\partial F}{\partial \theta} = -2(x - x_c)\frac{dx_c}{d\theta} - 2(y - y_c)\frac{dy_c}{d\theta} = 0, \tag{5.44}$$

where, from Eqs. (5.40) and (5.41),

$$\frac{dx_c}{d\theta} = -(r_b + r_f)\cos(\theta + \beta) - s\cos\theta - \frac{ds}{d\theta}\sin\theta \tag{5.45}$$

and

$$\frac{dy_c}{d\theta} = -(r_b + r_f)\sin(\theta + \beta) - s\sin\theta + \frac{ds}{d\theta}\cos\theta. \tag{5.46}$$

Solving Eqs. (5.39) and (5.44) simultaneously gives the coordinates of the cam profile:

$$x = x_c \pm r_f\left(\frac{dy_c}{d\theta}\right)\left[\left(\frac{dx_c}{d\theta}\right)^2 + \left(\frac{dy_c}{d\theta}\right)^2\right]^{-1/2}; \tag{5.47}$$

$$y = y_c \mp r_f\left(\frac{dx_c}{d\theta}\right)\left[\left(\frac{dx_c}{d\theta}\right)^2 + \left(\frac{dy_c}{d\theta}\right)^2\right]^{-1/2}. \tag{5.48}$$

Note the plus-or-minus sign in Eqs. (5.47) and (5.48). This reflects the fact that there are two envelopes: an inner profile (shown in Figure 5.19) and an outer profile. Also, observe that the plus sign in Eq. (5.47) goes with the minus sign in Eq. (5.48) and vice versa.

Pressure Angle

A mathematical expression can also be derived for the pressure angle ϕ. (See Figure 5.19.) Recall that the pressure angle is defined as the angle between the common normal at the cam–follower contact point and the line of travel of the follower. The common normal is the straight line passing through the contact point $Q(x, y)$ [see Eqs. (5.47) and (5.48)] and the roller center $c(x_c, y_c)$ [see Eqs. (5.40) and (5.41)]. The angle ψ that the common normal makes with the x direction is given by

$$\psi = \arctan\left(\frac{y_c - y}{x_c - x}\right).$$

Angle ψ can also be expressed in terms of the pressure angle ϕ and the cam angle θ as

$$\psi = \frac{\pi}{2} + \theta - \phi.$$

Therefore, equating the two expressions for ψ, we find that the pressure angle is

$$\phi = \frac{\pi}{2} + \theta - \arctan\left(\frac{y_c - y}{x_c - x}\right). \tag{5.49a}$$

This equation is based on the inner-envelope cam profile coordinates. The pressure angle for the outer envelope is equal to that for the inner envelope. As discussed earlier, the pressure angle is an important design characteristic in cam-and-follower systems.

A useful relationship between the pressure angle ϕ and the cam dimensions can be derived from Eq. (5.49a). Rewriting that equation, we have

$$\frac{y_c - y}{x_c - x} = \tan\left[\frac{\pi}{2} + (\theta - \phi)\right] = -\cot(\theta - \phi) = \frac{-1}{\tan(\theta - \phi)},$$

from which it follows that

$$\tan(\theta - \phi) = -\left(\frac{x - x_c}{y - y_c}\right).$$

Substituting Eqs. (5.47) and (5.48) for

$$\tan(\theta - \phi) = -\left(\frac{\pm y'_c}{\mp x'_c}\right) = \frac{y'_c}{x'_c},$$

where the prime notation denotes differentiation with respect to angle θ (i.e., $x'_c = dx_c/d\theta$, etc.). The derivatives x'_c and y'_c are given by Eqs. (5.45) and (5.46), and upon substitution, we obtain

$$\tan(\theta - \phi) = \frac{\sin(\theta - \phi)}{\cos(\theta - \phi)} = \frac{(r_b + r_f)\sin(\theta + \beta) + s\sin\theta - s'\cos\theta}{(r_b + r_f)\cos(\theta + \beta) + s\cos\theta + s'\sin\theta}.$$

Employing trigonometric identities, this equation reduces to the form

$$\frac{\sin\phi}{\cos\phi} = \tan\phi = \frac{s' - (r_b + r_f)\sin\beta}{s + (r_b + r_f)\cos\beta}.$$

But from Eq. (5.42),

$$\sin\beta = \frac{e}{r_b + r_f},$$

and therefore,

$$\cos\beta = \sqrt{1 - \sin^2\beta} = \sqrt{1 - \left(\frac{e}{r_b + r_f}\right)^2}.$$

Finally, making the latter two substitutions, we have the desired relationship:

$$\tan\phi = \frac{s' - e}{s + \sqrt{(r_b + r_f)^2 - e^2}}. \qquad (5.49b)$$

Given a required follower displacement function s and a limit on how large the pressure angle ϕ can be, Eq. (5.49b) can be used to size the cam base circle, the roller radius, and the follower offset. Note that the equation also applies to the case of zero offset ($e = 0$).

The analytical cam synthesis equations derived in this section for a translating roller follower, as well as those equations derived for other types of follower, are most effectively implemented in design practice through the use of a digital computer or a programmable calculator. The intention of the examples that follow is to illustrate the type of computation that would be performed at a relatively large number of positions in such a process.

SAMPLE PROBLEM 5.9

Disk Cam with a Translating Roller Follower

One segment of a prescribed cam-and-follower motion calls for a lift of 30 mm with cycloidal motion during cam rotation from $\theta = 0$ to $\theta = 90°$. The disk cam base-circle radius is 40 mm. The follower is a translating roller follower with a roller radius of 10 mm and an offset of 20 mm. Determine the profile coordinates and the pressure angle corresponding to a cam angle of $\theta = 60°$.

Solution. Applying the equation for cycloidal motion, we have

$$s = \left(\frac{h}{\alpha}\right)\theta - \left(\frac{h}{2\pi}\right)\sin\left(\frac{2\pi\theta}{\alpha}\right),$$

where $h = 30$ mm and $\alpha = \pi/2$ rad, leading to the relationship

$$s = \left(\frac{60}{\pi}\right)\theta - \left(\frac{15}{\pi}\right)\sin 4\theta.$$

Differentiating with respect to θ yields

$$s' = \frac{ds}{d\theta} = \frac{60}{\pi} - \left(\frac{60}{\pi}\right)\cos 4\theta.$$

When $\theta = 60° = \pi/3$ rad,

$$s = \left(\frac{60}{\pi}\right)\left(\frac{\pi}{3}\right) - \left(\frac{15}{\pi}\right)\sin\left(\frac{4\pi}{3}\right) = 24.13 \text{ mm},$$

or

$$s' = \left(\frac{60}{\pi}\right) - \left(\frac{60}{\pi}\right)\cos\left(\frac{4\pi}{3}\right) = 28.65 \text{ mm/rad}.$$

From Eq. (5.42),

$$\beta = \arcsin\left(\frac{20}{40 + 10}\right) = \arcsin(0.4) = 23.6°.$$

Now the quantities x_c, y_c, x'_c, and y'_c can be determined from Eqs. (5.40), (5.41), (5.45), and (5.46), respectively:

$$x_c = -(40 + 10)\sin(60° + 23.6°) - 24.13 \sin 60° = -70.59 \text{ mm};$$

$$y_c = (40 + 10)\cos(60° + 23.6°) + 24.13 \cos 60° = 17.64 \text{ mm};$$

$$x'_c = -(40 + 10)\cos(60° + 23.6°) - 24.13\cos 60°$$
$$- 28.65\sin 60° = -42.45 \text{ mm/rad};$$
$$y'_c = -(40 + 10)\sin(60° + 23.6°) - 24.13\sin 60°$$
$$= 28.65\cos 60° = -56.26 \text{ mm/rad}.$$

Substituting these values into Eqs. (5.47) and (5.48), we have the profile coordinates:

$$x = -70.59 \pm 10(-56.26)[(-42.45)^2 + (-56.26)^2]^{-1/2}$$
$$= -78.57, -62.61$$
$$y = 17.64 \mp 10(-42.45)[(-42.45)^2 + (-56.26)^2]^{1/2} = 23.66, 11.62.$$

Inspection of these results shows that the coordinates of the contact point on the inner profile corresponding to this cam angle ($\theta = 60°$) are

$$x = -62.61 \text{ mm} \quad \text{and} \quad y = 11.62 \text{ mm},$$

and the outer-profile coordinates are

$$x = -78.57 \text{ mm} \quad \text{and} \quad y = 23.66 \text{ mm}.$$

The complete cam profile can be generated by repeating the foregoing procedure over the total range of angle θ.

The pressure angle is obtained from either Eq. (5.49a) or Eq. (5.49b). From the first of these equations,

$$\phi = 90° + 60° - \arctan\left[\frac{17.64 - 11.62}{-70.59 - (-62.61)}\right]$$
$$= 150° - \arctan\left(\frac{6.02}{-7.98}\right) = 150° - 143.0° = 7.0°,$$

where the quadrant of the arctangent function was determined from the signs of the numerator and denominator. As a check, using Eq. (5.49b), we find that

$$\tan\phi = \frac{28.65 - 20}{24.13 + \sqrt{(40 + 10)^2 - (20)^2}} = 0.124,$$

so that

$$\phi = 7.0°.$$

This pressure angle may be acceptable. However, the pressure angle varies with position, so the value is 7.0° probably not the maximum that occurs during the overall motion.

SAMPLE PROBLEM 5.10

Translating Roller Follower with Zero Offset

Repeat Sample Problem 5.9, but with a follower offset of zero (i.e., the follower is a radial follower).

Solution. As before, at $\theta = 60°$ the values of the displacement and its derivative are $s = 24.13$ mm and $s' = 28.65$ mm/rad. But now, $\beta = 0$, because $e = 0$. Therefore, from

Eqs. (5.40), (5.41), (5.45), and (5.46),

$$x_c = -50\sin 60° - 24.13\sin 60° = -64.20 \text{ mm},$$
$$y_c = 50\cos 60° + 24.13\cos 60° = 37.06 \text{ mm},$$
$$x_c' = -50\cos 60° - 24.13\cos 60° - 28.65\sin 60° = -61.87 \text{ mm/rad},$$

and

$$y_c' = -50\sin 60° - 24.13\sin 60° + 28.65\cos 60° = -49.88 \text{ mm/rad},$$

This leads to the following profile coordinate values from Eqs. (5.47) and (5.48):

$$x = -64.20 \pm 10(-49.88)[(-61.87)^2 + (-49.88)^2]^{-1/2}$$
$$= -70.48, -57.92;$$
$$y = 37.06 \mp 10(-61.87)[(-61.87)^2 + (-49.88)^2]^{-1/2} = 44.84, 29.27.$$

The inner-profile coordinates are $x = -57.92$ mm and $y = 29.27$ mm, and the outer-profile coordinates are $x = -70.48$ mm and $y = 44.84$ mm.

The pressure angle is calculated from Eq. (5.49b):

$$\tan\phi = \frac{28.65 - 0}{24.13 + \sqrt{(40 + 10)^2 - (0)^2}} = 0.386,$$

so that

$$\phi = 21.1°.$$

Note that, whereas the value of the pressure angle for the offset case ($\phi = 7.0°$) is less than 21.1° just obtained, it should be realized that there may be other parts of the cam mechanism cycle wherein the offset arrangement has higher pressure angle values than the radial arrangement. However, the offset configuration may be designed to utilize the pressure angle advantage during that portion of the cycle when loads are large and, in turn, when side forces in the follower guideway are most critical (especially during the follower lift). Note also that the follower may be offset either to the left, as in Figure 5.19, or to the right, in which case e would have a negative value.

Disk Cam with Rotating Flat-Faced Follower

The oscillating, or pivoted, flat-faced follower is shown in Figure 5.20, where the various parameters are defined. The family of follower positions is a set of straight lines.

The equation for the family of lines that determine the envelope is the same as Eq. (5.30):

$$y = mx + b.$$

From Figure 5.20, the slope is given by

$$m = \tan(\theta - \beta - \gamma), \tag{5.50}$$

where γ is the angular displacement of the follower and $\beta = \arcsin\left(\dfrac{r_b + e}{r_c}\right)$

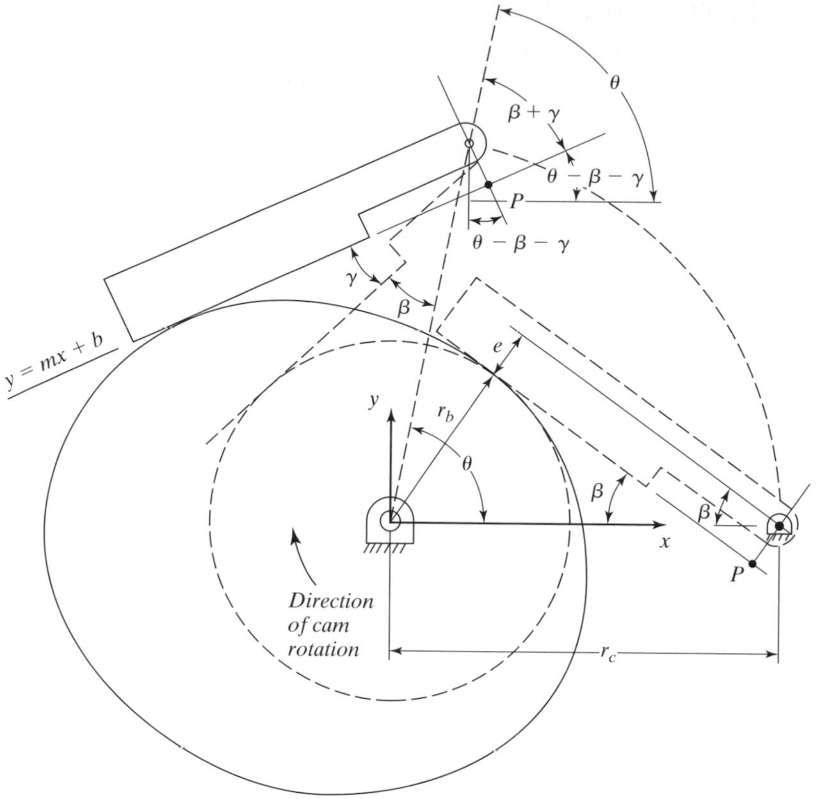

FIGURE 5.20 A disk cam with a rotating flat-faced follower. The motion of the follower consists of rotation γ during cam rotation θ. Angle β is a function of the base-circle radius r_b, the distance r_c between the centers of rotation of the cam and follower, and the perpendicular distance e from the follower pivot to point P on the extension of the follower face.

is the initial follower angle. In a typical design situation, follower displacement γ would be a prescribed function of cam angle θ. The coordinates of point P on the extension of the follower face are

$$x = r_c \cos\theta + e\sin(\theta - \beta - \gamma) \tag{5.51}$$

and

$$y = r_c \sin\theta - e\cos(\theta - \beta - \gamma). \tag{5.52}$$

Substituting Eqs. (5.50), (5.51), and (5.52) into Eq. (5.30) and solving for b yields

$$b = r_c \sin\theta - e\cos A - (r_c\cos\theta + e\sin A)\tan A,$$

where

$$A = \theta - \beta - \gamma.$$

Then

$$F(x, y, \theta) = y + (r_c \cos\theta + e \sin A - x) \tan A - r_c \sin\theta + e \cos A$$
$$= 0. \tag{5.53}$$

The partial-derivative expression is

$$\frac{\partial F}{\partial \theta} = \left(-r_c \sin\theta + e \cos A \cdot \frac{dA}{d\theta}\right) \tan A + (r_c \cos\theta + e \sin A - x)\sec^2 A \cdot \frac{dA}{d\theta}$$
$$- r_c \cos\theta - e \sin A \cdot \frac{dA}{d\theta}, \tag{5.54}$$

where

$$\frac{dA}{d\theta} - 1 - \frac{d\gamma}{d\theta}.$$

Solving Eqs. (5.53) and (5.54) simultaneously gives the following profile coordinates:

$$x = e \sin A + r_c\left[\cos\theta - \frac{\cos A \cos(\theta - A)}{dA/d\theta}\right]; \tag{5.55}$$

$$y = -e \cos A + r_c\left[\sin\theta - \frac{\sin A \cos(\theta - A)}{dA/d\theta}\right]. \tag{5.56}$$

Disk Cam with Rotating Roller Follower

The family of circles for the disk cam with a rotating roller follower is expressed by Eq. (5.39):

$$F(x, y, \theta) = (x - x_c)^2 + (y - y_c)^2 - r_f^2 = 0.$$

The parameters are defined as shown in Figure 5.21: r_a is the length of the follower arm, r_f is the roller radius, r_c is the distance between the center of cam rotation and the follower pivot, and γ is the follower angular displacement, which is a prescribed function of cam angle θ. From the figure, it can be seen that the coordinates of the center c of the roller are

$$x = r_c \cos\theta - r_a \cos(\theta - \beta - \gamma) \tag{5.57}$$

and

$$y = r_c \sin\theta - r_a \sin(\theta - \beta - \gamma), \tag{5.58}$$

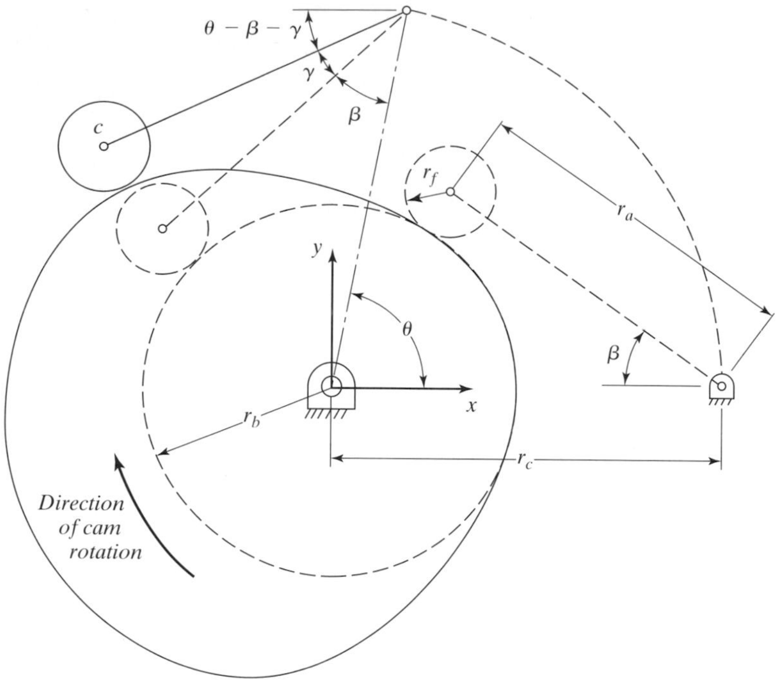

FIGURE 5.21 A disk cam with a rotating roller follower. The motion of the follower consists of rotation γ during cam rotation θ. Initial follower angle β is a function of the base-circle radius r_b, the roller radius r_f, the center distance r_c between the cam and follower pivots, and the length r_a of the follower arm.

where the initial follower angle β is determined by the law of cosines, viz.,

$$\beta = \arccos\left[\frac{r_a^2 + r_c^2 - (r_b + r_f)^2}{2r_a r_c}\right].$$

The solution parallels that for the translating roller follower given earlier, and the cam profile coordinates are

$$x = x_c \pm r_f\left(\frac{dy_c}{d\theta}\right)\left[\left(\frac{dx_c}{d\theta}\right)^2 + \left(\frac{dy_c}{d\theta}\right)^2\right]^{-1/2}$$

and

$$y = y_c \mp r_f\left(\frac{dx_c}{d\theta}\right)\left[\left(\frac{dx_c}{d\theta}\right)^2 + \left(\frac{dy_c}{d\theta}\right)^2\right]^{-1/2},$$

where x_c and y_c are given by Eqs. (5.57) and (5.58), respectively, and

$$\frac{dx_c}{d\theta} = -r_c \sin\theta + r_a\left(1 - \frac{d\gamma}{d\theta}\right)\sin(\theta - \beta - \gamma) \tag{5.59}$$

and

$$\frac{dy_c}{d\theta} = r_c \cos\theta - r_a\left(1 - \frac{d\gamma}{d\theta}\right)\cos\,(\theta - \beta - \gamma). \tag{5.60}$$

As in the previous case of a roller follower, there are two envelopes, designated by the plus and minus signs in the coordinate equations, representing inner and outer cam profiles.

Cam Curvature

Another important factor affecting cam size and performance is the cam curvature. If not limited by the pressure angle or some other consideration, the minimum size that a cam can have for a given application will be dictated by the cam curvature. As one attempts to make the base-circle radius, and therefore the cam, smaller, both the graphical and analytical approaches may show the presence of cusps in the cam profile. Obviously, such a cam will not perform satisfactorily. However, there are other less obvious situations in which the curvature can adversely affect the cam's performance, and these situations can be identified more readily by the analytical approach.

From calculus, the parametric expression for the radius of curvature of a curve confined to the xy-plane is

$$\rho = \frac{[(dx/d\theta)^2 + (dy/d\theta)^2]^{3/2}}{(dx/d\theta)(d^2y/d\theta^2) - (dy/d\theta)(d^2x/d\theta^2)}. \tag{5.61}$$

The interpretation of the sign of Eq. (5.61) is as follows: In moving along the curve in the direction corresponding to increasing values of the parameter θ, if the sign of ρ is positive, then the center of curvature is along the perpendicular to the curve on the left side, whereas if the sign of ρ is negative, then the center of curvature of the curve is to the right. A straight-line portion of a curve has an infinite radius of curvature, and a cusp has a radius of curvature equal to zero. A change of sign for ρ indicates a transition from a convex portion of the curve to a concave portion, or vice versa. (See Figure 5.22.) Equation (5.61) can be utilized in examining the cam curvature for any of the types of follower that have been considered. The sections that follow show the application of the equation to translating follower systems.

Translating Flat-Faced Follower

The radius of curvature of the cam profile in a translating flat-faced follower system is obtained by substituting Eqs. (5.35) and (5.36) into Eq. (5.61). The various derivatives are as follows: From Eq. (5.35),

$$\frac{dx}{d\theta} = -\left(r_b + s + \frac{d^2s}{d\theta^2}\right)\cos\theta$$

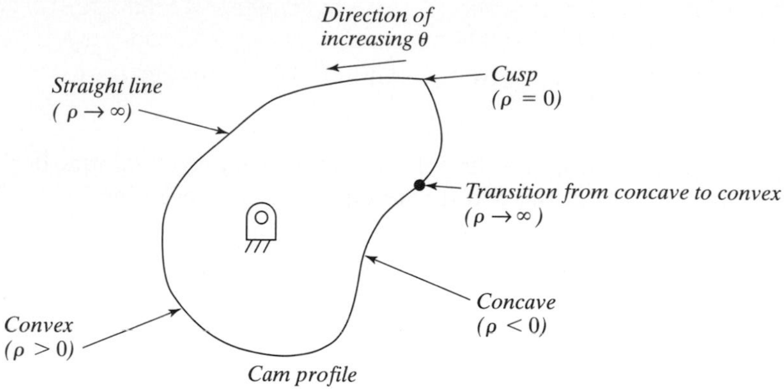

FIGURE 5.22 Various cases of the radius of curvature, ρ, for a cam profile.

and

$$\frac{d^2x}{d\theta^2} = \left(r_b + s + \frac{d^2s}{d\theta^2}\right)\sin\theta - \left(\frac{ds}{d\theta} + \frac{d^3s}{d\theta^3}\right)\cos\theta.$$

From Eq. (5.36),

$$\frac{dy}{d\theta} = -\left(r_b + s + \frac{d^2s}{d\theta^2}\right)\sin\theta$$

and

$$\frac{d^2y}{d\theta^2} = -\left(r_b + s + \frac{d^2s}{d\theta^2}\right)\cos\theta - \left(\frac{ds}{d\theta} + \frac{d^3s}{d\theta^3}\right)\sin\theta.$$

Substituting into Eq. (5.61) yields

$$\rho = r_b + s + \frac{d^2s}{d\theta^2}. \tag{5.62}$$

It can be seen from Eq. (5.62) that the radius of curvature of the cam is dependent on the base-circle radius and the prescribed follower motion. Figure 5.22 shows a cam profile depicting various cases of ρ, as given by Eq. (5.62).

Obviously, cusps are to be avoided; in other words, the condition $\rho = 0$ should not occur. Furthermore, the flat-faced follower will not operate properly on a concave portion of a cam profile. Therefore, from Eq. (5.62), the cam should be designed so that the following inequality is maintained at all points on the profile:

$$r_b + s + \frac{d^2s}{d\theta^2} > 0. \tag{5.63}$$

Given the desired follower motion s, the inequality of Eq. (5.63) can be used to determine an acceptable value for the base-circle radius. In this way, much of the trial-and-error process inherent in graphical cam layout can be avoided.

SAMPLE PROBLEM 5.11

Cam Curvature in a Flat-Faced Follower System

Determine the minimum allowable base-circle radius based on curvature for the cam design of Sample Problem 5.8. Recall that the motion requirements for the translating flat-faced follower in that problem were a dwell during 30° of cam rotation, a 2-inch rise with parabolic motion during the next 150° of rotation, a second dwell during the next 60° of rotation, and a 2-inch return with simple harmonic motion during the remaining 120° of the cam rotation cycle.

Solution. We will determine the minimum value of the quantity $Q = s + s''$ (where the prime notation refers to differentiation with respect to θ) over the entire cam cycle and then use the inequality of Eq. (5.63) to size the base circle. This will be accomplished by analyzing the four prescribed motion segments separately.

First we consider the dwell. Here, $s = s'' = 0$, and therefore, $Q = 0$ throughout this segment. A dwell period corresponds to a circular cam profile arc with a constant radius of curvature, which here is equal to the base-circle radius.

Next, we examine the parabolic rise. Note that the designation of a particular cam position as the zero-angle position ($\theta = 0$) is arbitrary. For convenience in analyzing this motion segment, we will assume that $\theta = 0$ corresponds to the beginning of the rise. For $h = 2$ in and $\alpha = 5\pi/6$ rad (150 degrees) for the first half of the rise,

$$s = \left(\frac{2h}{\alpha^2}\right)\theta^2 = \left[\frac{2(2)}{(5\pi/6)^2}\right]\theta^2 = 0.584\theta^2,$$

$$s' = 1.17\theta,$$

$$s'' = 1.17,$$

and

$$Q = 0.584\theta^2 + 1.17.$$

Obviously, Q will be greater than zero within the first half of the parabolic rise, and we need not pursue this case any further.

For the second half of the rise, we have

$$s = h\left[-1 + \left(\frac{4}{\alpha}\right)\theta - \left(\frac{2}{\alpha^2}\right)\theta^2\right] = -2 + 3.06\theta - 0.584\theta^2,$$

$$s' = 3.06 - 1.17\theta,$$

$$s'' = -1.17,$$

and

$$Q = -3.17 + 3.06\theta - 0.584\theta^2.$$

Examining the associated range of θ ($\theta = 1.31$ to $\theta = 2.62$), we see that the minimum value of Q occurs at $\theta = 1.31$ and is

$$Q = -3.17 + 3.06(1.31) - 0.584(1.31)^2 = -0.164.$$

Now we look at the second dwell. As with the first dwell, Q is constant within this range, with a value of $Q = s + s'' = 2 + 0 = 2$.

Finally, we consider the simple harmonic return. Once again, for convenience, we will assume that a cam angle of $\theta = 0$ corresponds to the beginning of the return motion. The follower position is $s = h$ at $\theta = 0$ and $s = 0$ at $\theta = \alpha$.

Thus,

$$s = h - \left[\frac{h}{2} - \left(\frac{h}{2}\right)\cos\left(\frac{\pi}{\alpha}\theta\right)\right] = \frac{h}{2} + \left(\frac{h}{2}\right)\cos\left(\frac{\pi}{\alpha}\theta\right).$$

Substituting values ($h = 2$ and $\alpha = 2\pi/3$, or $120°$) and differentiating, we have

$$s = 1 + \cos 1.5\theta,$$
$$s' = -1.5\sin 1.5\theta,$$
$$s'' = -2.25\cos 1.5\theta,$$

and

$$Q = s + s'' = 1 - 1.25\cos 1.5\theta.$$

By inspection, the minimum value of Q for this segment occurs at $\theta = 0$ and is

$$Q = 1 - 1.25 = -0.25.$$

Therefore, the minimum value of Q over the entire motion is (-0.25), and from the inequality of Eq. (5.63), we obtain

$$r_b + (-0.25) > 0,$$

which indicates that the minimum limit for the base-circle radius is $r_b = 0.25$ in.

Translating Roller Follower

Generally, the curvature of the pitch curve is analyzed in considering roller followers. The equation for the radius of curvature of the pitch curve is easier to obtain than that of the cam profile, and, as will be seen, the necessary design conditions can be expressed in terms of the pitch curve. The radius of curvature of the pitch curve of a translating roller follower can be determined from Eqs. (5.40), (5.41), and (5.61).

Consider the special case of a radial follower whose offset $e = 0$. Then, from Eq. (5.42), $\beta = 0$, and the pitch curve coordinates, Eqs. (5.40) and (5.41), are

$$x_c = -(r_b + r_f + s)\sin\theta$$

and

$$y_c = (r_b + r_f + s)\cos\theta.$$

The derivatives are

$$\frac{dx_c}{d\theta} = -(r_b + r_f + s)\cos\theta - \frac{ds}{d\theta}\sin\theta,$$

$$\frac{dy_c}{d\theta} = -(r_b + r_f + s)\sin\theta + \frac{ds}{d\theta}\cos\theta,$$

$$\frac{d^2x_c}{d\theta^2} = \left(r_b + r_f + s - \frac{d^2s}{d\theta^2}\right)\sin\theta - 2\frac{ds}{d\theta}\cos\theta,$$

and

$$\frac{d^2y_c}{d\theta^2} = -\left(r_b + r_f + s - \frac{d^2s}{d\theta^2}\right)\cos\theta - 2\frac{ds}{d\theta}\sin\theta.$$

Substituting into Eq. (5.61), we have the radius of curvature for a radial roller follower:

$$\rho_p = \frac{[(r_b + r_f + s)^2 + (ds/d\theta)^2]^{3/2}}{(r_b + r_f + s)^2 + 2(ds/d\theta)^2 - (r_b + r_f + s)^2(d^2s/d\theta^2)}. \tag{5.64}$$

Here, ρ_p is the radius of curvature of the pitch curve, where a positive value refers to a convex portion of the curve and a negative value refers to a concave portion, based on the previous definition of parameter θ.

Roller followers may move along concave cam profiles as well as convex cam profiles. The two cases are depicted in Figure 5.23. For the convex case, the absolute value of the radius of curvature, ρ, of the cam profile is

$$|\rho| = \rho_P - r_f.$$

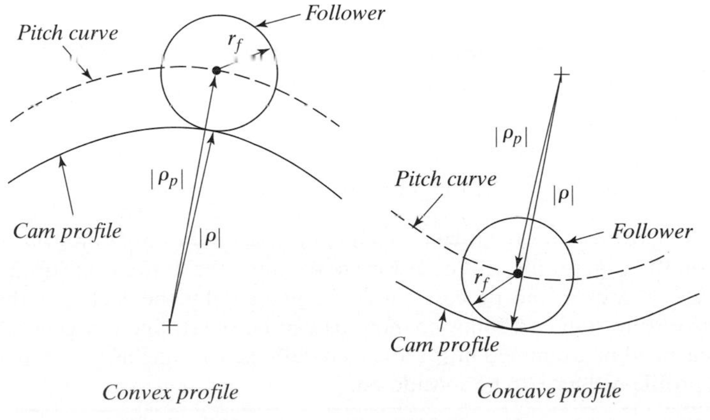

FIGURE 5.23 A roller follower can operate on both convex and concave portions of cam profiles.

The cam will become pointed (i.e., there will be a cusp) when

$$|\rho| = \rho_p - r_f = 0.$$

SAMPLE PROBLEM 5.12

Cam Curvature in a Roller Follower System

Evaluate the curvature of the disk cam with a radial roller follower of Sample Problem 5.10 for the position corresponding to a cam angle of $\theta = 60°$.

Solution. Recall from Sample Problems 5.9 and 5.10 that $r_b = 40$ mm, $r_f = 10$ mm, and the follower displacement function is

$$s = \left(\frac{60}{\pi}\right)\theta - \left(\frac{15}{\pi}\right)\sin 4\theta.$$

Differentiating this function twice with respect to angle θ, we have

$$\frac{ds}{d\theta} = \left(\frac{60}{\pi}\right) - \left(\frac{60}{\pi}\right)\cos 4\theta$$

and

$$\frac{d^2s}{d\theta^2} = \left(\frac{240}{\pi}\right)\sin 4\theta.$$

Evaluating these functions at $\theta = \pi/3$ rad yields

$$s = 24.13, \quad \frac{ds}{d\theta} = 28.65, \text{ and } \frac{d^2s}{d\theta^2} = -66.16.$$

Summing terms produces

$$(r_b + r_f + s) = (40 + 10 + 24.13) = 74.13.$$

Substituting into Eq. (5.64), we see that the radius of curvature of the pitch curve at this point is

$$\rho_p = \frac{[(74.13)^2 + (28.65)^2]}{(74.13)^2 + 2(28.65)^2 - (74.13)(-66.16)} = 41.69 \text{ mm}.$$

Thus, at the location in question, the cam profile is convex. Of course, since the curvature will, in general, vary with position on the cam profile, it is necessary to examine the entire cam surface for adverse curvature effects. However, the process can be expedited by the fact that the extreme curvature positions during standard follower motions can be determined in general, and then only these positions need be examined in specific cam designs. This means that only a few points on the total cam profile will have to be considered.

5.8 POSITIVE-MOTION CAMS

The various types of cam discussed up to this point depend on the force of gravity or a spring force to maintain contact with the follower during the return stroke. In many applications, it is necessary for the cam to exert positive control over the follower during the return as well as during the rise. In this section, some of the more common types of positive-motion cams are discussed.

Face Cam

One method of achieving positive motion is to cut a groove into the face of the cam. The roller follower then rides in the groove. During the rise, the inner surface of the groove (the side of the groove nearest the cam axis) causes the follower to move up, while on the return stroke, the outer surface of the groove forces the follower down. This type of cam is known as a *face cam*.

Constant-Breadth Cams

A constant-breadth cam-and-follower system is designed so that the cam surface is always in contact with two follower surfaces. That is, the follower "boxes in" the cam. Figure 5.24 shows a lawn sprinkler that employs a constant-breadth cam. A water turbine and reduction gears drive the cam. The follower consists of a slotted link with two projections contacting the cam. Note that the cam follower rotates and translates. A guide at the cam rotation axis maintains the relationship between the slotted link and cam. Finally, the slotted link drives the output crank, causing the sprinkler bar to oscillate.

The manual adjustment knob effectively changes output crank length, so that the user has the option of watering a large or small section of lawn, or watering the lawn on only one side of the sprinkler. Can you show that the linkage has one degree of freedom (once the manual adjustment is set)? Compare this design with a sprinkler based on a

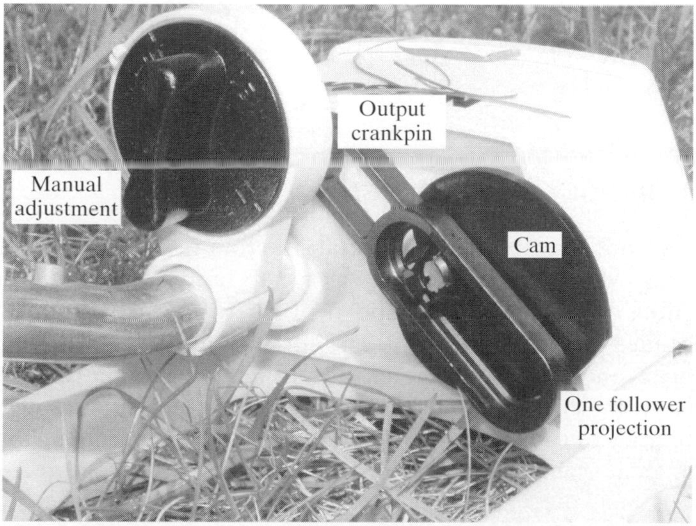

FIGURE 5.24 A lawn sprinkler driven by a constant-breadth cam.

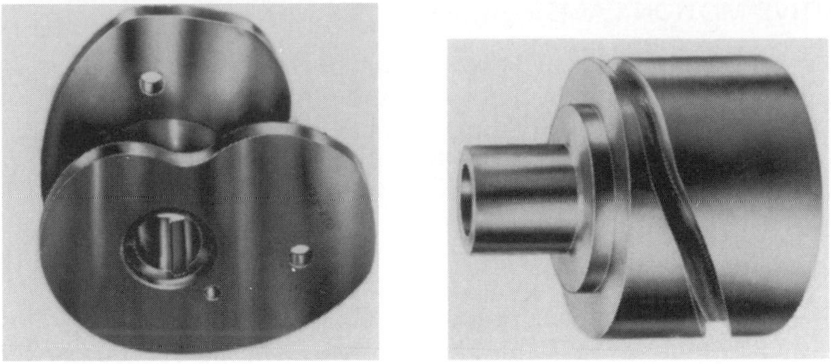

(a) (b)

FIGURE 5.25 (a) A conjugate disk cam. (b) A cylindrical barrel cam. (*Source:* Commercial Cam Co.)

four-bar linkage (Figure 1.42). Jensen (1987) discusses design techniques for constant-breadth cams (also called constant-diameter cams).

Conjugate Cams

A conjugate-cam system uses two cams (on the same camshaft) and two rigidly connected followers to ensure positive motion. (See Figure 5.25a.) Reeve (1995) illustrates design options for conjugate cams, including methods for reducing backlash. (For a definition of backlash, see Section 5.9.)

Cylindrical Cams

A cylindrical, or barrel, cam (see Figure 5.25b) is used to drive a translating follower, which moves parallel to the axis of the cam. A groove cut into the side of the cylinder provides the path for the follower. Cylindrical cams are positive-motion cams, except for the type known as end cams.

5.9 PRACTICAL CONSIDERATIONS IN CAM DESIGN

Practical considerations usually are the primary factors in the decision a designer must make with regard to the type of cam system to be used to solve a particular problem. Limitations of space and the speed of operation are perhaps the most common factors that govern the decision. The force required to keep the follower and cam in contact is another important consideration. The system required may be a gravity type or a spring type. Another important factor the designer must take into account is the accuracy of the machining operations used in manufacturing the cam. Machining errors may cause kinematic variations from the required operating conditions.

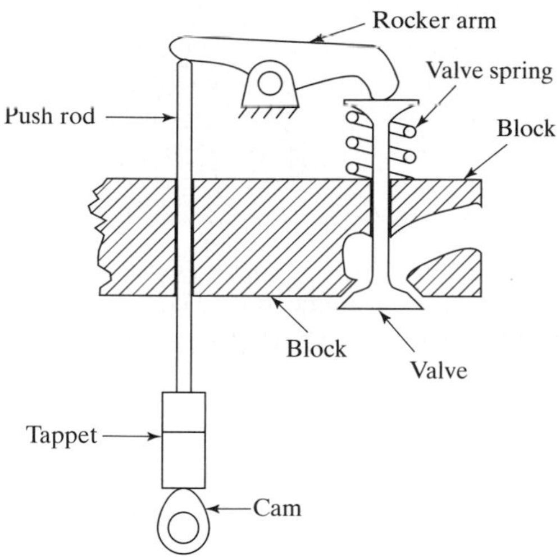

FIGURE 5.26 Valve train schematic (not to scale).

With the need for high-speed cams, the effect of vibrations on a cam's performance has also become more important. An investigation into the vibration of a cam requires information concerning the elasticity of the system, and determining the elastic response of a cam system is an extremely difficult task. It is very often desirable to construct a prototype of the cam system. As a result of tests on the prototype, modifications can then be made to ensure a satisfactory solution to the given problem.

Tolerances, Wear, Temperature Effects, and Backlash

Cam systems, like books, professors and students are not error-free. We cannot specify a dimension as "20 ± 0 mm". A nonzero tolerance must be specified or implied. Add wear and thermal expansion or contraction, and the result is backlash. In cam design, the term *backlash* or *lash* is used to refer to accumulated tolerance, wear and temperature effects, or the shock loading due to these effects.

Backlash is of particular importance in automotive valve trains that are subject to large temperature changes and operate at high speed. Figures 5.1e and 5.2a show parts of reciprocating engines, including valve trains. Figure 5.26 is a schematic of a valve train with the valve open. If we try to adjust the system so that the valve closes exactly at end of the return part of the cam cycle, backlash may result in:

- the valve staying partly open
- or shock loading because the valve closes before the end of the return.

SAMPLE PROBLEM 5.13

Backlash in a Valve Train

A 3–4–5 polynomial cam is designed for a 20 mm rise, and rotates at 1800 rpm. The design rise and return intervals are each 105° and there is a 25° top dwell.

The cam-valve train system is intended to operate a valve with approximately the same motion as the tappet (follower). As a result of tolerances, wear, and temperature changes, there is an error equivalent to 10% of the rise, causing the valve to close early. What effect will this have?

Solution. A 3–4–5 cam system was analyzed previously. We can use the same basic equations, but subtract the backlash from the position equation. Approximate results are shown in Figure 5.27. The valve is at rest until the backlash is taken up. We assume that the valve velocity is zero up to that instant, using the IF-statement

$$V_v = \text{if}\,[S(\theta) \geq 0, V(\theta), 0],$$

which gives the design velocity for a displacement greater than or equal to zero and zero otherwise. There is a jump in velocity that appears to occur in zero time. But, of course, acceleration cannot be infinite. The elasticity of the system moderates the acceleration a bit, but loads are still severe.

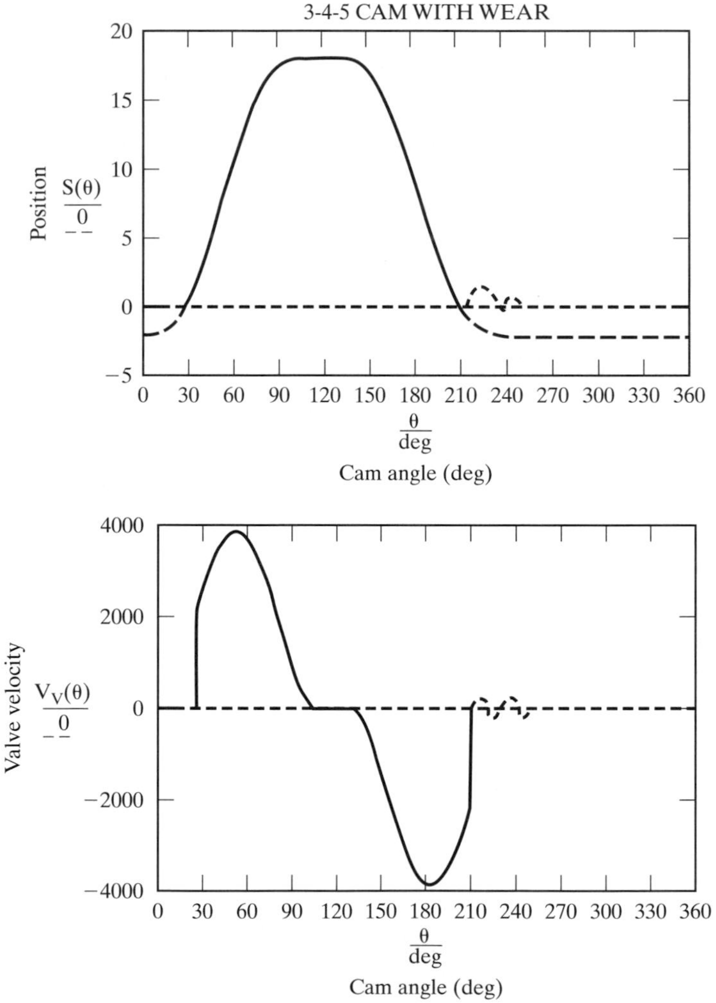

FIGURE 5.27 The effect of backlash in a valve train.

Deceleration is severe when the valve strikes the valve seat, again causing severe shock loading. The dots at the end of the return are a "guesstimate" of the valve bounce due to elastic collisions. Shock loading will cause excessive wear and premature failure. Valve bounce means that the valve has not closed properly.

5.10 REDUCING SHOCK LOADING

Theoretically, the follower velocity at the beginning of the rise and at the end of the return is zero for good high-speed cams (including the 3–4–5 polynomial cam). However, actual valve train backlash may cause a valve to seat at high velocity.

In the previous sample problem, the backlash was excessive. Our first task would be to improve the valve adjustment. Modifying the cam profile to produce lower velocities near the beginning of the rise and near the end of the return is another possibility. Reeve (1995) describes a seventh-order polynomial cam law designed for low-impact velocity in high-speed mechanisms in which backlash cannot be avoided. Be sure to check any proposed equations for your specific application and estimate of backlash.

Hydraulic Tappets to Reduce Backlash

Manual valve adjustment is difficult and expensive in automotive engines. Hydraulic tappets are designed to compensate for changes in valve train length due to temperature and wear. The hydraulic tappet is a cam follower that consists of a cylinder, a plunger, an oil chamber, a check valve, and a spring. When the valve is off its seat, the load is carried by a column of oil, which acts, for a moment, as a rigid link. The tappet is designed with a predetermined slight oil leakage to permit the valve to seat, even if the length of the valve train increases. If the length decreases, the plunger spring keeps the parts in contact, increasing the volume under the plunger. Oil is then fed through the check valve, maintaining the correct valve train length.

SUMMARY

There are hundreds of varieties of cam-and-follower systems. Some of the most common types of follower used with disk cams are translating roller followers, translating flat-face followers, rotating roller followers, and rotating flat-face followers.

If we specify the desired follower position as a function of the cam angle, then we can generate a prototype cam using a numerically controlled (NC) or computer numerically controlled (CNC) machine. There are limitations, of course. If the cam is to operate at high speed, smooth operation is critical. In particular, follower velocity and acceleration discontinuities should be avoided. We even consider jerk, the derivative of acceleration with respect to time.

Cams that produce cycloidal motion or higher order polynomial motion in the follower are generally satisfactory for high-speed operation. Harmonic (sinusoidal) motion is used for high-speed operation only if there are no dwells.

A cam profile may be constructed graphically or may be constructed analytically using the theory of envelopes. A profile so constructed is not used to generate a cam, but is useful for checking for undercuts and checking the pressure angle.

Since we are interested in the follower position, velocity, acceleration, and jerk, plots of motion characteristics are important. The Heaviside step function and/or an interval function can be used in describing the follower rise, return, and dwell(s).

Cams can be generated with very close tolerances. Nevertheless, wear, temperature effects, and tolerances in other parts can result in shock loading. Hydraulic tappets are used in valve trains to compensate for dimensional changes that cause shock loading.

A Few Review Items

- Make a rough sketch of the position, velocity, and acceleration vs. time characteristics for a reciprocating flat-face follower driven by an eccentric circular cam. Are the curves smooth?
- Make a rough sketch of the position, velocity, and acceleration vs. cam angle characteristics for a rise–dwell–return–dwell (RDRD) cam-and-follower system. Select sinusoidal follower motion. The rise and return intervals each correspond to 1.5 radians of cam rotation, and there is a top dwell corresponding to 0.5 radian of cam rotation. Does the acceleration plot suggest potential problems for high-speed operation?
- Can selecting a different form of motion eliminate the problems identified in the previous item? Write the rise equations for two satisfactory forms of motion.
- Suppose we eliminate the top dwell. Write a third equation, which is useful for a rise–return–dwell (RRD) cam.
- Make a rough sketch of a cam and radial reciprocating roller follower. The camshaft rotates clockwise. Orient the cam so that the pressure angle appears to be a maximum. Can you offset the follower to improve the pressure angle? Suppose the direction of camshaft rotation is sometimes clockwise and sometimes counterclockwise. Would offsetting the follower still be a good idea?
- Write an interval function that can be used to turn another function on at $\theta = \theta_a$ and off at $\theta = \theta_b$. Use the programming language or software form that you prefer.
- Identify methods for reducing backlash or mitigating its effects in valve trains or other cam-and-follower systems.

PROBLEMS

Some of the problems that follow are based on follower motion that is unsatisfactory for high-speed cams. This will be apparent if you calculate the follower acceleration and jerk.

5.1 A follower rises 50 mm in 120° with constant velocity, dwells for 60°, returns in 120° with constant velocity, and dwells for 60°. Draw the follower displacement diagram.

5.2 A follower rises 50 mm in 120° with simple harmonic motion, dwells for 60°, returns in 120° with simple harmonic motion, and dwells for 60°. Draw the follower displacement diagram.

5.3 A follower rises 50 mm in 120° with cycloidal motion, dwells for 60°, returns in 120° with cycloidal motion, and dwells for 60°. Write a computer program to calculate and plot the displacement diagram.

5.4 A follower rises 50 mm in 120° with parabolic motion, dwells for 60°, returns in 120° with parabolic motion, and dwells for 60°. Write a computer program to calculate and plot the displacement diagram.

5.5 A follower rises 2 in in 120° with constant velocity, returns 1 in in 90° with constant velocity, dwells for 60°, and returns 1 in in 90° with constant velocity. Draw the displacement diagram.

5.6 A follower rises 1 in in 60° with simple harmonic motion, dwells for 30°, rises another 1 in in 60° with simple harmonic motion, dwells for 30°, returns in 150° with simple harmonic motion, and dwells for the remaining 30°. Draw the displacement diagram.

5.7 A follower rises 2 in in 210° and returns in 150°, both with cycloidal motion. Write a computer program to calculate and plot the displacement diagram.

5.8 A follower rises $\frac{1}{2}$ in in 30° with constant acceleration, rises 1 in in 30° with constant velocity, rises $\frac{1}{2}$ in in 30° with constant deceleration, dwells for 30°, returns 2 in in 180° with parabolic motion, and dwells for 60°. Write a computer program to calculate and plot the displacement diagram.

5.9 A follower rises 2 in in 90° with simple harmonic motion, dwells for 45°, returns in 180° with parabolic motion, and dwells for 45°. Write a computer program to calculate and plot the displacement diagram.

5.10 A follower rises 2 in in 150° with constant velocity, dwells for 30°, returns in 150° with cycloidal motion, and dwells for 30°. Write a computer program to calculate and plot the displacement diagram.

5.11 A follower rises 40 mm in 150° with simple harmonic motion, dwells for 90°, and returns 40 mm in 120° with cycloidal motion. Write a computer program to calculate and plot the displacement diagram.

5.12 A follower rises 40 mm in 120° with parabolic motion, returns 20 mm in 90° with simple harmonic motion, dwells for 30°, returns the final 20 mm in 90° with cycloidal motion, and dwells for 30°. Write a computer program to calculate and plot the displacement diagram.

For Problems 5.13 through 5.16

Graphically lay out the profile of a disk cam for clockwise rotation of the cam. The base-circle diameter of the cam is to be 100 mm. For those problems involving roller followers, the roller diameter is to be 25 mm.

5.13 Lay out the cam described in Problem 5.1. Use a translating radial roller follower.

5.14 Lay out the cam described in Problem 5.2. Use a translating radial roller follower. Determine the maximum pressure angle.

5.15 Lay out the cam described in Problem 5.2. Use a translating flat-faced follower.

5.16 Lay out the cam described in Problem 5.2. Use a translating roller follower with an offset of 25 mm to the right of the center of the camshaft. Determine the maximum pressure angle.

5.17 /rite a computer program to generate the disk cam for the motion described in Problem 5.3. Use a translating radial roller follower. The cam rotation is to be clockwise. The base-circle diameter is 100 mm, and the roller diameter is 25 mm. Determine the maximum pressure angle.

5.18 Write a computer program to generate the disk cam for the motion described in Problem 5.4. Use a translating radial roller follower. The cam rotation is to be clockwise. The base-circle diameter is 100 mm, and the roller diameter is 25 mm. Determine the maximum pressure angle.

For Problems 5.19 through 5.23

Write a computer program to generate the profile of a disk cam for clockwise rotation of the cam. The base-circle diameter of the cam is to be 4 in. For those problems involving roller followers, the roller diameter is to be 1 in.

5.19 Lay out the cam described in Problem 5.5. Use a translating radial roller follower.

5.20 Lay out the cam described in Problem 5.6. Use a translating roller follower with an offset of 1 in to the right of the center of the camshaft. Determine the maximum pressure angle.

5.21 Lay out the cam described in Problem 5.7. Use a translating flat-faced follower.

5.22 Lay out the cam described in Problem 5.8. Use a translating flat-faced follower with an offset of 1 in to the right of the center of the camshaft.

5.23 Lay out the cam described in Problem 5.9. Use a translating flat-faced follower.

5.24 Design a cam for an oscillating, pivoted flat-faced follower (like that shown in Figure P5.1) to provide the following sequence of motion: the follower rotates clockwise for 15° with simple harmonic motion in 150° of cam rotation, dwells for 30°, returns with cycloidal motion in 150° of cam rotation, and dwells for 30°.

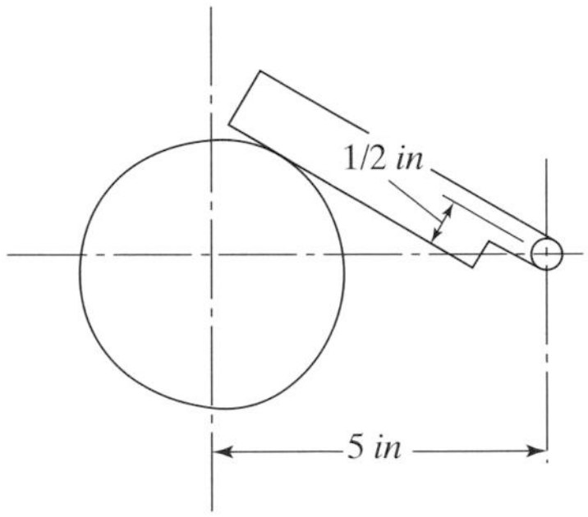

Base circle diam. = 4 in FIGURE P5.1

5.25 Design a cam for a pivoted roller follower (like that shown in Figure P5.2) to provide the following sequence of motion: The follower rotates clockwise 20° with simple harmonic motion in 150° of cam rotation, dwells for 90°, and returns 20° with parabolic motion in 120° of cam rotation.

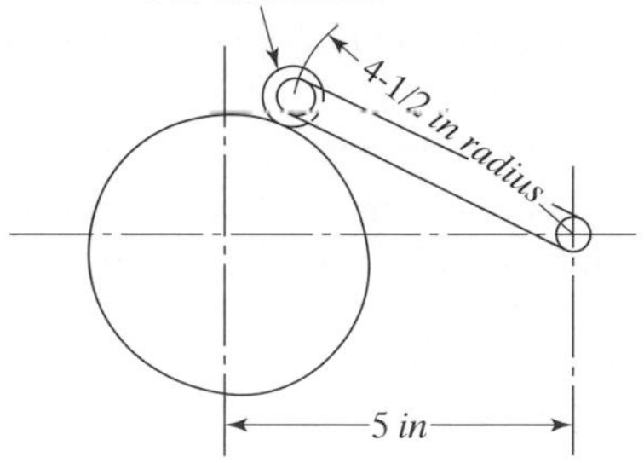

FIGURE P5.2

For Problems 5.26 through 5.31

A follower rises 4 *in in* 180° *of cam rotation and then returns during the next* 180° *of cam rotation. The cam rotates at* 60 *rev/min. For the given follower motions, determine: (a) mathematical expressions for the displacement, velocity, acceleration, and jerk of the follower; (b) the magnitudes and locations of maximum velocity and maximum acceleration of the follower; (c) the follower displacement, velocity, acceleration, and jerk when the cam angle is* 120°*; and (d) the follower displacement, velocity, acceleration, and jerk when the follower displacement is 3 in during the rise segment of the motion.*

5.26 Uniform motion.

5.27 Modified uniform motion (with constant velocity from 45° to 135° and from 225° to 315°).

5.28 Simple harmonic motion.

5.29 Cycloidal motion.

5.30 Parabolic motion.

5.31 3–4–5 polynomial motion.

For Problems 5.32 through 5.36

A follower rises 50 *mm in* 120° *of cam rotation, dwells for* 60°*, returns in* 120°*, and dwells for* 60°*. The cam rotational speed is* 60 *rev/min. For the given follower motions, determine: (a) mathematical expressions for the displacement, velocity, acceleration, and jerk of the follower and (b) the magnitudes and locations of maximum velocity and maximum acceleration of the follower.*

5.32 Uniform motion.

5.33 Simple harmonic motion.

5.34 Cycloidal motion.

5.35 Parabolic motion.

5.36 3–4–5 polynomial motion.

5.37 A 3–4–5 polynomial cam imparts the following motion to a follower: a rise of 22 mm in 80° of cam rotation, a dwell for 70° of cam rotation, and a return of 22 mm in 80° of cam rotation. Find the maximum follower velocity. The cam speed is 3200 rev/min.

5.38 A cam follower rises 2 inches in 180° of cam rotation. The constant acceleration for the first part of the rise is three times as great as the constant deceleration for the second part of the lift period. If the cam is rotating at 300 rev/min, determine the value of the acceleration.

5.39 A disk cam rotates at 375 rev/min. The follower rises $\frac{3}{4}$ in with constant acceleration in 80° of cam rotation and then rises an additional $\frac{3}{4}$ in with constant deceleration in the next 80° of cam rotation. Find the acceleration and maximum velocity of the follower.

5.40 The follower of a disk cam rises $\frac{1}{2}$ in with constant acceleration, rises an additional $\frac{3}{4}$ in with constant velocity, and then rises an additional $\frac{1}{2}$ in with constant deceleration. The cam rotates at 200 rev/min, and the follower has a maximum velocity of 35 in/s. Calculate the acceleration and deceleration of the follower. How many degrees has the cam rotated while this motion is being performed?

5.41 A disk cam is to give its follower a rise through a travel distance h during a cam rotation α. This motion is preceded by a dwell and is followed by a dwell. Derive the displacement, velocity, acceleration, and jerk expressions that will satisfy the boundary conditions for all four of these characteristics at the beginning and end of the travel distance.

For Problems 5.42 through 5.46

Write a computer program to calculate the displacement, velocity, acceleration, and jerk of a follower for the types of motion given. During the motion segment, the cam rotates from angle ANG1 to angle ANG2, and the follower moves from position S 1 to position S2 = S1 + H. Note that the travel distance H may be positive for a rise motion, negative for a return motion, and zero for a dwell. Also, the follower motion may be either translation or rotation. Check out the program for the case of a follower that is to have a translational displacement from 1 *to* 3 *in during a cam rotation from* 30° *to* 150°.

5.42 Uniform motion.

5.43 Simple harmonic motion.

5.44 Cycloidal motion.

5.45 Parabolic motion.

5.46 3–4–5 polynomial motion.

5.47 A family of circles has centers along a straight line passing through the origin of an xy-coordinate system at a clockwise angle of 45° from the x-axis. The radius of each circle is equal to one-half of the distance from the origin to the center of the circle. Determine the envelope for this family of curves

(a) as a function relating x and y and

(b) in parametric form, with x and y as functions of a parameter λ.

5.48 A translating flat-face follower is to move through a distance h with cycloidal motion during 180° of clockwise cam rotation. Determine expressions for the x- and y-coordinates of that portion of a cam profile that will produce this motion. For a travel of 50 mm and a base-circle radius of 100 mm, calculate and plot the cam profile. Based on this portion of the cam design, what is the minimum allowable width of the follower face?

5.49 Repeat Problem 5.48 for uniform motion.

5.50 Repeat Problem 5.48 for parabolic motion.

5.51 Design a disk cam to produce the following motion of a translating radial roller follower: a rise through a distance of 50 mm with simple harmonic motion during 180° of rotation, followed by a return, also with simple harmonic motion, during the remaining 180° of cam rotation. The base-circle radius is 100 mm and the roller radius is 25 mm. Determine the pressure angle corresponding to cam angles of 30°, 60°, and 90°.

5.52 Determine an expression for the curvature of the cam profile segment synthesized in Problem **5.48**. Find the minimum and maximum curvature values.

5.53 Determine an expression for the curvature of the cam profile segment synthesized in Problem **5.49**. Find the minimum and maximum curvature values.

5.54 Determine an expression for the curvature of the cam profile segment synthesized in Problem **5.50**. Find the minimum and maximum curvature values.

5.55 Determine an expression for the curvature of the cam profile segment synthesized in Problem **5.51**. Find the minimum and maximum curvature values.

5.56 Suppose we need a cam with an 18-mm rise. The rise and return intervals are both 105°, and the top dwell is 25°. The cam rotation speed will be 1000 rpm. Plot the follower position, velocity, acceleration, and jerk for the full range of motion. Check your results. Select cycloidal motion in your design.

5.57 A cam is needed to produce a 22-mm follower rise. The cam angular velocity will be 80 rad/s. The rise and return intervals are both 100°, and the top dwell is 20°. Plot the follower position, velocity, acceleration, and jerk for the full range of motion. Check your results. Select cycloidal motion in your design.

5.58 Suppose we need a cam with a 10-mm rise. The rise and return intervals are both 120°, and the top dwell is 45°. The cam rotation speed will be 400 rpm. Plot the follower position, velocity, acceleration, and jerk for the full range of motion. Check your results. Select cycloidal motion in your design.

5.59 Design a cam for a 12.5-mm rise. The rise and return intervals are both 125°, and the top dwell is 15°. The cam rotation speed will be 440 rpm. Find the maximum acceleration. Plot the follower position, velocity, acceleration, and jerk for the full range of motion. Are the boundary conditions satisfied? Check your results. Use 3–4–5 polynomial motion in your design.

5.60 Design a cam for a 20-mm rise. The rise and return intervals are both 95°, and the top dwell is 40°. The cam angular velocity will be 90 rad/s. Find the maximum acceleration. Plot the follower position, velocity, acceleration, and jerk for the full range of motion. Are the boundary conditions satisfied? Check your results. Use 3–4–5 polynomial motion in your design.

5.61 Design a cam for a 15-mm rise. The rise and return intervals are both 100°, and the top dwell is 45°. The cam rotation speed will be 1800 rpm. Find the maximum acceleration. Plot the follower position, velocity, acceleration, and jerk for the full range of motion. Are the boundary conditions satisfied? Check your results. Use 3–4–5 polynomial motion in your design.

5.62 Suppose we need a cam for a rise–return–dwell (RRD) application. The position, velocity, and acceleration should be zero at the beginning of the rise h, and the velocity should be zero at the end of the rise. A cam described by the follower displacement equation

$$s = h[4x^3 - 3x^4]$$

is proposed. Plot the position, velocity, and acceleration to see whether the boundary conditions are satisfied. Compare this 3–4 polynomial cam with a 3–4–5 polynomial cam. Normalize the results by using $h = \alpha = \omega = 1$.

5.63 Suppose we need a cam to meet the following follower-motion requirements: beginning of rise and end of return: $s = v = a = j = 0$; end of rise: $s = h$; $v = j = 0$ (there is no top dwell; the acceleration is unspecified at end of the rise). Show that an eighth-order polynomial cam can satisfy these requirements. Find the arbitrary constants. Plot the follower motion for a 10-mm rise. The cam angular velocity is 100 rad/s, and the rise and return intervals are each 125°. The follower is to dwell for the remainder of the cycle.

5.64 Suppose we need a cam to meet the following follower-motion requirements: beginning of rise and end of return: $s = v = a = j = 0$; end of rise: $s = h$; $v = j = 0$ (there is no top dwell; the acceleration is unspecified at end of the rise). Show that an eighth-order polynomial cam can satisfy these requirements. Find the arbitrary constants. Plot the follower motion for a 20-mm rise. The cam speed is 880 rpm, and the rise and return intervals are each 105°. The follower is to dwell for the remainder of the cycle.

5.65 Suppose we need a cam to meet the following follower-motion requirements: beginning of rise and end of return: $s = v = a = j = 0$; end of rise: $s = h$; $v = j = 0$ (there is no top dwell; the acceleration is unspecified at end of the rise). Show that an eighth-order polynomial cam can satisfy these requirements. Find the arbitrary constants. Plot the follower motion for an 18 mm rise. The cam speed is 400 rpm, and the rise and return intervals are each 95°. The follower is to dwell for the remainder of the cycle.

5.66 A 3–4–5 polynomial cam is designed for a 14.5-mm rise and rotates at 1000 rpm. The design rise and return intervals are each 85°, and there is a 20° top dwell. The cam is intended to operate a valve, with essentially the same motion as the tappet (the cam follower). As a result of tolerances, wear, and temperature changes, there is an error equivalent to 10% of the rise, causing the valve to close early. What effect will this error have? Show probable position and velocity plots.

5.67 A 3–4–5 polynomial cam is designed to rotate at 2000 rpm and provide an 18-mm rise. The design rise and return intervals are each 120°, and there is a 45° top dwell. The cam is intended to operate a valve, with essentially the same motion as the tappet (the cam follower). As a result of tolerances, wear, and temperature changes, there is an error equivalent to 10% of the rise, causing the valve to close early. What effect will this error have? Show probable position and velocity plots.

BIBLIOGRAPHY AND REFERENCES

Chen, F. Y., *Mechanics and Design of Cam Mechanisms*, Pergamon, New York, 1982.

Jensen, Preben W. *Cam Design and Manufacture*, Marcel Dekker, New York, 2d ed., 1987.

Rothbert, H. A., *Cams*, Wiley, New York, 1956.

Reeve, John, *Cams for Industry*, Mechanical Engineering Publications, London, 1995.

Stone, Richard, *Introduction to Internal Combustion Engines*, Society of Automotive Engineers, Warrendale, PA, 2d ed 1993.

Tesar, D., and G. K. Matthew, *The Dynamic Synthesis, Analysis, and Design of Modeled Cam Systems*, Lexington Books, Lexington, MA, 1976.

Note also the Internet references listed under the headings of **Cams and part handlers**, and **Automotive drive components** found at the end of Chapter 1. You will find many more commerical sites related to cams by searching the Internet yourself.

CHAPTER 6

Spur Gears: Design and Analysis

Spur gears are used to transmit power between parallel shafts and to change rotation speeds. Speed ratios are precise, and gear systems can be designed for high power; however, shafts that carry gears must be located precisely. When precise location cannot be maintained, V-belt or chain drives are sometimes substituted for gear drives.

Concepts and Definitions You Will Learn and Apply When Studying This Chapter

- Speed ratios of gear sets
- Pitch circle and circular pitch
- Addendum and dedendum; standard and stub teeth
- Module and diametral pitch
- Pressure angle and line of action
- Backlash
- Base circle and properties of an involute
- Design of gear sets for adequate contact ratio
- Design and selection of gears to avoid interference; the relationship between pressure angle, minimum number of teeth, and interference
- Forces on gear teeth

If you plan to design or select gears for high loads and critical applications, you will need additional study and references beyond the scope of this text. Authoritative references related to gear inspection, gear design for wear and bending stress, and gear failure modes are available from the American Gear Manufacturers Association (1980, 1982, 1988).

6.1 BASIC CONSIDERATIONS

This chapter is devoted to a discussion of the power-transmitting machine element known as the *spur gear*. The terminology, kinematics, and force analysis of spur gears will be presented.

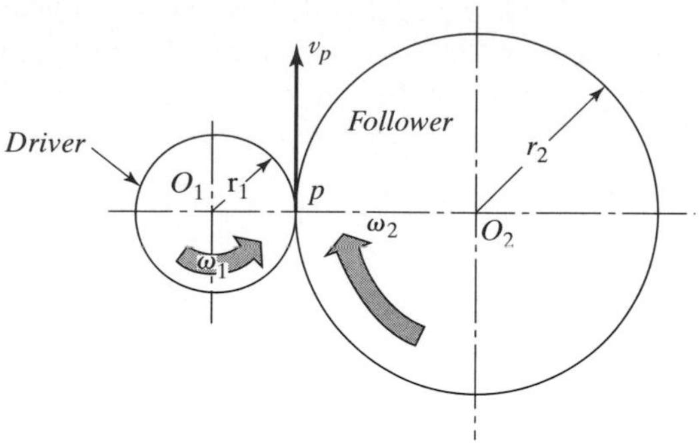

FIGURE 6.1 Two friction wheels (external cylinders).

A variety of machine elements can transmit power from one shaft to another. A pair of friction wheels (rolling cylinders) is shown in Figure 6.1, where wheel 1 is the driver and wheel 2 is the follower.

The force that cylinder 1 can transmit to cylinder 2 depends on the friction that can be developed between the two cylinders. Assuming that the frictional resistance between the two wheels is sufficiently large to prevent slipping of one cylinder relative to the other, the following kinematic relationship holds:

$$v_P = r_1\omega_1 = r_2\omega_2,$$

or

$$\frac{\omega_1}{\omega_2} = \frac{r_2}{r_1}. \tag{6.1}$$

Here, v_p is the instantaneous velocity of the point of contact, ω is the angular velocity, and r is the radius of the cylinder.

Equation (6.1) indicates that the ratio of the angular speeds of the cylinders is inversely proportional to the ratio of their radii. Another important fact to be observed, this time from Figure 6.1, is that the rotations of the cylinders are in opposite directions. (Wheel 1 rotates counterclockwise, while wheel 2 rotates clockwise.)

The cylinders shown in Figure 6.1 are external to each other. Figure 6.2 shows a similar situation, except that one of the cylinders is internal to the other. The only difference between the external and internal cylinder pairs is that the direction of rotation for both cylinders of the internal pair is the same. Thus, in Figure 6.2, cylinder 1 and cylinder 2 are both rotating counterclockwise.

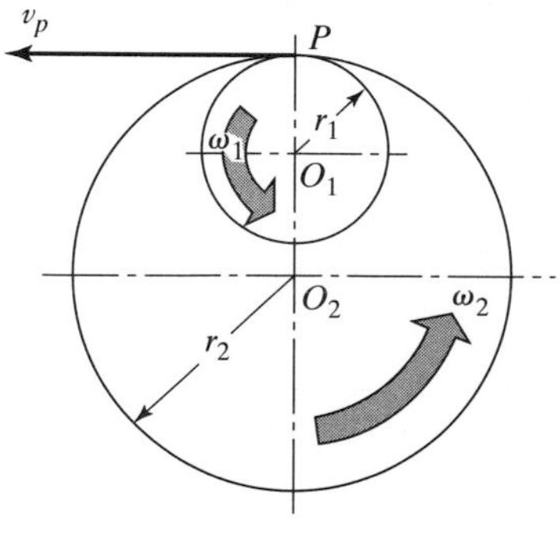

FIGURE 6.2 Two friction wheels (internal contact).

The big disadvantage of friction wheels is the possibility that slipping may occur between the cylinders. Therefore, when exact angular velocity ratios are required or a constant-phase relationship must be maintained between the driver and the driven shaft, gears are commonly used.

6.2 GEAR TYPES

There are several types of gears in common use. Among the more important types are the following:

1. Spur gears (Figure 6.3a, b, c) and *helical gears* (Figure 6.4a and b) are used when the driver and follower shafts are parallel to each other.

(a) (b) (c)

FIGURE 6.3 (a) A set of external spur gears. This type of gear is easily identified by its straight teeth, which are parallel to the gear axis. (*Source*: Boston Gear Works.) (b) An internal spur gear. This type of gear permits a closer positioning of the gear shafts. (*Source:* Richmond Gear, Wallace Murray Corporation.) (c) A spur gear and rack set. (*Source:* Browning Manufacturing Company.)

(a) (b)

FIGURE 6.4 (a) Helical gears. These gears provide less shock and offer smoother, quieter operation than do straight spur gears. (*Source:* Browning Manufacturing Company.) (b) Herringbone gear, or double helical gear. In some cases, the presence of end thrust inherent in helical gears is undesirable. Gears with opposing helices neutralize the end thrust of each helix. (*Source:* Horsburgh & Scott Company.)

2. Bevel gears (Figure 6.5) are used when the shaft axes intersect.
3. Worm gears (Figure 6.6) and crossed helical gears (Figure 6.7) are employed when the shaft axes are nonintersecting and nonparallel.

Gears other than spur gears will be considered in Chapter 7.

6.3 SPUR GEAR TERMINOLOGY

A spur gear can be visualized as a right circular cylinder that has teeth cut on its circumference parallel to the axis of the cylinder. Its design is the least complicated of gear designs. For this reason, the spur gear offers a convenient starting point for the study of gears, since the terms introduced will also apply to more complex gears discussed in the next chapter. When two gears are in mesh, it is customary to refer to the smaller as the *pinion* and the larger as the *gear*.

The following terms are in common use (see also Figure 6.8):

Pitch circle. The circle on a gear that corresponds to the contact surface of a friction wheel. Thus, for two gears in contact, the respective pitch circles can be

FIGURE 6.5 Bevel gears are used to provide end effector rotation about two axes.

imagined to roll on each other in the same manner as the circles of two friction wheels in contact. A gear may be thought of as similar to a friction cylinder, with the face width of the gear equal to the length of the cylinder and the diameter of the pitch circle of the gear equal to the diameter of the cylinder.

Addendum circle. The circle circumscribing the gear.

Addendum. The radial distance from the pitch circle to the addendum circle.

Root or dedendum circle. The circle drawn through the bottom of the gear teeth.

Dedendum. The radial distance from the pitch circle to the root circle.

Clearance circle. The largest circle centered at the gear center that is not penetrated by the teeth of the mating gear.

FIGURE 6.6 A worm and worm gear set. The worm gear is a special helical gear used for large reductions in speed. (*Source:* Cleveland Gear Company.)

FIGURE 6.7 Crossed helical gears are used with shafts that are nonparallel and nonintersecting. (*Source:* Browning Manufacturing Company.)

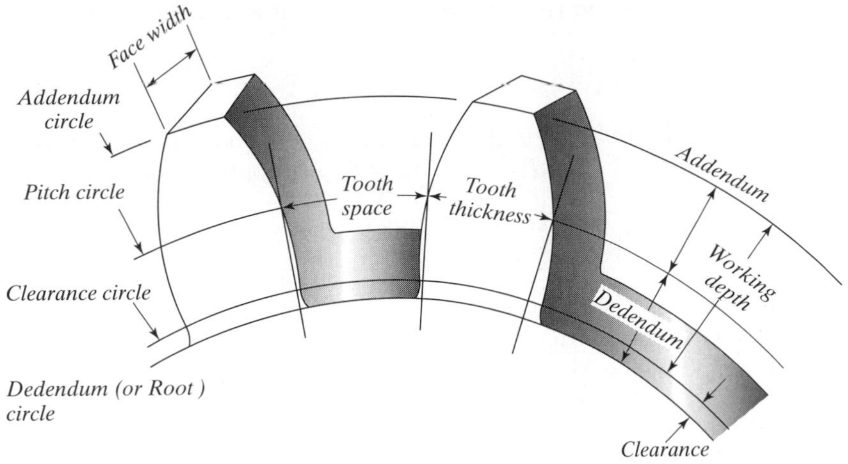

FIGURE 6.8 Spur gear nomenclature. This figure illustrates some of the more important terms and dimensions associated with spur gears.

Clearance. The radial distance from the clearance circle to the root circle. Since the clearance is also equal to the distance between the root of the tooth and the top of the tooth of the mating gear, it can also be defined as the difference between the dedendum of one gear and the addendum of the mating gear.

Whole depth. The radial distance between the addendum and dedendum circles.

Working depth. The radial distance between the addendum and clearance circles. The working depth is also equal to the sum of the addendums of the two meshing gears.

Circular pitch. The circular pitch p is the sum of the tooth width and the tooth space. It is the arc distance measured along the pitch circle from a point on one tooth to the corresponding point on the adjacent tooth of the gear. Therefore,

$$p = \frac{\pi d}{N}, \tag{6.2}$$

where d is the diameter of the pitch circle in inches or millimeters and N is the number of teeth of the gear. Accordingly, the circular pitch is equal to the circumference of the pitch circle divided by the number of teeth.

Diametral pitch. The number of teeth of a gear divided by the diameter of the pitch circle in inches. Thus,

$$P = \frac{N}{d}, \tag{6.3}$$

where P is the diametral pitch.

A simple relationship between the circular and diametral pitches is

$$Pp = \pi. \tag{6.4}$$

Module. In SI units, the module m is used to express the gear tooth size, rather than the diametral pitch P used in the U.S. customary system. The module is defined as

$$m = \frac{d}{N}, \tag{6.5}$$

where d and m have units of millimeters. Clearly, the module is the reciprocal of the diametral pitch; keep in mind, however, that the diametral pitch is a function of inches and the module is a function of millimeters.

The circular pitch, in millimeters, is

$$p = \frac{\pi d}{N} \tag{6.6}$$

or

$$p = \pi m. \tag{6.7}$$

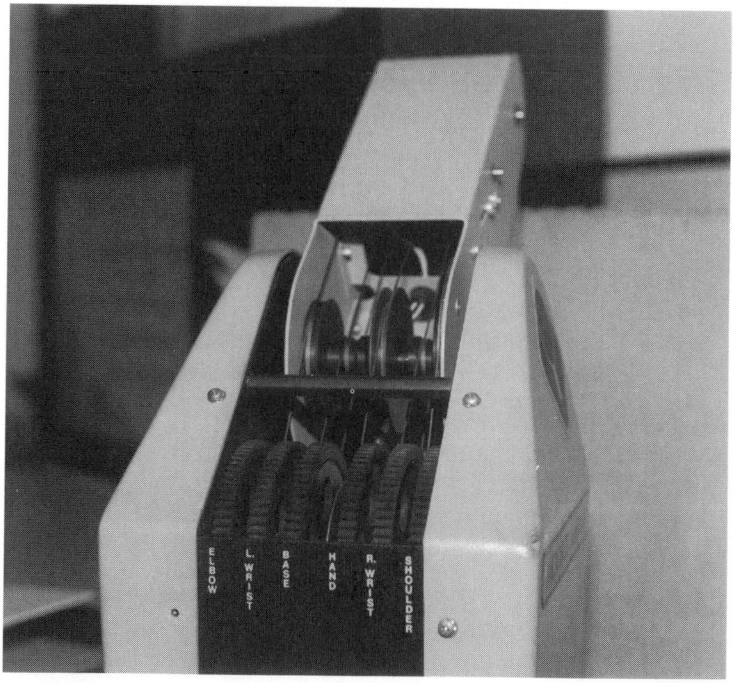

FIGURE 6.9 Plastic spur gears are among the components of this light-duty teaching robot.

A pair of meshing gears must have the same circular pitch. Thus, they must have the same module or the same diametral pitch. Heavy-duty applications call for gears with a large module. Plastic gears with a small module are used in the light-duty teaching robot shown in Figure 6.9.

SAMPLE PROBLEM 6.1

Spur Gear Properties (Using Diametral Pitch)

A spur gear, with 32 teeth and a diametral pitch of 4 is rotating at 400 rev/min. Determine its circular pitch and its pitch-line velocity.

Solution. Since we know the diametral pitch, the circular pitch can be obtained directly from Eq. (6.4):

$$p = \frac{\pi}{P} = \frac{\pi}{4} = 0.7854 \text{ in.}$$

To find the pitch-line velocity (equal to $r\omega$), we will first have to find the pitch diameter of the gear. From Eq. (6.3),

$$d = \frac{N}{P} = \frac{32}{4} = 8 \text{ in.}$$

Converting the angular velocity of the gear from revolutions per minute to radians per second, we have

$$\omega = 400 \text{ rev/min}\left(\frac{2\pi \text{ rad}}{1 \text{ rev}}\right)\left(\frac{1 \text{ min}}{60 \text{ s}}\right) = 41.9 \text{ rad/s.}$$

Finally, the pitch-line velocity is the product of the pitch-circle radius and the angular velocity of the gear [see Eq. (6.1)]:

$$v_p = r\omega = \frac{d\omega}{2} = \left(\frac{8}{2}\text{ in}\right)(41.9 \text{ rad/s}) = 167.6 \text{ in/s.}$$

SAMPLE PROBLEM 6.2

Spur Gear Properties (Using Module)

Repeat Sample Problem 6.1 for a gear manufactured with a module of 1.5 mm rather than a diametral pitch of 4.

Solution. From Eq. (6.7),

$$p = \pi m = \pi(1.5) = 4.71 \text{ mm}.$$

From Eq. (6.5),

$$d = mN = 1.5(32) = 48 \text{ mm}.$$

Finally,

$$v_p = r\omega = \left(\frac{48}{2} \text{ mm}\right)(41.9 \text{ rad/s}) = 1{,}005.6 \text{ mm/s}.$$

Backlash

If tooth spaces were exactly equal to tooth thicknesses, it would be extremely difficult for the gears to mesh. Any inaccuracies in manufacturing would cause the gears to jam. It is also very often necessary to lubricate gears. For these reasons, space must be provided between the meshing teeth. This is accomplished by making the tooth thickness less than the tooth space. The difference between tooth space and tooth thickness is known as *backlash*.

Backlash, which is measured on the pitch circle, is then equal to the distance between the nondriving side of a tooth and the side of the corresponding tooth of the meshing gear. If one of a pair of meshing gears is held stationary, the amount of backlash is then proportional to the angle the other gear can be rotated through. Figure 6.10 shows the backlash between two gears.

The cutting tool used to manufacture gears can be set further into the gear blank, thus decreasing the tooth thickness and increasing the tooth space. This is the most common method of providing backlash for gears. Slight variations in backlash can also be obtained by changing the distance between gear centers.

It should be emphasized that, while some backlash is necessary, too much backlash can result in large shock loads. Excessive backlash will also result in inaccurate gear motion.

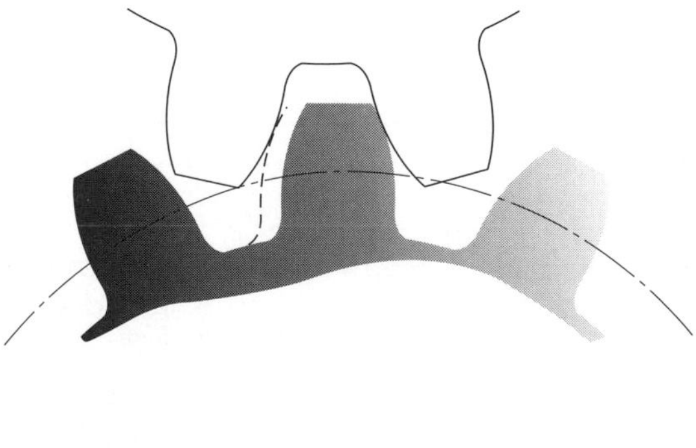

FIGURE 6.10 Backlash.

6.4 FUNDAMENTAL LAW OF GEARING

An important reason for the use of gears is to maintain a constant angular-velocity ratio. The fundamental law of gearing states the condition that the gear tooth profiles must satisfy in order to maintain a constant angular-velocity ratio. The law may be stated as follows: *The shape (profile) of the teeth of a gear must be such that the common normal at the point of contact between two teeth always passes through a fixed point on the line of centers of the gears. The fixed point is called the pitch point.* When the fundamental law is satisfied, the gears in mesh are said to produce conjugate action.

In Figure 6.11, O_1 and O_2 are the centers of the two gears in mesh, r_1 and r_2 are the radii of the pitch circles, P is the pitch point, and A is the point at which the gears are in contact.

Before proceeding with the discussion of conjugate action, we define the velocity ratio as the angular speed (ω) of the *follower* (driven gear), divided by the angular speed of the driving gear. However, the ratio can also be defined in terms of revolutions per minute, pitch radii, and the number of gear teeth.

In the following equation, the subscript 1 refers to the driver and the subscript 2 refers to the follower or driven gear:

$$r_v = \frac{\omega_2}{\omega_1} = \frac{n_2}{n_1} = \frac{r_1}{r_2} = \frac{N_1}{N_2}. \tag{6.8}$$

In this equation, r_v = velocity ratio,
ω = angular velocity (rad/s),
n = angular velocity (rev/min),
r = pitch-circle radius, and
N = number of teeth.

If a pair of spur gears is to mesh and operate properly, the following characteristics must be common to the two gears:

- circular pitch (mm or in).
- diametral pitch (number of teeth divided by diameter in inches).

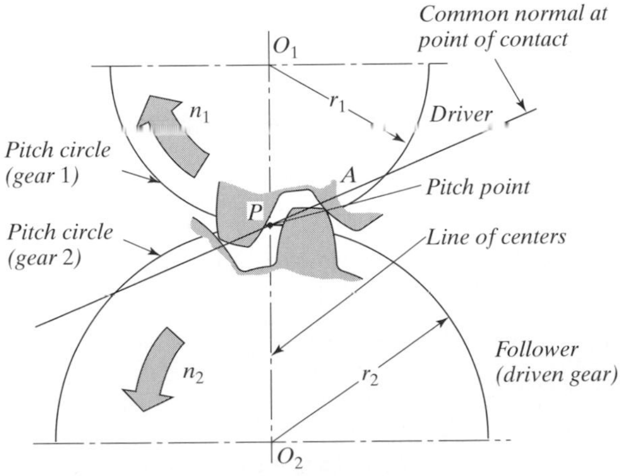

FIGURE 6.11 Two gears in mesh, with the pitch point at P. The meshing gear teeth are shown in contact at point A. The circles centered at points O_1 and O_2 and passing through the pitch point are the pitch circles.

- module (diameter in mm divided by number of teeth).
- pressure angle (degrees or radians).

When we select standard gears, the addendums and dedendums are common to the gears as well, but we sometimes use nonstandard gears with unequal addendums for special applications. In addition, there are other restrictions. Note that the velocity ratio equation applies to some other gear pairs, including helical gears on parallel shafts. Usually, the pinion (the smaller gear) drives the (larger) gear, producing a reduction in speed. (*Caution*: A pair of external gears rotate in opposite directions. If you are identifying counterclockwise and clockwise rotation by plus and minus signs, then use $\omega_2/\omega_1 = -N_1/N_2$ for a pair of external gears.)

The following partial set of rules may come in handy:

Some rules for the speed ratio of a pair of gears

The speed ratio of a pair of gears is equal to:	Applies to:	but not to:
the inverse of the ratio of the pitch radii	spur gears and helical gears on parallel shafts	planetary trains, worm drives, crossed helical gears, etc.
the inverse of the ratio of pitch diameters	spur gears and helical gears on parallel shafts	planetary trains, worm drives, crossed helical gears, etc.
the inverse of the ratio of tooth numbers	almost all gears	planetary trains

SAMPLE PROBLEM 6.3

Analysis of a Spur Gearset

Two spur gears have a velocity ratio of $\frac{1}{4}$. The driven gear has a module of 6 mm, possesses 96 teeth, and rotates at 500 rev/min. Determine the number of revolutions per minute of the driver, the number of teeth of the driver, and the pitch-line velocity.

Solution. The angular velocity (in revolutions per minute) is obtained directly from the velocity ratio, Eq. (6.8):

$$r_v = \frac{n_2}{n_1}.$$

Thus,

$$n_1 = \frac{n_2}{r_v} = \frac{500}{1/4} = 2000 \text{ rev/min}.$$

The number of teeth on the driver also follows directly from the velocity ratio:

$$r_v = \frac{N_1}{N_2},$$

or

$$N_1 = r_v N_2 = \frac{1}{4}(96) = 24 \text{ teeth}.$$

The pitch-line velocity is given by Eq. (6.1), the product of the pitch-circle radius and the angular velocity, just as in Sample Problem 6.2. We have

$$r_2 = \frac{d_2}{2} = \frac{mN_2}{2} = \frac{6(96)}{2} = 288 \text{ mm}$$

and

$$\omega_2 = n_2\left(\frac{2\pi}{60}\right) = 500 \text{ rev/min}\left(\frac{2\pi \text{ rad}}{1 \text{ rev}}\right)\left(\frac{1 \text{ min}}{60 \text{ s}}\right) = 52.3 \text{ rad/s},$$

so that

$$v_p = r_2\omega_2 = 288 \text{ mm } (52.3 \text{ rad/s}) = 15{,}062 \text{ mm/s}.$$

As a check, since $v_p = r_1\,\omega_1$ and

$$r_v = \frac{r_1}{r_2} = \frac{\omega_2}{\omega_1},$$

it follows that

$$r_1 = \frac{1}{4}(288 \text{ mm}) = 72 \text{ mm},$$

$$\omega_1 = \frac{\omega_2}{r_v} = \frac{52.3}{1/4} = 209.2 \text{ rad/s},$$

and

$$v_p = r_1\omega_1 = 72 \text{ mm } (209.2 \text{ rad/s}) = 15{,}062 \text{ mm/s}.$$

The number of teeth must be an integer. If our calculations suggest a non-integer number of teeth, we may relax the speed ratio requirement and adjust the calculated number of teeth upward or downward. If the speed ratio is critical, then we can try various combinations of pinion and gear teeth to get it right. Sometimes, a double reduction works, or a planetary train may be required. Irrational ratios like $n_2/n_1 = 1/\pi$ or $n_2/n_1 = 0.5^{-2}$ cannot be obtained precisely.

Conjugate Action and the Involute Curve

Figure 6.12 is a magnified view of the point of contact of two gears. Line *tt* is tangent to each of the two teeth at the point of contact A. Line *nn* is perpendicular to *tt* and is the common normal at the point of contact A. In order for the fundamental law to be satisfied, *nn* must always pass through a fixed point P on the line of centers.

As the gears continue to rotate, other points on the teeth will come into contact. However, for each successive point of contact, the common normal at that point must

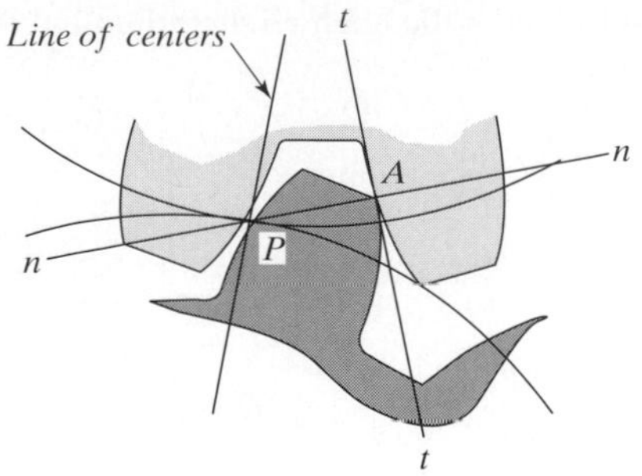

FIGURE 6.12 Two gears in contact at point A. The tangent drawn at A is tt, while the normal, nn, at point A passes through the pitch point P.

continue to pass through the fixed point P in order for conjugate action to take place (to maintain a constant angular-velocity ratio).

When gear profiles are cut in such a way as to produce conjugate action, the curves are known as *conjugate curves*. Most gears are cut using the involute curve to obtain conjugate action.

The Base Circle

Consider a cylinder with a string wrapped around it. An *involute curve* is the curve traced out by a point on the string as the string is unwrapped from the cylinder. In gear terminology, the cylinder around which the string is wrapped is known as the *base circle*. To better understand what the involute curve looks like, consider Figure 6.13a. The base circle, of radius r_b, has a string wrapped around it. Point A is the point on the end of the string, while point B is the corresponding point on the circle at which the string leaves the circle.

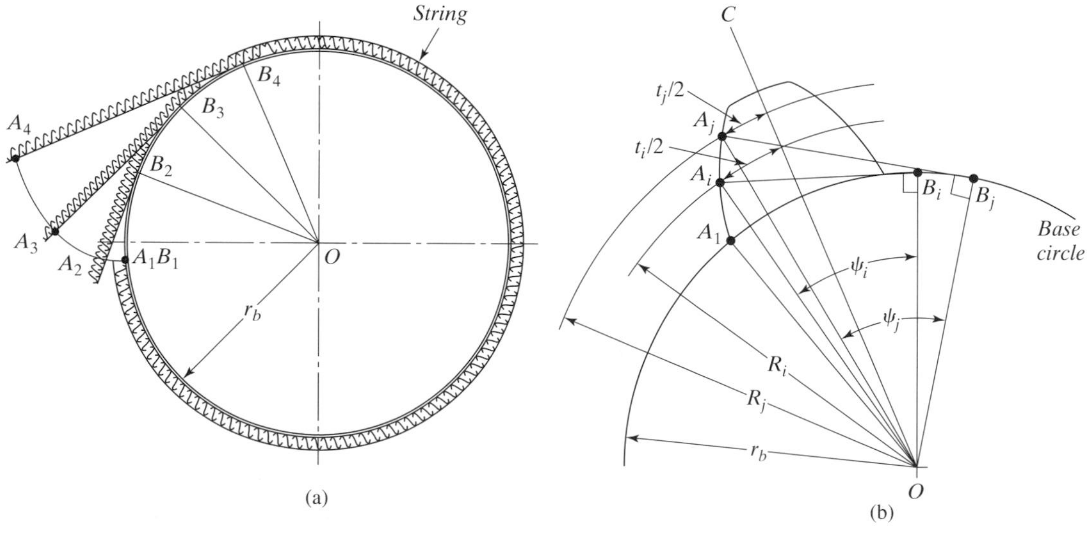

FIGURE 6.13 (a) Involute curve generation. As a string is unwrapped from a cylinder, the curve traced by a point on the string is an involute curve (A_1, A_2, A_3, A_4). The tangents to the base circle are the instantaneous radii of curvature of the involute. (b) A gear tooth with an involute profile.

Initially, while the string is still completely wrapped around the base circle, points A and B coincide. This condition is represented by points A_1 and B_1. As the string is unwrapped, point A moves to position A_2, while B_2 is the point at which the string leaves the base circle. As the string is further unwrapped, the positions A_3, B_3, A_4, and B_4 are likewise determined. The curve then drawn through points A_1, A_2, A_3, and A_4 is the involute curve. A special case is the rack, for which the involute tooth profile is a straight line.

An important property possessed by the involute curve is that a normal drawn to it is tangent to the base circle. Referring again to Figure 6.13a, we see that OB_1, OB_2, OB_3, and OB_4 are radii of the base circle, while A_2B_2, A_3B_3, and A_4B_4 are tangent to the base circle and perpendicular to the radii. The distance A_2B_2 is the radius of curvature of the involute at that instant, since point A_2 is rotating about point B_2 at that same instant. Similar reasoning shows that A_3B_3 and A_4B_4 are also instantaneous radii of curvature. Clearly, then, the radius of curvature of an involute curve is continuously varying. But, since at any given point on a curve, the radius of curvature is normal to the curve, lines A_2B_2, A_3B_3, and A_4B_4 are normals drawn to the involute curve and are also tangent to the base circle. Thus, the earlier statement that a normal to the involute curve is tangent to the base circle is correct.

From the involute geometry depicted in Figure 6.13a, a useful equation can be derived relating the tooth thicknesses at any two arbitrary radial positions on an involute gear tooth. Figure 6.13b shows a gear tooth that has thicknesses t_i and t_j at radial locations R_i and R_j, respectively. The tooth thickness is measured as an arc length along a circle that is centered at the gear center. Thus, from the figure,

$$\frac{t_i}{2} = R_i|\angle A_iOC|$$

and

$$\frac{t_j}{2} = R_j|\angle A_jOC|,$$

where $\angle A_iOC$ is the angle corresponding to arc($t_i/2$) and $\angle A_jOC$ is the angle corresponding to arc($t_j/2$). Combining these equations, we have

$$\frac{t_i}{2R_i} - \frac{t_j}{2R_j} = \angle A_iOC - \angle A_jOC = \angle A_iOA_j.$$

But

$$\angle A_iOA_j = \angle A_1OA_j - \angle A_1OA_i.$$

Therefore,

$$\frac{t_i}{2R_i} - \frac{t_j}{2R_j} = \angle A_1OA_j - \angle A_1OA_i.$$

Now we wish to determine alternative expressions for the angles $\angle A_1OA_j$ and $\angle A_1OA_i$. This can be accomplished by using the involute properties described earlier. In particular, we see from Figure 6.13b that

$$\angle A_1OA_j = \angle A_1OB_j - \angle A_jOB_j = \angle A_1OB_j - \psi_j,$$

where the angle ψ_j is referred to as the *involute angle* and is defined in the figure. Also,

$$\angle A_1OB_j = \frac{\text{arc } A_jB_j}{r_b}.$$

Now recall that, based on the string analogy, the arc length A_1B_j is equal to the distance A_jB_i. (See Figure 6.13a.) Therefore,

$$\angle A_1OB_j = \frac{A_jB_j}{r_b}.$$

But from Figure 6.13b,

$$\tan \psi_j = \frac{A_jB_j}{r_b},$$

which leads to

$$\angle A_1OB_j = \tan \psi_j.$$

Combining equations, we have

$$\angle A_1OA_j = \tan \psi_j - \psi_j = \text{inv } \psi_j,$$

where inv ψ is the involute function. (Note that the angle ψ must be expressed in radians.) In a similar manner,

$$\angle A_1OA_i = \tan \psi_i - \psi_i = \text{inv } \psi_i.$$

Finally, combining equations, we obtain the desired relationship between the tooth thicknesses, viz.,

$$\frac{t_i}{2R_i} - \frac{t_j}{2R_j} = \text{inv } \psi_j - \text{inv } \psi_i, \tag{6.9a}$$

where, from Figure 6.13b,

$$\cos \psi_i = \frac{r_b}{R_i} \tag{6.9b}$$

and

$$\cos \psi_j = \frac{r_b}{R_j}. \tag{6.9c}$$

Equations (6.9a) through (6.9c) can be used to determine the tooth thickness at any location in terms of that at another location—for example, the pitch circle.

We can now consider the action that occurs when two gear teeth, cut with involute curve profiles, are in contact.

In Figure 6.14, gear 1 is the driver and is rotating clockwise, while gear 2, the follower, rotates counterclockwise. The distance C is called the *center distance* and represents the spacing between the centers of the shafts upon which the gears are mounted. The following equation may be used to determine the center distance:

$$c = \frac{d_1 + d_2}{2}. \tag{6.10}^*$$

In this equation, d_1 and d_2 are the diameters of the pitch circles. Suppose now that the center distance between two shafts and the speed ratio are specified. Then, with that

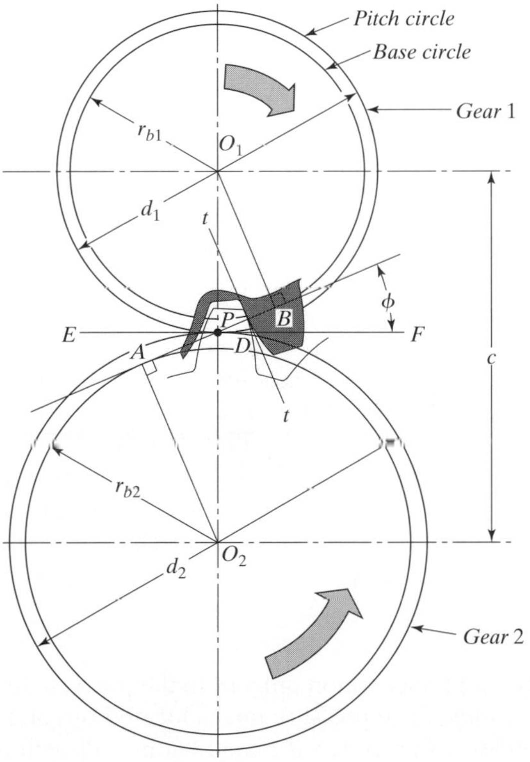

FIGURE 6.14 Two gears in contact. The center distance c is equal to one-half the sum of the pitch diameters. As the gears continue to rotate, the other points of contact must have their common normal passing through the pitch point P. For involute profiles, all contact points lie on line AB, called the *line of action*. Line AB is also called the *pressure line*, and ϕ is referred to as the *pressure angle*.

information, it is possible to determine the required pitch diameters. Since $d = N/P$ (inches) and $d = mN$ (millimeters), we have

$$c = \frac{1}{2P}(N_1 + N_2), \tag{6.11a)*}$$

where c is in inches, and

$$c = \frac{m(N_1 + N_2)}{2}. \tag{6.11b)*}$$

where c is in millimeters.

* Equations (6.10), (6.11a), and (6.11b) apply to a pair of external gears.

The following equations apply if gear number 2 is an *internal gear*:

$$c = (d_2 - d_1)/2, \tag{6.11c}$$

$$c = (N_2 - N_1)/(2P), \tag{6.11d}$$

and

$$c = m(N_2 - N_1)/2. \tag{6.11e}$$

The Line of Action

When two curves are in contact at a point, they must have the same tangent and normal at that point (Figure 6.12). In Figure 6.14, point D is the point of contact between the two involute curves to which the teeth of gears 1 and 2 have been cut. The common tangent to the tooth surfaces is tt, while line AB, which is perpendicular to tt at the point of contact, is the common normal.

According to the fundamental law of gearing, the common normal must pass through a fixed point on the line of centers, O_1O_2. But according to the properties of the involute curve discussed previously, the normal to each of the two involutes at the point of contact is tangent to the respective generating, or base, circle. It follows that the common normal must be simultaneously tangent to both of the base circles and therefore is a unique line. This means that, as the gears continue to rotate and other points become contact points, these contact points will always lie on line AB, which will always be the common normal. Therefore, the intersection P of the common normal and the line of centers is a fixed point, and the gears have conjugate action.

The line AB is often called the *line of action* because the contact points of two gears in mesh must lie along it. The force that one gear tooth exerts on the tooth of the meshing gear acts along the common normal, which is also line AB. Therefore, the *pressure line* is another name commonly given to line AB.

The Pressure Angle

The angle ϕ between the pressure line AB and the common tangent to the pitch circles EF in Figure 6.14 is known as the *pressure angle*. The pressure line is located by rotating the common tangent to the pitch circles, EF, through the angle ϕ in a direction opposite the direction of rotation of the driver. Referring again to Figure 6.14, since gear 1, the driver, is rotating clockwise, we can locate the pressure line AB by rotating the common tangent EF counterclockwise through the angle ϕ.

While gears may be manufactured with a wide range of pressure angles, most gears are made with standard angles of $14\frac{1}{2}°$, 20°, or 25°. Although gears are designated by their pressure angle, it must be emphasized that the actual pressure angle between two gears in contact may differ from the designated value. Changes in center distance c will result in corresponding changes of the actual pressure angles. In other words, two nominally 20° gears actually may be given a slightly larger pressure angle by increasing their center distance.

Later, in Sample Problem 6.6, we shall explain the difference between designated and actual pressure angles. However, at this point, a diagram can illustrate the effect of increasing the center distance. In Figure 6.15a, gears 1 and 2 have their centers at O_1

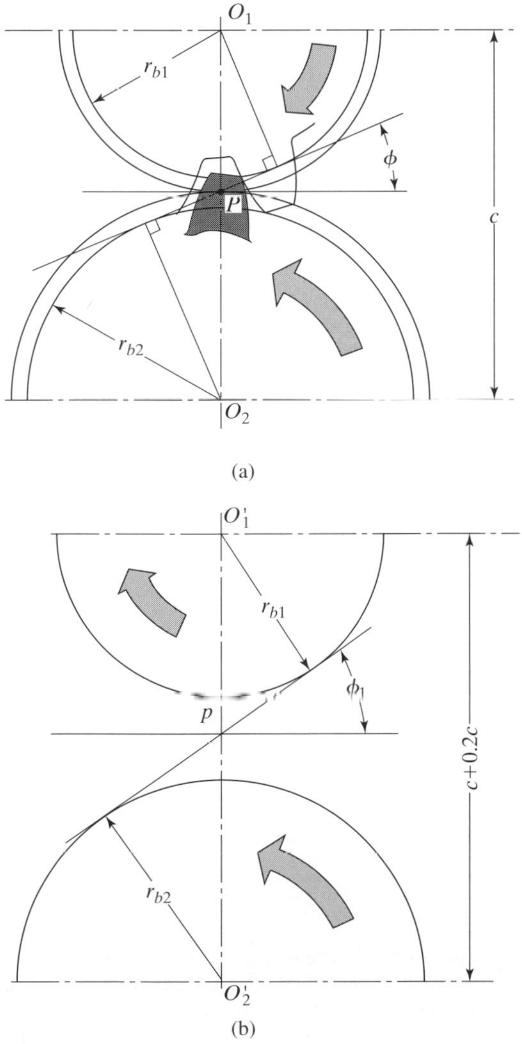

FIGURE 6.15 (a) Two gears in mesh showing center distance c, a pressure angle ϕ, and base-circle radii r_{b_1} and r_{b_2}. (b) The center distance of the gears in Figure 6.15a is shown increased. The base radii remain unchanged; however, the pressure angle is now increased to ϕ_1.

and O_2. The pressure angle is ϕ, and the base-circle radii are r_{b_1} and r_{b_2}. The center distance is then increased so that the centers of the gears are at O_1' and O_2'. Figure 6.15b shows the new situation with the increased center distance and unchanged base-circle radii. As in Figure 6.15a, the pressure line is drawn tangent to the base circles and through the pitch point. The pressure angle is now increased (shown much larger than would normally be the case, for purposes of illustration). This large change in center distance could result in excessive backlash, and the contact ratio (considered in a later section) might now be unsatisfactory.

To obtain a better understanding of gear tooth action, consider Figure 6.16. Two gears, 1 and 2, are shown in mesh. The pitch radii r_1 and r_2, as well as the base-circle radii r_{b_1} and r_{b_2}, are shown. The base circles were determined by drawing circles tangent to the pressure line AB. Therefore, the radii r_{b_1} and r_{b_2} are perpendicular to the pressure line at points D and C, respectively.

By considering the right triangles O_1PD and O_2PC, a simple relationship between the pitch circle radius and the base circle radius is seen to exist:

$$\cos\phi = \frac{r_{b_1}}{r_1} = \frac{r_{b_2}}{r_2} \text{ or } r_b = r\cos\phi. \tag{6.12}$$

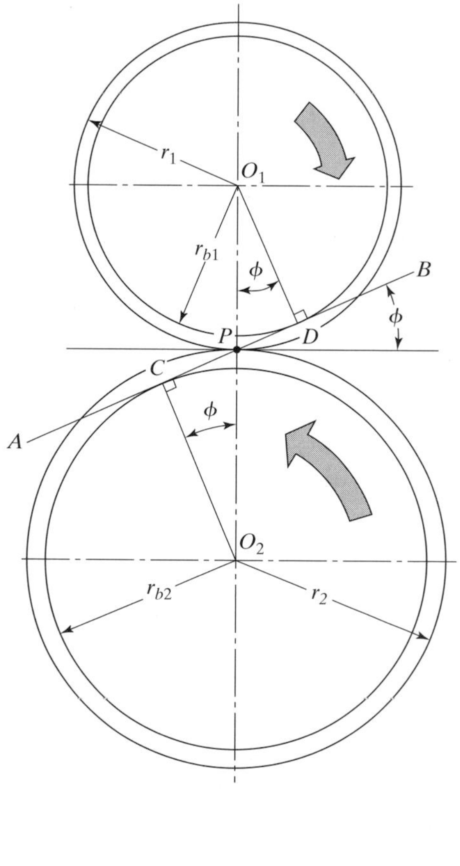

FIGURE 6.16 The relationship between the base-circle radii, the pressure angle, and the pitch circle radii: $r_b = r\cos\phi$.

SAMPLE PROBLEM 6.4

Analysis of a Gear Pair (based on diametral pitch)

Two meshing spur gears with a diametral pitch of 4, a pressure angle of 20°, and a velocity ratio of $\frac{1}{4}$ have their centers 15 in apart. Determine the number of teeth on the driver (pinion) and the base-circle radius of the gear.

Solution. As usual, subscript 1 will refer to the pinion and subscript 2 will refer to the gear. From the velocity ratio, we know that

$$r_v = \frac{n_2}{n_1} = \frac{N_1}{N_2} = \frac{1}{4},$$

or

$$N_2 = 4N_1.$$

We can set up a second equation in N_1 and N_2 (giving us two equations in two unknowns) and determine the values of those variables. Knowing the diametral pitch P and the distance between centers c, we can use Eq. 6.11a to obtain

$$15 = \frac{1}{2 \times 4}(N_1 + N_2).$$

But $N_2 = 4N_1$; therefore,

$$120 = N_1 + 4N_1 = 5N_1$$

and it follows that

$$N_1 - 24 \text{ teeth.}$$

To solve for the base-circle radius of the gear, we simply solve for r_1 and then use the velocity ratio and pressure angle formulas to find r_{b_2}. Thus,

$$d_1 = \frac{N_1}{P} = \frac{24}{4} = 6 \text{ in},$$

so that $r_1 = 3$ in and

$$r_2 = \frac{r_1}{n_2/n_1} = \frac{3}{1/4} = 12 \text{ in.}$$

Finally, from Eq. (6.12), we find the base-circle radius

$$r_{b_2} = r_2 \cos\phi = 12 \cos 20° = 11.276 \text{ in.}$$

SAMPLE PROBLEM 6.5

Analysis of a Gear Pair (Based on Module)

Repeat Sample Problem 6.4 for a module of 3 mm and a center distance of 180 mm.

Solution. From the velocity ratio,

$$r_v = \frac{N_1}{N_2} = \frac{1}{4}.$$

Therefore,

$$N_2 = 4N_1.$$

From Eq. (6.11b),

$$180 = \frac{3(N_1 + 4N_1)}{2}.$$

Hence, $N_1 = 24$ teeth and $N_2 = 96$ teeth.
From Eq. (6.5),

$$d_2 = mN_2 = 3(96) = 288 \text{ mm}$$

and

$$r_2 = \frac{d_2}{2} = \frac{288}{2} = 144 \text{ mm}.$$

Thus, from Eq. (6.12),

$$r_{b_2} = r_2 \cos\phi = 144 \cos 20° = 135.3 \text{ mm}.$$

SAMPLE PROBLEM 6.6

Determination of Pressure Angle

Two 20° gears have a diametral pitch of 4. The pinion has 28 teeth, while the gear has 56 teeth. Determine the center distance for an actual pressure angle of 20°. What is the actual pressure angle if the center distance is increased by 0.2 in?

Solution. For the data given, the center distance is found by using Eq. (6.3) to obtain the pitch-circle diameters and then using values obtained in Eq. (6.10):

$$d_1 = \frac{N_1}{P} = \frac{28}{4} = 7 \text{ in};$$

$$d_2 = \frac{N_2}{P} = \frac{56}{4} = 14 \text{ in}.$$

Thus,

$$c = \frac{1}{2}(d_1 + d_2) = \frac{1}{2}(7 + 14) = 10.5 \text{ in.}$$

The base-circle radius was determined when the gears were cut, and changing the center distance does not change that radius. *However, increasing the center distance does increase the pitch radius, which in turn results in a larger pressure angle.* To find the actual pressure angle resulting from an increased center distance, we must first find the base-circle radius and the new pitch radius. From Eq. (6.12),

$$r_{b_1} = r_1 \cos\phi = 3.5 \cos 20° = 3.29 \text{ in.}$$

The new center distance c′ is

$$10.5 + 0.2 = 10.7 \text{ in.}$$

The new pitch radii, r_1' and r_2', although changed numerically, will maintain the same proportion held by the original pitch radii. Thus, from Eq. (6.8),

$$r_v = \frac{N_1}{N_2} = \frac{r_1}{r_2} = \frac{r_1'}{r_2'} = \frac{1}{2},$$

or

$$r_2' = 2r_1'.$$

But

$$c' = r_1' + r_2' = 10.7.$$

Therefore,

$$r_1' + 2r_1' = 10.7,$$

or

$$r_1' = 3.57 \text{ in.}$$

Since the base-circle radius does not change, we can finally calculate the new pressure angle using Eq. (6.12):

$$r_{b_1} = r_1' \cos\phi,$$

so

$$\cos\phi = \frac{r_{b_1}}{r_1'} = \frac{3.29}{3.57} = 0.922,$$

or

$$\phi = 22.8°.$$

Contact Length

As mentioned earlier, all the points of contact between two gear teeth with involute profiles lie along the pressure line. The initial contact between two teeth will occur when the tip of the driven gear tooth contacts the driver tooth. The final contact occurs when the driven gear tooth contacts the tip of the driver tooth. Another way of describing the interval of contact is to say that initial contact occurs where the addendum circle of the driven gear intersects the pressure line and final contact occurs at the point where the addendum circle of the driver intersects the pressure line.

Figure 6.17 illustrates the important points of the preceding discussion. Initial contact occurs at point E, which is the intersection of the addendum circle of the driven gear and the pressure line. Point F, the intersection of the addendum circle of the driver and the pressure line, is the final contact point.

The distance between points E and F is known as the length of the line of action, or the *contact length*.

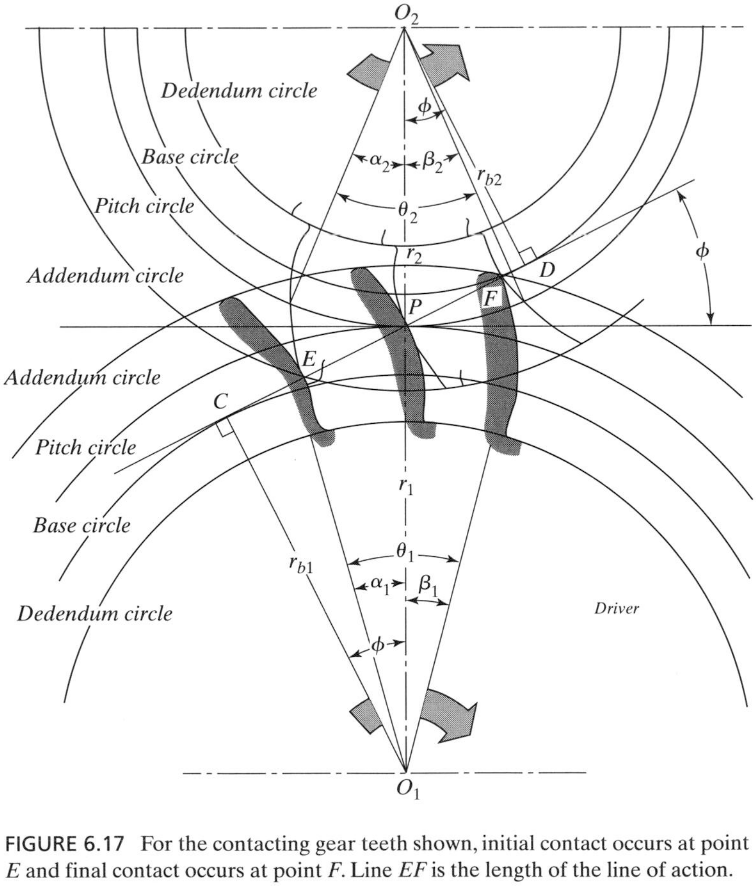

FIGURE 6.17 For the contacting gear teeth shown, initial contact occurs at point E and final contact occurs at point F. Line EF is the length of the line of action. Angles θ_1 and θ_2 are the angles of action, α_1 and α_2 are the angles of approach, and β_1 and β_2 are the angles of recess. The angle of action is equal to the sum of the angle of approach and the angle of recess.

The angles θ_1 and θ_2 shown in the figure are known as the *angles of action*—the angles turned through by the driver and follower gears, respectively, during the interval of contact between a pair of gear teeth. The *arc of action* is the arc, measured on the pitch circle, turned through by a gear as a pair of meshing gear teeth go from initial to final contact. Note that θ_1 and θ_2 subtend the arcs of action.

Angles α_1 and α_2 are known as *angles of approach*. The angle of approach is the angle turned through by a gear from the instant of initial tooth contact until the same pair of teeth contact at the pitch point.

The *angle of recess* is defined as the angle turned through by a gear while the contact between the teeth goes from the pitch point to the point of final contact. In Figure 6.17, β_1 and β_2 are the angles of recess. It should be clear from the definitions as well as from the illustration that the angle of action is equal to the sum of the angle of approach and the angle of recess.

The angles of approach and angles of recess could be measured from points where tooth profiles cross the pitch circle. This requires a time-consuming construction, however. Since the bottom line is the ***contact ratio***, the analytical method in the next section will usually be preferred.

Contact Ratio

The circular pitch, as defined in an earlier section, is equal to the distance, measured on the pitch circle, between corresponding points of adjacent teeth. Let γ be the angle determined by the circular pitch AB, as shown in Figure 6.18. The angle γ is known as the *pitch angle*. We now define the *contact ratio* as *the angle of action divided by the pitch angle*, or

$$\text{Contact ratio} = \frac{\theta}{\gamma} = \frac{\alpha + \beta}{\gamma}. \tag{6.13}$$

If the contact ratio were equal to unity, Eq. (6.13) would indicate that the angle of action is equal to the pitch angle. A contact ratio of unity means that one pair of teeth are in contact at all times. If the contact ratio were less than unity, there would be an

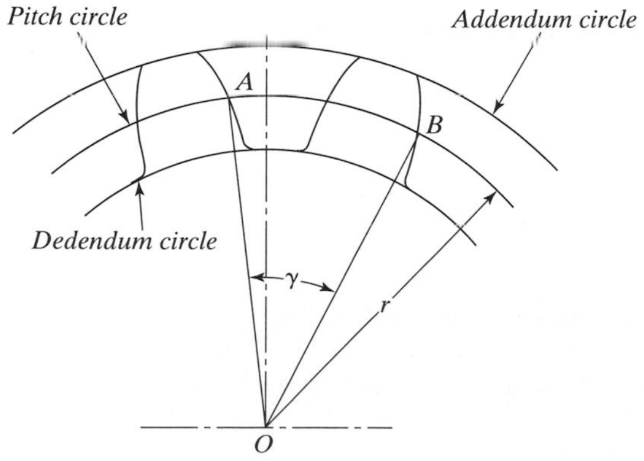

FIGURE 6.18 The pitch angle γ is the angular equivalent of the circular pitch.

interval during which *no* teeth would be in contact. Gears are usually designed with a contact ratio of at least 1.2. Higher values should result in smoother, quieter operation. For a contact ratio of 1.2, one pair of teeth are always in contact, and two pairs of teeth are in contact 20 percent of the time. Therefore, the contact ratio is also commonly defined as the average number of tooth pairs in contact.

The contact ratio can be defined in still another way. However, in order to understand this definition, we must first define the base pitch. *The base pitch* p_b *is equal to the distance between corresponding points of adjacent teeth measured on the base circle.* Mathematically,

$$p_b = \frac{\pi d_b}{N}, \tag{6.14}$$

where d_b is the diameter of the base circle and N is the number of teeth. Since $r_b = r \cos \phi$, it follows that $d_b = d \cos \phi$. Therefore,

$$p_b = \frac{\pi d \cos \phi}{N}.$$

But since $P = N/d$, we have

$$p_b = \frac{\pi \cos \phi}{P}.$$

Also, $p = \pi/P$; hence,

$$p_b = p \cos \phi, \tag{6.15}$$

or, SI units,

$$p_b = m \, \pi \cos \phi.$$

The contact ratio can now also be defined as the ratio of the length of action divided by the base pitch. Referring again to Figure 6.17, we see that the line of action is equal to the distance between points E and F. This definition is valid, since the length of action, EF, is the same distance the base circle rolls through as the involute is being generated. Therefore,

$$\text{Contact ratio} = \frac{EF}{p_b}. \tag{6.16}$$

The length of action can be computed from the formulas

$$EF = EP + PF, \; EP = ED - PD, \text{ and } PF = CF - CP.$$

To obtain EF, we must determine component lengths EP and PF. In right triangle O_2DE, length O_2E is equal to the sum of the pitch radius and the addendum of driven gear 2:

$$O_2E = r_2 + a_2.$$

O_2D is the base radius of gear 2 and is given by

$$O_2D = r_{b_2} = r_2 \cos \phi.$$

We can therefore determine length ED by using the Pythagorean theorem:

$$\begin{aligned} ED &= \sqrt{(O_2E)^2 - (O_2D)^2} \\ &= \sqrt{(r_2 + a_2)^2 - r_2^2 \cos^2 \phi}. \end{aligned}$$

Also,

$$PD = r_2 \sin\phi.$$

Therefore,

$$EP = ED - PD = \sqrt{(r_2 + a_2)^2 - r_2^2 \cos^2 \phi} - r_2 \sin \phi.$$

The distance PF is found in a similar manner. In triangle O_1CF, length O_1F equals the sum of the pitch radius and the addendum of the driver:

$$O_1F = r_1 + a_1.$$

The base radius of the driver is

$$O_1C = r_{b_1} = r_1 \cos \phi.$$

As with ED, CF is found by using the Pythagorean theorem:

$$\begin{aligned} CF &= \sqrt{(O_1F)^2 - (O_1C)^2} \\ &= \sqrt{(r_1 + a_1)^2 - r_1^2 \cos^2 \phi}. \end{aligned}$$

Also,

$$CP = r_1 \sin \phi.$$

Therefore,

$$PF = CF - CP = \sqrt{(r_1 + a_1)^2 - r_1^2 \cos^2 \phi} - r_1 \sin \phi.$$

Finally, we obtain the length of the line of action solely in terms of the pitch radii, the addendums, and the common pressure angle:

$$\begin{aligned} EF &= EP + PF \\ &= \sqrt{(r_2 + a_2)^2 - r_2^2 \cos^2\phi} - r_2 \sin\phi \\ &\quad + \sqrt{(r_1 + a_1)^2 - r_1^2 \cos^2\phi} - r_1 \sin\phi. \end{aligned}$$

Thus, the formula for the contact ratio, Eq. (6.16), becomes

$$\text{Contact ratio} = \frac{\sqrt{(r_2 + a_2)^2 - r_2^2 \cos^2\phi} - r_2 \sin\phi}{p_b} + \frac{\sqrt{(r_1 + a_1)^2 - r_1^2 \cos^2\phi} - r_1 \sin\phi}{p_b}. \qquad (6.17)$$

SAMPLE PROBLEM 6.7

Determination of Contact Ratio

Two 25° full-depth spur gears have a velocity ratio of 1/3. The diametral pitch is 5 and the pinion has 20 teeth. Determine the contact ratio (number of teeth in contact) of the gears. (The formula for the addendum of a 25° full-depth gear is given in Table 6.1 in Section 6.6.) as $a = 1/P$.

Solution. Before using Eq. (6.17) to determine the contact ratio, we will have to find the following values: r_1, r_2, a_1, a_2, p_b, $\sin\phi$, and $\cos\phi$. Thus, from Eq. (6.3),

$$d_1 = \frac{N_1}{P} = \frac{20}{5} = 4 \text{ in, or } r_1 = 2 \text{ in,}$$

and from the velocity ratio equation,

$$r_v = \frac{n_2}{n_1} = \frac{r_1}{r_2} = \frac{1}{3}, \text{ or } r_2 = 6 \text{ in.}$$

TABLE 6.1 Standard Gear Profiles

System	Addendum	Dedendum	Clearance	Whole depth
$14\frac{1}{2}°$ full-depth involute	$\frac{1}{P}$	$\frac{1.157}{P}$	$\frac{0.157}{P}$	$\frac{2.157}{P}$
20° full-depth involute (coarse pitch)	$\frac{1}{P}$	$\frac{1.25}{P}$	$\frac{0.25}{P}$	$\frac{2.25}{P}$
20° full-depth involute (fine pitch)	$\frac{1}{P}$	$\frac{1.2}{P} + 0.002$ in	$\frac{0.2}{P} + 0.002$ in	$\frac{2.2}{P} + 0.002$ in
20° stub-tooth involute	$\frac{0.8}{P}$	$\frac{1}{P}$	$\frac{0.2}{P}$	$\frac{1.8}{P}$
25° full-depth involute	$\frac{1}{P}$	$\frac{1.25}{P}$	$\frac{0.25}{P}$	$\frac{2.25}{P}$
20° full-depth metric standard	m	$1.25m$	$0.25m$	$2.25m$

The addendum is given by

$$a_1 = a_2 = \frac{1}{P} = \frac{1}{5} = 0.2 \text{ in.}$$

The last of the unknowns is the base pitch. From Eq. (6.4),

$$p = \frac{\pi}{P} = \frac{\pi}{5},$$

and from Eq. (6.15),

$$p_b = p \cos \phi = \frac{\pi}{5} \cos 25^\circ = 0.569.$$

Thus, the contact ratio formula, Eq. (6.17), is solved with the values just determined:

$$\begin{aligned}
\text{Contact ratio} &= \frac{\sqrt{(r_2 + a_2)^2 - r_2^2 \cos^2 \phi} - r_2 \sin \phi}{p_b} \\
&\quad + \frac{\sqrt{(r_1 + a_1)^2 - r_1^2 \cos^2 \phi} - r_1 \sin \phi}{p_b} \\
&= \frac{\sqrt{(6 + 0.2)^2 - 6^2(0.906)^2} - 6(0.423)}{0.569} \\
&\quad + \frac{\sqrt{(2 + 0.2)^2 - 2^2(0.906)^2} - 2(0.423)}{0.569} = 1.48.
\end{aligned}$$

6.5 INTERNAL GEARS

The use of an internal gear is highly desirable in many applications, (e.g., epicyclic gear trains, to be discussed in a later chapter). An internal gear has its teeth cut on the inside of the rim rather than on the outside.

Figure 6.19 shows a typical internal gear in mesh with an external pinion. The important terms and dimensions associated with internal gears are illustrated. As can be seen from the illustration, the directions of rotation for an internal and external gear in mesh are the same, whereas two external gears in contact have opposite directions of rotation.

Since the internal gear has a concave tooth profile, while the external gear's tooth profile is convex, the surface contact between the gears is increased, thus decreasing the contact stress. The center distance between the gears is less, thus making a more compact arrangement than that for external gear sets. Internal–external gear sets also have a greater number of teeth in contact, resulting in smoother and quieter operation. However, the manufacture of internal gears presents some unique problems.

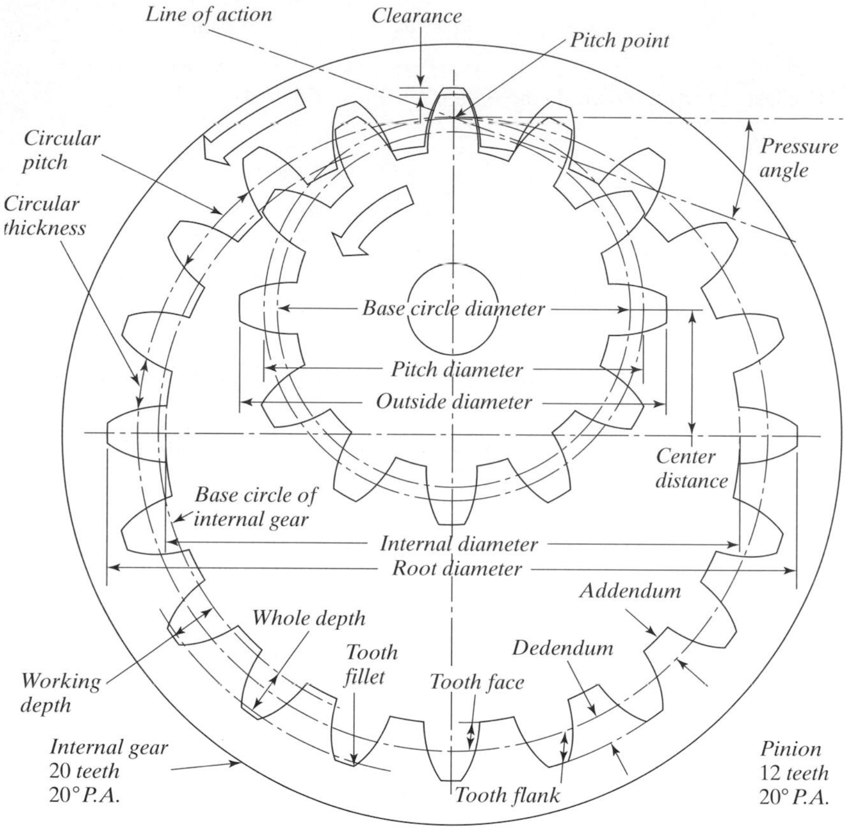

FIGURE 6.19 Nomenclature for internal gears.

6.6 STANDARD GEARS

It is economically desirable to standardize gears so that they can be interchanged. Standard tooth systems have specified values for the addendum, dedendum, clearance, and tooth thickness. Mating gears should have the same pressure angle, the same module or diametral pitch, and (usually) the same addendum and dedendum.

Table 6.1 lists some of the most commonly used standard gear profile systems. The standard gear profiles given in the table can be expressed in SI units by replacing P by $1/m$. For example, for a 20° full-depth involute (coarse pitch), the addendum $= m$, the dedendum $= 1.25m$, and so on.

The following are commonly used values of diametral pitch: coarse pitch, $1, 1\frac{1}{4}, 1\frac{1}{2}, 1\frac{3}{4}, 2, 2\frac{1}{4}, 2\frac{1}{2}, 2\frac{3}{4}, 3, 3\frac{1}{2}$, 4, 5, 6, 7, 8, 9, 10, 12, 14, 16, and 18; and fine pitch, 20, 22, 24, 26, 32, 40, 48, 64, 72, 80, 96, and 120. Some of the standard values for metric module are 0.3, 0.5, 1, 1.25, 1.5, 2, 2.5, 3, 4, 5, 6, 8, 10, 12, 16, 20, 25, 32, 40, 50. Obviously, standard values of module do not correspond with standard values of diametral pitch.

Therefore, the two systems are not interchangeable. Many manufacturers and suppliers stock only $14\frac{1}{2}^\circ$ and 20° pressure angle full-depth gears.

SAMPLE PROBLEM 6.8

Standard Gear Calculations

A 20° full-depth spur gear has 35 teeth and a diametral pitch of 5. Determine the addendum-circle diameter, the dedendum-circle diameter, and the working depth.

Solution. The addendum- and dedendum-circle radii are found by adding the addendum to, and subtracting the dedendum from, the pitch-circle radius, respectively. Thus, we must find the pitch radius. From Eq. (6.3), the pitch-circle diameter is

$$d = \frac{N}{P} = \frac{35}{5} = 7 \text{ in.}$$

From Table 6.1, we obtain the expression for the addendum of the gear profile system being considered:

$$\text{Addendum} = \frac{1}{P} = \frac{1}{5} = 0.2 \text{ in.}$$

The addendum-circle diameter is

$$d_a = d + 2(\text{addendum}) = 7 + 2(0.2) = 7.4 \text{ in.}$$

Similarly, the dedendum is given by

$$\text{Dedendum} = \frac{1.25}{P} = \frac{1.25}{5} = 0.25 \text{ in,}$$

and the dedendum-circle diameter is

$$d_d = d - 2(\text{dedendum}) = 7 - 2(0.25) = 6.5 \text{ in}$$

Also,

$$\text{Working depth} = \text{whole depth} - \text{clearance.}$$

The clearance, from Table 6.1, is

$$\text{Clearance} = \frac{0.25}{P} = \frac{0.25}{5} = 0.05 \text{ in.}$$

Or, alternatively,

$$\text{Clearance} = \text{dedendum} - \text{addendum} = 0.25 - 0.20 = 0.05 \text{ in}$$

and

$$\text{Whole depth} = \text{addendum} + \text{dedendum} = 0.20 + 0.25 = 0.45 \text{ in.}$$

Finally,

$$\text{working depth} = \text{whole depth} - \text{clearance} = 0.45 - 0.05 = 0.40 \text{ in.}$$
$$\text{Or, working depth} = 2 \times \text{addendum} = 2(0.2) = 0.4 \text{ in.}$$

Contact Ratio in Terms of Tooth Numbers (Standard Gears)

"Designing smart" often involves using standard components. Standard "off-the-shelf" gears cost less than nonstandard gears, and standard gear specifications are set to provide an adequate contact ratio for most applications.

We can see what influences the contact ratio by rewriting the equation. For standard full-depth teeth, we substitute

$$r = Nm/2 = N/(2P),$$
$$a = m = 1/P,$$
$$p = \pi m = \pi/P,$$

and

$$p_b = p \cos \phi,$$

where r = pitch radius (mm or in),
N = number of teeth,
m = module (mm),
P = diametral pitch (teeth/in),
a = addendum (mm or in),
p = circular pitch (mm or in), and
ϕ = pressure angle.

The result, given in the following sample problem, is a contact ratio equation in terms of tooth numbers and the pressure angle.

SAMPLE PROBLEM 6.9

A Three-dimensional Bar Chart of the Contact Ratio

How is the contact ratio related to tooth numbers in a gear pair? Make a three-dimensional bar chart of the contact ratio vs. tooth numbers.

As a design decision, we select 20° pressure angle standard full-depth gears.

Solution summary. We begin with the contact ratio equation in terms of pitch radii and make the substitutions just suggested . The result is an equation in terms of tooth numbers that applies to metric or customary U.S. standard gears of any module or diametral pitch. Gears with even numbers of teeth between 14 and 34 are considered. The results are formed into a matrix and plotted as a three-dimensional bar chart in Figure 6.20. The contact ratio ranges from about 1.46 for two 14-tooth gears to 1.68 for two 34-tooth gears. If a 14-tooth 20° pressure angle gear meshes with a gear with more than 26 teeth, there is interference. As a result, a few of the plotted values are not valid and are marked with an X. We will look into the problem of interference later.

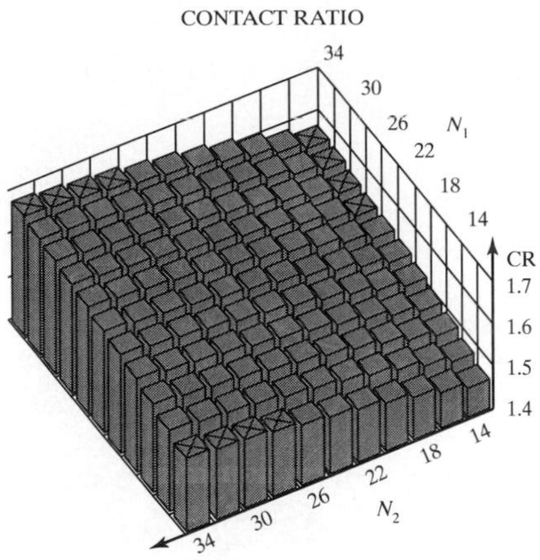

FIGURE 6.20 Contact ratio vs. number of pinion and gear teeth.

Detailed solution.

Contact Ratio: Full-Depth Teeth

$$CR(N_1, N_2, \phi) := \frac{[(N_1 + 2)^2 - (N_1 \cdot \cos(\phi))^2]^{\frac{1}{2}} + [(N_2 + 2)^2 - (N_2 \cdot \cos(\phi))^2]^{\frac{1}{2}} - (N_1 + N_2) \cdot \sin(\phi)}{2\pi \cdot \cos(\phi)}$$

$i := 0..10 \qquad j := 0..10$

$N_i := 2 \cdot i + 14 \quad N_j := 2 \cdot j + 14$

$M_{i,j} := CR(N_i, N_j, 20 \cdot deg) - 1.4 \qquad CR(14, 14.20 \cdot deg) = 1.463 \qquad CR(34, 34.20 \cdot deg) = 1.681$

6.7 GEAR MANUFACTURE

Gears are usually manufactured by milling, generating, or molding. In the milling method of manufacture, the milling cutter is shaped so as to conform to the shape of the space between the teeth. The cutter is then moved across the face of the gear blank, thus cutting out a space between teeth. The blank is then automatically rotated until the next space to be cut lines up with the cutter. This process is continued until all the spaces have been cut out, completely forming the gear. Figure 6.21a shows a typical milling cutter.

The disadvantage of the milling cutter is that a different cutter must be used not only for different pitches, but also for different numbers of teeth. Gear manufacturers usually shape milling cutters so that they are correct for the gear with the smallest number of teeth, in each of eight ranges of tooth numbers, for a given pitch. This means that when gears having a greater number of teeth are cut with a particular milling cutter, an error in the tooth profile results. The error increases toward the high end of each range of tooth numbers, but is acceptable for most applications.

Gears to be used for high-speed, high-load applications are not cut accurately enough by the milling process. Instead, the generating method should be used whenever high accuracy is required.

(a)

(b)

FIGURE 6.21 (a) Like all milling cutters, the milling cutter shown, is accurate only for a specific gear with a fixed pitch and a fixed number of teeth. (*Source:* Horsburgh & Scott Company.) (b) Hobbing a gear. This method of gear production is more accurate than milling, since a given hobbing cutter can cut gears of various tooth numbers. with equal accuracy. (*Source:* Horsburgh & Scott Company.)

The generating process of gear cutting entails the use of either a hob or a shaper. A hob, the cutting tool used in the hobbing process, is shown in Figure 6.21b. Cutting is accomplished by moving the hob across the gear blank as both the gear blank and the hob are rotated.

A second method of generating gears is by shaping. The cutting tool used in the shaping method is either a rack cutter or a pinion cutter. The rack cutter, which has teeth with straight sides, has its addendum made equal to the dedendum of the gear being cut. The angular orientation of the side of the rack tooth is equal to the pressure angle ϕ of the gear to be cut. Figure 6.22 shows a gear blank and rack cutter. Cutting is

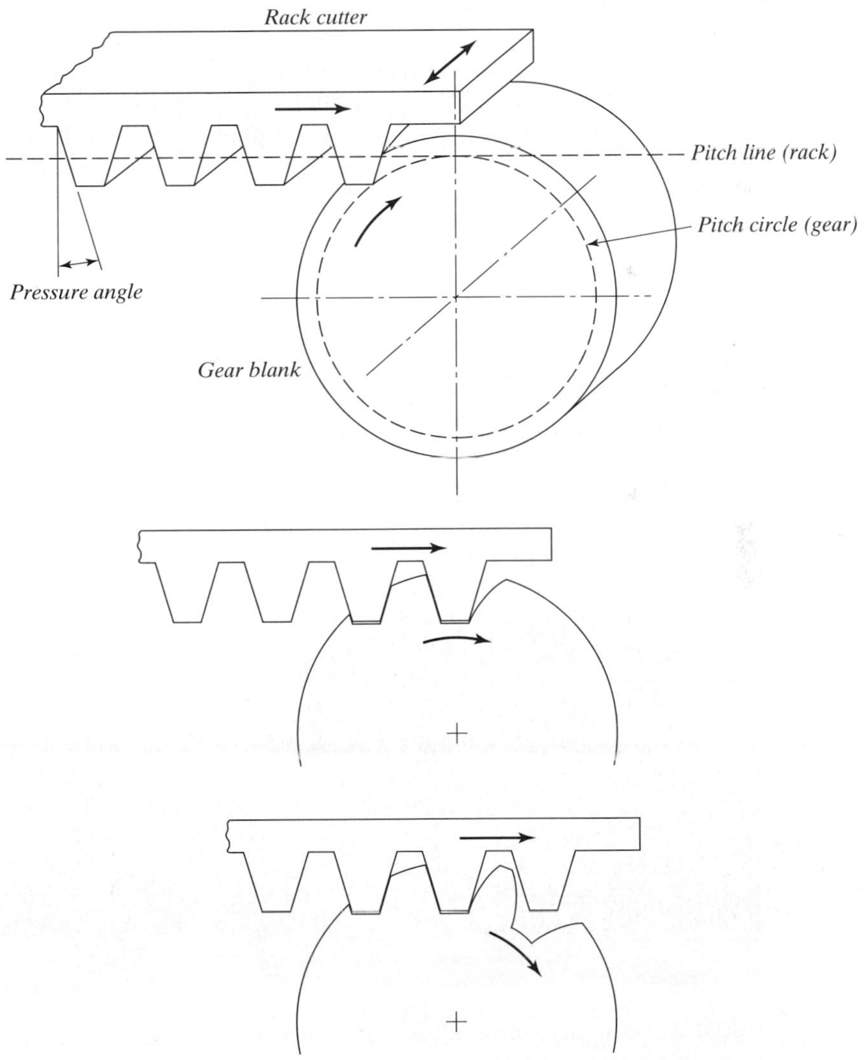

FIGURE 6.22 Generation of involute gear teeth with a rack cutter. The pitch line of the rack is tangent to the pitch circle of the gear being cut.

started when the gear blank has been moved into the cutter until the pitch circle of the gear blank is tangent to the pitch line of the rack cutter. The cutter is then given a reciprocating motion across the face of the gear blank as the gear blank slowly rotates and the rack translates. The cutting of the space between two teeth is not accomplished in one pass, but rather, requires several passes of the cutter parallel to the gear axis. The cutting and rolling action is continued until the end of the rack is reached, at which point the gear blank and the cutting rack are repositioned and the rolling and cutting action is continued until all the teeth on the gear have been cut.

The pinion cutter, as the name implies, is in the form of a gear rather than a rack. (See Figure 6.23.) The cutting operation for the pinion cutter is basically the same as that for the rack cutter; both are shaper operations in which the cutting tool passes back and forth across the rotating gear blank. The pinion cutter is the tool used to cut internal gears.

Two principal advantages to using generating cutters rather than milling cutters are that a much higher degree of accuracy can be obtained in the cutting process and a single cutter can be used to cut gears with any number of teeth of the same pitch.

The third general method of manufacturing gears is molding. Injection molding and die casting are used when a large number of gears is required. Injection molding is employed when the material is a plastic, while die casting is often the process utilized for metals such as brass and aluminum.

FIGURE 6.23 Generation of a spur gear with a pinion cutter. (*Source:* Fellows Gear Shaper Company.)

Powder metallurgy (P/M) processes are also used to form gears. Metal powders are compressed in a die and then sintered or heated in a controlled-atmosphere furnace. P/M processes can produce complex shapes (e.g., components of compound planetary gear drives).

A drawing process is used to produce steel, stainless steel, aluminum, bronze, and brass gears of small diameter (up to 1 in). The material is drawn through a series of dies to form the teeth. The final product, called *pinion wire*, is cut to form gears with the desired face width.

Gears intended for high-speed, high-load applications often require a finishing process. One method of finishing is shaving, which results in the removal of small amounts of the surface of the gear. Another popular method used to finish gears is grinding. In this method, a form grinder or a grinding wheel is used to obtain a high degree of accuracy. Of the two methods, grinding produces the more accurate finish. Other finishing methods are honing, lapping, and burnishing.

Gear tolerances are specified according to a *quality number*, where high quality numbers imply small tolerances (high precision). High quality numbers are often specified for high-speed operation and for critical applications such as aircraft guidance systems and space navigation systems. Low quality numbers are specified for many agricultural machinery and construction equipment applications. The American Gear Manufacturers Association (1988) covers gear classification and inspection in detail.

6.8 SLIDING ACTION OF GEAR TEETH

In our earlier discussion of friction wheels, we let the motion between the cylinders be pure rolling, except when the transmitted force was large enough to cause sliding. For gears, however, pure rolling motion occurs only when the contact point between gear teeth is at the pitch point (for spur gears, bevel gears, and helical gears on parallel shafts). Every other point of contact along the line of action results in sliding of one tooth on the other. Relative sliding velocity is of importance in the design of gears because sliding is one of the factors contributing to gear tooth wear.

Figure 6.24 shows two teeth in contact at the pitch point. The pitch-line velocity, v_p is given by

$$v_p = r_1\omega_1 = r_2\omega_2.$$

Now, consider velocity components normal and tangential to the tooth surfaces. Since the coincident contact points. P_1 and P_2 have the same velocity, it follows that the normal and tangential components of velocity will also be equal; that is, $v_1^n = v_2^n$ and $v_1^t = v_2^t$. The identical normal components represent the velocity of the gears along the line of action. Recall from Chapter 3 that the difference in the tangential components of velocity of the two gears represents the relative motion, or the sliding. Since in this case the tangential components of velocity are equal, there is no relative motion and therefore no sliding velocity. The motion is pure rolling at the instant in question.

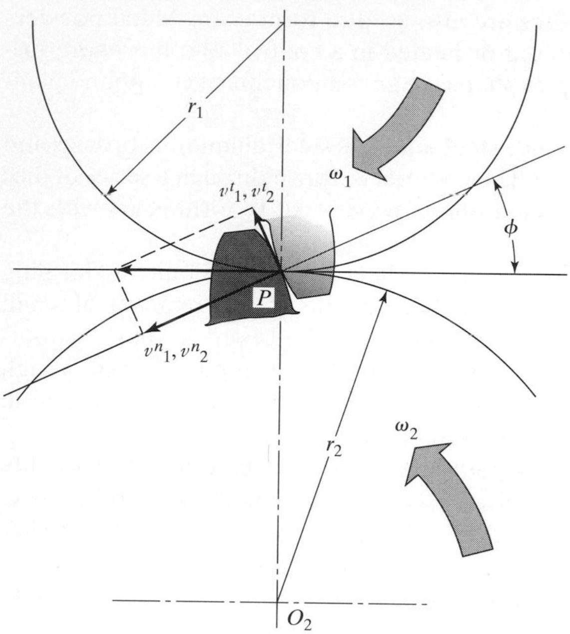

FIGURE 6.24 The contact between the gears at the instant shown is at the pitch point. The pitch-line velocity is identical for P_1 and P_2. Since normal and tangential velocity components are also identical, there can be no relative, or sliding, motion. At this point, the motion is pure rolling.

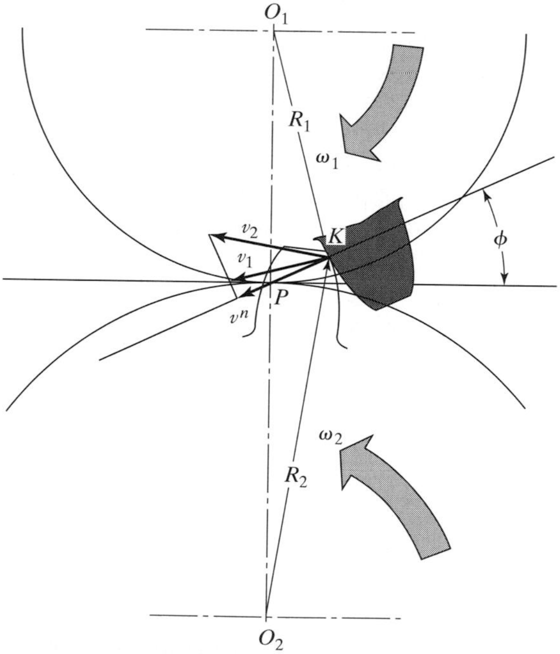

FIGURE 6.25 Unlike the contact point in Figure 6.24, the contact point K does not coincide with the pitch point P. In this case, the tangential velocities are unequal, and sliding of one gear on the other must occur.

Figure 6.25 shows two teeth in contact at a point K on the line of action other than the pitch point. The velocities v_1 and v_2 are again given by $v_1 = R_1\omega_1$ and $v_2 = R_2\omega_2$, respectively, in the directions shown. (Note that R_1 and R_2 are not pitch radii.) The components of these velocities acting along the line of action are equal and are shown in the figure as v^n. Note that if the components along the line of action were unequal, the gear teeth would be separating. Components of v_1 and v_2 in the tangential direction are unequal, and therefore, one tooth must slide on the other.

The greater as the distance of the point of contact between the teeth from the pitch point, the greater the sliding velocity becomes. The sliding velocity will therefore be maximum at the points where the teeth first come into contact and where they leave contact. The sliding velocity can be determined by any of the procedures discussed in Chapter 3.

6.9 INTERFERENCE (EXTERNAL GEARS)

Involute gear teeth have involute profiles between the base circle and the addendum circle. If the dedendum circle lies inside the base circle, the portion of the tooth between the base circle and the dedendum circle will not be an involute. Instead, that portion of the tooth profile may be a radial line with a fillet where it joins the base circle, a true clearance curve, or a circular arc. For this reason, if contact between two gears occurs below the base circle of one of the gears, interference is said to occur. The contact is then between two nonconjugate curves, and the fundamental law of gearing will be broken.

In other words, interference occurs whenever the addendum circle of a gear intersects the line of action beyond the *interference point*, which is the point where the line of action is tangent to the base circle of the other gear.

To better understand interference, consider Figure 6.26. Point B is the intersection of the addendum circle of the gear with the line of action. Points C and D are the points where the base circles are tangent to the line of action. Interference will occur, since point B lies outside line segment CD. The interfering portion of the tooth of gear 2 is shown as the shaded area in the figure 6.26.

To avoid interference, the size of the addendum circle diameter must be limited. Referring again to Figure 6.26, we see that the maximum radii that the addendum circles may have to avoid interference are O_1C and O_2D. From right triangle O_1CD, we observe the following relationships:

$$O_1C = r_{a_1}(\text{radius of addendum circle}) = \sqrt{(O_1D)^2 + (CD)^2};$$
$$O_1D = O_1P\cos\phi = r_1\cos\phi = r_{b_1};$$
$$PD = O_1P\sin\phi;$$
$$CP = O_2P\sin\phi;$$
$$CD = CP + PD = (O_1P + O_2P)\sin\phi.$$

Since the distance between gear centers is

$$c = O_1P + O_2P,$$

the expression for CD becomes

$$CD = c\sin\phi.$$

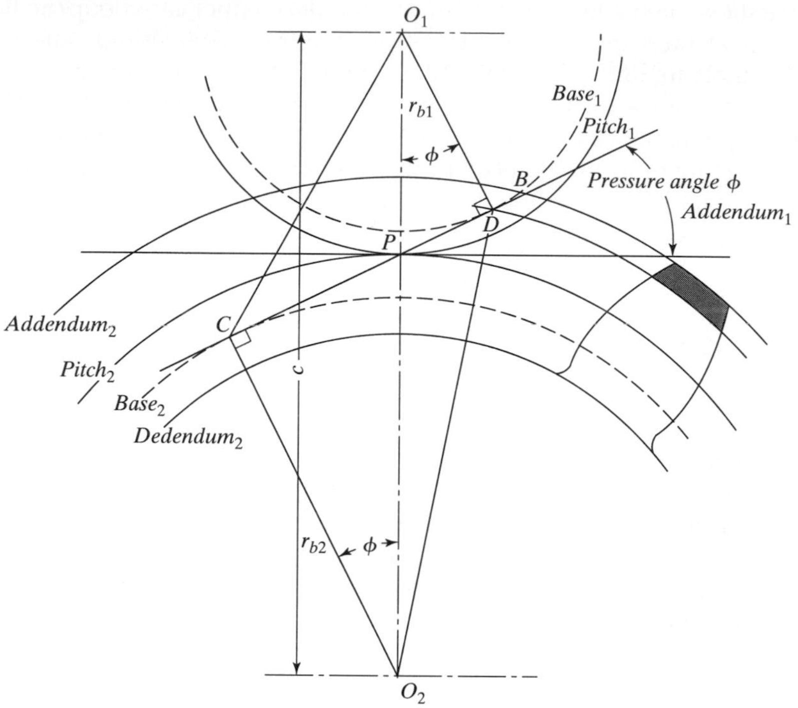

FIGURE 6.26 Points C and D are referred to as interference points. The part of the tooth that would have to be removed in order to prevent interference is shown shaded.

The expression for the radius of the addendum circle of gear 1 then becomes

$$O_1C = \sqrt{r_1^2 \cos^2\phi + c^2 \sin^2\phi},$$

or

$$r_{a_1}(\max) = \sqrt{r_1^2 \cos^2\phi + c^2 \sin^2\phi} = \sqrt{r_{b_1}^2 + c^2 \sin^2\phi}. \quad (6.18)$$

Similarly, for the other gear,

$$r_{a_2}(\max) = \sqrt{r_2^2 \cos^2\phi + c^2 \sin^2\phi} = \sqrt{r_{b_2}^2 + c^2 \sin^2\phi}. \quad (6.19)$$

If the addendum circle-radius exceeds the calculated value, interference will occur. If it is equal to or less than the calculated value, no interference will occur. However, clearance must still be provided for normal operation.

SAMPLE PROBLEM 6.10

Detection of Gear Interference

Two standard 20° full-depth gears have a module of 8 mm. The larger gear has 30 teeth, while the pinion has 15 teeth. Will the gear interfere with the pinion?

Solution. Equation (6.19) will indicate the maximum permissible addendum-circle radius for the gear in this problem. We must therefore calculate the base-circle radius for the gear, r_{b_2}, and the center distance c in order to use the formula. Thus, from Eq. (6.5),

$$d_2 = mN_2 = 8(30) = 240 \text{ mm}$$

and

$$d_1 = mN_1 = 8(15) = 120 \text{ mm.}$$

Hence,

$$c = \frac{1}{2}(d_1 + d_2) = \frac{1}{2}(120 + 240) = 180 \text{ mm.}$$

The radius of the base circle of the gear is therefore

$$r_{b_2} = r_2 \cos\phi = \frac{240}{2}\cos 20^\circ = 112.8 \text{ mm.}$$

We can now determine the maximum permissible addendum radius. By comparing the result with the actual value as determined from Table 6.1, we find whether interference exists in this case. Maximum permissible addendum radius

$$\begin{aligned} r_{a_2}(\text{max}) &= \sqrt{r_{b_2}^2 + c^2 \sin^2\phi} \\ &= \sqrt{(112.8)^2 + (180)^2(0.342)^2} = 128.5 \text{ mm.} \end{aligned}$$

From Table 6.1, for the 20° full-depth gear,

$$a_2 = m = 8 \text{ mm,}$$

and the addendum-circle radius is the sum of the pitch radius and the addendum, or

$$r_{a_2} = r_2 + a_2 = 120 + 8 = 128 \text{ mm.}$$

Comparing the maximum permissible addendum radius with the actual value, we find that there is no interference, since the actual addendum circle radius is slightly less than the maximum allowable addendum circle radius. If r_{a_2} had been greater than r_{a_2} (max), interference would have occurred.

Note that, for a pair of standard gears, if the addendum of the larger gear does not interfere with the smaller gear, then it follows that the addendum of the smaller gear will not interfere with the larger gear. That is, if r_{a_2} is less than r_{a_2} (max), then r_{a_1} will be less than r_{a_1} (max), where the subscript 1 designates the smaller of the two gears. This relationship can be illustrated by considering the preceding example again, in which it was determined that

$$r_{a_2} = 128 \text{ mm and } r_{a_2}(\text{max}) = 128.5 \text{ mm,}$$

and therefore, no interference occurs, although the condition is close to interference. Examining the smaller gear, we have

$$r_{a_1} = r_1 + a_1 = 60 + 8 = 68 \text{ mm}$$

and

$$\begin{aligned} r_{a_1}(\max) &= \sqrt{r_1^2 \cos^2\phi + c^2 \sin^2\phi} \\ &= \sqrt{(60)^2(0.940)^2 + (180)^2(0.342)^2} = 83.5 \text{ mm}. \end{aligned}$$

As can be seen, the actual addendum circle radius of 68 mm for the pinion is well within the limit of 83.5 mm.

The following limiting condition based on Eq. (6.19) can be used to determine the number of gear teeth necessary to avoid interference:

$$(r_2 + a_2)^2 \le r_2^2 \cos^2\phi + c^2 \sin^2\phi$$

Again, the subscript 2 represents the larger of the two meshing gears. Substituting the expressions

$$r_2 = \frac{N_2}{2P}, \; a_2 = \frac{k}{P}, \text{ and } c = \frac{1}{2P}(N_1 + N_2)$$

leads to

$$\left(\frac{N_2}{2P} + \frac{k}{P}\right)^2 \le \left(\frac{N_2}{2P}\right)^2 \cos^2\phi + \left(\frac{N_1 + N_2}{2P}\right)^2 \sin^2\phi,$$

where $k = 1$ for standard full-depth gears. Factoring P from the preceding inequality and simplifying yields

$$4kN_2 + 4k^2 \le N_1^2 \sin^2\phi + 2N_1N_2 \sin^2\phi. \tag{6.20}$$

For any given pressure angle ϕ and addendum constant k, this inequality can be used to determine the necessary number of teeth on one of the gears in terms of the number of teeth on the other gear in order to avoid interference. For example, rearranging terms, we have

$$4k - 2N_1 \sin^2\phi \le \frac{N_1^2 \sin^2\phi - 4k^2}{N_2}.$$

If we now consider the extreme case where N_2 approaches infinity, (i.e., the gear becomes a rack), then

$$4k - 2N_1 \sin^2 \phi \le 0,$$

or

$$N_1 \ge \frac{2k}{\sin^2 \phi},$$

for noninterference with a rack. The equation

$$N_1 = \frac{2k}{\sin^2 \phi} \tag{6.21}$$

defines the minimum number of teeth that a pinion can have without interference occurring when the pinion meshes with a rack. The following values of N_1 for standard tooth forms are obtained from this equation, where the fractional value calculated is rounded to the next highest integer:

$14\frac{1}{2}°$ full depth: $N_1 = 32$;
20° full depth: $N_1 = 18$;
20° stub: $N_1 = 14$ ($k = 0.8$);
25° full depth: $N_1 = 12$.

These values of N_1 satisfy the general inequality condition of Eq. (6.20) not only for $N_2 \to \infty$, but also for any value of $N_2 \ge N_1$. However, the values listed are no longer the minimum values required. That is, in some cases smaller numbers of teeth can be used if the pinion meshes with a gear other than a rack.

Minimum Number of Pinion Teeth Required to Mesh with an External Gear

Let us check the last statement in the preceding paragraph if, for example, 25° pressure angle full-depth teeth are specified, can a pinion with less than 12 teeth mesh with an external gear? We can set $k = 1$ for full-depth teeth replace "$\le$" with "$=$" in the interference inequality, and solve for the minimum number of pinion teeth in terms of the number of gear teeth. Software with symbolic capability makes the solution easier. If the result is not a whole number, we go the next higher integer.

SAMPLE PROBLEM 6.11

Minimum Number of Teeth to Avoid Interference

Suppose we plan to use gears with up to 80 teeth. Plot the minimum pinion size vs. the gear size.

Decision. We will consider standard full-depth teeth with pressure angle $\phi = 14.5°$, 20°, and 25°.

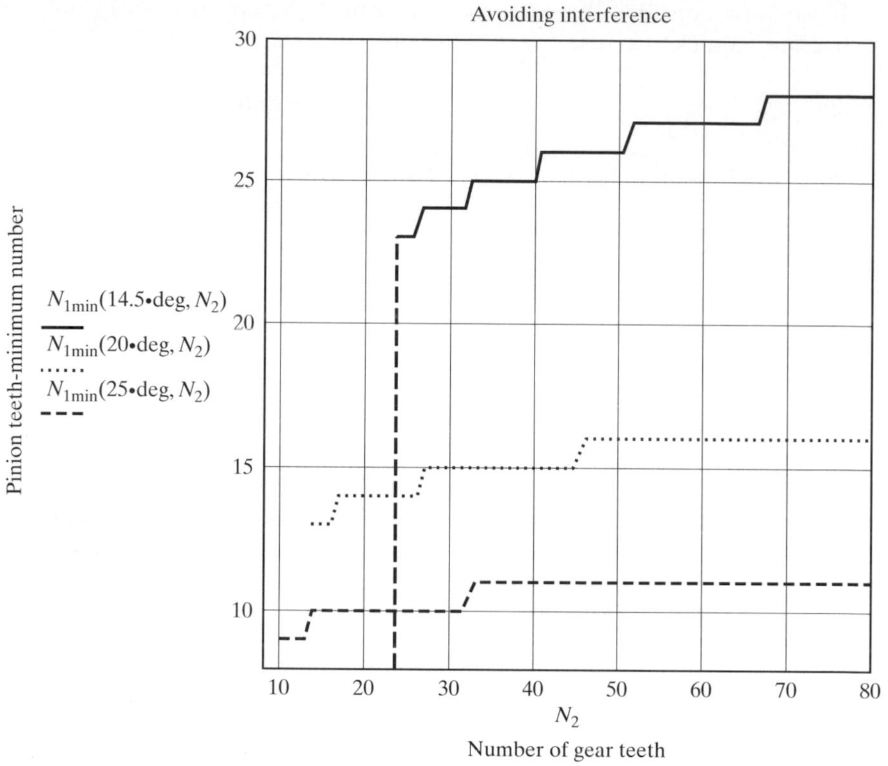

FIGURE 6.27 Minimum number of gear and pinion teeth required to avoid interference.

Solution summary. Software with symbolic capability is used to help us solve for N_1(minimum), as outlined in the paragraph preceding this sample problem. We select the second (positive) root and consider gears with 8 to 80 teeth. The interference equation assumes that $N_2 \geq N_1$; we reject other solutions. Plotted results (Figure 6.27) show the minimum number of gear and pinion teeth required to avoid interference. For example, if we select a 25° pressure angle, a pinion with at least 10 teeth can mesh with a gear that has 14 to 32 teeth.

Solution details.

Minimum Number Of Teeth (Full-Depth)

Pinion teeth N_1; Gear teeth N_2 Solve for minimum number of pinion teeth;

$$4 - 2N_1(\sin(\phi))^2 = \frac{(N_1\sin(\phi))^2 - 4}{N_2} \text{ solve,}$$

$$N_1 \rightarrow \begin{bmatrix} \frac{-1}{2} \cdot \frac{N_2}{\sin(\phi)^2} \cdot \left(2 \cdot \sin(\phi)^2 + 2 \cdot \frac{\sin(\phi)}{N_2} \cdot \sqrt{\sin(\phi)^2 \cdot {N_2}^2 + 4 \cdot N_2}\right) \\ \frac{-1}{2} \cdot \frac{N_2}{\sin(\phi)^2} \cdot \left(2 \cdot \sin(\phi)^2 - 2 \cdot \frac{\sin(\phi)}{N_2} \cdot \sqrt{\sin(\phi)^2 \cdot N_2^2 + 4 \cdot N_2}\right) \end{bmatrix}$$

$$N_1(\phi \cdot N_2) := \text{ceil}\left[\frac{-1}{2} \cdot \frac{N_2}{\sin(\phi)^2} \cdot \left(2 \cdot \sin(\phi)^2 - 2 \cdot \frac{\sin(\phi)}{N_2} \cdot \sqrt{\sin(\phi)^2 \cdot N_2^2 + 4 \cdot N_2 + 4}\right)\right]$$

Do not plot if $N_1 > N_2$ $\quad N_{1\,min}(\phi, N_2) := \text{if}(N_1(\phi, N_2) < N_2 \cdot N_1(\phi, N_2), 0)$

Contact ratio and interference are important measures of performance for gears. If interference is predicted, we must revise the design. In an earlier section, we made a three-dimensional bar chart of the contact ratio, but X'd out some gear combinations due to interference. This is explained in the next sample problem.

SAMPLE PROBLEM 6.12

Contact Ratio and Interference

A 20°-pressure-angle, 14-tooth pinion is to be used in a gearbox. The meshing gear may have from 14 to 35 teeth, depending on the speed ratio requested by the customer. Find the contact ratios.

Solution summary. We will use the contact ratio and interference equations derived previously. Contact ratios (CRs) are calculated and tabulated for the stated range of gears; however, the pinion must have a minimum of 15 teeth to mesh with a gear of 27 or more teeth. Thus, contact ratio values are valid only if the gear has 26 teeth or less. The valid range of $CR(N_1, N_2, \phi)$ is $CR(14, 14, 20°) = 1.463$ to $CR(14, 26, 20°) = 1.542$. If we had selected a 15-tooth pinion, it could mesh with gears of up to 100 teeth or so. The solution is also used to simulate a rack by putting in a large number (10^{10}) of gear teeth. Eighteen pinion teeth are required; the result checks previous calculations.

Solution details.

Contact Ratio and Interference: Full-Depth Teeth

Pinion teeth $N_1 := 14$ Gear teeth $N_2 := 14 \ldots 35$. Pressure angle $\phi = 20.\text{deg}$.

Contact Ratio

$$CR(N_1, N_2, \phi) := \frac{[(N_1 + 2)^2 - (N_1 \cdot \cos(\phi))^2]^{\frac{1}{2}} + [(N_2 + 2)^2 - (N_2 \cdot \cos(\phi))^2]^{\frac{1}{2}} - (N_1 + N_2) \cdot \sin(\phi)}{2 \cdot \pi \cdot \cos(\phi)}$$

Minimum Number of Teeth to Avoid Interference

Round NN_1 upward:

$$NN_1(\phi, N_2) := \text{ceil}\left[\frac{-1}{2} \cdot \frac{N_2}{\sin(\phi)^2} \cdot \left(2 \cdot \sin(\phi)^2 - 2 \cdot \frac{\sin(\phi)}{N_2} \cdot \sqrt{\sin(\phi)^2 \cdot N_2^2 + 4 \cdot N_2 + 4}\right)\right]$$

Not applicable if $N_1 > N_2$ $\quad N_{1\,min}(\phi, N_1, N_2) := \text{if}(N_1 \le N_2, NN_1(\phi, N_2), \text{"NA"})$

N_2	$CR(N_1, N_2, \phi)$	$N_{1\,min}(\phi, N_1, N_2)$
14	1.463	13
15	1.472	13
16	1.481	13
17	1.489	13
18	1.496	14
19	1.503	14
20	1.51	14
21	1.516	14
22	1.522	14
23	1.527	14
24	1.532	14
25	1.537	14
26	1.542	14
27	1.546	15
28	1.55	15
29	1.554	15
30	1.558	15
31	1.562	15
32	1.565	15
33	1.568	15
34	1.572	15
35	1.475	15

If gear 2 is a rack:

$N_{1\,min}(\phi, N_1, 10^{10}) = 18$

The preceding equations and results apply to a pair of external gears, or a rack and pinion. Internal gears (ring gears) present additional problems; the equations do not apply if one of the gears is an internal gear. Interference may occur in an internal gear set if the two gears are close in size. A suggested guideline is

$$N_I - N_P \geq 15 \quad \text{for} \quad \phi = 14.5$$

and

$$N_I - N_P \geq 12 \quad \text{for} \quad \phi = 20,$$

where N_I = number of internal gear teeth,
N_P = number of pinion teeth, and
ϕ = pressure angle.

Undercutting

In Section 6.7, the generating gear cutters (rack and pinion) were discussed. During cutting, the cutter and gear blank act in a manner similar to two meshing gears. It is therefore possible for interference to occur. However, since one of the elements is a cutting

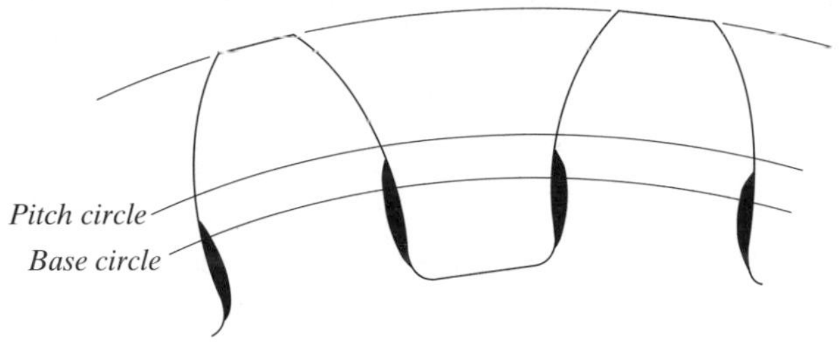

FIGURE 6.28 When gears are undercut, the shaded portions of the teeth are removed. Since a part of the profile that lies above the base circle is removed in the process, the result is a reduced length of contact.

tool, the portion of the gear that would be interfering is cut away. A gear that has had material removed in this manner is said to be *undercut*.

Figure 6.28 shows undercut gear teeth, with the undercut portion shaded. As can be seen, undercutting also removes a portion of the tooth profile above the base circle. This is a definite disadvantage, since the length of contact is reduced, which in turn decreases the contact ratio and results in rougher and noisier gear action, because fewer teeth are in contact. Moreover, since undercutting removes material from the base of the tooth, it also may seriously weaken the tooth.

If undercutting is permitted, the minimum recommended number of teeth is 16 for 14.5°-pressure-angle gears and 13 for 20°-pressure-angle gears.

Stub Teeth

To reduce the amount of interference, methods other than undercutting are also available. For example, interference can be eliminated if the height of the tooth is reduced by cutting off a portion of its tip. Interference occurs when the tip of one gear is in contact below the base circle of the mating gear. Removing a portion of the tip of the tooth will therefore prevent contact below the base circle. This type of gear is called a *stub-tooth gear*. We must recalculate and evaluate the contact ratio.

The Effect of the Pressure Angle

Increasing the pressure angle will also decrease the problem of interference. A larger pressure angle decreases the diameter of the base circle and thus increases the involute portion of the tooth profile. But larger pressure angles increase shaft loading as well.

An equation for determining the minimum pressure angle for which no interference will be present can be derived from Eq. (6.19). We have

$$r_a(\max) = \sqrt{r_b^2 + c^2 \sin^2 \phi}$$

and

$$r_a^2(\max) = r_b^2 + c^2 \sin^2 \phi,$$

and since $r_b = r\sin\phi$,

$$r_a^2(\max) = r^2\cos^2\phi + c^2\sin^2\phi.$$

But $\cos^2\phi + \sin^2\phi = 1$; therefore,

$$r_a^2(\max) = r^2(1 - \sin^2\phi) + c^2\sin^2\phi = r^2 + (c^2 - r^2)\sin^2\phi.$$

Hence,

$$\sin^2\phi = \frac{r_a^2(\max) - r^2}{c^2 - r^2},$$

so that

$$\sin\phi = \sqrt{\frac{r_a^2(\max) - r^2}{c^2 - r^2}}. \tag{6.22}$$

In employing Eq. (6.22), the standard value for the addendum is to be used in determining r_a (max). Therefore,

$$\sin\phi_{\min} = \sqrt{\frac{r_a^2 - r^2}{c^2 - r^2}}, \tag{6.23}$$

where r_a and r refer to the larger gear. Of course, the value for the pressure angle ϕ must still provide an acceptable contact ratio.

SAMPLE PROBLEM 6.13

Minimum Pressure Angle to Avoid Interference

We plan to design a line of gear sets. Find the minimum allowable pressure angle.

Design decisions. We will use pinions with 9 to 14 teeth and gears with 9 to 60 teeth. We will design for full-depth addendums, and determine the minimum pressure angle necessary to avoid interference.

Solution summary. The equation for the pressure angle in the preceding paragraph is rewritten in terms of tooth numbers and module. The larger gear of the pair is the offender when it comes to interference.

Assuming that $N_2 \geq N_1$, we substitute

$$r = m\,N_2/2,$$
$$r_a = m\,N_2/2 + m,$$

and

$$c = (N_1 + N_2)/(2\,m).$$

We then simplify to get an equation for the minimum pressure angle in terms of tooth numbers only. An IF-statement is used to call for a revised equation for combinations where in $N_1 > N_2$. The result is then plotted for each possible combination in the specified range. (See Figure 6.29.)

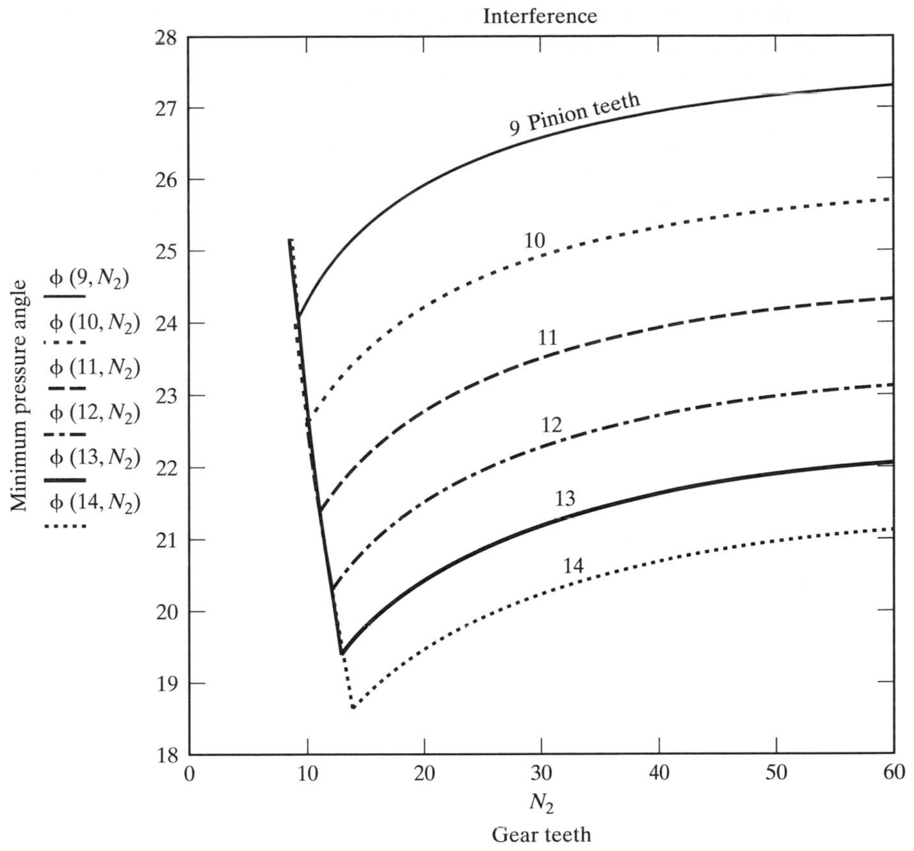

FIGURE 6.29 Minimum pressure angle required to avoid interference vs. number of gear and pinion teeth (based on full-depth teeth).

Note that the final result would be the same if we had used the customary U.S. form instead of the metric form, substituting

$$r = N_2/(2P), \quad \text{etc.}$$

Solution details.

Interference

Minimum pressure angle based on interference (full-depth teeth)

Pinion teeth N_1. Gear teeth $N_2 := 9,10\ldots60$.

$$\phi(N_1,N_2) := \text{if}\left[N_2 \geq N_1, \frac{\text{asin}\left[2\left(\frac{N_2+1}{2\cdot N_1\cdot N_2+N_1^2}\right)^{\frac{1}{2}}\right]}{\text{deg}}, \frac{\text{asin}\left[2\left(\frac{N_1+1}{2\cdot N_1\cdot N_2+N_2^2}\right)^{\frac{1}{2}}\right]}{\text{deg}}\right]$$

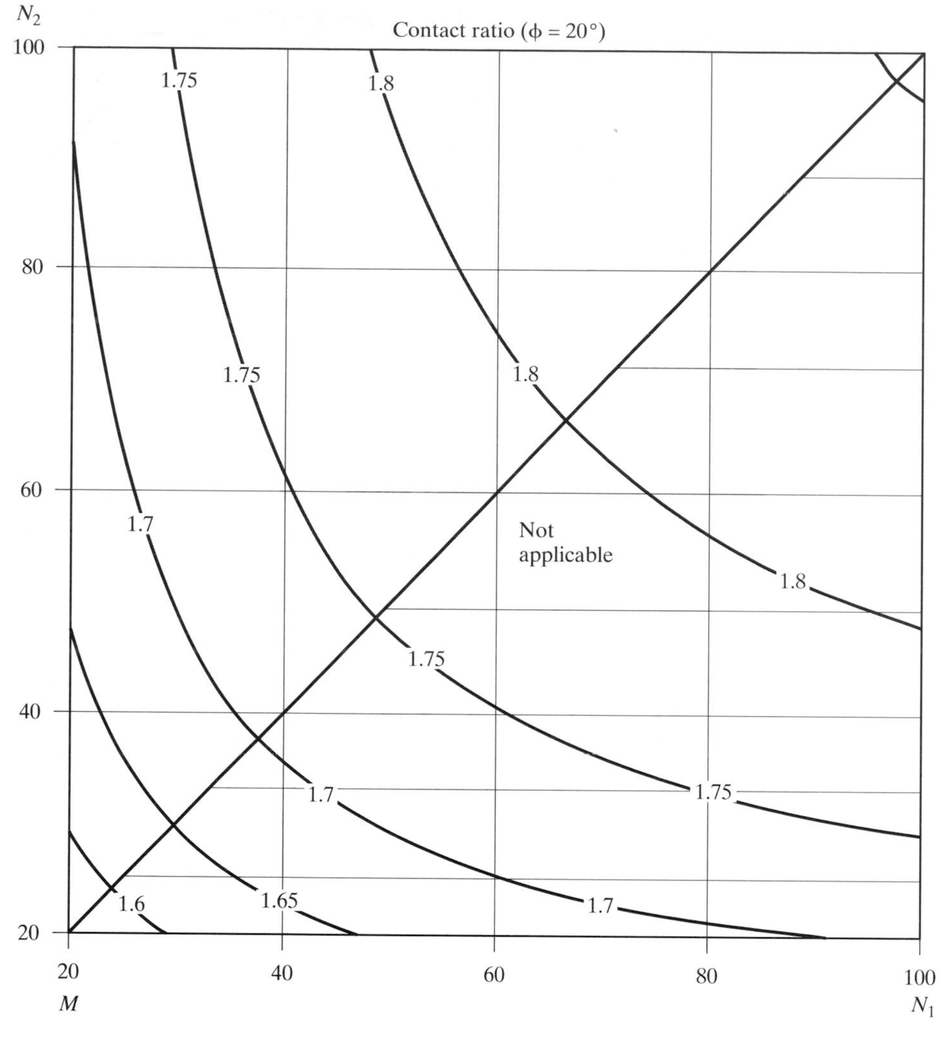

FIGURE 6.31 Contour plot of contact ratios for various gear and pinion tooth number combinations (based on a pressure angle of 20°).

Eliminating Interference by Using Unequal Addendums

The final method commonly used to eliminate interference is to cut the gears with unequal addendum and dedendum teeth. This is accomplished by increasing the addendum of the driver while decreasing its dedendum. The mating gear is then cut with a decreased addendum and an increased dedendum. The result of this procedure is to increase the length of action for which involute action is obtainable. Gears of this type are usually called *long-and-short addendum teeth gears*.

The disadvantage of cutting gears with unequal addendum and dedendum teeth is an increase in the cost of the gear and the fact that gears cut in this manner are non-interchangeable. Such gears are also known as *nonstandard gears*.

Finally, it should be clear from the previous discussion that the method used to eliminate interference depends on the application at which the gear is aimed.

6.10 GEAR TOOTH FORCES

Aside from the kinematic considerations already discussed, the forces and torques acting on spur gears are of vital importance to the designer. As was discussed earlier, the normal force one gear exerts upon another always lies along the line of action between them (also known as the pressure line).

In Figure 6.32, with gear 1 the driver, the normal force F_n exerted by gear 1 on gear 2 is shown. Keep in mind that no matter where on the line of action the two teeth are in contact, the normal force will always pass through the pitch point P. At the pitch point, the normal force can be resolved into the two components F_t, the tangential force, and F_r, the radial force. Thus

$$F_t = F_n \cos \phi \tag{6.24}$$

and

$$F_r = F_n \sin \phi, \tag{6.25}$$

or

$$F_r = F_t \tan \phi, \tag{6.26}$$

where ϕ is the pressure angle and the forces are expressed in pounds in the English system or in newtons in the SI system.

Clearly, the normal force exerted by gear 2 on gear 1 would act along the pressure line, but in the direction opposite to that shown in the figure.

The radial force on external spur gears always acts inward toward the center of the gear. On the other hand, the radial force acting on internal spur gears always acts

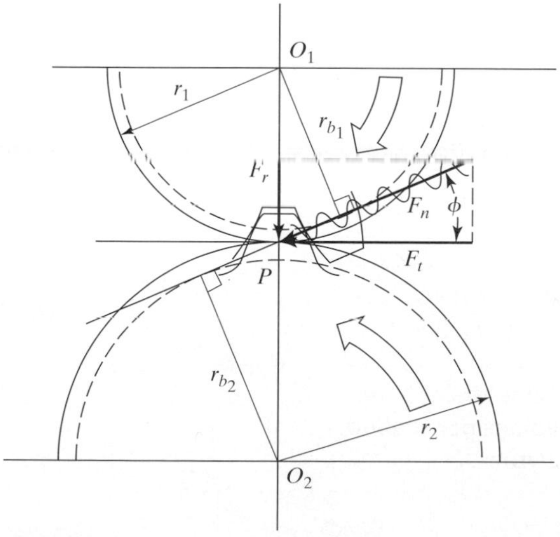

FIGURE 6.32 The normal force F_n exerted by gear 1 on gear 2 is shown, as well as its two components: the tangential force F_t and the radial force F_r. Gear 1 is the driver.

outward from the center of rotation. Obviously, the radial force tends to move the gear out of contact with its meshing gear.

The direction of the normal force along the pressure line is determined by the directi n of rotation of the driver. Follow the direction of rotation of the driver along the pressure line to determine the direction of the normal force on the driven gear. Clearly, the normal force exerted by the driven gear on the driver acts in the opposite direction.

The torque that is produced about the center of a gear is given by

$$T = F_n\left(\frac{d_b}{2}\right) = F_n\left(\frac{d}{2}\right)\cos\phi = F_t\left(\frac{d}{2}\right), \tag{6.27}$$

where F_n = normal force,
F_t = tangential force,
d = pitch circle diameter,
d_b = base-circle diameter, and
T = torque.

The power that can be transmitted from one gear to another is determined from

$$hp = \frac{Tn}{63{,}025}, \tag{6.28}$$

or

$$hp = \frac{F_t v_p}{33{,}000} \tag{6.29}$$

where, in the English system,

T = torque in inch-pounds,
n = revolutions per minute,
F_t = tangential force in pounds,
v_p = pitch-line velocity in feet per minute, and
hp = power in units of horsepower.

In the SI system, power is

$$kW = \frac{T\omega}{1{,}000{,}000}, \tag{6.30}$$

or

$$kW = \frac{F_t v_p}{1{,}000{,}000}, \tag{6.31}$$

where T = torque in newton-millimeters,
ω = rotational speed in radians per second,
F_t = tangential force in newtons,

v_p = pitch-line velocity in millimeters per second, and
kW = power in kilowatts.

Finally, it should be noted that the efficiency of power transmission from one gear to another is not 100 percent. Generally, we can expectabout a 1- or 2-percent power loss at maximum transmitted power for spur gears. Unless stated otherwise, we will assume, when doing problems, that there is no power loss.

SAMPLE PROBLEM 6.15

Force Analysis of Spur Gears

For the three gears shown in Figure 6.33, gear 1, the driver, rotates at 1000 rev/min clockwise and delivers 30 kW of power. Gear 1 has a module of 10 mm, a pressure angle of 20°, and 35 teeth, while gear 2 has 45 teeth and gear 3 has 60 teeth.

(a) What is the distance between A and C?
(b) What is the speed ratio between gears 1 and 3?
(c) If gear 2 were not in the train, what would be the speed ratio between gears 1 and 3? On the basis of this answer, what purpose does the idler gear, gear 2, serve?
(d) Draw and completely label the free-body force diagram of each gear.

Solution. (a) From Eq. (6.5),

$$d_1 = mN_1 = 10(35) = 350 \text{ mm},$$
$$d_2 = mN_2 = 10(45) = 450 \text{ mm},$$

and

$$d_3 = mN_3 = 10(60) = 600 \text{ mm}.$$

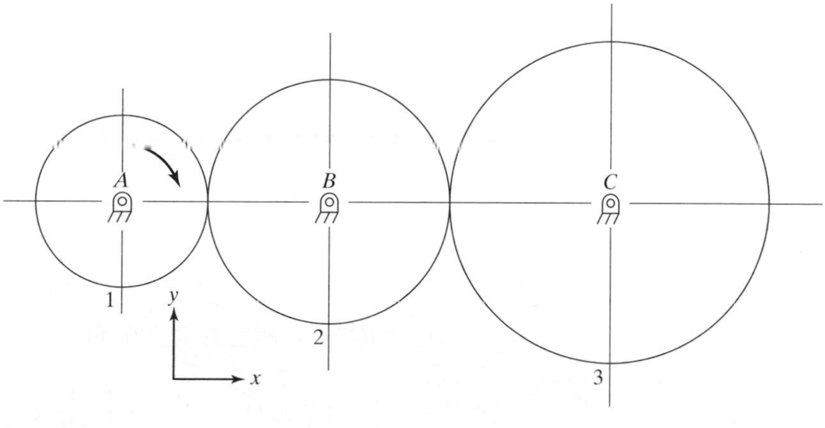

FIGURE 6.33 A gear train consisting of three spur gears.

By Eq. (6.10),

$$c_{AB} = \frac{1}{2}(d_1 + d_2) = \frac{1}{2}(350 + 450) = 400 \text{ mm}$$

and

$$c_{BC} = \frac{1}{2}(d_2 + d_3) = \frac{1}{2}(450 + 600) = 525 \text{ mm}.$$

Therefore,

$$c_{AC} = c_{AB} + c_{BC} = 400 + 525 = 925 \text{ mm}.$$

Check: Upon considering the figure, it is clear that the distance between shafts A and C is

$$r_1 + r_2 + r_2 + r_3 = \frac{350}{2} + \frac{450}{2} + \frac{450}{2} + \frac{600}{2} = 925 \text{ mm}.$$

(b) From Eq. (6.8),

$$r_{v_{1\text{-}3}} = \frac{N_1}{N_2} \times \frac{N_2}{N_3} = \frac{35}{45} \times \frac{45}{60} = 0.583.$$

(c) Without gear 2,

$$r_v = \frac{N_1}{N_3} = \frac{35}{60} = 0.583.$$

Obviously, gear 2 does not affect the speed ratio between gears 1 and 3. However, with gear 2 in the system, gear 3 rotates clockwise. Without gear 2, gear 3 would rotate counterclockwise. Therefore, the purpose in introducing an idler gear between the driver and driven gears is to permit the driver and driven gears to have the same direction of rotation. (d) The angular velocity of gear 1 is

$$\omega_1 = 1000 \text{ rev/min}\left(\frac{2\pi \text{ rad}}{1 \text{ rev}}\right)\left(\frac{1 \text{ min}}{60 \text{ s}}\right) = 104.7 \text{ rad/s}.$$

From Eq. (6.1),

$$v_p = r_1\omega_1 = \frac{350}{2} \text{ mm}(104.7 \text{ rad/s}) = 18{,}326 \text{ mm/s}.$$

By Eq. (6.31),

$$F_{t_{1\text{-}2}} = \frac{kW(1{,}000{,}000)}{v_p} = \frac{30(1{,}000{,}000)}{18{,}322} = 1637 \text{ N}.$$

From Eq. (6.27),

$$T_1 = F_{t_{1-2}}\left(\frac{d_1}{2}\right) = 1637\text{ N}\cdot\left(\frac{350}{2}\text{ mm}\right) = 286{,}500\text{ N}\cdot\text{mm} = 286.5\text{ N}\cdot\text{m}.$$

By Eq. (6.26),

$$F_{r_{1-2}} = F_{t_{1-2}}\tan\phi = 1637\tan 20^\circ = 1637(0.364) = 595.9\text{ N}.$$

The pressure line is located by rotating the common tangent through the pressure angle in a direction opposite that of rotation of the driver. The force F_{A_x}, which is equal in magnitude to force $F_{r_{1-2}}$, and the force F_{A_y}, which is equal in magnitude to force $F_{t_{1-2}}$, are the forces that shaft A exerts on gear 1. The torque T_A, which is equal in magnitude to torque T_1, is the torque the shaft exerts on gear 1. Figure 6.34 shows all of these vectors, with their proper directions, on a free-body diagram of gear 1.

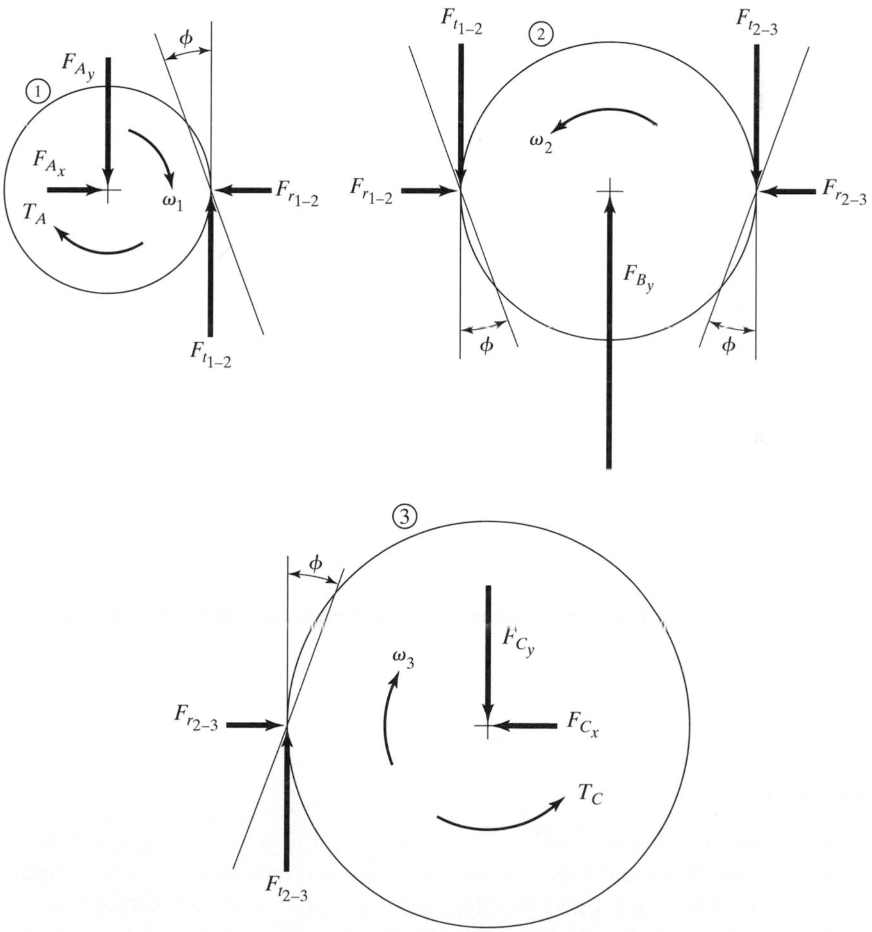

FIGURE 6.34 Free-body force diagrams of the gears in the gear train of Figure 6.29 and Sample Problem 6.10.

From Newton's third law (action and reaction), and since the net torque on gear 2 must be zero, it follows that

$$F_{t_{2\text{-}3}} = F_{t_{1\text{-}2}} \text{ and } F_{r_{2\text{-}3}} = F_{r_{1\text{-}2}}.$$

(See Figure 6.30.) And for gear 3,

$$F_{C_y} = F_{t_{2\text{-}3}} = 1{,}637 \text{ N}$$
$$F_{C_x} = F_{r_{2\text{-}3}} = 595.9 \text{ N},$$

and

$$T_C = T_3 = F_{t_{2\text{-}3}}\left(\frac{d_3}{2}\right) = 1637 \text{ N}\cdot\left(\frac{600}{2}\text{ mm}\right) = 491{,}100 \text{ N}\cdot\text{mm} = 491.1 \text{ N}\cdot\text{m}.$$

Check: From Eq. (6.8),

$$\frac{\omega_3}{\omega_1} = \frac{N_1}{N_3}.$$

Therefore,

$$\omega_3 = \omega_1 + \left(\frac{N_1}{N_3}\right) = 104.7 \text{ rad/s}\left(\frac{35}{60}\right) = 61.1 \text{ rad/s}$$

and

$$v_p = r_3\omega_3 = \frac{600}{2}\text{ mm}(61.1 \text{ rad/s}) = 18{,}326 \text{ mm/s}.$$

Using Eq. (6.31), we have

$$F_{t_{2\text{-}3}} = \frac{\text{kW}(1{,}000{,}000)}{v_p} = \frac{30(1{,}000{,}000)}{18{,}330} = 1{,}637 \text{ N}.$$

The complete force diagrams of the individual gears are shown in Figure 6.34. Note that each gear satisfies force and moment equilibrium conditions. Therefore, $F_{B_y} = F_{t_{1\text{-}2}} + F_{t_{2\text{-}3}} = 2F_{t_{1\text{-}2}}$ and $F_{B_x} = 0$, where F_{B_x} and F_{B_y} are the bearing forces on idler gear 2 at shaft B. Torque T_c represents the load on shaft C, against which the gear train is driving.

6.11 GEAR TOOTH FAILURE

Gear tooth failures usually result from bending or wear. Fatigue loading is an important factor. We usually specify face width and module or diametral pitch on the basis of tooth loading, pitch-line velocity, quality number, and related factors. Determining the face width, module, and diametral pitch for given loading scenarios is discussed by American Gear Manufacturers Association (1982) and Wilson (1997).

The consequences of a potential failure also affect our design. Will a tooth failure be a minor inconvenience? Or will there be a costly interruption in production? Or will failure cause possible injury or loss of life? Critical applications sometimes call for extensive testing. If our tentative design is inadequate, we can examine the failure mode as an aid to redesign. American Gear Manufacturers Association (1980) is an excellent reference for this purpose.

6.12 GEAR MATERIALS

Common gear materials include steel, stainless steel, cast iron, bronze, nylon, and phenolics. Steel with 40 points of carbon (0.40% carbon) is often selected for good fracture strength, toughness, impact resistance, and fatigue strength. Gear teeth can be heat treated to improve their resistance to wear. Typical hardened teeth have a Rockwell C-scale hardness of 45 R_c to over 60 R_c. Stainless steel is used when resistance to atmospheric corrosion is required. Food-industry specifications sometimes specify stainless steel. Cast iron is often selected for large gears. Cast iron has good wear characteristics and excellent machinability. Bronze is selected for applications with high sliding velocities, particularly for worm gears. When weight must be minimized, aluminum gears are used. Nylon gears are also light, have a low coefficient of friction, and are flexible and capable of absorbing shock loads. Gears are also made of laminated phenolic fiber, a non metallic material that provides electrical insulation.

SUMMARY

When two gears mesh, the smaller gear is usually called a pinion. When we use gears to reduce speed (the usual case), the pinion is the drive gear. The speed ratio of a pair of gears (other than planetary gears) is the inverse of the ratio of the numbers of teeth. Standard gears are based on the involute form and made with a 14.5°, 20°, or 25° pressure angle. Gear tooth size is based on the module (mm) or the diametral pitch (teeth per inch of diameter). For a pair of gears to mesh, the two gears must have the same pressure angle and the same module or diametral pitch. The contact ratio of a pair of gears is the average number of pairs of teeth in contact. Gears are usually designed with a contact ratio of at least 1.2. Higher values result in smoother, quieter operation.

Gears are designed to avoid interference. Gears with a 14.5° pressure angle require up to 32 teeth to avoid interference. Pressure angles of 20° and 25° allow for smaller tooth numbers. The resultant force on a gear tooth is broken into two components: the tangential force F_t and the normal force F_r. The tangential force is proportional to the power transmitted and inversely proportional to the rotational speed.

A Few Review Items

- A 20°-pressure-angle gear is to be cut with 21 full-depth teeth and a module of 8 mm. Calculate the radii of the pitch circle, addendum circle, base circle, and dedendum circle.

- Sketch and label the pitch circle, addendum circle, base circle, and dedendum circle of the gear just described. Sketch two adjacent teeth, approximately to scale. Show the circular pitch. Show the line of action if this 21-tooth gear rotates counterclockwise and drives another gear directly above it.
- Can a pair of gears produce an output-to-input speed ratio of exactly $1:\pi$? Select a pair of gears that will approximate this ratio.
- A 1:2 speed reducer is proposed with an 18-tooth pinion and 36-tooth gear. Will 14.5°-pressure-angle gears be OK? 20°? 25°?
- Suppose you can only cut 14.5°-pressure-angle gears. Do you have an alternative plan to produce the required speed ratio in the preceding problem?
- Comment on the expected performance of a pairs of gears with the following values of contact ratio: CR = 0.95, CR = 1.01, CR = 1.4, and CR = 2.2.
- Write an equation that relates power transmitted kW, torque (N·mm), and speed (rpm).

PROBLEMS

6.1 For what reasons are gears preferred over friction drives?

6.2 State the fundamental law of gearing.

6.3 Define the following terms:

(a) Pitch circle
(b) Diametral pitch
(c) Circular pitch
(d) Pitch point
(e) Addendum
(f) Dedendum
(g) Backlash
(h) Base circle
(i) Pressure angle
(j) Angle of approach
(k) Angle of recess
(l) Contact ratio
(m) Interference
(n) Speed ratio
(o) Module

6.4 What are the methods used to eliminate interference?

6.5 What is the pitch diameter of a 40-tooth spur gear having a circular pitch of 1.5708 in?

6.6 How many revolutions per minute is a spur gear turning at if it has 28 teeth, a circular pitch of 0.7854 in, and a pitch-line velocity of 12 ft/s?

6.7 How many revolutions per minute is a spur gear turning at if it has a module of 2 mm, 40 teeth, and a pitch-line velocity of 2500 mm/s?

6.8 A spur gear having 35 teeth is rotating at 350 rev/min and is to drive another spur gear at 520 rev/min.

(a) What is the value of the velocity ratio?

(b) How many teeth must the second gear have?

6.9 An external 20°, full-depth spur gear has a diametral pitch of 3. The spur gear drives an internal gear with 75 teeth to produce a velocity ratio of $\frac{1}{3}$. Determine the center distance.

6.10 A standard 20°, full-depth spur gear has 24 teeth and a circular pitch of 0.7854 in. Determine **(a)** the working depth, **(b)** the base circle diameter, **(c)** the outside diameter, **(d)** the tooth thickness at the base circle, and **(e)** the tooth thickness at the outside diameter.

6.11 A spur gear is rotating at 300 rev/min and is in mesh with a second spur gear having 60 teeth and a diametral pitch of 4. The velocity ratio of the pair of meshing gears is $\frac{1}{3}$. What is the magnitude of the pitch-line velocity?

6.12 Two meshing spur gears have a diametral pitch of 4, a velocity ratio of $\frac{1}{5}$, and a center distance of 15 in. How many teeth do the gears have?

6.13 Two meshing spur gears have a module of 1.25 mm, a center distance of 87.5 mm, and a velocity ratio of 0.4. How many teeth do the gears have? If the pinion speed is 1000 rev/min, what is the pitch-line velocity?

6.14 A pair of spur gears has a circular pitch of 15.708 mm, a pitch-line velocity of 4000 mm/s, and a center distance of 350 mm. The larger gear has 84 more teeth than the smaller gear. Determine the rotational speeds of the gears in revolutions per minute.

6.15 A pinion has 32 teeth, a diametral pitch of 4, and a 20° pressure angle. The driven gear is such that the velocity ratio is $\frac{1}{3}$.

(a) What is the center distance?

(b) What is the base-circle radius of the driven gear?

6.16 A standard pinion has 30 teeth, a module of 6 mm, and a 20° pressure angle. The velocity ratio is 0.3.

(a) What is the center distance?

(b) What are the base-circle radii?

(c) What are the tooth thicknesses at the respective base circles?

6.17 Two meshing 20° spur gears have a diametral pitch of 5. The pinion has 35 teeth and the center distance is 15 in.

(a) How many teeth does the driven gear have?

(b) To what value should the center distance be increased in order for the actual pressure angle to become 24°?

6.18 Two meshing 20° spur gears have a module of 5 mm. The pinion has 32 teeth and the center distance is 200 mm. To what value should the center distance be increased in order for the actual pressure angle to become 23°?

6.19 Two meshing standard full-depth spur gears have 20° pressure angles, addendums of $\frac{1}{3}$ in, and a velocity ratio of $\frac{1}{2}$. The pinion has 24 teeth. Calculate the contact ratio.

6.20 Two meshing standard full-depth spur gears have 25° pressure angles. The pinion has 32 teeth and a base pitch of 0.712 in, while the gear has 48 teeth. How many teeth are in contact?

6.21 A pair of standard 20° full-depth spur gears having a diametral pitch of 6 are to produce a speed ratio of $\frac{1}{3}$. What is the minimum allowable center distance if neither gear can have fewer than 20 teeth and the contact ratio cannot be less than 1.7?

6.22 Two meshing standard 20° stub spur gears have a module of 5 mm and a center distance of 200 mm. The pinion has 32 teeth. Calculate the contact ratio.

6.23 Two meshing standard 20° full-depth spur gears have a module of 4 mm and a speed ratio of $\frac{1}{3}$. The gear has 60 teeth. Calculate the contact ratio.

6.24 Two meshing standard 20° full-depth spur gears have a module of 5 mm and a center distance of 200 mm. The pinion has 32 teeth. If the pinion speed is 1,000 rev/min, calculate the extreme values of the relative sliding velocity of the gear teeth.

6.25 A pinion has 32 teeth, a diametral pitch of 4, and a 25° pressure angle. The velocity ratio is $\frac{1}{2}$, and the pinion speed is 200 rev/min. Calculate the extreme values of the relative sliding velocity of the gear teeth. The gears are standard.

6.26 Two meshing standard 20° full-depth gears have a diametral pitch of 3, the pinion has 12 teeth, and the velocity ratio is to be $\frac{1}{2}$. How much of the addenda must be removed in order to prevent interference?

6.27 The data for this problem are the same as for Problem 6.26. What must the pressure angle be in order to prevent interference?

6.28 A 25° full-depth spur gear with 45 teeth has an addendum of $\frac{1}{3}$ in. What is the minimum number of teeth the pinion may have without interference occurring?

6.29 In Figure P6.1, two base circles are shown. Gear 1 drives.

(a) Show the line of action.

(b) Label interference points I_1 and I_2 and pitch point P.

(c) Show the maximum permissible addendums on gears 1 and 2 without interference.

(d) Find the contact ratio with maximum addendums.

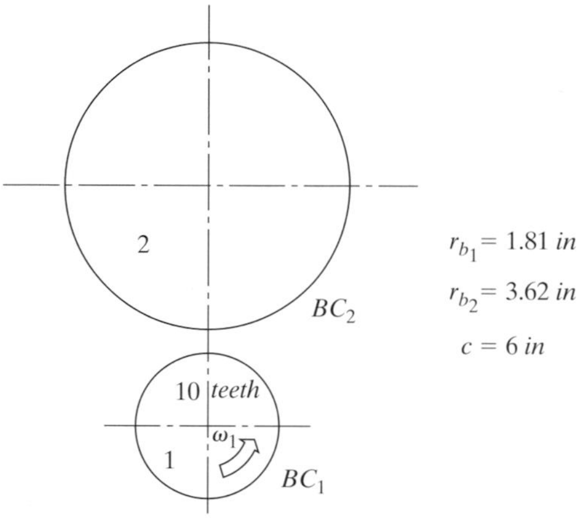

FIGURE P6.1

6.30 A spur gear with 32 teeth and a diametral pitch of 4 is meshed with a second gear having a pitch diameter of 16 in. Both gears are external standard full-depth $14\frac{1}{2}°$ involute gears. Determine **(a)** the number of teeth on the second gear, **(b)** the standard addendum, and **(c)** the maximum addendum for gear 2 for which no interference will occur.

6.31 Two meshing standard 20° full-depth gears have a module of 3 mm. The pinion has 13 teeth and the velocity ratio is $\frac{1}{2}$. Determine whether the gears interfere. If so, how would you change the addendum radii in order to prevent interference?

6.32 A standard 20° full-depth spur gear has 40 teeth and a module of 8 mm. What is the minimum number of teeth that a meshing pinion may have without interference occurring?

6.33 A standard 20° full-depth pinion has 15 teeth and a module of 2 mm. What is the maximum number of teeth that a meshing gear may have without interference occurring?

6.34 Two meshing standard 20° full-depth spur gears with a module of 4 mm have 14 teeth and 45 teeth, respectively. What must be the actual operating pressure angle, obtained by increasing the center distance, in order to avoid interference?

6.35 A shaft rotating at 2000 rev/min has a 20-tooth, 5-diametral-pitch pinion gear keyed to it. The pinion meshes with another spur gear whose center is 6 in from the centerline of the first shaft. Compounded with the gear is a double-threaded left-hand worm that drives a 56-tooth worm wheel keyed to the shaft of an 8-in-diameter hoisting drum. Calculate the distance a load attached to a cable wrapped around the drum moves through in 1 min.

6.36 Find the maximum addendum of a $14\frac{1}{2}°$ pressure angle rack that is to mesh with a 6-in-diameter pinion. Draw the gears and label the rack addendum A.

6.37 In Problem 6.36, assume that the pinion has 15 full-depth teeth. (Note that addendum $= 1/P$ for pinion only.)

(a) Label the beginning and end of contact with the pinion driving clockwise.

(b) Determine the contact ratio.

(c) Is the contact ratio adequate?

6.38 Find the minimum pressure angle required for a rack to mesh with a 6-in-diameter pinion with 15 teeth if both the rack and pinion are to have full-depth teeth. (Addendum $= 1/P$.)

6.39 Two spur gears, an external pinion, and an internal gear have a diametral pitch of 4. The center distance is 10 in and the speed ratio is $\frac{1}{5}$. How many teeth do the gears have? If the pinion speed is 250 rev/min, what is the pitch-line velocity?

6.40 The gear set shown in Figure P6.2 consists of three 20° full-depth spur gears with a diametral pitch of 4 and the following tooth numbers: $N_1 = 20$, $N_2 = 60$, and $N_3 = 40$. Gear 2 is an idler. The unit transmits 10 hp with the driveshaft to gear 1 rotating at 800 rev/min counterclockwise. Determine the following:

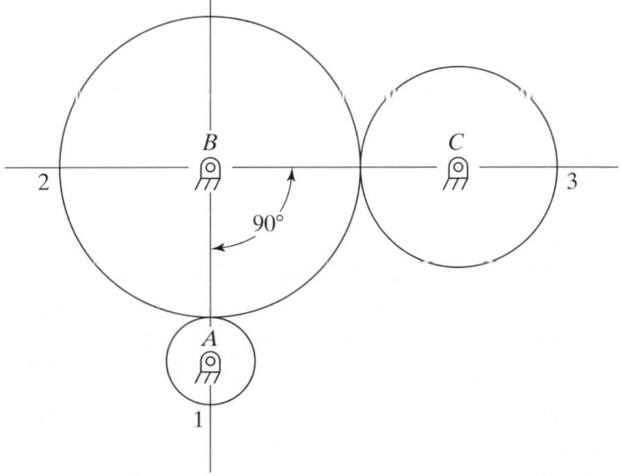

FIGURE P6.2

(a) The center distances.

(b) The rotational velocity (magnitude and direction) of each gear.

(c) The pitch-line velocity.

(d) The torque transmitted to each shaft.

(e) All forces (magnitude and direction) on each gear.

6.41 The gear set shown in Figure P6.3 consists of three 20° full-depth spur gears with a module of 8 mm and the following tooth numbers: $N_1 = 20$, $N_2 = 40$, and $N_3 = 100$. Gear 2 is an idler. The unit transmits 20 kW at a rotational speed of 1000 rev/min of gear 1 in the clockwise direction. Determine the following:

(a) The center distances

(b) The rotational velocity (magnitude and direction) of each gear.

(c) The torque transmitted to each shaft.

(d) All forces (magnitude and direction) on each gear.

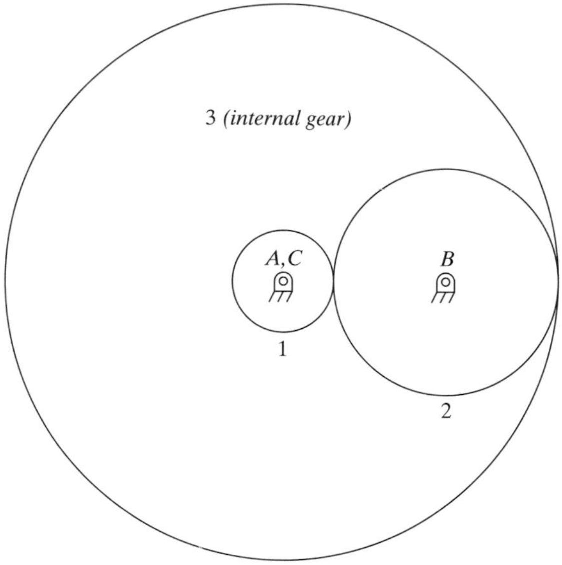

FIGURE P6.3

6.42 A double-reduction gear set has the configuration shown in Figure P6.4, where gear 1 is rigidly attached to shaft A, gears 2 and 3 are rigidly fastened to shaft B, and gear 4 is fastened to shaft C. Gears 1 and 2 have 24 teeth and 60 teeth, respectively, and are 25° full-depth spur gears of diametral pitch 5. Gears 3 and 4 have 20 teeth and 60 teeth, respectively, and are 20° full-depth spur gears of diametral pitch 4. The system transmits 20 hp at a rotational speed of 1000 rev/min of gear 1 in the clockwise direction. Find the speed and the torque transmitted by each shaft. Determine the forces acting on each shaft.

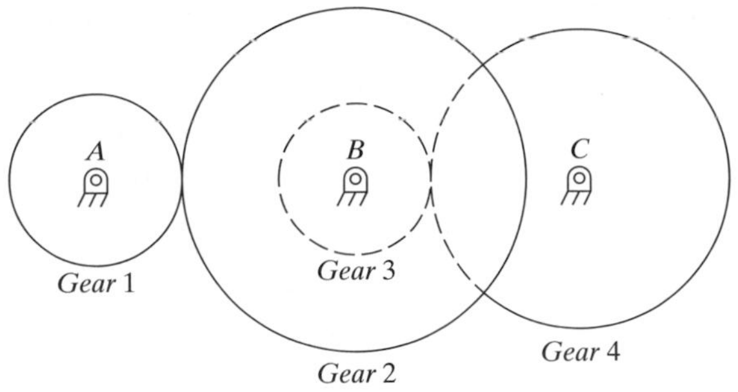

FIGURE P6.4

6.43 A double-reduction gear set has the configuration shown in Figure P6.4, where gear 1 is rigidly attached to shaft A, gears 2 and 3 are rigidly fastened to shaft B, and gear 4 is fastened to shaft C. Gears 1 and 2 have 32 teeth and 80 teeth, respectively, and are 20° full-depth spur gears with a module of 4 mm. Gears 3 and 4 have 24 teeth and 50 teeth, respectively, and are 25° full-depth spur gears with a module of 6 mm. The system transmits 30 kW at a rotational speed of 1200 rev/min of gear 1 in the counterclockwise direction. Find the speed and the torque transmitted by each shaft. Determine the forces acting on each shaft.

6.44 A gear set is to be designed with 14.5°-pressure-angle gears. Construct a three-dimensional bar graph relating the contact ratio to tooth numbers. Consider gears with 24 to 44 teeth. Calculate the minimum and maximum contact ratio for this range of gear pairs. *Note*: Values for a few of the combinations may not be valid due to interference. (Interference will be considered in other problems.)

6.45 A gear set is to be designed with 25°-pressure-angle gears. Construct a three-dimensional bar graph relating the contact ratio to tooth numbers. Consider gears with 12 to 32 teeth. Calculate the minimum contact ratio for this range of gear pairs.

6.46 A gear set is to be designed with 20°-pressure-angle gears. Construct a three-dimensional bar graph relating the contact ratio to tooth numbers. Consider gears with 15 to 65 teeth. Calculate the minimum and maximum contact ratio for this range of gear pairs. *Note*: Values for a few of the combinations may not be valid due to interference. (Interference will be considered in other problems.)

6.47 Suppose we plan to use a wide range of gears with possibly up to 1000 teeth. Plot the minimum pinion size vs. the gear size. Consider standard full-depth teeth with pressure angle $\phi = 14.5°, 20°$, and 25°. (*Suggestion:* Use a logarithmic scale for N_2.)

6.48 Design a series of gearboxes with output-to-input speed ratios of 1:1, 1:1.1, 1:1.2, ..., to about 1:3.5. Find the contact ratio for each pair in the series. *Design decision:* Try a 10-tooth 25°-pressure-angle pinion for each pair. Do you have to reject any designs due to interference? Check the minimum allowable number of pinion teeth.

6.49 Design a series of gearboxes with output-to-input speed ratios of 1:1, 25:26, 25:27, ..., to about 5:9. Find the contact ratio for each pair in the series. *Design decision:* Try a 25-tooth 14.5°-pressure-angle pinion for each. Do you have to reject any designs due to interference? Check the minimum allowable number of pinion teeth.

6.50 Design a series of gearboxes with output-to-input speed ratios of about 1:1 to about 1:4. Design decision: Try a 15-tooth 20°-pressure-angle pinion for each pair, and use gears with even tooth numbers. Find the contact ratio for each pair in the series. Do you have to reject any designs due to interference? Check the minimum allowable number of pinion teeth.

6.51 Suppose we plan to design a line of gearboxes to produce a 1:1 ratio and various speed reduction ratios. If necessary, we will consider nonstandard pressure angles. We will try sets with 16 to 21 pinion teeth and up to 100 gear teeth. Plot the minimum pressure angle for full-depth teeth. Be sure to avoid interference.

6.52 Suppose we plan to design a line of gearboxes to produce a 1:1 ratio and various speed reduction ratios. If necessary, we will consider nonstandard pressure angles. We will try sets with 5 to 10 pinion teeth and up to 60 gear teeth. Plot the minimum pressure angle for full-depth teeth. Be sure to avoid interference.

6.53 Suppose we plan to design a line of gearboxes to produce a 1:1 ratio and various speed reduction ratios. If necessary, we will consider nonstandard pressure angles. We will try sets with 6, 8, 10, 15, and 25 pinion teeth and up to 50 gear teeth. Plot the minimum pressure angle for full-depth teeth. Be sure to avoid interference.

6.54 Make a contour plot of the contact ratio for 25°-pressure-angle pinions and gears with 20 to 100 teeth. Reject combinations (if any) that interfere.

BIBLIOGRAPHY AND REFERENCES

American Gear Manufacturers Association, *American National Standard: Gear Classification and Inspection Handbook*, ANSI/AGMA 2000-A88, American Gear Manufacturers Association, Alexandria VA, March 1988.

American Gear Manufacturers Association, *AGMA Standard for Rating the Pitting Resistance and Bending Strength of Spur and Helical Involute Gear Teeth*, AGMA 218.01, American Gear Manufacturers Association, Alexandria VA, Dec. 1982.

American Gear Manufacturers Association, *AGMA Standard Nomenclature of Gear Tooth Failure Modes*, ANSI/AGMA 110.04, American Gear Manufacturers Association, Alexandria VA, Aug. 1980.

Colbourne, J. R., *The Geometry of Involute Gears*, Springer, New York, 1987.

Dudley, D. W. (ed.), *Gear Handbook*, McGraw-Hill, New York, 1962.

Stock Drive Products, *Design and Application of Small Standardized Components*, Data Book 757, Vol. 2, Stock Drive Products, New Hyde Park, NY, 1983.

Wilson, C. E., *Computer Integrated Machine Design*, Prentice Hall, Upper Saddle River, NJ, 1997.

Note also the Internet references listed under the heading of **Gears** found at the end of Chapter 1. You will find many more commercial sites related to the gear industry by searching the Internet yourself.

CHAPTER 7

Helical, Worm, and Bevel Gears: Design and Analysis

While the previous chapter was devoted exclusively to spur gears, they are by no means the only type of gears in common use. This chapter is devoted to a discussion of the more important types of gears other than spur gears. Figures 7.1a and b show some of the gears that will be discussed.

Concepts You Will Learn and Apply When Studying This Chapter

- Reasons for selecting helical gears over spur gears
- Helical gear descriptors, including the helix angle, normal pitch, and normal pressure angle
- Determination of normal, tangential, and thrust forces on helical gear pairs
- Selection of gear sets to transmit power between nonparallel shafts and selection of gear sets for large speed reductions
- Crossed helical gears: their geometry, center distance, and velocity ratio
- Worm gear sets: lead, velocity ratio, and forces
- Bevel gears: pitch angle, the effect of inboard and overhung mounting, velocity ratio, and forces
- Gears for special applications

7.1 HELICAL GEARS ON PARALLEL SHAFTS

The teeth on a helical gear are cut at an angle ψ called the *helix angle*. Although there is no set standard, "off-the-shelf" gears are often available with helix angles of 15°, 21.5°, and 45°. Other helix angles are used in gear clusters to balance thrust loads (more on that later).

(a)

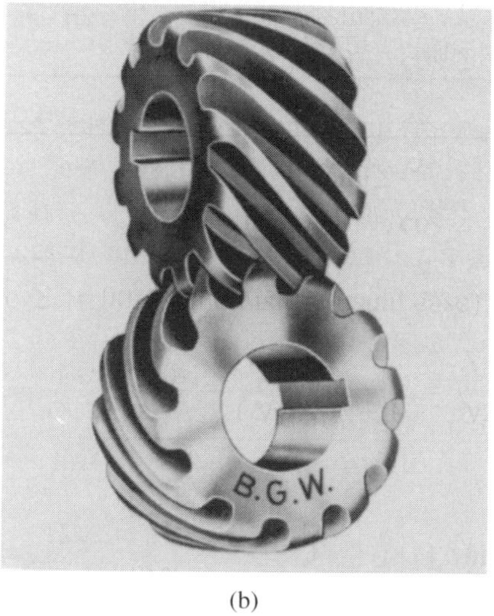

(b)

FIGURE 7.1 (a) A gear system consisting of helical gears (left foreground) and a worm gear set. (*Source:* Horsburgh & Scott Company.) (b) A pair of crossed helical gears. (*Source:* Boston Gear Works.)

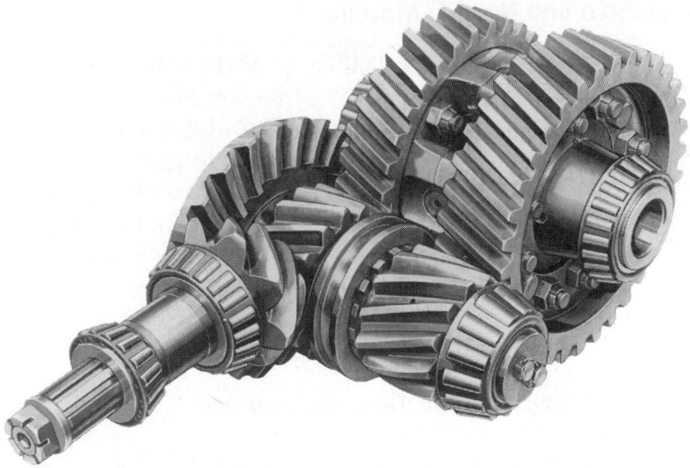

FIGURE 7.2 A speed reducer employing hypoid and helical gears.
(*Source:* Rockwell International.)

The speed reducer shown in Figure 7.2 employs a pair of hypoid gears, two left-hand helical pinions, and two right-hand helical gears. Can you see how a right-hand helical gear resembles a right-hand screw?

In order for two helical gears on parallel shafts to mesh, they must have equal helix angles, but be of different handedness. For example, if the driver is right-handed, the driven gear, or follower, must be left-handed.

Helical Gear Tooth Contact

Spur gears have an initial line contact, with the result that the impact (shock) which occurs when two teeth come into contact is much larger than for helical gears. The initial contact between two helical gear teeth is a point. As the motion continues, the contact between the teeth becomes a line.

The gradual engagement of helical gear teeth permits a larger load transmission, smoother operation, and quieter transmission of power, compared with spur gears of similar size. For these reasons, helical gears are often preferred over spur gears, even though they are usually more expensive and more difficult to manufacture.

Helical Gear Terminology and Geometry

The terminology used for helical gears is similar to that used for spur gears. In fact, most of the relationships developed for spur gears are equally applicable to helical gears on parallel shafts. Several additional terms are necessary, however.

Normal Pitch and Normal Module

The circular pitch p is defined, as for spur gears, as the distance between corresponding points on adjacent teeth, as measured on the pitch surface. However, the pitch of a helical gear can be measured in two different ways. The *transverse* circular pitch p is measured along the pitch circle, just as for spur gears. The *normal* circular pitch p^n is measured normal to the helix of the gear, as shown in Figure 7.3. Note that the sketch shows the pitch surface; the addendums of the teeth have been "cut off".

The diametral pitch P, just as for spur gears, is given by the formula

$$P = \frac{N}{d}, \tag{7.1}$$

where N is the number of teeth on the gear and d is the diameter of the pitch circle in inches.

As before, in SI units, the module m is used to express gear tooth size, rather than the diametral pitch P employed in the English system. As was the case for spur gears,

$$m = \frac{d}{N}, \tag{7.2}$$

where d is expressed in millimeters and m is the module in the transverse plane.

The relationship between the normal diametral pitch and the normal circular pitch is identical to that between the diametral pitch and the circular pitch given in the previous chapter:

$$P = \frac{\pi}{p} \quad \text{and} \quad P^n = \frac{\pi}{p^n}. \tag{7.3}$$

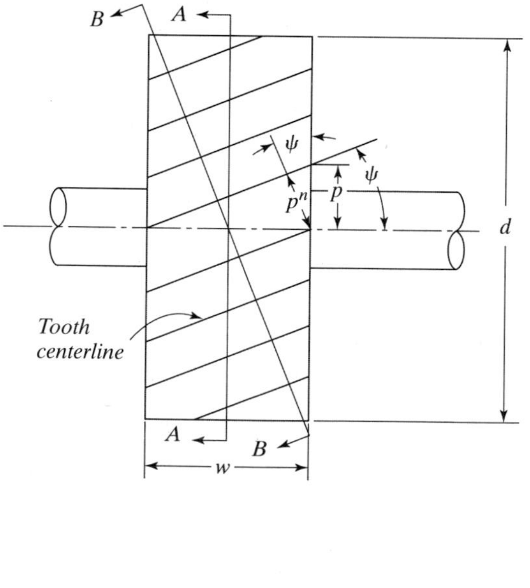

FIGURE 7.3 Helical gear terms: The diagram shows the helix angle ψ, the pitch diameter d, and the gear width w. The transverse circular pitch p, and the normal circular pitch p^n are measured on the pitch surface.

In the SI system,

$$p = \pi m \quad \text{and} \quad p^n = \pi m^n, \tag{7.4}$$

where m^n is the module in the normal plane. The normal pitch of a helical gear is an important dimension, because it becomes the circular pitch of the hob cutter used to manufacture the gear. When the cutting is done, instead, by a gear shaper, the transverse circular pitch of the gear becomes the circular pitch of the cutter. It can thus be seen why both pitches are consequential.

As seen from Figure 7.3, the relationship between the transverse and normal circular pitch is

$$p^n = p \cos \psi. \tag{7.5}$$

Substituting the expression for the circular pitch employed in the last chapter ($p = \pi d/N$), we can also derive an expression for the pitch diameter in terms of the number of teeth N, the *normal* circular pitch p^n, and the helix angle ψ:

$$d = \frac{pN}{\pi} = \frac{p^n N}{\pi \cos \psi}. \tag{7.6}$$

From Eqs. (7.3) and (7.6), we can express the *normal diametral pitch* in terms of the diametral pitch P and the helix angle:

$$P^n = \frac{\pi}{p^n} = \frac{\pi}{p \cos \psi} = \frac{P}{\cos \psi}. \tag{7.7}$$

In SI units,

$$\cos \psi = \frac{m^n}{m}. \tag{7.8}$$

Having derived the foregoing equations from Figure 7.3, we may now determine most of the important dimensions of the helical gear.

SAMPLE PROBLEM 7.1

Properties of a Helical Gear in Terms of Module

A 30-tooth helical gear with a 25° helix angle has a module of 10 mm. Determine the pitch diameter, the normal module, and the normal and transverse circular pitches.

Solution. The pitch diameter is obtained directly from Eq. (7.2):

$$d = mN = 10(30) = 300 \text{ mm}.$$

From Eq. (7.8),

$$m^n = m \cos \psi = 10 \cos 25° = 9.063 \text{ mm}.$$

From Eq. (7.4),

$$p = \pi m = \pi(10) = 31.42 \text{ mm}$$

and

$$p^n = \pi m^n = \pi(9.063) = 28.47 \text{ mm}.$$

Check: Using Eq. (7.5), we have

$$p^n = p \cos \psi = 31.42 \cos 25° = 28.47 \text{ mm}.$$

SAMPLE PROBLEM 7.2

Properties of a Helical Gear in Terms of Diametral Pitch

A helical gear has 25 teeth, a helix angle of 25°, and a transverse circular pitch of $\pi/5$ in. Determine the pitch diameter, the diametral pitch, and the normal circular and diametral pitches.

Solution. The pitch diameter is directly obtained from Eq. (7.6):

$$d = \frac{pN}{\pi} = \frac{(\pi/5)(25)}{\pi} = 5 \text{ in}.$$

The diametral pitch P is given by Eq. (7.3):

$$P = \frac{\pi}{p} = \frac{\pi}{\pi/5} = 5.$$

The normal diametral pitch and the normal circular pitch are:

$$P^n = \frac{P}{\cos \psi} = \frac{5}{\cos 25°} = 5.52$$

and

$$p^n = \frac{\pi}{P^n} = \frac{\pi}{5.52} = 0.569 \text{ in}.$$

Pressure Angle

If we view a helical gear along the shaft axis (section AA in Figure 7.3), we "see" the *transverse pressure angle* ϕ. However, when the gear is viewed in the normal plane (section BB), we have the *normal pressure angle* ϕ^n. The two are related to the helix angle by the equation

$$\cos \psi = \frac{\tan \phi^n}{\tan \phi}. \tag{7.9}$$

It is clear that the transverse pressure angle must always be larger than the normal pressure angle.

Center Distance

Some manufacturers standardize on the transverse diametral pitch, module, and pressure angle. Others standardize on the normal diametral pitch, normal module, and normal pressure angle. Some methods of generating gears make the latter system practical. How can we determine the distance between shaft centers in terms of numbers of gear teeth? We should be able to do this using customary U.S. and metric descriptors. Let us start with the pitch diameter.

From the formulas for the pitch diameter and the center distance derived in the last chapter, we obtain

$$d_1 = \frac{N_1 p}{\pi}, \quad d_2 = \frac{N_2 p}{\pi}, \tag{7.10}$$

and

$$c = \frac{d_1 + d_2}{2} = \frac{p}{2\pi}(N_1 + N_2) = \frac{N_1 + N_2}{2P}. \tag{7.11}$$

But $p^n = p \cos \psi$; therefore,

$$c = \frac{p^n(N_1 + N_2)}{2\pi \cos \psi}, \tag{7.12}$$

and since $p^n P^n = \pi$, another expression for the center distance is

$$c = \frac{N_1 + N_2}{2P^n \cos \psi}. \tag{7.13}$$

In the SI system,

$d_1 = mN_1, \qquad d_2 = mN_2,$

and

$$c = \frac{d_1 + d_2}{2} = \frac{m(N_1 + N_2)}{2} = \frac{m^n(N_1 + N_2)}{2\cos \psi}. \tag{7.14}$$

As an example, suppose the gears that are to be designed are to be mounted on shafts 10 in apart. The center distance is thus seen to be fixed. If the normal pitch and helix angle are chosen first, the formulas may be used to determine the sum of N_1 and N_2. This value, together with the speed ratio, will enable the designer to determine the appropriate values for N_1 and N_2.

SAMPLE PROBLEM 7.3

Analysis of a Helical Gear Pair

A pair of meshing helical gears has a normal pressure angle of 20°, a diametral pitch of 5, and a normal circular pitch of 0.55 in. The driver has 18 teeth and the follower has 36 teeth. Determine the pressure angle ϕ and the center distance c.

Solution. From Eq. (7.3) and the given data, the circular pitch is

$$p = \frac{\pi}{P} = \frac{\pi}{5} = 0.628 \text{ in.}$$

The helix angle is given by

$$\cos\psi = \frac{p^n}{p} = \frac{0.55}{0.628} = 0.876, \quad \text{or} \quad \psi = 28.8°$$

and the pressure angle is

$$\tan\phi = \frac{\tan\phi^n}{\cos\psi} = \frac{0.364}{0.876} = 0.416.$$

Hence,

$$\phi = 22.6°.$$

Finally, the center distance is found by using Eq. (7.12):

$$c = \frac{p^n(N_1 + N_2)}{2\pi\cos\psi} = \frac{0.55(18 + 36)}{2\pi(0.876)} = 5.4 \text{ in.}$$

Face Width

The face width w is the thickness of a gear measured parallel to the shaft axis. A common range for face width is

$$m \cdot 6 \le w \le m \cdot 8,$$

or

$$6/P \le w \le 8/P, \tag{7.15}$$

where m = module (mm),
P = diametral pitch (in^{-1}),

and

w = face width (mm or in).

Axial Pitch

Circular pitch and axial pitch are both measured from one tooth face to the next corresponding face. Circular pitch is an arc distance on the pitch circle in the plane of rotation; axial pitch is measured parallel to the shaft axis. We calculate axial pitch from

$$p_x = p/\tan\psi, \tag{7.16}$$

where p_x = axial pitch (mm or in),
p = circular pitch (in or mm) $= \pi m = \pi/P$,

and

ψ = helix angle.

Contact Ratio and Axial Contact Ratio (Axial Overlap)

The axial contact ratio (axial overlap) accounts for the advantage of helical gears over straight spur gears. The axial overlap of teeth ensures smoother, quieter operation. The contribution to the contact ratio is the face width divided by the axial pitch:

$$\mathrm{CR}_x = w/p_x. \tag{7.17}$$

The total contact ratio—the average number of pairs of helical gear teeth in contact—is

$$\mathrm{CR} = \mathrm{CR}_x + \mathrm{CR}_{spur}, \tag{7.18}$$

where CR_{spur} = contact ratio computed for a pair of spur gears with the same number of teeth and the same tooth form. We usually try to set $\mathrm{CR}_x \geq 1$—that is, a minimum face width equal to the axial pitch. This recommendation applies to moderate and large helix angles; it is not practical, however, for small helix angles, because the face width would be too large and loads would not be distributed across the tooth.

Velocity Ratio of Helical Gears on Parallel Shafts

If we view a pair of helical gears along the shaft axes, we see that the speed ratio is the same as for straight spur gears. That is,

$$n_2/n_1 = \omega_2/\omega_1 = N_1/N_2 = d_1/d_2 = r_1/r_2,$$

where n = shaft speed (rpm),
ω = angular velocity (rad/s),

N = number of teeth,
d = pitch diameter (mm or in),

and

r = pitch radius (mm or in).

The speed ratio equals the inverse of the tooth numbers *and* the inverse of the diameters or radii for spur gears and for helical gears on parallel shafts. This relationship does not apply to planetary trains. Speed ratio equals the inverse of the tooth numbers for all *non*planetary gear pairs. But, we cannot use the diameter or radius ratios for worm drives and other helical gears on nonparallel shafts.

Helical Gear Forces

As in all direct-contact mechanisms, the force that one helical gear exerts on its meshing gear acts normal (perpendicularly) to the contacting surfaces if friction forces are neglected. This normal force can be resolved into tangential and radial components, as was the case with spur gear forces. However, for helical gears, there is a third component, known as the axial or thrust load. This force acts parallel to the axis of the shaft the gear is mounted on. Figure 7.4 shows the forces acting on a helical gear. Let us assume that the gear depicted is the driver and is rotating in the counterclockwise direction. The forces shown in the figure are those exerted on the driver by the driven gear.

From the figure, the magnitudes of the components of the normal force F_n are

$$F_t = \text{tangential force} = F_n \cos \phi^n \cos \psi, \tag{7.19}$$

$$F_r = \text{radial force} = F_n \sin \phi^n = F_t \tan \phi, \tag{7.20}$$

and

$$F_a = \text{thrust or axial load} = F_n \cos \phi^n \sin \psi = F_t \tan \psi. \tag{7.21}$$

The direction of the thrust load is determined by the right- or left-hand rule (depending on the hand of the driver), applied to the driver: *The fingers point in the direction of rotation; then the thumb points in the direction of the thrust load.* The *driven* thrust load is then *opposite* the direction of the thrust load on the driver. Actually, the thrust force is applied at the pitch circle, leading to a combination of a thrust force and a thrust moment at the shaft center.

The equations for torque and power are the same as those used for spur gears, namely,

$$T = F_t\left(\frac{d}{2}\right) \tag{7.22}$$

and

$$hp = \frac{F_t v_P}{33{,}000} = \frac{Tn}{63{,}025}, \tag{7.23}$$

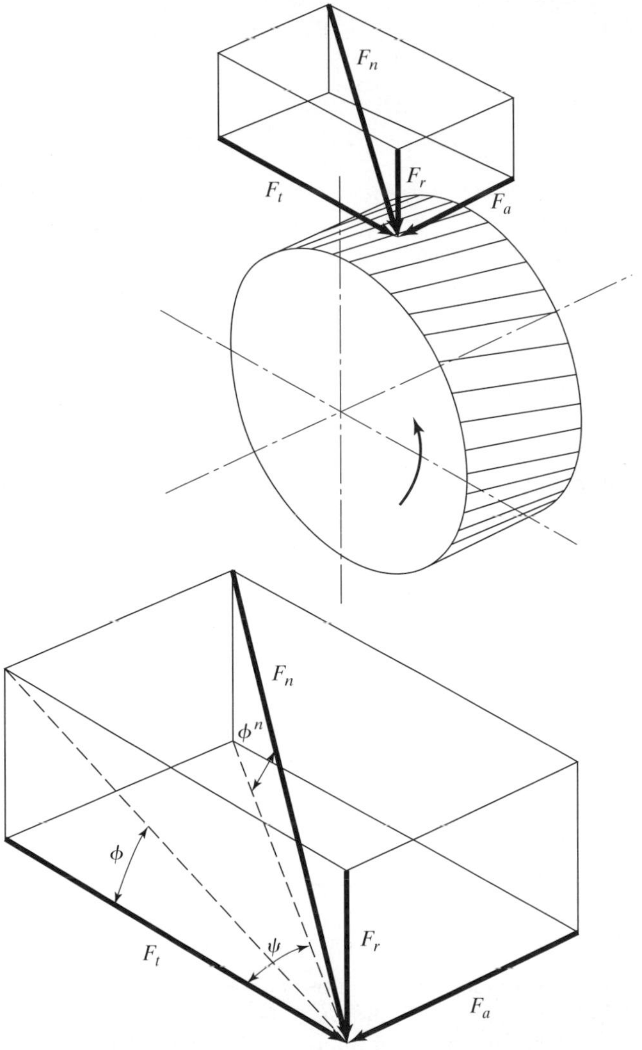

FIGURE 7.4 The normal force F_n acting on a helical gear consists of three components: tangential force F_t, radial force F_r, and axial or thrust force F_a.

where hp = horsepower, F_t = tangential force (pounds), v_P = pitch velocity (feet per minute), T = torque (inch · pounds), and n = speed (revolutions per minute). In the SI system,

$$kW = \frac{F_t v_P}{1{,}000{,}000} = \frac{T\omega}{1{,}000{,}000}, \tag{7.24}$$

where kW = power (kilowatts), F_t = tangential force (newtons), v_P = pitch velocity (millimeters per second), T = torque (newton · millimeters), and ω = angular velocity (radians per second).

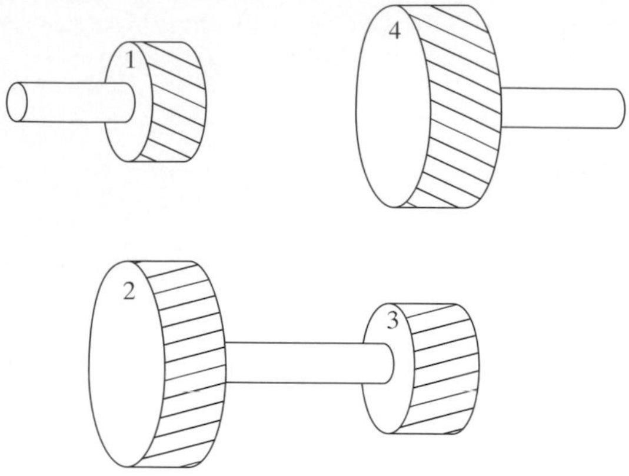

FIGURE 7.5 Reverted gear train.

Balancing Thrust in Countershafts

Suppose we need to reduce the shaft speed by a factor of 10 or more. A single pair of spur or helical gears might not be practical, due to the size of the larger gear. In that case, a two-step reduction can be used. Figure 7.5 is an exploded schematic of one such design, a reverted gear train. Input and output shafts lie on the same centerline. This general scheme is also practical for speed changers (about which we shall have more to say in the next chapter).

Let the input shaft drive gear 1 in the figure, and let gear 4 drive the output shaft. The shaft of gears 2 and 3 is the countershaft. (No bearings are shown, and the sketch is not to scale). It is possible to balance thrust loads so that there is no need for a thrust bearing on the countershaft.

SAMPLE PROBLEM 7.4

Design of a Reverted Gear Train with Thrust Balancing

We need a line of speed reducers with output-to-input speed ratios of about $\frac{1}{30}$, $\frac{1}{20}$, $\frac{1}{15}$, and $\frac{2}{15}$. The input speed will be 3000 rpm, and 1.5 kW of mechanical power is to be transmitted. Design the reducers and examine the geometry and loading. Check for interference and check the contact ratio.

Design decisions. We will design a reverted gear train similar to that shown in the figure. The actual train will be more compact to reduce bending loads. To reduce cost, gears 1 and 2 will be selected from commercial stock, with a module of 2.5 mm and a 45° helix angle. Gear 1 will have a right-handed helix and will rotate counterclockwise; gear 2 will have a left-handed helix. Gears 3 and 4, which are expected to have higher tooth loads, will have a module of 3 mm. We will specify a 20° transverse pressure angle and specify a face width of 8 times the module for all gears. We will try to balance thrust forces on the countershaft.

Solution summary. Sixteen-tooth gears will be selected for gears 1 and 3. Since the input and output shafts are collinear in a reverted train, the center distance of gears 1 and 2 equals the center distance of gears 3 and 4. In terms of tooth numbers, the center distance requirement is equivalent to

$$(N_1 + N_2)m_1 = (N_3 + N_4)m_3.$$

This requirement limits the tooth number selection. To get whole numbers for gear 4's teeth, we can set $N_2 = 20, 26, \ldots, 104$ teeth. We then find pitch radii

$$r = m\,N/2.$$

The speed ratios are the inverse of tooth numbers. The countershaft speed can be found from

$$n_2 = n_3 = n_1 N_1/N_2$$

and the output speed from

$$n_4 = n_3 N_3/N_4 = n_1 N_1 N_3/(N_2\,N_4).$$

The pitch-line velocity (i.e., the tangential speed at the pitch circle), is given by

$$v = \omega r,$$

where $\omega = \pi\,n/30.$

The torque applied to gear 1, the driver, is found from the power transmitted and angular velocity. The tangential force on gear 1 is the torque divided by the pitch radius. The tangential force on gear 2 is equal and opposite. Radial and thrust forces are found from the tangential force, the pressure angle, and the helix angle. Force and torque directions are given in the detailed solution and shown in Figure 7.6. The figure shows partial free-body diagrams, with the countershaft broken between gears 2 and 3. (Bearings and bearing reactions are not shown.)

The tangential, radial, and thrust loads on the gears produce lateral and thrust forces and moments on the shafts that must be considered in selecting bearings. However, we can design for a zero net thrust on the countershaft. The torque on gear 3 is equal and opposite to the torque on gear 2, which leads to the equations

$$F_{t3}\,r_3 = F_{t2}\,r_2$$

and

$$(\tan\psi_2)/r_2 = (\tan\psi_3)/r_3.$$

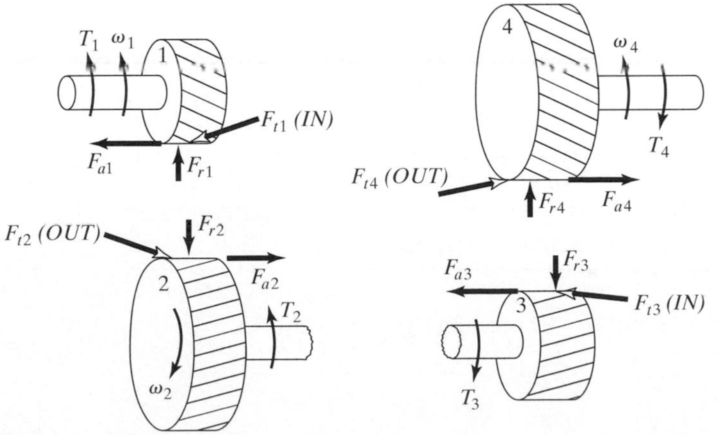

FIGURE 7.6 Reverted gear train forces and torques.

The latter equation can be used to find the helix angle for gears 3 and 4 that will balance the countershaft thrust.

The tabulated results indicate that the required ratios are approximated if we use $N_2 = 98$, 80, 68, and 50 teeth in our line of reducers. An interference check shows that gear 2 cannot have more than 101 teeth without interference. This is no problem, since $N_2 = 98$ for an approximate speed ratio of 1/30. The contact ratio for both meshing pairs is greater than 2.0 for all speed ratios; hence, we can expect smooth, quiet operation. Smaller helix angles on some gears result in lower values of the axial contact ratio (axial overlap). The contact ratio for $N_2 = 104$ teeth is arbitrarily given a value of zero, because we do not want to undercut the teeth or try some other remedy for interference.

There are dozens of chances for errors in a complicated problem like this. But the computer makes checking easier. Does the output power equal the input power if friction is neglected? Is the center distance the same for both gear pairs? Sketches help, too. Have we violated Newton's first law (the one about equilibrium)? You need bearing reactions to check this fully. Does Newton's third law apply (action and reaction)? Check force directions on meshing gears.

Detailed solution.

Helical Gears on Parallel Shafts (Units: mm, N, rad, s, kW)

Transmitted power (kW) $P_{kw} := 1.5$ Module (mm) $m_1 := 2.5$ $m_3 := 3$

Face width, gears 1 and 2: $w_1 := 8\, m_1$ $w_1 = 20$ 3 and 4: $w_3 := 8\, m_3$ $w_3 = 24$

Helix angles $\psi_1 := 45 \cdot \text{deg}$ RH $\psi_2 := 45 \cdot \text{deg}$

Hand of gears 2 and 3 opposite gear 1; gear 4 same as gear 1

Input speed (rpm) $n_1 := 3000$ counterclockwise

Angular velocity (rad/s) $\omega_1 := n_1 \cdot \dfrac{\pi}{30}$ $\omega_1 = 314.159$

Transverse pressure angle $\phi := 20 \cdot \text{deg}$ $\phi = 0.349$ rad

Tooth numbers $N_1 := 16$ $N_2 := 20, 26..\, 104$ $N_3 := 16$

Collinear input and output shafts require $N_4(N_2) := (N_1 + N_2)\dfrac{m_1}{m_2} - N_3$

Pitch radii $r_1 := \dfrac{m_1 \cdot N_1}{2}$ $r_1 = 20$ $r_2(N_2) := \dfrac{m_1 \cdot N_2}{2}$

$r_3 := \dfrac{m_3 \cdot N_3}{2}$ $r_3 = 24$ $r_4(N_2) := \dfrac{m_3 \cdot N_4(N_2)}{2}$

Center distance $c(N_2) := r_1 + r_2(N_2)$ Check $cc(N_2) := r_3 + r_4(N_2)$

Rotation speed (rpm)

Gears 2 and 3: $n_2(N_2) := n_1 \cdot \frac{N_1}{N_2}$ $\omega_2(N_2) := n_2(N_2)\frac{\pi}{30}$ Opposite input direction

Output $n_4(N_2) := n_1 \cdot \frac{N_1 \cdot N_3}{N_2 \cdot N_4(N_2)}$ $\omega_4(N_2) := n_4(N_2) \cdot \frac{\pi}{30}$ Same direction as input

Pitch-line velocity (mm/s) gears 1 and 2 $v_1 := \omega_1 \cdot r_1$ $v_1 = 6.283 \cdot 10^3$

gears 3 and 4 $v_3(N_2) := \omega_2(N_2) \cdot r_3$

Torque (N · mm) for free body $T_1 := \frac{10^6 \cdot P_{kw}}{\omega_1}$ $T_1 = 4.775 \cdot 10^3$ Same direction as rotation

Tangential force $F_{t1} := \frac{T_1}{r_1}$ $F_{t1} = 238.732$ opposes rotation

$F_{t2} := F_{t1}$ opposite tangential force on gear 1

Radial force $F_{r1} := F_{t1} \cdot \tan(\phi)$ $F_{r1} = 86.891$

$F_{r2} := F_{r1}$ Radial forces directed toward gear center

Thrust force at gear tooth $F_{a1} := F_{t1} \cdot \tan(\psi_1)$ $F_{a1} = 238.732$

For a right-hand helical driver turning counterclockwise, thrust at tooth is toward observer $F_{a2} := F_{a1}$, but direction is opposite.

Determine helix angle for gear 3 so that countershaft thrust is balanced. Make all thrust forces equal. Gears 2 and 3 must have same hand. Thrust on gear 3 opposite thrust on gear 2; thrust on gear 4 same direction as thrust on gear 2.

$$\psi_3(N_2) := \mathrm{atan}\left(\tan(\psi_1) \cdot \frac{r_3}{r_2(N_2)}\right)$$

Normal module Gears 1 and 2 $m_{n1} := m_1 \cdot \cos(\psi_1)$ $m_{n1} = 1.768$

Gears 3 and 4 $m_{n3}(N_2) := m_3 \cdot \cos(\psi_3(N_2))$

Normal pressure angle Gears 1 and 2

$$\phi_{n1} := \mathrm{atan}(\tan(\phi) \cdot \cos(\psi_1)) \qquad \frac{\phi_{n1}}{\deg} = 14.433 \deg$$

Gears 3 and 4 $\phi_{n3}(N_2) := \mathrm{atan}(\tan(\phi) \cdot \cos(\psi_3(N_2)))$

Torque (N · mm) for free body $T_2(N_2) := F_{t2} \cdot r_2(N_2)$ opposes rotation

$T_3(N_2) := T_2(N_2)$, but opposite direction

Tangential force $F_{t3}(N_2) := \frac{T_3(N_2)}{r_3}$ opposes rotation

$F_{t4}(N_2) := F_{t3}(N_2)$ opposite tangential force on gear 3

Radial force: gears 3 and 4 $F_{r3}(N_2) := F_{t3}(N_2) \cdot \tan(\phi)$ toward gear center

Thrust force-gears 3 and 4 $F_{a3}(N_2) := F_{t3}(N_2) \cdot \tan(\psi_3(N_2))$ $F_{a3}(50) = 238.732$

Torque (N · mm) for free body $T_4(N_2) := F_{t4}(N_2) \cdot r_4(N_2)$ Reaction torque opposes rotation

Output power check (neglects friction) $P_0(N_2) := T_4(N_2) \cdot \omega_4(N_2) \cdot 10^{-6}$

Circular pitch: gears 1 and 2 $p_1 := \pi m_1$ $p_1 = 7.854$

gears 3 and 4 $p_3 := \pi m_3$ $p_3 = 9.425$

Axial pitch: gears 1 and 2 $p_{x1} := \dfrac{p_1}{\tan(\psi_1)}$ $p_{x1} = 7.854$

gears 3 and 4 $p_{x3}(N_1) := \dfrac{p_3}{\tan(\psi_3(N_2))}$

Axial overlap due to helix angle: gears 1 and 2: gears 3 and 4:

$$CR_{x1} := \frac{w_1}{p_{x1}} \quad CR_{x1} = 2.546 \qquad CR_{x3}(N_2) := \frac{w_3}{p_{x3}(N_2)}$$

Check interference. Pinion teeth N_1 Gear teeth N_2 Solve for maximum number of gear teeth

$$4 - 2N_1 \cdot (\sin(\phi))^2 = \frac{(N_1 \cdot \sin(\phi))^2 - 4}{N_{2\max}} \text{ solve, } N_{2\max} \rightarrow \frac{-(-256 \cdot \sin(20 \cdot \deg)^2 + 4)}{(4 - 32 \cdot \sin(20 \cdot \deg)^2)}$$

Drop fractional teeth $N_{2\max} := \text{floor}\left[\dfrac{-(-256 \cdot \sin(20 \cdot \deg)^2 + 4)}{(4 - 32 \cdot \sin(20 \cdot \deg)^2)}\right]$ $N_{2\max} = 101$

$$\text{Int}(N_2) := \text{if}\,(N_2 > N_{2\max}, 0, 1) \quad \text{Int3}(N_2) := \text{if}\,(N_4(N_2) > N_{2\max}, 0, 1)$$

Profile contact ratio: gears 1 and 2
$CR_p(N_2) :=$

$$\frac{\left[(N_1 + 2)^2 - (N_1 \cdot \cos(\phi))^2\right]^{\frac{1}{2}} + \left[(N_2 + 2)^2 - (N_2 \cdot \cos(\phi))^2\right]^{\frac{1}{2}} - (N_1 + N_2) \cdot \sin(\phi)}{2\pi \cdot \cos(\phi)}$$

Total contact ratio (call it zero if interference occurs) $CR_1(N_2) := (CR_{x1} + CR_p(N_2)) \cdot \text{Int}(N_2)$

$$CR_{p3a}(N_2) := \frac{\left[(N_3 + 2)^2 - (N_3 \cdot \cos(\phi))^2\right]^{\frac{1}{2}} + \left[(N_4(N_2) + 2)^2 - (N_4(N_2) \cdot \cos(\phi))^2\right]^{\frac{1}{2}}}{2\pi \cdot \cos(\phi)}$$

$$CR_{p3}(N_2) := \frac{-(N_3 + N_4(N_2)) \cdot \sin(\phi)}{2\pi \cdot \cos(\phi)} + CR_{p3a}(N_2)$$

$$CR_3(N_2) := (CR_{x3}(N_2) + CR_{p3}(N_2)) \cdot \text{Int3}(N_2)$$

N_2	$N_4(N_2)$	$r_2(N_2)$	$r_4(N_2)$	$c(N_2)$	$n_2(N_2)$	$n_4(N_2)$	$v_3(N_2)$	$\frac{\psi_3(N_2)}{deg}$	$m_{n3}(N_2)$	$\frac{\phi_{n3}(N_2)}{deg}$
20	14	25	21	45	2400	2742.9	6031.9	43.831	2.164	14.712
26	19	32.5	28.5	52.5	1846.2	1554.7	4639.9	36.444	2.413	16.319
32	24	40	36	60	1500	1000	3769.9	30.964	2.572	17.333
38	29	47.5	43.5	67.5	1263.2	696.9	3174.7	26.806	2.678	17.997
44	34	55	51	75	1090.9	513.4	2741.8	23.575	2.75	18.448
50	39	62.5	58.5	82.5	960	393.8	2412.7	21.007	2.801	18.767
56	44	70	66	90	857.1	311.7	2154.2	18.925	2.838	18.998
62	49	77.5	73.5	97.5	774.2	252.8	1945.8	17.207	2.866	19.172
68	54	85	81	105	705.9	209.2	1774.1	15.767	2.887	19.304
74	59	92.5	88.5	112.5	648.6	175.9	1630.2	14.545	2.904	19.408
80	64	100	96	120	600	150	1508	13.496	2.917	19.49
86	69	107.5	103.5	127.5	558.1	129.4	1402.8	12.585	2.928	19.556
92	74	115	111	135	521.7	112.8	1311.3	11.788	2.937	19.611
98	79	122.5	118.5	142.5	489.8	99.2	1231	11.085	2.944	19.656
104	84	130	126	150	461.5	87.9	1160	10.46	2.95	19.693

N_2	$T_2(N_2)$	$T_3(N_2)$	$F_{t3}(N_2)$	$F_{r3}(N_2)$	$T_4(N_2)$	$P_0(N_2)$	$CR_1(N_2)$	$CR_3(N_2)$
20	5968.3	5968.3	248.7	90.512	5222.3	1.5	4.074	3.925
26	7758.8	7758.8	323.3	117.666	9213.6	1.5	4.106	3.402
32	9549.3	9549.3	397.9	144.819	14323.9	1.5	4.13	3.078
38	11339.8	11339.8	472.5	171.973	20553.4	1.5	4.148	2.859
44	13130.3	13130.3	547.1	199.126	27901.9	1.5	4.162	2.701
50	14920.8	14920.8	621.7	226.28	36369.4	1.5	4.173	2.581
56	16711.3	16711.3	696.3	253.434	45956	1.5	4.183	2.488
62	18501.8	18501.8	770.9	280.587	56661.6	1.5	4.191	2.414
68	20292.3	20292.3	845.5	307.741	68486.4	1.5	4.198	2.352
74	22082.7	22082.7	920.1	334.894	81430.1	1.5	4.204	2.301
80	23873.2	23873.2	994.7	362.048	95493	1.5	4.209	2.258
86	25663.7	25663.7	1069.3	389.201	110674.9	1.5	4.213	2.221
92	27454.2	27454.2	1143.9	416.355	126975.8	1.5	4.217	2.188
98	29244.7	29244.7	1218.5	443.509	144395.8	1.5	4.221	2.16
104	31035.2	31035.2	1293.1	470.662	162934.9	1.5	0	2.135

Eliminating Thrust with Herringbone Gears

A herringbone gear can be thought of as a helical gear with half of its face cut right-handed and the other half cut left-handed. Thus, the thrust loads generated by the left-hand and right-hand teeth cancel each other. The speed changer shown in Figure 7.7 utilizes two pairs of herringbone gears. Some herringbone gears are cut with a space between the right-hand and left-hand teeth for ease of manufacture. Herringbone gear sets are designed with axial play to allow the gears to mesh without binding.

FIGURE 7.7 A speed changer that utilizes four herringbone gears.
(*Source:* Horsburgh & Scott Company.)

7.2 CROSSED HELICAL GEARS

Helical gears can also be used when power is to be transmitted from one shaft to another, nonparallel, nonintersecting shaft. The two helical gears are collectively referred to as *crossed helical gears*. Any helical gear can be used as a crossed helical gear; a helical gear becomes a crossed helical gear when it is meshed with another helical gear whose shaft is nonparallel and nonintersecting with the shaft of the first gear. A typical set of crossed helical gears is illustrated in Figure 7.8.

While helical gears on parallel shafts must be of opposite hand, two crossed helical gears usually have the *same* hand. And while helical gears on parallel shafts must have identical helix angles, the helix angles for crossed helical gears do *not* have to be equal.

Theoretically, helical gears on crossed shafts make point contact, while spur gears make line contact. Helical gears on parallel shafts also make line contact. The areas of line and point contact are not mathematical lines and points of zero area. Rather, the point of contact may be about the size of the period at the end of this sentence for helical gears on crossed shafts and like a pencil line for properly aligned spur gears or helical gears on parallel shafts. For the same load, point

FIGURE 7.8 A pair of crossed helical gears, used when the shafts are not parallel. Usually, crossed helical gears have the same handedness. (*Source:* Richmond Gear, Wallace Murray Corporation.)

contact results in higher stresses than line contact. Design implications may include the following:

- Rejection of crossed helical gears for heavily loaded drives
- Rejection of crossed helical gears when failure could have severe consequences
- Specifying adequate lubrication and frequent inspection when the design includes crossed helical gears

In specifying pitch, the normal pitch rather than the transverse pitch is usually referred to for crossed helical gears. The reason for this is that, while the normal pitches for meshing helical gears must be equal, the transverse pitches will be unequal if the helix angles are unequal.

Crossed Helical Gear Geometry

Some of the more important relationships involving crossed helical gears can be obtained by considering Figure 7.9. The two helical gears shown have different helix angles ψ_1 and ψ_2. Both gears are right-handed, and Σ is the angle between the shafts. For crossed helical gears, the angle between the shafts always equals the sum or difference of the helix angles of the two gears. As can be seen from the figure,

$$\Sigma = \psi_1 + \psi_2. \tag{7.25a}$$

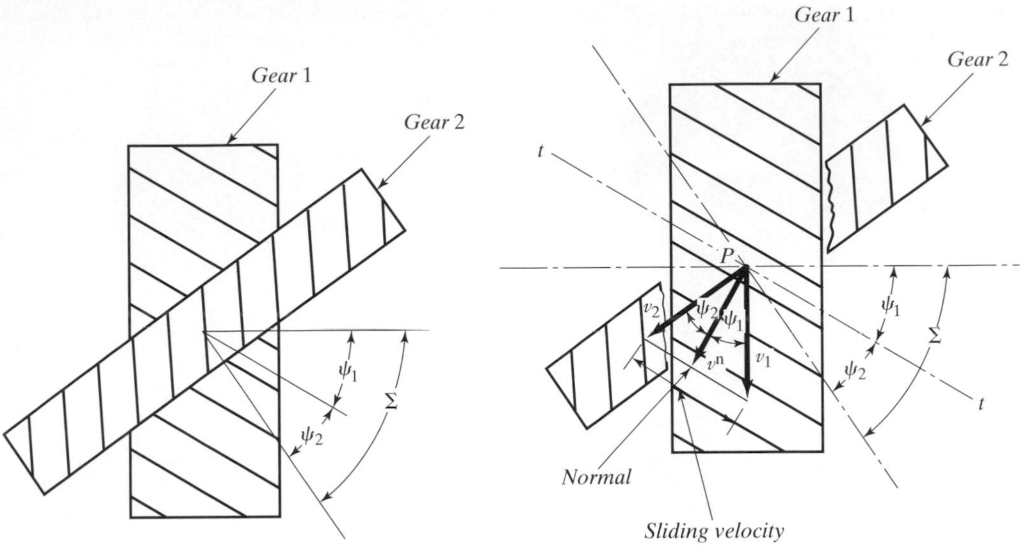

FIGURE 7.9 The two crossed helical gears are of the same hand, but have different helix angles. The shaft angle Σ is equal to the sum of the helix angles. The normal components of the pitch-line velocities are equal, but a sliding velocity exists.

If the gears were of opposite hand, the shaft angle would instead be

$$\Sigma = \psi_1 - \psi_2. \tag{7.25b}$$

Pitch line velocities v_1 and v_2 are shown in the figure. The normal components of v_1 and v_2 must be equal, are perpendicular to the axis t–t (the axis tangent to the teeth in contact), and are both labeled v^n. Also as shown in the figure, a sliding velocity exists for crossed helical gears, even when they contact at the pitch point.

Center Distance

A procedure similar to that used to find the center distance for helical gears on parallel shafts, via Eqs. (7.12) through (7.14), can be followed to obtain the formulas for the center distance between two crossed helical gears. Thus the pitch diameters are again given by

$$d_1 = \frac{N_1 p_1}{\pi} = \frac{N_1 p^n}{\pi \cos \psi_1}$$

and

$$d_2 = \frac{N_2 p_2}{\pi} = \frac{N_2 p^n}{\pi \cos \psi_2},$$

where the normal circular pitches of the two meshing gears must be equal. The center distance is given by

$$c = \frac{d_1 + d_2}{2}.$$

Substituting the expressions for the pitch diameters into the formula for the center distance, we obtain

$$c = \frac{p^n}{2\pi}\left(\frac{N_1}{\cos\psi_1} + \frac{N_2}{\cos\psi_2}\right). \tag{7.26}$$

Since $p^n P^n = \pi$, another form for the center distance is

$$c = \frac{1}{2P^n}\left(\frac{N_1}{\cos\psi_1} + \frac{N_2}{\cos\psi_2}\right), \tag{7.27}$$

while in the SI system,

$$c = \frac{m^n}{2}\left(\frac{N_1}{\cos\psi_1} + \frac{N_2}{\cos\psi_2}\right). \tag{7.28}$$

Velocity Ratio of Crossed Helical Gears

The output-to-input angular-velocity ratio for crossed helical gears is the inverse of the tooth number ratio; that is,

$$r_v = n_2/n_1 = \omega_2/\omega_1 = N_1/N_2, \tag{7.29}$$

the same equation that applies to all nonplanetary gears. We cannot use the inverse ratio of the diameters unless the helix angles are equal. A sketch showing a pair of teeth in mesh makes it easy to find the direction of the output shaft rotation.

Torque, Tangential Force, Radial Force, and Thrust. Suppose we need to find the tooth loads on a pair of crossed helical gears. Forces are applied at the location of tooth engagement. Torque, rotation speed, and power are related by the same equations used for spur gears and helical gears on parallel shafts. Once we have found the driver torque, the tangential force can be found from the relationship

$$F_{t1} = T_1/r_1, \tag{7.30}$$

where F_{t1} = driver tangential force,
T_1 = driver torque, and
r_1 = driver pitch radius.

The radial force on the driver is

$$F_{r1} = F_{t1} \tan \phi,$$

and the thrust on the driver is

$$F_{a1} = F_{t1} \tan \psi_1.$$

In general, the tangential force on the driven gear will not equal the tangential force on the driver. The *normal* forces on the driver and the driven gear, however, *are equal and opposite. If the input and output shafts are perpendicular*, the following relationships hold:

- The tangential force on the driver is equal to and opposite the thrust force on the driven gear.
- The thrust force on the driver is equal to and opposite the tangential force on the driven gear.
- The radial force on the driver is equal to and opposite the radial force on the driven gear.

Think of a right-hand screw to get the *thrust directions*. Thrust force directions at a driving gear tooth are as follows:

- If a right-hand helical drive gear turns clockwise, the thrust force is away from the observer.
- If a right-hand helical drive gear turns counterclockwise, the thrust force is toward the observer.
- If a left-hand helical drive gear turns clockwise, the thrust force is toward the observer.
- If a left-hand helical drive gear turns counterclockwise, the thrust force is away from the observer.

And remember the law of action and reaction to get forces on the driven gear.

SAMPLE PROBLEM 7.5

Analysis of Crossed Helical Gears

Two crossed helical gears have a normal module of 15 mm. The driver has 20 teeth and a helix angle of 20°. The angle between the shafts of the driver and follower is 50°, and the velocity ratio is $\frac{1}{2}$. The driver and the follower are both right handed. Determine the center distance.

Solution. The helix angle of the follower is found by using the known helix angle of the driver, the shaft angle, and Eq. (7.25a):

$$\psi_2 = \sum - \psi_1 = 50° - 20° = 30°.$$

The number of teeth on the follower is found by using the number of teeth on the driver and the velocity ratio formula:

$$N_2 = \frac{N_1}{r_v} = \frac{20}{1/2} = 40 \text{ teeth}.$$

Finally, the center distance for a pair of crossed helical gears is found by using Eq. (7.28):

$$c = \frac{m^n}{2}\left(\frac{N_1}{\cos\psi_1} + \frac{N_2}{\cos\psi_2}\right) = \frac{15 \text{ mm}}{2}\left(\frac{20}{\cos 20°} + \frac{40}{\cos 30°}\right) = 506 \text{ mm}.$$

7.3 WORM GEARS

If large speed reduction ratios are necessary between nonparallel shafts, crossed helical gears with a small driver and large follower can be used. However, the magnitude of the load that can be transmitted by these gears is limited. A better solution to the problem is the use of a worm and worm gear. Note, however, that worm gear sets can be considered a special case of crossed helical gears.

In Figure 7.10, a typical worm gear set is shown. Clearly, the worm is similar to a screw. In fact, the teeth on a worm are often spoken of as threads. The worm gear, sometimes called a worm wheel, is a helical gear.

If an ordinary cylindrical helical gear is used, there is "point contact". However, worm gears are usually cut with a concave rather than a straight width. (See Figure 7.10.) This results in the worm gear partially enclosing the worm, thus giving "line contact". Such a set, which is called a single enveloping worm gear set, can transmit much more power. If the worm is also manufactured with its length concave rather than straight, the worm teeth will partially enclose the gear teeth, as well as the gear teeth partially enclosing the worm teeth. Such a gear set, shown in Figure 7.11, is known as a double-enveloping worm gear set and will provide still more contact between gears, thus permitting even greater transmission of power.

Alignment is extremely important for proper operation of worm gear sets. For single-enveloping sets, the worm gear must be mounted accurately, while for double-enveloping sets, both the worm and the worm gear must be mounted accurately.

FIGURE 7.10 A cylindrical worm gear set. This gear set is single-enveloping. (*Source:* Horsburgh & Scott Company.)

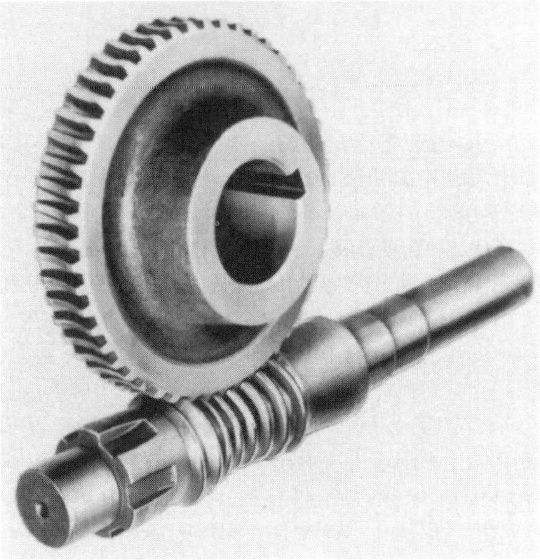

FIGURE 7.11 A double-enveloping worm gear set. The worm body is concave, so that the worm encloses, in addition to being enclosed by, the gear. (*Source:* Excello Corporation.)

Worm Gear Terminology and Geometry

The *axial pitch* of a worm is equal to the distance between corresponding points on adjacent threads, measured along the axis of the worm. (See Figure 7.12a.) The axial pitch of the worm and the circular pitch of the gear are equal in magnitude if the shaft axes are 90° apart. *We will consider worm drives with perpendicular shafts only*. Other shaft angles are extremely rare (except in textbooks).

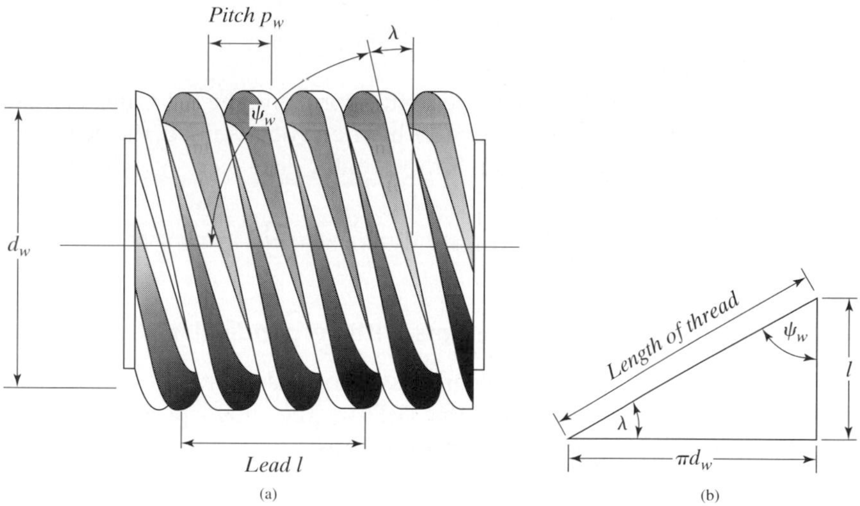

FIGURE 7.12 (a) The relationship between the pitch and lead for a worm with a triple thread. (b) One tooth of a worm is shown unwrapped to illustrate the relationship between the lead, the lead angle, the pitch diameter, and the helix angle. Note that the lead and helix angles are complementary.

Lead. The lead of a worm is equal to the apparent axial distance that a thread advances in one revolution of the worm. For a single-threaded worm (a worm with one tooth), the lead is equal to the pitch. A double-threaded worm (a worm with two teeth), has a lead equal to twice the pitch. The worm of Figure 7.12a has a triple thread and thus a lead equal to three times the pitch. The lead and the pitch of a worm are thus related by the equation.

$$l = p_w N_w, \tag{7.31}$$

where l = lead,
p_w = axial pitch of the worm, and
N_w = number of teeth on the worm.

In the SI system,

$$l = \pi m_w N_w, \tag{7.32}$$

where m_w is the module of the worm. The triangle shown in Figure 7.12b represents the unwrapping of one tooth of the worm illustrated in Figure 7.12a. From Figure 7.12b, we have

$$\tan \lambda = \frac{l}{\pi d_w}. \tag{7.33}$$

Since the lead angle λ and the helix angle ψ_w of the worm are complementary to each other, Eq. (7.33) can be rewritten as

$$\cot \psi_w = \frac{l}{\pi d_w}. \tag{7.34}$$

For shafts 90° apart, the lead angle of the worm and the helix angle of the gear are equal.

Once the worm pitch diameter has been determined, the gear pitch diameter can be found by using the equation for the center distance:

$$c = \frac{d_w + d_g}{2}. \tag{7.35}$$

Velocity Ratio of Worm Gear Sets with Perpendicular Shafts

The velocity ratio for worm gear sets is derived as was the velocity ratio for crossed helical gears. In the derivation that follows, the subscript w refers to the worm, while the subscript g refers to the worm gear. Thus,

$$r_v = \frac{\omega\,(\text{follower})}{\omega\,(\text{driver})} = \frac{\omega_g}{\omega_w} = \frac{N_w}{N_g}. \tag{7.36}$$

From Eq. (7.31), the number of teeth on the worm is given by

$$N_w = \frac{l}{p_w}.$$

The number of teeth on the worm gear is given by

$$N_g = P_g d_g.$$

But $P_g = \pi/p_g$; therefore,

$$N_g = \frac{\pi d_g}{p_g}.$$

Substituting the preceding expressions for N_w and N_g into the expression for the velocity ratio, we obtain

$$r_v = \frac{N_w}{N_g} = \frac{l p_g}{p_w \pi d_g}.$$

Since the shafts are perpendicular, the axial pitch of the worm, p_w, and the circular pitch of the gear, p_g, are equal. then the velocity ratio for the worm gear set becomes

$$\frac{\omega_g}{\omega_w} = r_v = \frac{l}{\pi d_g}. \tag{7.37}$$

In most worm gear sets, the worm is the driver. The set is therefore a speed-reduction unit. If the lead angle of the worm is greater than 11° or so, it is possible for the gear to be the driver, thus making the set a speed-increasing unit. Whether a given set is reversible or not depends on how much frictional force exists between the worm and the gear. Gear sets that are irreversible are usually referred to as *self-locking*. Small lead angles (less than 5°) usually result in irreversible gear sets.

In some applications, a self-locking gear set is a distinct advantage. For example, in a hoisting-machine application, a self-locking gear set would be an advantage because of the braking action it provides. The designer, however, must be certain that the braking capacity of a gear set is sufficient to perform satisfactorily as a self-locking unit. Toward that end, he or she should include a secondary braking device to ensure safety.

SAMPLE PROBLEM 7.6

Analysis of a Worm Gear Set

A quadruple-threaded worm has an axial pitch of 1 in and a pitch diameter of 2 in. The worm drives a gear having 42 teeth. Determine the lead angle of the worm and the center distance between worm and gear.

Solution. Since we know that the worm has four threads ($N_w = 4$) and that the axial pitch is 1 in ($p_w = 1$), we can easily determine the lead of the worm from Eq. (7.31):

$$l = p_w N_w = 1(4) = 4 \text{ in.}$$

Substituting this value into Eq. (7.33), we can determine the lead angle λ of the worm: We have

$$\tan \lambda = \frac{l}{\pi d_w} = \frac{4}{\pi(2)} = 0.637,$$

so that

$$\lambda = 32.5°.$$

We can now determine the center distance. Since we must have the pitch diameters of the gear and worm in order to use the formula for the center distance [Eq. (7.35)], we must first determine the pitch diameter of the gear. Equation (7.37) gives us an expression containing this unknown. The value of the velocity ratio can be found by using Eq. (7.36):

$$r_v = \frac{N_w}{N_g} = \frac{4}{42} = 0.095.$$

From Eq. (7.37), we have the pitch diameter

$$d_g = \frac{l}{\pi r_v} = \frac{4}{\pi(0.095)} = 13.4 \text{ in.}$$

Finally, having found the pitch diameter of the gear d_g, we can use Eq. (7.35) to find the center distance:

$$c = \frac{d_w + d_g}{2} = \frac{2 + 13.4}{2} = 7.7 \text{ in.}$$

Forces in Worm Gear Sets

The forces acting on worms and worm gears are the same as those acting on helical gears, except that the shaft axes are 90° apart in almost all practical applications. The free-body diagrams of a worm and worm gear shown in Figure 7.13 depict the relationships of the forces for a normal force F_n exerted between the worm and gear.

First, considering the worm gear and using Eqs. (7.19) through (7.21), we see that the components of the resultant force F_n consist of a tangential force

$$F_{tg} = F_n \cos \phi^n \cos \psi_g,$$

a radial force

$$F_{rg} = F_n \sin \phi^n = F_{tg} \tan \phi,$$

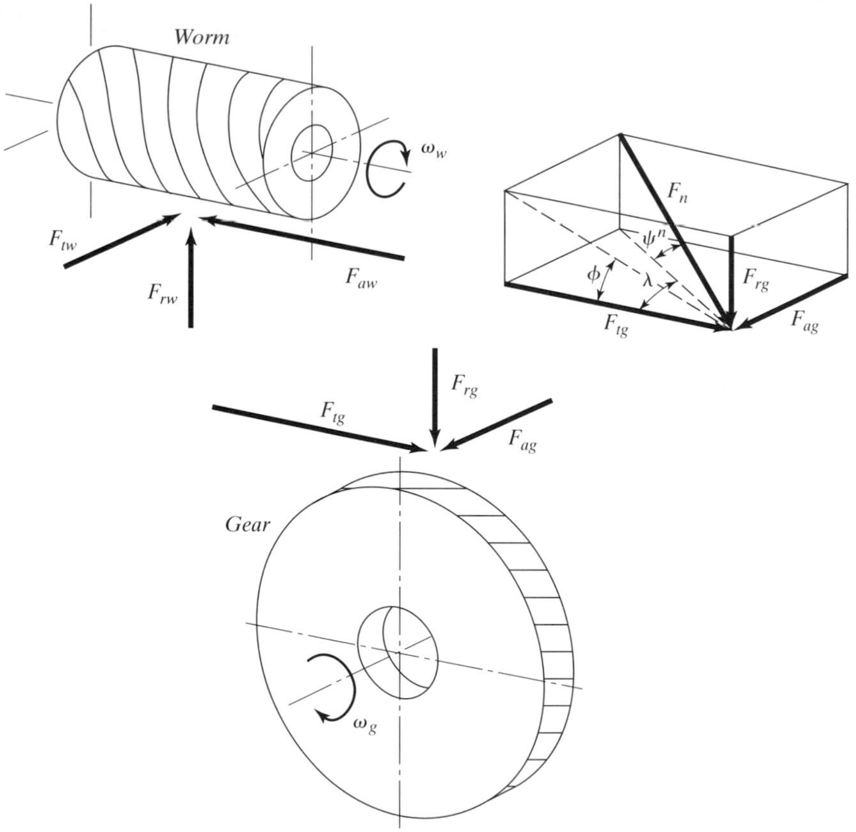

FIGURE 7.13 Forces exerted on a worm and an engaged gear. Both gears are righthanded, with the worm driving in the direction shown. The force diagram shows the breakdown into components of the resultant normal force F_n on the gear. Note that $\lambda = \psi_g$.

and an axial or thrust force

$$F_{ag} = F_n \cos \phi^n \sin \psi_g = F_{tg} \tan \psi_g,$$

where ϕ^n and ϕ are the normal and transverse pressure angles, respectively, and ψ_g is the helix angle of the worm gear. Since the helix angle of the gear is equal to the lead angle of the worm (for 90° shafts), these forces can be expressed as

$$F_{tg} = F_n \cos \phi^n \cos \lambda,$$

$$F_{rg} = F_n \sin \phi^n = F_{tg} \tan \phi,$$

and

$$F_{ag} = F_n \cos \phi^n \sin \lambda = F_{tg} \tan \lambda.$$

Next, from an examination of the free body of the worm in Figure 7.13, it is clear that the magnitude of the axial or thrust force F_{aw}, on the worm is equal to that of the tangential gear force F_{tg} (the directions are opposite). The magnitude of the tangential force F_{tw} on the worm is equal to that of the axial gear force F_{ag}, and the magnitudes of the two radial force components F_{rw} and F_{rg} are equal.

The relationships of these forces to the shaft torques about the respective axes of rotation are

$$T_w = F_{tw} r_w$$

and

$$T_g = F_{tg} r_g,$$

where r_w and r_g are the pitch radii of the worm and the gear respectively. Note that, as pointed out earlier, the forces F_{tw} and F_{tg} are not equal, as they are for a pair of spur gears or a pair of helical gears on parallel shafts. Instead, observing that $F_{tw} = F_{ag}$, we have

$$T_w = F_{ag} r_w = F_{tg} \tan \lambda \, r_w,$$

which leads to the following equation for the torque ratio:

$$\frac{T_g}{T_w} = \frac{F_{tg} r_g}{F_{tg} \tan \lambda r_w} = \frac{1}{\tan \lambda}\left(\frac{r_g}{r_w}\right) = \frac{2\pi r_w}{l}\left(\frac{r_g}{r_w}\right) = \frac{\pi d_g}{l} = \frac{1}{r_v}.$$

This relationship follows from the fact that, in the absence of friction and other losses, the power going into a gear set ($T_w\omega_w$) is equal to the power coming out of the gear set ($T_g\omega_g$). The torque ratio that is derived therefore assumes a power transmission efficiency of 100 percent. (Actually, efficiencies of considerably less than 100 percent can occur in worm gear units.)

7.4 BEVEL GEARS

When power is to be transmitted between two shafts that intersect, the type of gear usually used is a bevel gear. The pitch surfaces of two mating bevel gears are rolling cones, rather than the rolling cylinders that two mating spur gears have. Figure 7.14 shows a typical pair of meshing bevel gears. While the shafts that bevel gears are mounted on are usually 90° apart, in certain applications the shaft angle is greater or less than 90°.

Bevel Gear Terminology and Geometry

Some of the more common terms used in bevel gearing are illustrated in Figure 7.15. As seen in the illustration, the tooth size decreases along the face width as the apex of the pitch cone is approached. The pressure angle for most straight bevel gears is 20°.

The definitions of the terms shown in the figure are as follows:

Pitch cone: the geometric shape of bevel gears, based on equivalent rolling contact.

Apex of pitch cone: the intersection of the elements making up the pitch cone.

Cone distance: the slant height of the pitch cone—in other words, the length of a pitch cone element.

Face cone: the cone formed by the elements passing through the top of the teeth and the apex.

Root cone: the cone formed by the elements passing through the bottom of the teeth and the apex.

FIGURE 7.14 A pair of bevel gears. The pitch surfaces for the two gears are rolling cones. (*Source:* Browning Manufacturing Company.)

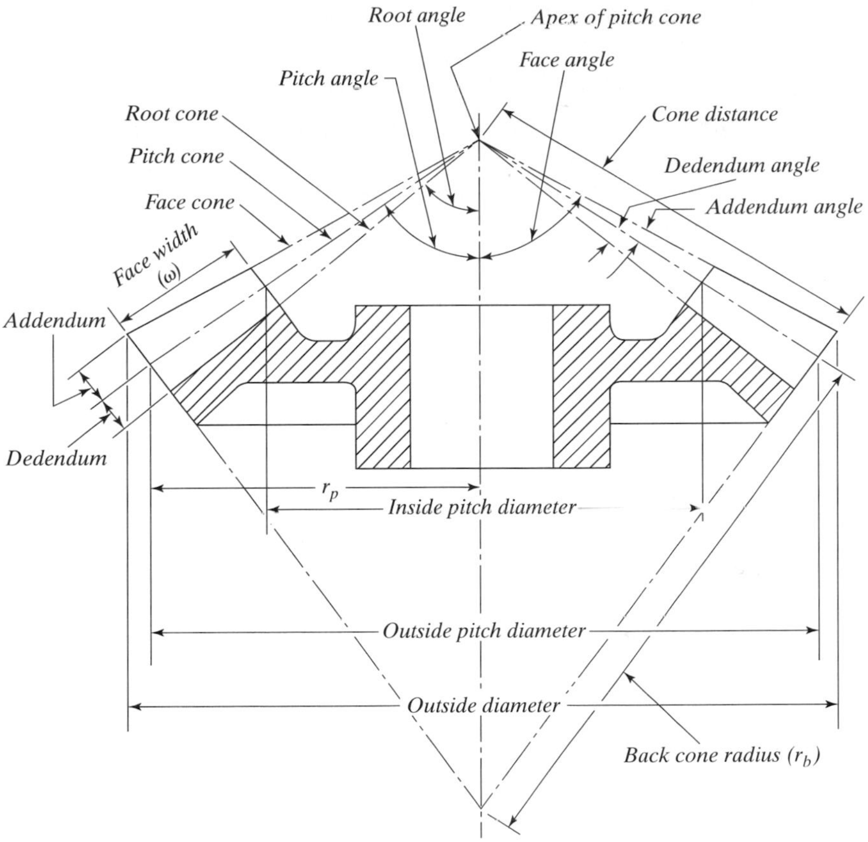

FIGURE 7.15 Terminology associated with bevel gears.

Face angle: the angle between an element of the face cone and the axis of the gear.

Pitch angle: the angle between an element of the pitch cone and the axis of the gear.

Root angle: the angle between an element of the root cone and the axis of the gear.

Face width: the width of a tooth.

Addendum: the distance from the pitch cone to the face cone, measured on the outside of the tooth.

Dedendum: the distance from the pitch cone to the root cone, measured on the outside of the tooth.

Addendum angle: the angle between an element on the pitch cone and an element on the face cone.

Dedendum angle: the angle between an element on the pitch cone and an element on the root cone.

Inside pitch diameter: the pitch diameter measured on the inside of the tooth.

Outside pitch diameter: the pitch diameter measured on the outside of the tooth.

Back cone: the cone formed by elements perpendicular to the pitch cone elements at the outside of the teeth.

Back cone radius: the length r_b of a back cone element.

Since most bevel gears are mounted on intersecting shafts, at least one is usually mounted *outboard*. That is, one gear is mounted on the cantilevered end of a shaft. Because of the outboard mounting, the deflection of the shaft where the gear is attached may be rather large. This could result in the teeth at the small end moving out of mesh. The load would thus be unequally distributed, with the larger ends of the teeth taking most of the load. To mitigate this effect, the tooth face width is usually made no greater than $\frac{1}{3}$ of the cone distance.

Classifying Bevel Gears by Pitch Angle

Bevel gears are usually classified according to their pitch angle. A bevel gear having a pitch angle of 90° and a plane for its pitch surface is known as a *crown gear*. Figure 7.16 shows such a gear.

When the pitch angle of a bevel gear exceeds 90°, the gear is called an *internal bevel gear*. Like the gear shown in Figure 7.17 internal bevel gears, cannot have pitch angles very much greater than 90° because of the problems incurred in manufacturing such gears. In fact, these manufacturing difficulties are the main reason internal bevel gears are rarely used.

Bevel gears with pitch angles less than 90° are the type most commonly used. Figure 7.15 illustrates this kind of external bevel gear.

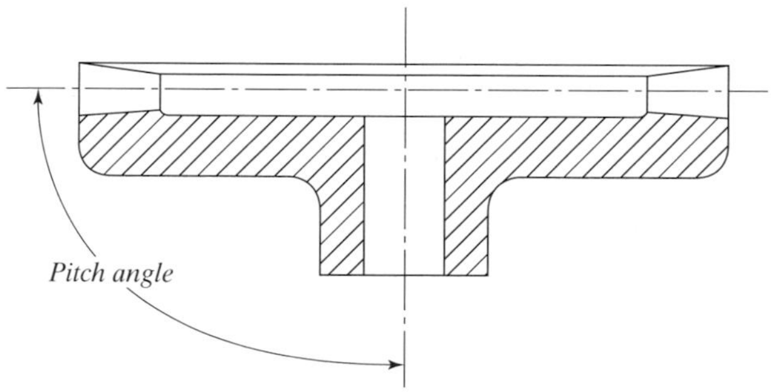

FIGURE 7.16 A crown gear is a bevel gear with a pitch angle of 90°. The entire pitch surface lies in a single plane perpendicular to the gear axis.

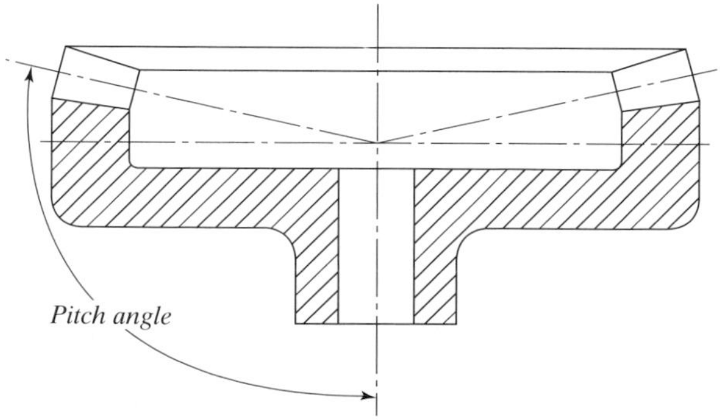

FIGURE 7.17 An internal bevel gear is a bevel gear with a pitch angle greater than 90°.

When two meshing bevel gears have a shaft angle of 90° and have the same number of teeth, they are called *miter gears*. Miter gears have a speed ratio of unity. Each of the two gears has a 45° pitch angle.

Velocity Ratio of Bevel Gears

The velocity ratio of bevel gears is given by the same expressions used to determine the velocity ratio of spur gears, where the subscripts 1 and 2 refer to the driver and follower:

$$r_v = \frac{\omega_2}{\omega_1} = \frac{r_1}{r_2} = \frac{N_1}{N_2}. \tag{7.39}$$

Here, r is the pitch-circle radius and N is the number of teeth.

At this point, it is desirable to derive some relationships between numbers of teeth and pitch angles for bevel gears. In Figure 7.18, which shows the pitch cones of two external bevel gears in mesh, Σ is the shaft angle, γ and Γ are the pitch angles, and r_p and r_g are the pitch radii for the pinion and gear, respectively. From Figure 7.18,

$$\Sigma = \Gamma + \gamma,$$

and

$$\sin\Gamma = \frac{r_g}{OP}, \quad \sin\gamma = \frac{r_p}{OP},$$

$$OP = \frac{r_g}{\sin\Gamma} = \frac{r_p}{\sin\gamma}.$$

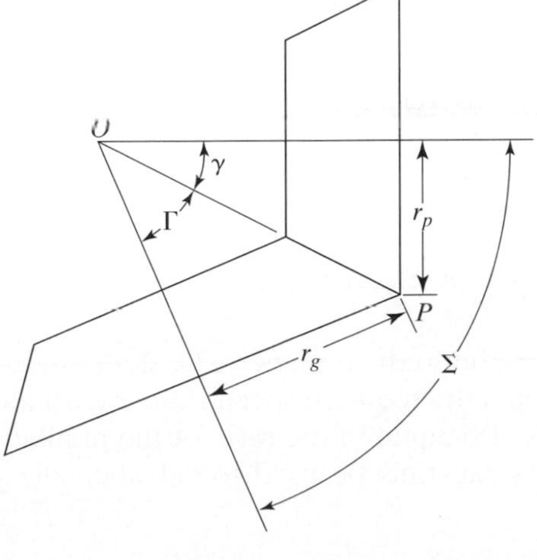

FIGURE 7.18 Two meshing external bevel gears, illustrating the relationships among the shaft angle, the pitch angles, and the pitch radii.

Thus,

$$\sin\Gamma = \frac{r_g}{r_p}\sin\gamma = \frac{r_g}{r_p}\sin(\Sigma - \Gamma)$$

$$= \frac{r_g}{r_p}(\sin\Sigma\cos\Gamma - \cos\Sigma\sin\Gamma).$$

Dividing this expression for $\sin\Gamma$ by $\cos\Gamma$, we obtain

$$\frac{\sin\Gamma}{\cos\Gamma} = \tan\Gamma = \frac{r_g}{r_p}(\sin\Sigma - \cos\Sigma\tan\Gamma),$$

or

$$\tan\Gamma = \frac{(r_g/r_p)\sin\Sigma}{1 + (r_g/r_p)\cos\Sigma} = \frac{\sin\Sigma}{(r_p/r_g) + \cos\Sigma}.$$

At this point, we can utilize our original velocity ratio formula to arrive at the desired formula involving both the pitch angles and the numbers of teeth for both bevel gears. From Eq. (7.39),

$$\frac{r_p}{r_g} = \frac{N_p}{N_g}.$$

Therefore,

$$\tan\Gamma = \frac{\sin\Sigma}{(N_p/N_g) + \cos\Sigma}. \tag{7.40}$$

Similarly,

$$\tan\gamma = \frac{\sin\Sigma}{(N_g/N_p) + \cos\Sigma}. \tag{7.41}$$

Finally, for shaft angle $\Sigma = 90°$ (which is the usual case), we obtain

$$\tan\Gamma = \frac{1}{(N_p/N_g) + 0} = \frac{N_g}{N_p} \tag{7.42}$$

and

$$\tan\gamma = \frac{1}{(N_g/N_p) + 0} = \frac{N_p}{N_g}. \tag{7.43}$$

The formulas just derived are quite useful to the designer. The shaft angle, or angle at which one shaft intersects the other, and the required speed ratio are usually known to the designer. Since the speed ratio is also equal to the ratio of the number of teeth, it should be obvious that the formulas can thus be used to calculate the pitch angle required for each gear.

SAMPLE PROBLEM 7.7

Analysis of Bevel Gears

A pair of straight-toothed bevel gears is mounted on shafts that intersect each other at an angle of 70°. The velocity ratio of the gears is $\frac{1}{2}$. Determine the pitch angles of the gears.

Solution. Knowing the velocity ratio, we can use Eqs. (7.40) and (7.41) to find the pitch angles for bevel gears with a shaft angle other than 90°. Thus,

$$\tan \Gamma = \frac{\sin \Sigma}{(N_p/N_g) + \cos \Sigma},$$

and since $N_p/N_g = r_p/r_g = \frac{1}{2}$,

$$\tan \Gamma = \frac{\sin 70°}{1/2 + \cos 70°} = \frac{0.940}{0.842} = 1.12,$$

or

$$\Gamma = 48.2°.$$

Similarly,

$$\tan \gamma = \frac{\sin \Sigma}{(N_g/N_p) + \cos \Sigma} = \frac{0.940}{2.342} = 0.401,$$

so that

$$\gamma = 21.8°.$$

Other Types of Bevel Gears

There are a number of other types of bevel gears in addition to straight-toothed gears. Spiral bevel gears (Figure 7.19), are used with high-speed, high-load applications. For these gears, the transmission of power is much smoother than for straight bevel gears, since there is gradual tooth contact in addition to more teeth being in contact at any instant.

While a wide variety of spiral angles are used, 35° is the most common. The spiral angle is the angle of the spiral relative to the axis of the gear, as measured at the middle of the face width of the tooth.

Hypoid gears are used on nonintersecting shafts (Figure 7.20). Large reductions in speed are possible with hypoid gears, which, incidentally, helped solve a problem for designers of rear-wheel-drive automobiles (before the popularity of the front-wheel drive). A hypoid pinion on the drive shaft is used to drive a large hypoid gear, which in turn drives the differential case. The drive shaft is mounted below the center of the large hypoid gear. A design with straight bevel gears results in a higher drive shaft, and loss of clear floor space in the passenger compartment.

FIGURE 7.19 Spiral bevel gears feature a spiral rather than a helical tooth design. However, for simplification of the manufacture of spiral bevel gears, the tooth is usually made circular instead of spiral. (*Source:* Arrow Gear Company.)

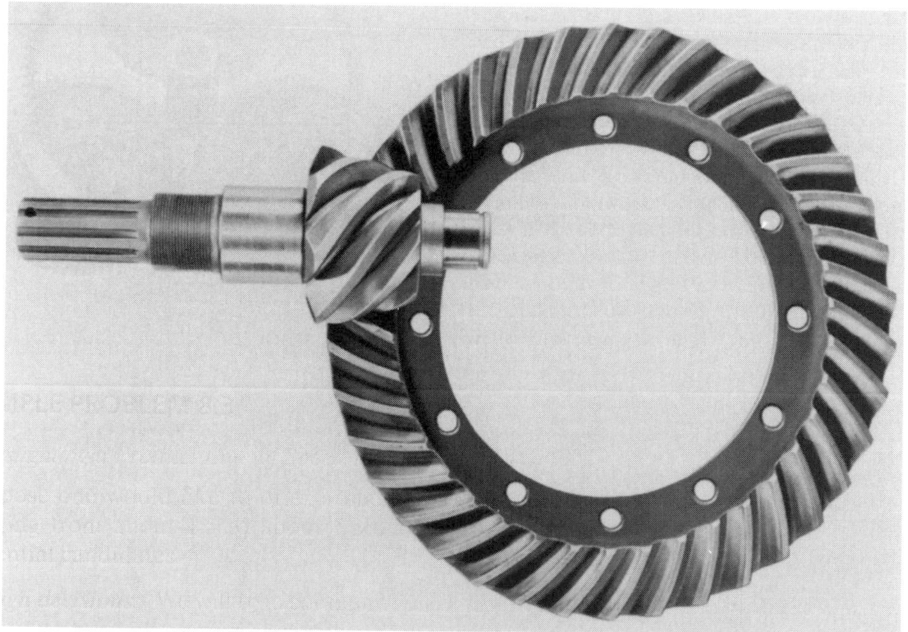

FIGURE 7.20 Hypoid gears are spiral bevel gears designed to operate on nonintersecting shafts. (*Source:* Richmond Gear, Wallace Murray Corporation.)

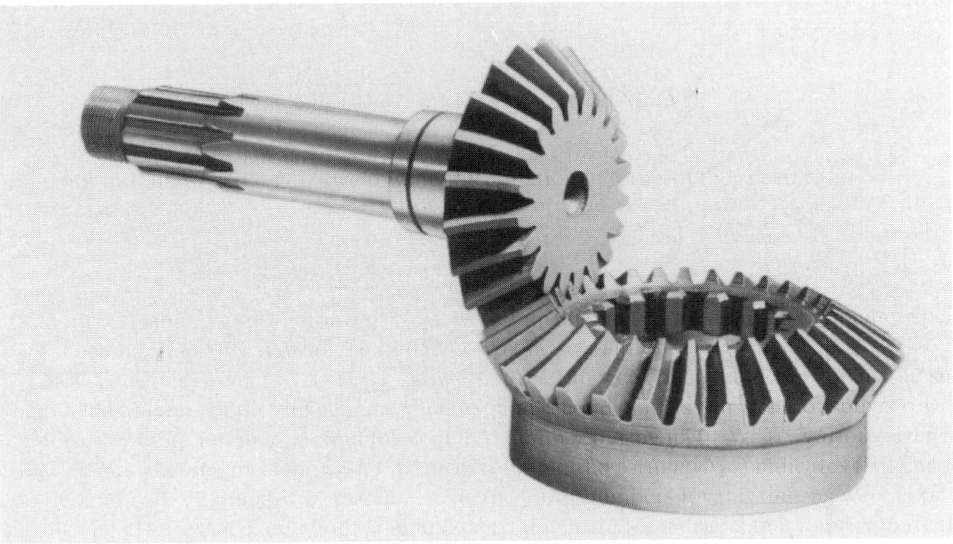

FIGURE 7.21 Zerol bevel gears are spiral bevel gears with a zero spiral angle. (*Source:* Richmond Gear, Wallace Murray Corporation.)

Zerol bevel gears, shown in Figure 7.21, are the same as spiral gears, except that the spiral angle is zero. Zerol gears are used when it is desirable to reduce the thrust loads that occur when spiral bevel gears are used.

While there are still more types of bevel gears, those which have been discussed are the types most often used. Gear manufacturers' catalogs should be consulted when information about other specialized gears is desired.

Forces on Straight Bevel Gears with 90° Shaft Angles

Let us consider the forces exerted by a bevel gear on its mating gear. (See Figure 7.22.) We will draw the force diagram for one of the gears, assuming that we have straight-toothed bevel gears mounted on shafts that are 90° apart. Another reasonable assumption we will make is that the resultant tooth load acts at the center of the tooth. In other words, the force acts at the mean pitch radius

$$r_m = \frac{d_i + d_o}{4},$$

where d_i is the inside pitch diameter and d_o is the outside pitch diameter of the gear. (See Figure 7.15.)

In Figure 7.22, the force F_n is the normal force exerted by the driven gear on the driving pinion teeth. The driving pinion rotates counterclockwise when viewed from the left. A force of equal magnitude and opposite direction will act on the driven gear. If friction is neglected, these are the resultant gear mesh forces on the individual gears.

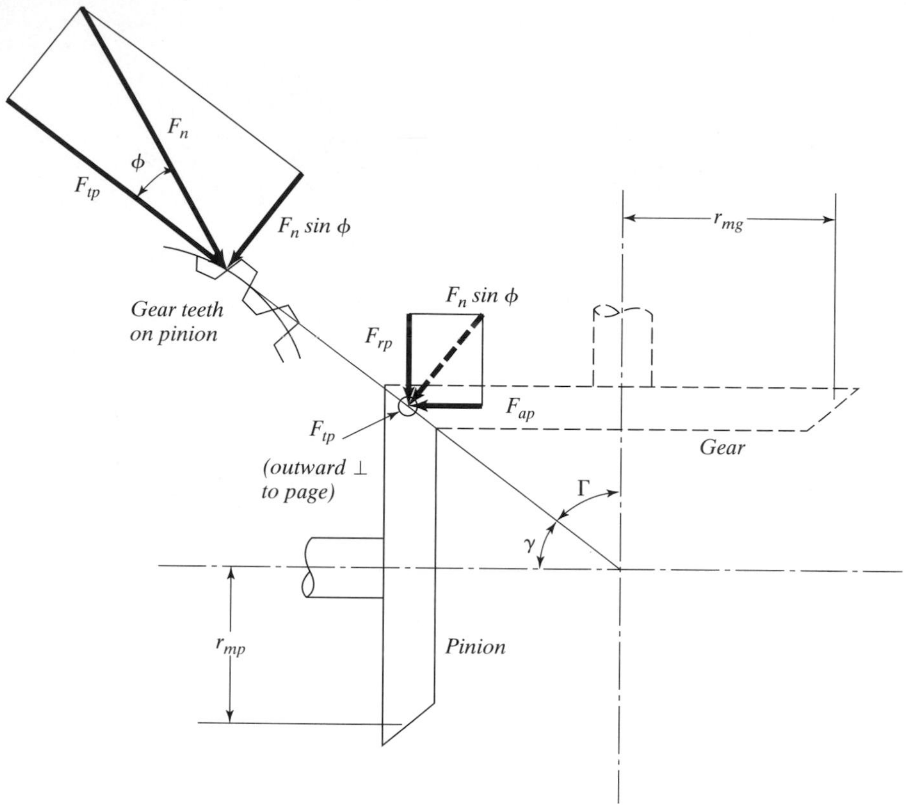

FIGURE 7.22 Tooth loads on straight-toothed bevel gears. The forces shown are those acting on the driving pinion, rotating in the direction shown. Equal and opposite forces act on the driven gear.

From Figure 7.22, for a pressure angle ϕ, the component forces on the pinion are a tangential force

$$F_{tp} = F_n \cos\phi, \tag{7.44}$$

a radial force

$$F_{rp} = F_n \sin\phi \cos\gamma = F_{tp} \tan\phi \cos\gamma, \tag{7.45}$$

and an axial force

$$F_{ap} = F_n \sin\phi \sin\gamma = F_{tp} \tan\phi \sin\gamma. \tag{7.46}$$

For the gear, the tangential, radial, and axial force components are, respectively,

$$F_{tg} = F_n \cos\phi = F_{tp}, \tag{7.47}$$

$$F_{rg} = F_n \sin\phi \cos\Gamma = F_{tg} \tan\phi \cos\Gamma, \tag{7.48}$$

and

$$F_{ag} = F_n \sin \phi \sin \Gamma = F_{tg} \tan \phi \sin \Gamma. \tag{7.49}$$

It should be obvious from Figure 7.22 that the radial force component for the pinion is equal in magnitude to the thrust or axial component for the gear, while the thrust component for the pinion is equal in magnitude to the radial component for the gear. Therefore, for shafts that are 90° apart,

$$F_{ap} = F_{rg} = F_n \sin \phi \sin \gamma = F_n \sin \phi \cos \Gamma$$

and

$$F_{rp} = F_{ag} = F_n \sin \phi \cos \gamma = F_n \sin \phi \sin \Gamma.$$

These equations follows from the fact that

$$\gamma + \Gamma = 90^\circ,$$

and therefore,

$$\sin \gamma = \sin(90^\circ - \Gamma) = \cos \Gamma$$

and

$$\cos \gamma = \cos (90^\circ - \Gamma) = \sin \Gamma.$$

Finally, the gear tooth forces produce the following torques on the pinion and gear:

$$T_p = F_{tp}\, r_{mp} \tag{7.50}$$

and

$$T_g = F_{tg}\, r_{mg} = F_{tp}\, r_{mg}. \tag{7.51}$$

Here, T_p is the torque acting on the pinion and T_g is the torque acting on the gear. The preceding force and torque expressions can be used to determine the power transmitted by a bevel gear set.

SUMMARY

Almost all gear sets are used to reduce speed. Except for planetary trains, the speed ratio of any pair of gears equals the inverse of the tooth number ratio. The speed ratio of helical gears on parallel shafts is also given by the inverse of the pitch diameter ratio.

The contact ratio of a pair of helical gears on parallel shafts is higher than that of equivalent spur gears. Helical gears are often selected to replace spur gears when smooth, quiet operation is required. However, helical gears introduce thrust loads, a problem that is not present when we specify straight spur gears.

Bevel gears are used to turn a corner—for example, if we want to drive a vertical shaft from a horizontal shaft. Crossed helical gears are used to transmit power between non-intersecting, nonparallel shafts.

Worm drives are a special case of crossed helical gears. They are usually used for large reductions in speed. For example, a single-toothed worm driving a 50-tooth gear has an output-to-input speed ratio of 1 to 50. The output-to-input torque ratio would be 50 to 1 without friction. However, sliding of the worm teeth on the gear teeth results in substantial friction loss.

When the required speed and transmitted power are specified, we can find the torque. The tangential force on a gear tooth equals the torque divided by the pitch radius. The tangential force on a drive gear opposes rotation. The radial force and thrust force are then found from the tangential force, pressure angle, and helix angle. Free-body diagrams using an exploded view of the gear pair are useful in this regard. The diagrams would show, for example, that the thrust force on a worm is balanced by the tangential force on a worm gear.

A Few Review Items

- Sketch a pair of helical gears on parallel shafts. (Shown an exploded view.) Identify the driver and driven gears, rotation directions, helix angles, and hand. Show the torque, tangential force, radial force, and thrust on each. Is there a torque balance? Have you violated Newton's third law?
- A proposed spur gear pair has a contact ratio of 1.2. You decide to replace it with a pair of helical gears to obtain a contact ratio of at least 2.2. How would you determine the minimum face width to reach this goal?
- Why would you be reluctant to specify crossed helical gears for high-power applications?
- What is the significance of the normal diametral pitch and normal module in designing crossed helical gears?
- Suppose we want to balance thrust on a countershaft. Relate the helix angle to the pitch radius. Will you specify the same hand for both gears on that shaft?
- Relate the lead angle of a worm to the helix angle of a meshing worm gear.
- Suppose you need an output-to-input speed ratio of 1 to 80. What type of gearing would you choose? Specify the numbers of teeth.
- Suppose you need a pair of gears to operate on perpendicular shafts with intersecting centerlines. The output-to-input speed ratio is to be 1:2.5. Select the type of gear. Sketch the gears (in an exploded view). Show the directions of the tooth force and torque on the input and output gears.
- Suppose is necessary to stir four vats of chemicals. Can you design a system that utilizes a single motor driving a horizontal shaft which, in turn, drives four stirrers? What type of gears have you chosen?

- Can you suggest an alternative design with a different type of gear?
- Suppose you hear some complaints that a speed changer utilizing spur gears is too noisy. Suggest an alternative design.
- Identify the conditions, in terms of friction and lead angle, that result in a self-locking worm gear set.
- Give two potential applications for a self-locking worm gear set.
- Are there situations in which the self-locking feature would be undesirable?

PROBLEMS

7.1 Two helical gears mounted on parallel shafts are in mesh. The gears have a diametral pitch of 5, a 40° helix angle, and 20 and 30 teeth, respectively. Calculate the normal circular pitch.

7.2 Two meshing helical gears have a 25° transverse pressure angle and a 20° normal pressure angle. The gears have a diametral pitch of 10 and possess 15 and 45 teeth, respectively. Determine the center distance and helix angle.

7.3 Two parallel shafts are spaced 5 in apart. A pair of helical gears are to be selected to provide a velocity ratio of about $\frac{1}{2}$. The normal diametral pitch is to be 6, the normal pressure angle is to be 20°, and the gears are to have at least 20 teeth. Determine the number of teeth for the gears and the transverse pressure angle (There are many possible solutions.)

7.4 A helical pinion has a normal pressure angle of 20°, and a transverse pressure angle of 25°. The pinion rotates at 2000 rev/min and is to drive a meshing helical gear so that the speed ratio is $\frac{1}{4}$. The centers of the shafts are 10 in apart. Determine the normal diametral pitch and the pitch diameters if the pinion has 20 teeth.

7.5 For the gears shown in Figure P7.1, show the rotation direction of each gear and also indicate the direction of the thrust load for each gear if shaft C is the input and rotates counterclockwise as observed from the left.

7.6 Two meshing helical gears on parallel shafts have a normal pressure angle of 20° and a transverse pressure angle of 23°. The normal circular pitch is 0.6 in. If the speed ratio is to be 0.4, determine the number of teeth for each gear. The center distance is 8 in.

7.7 Two meshing helical gears are mounted on parallel shafts that have rotational speeds of 1000 and 400 rev/min. The helix angle is 30° and the center distance is 252 mm. The gears have a module of 6 mm. Determine the normal circular pitch and the transverse circular pitch. Also, determine the number of teeth on each gear.

7.8 Two helical gears on parallel shafts have a normal circular pitch of 15 mm and a pitch-line velocity of 4500 mm/s. If the rotational speed of the pinion is 800 rev/min and the number of pinion teeth is 20, what must be the helix angle?

7.9 Two helical gears on parallel shafts have 30 teeth and 60 teeth and a normal module of 5 mm. The normal and transverse pressure angles are 20° and 24°, respectively. Determine the center distance.

7.10 Two helical gears on parallel shafts have a helix angle of 20°, a normal pressure angle of 25°, and a normal diametral pitch of 4. The numbers of teeth are 30 and 50. The pinion rotates at 800 rev/min and the gear set transmits 100 hp. Determine the tangential, radial, and axial gear tooth loads. Show these forces on a sketch of the gears. The pinion is left handed and rotates counterclockwise.

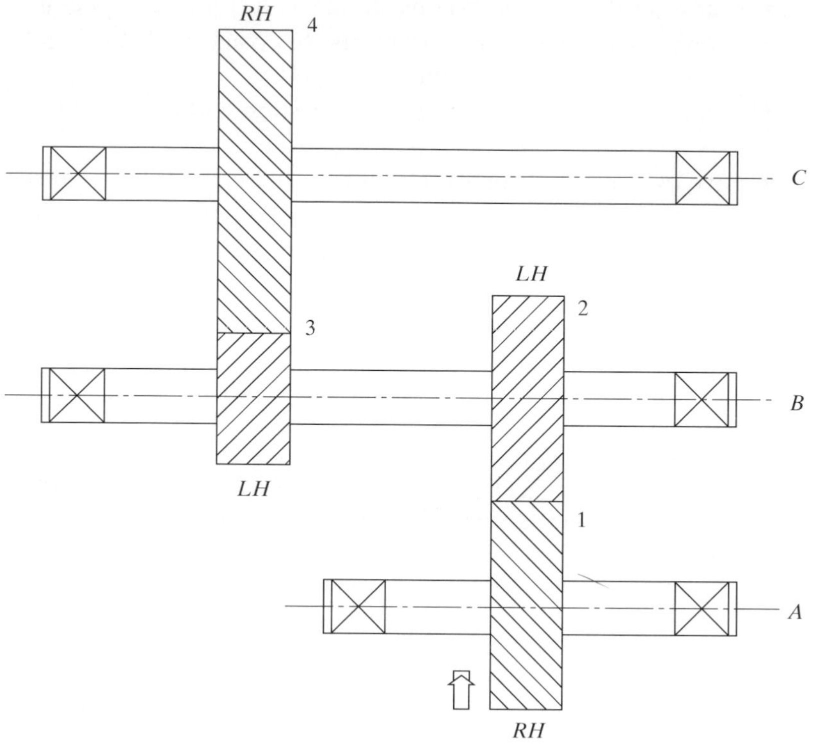

FIGURE P7.1

7.11 Two helical gears on parallel shafts have a normal pressure angle of 20° and a normal module of 6 mm. The center distance is 200 mm and the tooth numbers are 20 and 40. The gear set transmits 50 kW at a pinion speed of 1200 rev/min. Determine the tangential, radial, and thrust loads on the gear teeth, and show these forces on a sketch of the gears. The pinion is right handed and rotates clockwise.

7.12 The double-reduction helical gear train in Figure P7.1 employs helical gears with a helix angle of 30° and a normal pressure angle of 20°. The normal module is 8 mm and the numbers of teeth are $N_1 = 30$, $N_2 = 40$, $N_3 = 20$, and $N_4 = 50$. The pinion has a rotational speed of 1000 rev/min and the power transmitted is 20 kW. Determine the torque carried by each shaft, the magnitude and direction of the thrust force on each shaft, and the resultant gear load on each gear.

7.13 Two helical gears of the same hand are used to connect two shafts that are 90° apart. The smaller gear has 24 teeth and a helix angle of 35°. Determine the center distance between the shafts if the speed ratio is $\frac{1}{2}$. The normal circular pitch is 0.7854 in.

7.14 Repeat Problem 7.13, but assume that the gears are of opposite hand and that the shaft angle is 10°.

7.15 Two left-hand helical gears having the same helix angle are used to connect two shafts 60° apart. The velocity ratio is to be 0.4 and the gears have a normal diametral pitch of 4. If the center distance is to be about 12 in, determine the numbers of teeth for each gear.

7.16 Two right-hand helical gears with a normal module of 4 mm connect two shafts that are 60° apart. The pinion has 32 teeth and the velocity ratio is $\frac{1}{2}$. The center distance is 220 mm. Determine the helix angles of the two gears.

7.17 Determine the pitch diameters of a worm gear set having a velocity ratio of 0.1 and a center distance of $2\frac{1}{2}$ in, if the worm has three teeth and a lead of 1 in.

7.18 A worm gear set has a velocity ratio of 0.05. The worm has two teeth, a lead of 3 in, and a pitch diameter of $1\frac{1}{2}$ in. Determine the helix angle and pitch diameter of the worm gear.

7.19 A worm gear set is to have a velocity ratio of 0.05. The worm has three teeth, a lead angle of 20°, and a pitch of 1.5 in. Determine the center distance.

7.20 A worm gear set has a velocity ratio of 0.04. Find the center distance if the worm has three teeth, a pitch diameter of 2.5 in, and an axial pitch of 0.5 in.

7.21 A worm gear set has a speed ratio of 0.05, a lead angle of 20°, and a center distance of 10 in. Determine the pitch diameters.

7.22 A single-threaded worm has an axial pitch of 20 mm and a pitch diameter of 50 mm. The worm rotates at 500 rev/min and drives a gear having 40 teeth and a 25° transverse pressure angle. The power transmitted is 0.5 kW. Determine **(a)** the lead angle of the worm, **(b)** the center distance between worm and gear, and **(c)** the tangential, radial, and thrust forces on the worm gear.

7.23 For the gear train of Figure 7.1a, assume that the small helical gear is the input and rotates counterclockwise at 1,000 rev/min. The power transmitted is 1.0 kW. The helical gears on parallel shafts have a 20° normal pressure angle, a 20° helix angle, a center distance of 120 mm, and 24 and 48 teeth, respectively. The worm is single threaded with a pitch of 20 mm, and the worm gear has 40 teeth and a 25° transverse pressure angle. The center distance of the worm gear set is 150 mm. Determine the torque and thrust load for each shaft, and sketch the intermediate shaft, showing the gear force components acting on it.

7.24 Two bevel gears are to be used to connect two shafts that are 90° apart. The pinion has 18 teeth and a diametral pitch of 6. If the velocity ratio is to be 0.4, determine **(a)** the pitch angles, **(b)** the back cone radii.

7.25 Repeat Problem 7.24 for two shafts that are 60° apart.

7.26 Repeat Problem 7.24 for two shafts that are 110° apart.

7.27 Two 20° straight bevel gears have a diametral pitch of 4, and 24 and 48 teeth, respectively. The tooth face width is 2 in. The pinion rotates at 1000 rev/min and transmits 50 hp. The shafts are at 90°. Determine the components of the gear tooth force and show these on a sketch of the gears.

7.28 Suppose we need a line of speed reducers with output-to-input speed ratios of about 1/25 to about 1/6. The input speed will be 2400 rpm, and 2 kW of mechanical power are to be transmitted. Design the reducers and examine the geometry and loading. Check for interference and check the contact ratio. Among our design decisions will be the following: We will design a reverted gear train similar to the one in Figure 7.5. The actual train will be designed for bearings and will be made compact to reduce bending loads. Gears 1 and 2 will be selected with a module of 3 mm and a 40° helix angle. Gear 1 will have a right-hand helix and rotate counterclockwise; gear 2 will have a left-hand helix. Gears 3 and 4, which are expected to have higher tooth loads, will have a module of 4 mm. We will specify a 20° transverse pressure angle and a face width of eight times the module for all gears. We will try to balance thrust forces on the countershaft. Sixteen-tooth gears will be selected for gears 1 and 3.

7.29 Suppose we need a line of speed reducers with output speeds of about 120 to 600 rpm. The input speed will be 3600 rpm, and 1.2 kW of mechanical power is to be transmitted. Design the reducers and examine the geometry and loading. Check for interference and check the contact ratio. Among our design decisions will be the following: We will design a reverted gear train similar to the one in Figure 7.5. Gears 1 and 2 will be selected with a module of 1.5 mm and a 45° helix angle. Gear 1 will have a right-hand helix and rotate counterclockwise; gear 2 will have a left-hand helix. Gears 3 and 4, which are expected to have higher tooth loads, will have a module of 2 mm. We will specify a 20° transverse pressure angle and a face width of eight times the module for all gears. We will try to balance thrust forces on the countershaft. Sixteen-tooth gears will be selected for gears 1 and 3.

7.30 Suppose we need speed reducers with output speeds of about 100 to 160 rpm. The input speed will be 4000 rpm, and 1.2 kW of mechanical power is to be transmitted. Design the reducers and examine the geometry and loading. Check for interference and check the contact ratio. Among our design decisions will be the following: We will design a reverted gear train similar to Figure 7.5. Gears 1 and 2 will be selected with a 45° helix angle. Gear 1 will have a right-hand helix and rotate counterclockwise; gear 2 will have a left-hand helix. All gears will have a module of 2 mm, a 20° transverse pressure angle, and a face width of eight times the module. We will try to balance thrust forces on the countershaft. Sixteen-tooth gears will be selected for gears 1 and 3.

BIBLIOGRAPHY AND REFERENCES

American Gear Manufacturers Association, *American National Standard—Gear Classification and Inspection Handbook*, ANSI/AGMA 2000-A88, American Gear Manufacturers Association, Alexandria VA, March 1988.

American Gear Manufacturers Association, *AGMA Standard for Rating the Pitting Resistance and Bending Strength of Spur and Helical Involute Gear Teeth*, AGMA 218.01, American Gear Manufacturers Association, Alexandria VA, Dec. 1982.

American Gear Manufacturers Association, *AGMA Standard Nomenclature of Gear Tooth Failure Modes*, ANSI/AGMA 110.04, American Gear Manufacturers Association, Alexandria VA, Aug. 1980(a).

American Gear Manufacturers Association, *AGMA Gear Handbook— Gear Classification, Materials and Measuring Methods for Bevel, Hypoid, Fine Pitch Wormgearing and Racks Only as Unassembled Gears*, AGMA 390.03a, American Gear Manufacturers Association, Alexandria VA, Aug. 1980(b).

Boston Gear, *Gears and Shaft Accessories Catalog*, Boston Gear Div., Imo Industries, Inc., Quincy, MA, 1992.

CHAPTER 8

Drive Trains: Design and Analysis

Concepts You Will Learn and Apply When Studying This Chapter

- Design of speed changers and gearing to reverse direction
- Transmissions with axial shifting
- Design of simple and compound planetary gear trains, using the formula and superposition methods
- Design of balanced planetary trains; load sharing and elimination of lateral shaft loads
- Design of differential drives for high reduction ratios
- The free-floating transmission
- Planetary transmissions for changing speed
- Bevel-gear differentials and other trains with more than one input
- Fixed-ratio and variable-speed chain drives
- Fixed-ratio and variable-speed belt drives
- Other friction-drive speed changers
- Flexible spline drives and impulse drives
- How we select drive train components
- Forces, torques, and transmitted power
- Gear train diagnostics; tooth error and tooth meshing frequencies

8.1 INTRODUCTION

Most electric motors, internal-combustion engines, and turbines operate efficiently and produce maximum power at high rotating speeds—speeds much higher than the optimum speeds for operating machinery. For this reason, gear trains and other speed reducers are commonly used with industrial and domestic engines. In this chapter, various drive trains will be examined in terms of their operation—that is, the manner in which they effect a reduction in speed.

Some gear trains permit us to change output speed even though the input speed remains constant. A pair of gears may be removed from the train and replaced by a pair having a different speed ratio. When the change in speed ratio is required only occasionally, a design of this type is satisfactory. For rapid or frequent speed ratio changes, pairs of gears having different ratios are engaged by shifting the location of the gears themselves and by employing bands and clutches within the transmission. Other continuously variable (stepless) transmissions employing belts, chains, and friction drives are available as well.

8.2 VELOCITY RATIOS FOR SPUR AND HELICAL GEAR TRAINS

Spur Gear Trains

We know from the preceding chapters that, for any pair of gears with fixed centers, the angular-velocity ratio is given by

$$\left|\frac{n_2}{n_1}\right| = \frac{N_1}{N_2}, \tag{8.1}$$

where $|n_2/n_1|$ is the absolute value of the ratio of the speeds of rotation and N_1 and N_2 represent the number of teeth in each gear. Since the diametral pitch P (the number of teeth per inch of pitch diameter) must also be equal in the case of meshing gears on parallel shafts, pitch diameters d_1 and d_2 must be proportional, respectively, to N_1 and N_2:

$$P = \frac{N_1}{d_2} = \frac{N_2}{d_2} \text{ and } \frac{N_1}{N_2} = \frac{d_1}{d_2}.$$

For straight spur gears, this equation may be written as

$$\frac{n_2}{n_1} = -\frac{N_1}{N_2} = -\frac{d_1}{d_2}. \tag{8.2}$$

The minus sign refers to the fact that the direction of rotation changes for any pair of external gears. If one of the gears in the pair is a ring gear (an internal gear), then

$$\frac{n_2}{n_1} = \frac{N_1}{N_2} = \frac{d_1}{d_2}, \tag{8.3}$$

and there is no change in the direction of rotation.

The result is the same if the gears conform to standard metric sizes, where in the module m is defined as the pitch diameter (in millimeters) divided by the number of teeth. The module is common to a pair of meshing gears, from which it follows that

$$m = \frac{d_1}{N_1} = \frac{d_2}{N_2},$$

leading to Eqs. (8.2) and (8.3).

Helical Gear Trains

Helical gears are often used in place of spur gears to reduce vibration and noise levels. A pair of helical gears on *parallel shafts* have the *same* helix angle, but one gear is right-handed and the other left-handed. The hand is defined as it is for screw threads. Also, as in the case of spur gears, the speed ratio for helical gears on parallel shafts is $n_2/n_1 = -N_1/N_2 = -d_1/d_2$, where N is again the number of teeth and d is the pitch diameter.

In the case of *crossed* helical gears (gears on nonparallel shafts), the velocity ratio is still equal to the inverse ratio of the number of teeth: $|n_2/n_1| = N_1/N_2$. While helical gears on parallel shafts must have equal helix angles (of opposite hand), *crossed* helical gears usually have *unequal* helix angles (usually the same hand). This means that the velocity ratio for crossed helical gears is, in general, *not* equal to the ratio of the pitch diameters. Generally, the velocity ratio for helical gears is given by

$$\left|\frac{n_2}{n_1}\right| = \frac{d_1 \cos \psi_1}{d_2 \cos \psi_2}, \tag{8.4}$$

where d_1 and d_2 are the pitch diameters and ψ_1 and ψ_2 are the respective helix angles. (See the section on crossed helical gears in Chapter 7.)

Worm Drives

A worm drive is a special case of a pair of helical gears on crossed shafts. Most worm drives have a 90° shaft angle. In that case, using subscript 1 for the worm and subscript 2 for the worm gear, we have

$$\frac{n_2}{n_1} = \frac{d_1 \cos \psi_1}{d_2 \sin \psi_1} = \frac{d_1/d_2}{\tan \psi_1} = \left(\frac{d_1}{d_2}\right)\tan \lambda,$$

where λ is the lead angle of the worm.

Helical gears may be designed to transmit power between a pair of shafts at any angle to one another, as long as the shaft centerlines do not intersect. Bevel gears may be used to transmit power between shafts that have intersecting centerlines. Note that Eqs. (8.1) and (8.4) apply to any pair of helical gears (except in planetary trains), while Eqs. (8.2) and (8.3) require that the gears be on parallel shafts.

In the case of a worm drive, the direction of worm gear rotation may be determined by analogy to the screw and nut: If a right-hand worm turns clockwise, the worm gear teeth in contact with the worm move toward the observer. In solving gear train problems, the direction of rotation should be marked on each gear.

Idlers

An idler may be described as a gear placed between, and meshing with, the input and output gears. Its purpose is to reverse the direction of the output. Thus, an idler gear affects the sign of the angular-velocity ratio. For the gear train of Figure 8.1,

$$\frac{n_2}{n_1} = -\frac{N_1}{N_2} \text{ and } \frac{n_3}{n_2} = -\frac{N_2}{N_3}.$$

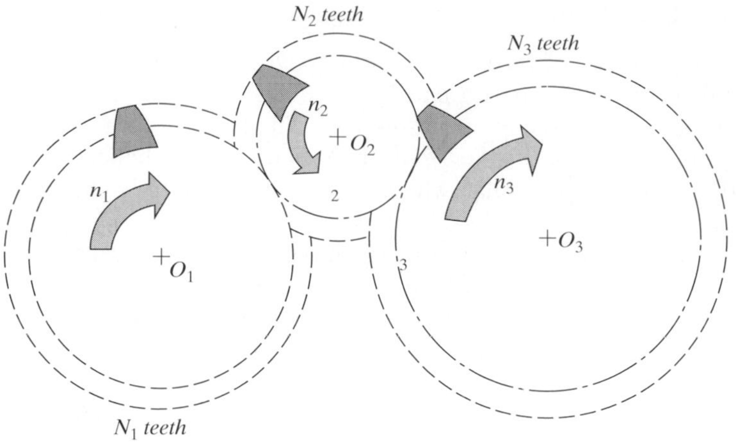

FIGURE 8.1 Gear train with idler. The idler affects the direction of rotation, but not the numerical value of the speed ratio.

Multiplying the first equation by the second, we obtain

$$\left(\frac{n_3}{n_2}\right)\left(\frac{n_2}{n_1}\right) = \left(-\frac{N_2}{N_3}\right)\left(-\frac{N_1}{N_2}\right),$$

which reduces to

$$\frac{n_3}{n_1} = \frac{N_1}{N_3}. \tag{8.5a}$$

The number of teeth in gear 2, the idler, does not affect the velocity ratio of the train. However, the idler does affect the direction of rotation of the output gear and, of course, takes up space. Since gear 2 meshes with both gears 1 and 3, all three gears must have the same diametral pitch P or the same module m. Thus, for spur gears, we may also write

$$\frac{n_3}{n_1} = \frac{d_1}{d_3}, \tag{8.5b}$$

since $P = N_1/d_1 = N_3/d_3$ and $m = d_1/N_1 = d_3/N_3$.

Occasionally, several idlers are used to transmit power between shafts that are too far apart for the use of a single pair of gears. With an odd number of idlers, the driver and driven shafts rotate in the same direction. With an even number of idlers, one shaft turns clockwise and the other counterclockwise.

Reversing Direction

The previous section indicates that the introduction of an idler in a spur gear train changes the direction of velocity of the output. Both the driver and the driven gear

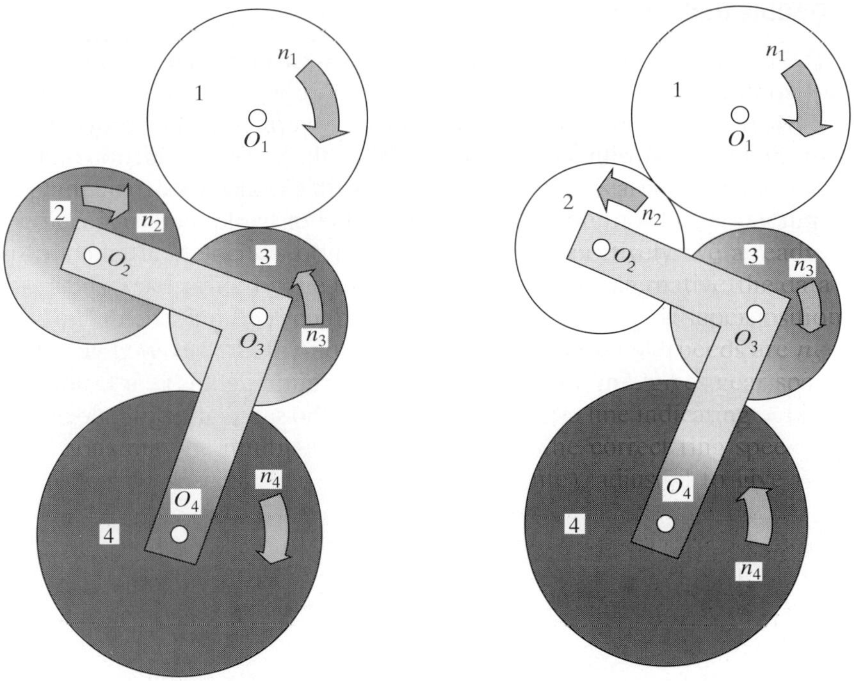

FIGURE 8.2 A reversing gearbox power train. In the position shown at the *left*, there is one idler in the train, and the input and output shafts turn in the same direction. There are two idlers in the train when the arm is moved to the position shown at the *right*, and the direction of the output shaft is reversed.

rotate in the same direction with one idler in a simple train. When two idlers are inserted in the train, the driver and the driven gear rotate in opposite directions. In Figure 8.2, gears 1 and 4 have fixed centers O_1 and O_4, while gears 2 and 3 rotate in bearings that are held in an arm. Reversing trains of this type have been used in lathes. When the arm is fixed in the position shown at the *left* of the figure, the velocity ratio is given by

$$\frac{n_4}{n_1} = \frac{N_1}{N_4},$$

where n refers to the rotation speed and N denotes the number of teeth.

When the arm is rotated about O_4 to the position shown at the right the velocity ratio becomes

$$\frac{n_4}{n_1} = -\frac{N_1}{N_4}.$$

This type of train is satisfactory for occasional changing of direction, but unsatisfactory for frequent changes, because the gears do not always correctly engage when shifted. In some cases, the gears must be manipulated by hand before they will fully mesh.

Double Reductions

The minimum number of teeth that a spur gear may have is limited by considerations of contact ratio and interference. While there is no theoretical *maximum* number of teeth, practical considerations like cost and overall size may prevent the designer from specifying a gear with more than, say, 100 teeth. A speed reduction on the order of 100 : 1 can be accomplished in two to four steps with a transmission requiring as many pairs of spur gears and considerable space. A double or triple reduction is thus used in preference to a single pair of gears in cases when the required reduction in speed is so great that the output gear of a single pair of spur gears would have to be unreasonably large. Worm drives are another alternative for large speed reductions.

The double reduction of Figure 8.3a is called a *reverted* gear train, because the output shaft is in line with the input shaft. Examining this gear train, we see that gears 2 and 3 are keyed to the same shaft and have the same angular velocity, viz.,

$$n_3 = n_2 = -n_1\left(\frac{N_1}{N_2}\right),$$

and that

$$n_4 = -n_3\left(\frac{N_3}{N_4}\right),$$

from which the ratio of output speed to input speed becomes

$$\frac{n_4}{n_1} = \frac{N_1 N_3}{N_2 N_4}. \tag{8.6a}$$

Since gears 1 and 2 must have the same diametral pitch, and since gears 3 and 4 must also have the same diametral pitch, Eq. (8.6a) may be rewritten as

$$\frac{n_4}{n_1} = \frac{d_1 d_3}{d_2 d_4}. \tag{8.6b}$$

Note that gear 1 drives gear 2 and that gear 3 drives gear 4. By adding more pairs of gears and examining the results, we find the general relationship to be

$$\left|\frac{n_{\text{output}}}{n_{\text{input}}}\right| = \frac{\text{product of driving gear teeth}}{\text{product of driven gear teeth}}, \tag{8.7}$$

which applies to all gear trains in which the shaft centers are fixed in space. The torque-arm speed reducer shown in Figure 8.3b is a practical example of such a double reduction.

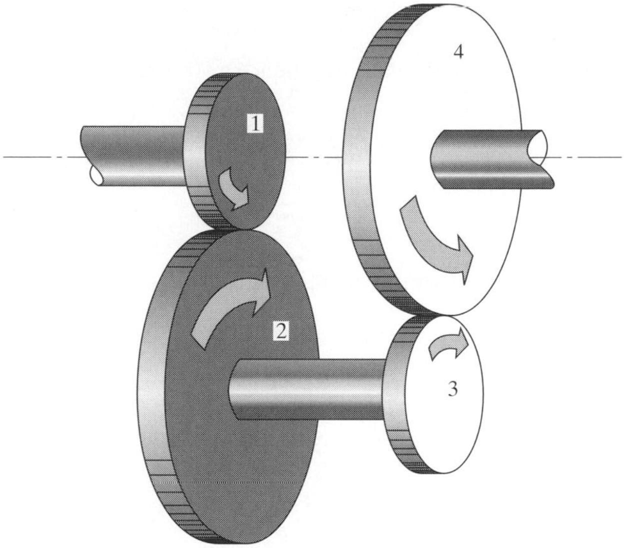

(a)

(b)

FIGURE 8.3 (a) A reverted gear train. The input and output shafts have the same centerline. The speed ratio equals the product of the tooth numbers of the driving gears divided by the product of the tooth numbers of the driven gears. (b) Torque-arm speed reducer. This double reduction is available with 1:15 and 1:25 output-to-input speed ratios. The helical involute teeth have an ellipsoid form, slightly narrower at the ends for a more even load distribution. (*Source:* Reliance Electric Company.)

SAMPLE PROBLEM 8.1

Speed Reducer

Figure 8.4 shows a gear train that is to produce a 50 : 1 reduction in speed. The following data are given:

	Gear 1, Worm	*Gear 2 Worm Gear*	*Gear 3, Straight Bevel*	*Gear 4, Straight Bevel*
Tooth numbers	2		20	40
rev/min	1,000			20

Find the number of worm gear teeth. Find the speeds and directions of the gears.

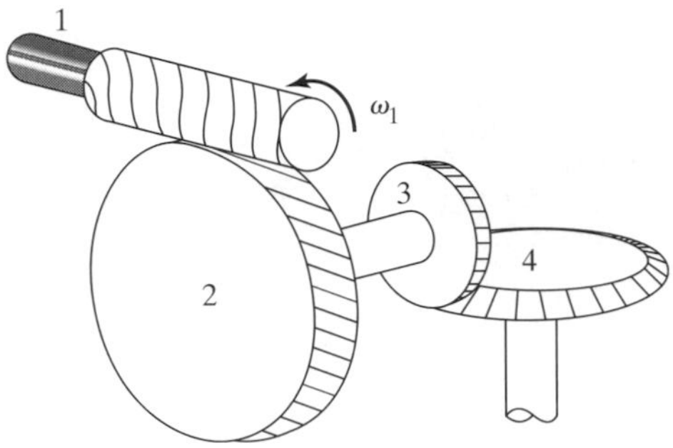

FIGURE 8.4 Speed reducer.

Solution. Using Eq. (8.7), we have $|n_{output}/n_{input}|$ = (product of driving gear teeth)/(product of driven gear teeth), from which it follows that $20/1000 = 2 \times 20/(N_2 \times 40)$, yielding $N_2 = 50$ teeth. Using Eq. (8.1) yields $|n_2/n_1| = N_1/N_2$, or $n_2/1000 = 2/50$, from which we obtain $n_2 = n_3 = 40$ rev/min. Gears 2 and 3 turn counterclockwise and gear 4 turns counterclockwise, as viewed from above in Figure 8.4.

8.3 SPEED RATIO CHANGE

Idlers may also be used to permit the changing of speed ratios. An arm holds idler gear 2 to drive gear 1 in Figure 8.5. Gear 1 is keyed, or splined, to the input shaft and turns with it. Arm *A* does not rotate with the input shaft and gear 1 but is connected to gear 1, by a sleeve. The arm is moved axially along the input shaft with gear 1 when a new speed ratio is required. The speed ratios available are

$$\frac{n_{output}}{n_{input}} = \frac{N_1}{N_3}, \frac{N_1}{N_4}, \frac{N_1}{N_5}, \ldots.$$

A typical industrial lathe may have a "cone" of as many as 12 gears on the output shaft of a train similar to that shown in Figure 8.5. This train is partly responsible for the wide variation of feeds that are available. (*Feed* refers to the movement of the cutting tool along or into the workpiece.) With two other speed selectors, the total number of feeds available, and therefore the number of different pitches of screw threads that can be cut, is 48. Since thread cutting requires a high degree of accuracy, a gear drive, which provides precise speed ratios, is ideal for this application.

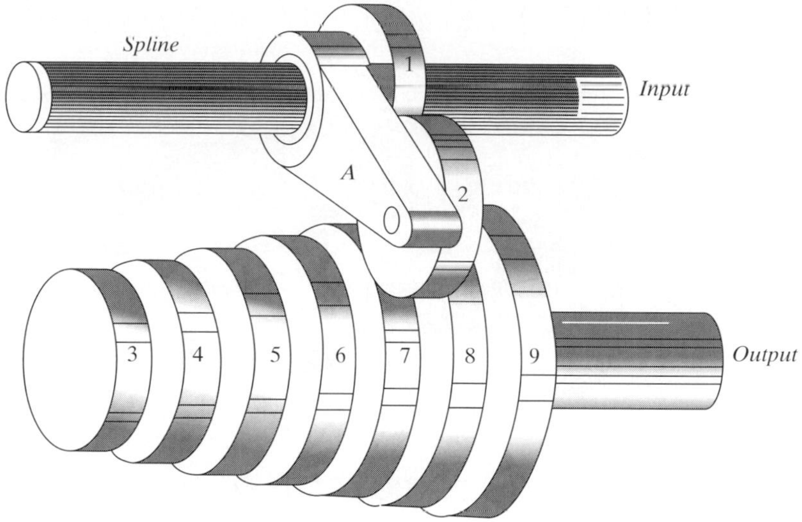

FIGURE 8.5 Speed changer employing an idler. Gear 1 turns the input shaft. Arm A holds idler gear 2 in contact with gear 1 and one of the gears on the output shaft. Arm A is fixed in space except when changing gears.

The operator of a lathe or other machine tool may find it quite satisfactory to interrupt production in order to change speed ratio. Practical, efficient, and safe vehicle operation, however, requires a smooth, quick transition from one speed ratio to another without completely stopping the machine; different types of transmissions are therefore necessary.

Transmissions with Axial Shifting

Automotive transmissions include gear trains with axial shifting, fluid drive units, planetary gear trains, and combinations of these. The transmission shown in Figure 8.6a is called a *three-speed transmission*, even though it offers a reverse speed ratio in addition to three forward speed ratios and a neutral position. The axial distance between gears in the sketch has been exaggerated for clarity; a typical transmission (like the one shown in Figure 8.6b) would be more compact.

In Figure 8.6a, shafts A, B, and D and the shaft of idler gear 6 turn in bearings mounted in the transmission housing, but bearings and parts of the shafts have been omitted from the sketch. Gear 1 is an integral part of input shaft A. Output shaft D is not directly connected to the input, but power may be transmitted to it through clutch C or through gears on countershaft B. In the position shown, no torque is transmitted between the input and output shafts, because gear 7 is not rigidly connected to output shaft D, but turns freely on it. This position is called *neutral*.

Clutch C and gear 1 have cone-shaped internal and external mating faces, respectively. Clutch C and the end of the output shaft on which it rides are splined so that the two turn together, even though C may be moved axially on the shaft. When C is made

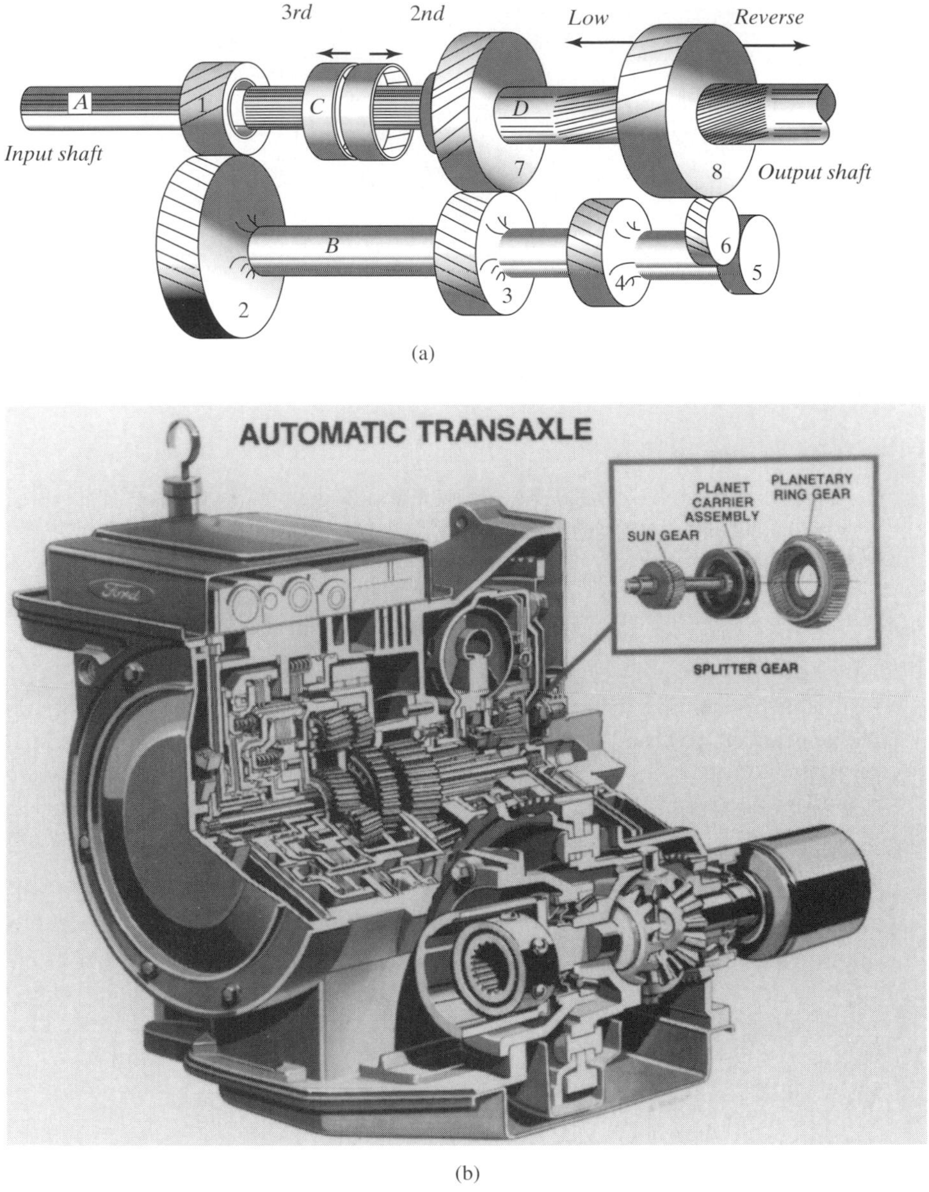

FIGURE 8.6 (a) A three-speed transmission. (b) Automatic transaxle. A three-speed automatic transaxle designed for certain front-wheel-drive automobiles. A "splitter" gear set within the torque converter improves efficiency. In third (or drive) gear, 93 percent of the torque is transmitted mechanically, as in a manual transmission. Only 7 percent is transmitted through the torque converter, where slippage reduces efficiency. (*Source:* Ford Motor Company.)

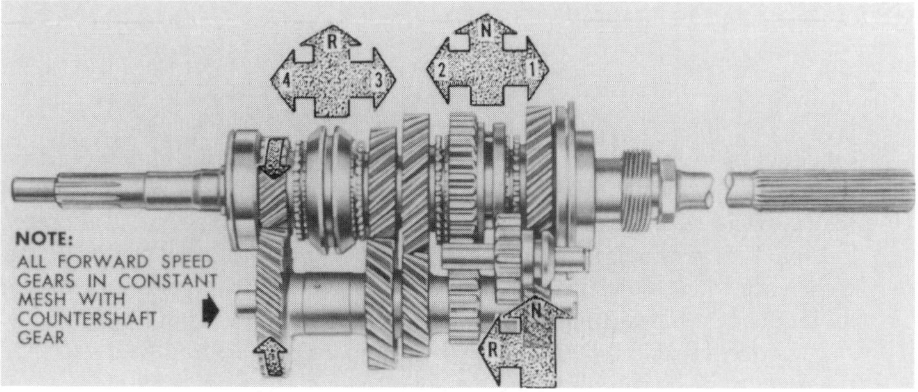

FIGURE 8.6 (c) A four-speed transmission. (*Source:* Ford Motor Company.)

to contact the conical face of gear 1, the two begin to move at the same speed. In addition to the conical friction faces, C and gear 1 have internal and external mating "teeth" that ensure a positive drive after the initial contact synchronizes the input and output shaft speeds. The synchronizer teeth are smaller than the gear teeth, since all of the synchronizer teeth are engaged at once. The resulting direct-drive, or one-to-one, speed ratio is called *high* or *third gear* in the three-speed transmission.

The engine is not directly connected to the input shaft, but a disk clutch (not shown) is installed between the two. Synchronizing the input and output shafts is essentially independent of the engine speed when the disk clutch is disengaged. The word *clutch* ordinarily refers to the disk clutch, while clutch C, called a *synchromesh clutch*, is shifted by a fork that rides in an annular groove.

We see in Figure 8.6a that gears 1 and 2, gears 3 and 7, and gears 5 and 6 mesh at all times. Gears 2, 3, 4, and 5 are integral parts of countershaft B. Therefore, all gears except gear 8 turn at all times when the input shaft is in motion. One face of gear 7, like gear 1, mates with clutch C. Since output shaft D and the internal surface of gear 7 are smooth where they contact, clutch C must engage the clutch face of gear 7 when power is transmitted through that gear. The effect is a reverted train of gears 1, 2, 3, and 7 or, symbolically, the path of power transmission from input to output is A–1–2–3–7–C–D. The output-to-input speed ratio is given by

$$\frac{n_D}{n_A} = \frac{N_1 N_3}{N_2 N_7},$$

where N_1 is the number of teeth in gear 1, and so on. Ratios N_1/N_2 and N_3/N_7 are both less than unity; hence, there is a reduction in speed. This position is called *second gear*. The disk clutch is disengaged when shifting, so that the speed of gear 7 can be synchronized with the output shaft speed.

Gear 4 is made smaller than gear 3; consequently, engaging gear 8 with gear 4 produces an even lower speed ratio called *low* or *first gear*, symbolically noted as

A–1–2–4–8–D. The speed ratio is given by

$$\frac{n_D}{n_A} = \frac{N_1 N_4}{N_2 N_8}.$$

Of course, the shifting mechanism must be designed so that clutch C is first disengaged from gear 7, or we would simultaneously have two different speed ratios. (Actually, the result would be a locked gear train or a broken transmission.) If there is no synchromesh in first gear, it is best to shift when the output and input shafts are stationary. This is the case when the vehicle is stationary and the disk clutch has been disengaged a few moments before shifting. Shifting of gear 8 along the splined portion of the output shaft is accomplished by a fork that rides in a grooved ring (not shown). The ends of the teeth of gears 4, 6, and 8 are rounded to facilitate engagement.

Gear 5 is made slightly smaller than gear 4 so that gear 8 cannot mesh with it, but may be shifted to mesh with idler gear 6. As we have just seen, for first, second, and third gears, the ratios were positive; the output shaft turned in the same direction as the input shaft. But when engaged with gear 5, idler gear 6 causes an odd number of changes in direction, and the output-to-input speed ratio is given by

$$\frac{n_D}{n_A} = -\frac{N_1 N_5}{N_2 N_8}.$$

This arrangement is *reverse gear*, represented symbolically by A–1–2–5–6–8–D. When one shifts into reverse, as when shifting to first gear, the disk clutch is disengaged, and both the input and the output shafts are stationary. Except for reverse and first gear ratios, gear 8 must turn freely, engaging neither gear 4 nor gear 6.

Helical gears are often selected for transmissions because of their greater strength and smoother, quieter operation. Meshing helical gears on parallel shafts are of opposite hand with a right-hand helical gear resembling a right-hand screw. If gear 1 of Figure 8.6a is a left hand gear, then gear 2 must be right hand gear. If gear 1 turns counterclockwise (as seen from the right), creating a thrust to the left on gear 1, then there is a thrust to the right on the countershaft at gear 2. In second gear, gear 3 will have a balancing thrust (to the left) if it is a right-hand helix. Gear 4 is also a right-hand helix, so that countershaft thrust is balanced in first gear. Since gear 8 meshes with both gears 4 and 6, gear 6 must be right hand, making gear 5 left hand. Thus, thrust is not balanced when in reverse. Finally, all speed ratios and paths of power transmission for the transmission of Figure 8.6a are summarized in Table 8.1. Automatic transmissions, including transaxles, employ several gear trains. Figure 8.6b shows a transaxle with a torque converter, differential and planetary gear trains. Methods of analyzing planetary gear trains and differentials are discussed in the next section.

Automotive transmissions with four or more forward speed ratios are available, some including synchromesh in all of the forward gears. (See Figure 8.6c.) While there are many innovations among manufacturers, the basic principles are often the same. Although the preceding discussion centers around reducing speed, in a few instances it is desirable to increase speed. Kinematically, the equations apply to increases, as well as reductions, in speed. Friction losses, however, make large increases in speed impossible.

TABLE 8.1 Position, Path, and Speed Ratios for Three-Speed Transmission of Figure 8.6a

Gear	Position of synchromesh clutch, C	Position of gear 8	Path of transmitted power	Output-to-input speed ratio n_D/n_A
Neutral	Center	Center	—	—
Third (high)	Left	Center	A–1–C–D	+1
Second	Right	Center	A–1–2–3–7–C–D	$+\frac{N_1N_3}{N_2N_7}$
First (low)	Center	Left	A–1–2–4–8–D	$+\frac{N_1N_4}{N_2N_8}$
Reverse	Center	Right	A–1–2–5–6–8–D	$-\frac{N_1N_5}{N_2N_8}$

Designing for a Particular Speed Ratio

For many applications, it is necessary to have a particular relationship between the output and input speed. When the required relationship is a ratio of small whole numbers, say, one-half or five-sevenths, we have a wide selection of satisfactory pairs of gears to choose from. However, some speed ratios are impossible to obtain exactly, and others may be impractical to obtain exactly. An example of a speed ratio that cannot be obtained with gears is the square root of two, an irrational number. An example of a speed ratio that is difficult to obtain exactly (from a practical standpoint) is 503/2003, the ratio of two prime numbers.

In the first case, we cannot express the desired ratio as a fraction made up of whole numbers and thus cannot select a corresponding set of gear tooth numbers. The second case involves a pair of numbers, neither of which can be factored. An exact solution involving a pair of gears with 503 and 2003 teeth might be prohibitively expensive. Either problem, however, may be solved if a small variation from the desired ratio is permitted.

In some instances, a table of factors may be used to advantage. Suppose, for example, we needed the exact speed ratio $n_o/n_i = 1501/1500$, where n_o and n_i represent the output and input speeds, respectively. For a gear train similar to that of Figure 8.3a, $n_4/n_1 = +N_1N_3/(N_2N_4)$ leading us to try $N_1N_3 = 1501$ and $N_2N_4 = 1500$. Fortunately, both numbers are factorable: $1501 = 19 \times 79$ and $1500 = 2^3 \times 3 \times 5^3$. Letting $N_1 = 19$, $N_3 = 79$, $N_2 = 2^2 \times 5 = 20$, and $N_4 = 3 \times 5^2 = 75$, we have the desired ratio exactly. When the desired ratio consists of a pair of large numbers that cannot be factored to give reasonable gear sizes, we may be forced to try a more complicated and expensive solution or an approximate solution.

8.4 PLANETARY GEAR TRAINS

Gear sets of the type shown in Figure 8.7a through l are called *epicyclic*, or *planetary*, gear trains. In planetary trains, one or more gears are carried on a rotating planet carrier,

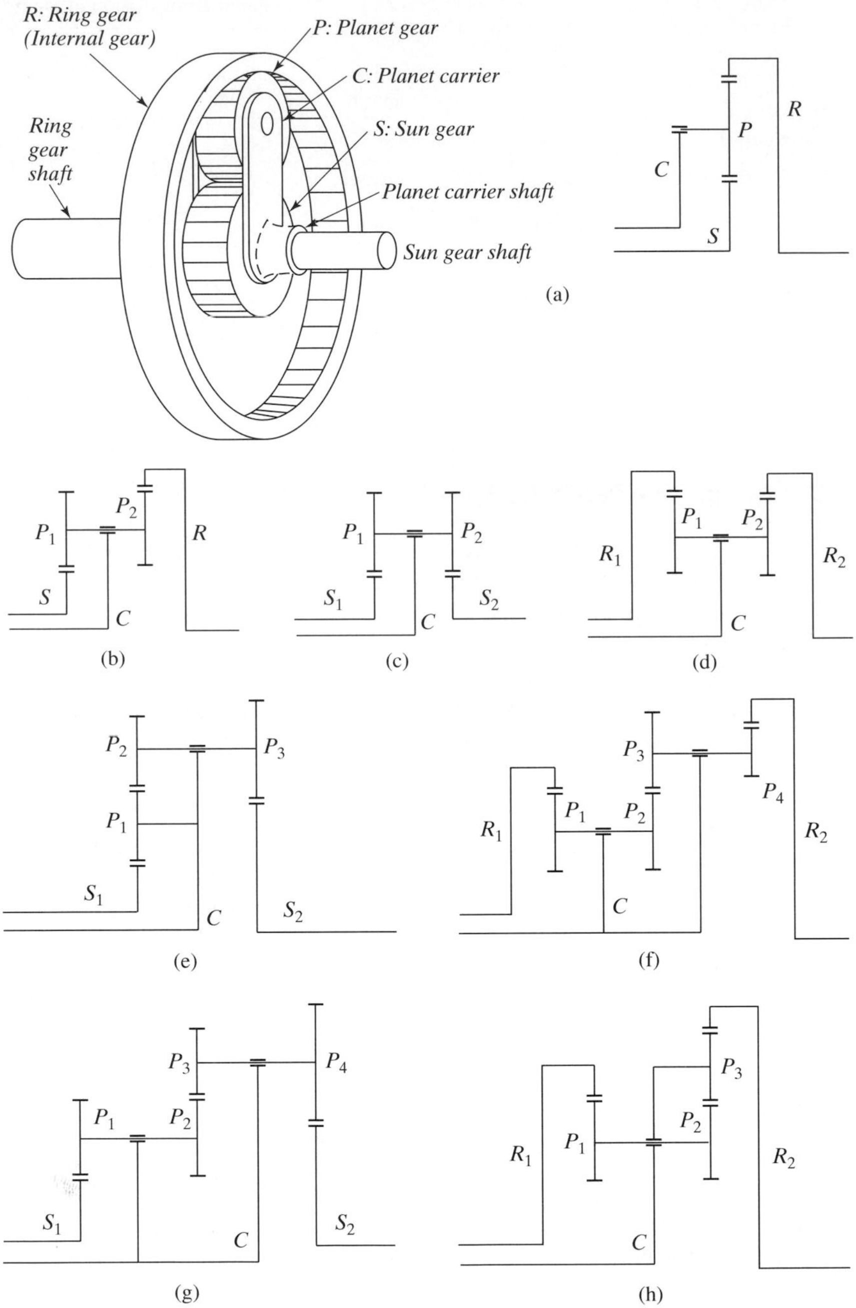

FIGURE 8.7 Types of planetary gear train identified by Lévai. Part a includes a key to the skeleton diagrams.

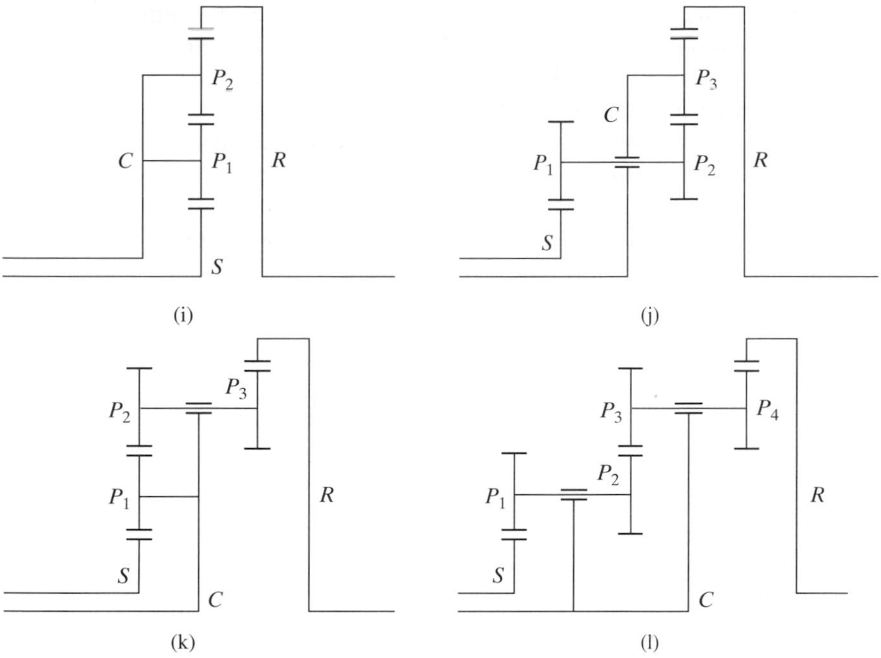

FIGURE 8.7 *(continued)*

rather than on a shaft that rotates on a fixed axis. Several types of gear trains may be shifted manually to obtain greater or lesser reductions in speed. The shifting process, however, is difficult to accomplish automatically with gears that rotate about fixed centers. On the other hand, planetary gear trains are readily adapted to automatic control. Some planetary gear trains are designed to change ratios simply by using electrically or hydraulically operated band brakes to keep one or more of the gears stationary. Other planetary trains operating with fixed gear ratios are selected for their compact design and high efficiency.

Lévai (1966) identified 12 possible variations of planetary trains. These are shown in Figure 8.7a through l, using simplified schematics. The simple planetary train shown in skeleton form in Figure 8.7a has hundreds of applications, including automatic transmissions for automobiles and wheel drives for loaders and scrapers. Motorized wheels designed for heavy-service vehicles utilize a planetary train based on part b of the figure. Extreme reductions in speed are possible with gear drives based on Figure 8.7c. The configuration of Figure 8.7i is used in automotive final drives, replacing a bevel-gear differential. Some planetary train variations shown in the figure have no commercial applications at this time. Other possible variations include planetary trains in which the planet carrier rotates freely, serving as neither input nor output of the mechanism.

The simple planetary gear train sketched in Figure 8.7a consists of a sun gear (S) in the center, a planet gear (P), a planet carrier or arm (C), and an internal, or ring, gear (R). The sun gear, ring gear, and planet carrier all rotate about the same axis. The planet gear is mounted on a shaft that turns in a bearing in the planet carrier; the planet gear meshes with both the sun gear and the ring gear. Real gear trains are designed with three or four planets held in a carrier that encircles the sun. Since the planets do not rotate about a fixed center, some of the rules developed for gears rotating about fixed centers must be reexamined.

Formula Method (Train Value Formulation Method) for Solving Planetary Trains

We noted earlier that, for gears in which the shaft centers are fixed in space, the train value is given by

$$\left|\frac{n_{\text{output}}}{n_{\text{input}}}\right| = \frac{\text{product of driving gear teeth}}{\text{product of driven gear teeth}}.$$

The foregoing equation does not directly apply to planetary trains, but we will use the train value to aid in solving planetary train problems.

If the actual speeds of two of the gears in a planetary train are known, we arbitrarily designate one of them as the input gear and the other as the output gear. We then find the train value with the carrier fixed:

$$r^* = (\pm)\frac{\text{product of driving gear teeth}}{\text{product of driven gear teeth}}. \tag{8.8}$$

Equation (8.8) is based on our arbitrary designation of input and output gears. The asterisk indicates that the planet carrier does not rotate. The sign of r^* is positive if the arbitrarily designated input and output gears would rotate in the same direction with the carrier fixed. Of course, we must correct for planet carrier rotation.

Let gears x and y be designated as input and output gears, respectively. With the concept of relative velocity, n_x^*, the speed of gear x relative to carrier C, is given by the difference in rotation speeds between the two:

$$n_x^* = n_x - n_c.$$

Here, n_x and n_c are, respectively, the actual speed of gear x and the actual speed of carrier C. Likewise, for gear y, the speed relative to the carrier is

$$n_y^* = n_y - n_c,$$

from which the train value with the carrier fixed is

$$r^* = \frac{n_y^*}{n_x^*} = \frac{n_y - n_c}{n_x - n_c}. \tag{8.9}$$

Equations (8.8) and (8.9) are the basis for the formula method of solving for planetary train speed ratios. To apply these equations, we first designate gears x and y. The designation is arbitrary, except that we would designate gears whose speed we knew or wished to know. Planet gears would not ordinarily be so designated. Let the number of teeth in each gear be given. The fixed carrier train ratio r^* is determined as if x and y were input and output gears, respectively. Then, with Eq. (8.9), if two of the speeds n_x, n_y, and n_c are given, the third can be found.

Suppose, for example, that the speeds n_S and n_R are given for the sun gear and ring gear, respectively, in the simple planetary train shown in Figure 8.7a. Let the sun and ring gears be arbitrarily designated as input and output gears, respectively. Then,

$$n_x = n_S \quad \text{and} \quad n_y = n_R.$$

From Eq. (8.8), noting that the direction of rotation changes when two external gears mesh, but does not change when an external gear meshes with an internal gear, we obtain

$$r^* = \left(\frac{-N_S}{N_P}\right)\left(\frac{+N_P}{N_R}\right) = -\frac{N_S}{N_R}.$$

From Eq. (8.9),

$$r^* = \frac{n_R - n_C}{n_S - n_C}.$$

Combining the two results, we obtain, for the train shown in Figure 8.7a,

$$\frac{n_R - n_C}{n_S - n_C} = -\frac{N_S}{N_R}.$$

SAMPLE PROBLEM 8.2

Planetary Train Analyzed by the Formula Method

For the planetary train of Figure 8.7a, let the tooth numbers be $N_S = 40$, $N_P = 20$, and $N_R = 80$. Find the speed of the planet carrier if the sun gear rotates counterclockwise at 100 rev/min and the ring gear clockwise at 300 rev/min. Use the formula method.

Solution. If we arbitrarily select the sun as the input gear and the ring as the output gear, the train value [Eq. (8.8)] is

$$r^* = -\frac{N_S}{N_R} = -0.5.$$

Using Eq. (8.9), where $n_x = -100$ and $n_y = 300$, we get

$$r^* = \frac{300 - n_C}{-100 - n_C}.$$

If we equate the two expressions for r^*, the result is $n_c = 166.7$ rev/min clockwise.

In some cases, the formula method must be applied in two steps. For the planetary speed reducer of Figure 8.9 appearing later in this chapter, for example, we may let $n_x = n_S$ and $n_y = n_{R_1}$ in Eq. (8.9) to find n_C if n_S and n_{R_1} are known. Then, using the value of n_C we have just found and letting $n_x = n_S$ and $n_y = n_R$ in Eq. (8.9), we can find n_{R_2}.

Input and Output Shafts

In most cases, gear trains are used to obtain a reduction in speed. Other applications include a reversal in direction, a differential effect, an increase in speed, and even a one-to-one, input-to-output relationship. In the skeleton diagrams of 12 types of planetary train identified by Lévai, (Figures 8.7a to l), carrier, sun gear, and ring gear shafts serve as potential input and output shafts. In Figure 8.7a, for example, the ring gear may be fixed by a band brake. Then, if the sun gear shaft is used as the input and the carrier shaft as output, there will be a reduction in speed. If the band brake is released, allowing the ring gear to rotate freely, and if a clutch engages the sun gear to the carrier, a one-to-one speed ratio results. If, instead, the sun gear in Figure 8.7a is fixed, the ring gear shaft can be used as the input and the planet carrier shaft as output. Considering the planetary trains shown, with different combinations of gear tooth numbers, the number of possible output-to-input speed ratios is virtually limitless.

Tabular Analysis (Superposition): An Alternative Method for Analyzing of Planetary Trains

As we have observed, the rotation of the planet carrier complicates the problem of determining gear speeds in a planetary train. However, by the simple device of calculating rotation *relative to the carrier* and combining it with the rotation of the entire train turning as a unit, we can find velocities in two steps. If the planet carrier is kept stationary so that the centers of all gears are fixed, the gear speed ratios equal the inverse of the ratios of the tooth numbers.

If the sun or ring gear is actually fixed, the constraint is (theoretically) temporarily relaxed, and that gear is given one turn. The effect on the entire train is calculated. Of course, since the net rotation of the fixed sun or ring gear must be zero, a compensating rotation must take place to correct its position. That compensating motion is one rotation of the entire locked train *in the opposite direction*.

As an example, consider the planetary train of Figure 8.8, in which the sun, planet, and ring gears have, respectively, 40, 20, and 80 teeth, with the sun gear fixed. The ring gear is the driver and rotates at 300 rev/min clockwise. As a first step in the solution, let the entire train be locked together and given one clockwise rotation. Then sun and ring gears and the planet carrier will each have turned through one clockwise rotation about their common center of rotation. This motion, though, violates the requirement that the sun gear be fixed. In the second step, the sun gear will be given one counterclockwise rotation, yielding a net sun gear motion of zero. While the sun gear is rotated counterclockwise, the carrier will be fixed so that all gears rotate on fixed centers. The results up to this point are tabulated by denoting clockwise rotations as positive (+) and counterclockwise rotations as negative (−). Thus, we are able to construct Table 8.2.

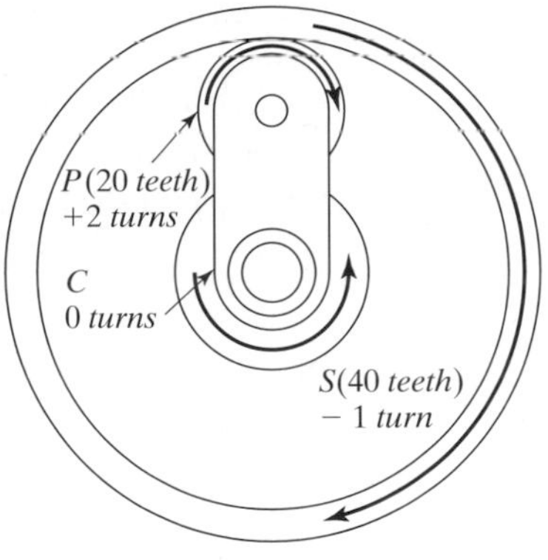

FIGURE 8.8 Planetary train. This figure corresponds to step 2 in Table 8.2.

TABLE 8.2 Superposition Method for Solving Planetary Train Speed Ratios with Sun Gear Fixed (See Figure 8.8)

Gear	Sun	Planet	Ring	Planet carrier
No. of teeth	40	20	80	
Step 1: rotations with train locked	+1	+1	+1	+1
Step 2: rotations with planet carrier fixed	−1	+2	+0.5	0
Total rotations	0	+3	+1.5	+1
Speed (rev/min)	0	600 cw	300 cw	200 cw

As we have noted, the planet carrier will be fixed as we complete the problem. One counterclockwise rotation of the sun gear results in

$$n_p = n_s\left(-\frac{N_s}{N_p}\right) = -1\left(-\frac{40}{20}\right) = +2$$

rotations of the planet and

$$n_R = n_p\left(\frac{N_p}{N_R}\right) = +2\left(+\frac{20}{80}\right) = +0.5$$

rotations of the ring, as shown in Figure 8.8. The planet carrier is given zero rotations. Including these values in Table 8.2 and adding each column, we obtain the total number of rotations. We see that the ring gear makes 1.5 turns for each turn of the planet carrier. Hence, if the ring gear is on the input shaft and the planet carrier on the output shaft, the ratio of output to input is given by $n_o/n_i = 1/1.5 = +2/3$. The actual rotation speeds bear the same relationship. Thus, we divide 300 rev/min, the given speed of the ring, by 1.5, the total rotations, to obtain a factor of 200. Multiplying each figure in the total-rotations row by 200, we obtain the actual speeds in revolutions per minute. The reader will observe that, in the second step of the calculations, the planet acts as an idler and affects only the sign of the speed ratio. Thus, it is not necessary to compute rotations of the planet to obtain output-to-input speed ratios.

Using the data from this example in the formula method, we find that the train value is $r^* = -0.5$. Selecting the sun and ring as the arbitrary input and output gears, respectively (although the ring is the actual input gear and the carrier is the actual output), we have

$$r^* = \frac{300 - n_C}{0 - n_C},$$

from which it follows that $n_C = 200$ rev/min (clockwise).

As a second example, let the ring gear in Figure 8.8 be held stationary while the planet carrier, sun, and planet gears are permitted to rotate. In order that the results be more general, the number of teeth in the sun, planet, and ring gears will be represented, respectively, by N_S, N_P, and N_R. In tabulating the solution in Table 8.3, the first step is identical with the previous example. In this case, however, the net rotation of the ring gear must be zero. To accomplish this, the ring gear is given one counterclockwise rotation in the second step of the table, while the planet carrier remains fixed. Noting that the planet gear acts as an idler in the second step, we obtain $+N_R/N_S$ rotations of the sun gear for -1 rotation of the ring gear. Tabulating these values and adding as before, we obtain the total number of rotations.

The last line of the Table 8.3 indicates that, with the ring gear fixed, the ratio of sun gear speed to planet carrier speed is

$$\frac{n_S}{n_C} = 1 + \frac{N_R}{N_S}.$$

The speed relationship obtained for this train by the formula method is

$$\frac{n_R - n_C}{n_S - n_C} = -\frac{N_S}{N_R}.$$

TABLE 8.3 Speed Ratios for a Planetary Train with the Ring Fixed (See Figure 8.8)

Gear	Sun	Planet	Ring	Planet carrier
No. of teeth	N_S	N_P	N_R	
Step 1: rotations with train locked	$+1$	$+1$	$+1$	$+1$
Step 2: rotations with planet carrier fixed	$+\frac{N_R}{N_S}$	$-\frac{N_R}{N_P}$	-1	0
Total rotations	$1 + \frac{N_R}{N_S}$	$1 - \frac{N_R}{N_P}$	0	$+1$

With the ring gear fixed, we have

$$\frac{0 - n_C}{n_S - n_C} = -\frac{N_S}{N_R},$$

from which we obtain the same result:

$$\frac{n_S}{n_C} = 1 + \frac{N_R}{N_S}.$$

Repeating the example with the sun gear fixed, but using symbols to represent the actual numbers of teeth, we find that the ratio of ring gear speed to planet carrier speed becomes

$$\frac{n_R}{n_C} = 1 + \frac{N_S}{N_R}.$$

If the gear train operates instead with the planet carrier stationary, it no longer acts as a planetary train. Then, the ratio of sun gear speed to ring gear speed becomes $-N_R/N_S$ to 1, by inspection.

Compound Planetary Trains

A gear train that may be designed for extremely low ratios of output to input speed employs *two* planet gears and *two* ring gears. The reducer of Figure 8.9 is made up of a sun gear S, two ring gears R_1 and R_2, and two planet gears P_1 and P_2. The planets rotate at the same speed. They are held in the common planet carrier C, which is free to rotate. The input is the sun gear, and the ring gear R_2 is the output. Ring gear R_1 is fixed.

As in the previous examples, we begin by rotating the entire train one turn clockwise (step 1, Table 8.4). In the second step, the planet carrier is fixed, while ring gear R_1 is returned to its original position by one counterclockwise turn. The number of rotations of each gear is entered in Table 8.4, noting that both planets turn at the same speed. From the sum of the motions of steps 1 and 2, it is seen that the ratio of output to input speed is given by

$$\frac{n_{R_2}}{n_S} = \frac{1 - [N_{R_1}N_{P_2}/(N_{P_1}N_{R_2})]}{1 + (N_{R_1}/N_S)}. \tag{8.10}$$

If the gears are chosen so that the value of the term $N_{R_1}N_{P_2}/(N_{P_1}N_{R_2})$ is very near to unity, the output speed will be very low, making a very high output *torque* available. Of course, if the term exactly equals unity, the reducer will be useless, since the output shaft will not turn.

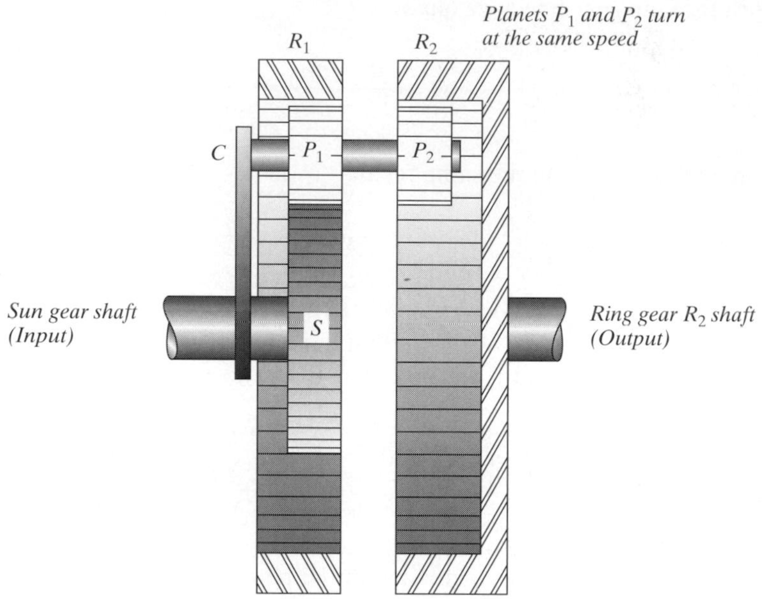

FIGURE 8.9 Planetary speed reducer. This planetary train employs two planets and two ring gears.

TABLE 8.4 Speed Ratios for a Compound Planetary Speed Reducer with Ring Gear R_1 Fixed (See Figure 8.9)

Gear	S	P_1	P_2	R_1	R_2	C
No. of teeth	N_S	N_{P_1}	N_{P_2}	N_{R_1}	N_{R_2}	
Step 1: rotations with train locked	+1	+1	+1	+1	+1	+1
Step 2: rotations with planet carrier fixed	$+\frac{N_{R_1}}{N_S}$	$-\frac{N_{R_1}}{N_{P_1}}$	$-\frac{N_{R_1}}{N_{P_1}}$	−1	$-\frac{N_{R_1}}{N_{P_1}}\frac{N_{P_2}}{N_{R_2}}$	0
Total number of rotations	$1+\frac{N_{R_1}}{N_S}$			0	$1-\frac{N_{R_1}}{N_{P_1}}\frac{N_{P_2}}{N_{R_2}}$	+1

Balanced Planetary Trains

In the preceding examples, only one planet gear was shown meshing with the ring gear. Kinematically, this is sufficient, but balancing gear tooth loads and inertial forces requires two or more planets meshing with each ring gear. (See Figures 8.10, 8.11a, and 8.11b.) The planetary gear train of Figure 8.8, for example, can be redesigned with four planets. The planet shafts are mounted in bearings in a planet carrier equivalent to the

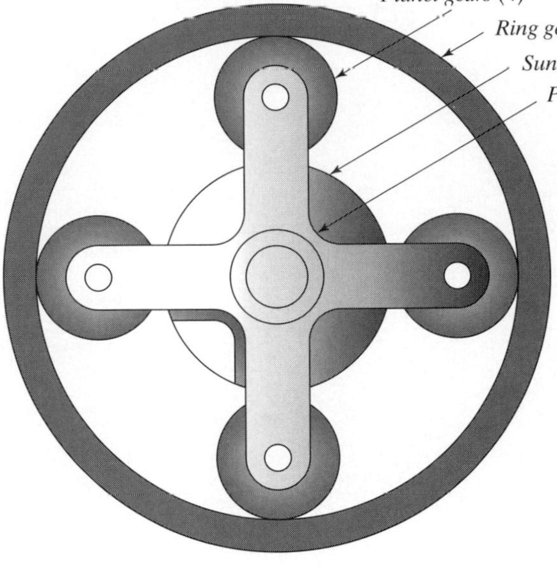

FIGURE 8.10 Planetary train with four planets. Forces are more readily balanced in a planetary train with three or four planets. Kinematically, there is no difference; the speed ratio remains the same whether one or several planets are used.

arrangement in Figure 8.10. Kinematically, the gear train of that figure is identical to the train shown in Figure 8.8.

For a simple train of the type shown in Figures 8.8 and 8.10, the pitch diameter of the ring gear is obviously equal to the sum of the pitch diameter of the sun gear plus twice the pitch diameter of the planet gear. Since the metric module or the diametral pitch must be the same on all of the gears in order that they mesh, tooth numbers are related by the equation

$$N_R = N_S + 2N_P, \tag{8.11a}$$

where subscripts R, S, and P refer to the ring, sun, and planet, respectively. When several equally spaced planet gears are to be used, as in Figure 8.10, the designer must ensure that it is possible to assemble the train. For example, if $N_S = 25$, $N_P = 20$, and $N_R = 65$, Eq. (8.11) is satisfied. However, a layout will show that four planets cannot be equally spaced in the train. Equal spacing is possible only if

$$\frac{N_S + N_R}{\#_P} = \text{an integer}, \tag{8.11b}$$

where $\#_P$ = the number of planets, which in turn is limited by the requirement that there be clearance between addendum circles of adjacent planets. Figure 8.11a illustrates a balanced train employing three planets.

(a)

(b)

FIGURE 8.11 (a) Simple planetary train. A train typical of the wheel planetaries used in drive axles of loaders and scrapers. Although this drive with three planets is kinematically equivalent to a drive with only one planet, the configuration shown balances the loading and has greater capacity. The planet carrier is not shown. (*Source:* Fairfield Manufacturing Company, Inc.) (b) Compound planetary train. This assembly with two sets of planet gears is used in a motorized wheel. The sun gear is driven by the shaft of an electric motor rated at 400 hp. The outside diameter of the ring gear is over 39 in, and the unit is designed for heavy-duty service. (*Source:* Fairfield Manufacturing Company, Inc.)

SAMPLE PROBLEM 8.3

Investigating a Product Line of Speed Reducers

Suppose one would like to offer a line of speed reducers with output-to-input speed ratios ranging from about 1:5 to about 1:30. Typical input speeds will be 4000 rpm. Investigate potential candidates.

Design decisions. We will try a configuration similar to that in Figure 8.11b. The sun gear will be the input, the planet carrier (not shown) will be the output, and the ring will be kept stationary. We will use four sets of planets to share the load and balance the train. The two planets in each set rotate as a unit. We will try 18 teeth in the sun gear and 18 or more teeth in the planets meshing with it. The planets meshing with the ring will have 20 teeth. We will use straight spur gears, all with a 2.5-mm module.

Solution summary. We will sit back and let the computer do most of the work. Using the formula method, we obtain

$$(n_R - n_C)/(n_S - n_C) = -N_S \cdot N_{P2}/(N_{P1}\, N_R).$$

Noting that the ring speed is zero, we rewrite the equation in terms of the speed ratio

$$n_{CS} \equiv n_C/n_S$$

and call for a symbolic solution for n_{CS}. Next, we set up an equation containing the planet speed and let the computer solve it. Can you find a more concise solution?

Now consider geometry. A sketch shows that the ring gear pitch diameter equals the sum of the pitch diameters of the sun gear and the first and second planets. This relationship is rewritten in terms of tooth numbers to find the required number of ring teeth. We try various sizes for the first planet (the one meshing with the sun). Note that the form $n_c(N_{P1})$ indicates that carrier speed depends on the number of planet one teeth. Tooth numbers for the ring, as well as pitch radii, and speeds, are calculated and tabulated. The table includes reducers that approximate the required ratios, and we can select from many in-between values for our line of speed reducers. The selections should be checked to see that the addendums of the planets do not interfere. We should also check for the required position of one planet relative to the other on the shaft.

Solution details. With the sun gear as input and the carrier as output, the sun drives the first planet, and the second planet on the same shaft contacts the fixed ring.

Output/input speed ratio $n_{CS} = \dfrac{n_C}{n_S}$

Formula method for speed ratio:

$$\frac{n_{CS}}{1 - n_{CS}} = \frac{N_S \cdot N_{P2}}{N_{P1} \cdot N_R} \text{ solve, } n_{CS} \rightarrow -N_S \cdot \frac{N_{P2}}{(-N_{P1} \cdot N_R - N_S \cdot N_{P2})}$$

Planet speed: $\dfrac{n_{P1} - n_C}{n_S - n_C} = \dfrac{-N_S}{N_{P1}}$ solve, $n_{P1} \rightarrow -\left[\dfrac{-1}{(n_S - n_C)} \cdot n_C + \dfrac{N_S}{N_{P1}}\right] \cdot (n_S - n_C)$

Module for all gears (mm) $m := 2.5$

Number of teeth in sun: $N_S := 18$ First planet: $N_{P1} := 18, 20 .. 90$

second planet: $N_{P2} := 20$ Ring: $N_R(N_{P1}) := N_S + N_{P1} + N_{P2}$

Pitch radii

$$r_S := N_S \cdot \frac{m}{2} \quad r_S = 22.5 \quad r_{P1}(N_{P1}) := N_{P1} \cdot \frac{m}{2}$$

$$r_{P2} := N_{P2} \cdot \frac{m}{2} \quad r_{P2} = 25 \quad r_R(N_{P1}) := N_R(N_{P1}) \cdot \frac{m}{2}$$

Calculate speed ratio $n_{CS}(N_{P1}) := N_S \dfrac{N_{P2}}{(N_{P1} \cdot N_R(N_{P1}) + N_S \cdot N_{P2})}$

Number of planet pairs: $P_{num} := 4$

Input speed (rpm CCW) $n_S := 4000$ Angular velocity (rad/s) $\omega_S := \dfrac{\pi \cdot n_S}{30}$ $\omega_S = 418.879$

Output speed $n_C(N_{P1}) := n_S \cdot n_{CS}(N_{P1})$

Calculate planet speed: $n_{P1}(N_{P1}) := -\left[\dfrac{-1}{(n_S - n_C(N_{P1}))} \cdot n_C(N_{P1}) + \dfrac{N_S}{N_{P1}}\right] \cdot (n_S - n_C(N_{P1}))$

N_{P1}	$r_{P1}(N_{P1})$	$N_R(N_{P1})$	$r_R(N_{P1})$	$n_{CS}(N_{P1})$	$n_{P1}(N_{P1})$	$n_C(N_{P1})$
18	22.5	56	70	0.263	−1,894.737	1,052.632
20	25	58	72.5	0.237	−1,800	947.368
22	27.5	60	75	0.214	−1,714.286	857.143
24	30	62	77.5	0.195	−1,636.364	779.221
26	32.5	64	80	0.178	−1,565.217	711.462
28	35	66	82.5	0.163	−1,500	652.174
30	37.5	68	85	0.15	−1,440	600
32	40	70	87.5	0.138	−1,384.615	553.846
34	42.5	72	90	0.128	−1,333.333	512.821
36	45	74	92.5	0.119	−1,285.714	476.19
38	47.5	76	95	0.111	−1,241.379	443.35
40	50	78	97.5	0.103	−1,200	413.793
42	52.5	80	100	0.097	−1,161.29	387.097
44	55	82	102.5	0.091	−1,125	362.903
46	57.5	84	105	0.085	−1,090.909	340.909
48	60	86	107.5	0.08	−1,058.824	320.856
50	62.5	88	110	0.076	−1,028.571	302.521
52	65	90	112.5	0.071	−1,000	285.714
54	67.5	92	115	0.068	−972.973	270.27
56	70	94	117.5	0.064	−947.368	256.046
58	72.5	96	120	0.061	−923.077	242.915
60	75	98	122.5	0.058	−900	230.769
62	77.5	100	125	0.055	−878.049	219.512
64	80	102	127.5	0.052	−857.143	209.059
66	82.5	104	130	0.05	−837.209	199.336
68	85	106	132.5	0.048	−818.182	190.275
70	87.5	108	135	0.045	−800	181.818
72	90	110	137.5	0.043	−782.609	173.913
74	92.5	112	140	0.042	−765.957	166.512
76	95	114	142.5	0.04	−750	159.574
78	97.5	116	145	0.038	−734.694	153.061
80	100	118	147.5	0.037	−720	146.939
82	102.5	120	150	0.035	−705.882	141.176
84	105	122	152.5	0.034	−692.308	135.747
86	107.5	124	155	0.033	−679.245	130.624
88	110	126	157.5	0.031	−666.667	125.786
90	112.5	128	160	0.03	−654.545	121.212

Planetary Differential Drives

An alternative method of obtaining low ratios of output to input speed employs two sun gears and two planet gears as shown in Figure 8.12. The planet carrier C is keyed to the input shaft, which goes through the center of sun gear S_1, which is fixed. Sun gear S_2 is keyed to the output shaft. Planet gears P_1 and P_2 are both keyed to the same shaft, which turns freely in the planet carrier. Gears, P_1 and P_2 in the schematic represent three or four pairs of planets, equally spaced in a planet carrier. Ring gears are not used in this speed reducer.

Speed ratios are again found by rotating the entire locked gear train and then correcting the position of sun gear S_1 while the planet carrier remains stationary. The result, given in Table 8.5, is an output-to-input speed ratio:

$$\frac{n_o}{n_i} = \frac{n_{S_2}}{n_C} = 1 - \frac{N_{S_1}N_{P_2}}{N_{P_1}N_{S_2}}. \tag{8.12}$$

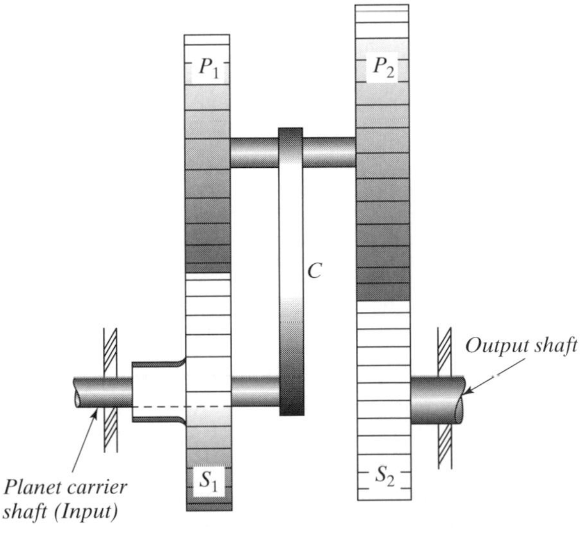

FIGURE 8.12 Planetary differential drive.

TABLE 8.5 Speed Ratios for a Planetary Differential Drive with Sun Gear S_1 Fixed (See Figure 8.12)

Gear	S_1	P_1	P_2	S_2	C
No. of teeth	N_{S_1}	N_{P_1}	N_{P_2}	N_{S_2}	
Step 1: rotations with train locked	+1	+1	+1	+1	+1
Step 2: rotations with planet carrier fixed	−1	$+\frac{N_{S_1}}{N_{P_1}}$	$+\frac{N_{S_1}}{N_{P_1}}$	$-\frac{N_{S_1}N_{P_2}}{N_{P_1}N_{S_2}}$	0
Total rotations	0			$1 - \frac{N_{S_1}N_{P_2}}{N_{P_1}N_{S_2}}$	1

This result may be checked by the formula method. Considering rotations relative to the carrier, we have

$$\frac{n^*_{S_2}}{n^*_{S_1}} = \frac{N_{S_1}N_{P_2}}{N_{P_1}N_{S_2}} = \frac{n_{S_2} - n_C}{n_{S_1} - n_C}.$$

If sun gear S_1 is fixed, then

$$\frac{N_{S_1}N_{P_2}}{N_{P_1}N_{S_2}} = \frac{n_{S_2} - n_C}{0 - n_C},$$

from which it follows that

$$\frac{n_{S_2}}{n_C} = 1 - \frac{N_{S_1}N_{P_2}}{N_{P_1}N_{S_2}},$$

as before.

Gear sizes can be specified to produce a wide variety of output-to-input ratios. By selecting gears so that the fraction $N_{S_1}N_{P_2}/(N_{P_1}N_{S_1})$ is approximately (but not exactly) unity, we obtain very great speed reductions. For example, let

$$N_{S_1} = 102, \qquad N_{P_2} = 49,$$
$$N_{P_1} = 50, \quad \text{and} \quad N_{S_2} = 100.$$

Then, from Eq. (8.12),

$$\frac{n_o}{n_i} = 1 - \frac{4998}{5000} = +\frac{1}{2500}.$$

In considering the preceding example, the reader may come to an extraordinary conclusion. According to the calculations, if the shaft of S_2 is given one turn, the planet carrier C will make 2500 revolutions. In an actual gear train possessing this speed ratio, however, friction would prevent gear S_2 from driving the train, since a small friction torque on the planet carrier shaft would be so greatly magnified. As is also the case with most worm drives, this particular gear train may be used only to reduce speed. Commercially available planetary trains of that variety may be employed to *step up* speed by using sun gear S_2 for the input and the planet carrier C for the output. Ordinarily, n_o/n_i is limited to about 10 or 20.

A planetary speed reducer of the type sketched in Figure 8.12 is very flexible; most output-to-input ratios may be approximated simply by selecting the appropriate gear sizes. Exact ratios, however, are not always obtainable. Suppose, for example, that a speed ratio $n_o/n_i = 1/2000$ is called for, where n_o is the speed of S_2, the output, and n_i is the speed of C, the input. Instead of using a trial-and-error method to select a suitable combination of gears, we will examine Eq. (8.12). If the term

$$\frac{N_{S_1}N_{P_2}}{N_{P_1}N_{S_2}} = \frac{1999}{2000},$$

then

$$\frac{n_o}{n_i} = \frac{1}{2000} \text{ (exactly)}.$$

Let us, therefore, try to specify the gear sizes so that $N_{S_1} \times N_{P_2} = 1999$ and $N_{P_1} \times N_{S_2} = 2000$. Referring to a table of factors and primes or using software with a *factor* command, we see that $2000 = 2^4 \times 5^3 = 2 \times 2 \times 2 \times 2 \times 5 \times 5 \times 5$, which suggests that $N_{P_1} = 5 \times 5 = 25$ and $N_{S_2} = 2 \times 2 \times 2 \times 2 \times 5 = 80$. Unfortunately, 1999 has no factors except 1 and 1999; it is a prime number.

Since we cannot manufacture a pair of gears with 1 and 1999 teeth for this application, we refer again to the table of factors and primes. Noting that $2001 = 3 \times 23 \times 29$, we may specify $N_{S_1} = 3 \times 23 = 69$ and $N_{P_2} = 29$, so that $N_{S_1} \times N_{P_2} = 2001$. Then, from Eq. (8.12), $n_o/n_i = 1 - 2001/2000 = -1/2000$. The minus sign indicates that the output and input shafts rotate in opposite directions. If this is objectionable, an additional gear may be added to the train to change the direction of the output.

Alternatively, the desired speed ratio of 1/2000 may be approximated by using the same factors, except that in this case

$$N_{P_1} = 29, \qquad N_{P_2} = 25,$$
$$N_{S_2} = 69, \quad \text{and} \quad N_{S_1} = 80.$$

From Eq. (8.12), the result is

$$\frac{n_o}{n_i} = 1 - \frac{80 \times 25}{29 \times 69} = +\frac{1}{2001},$$

a value that in most cases would be close enough to be acceptable.

Manufacturing difficulties are sometimes encountered with reverted gear trains. For the gear train of Figure 8.12, the distance between the centers of gears S_1 and P_1 is the same as that between S_2 and P_2, since S_1 and S_2 turn about the same axis. Therefore, the pitch diameters are related by the equation

$$d_{S_1} + d_{P_1} = d_{S_2} + d_{P_2}.$$

(Recall that gears which mesh must have the same diametral pitch or module.) If the loading is such that a diametral pitch of 10 teeth per inch of diameter is satisfactory for gears S_2 and P_2 in the previous example, then

$$d_{S_1} + d_{P_1} = d_{S_2} + d_{P_2} = \frac{25}{10} + \frac{69}{10} = 9.4 \text{ in.}$$

The diametral pitch of gears S_1 and P_1 will be

$$P = \frac{N_{S_1} + N_{P_1}}{d_{S_1} + d_{P_1}} = \frac{80 + 29}{9.4} = 11\frac{56}{94}$$

teeth per inch of diameter. Since this value is not a standard diametral pitch, special cutters must be manufactured, increasing the cost of the gear train. On the other hand, the use of only standard gears limits the number of different speed ratios that may be obtained. The planetary differential drive shown in Figures 8.13a and b offers a wide range of speed ratios, due to a broad selection of available component gears.

(a)

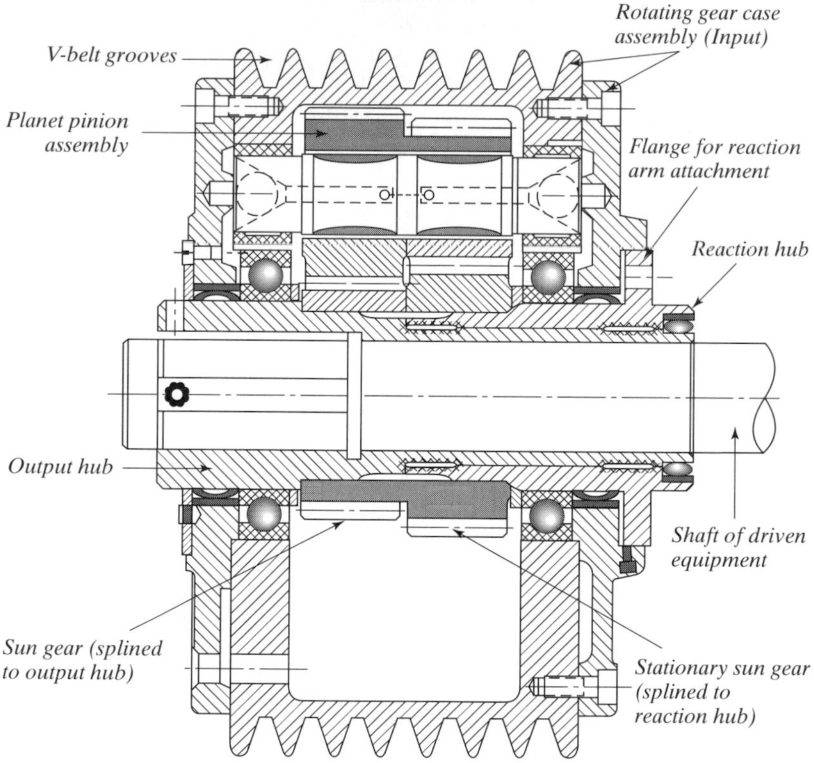

(b)

FIGURE 8.13 (a) Planetary differential drive. Mounted within the driven pulley and directly on the shaft of the driven equipment, this speed reducer requires little additional space. A wide range of speed ratios and output torque capacities results from changes in tooth number. The available output-to-input speed ratios range from 1:1.79 to 1:482, with the direction of the output opposite in some cases. (*Source:* Airborne Accessories Corporation.) (b) Sectional view of planetary differential drive. The driven pulley of a V-belt drive forms the rotating gear case of the compound planetary drive. The gear case carries the planet assemblies, each consisting of two helical planet gears splined to the same shaft. Power is transmitted to a helical sun gear on the output shaft through differential action between the planets and a fixed helical sun gear.

Nonstandard gears may also result when metric sizes are used. Suppose, for example, we require an output speed of $\frac{1}{16}$ rev/min for an input speed of 100 rev/min. Let us select a transmission similar to that shown in Figure 8.12, using carrier C as input and sun gear S_2 as output, with sun gear S_1 fixed. Referring to Table 8.5, we see that

$$1 - \frac{N_{S_1}N_{P_2}}{N_{P_1}N_{S_2}} = 0.000625.$$

After several attempts, we find that one possible combination of spur gears is as follows:

$$N_{S_1} = 82, \qquad N_{S_2} = 80,$$
$$N_{P_1} = 40, \text{ and } N_{P_2} = 39.$$

The speeds are

$$n_{S_1} = 0, \quad n_{P_1} = n_{P_2} = 305 \text{ rev/min (cw)},$$
$$n_{S_2} = 0.0625 \text{ rev/min cw}, \quad \text{and} \quad n_C = 100 \text{ rev/min cw}.$$

If we use a standard metric module $m_1 = 4$ for gears S_1 and P_1, the center distance between the shafts is

$$c = \frac{d_{S_1} + d_{P_1}}{2} = \frac{m_1(N_{S_1} + N_{P_1})}{2} = 244 \text{ mm}.$$

Then, the metric module m_2 for gears P_2 and S_2 is given by

$$m_2 = \frac{2c}{N_{S_2} + N_{P_2}} = 4.1008,$$

which is not a standard value.

Tandem Planetary Trains

For most planetary trains, there is no loss in generality when speed ratios are analyzed by giving the entire train one rotation and then superimposing a correcting rotation with the planet carrier fixed. When the gear train has more than one planet carrier, however, this procedure, if applied to the train as a whole, would arbitrarily restrict the planet carriers to the same speed—a result that may contradict the constraints of the actual problem. One method of solving tandem planetary trains is to divide the train at some convenient point and solve for the speed ratios for each half. The speed ratio for the entire train is then the product of the separate speed ratios.

Figure 8.14 illustrates a tandem planetary train that divides conveniently between R_2 and R_3. Following the usual procedure for the left side of the train, which

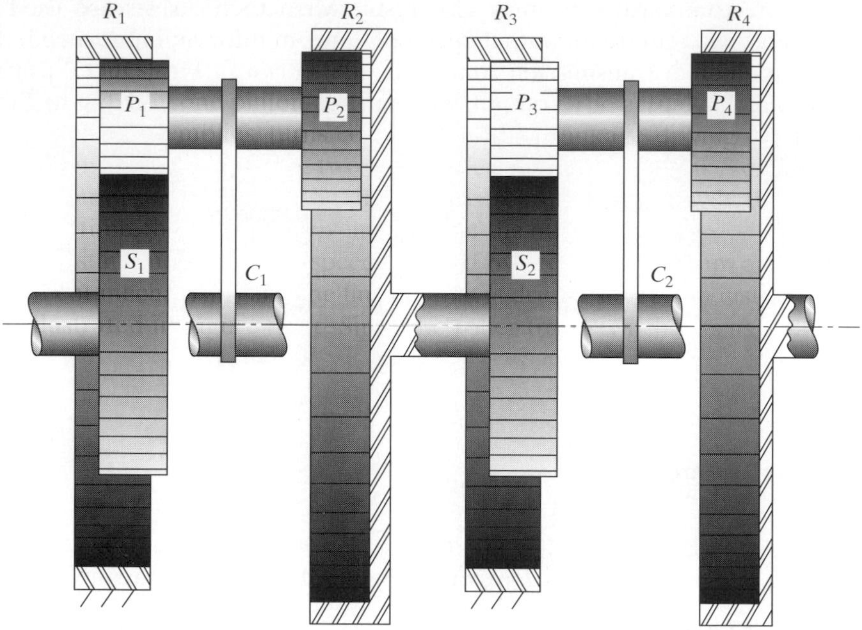

FIGURE 8.14 Train with two planet carriers. With two independent planet carriers in a train, the reader should assume that they do *not* rotate at the same speed. To solve for the overall speed ratio, one can solve for the speed ratio of each half of this train separately, thus correctly using the superposition method. The speed ratio for the entire train is then the *product* of the ratios of the two halves.

includes S_1, P_1, P_2, R_1, R_2, and C_1, we obtain

$$\frac{n_{R_2}}{n_{S_1}} = \frac{1 - [N_{P_2}N_{R_1}/(N_{P_1}N_{R_2})]}{1 + (N_{R_1}/N_{S_1})}.$$

Similarly, for the last half of the train,

$$\frac{n_{R_4}}{n_{S_2}} = \frac{1 - [N_{R_3}N_{P_4}/(N_{P_3}N_{R_4})]}{1 + (N_{R_3}/N_{S_2})}.$$

Since $n_{R_2} = n_{S_2}$, the output-to-input ratio for the entire train is the product of the foregoing ratios:

$$\frac{n_o}{n_i} = \frac{n_{R_2}}{n_{S_1}} \cdot \frac{n_{R_4}}{n_{S_2}}.$$

Using factors and primes as in a previous example, we may demonstrate the flexibility and the extreme speed reductions available with this type of gear train. For example, letting

$$N_{R_1} = N_{R_3} = 87 \text{ teeth},$$
$$N_{P_1} = N_{P_3} = 20 \text{ teeth},$$
$$N_{S_1} = N_{S_2} = 47 \text{ teeth},$$
$$N_{P_2} = N_{P_4} = 23 \text{ teeth},$$

and

$$N_{R_2} = N_{R_4} = 100 \text{ teeth},$$

we find that the speed ratio of each part of the gear train separately is

$$\frac{n_{R_4}}{n_{S_2}} = \frac{n_{R_2}}{n_{S_1}} = -\frac{47}{268{,}000}.$$

The product of the two ratios is the output-to-input speed ratio:

$$\frac{n_o}{n_i} = +30.756 \times 10^{-9} = \frac{+1}{32{,}514{,}260}.$$

The preceding set of values is only an illustration of the planetary train's potential for speed reduction. We might never need such extreme ratios in practice, but trains of this type are commonly used for speed reductions from one thousand to several thousand.

Free-Floating Planetary Transmission

The free-floating planetary transmission (Figures 8.15a and b) consists of a sun gear S (the input) which drives five planetary spindles. Each spindle has three planets: P_1, P_2, and P_3. Ring gear R_1 is the output gear, and ring gear R_2 is fixed. There is no planet carrier as such; the planets are constrained by the gear meshes and the spindles, which roll on cylindrical rings. All planets rotate at the same speed. All forces and reactions are transmitted through the gear meshes and through rolling cylinders. The planet gears are so spaced that tangential gear tooth forces keep the planet spindles in equilibrium. Tooth separating and centifugal forces are balanced out by the cylindrical rings.

Using the superposition method, we may find relative speeds for the free-planet transmission. Answers are given in Table 8.6 in terms of tooth numbers N_S, N_{P_1}, and so on, where C refers to motion of the planet spindle centers about the centerline of the transmission, since there is no planet carrier.

Changing Speed Ratios with Planetary Transmissions

When speed ratios must be changed smoothly and rapidly, planetary transmissions are often selected. Automotive requirements, for example, may call for a gear train with two

(a)

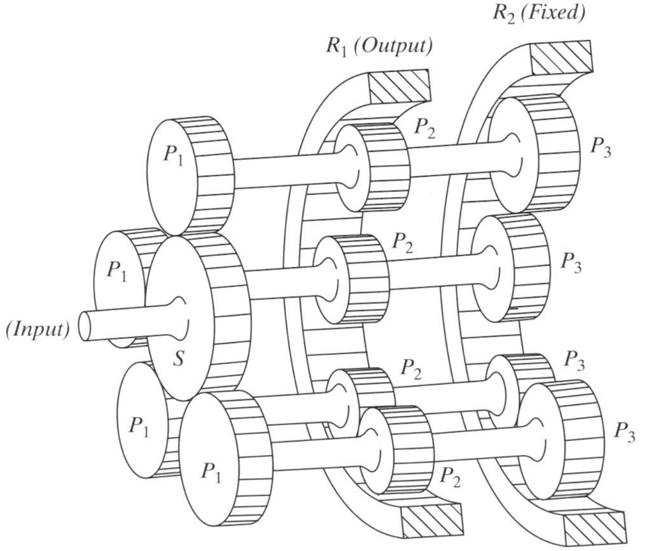

(b)

FIGURE 8.15 (a) Free-floating planetary transmission with fixed ring gear removed to show details of planet spindles. (*Source:* The Free-Floating Planetary Transmission is a proprietary concept of the Curtiss-Wright Corp., Wood-Ridge, NJ, and is covered by a number of patents, including No. 3,540,311 and others pending.) (b) Schematic of free-floating planetary transmission, showing four of five planetary spindles.

TABLE 8.6 Free-Floating Planetary Transmission

	Input		Output	Stationary	
Gear	S	$P_1P_2P_3$	R_1	R_2	C
Number of teeth	N_S	$N_{P_1}\ N_{P_2}\ N_{P_3}$	N_{R_1}	N_{R_2}	—
1: Train locked together	+1	+1	+1	+1	+1
2: Correct stationary gear with spindle centers fixed	$+\dfrac{N_{R_2}N_{P_1}}{N_{P_3}N_S}$	$-\dfrac{N_{R_2}}{N_{P_3}}$	$-\dfrac{N_{R_2}N_{P_2}}{N_{P_3}N_{R_1}}$	−1	0
Σ (true relative speed)	$1+\dfrac{N_{R_2}N_{P_1}}{N_{P_3}N_S}$	$1-\dfrac{N_{R_2}}{N_{P_3}}$	$1-\dfrac{N_{R_2}N_{P_2}}{N_{P_3}N_{R_1}}$	0	+1

or more positive (forward) speed ratios and one negative (reverse) speed ratio. The speed changes can be accomplished by clutches and brakes in a planetary transmission.

Figure 8.16 illustrates, schematically, such a transmission. With the gears as shown, there is no output since both sun gear S_2 and ring gear R are free to turn. Gear P_1 meshes only with S_1 and P_2. Thus, carrier C could remain stationary, even though the input shaft is turning.

If sun gear S_2 is locked to the input shaft by engaging a clutch (not shown), a direct drive is obtained. Since the two sun gears rotate at the same speed (and in the same direction), the planet gears are, in effect, locked to the sun gears. The planets are also locked to each other, because the sun gears tend to rotate the meshed planets in the same direction (preventing any relative motion). Any motion of planets P_1 and P_2

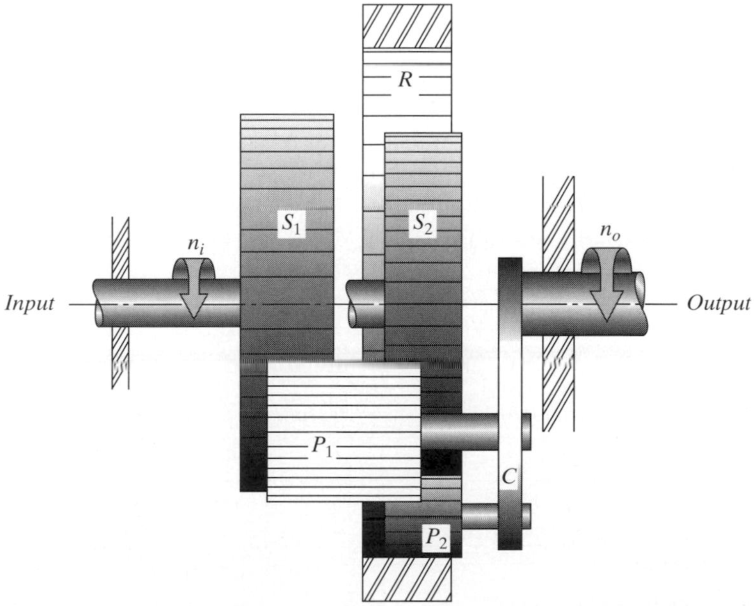

FIGURE 8.16 Planetary transmission. Planets P_1 and P_2 are carried by the *common* planet carrier. P_1 meshes only with S_1 and P_2. When both S_2 and R are free to turn, there is no output; the transmission is in neutral. S_2 may be locked to the input shaft by engaging a clutch, giving a one-to-one ratio of output to input. (See Tables 8.7 and 8.8 for other ratios.) (*Note:* Gear positions in an actual transmission do not correspond with those shown in this schematic.)

TABLE 8.7 Speed Ratios for the Planetary Transmission of Figure 8.16 with Sun Gear S_2 Fixed

Gear	S_1	P_1	P_2	S_2	R	C
No. of teeth	N_{S_1}	N_{P_1}	N_{P_2}	N_{S_2}	N_R	
Step 1: rotations with train locked	+1	+1	+1	+1	+1	+1
Step 2: rotations with planet carrier fixed	$+\dfrac{N_{S_2}}{N_{S_1}}$	—	—	−1	—	0
Total number of rotations	$1+\dfrac{N_{S_2}}{N_{S_1}}$	—	—	0	—	+1

relative to the sun gears would be incompatible with the fact that both planets are held in the same carrier. Thus, when S_2 is locked to the input shaft, all gears lock together and turn as a unit. The ratio of output to input speed then becomes $n_o/n_i = 1$.

To fix either the ring gear or sun gear S_2 to the frame, band brakes may be used. The brakes would engage the outer surface of the ring gear or a drum attached to sun gear S_2. Both bands and drums are omitted from the figure so that we may examine the kinematics of the problem without additional complications. The effect of locking sun gear S_2 to the frame is shown in Table 8.7. In the first step, the entire train is given one clockwise rotation, just as in each of the previous examples. In the second step, the carrier is fixed, while sun gear S_2 is given one counterclockwise rotation so that it has zero net motion.

By adding steps 1 and 2 in the table, we obtain an output-to-input speed reduction of

$$\frac{n_o}{n_i} = \frac{1}{1 + (N_{S_2}/N_{S_1})} \tag{8.13}$$

when gear S_2 is fixed.

When the ring gear is free to rotate (Figure 8.16), it serves no purpose in the transmission. When the ring is fixed and gear S_2 is free to turn, however, a negative or reverse ratio is obtained. The results are shown in Table 8.8. which is obtained by proceeding as in Table 8.7, except that in this case ring gear R is given the zero net rotation.

In this case, sun gear S_2 turns freely and serves no useful purpose. The output-to-input speed ratio is

$$\frac{n_o}{n_i} = \frac{1}{1 - (N_R/N_{S_1})} \tag{8.14}$$

TABLE 8.8 Speed Ratios for the Planetary Transmission of Figure 8.16 with Ring Gear Fixed

Gear	S_1	P_1	P_2	S_2	R	C
No. of teeth	N_{S_1}	N_{P_1}	N_{P_2}	N_{S_2}	N_R	
Step 1: rotations with train locked	+1	+1	+1	+1	+1	+1
Step 2: rotations with planet carrier fixed	$-\dfrac{N_R}{N_{S_1}}$	—	—	—	−1	0
Total rotations	$1-\dfrac{N_R}{N_{S_1}}$	—	—	—	0	1

with the ring fixed. Since N_R must be greater than N_{S_1}, their ratio is always negative, indicating that the direction of rotation of the output is opposite that of the input.

Planetary Trains with More than One Input

In some planetary gear train applications, *none* of the gears are fixed. As an example, it might be necessary to find the angular velocity n_C of the planet carrier of Figure 8.8, given n_S and n_R the velocities of the sun and ring, respectively. We already solved a problem of this type earlier by the formula method. As an alternative, the data already calculated in Tables 8.2 and 8.3 enable us to solve the problem by superposition.

Let the sun have 40 teeth and the ring 80 teeth. The given speeds are $n_S = +900$ rev/min and $n_R = +1500$ rev/min (both clockwise). Table 8.2 gives gear speeds when the sun is fixed. Since these speeds are only relative, the line indicating the total number of rotations may be multiplied by 1000 to give the correct ring speed. We have, then, the following speeds (in revolutions per minute), adjusted to give the correct speed for n_R:

$$n_S = 0;$$
$$n_R = 1.5 \times 1000 = 1500;$$
$$n_C = 1 \times 1000 = 1000.$$

Next, the values of N_S and N_R are substituted into the last line of Table 8.3, and the result is multiplied by 300 to give the correct sun gear speed when the ring is fixed. We obtain

$$n_S = 3 \times 300 = 900,$$
$$n_R = 0,$$

and

$$n_C = 1 \times 300 = 300.$$

Adding the values for the case where the sun gear is fixed to the values for the case where the ring is fixed, we get

$$n_S = 0 + 900 = +900,$$

$$n_R = 1500 + 0 = +1500,$$

and

$$n_C = 1000 + 300 = +1300.$$

The method that follows avoids making use of previous results. We begin the problem by rotating the entire train $+v$ revolutions (where the value of v is to be found later), instead of $+1$ revolutions as done previously. Thus, we have step 1 of Table 8.9. In step 2, sun gear S is given $+w$ rotations with the carrier fixed. The value of w is also

TABLE 8.9 Speed Ratios for the Planetary Train of Figure 8.8 with No Fixed Gears

Gear	Sun	Planet	Ring	Carrier
No. of teeth	40	20	80	
Step 1: rotations with train locked	$+v$	$+v$	$+v$	$+v$
Step 2: rotations with carrier fixed	$+w$	$-\left(\frac{N_S}{N_P}\right)w$	$-\left(\frac{N_S}{N_R}\right)w$	0
Total rotations	$v + w$	$v - \left(\frac{N_S}{N_P}\right)w$	$v - \left(\frac{N_S}{N_R}\right)w$	v

unknown at this time. Noting the gear ratio, we obtain $(-N_S/N_R) \times w$ rotations of the ring gear in step 2. The last line of the table is the sum of steps 1 and 2, from which we get

$$n_S = v + w,$$

$$n_R = v - \left(\frac{N_S}{N_R}\right)w,$$

and

$$n_C = v.$$

Eliminating v and w from the preceding equations, we obtain

$$n_R = n_C - \left(\frac{N_S}{N_R}\right)(n_S - n_C), \tag{8.15}$$

from which it follows that

$$n_C = \frac{n_R + n_S(N_S/N_R)}{1 + (N_S/N_R)} \tag{8.16}$$

and

$$n_S = n_C + \left(\frac{N_R}{N_S}\right)(n_C - n_R). \tag{8.17}$$

These three equations describe the motion of the planetary train shown in Figures 8.7a and 8.8, *for any input or output given in terms of angular velocity*. When the data of the foregoing problems are substituted into the equation for n_C, we again obtain $n_C = 130$ rev/min clockwise.

Differentials and Phase Shifters

While engineering problem solving often suggests the use of electronic calculators and computers, there are many specialized applications for mechanical computing devices. Automatic control of a certain process may depend on the addition of two input functions. Mechanically computed products and sums of signals from sensing devices have been used to guide aircraft.

The tabular method may be employed with a modification to incorporate two *angular* inputs. (See Figure 8.17.) In the first step (Table 8.10), the entire locked train is given $+v$ rotations. In the second step, the planet carrier is fixed while gear S_1 is given $+w$ rotations. If S_1 and S_2 have the same number of teeth, then S_2 makes $-w$ rotations. The results are given in Table 8.10, where x and y represent the actual total number of rotations of S_1 and S_2, respectively.

On adding the values of x and y, we see that $x + y = 2v$; that is,

$$v = \frac{x + y}{2} \quad \text{or} \quad n_C = \frac{n_{S_1} + n_{S_2}}{2}. \tag{8.18}$$

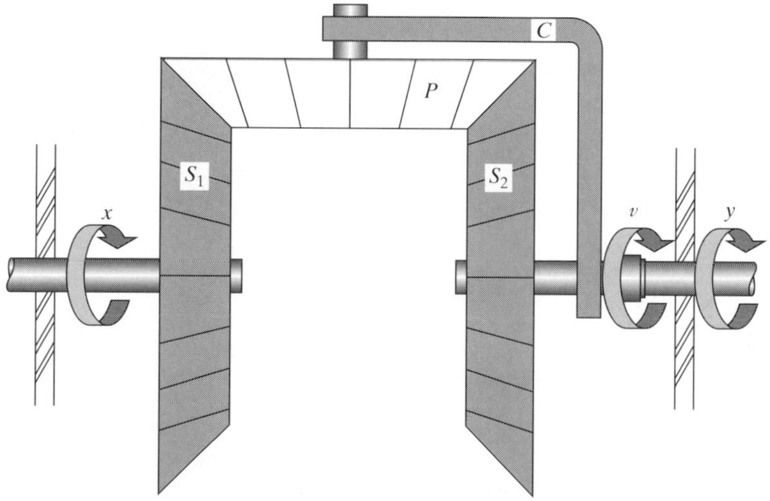

FIGURE 8.17 When x and y are given as *angular* velocities of S_1 and S_2, respectively, $v = (x + y)/2$ is the angular velocity of the planet carrier. Table 8.10 illustrates how this result may be obtained by using a variation of the superposition method.

TABLE 8.10 Speed Ratio for the Bevel Gear Differential of Figure 8.17 with Two Inputs

Gear	S_1	S_2	C
Step 1: rotations with train locked	v	v	v
Step 2: rotations with planet carrier fixed	$+w$	$-w$	0
Total rotations	$x = v + w$	$y = v - w$	v

(a)

(b)

FIGURE 8.18 (a) The large bevel gears that act as sun gears are mounted on bearings. Each of the sun gears may be directly attached to a spur gear through which input motion is transmitted. The planet bevel gear drives the junction block, which is rigidly connected to the shaft through the sun gears. (*Source:* Precision Industrial Components Corporation, Wells-Benrus Corporation.) (b) When this mechanism is used to *add* or *subtract*, the shafts at the right and left foreground act as inputs (to the pair of gears driving the bevel sun gears), and the planetary bevel gear cross arm drives the output shaft. When the device is used as a *phase shifter*, the shafts at the right and left foreground become the input and output, respectively. Rotation of the junction block changes the relative position (phase) between input and output. (*Source:* Precision Industrial Components Corporation. Wells-Benrus Corporation.)

In this mechanism, x, y, and v may represent angular displacement, angular velocity, or angular acceleration. Figure 8.18a illustrates a miniature bevel gear differential that can also be used as a phase shifter.

If the sum of the two inputs x and y is required, it is necessary only to double the output (the rotation of the planet carrier shaft). This may be accomplished simply by a change in scale or by using gears to effect a two-to-one increase in speed. The value of the function $v = Ax + By$, where A and B are constants, is obtained by additional pairs of gears driving the input shafts. (See Figure 8.18b.) Since the direction of rotation changes with each meshing pair of external spur gears, idlers may be used to obtain the correct sign.

When an automobile is making a turn, the wheels on the outside of the turn travel farther than the wheels on the inside. A differential allows the wheels to rotate

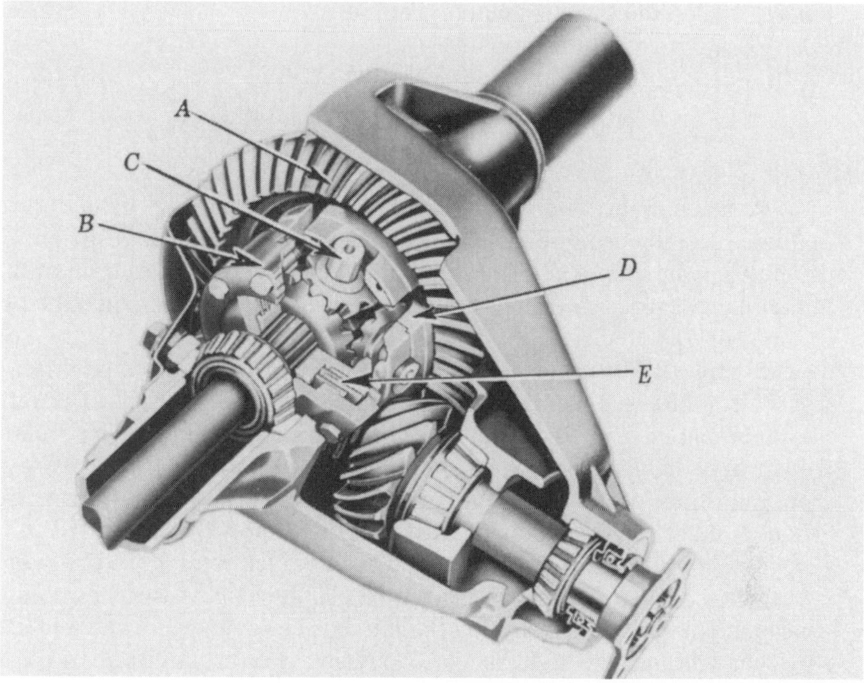

FIGURE 8.19 Limited-slip automotive differential. Engine power is transmitted to the large hypoid gear (A), which drives the differential case (planet carrier B). The driving force moves cross pins (C) up cam surfaces (D), engaging disk clutches (E). This type of differential *applies the greatest amount of torque to the wheel with the most traction* to prevent the other wheel from spinning on ice or snow. (*Source:* Dana Corporation.)

at different velocities so that the tires are not required to drag along the road during a turn. A typical automotive differential is similar to the sketch in Figure 8.17, except that *several* bevel planet gears are carried on the same planet carrier, which is driven by a pair of hypoid gears (Figure 8.19). Kinematically, there is no difference in the differential itself. The input to the planet carrier (Figure 8.17) is transmitted through the planet P to the axles represented by the shafts of S_1 and S_2. The sum of the axle speeds is equal to twice the planet carrier speed: $x + y = 2v$. On a straight, dry road, $x = y = v$. During turning, x does not equal y; however, the average of x and y is v. If there is no provision for positive traction, the torque delivered to both axles is equal. Then, with one wheel on ice and the other on dry concrete, the tire on the ice may turn freely at a speed of $2v$ while the other tire does not turn at all. This is the price we pay to avoid excessive tire wear. Such an obvious disadvantage can be remedied by devices that deliver most of the torque to the wheel that is *not* slipping. (See Figure 8.19.)

Using the formula method to describe the motion of a bevel gear differential, we have

$$\frac{n^*_{S_2}}{n^*_{S_1}} = \frac{n_{S_2} - n_C}{n_{S_1} - n_C} = -\frac{N_{S_1}}{N_{S_2}}.$$

For $N_{S_1} = N_{S_2}$, the carrier speed is given by

$$n_C = \frac{n_{S_1} + n_{S_2}}{2},$$

as in Eq. (8.18).

A mechanism kinematically equivalent to the bevel gear differential may be manufactured by using spur gears. While one planet gear is sufficient in a bevel gear differential, the spur gear differential of Figure 8.20 requires a train of two planets held in the planet carrier. The number of teeth in sun gear S_1 will equal the number of teeth in sun gear S_2.

Figure 8.21 illustrates the use of a different type of spur gear differential in a final drive assembly. This configuration is equivalent to Figure 8.7i. The differential drive pinion is splined directly to an automatic transmission. In this particular application, a front-wheel drive, a pair of spiral bevel gears was chosen to avoid offsetting the pinion from the center of the planetary train. A hypoid gear is used in rear-wheel-drive vehicles to make it possible to locate the driveshaft below the axle.

The spiral bevel gears produce a speed reduction of 3.21 to one, and when the car is traveling in a straight path, the entire planetary train turns as a unit. A right turn causes the planet carrier (driving the left axle) to overspeed, and a left turn causes the sun gear (driving the right axle) to overspeed. During either turn, relative motion between the ring, planets, and sun provide the differential action. The planet gears must be in pairs, one contacting the ring and the other the sun, for proper rotation direction. The number of teeth in the gears are as follows: sun, 36; inner planets, 16; outer planets, 16; internal ring, 72; large spiral bevel, 45; and spiral bevel pinion, 14. So that the axles are flexible, Rzeppa-type, constant-velocity universal joints are used in the drive assembly.

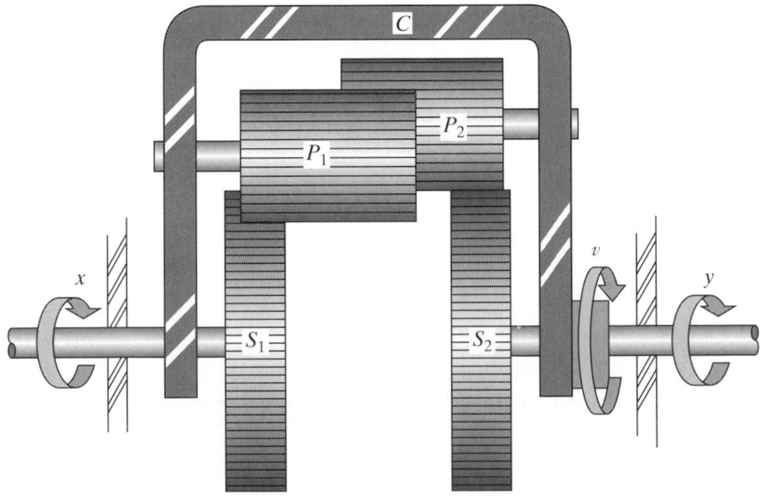

FIGURE 8.20 Spur gear differential.

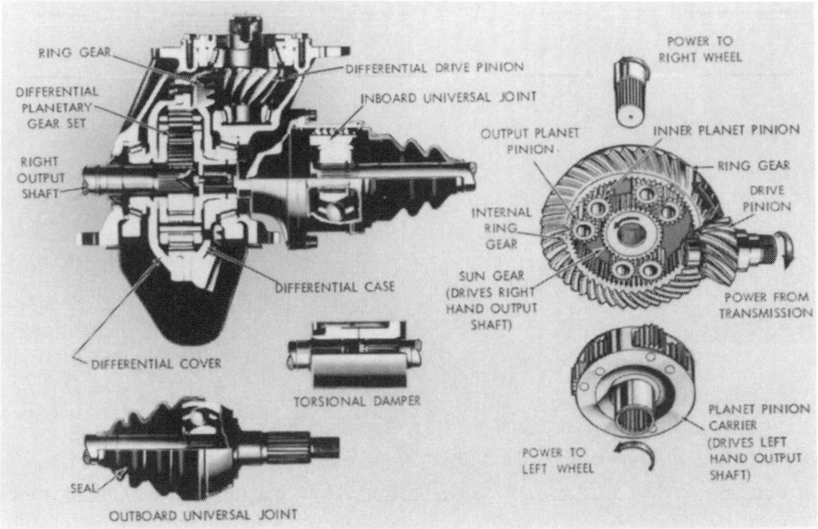

FIGURE 8.21 An automotive final drive assembly. (*Source:* General Motors Corporation.)

8.5 OTHER DRIVE TRAIN ELEMENTS

Chain Drives

Chain drives, like gear drives, result in definite output-to-input speed ratios and have high-power transmission capacity—over 1000 hp for some designs operating at high speeds. Unlike the situation with gear trains, however, the distance between shafts is not a critical factor and may even fluctuate without adverse effects during the operation of a chain drive. The chain length is changed simply by adding or removing a link.

Due to chordal action, speed does fluctuate slightly in chain drives. This effect is most significant when sprockets with a small number of teeth are used. Chain drives tend to be noisy, because of the impact of the links with sprocket teeth. An inverted-tooth chain (sometimes called a silent chain) is usually quieter than a roller chain of the same capacity. Chain drives should be guarded to prevent injury.

A roller chain consists of side plates and pin and bushing joints that mesh with toothed sprockets. Roller chain sprockets have modified involute teeth. Unwrapping the chain from the sprocket tooth resembles involute generation, although the involute form is modified by the nonzero roller diameter. Most high-speed chain drives use a roller-type chain or an inverted-tooth type made up of pin-connected, inward-facing teeth. (See Figure 8.22.) For maximum chain life, both sprockets are vertical and in the same plane. Although there is little restriction on center distances, the chain should wrap around at least 120° on both sprockets. The chain tension is adjusted if necessary, by moving one of the shafts or by using an idler sprocket.

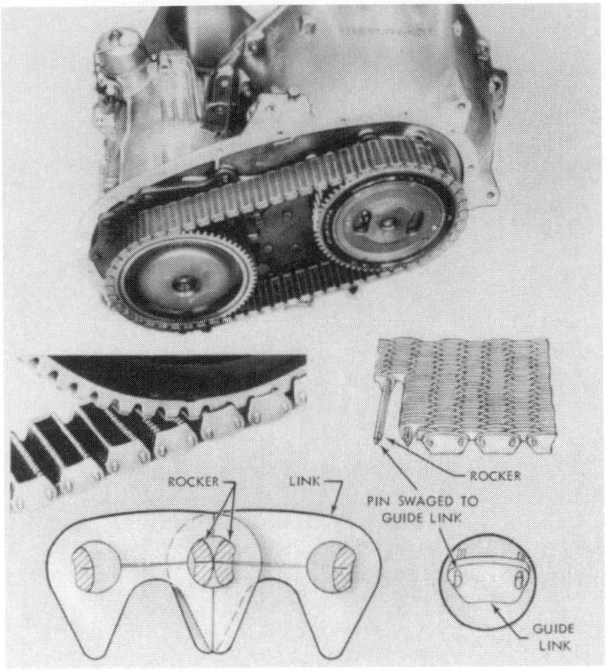

FIGURE 8.22 Automotive chain drive. This inverted-tooth chain transmits power from a torque converter (*left*) to a planetary transmission (*right*). It was selected in preference to a gear drive, which would have required an idler. The links are designed for smooth engagement with the sprockets. Although timing chains are common in automobiles, the development of a drive chain posed several problems, including excessive noise. The manufacturer arrived at the design shown after considerable analytical and experimental study, including the use of high-speed motion pictures. (*Source:* General Motors Corporation.)

For instrument drives and other very light-load, low-speed applications, bead chains (Figure 8.23a) may be used. Sprockets need not be in the same plane if idlers are provided where it is necessary to maintain chain-to-sprocket contact. The complex drive shown in Figure 8.23b employs both bead chain and roller chain drives.

Sprocket speeds are inversely proportional to the number of teeth in each sprocket (or the number of pockets or indentations in bead chain sprockets). Both sprockets rotate in the same direction. The speed ratio is given by

$$\frac{n_2}{n_1} = +\frac{N_1}{N_2},$$

where n_1 and n_2 are, respectively, the driver and driven sprocket speeds and N_1 and N_2 are, respectively, the numbers of teeth in the driver and driven sprockets. Idler sprockets and the center distance do not affect speed. When bead chains are used, the chain may be crossed between sprockets to reverse the direction of rotation.

Variable-Speed Chain Drives

Changing the speed ratio is generally *not* practical with high-speed or high-load roller chain drives, except by using two or more chains and sprocket pairs, with clutches to engage one set at a time. For low loads and low speeds, a movable idler sprocket may be used to shift the chain from one sprocket to another without interrupting the driving. One application is a bicycle *derailleur* type of transmission having two driving sprockets

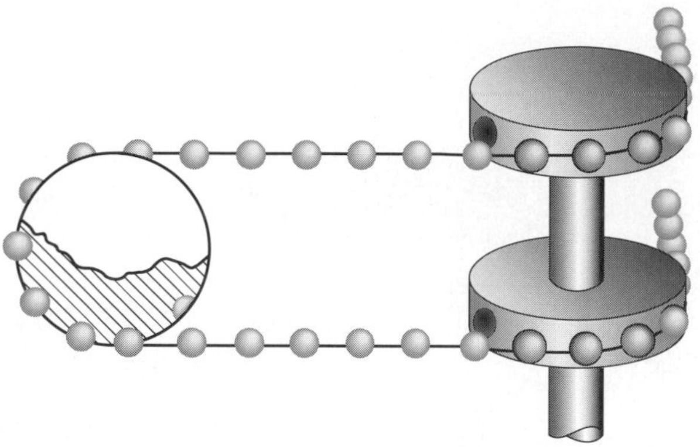

(a)

(b)

FIGURE 8.23 (a) Bead chain drive. Shown is a close-tolerance bead chain on pocketed sprockets for a positive, low-speed drive. The beads swivel on pins so that sprockets need not be in the same plane. (b) Instrument dial drive. Both bead chains and roller chains are used in this drive, assuring precise speed ratios with maximum flexibility. (*Source:* Voland Corporation.)

with different numbers of teeth and a "cone" of five driven sprockets at the rear wheel with graduated tooth numbers. Thus, 10 sprocket combinations are available, yielding 10 different speed ratios. Drives of this type are also available with additional sprockets, producing 12, 15, 18, or 21 different speed ratios. For most of the ratios, the sprockets and the chain are not exactly in line (i.e., the sprockets in use are not in the same plane), but in this application, loads are light enough so that chain wear due to misalignment is slight.

A variable-speed chain drive that provides essentially stepless variation in the speed ratio is shown in Figure 8.24. The effective sheave diameter is changed by the spacing between the sides of the grooved sheaves. The chain has floating laminations that conform to the grooves. Thus, the output-to-input speed ratio is related to the effective sheave diameter by Eq. (8.3): $n_2/n_1 = d_1/d_2$.

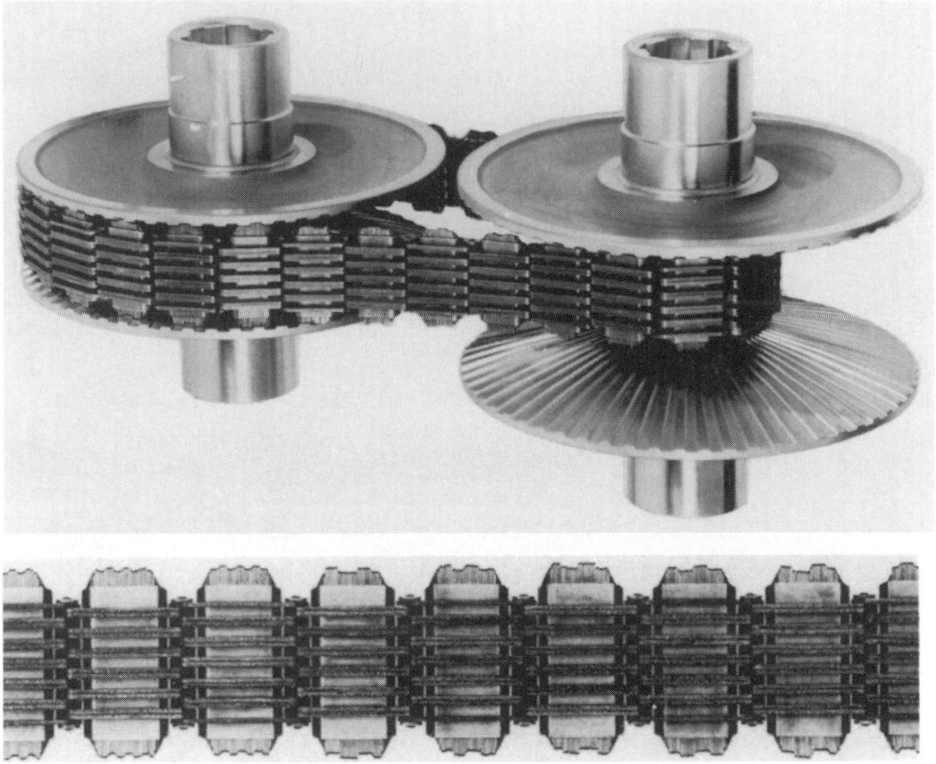

FIGURE 8.24 Variable-speed chain drive. This drive provides essentially stepless variation in the speed ratio with a positive drive. The effective sheave (driving surface) diameter is changed by increasing or decreasing the spacing between the sides of the grooved sheaves. The chain has floating laminations that conform to the grooves; thus, the output-to-input speed ratio depends on the effective sheave diameters only. The speed ratio is independent of the number of grooves in the sheaves. (*Source:* FMC Corporation.)

Belt Drives

While the low cost of belt drives is the most common reason for selecting them over other means of power transmission, belts have other positive features. Belts absorb shock that might otherwise be transmitted from the driven shaft back to the driver and prevent vibrations in the driver from being transmitted to the driven shaft, thus reducing noise and damage from these sources.

In addition, a belt may act as a clutch, disengaging the driver and driven shafts when loose and engaging them when tight. Belt tightening may be accomplished by moving either the driver or the driven shaft, or if both shafts are fixed, a movable idler pulley may be used to tighten the belt. This method is used to engage some power lawn mower drives to the gasoline engine and to drive punch presses and shears, wherein a belt is momentarily engaged to a large flywheel for the working stroke of the machine. An idler is the less desirable of the two methods from the standpoint of belt wear; if used, it should be as large as practical and, preferably, ride on the inside of the slack span of the belt.

V Belt Drives

V belts, the most common type of belts, are made of rubber reinforced with load-carrying cords. They operate in V-grooved pulleys or sheaves. The belt sidewalls wedge into the pulley; the belt does not ride in the bottom of the groove. The effect of this wedging action is to improve driving traction by more than tripling the friction force over what it would be if the pulleys were flat. The output-to-input speed ratio is given by Eq. (8.3), viz.,

$$\frac{n_2}{n_1} = \frac{d_1}{d_2},$$

where d_1 and d_2 refer, respectively, to the driver and driven pulley pitch diameters—the diameters of the neutral surface of the belt riding in the pulley. (The neutral surface of the belt is the surface that is neither compressed nor extended due to bending around the pulleys.)

Since the belt tension varies as each element of the belt turns about a pulley, there is some slippage; therefore, speed ratios are not precise, except with specially designed positive-drive belts and pulleys. If the belt does not cross itself, the driver and driven pulleys turn in the same direction. For the best results, the pulleys should be as large as practical and the belt should operate in a single vertical plane. Heavy loads may be carried by a multigrooved V pulley with a single-backed multiple-V belt.

Variable-Speed Belt Drives

To change the speed ratio occasionally on light machinery, a "cone" of graduated-diameter V pulleys is matched with a similar "cone" (with the large pulley on one opposite the small pulley on the other). The belt is manually transferred from one pair of pulleys to the other when the machine is stopped.

A pair of variable belt-drive pulleys, similar in appearance and operation to the variable-speed chain drive pulleys shown in Figure 8.24, may be used to change speed ratios on moving machinery without interrupting power. The sides of one pulley are moved apart to decrease its pitch diameter, while the sides of the opposite pulley are moved together to increase its pitch diameter. The output-to-input speed ratio again equals the inverse ratio of the pitch diameters and may be continuously varied. Each variable pulley may change diameter by as much as a factor of three or four.

The variable pulleys may be varied manually, pneumatically, or hydraulically. The change in speed ratio may be entirely automatic in response to a speed governor on the output shaft, output torque, engine vacuum, or some combination of them. (See Figure 8.25.)

Some variable-speed drives use a fixed-pitch-diameter pulley opposite a *spring-loaded pulley* like that shown in Figure 8.26. The sides of the spring-loaded pulley are separated simply by increasing the distance between the driver and driven shafts. A cam may be incorporated into the variable pulley to prevent separation of the sides when the driven load increases. The cam is visible on the left side of the pulley shown in Figure 8.26.

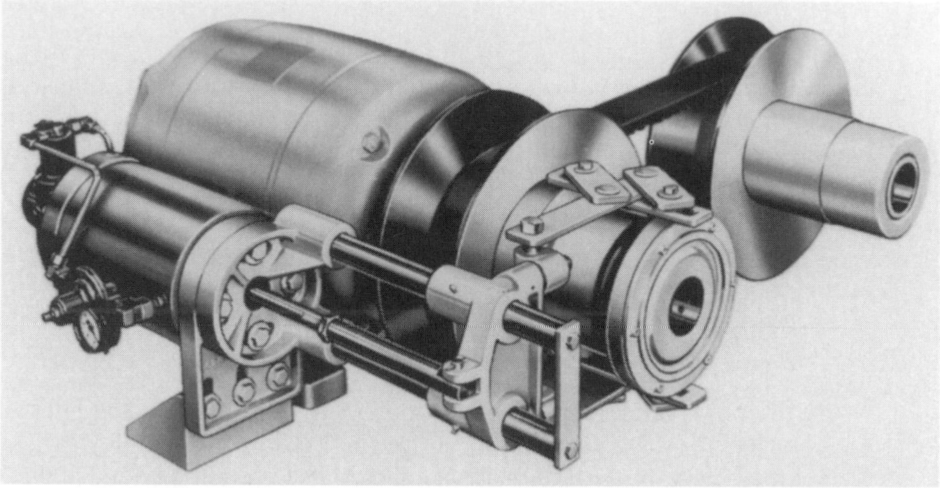

FIGURE 8.25 Pneumatically controlled belt drive. The pneumatic control regulator at the left responds to pressure signals, continuously varying the speed of the drive for process control. Speeds may be varied or may be adjusted to remain constant, or a predetermined cycle of variation in speed may be programmed. The linkage simultaneously moves both halves of the variable-pitch pulley (*foreground*) in order to keep the belt aligned as speeds are changed. Units are available with horsepowers to 30 and speed ranges to 10:1. (*Source:* Lewellen Manufacturing Corporation.)

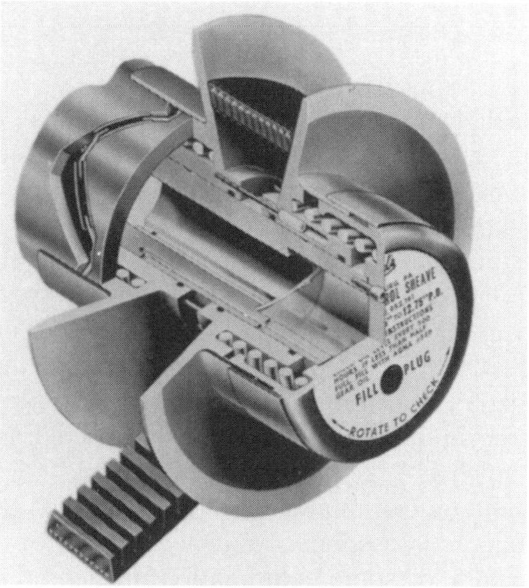

FIGURE 8.26 Variable-pitch pulley. This smooth-faced pulley is part of a variable-speed *belt* drive. When the distance between pulley centers is increased, the spring-loaded sides of the variable pulley separate to decrease its effective diameter. A cam at the left side of the pulley increases pressure on the pulley sides with increases in torque, compensating for the tendency of the pulley to separate at an increased load. (*Source:* T. B. Wood's Sons Company.)

Countershaft Pulleys

If the distance between the driver and driven shafts is fixed, a variable double pulley may be used on a countershaft, as shown in Figure 8.27a. By moving the countershaft toward either the driver or the driven shaft, the pitch diameters of the intervening double pulley change simultaneously, one increasing while the other decreases. This action gives a wide range of speed ratios.

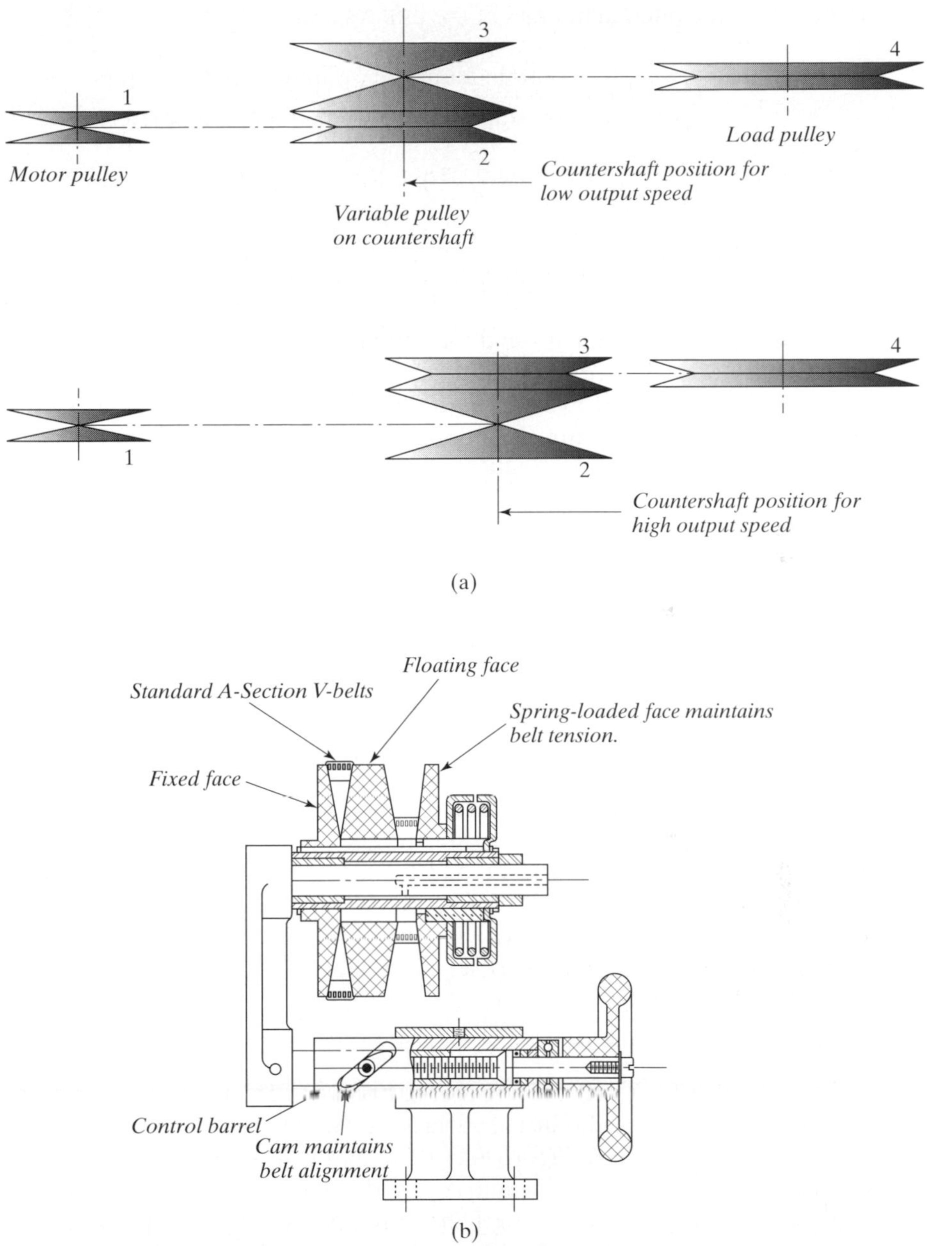

FIGURE 8.27 (a) Variable-speed belt drive. When the countershaft is moved toward the load pulley, the spring-loaded faces of the countershaft double pulley change the effective pitch diameters d_2 and d_3. Pitch diameter d_3 increases, while d_2 decreases. (*Source:* Speed Selector, Inc.) (b) A sectional view of a spring-loaded, cam-aligned, countershaft double pulley for a compound variable-speed belt drive. (*Source:* Speed Selector, Inc.)

When one or more countershafts are used, the output-to-input speed ratio again depends on the pitch diameters of the pulleys and is given by

$$\left|\frac{n_o}{n_i}\right| = \frac{\text{product of driving-pulley pitch diameters}}{\text{product of driven-pulley pitch diameters}}. \tag{8.19}$$

In this case, output-to-input speed ratio is given by

$$\frac{n_4}{n_1} = \frac{d_1 d_3}{d_2 d_4},$$

where d is the pitch diameter and the subscripts 1 to 4 refer to the driver pulley, countershaft driven pulley, countershaft driver pulley, and driven pulley, in that order. The two variable diameters are d_2 and d_3. Figure 8.27b is a sectional view of a spring-loaded, cam-aligned, countershaft double pulley.

Flat Belts

When a single steam engine was used to drive dozens of separate pieces of factory machinery, power was transmitted by large flat belts turning pulleys on overhead countershafts. Belts were shifted, removed, and replaced by skilled operators who worked while the shafts were turning. While far less prominent now, flat belts have many modern, specialized applications. Light fabric belts and flat rubber-covered fabric belts are used at speeds up to 10,000 ft/min and higher—speeds that are well above the operating range of ordinary V-belt drives. When used on balancing machine drives, lightweight flat belts do not significantly affect the rotating mass, and in many high-speed applications, such belts are least likely to excite and transmit vibration.

The output-to-input speed ratio for flat belts is again given by $n_2/n_1 = d_1/d_2$, where pitch diameter d is measured to the center of the belt and is equal to the pulley diameter plus one thickness of belt. As before, subscripts 1 and 2 refer to the driver and driven pulleys, respectively. Either the driver or the driven pulley of a flat belt drive is usually crowned to prevent the belt from riding off.

Positive-Drive Belts

A special type of flat belt called a *timing belt* is used as a positive drive, ensuring exact speed ratios. Teeth on the inside surface of the belt mesh with a grooved pulley, and there is no slippage. (See Figure 8.28.) The output-to-input speed ratio is given by Eq. (8.1), $n_2/n_1 = N_1/N_2$, where N_1 and N_2 are the *number of grooves* in the input and output pulleys, respectively. Timing belts may operate at high speeds, and they require only light tensioning, since the drive does not depend on friction.

A patented drive system for instruments and other light-duty applications uses a reinforced plastic belt with side lugs (Figure 8.29), which meshes with a gearlike pulley that has a central groove to accommodate the belt, but that may be meshed with a spur gear. As with timing belts, the output-to-input speed ratio is equal to the inverse ratio of the numbers of pulley teeth.

FIGURE 8.28 Timing belt drive. Teeth on the inside surface of the belt mesh with grooves in the pulleys. The speed ratio depends on the number of grooves on the input and output pulleys. (*Source:* T. B. Wood's Sons Company.)

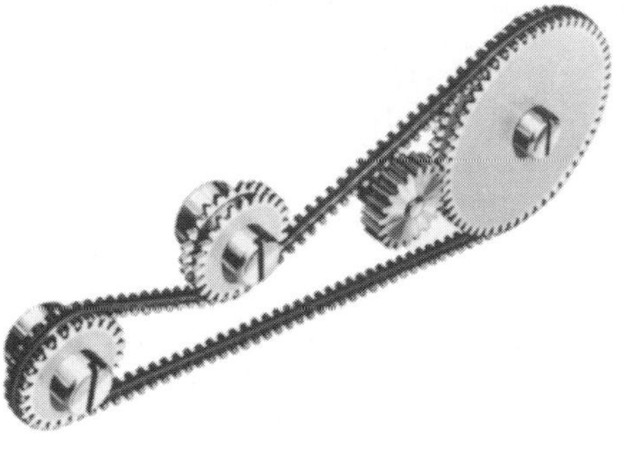

FIGURE 8.29 Positive-drive belt used with geared pulleys. The reinforced plastic belt was originally developed as a silent drive for sound and recording systems. The geared pulleys may be directly meshed with spur gears, eliminating intermediate components. An idler is usually used as shown to take up the slack. The flexibility of this type of belt allows for considerable shaft misalignment, and, with idlers or guides, the belt may even be used between pulleys on perpendicular intersecting shafts. (*Source:* Precision Industrial Components Corporation, Wells-Benrus Corporation.)

Friction Disk Drives

For certain low-torque applications, disks may be used to transmit power from one shaft to another. (See Figure 8.30.) When two disks on parallel shafts make contact at their outer edges, the output-to-input speed ratio is equal to the inverse ratio of the disk diameters. For any number of disks, the output-to-input speed ratio is given by

$$\left|\frac{n_o}{n_i}\right| = \frac{\text{product of driving - disk diameters}}{\text{product of driven - disk diameters}}.$$

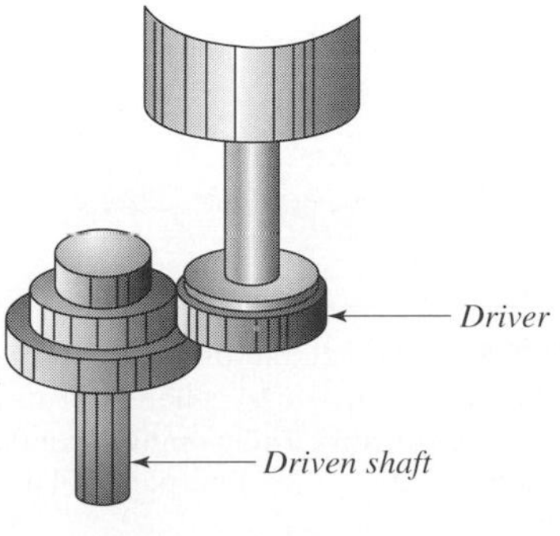

FIGURE 8.30 Friction drive. For this type of drive, the *speed ratio equals the inverse ratio of the disk diameters*. The drive is shifted to a new position to change ratios, and no clutch is required.

To maintain adequate frictional force between the disks, alternate disks might have rubber-covered driving surfaces.

Speed may be changed in discrete steps by using a stepped set of disks on the driven shaft and changing the location of the driving disk to mesh with the driven disk, giving the desired speed. Alternatively, disk speed changers and reversers may employ an idler, as is done with simple geared speed changers and reversing gear trains. Possible slippage between disks makes this type of drive unsuitable for most instrument applications, but permits a simple and inexpensive drive system for low-torque applications, eliminating the need for a clutch. The speed ratio may be changed by shifting disks from one location to another without stopping the drive, since there is no problem of meshing teeth.

Variable-Speed Wheel-and-Disk Drives

A wheel-and-disk type of transmission provides stepless speed changing and reversing. The wheel (1) of Figure 8.31 traces a path of radius r_2 on the disk (2), and the output-to-input speed ratio is given by

$$\frac{n_2}{n_1} = \frac{r_1}{r_2},$$

where r_1 is the radius of the driver and r_2 is the distance from the axis of the driven disk to the point of contact of the wheel and disk. When the wheel contacts the disk to the left of the disk center, the output direction is reversed.

An alternative transmission employs two disks and an idler wheel. Only the idler wheel is shifted while the input and output shafts rotate in place, as in Figure 8.32. Since the velocity ratio is unaffected by the idler diameter, the speed ratio is again given by $n_2/n_1 = r_1/r_2$, where r_1 and r_2 vary simultaneously when the idler is shifted.

The drive shown in the figure is not reversible, but by simply moving the input and output shafts closer together, or by increasing the output disk diameter, one could design a drive with same-direction, zero, or reverse output. When the idler is shifted to the center of the input disk, it does not turn, and the output speed is zero. When the idler is shifted farther to the left, the output is reversed.

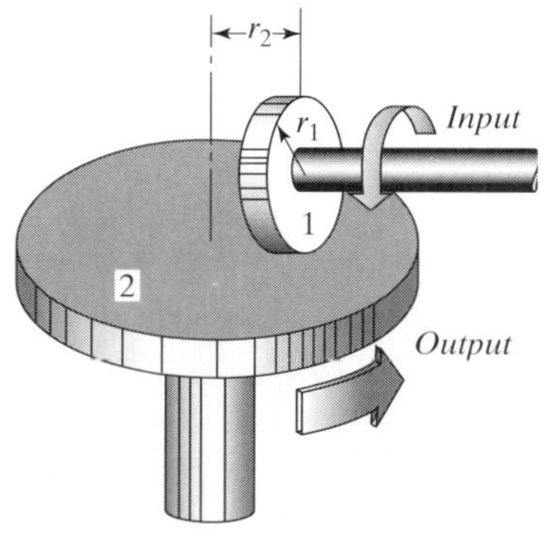

FIGURE 8.31 Wheel-and-disk drive. The output-to-input speed ratio is $n_2/n_1 = r_1/r_2$, where r_2 is measured from the centerline of the disk to the point of contact. The input and output shafts are kept perpendicular and in the same plane.

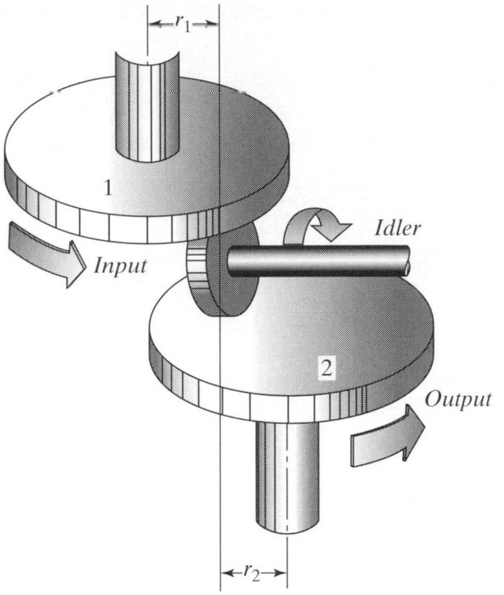

FIGURE 8.32 Friction disk drive with idler wheel. As in the Figure 8.31, the output-to-input speed is given by $n_2/n_1 = r_1/r_2$, where both radii are measured from the respective disk centerlines to the points of contact. The input and output shafts are parallel, and the axes of all three shafts lie in the same vertical plane. The idler shaft is perpendicular to the other two and is moved axially to change speed.

Variable-Speed Cone-Roller Disk Drives

An alternative speed changer design uses one or more *cone rollers*, shifted by an adjusting screw when the unit is running (Figure 8.33a). Cam surfaces on the split output shaft respond to high output torque by increasing the normal force between the roller and disks to ensure adequate driving traction. When the load decreases, the normal force is decreased so that wear is not excessive. A disk drive with four cone rollers is diagrammed in Figure 8.33b.

Variable-Speed Ball-and-Disk Drives

Friction drives require relatively high normal forces wherever the driver and driven parts come into contact. Optimum conditions for shifting, however, include *low* normal forces for drives of the type shown in Figures 8.31 and 8.32, neither of which would be satisfactory if changes in speed were frequent or continuous.

The ball-and-disk drive of Figure 8.34, however, does not suffer extensive wear when the speed is continuously varied, because the two balls in the cage *roll* on one another and on the input disk and output shaft when the balls are shifted. (Note that a single ball could not roll on both surfaces as radius r_1 was changed.) As with the previously discussed drives, the equation

$$\frac{n_2}{n_1} = \frac{r_1}{r_2}$$

can also be used to find the output-to-input speed ratio for the ball-and-disk drive. When the ball cage is shifted axially to the center of the input disk, there is no output motion, but there is still no significant wear on the drive members. Shifting farther to the left results in an output in the opposite direction from that shown in the figure.

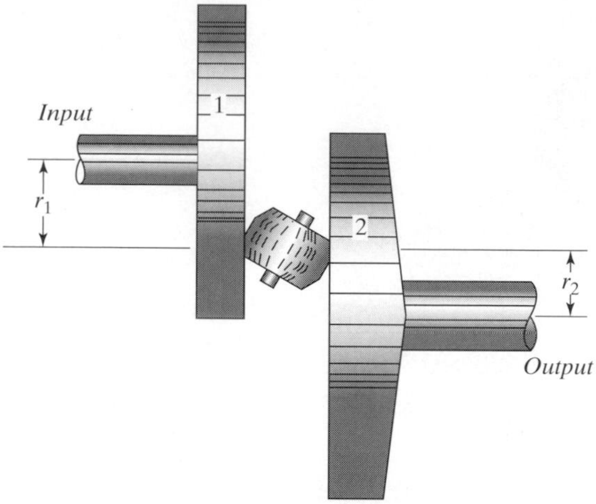

(a)

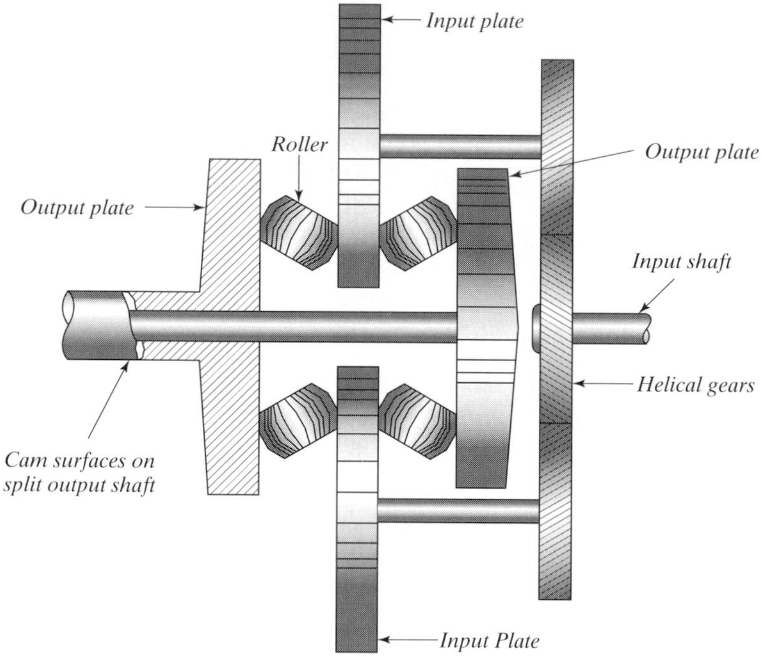

(b)

FIGURE 8.33 (a) Friction disk drive with cone roller. The cone roller is shifted by means of an adjusting screw to change the radius of the rolling paths on disks 1 and 2. As with the other disk drives, the output-to-input speed ratio is given by $n_2/n_1 = r_1/r_2$. (b) Disk drive with four cone rollers. With the use of four adjustable rollers instead of one, the maximum power capacity is increased. (*Source:* Michigan Tool Co.)

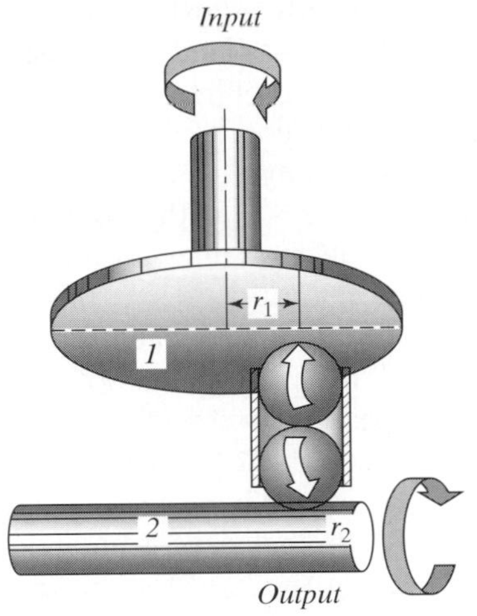

FIGURE 8.34 Ball-and-disk drive. The output is zero when the ball cage is at the disk center. The direction reverses when the ball cage is moved past (to the left of) the disk center.

Wheel-and-disk and ball-and-disk drives may be used as multipliers. From the familiar speed ratio equation, we have

$$\theta_2 = \left(\frac{r_1}{r_2}\right)\theta_1,$$

where θ_1 and θ_2 are the input and output angular displacement, respectively (measured in degrees, radians, or revolutions, with both θ_1 and θ_2 in the same units). To multiply two numbers by using the ball-and-disk drive, we adjust r_1, the ball cage position, so that one of the numbers is represented by r_1/r_2, where r_2 is the radius of the output shaft. The driver is then given a rotation θ_1 representing the other number, and the product is given by θ_2, the rotation of the output shaft.

When r_1 is varied as a function of θ_1, θ_2 is the integral of r_1/r_2 with respect to θ_1. Thus, the ball-and-disk drive may be used as an integrator. (See Figure 8.35.) For

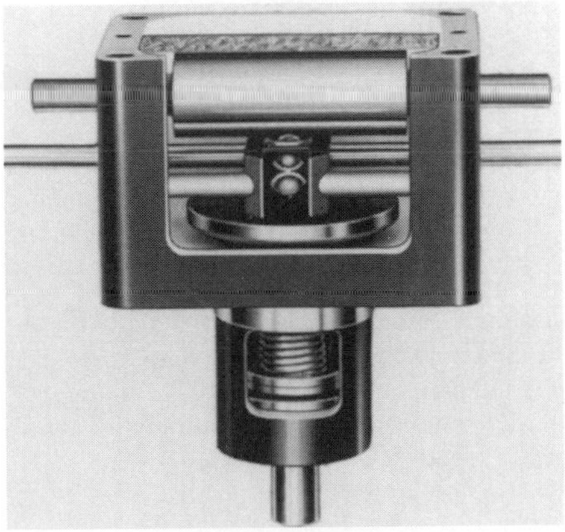

FIGURE 8.35 Cutaway view of a ball-and-disk integrator. (*Source:* The Singer Company, Librascope Division.)

example, let the ball cage position be automatically adjusted so that r_1/r_2 is proportional to the flow of liquid through a meter. Disk 1 is turned at a constant rate, and the rotation of shaft 2 is proportional to the total amount of liquid that passed through the meter during the measuring period. Measurements could, of course, be scaled to keep both input and output values within a reasonable range.

Now, consider the friction disk drive of Figure 8.32 with two offset disks, but with the idler wheel replaced instead by a single ball restrained in a cage. The variable-speed, ball-and-disk drive sketched in Figure 8.36 operates on this principle, except that, instead of a single ball, a cage of several steel balls rotates about its own center between the steel input and output disks. Each ball contacts both input and output disks at all times, permitting greater tractive forces between the two offset disks.

As far as input and output speeds are concerned, the rotating ball cage is kinematically equivalent to a single ball at the cage axis. When the ball cage axis coincides with the input axis, the output speed is zero. As the ball cage is moved toward the output shaft axis, the output speed increases until, at the limit for this design, the output speed is 1.5 times the input speed. Speed ratios are changed when the unit is running. When the recommended torque is not exceeded, there is pure rolling without sliding between the balls and disks, except during speed changes. In general, the ratio of the output to the input speed is given by $n_2/n_1 = r_1/r_2$, where r_1 and r_2 are defined in Figure 8.36.

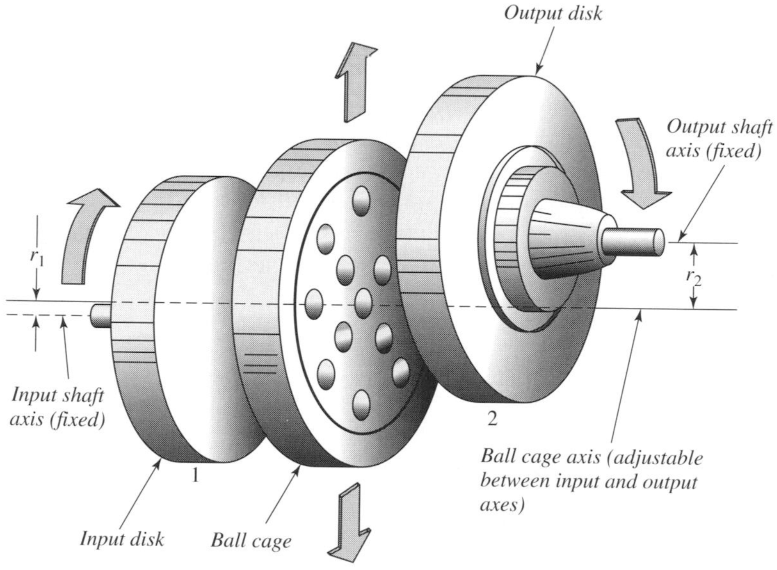

FIGURE 8.36 The Graham *ball-and-disk* drive employs a number of balls held in a rotating cage. The balls contact both the input and output disks. The output-to-input ratio is again given by $n_2/n_1 = r_1/r_2$ and may vary from 0 to 1.5. The output speed is increased as the ball cage centerline is moved toward the output disk axis. (*Source:* Graham Company.)

The angular velocity of the cage is half the sum of the input and output speeds:

$$n_{cage} = \frac{n_1}{2}\left(\frac{r_1}{r_2} + 1\right).$$

Variable-Speed Axle-Mounted Ball-and-Disk Drives

Another compact metal-to-metal friction drive employs axle-mounted balls that transmit power between beveled disks, as shown in Figure 8.37. The ball axles are tilted uniformly by a slotted iris plate (acting as a set of cams) when it is desired to change speed ratios. (See Figure 8.38a.) In analyzing this drive kinematically, we need consider only one ball and observe effective radii of the driving and driven paths. When the ball axle is parallel to the input and output shafts, as shown in the figure, the distances r_1 and r_2 (from the ball axle to the points of contact of the ball with the beveled input and output disks) are equal. As the ball axle tilts either up or down (Figures 8.38b and c), the radii r_1 and r_2 change simultaneously. Using the relationship for the output-to-input speed ratio, viz.,

$$\frac{n_o}{n_i} = \frac{\text{product of effective driving member radii}}{\text{product of effective driven member radii}},$$

we obtain

$$\frac{n_o}{n_i} = \frac{r_i r_2}{r_1 r_o} = \frac{r_2}{r_1}$$

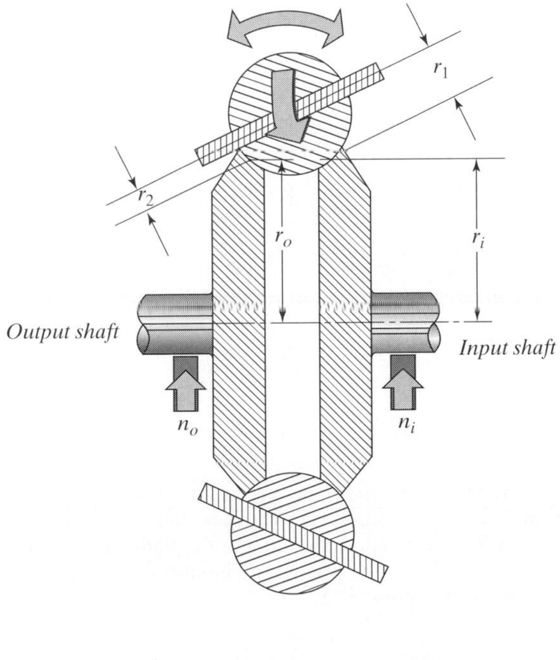

FIGURE 8.37 Axle-mounted ball-and-disk drive. Since the output and input disk radii are equal ($r_o = r_i$) the output-to-input speed is given by $n_o/n_i = r_2/r_1$. A change in the angle of the ball axle simultaneously changes r_1 and r_2, the distances from the ball axle to the point of contact of the ball surface with the beveled faces of the input and output disks, respectively. (*Source:* Cleveland Gear Company, subsidiary of Vesper Corporation.)

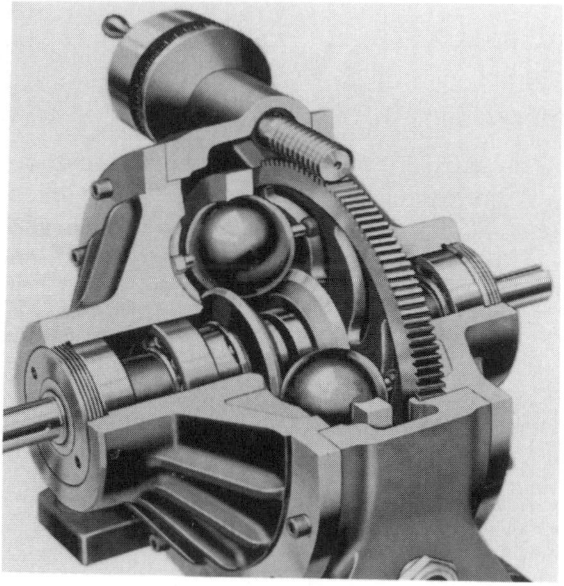

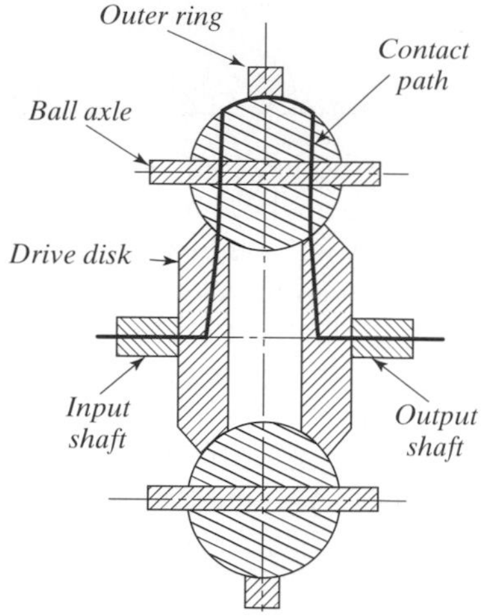

1:1 *Ratio*

(a)

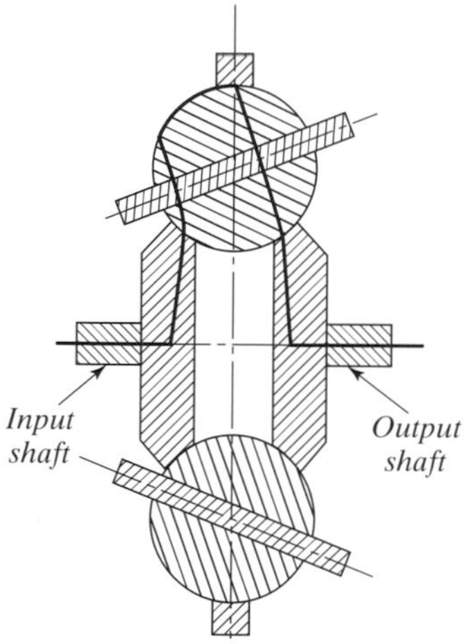

1:3 *Increase*

(b)

FIGURE 8.38 (a) With the ball axles in the position shown, the input speed equals the output speed. (*Source:* Cleveland Gear Company, subsidiary of Vesper Corporation.) (b) The precision speed adjustment device (worm drive) at the top of the transmission is manually adjusted. The iris plate tilts the ball axles, simultaneously changing r_1 and r_2 as shown in the insert. The effect is a 1:3 input-to-output ratio. (*Source:* Cleveland Gear Company, subsidiary of Vesper Corporation.)

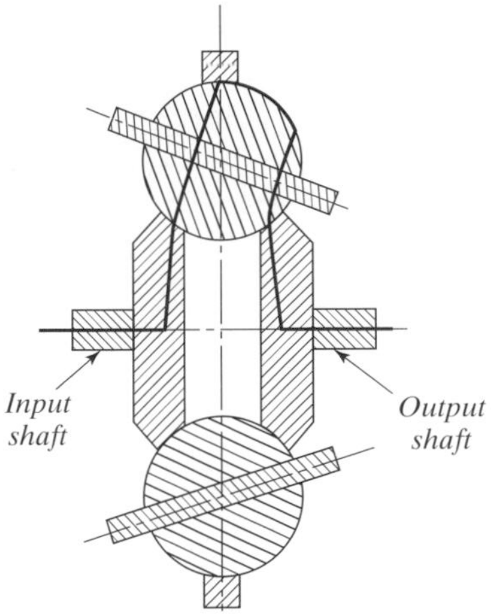

3:1 *Decrease*

(c)

FIGURE 8.38 (c) With the axles tilted to the position shown, the input-to-output ratio becomes 3:1. For automatic process control or remote speed adjustment, the handwheel drive may be replaced by an electric or a pneumatic actuator. The ring that encircles the balls balances radial forces, but runs freely and does not affect the speed ratio. (*Source:* Cleveland Gear Company, subsidiary of Vesper Corporation.)

after noting that the effective radii of the input and output disks are equal. Reversing is not possible with this mechanism alone, but the speed ratio may be varied from $\frac{1}{3}$ to 3 while the unit is running.

Variable-Speed Roller–Torus Drives

A variation of the axle-mounted ball-and-disk drive is the roller–torus drive shown in Figure 8.39a. This drive employs two tiltable rollers within a split torus. The roller axes are shifted equally by gearing, which includes a worm drive to provide fine speed adjustment. (See Figure 8.39b.) The rollers act as idlers between the input and output tori. The output-to-input speed ratio is given by

$$\frac{n_2}{n_1} = -\frac{r_1}{r_2},$$

where r_1 and r_2 are measured from the driveshaft centerline to points of contact between the roller and the input torus and output torus, respectively. The minus sign indicates that the direction of rotation of the output is opposite that of the input. The roller–torus drive by itself cannot have a zero output and cannot be shifted to change the direction of the output.

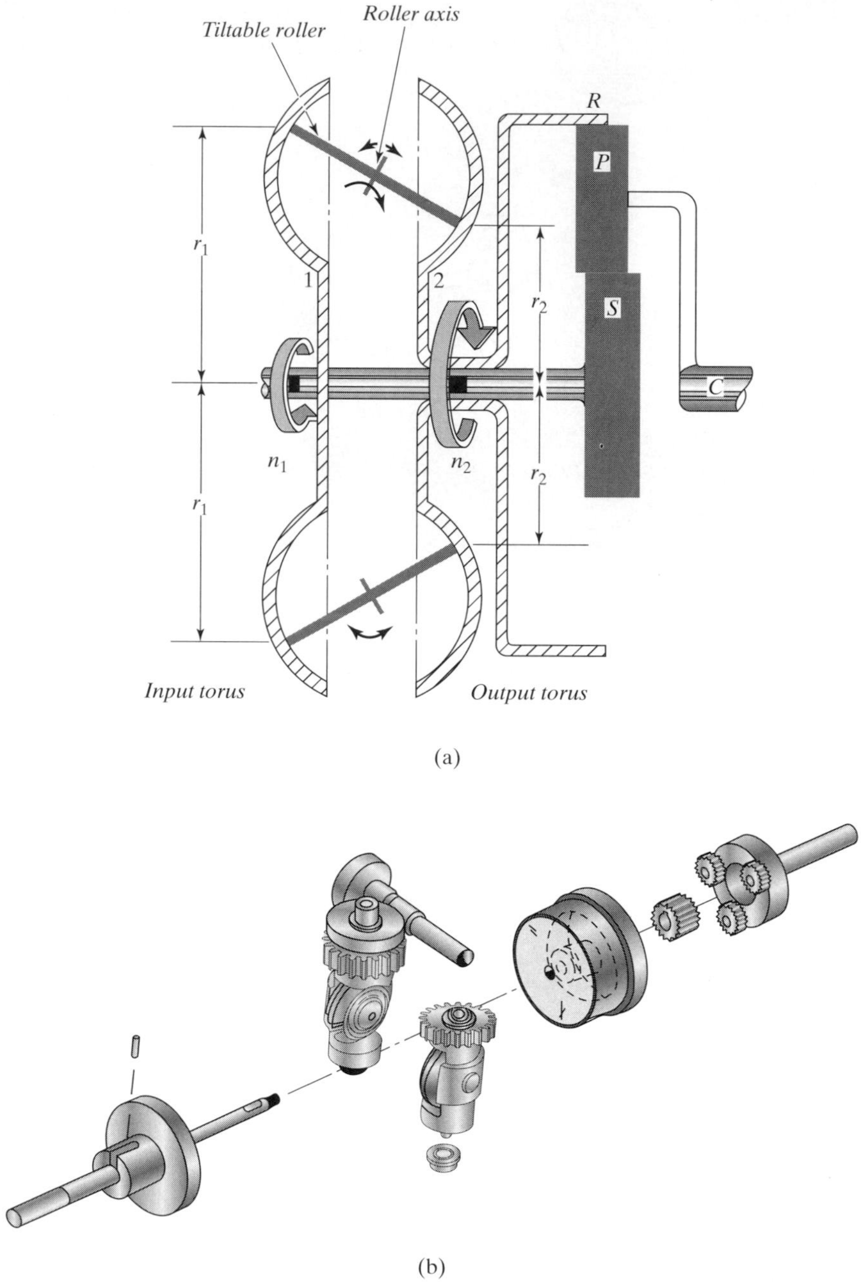

FIGURE 8.39 (a) Schematic of roller–torus drive. The roller axes are shifted to change r_1 and r_2. The input torus drives the sun gear S; the output torus drives the ring R. (*Source:* Metron Instruments, Inc.) (b) Roller–torus drive, shown combined with a planetary gear transmission in a low-power application. (*Source:* Metron Instruments, Inc.)

Friction drives and gearing are often combined to provide greater flexibility. Consider a simple planetary gear train with no fixed gears. From our earlier discussion of planetary trains with more than one input, the planet carrier speed is given by Eq. (8.16):

$$n_C = \frac{n_R + n_S(N_S/N_R)}{1 + (N_S/N_R)}.$$

Here, n represents speed in revolutions per minute (or any other unit, provided that we are consistent), N represents numbers of teeth, and subscripts C, R, and S refer to the carrier, ring, and sun, respectively. Let the output torus of a roller–torus drive be directly connected to the ring gear of the planetary train, and let the input torus shaft pass through the output torus to drive the sun gear. Then the planet carrier may be used as output of the combined transmission, so that a zero speed and a reverse are possible. Recall that, for all roller positions, the output torus rotates in a direction opposite that of the input torus. Referring to the roller–torus equation $n_2/n_1 = -r_1/r_2$ and to the previous equation for n_C, we have the ratio of the output to the input speed for the combined transmission:

$$\frac{n_C}{n_1} = \frac{-(r_1/r_2) + (N_S/N_R)}{1 + (N_S/N_R)}.$$

By rotating the rollers until r_1/r_2 is less than the fraction N_S/N_R, we obtain output and input in the same direction (with a considerable reduction in speed). The roller adjustment at which $r_1/r_2 = N_S/N_R$ results in a zero output speed. Most of the range of the transmission is obtained with r_1/r_2 greater than N_S/N_R, resulting in an output direction opposite that of the input.

Planetary Traction Drives

The analysis of planetary traction drives differs little from the analysis of planetary gear trains. The same equations are applicable, except that ratios of gear teeth are replaced by ratios of rolling path radii.

Planetary Cone Transmission

One variable-speed planetary friction drive employs *cone* planets that make contact with a variable-position reaction ring (Figure 8.40a). The drive motor may be directly connected to the planet carrier. A set of planet pinions (integral with the planet cones) drives a ring gear, which serves as the output. At most input speeds, inertial forces hold the cones against the friction ring, but if the input speed is low, auxiliary springs are used to ensure contact. For a change of speed ratio, the reaction ring is shifted axially (Figure 8.40b) to contact the planet cones at a different location (changing the effective planet radius r_{P_1}). Speeds may be changed when the unit is operating if the cone planets are spring loaded. The change in speed is stepless.

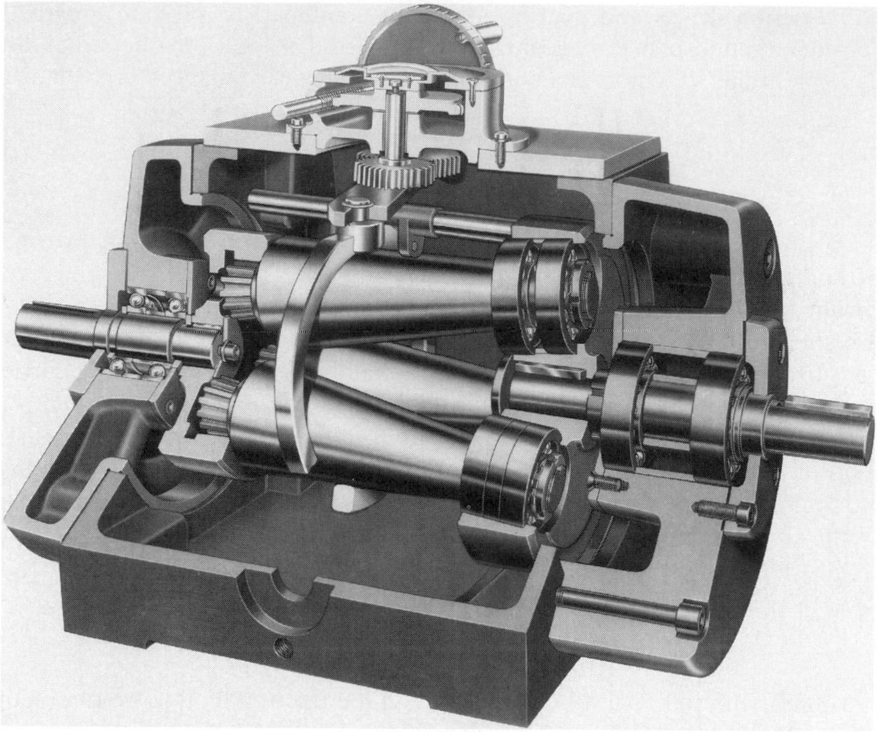

(a)

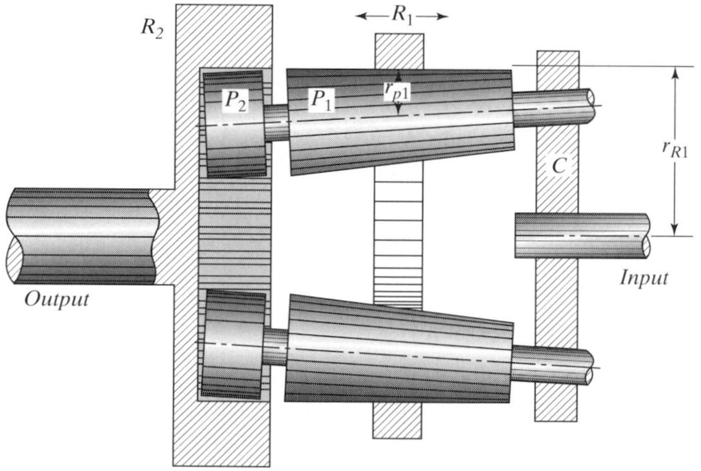

(b)

FIGURE 8.40 (a) Cutaway of planetary cone transmission, showing speed control mechanism. The mechanism at the top of the transmission permits fine adjustment of the reaction ring position and a wide range of speed ratios through zero and reverse speeds. Most models of this transmission depend on centrifugal force to hold the cone rollers against the ring. This model, however, is designed with a spring-loading assembly in the center to ensure traction between the cones and the ring, even at low speeds. (*Source:* Graham Company.) (b) Schematic of the planetary cone transmission.

TABLE 8.11 Speed Ratios for a Planetary Cone Transmission with the Reaction Ring R_1 Fixed (See Figure 8.40b)

Component	C	R_1	P_1	P_2	R_2
Step 1: rotations with train locked	+1	+1	+1	+1	+1
Step 2: rotations with carrier fixed [R_1 given (−1) rotations]	0	−1	$-\dfrac{r_{R_1}}{r_{P_1}}$	$-\dfrac{r_{R_1}}{r_{P_1}}$	$-\left(\dfrac{r_{R_1}}{r_{P_1}}\right)\left(\dfrac{N_{P_2}}{N_{R_2}}\right)$
Total rotations (R_1 fixed)	+1	0			$1-\dfrac{r_{R_1}N_{P_2}}{r_{P_1}N_{R_2}}$

Superposition will be used to solve for relative speeds. In the first step of Table 8.11, the entire transmission is given one positive rotation. In step 2, the reaction ring rotation is corrected by giving it one negative rotation with the carrier fixed. The effect is calculated for the cone planet and planet pinion (which turn as a unit) and for the ring gear. Adding steps 1 and 2 of the table, we have the required rotation of the reaction ring (zero) and one rotation of the planet carrier. The output-to-input speed ratio is given by the total number of rotations of the ring gear divided by the carrier rotations:

$$\frac{n_{R_2}}{n_C} = 1 - \frac{r_{R_1}N_{P_2}}{r_{P_1}N_{R_2}}. \tag{8.20}$$

In one model, when the reaction ring (also called the control ring) is adjusted to approximately the middle of its range, the ratio of the reaction ring radius to the cone radius is equal to the ratio of the ring gear teeth to the pinion teeth (N_{R_2}/N_{P_2}). Then, from Eq. (8.20), the output speed is zero. As we move the reaction ring to the left, r_{P_1} increases and the output speed is in the same direction as the input speed until, at its maximum value, $n_{R_2}/n_C = \frac{1}{5}$. Moving the control ring to the right of the zero output position gives us an output rotation opposite that of the input until we reach a limiting value of $n_{R_2}/n_C = -\frac{1}{5}$.

Impulse Drives and Flexible-Spline Drives

A gear-and-linkage impulse drive driven by an eccentric is shown in Figure 8.41a. This particular drive is no longer commercially available. The sketch in Figure 8.41b is incomplete, depicting only one of the three pinion-and-linkage assemblies of the unit, but we will first examine the part of the drive shown before considering the entire mechanism.

The eccentric drives a roller follower pinned to link 1. The follower in turn drives links 2 and 3. Link 3 turns a pinion counterclockwise through a one-way clutch, and the pinion drives the output gear. The cam and follower will cause link 3 to oscillate through the same angle whether the input cam turns clockwise or counterclockwise. Thus, the orientation of the one-way clutch determines the directions the pinion and output gears will turn. In this case, the output is clockwise regardless of whether the input is clockwise or counterclockwise.

When links 2 and 3 are nearly perpendicular, the oscillation of link 3 is greatest and the output speed is maximal. The pivot point of link 1 may be moved to make the angle between links 2 and 3 quite small, so that link 3 oscillates through a small angle and the output speed is reduced.

The output shaft would be stationary for about half of each input shaft revolution if the drive had only one pinion-and-linkage assembly. The actual drive (Figure 8.41a), with three such assemblies, turns continuously, each linkage driving one-third of the time. For a given linkage, during two-thirds of each input cycle, link 3 either is rotating in a direction opposite that of the pinion or is rotating more slowly than the pinion. During that time, the pinion is driven by one of the other linkages.

The instantaneous output speed is found by sketching the cam and follower (or its equivalent mechanism) and the linkage that is driving at a given instant. The procedure is illustrated in Figure 8.41c. Since the roller track, which acts as a cam, and the

(a)

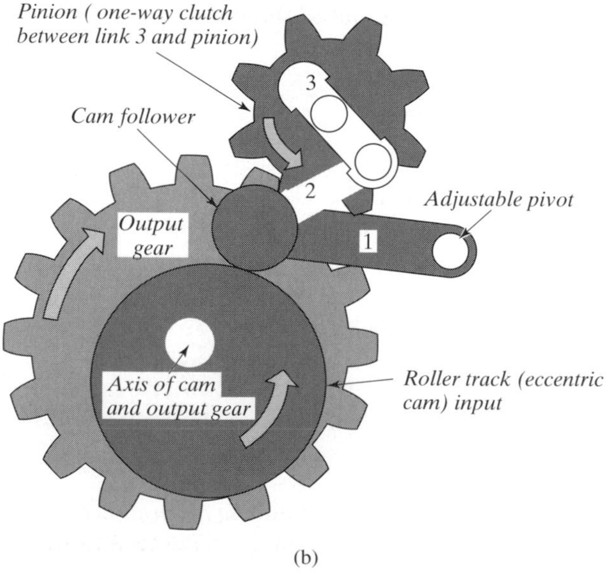

(b)

FIGURE 8.41 (a) Cutaway view of the gear-and-linkage impulse drive. (*Source:* Morse Chain Division, Borg Warner.) (b) Schematic of the gear-and-linkage impulse drive (showing only one of the three pinion-and-linkage assemblies).

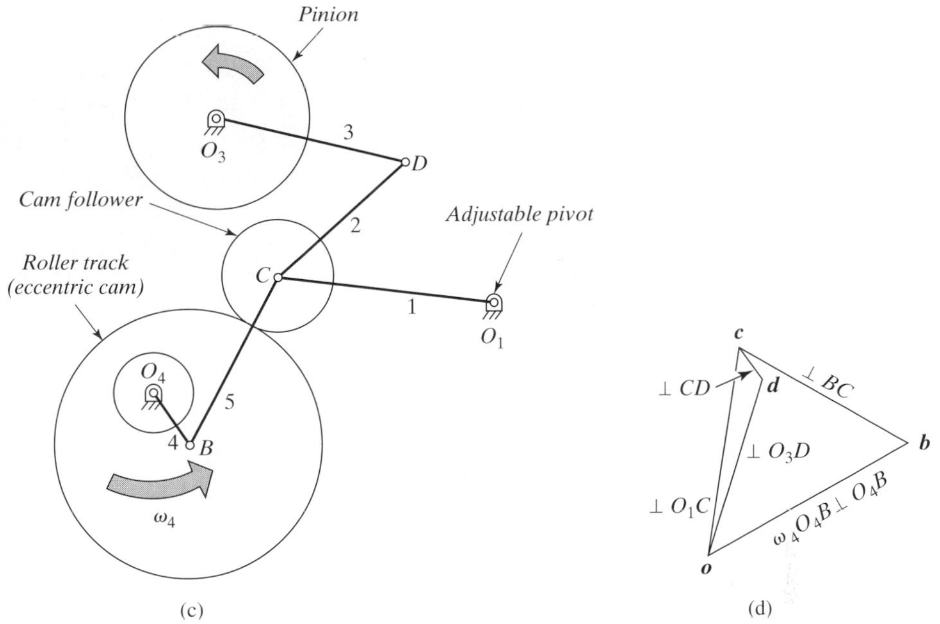

FIGURE 8.41 (c) Equivalent-linkage drawing for the pinion-and-linkage assembly, showing the principle of operation of the mechanism. (d) The velocity polygon for the pinion-and-linkage assembly.

cam follower are circular, the distance from roller track center B to cam follower center C is fixed and will be considered rigid link 5. The axis of rotation of the roller track is labeled O_4, and O_4B is designated as link 4. For a specified position of adjustable pivot O_1, we form a four-bar linkage with links 4, 5, and 1 and the frame. Given ω_4, the angular velocity of the roller track, we may draw velocity polygon ***obc*** by using the methods of Chapter 3. (See Figure 8.41d.) We then note that links 1, 2, and 3 and the frame also form a four-bar linkage, permitting us to complete the polygon with velocity point ***d***. The linkage shown is obviously driving the transmission at this instant (through a one-way clutch whereby link 3 drives the pinion counterclockwise). Thus, the pinion angular velocity is given by $\omega_3 = od/O_3D$.

The angular velocity of the output gear is related to the angular velocity of link 3 by the negative ratio of the number of pinion teeth to the number of output gear teeth. The same roller track drives the three identical linkages; thus, the output motion is repeated every 120° of cam rotation.

If it is not obvious which of two linkages is driving the output gear at a particular instant, the problem is solved for both linkages: The linkage giving the highest speed is the actual driver. It is seen that the output is pulsating, making a drive of this type practical only for relatively low speeds.

Flexible-Spline Drives

The *flexible-spline drive* of Figure 8.42a operates on a unique principle, employing a wave generator to deflect a flexible external spline. The flexible spline meshes with a slightly

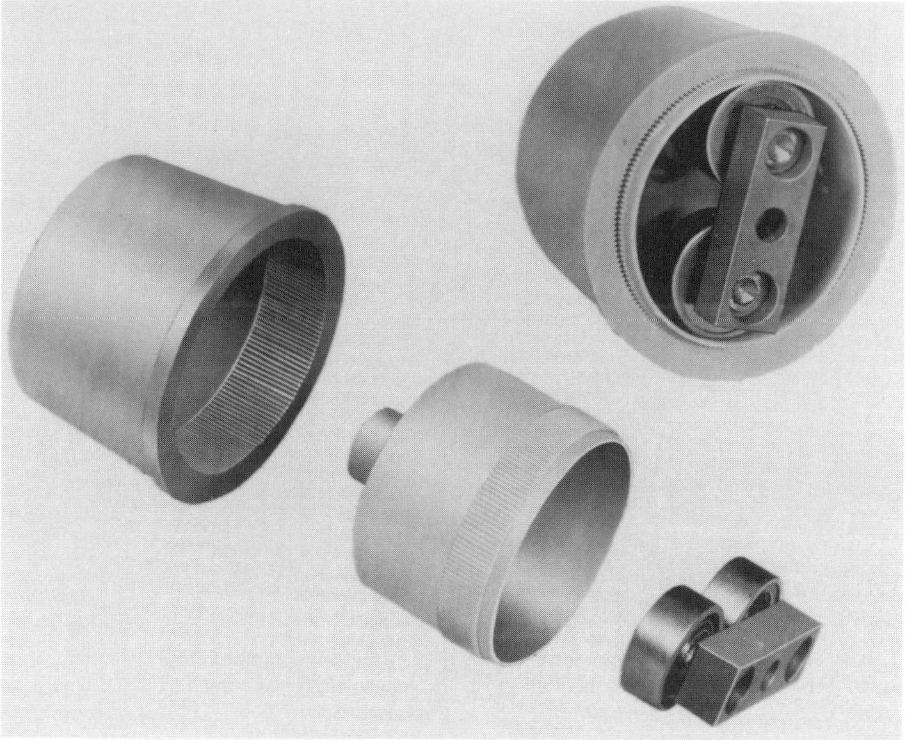

(a)

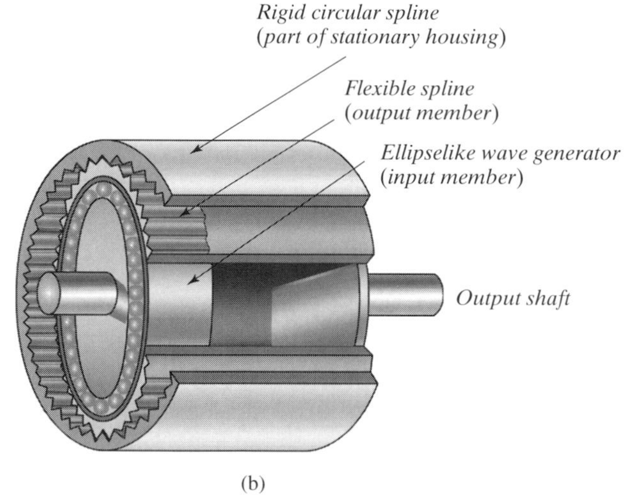

(b)

FIGURE 8.42 (a) Components of the flexible-spline drive are a housing with an integral 162-tooth circular spline (fixed), a plastic 160-tooth flexible spline (the output member), and a two-lobe ball-bearing wave generator (mounted on the output shaft). Other flexible spline drives are available for heavy-duty and high-precision applications. A combination of two rollers or an ellipselike cam is used as the wave generator for drives of this type. (*Source:* Harmonic Drive Division, USM Corporation.) (b) The wave generator makes the flexible spline "walk" around the inside of the stationary circular spline.

larger, rigid, stationary member having an internal circular spline. The major components are shown in exploded view in the figure. The wave generator, usually an ellipselike cam, is ordinarily the input to the drive. The flexible spline serves as the output, rotating in a direction opposite that of the input. Both splines have the same circular pitch, but the stationary spline typically has two or four more teeth than the flexible spline.

The wave generator causes the flexible spline to "walk" around the inside of the stationary spline. Since the splines mesh tooth for tooth, when the wave generator turns one rotation clockwise, the flexible spline will advance slightly counterclockwise. The difference in pitch circumference between the splines results in a tangential motion with a magnitude of two or four circular pitches. In general, then, there are

$$\frac{N_f - N_s}{N_f}$$

rotations of the flexible spline (output) for each wave generator rotation (input), where N_f is the number of flexible-spline teeth and N_s is the number of stationary internal-spline teeth. (See Figure 8.42b.)

The speed ratio can also be obtained by employing the superposition method used previously for planetary drives. In Table 8.12, the locked train is given one rotation for step 1. In step 2, the rigid, internal spline is given one negative rotation with the wave generator fixed. The two steps are then added as shown in the table.

The wave generator of the flexible-spline drive has two or more equally spaced lobes or rollers; the two splines mate where the lobes or rollers press the flexible-spline unit against the rigid splines. The difference in tooth numbers, $N_s - N_f$, is a whole-number multiple of the number of lobes. An output-to-input speed ratio of $-\frac{1}{80}$, for example, is obtained by using a two-roller or two-lobe wave generator input and a flexible spline of 160 teeth meshing with a 162-tooth, fixed internal spline. If the output and input directions are to be the same, the wave generator is again used as the input member, but the output is taken from the rigid circular spline with the flexible spline fixed. Using the data from the preceding example, we find that this condition would yield an output-to-input ratio of

$$\frac{N_s - N_f}{N_s} = +\frac{1}{81}.$$

TABLE 8.12 Speed Ratios for the Flexible-Spline Drive (See Figure 8.42a)

Component	Flexible spline	Circular spline	Wave generator
Step 1: rotations with train locked	+1	+1	+1
Step 2: rotations with wave generator fixed	$-\frac{N_s}{N_f}$	−1	0
Total number of rotations	$1 - \frac{N_s}{N_f}$	0	+1

Speed can be reduced to very low values with correspondingly great increases in output torque with compact one- or two-stage flexible-spline drives. For this reason, such drives have been used in aircraft and space applications, including vertical-takeoff aircraft, short-takeoff-and-landing aircraft, and a meteor-detection satellite. In this last application, a two-stage, 3-by-4-in flexible-spline drive provides a 1-to-3840 reduction in speed to enable a secondary drive to extend 15-by-48-ft detector "wings" (while the satellite is orbiting).

Selection of Drive Train Components

A given set of design requirements might be satisfied by more than one of the gear train configurations, packaged speed reducers, or drive train components considered in this chapter. A few of the most important design objectives will be discussed as an aid to making an optimum selection.

Shaft Geometry Considerations. One of the first requirements might involve the *arrangement* of the input and output shafts. If the shaft centerlines intersect as in an outboard motor drive, bevel gears might be used. For nonintersecting perpendicular shafts, as in a rear-wheel-drive automobile, for example, hypoid gears are used. Power may also be transmitted between nonintersecting perpendicular shafts by a worm and worm gear if a considerable reduction in speed is desired. If the shaft centerlines do not intersect and are neither parallel nor perpendicular, crossed helical gears, bead chain, and some types of belt drives are used. When the design requires parallel or collinear input and output axes, packaged speed reducers, belt-and-chain drives, and spur and parallel helical gear trains may be considered.

Speed Ratio Considerations. If speed must be reduced considerably, worm and worm gear drives and some compound planetary drives are used. Rapid *speed ratio changing* (in discrete steps) suggests the use of clutches engaging and disengaging spur gears or helical gears on parallel shafts. An alternative method employs a planetary train with band brakes and clutches to lock certain elements.

When speed must be continuously variable (i.e., if stepless variation is required), variable-speed belt drives may be employed. For low-power applications (e.g., instrument-type applications), disk and wheel-disk drives are used. Some disk, roller-and-disk, and ball-and-disk drives are variable through zero output speed to negative speed ratios.

Control Considerations. The method of *speed control* is an important factor in selecting a transmission. Spur and helical gear trains are ordinarily shifted manually through mechanical linkages, but changes in speed may be actuated or assisted pneumatically, hydraulically, or electrically. The bands and clutches in most planetary trains are automatically actuated by speed-sensitive and load-sensitive controls.

Load (Power) Considerations. Bead chains and plastic belts with lugs are limited to very low-torque applications (chart drives, etc.). *High-torque or -power requirements* preclude the selection of these drives and most traction drives for many industrial applications. Drives available for high horsepower include chains, belts, and gears.

The most severe power transmission and torque requirements are met by various forms of gearing. A single mesh using high-speed double-helical gears has been employed for loads as high as 22,000 to 30,000 kW (30,000 to 40,000 hp). Two or more input pinions are used (as in ship propulsion units) for drives with higher horsepower.

Precision-Drive Considerations

A requirement for *precision* in the transfer of motion is usually met by gear or power screw drives. Most high-precision instrument drives employ spur gear trains, but for high-velocity applications, helical gears are used. When a considerable reduction in speed is required, a single precision worm and worm gear pair may be used.

Sources of positional errors in gear drives include errors in tooth form and location; gear and shaft eccentricity, misalignment, and deflection under load; thermal expansion and contraction; and looseness and deflection in couplings, bearings, and other train components. When the output of a train must accurately reflect the input motion, the designer may use one or more of the following methods to control error:

1. Select precision components.
2. Use as few pairs of gears as possible to reduce integrated errors.
3. Use larger diameter or shorter shafting and rigid supports to limit deflection under a load.
4. Replace alternate gears in the train with split antibacklash gears. The split gear is spring loaded so that the halves rotate slightly with respect to one another and eliminate play.
5. Attach one gear to a floating shaft that moves slightly to eliminate play between meshing teeth.

In multistage speed reducers, angular error from each stage is reduced by the succeeding stages. If the final (output) stage has a high reduction ratio, precision components may be required in that stage only.

Noise, Shock, and Vibration Considerations

Noise, shock, and vibration effects are usually reduced by nonmetallic drives. Belt drives absorb vibration and shock, rather than transmitting these effects between the driver and driven shaft. For low loads, nylon or fiber gears are used to reduce noise.

Noise and vibration result as the load is transferred from one pair of gear teeth to another. The load transfer in helical gears on parallel shafts is smoother, and more pairs of teeth are in contact at a given time, than for similar spur gears. For this reason, helical gears are often chosen for high-power, high-speed drives.

Efficiency Considerations

Efficiency is an important factor in continuously operated drive trains. Over a period of years, losses of a few percent of transmitted power can be expensive. Low efficiency also may exact additional penalties, since friction losses represent (1) *heat*, which must be dissipated, and (2) additional fuel *weight* to be carried (a problem in surface vehicle design as well as in that of aircraft and space vehicles).

Spur gears, helical gears on parallel shafts, and bevel gears have efficiencies of 98 to 99 percent at rated power (for each mesh). If lubrication is poor or if the drive operates at lower-than-rated power, the efficiency will be lower. Efficiency is considerably reduced in pairs of gears with a high sliding velocity; with high-reduction worm drives, losses may exceed 50 percent of the power transmitted. Belt drive efficiencies are typically about 95 percent for single V belts. Traction drives have efficiencies from 65 to 90 percent (lower at very low output speeds).

Miscellaneous Considerations

Some other major selection criteria are *size, weight, reliability, and cost*. On a cost-per-transmitted-horsepower basis, belt drives usually have the advantage. If speed is to be reduced by a large factor, a worm and worm-wheel drive may be selected on the basis of its compactness and low cost, compared with more complicated drive trains. Reliability is an important characteristic of gear trains, particularly spur gears and helical gears on parallel shafts. Planetary gear trains offer a compact package, taking up little space in an axial direction, even for high-power drives.

8.6 FORCES, TORQUES, AND TRANSMITTED POWER IN GEAR TRAINS

The torque T on a gear is given by the tangential force F_t multiplied by the pitch radius r of the gear. If a gear makes contact with two or more other gears (e.g., as in a planetary train), each contact contributes to the torque. The net torque on an idler gear due to gear tooth contacts is zero.

The radial force on a gear is given by

$$F_r = F_t \tan \phi, \tag{8.23}$$

where ϕ is the pressure angle. The radial force is directed toward the center of a gear, except for a ring gear, in which case the radial force is directed outward. Axial forces, as well as radial and tangential forces, occur in helical gears (including worms and worm gears), bevel gears, and hypoid gears.

One of the reasons for selecting balanced planetary trains over other drive trains is that the former eliminate shaft bending loads due to unbalanced radial and tangential forces on the sun and ring gears. Another advantage is the reduction in tooth loads for a given torque capacity due to the presence of two or more planet gears.

Noting that 1 W = 1 N · m/s, we observe that the transmitted power is given by

$$P(\text{kW}) = T(\text{N} \cdot \text{mm})\omega(\text{rad/s}) \times 10^{-6}. \tag{8.24}$$

The equivalent expression in customary U.S. units is

$$P(\text{hp}) = \frac{T(\text{in} \cdot \text{lb})n(\text{rad/min})}{63{,}025}. \tag{8.25}$$

SAMPLE PROBLEM 8.4

Forces, Torques, and Transmitted Power in a Simple Planetary Train

Figure 8.43 shows a balanced planetary train. Spur gears of module $2(d/N_t = 2\text{ mm})$ and $\phi = 20°$ are used. The sun gear is the input gear and the carrier is the output. The ring gear is fixed. We have

$$N_S = 36, \quad N_P = 32, \quad \omega_S = 400 \text{ rad/s, and } P = 14.4 \text{ kW}.$$

Determine the angular velocities. Tabulate the results. Determine all forces and torques. Draw free-body diagrams. Do the external torques balance? Does the input power equal the output power? Do the forces balance?

Solution. For this planetary train,

$$d_R = d_S + 2d_P,$$

from which

$$N_R = N_S + 2N_P = 36 + 2 \times 32 = 100 \text{ teeth}.$$

For a module $m = 2$ mm, the pitch radius of each gear is given by

$$r_P = \frac{1}{2}mN_t = \frac{1}{2} \times 2 \times N_t.$$

The radius of the carrier is

$$r_C = r_S + r_P = 36 + 32 = 68 \text{ mm}.$$

Rotation speeds are obtained by the superposition method, as shown in Table 8.13.

Free-body diagrams are the key to the remainder of the problem. The torque on the sun gear shaft is

$$T_S = \frac{10^6 P}{\omega} = \frac{10^6 \times 14.4 \text{ kW}}{400 \text{ rad/s}} = 36{,}000 \text{ N} \cdot \text{mm}.$$

This torque is balanced by tangential forces at each of the four planets:

$$T_S = 4F_t r_S.$$

TABLE 8.13 Speeds, Torques, and Transmitted Power in a Planetary Train

	Equation/ procedure	Gear S (input)	P	R (fixed)	C (output)
N_t	—	36	32	100	—
r_{pitch}(mm)	$r_p = \_mN_t$	36	32	100	$r_c = 68$
1	Lock train	+1	+1	+1	+1
2	Correct R, C fixed	$+\frac{100}{36}$	$-\frac{100}{32}$	−1	0
Sum	1 + 2	+3.7777	−2.125	0	+1
ω (rad/s)	$\frac{400}{3.777} \times$ sum	400 cw	225 ccw	0	105.88 cw
Shaft T (N·mm)	From free body	36,000 cw	0	100,000 cw	136,000
Shaft P (kW)	$10^{-6}T\omega$	14.4	0	0	14.4

From this equation, the tangential force is

$$F_t = \frac{T_S}{4r_S} = \frac{36{,}000}{4 \times 36} = 250 \text{ N},$$

and the radial force is

$$F_r = F_t \tan \phi = 250 \tan 20° = 91 \text{ N},$$

as shown. The 250-N tangential force and 91-N radial force are each applied at four locations on the sun gear, two locations on each of the four planet gears, and four locations on the ring gear, as shown on the free-body diagrams in Figure 8.43. The direction of the tangential forces on the sun and the planet, where they make contact, are opposite. The planet gear in this train acts as an idler. Thus, planet torque $T_P = 0$, enabling us to obtain the magnitude of the tangential force on the planet where that force contacts the ring gear. The tangential force on the ring is equal and opposite.

A reaction torque is required on the ring gear to prevent it from turning. This torque is given by

$$T_R = 4F_t r_R = 4 \times 250 \times 100 = 100{,}000 \text{ N} \cdot \text{mm cw}.$$

The reaction torque may be provided by bolting the ring in place if the transmission is designed for a fixed speed ratio. If speed changing is required, the reaction torque may be provided by a band brake about the outside of the ring.

Note that forces are balanced in the horizontal and vertical directions on the sun and ring gears (and on the planet carrier, as will be seen later). Thus, these elements of the gear train do not cause bending moments on their respective shafts. For the planet shown (in the top position at this instant), the two tangential forces must be balanced by a horizontal force of $2F_t = 2 \times 250 = 500$ N to the left. Hence, each planet causes an equal and opposite force on the planet carrier. Torque equilibrium on the planet carrier requires, that

$$T_C = 4 \times 2F_t r_C = 4 \times 2 \times 250 \times 68 = 136{,}000 \text{ N} \cdot \text{mm ccw}$$

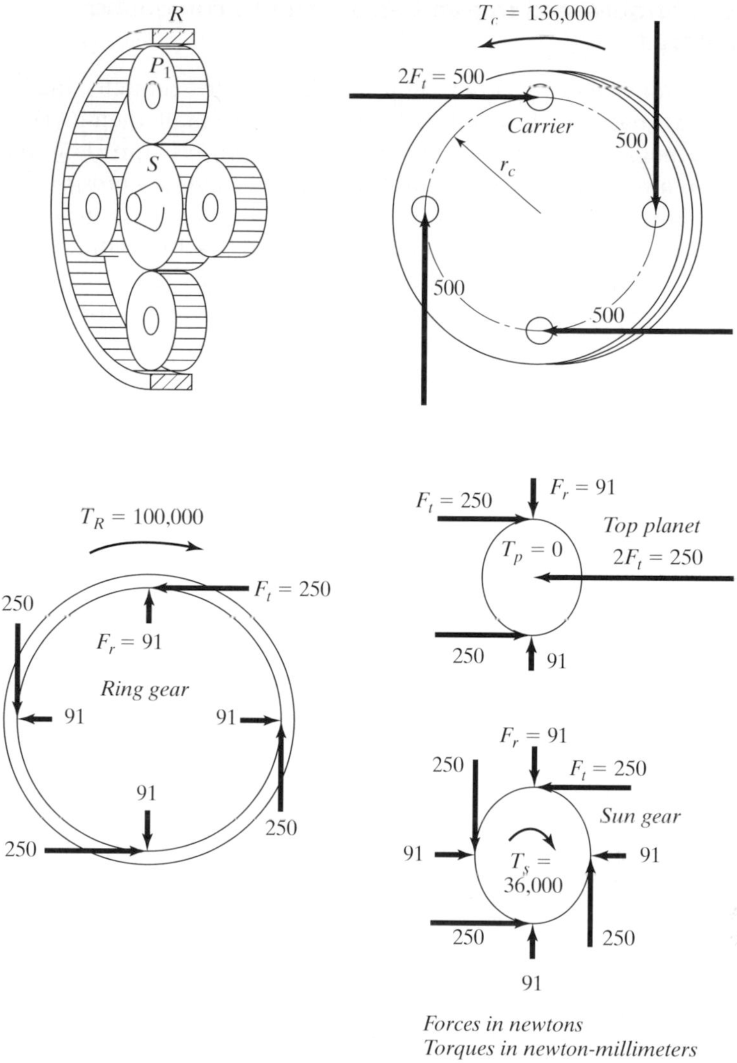

FIGURE 8.43 Forces and torques in a planetary train.

on the free-body diagram of the carrier. Of course, the carrier produces a 136,000 N · mm clockwise torque on the output shaft, which turns clockwise.

External torques on the transmission as a whole are in balance, since

$$T_S + T_C + T_R = 36{,}000 - 136{,}000 + 100{,}000 = 0.$$

Note that friction losses have been neglected. The output power is given by

$$P_C = 10^{-6} T_C \omega_C = 10^{-6} \times 136{,}000 \text{ N} \cdot \text{mm} \times 105.88 \text{ rad/s} = 14.4 \text{ kW}.$$

Forces, torques, and transmitted power in a compound planetary train

Compound planetary trains require a little more thought and care. Be sure to use free-body diagrams. If angular accelerations are small, then inertial forces and torques may be neglected. And one should apply Newton's third law: *When a body exerts a force on a second body, the second exerts an equal and opposite force on the first.* Insert checks into your work; test your results. Do the external torques balance? Remember to include reaction torques on fixed gears. Does the output power equal the input power? The answer is *yes* if friction losses are neglected.

If angular accelerations are significant, then inertial forces and torques cannot be neglected. Reverse-effective forces and torques are added to the free-body diagrams; that is, d'Alembert's principle applies. (Problems employing d'Alembert's principle are solved in Chapter 10.)

SAMPLE PROBLEM 8.5

Design of a Line of Compound Planetary Trains; Determination of Forces and Torques

We want to offer a series of speed reducers with fixed speed ratios. The design specifications call for an input speed of 3000 rpm and output speeds ranging from about 85 rpm to about 600 rpm. Design the reducers and tabulate forces and torques when 2.4 kW of mechanical power is transmitted.

Design decisions. We will use a compound planetary train similar to that shown in Figure 8.44 with three sets of planets. The sun gear drives, and contacts planet 1 of each set. Both planets of each set rotate together. The planet carrier drives the output shaft. Planet 2 of each set contacts the fixed ring gear. All gears are straight spur gears with a 3-mm module and 20° pressure angle. We will select a 19-tooth sun gear; planets 1 will each have 20 or more teeth, and planets 2 will have 18 teeth each.

Solution summary. We will specify 20, 24, 28, ..., 92 teeth for planet 1. For this configuration, the number of teeth in the ring gear equals the sum of the teeth in the sun and planets 1 and 2. The formula method is used to determine speeds.

The input torque is determined from the input power and the angular velocity. Each planet 1 contributes one-third of torque T_S. Tangential forces oppose the torque on the sun gear, as shown in the figure. The tangential force on each planet 1 is equal to and opposite that on the sun. There is no net torque on a planet pair. We use this condition to find the tangential force on planet 2. A planet pair that is (instantaneously) in the top position is shown in the sketch. Tangential forces on both planets in each pair contribute to the planet carrier torque. Note that the load reaction torque is shown on the free-body diagram of the planet carrier; the torque on the output shaft is in the same direction as the input torque.

Do you know where the additional torque comes from? We apply a torque to the stationary ring gear. In some transmissions, this is done with a band brake. The ring gear can be bolted in place in a transmission designed for a single speed ratio. The ring gear torque must balance the torque contributions due to the tangential forces from each planet 2.

The sum of the external torques applied to the transmission (the input torque plus the output reaction torque plus the torque applied to the ring gear) is zero. We check this equation by comparing the carrier torque with the sum of the sun torque and ring torque. Our calculations show that the output power is equal to the input power. (In a real transmission, there will be some friction loss; we neglected it here.)

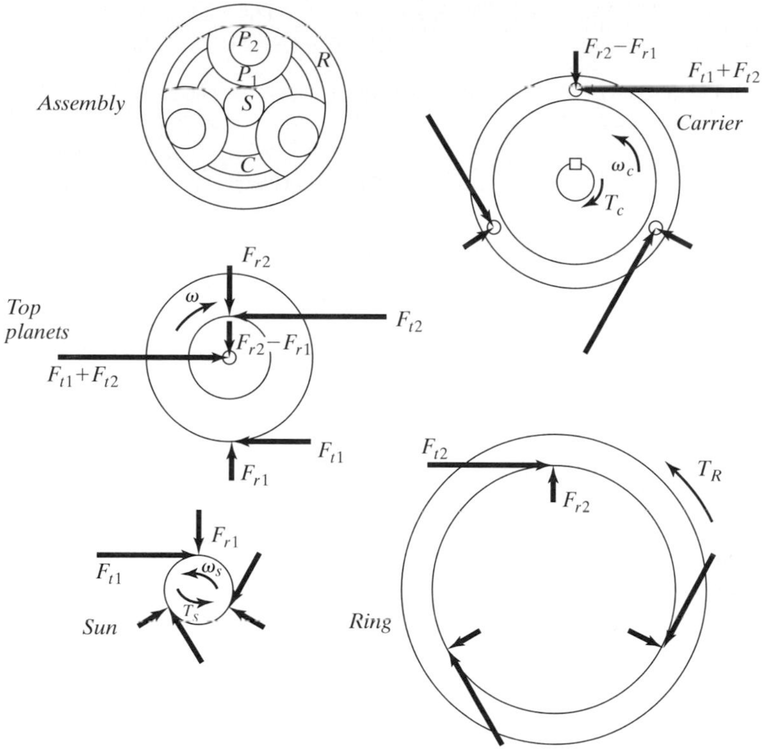

FIGURE 8.44 Compound planetary train.

Detailed calculations. Use the sun gear as input and the carrier as output. The sun drives the first planet. The second planet on the same shaft contacts the fixed ring

Input speed (rpm CCW) $n_S := 3000$ Angular velocity (rad/s) $\omega_S := \dfrac{\pi \cdot n_S}{30}$ $\omega_S = 314.2$

Module for all gears (mm) $m := 3$ Pressure angle $\phi := 20 \cdot \text{deg}$

Number of teeth in sun: $N_S := 19$ First planet: $N_{P1} := 20, 24 .. 92$

Second planet: $N_{P2} := 18$ Ring: $N_R(N_{P1}) := N_S + N_{P1} + N_{P2}$

Pitch radii of gears: $r_S := N_S \cdot \dfrac{m}{2}$ $r_S = 28.5$ $r_{P1}(N_{P1}) := N_{P1} \cdot \dfrac{m}{2}$

$r_{P2} := N_{P2} \cdot \dfrac{m}{2}$ $r_{P2} = 27$ $r_R(N_{P1}) := N_R(N_{P1}) \cdot \dfrac{m}{2}$

Planet carrier radius: $r_C(N_{P1}) := r_S + r_{P1}(N_{P1})$

Number of planet pairs: $p_{num} := 3$ Transmitted power $P_{kw} := 2.4$

Output/input speed ratio $n_{CS} = \dfrac{n_C}{n_S}$

Formula method for speed ratio: $\dfrac{n_{CS}}{1 - n_{CS}} = \dfrac{N_S \cdot N_{P2}}{N_{P1} \cdot N_R}$ solve, $n_{CS} \rightarrow \dfrac{-342}{(-N_{P1} \cdot N_R - 342)}$

Planet speed: $\dfrac{n_{P1} - n_C}{n_S - n_C} = \dfrac{-N_S}{N_{P1}}$ solve, $n_{P1} \rightarrow -\left[\dfrac{-1}{(3000 - n_C)} \cdot n_C + \dfrac{19}{N_{P1}}\right] \cdot (3000 - n_C)$

Calculate speed ratio $n_{CS}(N_{P1}) := N_S \cdot \dfrac{N_{P2}}{(N_{P1} \cdot N_R(N_{P1}) + N_S \cdot N_{P2})}$

Output speed $n_C(N_{P1}) := n_S \cdot n_{CS}(N_{P1}) \quad \omega_C(N_{P1}) := n_C(N_{P1}) \cdot \dfrac{\pi}{30}$

Calculate planet speed: $n_{P1}(N_{P1}) := -\left[\dfrac{-1}{(n_S - n_C(N_{P1}))} \cdot n_C(N_{P1}) + \dfrac{N_S}{N_{P1}}\right] \cdot (n_S - n_C(N_{P1}))$

Torque on sun (N mm) $T_S := 10^6 \dfrac{P_{kw}}{\omega_S} \quad T_S = 7639.4$

Tangential forces on sun and first planet (N) $F_{t1} := \dfrac{T_S}{r_S \cdot P_{num}} \quad F_{t1} = 89.4$

Radial forces on sun and first planet (N) $F_{r1} := F_{t1} \cdot \tan(\phi) \quad F_{r1} = 32.5$

Planet torque balance, tangential forces on second planet and ring

$$T_P(N_{P1}) := F_{t1} \cdot r_{P1}(N_{P1}) \quad F_{t2}(N_{P1}) := \frac{F_{t1} \cdot r_{P1}(N_{P1})}{r_{P2}}$$

Radial forces on second planet and ring (N) $F_{r2}(N_{P1}) := F_{t2}(N_{P1}) \cdot \tan(\phi)$

Reaction torque on ring $T_R(N_{P1}) := F_{t2}(N_{P1}) \cdot r_R(N_{P1}) \cdot P_{num}$

Torque on carrier $T_C(N_{P1}) := (F_{t1} + F_{t2}(N_{P1})) \cdot r_C(N_{P1}) \cdot P_{num}$

Check torque $T_{Ccheck}(N_{P1}) := T_S + T_R(N_{P1})$

Output power check (kW, neglecting friction losses)

$$P_{out}(N_{P1}) := T_C(N_{P1}) \cdot \omega_C(N_{P1}) \cdot 10^{-6} \quad P_{out}(30) = 2.4$$

N_{P1}	$r_{P1}(N_{P1})$	$N_R(N_{P1})$	$r_R(N_{P1})$	$n_{P1}(N_{P1})$	$n_C(N_{P1})$
20	30	57	85.5	−1,500	692.3
24	36	61	91.5	−1,357.1	568.1
28	42	65	97.5	−1,239.1	474.6
32	48	69	103.5	−1,140	402.4
36	54	73	109.5	−1,055.6	345.5
40	60	77	115.5	−982.8	299.8
44	66	81	121.5	−919.4	262.7
48	72	85	127.5	−863.6	232
52	78	89	133.5	−814.3	206.4
56	84	93	139.5	−770.3	184.9
60	90	97	145.5	−730.8	166.5
64	96	101	151.5	−695.1	150.7
68	102	105	157.5	−662.8	137.1
72	108	109	163.5	−633.3	125.3
76	114	113	169.5	−606.4	114.9
80	120	117	175.5	−581.6	105.8
84	126	121	181.5	−558.8	97.7
88	132	125	187.5	−537.7	90.5
92	138	129	193.5	−518.2	84

N_{P1}	$F_{t2}(N_{P1})$	$F_{r2}(N_{P1})$	$T_R(N_{P1})$	$T_C(N_{P1})$	$T_{Check}(N_{P1})$	$P_{out}(N_{P1})$
20	99.3	36.1	25,464.8	33,104.2	33,104.2	2.4
24	119.1	43.4	32,702.2	40,341.6	40,341.6	2.4
28	139	50.6	40,654.3	48,293.8	48,293.8	2.4
32	158.8	57.8	49,321.3	56,960.7	56,960.7	2.4
36	178.7	65	58,703	66,342.5	66,342.5	2.4
40	198.6	72.3	68,799.6	76,439	76,439	2.4
44	218.4	79.5	79,611	87,250.4	87,250.4	2.4
48	238.3	86.7	91,137.1	98,776.6	98,776.6	2.4
52	258.1	93.9	103,378.1	111,017.6	111,017.6	2.4
56	278	101.2	116,333.9	123,973.3	123,973.3	2.4
60	297.8	108.4	130,004.5	137,643.9	137,643.9	2.4
64	317.7	115.6	144,389.8	152,029.3	152,029.3	2.4
68	337.5	122.9	159,490	167,129.4	167,129.4	2.4
72	357.4	130.1	175,305	182,944.4	182,944.4	2.4
76	377.3	137.3	191,834.8	199,474.2	199,474.2	2.4
80	397.1	144.5	209,079.3	216,718.8	216,718.8	2.4
84	417	151.8	227,038.7	234,678.2	234,678.2	2.4
88	436.8	159	245,712.9	253,352.3	253,352.3	2.4
92	456.7	166.2	265,101.9	272,741.3	272,741.3	2.4

8.7 SPREADSHEETS APPLIED TO THE DESIGN OF GEAR TRAINS

The design of a gear train may require many trials in order to meet a given set of specifications. The use of a computer permits a designer to consider several different combinations in attempting to optimize a design. A number of computer programs are available for gear train design. Electronic spreadsheets are also ideally suited to the task.

The superposition method of planetary gear train analysis fits easily into the spreadsheet format. Calculations should be checked by a second method, though, whenever practical. The formula method for analyzing planetary trains can be coded into the spreadsheet as a check.

SAMPLE PROBLEM 8.6

Design of a Two-Speed Transmission

Design a transmission with output-to-input speed ratios of approximately 1:3.5 and 1:1.

Solution. There are many possible solutions. Since we have just considered planetary transmissions, let us attempt to use a simple planetary transmission of the type shown in Figure 8.43. We will try sun gears with 24 to 26 teeth and planet gears with 17 to 19 teeth, using either three or four planet gears (18 possible combinations). The number of teeth in the ring gear is given by

$$N_R = N_S + 2N_P.$$

(The solution will be outlined using symbols for convenience, whereas spreadsheet formulas are coded in terms of cell addresses.)

We want to consider only balanced planetary trains, in order to avoid lateral loads on the sun and carrier shafts. The train can be balanced with equally spaced planet gears if

$$(N_S + N_R)/(\text{the number of planets}) = \text{an integer.}$$

An IF statement and modulo (remainder) function may be used. The test for a balanced train is based on the statement

$$\text{IF}\{\text{MODULO}[(N_S + N_R)/(\text{number of planets})] < 10^{-6}, \text{"yes"}, \text{"no"}\},$$

which returns "yes" if the remainder is zero (within a small round-off error). If the test returns "no," the train cannot be assembled with equally spaced planets, and we reject that combination of gears.

Let the sun gear be integral with the input shaft, and let the planet carrier shaft be the output. The 1:1 speed ratio requirement can be met by incorporating a clutch into the system. If the ring gear is free to rotate and the clutch engages the sun gear to the planet carrier, then the input and output speeds are equal.

A speed reduction is possible if the clutch is disengaged and the ring gear is held fixed. One way to stop the ring gear from rotating is to tighten a band brake on the outer surface of the gear.

Combining two steps in the superposition method, for one rotation of the carrier and zero rotations of the ring gear, we have the following results:

$$n_P/n_C = 1 - N_R/N_S$$

and

$$n_S/n_C = 1 + N_R/N_S.$$

Speed ratios are calculated for various combinations of gears.

If the speed ratios are divided by n_s/n_c, we obtain the ratios n/n_s. The result may also be checked by the formula method, from which we obtain

$$n_C/n_S = N_S/[N_S + N_R].$$

A spreadsheet program was used to obtain the results shown in Table 8.14. It can be seen that 7 of the 18 combinations tested form balanced trains. The speed ratio n_c/n_s is precisely 1:3.5 with 24, 18, and 60 teeth in the sun, planet, and ring gears, respectively. This combination will be selected as our tentative design. Note that either three or four planet gears may be used.

If we wish to calculate shaft torques and gear tooth loading, the spreadsheet may be expanded for that purpose. If it is decided to evaluate a change in any design parameter, the spreadsheet program automatically recalculates all values related to the change. It is again recommended that checks be incorporated to detect programming errors.

TABLE 8.14 Planetary Gear Train with Stationary Ring, Superposition Method (Check by formula method)

					Planets	
					3	4
	Sun	Planet	Ring	Carrier	Balanced	
Teeth	24	17	58		No	No
Speed $n/n(c)$	3.416667	−2.41176	0	1		
Speed $n/n(s)$	1	−.705882	0	.2926829		
Speed check	1		0	.2926829		
Teeth	24	18	60		Yes	Yes
Speed $n/n(c)$	3.5	−2.33333	0	1		
Speed $n/n(s)$	1	−.666667	0	.2857143		
Speed check	1		0	.2857143		
Teeth	24	19	62		No	No
Speed $n/n(c)$	3.583333	−2.26316	0	1		
Speed $n/n(s)$	1	−.631579	0	.2790698		
Speed check	1		0	.2790698		
Teeth	25	17	59		Yes	Yes
Speed $n/n(c)$	3.36	−2.47059	0	1		
Speed $n/n(s)$	1	−.735294	0	.297619		
Speed check	1		0	.297619		
Teeth	25	18	61		No	No
Speed $n/n(c)$	3.44	−2.38889	0	1		
Speed $n/n(s)$	1	−.694444	0	.2906977		
Speed check	1		0	.2906977		
Teeth	25	19	63		No	Yes
Speed $n/n(c)$	3.52	−2.31579	0	1		
Speed $n/n(s)$	1	−.657895	0	.2840909		
Speed check	1		0	.2840909		
Teeth	26	17	60		No	No
Speed $n/n(c)$	3.307692	−2.52941	0	1		
Speed $n/n(s)$	1	−.764706	0	.3023256		
Speed check	1		0	.3023256		
Teeth	26	18	62		No	Yes
Speed $n/n(c)$	3.384615	−2.44444	0	1		
Speed $n/n(s)$	1	−.722222	0	.2954545		
Speed check	1		0	.2954545		
Teeth	26	19	64		Yes	No
Speed $n/n(c)$	3.461538	−2.36842	0	1		
Speed $n/n(s)$	1	−.684211	0	.2888889		
Speed check	1		0	.2888889		

EXAMPLE PROBLEM 8.7

Power, Torque, and Tooth Loading in a Tentative Transmission Design

Design a planetary transmission to transmit 12 kW of mechanical power with an input speed of 120 rad/s and a speed reduction of 1:3.75.

Solution. A simple planetary train will be selected for this application. Let the sun gear be the input and the carrier the output, and let the ring gear be held stationary. After investigating a number of combinations, it was found that 24, 21, and 66 teeth in the sun, planet, and ring gears, respectively, produce the required ratio. The results are shown in Table 8.15. Three (but not four) equally spaced planet gears may be used. Design decisions include the selection of a 20° pressure angle and a module of $m = 5$ mm. Gear pitch radii are given by

$$r = mN/2.$$

The radius of the planet carrier, measured at the planet centers, is

$$r_C = r_S + r_P.$$

Speed is calculated relative to the carrier speed, using the superposition method. We see that $n_s/n_c = 3.75$, as required. Multiplying this row by 120/3.75, we obtain the actual speed of each gear when $\omega_s = 120$ rad/s. The result is checked by the formula method. The torque (N · mm) on the input shaft is given by

$$T_S = 10^6 P/\omega_S,$$

where P = power (kW). Since as the design calls for three planet gears, the tangential force is given by

$$F_t = T_S/[3r_S].$$

The radial force is

$$F_r = F_t \tan\phi,$$

where $\phi = 20°$, the pressure angle.

TABLE 8.15 Planetary Gear Train with Stationary Ring, Module: 5, Power: 12 kW, Pressure angle 20°. Superposition method (Check by formula method; shaft torque is in N · mm)

					Planets	
					3	4
	Sun	Planet	Ring	Carrier	Balanced?	
Teeth	24	21	66		Yes	No
Radius(mm)	60	52.5	165	112.5		
Relative speed	3.75	−2.14286	0	1		
Speed (rad/s)	120	−68.5714	0	32		
Speed check	120		0	32		
Shaft torque	100,000	0	275,000	−375,000		
Tangential force (*N*)					555.5556	416.6667
Radial force (*N*)					202.2057	151.6543
Shaft power (kW)	12	0	0	−12		
Torque balance (N · mm)					0 OK	

The torque on the ring gear is given by

$$T_R = 3F_t r_R$$

and the torque on the carrier by

$$T_C = 2 \times 3F_t r_C.$$

Free-body sketches should be used to check directions. Note that the torques in a free-body diagram of the assembly balance. Of course, there are many other aspects to the design of a transmission, including considerations of strength and wear.

8.8 GEAR TRAIN DIAGNOSTICS BASED ON NOISE AND VIBRATION FREQUENCIES

Some machines produce objectionable noise and vibration. Can we track down and correct the problem? Diagnosing noise and vibration problems is difficult because they have many sources. However, diagnostics can be rewarding. It may be possible to reduce one's annoyance and risk of hearing loss due to noise. Sometimes a serious problem is detected before it causes a catastrophic failure. We may find that gears are to blame, or we may be able to rule out gears as the cause of noise and vibration problems.

Noise and vibration frequencies. Noise and vibration frequencies are measured in hertz (Hz). The range of audible sound is roughly

$$20 \leq \mathrm{f} \leq 20{,}000 \text{ Hz},$$

where noise is unwanted sound and the terms *noise* and *sound* are used interchangeably,

$$f = \text{frequency (Hz), and } 1 \text{ Hz} = 1 \text{ cycle/s} = 60 \text{ cycles/minute.}$$

Interpreting noise and vibration measurements. Noise and vibration measurement techniques are beyond the scope of this book. But we discuss the subject briefly because you may sometime be asked to interpret noise or vibration measurements.

Sound levels are defined by

$$L = 10 \log [p_{\text{rms}}^2 / p_{\text{ref}}^2],$$

where
L = sound level in decibels (dB)
log = common (base-10) logarithm,
p_{rms} = root-mean-square sound pressure, and
p_{ref} = reference pressure = $2 \cdot 10^{-5}$ Pa.

Vibration is usually measured with an accelerometer; common units of measurement are m/s^2, in/s^2, g's (where 1 g = the acceleration due to gravity), and decibels (with various reference levels).

Spectral analysis involves breaking down a noise or vibration signal into frequency components. A fast-fourier-transform (FFT) analyzer is often used to produce a plot of sound or vibration level against frequency. Sometimes we find distinct peaks in the plots. The equations that follow may help us interpret those peaks and lead to the source of the noise or vibration.

Shaft-speed-related noise and vibration. The fundamental noise or vibration frequency due to shaft imbalance is given by

$$f_{\text{shaft imbalance}} = n/60,$$

where f = fundamental frequency (Hz) and
n = speed (rpm) of the shaft in question.

If this frequency is prominent, we could consider balancing the offending gear and shaft. (But note the next paragraph.)

Tooth-error frequencies. Consider a single tooth that is damaged or poorly cut. If that tooth is on one of a pair of nonplanetary gears, the result is the same as before, viz.,

$$f_{\text{tooth error}} = n/60,$$

where n = speed (rpm) of the gear with the tooth error.

Tooth-meshing frequencies. Suppose we could manufacture perfect error-free gears. (We cannot, of course.) There would still be shock loading due to tooth deflection. Figure 8.45 is a representation showing tooth centerlines. Gear 1 drives, and at this instant, teeth b and c contact teeth B and C on gear 2. The deflection of these teeth is exaggerated in the sketch. As a result of the elastic deflection, tooth d, which is about to come into mesh, leads its theoretical position. Tooth D lags. The result is shock loading each time gear teeth come into contact. The fundamental tooth meshing frequency is

$$f_{\text{tooth meshing}} = nN/60,$$

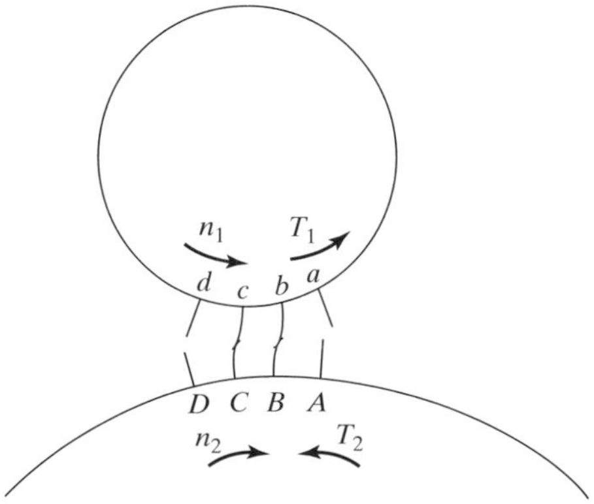

FIGURE 8.45 Tooth-meshing frequencies

where n = speed (rpm) of one of the gears and
N = number of teeth on the same gear

If this frequency is a problem, we might consider improving the contact ratio. The usual solution is to redesign with helical gears instead of spur gears.

Planetary trains. The foregoing results apply to a pair of gears that rotate on shafts with fixed centerlines. What changes do we need when diagnosing a noise or vibration problem in planetary gears? Think of the problem this way:

- The sun and planet carrier shafts rotate on fixed centerlines, so there is no change in the sun and carrier shaft imbalance equations.
- The planets rotate within the planet carrier. Rewrite the shaft imbalance equation for planet gears to take into account the speed of the planet with respect to the carrier.
- The number of tooth meshes depends on motion relative to the carrier. Suppose you could stand on the planet carrier and count the number of tooth contacts. The frequency of contacts between the sun and a single planet gear is given by

$$(n_S - n_C)N_S/60.$$

- Tooth error frequencies depend on motion of a gear with respect to the carrier and, the for sun and ring gears, the number of planets. If a given planet contacts both the sun and the ring gear (as in a simple planetary train), the tooth error frequency is

$$2(n_P - n_C)/60.$$

Sideband frequencies. Sometimes tooth-meshing frequencies combine with other sources of noise and vibration to produce sideband frequencies, which may appear in a frequency analysis as

$$f_{\text{sideband}} = f_{\text{tooth meshing}} \pm f_{\text{shaft imbalance}}$$

or

$$f_{\text{sideband}} = f_{\text{tooth meshing}} \pm f_{\text{tooth error}}.$$

Harmonics. Harmonics are integer multiples of the fundamental frequencies. The shock loading of tooth meshing, for example, does not produce a pure tone. The result is the fundamental frequency just calculated, together with harmonics. For fundamental frequency f_1, we might have significant harmonics of frequency $2f_1$ and $3f_1$.

In dealing with planetary trains, there is another complication: If the sun gear meshes with three planet gears, are the shock loads of tooth meshing in phase or out of phase? A significant noise or vibration level might occur at three times the tooth meshing frequency. A significant noise or vibration level might occur at twice the tooth meshing frequency for an idler gear (depending on relative shaft position).

SAMPLE PROBLEM 8.8

Noise and Vibration Frequencies of a Compound Planetary Train

Suppose we are informed that our compound planetary trains may be responsible for noise and vibration problems. The planetary train is similar to that shown in Figure 8.44. A 17-toothed sun gear, the input, rotates at 7,500 rpm, driving planet gears that may have from 60 to 78 teeth, depending on customer requirements. Seventeen-toothed planet gears are mounted on the same shaft with each large planet. They mesh with a fixed ring gear. There are three equally spaced pairs of such planets. Find the fundamental frequencies due to the gear trains so that they can be compared with the results of measured noise and vibration spectra. This makes it possible to propose corrective action or to rule out the planetary trains as a cause of the noise and vibration.

Solution summary. Check the geometry constraints of the gear train. The ring gear size and the speeds of the planet gears and carrier depend on the number of teeth in planet 1. The formula method is used to calculate speeds, but you may prefer the tabular (superposition) method. Fundamental frequencies are calculated for shaft imbalance, tooth meshing, and tooth error. The fundamental frequency for input shaft imbalance is 125 Hz. The other values are tabulated for each gear train option. Sideband frequencies and harmonics can be calculated from the tabulated values.

Detailed calculations. Use the sun gear as input and the carrier as output. The sun drives the first planet, and the second planet on the same shaft contacts the fixed ring.

Input speed (rpm CCW) $n_S := 7500$ Angular velocity (rad/s) $\omega_S := \frac{\pi \cdot n_S}{30}$ $\omega_S = 785.4$

Module for all gears (mm) $m := 1.5$ Pressure angle $\phi := 20 \cdot deg$

Number of teeth in sun: $N_S := 17$ First planet: $N_{P1} := 60..78$

Second planet: $N_{P2} := 17$ Ring: $N_R(N_{P1}) := N_S + N_{P1} + N_{P2}$

Pitch radii of gears:

$$r_S := N_S \cdot \frac{m}{2} \quad r_S = 12.8 \qquad r_{P1}(N_{P1}) := N_{P1} \cdot \frac{m}{2}$$

$$r_{P2} := N_{P2} \cdot \frac{m}{2} \quad r_{P2} = 12.8 \qquad r_R(N_{P1}) := N_R(N_{P1}) \cdot \frac{m}{2}$$

Planet carrier radius: $r_C(N_{P1}) := r_S + r_{P1}(N_{P1})$

Number of planet pairs: $P_{num} := 3$

Output/input speed ratio $n_{CS} = \frac{n_C}{n_S}$

Formula method for speed ratio: $\frac{n_{CS}}{1 - n_{CS}} = \frac{N_S \cdot N_{P2}}{N_{P1} \cdot N_R}$ solve, $n_{CS} \rightarrow \frac{-289}{(-N_{P1} \cdot N_R - 289)}$

Planet speed: $\frac{n_{P1} - n_C}{n_S - n_C} = \frac{-N_S}{N_{P1}}$ solve, $n_{P1} \rightarrow -\left[\frac{-1}{(7500 - n_C)} \cdot n_C + \frac{17}{N_{P1}}\right] \cdot (7500 - n_C)$

Calculate speed ratio $n_{CS}(N_{P1}) := N_S \cdot \frac{N_{P2}}{(N_{P1} \cdot N_R(N_{P1}) + N_S \cdot N_{P2})}$

Output speed $n_C(N_{P1}) := n_S \cdot n_{CS}(N_{P1})$ $\omega_C(N_{P1}) := n_C(N_{P1}) \cdot \frac{\pi}{30}$

Calculate planet speed: $n_{P1}(N_{P1}) := -\left[\frac{-1}{(n_S - n_C(N_{P1}))} \cdot n_C(N_{P1}) + \frac{N_S}{N_{P1}}\right] \cdot (n_S - n_C(N_{P1}))$

Possible noise and vibration frequencies (fundamental frequencies, Hz)

Sun shaft $f_S := \frac{n_S}{60}$ $f_S = 125$

Planet carrier shaft $f_C(N_{P1}) := \frac{n_C(N_{P1})}{60}$

Planet shaft $f_P(N_{P1}) := \frac{|n_{P1}(N_{P1}) - n_C(N_{P1})|}{60}$

Tooth-meshing frequencies

Sun and planet $f_{SP}(N_{P1}) := \frac{(n_S - n_C(N_{P1})) \cdot N_S}{60}$

Planet and ring $f_{PR}(N_{P1}) := \frac{|n_{P1}(N_{P1}) - n_C(N_{P1})| \cdot N_{P2}}{60}$

Check $f_{PRcheck}(N_{P1}) := \frac{(n_C(N_{P1}) \cdot N_R(N_{P1}))}{60}$

Tooth error frequencies

Sun $f_{SE}(N_{P1}) := \frac{(n_S - n_C(n_{P1})) \cdot P_{num}}{60}$

Either planet $f_{PE}(N_{P1}) := \frac{|n_{P1}(N_{P1}) - n_C(N_{P1})|}{60}$

Ring $f_{RE}(N_{P1}) := \frac{n_C(N_{P1}) P_{num}}{60}$

N_{P1}	$r_{P1}(N_{P1})$	$N_R(N_{P1})$	$r_R(N_{P1})$	$n_{P1}(N_{P1})$	$n_C(N_{P1})$	$f_C(N_{P1})$	$f_P(N_{P1})$
60	45	94	70.5	−1,655.8	365.6	6.1	33.7
61	45.8	95	71.3	−1,634.6	356.3	5.9	33.2
62	46.5	96	72	−1,613.9	347.3	5.8	32.7
63	47.3	97	72.8	−1,593.8	338.7	5.6	32.2
64	48	98	73.5	−1,574.1	330.4	5.5	31.7
65	48.8	99	74.3	−1,554.9	322.4	5.4	31.3
66	49.5	100	75	−1,536.1	314.6	5.2	30.8
67	50.3	101	75.8	−1,517.9	307.2	5.1	30.4
68	51	102	76.5	−1,500	300	5	30
69	51.8	103	77.3	−1,482.6	293.1	4.9	29.6
70	52.5	104	78	−1,465.5	286.4	4.8	29.2
71	53.3	105	78.8	−1,448.9	279.9	4.7	28.8
72	54	106	79.5	−1,432.6	273.6	4.6	28.4
73	54.8	107	80.3	−1,416.7	267.6	4.5	28.1
74	55.5	108	81	−1,401.1	261.7	4.4	27.7
75	56.3	109	81.8	−1,385.9	256.1	4.3	27.4
76	57	110	82.5	−1,371	250.6	4.2	27
77	57.8	111	83.3	−1,356.4	245.3	4.1	26.7
78	58.5	112	84	−1,342.1	240.2	4	26.4

N_{P1}	$f_{SP}(N_{P1})$	$f_{PR}(N_{P1})$	$f_{PRcheck}(N_{P1})$	$f_{SE}(N_{P1})$	$f_{PE}(N_{P1})$	$f_{RE}(N_{P1})$
60	2,021.4	572.7	572.7	356.7	33.7	18.3
61	2,024.1	564.1	564.1	357.2	33.2	17.8
62	2,026.6	555.7	555.7	357.6	32.7	17.4
63	2,029	547.5	547.5	358.1	32.2	16.9
64	2,031.4	539.6	539.6	358.5	31.7	16.5
65	2,033.7	531.9	531.9	358.9	31.3	16.1
66	2,035.9	524.4	524.4	359.3	30.8	15.7
67	2,038	517.1	517.1	359.6	30.4	15.4
68	2,040	510	510	360	30	15
69	2,042	503.1	503.1	360.3	29.6	14.7
70	2,043.9	496.4	496.4	360.7	29.2	14.3
71	2,045.7	489.8	489.8	361	28.8	14
72	2,047.5	483.4	483.4	361.3	28.4	13.7
73	2,049.2	477.2	477.2	361.6	28.1	13.4
74	2,050.8	471.1	471.1	361.9	27.7	13.1
75	2,052.4	465.2	465.2	362.2	27.4	12.8
76	2,054	459.4	459.4	362.5	27	12.5
77	2,055.5	453.8	453.8	362.7	26.7	12.3
78	2,057	448.3	448.3	363	26.4	12

SUMMARY

The output-to-input speed ratio of a nonplanetary train is given by the product of the driving gear teeth divided by the product of the driven gear teeth. Idlers serving as both driving and driven gears change the direction of rotation, but do not affect the speed ratio.

We can obtain a wide range of speed ratios with planetary gear trains, which can be designed with clutches and brakes to provide smooth, rapid changes of the speed ratio. The formula method and the tabular (superposition) method are used to analyze planetary trains. In both methods, we consider *motion relative to the planet carrier* and *motion of the planet carrier*. The formula method is superior if there are two inputs. It also works well if we want a computer to produce a symbolic solution. The tabular method helps keep everything in order. It is convenient if we want to include tooth numbers, pitch radii, forces, torques, and transmitted power in tabular form.

Simple planetary trains almost always incorporate three or four planets to reduce tooth loading and to eliminate bending loads on the input and output shafts. This imposes special geometry requirements. Some planetary trains can be balanced with three equally spaced planets, some with four. The design of compound planetary trains also is controlled by geometry requirements.

Differentials and phase shifters can be designed with spur gears, but bevel gears are more commonly used. The analysis of differentials and phase shifters is similar to that used for other planetary trains. A sketch helps to get the directions right.

Gear-drive speed changers are limited to a few discrete speed ratios. Chain drives and belt drives are commonly used to transmit power. Variable-pitch sheaves on special belt and chain drives can provide smooth, stepless changes in speed. These drives and various friction drives are not restricted to discrete speed ratios.

Free-body diagrams are essential in analyzing forces and torques in drive trains. When a body exerts a force on a second body, the second exerts an equal and opposite force on the first. We can often detect errors by trying to balance external torques on a speed reducer and by comparing input and output power.

Shaft imbalance, gear tooth errors, and tooth meshing produce noise and vibration. We can predict the noise and vibration frequencies due to each of these factors and compare those frequencies with noise and vibration measurements. Such comparisons are helpful for diagnostic purposes and as an aid to redesign.

A Few Review Items

- Sketch two designs for reverse gears (i.e., two schemes for changing the direction of output shaft rotation).
- Suppose you plan to design a simple planetary train with a 20-tooth sun gear, three planet gears in a planet carrier, and a ring gear. Can you use planets with 20 teeth each? What do you recommend?
- Describe the function of an automobile differential. Does its operation pose any problems? Consider the consequences of eliminating the differential.
- Identify a set of requirements that would suggest selecting one of the following drive train systems over another: reverted gear trains, simple planetary gear trains, V-belt drives, toothed (timing) belts, inverted-tooth (silent) chain drives, and friction drives. Consider capacity, complexity, shaft loading, speed ratio precision, input and output shaft location requirements, and other positive and negative features.
- A clockwise torque T_S is applied to the sun (input) shaft of a simple planetary train. The planet carrier *applies* a clockwise torque of $4 \cdot T_S$ *to* the output shaft. A band brake holds the ring gear stationary. What is the brake torque? In what direction?
- Suppose there are four planet gears in the transmission just described. Find tangential and radial forces in terms of the sun gear pitch radius, etc. Sketch a free-body diagram of the sun gear. Do the forces and torques balance? Sketch a free-body diagram of the planet carrier. Do the forces and torques balance?
- Suppose someone has measured noise and vibration due to a gear train and finds that significant noise and vibration levels occur at tooth-meshing frequencies. Does this result indicate a design error or a manufacturing error? Do you have any suggestions for reducing the noise and vibration?
- Suppose there is a defect on the inner race of a ball bearing. Can you relate the resulting noise and vibration frequency to the ball bearing's rotation speed? (*Suggestion*: Use radii of the rolling paths instead of tooth numbers in the equations for a planetary train.)

PROBLEMS

8.1 In Figure 8.3a (a reverted gear train), input shaft 1 rotates at 2000 rev/min, $N_1 = 20$, and $N_4 = 40$. Find the highest and lowest output speeds obtainable if the tooth numbers for gears 2 and 3 are to be no fewer than 20 and no more than 60. (Note that shafts 1 and 4 lie on the same centerline.) All gears have the same module.

8.2 In Figure P8.1, find the speed and direction of rotation of gear 4. Gear 1 rotates at 1000 rev/min, as shown.

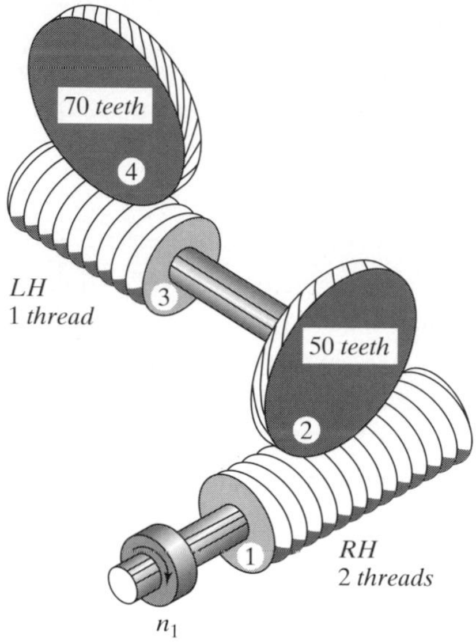

FIGURE P8.1

8.3 Specify the gears for a speed changer similar to that shown in Figure 8.5 (a speed changer employing an idler), having available output-to-input speed ratios of 0.8, 0.75, 0.6, 0.5, and 0.4. The smallest gear cannot have fewer than 20 teeth.

8.4 Sketch a transmission similar to that shown in Figure 8.6a (a three-speed transmission). A synchromesh clutch is to be used in all gears. For the transmission you have sketched, specify tooth numbers to produce (approximately) the following output-to-input speed ratios: $n_O/n_i = +1, +0.5, +0.25$, and -0.25. No gear may have fewer than 18 teeth.

8.5 Specify the gears required to obtain an output-to-input speed ratio of exactly 1131/2000 with a gear train similar to that shown in Figure 8.3a (a reverted gear train). Use gears of not fewer than 20 nor more than 50 teeth.

8.6 In Figure P8.2, tooth numbers are $N_p = 30$, $N_{S_1} = 50$, $N_{S_2} = 49$, and $N_{S_3} = 51$. Gear S_1 is fixed, and the carrier speed is 100 rev/min clockwise. Find the speed and direction of rotation of S_2 and S_3, which are mounted on separate shafts. Use the tabular method. (*Note*: If the three sun gears have the same diameter, their pitches will be slightly different, and the drive will not, theoretically, be precise.)

8.7 In Figure P8.3, assume that the sun gear is fixed. Find the speed of each gear by the tabular method if the planet carrier speed n_c is 200 rev/min clockwise. Give your answer in terms of tooth numbers N_S, N_{P_1}, and so on.

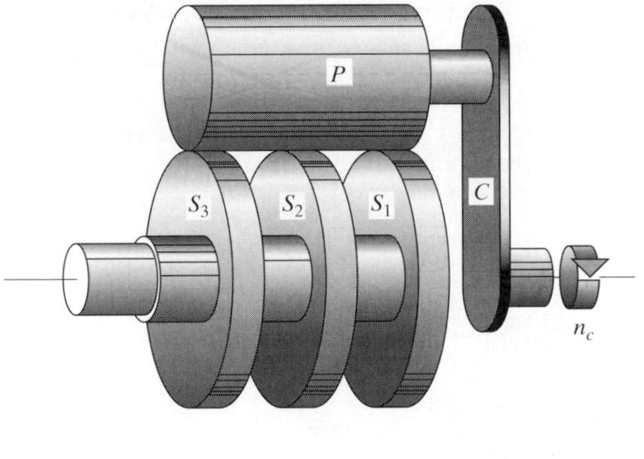

FIGURE P8.2

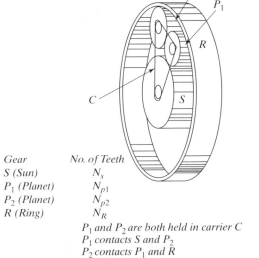

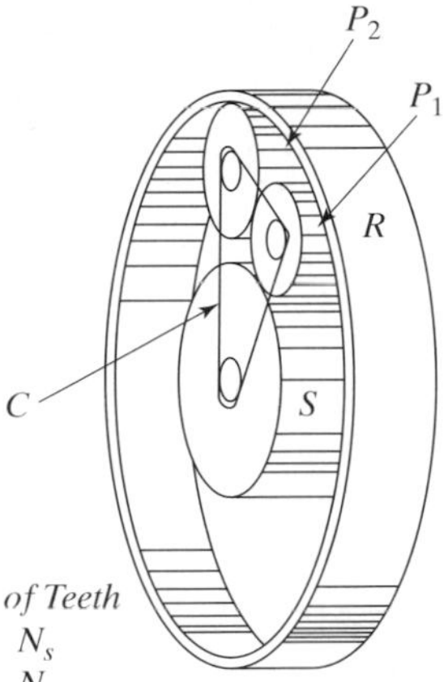

Gear	*No. of Teeth*
S (Sun)	N_s
P_1 *(Planet)*	N_{p1}
P_2 *(Planet)*	N_{p2}
R (Ring)	N_R

P_1 and P_2 are both held in carrier C
P_1 contacts S and P_2
P_2 contacts P_1 and R

FIGURE P8.3

8.8 Repeat Problem 8.7 for the case where the ring gear is fixed.

8.9 In Figure 8.9 (a planetary speed reducer), ring gear R_1 is fixed. The sun gear rotates at 100 rev/min clockwise.

(a) Find the output speed if the tooth numbers are $N_{R_1} = 95$, $N_{P_2} = 21$, $N_{P_1} = 20$, $N_{R_2} = 100$, and $N_S = 55$.

(b) Let R_2 serve as the input gear rotating at 100 rev/min, and let S serve as the output gear. Will a gear train actually operate in this manner?

8.10 Using a speed reducer similar to that shown in Figure 8.10 (a planetary train with four planets), and with the ring gear fixed, obtain an output speed of approximately 1000 rev/min with an input speed of 3500 rev/min.

(a) Specify the tooth numbers, letting the smallest gear have at least 20 teeth.

(b) Determine the exact output speed for the gears you selected.

(c) If the planets are to be equally spaced, how many will the design call for?

8.11 In Figure P8.4, carrier C rotates at 1000 rev/min clockwise and sun gear S_1 is fixed. Tooth numbers are $N_{S_1} = 51$, $N_{P_1} = 20$, $N_{P_2} = 19$, and $N_{S_2} = 50$. Find the speed and direction of rotation for each gear, using the tabular method.

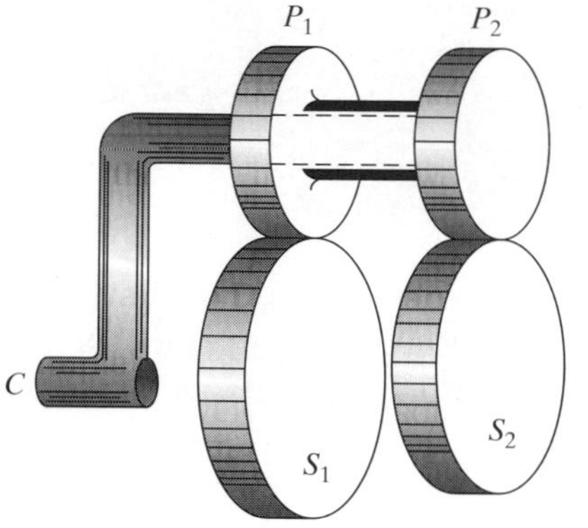

FIGURE P8.4

8.12 Repeat Problem 8.11 for the case where sun gear S_2 is fixed.

8.13 In Figure P8.5, tooth numbers are $N_{S_1} = N_{S_3} = 101$, $N_{S_2} = N_{S_4} = 100$, $N_{P_1} = N_{P_3} = 100$, and $N_{P_2} = N_{P_4} = 99$. Sun gears S_1 and S_3 are fixed. If the input shaft turns at 60 rev/min, how long will it take for one revolution of the output shaft?

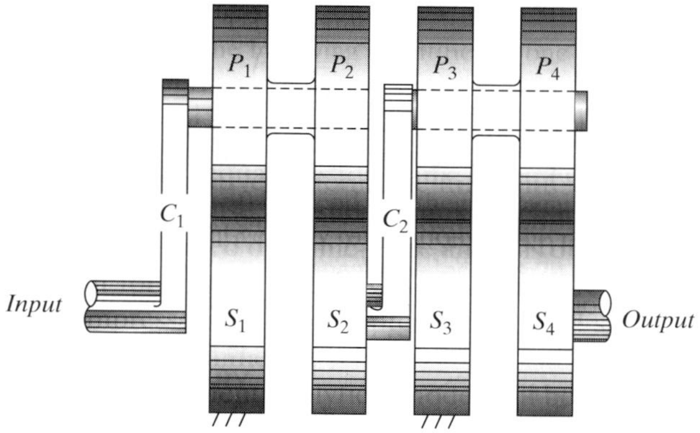

FIGURE P8.5

8.14 In Figure P8.3, $n_S = 400$ rev/min clockwise and $n_R = 300$ rev/min clockwise. Let $N_S = 40$ teeth and $N_R = 100$ teeth. Find the carrier speed.

8.15 In Figure P8.4, let speeds n_C and n_{S_1} be given. Find n_{S_2} in terms of speeds n_C and n_{S_1} and tooth numbers N_{S_1}, N_{P_1}, N_{P_2}, and N_{S_2}.

Problems 8.16 through 8.49 Refer to the Simplified Schematics of Figure 8.7

Find carrier speed n_C in terms of input speed n_i and tooth numbers N_S, N_P, and so on. Use the formula method to solve Problems 8.16 *through* 8.33.

8.16 Use Figure 8.7b, input gear S, and fixed gear R.
8.17 Use Figure 8.7c, input gear S_1, and fixed gear S_2.
8.18 Use Figure 8.7d, input gear R_1, and fixed gear R_2.
8.19 Use Figure 8.7f, input gear R_1, and fixed gear R_2.
8.20 Use Figure 8.7g, input gear S_1, and fixed gear S_2.
8.21 Use Figure 8.7h, input gear R_1, and fixed gear R_2.
8.22 Use Figure 8.7i, input gear S, and fixed gear R.
8.23 Use Figure 8.7j, input gear S, and fixed gear R.
8.24 Use Figure 8.7k, input gear S, and fixed gear R.
8.25 Use Figure 8.7l, input gear S, and fixed gear R.
8.26 Use Figure 8.7b, input gear R, and fixed gear S.
8.27 Use Figure 8.7e, input gear S_1, and fixed gear S_2.
8.28 Use Figure 8.7e, input gear S_2, and fixed gear S_1.
8.29 Use Figure 8.7h, input gear R_2, and fixed gear R_1.
8.30 Use Figure 8.7i, input gear R, and fixed gear S.
8.31 Use Figure 8.7j, input gear R, and fixed gear S.
8.32 Use Figure 8.7k, input gear R, and fixed gear S.
8.33 Use Figure 8.7l, input gear R, and fixed gear S.

Use the tabular (superposition) method to solve problems 8.34 *through* 8.49.

8.34 Use Figure 8.7b. The input gear is S and the fixed gear R.
8.35 Use Figure 8.7c. The input gear is S_1 and the fixed gear S_2.
8.36 Use Figure 8.7d. The input gear is R_1 and the fixed gear R_2.
8.37 Use Figure 8.7f. The input gear is R_1 and the fixed gear R_2.
8.38 Use Figure 8.7g. The input gear is S_1 and the fixed gear S_2.
8.39 Use Figure 8.7h. The input gear is R_1 and the fixed gear R_2.
8.40 Use Figure 8.7i. The input gear is S and the fixed gear R.
8.41 Use Figure 8.7j. The input gear is S and the fixed gear R.
8.42 Use Figure 8.7k. The input gear is S and the fixed gear R.
8.43 Use Figure 8.7l. The input gear is S and the fixed gear R.
8.44 Use Figure 8.7b. The input gear is R and the fixed gear S.
8.45 Use Figure 8.7h. The input gear is R_2 and the fixed gear R_1.
8.46 Use Figure 8.7i. The input gear is R and the fixed gear S.
8.47 Use Figure 8.7j. The input gear is R and the fixed gear S.
8.48 Use Figure 8.7k. The input gear is R and the fixed gear S.
8.49 Use Figure 8.7l. The input gear is R and the fixed gear S.
8.50 Design a gear train that produces one rotation of the output shaft in approximately 16 minutes, for an input speed of 100 rev/min. Use a planetary train similar to that shown in Figure 8.7c, with the center distance between the planet shaft and the input and output shafts = 244 mm. Find the speeds of all gears and the modules (which may not be standard for this problem). More than one trial solution may be required.
8.51 **(a)** Design a gear train similar to that of Figure P8.6 so that ω_C = 100 rad/s ccw for ω_S = 300 rad/s cw. Let the sun gear diameter be 80 mm and the module 4 mm. Give the possible range of planet sizes. Find the speed of each gear after selecting planets.
(b) Check the speed ratio using the formula method.

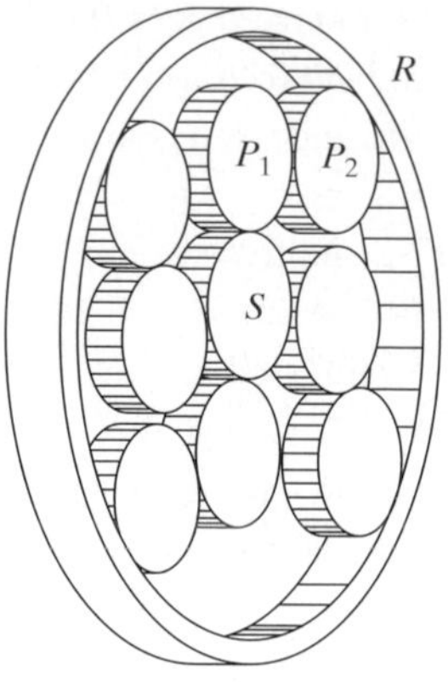

FIGURE P8.6

8.52 In the planetary train shown in Figure P8.7, the input gear is sun S_1 and the output gear is sun S_2. Ring gear R is fixed. Planets P_1 and P_2 rotate at the same speed. Find the output-to-input speed ratio n_{S_2}/n_{S_1} by the superposition method (in terms of tooth numbers N_{S_1}, etc.).

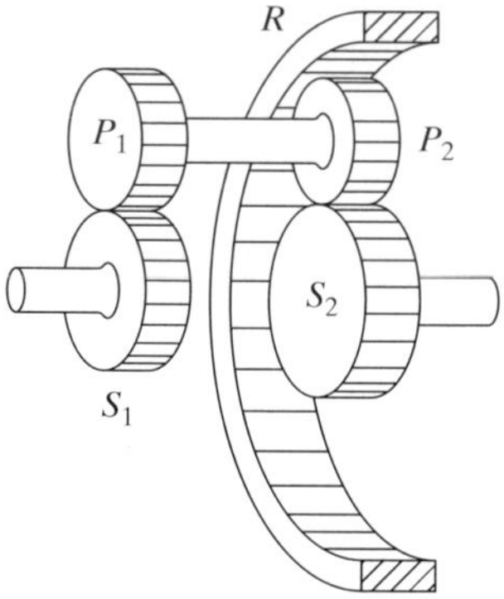

FIGURE P8.7

8.53 In the planetary train of Figure P8.7, let $d_{S_1} = 100$, $d_{P_1} = 100$, $d_{P_2} = 95$, $d_{S_2} = 105$, $d_R = 295$ mm, and the module $m = 5$ mm. Find n_{S_2}/n_{S_1} by the formula method if the ring gear is stationary. (*Suggestion*: Find $n_R^*/n_{S_1}^*$ to obtain n_C/n_{S_1}. Then find $n_{S_2}^*/n_{S_1}^*$.)

8.54 Repeat Problem 8.53, using the superposition method.

8.55 Suppose a gear train was designed similar to the transmission shown in Figure 8.15, but with $d_S = 60$, $d_{P_1} = 60$, $d_{P_2} = 40$, $d_{P_3} = 44$, $d_{R_1} = 160$, $d_{R_2} = 164$ mm, and $m = 2$ mm for all gears. Find speed ratio ω_{R_1}/ω_S if ring gear R_2 is fixed. Use the formula method (in two steps).

8.56 Repeat Problem 8.55 by the tabular method.

8.57 In the planetary train of Figure 8.10, $m = 2$ mm, $d_S = 36$, $d_P = 32$, and $d_R = 100$ mm. Find the speed of all gears if $\omega_S = 400$ rad/s cw and $\omega_R = 0$. Use the tabular method.

8.58 Repeat Problem 8.57 by the formula method.

8.59 In Figure 8.17 (a bevel gear differential), let the planet P have 20 teeth and both sun gears 30 teeth. Gear S_1 drives the left rear axle and S_2 the right rear axle of a vehicle making a right turn at 20 mi/h. The 26-in-diameter tires are 56 in apart (from center to center). The right wheel rolls in a 30-ft-radius path. Find the speed of the carrier, the speed of each sun gear, and the speed of the planet with respect to the carrier. Use the tabular method.

8.60 Repeat Problem 8.59, using the formula method.

8.61 In Figure P8.8, tooth numbers are $N_{S_1} = 35$, $N_P = 22$, and $N_{S_2} = 25$.

(a) If gear S_1 makes 40 rotations clockwise and S_2 makes 15 rotations counterclockwise, find the angular displacement of the carrier by the tabular method. How many rotations does the planet make about its own axis?

(b) Represent the motion of S_1 and S_2 by x and y, respectively. Write an expression for the carrier motion in terms of x and y.

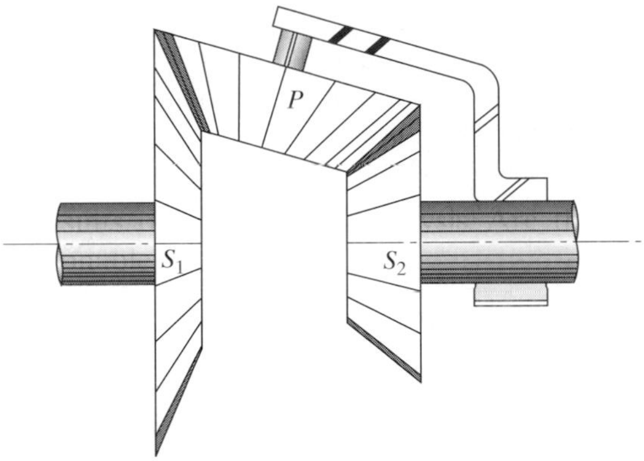

FIGURE P8.8

8.62 Repeat Problem 8.61 by the formula method.

8.63 In Figure 8.27a (a variable-speed belt drive), the pulley pitch diameters are $d_1 = 4$ in and $d_4 = 10$ in. Both countershaft pulleys are to have a minimum pitch diameter of 5 in. Determine the range of d_2 and d_3 so that the output speed may be varied from 400 to 1400 rev/min with a motor speed of 1800 rev/min.

8.64 Repeat Problem 8.63 with $d_1 = 80$ mm, $d_4 = 200$ mm, and the minimum values of d_2 and $d_3 = 100$ mm.

8.65 In Figure 8.32 (a disk drive with an idler wheel), the input shaft rotates at 90 rev/min counterclockwise. The input and output shaft centerlines are 2 in apart. Find idler positions for output speeds of 16, $33\frac{1}{3}$, 45, and 78 rev/min counterclockwise.

8.66 Repeat Problem 8.65 if the input and output shaft centerlines are 38 mm apart.

8.67 Design a variable-speed drive similar to that of Figure 8.31 (a wheel-disk drive) to produce output speeds from 2000 rev/min through 0 to 2000 rev/min in the reverse direction with a constant input speed of 1000 rev/min. (*Hint*: Disk 2 may be used as input and disk 1 as output.)

8.68 In Figure 8.40b (a planetary cone transmission), let the ring gear have 45 teeth and the planets 15 teeth. If the reaction ring has an inside diameter of 5 in, what would the required cone diameters be for output speeds ranging from 450 rev/min clockwise to 180 rev/min counterclockwise, with 1800 rev/min input clockwise. Give cone diameters at points of contact for extreme positions of the reaction ring. (*Note*: The data for this problem do not represent actual dimensions of any commercially available transmission.)

8.69 Repeat Problem 8.68 for a planet ring of 200 mm inside diameter.

8.70 A reverted gear train is to be designed similar to that of Figure 8.3a, with input and output shafts collinear. The distance from the input shaft (gear 1) to the countershaft (gears 2 and 3) is 112.5 mm. Module $m = 5$ mm. Find gear diameters for minimum and maximum values of the speed ratio if no gear is to have fewer than 20 teeth.

8.71 How many different speed ratios are possible in Problem 8.70? Calculate them.

8.72 Calculate the ratio of crankshaft speed to rotor speed for the rotating combustion engine (Wankel engine) described in Chapter 1. Note that the sun gear is fixed, the crankshaft is equivalent to the carrier, and the internal gear integral with the rotor acts as a planet. $N_P/N_S = 1.5$. Use the tabular method.

8.73 Repeat Problem 8.72 by the formula method.

8.74 A planetary gear train similar to that shown in Figure 8.43, but with only two planets, transmits 20 hp. Input is to the sun gear and the carrier drives the load. The ring gear is fixed. The sun gear has 25 teeth and the planets 20. The diametral pitch is 5 and the pressure angle 20°. The sun gear rotates at 630 rev/min cw. Find speeds and draw free-body diagrams of each gear and the carrier.

8.75 Refer to Figure 8.12. Let $n_C = 1960$ rev/min cw, $N_{S_1} = 25$, $N_{P_2} = 39$, and $N_{S_2} = 49$. Find the required number of teeth in P_1 so that $n_{S_2} = 10$ rev/min cw. The module may be nonstandard.

8.76 Refer to Figure P8.9. The sun gear is fixed and has a radius of 20 mm. The planets each have 20 teeth, a 25° pressure angle and a 10-mm radius. A clockwise torque of 12,000 N · mm is applied to the ring (the input gear), which rotates at 200 rev/min clockwise. Determine speeds. Show free-body diagrams of each gear and determine the power transmitted. The planet carrier is the transmission output.

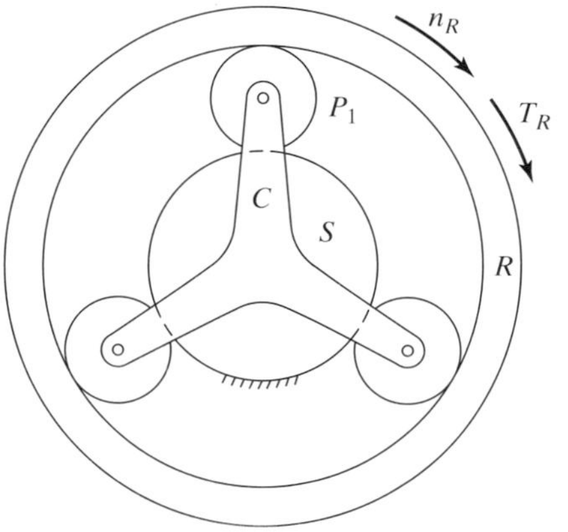

FIGURE P8.9

8.77 Design a two-speed transmission with output-to-input speed ratios of approximately 1:4.2 and 1:1. Many possible designs will satisfy this requirement. (*Suggestions*: Use a computer to evaluate possible solutions. Consider a simple planetary train with 18 to 20 sun gear teeth and 19 to 21 planet gear teeth. Use three or four planet gears.)

8.78 Design a two-speed transmission with speed ratios of approximately 1:3.8 and 1:1. Many possible designs will satisfy this requirement. (*Suggestions*: Use a computer to evaluate possible solutions. Consider a simple planetary train with 21 to 23 sun gear teeth and 20 to 22 planet gear teeth. Use three or four planet gears.)

8.79 Design a two-speed transmission with output-to-input speed ratios of approximately 3:10 and 1:1. Many possible designs will satisfy this requirement. (*Suggestions*: Use a computer to evaluate possible solutions. Consider a simple planetary train with 18 to 26 sun gear teeth and 18 planet gear teeth. Use three or four planet gears.)

8.80 A transmission is to be designed to transmit 2.8 kW of mechanical power. Speed is to be reduced from 220 rad/s to about 58 rad/s. A 1:1 speed ratio should be available as well. Evaluate the kinematic and dynamic aspects of a simple planetary transmission with 20 sun gear teeth, 18 planet gear teeth, a 20° pressure angle, and a module of 1.5. Find

(a) The number of ring gear teeth.
(b) The actual output speed.
(c) The number of planets that will result in a balanced train.
(d) The pitch radii of the gears.
(e) The shaft torques.
(f) The tangential force at each gear mesh.
(g) The radial force at each gear mesh.

Check the torque balance.

8.81 A transmission is to be designed to transmit 12 kW of mechanical power. Speed is to be reduced from 400 rad/s to about 107 rad/s. A 1:1 speed ratio should be available as well. Evaluate the kinematic and dynamic aspects of a simple planetary transmission with 24 sun gear teeth, 21 planet gear teeth, a 20° pressure angle, and a module of 5. Find

(a) The number of ring gear teeth.
(b) The actual output speed.
(c) The number of planets that will result in a balanced train.
(d) The pitch radii of the gears.
(e) The shaft torques.
(f) The tangential force at each gear mesh.
(g) The radial force at each gear mesh.

Check the torque balance.

8.82 A transmission is to be designed to transmit 3 kW of mechanical power. Speed is to be reduced from 400 rad/s to about 105 rad/s. A 1:1 speed ratio should be available as well. Evaluate the kinematic and dynamic aspects of a simple planetary transmission with 22 sun gear teeth, 20 planet gear teeth, a 20° pressure angle, and a module of 2.5. Find

(a) The number of ring gear teeth.
(b) The actual output speed.
(c) The number of planets that will result in a balanced train.
(d) The pitch radii of the gears.
(e) The shaft torques.
(f) The tangential force at each gear mesh.
(g) The radial force at each gear mesh.

Check the torque balance.

8.83 A transmission is to be designed to transmit 1.75 kW of mechanical power. Speed is to be reduced from 400 rad/s to about 100 rad/s. A 1:1 speed ratio should be available as well. Evaluate the kinematic and dynamic aspects of a simple planetary transmission with 20 sun gear teeth, 20 planet gear teeth, a 20° pressure angle, and a module of 1.5. Find

(a) The number of ring gear teeth.
(b) The actual output speed.
(c) The number of planets that will result in a balanced train.
(d) The pitch radii of the gears.
(e) The shaft torques.
(f) The tangential force at each gear mesh.
(g) The radial force at each gear mesh.

Check the torque balance.

8.84 Suppose we would like to offer a line of speed reducers with output-to-input speed ratios ranging from about 1:4 to about 1:17. Typical input speeds will be 2400 rpm. Investigate candidates for the job. Among our design decisions, let us try a configuration similar to that in Figure 8.11b. The sun gear will be the input, the planet carrier (not shown) will be the output, and the ring will be kept stationary. We will use three sets of planets to share the load and balance the train. The two planets in each set rotate as a unit. We will try 17 teeth in the sun gear and 17 or more teeth in the planets meshing with it. The planets meshing with the ring will have 17 teeth also. We will use straight spur gears, all with a 1.5-mm module. Calculate and tabulate the number of ring teeth, as well as the pitch radii, speed ratios, and speeds for various possible configurations.

8.85 Suppose we plan to design a line of speed reducers for an input speed of 4000 rpm and output speeds ranging from 250 rpm to 1000 rpm. Among our design decisions; let us try a compound planetary train similar to that in Figure 8.11b. The sun gear will be the input, the planet carrier (not shown) will be the output, and the ring will be kept stationary. We will use four sets of planets to share the load and balance the train. The two planets in each set rotate as a unit. We will try 19 teeth in the sun gear and 18 or more teeth in the planets meshing with it. The planets meshing with the ring will have 19 teeth. We will use straight spur gears, all with a 2-mm module. Calculate and tabulate the number of ring gear teeth, as well as the pitch radii, speed ratios, and speeds for possible speed reducer designs.

8.86 Suppose we plan to design a line of speed reducers for an input speed of 1800 rpm and output speeds ranging from about 75 to 450 rpm. Among our design decisions, let us try a compound planetary train similar to that in Figure 8.11b. The sun gear will be the input, the planet carrier (not shown) will be the output, and the ring will be kept stationary. We will use three sets of planets to share the load and balance the train. The two planets in each set rotate as a unit. We will try 20 teeth in the sun gear and 20 or more teeth in the planets meshing with it. The planets meshing with the ring will have 20 teeth also. We will use straight spur gears, all with a 3-mm module. Calculate and tabulate the number of ring gear teeth, as well as the pitch radii, speed ratios, and speeds for possible speed reducer designs.

8.87 Suppose we want to offer a series of speed reducers with fixed speed ratios. The design specifications call for an input speed of 5000 rpm and output speeds ranging from about 150 rpm to about 1,200 rpm. Design the reducers and tabulate the forces and torques when 6 kW of mechanical power are transmitted. Check the output power and the external torque equilibrium. Among our design decisions, let us use a compound planetary train similar to that shown in Figure 8.44, with three sets of planets. The sun gear drives and contacts planet 1 of each set. Both planets of each set rotate together. The planet carrier drives the output shaft. Planet 2 of each set contacts the fixed ring gear. All gears are straight spur gears with a 4-mm module and 20° pressure angle. We will select an 18-tooth sun gear; planet 1 of each set will each have 18 or more teeth, and planet 2 of each set will have 19 teeth.

8.88 Suppose we need a series of speed reducers with fixed speed ratios. The design specifications call for an input speed of 1760 rpm and output speeds ranging from about 60 rpm to about 430 rpm. Design the reducers and tabulate forces and torques when 1.5 kW of mechanical power are transmitted. Check the output power and the check external torque equilibrium. Among our design decisions, let us use a compound planetary train similar to that shown in Figure 8.44, with four sets of planets. The sun gear drives and contacts planet 1 of each set. Both planets of each set rotate together. The planet carrier drives the output shaft. Planet 2 of each set contacts the fixed ring gear. All gears are straight spur gears with a 1.75-mm module and 20° pressure angle. We will select a 20-tooth sun gear; planet 1 of each set will each have 20 or more teeth, and planet 2 of each set will have 19 teeth.

8.89 Suppose we need a series of speed reducers with fixed input-to-output speed ratios ranging from about $\frac{1}{32}$ to about $\frac{1}{4}$. The design specifications call for an input angular velocity of 200 rad/s. Design the reducers and tabulate forces and torques when 3.5 kW of mechanical power are transmitted. Check the output power and the external torque equilibrium. Among our design decisions, let us use a compound planetary train similar to that shown in Figure 8.44, with three sets of planets. The sun gear drives and contacts planet 1 of each set. Both planets of each set rotate together. The planet carrier drives the output shaft. Planet 2 of each set contacts the fixed ring gear. All gears are straight spur gears with a 2.5-mm module and 20° pressure angle. We will select a 19-tooth sun gear; plancts 1 of each set will each have 18 or more teeth, and planet 2 of each set will have 20 teeth.

8.90 A customer insists that our compound planetary trains cause noise and vibration problems. The planetary trains are similar to that shown in Figure 8.44. A 20-tooth sun gear, the input, rotates at 2250 rpm and drives planet gears that may have 20, 24, 28, ..., 90 teeth, depending on customer requirements. Nineteen-tooth planet gears are mounted on the same shaft with the planets that mesh with the sun gear. These planet gears mesh with a fixed ring gear. There are three equally spaced pairs of such planets. Check the geometry constraints of the gear train. The ring gear size and the speeds of the planet gears and carrier depend on the number of teeth in planet 1. Use the formula method or the tabular (superposition) method to calculate speeds. Calculate and tabulate fundamental frequencies for shaft imbalance, tooth meshing, and tooth error. This makes it possible to propose corrective action or to rule out the planetary trains as a cause of the noise and vibration problems.

8.91 Suppose we have a noise and vibration problem with a series of compound planetary trains similar to those shown in Figure 8.44. A 19-tooth sun gear, the input, rotates at 8000 rpm and drives planet gears that may have 18, 22, 26, ..., 86 teeth, depending on customer requirements. Nineteen-tooth planet gears are mounted on the same shaft with the planets that mesh with the sun gear. These planet gears mesh with a fixed ring gear. There are four equally spaced pairs of such planets. Check the geometry constraints of the gear train. The ring gear size and the speeds of the planet gears and carrier depend on the number of teeth in planet 1. Use the formula method or the tabular (superposition) method to calculate speeds. Calculate and tabulate fundamental frequencies for shaft imbalance, tooth meshing, and tooth error. This makes it possible to propose corrective action.

8.92 Suppose a customer insists that our compound planetary trains cause noise and vibration problems. The planetary trains are similar to those shown in Figure 8.44. A 17-tooth sun gear, the input, rotates at 10,000 rpm and drives planet gears that may have $30, 32, 34, \ldots, 64$ teeth, depending on customer requirements. Eighteen-tooth planet gears are mounted on the same shaft with the planets that mesh with the sun gear. These planet gears mesh with a fixed ring gear. There are three equally spaced pairs of such planets. Check the geometry constraints of the gear train. The ring gear size and the speeds of the planet gears and carrier depend on the number of teeth in planet 1. Use the formula method or the tabular (superposition) method to calculate speeds. Calculate and tabulate fundamental frequencies for shaft imbalance, tooth meshing, and tooth error. This makes it possible to propose corrective action or to rule out the planetary trains as a cause of the noise and vibration problems.

PROJECTS

8.1 Design a light-duty transmission with six forward speed ratios and six reverse speed ratios. Output-to-input speed ratios are to range from 1.5:1 to 0.1:1 and −0.1:1 to −1:1.

8.2 Design a light-duty transmission with continuously variable speed ratios. Output-to-input speed ratios are to range from 2:1 to −0.75:1.

BIBLIOGRAPHY AND REFERENCES

1. Adams, A. E., *Plastics Gearing: Selection and Application*, Marcel Dekker, New York, 1986.
2. Chironis, N. P. (ed.), *Gear Design and Application*, McGraw-Hill, New York, 1967.
3. Coy, J. J., D. Townsend, and E. Zaretsky, *Gearing*, NASA Scientific and Technical Branch, Springfield, VA, 1985.
4. Drago, R. J., *Fundamentals of Gear Design*, Butterworths, Boston, 1988.
5. Dudley, D. W., *Handbook of Practical Gear Design*, McGraw-Hill, New York, 1984.
6. Dyson, A., *A General Theory of the Kinematics and Geometry of Gears in Three Dimensions*, Clarendon Press, Oxford, 1969.
7. Lévai, Zoltan, *Theory of Epicyclic Gears and Epicyclic Change-Speed Gearing*, Technical University of Building, Civil and Transport Engineering, Budapest, 1966.
8. Selby, S. M., and R. C. Weast (eds.), "Table of Factors and Primes," in *Handbook of Chemistry and Physics*, 47th ed., Chemical Rubber, Cleveland, 1967, pp. A167–A176.
9. White, G., "Epicyclic Gears Applied to Early Steam Engines," *Mechanism and Machine Theory*, Penton, Cambridge, vol. 23, no. 1, 1988, pp. 25–38.

Note also the Internet references listed under the heading of **Gears** found at the end of Chapter 1. You will find may more illustrations of gear trains by searching the Internet yourself.

CHAPTER 9

Static-Force Analysis

Concepts You Will Learn and Apply When Studying This Chapter

- Common assumptions we make while performing static force analysis; limitations of static force analysis.
- Review of basic principles: force and torque vectors, free-body diagrams, couples, static equilibrium, and superposition of forces.
- Graphical force analysis for insight into static force analysis of linkages and for checking analytical work.

  Two- and three-force members.

  Links with two forces and a couple.
- Graphical force analysis of the slider-crank mechanism and four-bar linkage.
  An application of superposition.
- Graphical force analysis of more complex linkages.
- Analytical statics.
  Equilibrium equations.
- Analytical statics applied to construction machinery and slider-crank and four-bar linkages.
- Analyzing linkages by the method of virtual work.
- Dealing with friction in mechanisms.

9.1 INTRODUCTION

A machine is a device that performs work and thereby transmits energy by means of mechanical force from a power source to a driven load. In the design of machine mechanisms, it is necessary to know the manner in which forces are transmitted from the input to the output, so that the components of the machine can be properly sized to withstand the stresses that are developed. If the members are not designed to be strong enough, then the machine will fail during its operation; if, on the other hand, the machine is designed to have much more strength than is required, then it may not be competitive with other machines in terms of cost, weight, size, power requirements, or other criteria.

During the course of a force analysis, various assumptions must be made that reflect the specific characteristics of the particular system being investigated. These assumptions should be verified as the design proceeds. A major assumption concerns dynamic, or inertial, forces. All machines have mass, and if parts of a machine are accelerating, inertial forces will be associated with this motion. If the magnitudes of these inertial forces are small relative to externally applied loads, then they can be neglected in the force analysis. Such an analysis is referred to as static-force analysis and is the topic of this chapter.

As an example, during the normal operation of a front-end loader, such as that shown later in the chapter in Figure 9.13a, the bucket load and static weight loads may far exceed any dynamic loads due to accelerating masses, and a static-force analysis may be justified. An analysis that includes inertial effects is called a *dynamic-force analysis* and will be discussed in the next chapter. An example of an application requiring a dynamic-force analysis is the design of an automatic sewing machine, wherein, due to high operating speeds, the inertial forces may be greater than the external loads on the machine.

Another assumption deals with the rigidity of the machine components. No material is truly rigid, and all materials experience significant deformation if the forces acting on them, either external or inertial in nature, are great enough. In this chapter and the next, we assume that deformations are so small as to be negligible, and therefore, the members will be treated as though they are rigid. The subject of mechanical vibrations, which is beyond the scope of this book, considers the flexibility of machine components and the resulting effects on machine behavior.

A third major assumption that is often made is that friction effects are negligible. Friction is inherent in all devices, and its degree is dependent upon many factors, including types of bearings, lubrication, loads, environmental conditions, and so on. Friction will be neglected in the first few sections of this chapter; we introduce the subject in Section 9.5.

In addition to the preceding assumptions, other assumptions may be necessary, and some of these will be addressed at various points throughout the chapter.

The first part of the chapter is a review of general force analysis principles and will also establish some of the conventions and terminology to be used in succeeding sections. The remainder of the chapter will present both graphical and analytical methods for the static-force analysis of machines.

9.2 BASIC PRINCIPLES OF FORCE ANALYSIS

In this section, we review the important concepts of force and torque as vector quantities, free-body diagrams, equilibrium, and superposition.

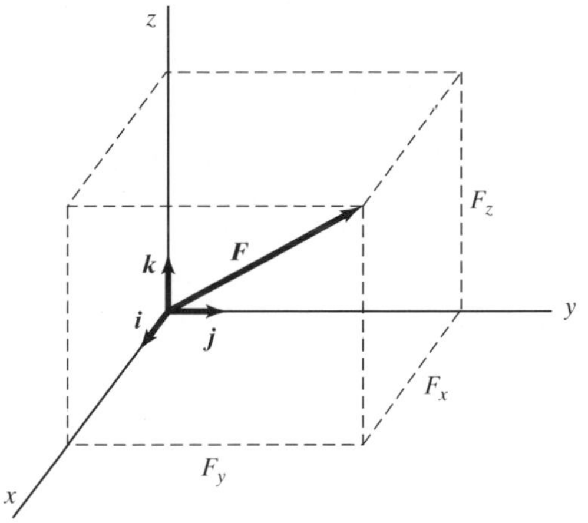

FIGURE 9.1 A force vector $\boldsymbol{F}$ in a Cartesian coordinate system.

Force and Torque

A force is a vector quantity, with magnitude, a direction or line of action, and a sense. Figure 9.1 shows a force vector $\boldsymbol{F}$, which can be expressed in terms of Cartesian coordinates as

$$\boldsymbol{F} = F_x\boldsymbol{i} + F_y\boldsymbol{j} + F_z\boldsymbol{k}, \tag{9.1}$$

where F_x, F_y, and F_z are the components of the force in the x, y, and z directions, respectively, with these directions represented in turn by unit vectors $\boldsymbol{i}$, $\boldsymbol{j}$, and $\boldsymbol{k}$. The resultant force $\boldsymbol{F}$ of two forces $\boldsymbol{F}_1$ and $\boldsymbol{F}_2$ is the vector sum of those forces. This is expressed graphically in Figure 9.2 and mathematically as:

$$\boldsymbol{F} = \boldsymbol{F}_1 + \boldsymbol{F}_2 = (F_{1x} + F_{2x})\boldsymbol{i} + (F_{1y} + F_{2y})\boldsymbol{j} + (F_{1z} + F_{2z})\boldsymbol{k}, \tag{9.2}$$

where F_{1x} is the x component of force $\boldsymbol{F}_1$, and so on.

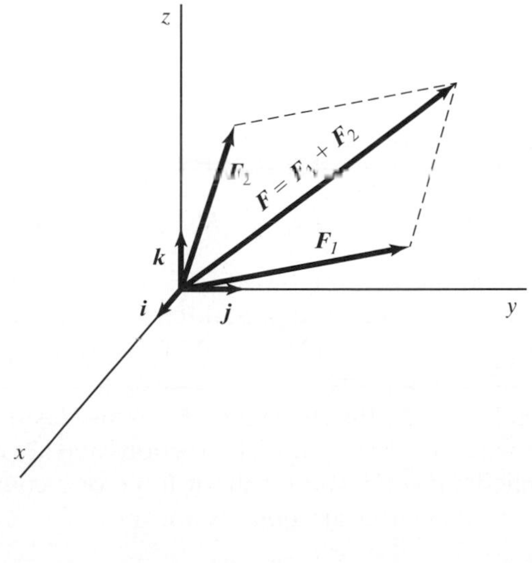

FIGURE 9.2 The resultant force $\boldsymbol{F}$ of two forces $\boldsymbol{F}_1$ and $\boldsymbol{F}_2$.

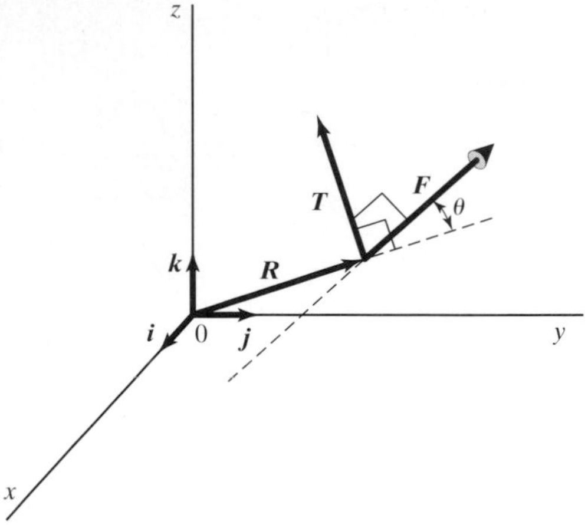

FIGURE 9.3 Torque $\boldsymbol{T}$ is the moment of force $\boldsymbol{F}$ about point 0. Vector $\boldsymbol{R}$ locates the line of action of the force relative to point 0.

A torque, or moment, $\boldsymbol{T}$, is defined as the moment of a force about a point and is a vector quantity, too. Using the vector cross-product notation, we have

$$\boldsymbol{T} = \boldsymbol{R} \times \boldsymbol{F}, \tag{9.3}$$

where $\boldsymbol{R}$ is a position vector directed from the point about which the moment is taken to any point on the line of action of force $\boldsymbol{F}$. (See Figure 9.3.) The magnitude of $\boldsymbol{T}$ is

$$T = RF|\sin\theta|,$$

where θ is the angle between vectors $\boldsymbol{R}$ and $\boldsymbol{F}$, and R and F are the magnitudes of the vectors. The direction of $\boldsymbol{T}$ is perpendicular to the plane containing $\boldsymbol{R}$ and $\boldsymbol{F}$, and the sense of $\boldsymbol{T}$ is given by the right-hand rule. Alternatively, in determinant form,

$$\boldsymbol{T} = \begin{vmatrix} \boldsymbol{i} & \boldsymbol{j} & \boldsymbol{k} \\ R_x & R_y & R_x \\ F_x & F_y & F_z \end{vmatrix}$$

$$= (R_yF_z - R_zF_y)\boldsymbol{i} + (R_zF_x - R_xF_z)\boldsymbol{j} + (R_xF_y - R_yF_x)\boldsymbol{k}. \tag{9.4}$$

An infinite number of combinations of a force vector $\boldsymbol{F}$ and a moment arm vector $\boldsymbol{R}$ exist that will produce the same moment $\boldsymbol{T}$; that is, different values of vectors $\boldsymbol{R}$ and $\boldsymbol{F}$ can lead to the same cross product as given by Eq. (9.3). The resultant of two or more moments is the vector sum of the moments.

Figure 9.4 shows two forces $\boldsymbol{F}_1$ and $\boldsymbol{F}_2$ that have equal magnitudes but different lines of action. Furthermore, the two forces have parallel directions and are of opposite sense. Such a pair of forces is called a *couple*. The resultant force of a couple is zero. However, the resultant moment about an arbitrary point is not zero.

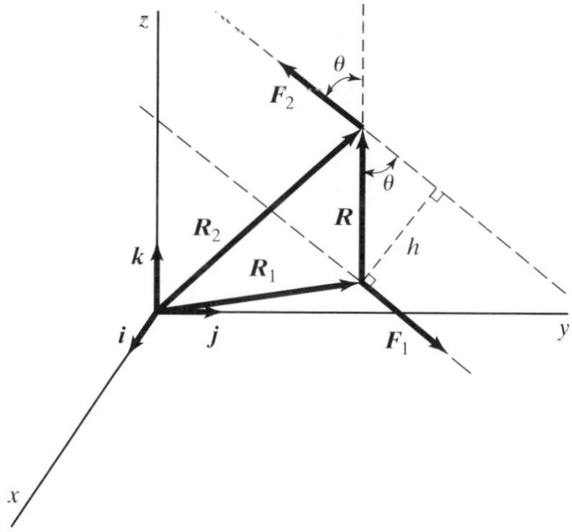

FIGURE 9.4 Forces F_1 and F_2 form a couple, which has zero resultant force, but a nonzero resultant moment.

For example, if moments about the origin in Figure 9.4 are summed, the resultant moment $\boldsymbol{T}$ is

$$\boldsymbol{T} = \boldsymbol{R}_1 \times \boldsymbol{F}_1 + \boldsymbol{R}_2 \times \boldsymbol{F}_2.$$

But $\boldsymbol{F}_1 = -\boldsymbol{F}_2$; therefore,

$$\boldsymbol{T} = \boldsymbol{R}_1 \times (-\boldsymbol{F}_2) + \boldsymbol{R}_2 \times \boldsymbol{F}_2 = (\boldsymbol{R}_2 - \boldsymbol{R}_1) \times \boldsymbol{F}_2 = \boldsymbol{R} \times \boldsymbol{F}_2, \tag{9.5}$$

where $\boldsymbol{R} = \boldsymbol{R}_2 - \boldsymbol{R}_1$ is a vector from any point on the line of action of $\boldsymbol{F}_1$ to any point on the line of action of $\boldsymbol{F}_2$. The direction of the torque is perpendicular to the plane of the couple, and the magnitude of the torque is given by

$$T = RF_2|\sin\theta| = hF_2, \tag{9.6}$$

where $h = R|\sin\theta|$ is the perpendicular distance between the lines of action. It can be seen that the resultant moment of a couple, given by Eq. (9.5), is independent of the point about which moments are taken. Conversely, the moment of a couple about a particular point is independent of the position of the couple relative to the point. For these reasons, a couple is sometimes referred to as a *pure moment* or *pure torque*. As will be seen, the concept of a couple is very useful in force analysis applications.

Free-Body Diagrams

Free-body diagrams are extremely important and useful in force analysis. A free-body diagram is a sketch or drawing of part or all of a system, isolated in order to determine the nature of the forces acting on that body. Sometimes a free-body diagram may take the form of a mental picture; however, actual sketches are strongly recommended, especially for complex mechanical systems.

Generally, the first step (and one of the most important) in a successful force analysis is the identification of the free bodies to be used. Figures 9.5b through 9.5e

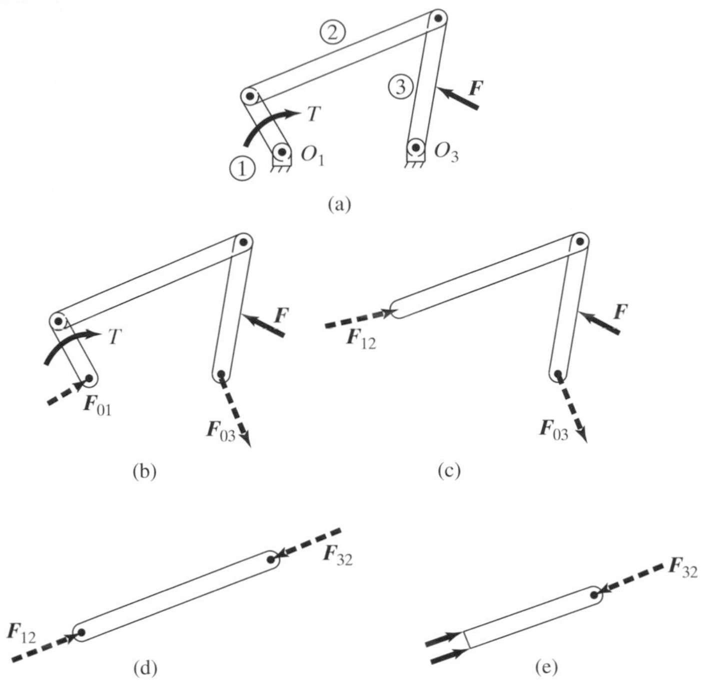

FIGURE 9.5 (a) A four-bar linkage. (b) Free-body diagram of the three moving links. (c) Free-body diagram of two connected links. (d) Free-body diagram of a single link. (e) Free-body diagram of part of a link.

give examples of various free bodies that might be considered in the analysis of the four-bar linkage shown in Figure 9.5a. In Figure 9.5b, the free body consists of the three moving members isolated from the frame; here, the forces acting on the free body include a driving force or torque, external loads, and the forces transmitted from the frame at bearings O_1 and O_3. The force convention is defined as follows: $\boldsymbol{F}_{ij}$ represents the force exerted by member i on member j. This convention will be used throughout the text. Figure 9.5c is a free-body diagram of two links acted upon by the forces transmitted from adjoining links as well as other applied loads. Probably the most commonly used form of a free-body diagram is that of a single link. (See Figure 9.5d.) Most force analyses can be accomplished by examining each of the individual members that make up the system. Such an approach leads to the determination of all of the bearing forces between members, as well as the required input force or torque for a given output load or set of loads. For investigating internal forces or stresses in members, free bodies consisting of portions of members, as in Figure 9.5e, are useful.

Static Equilibrium

For a free body in static equilibrium, the vector sum of all forces acting on the body must be zero, and the vector sum of all moments about any arbitrary point must also be zero. These conditions can be expressed mathematically as

$$\sum \boldsymbol{F} = 0 \tag{9.7a}$$

and

$$\sum \boldsymbol{T} = 0 \tag{9.7b}$$

Since each of these vector equations represents three scalar equations, a total of six independent scalar conditions must be satisfied for the general case of equilibrium under three-dimensional loading.

In many situations, the loading is essentially planar, and the forces can be described by two-dimensional vectors. If the xy-plane designates the plane of loading, then the applicable form of Eqs. (9.7a) and (9.7b) is

$$\sum F_x = 0, \tag{9.8a}$$

$$\sum F_y = 0, \tag{9.8b}$$

and

$$\sum T_z = 0. \tag{9.8c}$$

Equations (9.8a) through (9.8c) are three scalar equations which state that, for the case of two-dimensional xy loading, the summations of forces in the x and y directions must individually equal zero, and the summation of moments about any arbitrary point in the plane must also equal zero. The remainder of this chapter deals with two-dimensional force analysis. A common example of three-dimensional forces is gear forces, which were discussed in Chapter 7.

Superposition

The principle of superposition of forces is an extremely useful concept, particularly in graphical force analysis. Basically, the principle states that, for linear systems, the net effect of multiple loads on a system is equal to the superposition (i.e., vector summation) of the effects of the individual loads considered one at a time. Physically, linearity refers to a direct proportionality between input force and output force. Its mathematical characteristics will be discussed in the section, on analytical force analysis. Generally, in the absence of coulomb or dry friction, most mechanisms are linear for force analysis purposes, despite the fact that many of these mechanisms exhibit nonlinear motions. Examples and further discussion in later sections will demonstrate the application of the principle of superposition.

9.3 GRAPHICAL FORCE ANALYSIS

Graphical force analysis employs scaled free-body diagrams and vector graphics in the determination of unknown machine forces. The graphical approach is best suited for planar force systems. Since forces are normally not constant during the motion of a machine, analyses may be required for a number of mechanism positions; however, in many cases, critical maximum-force positions can be identified and graphical analyses performed for those positions only. An important advantage of the graphical approach is that it provides a useful insight into the nature of the forces in the physical system.

The approach, however, suffers from disadvantages related to accuracy and time. As is true of any graphical procedure, the results are susceptible to drawing and measurement errors. Further, a great amount of graphics time and effort can be expended in the iterative design of a machine mechanism for which fairly thorough knowledge of force–time relationships is required. In recent years, the physical insight of the graphics approach and the speed and accuracy inherent in the computer-based analytical approach have been brought together through computer graphics systems, which have proved to be highly effective engineering design tools.

A few special types of member loadings are repeatedly encountered in the force analysis of mechanisms: a member subjected to two forces, a member subjected to three forces, and a member subjected to two forces and a couple. Other loading cases with more forces can usually be reduced to one of these situations by combining known individual forces into equivalent resultant forces. These special cases will be considered in the paragraphs that follow, before proceeding to the graphical analysis of complete mechanisms.

Analysis of a Two-Force Member

A *member subjected to two forces is in equilibrium if and only if the two forces (1) have the same magnitude, (2) act along the same line, and (3) are opposite in sense*. Figure 9.6a shows a free-body diagram of a member acted upon by forces $\boldsymbol{F}_1$ and $\boldsymbol{F}_2$, where the points of application of these forces are A and B. For equilibrium, the directions of $\boldsymbol{F}_1$ and $\boldsymbol{F}_2$ must be along line AB, and $\boldsymbol{F}_1$ must equal $(-\boldsymbol{F}_2)$. The graphical vector addition of forces $\boldsymbol{F}_1$ and $\boldsymbol{F}_2$ is shown in Figure 9.6b, and, obviously, the resultant net force on the member is zero when $\boldsymbol{F}_1 = -\boldsymbol{F}_2$. The resultant moment about any point will also be zero, as can be seen from inspection or by the application of Eq. (9.5).

Thus, if the load application points for a two-force member are known, the line of action of the forces is defined, and if the magnitude and sense of one of the forces are known, then the magnitude and sense of the other force can immediately be determined. Such a member will be in either tension or compression. [*Caution:* When a slender link is in compression, it may buckle. Machine design and stress analysis texts can help us design compression members and avoid buckling (elastic stability failures).]

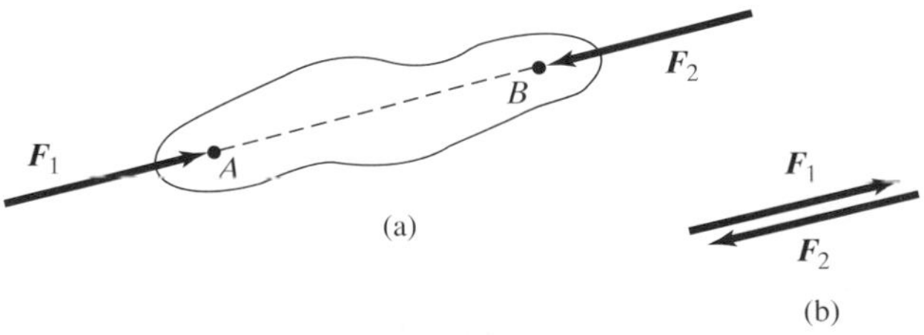

FIGURE 9.6 (a) A two-force member. The resultant force and the resultant moment both equal zero. (b) Force summation for a two-force member.

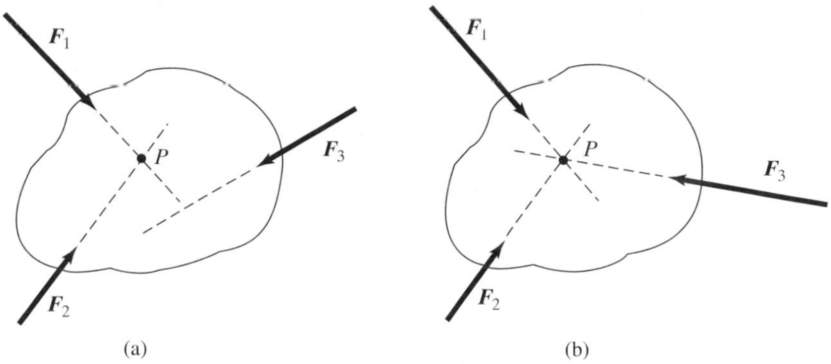

FIGURE 9.7 (a) The three forces on the member do not intersect at a common point, and there is a net resultant moment. (b) The three forces intersect at the same point $\boldsymbol{P}$, called the *concurrency point*, and the net moment is zero.

Analysis of a Three-Force Member

A *member subjected to three forces is in equilibrium if and only if (1) the resultant of the three forces is zero and (2) the lines of action of the forces all intersect at the same point.* The first condition guarantees equilibrium of forces, while the second guarantees equilibrium of moments. The second condition can be understood by considering the case when it is not satisfied. (See Figure 9.7a.) If moments are summed about point P, the intersection of forces $\boldsymbol{F}_1$ and $\boldsymbol{F}_2$, then the moments of these forces will be zero, but $\boldsymbol{F}_3$ will produce a nonzero moment, resulting in a nonzero net moment on the member. In contrast, if the line of action of force $\boldsymbol{F}_3$ also passes through point P (Figure 9.7b), the net moment will be zero. This common point of intersection of the three forces is called the *point of concurrency*.

A typical situation encountered is that in which one of the forces, $\boldsymbol{F}_1$, is known completely (i.e., its magnitude and direction are known), a second force, $\boldsymbol{F}_2$, has a known direction but an unknown magnitude, and a third force, $\boldsymbol{F}_3$, has an unknown magnitude and direction. The graphical solution of this case is depicted in Figure 9.8a through c. First, the free-body diagram is drawn to a convenient scale, and the points of application of the three forces are identified—points A, B, and C in this case. Next, the known force $\boldsymbol{F}_1$ is drawn on the diagram with the proper direction and a suitable magnitude scale. The direction of force $\boldsymbol{F}_2$ is then drawn, and the intersection of this line with an extension of the line of action of force $\boldsymbol{F}_1$ is the concurrency point P. For equilibrium, the line of action of force $\boldsymbol{F}_3$ must pass through points C and P and is therefore as shown in Figure 9.8a.

The force equilibrium condition states that

$$\boldsymbol{F}_1 + \boldsymbol{F}_2 + \boldsymbol{F}_3 = 0.$$

Since the directions of all three forces are now known and the magnitude of $\boldsymbol{F}_1$ was given, this equation can be solved for the remaining two magnitudes. A graphical solution follows from the fact that the three forces must form a closed vector loop, called a

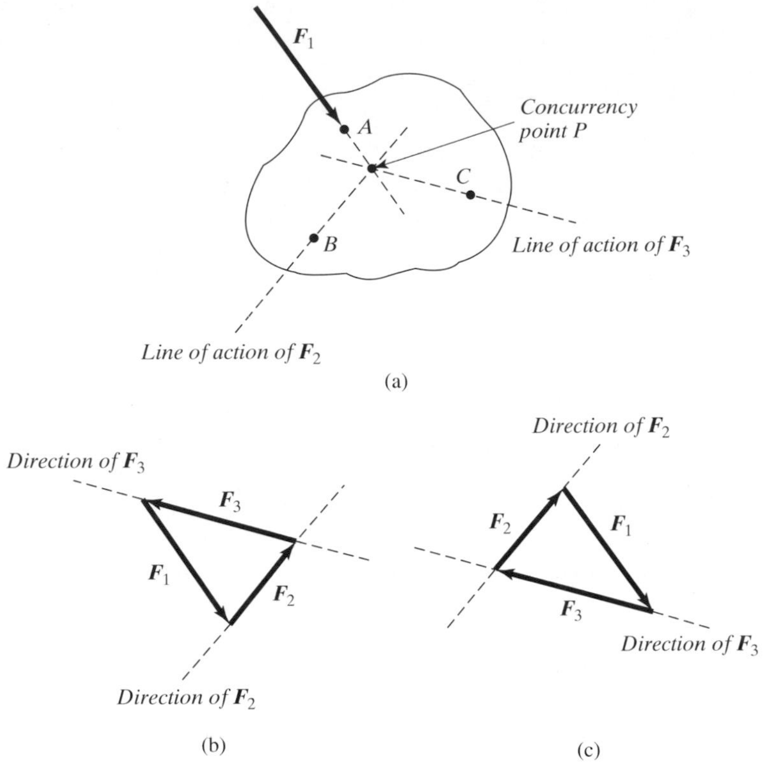

FIGURE 9.8 (a) Graphical force analysis of a three-force member. (b) Force polygon for the three-force member. (c) An equivalent force polygon for the three-force member.

force polygon. The procedure is shown in Figure 9.8b. Vector $\boldsymbol{F}_1$ is redrawn. From the head of this vector, a line is drawn in the direction of force $\boldsymbol{F}_2$, and from the tail, a line is drawn parallel to $\boldsymbol{F}_3$. The intersection of these lines closes the vector loop and determines the magnitudes of forces $\boldsymbol{F}_2$ and $\boldsymbol{F}_3$. Note that the same solution is obtained if, instead, a line parallel to $\boldsymbol{F}_3$ is drawn from the head of $\boldsymbol{F}_1$ and a line parallel to $\boldsymbol{F}_2$ is drawn from the tail of $\boldsymbol{F}_1$. (See Figure 9.8c.) This is so because vector addition is commutative, and therefore, both force polygons are equivalent to the foregoing vector equation.

It is important to remember that, by the definition of vector addition, the force polygon corresponding to the general force equation

$$\sum \boldsymbol{F} = 0$$

will have adjacent vectors connected head to tail. This principle is used in identifying the sense of forces $\boldsymbol{F}_2$ and $\boldsymbol{F}_3$ in Figures 9.8b and c. Also, if the lines of action of $\boldsymbol{F}_1$ and

$\boldsymbol{F}_2$ are parallel, then the point of concurrency is at infinity and the third force $\boldsymbol{F}_3$ must be parallel to the other two. In this case, the force polygon collapses to a straight line.

Analysis of a Member with Two Forces and a Couple

In performing force analyses, it is imperative that we know the nature of the forces that drive a system or that act as loads on the system. Only by knowing where these forces act and how they act can we proceed with a complete force analysis of the system.

This point can be illustrated by considering two ways by which the input crank of a four-bar linkage can be driven. (See Figures 9.9a and b.) In Figure 9.9a, the bell crank is driven by a hydraulic cylinder attached at the point shown. In this case, the crank is a three-force member, and the analysis proceeds according to the preceding section. If, on the other hand, the crank is driven by a shaft with a direct connection to an electric motor, as shown in Figure 9.9b, then the torque applied by the shaft takes the form of a pure torque, and the crank is subjected to two forces plus an input couple. Both of these drive systems can be designed to produce the same torque about pivot O, but the forces acting on the crank will differ in the two cases.

For equilibrium of a member subjected to two forces $\boldsymbol{F}_1$ *and* $\boldsymbol{F}_2$ *plus an applied couple, forces* $\boldsymbol{F}_1$ *and* $\boldsymbol{F}_2$ *must form a couple that is equal and opposite to the applied couple.* Hence, if the magnitude and direction, of force $\boldsymbol{F}_1$ are known, then force $\boldsymbol{F}_2$ will be equal in magnitude, parallel in direction, and opposite in sense, and the moment of the applied couple must be equal and opposite to the moment of couple $\boldsymbol{F}_1$, $\boldsymbol{F}_2$. This is illustrated in Figure 9.9b, in which the magnitude of couple $\boldsymbol{T}$ is equal to the product $hF_1 = hF_2$, where F_1 and F_2 are the magnitudes of forces $\boldsymbol{F}_1$ and $\boldsymbol{F}_2$, respectively.

Graphical Force Analysis of the Slider-Crank Mechanism

The slider-crank mechanism finds extensive application in reciprocating compressors, piston engines, presses, toggle devices, and other machines in which force characteristics are important. The force analysis of this mechanism employs most of the principles described in previous sections, as demonstrated by Sample Problem 9.1.

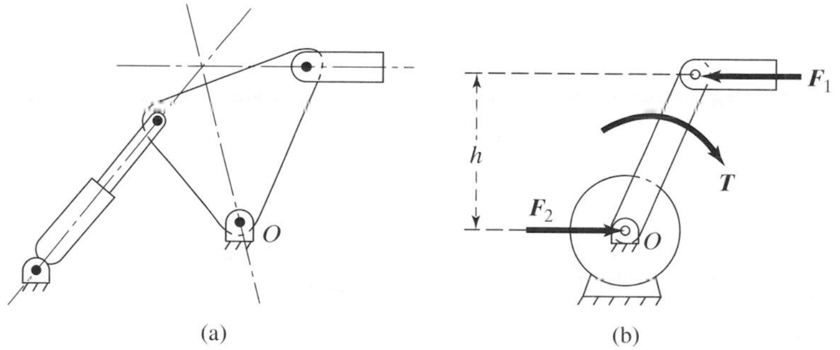

FIGURE 9.9 (a) The crank is driven by a piston and cylinder and is a three-force member. (b) The crank is driven by an electric motor and is subjected to two forces and a pure torque.

SAMPLE PROBLEM 9.1

Static-Force Analysis of a Slider-Crank Mechanism

Consider the slider-crank linkage shown in Figure 9.10a, representing a compressor, operating at so low a speed that inertial effects are negligible. It is assumed that gravity forces also are small compared with other forces and that all forces lie in the same plane. The dimensions are $OB = 30$ mm and $BC = 70$ mm. We wish to find the required crankshaft torque $\boldsymbol{T}$ and the bearing forces for a total gas pressure force $\boldsymbol{P} = 40$ N at the instant when the crank angle $\phi = 45°$.

Solution. The graphical analysis is shown in Figure 9.10b. First, consider connecting rod 2. In the absence of gravity and inertial forces, this link is acted on by two forces only, at pins B and C. These pins are assumed to be frictionless and, therefore, transmit no torque. Thus, link 2 is a two-force member loaded at each end as shown. Forces $\boldsymbol{F}_{12}$ and $\boldsymbol{F}_{32}$ lie along the link, producing zero net moment, and must be equal and opposite for equilibrium of the link. At this point, the magnitudes and senses of these forces are unknown.

Next, examine piston 3, which is a three-force member. The pressure force $\boldsymbol{P}$ is completely known and is assumed to act through the center of the piston (i.e., the pressure distribution on the

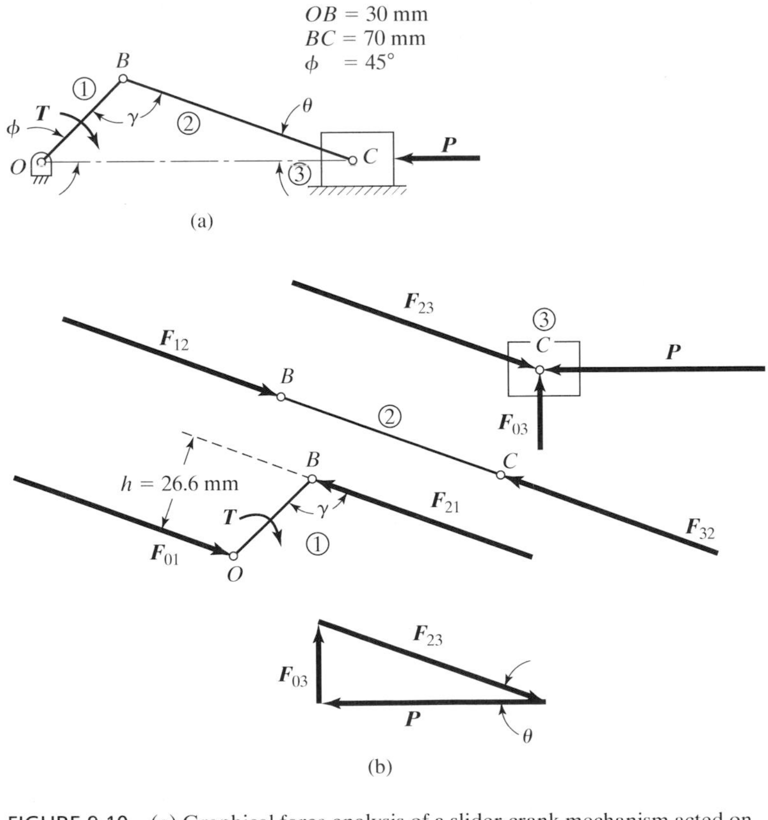

FIGURE 9.10 (a) Graphical force analysis of a slider-crank mechanism acted on by piston force $\boldsymbol{P}$ and crank torque $\boldsymbol{T}$. (b) Static force balances for the three moving links, each considered as a free body.

piston face is assumed to be symmetric). From Newton's third law, which states that for every action there is an equal and opposite reaction, it follows that $\boldsymbol{F}_{23} = -\boldsymbol{F}_{32}$, and the direction of $\boldsymbol{F}_{23}$ is therefore known. In the absence of friction, the force of the cylinder on the piston, $\boldsymbol{F}_{03}$, is perpendicular to the cylinder wall, and it also must pass through the concurrency point, which is the wrist pin C. Now knowing the directions of the forces, we can construct the force polygon for member 3 (Figure 9.10b). Scaling from this diagram, we see that the contact force between the cylinder and piston is $F_{03} = 12.7$ N, acting upward, and the magnitude of the bearing force at C is $F_{23} = F_{32} = 42.0$ N. This is also the bearing force at crankpin B, because $\boldsymbol{F}_{12} = -\boldsymbol{F}_{32}$. Further, the directions of the forces on the connecting rod shown in the figure are correct, and the link is in compression.

Finally, crank 1 is subjected to two forces and a pure torque $\boldsymbol{T}$. The force at B is $\boldsymbol{F}_{21} = -\boldsymbol{F}_{12}$ and is now known. For force equilibrium, $\boldsymbol{F}_{01} = -\boldsymbol{F}_{21}$, as shown on the free-body diagram of link 1. However, these forces are not collinear, and for equilibrium, the moment of this couple must be balanced by torque $\boldsymbol{T}$. Thus, the required torque is clockwise and has magnitude

$$T = F_{21}h = (42.0\ \text{N})(26.6\ \text{mm}) = 1120\ \text{N}\cdot\text{mm} = 1.12\ \text{N}\cdot\text{m}.$$

It should be emphasized that this is the torque required for static equilibrium in the position shown in Figure 9.10a. If information about the torque is needed for a complete compression cycle, then the analysis must be repeated at other crank positions throughout the cycle. In general, the torque will vary with position.

Graphical Force Analysis of the Four-Bar Linkage

The force analysis of the four-bar linkage proceeds in much the same manner as that of the slider-crank mechanism. However, in the next example, we consider the case of external forces on both the coupler and follower links and utilize the principle of superposition.

SAMPLE PROBLEM 9.2

Static-Force Analysis of a Four-Bar Linkage

The link lengths for the four-bar linkage of Figure 9.11a are given in the figure. In the position shown, coupler link 2 is subjected to force $\boldsymbol{F}_2$ of magnitude 47 N, and follower link 3 is subjected to force $\boldsymbol{F}_3$ of magnitude 30 N. Determine the shaft torque $\boldsymbol{T}_1$ on input link 1 and the bearing loads for static equilibrium.

Solution. As shown in Figure 9.11a, the solution of the stated problem can be obtained by superposition of the solutions of subproblems I and II. In subproblem I, force $\boldsymbol{F}_3$ is neglected, and in subproblem II, force $\boldsymbol{F}_2$ is neglected. This process facilitates the solution by dividing a more difficult problem into two simpler ones.

The analysis of subproblem I is shown in Figure 9.11b, with quantities designated by superscript I. Here, member 3 is a two-force member, because force $\boldsymbol{F}_3$ is neglected. The direction of forces $\boldsymbol{F}^{\text{I}}_{23}$ and $\boldsymbol{F}^{\text{I}}_{03}$ are as shown, and the forces are equal and opposite. (Note that the magnitudes and senses of these forces are as yet unknown.) This information allows the analysis of member 2, which is a three-force member with completely known force $\boldsymbol{F}^{\text{I}}_{2}$, known directions for $\boldsymbol{F}^{\text{I}}_{32}$, and, using the concurrency point, known direction for $\boldsymbol{F}^{\text{I}}_{12}$. Scaling from the force polygon,

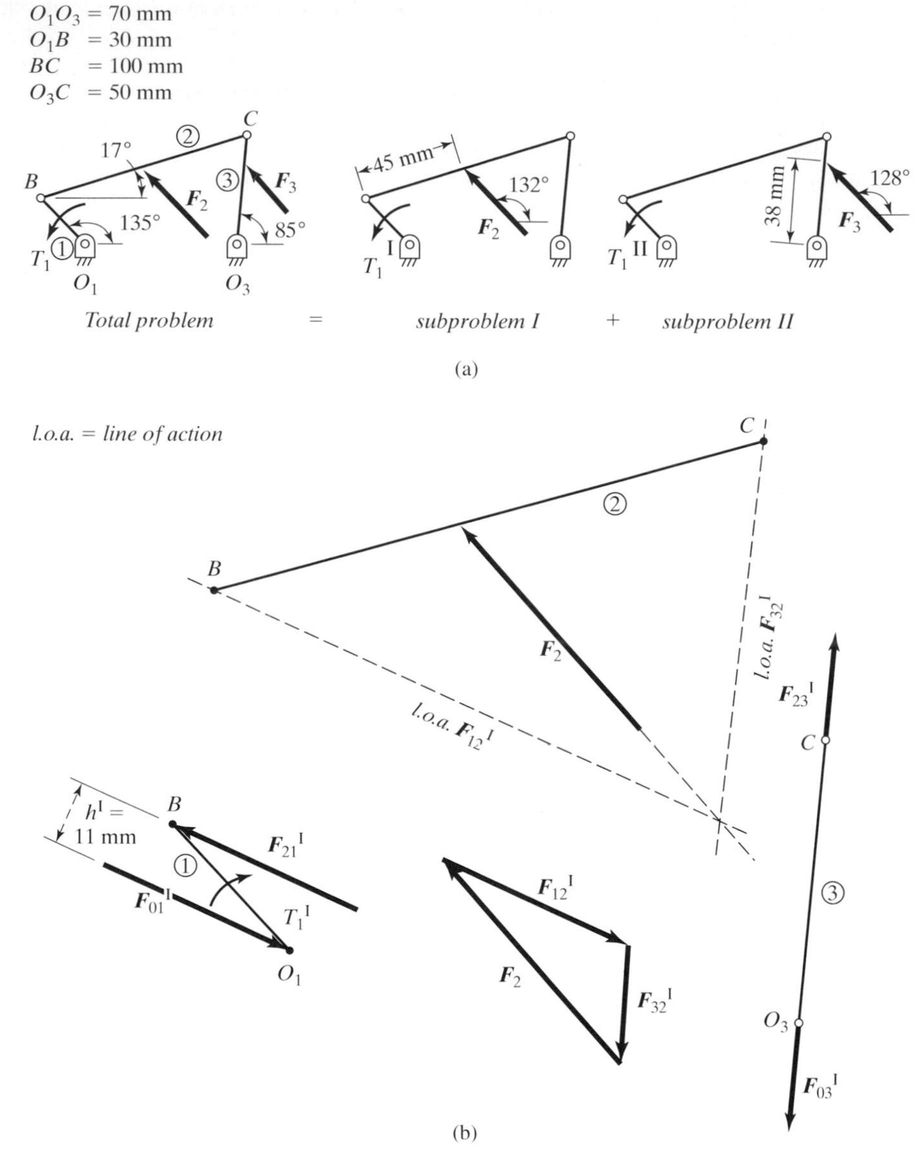

FIGURE 9.11 (a) Graphical force analysis of a four-bar linkage, utilizing the principle of superposition. (b) The solution of subproblem I.

we determine the following magnitudes for the forces (the directions of the forces are shown in Figure 9.11b):

$$F_{32}^{I} = F_{23}^{I} = F_{03}^{I} = 21 \text{ N};$$

$$F_{12}^{I} = F_{21}^{I} = 36 \text{ N}.$$

Link 1 is subjected to two forces and couple T_1^{I}, and, for equilibrium,

$$F_{01}^{\mathrm{I}} = -F_{21}^{\mathrm{I}}$$

and

$$T_1^{\mathrm{I}} = F_{21}^{\mathrm{I}} h^{\mathrm{I}} = (36\ \mathrm{N})(11\ \mathrm{mm}) = 396\ \mathrm{N \cdot mm\ cw.}$$

The analysis of subproblem II is similar and is shown in Figure 9.11c, where superscript II is used. In this case, link 2 is a two-force member and link 3 is a three-force member, and the following results are obtained:

$$F_{03}^{\mathrm{II}} = 29\ \mathrm{N},$$
$$F_{23}^{\mathrm{II}} = F_{21}^{\mathrm{II}} = F_{01}^{\mathrm{II}} = 17\ \mathrm{N},$$

and

$$T_1^{\mathrm{II}} = F_{21}^{\mathrm{II}} h^{\mathrm{II}} = (17\ \mathrm{N})(26\ \mathrm{mm}) = 442\ \mathrm{N \cdot mm\ cw.}$$

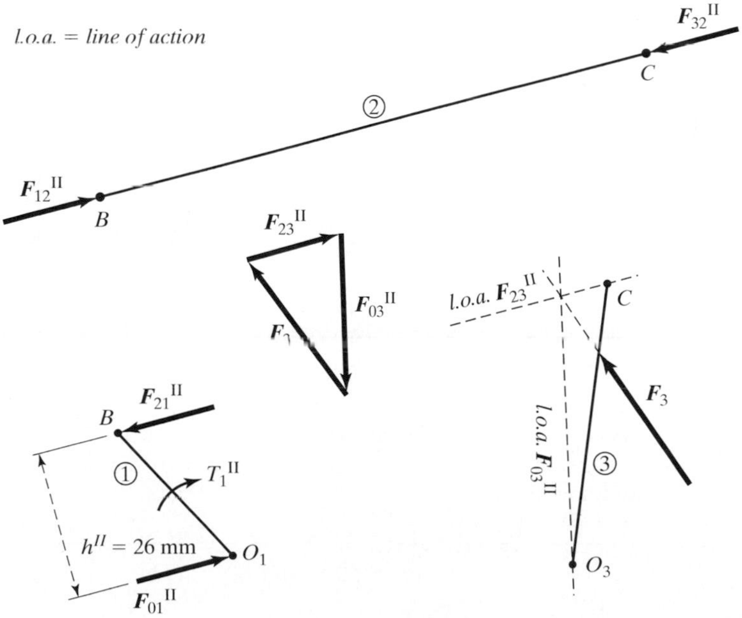

FIGURE 9.11 (c) The solution of subproblem II.

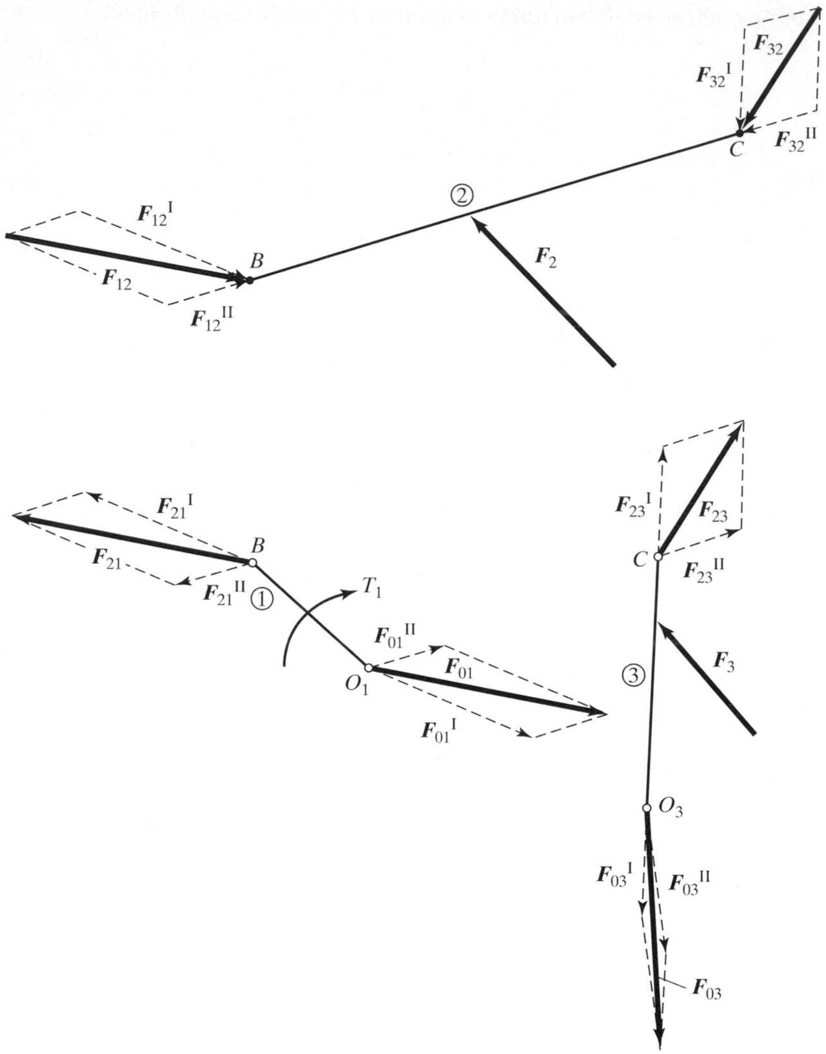

FIGURE 9.11 (d) The solutions combine to give the total solution.

The superposition of the results of Figures 9.11b and c is shown in Figure 9.11d. The results must be added vectorially, as shown. By scaling from the free-body diagrams, the overall bearing force magnitudes are found to be

$$F_{01} = 50\text{ N}, \qquad F_{23} = 31\text{ N},$$
$$F_{12} = 50\text{ N, and } F_{03} = 49\text{ N},$$

and the net crankshaft torque is

$$T_1 = T_1^{\text{I}} + T_1^{\text{II}} = 838\text{ N}\cdot\text{mm cw.}$$

The directions of the bearing forces are as shown in the figure. These resultant quantities represent the actual mechanism forces.

It can be seen from the analysis that the effect of the superposition principle in this example was to create subproblems containing two-force members, from which the separate analyses could begin. If one were to attempt a graphical analysis of the original problem without superposition, one would find that not enough intuitive information about the forces was given to analyze three-force members 2 and 3, because none of the bearing force directions can be determined by inspection.

Graphical Force Analysis of Complex Linkages

In this section, an example is presented involving the static-force analysis of a mechanical system that is somewhat more complex than the previous cases considered. This example will demonstrate that, although each force analysis problem has its own special characteristics, the solution procedure for a broad range of mechanisms is essentially unchanged, relying on the basic force analysis groundwork that has been developed.

SAMPLE PROBLEM 9.3

Force Analysis of a Door Mechanism

Figure 9.12a shows the plan view of a fourfold industrial door. The door, which opens at the center, has four panels, two of which fold to the left side and two to the right side. The door is shown in the closed position, with the open position of the panels inserted as dashed lines for reference. The figure also shows the electrically powered operating system mounted above the doors and consisting of two symmetric linkages driven by the same motor.

Figure 9.12b is a schematic drawing of the right half of the system, drawn to scale for an intermediate position between the open- and closed-door positions. Including the door frame and the two door panels as links, the mechanism is an eight-bar linkage. Member 0 is the frame, and members 1 and 2 are the door panels hinged together at point *B*. Slider 3 is pinned to panel 2 and moves along a fixed track as the doors open and close. Power is transmitted to the door by means of drive arm 7, connecting rod 6, and links 4 and 5.

Member 4 is connected to door panel 1 at point *E*, and member 5 is connected to the door frame at point *D*.

For the position shown in Figure 9.12b, determine the required shaft torque on drive arm 7 for static equilibrium against applied load F_2, which has a magnitude of 1000 N and acts on door panel 2 as shown.

Solution. It is assumed that inertial forces and friction effects are negligible. A planar force analysis will be performed that takes into account those forces which act in planes parallel to that shown in Figure 9.12b. Gravity loads, which act perpendicular to these planes, are not included in the analysis.

The graphical analysis, presented in Figure 9.12c, starts with slider 3, which is a two-force member. Since friction is neglected, these forces must act perpendicular to the guide track, thus establishing the directions of forces F_{03} and F_{23}. Door panel 2 is a three-force member with known force F_2 and known direction of force F_{32}. From this information, the concurrency point can be found and the force polygon constructed, yielding the following force magnitudes:

$$F_{32} = 420 \text{ N}; \quad F_{12} = 730 \text{ N}.$$

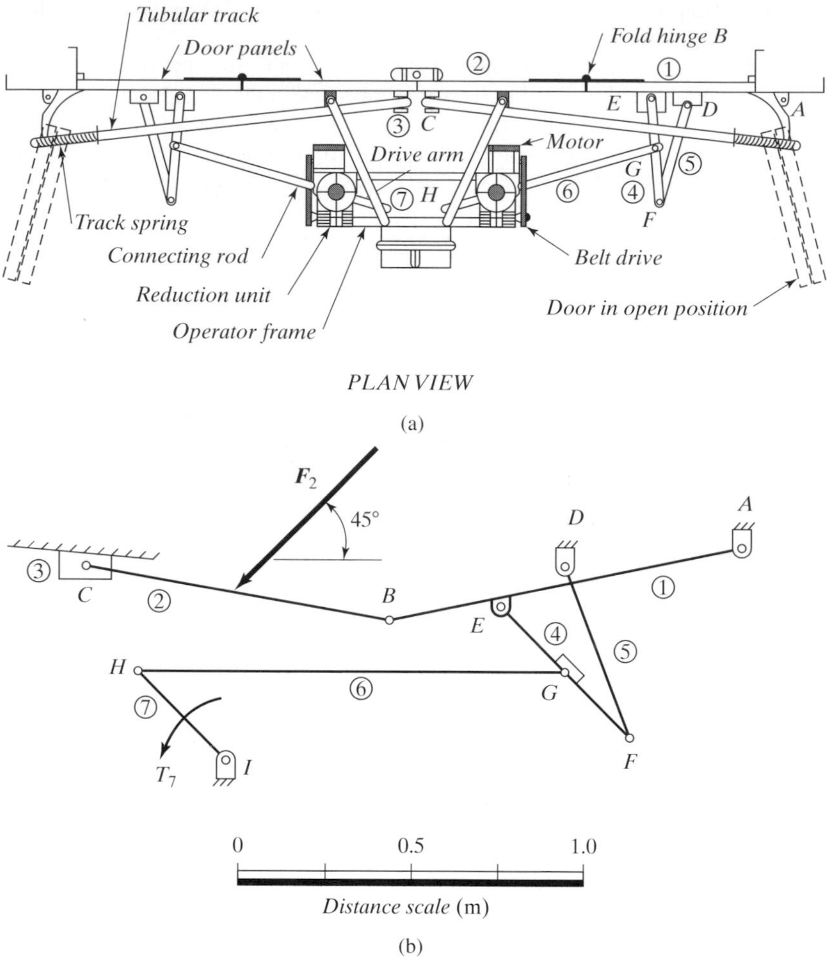

FIGURE 9.12 (a) An industrial door mechanism. (*Source:* Electric Power Door Company.) (b) Schematic linkage diagram.

Member 1 is also a three-force member, acted on by force $\boldsymbol{F}_{21}$, which is now completely known, and the forces $\boldsymbol{F}_{01}$ and $\boldsymbol{F}_{41}$, both of which have unknown direction and magnitude. In order for the analysis of this member to be completed, the direction of either $\boldsymbol{F}_{01}$ or $\boldsymbol{F}_{41}$ must be determined.

The direction of $\boldsymbol{F}_{41}$ can be found by considering link 4, which is another three-force member, acted upon by force $\boldsymbol{F}_{14}$ from link 1 at point E, force $\boldsymbol{F}_{54}$ from link 5 at point F, and force $\boldsymbol{F}_{64}$ from link 6 at point G. Since links 5 and 6 are two-force members, the lines of action of forces $\boldsymbol{F}_{54}$ and $\boldsymbol{F}_{64}$ are along the respective links, and the intersection of these lines is the concurrency point for member 4. (See Figure 9.12c.) This analysis leads to the line of action for $\boldsymbol{F}_{14}$ and, in turn, the direction of $\boldsymbol{F}_{41}$.

The force polygon can now be constructed for member 1, as shown in Figure 9.12c, yielding the following force magnitudes:

$$F_{01} = 960 \text{ N}; \quad F_{41} = 350 \text{ N}.$$

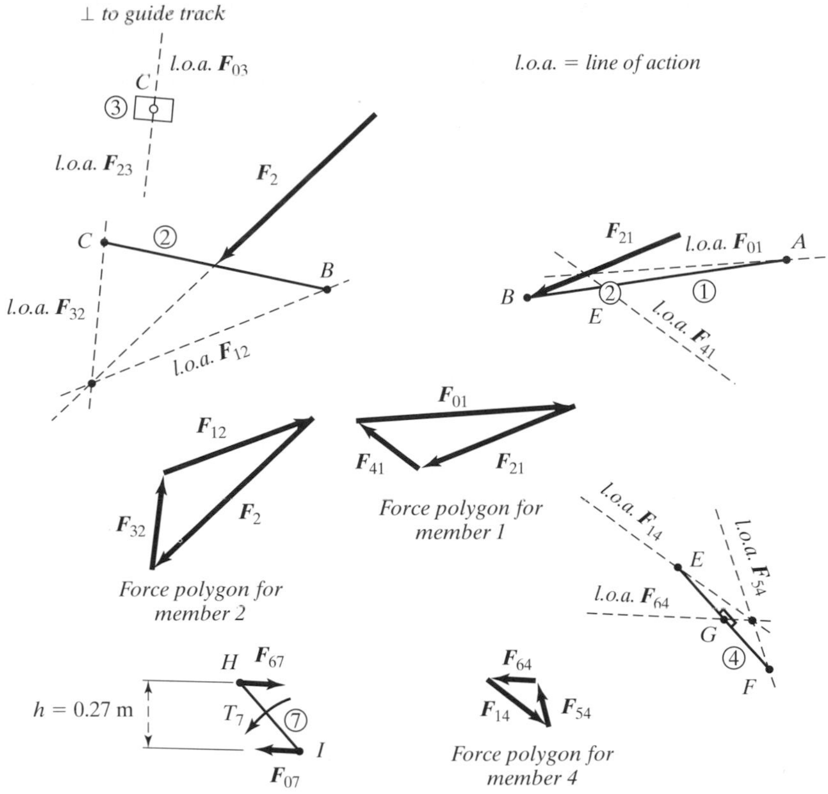

FIGURE 9.12 (c) Graphical force analysis of an industrial door mechanism.

Next, the polygon is constructed for member 4 (see Figure 9.12c), from which we obtain

$$F_{54} = 210 \text{ N} \quad \text{and} \quad F_{64} = 220 \text{ N}.$$

Finally, member 7 is acted upon by two forces—known force $\boldsymbol{F}_{67}$ and equal and opposite force $\boldsymbol{F}_{07}$ (see Figure 9.12c)—and shaft torque $\boldsymbol{T}_7$, which must be equal and opposite to the moment of the couple, $\boldsymbol{F}_{67}$, $\boldsymbol{F}_{07}$. Therefore, the torque is counterclockwise and, scaling the moment arm from Figure 9.12b, we find that the magnitude of the torque is

$$T_7 = hF_{67} = (0.27 \text{ m})(220 \text{ N}) = 59.4 \text{ N} \cdot \text{m ccw}.$$

If the other half of the door system is loaded symmetrically, a total torque double that just found would be required. From this and knowledge of the speed reduction unit, the necessary motor torque can be found.

Similar analyses can be performed throughout the range of motion of the mechanism in order to size components for proper operation under various loading conditions, such as wind loads, which would be represented by external loads on both door panels.

9.4 ANALYTICAL STATICS

Analytical methods for investigating static and dynamic forces in machines employ mathematical models that are solved either (1) for unknown forces and torques associated with a known motion of the mechanism, or (2) for an unknown motion of a given mechanism resulting from known driving forces or torques. This text deals almost exclusively with the former; however, a brief discussion of the latter appears in Chapter 10. There are two approaches to formulating mathematical models, one based on force and moment equilibrium and the other is based on energy principles. Methods utilizing force and moment equilibrium equations parallel very closely the graphical method that has been presented. Both rely heavily on free-body diagrams, but the graphical force polygons are replaced in the analytical approach by equivalent vector equations. Energy methods utilize the principle of conservation of energy; one of the best-known such examples is the method of virtual work.

The mathematical basis of the analytical approach lends itself well to a computer implementation. Solutions can be obtained quickly and accurately for many positions of a mechanism, and the computer is particularly useful in design situations in which many variations of the mechanism are to be considered. The computer facilitates design optimization, wherein those values of design parameters are determined such that selected performance criteria are optimized. The designer may choose to write his or her own computer program for analysis or apply one of a number of general-purpose programs that are available.

The next sections introduce some of the basic theory involved in analytical methods uesed in static-force analysis.

Static-Equilibrium Equations

The mathematical conditions for static equilibrium of a body were stated in Eqs. (9.7a) and (9.7b), that is,

$$\sum \boldsymbol{F} = 0$$

and

$$\sum \boldsymbol{T} = 0.$$

The detailed mathematical expression of these equations can take many forms, depending on the vector representation used, and, for example, may involve Cartesian vectors or complex-number vectors, fixed or moving coordinate systems, and so on.

Employing Cartesian vectors referenced to a fixed x, y, z coordinate frame, we see that the component forms of Eqs. (9.7a) and (9.7b) become

$$\sum F_x = 0, \tag{9.9a}$$

$$\sum F_y = 0, \tag{9.9b}$$

$$\sum F_z = 0, \tag{9.9c}$$

$$\sum T_x = 0, \tag{9.9d}$$

$$\sum T_y = 0, \tag{9.9e}$$

and

$$\sum T_z = 0, \tag{9.9f}$$

which together state that the net forces on a body in the x, y, and z directions must be zero and the net moments on the body about any three axes parallel to the x, y, and z directions must be zero. For two-dimensional force problems in the xy-plane, Eqs. (9.9a) through (9.9f) reduce to the three conditions of Eqs. (9.8a) through (9.8c) presented earlier in the chapter:

$$\sum F_x = 0;$$

$$\sum F_y = 0;$$

$$\sum T_z = 0.$$

For determinate force systems, Eqs. (9.9a) through (9.9f) will yield solutions to spatial force problems, and Eqs. (9.8a) through (9.8c) will yield solutions to planar force problems.

SAMPLE PROBLEM 9.4

Force Analysis of a Front-End Loader

Figure 9.13a is a photograph of a front-end loader showing the linkage arrangement for the boom mechanism. The boom is actuated by two hydraulic cylinders, one on each side of the machine, and the bucket is pivoted relative to the boom by a third hydraulic cylinder. Neglecting member weights and friction effects, determine the cylinder force required for static equilibrium of the boom in the position shown under a total bucket load of 4000 N.

Solution. In a thorough design analysis of the loader, the member weights would also be considered; they are neglected here in order to simplify the example. Also, it is assumed that the bucket load is evenly distributed between the two sides of the loader. Therefore, we will consider just one side under a vertical bucket load P having a magnitude of 2000 N.

FIGURE 9.13 (a) A front-end loader. (*Source:* Sperry New Holland Company.)

Figure 9.13b is a drawing of the boom mechanism in the analysis position showing the force P and various dimensions and angular orientations. The xy coordinate system has been selected, with x horizontal and y vertical.

A free body consisting of the bucket and boom is shown in Figure 9.13c, and, as indicated, four forces act on the body. These forces, which are identified in x and y component form, are the vertical bucket load $\boldsymbol{P}$, the force $\boldsymbol{F}_{12}$ from member 1, the force $\boldsymbol{F}_{32}$ from member 3, and the cylinder force $\boldsymbol{F}_c$. Applying Eqs. (9.8a) and (9.8b), we have

$$F_{cx} + F_{12x} + F_{32x} = 0 \tag{9.10a}$$

and

$$F_{cy} + F_{12y} + F_{32y} - P = 0, \tag{9.10b}$$

and summing moments about point O, we find that Eq. (9.8c) becomes

$$2.84P - 0.71F_{32x} - 0.90F_{cx} - 0.22F_{cy} = 0. \tag{9.10c}$$

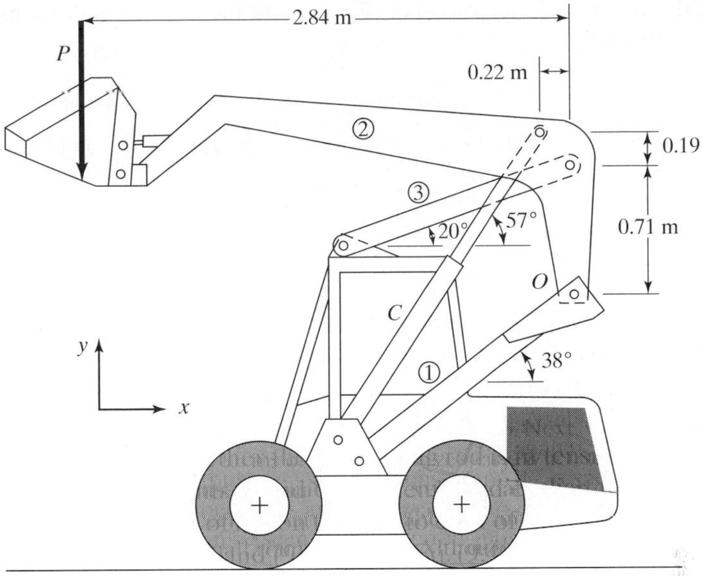

(b)

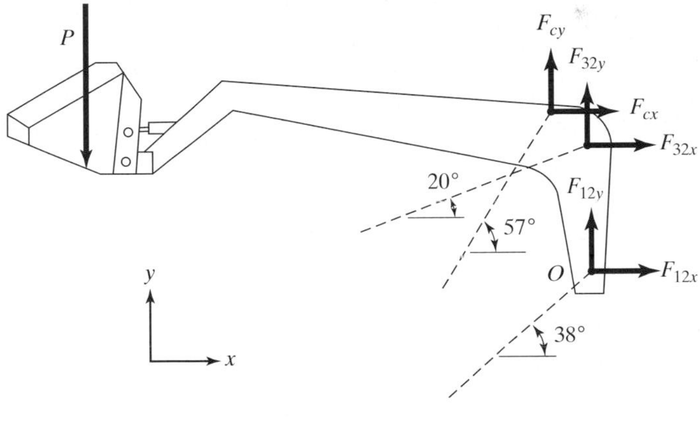

(c)

FIGURE 9.13 (b) Dimensions of the front-end loader. (c) Free body diagram of the bucket and boom.

Equations (9.10a) through (9.10c) are a system of three equations in six unknowns: F_{cx}, F_{cy}, F_{12x}, F_{12y}, F_{32x}, and F_{32y}. However, links 1 and 3 and the hydraulic cylinder are two-force members, and therefore, the directions of the forces that they exert on the boom will be along the links. Thus, the six unknowns can be expressed in terms of three unknowns as:

$$F_{cx} = F_c \cos(57°), \qquad F_{cy} = F_c \sin(57°),$$
$$F_{12x} = F_{12} \cos(38°), \qquad F_{12y} = F_{12} \sin(38°),$$
$$F_{32x} = F_{32} \cos(20°), \text{ and } F_{32y} = F_{32} \sin(20°),$$

where the angular orientations of the links are for the position under consideration. Substituting into Eqs. (9.10a) through (9.10c) we have

$$(0.545)\,F_c + (0.788)\,F_{12} + (0.940)\,F_{32} = 0,$$

$$(0.839)\,F_c + (0.616)\,F_{12} + (0.342)\,F_{32} - 2{,}000 = 0,$$

and

$$5{,}680 - (0.667)\,F_{32} - (0.675)\,F_c = 0.$$

Solving these equations for the three unknowns, we obtain

$$F_c = 6{,}595\text{ N},\ F_{12} = -6{,}758\text{ N, and } F_{32} = 1{,}842\text{ N},$$

or, in component form,

$$F_{cx} = 3{,}594\text{ N},\quad F_{cy} = 5533\text{ N},$$

$$F_{12x} = -5{,}325\text{ N},\quad F_{12y} = -4{,}163\text{ N},$$

$$F_{32x} = 1{,}731\text{ N, and } F_{32y} = 630\text{ N}.$$

The signs indicate whether the force components are in the positive or negative coordinate directions. Thus, the actual directions of the components of forces $\boldsymbol{F}_c$ and $\boldsymbol{F}_{32}$ are as shown in Figure 9.13c, whereas the components of $\boldsymbol{F}_{12}$ act in the negative coordinate directions. This means that member 3 and the cylinder are acted on by compressive forces; whereas member 1 is in tension for the position analyzed.

Analytical Solution for the Slider-Crank Mechanism

Because of its extensive use, the slider-crank mechanism deserves special attention. In the next chapter, a detailed dynamic-force analysis of this mechanism will be presented that will account for inertial effects, which are usually significant in machines such as engines and compressors. In this chapter, a graphical static-force analysis has already been presented to determine the relationship between piston force and crank torque. In this section, an equivalent analytical model will be developed.

An in-line slider-crank mechanism is shown in Figure 9.14a with crank length r, connecting rod length ℓ, and piston force $\boldsymbol{P}$. A mathematical expression is sought relating force P to the crankshaft torque T required for equilibrium. This expression will be a function of the position of the mechanism, as given by crank angle ϕ.

Free bodies of the moving links are shown in Figure 9.14b. Connecting rod 2 is a two-force member, and therefore, the force it exerts on piston 3, $\boldsymbol{F}_{23}$, will act at the angle θ of the connecting rod. Summing the forces on the piston in the x direction, we have

$$F_{23}\cos\theta = P,$$

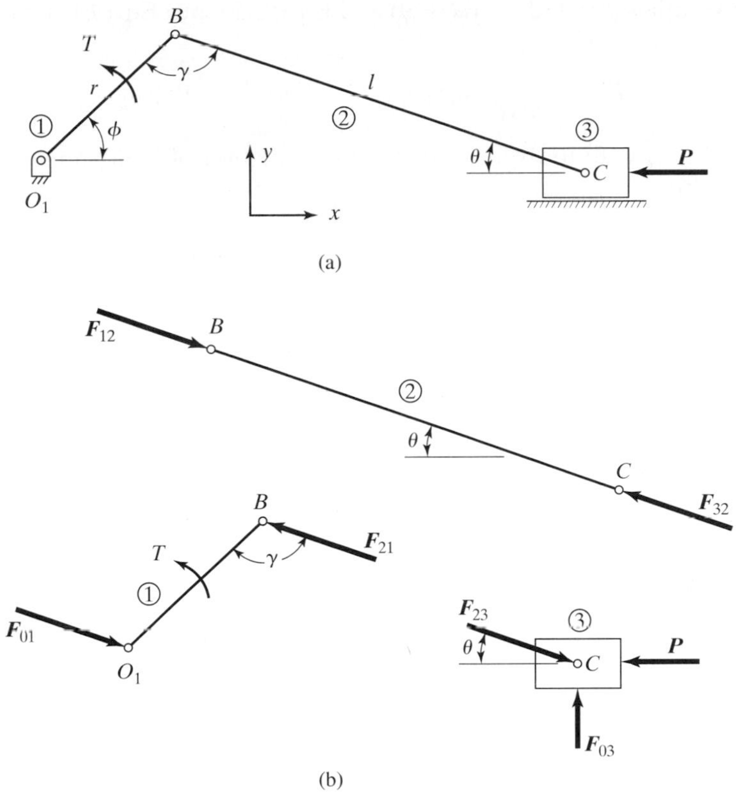

FIGURE 9.14 (a) An in-line slider-crank linkage. (b) Free-body diagrams of the moving members, employed in an analytical solution for torque T as a function of piston force $\boldsymbol{P}$.

or

$$F_{23} = \frac{P}{\cos\theta}. \tag{9.11}$$

Again, note that the connecting rod is a two-force member. Force F_{21} is equal and opposite force F_{23}. The magnitude of the force at the end of the crank is given by

$$F_{21} = \frac{P}{\cos\theta}. \tag{9.12}$$

Summing moments about point O_1, we have

$$T = -F_{21}\, r \sin\gamma \tag{9.13}$$

A negative sign is used to denote a clockwise torque, which will occur for the force convention shown (which assumes that a positive piston force P acts to the left).

Substituting $\gamma = 180° - (\phi + \theta)$ and Eq. (9.12) into Eq. (9.13), we have

$$T = -\frac{Pr}{\cos\theta}\sin[180° - (\phi + \theta)] = -Pr(\sin\phi + \cos\phi\tan\theta).$$

Finally, we wish to express angle θ as a function of crank angle ϕ. From the linkage geometry,

$$\sin\theta = \frac{r}{\ell}\sin\phi$$

and

$$\cos\theta = \sqrt{1 - \sin^2\theta} = \sqrt{1 - \left(\frac{r}{\ell}\sin\phi\right)^2}.$$

Substitution yields

$$T = -Pr\sin\phi\left(1 + \frac{r\cos\phi}{\sqrt{\ell^2 - r^2\sin^2\phi}}\right). \tag{9.14}$$

As can be seen, even if force $\boldsymbol{P}$ is constant, torque $\boldsymbol{T}$ will vary as the linkage orientation changes. Of course, in engines and compressors, $\boldsymbol{P}$ (the force due to cylinder pressure) will also vary with position. This can be accounted for in Eq. (9.14) by expressing $\boldsymbol{P}$ as a function of ϕ. The torque will be zero when $\phi = 0$ or 180°, corresponding, respectively, to the top and bottom dead-center positions of the mechanism. An engine can lock in these positions under force $\boldsymbol{P}$, unless acted upon by other forces, such as inertial effects, or the torque from other cylinders.

SAMPLE PROBLEM 9.5

Analysis of a Slider-Crank Mechanism

Calculate the torque required for static equilibrium of an in-line slider-crank mechanism in the position when crank angle $\phi = 45°$. The dimensions are $r = 30$ mm and $l = 70$ mm, and the piston force is $P = 40$ N.

Solution. Substituting into Eq. (9.14), we have

$$T = -(40)(30)\sin(45°)\left[1 + \frac{(30)\cos(45°)}{\sqrt{(70)^2 - (30)^2\sin^2(45°)}}\right]$$

$$= -1119 \text{ N} \cdot \text{mm}.$$

This result agrees with that determined by graphical solution in Sample Problem 9.1.

Detailed Force Analysis of a Linkage Using Vector Methods

After proposing a linkage design, we analyze the design through a full cycle of motion.

Link Positions. Graphical methods are out of the question for a detailed analysis (except for checking one or two positions). But we can choose complex-number methods, analytical vector methods, or simulation software. Suppose we select vectors, which may seem complicated, but they certainly keep everything in order, and we can use them later to find velocities, forces, and torques. One possible procedure is as follows:

- Sketch the proposed linkage design.
- Define the rectangular unit vectors, and describe each link as a vector.
- Identify angles in standard form.
- See "Solution of Planar Vector Equations" and "Displacement Analysis of Planar Linkages: Analytical Vector Methods" in Chapter 2.
- Select the solution that fits the mechanism and the given data.
- Solve for the unknown vectors.
- Check the vector closure equation (for at least one arbitrary position). Do the links form a closed loop? Does the sum of the link vectors equal zero?
- Check the transmission angle if applicable. This step may provide redesign clues if we are not happy with the force analysis that follows.
- Plot the results of the position analysis. Are the curves continuous? Are the results reasonable?

Velocities and Accelerations. A velocity analysis may not be necessary, but it helps to check the design. You could do the following:

- Write the vector velocity equations for the linkage considered. Use the cross-product form (as in Chapter 3).
- Find the unknown velocities and angular velocities.
- Repeat for accelerations if desired. Refer to Chapter 4.
- Check at least one position, using graphical methods, complex-number methods, or numerical differentiation.

Force Analysis. Are we dealing with a pump, a compressor, or a materials-handling device? Applied forces are likely to be given as a function of time or a function of link position. You may be able to find linkage forces as follows:

- Estimate component masses and the magnitude of inertial forces and torques. If inertial effects are small compared with applied forces and torques, then static analysis is valid.
- Construct free-body diagrams of each link. A few minutes spent drawing such a diagram will be repaid as you avoid errors in the analysis.
- Identify known and unknown forces and torques. If there is a piston, be sure to include the lateral force that the cylinder exerts on the piston.
- Write the force and moment equilibrium equations.
- Look for easy solutions. If you start in the right place, the equations may be uncoupled. Otherwise you may have to solve simultaneous equations.
- If a connecting rod or coupler is a two-force member, then the force in that body lies along its axis.

- Use Newton's third law: *If one link exerts a force on a second link, the second link exerts an equal and opposite force on the first.* Solve one link; move along to solve the next link.
- If the force at the head of a link vector has the same sense as the vector itself, the link is in tension at that time. The link is in compression if the force and the vector oppose each other.

Torque. If you are designing a pump, compressor, or crusher, you probably know the required output forces. You must find the required input torque. If you are designing an engine, and gas pressure is known in terms of piston position, then you can predict the output torque. Try these steps:

- Calculate link forces.
- A force vector **F** at the head of a link vector **r** has a moment $\mathbf{r} \times \mathbf{F}$. That moment is balanced by a torque $\mathbf{T} = -\mathbf{r} \times \mathbf{F}$ at the other end of the link.
- Consider a free-body diagram of the linkage as a whole. Do the external forces and torques balance?
- Plot forces and torques against crank position.
- You may also want to plot forces and torques against piston position. These plots will be continuous curves.
- Examine both sets of plots. Redesign the linkage if it does not meet your needs.

Static-Force Analysis as a First Approximation

Consider a moving component in a linkage. Suppose the product of mass and acceleration may be significant in comparison with applied forces. Or suppose the product of mass moment of inertia and angular acceleration may be significant compared with applied torques. Then, a designer must consider inertial effects. The difficulty is twofold:

- A designer needs to know inertial forces and inertial torques in order to specify the magnitudes of components, but
- Inertial effects depend on the size and mass of components that are yet to be designed.

Static-force analysis will at least provide a first approximation of loading on a linkage, even before we know the final masses. If it appears that inertial forces will be significant compared with static forces, then we must use dynamic-force analysis (to be examined in the next chapter).

SAMPLE PROBLEM 9.6

Detailed Analysis of a Proposed Compressor Design

Design a single-cylinder compressor to operate at 30 rpm and compress air to about 3.25 MPa (gage pressure). Analyze positions, forces and torques through a full cycle of operation. Plot the results.

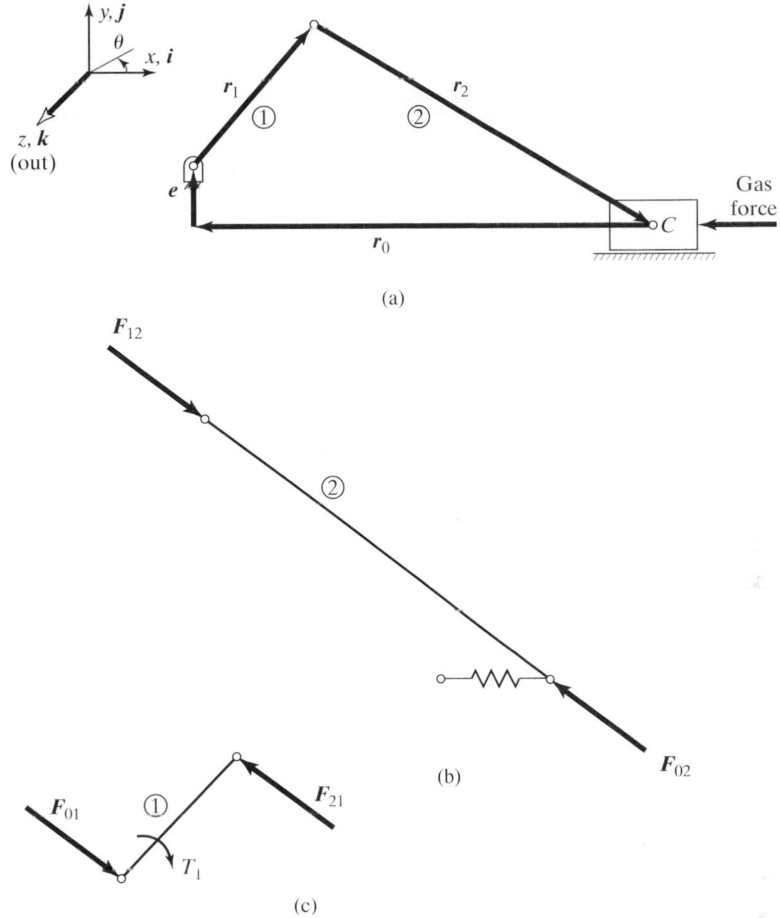

FIGURE 9.15 Analysis of a proposed compressor design. (a) Linkage sketched as a closed vector loop. (b) Connecting rod forces. (c) Crank forces and torque.

Design decisions. An 80-mm-diameter cylinder will be used. We will try an offset slider-crank linkage with 75-mm crank length, 135-mm connecting-rod length, and 10-mm offset. (See Figure 9.15.) The offset is not typical of this application; it is included only to show calculation methods for the more general case. Linkage proportions are usually based on the designer's experience with similar applications. A static-force analysis will enable the designer to tentatively specify the magnitudes and weights of components. If inertial effects appear significant compared to static forces and torques, then a dynamic-force analysis should follow.

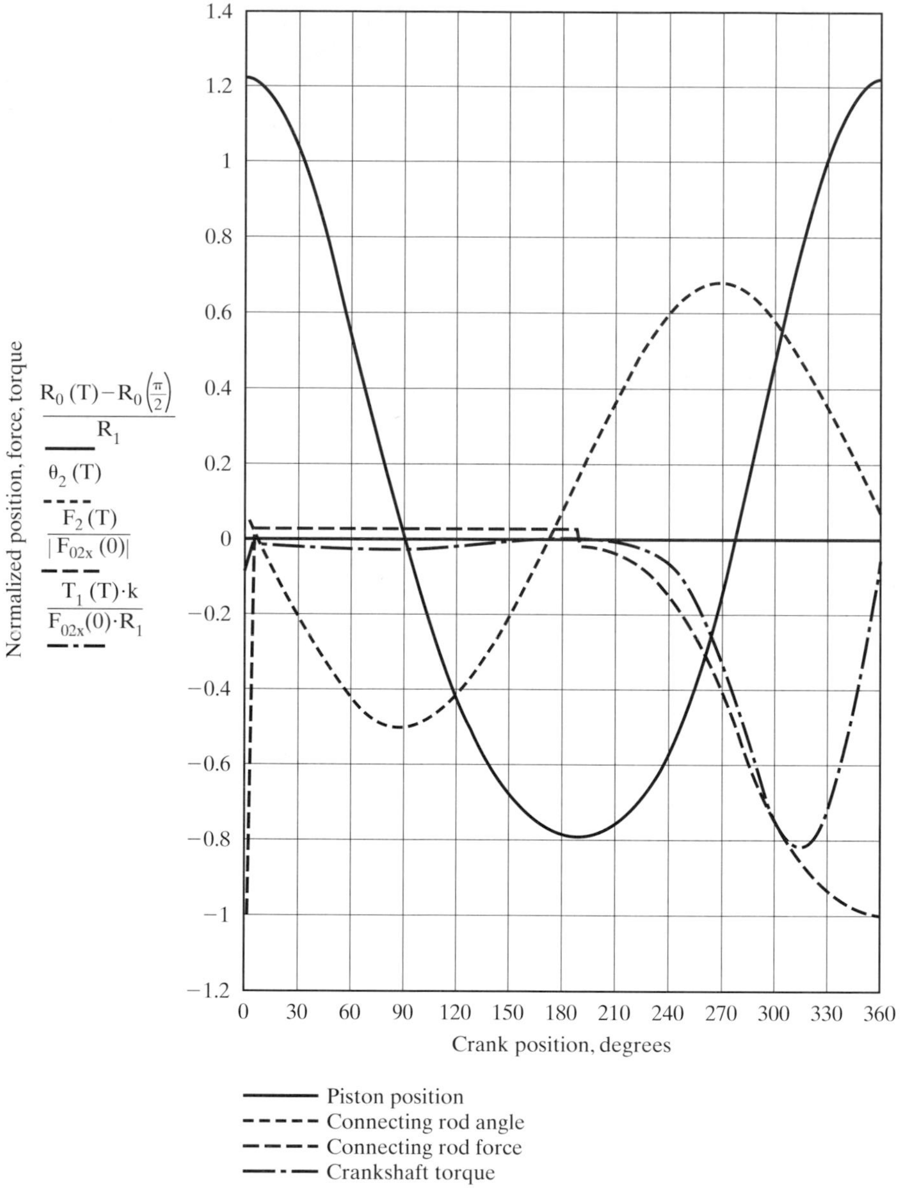

FIGURE 9.15 (d) Piston position*, connecting-rod angle (rad), connecting-rod force*, and crankshaft torque* plotted against crank position (°).
*Normalized.

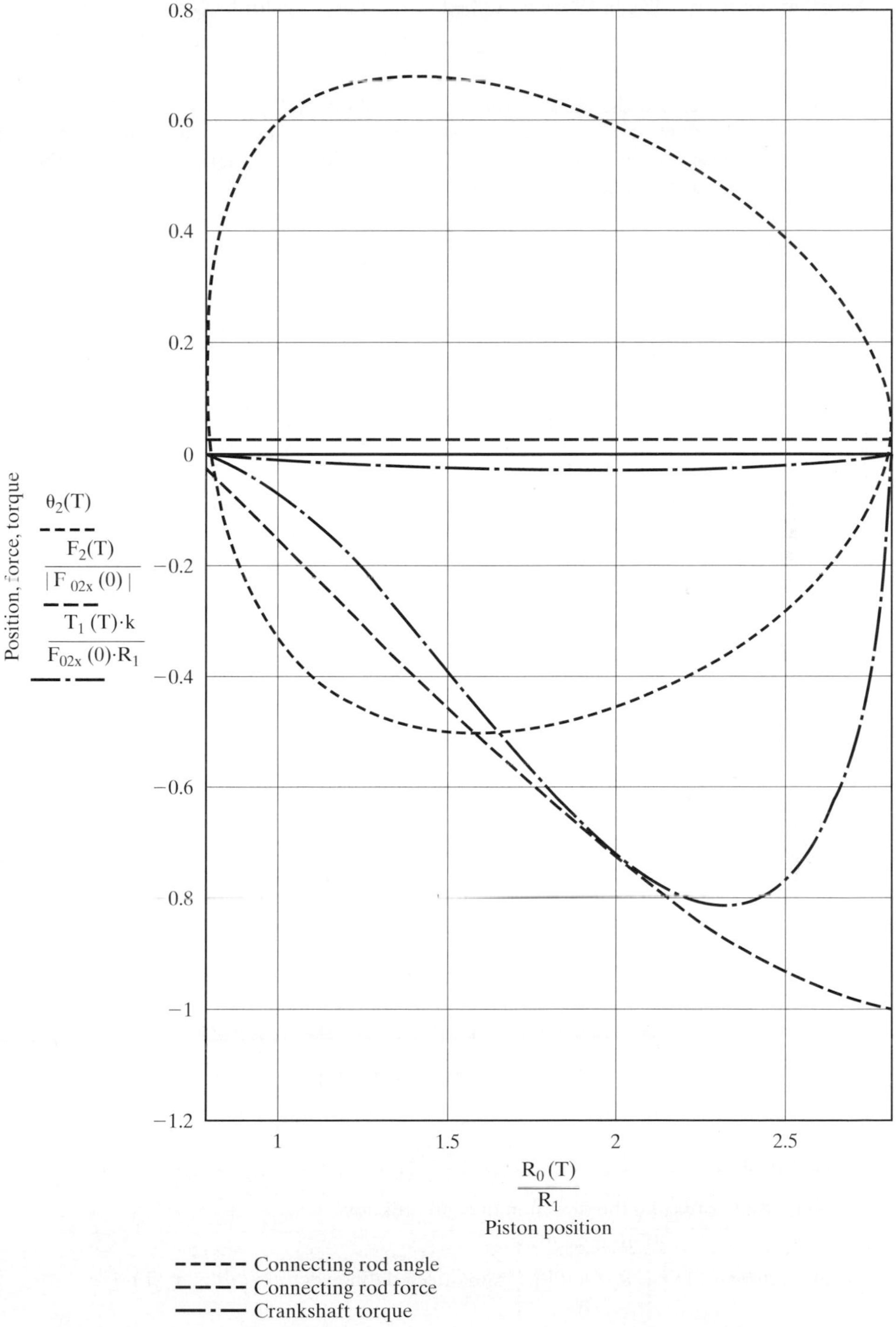

FIGURE 9.15 (e) Connecting-rod angle (rad), connecting-rod force*, and crankshaft torque* plotted against piston position*.
*Normalized.

Solution summary. The links are described by the closed vector loop

$$\boldsymbol{r}_3 + \boldsymbol{r}_2 + \boldsymbol{r}_0 = 0,$$

where $\boldsymbol{r}_3 = \boldsymbol{r}_1 + \boldsymbol{e}$, and the crank position is identified as $\boldsymbol{T}$ (rad) for convenience. Analytical vector methods are used to complete the position analysis. Checking the vector loop, we find the closure error insignificant. Vector methods are used to find the angular velocity of the connecting rod and the piston velocity. Numerical differentiation is used to check the results for one position; the results agree.

Maximum air pressure causes a piston force of

$$-\ p\ A_{\text{cyinder}} = -3.25 \cdot \pi\ 40^2 = -16{,}336\ \text{N}.$$

This gas force on the piston is modeled with a spring that is "connected" for only part of the cycle (when $V_C \geq 0$). The spring force is

$$F_{\text{spring}} = -K[R_0 - (R_2 - R_1)],$$

where K = spring rate = 110 N/mm.

An if-statement is used to turn off the spring when a valve discharges the compressed air. An additional force that opposes the motion of the piston models friction effects with a constant force of $0.05R_1\ K$. The gas force and friction force combine to a produce total horizontal force F_{02x} on the wrist pin.

Detailed calculations.

Compressor static-force analysis using vector methods directly. Units: N, mm, sec, rad

Vector equation $\mathbf{r}_3 + \mathbf{r}_2 + \mathbf{r}_0 = 0$ where $\mathbf{r}_3 = \mathbf{r}_1 + \mathbf{e}$

Let $\mathrm{T} = \theta_1$ = crank position (radians) $\mathrm{T} := 0, \dfrac{\pi}{72} .. 2\pi$

Given: Clockwise configuration

Crank length $\mathrm{R}_1 := 75$ Connecting rod length $\mathrm{R}_2 := 135$

Offset $\mathrm{e} := \begin{bmatrix} 0 \\ -10 \\ 0 \end{bmatrix}$ $\mathrm{r_o}$ unit vector $\mathrm{r}_{0\mathrm{u}} := \begin{bmatrix} -1 \\ 0 \\ 0 \end{bmatrix}$ Crank speed rpm $\mathrm{n}_1 := 30$

Angular velocity rad/s $\omega_1 := \dfrac{\pi \cdot \mathrm{n}_1}{30}$ $\omega_1 = 3.142$

Rectangular unit vectors $\mathrm{i} := \begin{bmatrix} 1 \\ 0 \\ 0 \end{bmatrix}$ $\mathrm{j} := \begin{bmatrix} 0 \\ 1 \\ 0 \end{bmatrix}$ $\mathrm{k} := \begin{bmatrix} 0 \\ 0 \\ 1 \end{bmatrix}$

Angular acceleration $\alpha_1 := 0$

Position analysis

The magnitude of r_0 and the direction of r_2 are unknown.

Crank vector $\mathrm{r}_1(\mathrm{T}) := \begin{bmatrix} \mathrm{R}_1 \cdot \cos(\mathrm{T}) \\ \mathrm{R}_1 \cdot \sin(\mathrm{T}) \\ 0 \end{bmatrix}$ Add offset, define vector $\mathrm{r}_3(\mathrm{T}) := \mathrm{r}_1(\mathrm{T}) + \mathrm{e}$

Slider position vector $\mathrm{r}_0(\mathrm{T}) := \left[-\mathrm{r}_3(\mathrm{T}) \cdot \mathrm{r}_{0\mathrm{u}} + \left[\mathrm{R}_2{}^2 - \left[\mathrm{r}_3(\mathrm{T}) \cdot (\mathrm{r}_{0\mathrm{u}} \times \mathrm{k})\right]^2\right]^{\frac{1}{2}}\right] \cdot \mathrm{r}_{0\mathrm{u}}$

Scalar for plotting $R_0(T) := -r_0(T) \cdot i$ (+ to right) $R_0(0) = 209.629$

For convenience, define $A(T) := r_3(T) \cdot (r_{0u} \times k)$ $R_0(\pi) = 59.629$

Connecting rod vector $r_2(T) := -A(T) \cdot (r_{0u} \times k) - (R_2^2 - A(T)^2)^{\frac{1}{2}} \cdot r_{0u}$

Angular position $q(T) :=$ angle $(r_2(T)_0, r_2(T)_1)$

for plotting $\theta_2(T) := \text{if}(q(T) \le \pi, q(T), q(T) - 2\pi)$, $\dfrac{\theta_2(60 \cdot \text{deg})}{\text{deg}} = -24.02$ deg

Check vector closure $e + r_0(1) + r_1(1) + r_2(1) = \begin{bmatrix} 1.421 \cdot 10^{-14} \\ 0 \\ 0 \end{bmatrix}$

Velocity analysis

Connecting-rod angular velocity $\omega_2(T) := \dfrac{-\omega_1 \cdot (r_1(T) \cdot i)}{r_2(T) \cdot i}$

Slider velocity vector (positive to right) $v_C(T) := \omega_1 \cdot (k \times r_1(T)) + \omega_2(T) \cdot (k \times r_2(T))$

scalar $V_C(T) := v_C(T) \cdot i$

Numerical differentiation $V_{Cn}(T) := \omega_1 \cdot \left(\dfrac{d}{dT} R_0(T)\right)$ $\omega_{2n}(T) := \omega_1 \cdot \left(\dfrac{d}{dT} \theta_2(T)\right)$

Check: $V_C(1) = -252.743$ $V_{Cn}(1) = -252.743$ mm/s

$\omega_2(1) = -1.026$ $\omega_{2n}(1) = -1.026$ rad/s

Force analysis

Model gas forces with spring. Spring force opposes motion of piston

Spring rate N/mm $K := 110$ Add an additional force to approximate friction effects:

$$F_{02x}(T) := \text{if}\,[V_C(T) \ge 0, -K \cdot [R_0(T) - (R_2 - R_1) + .05 \cdot R_1], .05 \cdot R_1 \cdot K]$$

$$F_{02x}(0) = -16{,}871.703 \quad F_{02x}(.25 \cdot \pi) = 412.5 \quad F_{02x}(1.75 \cdot \pi) = -12{,}778.049$$

Reaction force due to cylinder

Connecting rod is a two-force member $F_{02y}(T) := F_{02x}(T) \cdot \tan(\theta_2(T))$

$$F_{02y}(1) = -176.515 \quad F_{02y}(1.75 \cdot \pi) = -6746.789$$

Force vector on connecting rod at piston

$$F_{02}(T) := \begin{bmatrix} F_{02x}(T) \\ F_{02y}(T) \\ 0 \end{bmatrix} \quad F_{02}(0) = \begin{bmatrix} -16{,}871.703 \\ -1253.199 \\ 0 \end{bmatrix}$$

If force vector has same sense as connecting-rod vector, connecting rod is in tension (+); If vectors oppose, compression (−)

Total force on link 2 $F_2(T) := F_{02}(T) \cdot \dfrac{r_2(T)}{R_2}$ $F_2(1.75 \cdot \pi) = -14{,}449.833$

$$F_2(0) = -16{,}918.182 \quad F_2(\pi) = 413.636$$

Check magnitude $F_{2m}(T) := (F_{02x}(T)^2 + F_{02y}(T)^2)^{\frac{1}{2}}$ $F_{2m}(1.75 \cdot \pi) = 14449.833$

For two-force connecting rod, force from piston is applied at crankpin; the torque is $R \times F$, and the torque applied by the crankshaft is

$$T_1(T) := -r_1(T) \times F_{02}(T) \quad r_1(1.75 \cdot \pi) = \begin{bmatrix} 53.033 \\ -53.033 \\ 0 \end{bmatrix} \qquad T_1(1.75 \cdot \pi) = \begin{bmatrix} 0 \\ 0 \\ 1.035 \cdot 10^6 \end{bmatrix}$$

Check torque based on external forces

$$T_{1c}(T) := -(F_{02y}(T) \cdot R_0(T) + F_{02x}(T) \cdot e \cdot j) \quad T_{1c}(1.75 \cdot \pi) = 1.035 \cdot 10^6$$

Analytical Solution for the Four-Bar Linkage

In this section, a generalized analysis of the four-bar linkage will be presented. The equations to be derived are applicable to a wide variety of static-force situations, and in the next chapter it will be shown that they also apply to dynamic-force analysis. In addition, the same procedure with essentially the same equations can be used to analyze more complex mechanisms.

A four-bar linkage is shown in Figure 9.16a, with the link lengths designated by ℓ_i and angular positions represented by angles ϕ_i, where $i = 1,2,3$. The distances r_i, with $i = 1,2,3$, locate the points of intersection of the lines of action of applied forces $\boldsymbol{F}_1$, $\boldsymbol{F}_2$, and $\boldsymbol{F}_3$ with the respective links. In addition to these loads, it is assumed that each link is acted upon by an externally applied couple C_i and input link 1 is driven by shaft torque T. The sign convention to be used for torques is that counterclockwise torques are positive and clockwise torques are negative. Employing this convention, we treat torques, which in the strict sense are vectors perpendicular to the xy-plane, as scalars. The forces $\boldsymbol{F}_i$ and couples C_i are assumed to be known quantities. Most typical forms of external loading can be represented by some combination of these forces and couples. Torque T will be treated as an unknown input required for equilibrium of the mechanism in the given position under specified loads.

Free-body diagrams of the three moving links are drawn in Figure 9.16b, which shows the bearing forces as well as the loads just defined. The location and orientation of the xy coordinate system are arbitrary. Angles ϕ_i are measured counterclockwise from the positive x direction. All forces are expressed in terms of x and y components.

Since the mechanism is planar, a maximum of three independent equilibrium equations can be written for each link considered as a free body. Beginning with link 3 and summing forces in the x and y directions and moments about pivot O_3, we have

$$F_{03x} + F_{23x} + F_{3x} = 0, \tag{9.15a}$$

$$F_{03y} + F_{23y} + F_{3y} = 0, \tag{9.15b}$$

and

$$F_{23y}\,\ell_3 \cos\phi_3 - F_{23x}\,\ell_3 \sin\phi_3 + F_{3y}\,r_3 \cos\phi_3 - F_{3x} r_3 \sin\phi_3 + C_3 = 0. \tag{9.15c}$$

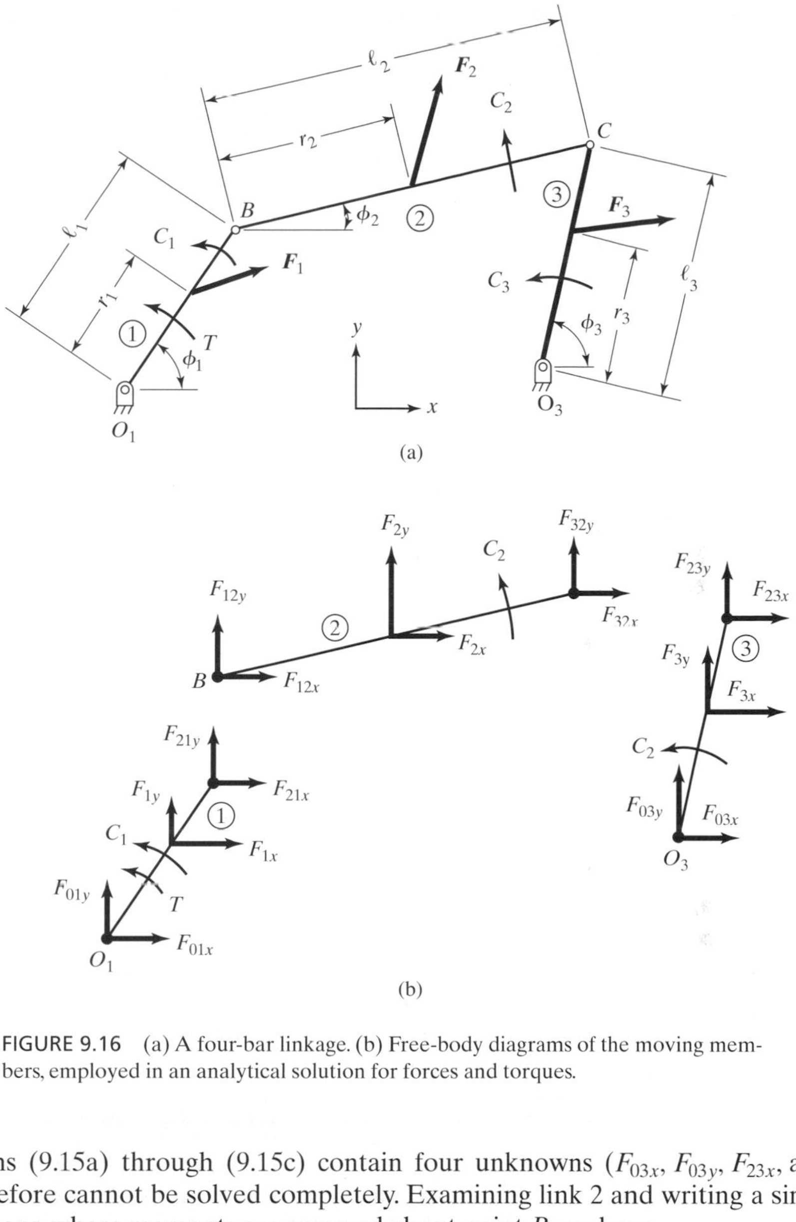

FIGURE 9.16 (a) A four-bar linkage. (b) Free-body diagrams of the moving members, employed in an analytical solution for forces and torques.

Equations (9.15a) through (9.15c) contain four unknowns (F_{03x}, F_{03y}, F_{23x}, and F_{23y}) and therefore cannot be solved completely. Examining link 2 and writing a similar set of equations, where moments are summed about point B, we have

$$F_{12x} + F_{32x} + F_{2x} = 0,$$
$$F_{12y} + F_{32y} + F_{2y} = 0,$$

and

$$F_{32y}\,\ell_2 \cos\phi_2 - F_{32x}\,\ell_2 \sin\phi_2 + F_{2y}\,r_2 \cos\phi_2 - F_{2x} r_2 \sin\phi_2 + C_2 = 0.$$

These three equations appear to introduce four new unknowns: F_{12x}, F_{12y}, F_{32x}, and F_{32y}. However, from Newton's third law,

$$F_{32x} = -F_{23x}$$

and

$$F_{32y} = -F_{23y}.$$

Substituting these relationships into the equilibrium equations for link 2, we get

$$F_{12x} + F_{23x} + F_{2x} = 0, \tag{9.16a}$$

$$F_{12y} + F_{23y} + F_{2y} = 0, \tag{9.16b}$$

and

$$-F_{23y}\,\ell_2 \cos\phi_2 - F_{23x}\,\ell_2 \sin\phi_2 + F_{2y}\,r_2 \cos\phi_2 - F_{2x}\,r_2 \sin\phi_2 + C_2 = 0. \tag{9.16c}$$

Now, Eqs. (9.15a) through (9.16c) are a system of six equations in six unknowns: F_{12x}, F_{12y}, F_{23x}, F_{23y}, F_{03x}, and F_{03y}. The solution of these equations is simplified by observing that Eqs. (9.15c) and (9.16c) contain only two of the unknowns: F_{23x} and F_{23y}. Rearranging those equations, we have

$$a_{11}F_{23x} + a_{12}F_{23y} = b_1 \tag{9.17a}$$

and

$$a_{21}F_{23x} + a_{22}F_{23y} = b_2, \tag{9.17b}$$

where

$$\begin{aligned}
a_{11} &= -\ell_3 \sin\phi_3,\\
a_{12} &= -\ell_3 \cos\phi_3,\\
a_{21} &= \ell_2 \sin\phi_2,\\
a_{22} &= -\ell_2 \cos\phi_2,\\
b_1 &= F_{3x} r_3 \sin\phi_3 - F_{3y}\,r_3 \cos\phi_3 - C_3,
\end{aligned}$$

and

$$b_2 = F_{2x}\,r_2 \sin\phi_2 - F_{2y} r_2 \cos\phi_2 - C_2.$$

Solving, we obtain

$$F_{23x} = \frac{a_{22}\,b_1 - a_{12}\,b_2}{a_{11}\,a_{22} - a_{12}\,a_{21}} \tag{9.18a}$$

and

$$F_{23y} = \frac{a_{11}\,b_2 - a_{21}\,b_1}{a_{11}\,a_{22} - a_{12}\,a_{21}}. \tag{9.18b}$$

Returning to Eqs. (9.15a), (9.15b), (9.16a), and (9.16b), we can determine the other four unknown bearing force components as follows:

$$F_{03x} = F_{23x} - F_{3x}, \tag{9.19a}$$

$$F_{03y} = -F_{23y} - F_{3y}, \tag{9.19b}$$

$$F_{12x} = F_{23x} - F_{2x}, \tag{9.20a}$$

and

$$F_{12y} = F_{23y} - F_{2y} \tag{9.20b}$$

Negative values for any of the quantities indicate that their directions are actually in the negative coordinate directions, opposite to the directions shown in Figure 9.16b.

Proceeding to member 1, the equilibrium equations are slightly different due to the presence of torque T (moments are summed about pivot O_1):

$$F_{01x} + F_{21x} + F_{1x} = 0;$$

$$F_{01y} + F_{21y} + F_{1y} = 0;$$

$$T + F_{21y}\,\ell_1 \cos\phi_1 - F_{21x}\,\ell_1 \sin\phi_1 + F_{1y}\,r_1 \cos\phi_1 - F_{1x}\,r_1 \sin\phi_1 + C_1 = 0.$$

Substituting $F_{21x} = -F_{12x}$ and $F_{21y} = -F_{12y}$ and rearranging terms, we solve these equations for F_{01x}, F_{01y}, and T:

$$F_{01x} = F_{12x} - F_{1x}; \tag{9.21a}$$

$$F_{01y} = F_{12y} - F_{1y}; \tag{9.21b}$$

$$T = F_{12y}\ell_1 \cos\phi_1 - F_{12x}\ell_1 \sin\phi_1 - F_{1y}r_1 \cos\phi_1 + F_{1x}r_1 \sin\phi_1 - C_1. \tag{9.22}$$

This completes the analysis of the four-bar linkage, in which we already determined the x and y components of the four bearing forces, Eqs. (9.18a) through (9.21b), and the required input torque given by Eq. (9.22).

SAMPLE PROBLEM 9.7

Analysis of a Four-Bar Linkage

Solve Sample Problem 9.2 by the analytical method.

Solution. The four-bar linkage of Figure 9.11 is redrawn in Figure 9.17, showing the various dimensions and forces F_2 and F_3 in component form. The following information may be determined from the figure:

$$\ell_1 = 30 \text{ mm}; \quad \ell_2 = 100 \text{ mm}; \quad \ell_3 = 50 \text{ mm};$$

$$\phi_1 = 135°; \quad \phi_2 = 17°; \quad \phi_3 = 85°;$$

$$r_1 = 0; \quad r_2 = 45 \text{ mm}; \quad r_3 = 38 \text{ mm};$$

$$F_{2x} = 47\cos(132°) = -31.4 \text{ N}; \quad F_{2y} = 47\sin(132°) = 34.9 \text{ N};$$

$$F_{3x} = 30\cos(128°) = -18.5 \text{ N}; \quad F_{3y} = 30\sin(128°) = 23.6 \text{ N};$$

$$F_{1x} = F_{1y} = C_1 = C_2 = C_3 = 0.$$

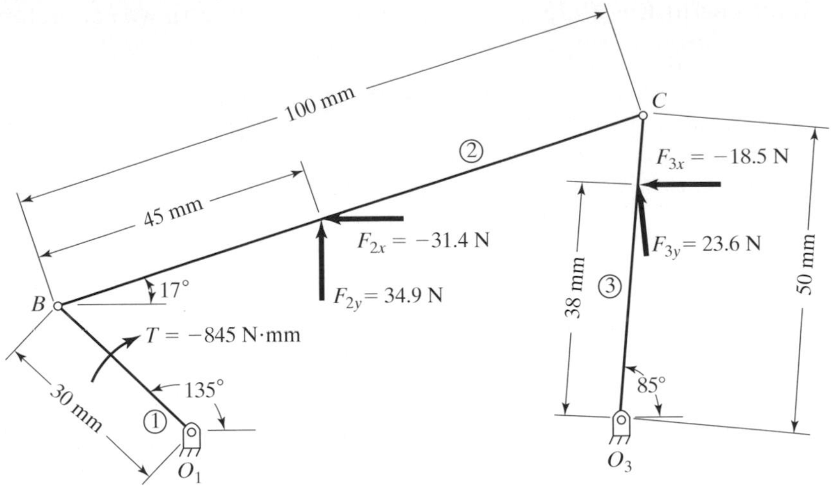

FIGURE 9.17 The four-bar linkage of Sample Problem 9.7.

From these data, the coefficients and right-hand terms in Eqs. (9.17a) and (9.17b) are calculated as follows:

$$a_{11} = -49.8; \quad a_{21} = 29.2; \quad b_1 = -778;$$
$$a_{12} = 4.36; \quad a_{22} = -95.6; \quad b_2 = -1910.$$

Then, from Eqs. (9.18a) through (9.21b),

$$F_{23x} = 17.8 \text{ N}, F_{23y} = 25.5 \text{ N},$$
$$F_{03x} = 0.64 \text{ N}, F_{03y} = -49.1 \text{ N}$$
$$F_{12x} = 49.2 \text{ N}, F_{12y} = -9.42 \text{ N},$$

and

$$F_{01x} = 49.2 \text{ N}, F_{01y} = -9.42 \text{ N},$$

and the magnitudes of these forces are

$$F_{23} = 31.1 \text{ N}, F_{12} = 50.2 \text{ N},$$

and

$$F_{03} = 49.1 \text{ N}, F_{01} = 50.2 \text{ N}.$$

Finally, the torque is calculated by Eq. (9.22):

$$T = -845 \text{ N} \cdot \text{mm}.$$

Recall that a negative torque is clockwise.

How closely does the calculated torque agree with a graphical solution to the same problem presented earlier?

Superposition

Sometimes it is convenient to divide a problem into two or more parts. For example, we can consider only the forces and torques (if any) acting on link 2 of the preceding four-bar linkage. Then we can consider only the forces and torques on link 3. Finally, we can combine the two steps in a procedure called *superposition*.

Each of the nine equations (i.e., three equilibrium equations for each of three members) that have been derived consists of a sum of multiples of the nine unknowns: $F_{23x}, F_{23y}, F_{03x}, F_{03y}, F_{12x}, F_{12y}, F_{01x}, F_{01y}$, and T. The coefficients of the unknowns in these equations do not depend on the applied forces, and there are no nonlinear terms in the unknowns or applied loads. Such a set of equations is said to be *linear*, and the principle of superposition applies. Recall that this principle states that the solution under a combined loads is equal to the sum of the solutions for the individual loads that combine to produce the total load.

For a linear set of equations, superposition can be demonstrated by examining Eqs. (9.17a) and (9.17b). The coefficients a_{11}, a_{12}, a_{21}, anda_{22} are functions of the linkage configuration only and are not functions of the loads on the linkage. Right-hand term b_1 is a linear function of the loading on member 3, and the term b_2 is a linear function of the loading on member 2. Consider the case where the forces on member 3 are zero; that is, $F_3 = C_3 = 0$ and, in turn, $b_1 = 0$. (We will refer to this case as subproblem I.) Then, Eqs. (9.17a) and (9.17b) become

$$a_{11}F^{\mathrm{I}}_{23x} + a_{12}F^{\mathrm{I}}_{23y} = 0 \tag{9.23a}$$

and

$$a_{21}F^{\mathrm{I}}_{23x} + a_{22}F^{\mathrm{I}}_{23y} = b_2, \tag{9.23b}$$

respectively, where F^{I}_{23x} and F^{I}_{23y} are the bearing forces resulting from this loading. Next, consider the case where the forces on member 2, and therefore b_2, are zero (subproblem II):

$$a_{11}F^{\mathrm{II}}_{23x} + a_{12}F^{\mathrm{II}}_{23y} = b_1 \tag{9.24a}$$

and

$$a_{21}F^{\mathrm{II}}_{23x} + a_{22}F^{\mathrm{II}}_{23y} = 0. \tag{9.24b}$$

Here, F^{II}_{23x}andF^{II}_{23y} are the bearing forces resulting from the loading on member 3 only. Combining the two sets of equations by adding Eq. (9.23a) to Eq. (9.24a) and Eq. (9.23b) to Eq. (9.24b) yields

$$a_{11}(F^{\mathrm{I}}_{23x} + F^{\mathrm{II}}_{23x}) + a_{12}(F^{\mathrm{I}}_{23y} + F^{\mathrm{II}}_{23y}) = b_1$$

and

$$a_{21}(F^{\mathrm{I}}_{23x} + F^{\mathrm{II}}_{23x}) + a_{22}(F^{\mathrm{I}}_{23y} + F^{\mathrm{II}}_{23y}) = b_2.$$

By comparison with Eqs. (9.17a) and (9.17b),

$$F_{23x} = F^{\mathrm{I}}_{23x} + F^{\mathrm{II}}_{23x}$$

and

$$F_{23y} = F^{\mathrm{I}}_{23y} + F^{\mathrm{II}}_{23y}.$$

Thus, superposition of the solutions to subproblems I and II leads to the total solution for force F_{23}. It can be shown that the other unknowns, including torque T, can also be found by superposition.

Detailed Analysis Using Simulation Software

A major strength of simulation softwarc is its ability to animate a linkage and track the motion of each part. If the animation of a proposed design does not look right to us, we can make design changes "on the spot." Loading can be represented by linear springs or by equations. Loading forces can be "turned on or off" during part of a cycle. If you are inexperienced in motion simulation, build your model one link at a time. Try animating the model after each step, to verify the effect of each constraint. It is advisable to review the results of a simulation carefully. Examine the motion of the linkage and the forces and torques on members for a full cycle of motion. Do the results approximate your expectations? Methods of static-force analysis are not restricted to bodies and parts that are stationary. If a linkage operates at low speed, then a static analysis can be used to check forces and torques for one or two positions.

Suppose your computer results are inconsistent with your independent analysis. Although software may contain errors, other sources of error are far more likely. You might begin by checking the input data. Vcrify thc properties and the geometry of individual links. Review the software manual and software "help" topics. Check the assembly of the bodies and the constraints in your model. Is the assembly actually consistent with your proposed design and consistent with the instructions in the software manual?

Suggestions for Building Complex Models

When you gain more skill in modeling, complex models will be within your grasp. But try a simplified model first; debug it and take approximate measurements from it. Knowledge Revolution (1996) recommends increasing the fidelity of a model gradually, verifying the behavior of the model at each step. This approach is faster than the everything-at-once approach because it saves debugging time. Also recommended is a modular method of modeling. Subcomponents are modeled in separate documents, tested as stand-alone linkages, and then incorporated into the main model.

SAMPLE PROBLEM 9.8

Using motion simulation software to aid in the design of a compressor

Suppose we need to supply air at a gage pressure of about 100,000 Pa. Propose and analyze a tentative design.

Design and modeling decisions. Let us try a tentative air compressor design based on an in-line slider-crank linkage with a 40-mm crank length, 64-mm connecting-rod length, and 78-mm-diameter piston. A motor rotating at 30 rpm will drive the compressor. We will simulate the air pressure with a linear spring that is active only during the compression stroke. The spring force on the piston will range from zero to 480 N during that stroke and will be zero during the other part of the cycle. Note that we are using only a rough approximation of actual conditions. For a more precise model, we would have to consider heat transfer as the air was compressed and know more about airflow through the valves. We want to find the required crank torque, the lateral force on the piston, and the forces on the connecting rod.

Solution. The linkage was simulated in Working Model™. (See Figure 9.18.) A linear spring with a spring rate of 6 N/mm is attached to the piston. The spring force is zero when the piston is moving to the left and goes from zero to 480 N when the piston is moving to the right.

Part a of the figure tracks the motion of each link and the center of the connecting rod as the crank rotates counterclockwise. The links are shown as narrow bars in the simulation. The crank, connecting rod, and piston in an actual linkage will not have the same proportions as the links used in the simulation.

F_Y is the vertical component of the connecting-rod force on the crank and the lateral force of the piston on the cylinder. Its maximum magnitude occurs at about 1.68 s into the cycle, corresponding to a crank angle of about 302°. The last-drawn linkage position in the simulation (which appears closest to the observer) results in the maximum magnitude for F_Y. The meters in part a show the spring tension, motor torque, and crankpin forces at that instant.

Part b of the figure shows the linkage position when the motor torque is at maximum magnitude. The meters show the maximum motor torque and the corresponding spring tension and crankpin forces. Part c of the figure shows the variation in the spring tension, motor torque, and crankpin forces during a full crank rotation.

Checking our results. At 30 rpm (2 s/rev), inertial forces should be negligible. If so, then the connecting rod is a two-force member. Measure the connecting-rod angle in parts a and b of the figure with a protractor, but correct for the proper quadrant.

Do the crankpin forces correspond to the angle of the connecting rod in each of the two positions shown? Does the maximum spring tension correspond to the maximum horizontal force on the crankpin? Now consider a free-body diagram of the entire linkage, and measure the distance from the wrist pin to the crankshaft in part b of the figure. Does the product of the lateral force on the cylinder and the distance from the wrist pin to the crankshaft balance the motor torque?

The Method of Virtual Work

The method of virtual work utilizes energy principles for force analysis. This approach offers advantages in certain types of analyses. For example, an entire mechanism can be examined as a whole, without the need for dividing it up into a number of free bodies. This leads directly to a relationship between input and output forces or torques without the need for an intermediate solution for bearing forces throughout the mechanism. The method is applicable to both static-force and dynamic-force analyses; we discuss its application to the former.

As the name indicates, the method of virtual work derives from the concept of work, which is defined as a force (or torque) acting through a displacement. In mathematical terms, work is the vector dot product of force and displacement; that is,

$$W = \boldsymbol{F} \cdot \boldsymbol{S}, \tag{9.25a}$$

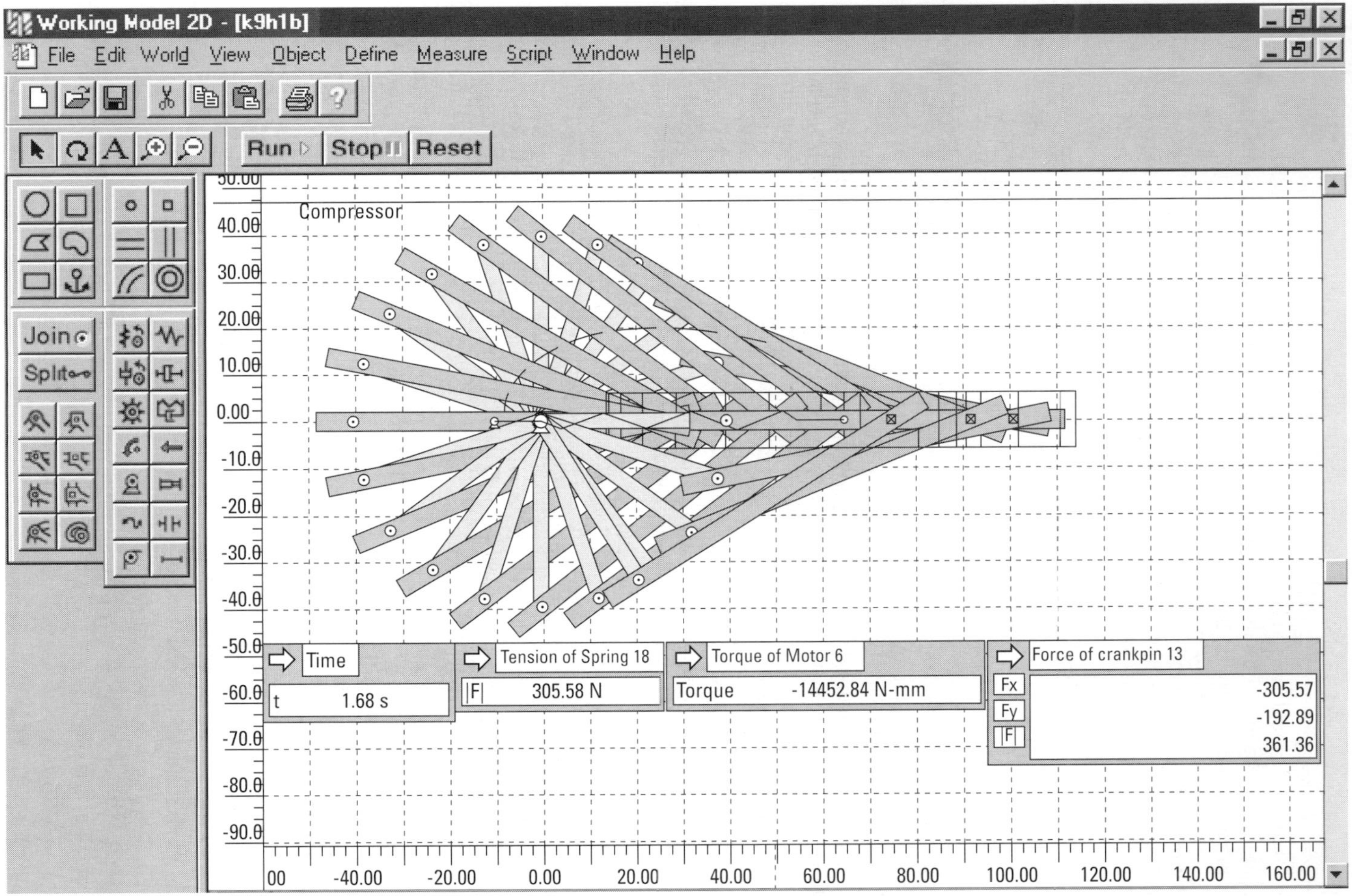

FIGURE 9.18 Using motion simulation software to analyze a tentative design: (a) Tracking the motion of each link.

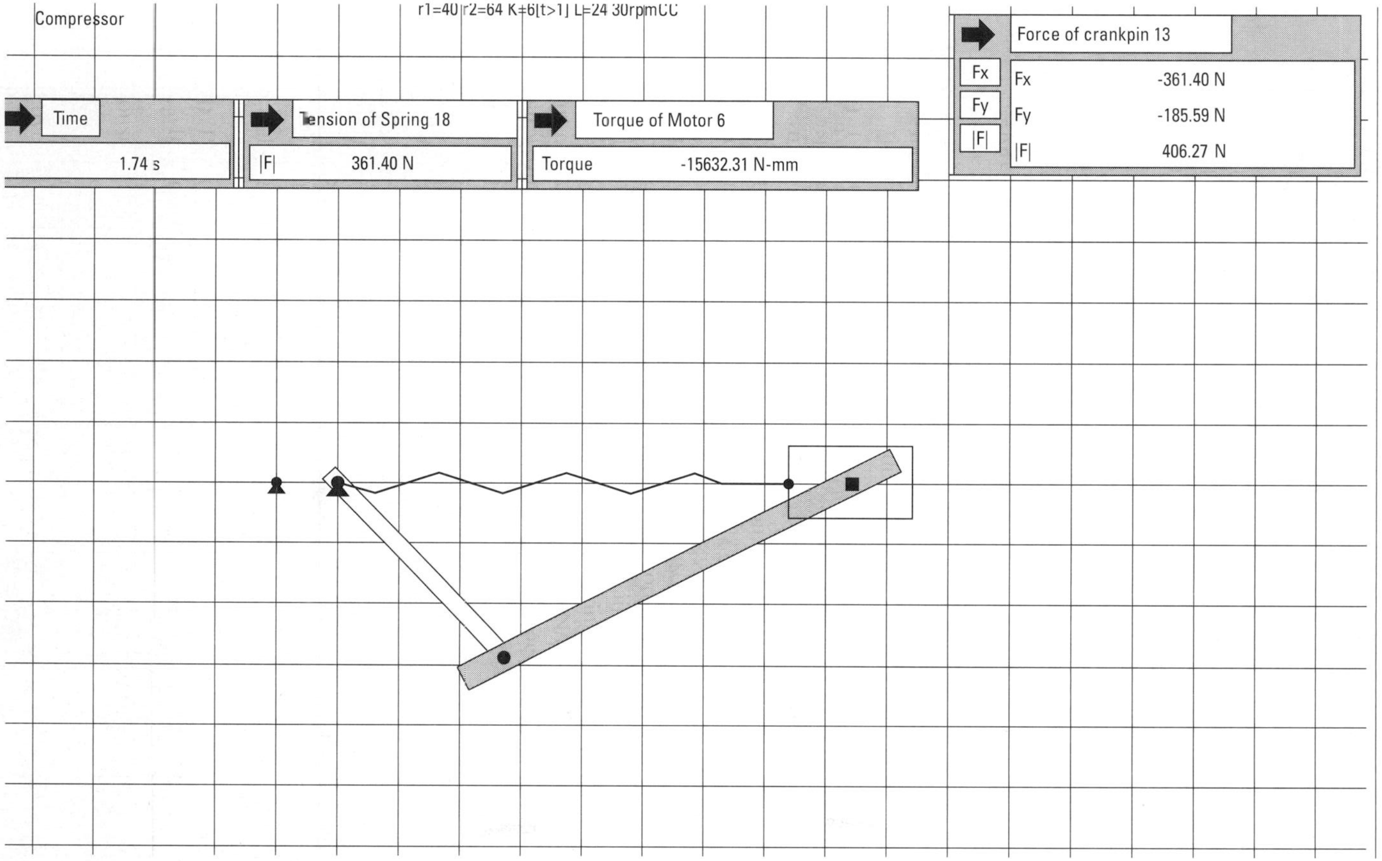

FIGURE 9.18 (b) Linkage position when motor torque is maximum.

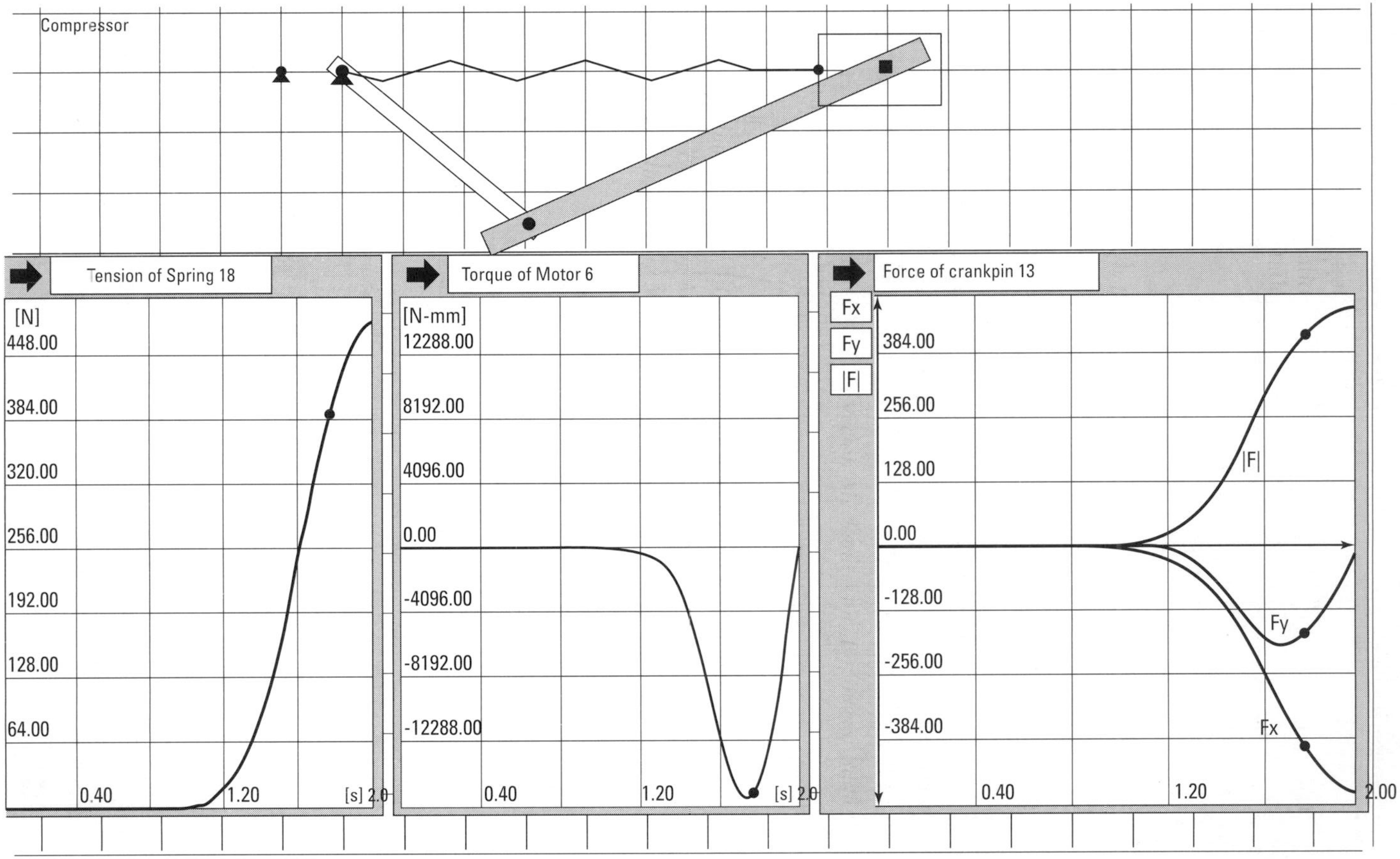

FIGURE 9.18 (c) Spring tension (simulating the effect of air pressure), motor torque, and crankpin force, all plotted against time.

where W is the work, $\boldsymbol{F}$ is the force, and $\boldsymbol{S}$ is the vector displacement at the point of application of the force. For rotational motion,

$$W = T\psi, \tag{9.25b}$$

where T is the torque and ψ is the angular displacement in a plane perpendicular to the torque. Since the result of the dot-product operation is a scalar, work is a scalar quantity. The units of work are those of energy.

The mathematical definition of work in Eq. (9.25a) provides some useful insight into the nature of work. In particular, no net work is performed in the following cases:

1. when there is no displacement at the point of application of the force.
2. When the force $\boldsymbol{F}$ is perpendicular to the displacement $\boldsymbol{S}$.
3. when equal, but opposite, forces act at the same point.

Consider these statements as they pertain to the slider-crank mechanism shown in Figure 9.19. The mechanism is acted on by external piston force $\boldsymbol{P}$ and external crankshaft torque T, as well as bearing forces. Bearing friction is negligible. We wish to determine the work performed as the mechanism travels through a small displacement from crank position ϕ, during which the angular displacement of the crank is $\delta\phi$ and the corresponding displacement of the piston is δx. (See Figure 9.19.) If we examine all forces acting on the mechanism, we see that the bearing force on the mechanism at crank pivot O_1 produces no work because there is no displacement of point O_1. Neglecting friction between the piston and cylinder, we see that the force of the cylinder on the piston is perpendicular to the piston displacement, and hence no work is performed by this force. No work is performed by internal forces either, including bearing forces at the crankpin and the wrist pin, because equal and opposite forces act at all internal points. Thus, the only work performed is that by torque T and force P:

$$W = T\delta\phi + P\delta x. \tag{9.26}$$

We assume here that torque T and force P are constant during the displacement of the mechanism. For small displacements, this is a reasonable approximation. If the force varies significantly during a displacement, then we must integrate find the work. Note that work is positive if the force acts in the same direction as the displacement and is negative if the force acts in a direction opposite that of the displacement. Both T and $\boldsymbol{P}$

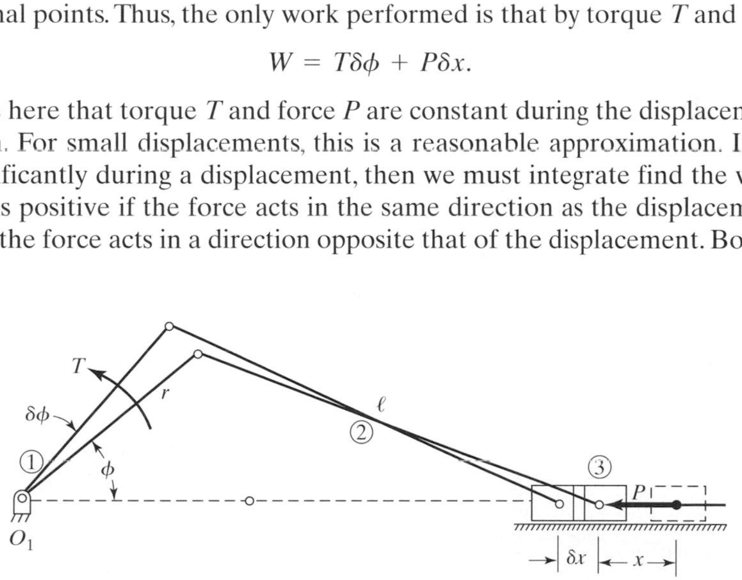

FIGURE 9.19 A slider-crank mechanism to be analyzed by the method of virtual work.

acting as indicated in Figure 9.19 would produce positive work during the displacement shown.

In a true statics problem, there is no displacement. We therefore introduce a quantity called a *virtual displacement*, which is defined as an imaginary infinitesimal displacement of the system that is consistent with the constraints on it. For example, the constraints on the slider-crank mechanism are that all members, including the frame, be rigid and that all joints maintain contact. Thus, Figure 9.19 depicts a virtual displacement of the mechanism, where $\delta\phi$ and δx are related by the kinematics of the rigid-membered linkage. *Virtual work* is defined as the work performed during a virtual displacement.

We now can state the *principle of virtual work* as it applies to equilibrium of mechanisms: *The work performed during a virtual displacement from equilibrium is equal to zero*. The interested student should consult a text on engineering mechanics for a complete derivation of this principle.

Let us now apply the principle to determine the torque T required in the slider-crank mechanism of Figure 9.19 for static equilibrium against applied force P for the mechanism position given by crank angle ϕ. For virtual displacements $\delta\phi$ and δx, it follows from Eq. (9.26) that

$$T\delta\phi + P\delta x = 0 \tag{9.27}$$

for equilibrium. Before solving this equation, we introduce one more characteristic of virtual displacements: They are assumed to take place during the same time interval dt. Dividing Eq. (9.27) by dt yields

$$T\frac{\delta\phi}{dt} + P\frac{\delta x}{dt} = 0,$$

or

$$T\dot{\phi} + P\dot{x} = 0, \tag{9.28}$$

where $\dot{\phi}$ and $\dot{x}$ are the instantaneous velocities of the crank and piston, respectively. Solving for T, we have

$$T = -\frac{\dot{x}}{\dot{\phi}}P. \tag{9.29}$$

Recall from Chapter 3 that

$$\dot{x} = r\dot{\phi}\sin\phi\left[1 + \frac{r\cos\phi}{\sqrt{\ell^2 - (r\sin\phi)^2}}\right],$$

and therefore,

$$T = -Pr\sin\phi\left[1 + \frac{r\cos\phi}{\sqrt{\ell^2 - (r\sin\phi)^2}}\right]. \tag{9.30}$$

The negative sign indicates that, for equilibrium, torque T must produce a negative amount of work in Eq. (9.27) equal to the positive amount produced by force P, and therefore, the torque must be clockwise in Figure 9.19.

It can be seen from this example that, for a mechanism in equilibrium under the action of two forces—an input driving force and an output load—the ratio of the magnitude of the input force to the magnitude of the output force equals the inverse ratio of the magnitudes of the corresponding velocities. Of course, the method can also be used to analyze mechanisms with multiple loads, in which case the summation of all of the virtual work performed by the individual forces and torques must equal zero for equilibrium. Gravity loadings can be treated like any other force. Furthermore, inertial forces can be included, so that the method can be employed in dynamic-force analysis.

9.5 FRICTION IN MECHANISMS

Whenever two connected members of a mechanism are in relative motion, friction occurs at the joint that connects them. The friction produces heat and wear, which may eventually lead to bearing failure. It can also adversely affect the motion response of the mechanism; for example, friction will slow the response of fast-action devices, such as mechanical circuit breakers for electrical transmission lines, and it may alter the synchronization of automated systems, such as multiple-input manipulators. In addition, friction can substantially increase the energy requirements of a machine.

The nature and amount of friction depend on the type of bearing employed. High-speed, high-load machinery is often designed with low-friction bearings: either rolling-contact bearings, utilizing balls or rollers to eliminate relative sliding, or thick-film bearings, in which the moving parts are separated by a layer of lubricant film. However, in many situations, direct physical contact between sliding members occurs. This type of friction is referred to as dry or Coulomb friction, which will be examined shortly.

Dry friction can occur for various reasons. For example, equipment may be poorly lubricated, or there may be loss of lubricant due to leakage. In other cases, the environment places severe restrictions on the use of lubrication. For example, lubricants are prohibited in outer-space applications and in certain food-processing applications; special self-lubricating bearing materials with low coefficients of friction have found successful application in these areas. Purely economic considerations sometimes rule against the use of low-friction bearings, resulting in machines or products in which dry-friction effects can be significant. Even in hydrodynamic thick-film bearings, direct surface contact will occur when the machine starts up, before the lubricant layer develops, resulting in Coulomb friction.

Two common types of joints in which sliding friction can be present are *prismatic or slider connections* and *revolute or journal bearings*. Figure 9.20 shows a sliding block moving relative to a flat surface, which may be moving or fixed. In general, the applied force on the slider consists of components parallel and perpendicular to its direction of motion. These components are shown as $\boldsymbol{P}$ and $\boldsymbol{N}$, respectively. The Coulomb friction force $\boldsymbol{F}$ is related to the normal force $\boldsymbol{N}$ between contacting surfaces, and its magnitude is given by

$$F = \mu N, \tag{9.31}$$

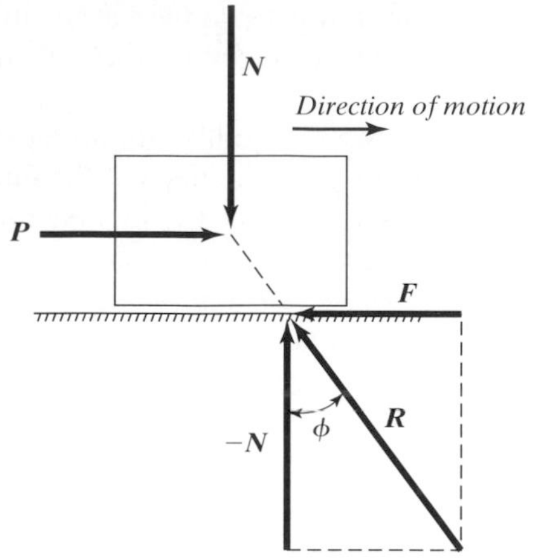

FIGURE 9.20 Dry friction in a translating bearing. Friction force F acting on the sliding block opposes the motion of the block relative to the contacting surface.

where μ is defined as the coefficient of sliding friction, which is a characteristic of both the contacting materials and the operating conditions. Although precise values for coefficients of friction are extremely difficult, if not impossible, to obtain, handbooks contain extensive lists of approximate values that are acceptable for most design purposes. The direction of the friction force is always such that the relative motion is opposed. In Figure 9.20, the block is assumed to be sliding to the right. The friction force F acting on the block is therefore directed to the left.

The resultant force $\boldsymbol{R}$ that the surface exerts on the block has components of magnitude N and $F = \mu N$, and therefore, the angle ϕ of this resultant from the normal direction is given by

$$\tan\phi = \frac{F}{N} = \frac{\mu N}{N} = \mu. \tag{9.32}$$

Angle ϕ is referred to as the *friction angle*, and it is evident that the direction of $\boldsymbol{R}$ is known once the coefficient of friction has been determined. As a limiting case, when there is no friction ($\mu = 0$), $\phi = 0$, and the resultant force is normal to the surfaces.

Friction in a journal bearing (see Figure 9.21) is essentially the same as that for a sliding block. Shown in the figure is a journal or pin of radius r attached to one member that rotates in a bearing or sleeve in another member. The center of the bearing is at point A, and the center of the journal is at point B; normally, the clearance in the joint is much less than that depicted in the figure, since large clearances can lead to serious impact problems. The two members instantaneously contact at point C, where there is a resultant force $\boldsymbol{R}$ of the bearing on the journal. This force consists of a compressive normal component $\boldsymbol{N}$ and a friction component $\boldsymbol{F}$, that opposes the sliding motion of the journal relative to the bearing.

As before, the two force components are related by the equation

$$F = \mu N,$$

where μ is the coefficient of friction, and the friction angle ϕ is given by

$$\tan\phi = \mu.$$

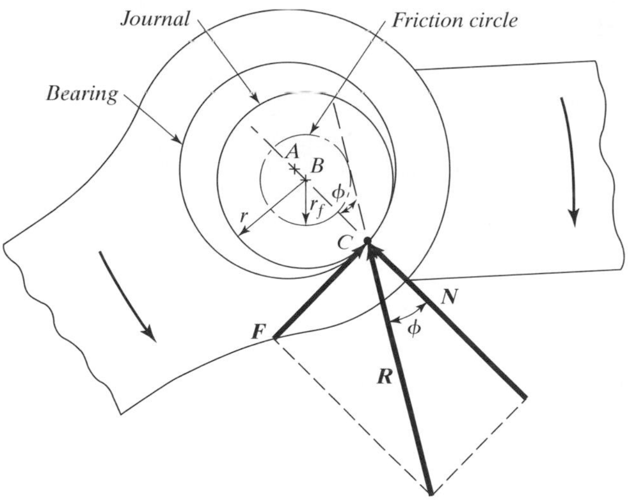

FIGURE 9.21 Dry friction in a journal bearing. The resultant force R acts tangent to the friction circle and includes a component F that opposes relative motion and a component N that acts normal to the contacting surfaces.

Thus, as long as the coefficient of friction is constant, the resultant force will always act at a fixed angle ϕ with respect to the common normal to the two surfaces. It follows that the line of action of resultant $\boldsymbol{R}$ will always be tangent to a circle with center at point B and radius

$$r_f = r \sin \phi. \tag{9.33}$$

(See Figure 9.21). This circle is called the *friction circle*, and with the following trigonometric identity, its radius can be expressed as

$$r_f = r \sin \phi = \frac{r \tan \phi}{\sqrt{1 + \tan^2 \phi}} = \frac{r\mu}{\sqrt{1 + \mu^2}}. \tag{9.34a}$$

For small values of μ, which is often the case, the radius can be approximated as

$$r_f = r \sin \phi \cong r \tan \phi = r\mu. \tag{9.34b}$$

In addition to being tangent to the friction circle, the resultant force R will be directed so as to produce a moment about journal center B that opposes the relative motion. As we shall demonstrated in the sections that follows, the friction circle is a useful concept for force analysis.

Graphical Solution for a Slider-Crank Mechanism Including Friction

We illustrate the material of the preceding section by investigating the slider-crank linkage shown in Figure 9.22a. The force analysis will consider the presence of friction at all four connections: the three turning connections and one sliding connection. A load $\boldsymbol{P}$ is applied to the piston and the various bearing forces and the required input torque for static equilibrium are to be determined. Crank 1 is rotating in the clockwise direction.

Figure 9.22b shows the free-body diagrams of the members. The friction circles are constructed for the three rotating joints by means of Eq. (9.34a) or Eq. (9.34b). The

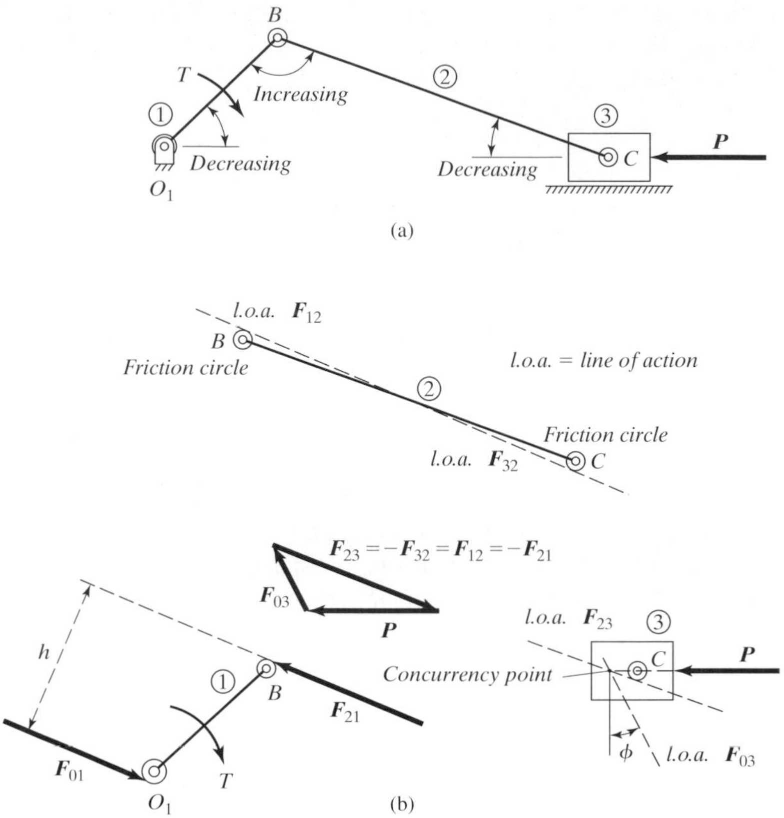

FIGURE 9.22 (a) Graphical force analysis of a slider-crank mechanism including friction. The crank rotates clockwise. (b) Free-body diagrams and force polygon.

forces at these joints must be tangent to the friction circles. Furthermore, the forces must act in such a way as to oppose relative motion at the joints. By inspection, the relative link motions are as shown in Figure 9.22a for clockwise crank rotation.

First, consider connecting rod 2. This link is a two-force member and obviously is in compression for the loading and position being analyzed. Therefore, the forces $\boldsymbol{F}_{12}$ and $\boldsymbol{F}_{32}$ must be collinear, with force $\boldsymbol{F}_{12}$ producing a clockwise moment about pin B opposing the counterclockwise rotation of link 2 relative to link 1. Force $\boldsymbol{F}_{32}$ produces a clockwise moment about pin C opposing the counterclockwise rotation of link 2 relative to piston 3. The line of action of the compressive forces, which is shown on the free body of link 2, will satisfy these conditions. Notice that four straight lines can be drawn tangent to the two friction circles, but that the one shown is the only one that will satisfy all of the conditions stated. Generally, some intuitive guessing or trial and error is necessary in properly locating lines of action.

Having found the line of action for member 2, we can now proceed with the analysis of the piston. Force $\boldsymbol{P}$ is known completely, and the line of action of force $\boldsymbol{F}_{23}$ is now known. Force $\boldsymbol{F}_{03}$, exerted by the frame on the piston, must pass through the concurrency point, given by the intersection of $\boldsymbol{P}$ and $\boldsymbol{F}_{23}$, and must act at the friction angle

ϕ with respect to the normal to the surfaces. Angle ϕ is measured as shown, taking into account a friction force to the left opposing the sliding of the block to the right. From this information, the force polygon is constructed, yielding forces $\boldsymbol{F}_{03}$ and $\boldsymbol{F}_{23}$ and, in turn, forces $\boldsymbol{F}_{32}$, $\boldsymbol{F}_{12}$, and $\boldsymbol{F}_{21}$.

Force $\boldsymbol{F}_{21}$ has been drawn on the free body of crank 1. Note that this force will produce a counterclockwise moment about the joint that opposes the clockwise rotation of link 1 relative to link 2. Force $\boldsymbol{F}_{01}$ must be equal in magnitude and opposite in direction to $\boldsymbol{F}_{21}$. It is drawn tangent to the friction circle at pin O_1 and is properly placed to oppose the clockwise rotation of the crank relative to the frame. Finally, the moment created by this couple must be opposed by the driving torque $T = hF_{21}$, the direction of which is clockwise, by inspection of the free-body diagram.

The graphical force analysis of other types of mechanisms with friction would proceed in a similar fashion. However, the more complex the mechanism is, the greater is the probability that a trial-and-error solution will be required in the determination of forces.

The performance of a machine with friction is evaluated relative to ideal, friction-free operation by means of a quantity called the *instantaneous efficiency*, which is defined as the ratio of the required input torque or force in the absence of friction to that in the presence of friction. For the slider-crank mechanism under consideration,

$$e = \frac{T_o}{T}, \tag{9.35}$$

where e is the instantaneous efficiency, T_o is the required input torque when there is no friction present, and T is the required input torque with friction included. For the case just analyzed, T is greater than T_o because the input driving torque must overcome friction from the mechanism as well as the piston force $\boldsymbol{P}$.

As the name implies, the instantaneous efficiency may vary with the position of the mechanism, as would be the case for the slider-crank mechanism. Also, for this mechanism, if one considers the case of counterclockwise rotation of crank 1, with the same applied piston force $\boldsymbol{P}$, the instantaneous efficiency will apparently be greater than unity. Looking at it another way, the piston force may more logically be thought of as the input in this situation, and for a given torque load T, the efficiency

$$e = \frac{P_o}{P}$$

will be less than unity.

Analytical Solution for a Slider-Crank Mechanism Including Friction

Some disadvantages are inherent in the analysis procedure of the previous section. First, friction circles may be much smaller than those utilized for illustrative purposes in the slider-crank example, which can lead to graphical inaccuracies. Second, since guesswork on the placement of forces is necessary in some analyses, the amount of

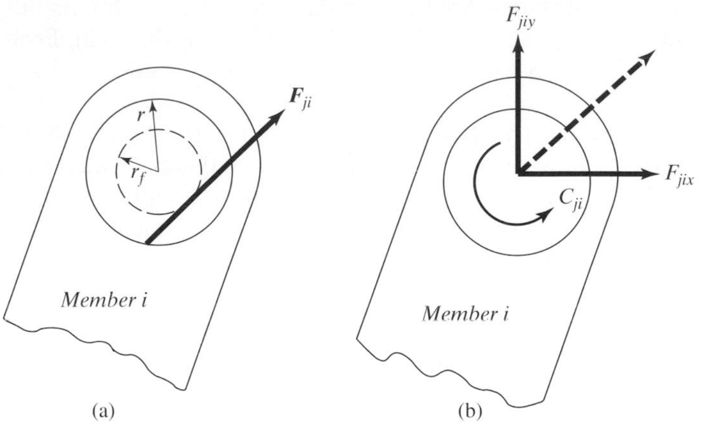

FIGURE 9.23 (a) The force transmitted at a journal bearing with friction. The diagram shows the actual line of action of the resultant bearing force. (b) An equivalent combination of a force and a torque.

graphical construction may be quite time consuming. An alternative is the analytical approach that follows, which can be computerized for a faster and more accurate solution.

Figure 9.23a shows a machine member i that is connected to a member j (not shown) by means of a journal bearing, of radius r and an assumed negligible clearance. The friction circle is shown, and bearing force $\boldsymbol{F}_{ji}$ of member j on member i acts along a direction tangent to the friction circle and is positioned so as to oppose the rotation of member i relative to member j.

In Figure 9.23b, the bearing force has been replaced by a combination of a force and a couple. The two force systems will be equivalent if they produce the same net force and the same net moment about any arbitrary point. The force, represented by perpendicular x and y components, has the same magnitude and direction as the original force, but has a line of action through the center of the joint. The couple C_{ji} accounts for the offset r_f of the original force and can be expressed as

$$C_{ji} = r_f|\boldsymbol{F}_{ji}|\text{sign}(\omega_j - \omega_i) \cong \mu r(F_{jix}^2 + F_{jiy}^2)^{1/2}\,\text{sign}(\omega_j - \omega_i). \tag{9.36}$$

The sign function is defined as

$$\text{sign}(\omega_j - \omega_i) = \begin{cases} -1 & \text{if } (\omega_j - \omega_i) < 0 \\ 0 & \text{if } (\omega_j - \omega_i) = 0^* \\ +1 & \text{if } (\omega_j - \omega_i) > 0 \end{cases} \tag{9.37}$$

where ω_i and ω_j are the angular velocities of members i and j, respectively. The difference in the angular velocities, $\omega_i - \omega_j$, is an indication of the motion of link i relative to link j. Therefore, the force system in Figure 9.23b has a net force equal to $\boldsymbol{F}_{ji}$ and a net moment about the bearing center that is equal in both magnitude and direction to that in Figure 9.23a, and the two systems are equivalent.

This representation of bearing friction eliminates any guesswork in the analysis and is well suited for implementation on a computer. For example, let us return to the

*For a more rigorous analysis, consider static friction and impending motion.

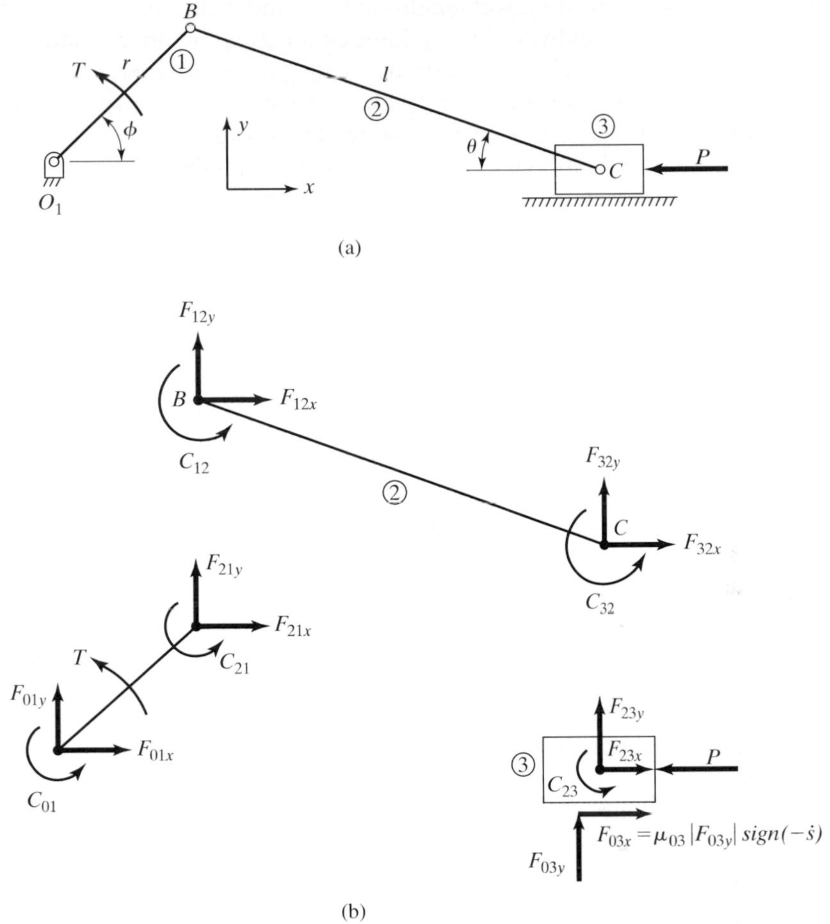

FIGURE 9.24 (a) A slider-crank mechanism. (b) Free-body diagrams to be used for an analytical solution including friction effects.

slider-crank mechanism of the previous section. The mechanism and free bodies are redrawn in Figures 9.24a and b with the new force representation. Summing forces on connecting rod 2 and moments about point B, we have

$$F_{12x} + F_{32x} = 0, \tag{9.38a}$$
$$F_{12y} + F_{32y} = 0, \tag{9.38b}$$

and

$$F_{32y}\ell\cos\theta + F_{32y}\ell\sin\theta + \mu_{32}r_{32}(F_{32x}^2 + F_{32y}^2)^{1/2}\,\text{sign}(\omega_3 - \omega_2)$$
$$+ \mu_{12}r_{12}(F_{12x}^2 + F_{12y}^2)^{1/2}\,\text{sign}(\omega_1 - \omega_2) = 0, \tag{9.38c}$$

where ℓ is the connecting-rod length and μ_{ij} and r_{ij} are the coefficient of friction and bearing radius, respectively, for the joint connecting members *i* and *j*. Note that, in this case, the angular velocity ω_3 of member 3, the piston, is zero.

Let us made a couple of observations. First, the equilibrium conditions have produced three equations in four unknowns. Therefore, more information, which can be obtained by examining the free bodies of other members, is necessary before the system of equations can be solved. For the piston in Figure 9.24b,

$$F_{03x} + F_{23x} = \mu_{03}|F_{03y}| \operatorname{sign}(-\dot{s}) - F_{32x} = P \tag{9.39a}$$

and

$$F_{03y} + F_{23y} = F_{03y} - F_{32y} = 0, \tag{9.39b}$$

where $\dot{s}$ is the piston velocity (assumed to be positive to the right) and $\operatorname{sign}(-\dot{s})$ is defined analogously to Eq. (9.37). Thus, when the piston is moving to the right, $\dot{s}$ will be positive and friction force F_{03x} will be negative, or directed towards the left. Equations (9.39a) and (9.39b) are two new equations introducing only one new unknown, F_{03y}, and the five equations can now be solved simultaneously.

The second observation to be made is that the equations are nonlinear in the unknowns, due to the presence of the square-root terms, which, by the way, disappear if the friction is zero. This nonlinearity has serious ramifications in that the principle of superposition no longer holds, and also, the solution procedure is much more difficult. In general, numerical techniques or approximation methods must be employed in solving the equations.

The solution is facilitated in the present case by the fact that the connecting rod is a two-force member. Solving Eqs. (9.38a) and (9.38b) for F_{12x} and F_{12y} and then substituting the resulting expressions into Eq. (9.38c) and factoring yields

$$F_{32y}\ell \cos\theta + F_{32x}\ell \sin\theta + k\sqrt{F_{32x}^2 + F_{32y}^2} = 0, \tag{9.40}$$

where

$$k = \mu_{12} r_{12} \operatorname{sign}(\omega_1 - \omega_2) + \mu_{32} r_{32} \operatorname{sign}(\omega_3 - \omega_2).$$

Next, we express the force components F_{32x} and F_{32y} in terms of polar coordinates as

$$F_{32x} = F_{32}\cos\psi \tag{9.41a}$$

and

$$F_{32y} = F_{32}\sin\psi, \tag{9.41b}$$

where F_{32} is the magnitude of the force transmitted from member 3 to member 2 and angle ψ is its argument with respect to the positive *x* axis. (See Figure 9.25.) The value

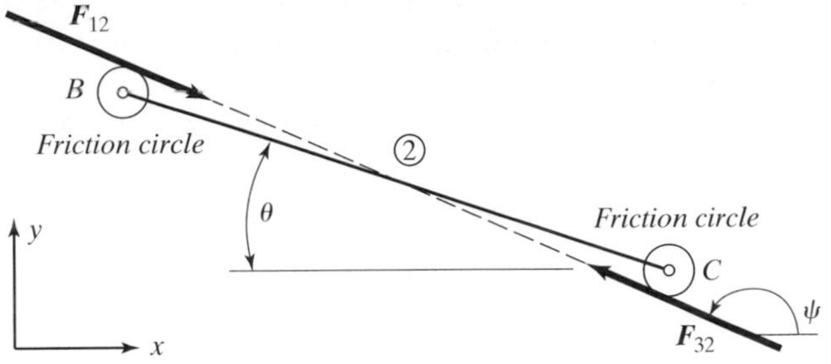

FIGURE 9.25 Free-body diagram of connecting rod 2, showing the orientation of the forces at bearings B and C.

of angle ψ will depend on whether the connecting rod is in tension or compression and on the relative motion at joints B and C. As mentioned earlier, there are four possible orientations of the force line of action tangent to each of the two friction circles.

Substituting Eqs. (9.41a) and (9.41b) into Eq. (9.40), we have

$$F_{32}\,\ell \sin\psi \cos\theta + F_{32}\,\ell \cos\psi \sin\theta + kF_{32} = 0,$$

or

$$\ell \sin(\psi + \theta) + k = 0.$$

Solving for angle ψ, we obtain

$$\psi = \arcsin\left(-\frac{k}{\ell}\right) - \theta. \tag{9.42}$$

In this equation, ψ has two solutions, one corresponding to compression of the connecting rod and the other corresponding to tension; note that the relative joint motions are accounted for by the sign functions in the expression for k. For example, when there is no friction, the solutions are

$$\psi = -\theta$$

and

$$\psi = \pi - \theta$$

Inspection of Figures 9.24a and 9.25 shows that the first of these solutions represents tension of the connecting rod, as would occur when force $\boldsymbol{P}$ on the piston is directed to the right. The second solution represents the situation depicted in Figure 9.24a with piston force $\boldsymbol{P}$ acting to the left and the connecting rod therefore in compression.

Equation (9.42) gives the argument of force $\boldsymbol{F}_{32}$; however, the magnitude of this force is as yet unknown. Combining Eqs. (9.39a) and (9.39b) results in

$$\mu_{03}|F_{32y}|\text{sign}(\dot{s}) + F_{32x} + P = 0,$$

where $\text{sign}(\dot{s}) = -\text{sign}(-\dot{s})$. Substituting Eqs. (9.41a) and (9.41b) into this equation yields,

$$\mu_{03}|F_{32y}|\text{sign}(\dot{s}) + F_{32}\cos\psi + P = 0,$$

or

$$F_{32} = \frac{-P}{\cos\psi + \mu_{03}|\sin\psi|\text{sign}(\dot{s})}. \tag{9.43}$$

Force $\boldsymbol{F}_{32}$ is now completely determined by Eqs. (9.42) and (9.43). The latter equation also serves as a check for the proper solution for angle ψ from Eq. (9.42). The correct value of ψ is that which will yield a positive value for the right side of Eq. (9.43), which represents the absolute value of force $\boldsymbol{F}_{32}$.

The various x and y force components can now be expressed as

$$F_{32x} = -F_{23x} = -F_{12x} = F_{21x} = F_{32}\cos\psi \tag{9.44a}$$

and

$$F_{32y} = -F_{23y} = -F_{12y} = F_{21y} = F_{03y} = F_{32}\sin\psi \tag{9.44b}$$

Once F_{21x} and F_{21y} have been determined, the analysis can be completed by solving of the following equations for crank 1 (see Figure 9.24b):

$$F_{01x} + F_{21x} = 0; \tag{9.45a}$$

$$F_{01y} + F_{21y} = 0; \tag{9.45b}$$

$$T + C_{01} + C_{21} + F_{21y}r\cos\phi - F_{21x}r\sin\phi = 0. \tag{9.45c}$$

In Eq. (9.45c), r is the crank length,

$$C_{01} = \mu_{21}r_{21}(F_{01x}^2 + F_{01y}^2)^{1/2}\text{sign}(-\omega_1),$$

and

$$C_{21} = \mu_{21}r_{21}(F_{21x}^2 + F_{21y}^2)^{1/2}\text{sign}(\omega_2 - \omega_1).$$

Note that the angular velocity of frame 0 is zero. The instantaneous efficiency can then be determined as before.

SAMPLE PROBLEM 9.9

Analysis of a Slider-Crank Mechanism with Friction

Determine the instantaneous efficiency of the slider-crank mechanism of Sample Problems 9.1 and 9.5. The mechanism has a crank length r of 30 mm and a connecting-rod length ℓ of 70 mm. The analysis is to be performed, as before, for a crank angle $\phi = 45°$. Recall that, in the absence of friction, the crankshaft torque for a piston force $P = 40$ N was found to be $T = -1119$ N · mm, with the negative sign indicating that the torque is clockwise. Assume that the coefficient of friction for all bearings is 0.1. The three journal bearings have radii of 10 mm, and the crank is rotating in the clockwise direction.

Solution. From the specified information, we have

$$r_{01} = r_{12} = r_{23} = 10\,\text{mm}$$

and

$$\mu_{03} = \mu_{01} = \mu_{12} = \mu_{23} = 0.1.$$

The sign functions for a complex mechanism would require a kinematic velocity analysis to determine relative joint velocities. In this case, the functions can be determined by inspection (for example, see Figure 9.22a):

$$\begin{aligned}
\text{sign}(\dot{s}) &= +1; \\
\text{sign}(\omega_3 - \omega_2) &= \text{sign}(-\omega_2) = -1; \\
\text{sign}(\omega_1 - \omega_2) &= -\text{sign}(\omega_2 - \omega_1) = -1; \\
\text{sign}(\omega_0 - \omega_1) &= \text{sign}(-\omega_1) = +1.
\end{aligned}$$

Angle θ (see Figure 9.24a) can be calculated from the following relationships developed earlier in the chapter:

$$\begin{aligned}
\sin\theta &= \frac{r}{\ell}\sin\phi = \frac{30}{70}\sin 45° = 0.303; \\
\theta &= 17.6°
\end{aligned}$$

Angle ψ, in Eq. (9.42) can now be determined as follows:

$$\begin{aligned}
k &= 0.1(10)(-1) + 0.1(10)(-1) = -2; \\
\psi &= \arcsin\left[-\frac{(-2)}{70}\right] - 17.6° = -15.9°, 160.7°.
\end{aligned}$$

Since the connecting rod is in compression, angle ψ must be in the second quadrant, indicating that $\psi = 160.7°$ is the correct value. Substituting into Eq. (9.43) yields

$$F_{32} = \frac{-40}{\cos(160.7°) + 0.1|\sin(160.7°)|(+1)} = 43.9\,\text{N}.$$

The positive value obtained for F_{32} confirms the choice of 160.7° for ψ. The force components, then, are

$$F_{32x} = -F_{23x} = -F_{12x} = F_{21x} = 43.9 \cos 160.7^\circ = -41.4\text{ N}$$

and

$$F_{32y} = -F_{23y} = -F_{12y} = F_{21y} = F_{03y} = 43.9 \sin 160.7^\circ = 14.5\text{ N}.$$

Noting that the magnitudes of forces $\boldsymbol{F}_{32}$, $\boldsymbol{F}_{12}$, and $\boldsymbol{F}_{01}$ are all equal, we can now compute the input crank torque from Eq. (9.45c):

$$\begin{aligned} T &= -0.1(10)(43.9)(+1) - 0.1(10)(43.9)(+1) \\ &\quad - (14.5)(30)\cos 45^\circ + (-41.4)(30)\sin 45^\circ \\ &= -1273\text{ N}\cdot\text{mm}. \end{aligned}$$

The instantaneous efficiency is thus

$$e = \frac{-1119}{-1273} = 0.88\ (88\text{ percent}).$$

9.6 FORCES IN GEAR AND CAM MECHANISMS

Up to now, this chapter has dealt exclusively with forces in linkage mechanisms. However, the general force analysis principles that have been presented are applicable to mechanisms of all types, including gears and cams.

Due to the special nature of gears, associated forces are described in Chapters 6, 7, and 8, where terminology specific to gears is defined. Helical, worm, and bevel gear systems are subject to three-dimensional loading, while forces on spur gears are essentially planar. Drawing free-body diagrams of individual gears usually makes the force analysis treatment of these machine components relatively straightforward.

The analysis methods and principles described earlier in the chapter can be applied to cam mechanisms, as sample problem 9.10 illustrates.

SAMPLE PROBLEM 9.10

Analysis of a Cam Mechanism

The cam in Figure 9.26a drives the four-bar linkage against an applied load $\boldsymbol{F}_3$ having a magnitude of 100 N. Determine the torque required on the camshaft for static equilibrium. Neglect friction.

Solution. Figure 9.26b shows free-body diagrams of the individual moving members. The solution will be carried out graphically. Before the force polygon for member 3 can be constructed, we need to determine the direction of the force at pin C. This is accomplished by considering member 2, because the directions of two of its three forces are already known: The force $\boldsymbol{F}_{42}$ at point D is directed along link 4, which is a two-force member, and, in the absence of friction, the

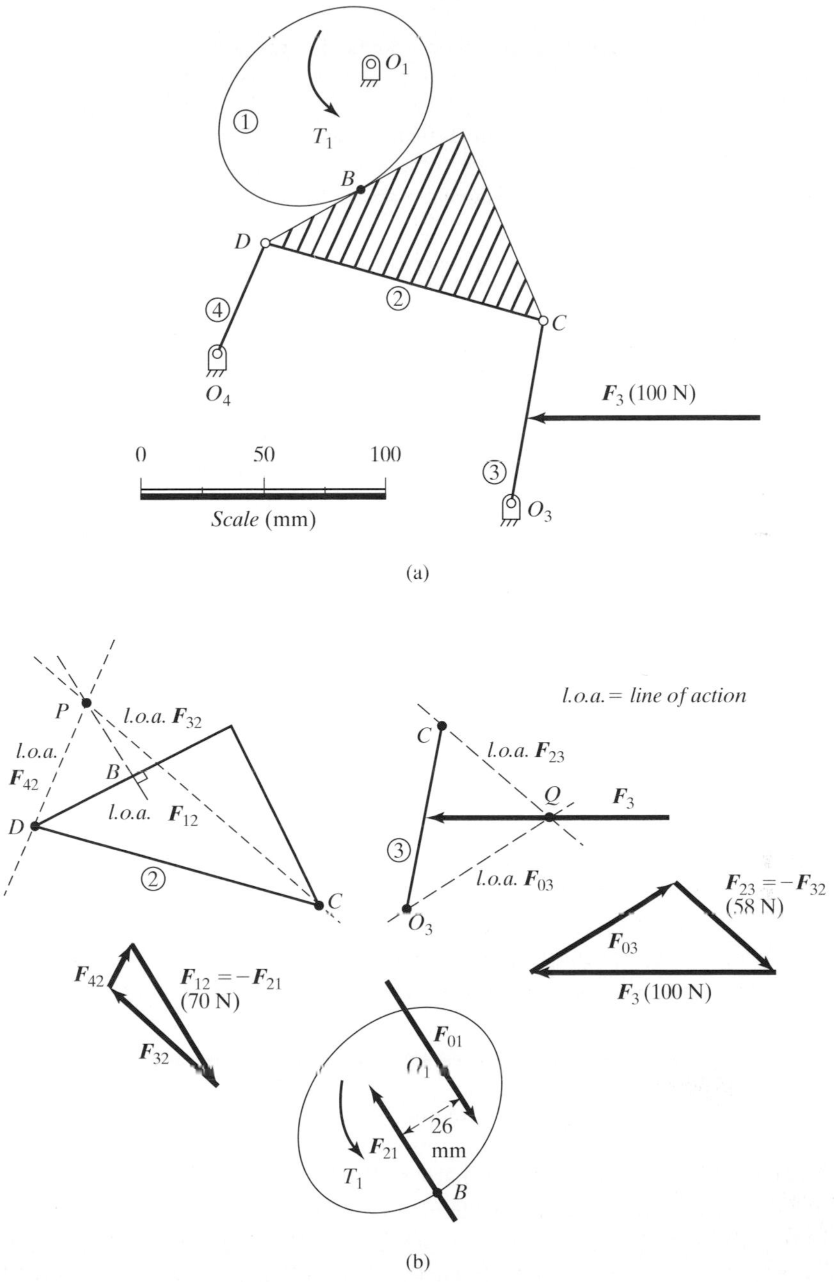

FIGURE 9.26 (a) Force analysis of a cam linkage system. Friction is neglected. (b) Free-body diagrams and force polygons.

contact force $\boldsymbol{F}_{12}$ of the cam acts along the common normal at point B. The intersection of these two lines of action is concurrency point P for member 2, and line PC is the direction of force $\boldsymbol{F}_{32}$ and, in turn, force $\boldsymbol{F}_{23}$.

Now the force polygon for member 3 can be determined, as shown in Figure 9.26b, by first establishing the location of concurrency point Q. If we measure the magnitude of force $\boldsymbol{F}_{32}$ from this polygon, the polygon for member 2 can then be constructed; this polygon is also shown in Figure 9.26b. Finally, considering cam 1, which is acted on by two forces and a torque, we now know force $\boldsymbol{F}_{21}$, and force $\boldsymbol{F}_{01}$ must be equal and opposite. Torque $\boldsymbol{T}_1$ must balance this couple, and measuring the perpendicular distance between these forces, we find that the magnitude is

$$T_1 = (70\,\text{N})(26\,\text{mm}) = 1{,}820\,\text{N} \cdot \text{mm}.$$

By inspection, the direction is counterclockwise.

As is true of other types of mechanism, friction effects may be significant in gears and cams, depending on factors such as loads, speeds, lubrication, and operating conditions. Efficiencies of less than 100 percent are tabulated for gear drives in gear design handbooks to account for friction losses, which can be substantial in units such as worm gear drives. In Sample Problem 9.11, the mechanism of Sample Problem 9.10 is considered again, this time with friction at the cam surface included.

SAMPLE PROBLEM 9.11

Analysis of a Cam Mechanism with Friction

The cam in Figure 9.27a rotates in the counterclockwise direction. Determine the camshaft torque $\boldsymbol{T}_1$ required for static equilibrium against applied load $\boldsymbol{F}_3$, which has a magnitude of 100 N. The coefficient of friction between cam 1 and coupler link 2 is 0.1, while friction in the linkage bearings and pin O_1 is negligible.

Solution. The solution procedure is similar to that of Sample Problem 9.10. The only difference is that the net force between members 1 and 2 does not act along the common normal in this case. Instead, it acts at an angle to the normal equal to the friction angle defined by Eq. (9.32):

$$\tan \phi = \mu = 0.1 \text{ or } \phi = 6°.$$

This angle is incorporated into the free-body diagrams of Figure 9.27b and must reflect the proper direction of the friction force. By inspection of Figure 9.27a, the sliding velocity $\boldsymbol{v}_{B_1B_2}$ will have the direction shown. This will therefore be the direction of the friction force acting on member 2, with the friction force on member 1 equal and opposite. Figure 9.27b shows the complete static-force analysis, from which torque $\boldsymbol{T}_1$ is found to have a magnitude of

$$T_1 = (64\,\text{N})\,(30\,\text{mm}) = 1{,}920\,\text{N} \cdot \text{mm}$$

and a counterclockwise direction. As expected, the presence of friction brings about increased torque requirements compared to with those of Sample Problem 9.10.

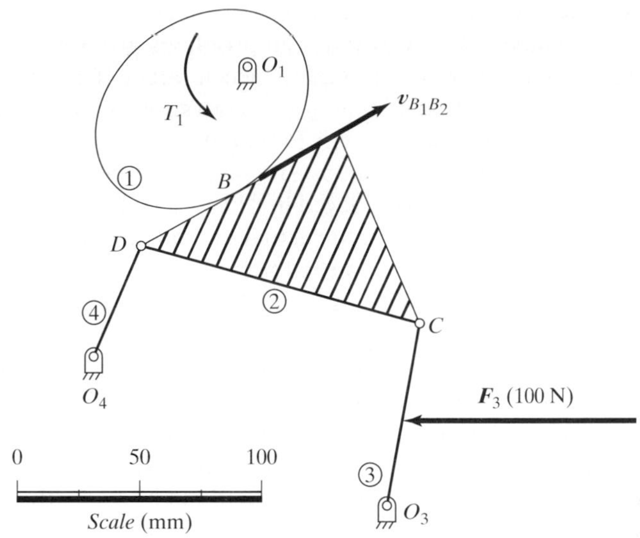

(a)

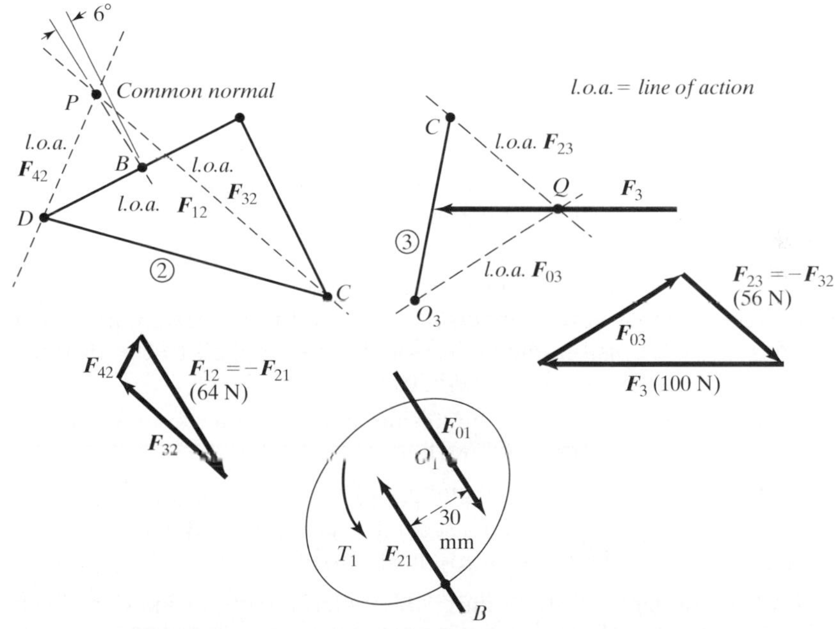

(b)

FIGURE 9.27 (a) Force analysis of a cam linkage system including friction at the cam surface. Friction at all other joints is neglected. (b) Free-body diagrams and force polygons.

SUMMARY

Academic problems in stress analysis and machine design often specify the load applied to a member. Real-world design problems are not so straightforward. Inertia effects are present whenever a mass is accelerated. If inertial forces are significant compared with applied loads, static-force analysis is inadequate. We then need a thorough kinematic and dynamic analysis to find forces in links, pins, bearings and fasteners. The same is true if inertial torques are significant compared with applied torques. (Dynamic-force analysis is considered in Chapter 10.)

Early in the design process, the actual shape and mass of each linkage component are often unspecified. If so, a static-force analysis may provide a reasonable first approximation of actual conditions. Using that analysis and other information, we may select tentative masses. We can then decide whether a dynamic-force analysis is required for accurate results. In some cases, it is also necessary to consider friction.

Tools commonly used for linkage design and force analysis of linkages include graphical methods, detailed computer analysis methods based on mathematics software, and motion simulation software. The method of virtual work based on energy principles is another alternative. Each method has advantages and disadvantages. Since errors can creep into any design or analysis, you may want to solve the same problem in two different ways.

If we need to find forces in a linkage for only one position, graphical methods may be considered. Graphical methods are also used to help formulate a solution in a detailed analysis of a linkage and to check a computer solution at an arbitrary position.

Are you comfortable with free-body diagrams? If not, spend a few minutes reviewing examples of such diagrams in this chapter or in an elementary engineering mechanics text. Here are a few concepts that you can use for static-force analysis of linkages:

- For a planar linkage, two force equilibrium equations and one torque equilibrium equation must be satisfied for each link.
- Suppose we consider the linkage as a whole; that is, we look at a free-body diagram of the assembled linkage. Then, external forces are in equilibrium and external torques and moments are in equilibrium. Examples of external forces on a pump or piston engine are fluid force applied to the piston and reaction forces on the piston and crankshaft.
- A two-force member is in equilibrium if and only if the two forces are equal and opposite and lie along the same line.
- The forces in a three-force member have a concurrency point.
- If one link exerts a force on a second link, the second exerts an equal and opposite force on the first (Newton's third law).

Linkage design usually requires a detailed computer analysis. Such an analysis begins with describing link positions. Vector methods are useful because they fit in well with force analysis methods. Be sure, however, to check for mechanism closure. Is the sum of the vector links zero? Check the transmission angle when designing a crank-rocker linkage. If the transmission angle falls outside of the generally accepted range, then the linkage may jam; the coupler may not be able to drive the output crank. After correcting any flaws in the tentative linkage design, redo the position analysis and follow up with velocity and acceleration analyses.

Before starting a detailed force analysis, make a rough estimate of maximum inertial forces and torques. If inertial effects are small compared with applied forces and torques, a static-force analysis will be reasonably accurate. Then

- Identify known and unknown forces and torques.
- If you are analyzing a piston engine, pump or compressor, be sure to include the lateral force that the cylinder exerts on the piston.
- Write force and moment equilibrium equations.
- Try to uncouple the equations.
- If you cannot uncouple the equations, set up the simultaneous equations in a form that is acceptable to the software you have chosen.
- Plot forces and torques as a function of crank angle.
- Interpret the results. Identify maximum values of tension and compression in links.
- Use your results to redesign the linkage if necessary.

Motion simulation software can be a powerful and efficient design tool, but it does not relieve the user of thinking. Instead, it reduces the labor of programming and allows more time for "what if" analysis and redesign based on analyzing various configurations.

A Few Review Items

- Identify a two-force member that is a component of a piston pump.
- Suppose the pump operates at high speed. Will static-force analysis apply? Is the component you identified still a two-force member?
- A piston, wrist pin, and connecting rod are assembled. Can you make a free-body diagram showing the external forces on this assembly?
- Sketch a crank-rocker linkage driven by the small crank. The large crank drives the output shaft. Sketch a free-body diagram of each link. Does your sketch violate Newton's third law? If so, make the necessary changes.
- Can you relate input torque and output torque to reaction forces on a four-bar linkage?

PROBLEMS

For Problems 9.1 through 9.17

Perform a graphical static-force analysis of the given mechanism. Construct the complete force polygon for determining bearing forces and the required input force or torque. Mechanism dimensions are given in the accompanying figures.

9.1 The applied piston load $\boldsymbol{P}$ on the in-line slider-crank mechanism of Figure P9.1 remains constant as angle ϕ varies. $\boldsymbol{P}$ has a magnitude of 500 N. Determine the required input torque $\boldsymbol{T}_1$ for static equilibrium at the following crank positions:

(a) $\phi = 45°$.

(b) $\phi = 135°$.

(c) $\phi = 270°$.

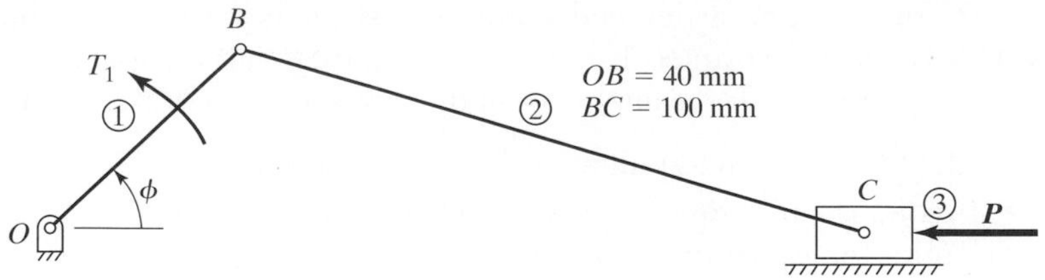

FIGURE P9.1

9.2 The applied piston load $\boldsymbol{P}$ on the offset slider-crank mechanism of Figure P9.2 remains constant as angle ϕ varies. $\boldsymbol{P}$ has a magnitude of 100 lb. Determine the required input torque $\boldsymbol{T}_1$ for static equilibrium at the following crank positions:

(a) $\phi = 45°$.
(b) $\phi = 135°$.
(c) $\phi = 270°$.
(d) $\phi = 315°$.

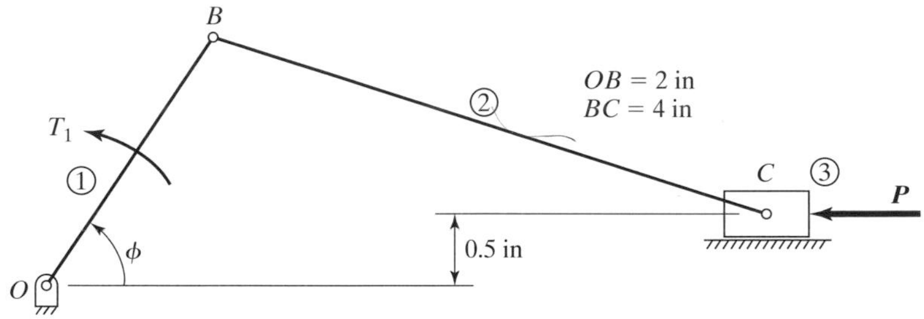

FIGURE P9.2

9.3 Determine the required input torque $\boldsymbol{T}_1$ for static equilibrium of the mechanism shown in Figure P9.3. Force $\boldsymbol{F}_2$ has a magnitude of 200 N.

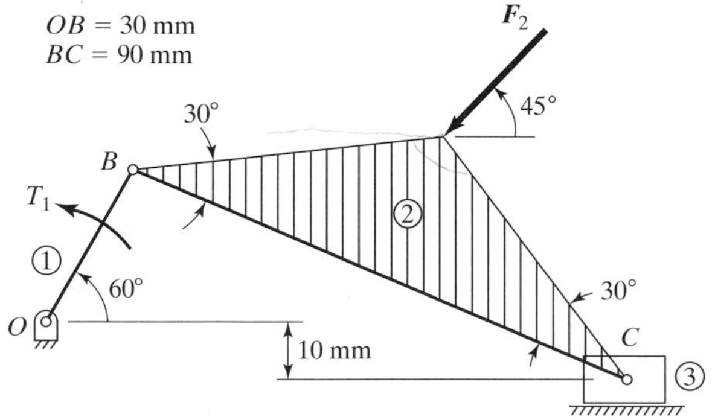

FIGURE P9.3

9.4 Determine the required input torque $\boldsymbol{T}_1$ for static equilibrium of the mechanism shown in Figure P9.4. Force $\boldsymbol{F}_2$ has a magnitude of 100 lb, and piston force $\boldsymbol{P}$ is 200 lb. Both forces act horizontally.

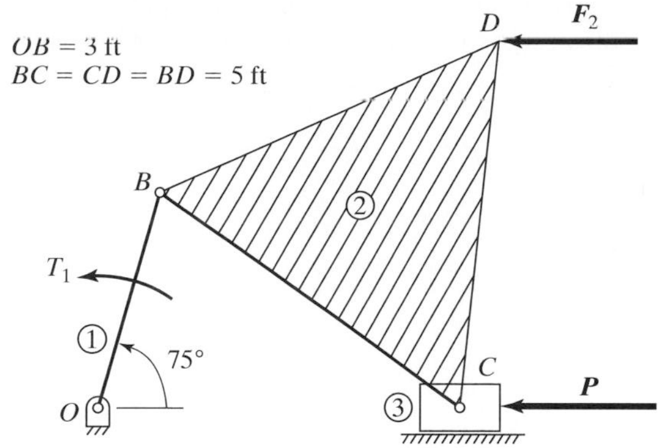

FIGURE P9.4

9.5 Determine the required input torque $\boldsymbol{T_1}$ for static equilibrium of the mechanism shown in Figure P9.5. Forces $\boldsymbol{F_2}$ and $\boldsymbol{F_3}$ have magnitudes of 50 N and 75 N, respectively. Force $\boldsymbol{F_2}$ acts in the horizontal direction.

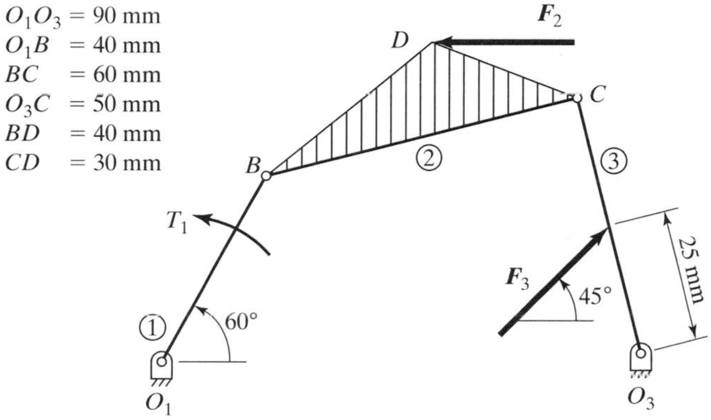

FIGURE P9.5

9.6 Determine the required input torque $\boldsymbol{T_1}$ for static equilibrium of the mechanism shown in Figure P9.6. Forces $\boldsymbol{F_2}$ and $\boldsymbol{F_3}$ have magnitudes of 20 lb and 10 lb, respectively. Force $\boldsymbol{F_3}$ acts in the horizontal direction.

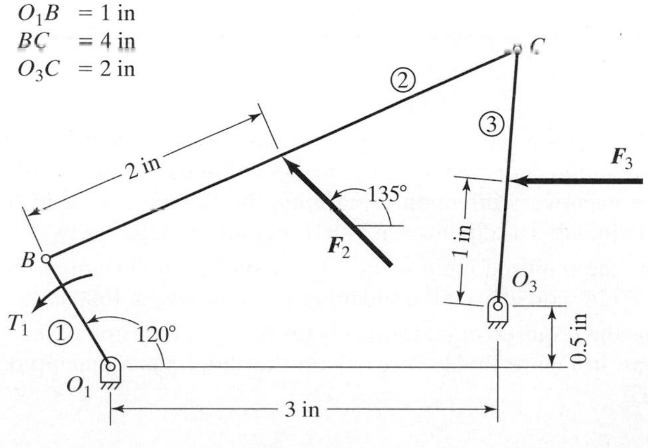

FIGURE P9.6

9.7 Determine the required input torque T_1 for static equilibrium of the mechanism shown in Figure P9.7. Torques T_2 and T_3 are pure torques, with magnitudes of $10 \, \text{N} \cdot \text{m}$ and $7 \, \text{N} \cdot \text{m}$, respectively.

$$O_1B = 80 \text{ mm}$$
$$BC = 160 \text{ mm}$$
$$O_3C = 100 \text{ mm}$$
$$O_1O_3 = 200 \text{ mm}$$

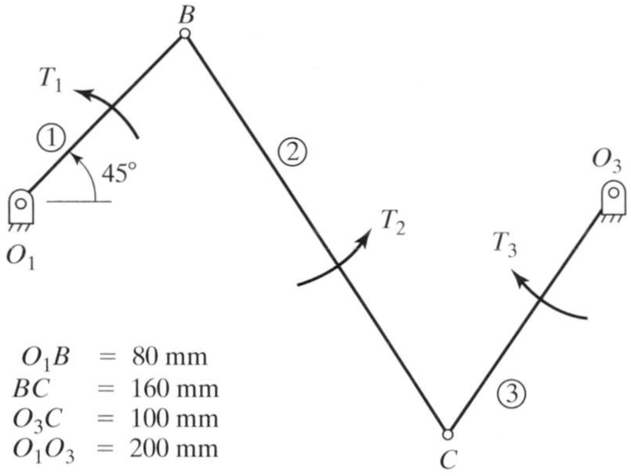

FIGURE P9.7

9.8 Determine the required cylinder gage pressure for static equilibrium of the mechanism shown in Figure P9.8. Torque T_3 has a magnitude of $1000 \, \text{in} \cdot \text{lb}$. The diameter of the piston is 1.5 in.

$$OA = 9 \text{ in}$$
$$O_1A = 6 \text{ in}$$
$$O_1B = 9 \text{ in}$$
$$BC = 15 \text{ in}$$
$$O_3C = 12 \text{ in}$$

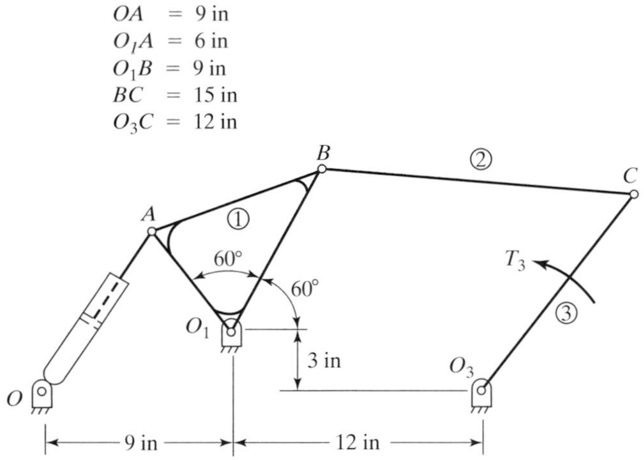

FIGURE P9.8

9.9 Figure P9.9 shows a four-bar linkage that has link masses of 7 kg, 15 kg, and 12 kg for members 1, 2, and 3, respectively. The centers of mass are at the link centers, and gravity acts in the negative y direction. Determine the required force in the horizontal spring attached to member 1 at point A for static equilibrium.

9.10 Determine the required input torque T_1 for static equilibrium of the mechanism shown in Figure P9.10. Force F_5 on the slider has a magnitude of 1000 lb.

9.11 Determine the required input torque T_2 for static equilibrium of the quick-return mechanism shown in Figure P9.11. Force F_5 on the slider has a magnitude of 800 lb. Angle θ equals 105°.

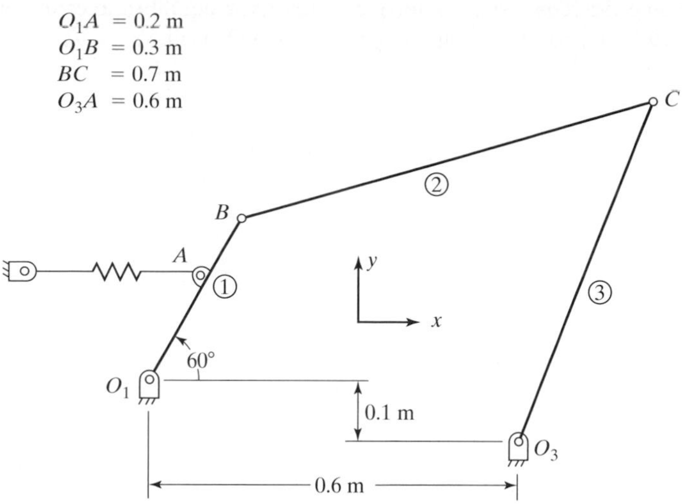

FIGURE P9.9

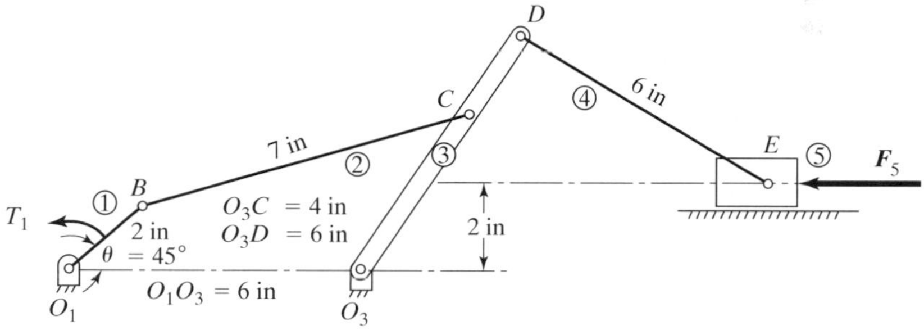

FIGURE P9.10

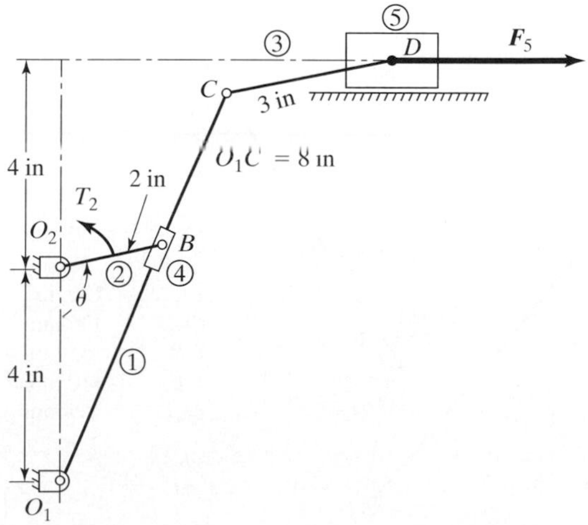

FIGURE P9.11

9.12 Determine the required input torque T_1 for static equilibrium of the mechanism shown in Figure P9.12. Torque T_5 has a magnitude of 600 in · lb.

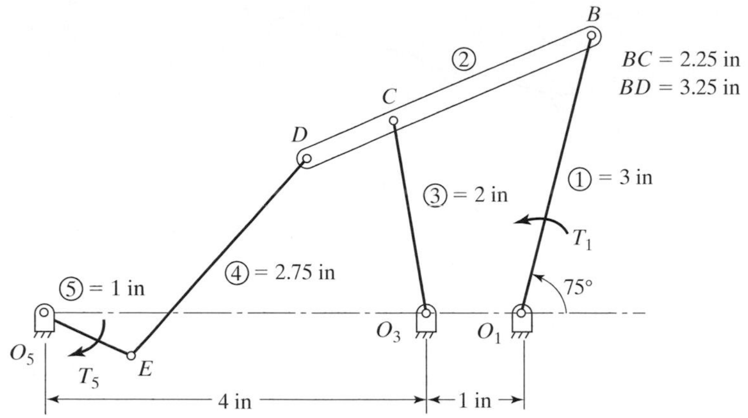

FIGURE P9.12

9.13 Figure P9.13 is a schematic diagram of a linkage similar to the variable-stroke pump shown in Figure 1.11. Determine the torque T_1 required for static equilibrium when piston load $\boldsymbol{P}$ has a magnitude of 500 N.

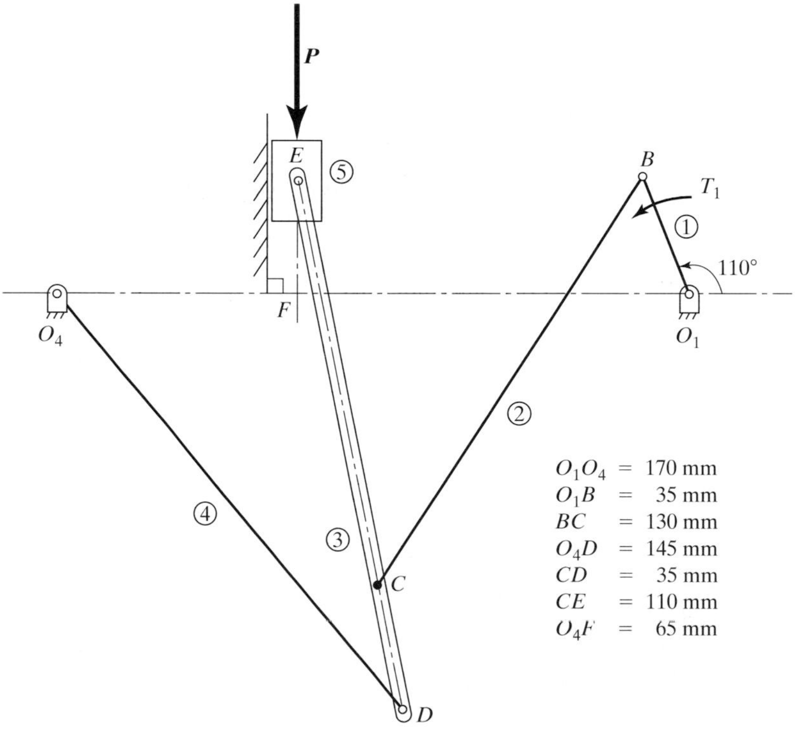

FIGURE P9.13

9.14 For the toggle mechanism shown in Figure P9.14, determine the force developed at slider 5 for an input torque T_1 of 100 in · lb clockwise, assuming that inertial forces and friction forces are negligible. Perform the analysis for the following crank positions:

(a) $\theta = 0$.

(b) $\theta = 30°$.

(c) $\theta = 60°$.

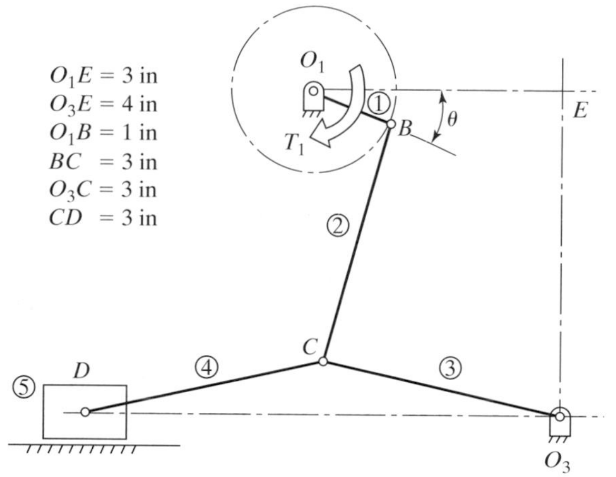

FIGURE P9.14

9.15 Determine the torque T_1 required for static equilibrium of the two-cylinder engine depicted in Figure P9.15. Piston forces F_4 and F_5 have magnitudes of 6,000 N and 2,000 N, respectively. Consider the case where cylinder V angle ψ is 90°, crank spacing θ is 90°, and crank angle ϕ is 160°.

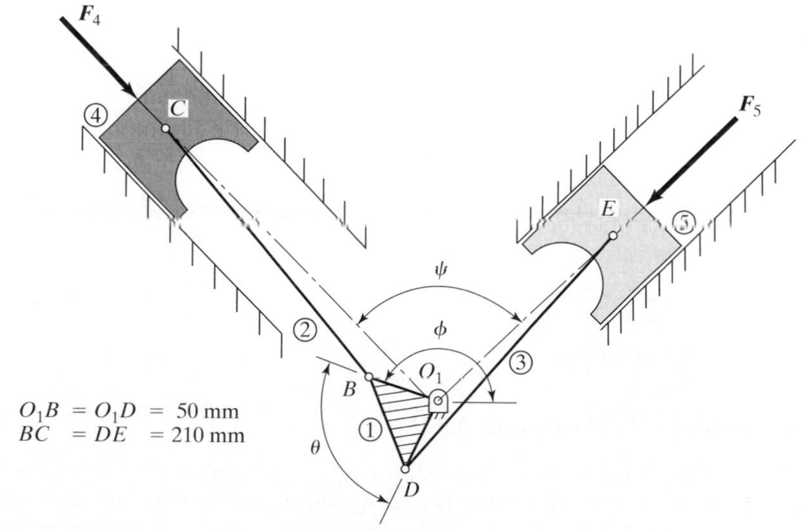

FIGURE P9.15

9.16 Determine torque T_1 required for static equilibrium of the two-cylinder engine depicted in Figure P9.16. Piston forces F_4 and F_5 have magnitudes of 3000 N and 7000 N, respectively. Consider the case where cylinder V angle ψ is 90° and crank angle ϕ is 20°.

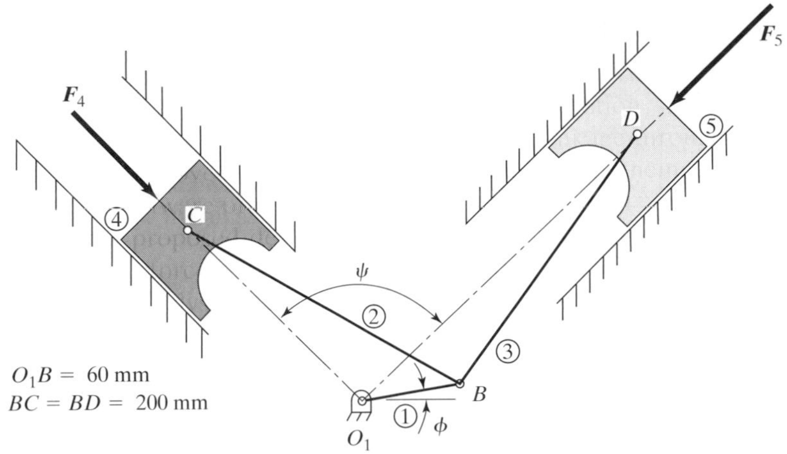

O_1B = 60 mm
$BC = BD$ = 200 mm

FIGURE P9.16

9.17 Determine the handle force P necessary to produce a jaw force Q of 100 lb on the lever wrench shown in Figure P9.17. AB = 3.0 in, BC = 1.2 in, CD = 1.3 in. Force P, force Q, and line BC are vertical. Angles are measured from the horizontal.

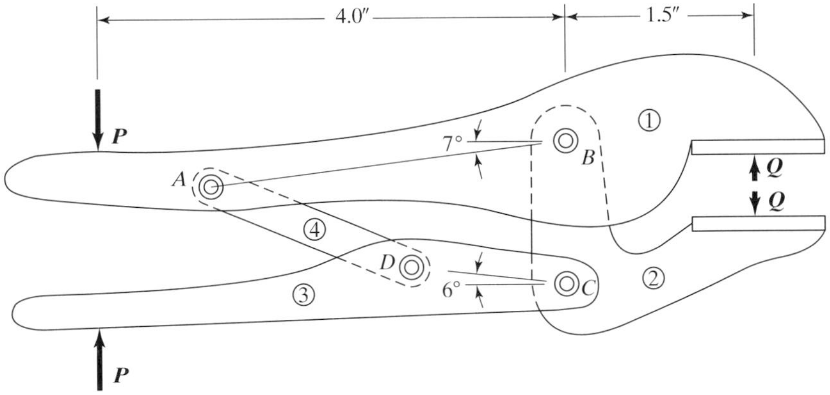

FIGURE P9.17

For Problems 9.18 through 9.34

Obtain an analytical static-force solution for the given mechanism, using the force and moment equilibrium approach. Determine all bearing forces and the required input force or torque at the specified position(s). Mechanism dimensions are given in the accompanying figures.

9.18 Analyze the mechanism of Problem 9.1.

9.19 Analyze the mechanism of Problem 9.2. Derive a general function relating input torque T_1 to piston force $\boldsymbol{P}$ and crank angle ϕ for an offset slider-crank mechanism with crank length r, connecting-rod length ℓ, and offset h. Calculate the input torque and bearing forces at the following crank positions of the mechanism of Figure P9.2:

(a) $\phi = 45°$.

(b) $\phi = 135°$.

(c) $\phi = 270°$.

(d) $\phi = 315°$.

9.20 Analyze the mechanism of Problem 9.3.

9.21 Analyze the mechanism of Problem 9.4.

9.22 Analyze the mechanism of Problem 9.5.

9.23 Analyze the mechanism of Problem 9.6.

9.24 Analyze the mechanism of Problem 9.7.

9.25 Analyze the mechanism of Problem 9.8.

9.26 Analyze the mechanism of Problem 9.9.

9.27 Analyze the mechanism of Problem 9.10.

9.28 Analyze the mechanism of Problem 9.11.

9.29 Analyze the mechanism of Problem 9.12.

9.30 Analyze the mechanism of Problem 9.13.

9.31 Analyze the mechanism of Problem 9.14.

9.32 Analyze the mechanism of Problem 9.15.

9.33 Analyze the mechanism of Problem 9.16.

9.34 Analyze the mechanism of Problem 9.17.

9.35 Analyze the mechanism of Problem 9.1 by the method of virtual work.

9.36 Analyze the mechanism of Problem 9.2 by the method of virtual work.

9.37 Analyze the mechanism of Problem 9.3 by the method of virtual work.

9.38 Analyze the mechanism of Problem 9.4 by the method of virtual work.

9.39 Analyze the mechanism of Problem 9.5 by the method of virtual work.

9.40 Analyze the mechanism of Problem 9.6 by the method of virtual work.

9.41 Analyze the mechanism of Problem 9.7 by the method of virtual work.

9.42 Analyze the mechanism of Problem 9.8 by the method of virtual work.

9.43 Analyze the mechanism of Problem 9.10 by the method of virtual work.

9.44 Analyze the mechanism of Problem 9.11 by the method of virtual work.

9.45 Analyze the mechanism of Problem 9.15 by the method of virtual work.

9.46 Analyze the mechanism of Problem 9.16 by the method of virtual work.

9.47 Perform a graphical force analysis of the mechanism of Problem 9.1, including friction effects. The coefficients of friction are 0.1 at bearing O, 0.15 at each of pins B and C, and 0.2 at the slider. The bearing radii are 20 mm at O, 15 mm at B, and 15 mm at C. The crank rotates clockwise. Determine the following for crank angle $\phi = 45°$:

(a) the torque $\boldsymbol{T_1}$ required for static equilibrium.

(b) the instantaneous efficiency.

9.48 Do Problem 9.47 by the analytical method.

9.49 Perform a graphical force analysis of the mechanism of Problem 9.2, including friction effects. The coefficients of friction are 0.15 at bearing O, 0.25 at bearing B, 0.1 at bearing C, and 0.2 at the slider. The bearing radii all equal 1.0 in. The crank rotates clockwise. Determine the following for crank angle $\phi = 45°$:

(a) the torque $\boldsymbol{T}_1$ required for static equilibrium.

(b) the instantaneous efficiency.

9.50 Do Problem 9.49 by the analytical method.

9.51 Perform a graphical force analysis of the mechanism of Problem 9.3, including friction effects. The radii of the friction circle are 4 mm at O, 5 mm at B, and 3 mm at C. The crank rotates clockwise. Determine the torque $\boldsymbol{T}_1$ required for static equilibrium. Neglect friction between the slider and the frame.

9.52 Perform a graphical force analysis, including friction effects, of the mechanism of Problem 9.5 with force $\boldsymbol{F}_2 = 0$. The coefficients of friction are 0.2 at each of bearings O_1 and O_2 and 0.15 at each of bearings B and C. The bearing radii are all equal to 15 mm. The crank rotates counterclockwise. Determine the torque $\boldsymbol{T}_1$ required for static equilibrium.

9.53 Perform a graphical force analysis, including friction of the mechanism of Problem 9.6 with force $\boldsymbol{F}_2 = 0$. The coefficients of friction all equal 0.2, and the bearing radii are 0.3 in at O_1 and 0.5 in at B, C, and O_3. The crank rotates clockwise. Determine the torque $\boldsymbol{T}_1$ required for static equilibrium.

9.54 Figure P9.18 shows a geared five-bar linkage, with the larger gear fixed to the frame and the smaller gear attached to link 2. The gears are spur gears with a pressure angle of 20° and pitch diameters of 6 in and 4 in. A force $\boldsymbol{F}_4$ of 200 lb is applied as shown at the midpoint of member 4. Determine the required torque $\boldsymbol{T}_1$ for static equilibrium.

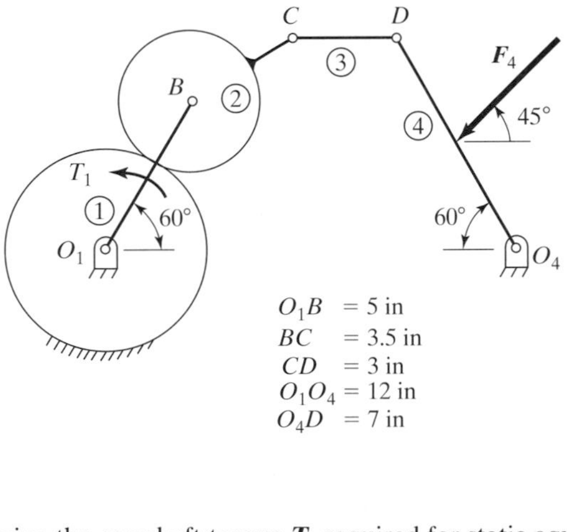

FIGURE P9.18

9.55 Determine the camshaft torque $\boldsymbol{T}_1$ required for static equilibrium of the cam-and-follower mechanism of Figure P9.19. Torque $\boldsymbol{T}_2$ has a magnitude of 50 in · lb and a counterclockwise direction. Neglect friction.

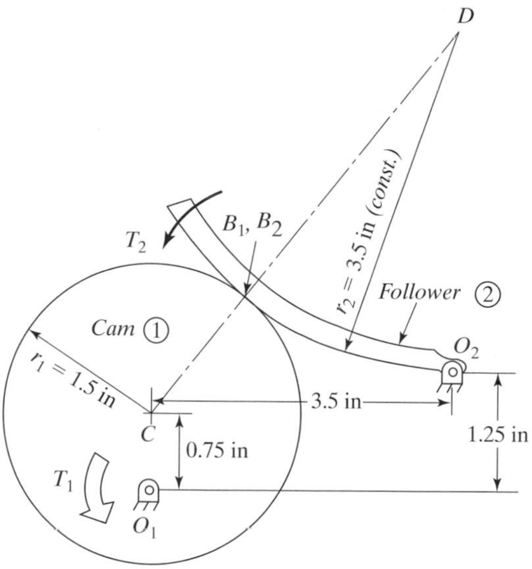

FIGURE P9.19 At the instant shown, point D, the center of curvature of the cam follower, lies directly above O_2, and point C lies directly above O_1.

9.56 Repeat Problem 9.55, including friction at the cam-follower interface. The coefficient of friction is 0.15, and the cam rotates clockwise.

9.57 Determine the camshaft torque $\boldsymbol{T}_1$ required for static equilibrium of the cam-and-follower mechanism of Figure P9.20. Force $\boldsymbol{F}_2$ on the follower has a magnitude of 20 lb. Neglect friction.

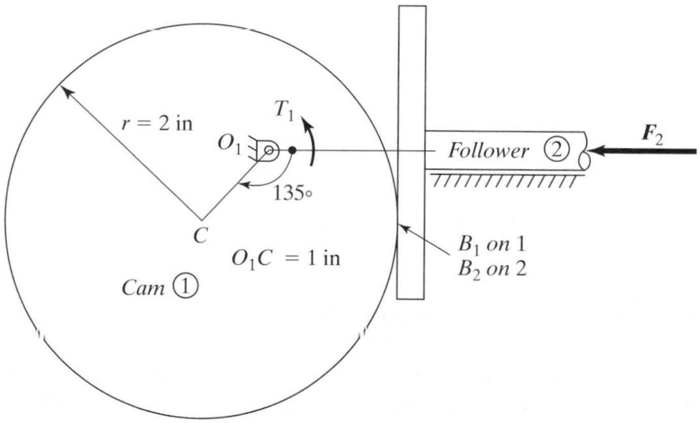

FIGURE P9.20

9.58 Repeat Problem 9.57, including friction at both the cam–follower interface and the follower–guideway interface. The coefficient of friction is 0.15, and the cam has counterclockwise rotation.

9.59 An in-line slider-crank mechanism with a 30-mm crank length and 70-mm connecting rod length operates at a crank speed of 20 rpm. There is a constant 40-N force on the slider (toward the crankshaft). Plot the slider position, connecting-rod position, connecting-rod force, and crankshaft torque against crank position. The same linkage was analyzed graphically and analytically for one crank position in sample problems 9.1 and 9.5. You may want to check your results against those problems.

9.60 Design a single-cylinder compressor to operate at 20 rpm. Analyze the compressor position, connecting-rod forces, and crankshaft torque through a full cycle of the machine's operation. Check the torque for one crankshaft position. Plot the results against the crank angle and piston position. Among your design decisions, try an in-line slider-crank linkage with crank length $R_1 = 50$ mm and connecting-rod length $R_2 = 110$ mm. Represent the axial piston force (N) by

$$\begin{cases} -100[R_0 - (R_2 - R_1) + 0.05R_1] & \text{for piston velocity} \geq 0 \text{ and} \\ 5R_1 \text{ otherwise,} & \end{cases}$$

where R_0 = wrist-pin position (mm), measured from the crankshaft.

9.61 Design a single-cylinder compressor to operate at 12 rpm. Analyze the compressor position, connecting-rod forces, and crankshaft torque through a full cycle of the machine's operation. Check the torque for one crankshaft position. Plot the results against the crank angle and piston position. Among your design decisions, try an in-line slider-crank linkage with crank length $R_1 = 65$ mm and connecting-rod length $R_2 = 144$ mm. Represent the axial piston force (N) by

$$\begin{cases} -200[R_0 - (R_2 - R_1) + 0.05R_1] & \text{for piston velocity} \geq 0 \text{ and} \\ 12R_1 \quad \text{for piston velocity} < 0, & \end{cases}$$

where R_0 = wrist-pin position (mm), measured from the crankshaft.

9.62 Consider an air compressor in the form of an offset slider crank linkage. The compressor will operate at a constant speed of 25 rpm. Analyze the compressor position, forces, and torques through a full cycle of the machine's operation. Plot the results against crank angle and piston position. Use the moments of external forces to check the torque for one crankshaft position. Among your design decisions, try crank length $R_1 = 60$ mm, connecting-rod length $R_2 = 130$ mm, and a 15-mm offset. The offset is not typical of this application. We estimate that the axial force on the piston due to air pressure and friction will be

$$\begin{cases} -K[R_0 - (R_2 - R_1) + 0.05R_1] & \text{when the piston is moving to the right and} \\ 0.05R_1K \quad \text{when the piston is moving to the left,} & \end{cases}$$

where $K = 150$ N/mm and R_0 = wrist-pin position (mm), measured from the crankshaft.

9.63 Consider an air compressor in the form of an offset slider-crank linkage. The compressor will operate at a constant speed of 30 rpm. Analyze the compressor position, forces, and torques through a full cycle of the machine's operation. Plot the results against the crank angle and piston position. Use the moments of external forces to check the torque for one crankshaft position. Among your design decisions, try crank length $R_1 = 75$ mm, connecting-rod length $R_2 = 155$ mm, and a 25-mm offset. We estimate that the axial force on the piston due to air pressure and friction will be

$$\begin{cases} -K[R_0 - (R_2 - R_1) + 0.05R_1] & \text{when piston velocity} \leq 0 \text{ and} \\ 0.06R_1K, \text{ when piston velocity} > 0 & \end{cases}$$

where R_0 = wrist-pin position (mm), measured from the crankshaft, and $K = 180$ N/mm.

9.64 Refer to the Sample Problem 9.8, and examine the effect of offsetting the crankshaft from the cylinder centerline. We need to supply air at a gage pressure of about 100,000 Pa. Among your design and modeling decisions, consider a tentative air compressor design based on an offset slider-crank linkage with a 40-mm crank length, 64-mm connecting-rod length, and 78-mm-diameter piston. Try offsetting the crankshaft 8 mm in the direction that will increase the time of the compression stroke. A motor rotating at a constant 30 rpm will drive the compressor. Simulate the air pressure with a linear spring that is active only during the compression stroke; the spring force will be zero during the other part of the cycle. The spring, with a spring rate of 6 N/mm, is attached to the piston. Note that we are using only a rough approximation of actual conditions. Find the required crank torque, the lateral force on the piston, and the forces on the connecting rod.

(a) Track the motion of the linkage components for one crank rotation.

(b) Find the maximum magnitude of F_Y, the vertical component of the connecting-rod force on the crank, and the lateral force of the piston on the cylinder. Show the linkage position, spring tension, and motor torque, as well as the other crankpin forces, at the instant that F_Y has maximum magnitude. Assume that inertial forces are small; that is, assume that a static-force analysis will give a reasonably accurate approximation of the actual conditions under which the machine operates.

(c) Find the maximum magnitude of the motor torque, $|\boldsymbol{T}|_{max}$. Show the crank position, spring tension, and crankpin forces at the instant that the motor torque has maximum magnitude.

(d) Plot the variation in the spring tension, motor torque, and crankpin forces during a full crank rotation.

9.65 Consider a tentative air compressor design based on an in-line slider-crank linkage with a 40-mm crank length and 76-mm connecting-rod length. A motor rotating at a constant 30 rpm will drive the compressor. Simulate the air pressure with a linear spring that is active only during the compression stroke; the spring force will be zero during the other part of the cycle. The spring, with a spring rate of 5 N/mm, is attached to the piston. Note that we are using only a rough approximation of the actual conditions. Find the required crank torque, the lateral force on the piston, and the forces on the connecting rod.

(a) Track the motion of the linkage components for one crank rotation.

(b) Find the maximum magnitude of F_Y, the vertical component of the connecting rod force on the crank, and the lateral force of the piston on the cylinder. Show the linkage position, spring tension, and motor torque, as well as the other crankpin forces, at the instant that F_Y has maximum magnitude. Ignore inertial forces.

(c) Find the maximum magnitude of the motor torque, $|\boldsymbol{T}|_{max}$. Show the crank position, spring tension, and crankpin forces at the instant that the motor torque has maximum magnitude

(d) Plot the variation in the spring tension, motor torque, and crankpin forces during a full crank rotation.

9.66 Consider a tentative air compressor design based on an in-line slider-crank linkage with a 19-mm crank length and 36-mm connecting rod length. A motor rotating at a constant 1 rad/s will drive the compressor. Simulate the air pressure with a linear spring that is active only during the compression stroke; the spring force will be zero during the other part of the cycle. The spring, with a spring rate of 10 N/mm, is attached to the piston. Note that we are using only a rough approximation of actual conditions. Find the required crank torque and the horizontal, vertical, and resultant forces that the connecting rod exerts on the wrist pin.

(a) Track the motion of the linkage components for one crank rotation.

(b) Plot the variation in the spring tension, motor torque, and wrist-pin forces during a full crank rotation.

PROJECTS

See Projects 1.1 through 1.6 and suggestions in Chapter 1.

Describe and plot representative forces and torques. Make use of computer software wherever practical. Check your results by a graphical method for at least one linkage position. Evaluate the linkage in terms of its performance requirements.

BIBLIOGRAPHY AND REFERENCES

Bedford, A., and W. Fowler, *Dynamics*, Addison-Wesley, Reading, MA, 1995.

Beer, F. P., and E. R. Johnston, Jr., *Vector Mechanics for Engineers: Statics and Dynamics*, McGraw-Hill, New York, 1984.

Knowledge Revolution, *Working Model 2D User's Manual*, Knowledge Revolution, San Mateo CA, 1996.

Mathsoft, *Mathcad 2000*™ *User's Guide*, Mathsoft, Inc., Cambridge MA, 1999.

Meriam, J. L., *Engineering Mechanics: Statics and Dynamics*, Wiley, New York, 1975.

Shames, I. H., *Engineering Mechanics: Statics and Dynamics*, Prentice-Hall, Upper Saddle River, NJ, 1980.

CHAPTER 10

Dynamic-Force Analysis

Methods and Concepts You Will Learn and Apply When Studying This Chapter

- Newton's second law of motion.
- Calculation of mass moment of inertia for linkage elements.
- Use of software to calculate mass, mass moment of inertia, and center of mass of complex shapes.
- Inertial forces and torques and d'Alembert's principle.
- Dynamic analysis of the four-bar linkage, using graphical and analytical methods.
- Application to equivalent linkages.
- Dynamic analysis of the slider-crank mechanism, using graphical and analytical methods.
- Verifying the results of a simulation.
- Dynamic analysis of compound linkages.
- Dynamic analysis of a linkage subject to a specified speed-related torque input.
- Applications involving inertial forces only and applications that include fluid forces.
- Balancing of machinery. Static balancing and dynamic balancing.
- Balancing of reciprocating machines. Single- and multicylinder machines and in-line-, opposed, and V engines.

10.1 INTRODUCTION

This chapter deals with dynamic forces in machines. Dynamic forces are associated with accelerating masses, and since virtually all machines contain accelerating parts, dynamic forces are always present.

Perhaps the simplest example of a dynamic force is the case of a mass mounted on a rod that is rotating about a fixed pivot with constant angular velocity. (See Figure 10.1a.) We know that in order for the mass to maintain a circular path of motion, the

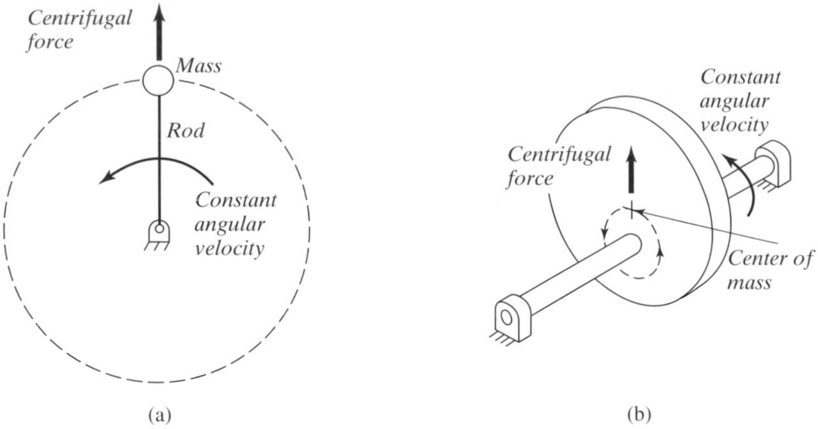

FIGURE 10.1 (a) A mass that is moving in a circular path undergoes centripetal acceleration, and there is a dynamic force, referred to as centrifugal force, associated with the acceleration. The centrifugal force is exerted by the mass on the rod, and is transmitted to the bearing. (b) The shaft is subject to centrifugal force because the center of mass of the rotor does not lie on the shaft enterline.

rod and, in turn, the pivot must provide a force acting radially inward (a centripetal force). As a result, the rod and, in turn, the pivot will be acted upon by a force directed radially outward. This force, commonly referred to as a *centrifugal force*, is a dynamic force because it results from the centripetal acceleration of the mass. This type of dynamic force is prevalent in machines. For example, a centrifugal force will act upon a rotating shaft or rotor if its center of mass does not lie exactly on the bearing centerline, as shown in Figure 10.1b. Discrepancies between the location of the center of mass and that of the center of rotation of an actual rotor can result from a number of factors, such as manufacturing inaccuracies, inhomogeneity of the material of the rotor and bowing or bending of the shaft.

In the previous chapter, it was pointed out that in certain applications dynamic forces are small compared with other forces acting on the system, and in those cases dynamic forces may be neglected in the analysis. However, in many situations the reverse is true, and dynamic forces are dominant or at least comparable in magnitude to external forces. The group of problems in this class continues to grow with the trend toward machines with higher and higher operating speeds, as, for example, in industrial manufacturing processes where the speed of a machine has a direct bearing on a worker's productivity. As new machines are designed to operate at higher speeds, dynamic-force effects become increasingly important. For example, rotors have been designed and built that run at rotational speeds in excess of 100,000 rev/min. Even the slightest eccentricity of the center of mass from the axis of rotation will lead to significant dynamic forces that, if not accounted for in the design process, may result in vibrations, noise, wear, or even failure of the machine.

10.2 NEWTON'S SECOND LAW OF MOTION

The basis for investigating dynamic forces in machines is Newton's second law of motion, which states that *a particle acted on by forces whose resultant is not zero will move in such a way that its acceleration will at any instant be proportional to the resultant force*. This law is expressed mathematically, for the special case of a particle with invariant mass, as

$$\boldsymbol{F} = m\boldsymbol{a}, \tag{10.1}$$

where $\boldsymbol{F}$ is the resultant force on the particle, m is the mass of the particle, and $\boldsymbol{a}$ is the acceleration of the particle. Newton's second law can be extended to a rigid body in the form

$$\boldsymbol{F}_e = m\boldsymbol{a}_G \tag{10.2a}$$

for linear motion or

$$\boldsymbol{T}_{eG} = \dot{\boldsymbol{H}} \tag{10.2b}$$

for rotational (uniform circular) motion.

Equation (10.2a) states that the resultant external force $\boldsymbol{F}_e$ on a rigid body is equal to the product of the mass of the body, m, and the acceleration of the center of mass, $\boldsymbol{a}_G$, where G designates the location of the center of mass. Equation (10.2b), which is the analogous statement for rotational motion, states that the resultant external moment on the body about its center of mass, $\boldsymbol{T}_{eG}$, is equal to the rate of change of the body's angular momentum $\boldsymbol{H}$ with respect to time. This statement is often expressed in the following component form, known as Euler's equations of motion for a rigid body:

$$T_{eG_x} = I_x\,\alpha_x + (I_z - I_y)\omega_y\omega_z; \tag{10.2c}$$
$$T_{eG_y} = I_y\,\alpha_y + (I_x - I_z)\omega_z\omega_x; \tag{10.2d}$$
$$T_{eG_z} = I_z\,\alpha_z + (I_y - I_x)\omega_x\omega_y. \tag{10.2e}$$

In this formulation, the xyz-axes are the principal axes of the body, with their origin at the center of mass, G, of the body.

The mass moment of inertia of a particle about a particular axis is defined as the product of the mass of the particle and the square of the distance of the particle from the axis. For a continuous body made up of an infinite number of mass particles, the moment of inertia, I, is the integral of all of the individual moments of inertia; that is,

$$I = \int r^2\,dm,$$

where dm is the mass of a particle and r is the distance of that particle from the axis. This mathematical definition can be used to compute mass moments of inertia of machine members, and tables are available in mechanics texts and handbooks that contain expressions for moments of inertia determined this way for standard shapes. However, experimental methods of determination are often used for complex shapes, such as connecting rods and cams. Note that the moment of inertia of a body depends on the particular reference axis, as well as on the shape of the body and the manner in which its mass is distributed.

In the SI system, mass moment of inertia is expressed in units of kilogram-meter squared ($kg \cdot m^2$).

The preferred customary U.S. unit for mass is the pound second squared per inch ($lb\text{-}s^2/in$), obtained by dividing the weight of a body by 386 in/s^2, the acceleration due to gravity. Thus, the customary U.S. unit for mass moment of inertia is the pound second squared inch ($lb \cdot s^2 \cdot in$).

For the special case of planar motion, angular momentum $\boldsymbol{H} = I_G \boldsymbol{\omega}$. Equation 10.2b can be stated as

$$\boldsymbol{T}_{eG} = I_G \boldsymbol{\alpha}, \tag{10.2f}$$

where vectors $\boldsymbol{T}_{eG}$, $\boldsymbol{\omega}$, and $\boldsymbol{\alpha}$ are perpendicular to the plane of motion of the body. The subscript G in Eq. (10.2f) indicates that the moment of inertia is with respect to an axis passing through the center of mass, G, and perpendicular to the plane of rotation of the body. Equations (10.2a) and (10.2f) are depicted in Figure 10.2a. Part b of the figure illustrates d'Alembert's principle (to be discussed later). Do you remember the ***parallel-axis theorem***? It is the relationship between the moment of inertia, I, of a body about any axis and the moment of inertia, I_G, about a parallel axis through the center of mass:

$$I = I_G + mR^2,$$

where m is the mass of the body and R is the perpendicular distance between the two axes.

An important observation from Eqs. (10.2a) and (10.2b) is based on the following two ways in which the equations can be interpreted:

1. The equations can be solved for the force and torque required to produce a known motion of a body.

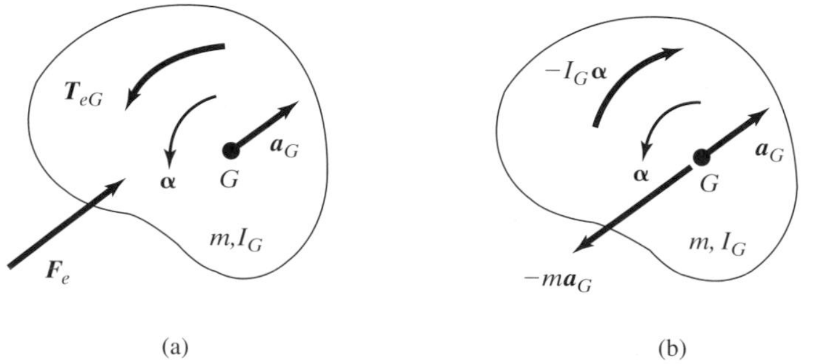

FIGURE 10.2 (a) A rigid body accelerating under the influence of external force $\boldsymbol{F}_e$ and external torque $\boldsymbol{T}_{eG}$. (b) The inertial force and the inertial torque corresponding to planar motion of a rigid body.

2. The equations can be solved for the motion of a body resulting from the application of known forces and torques.

Strictly speaking, the second approach is the more accurate analytical procedure in machinery dynamics, because we are more apt to have knowledge of the nature of applied loads and input forces (e.g., torque–speed characteristics for electric motors) than knowledge of the exact motion response of complex nonlinear mechanisms. However, the solution by this approach is more difficult than by the first approach.

In many design situations, it is common practice to utilize the first approach and to assume that the motion of the mechanism is equal to the ideal required motion for the specific application. The solution is then carried out for the resulting forces, and the information obtained thereby is used to size the members and bearings of the mechanism and to select an input power source. This is an iterative process, because the forces are dependent on the sizes of the members. Also, it is almost impossible to select design components so that the forces required for the ideal assumed motion are produced exactly. Therefore, it may be necessary to evaluate the mechanism design obtained in this way, either by construction and testing a physical prototype or by examining a computer model based on the second approach described; these procedures can be used to "fine-tune" the design by adjusting the final design parameters.

Most of the examples and problems of this chapter will be based on the approach of assumed motion of the mechanism. In the next section, it will be shown that dynamic-force analysis can be expressed in a form very similar to static-force analysis, so that solution techniques from Chapter 9 can be applied.

Calculation of Mass and Mass Moment of Inertia for Linkage Elements

Suppose we intend to construct mechanisms that include bodies shaped like those in Figure 10.3. The bodies have uniform thickness h and uniform mass density γ and will move in the plane of the figure. The center of mass (center of gravity) is shown for each body. We want to know the mass and mass moment of inertia of each body about an axis perpendicular to the figure and through the center of mass.

Consider, for example, a circular disk of radius R_0. The mass is

$$m = \gamma h \, \pi R_0^2.$$

We substitute

$$dm = \gamma h \cdot 2\pi r \, dr$$

in the equation for mass moment of inertia and integrate from

$$r = 0 \text{ to } r = R_0$$

Center of Mass

It is easy to locate the center of mass of regular shapes: a circle, square, or rectangle of uniform thickness and uniform mass. For an irregularly shaped body, the coordinates of the center of mass can be found from the integrals:

$$x_G = \int x\, dm / \int dm$$

and

$$y_G = \int y\, dm / \int dm,$$

where the origin of the coordinates is arbitrary,
dm is an element of mass,
$\int dm$ is the total mass of the body,
x and y are the coordinates of the element of mass,
integration takes place over the entire body, and
x_G and y_G are the coordinates of the center of mass of the body.

Locating the center of mass of a complex body can be a time-consuming calculation. If software (e.g., Working Model™) is used to find the center of mass and other properties, then we have more time for important design tasks, such as performing a dynamic analysis, interpretating results, and redesign.

10.3 D'ALEMBERT'S PRINCIPLE AND INERTIAL FORCES

An important principle, known as d'Alembert's principle, can be derived from Newton's second law. In words, d'Alembert's principle states that *the reverse-effective forces and torques and the external forces and torques on a body together are equivalent to static equilibrium. Transposing the right-hand terms of Eqs. (10.2a) and (10.2b) yields*

$$\boldsymbol{F}_e + (-m\boldsymbol{a}_G) = 0 \tag{10.3a}$$

and

$$\boldsymbol{T}_{eG} + (-\dot{\boldsymbol{H}}) = 0. \tag{10.3b}$$

The terms in parentheses in Eqs. (10.3a) and (10.3b) are called the reverse-effective *force* and the reverse-effective *torque*, respectively. (These quantities are also referred to as the inertial force and inertial torque.) Thus, we define the inertial force as

$$\boldsymbol{F}_i = -m\boldsymbol{a}_G \tag{10.4a}$$

which reflects the concept that a body resists any change in its velocity by an inertial force that is proportional to the mass of the body and its acceleration. The inertia force acts through the center of mass, G of the body. The inertial torque, or inertial couple, is given by

$$\boldsymbol{C}_i = -\dot{\boldsymbol{H}}, \tag{10.4b}$$

or, for the case of planar motion (see Eq. (10.2f)),

$$\boldsymbol{C}_i = -I_G\boldsymbol{\alpha}. \tag{10.4c}$$

As indicated, the inertial torque is a pure torque or couple. Figure 10.2b shows these quantities for the case of planar motion, where, from Eqs. (10.4a) and (10.4c), their directions are opposite to that of the accelerations. Substituting Eqs. (10.4a) and (10.4b) into Eqs. (10.3a) and (10.3b) leads to equations that are similar to those used in static-force analysis, viz.,

$$\sum \boldsymbol{F} = \sum \boldsymbol{F}_e + \boldsymbol{F}_i = 0 \tag{10.5a}$$

$$\sum \boldsymbol{T}_G = \sum \boldsymbol{T}_{eG} + \boldsymbol{C}_i = 0, \tag{10.5b}$$

where $\Sigma \boldsymbol{F}_e$ refers here to the summation of external forces and therefore is the resultant external force and $\Sigma \boldsymbol{T}_{eG}$ is the summation of external moments, or resultant external moment, about the center of mass, G. Thus, the dynamic-analysis problem is reduced in form to a static-force and -moment balance in which inertial effects are treated in the same manner as external forces and torques. In particular, for the case where the motion of the mechanism is assumed, the inertial forces and couples can be determined completely and thereafter treated as known mechanism loads. The graphical or analytical solution techniques described in Chapter 9 can then be employed.

Furthermore, d'Alembert's principle facilitates summing moments about any arbitrary point P in the body, as long as we remember that the moment due to inertial force $\boldsymbol{F}_i$ must be included in the summation. Hence,

$$\sum \boldsymbol{T}_P = \sum \boldsymbol{T}_{eP} + \boldsymbol{C}_i + \boldsymbol{R}_{PG} \times \boldsymbol{F}_i = 0, \tag{10.5c}$$

where $\Sigma \boldsymbol{T}_P$ is the summation of moments (including inertial moments) about point P, $\Sigma \boldsymbol{T}_{eP}$ is the summation of external moments about P, $\boldsymbol{C}_i$ is the inertial couple defined by Eq. (10.4b) or (Eq. 10.4c), $\boldsymbol{F}_i$ is the inertial force defined by Eq. (10.4a), and $\boldsymbol{R}_{PG}$ is a vector from point P to point G. It is clear that Eq. (10.5b) is the special case of Eq. (10.5c) in which point P is taken as the center of mass, G (i.e., $\boldsymbol{R}_{PG} = 0$).

For a body in plane motion in the xy-plane, with all external forces in that plane, Eqs. (10.5a) and (10.5b) become

$$\sum F_x = \sum F_{ex} + F_{ix} = \sum F_{ex} + (-ma_{Gx}) = 0, \qquad (10.6a)$$

$$\sum F_y = \sum F_{ey} + F_{iy} = \sum F_{ey} + (-ma_{Gy}) = 0, \qquad (10.6b)$$

$$\sum T_G = \sum T_{eG} + C_i = \sum T_{eG} + (-I_{G\alpha}) = 0, \qquad (10.6c)$$

where a_{Gx} and a_{Gy} are the x and y components of $\boldsymbol{a}_G$. Equations (10.6a) through (10.6c) are three scalar equations for which the sign convention for torques and angular accelerations is based on a right-hand xyz coordinate system; that is, counterclockwise is positive and clockwise is negative. The general summation of moments about arbitrary point P, Eq. (10.5c), becomes

$$\sum T_P = \sum T_{eP} + C_i + R_{PGx}F_{iy} - R_{PGy}F_{ix}$$

$$= \sum T_{eP} + (-I_G\alpha) + R_{PGx}(-ma_{Gy}) - R_{PG}(-ma_{Gx}) = 0, \qquad (10.6d)$$

where R_{PGx} and R_{PGy} are the x and y components of position vector $\boldsymbol{R}_{PG}$. This expression for dynamic moment equilibrium will be useful in the analyses to be presented in the rest of this chapter.

SAMPLE PROBLEM 10.2

Inertial Effects on a Rotating Body

A steel disk 0.3 m in radius and 0.0398 m thick with two blades is shown in Figure 10.4. Each blade is 1 m square by 5 mm thick, has a mass of 10 kg, and is subject to air resistance equal to $k \cdot v^2 \cdot A$, where $k = 0.025\ Ns^2/m^4$, v = velocity of the blade center, and A = blade area. An electric motor applying a torque of 25 N • m drives the assembly. Find the mass moment of inertia and the terminal angular velocity of the assembly. Find the angular acceleration and motor power at 20 rad/s angular velocity.

Calculation decision. We will use mathematics software in the solution.

Solution summary. The mass density of steel is about 8,000 kg/m^3. The table of masses and mass moments of inertia can be used to calculate the mass moment of inertia about an axis through the center of gravity of the disk and each vane. However, because the vanes rotate about the center of the disk, we must correct the mass moment of inertia of the vanes by adding the product of the mass of the vane and the square of the distance from the center of rotation to the center of each vane. We find the mass moment of inertia of the assembly (the disk and two vanes) to be about 18.5 kg • m^2.

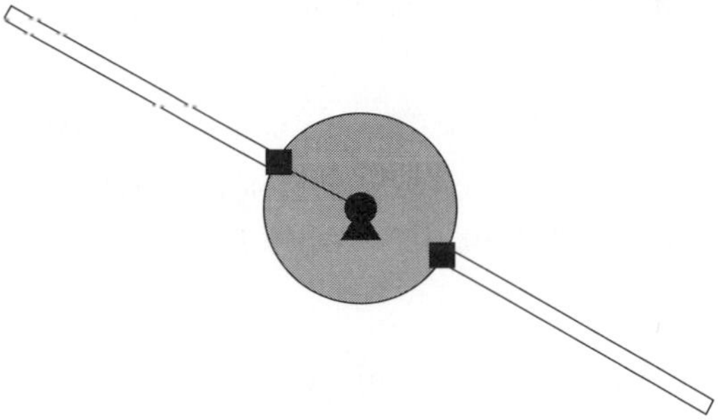

FIGURE 10.4 Inertial effects on a rotating body.

We now apply d'Alembert's principle, adding the applied torque, air resistance torque, and reverse-effective torque and then equating the sum to zero. Next, we rearrange the equation to find the angular acceleration in terms of the angular velocity. When the angular acceleration is zero, the assembly has reached its terminal angular velocity, about 31.25 rad/s. At the terminal angular velocity, the angular acceleration is zero, and the air resistance torque balances the applied torque, plus or minus a small rounding error. Although a numerical procedure is used in the detailed calculations, you may discover a closed-form solution for the terminal velocity.

The mechanical power (W) is given by the product of the angular velocity (rad/s) and the applied torque (N · m). The motor will, of course, be less than 100% efficient. Thus, the electric power requirement is somewhat higher than the calculated mechanical power.

Detailed calculation.

Inertia of rotating body units: mks

Circular disk Radius $R_0 := 0.3$ Thickness $h := 0.0398$

Material: steel, mass density $\gamma := 8000$

mass $m_c := \gamma \cdot h \cdot \pi \cdot R_0^2$ $m_c = 90.025$

mass moment of inertia $I_c := \dfrac{m_c \cdot R_0^2}{2}$ $I_c = 4.051$

Rectangular vane length $c := 1$ width $b := .005$ depth $h_v := 1$

mass $m_v := 10$

distance from center of disk to center of vane $x := 0.8$

mass moment of inertia

$$I_{v0} := \frac{m_v \cdot (b^2 + c^2)}{12} \qquad I_{v0} = 0.835$$

$$I_v := I_{v0} + m_v \cdot x^2 \qquad I_v = 7.235$$

Assembly:

mass moment of inertia $I := I_c + 2 \cdot I_v$ $I = 18.522$

air resistance at midpoint of one vane $k := 0.025$ $F_v(\omega) := k \cdot (\omega \cdot x)^2 \cdot h_v \cdot c$

total torque due to air resistance $T_a(\omega) := 2 \cdot F_v(\omega) \cdot x$

Motor: Torque $T := 25$ Power $P(\omega) := T \cdot \omega$

Angular momentum $H(\omega) := I \cdot \omega$

Angular acceleration $\alpha(\omega) := \dfrac{T - T_a(\omega)}{I}$

Terminal angular velocity Estimate: $\omega := 20$ $\omega_1 := \text{root}(\alpha(\omega), \omega)$ $\omega_1 = 31.25$

Force (air resistance) $F_v(\omega_1) = 15.625$ Torque $T_a(\omega_1) = 25.001$

Angular acceleration $\alpha(\omega_1) = -2.879 \cdot 10^{-5}$ Power $P(\omega_1) = 781.258$

Angular momentum of hub $I_c \cdot \omega_1 = 126.6$

Check for $\omega = 20$ $\omega_2 := 20$

Angular acceleration $\alpha(\omega_2) = 0.797$ Power $P(\omega_2) = 500$

Using Animation Software to Display Transient Effects

In the preceding sample problem, torque, angular acceleration, and power were expressed in terms of angular velocity. We sometimes need to know how these quantities vary with time. It is possible to work out the relationship mathematically. As an alternative, we can "work smart" (or be lazy) and let animation software solve the problem and plot the results.

SAMPLE PROBLEM 10.3

Transient Effects

Examine the motion of the assembly of Sample Problem 10.2.

a. Find the value of the following quantities at one-half second after starting the motor: mechanical power, and angular position, velocity, and acceleration of the assembly.

b. Plot the following variables against time: air resistance on one blade, angular velocity and angular acceleration of the assembly, and angular momentum of the disk.

Solution. The results at $t = 0.5$ second and the plots against time are shown in Figure 10.5. The plots agree with the terminal velocity calculations in the previous sample problem. In addition, we can find the time at which the angular velocity is 20 rad/s and the corresponding angular acceleration. Again the results agree with those of the previous example.

A Note on Fluid Resistance

The foregoing examples were used to illustrate dynamic effects, employing a very rough approximation of fluid behavior. It is assumed that the relationship between air resistance and speed is based on tests of similar assemblies. As an alternative, you can determine drag coefficients and integrate the torque contributions over the surface of the blade. If you study fluid mechanics, you will be able to develop a more precise model of fluid resistance.

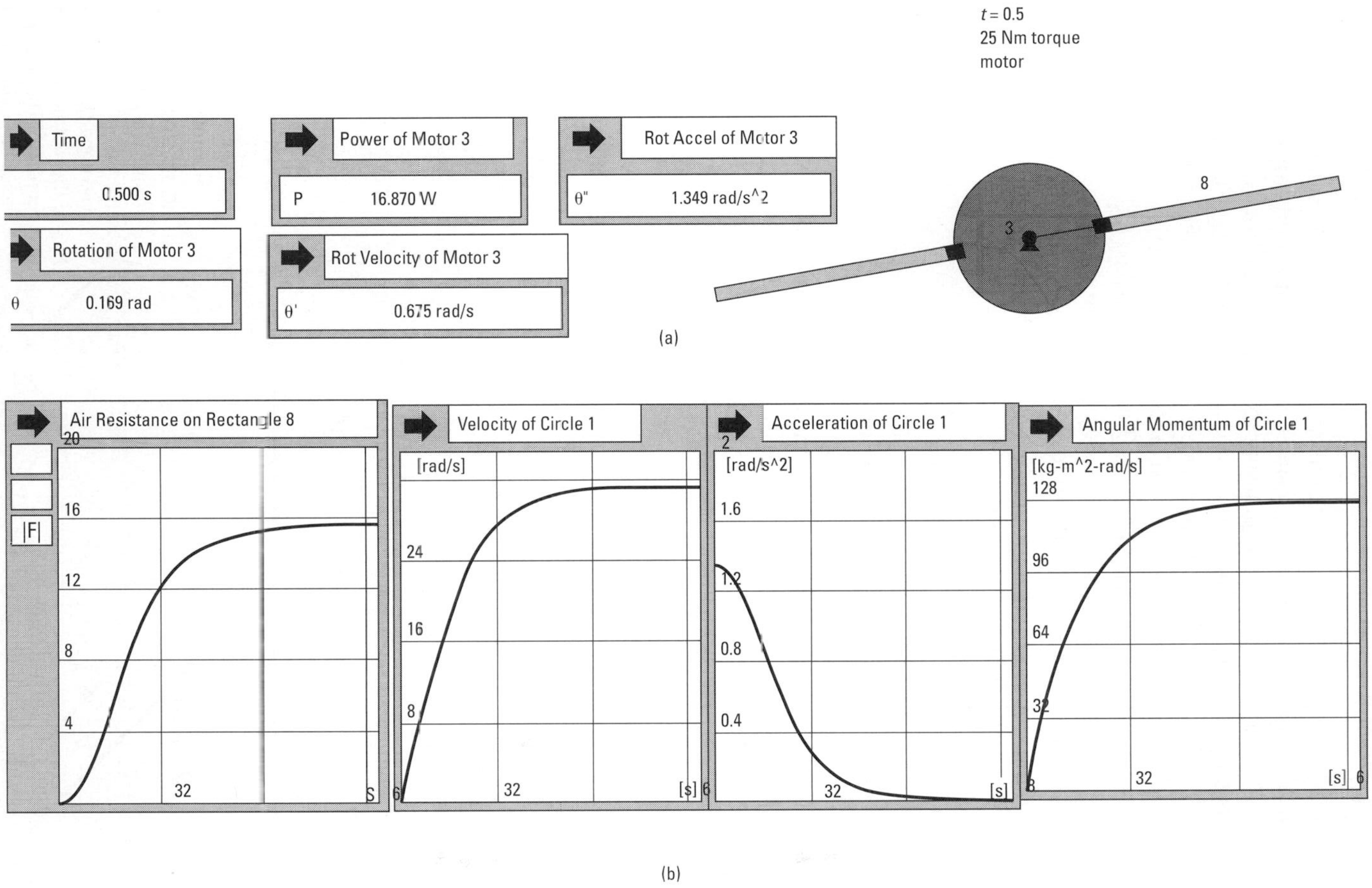

FIGURE 10.5 Inertial effects in Sample Problem 10.3.

Equivalent Offset Inertial Force

For purposes of graphical plane force analysis, it is convenient to define what is known as the equivalent offset inertial force, a single force that accounts for both translational inertia and rotational inertia corresponding to the plane motion of a rigid body. Figure 10.6a shows a rigid body, with planar motion represented by acceleration of the center of mass of the body, $\boldsymbol{a}_G$, and angular acceleration α. The inertial force and inertial torque associated with this motion are also shown. The inertial torque $(-I_G\alpha)$ can be expressed as a couple consisting of forces $\boldsymbol{Q}$ and $-\boldsymbol{Q}$, separated by perpendicular distance h, as shown in Figure 10.6b. The necessary conditions for the couple to be equivalent to the inertial torque are that the sense and magnitude of the two be the same. Therefore, in this case, the sense of the couple must be clockwise, and the magnitudes of $\boldsymbol{Q}$ and h must satisfy the relationship

$$|\boldsymbol{Q}h| = |I_G\alpha|.$$

Otherwise, the couple is arbitrary and an infinite number of possibilities will work. Furthermore, the couple can be placed anywhere in the plane.

Figure 10.6c shows a special case of the couple, where force vector $\boldsymbol{Q}$ is equal to $m\boldsymbol{a}_G$ and acts through the center of mass of the body. Force $-\boldsymbol{Q}$ must then be placed as shown to produce a clockwise sense at a distance

$$h = \frac{|I_G\alpha|}{|\boldsymbol{Q}|} = \frac{|I_G\alpha|}{|m\boldsymbol{a}_G|}. \tag{10.7}$$

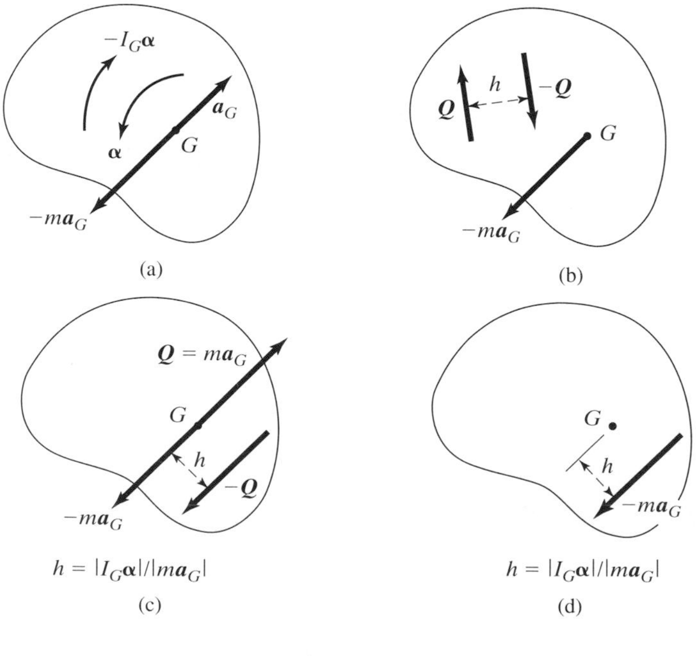

FIGURE 10.6 (a) Equivalent offset inertial force associated with planar motion of a rigid body. (b) Replacement of the inertial torque by a couple. (c) The strategic choice of a couple. (d) The single force is equivalent to the combination of a force and a torque in (a) .

Force $\boldsymbol{Q}$ will cancel with the inertia force $\boldsymbol{F}_i = -m\boldsymbol{a}_G$, leaving the single equivalent offset force shown in Figure 10.6d, which has the following characteristics:

1. The magnitude of the force is $|m\boldsymbol{a}_G|$.
2. The direction of the force is opposite to that of acceleration $\boldsymbol{a}_G$.
3. The perpendicular offset distance from the center of mass to the line of action of the force is given by Eq. (10.7).
4. The force is offset from the center of mass so as to produce a moment about the center of mass that is opposite in sense to acceleration α.

The usefulness of this approach for graphical force analysis will be demonstrated in the next section. Note, however, that the approach is usually unnecessary in analytical solutions, wherein Eqs. (10.6a) through (10.6d), including the original inertial force and inertial torque, can be applied directly.

10.4 DYNAMIC ANALYSIS OF THE FOUR-BAR LINKAGE

The analysis of a four-bar linkage will effectively illustrate most of the ideas that have been presented. Furthermore, the extension to other types of mechanisms should become clear from the analysis of the linkage.

SAMPLE PROBLEM 10.4

Dynamic-Force Analysis of a Four-Bar Linkage

In the four-bar linkage of Figure 10.7a, the dimensions are as shown in the figure, where G refers to center of mass and the mechanism has the following mass properties:

$$\begin{aligned} m_1 &= 0.1 \text{ kg} \quad I_{G1} = 20 \text{ kg}\cdot\text{mm}^2; \\ m_2 &= 0.2 \text{ kg} \quad I_{G2} = 400 \text{ kg}\cdot\text{mm}^2; \\ m_3 &= 0.3 \text{ kg} \quad I_{G3} = 400 \text{ kg}\cdot\text{mm}^2. \end{aligned}$$

Determine the instantaneous value of drive torque T required to produce an assumed motion given by input angular velocity $\omega_1 = 95$ rad/s counterclockwise and input angular acceleration $\alpha_1 = 0$ for the position shown in the figure. Neglect gravity and friction effects.

Solution. This problem falls into the first analysis category described in Section 10.2; that is, given the motion of the mechanism, determine the resulting bearing forces and the necessary input torque. Therefore, the first step in solving the problem is to determine the inertial forces and inertial torques. Thereafter, the problem can be treated as if it were a static-force analysis problem.

The mechanism can be analyzed kinematically by using the methods presented in Chapter 2, 3, and 4. Figure 10.7b shows a graphical analysis employing velocity and acceleration polygons. From the analysis, the following accelerations are determined:

$$\begin{aligned} \boldsymbol{a}_{G1} &= 0 \text{ (stationary center of mass)}; & \boldsymbol{\alpha}_1 &= 0 \text{ (given)}; \\ \boldsymbol{a}_{G2} &= 235{,}000\angle 312^\circ \text{ mm/s}^2; & \boldsymbol{\alpha}_2 &= 520 \text{ rad/s}^2 \text{ ccw}; \\ \boldsymbol{a}_{G3} &= 100{,}000\angle 308^\circ \text{ mm/s}^2; & \boldsymbol{\alpha}_3 &= 2740 \text{ rad/s}^2 \text{ cw}. \end{aligned}$$

Note that the angles of the acceleration vectors are measured counterclockwise from the positive x direction shown in the figure. From Eqs. (10.4a) and (10.4c), the inertial forces and inertial torques are

$$\mathbf{F}_{i1} = \mathbf{0},$$
$$\mathbf{F}_{i2} = -m_2\mathbf{a}_{G2} = 47{,}000\angle 132^\circ \text{ kg} \cdot \text{mm/s}^2 = 47\angle 132^\circ \text{ N},$$
$$\mathbf{F}_{i3} = -m_3\mathbf{a}_{G3} = 30{,}000\angle 128^\circ \text{ kg} \cdot \text{mm/s}^2 = 30\angle 128^\circ \text{ N},$$

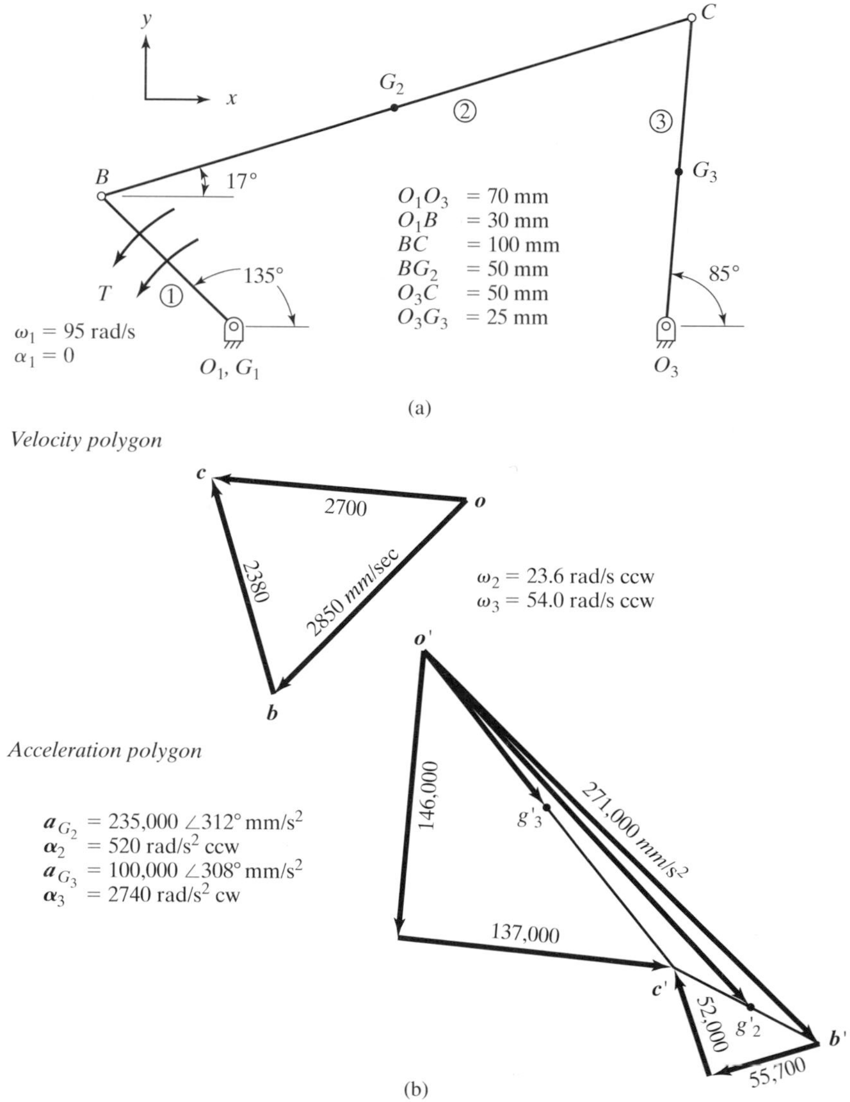

FIGURE 10.7 (a) The four-bar linkage of Sample Problem 10.4. (b) The velocity and acceleration analyses necessary for determining inertial forces and inertial torques.

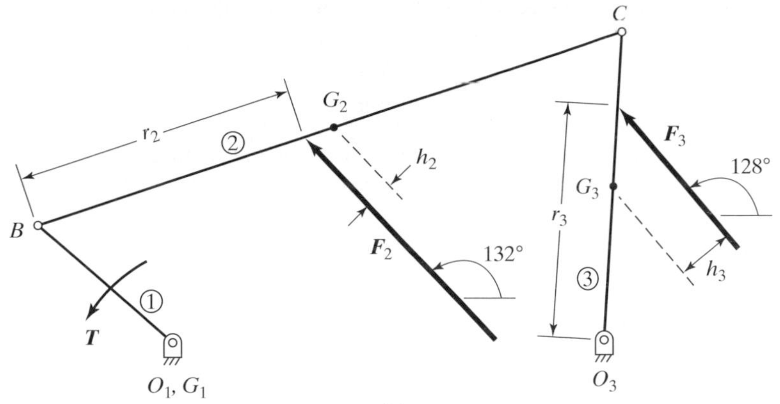

(c)

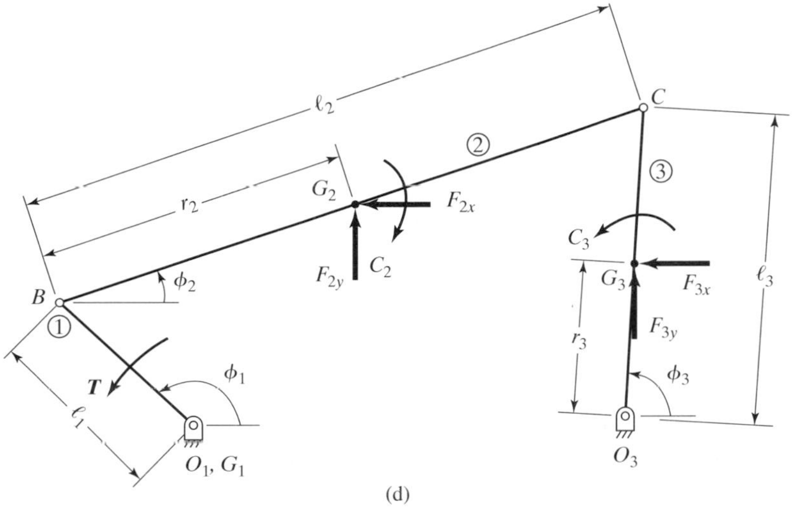

(d)

FIGURE 10.7 (c) Equivalent offset inertial forces for members 2 and 3.
(d) Combinations of inertial forces and inertial torques for members 2 and 3.

$$C_{i1} = 0,$$
$$C_{i2} = -I_{G2}\alpha_2 = 208{,}000 \text{ kg} \cdot \text{mm}^2/\text{s}^2 \text{ cw} = 208 \text{ N} \cdot \text{mm cw},$$

and

$$C_{i3} = -I_{G3}\alpha_2 = 274{,}000 \text{ kg} \cdot \text{mm}^2/\text{s}^2 \text{ ccw} = 274 \text{ N} \cdot \text{mm ccw}.$$

The inertial forces have lines of action through the respective centers of mass, and the inertial torques are pure torques.

Graphical Solution. To simplify the graphical force analysis, we will account for the inertial torques by introducing equivalent offset inertial forces. These forces are shown in Figure 10.7c, and their placement is determined according to the discussion of the previous section. For link 2, offset force $\boldsymbol{F}_2$ is equal and parallel to inertial force $\boldsymbol{F}_{i2}$. Therefore,

$$\boldsymbol{F}_2 = 47\angle 132° \text{ N}.$$

$\boldsymbol{F}_2$ is offset from the center of mass, G_2, by a perpendicular distance equal to

$$h_2 = \frac{|I_{G2}\boldsymbol{\alpha}_2|}{|m_2\boldsymbol{a}_{G2}|} = \frac{208}{47} = 4.43 \text{ mm},$$

measured to the left as shown in order to produce the required clockwise direction for the inertial moment about point G_2. In a similar manner, the equivalent offset inertial force for link 3 is

$$\boldsymbol{F}_3 = 30\angle 128° \text{ N},$$

at an offset distance

$$h_3 = \frac{|I_{G3}\boldsymbol{\alpha}_3|}{|m_3\boldsymbol{a}_{G3}|} = \frac{274}{30} = 9.13 \text{ mm},$$

measured to the right from G_3 in order to produce the necessary counterclockwise inertia moment about G_3. From the values of h_2 and h_3 and the angular relationships, the force positions r_2 and r_3 in Figure 10.7c are computed to be

$$r_2 = BG_2 - \frac{h_2}{\cos(132°-17°-90°)} = 45.1 \text{ mm}$$

and

$$r_3 = O_3G_3 + \frac{h_3}{\cos(90°-85°-128°)} = 38.4 \text{ mm}.$$

Now, we wish to perform a graphical force analysis for known forces $\boldsymbol{F}_2$ and $\boldsymbol{F}_3$. This has been done in Sample Problem 9.2, and the reader is referred to that analysis. The required input torque was found to be

$$T = 838 \text{ N} \cdot \text{mm cw},$$

and the bearing forces were also determined in that example. Free-body diagrams of the three moving members are shown in Figure 9.11d.

Analytical Solution. Having determined the equivalent offset inertia forces $\boldsymbol{F}_2$ and $\boldsymbol{F}_3$, the analytical solution could proceed according to Sample Problem 9.7, which examined the same problem. However, it is not necessary to convert to the offset force, and here we will carry out the analytical solution in terms of the original inertial forces and inertial couples.

Figure 10.7d shows the linkage with the inertial torques and the inertial forces in xy-coordinate form. We define the following quantities:

$$\begin{aligned}
\ell_1 &= 30 \text{ mm}; \quad \ell_2 = 100 \text{ mm}; \quad \ell_3 = 50 \text{ mm};\\
\phi_1 &= 135°; \quad \phi_2 = 17°; \quad \phi_3 = 85°;\\
r_1 &= 0; \quad r_2 = 50 \text{ mm}; \quad r_3 = 25 \text{ mm};\\
F_{2x} &= 47\cos(132°) = -31.4 \text{ N}; F_{2y} = 47\sin(132°) = 34.9 \text{ N};\\
F_{3x} &= 30\cos(128°) = -18.5 \text{ N}; F_{3y} = 30\sin(128°) = 23.6 \text{ N};\\
C_2 &= -208 \text{ N} \cdot \text{mm}; \quad C_3 = 274 \text{ N} \cdot \text{mm};\\
F_{1x} &= F_{1y} = C_1 = 0.
\end{aligned}$$

These data lead to approximately the same values as these found in Sample Problem 9.7 for the coefficients and right-hand terms in Eqs. (9.17a) and (9.17b), where the differences are due to round-off:

$$a_{11} = -49.8; \quad a_{21} = 29.2; \quad b_1 = -786;$$
$$a_{12} = 4.36; \quad a_{22} = -95.6; \quad b_2 = -1{,}920.$$

Then, from Eqs. (9.18a) through (9.22), we get

$$F_{23} = 31.3 \text{ N}, \quad F_{12} = 50.3 \text{ N},$$

and

$$F_{03} = 49.2 \text{ N}, \quad F_{01} = 50.3 \text{ N},$$

and it follows that

$$T = -185 \text{ N} \cdot \text{mm}.$$

Thus, it can be seen that the form of the general analytical solution of the four-bar linkage presented in Chapter 9 as a study of static-force analysis is equally well suited to dynamic-force analysis.

Before leaving this example, a couple of general comments should be made about it. First, the torque that is determined is the instantaneous value required for the prescribed motion, and the value will vary with position. Furthermore, for the position considered, the torque is opposite in direction to the angular velocity of the crank. This opposition can be explained by the fact that the inertia of the mechanism in the indicated position tends to accelerate the crank in the counterclockwise direction, and therefore, the required torque must be clockwise to maintain a constant angular speed. If a constant speed is to be maintained throughout the cycle of the mechanism's motion, there will be other positions of the mechanism for which the required torque will be counterclockwise. The second comment is that it may be impossible to find a mechanism actuator, such as an electric motor, that will supply the required torque-versus-position behavior. This problem can be alleviated, however, in the case of a "constant" rotational speed mechanism through the use of a *flywheel*, which is mounted on the input shaft and which produces a relatively large mass moment of inertia for crank 1. The flywheel can absorb torque and energy variations of the mechanism with minimal fluctuation in speed, thereby maintaining an essentially constant input speed. In such a case, the assumed-motion approach to dynamic-force analysis is appropriate.

SAMPLE PROBLEM 10.5

Dynamic-Force Analysis of an Air Pump

In Chapter 4, an acceleration analysis was carried out for a curved-wing air pump, utilizing an equivalent four-bar linkage. (See Sample Problem 4.13 and Figure 4.15.) The equivalent linkage considered in that example and the resulting acceleration polygon are reproduced here as Figures 10.8a and b. This linkage represents the drive crank (link 1), one of the four curved wings (link 2), and the housing (link 3), which constrains the motion of the tip of the wing (point C).

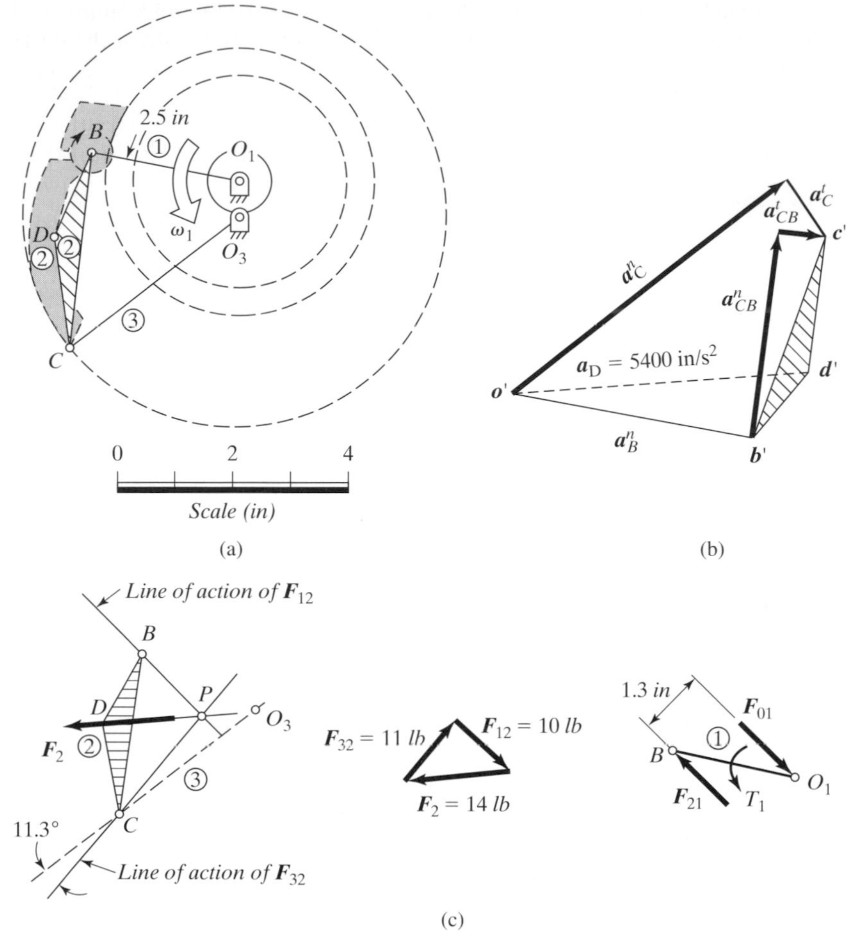

FIGURE 10.8 (a) Dynamic-force analysis of a curved-wing air pump. The diagram shows the equivalent four-bar linkage. (b) Acceleration polygon from Chapter 4. (c) The graphical force analysis.

Point D is the center of mass of the wing. The angular velocity of the shaft is constant at 400 rev/min counterclockwise. The acceleration of the center of mass of the wing at the instant shown was found to have a magnitude of 5400 in/s^2 and the direction shown in the acceleration polygon (Figure 10.8b). The instantaneous angular acceleration of the wing was determined to be 265 rad/s^2 counterclockwise.

Perform a dynamic-force analysis of the mechanism to determine (1) the restraining force exerted by the housing on the wing at point C at the instant shown in the figure and (2) the necessary drive torque on link 1 associated with the inertial force and torque of the wing. The weight of the wing is 1 lb and the moment of inertia about the wing's center of mass is 0.004 lb · s^2 · in. Furthermore, assume that the coefficient of sliding friction between the wing and the housing is 0.2. Neglect friction at pins O_1 and B. (As mentioned in Chapter 4, these dimensions may not correspond to an actual pump and are used here to illustrate general principles of force analysis only.)

Solution. The solution will be carried out graphically. First, we must determine the inertial forces and inertial torques. For link 1, we assume that the center of mass is coincident with the stationary point of rotation, O_1, since the member is symmetric (see Figure 4.15), and therefore, the associated inertial force is zero. Further, the inertial torque for member 1 is also zero, because the angular acceleration is zero. Since a moving member 3 does not actually exist, there are obviously no corresponding inertial terms for that member. Member 2 has an inertial force given by

$$\boldsymbol{F}_{i2} = -m_2\boldsymbol{a}_D$$

which has a magnitude of

$$F_{i2} = m_2 a_D = \left(\frac{w_2}{g}\right)a_D = \left(\frac{1 \text{ lb}}{386.4 \text{ in/s}^2}\right)(5400 \text{ in/s}^2) = 14.0 \text{ lb}$$

and a direction opposite to that of acceleration a_D with a line of action through point D. The inertial torque of member 2 has a magnitude equal to

$$|C_2| = |I_{G2}\alpha_2| = (0.004 \text{ lb}\cdot\text{s}^2\cdot\text{in})(265 \text{ rad/s}^2) = 1.06 \text{ lb}\cdot\text{in}$$

and a clockwise direction opposite to that of the angular acceleration.

Since the analysis will be graphical, we will make use of an equivalent offset inertial force F_2 for member 2. This force has a magnitude of 14.0 lb, a direction opposite to a_D, and a line of action that is displaced by the perpendicular distance

$$h_2 = \frac{I_{G2}\alpha_2}{m_2\alpha_D} = \frac{1.06}{14.0} = 0.076 \text{ in}$$

from the center of mass.

Inspection of Figure 10.8a shows that the offset should be downward from point D in order to produce a clockwise inertial torque. Note that the offset is negligible in this example. Figure 10.8c shows a free-body diagram of member 2 with the offset inertial force placed on the diagram to a suitable scale. The member is a three-force member with known force $\boldsymbol{F}_2$ and unknown forces at points B and C. We can, however, determine the direction of the force exerted by the housing at point C. In the absence of friction, it would act normal to the contacting surfaces—that is, along imaginary link 3 toward point O_3. With friction present, that force acts at an angle to the normal, given by Eq. (9.32):

$$\tan\phi = 0.2, \quad \text{or} \quad \phi = 11.3°.$$

The direction in which this angle is measured is based on the relative motion of the contacting surfaces and should produce a friction force that opposes that motion. In Figure 10.8a, point C on the wing moves counterclockwise relative to the housing: and therefore, the angle is laid off as shown in Figure 10.8c, thus producing a clockwise-acting friction force of the housing on the wing. This establishes the line of action of force $\boldsymbol{F}_{32}$ in Figure 10.8c.

Now, knowing the lines of action of forces $\boldsymbol{F}_2$ and $\boldsymbol{F}_{32}$, we can find the location of the point of concurrency, P, at the intersection of these lines and thereby obtain the line of action of the third force $\boldsymbol{F}_{12}$, applied at point B. Next, the force polygon for member 2 is constructed, as shown in Figure 10.8c. Finally, member 1 is subjected to two forces and a torque. (See the free-body diagram of Figure 10.8c.) Force $\boldsymbol{F}_{21}$ is equal and opposite to force $\boldsymbol{F}_{12}$, and force $\boldsymbol{F}_{01}$ is, in turn, equal and opposite to force $\boldsymbol{F}_{21}$.

We can now address the specific requirements of the problem.

From the force polygon of Figure 10.8c, the force $\boldsymbol{F}_{32}$ of the housing on the wing has a magnitude of approximately 11 lb and the direction shown. The components of this force are a normal force of magnitude 11 cos(11.3°) = 10.8 lb and a friction force of magnitude 11 sin(11.3°) = 2.2 lb.

The required shaft torque T_1 is equal in magnitude to the moment produced by the force couple $(\boldsymbol{F}_{21}, \boldsymbol{F}_{01})$. That is,

$$T_1 = (10 \text{ lb})(1.3 \text{ in}) = 13 \text{ lb} \cdot \text{in},$$

and, by inspection, the torque must be counterclockwise. Note that a complete force analysis of the pump would also consider the other wings and forces produced by air pressure.

10.5 DYNAMIC ANALYSIS OF THE SLIDER-CRANK MECHANISM

Dynamic forces are a very important consideration in the design of slider-crank mechanisms for use in machines such as internal-combustion engines and reciprocating compressors. A dynamic-force analysis of this mechanism can be carried out in exactly the same manner as for the four-bar linkage in the previous section. In accordance with such a process, a kinematic analysis is first performed, from which expressions are developed for the inertial force and inertial torque for each of the moving members. These quantities may then be converted to equivalent offset inertial forces for graphical analysis, or they may be retained in the form of forces and torques for an analytical solution, utilizing, in either case, the methods presented in Chapter 9. In fact, the analysis of the slider-crank mechanism is somewhat easier than that of the four-bar linkage, because there is no rotational motion and, in turn, no inertial torque for the piston or slider, which has translational motion only. The next few paragraphs describe an analytical approach in detail.

Figure 10.9a is a schematic diagram of a slider-crank mechanism, showing crank 1, connecting rod 2, and piston 3, all of which are assumed to be rigid. The locations of the center of mass are designated by letter G, and the members have masses m_i and moments of inertia I_{Gi}, for $i = 1, 2, 3$. The analysis that follows will consider the relationships of the inertial forces and torques to the bearing reactions and the drive torque on the crank, at an arbitrary position of the mechanism given by crank angle ϕ. Friction will be neglected.

Figure 10.9b shows free-body diagrams of the three moving members of the linkage. Applying the dynamic equilibrium conditions of Eqs. (10.4a) through (10.6d) to each member yields the required set of equations. For the piston (moment equation not included),

$$F_{23x} + (-m_3 \, a_{G3}) = 0 \tag{10.8a}$$

and

$$F_{03y} + F_{23y} = 0. \tag{10.8b}$$

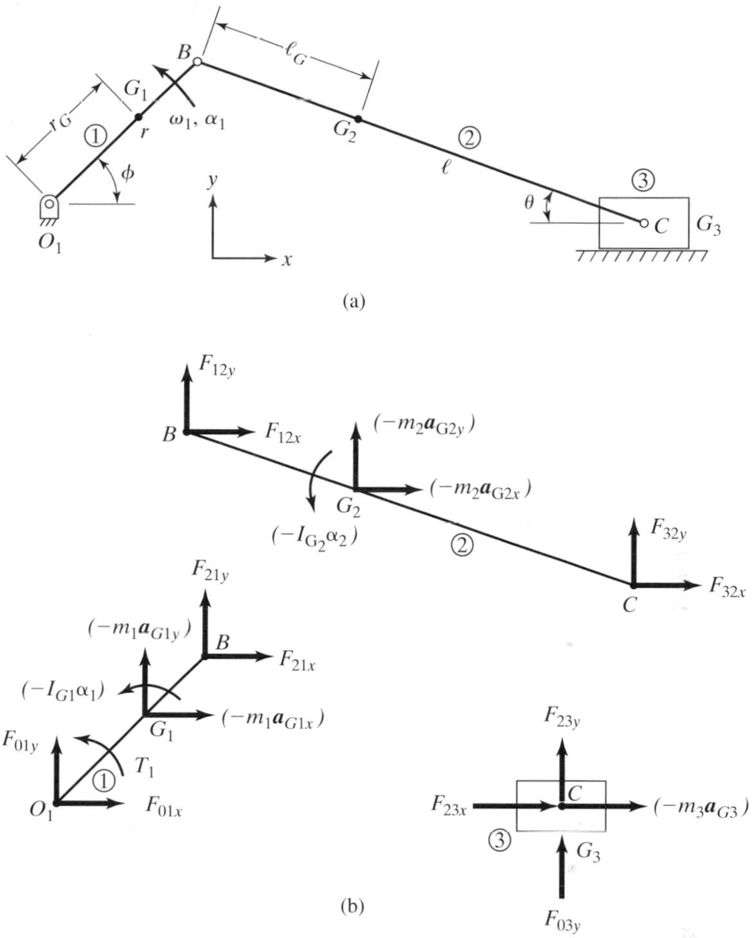

FIGURE 10.9 (a) Dynamic-force analysis of a slider-crank mechanism. (b) Free-body diagrams of the moving members.

For the connecting rod (with moments about point B),

$$F_{12x} + F_{32x} + (-m_2\, a_{G2x}) = 0, \tag{10.8c}$$
$$F_{12y} + F_{32y} + (-m_2\, a_{G2y}) = 0, \tag{10.8d}$$

and

$$F_{32x}\, \ell \sin\theta + F_{32y}\, \ell \cos\theta + (-m_2\, a_{G2x})\, \ell_G \sin\theta + (-m_2\, a_{G2y})\, \ell_G \cos\theta + (-I_{G2}\, \alpha_2) = 0. \tag{10.8e}$$

For the crank (with moments about point O_1),

$$F_{01x} + F_{21x} + (-m_1\, a_{G1x}) = 0, \tag{10.8f}$$
$$F_{01y} + F_{21y} + (-m_1\, a_{G1y}) = 0, \tag{10.8g}$$

and

$$T_1 - F_{21x} r \sin\phi + F_{21y} r \cos\phi - (-m_1 a_{G1x}) r_G \sin\phi$$
$$+ (-m_1 a_{G1y}) r_G \cos\phi + (-I_{G1}\alpha_1) = 0, \qquad (10.8h)$$

where T_1 is the input torque on the crank. This set of equations embodies both of the dynamic-force analysis approaches described in Section 10.2; however, its form is best suited for the case where the motion of the mechanism is known, as illustrated by the next sample problem.

SAMPLE PROBLEM 10.6

Dynamic-Force Analysis of a Slider-Crank Mechanism

Perform a dynamic-force analysis of the in-line slider-crank mechanism for which a kinematic acceleration analysis was carried out in Sample Problem 4.8. That mechanism has dimensions $r = 2$ in and $\ell = 3.76$ in and a known motion based on a crank angular velocity of $\omega_1 = +10$ rad/s and an angular acceleration of $\alpha_1 = +40$ rad/s^2 when the crank angle is $\phi = 70°$ (therefore, $\theta = 30°$). We assume, for the purpose of this example, that the locations of the center of mass are $r_G = 1$ in and $\ell_G = 1.88$ in and that the mass properties are $m_1 = 1$ lbm, $m_2 = 2$ lbm, $m_3 = 3$ lbm, $I_{G1} = 0.01$ lb · s^2 · in, and $I_{G2} = 0.02$ lb · s^2 · in.

Solution. From the previous acceleration analysis in Chapter 4, the following quantities are obtained:

$$a_{G1x} = -71.6 \text{ in/s}^2, \quad a_{G2y} = -80.0 \text{ in/s}^2,$$
$$a_{G1y} = -79.5 \text{ in/s}^3, \quad a_2 = 46 \text{ rad/s}^2,$$
$$a_1 = 40 \text{ rad/s}^2, \quad a_{G3} = -70 \text{ in/s}^2,$$
$$a_{G2x} = -106 \text{ in/s}^2.$$

The inertial force components and inertial torques are calculated as follows:

$$-m_1 a_{G1x} = -\left(\frac{1 \text{ lbm}}{386.4 \text{ lbm} \cdot \text{in/s}^2 \cdot \text{lb}}\right)(-71.6 \text{ in/s}^2) = 0.185 \text{ lb};$$

$$-m_1 a_{G1y} = -\left(\frac{1}{386.4}\right)(-79.5) = 0.206 \text{ lb};$$

$$-I_{G1}\alpha_1 = -(0.01 \text{ lb} \cdot \text{s}^2 \cdot \text{in})(40 \text{ rad/s}^2) = -0.4 \text{ lb} \cdot \text{in};$$

$$-m_2 a_{G2x} = -\left(\frac{2}{386.4}\right)(-106) = 0.549 \text{ lb};$$

$$-m_2 a_{G2y} = -\left(\frac{2}{386.4}\right)(-80.0) = 0.414 \text{ lb};$$

$$-I_{G2}\alpha_2 = -(0.02)(46) = -0.92 \text{ lb} \cdot \text{in};$$

$$-m_3 a_{G3} = -\left(\frac{3}{386.4}\right)(-70) = 0.543 \text{ lb}.$$

Substituting into Eqs. (10.8a) and (10.8e) yields

$$
\begin{aligned}
F_{23x} &= -0.543 = -F_{32x} \\
F_{32y} &= -F_{23y} \\
&= \frac{0.543(3.76)\sin 30^\circ - 0.549(1.88)\sin 30^\circ - 0.414(1.88)\cos 30^\circ + 0.92}{3.76\cos(30^\circ)} \\
&= -0.396 \text{ lb},
\end{aligned}
$$

and, from Eq. (10.8b),

$$F_{03y} = -0.396 \text{ lb}.$$

Using Eqs. (10.8c) and (10.8d), we have

$$F_{12x} = -F_{21x} = -0.543 - 0.549 = -1.09 \text{ lb}$$

and

$$F_{12y} = -F_{21y} = -0.396 - 0.414 - -0.018 \text{ lb}.$$

From Eqs. (10.8f) and (10.8g),

$$F_{01x} = -1.09 - 0.185 = -1.28 \text{ lb}$$

and

$$F_{01y} = -0.018 - 0.206 = -0.224 \text{ lb}.$$

Finally, substituting into Eq. (10.8h), we find the required input torque for the prescribed motion to be

$$
\begin{aligned}
T_1 &= (1.09)(2)\sin(70^\circ) - (0.018)(2)\cos(70^\circ) \\
&\quad + (0.185)(1)\sin(70^\circ) - (0.206)(1)\cos(70^\circ) + 0.4 \\
&= 2.54 \text{ lb} \cdot \text{in}.
\end{aligned}
$$

Thus, an instantaneous counterclockwise torque of 2.54 lb · in is required at the mechanism position considered.

Designers of slider-crank linkages for applications such as engines and compressors often use approximations, to simplify the analysis task. These approximations which deal with the mass distribution of the connecting rod and the acceleration of the piston, are the topics of the next several sections and will also be useful in a later section of this chapter on balancing reciprocating masses. In addition, we will analyze this mechanism from the two standpoints discussed earlier, namely, force analysis under assumed motion and motion analysis under assumed forces.

Using Motion-Simulation Software for Dynamic Force Analysis of Linkages

The analysis in Sample Problem 10.6 required substantial effort. Is there a better way to get the correct result? Mathematics software and motion simulation software should be considered when it is necessary to analyze a linkage through a full cycle of motion. And since improved design is the goal of analysis, mathematics software and motion simulation software offer another advantage: We can do "what if" analysis and find the effect of a proposed design change with a few keystrokes. Designers sometimes look at inertial forces alone, leaving out gas forces on the piston, etc. Then they use the result of this analysis to design and specify mounts for isolating vibration.

SAMPLE PROBLEM 10.7

Using Motion-Simulation Software for Dynamic Force Analysis of a Slider-Crank Linkage

A pump in the form of an in-line slider-crank linkage has a 51-mm crank length and 96-mm connecting-rod length. The crank, connecting rod, and piston masses are 0.45, 0.91, and 1.36 kg, respectively. Mass moments of inertia are $9.845 \cdot 10^{-5}$ and $7.008 \cdot 10^{-4}\ \mathrm{kg \cdot m^2}$ for the crank and connecting rod, respectively (about their centers of gravity). The crank angular velocity is a constant 100 rad/s. Plot the motor torque and linkage motion against time. Find the maximum motor torque and the corresponding values of power, linkage motion, and forces.

Decisions. We will examine inertial forces for the purpose of selecting vibration mounts. Fluid forces on the piston are not considered.

Solution. Working Model™ is used to animate the linkage and plot the results, beginning at the head-dead-center position. The maximum torque is 43.04 N · m at crank position $\theta_1 = 0.59$ rad and $t = 0.0059$ s. Note that discrete time steps are used; the accuracy of these values depends on the size of the step. (See Figure 10.10 for a detailed solution.)

Verifying the Results of a Simulation

We cannot rely totally on the results of mathematics software and motion simulation software. In addition, we should obtain plenty of output data and make simple checks of validity. Software errors are not unknown, but most discrepancies are the result of data entry errors. We can check velocity and acceleration at limiting positions and Newton's laws come in handy for checking forces and torques. If a connecting rod or coupler is temporarily given a very small mass and mass moment of inertia, then it approximates a two-force member. In that case, we can check whether the forces at either end balance and lie along the member. At high angular velocities, inertial effects are dominant and gravity forces are not important. For machines with heavy links moving at relatively low speeds, the weight of a link may be significant compared with the product of its mass and acceleration. When gravity forces are included in a simulation, the orientation of the linkage (relative to the vertical) is important.

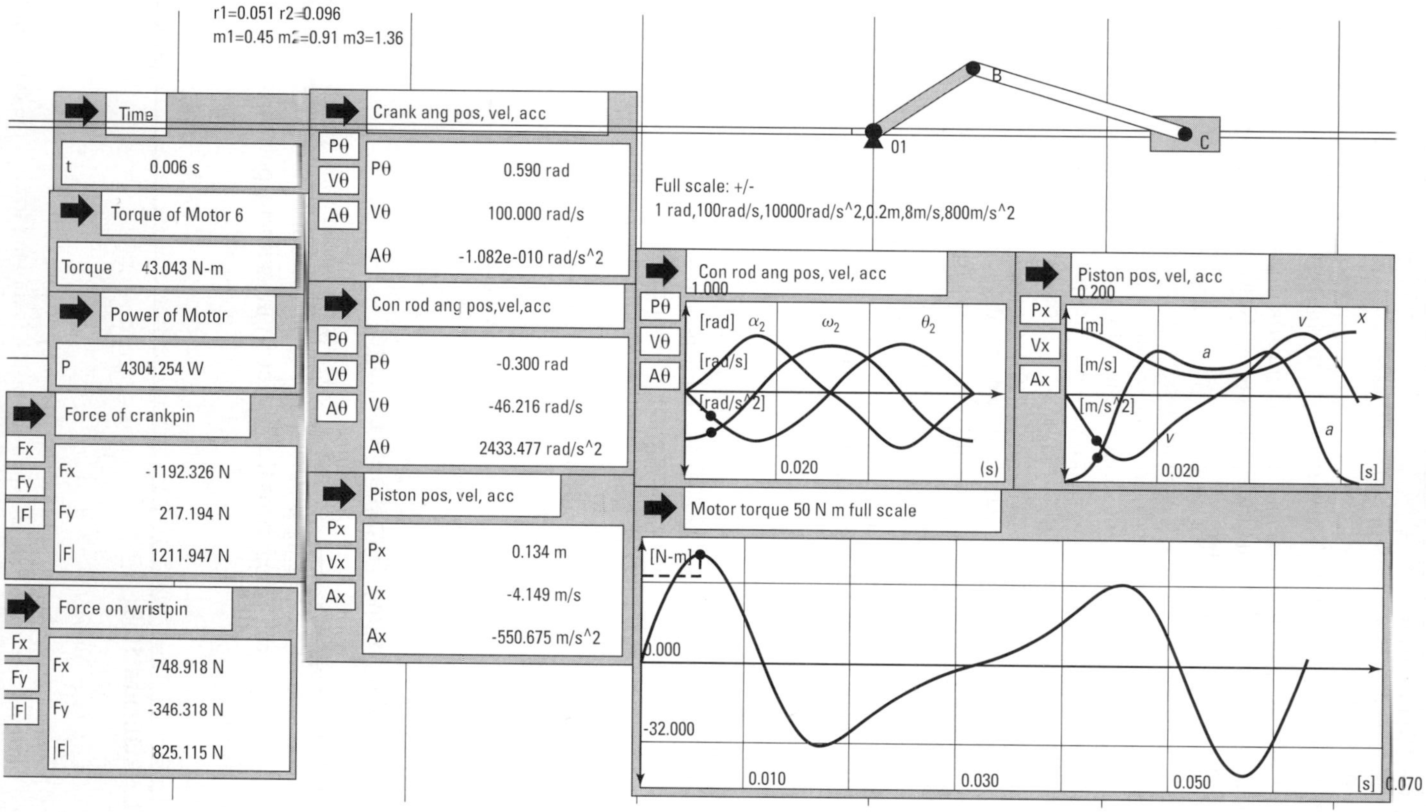

FIGURE 10.10 Using motion-simulation software for the dynamic force analysis of a slider-crank linkage.

SAMPLE PROBLEM 10.8

Checking a Simulation

Spot-check the results of Sample Problem 10.7.

Decisions. We can use known relationships from earlier chapters, particularly the values at limiting positions. We can also rerun the simulation, use Newton's laws, and begin a new simulation with a small connecting-rod mass.

Solution. The power transmitted by the motor equals the product of the torque and the angular velocity. The horizontal force on the crankpin (reverse effective force) equals the product of the piston mass and acceleration, with a change in sign. These calculations check exactly. For constant angular velocity of the crank, the crank position is proportional to time. We see that the curve of torque vs. time is antisymmetric about $\pi/100$ seconds, or crank position π radians. The net work over a full cycle is zero, because we considered only inertial effects. Checking the plots of piston position, velocity, and acceleration, we see that zero velocity corresponds to the extreme piston positions and that the greatest positive and negative velocities occur when the acceleration is zero.

A real machine could not go from zero to 100 rad/s crank speed instantaneously. Thus, accelerations at zero time are meaningless in this simulation. Let us go back to Chapter 4 to find the acceleration of the piston at top dead center. At the end of the first cycle, the acceleration should be

$$a_3 = -r_1\,\omega_1{}^2\,[1 + r_1/r_2] = -0.051 \cdot 100^2\,[1 + 0.051/0.096] = -780.9 \text{ m/s}^2,$$

where the subscripts 1, 2, and 3 refer to the crank, connecting rod, and piston, respectively. The closest point to a crank position of 2π radians essentially agrees with the calculated acceleration.

We could have checked the connecting-rod equilibrium with the use of d'Alembert's principle—that is, by including the reverse effective force and torque. Instead, the simulation is repeated with a connecting rod mass of 1 gram. The rod should then behave approximately as a two-force member. Forces on either end of the connecting rod (on the crankpin and wrist pin) are almost exactly equal and opposite. Finally, arctangent(F_y/F_x) checks the angular position of the connecting rod.

Fluid Forces and Dynamic Forces on a Linkage

Most real-world machines are intended to perform some task resulting in external forces or torques and requiring a net energy input. If a motor drives the crank of a slider-crank linkage, the integral of the power input with respect to time over a full cycle is positive.

SAMPLE PROBLEM 10.9

Analysis of a Linkage Subject to Fluid Forces

The piston of a pump is subject to fluid forces given by

$F_d = -60\, v_C^2$ when $v_C > 0$ (i.e, when the piston is moving toward the cylinder head) and
$F_d = 0$ otherwise,

where F_d = fluid force (N) resisting piston motion and
v_C = piston velocity (m/s).

The pump will have the form of an in-line slider-crank linkage with a 50-mm crank length and 90-mm connecting-rod length. The crank, connecting rod, and piston masses are 1.5, 1, and 2 kg, respectively. The mass moments of inertia are $3.156 \cdot 10^{-4}$ and $6.771 \cdot 10^{-4}$ kg $\cdot$ m^2 for the crank and connecting rod, respectively (about their centers of gravity). The crank angular velocity is a constant 100 rad/s. Plot the motor torque and linkage motion against time. Find the maximum motor power and the corresponding values of the torque, linkage motion, and forces.

Decisions. We will use Working Model™, inserting a damper that applies a force to the piston proportional to the square of its velocity.

Solution. The maximum motor power is 12.89 kW, occurring at 49.1 milliseconds, corresponding to crank position $\theta_1 = 4.91$ rad measured from the head-dead-center position. The actual input energy depends on the efficiency of the motor. For example, if we use an electric motor with 85% efficiency, the approximate electrical power load would peak at 12.89/0.85 = 15.2 kW.

Since the angular velocity is a constant 100 rad/s, we can multiply the time scale of the torque curve by 100 to get a curve of the crank torque vs. crank position. The area under that curve represents mechanical work in newton-meters = watt-seconds = joules. During the first half of the cycle, the torque is dependent on inertia of the parts only; the work input is followed by "coasting." During the second half of the cycle, the piston is opposed by fluid pressure, and there is net work input. The detailed solution is shown in Figure 10.11.

Compound Linkages

Motion simulation software can save considerable time when we are required to design and analyze multibar linkages. The time saved in analysis can be used for interpreting the results of the analysis and for redesigning the linkage on the basis of those results.

SAMPLE PROBLEM 10.10

Dynamic Analysis of a Multibar Linkage

Figure 10.12a shows a linkage made up of a four-bar linkage, a connecting rod, and a slider. At the initial position of the linkage, the crank angle = 0, and the coordinates of the points shown are given in mm as follows:

Motor 0_1: 0, 0	Fixed point 0_3: 140, 0
B: 52,0	D: 35.5, −81.6
C: 95.4, −90.6	Reentrant corner of coupler: 51.7, −71
Wrist-pin path: x, −50	E: 151.2, −50

The bodies have the following masses (kg) and mass moments of inertia (kg $\cdot$ mm^2), respectively:

Drive crank 0_1B: 0.08, 8.0	Coupler BDC: 0.20, 109.225
Driven crank 0_3C: 0.150, 60	Connecting rod DE: 0.18, 216.035
Piston E: 0.25, 30.9	

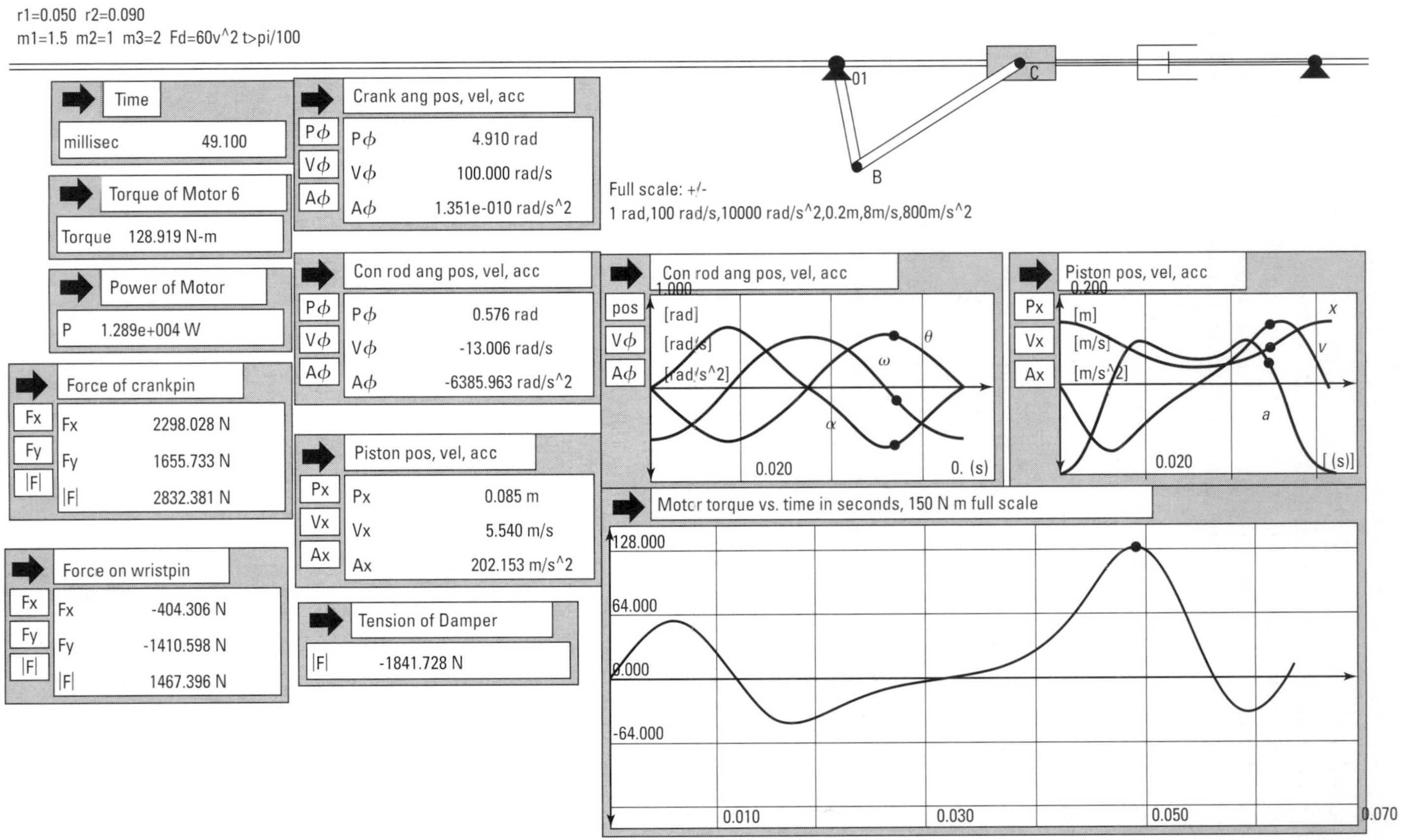

FIGURE 10.11 Fluid forces and dynamic forces on a linkage

a. Find the number of degrees of freedom of the linkage.

b. A motor drives the crank at a constant speed of 3000 rpm counterclockwise. Plot the piston position, velocity, and acceleration against time.

Determine the following for the instant that the magnitude of the piston acceleration is maximum: crankshaft position; piston position, velocity, and acceleration; and forces applied by the connecting rod to the wrist pin. Sketch the linkage for this instant.

c. Plot the horizontal, vertical, and resultant forces applied to the wrist pin by the connecting rod against time.

d. Is the force applied by the connecting rod to the wristpin in the x direction consistent with the piston acceleration?

Solution. **(a)** This is a planar linkage with six links, including the frame. There are six revolute joints and one sliding pair, for a total of seven pairs, each having one degree of freedom. The number of degrees of freedom for the linkage are given by the following equation (see Chapter 1):

$$\mathrm{DF}_{\text{planner}} = 3(n_L - n_J - 1) + \Sigma f_i = 3(6 - 7 - 1) + 7 = 1.$$

Thus, we can determine the linkage motion and inertial forces if the motion of one link is specified.

(b) Working Model™ was used to obtain the plots, sketch, and values shown in Figure 10.12b.

(c) See Part c of the figure.

(d) The force applied by the connecting rod to the wrist pin in the x direction is about -1687.75N. Applying d'Alembert's principle to horizontal forces on the piston yields

$$-1{,}687.75 - (0.25)(-6{,}751) = 0.$$

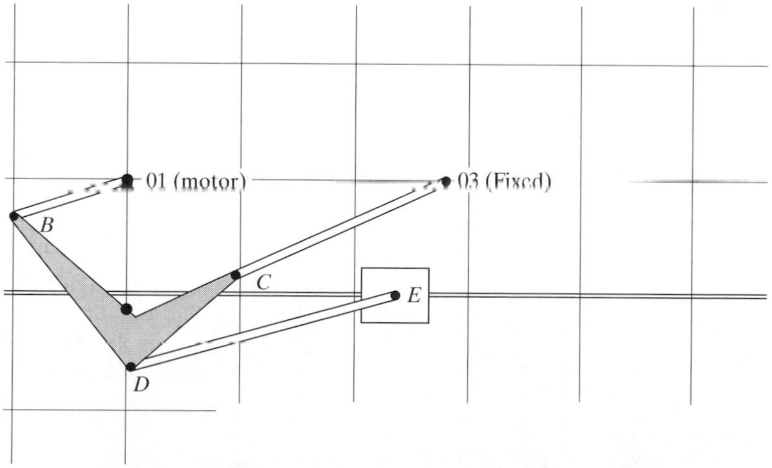

FIGURE 10.12 (a) Dynamic analysis of a compound linkage.

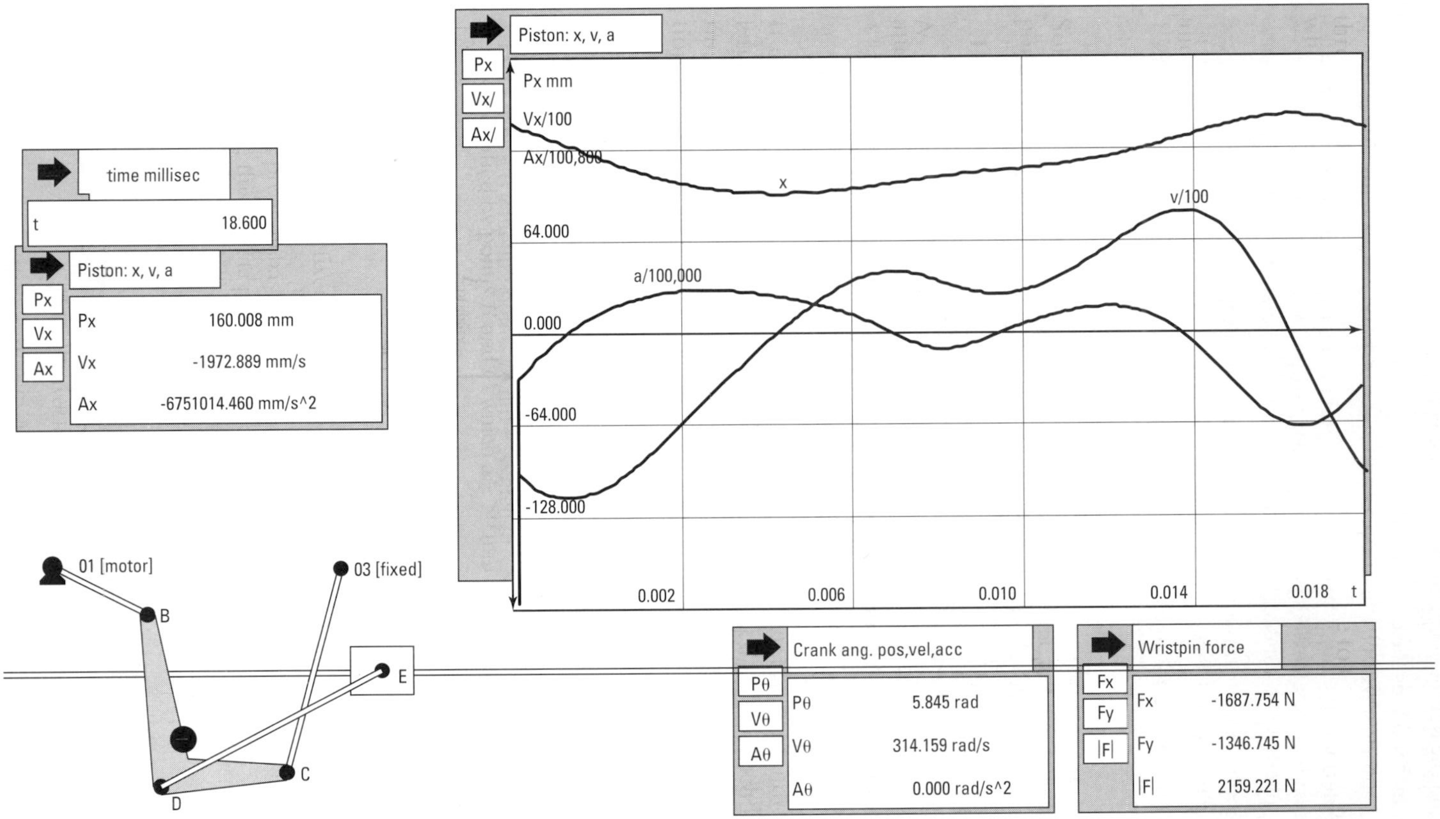

FIGURE 10.12 (b) Piston motion plotted against time. The linkage is sketched at the instant the magnitude of the piston acceleration is maximal; corresponding motion and force values are shown.

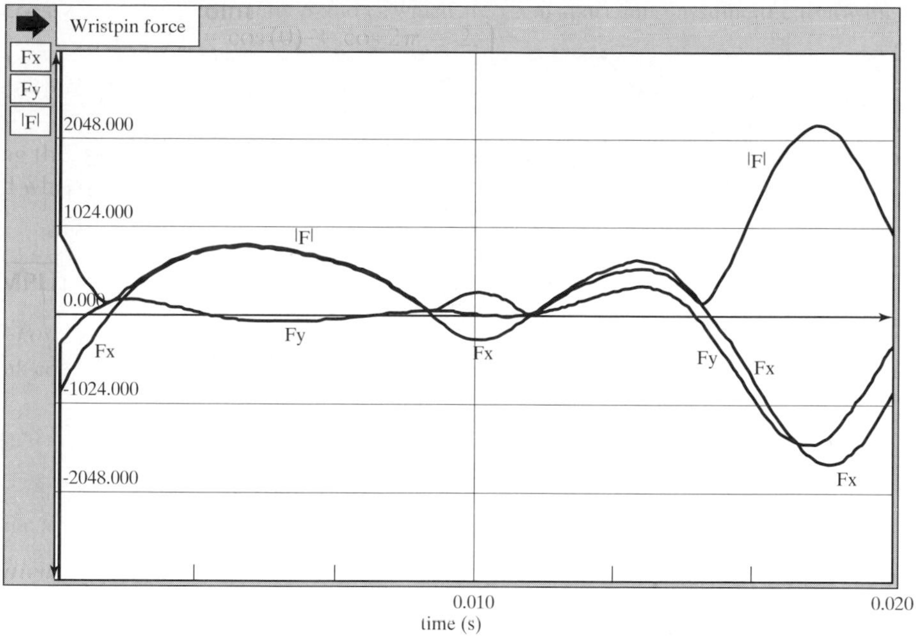

FIGURE 10.12 (c) The forces applied by the connectiong rod to the wrist pin, and the resultant force in the x and y directions are all plotted against time.

Equivalent Inertia

The slider-crank mechanism of Figure 10.9a is reproduced in Figure 10.13a. The crank and piston perform rotational motion and rectilinear motion, respectively, which are relatively simple motions to analyze. The connecting rod exhibits a more complex motion, except for the special points B (the crankpin) and C (the wrist pin). These points coincide with points on the crank and piston, and therefore, point B has a circular path and point C follows a straight-line path.

Because of the simplified kinematics for points B and C, we will examine the possibility of representing the mass distribution of the original connecting rod by an equivalent body consisting of two point masses m_B and m_C at points B and C, respectively, connected by a massless rigid rod. (See Figure 10.13.) The two members are equivalent for dynamic analysis purposes if the following three conditions are satisfied:

1. The centers of mass must be at the same location, point G_2.
2. The total masses must be equal.
3. The moments of inertia with respect to the center of mass must be equal.

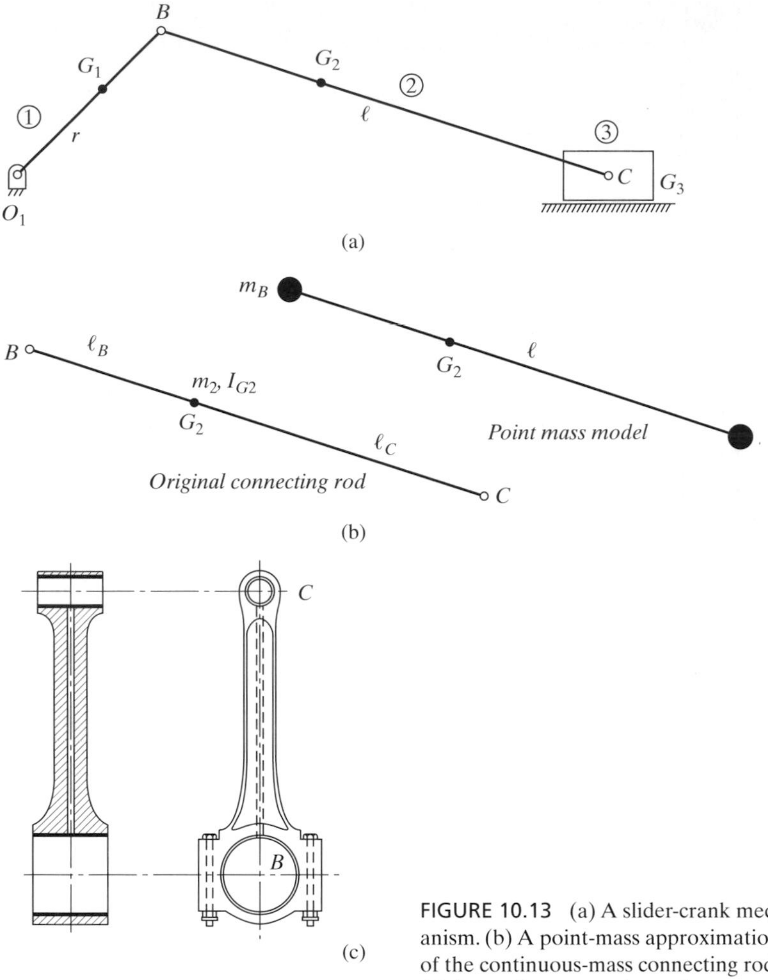

FIGURE 10.13 (a) A slider-crank mechanism. (b) A point-mass approximation of the continuous-mass connecting rod. (c) An engine connecting rod.

With reference to Figure 10.13b, these conditions can be expressed mathematically as

1. $m_B \ell_B = m_C \ell_C$,

2. $m_B + m_C = m_2$,

and

3. $m_B \ell_B^2 + m_C \ell_C^2 = I_{G2}$.

Since there are only two adjustable quantities (m_B and m_C) in these equations, not all three conditions will be satisfied in general. Therefore, the point-mass representation is an approximation of the original connecting rod. Using the first two equations to set values for masses m_B and m_C, we have

$$m_B = \left(\frac{\ell_C}{\ell_B + \ell_C} \right) m_2 = \left(\frac{\ell_C}{\ell} \right) m_2 \tag{10.9a}$$

and

$$m_C - \left(\frac{\ell_B}{\ell_B + \ell_C}\right) m_2 = \left(\frac{\ell_B}{\ell}\right) m_2. \tag{10.9b}$$

The accuracy of the approximation employing these point masses will depend on how closely the third condition is satisfied. There are many actual geometries for which such an approximation is quite satisfactory. For example, Figure 10.13c shows an engine connecting rod that can be modeled this way.

Approximate Dynamic Analysis Equations

In this section, we make use of the point-mass approximation of the previous section to derive equations for the dynamic analysis of the slider-crank mechanism. Figure 10.14a shows the mechanism and includes pertinent information. Note that the center of mass of the crank is located at fixed pivot O_1 for this analysis. As a result, the associated inertial force will be zero. The more general analysis of Eqs. (10.8a) through (10.8h) can be applied when the center of mass is located elsewhere.

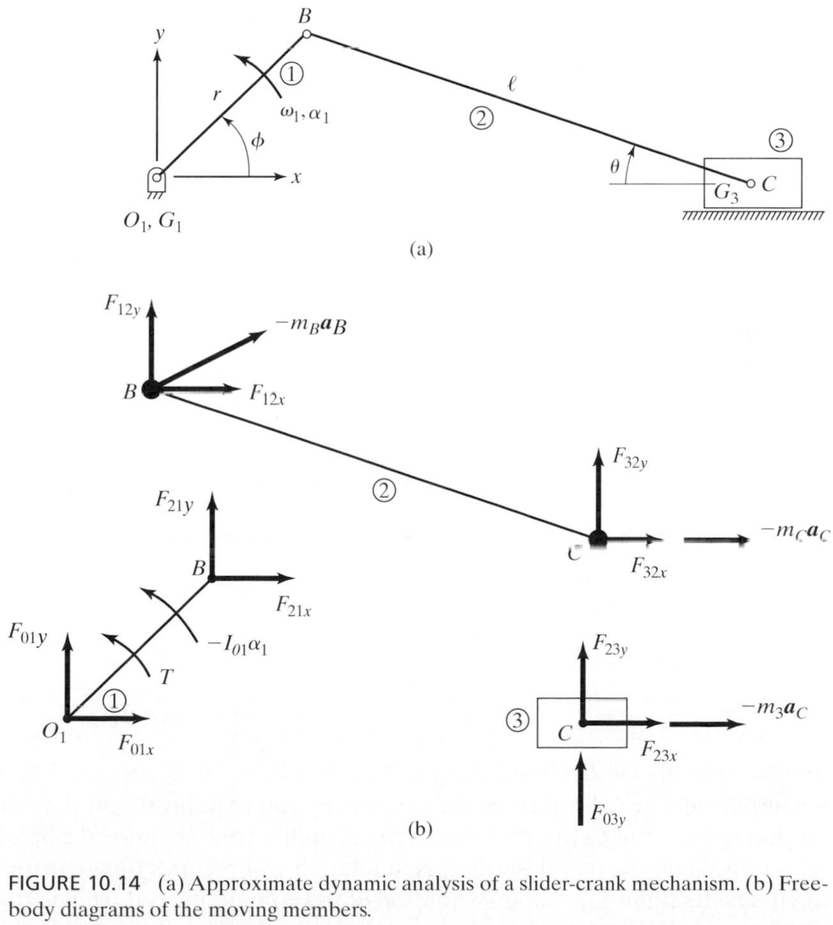

FIGURE 10.14 (a) Approximate dynamic analysis of a slider-crank mechanism. (b) Free-body diagrams of the moving members.

Figure 10.14b contains free-body diagrams of the three moving links, including inertial forces and torques. The piston is not acted upon by an inertial torque, because its angular acceleration is zero. The two inertial forces shown for the equivalent connecting rod represent the combined inertial force and inertial torque of the actual connecting rod. Bearing forces are shown in the figure in terms of x and y components. Expressing the inertial forces in component form, we have

$$-m_B\boldsymbol{a}_B = -m_B a_{Bx}\boldsymbol{i} - m_B a_{By}\boldsymbol{j}, \tag{10.10}$$

$$-m_C\boldsymbol{a}_C = -m_C a_{Cx}\boldsymbol{i}, \tag{10.11a}$$

and

$$-m_3\boldsymbol{a}_C = -m_3 a_{Cx}\boldsymbol{i}, \tag{10.11b}$$

where the accelerations $\boldsymbol{a}_B$ and $\boldsymbol{a}_C$ will, in general, be functions of the crank angular position, velocity, and acceleration and the linkage dimensions. Summing forces on the piston in the x and y directions yields

$$F_{23x} - m_3 a_{Cx} = 0$$
$$F_{03y} - F_{23y} = 0.$$

From the first of these equations,

$$F_{23x} = -F_{32x} = m_3 a_{Cx}. \tag{10.12}$$

Writing a moment equation for link 2 about point B, we obtain

$$(F_{32x} - m_C a_{Cx})\ell \sin\theta + F_{32y}\ell\cos\theta = 0,$$

from which it follows that

$$F_{32y} = -F_{23y} = F_{03y} = (m_C + m_3)\,a_{Cx}\tan\theta. \tag{10.13}$$

Summing forces on the connecting rod, we have

$$F_{12x} = -F_{21x} = F_{01x} = m_B a_{Bx} + (m_C + m_3)a_{Cx} \tag{10.14}$$

and

$$F_{12y} = -F_{21y} = F_{01y} = m_B a_{By} - (m_C + m_3)a_{Cx}\tan\theta \tag{10.15}$$

Finally, summing moments on the crank about pivot O_1 produces

$$T = I_{G1}\alpha_1 + F_{21x} r\sin\phi - F_{21y} r\cos\phi. \tag{10.16}$$

Notice that, in the determination of bearing forces, it is necessary to distinguish, for example, between the mass of the connecting rod at point C and the mass of the piston at that same point C. Similarly, mass m_B is on the connecting-rod side of crankpin B.

The analysis just described is similar to using Eqs. (10.8a) through (10.8h), but includes the point-mass approximation of the connecting rod. The next several sections will illustrate how the model, in conjunction with an approximation dealing with the kinematics of the mechanism, can be applied to a variety of operating conditions.

Dynamic-Force Analysis for an Assumed Motion of a Mechanism

Suppose we wish to determine the bearing forces and required crankshaft torque during a complete cycle of the mechanism for the case where the angular velocity ω_1 of the crank is assumed to be constant. We will substitute analytical acceleration expressions for this motion into the equations derived in the previous section. The result will be equations for the bearing forces and input torque as functions of the parameters (i.e., dimensions, masses, etc.) of the mechanism and the crank angle ϕ for constant crank speed ($\alpha_1 = 0$).

From Figure 10.14, the acceleration $\boldsymbol{a}_B$ is determined by differentiating the position vector

$$\boldsymbol{R}_B = r \cos \phi \boldsymbol{i} + r \sin \phi \boldsymbol{j}. \tag{10.17}$$

The velocity and acceleration so obtained, with ω_1 constant, are

$$\boldsymbol{v}_B = -r\omega_1 \sin \phi \boldsymbol{i} + r\omega_1 \cos \phi \boldsymbol{j} \tag{10.18}$$

and

$$\boldsymbol{a}_B = -r\omega_1^2 \cos \phi \boldsymbol{i} + r\omega_1^2 \sin \phi \boldsymbol{j}, \tag{10.19}$$

and, therefore, the associated inertial force is

$$-m_B \boldsymbol{a}_B = m_B r\omega_1^2(\cos \phi \boldsymbol{i} + \sin \phi \boldsymbol{j}). \tag{10.20}$$

This is a centrifugal force directed radially outward along the crank. Equations 10.14 through 10.16 indicate that torque T is independent of this inertial force.

From Figure 10.14a, the position vector for point C is

$$\boldsymbol{R}_C = (r \cos \phi + \ell \cos \theta)\boldsymbol{i}. \tag{10.21}$$

Angle θ can be expressed in terms of ϕ from the relationship

$$\ell \sin \theta = r \sin \phi, \tag{10.22}$$

which leads to

$$\cos \theta = \sqrt{1 - \sin^2 \theta} = \sqrt{1 - \left(\frac{r}{\ell} \sin \phi\right)^2}. \tag{10.23}$$

This expression can be substituted into the position vector and the result differentiated twice to obtain the exact acceleration. Alternatively, we introduce an approximate form that will reduce the mathematical complexity and that is quite satisfactory for crank-length-to-connecting-rod-length ratios commonly found in engines and compressors.

The following binomial series will be utilized:

$$\sqrt{1-s} = 1 - \frac{1}{2}s - \frac{1}{8}s^2 - \frac{1}{16}s^3 - \cdots. \tag{10.24}$$

This is an infinite series that holds for magnitudes of $s < 1$. Since r must be less than ℓ if the crank is to rotate completely, and since $\sin\phi$ cannot exceed unity, the series can be used in Eq. (10.23) with

$$s = \left(\frac{r}{\ell}\sin\phi\right)^2. \tag{10.25}$$

Furthermore, the square root may be approximated by a truncated partial sum of the infinite series. Using the first two terms of the series, we have

$$\cos\theta \cong 1 - \frac{1}{2}\left(\frac{r}{\ell}\right)^2 \sin^2\phi. \tag{10.26}$$

Because s is proportional to $(r/\ell)^2$, this approximation will be more accurate for smaller values of r/ℓ. For example, for $r/\ell \leq \frac{1}{4}$, the magnitude of the third term in Eq. (10.24) will be less than 0.0005, which is small enough to be neglected in most analyses.

Substituting the trigonometric identity

$$\sin^2\phi = \frac{1}{2} - \frac{1}{2}\cos 2\phi,$$

we find that Eq. (10.26) becomes

$$\cos\theta \cong 1 - \frac{1}{4}\left(\frac{r}{\ell}\right)^2 + \frac{1}{4}\left(\frac{r}{\ell}\right)^2 \cos 2\phi, \tag{10.27}$$

and it follows that

$$\boldsymbol{R}_C = \left(\ell - \frac{r^2}{4\ell} + r\cos\phi + \frac{r^2}{4\ell}\cos 2\phi\right)\boldsymbol{i}. \tag{10.28}$$

Differentiating twice with ω_1 held constant, we have

$$\boldsymbol{v}_C = \left(-r\omega_1\sin\phi - \frac{\omega_1 r^2}{2\ell}\sin 2\phi\right)\boldsymbol{i} \tag{10.29}$$

and

$$\boldsymbol{a}_C = \left(-r\omega_1^2\cos\phi - \frac{r^2\omega_1^2}{\ell}\cos 2\phi\right)\boldsymbol{i}. \tag{10.30}$$

The corresponding inertial forces are

$$-m_C\,\boldsymbol{a}_C = m_C\,r\omega_1^2\left(\cos\phi + \frac{r}{\ell}\cos 2\phi\right)\boldsymbol{i} \tag{10.31}$$

and

$$-m_3\,\boldsymbol{a}_C = m_3\,r\omega_1^2\left(\cos\phi + \frac{r}{\ell}\cos 2\phi\right)\boldsymbol{i}. \tag{10.32}$$

Substituting the expressions for the acceleration into the previous force analysis equations, Eqs. (10.12) through (10.16), leads to the following relationships for the bearing forces and crankshaft torque:

$$F_{32x} = m_3\,r\omega_1^2\left(\cos\phi + \frac{r}{\ell}\cos 2\phi\right), \tag{10.33}$$

$$F_{32y} = F_{03y} = \frac{-(m_C - m_3)r^2\omega_1^2\sin\phi\,[\cos\phi + (r/\ell)\cos 2\phi]}{\sqrt{\ell^2 - r^2\sin^2\phi}}, \tag{10.34}$$

$$F_{12x} = F_{01x} = -m_B\,r\omega_1^2\cos\phi - (m_C - m_3)\,r\omega_1^2\left(\cos\phi + \frac{r}{\ell}\cos 2\phi\right), \tag{10.35}$$

$$F_{12y} = F_{01y} = -m_B\,r\omega_1^2\sin\phi + \frac{(m_C + m_3)\,r^2\omega_1^2\sin\phi\,[\cos\phi + (r/\ell)\cos 2\phi]}{\sqrt{\ell^2 - r^2\sin^2\phi}}, \tag{10.36}$$

and

$$T = (m_C - m_3)r^2\omega_1^2\sin\phi\left(\cos\phi + \frac{r}{\ell}\cos 2\phi\right)\left(1 + \frac{r\cos\phi}{\sqrt{\ell^2 - r^2\sin^2\phi}}\right). \tag{10.37}$$

These are explicit relationships among forces, motion, and design parameters having to do with the mechanism that are useful to the design engineer who must properly size a mechanism for a given set of performance requirements.

In the absence of friction, the superposition principle can be applied to determine the combined effects of cylinder pressure forces and inertia of the mechanism. The force analysis for an external piston force has been presented in Chapter 9.

SAMPLE PROBLEM 10.11

Dynamic Analysis of a Slider-Crank Mechanism under Assumed Input Motion

An in-line slider-crank mechanism has a crank length $r = 0.04$ m, a connecting-rod length $\ell = 0.16$ m, and a reciprocating mass $(m_c + m_3) = 0.6$ kg. Determine the variation in torque required over a complete cycle in order to maintain a constant crank speed of 1000 rev/min.

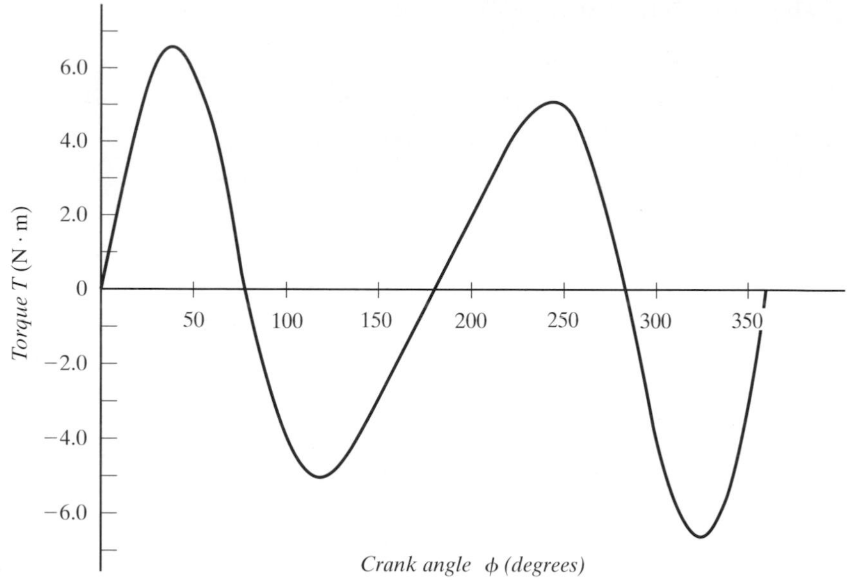

FIGURE 10.15 Input torque versus crank angle for the slider-crank mechanism of Sample Problem 10.11. The input crank is assumed to have a constant velocity of 1,000 rev/min.

Solution. The angular velocity ω_1 is

$$\omega_1 = \frac{1000 \text{ rev/min}}{60 \text{ s/min}} \cdot 2\pi \text{ rad/rev} = 104.72 \text{ rad/s}.$$

Substituting this value and the given data into Eq. (10.37), we find that the crank torque, as a function of crank angle ϕ, is

$$T = 10.53 \sin\phi\,(\cos\phi + 0.25\cos 2\phi)\left[1 + \frac{0.25\cos\phi}{\sqrt{1 - (0.25\sin\phi)^2}}\right] N \cdot m.$$

The torque is plotted as a function of ϕ in Figure 10.15.

Dynamic-Motion Analysis for an Assumed Input Torque

In this section, the dynamic analysis equations derived previously will be employed in dynamic-motion analysis, sometimes referred to as time-response analysis. Here, the motion of the mechanism is not assumed and is treated as an unknown. Instead, an assumption will be made about the input torque T representative of the behavior of a typical driving device. Once again, although a totally general analysis can be undertaken, the approximations that were described and that are typically valid, will be imposed to simplify the mathematical development somewhat.

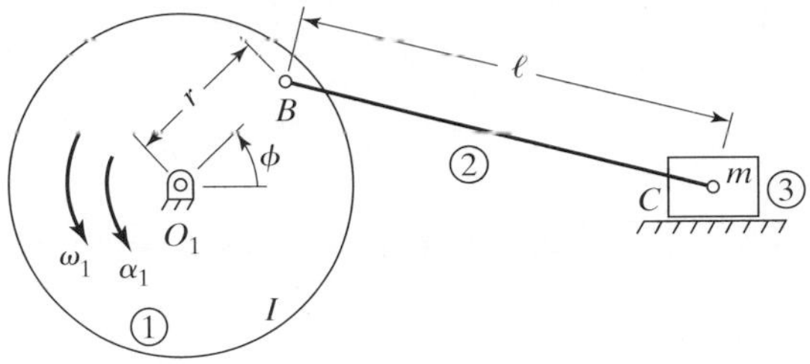

FIGURE 10.16 A slider-crank mechanism for which a dynamic-motion analysis is to be performed.

Consider the slider-crank mechanism shown in Figure 10.16. The inertial effects of the rotating part of the connecting-rod mass (mass m_B) can be nullified by counterweighting the crank; assuming that that has been done in this case, we set $m_B = 0$. The net mass moment of inertia of the crank will then be referred to as I. Further, let $m = m_C + m_3$. The mechanism is to be driven by an electric motor, governed by the following torque–speed equation, which is a function of the motor characteristics:

$$T = A - B\omega_1. \tag{10.38}$$

Here, A and B are motor constants. The relationship embodied in Eq. (10.38) which is typical of some DC motors, indicates that the input torque varies inversely with the crank speed; that is, the torque increases as ω_1 decreases and decreases as ω_1 increases. We wish to determine the motion of the mechanism when it is driven by such a motor.

Substituting Eqs. (10.14), (10.15), and (10.38) into Eq. (10.16), with $m_3 = 0$, we obtain

$$A - B\omega_1 = I\alpha_1 - mra_{Cx}\sin\phi - mra_{Cx}\cos\phi\tan\theta.$$

Rearranging terms yields

$$I\alpha_1 + B\omega_1 - mra_{Cx}(\sin\phi + \cos\phi\tan\theta) - A = 0. \tag{10.39}$$

Acceleration a_{Cx} is obtained by differentiating Eq. (10.29), where now $\alpha_1 \neq 0$:

$$\begin{aligned} a_{cx} &= -r\alpha_1\sin\phi - r\omega_1^2\cos\phi - \left(\frac{\alpha_1 r^2}{2\ell}\right)\sin 2\phi - \left(\frac{\omega_1^2 r^2}{\ell}\right)\cos 2\phi \\ &= -r\alpha_1\left[\sin\phi + \left(\frac{r}{2\ell}\right)\sin 2\phi\right] - r\omega_1^2\left[\cos\phi + \left(\frac{r}{\ell}\right)\cos 2\phi\right]. \end{aligned} \tag{10.40}$$

Substituting into Eq. (10.39) yields

$$K_1\frac{d^2\phi}{dt^2} + B\frac{d\phi}{dt} + K_2\left(\frac{d\phi}{dt}\right)^2 - A = 0, \tag{10.41}$$

where

$$\frac{d\phi}{dt} = \omega_1 \quad \frac{d^2\phi}{dt^2} = \alpha_1,$$

$$K_1 = I + mr^2\left[\sin\phi + \left(\frac{r}{2\ell}\right)\sin 2\phi\right](\sin\phi + \cos\phi\tan\theta),$$

$$K_2 = mr^2\left[\cos\phi + \left(\frac{r}{\ell}\right)\cos 2\phi\right](\sin\phi + \cos\phi\tan\theta),$$

and

$$\tan\theta = \frac{r\sin\phi}{\sqrt{\ell^2 - r^2\sin^2\phi}}.$$

Equation (10.41) is a nonlinear differential equation relating crank angle ϕ and time t. It can be solved by numerical methods.

Once the motion $\phi(t)$ and its derivatives $\omega_1(t)$ and $\alpha_1(t)$ have been calculated, the bearing forces can be determined from Eqs. (10.12) through (10.15). Although quantitative results are difficult to obtain, some qualitative information follows from inspection of Eq. (10.41). For example, solving for the second derivative of ϕ, we have

$$\frac{d^2\phi}{dt^2} = \frac{A - B\,(d\phi/dt) - K_2\,(d\phi/dt)^2}{K_1}. \tag{10.42}$$

As the moment of inertia, I, increases, denominator K_1 will grow large, and the magnitude of the acceleration will decrease. Therefore, for some suitably large value of I (such as that produced by a flywheel), α_1 will become negligibly small, and the mechanism can be assumed to be operating at constant angular speed ω_1. On the other hand, the inertia of the piston is represented by the second term in the equation for K_1, which varies between limiting values of zero and mr^2. Therefore, for a small moment of inertia, I, the acceleration will also vary and will have large values at certain crank positions, leading to a considerable fluctuation in the speed of the mechanism.

SAMPLE PROBLEM 10.12

Dynamic Analysis of a Slider-Crank Mechanism under an Assumed Input Torque

An in-line slider-crank mechanism has a crank length $r = 0.04$ m, a connecting-rod length $\ell = 0.16$ m, a reciprocating mass $m = 0.6$ kg, and a crank moment of inertia $I = 0.001\ \text{kg}\cdot\text{m}^2$. The mechanism is driven by a DC motor having the constants $A = 105\ N\cdot m$ and $B = 1.0\ \text{N}\cdot m\cdot s$. Determine how the crank velocity varies over a complete steady-state cycle of the mechanism.

Solution. Notice that this is the same mechanism as was studied in Sample Problem 10.11; however, a more realistic input actuator is now considered. Figure 10.17 shows a plot of the crank angular velocity versus angle ϕ for a complete 360° rotation of the crank. The response was obtained by numerically solving Eq. (10.41), starting from initial conditions of rest; that is, $\phi = d\phi/dt = 0$ when time $t = 0$. The data plotted in the figure are for the eighth cycle of rotation, by which time steady-state motion had been established.

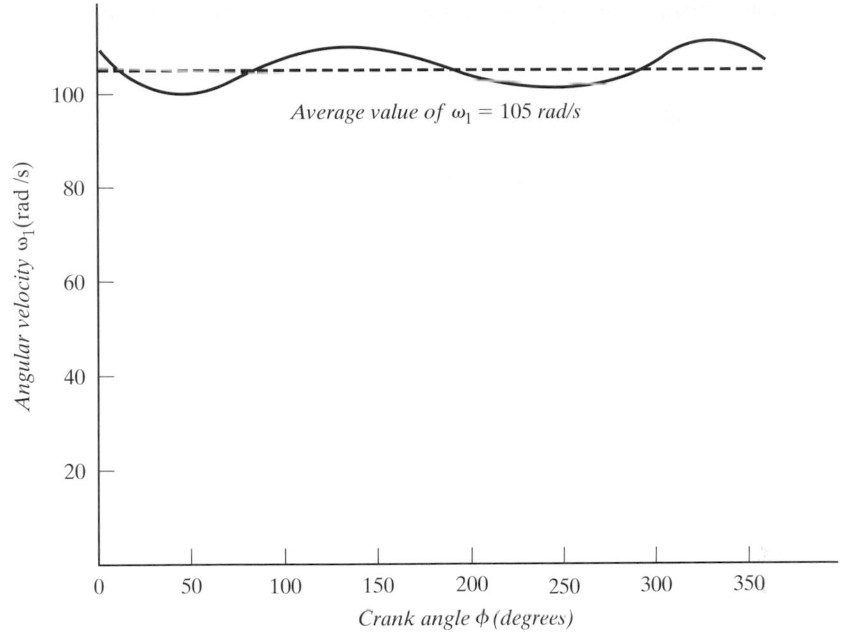

FIGURE 10.17 Crank angular velocity versus crank angle for the slider-crank mechanism of Sample Problem 10.12. The average velocity is 105 rad/s (approximately 1000 rev/min).

Linkage Motion in Response to a Specified Torque: Applying Motion Simulation Software to Solve for Motion and Forces in a Linkage Subject to Fluid Forces and Inertial Effects

The preceding analysis illustrates the difficulty in determining the motion of a linkage in response to a specified input torque. Motion simulation software can be a great help in a problem like this. A motor that applies a constant input torque would be unrealistic, because it would add energy to the system during each cycle, and the speed of the system could increase indefinitely. Motor characteristics can be modeled by an equation or by combining a torque motor with a torsional damper. The net motor power (W) is the product of the net torque($N \cdot m$) and the crank angular velocity (rad/s).

It is so easy to introduce errors into problems of this type. Measure all relevant variables. Check initial conditions; compare plots and other output data. Are the results consistent? Do they make sense? The output data may help you to redesign the linkage. For example, if the crankshaft speed varies over too wide a range, you might try using a higher crank mass moment of inertia.

SAMPLE PROBLEM 10.13

Applying Motion Simulation Software to Solve for Motion and Forces in a Linkage Subject to Fluid Forces and Inertial Effects

A motor driving a piston pump has torque characteristics approximated by

$$T = 60 - 0.005\,\omega_1^2,$$

where T = net torque (N · m) and
ω_1 = crank angular velocity.
The piston is subject to fluid forces given by
$F_d = -55v_C^2$ when $v_C > 0$ (i.e., when the piston is moving toward the cylinder head) and
$F_d = 0$ otherwise.
F_d is the fluid force (N) and
v_C is the velocity (m/s) of the piston.

The pump will have the form of an in-line slider-crank linkage with a 50-mm crank length and a 110-mm connecting-rod length. The crank, connecting-rod, and piston masses are 2, 1.25, and 1.36 kg, respectively. Mass moments of inertia are 0.005 and 0.001 kg · m^2 for the crank and connecting rod, respectively (about their centers of gravity). Plot the motor torque, motor power, and motion of the linkage against time. Find the maximum motor power and the corresponding values of torque, linkage motion, and forces.

Decisions. We will use Working Model™ to simulate the linkage, adding a torque motor and velocity-squared torsional damper. Fluid force will be simulated with a damper that opposes fluid motion when the piston velocity is positive (away from the crankshaft).

Solution. Figure 10.18 shows the linkage; plots of the crank angular position, velocity, and acceleration; plots of the piston position, velocity, and acceleration; and plots of the net motor torque and power, all against time. The time span is a bit more than two crankshaft revolutions. Values of the foregoing quantities and of the crankpin and wrist-pin forces are shown at the instant the power is approximately a maximum. The maximum power driving the linkage is about 2.53 kW.

Note that the crank starts out at zero angular velocity; the linkage is "getting up to speed" during the first revolution of the crankshaft, and the first crankshaft revolution takes longer than subsequent revolutions. After the first crankshaft revolution, we expect the plots to have a repetitive pattern. The intermittent piston load and piston and connecting-rod inertial forces cause a wide fluctuation in the crankshaft speed. We cannot eliminate this fluctuation, but increasing the crank mass moment of inertia can reduce it.

10.6 BALANCING OF MACHINERY

Up to this point, little mention has been made of the dynamic forces that are transmitted to the frame of a machine. And yet, some of the more serious problems encountered in high-speed machinery are the direct result of these forces. As can be seen from the analyses presented previously, the forces exerted on the frame by moving machine members will, in general, vary with time and will therefore impart vibratory motion to the frame. This vibration, together with the accompanying noise, can produce human discomfort, alter the desired machine performance, and adversely affect the structural integrity of the machine's foundation. Furthermore, these effects are intensified both by increased operating speeds, which lead to greater inertial forces, and by conditions of resonance (vibration at the natural frequency).

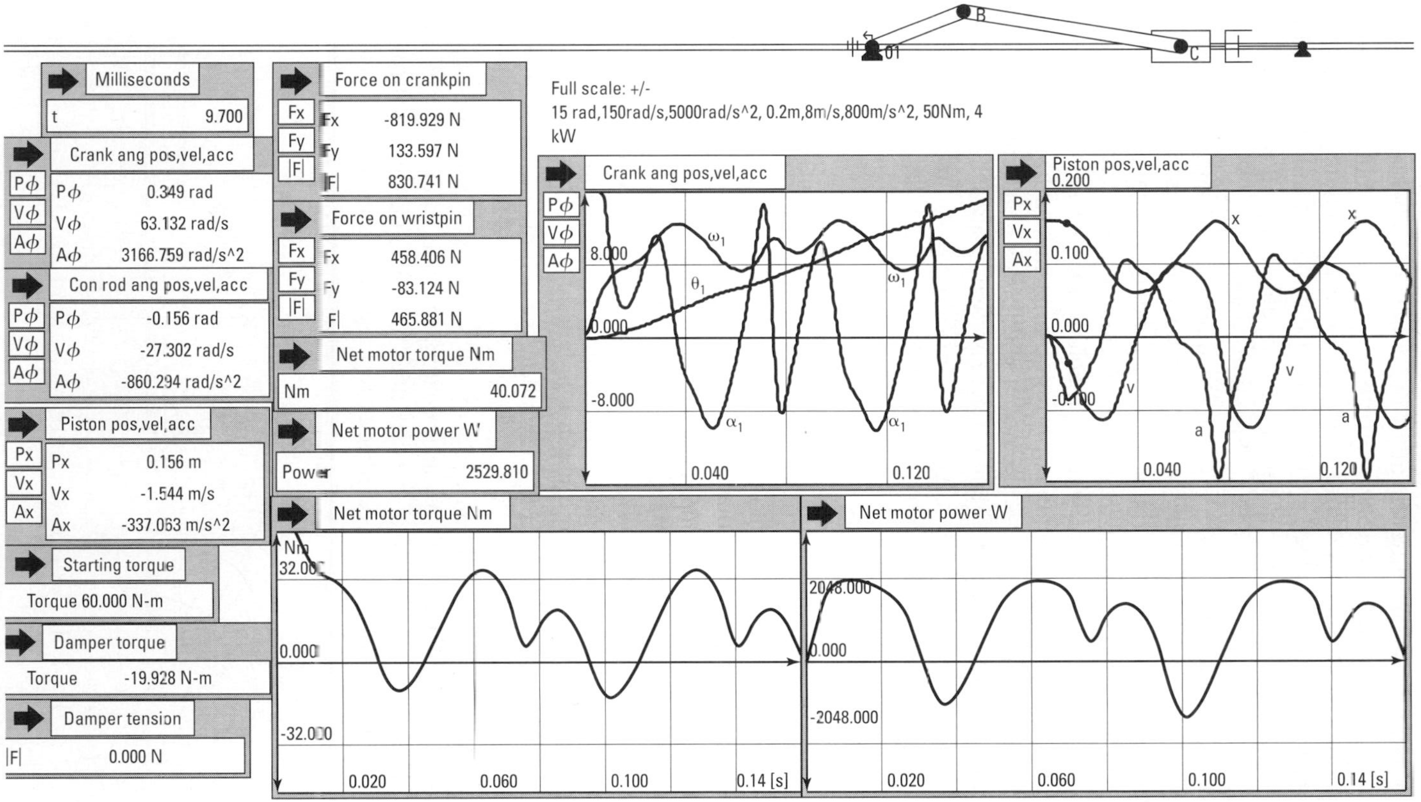

FIGURE 10.18 Applying motion simulation software to solve for motion and forces in a linkage subject to fluid forces and inertial effects.

A net unbalanced force acting on the frame of a machine or mechanism (i.e., the resultant of the forces transmitted at all of the connections between the machine and frame) is referred to as a *shaking force*. Likewise, a resultant unbalanced moment acting on the frame is called a *shaking moment*. Since the shaking force and shaking moment are unbalanced effects, they will cause the frame to vibrate, with the magnitude of the vibration dependent on the amount of unbalance. Clearly, then, an important design objective is to minimize machine unbalance.

The process of designing or modifying machinery in order to reduce unbalance to an acceptable level is called *balancing*. Since unbalance is caused in the first place by the inertial forces associated with the mass of a moving machine, the most common approach to balancing is to redistribute the machine's mass, accomplished by adding mass to, or removing mass from, various machine members. However, other techniques, involving springs and dampers or balancing mechanisms, are also used.

The method of balancing that is employed depends to a considerable extent on the type of unbalance present in the machine. The two basic types are rotating unbalance and reciprocating unbalance, which may occur separately or in combination. These two types of unbalance are considered in detail in the next few sections.

10.7 BALANCING OF RIGID ROTORS

In the introduction to this chapter, a dynamic force was presented that was a centrifugal force associated with a mass attached to a rotating rod. (See Figure 10.1a.) The same kind of force occurs in an eccentric rotor. (See Figure 10.1b.) In both cases, the bearing mounts on the machine frame were acted upon by a net unbalanced shaking force. This type of unbalance due to a rotating mass is referred to as *rotating unbalance*, and since virtually all machines contain rotating parts, that form of unbalance is very common. It occurs, for example, in such diverse applications as turbine rotors, engine crankshafts, washing machine drums, and window fans. Fortunately, rotating unbalance is relatively easy to deal with.

Figure 10.19 shows a rotor consisting of a disk of mass m attached to a rigid shaft with an assumed negligible mass. We also assume that the rotor has constant angular velocity ω. The center of mass, G, of the disk does not coincide with the bearing centerline AB, with the amount of this eccentricity represented by e. The rotor is acted upon

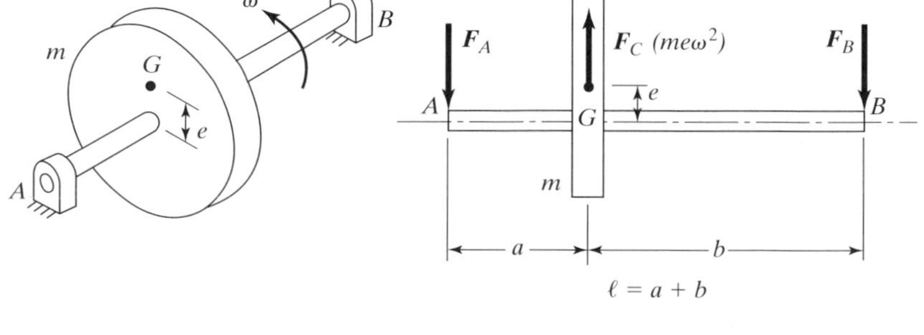

FIGURE 10.19 Static unbalance caused by an eccentric mass on a rotating shaft.

by the centrifugal force F_C, which has a magnitude equal to $me\omega^2$. Summing forces and moments on the rotor, we find that the bearing reactions on the shaft are

$$F_A = \left(\frac{b}{\ell}\right)me\omega^2$$

and

$$F_B = \left(\frac{a}{\ell}\right)me\omega^2,$$

which act in the directions shown in the figure. The net force on the frame will therefore be a force of magnitude

$$F_s = \left(\frac{a}{\ell}\right)me\omega^2 + \left(\frac{b}{\ell}\right)me\,\omega^2 = \left(\frac{a+b}{\ell}\right)me\omega^2 = me\omega^2$$

and with a direction that rotates with speed ω. Thus, the centrifugal force is transmitted directly to the frame. Notice that if the rotational speed is doubled, the shaking force is quadrupled.

The unbalance just described is fairly easy to detect. For example, if the shaft were mounted horizontally on knife-edge bearings, then, due to gravity, the rotor would always seek the static position with point G below the bearing centerline. Any rotating unbalance that can be detected in a static test is referred to as a *static unbalance*, which actually is somewhat of a misnomer in that we are concerned primarily with the dynamic effects caused by such an unbalance. Not only can a static unbalance be detected through a static test, but it can also be corrected through a static procedure. The bubble balance sometimes used for automobile wheels is an example; the device statically determines the location and amount of unbalance so that corrective counterweights can be attached in such a way that the combined center of mass is coincident with the bearing centerline (i.e., $e = 0$) and the resulting shaking force is zero.

Contrasted with the case of Figure 10.19 is the rotor of Figure 10.20, with two disks having masses m_1 and m_2 and eccentricities e_1 and e_2. Suppose that the centers of mass, G_1 and G_2, of the disks are 180° apart, as shown in the figure. Then, again assuming

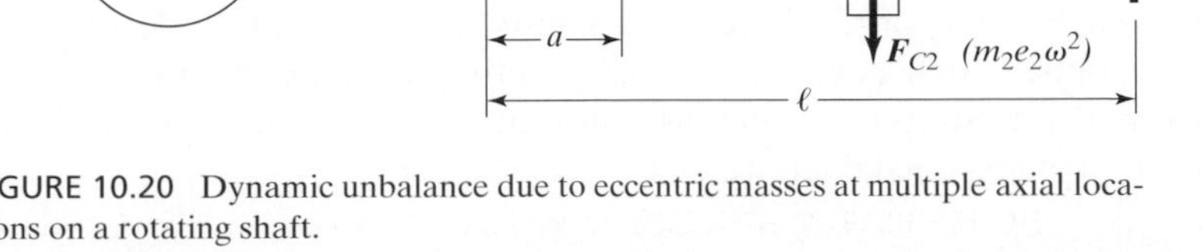

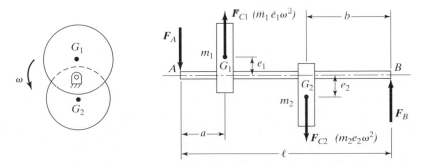

FIGURE 10.20 Dynamic unbalance due to eccentric masses at multiple axial locations on a rotating shaft.

constant angular velocity ω and a negligible mass for the shaft, we can determine the bearing reactions to be

$$F_A = \left(\frac{\omega^2}{\ell}\right)[m_1 e_1(\ell - a) - m_2 e_2 b]$$

and

$$F_B = \left(\frac{\omega^2}{\ell}\right)[m_2 e_2(\ell - b) - m_1 e_1 a].$$

The directions of F_A and F_B are based on the signs of these equations; if the sign is positive, then the force is oriented as shown in Figure 10.20; if the sign is negative, then the direction of the force is reversed. The forces rotate about the bearing centerline with angular speed ω. The magnitude of the shaking force is

$$F_S = |F_A - F_B| = \omega^2 |\, m_1 e_1 - m_2 e_2|,$$

and if $m_1 e_1 = m_2 e_2$, then the shaking force is zero. However, in this situation, even though the resultant force on the frame is zero, the individual bearing forces are nonzero, now having equal magnitudes given by

$$F_A = F_B = \left(\frac{m_1 e_1 \omega^2}{\ell}\right)[\ell - (a + b)],$$

but opposite directions, as shown in Figure 10.20. Thus, a resultant, or shaking, couple is still acting on the frame. In general, shaking couples occur when unbalanced masses are located at multiple axial positions on a rotor.

The unbalance shown in Figure 10.20 with $m_1 e_1 = m_2 e_2$ could not be detected in a static test; for example, the rotor would take random orientations in the gravity test described earlier. Such an unbalance can be detected only by means of a dynamic test in which the rotor is spinning. Any unbalance that can be detected through such a test is referred to as a *dynamic unbalance*. (By this definition, the static unbalance defined previously is also one form, but not the only form, of dynamic unbalance.) A static balancing procedure will not completely correct a dynamic unbalance, and a statically balanced rotor may perform very poorly under actual operating conditions. In such a situation, dynamic balancing procedures are required, wherein the rotor is driven at an arbitrary speed and bearing forces are measured. From this information, magnitudes and locations of corrective counterweights are determined.

In general, static unbalance is characterized by a net shaking force, and dynamic unbalance is characterized by a combination of a net shaking force and a net shaking couple. Dynamic unbalance is more apt to be significant in cases of rotors having their mass distributed over relatively large axial distances. For example, static balancing may be satisfactory for machine components such as automobile wheels or household window fans, which have short axial lengths, whereas dynamic balancing must be performed on equipment such as automotive crankshafts and multistage turbine rotors that have large axial lengths.

From the preceding discussion, general balancing procedures for sizing and positioning corrective masses on rotors are based on the following criteria:

1. For static balance, the shaking force must be zero.
2. For dynamic balance, the shaking force and the shaking moment must both be zero.

These procedures are illustrated in the next few sections.

Static Balancing

Consider the rigid rotor shown in Figure 10.21a. The rotor is assumed to be rotating with constant angular velocity ω. Unbalanced masses are depicted as point masses m at radial distances r and may represent a variety of actual rotating masses, including turbine or propeller blades, eccentric disks, crank throws, and so on. In this case, there are three masses, but there could be any number. It is assumed here that all of the masses lie in a single transverse plane at the same axial location along the shaft or close to the same plane. It will be shown that this arrangement can be balanced by a single counterbalance lying in that plane and represented by dashed lines in Figure 10.21b.

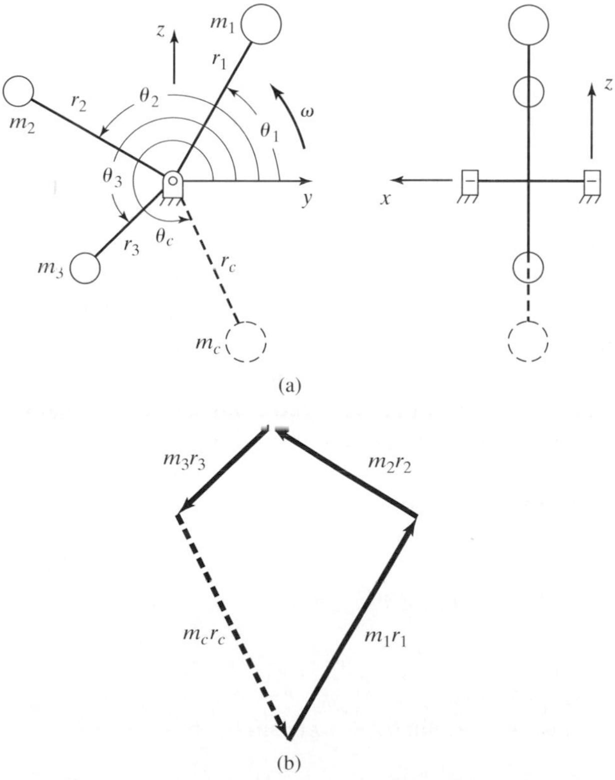

FIGURE 10.21 (a) A static unbalance can be eliminated by the addition of a single counterweight m_c at the proper radial distance r_c and angle θ_c. (b) Graphical determination of the counterweight size and location.

Each of the original masses produces a centrifugal force acting radially outward from the axis of rotation with a magnitude equal to $m_n r_n \omega^2$, where $n = 1, 2, 3$. The vector sum of these forces will be transmitted through the support bearings to the frame, resulting in a shaking force

$$\boldsymbol{F}_S = m_1\omega^2\boldsymbol{r}_1 + m_2\omega^2\boldsymbol{r}_2 + m_3\omega^2\boldsymbol{r}_3.$$

If this vector sum is zero, then the rotor is balanced. In general, though, that will not be the case, and we therefore introduce the counterweight of mass m_c shown in Figure 10.21a at radial distance $\boldsymbol{r}_c$. The magnitude and location of this counterweight are determined from the condition that the resultant inertial force must now be zero; that is,

$$m_1\omega^2\boldsymbol{r}_1 + m_2\omega^2\boldsymbol{r}_2 + m_3\omega^2\boldsymbol{r}_3 + m_c\omega^2\boldsymbol{r}_c = \boldsymbol{0}. \tag{10.43}$$

The quantity ω^2 can be factored out of Eq. (10.43), whereupon dividing by ω yields the following relationship for static balance:

$$m_1\boldsymbol{r}_1 + m_2\boldsymbol{r}_2 + m_3\boldsymbol{r}_3 + m_c\boldsymbol{r}_c = \boldsymbol{0}. \tag{10.44a}$$

This equation indicates that the combined center of mass must lie on the axis of rotation. In general, for N initial masses, the balancing condition is

$$\sum_{n=1}^{N} m_n\boldsymbol{r}_n + m_c\boldsymbol{r}_c = \boldsymbol{0}. \tag{10.44b}$$

Since all the vectors in Eq. (10.44a) lie in a plane that is parallel to the yz-plane in Figure 10.21a, that equation is a two-dimensional vector equation, and it therefore can be satisfied by the two parameters—that is, magnitude and direction—associated with the vector for the single counterweight m_c. The equation can be solved either graphically or mathematically. Figure 10.21b shows a graphical solution wherein vectors $m_1\boldsymbol{r}_1$, $m_2\boldsymbol{r}_2$, and $m_3\boldsymbol{r}_3$ are drawn in sequence to a suitable scale. The vector that closes this loop and that therefore satisfies Eq. (10.44a) is $m_c\boldsymbol{r}_c$. The direction of this vector identifies the angular orientation of the counterweight relative to that of the other masses, and the magnitude of the vector is the required amount of correction, $m_c\boldsymbol{r}_c$. Note that, because only the proper value of the product is required, either m_c or $\boldsymbol{r}_c$ can be selected arbitrarily. For example, if mass m_c is chosen, then the preceding solution determines vector location $\boldsymbol{r}_c$ of this counterweight.

Equation (10.44a) can also be solved mathematically by dividing it into y and z components. We then obtain

$$m_1r_1\cos\theta_1 + m_2r_2\cos\theta_2 + m_3r_3\cos\theta_3 + m_cr_c\cos\theta_c = 0$$

and

$$m_1r_1\sin\theta_1 + m_2r_2\sin\theta_2 + m_3r_3\sin\theta_3 + m_cr_c\sin\theta_c = 0,$$

where θ represents instantaneous angular orientation with respect to the y-axis. (See Figure 10.21a.) Solving for $m_c r_c$ and θ_c, we have

$$m_c r_c = [(m_1 r_1 \cos\theta_1 + m_2 r_2 \cos\theta_2 + m_3 r_3 \cos\theta_3)^2 + (m_1 r_1 \sin\theta_1 + m_2 r_2 \sin\theta_2 + m_3 r_3 \sin\theta_3)^2]^{1/2} \quad (10.45)$$

and

$$\theta_c = \arctan\left(\frac{-m_1 r_1 \sin\theta_1 - m_2 r_2 \sin\theta_2 - m_3 r_3 \sin\theta_3}{-m_1 r_1 \cos\theta_1 - m_2 r_2 \cos\theta_2 - m_3 r_3 \cos\theta_3}\right). \quad (10.46)$$

Note that the signs of the numerator and denominator of the arctan function in Eq. (10.46) will identify the proper quadrant for angle θ_c. Rotors are often balanced through the removal of mass, as by drilling holes, rather than by adding counterweights. This is accomplished in the foregoing procedure by specifying a negative correction mass. Therefore, the rotor in Figure 10.21a is also balanced by removing an amount of mass $-m_c$ at position $-r_c$.

SAMPLE PROBLEM 10.14

Static Balancing of a Rotor

The rotor of Figure 10.21a has the following properties:

$$m_1 = 3\text{ kg},\quad r_1 = 80\text{ mm},\quad \theta_1 = 60°,$$
$$m_2 = 2\text{ kg},\quad r_2 = 80\text{ mm},\quad \theta_2 = 150°,$$
$$m_3 = 2\text{ kg},\quad r_3 = 60\text{ mm, and } \theta_3 = 225°.$$

Determine the amount and location of the counterweight required for static balance.

Solution. Substituting the values given into Eq. (10.45), we have

$$m_c r_c = [(240\cos 60° + 160\cos 150° + 120\cos 225°)^2 + (240\sin 60° + 160\sin 150° + 120\sin 225°)^2]^{1/2}$$
$$= [(103.4)^2 + (203.0)^2]^{1/2} = 227.8\text{ kg}\cdot\text{mm}.$$

This product will result, for example, from a counterweight mass of 2.85 kg at a radial distance of 80 mm. The angular position of the counterweight is calculated from Eq. (10.46), viz.,

$$\theta_c = \arctan\left(\frac{-203.0}{+103.4}\right) = 297.0°,$$

where from the signs of the numerator and denominator in the argument of the arctan function indicate that the angle is in the fourth quadrant. Better still, use a two-argument function (e.g. angle (x, y) or $\arctan_2(x, y)$). A graphical solution of this example is sketched in Figure 10.21b.

Dynamic Balancing

Figure 10.22a shows a rotor with eccentric masses at multiple axial locations. As a result, the rotor experiences general dynamic unbalance. As in the preceding section, the case of three initial masses will be presented here; the results will then be generalized to any number of masses.

We will examine the possibility of completely balancing the given rotor through the addition of two countermasses m_{c1} and m_{c2}, placed in transverse planes at arbitrarily selected axial locations P and Q. For static balance, the sum of all the inertial forces must be zero, a condition that yields the following equations, which are similar to Eqs. (10.44a) and (10.44b):

$$m_1\mathbf{r}_1 + m_2\mathbf{r}_2 + m_3\mathbf{r}_3 + m_{c1}\mathbf{r}_{c1} + m_{c2}\mathbf{r}_{c2} = \mathbf{0}. \tag{10.47a}$$

For the general case of N original masses, we have

$$\sum_{n=1}^{N} m_n\mathbf{r}_n + m_{c1}\mathbf{r}_{c1} + m_{c2}\mathbf{r}_{c2} = \mathbf{0}. \tag{10.47b}$$

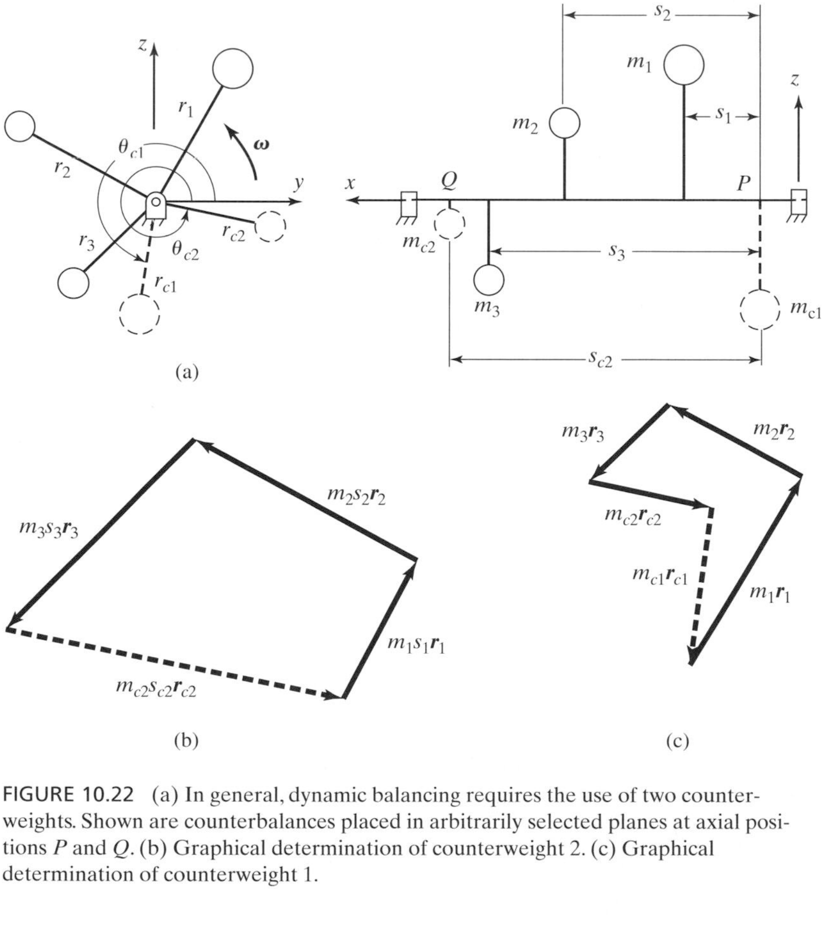

FIGURE 10.22 (a) In general, dynamic balancing requires the use of two counterweights. Shown are counterbalances placed in arbitrarily selected planes at axial positions P and Q. (b) Graphical determination of counterweight 2. (c) Graphical determination of counterweight 1.

However, a shaking couple will still exist if the inertial forces produce a net couple. Therefore, the additional condition for dynamic balance is that the sum of the moments of the inertia forces about any arbitrary point be zero. For convenience in determining the required counterbalances, we will take moments about point P, the axial location of counterweight 1, thereby eliminating this unknown counterweight from the moment equation. The axial distances of all other masses relative to point P are designated by symbol s in Figure 10.22a.

Recalling the definition for the moment of a force, Eq. (9.3), we find that the sum of the inertial force moments about point P is

$$(s_1\boldsymbol{i} \times m_1\omega^2\boldsymbol{r}_1) + (s_2\boldsymbol{i} \times m_2\omega^2\boldsymbol{r}_2) + (s_3\boldsymbol{i} \times m_3\omega^2\boldsymbol{r}_3) + (s_{c2}\boldsymbol{i} \times m_{c2}\omega^2\boldsymbol{r}_{c2}) = \mathbf{0}.$$

Factoring terms, we can rearrange this equation as

$$\boldsymbol{i} \times (m_1s_1\omega^2\boldsymbol{r}_1 + m_2s_2\omega^2\boldsymbol{r}_2 + m_3s_3\omega^2\boldsymbol{r}_3 + m_{c2}s_{c2}\omega^2\boldsymbol{r}_{c2}) = \mathbf{0}, \tag{10.48}$$

where vectors $\boldsymbol{r}_1$, $\boldsymbol{r}_2$, $\boldsymbol{r}_3$, and $\boldsymbol{r}_{c2}$ do not have $\boldsymbol{i}$ components. The only way Eq. (10.48) can be satisfied is if the second factor in the cross product is zero. Factoring out ω^2, which appears in each term, leads to the following condition for dynamic balance:

$$m_1s_1\boldsymbol{r}_1 + m_2s_2\boldsymbol{r}_2 + m_3s_3\boldsymbol{r}_3 + m_{c2}s_{c2}\boldsymbol{r}_{c2} = \mathbf{0}. \tag{10.49a}$$

Extended to the general case, Eq. (10.49a) yields

$$\sum_{n=1}^{N} m_ns_n\boldsymbol{r}_n + m_{c2}s_{c2}\boldsymbol{r}_{c2} = \mathbf{0}. \tag{10.49b}$$

Thus, Eq. (10.47a) or Eq. (10.47b), together with Eq. (10.49a) or Eq. (10.49b), constitutes the requirement for complete rotor balancing, and the pair of equations must be solved simultaneously for the necessary counterbalances. Thus, we have two-dimensional vector equations in two unknown vectors $m_{c1}\boldsymbol{r}_{c1}$ and $m_{c2}\boldsymbol{r}_{c2}$.

The graphical solution is shown in Figure 10.22b and c for the case of three initial masses. Equation (10.49a) is solved first (see Figure 10.22b) by drawing the first three vectors, which are completely known, to an appropriate scale. The vector that closes this polygon is $m_{c2}s_{c2}\boldsymbol{r}_{c2}$. The direction of this vector specifies the required angular orientation of the counterweight in the transverse plane at point Q, and the magnitude of the vector divided by the known distance s_{c2} is the required correction $m_{c2}\boldsymbol{r}_{c2}$. The counterweight in the plane at point P can now be determined from Eq. (10.47a), since vector $m_{c2}\boldsymbol{r}_{c2}$ is now known. (See Figure 10.22c.) The vector that closes this polygon is $m_{c1}\boldsymbol{r}_{c1}$, which identifies the direction and magnitude of the counterbalance.

The mathematical solution parallels the graphical approach. First, Eq. (10.49a) is divided into component form:

$$m_1s_1r_1\cos\theta_1 + m_2s_2r_2\cos\theta_2 + m_3s_3r_3\cos\theta_3 + m_{c2}s_{c2}r_{c2}\cos\theta_{c2} = 0;$$

$$m_1s_1r_1\sin\theta_1 + m_2s_2r_2\sin\theta_2 + m_3s_3r_3\sin\theta_3 + m_{c2}s_{c2}r_{c2}\sin\theta_{c2} = 0;$$

Solving for $m_{c2}r_{c2}$ and θ_{c2}, we have

$$m_{c2}r_{c2} = \frac{1}{s_{c2}}[(m_1s_1r_1\cos\theta_1 + m_2s_2r_2\cos\theta_2 + m_3s_3r_3\cos\theta_3)^2 + (m_1s_1r_1\sin\theta_1 + m_2s_2r_2\sin\theta_2 + m_3s_3r_3\sin\theta_3)^2]^{1/2} \quad (10.50)$$

and

$$\theta_{c2} = \arctan\left(\frac{-m_1s_1r_1\sin\theta_1 - m_2s_2r_2\sin\theta_2 - m_3s_3r_3\sin\theta_3}{-m_1s_1r_1\cos\theta_1 - m_2s_2r_2\cos\theta_2 - m_3s_3r_3\cos\theta_3}\right). \quad (10.51)$$

Next, Eq. (10.47a) is solved for $m_{c1}r_{c1}$ and θ_{c1}.

$$m_{c1}r_{c1} = [(m_1r_1\cos\theta_1 + m_2r_2\cos\theta_2 + m_3r_3\cos\theta_3 + m_{c2}r_{c2}\cos\theta_{c2})^2 + (m_1r_1\sin\theta_1 + m_2r_2\sin\theta_2 + m_3r_3\sin\theta_3 + m_{c2}r_{c2}\sin\theta_{c2})^2]^{1/2}; \quad (10.52)$$

$$\theta_{c1} = \arctan\left(\frac{-m_1r_1\sin\theta_1 - m_2r_2\sin\theta_2 - m_3r_3\sin\theta_3 - m_{c2}r_{c2}\sin\theta_{c2}}{-m_1r_1\cos\theta_1 - m_2r_2\cos\theta_2 - m_3r_3\cos\theta_3 - m_{c2}r_{c2}\cos\theta_{c2}}\right). \quad (10.53)$$

The signs of the numerator and denominator in the arctan functions of Eqs. (10.51) and (10.53) identify the correct quadrants of angles θ_{c2} and θ_{c1}.

SAMPLE PROBLEM 10.15

Dynamic Balancing of a Rotor

The rotor of Figure 10.22 has the following properties:

$$m_1 = 3\text{ kg},\quad r_1 = 80\text{ mm},\quad \theta_1 = 60^\circ,$$
$$m_2 = 2\text{ kg},\quad r_2 = 80\text{ mm},\quad \theta_2 = 150^\circ,$$

and

$$m_3 = 2\text{ kg},\quad r_3 = 60\text{ mm},\quad \theta_3 = 225^\circ.$$

The total axial length is 1000 mm between bearings. Counterweights are to be placed in planes that are 100 mm from each bearing. The axial distances in Figure 10.22a are then

$$s_1 = 200\text{ mm},\quad s_2 = 500\text{ mm},\quad s_3 = 700\text{ mm},\quad \text{and}\quad s_{c2} = 800\text{ mm}.$$

Determine the amounts and locations of the counterweights in planes P and Q required for complete balance.

Solution. From Eq. (10.50),

$$m_{c2}r_{c2} = \frac{1}{800}[(104{,}679)^2 + (22{,}172)^2]^{1/2} = 133.8\text{ kg}\cdot\text{mm}.$$

One combination that will produce this product is $m_{c2} = 2.23$ kg and $r_{c2} = 60$ mm. The angle is computed from Eq. (10.51):

$$\theta_{c2} = \arctan\left(\frac{22{,}172}{104{,}679}\right) = 348.0°.$$

Now, from Eqs. (10.52) and (10.53), $m_{c1}r_{c1}$ and θ_{c1} can be computed to be

$$m_{c1}r_{c1} = [(27.4)^2 + (175.3)^2]^{1/2} = 177.4 \text{ kg} \cdot \text{mm}$$

(e.g., 2.96 kg at 60 mm) and

$$\theta_{c1} = \arctan\left(\frac{-175.3}{-27.4}\right) = 261.1°.$$

The negative numerator and denominator indicate that θ_{c1} is in the third quadrant. The graphical solution of this example is shown to scale in Figures 10.22b and c.

A couple of observations follow from the previous discussion. First, the question arises as to whether a rotor can be completely balanced by means of a single counterbalance. Earlier, we saw that a single counterweight will suffice for the case of static unbalance alone. In general, however, a rotor cannot be completely balanced by one counterweight, because there are four scalar conditions to be satisfied, but only three design parameters: the axial location of the counterweight, the angular orientation θ_c, and the correction magnitude $m_c r_c$. One special case where a single counterbalance will work is the situation in which all the initial masses lie in a single plane containing the shaft axis (i.e., all the angular orientations θ are either equal or differ by 180°) and an initial static unbalance exists. In this case, Eqs. (10.47b) and (10.49b) will reduce to two scalar equations for the unknown counterweight.

We saw that, because two counterweights represent a total of six design parameters, two values can be selected arbitrarily: the axial locations P and Q in Figure 10.22a. Two different parameters could have been selected, of course, but an advantage of choosing the axial locations is that the counterweights can be placed near bearing supports in order to minimize the bending moments and resulting shaft deflection that they will produce. This fact leads to a second observation: Equations (10.44a), (10.44b), (10.47a), (10.47b), (10.49a), and (10.49b) are independent of shaft speed ω. This independence means that the rotor will be balanced at any speed for which the initial assumptions—particularly that dealing with the rigidity of the rotor—are valid. For a range of speeds, depending on the rotor material and size, deflections will be negligible and rigid-rotor balancing is satisfactory. However, as speeds are increased, eventually the flexibility of the shaft becomes significant.

Critical Speed. The *critical speed* of a shaft is the rotational speed at which severe vibration may occur, even with slight unbalance. Critical speed depends on the shaft elasticity and the mass of gears and other bodies integral with the shaft. The calculation of critical speed is discussed in a number of texts on vibration, including Dimarogonas (1996) and James et al. (1989).

10.8 BALANCING OF RECIPROCATING MACHINES

Another common type of machine unbalance is *reciprocating unbalance*, which is caused by the inertial forces associated with a translating mass. The effects of reciprocating unbalance are evident in machines such as piston engines and compressors. The balancing of reciprocating machines is more difficult than that of rotors, and in many cases complete balancing cannot be achieved by practical means.

Figure 10.23a shows a slider-crank mechanism with crank length r, connecting-rod length ℓ, and reciprocating mass m. Recall that m consists of the piston mass plus part of the connecting-rod mass based on a lumped-mass approximation. In what follows, it will be assumed that the angular velocity ω of the crank is constant. It will also be assumed that the rotating unbalance associated with the mechanism being examined is balanced by a counterweight mounted on the crank. This is common practice and reduces the inertial forces produced by the crank and the rotating part of the connecting-rod mass to such an extent that they may be neglected.

Figure 10.23b shows a free-body diagram of the frame. Based on the foregoing assumptions, the forces transmitted to the mechanism supports are obtained from Eqs. (10.34), (10.35), and (10.36), with $m_B = 0$, $m_C + m_3 = m$, and $\omega_1 = \omega$:

$$F_{10x} = mr\omega^2\left[\cos\phi + \left(\frac{r}{\ell}\right)\cos 2\phi\right]; \tag{10.54}$$

$$F_{10y} = \frac{-mr^2\omega^2 \sin\phi\left[\cos\phi + \left(\frac{r}{\ell}\right)\cos 2\phi\right]}{(\ell^2 - r^2\sin^2\phi)^{1/2}}; \tag{10.55}$$

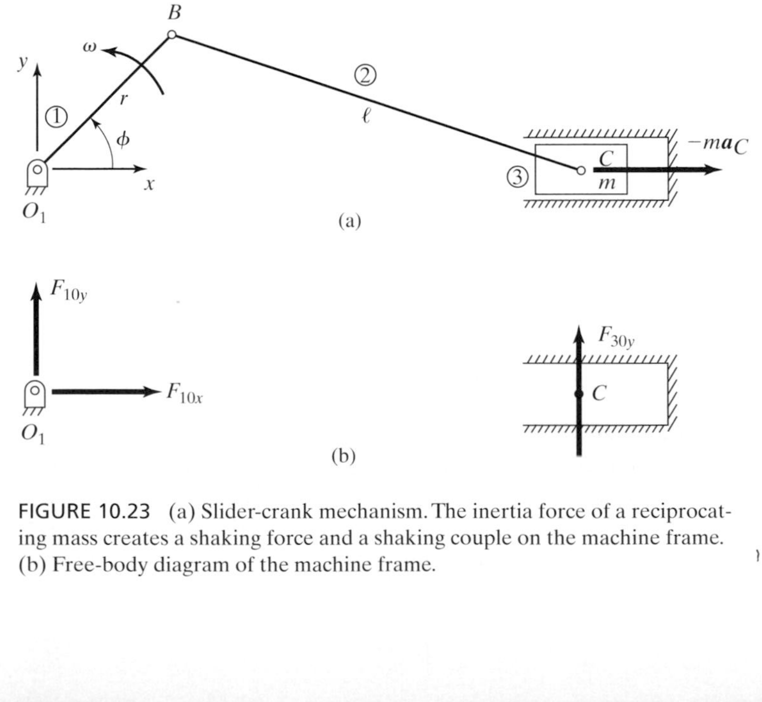

FIGURE 10.23 (a) Slider-crank mechanism. The inertia force of a reciprocating mass creates a shaking force and a shaking couple on the machine frame. (b) Free-body diagram of the machine frame.

$$F_{30y} = \frac{mr^2\omega^2 \sin\phi\left[\cos\phi + \left(\frac{r}{\ell}\right)\cos 2\phi\right]}{(\ell^2 - r^2\sin^2\phi)^{1/2}}. \tag{10.56}$$

The net effects of these forces are a shaking force F_{10x} and a shaking couple consisting of equal and opposite forces F_{10y} and F_{30y}. The shaking force produces translational vibration of the frame, whereas the shaking couple results in rotational vibration about an axis parallel to the crankshaft.

The largest, and probably the most critical, of the three forces just defined is F_{10x}. We will therefore restrict our attention to the balancing of this shaking force, which, upon substituton of $\phi = \omega t$, is rewritten as

$$F_s = mr\omega^2 \cos\omega t + mr\omega^2\left(\frac{r}{\ell}\right)\cos 2\omega t. \tag{10.57}$$

The shaking force has variable magnitude and sense, but its line of action is always along the cylinder centerline (i.e., line O_1C in Figure 10.23a; thus, the translational vibration induced will be in that direction. The first term in Eq. (10.57), which is the larger of the two terms, is called the *primary part* of the shaking force and has a frequency ω equal to the rotational frequency of the crank. The second term is referred to as the *secondary part* of the shaking force and has a frequency 2ω, twice that of the crank. In the sections that follow, we will explore ways of counteracting this shaking force.

Single-Cylinder Machines

One approach used to partially balance forces in single-cylinder engines and compressors is to add a rotating counterbalance to the crank. This counterweight supplements that described in the preceding section, which is used to counteract the rotating unbalance due to the crank mass and the rotating part of the connecting-rod mass. Figure 10.24a shows the mechanism of Figure 10.23a with a counterweight of mass m_c mounted on the crank at a radial distance r_c from main bearing O_1 and at an angular position equal to $\phi + 180°$. This mass will create a constant-magnitude centrifugal force at O_1 that rotates with speed ω. The total shaking force will then be the vector sum of the centrifugal force and the force of Eq. (10.57), as shown in Figure 10.24b. In terms of x and y unit vectors,

$$\boldsymbol{F}_s = \left[mr\omega^2\cos\phi + mr\omega^2\left(\frac{r}{\ell}\right)\cos 2\phi - m_c r_c\omega^2\cos\phi\right]\boldsymbol{i} - m_c r_c \omega^2 \sin\phi \boldsymbol{j} \tag{10.58}$$

Clearly, this counterweight cannot eliminate the shaking force entirely, because it introduces a nonzero y component, and the x component, though reduced, will not be identically equal to zero. However, by properly sizing the correction $m_c r_c$, the maximum magnitude of the shaking force can be reduced considerably.

Correction amounts typically used range from $m_c r_c = mr/2$ to $m_c r_c = 2mr/3$. For example, consider the case of $m_c r_c = 0.6mr$ for a mechanism with a ratio of crank

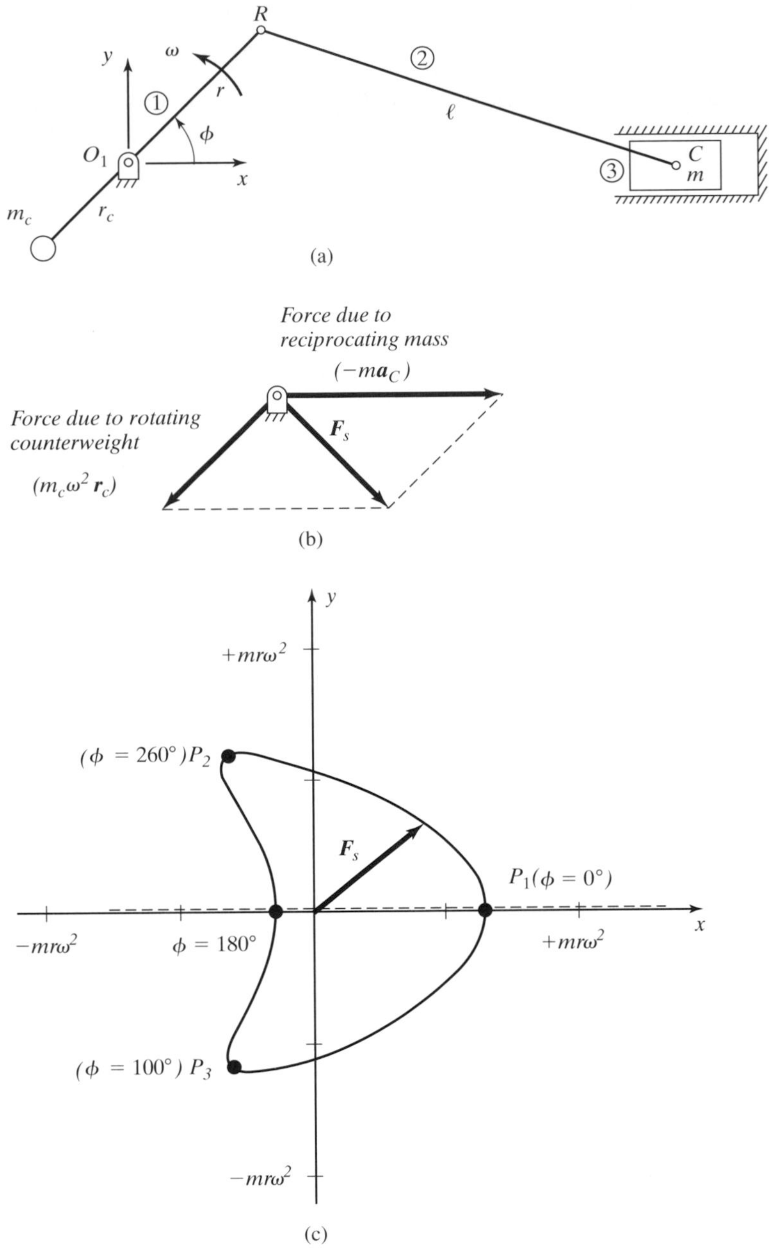

FIGURE 10.24 (a) Partial shaking-force balancing of a slider-crank mechanism by means of a rotating counterweight attached to the crank. (b) The net shaking force $\boldsymbol{F}_s$. (c) Polar plot of the shaking force.

length to connecting-rod length given by $r/\ell = 0.25$. The following expression for the shaking force results:

$$\boldsymbol{F}_s = (0.4mr\omega^2 \cos\phi + 0.25mr\omega^2 \cos 2\phi)\boldsymbol{i} - 0.6mr\omega^2 \sin\phi\,\boldsymbol{j}.$$

The magnitude of this force, as a function of crank angle ϕ, is

$$|\boldsymbol{F}_s| = mr\omega^2\sqrt{(0.4\cos\phi + 0.25\cos 2\phi)^2 + (0.6\sin\phi)^2}.$$

Figure 10.24c shows a polar plot of the shaking force, where each point on the curve defines the magnitude and direction of the force for a corresponding value of ϕ. The maximum magnitude of the shaking force is $|\boldsymbol{F}_s|_{\max} = 0.66mr\omega^2$ and occurs when ϕ equals 100° and 260°. Superimposed on Figure 10.24c is the initial shaking-force variation (dashed line) without the counterweight. The maximum shaking force is $|\boldsymbol{F}_s|_{\max} = 1.25mr\omega^2$ at $\phi = 0$. Thus, a 47-percent reduction in magnitude has been achieved through the addition of a rotating counterweight. The optimum size of the counterweight would be that which produces equal shaking force magnitudes at points P_1, P_2, and P_3 on the polar-force plot of Figure 10.24c. Examination of the figure shows that the correction used in this example is close to optimum, and therefore, little improvement beyond the 47-percent reduction could be obtained.

Multicylinder Machines

Many applications of the slider-crank mechanism in engines, pumps, and compressors involve the use of multiple mechanisms, which are designed to provide smoother flow of fluid or transmission of power than can be accomplished in a single-cylinder device. These multicylinder systems facilitate one of the more effective means of reducing the consequences of shaking forces. By a proper arrangement of the individual mechanisms, the shaking forces will partially, and perhaps totally, cancel one another. In the paragraphs that follow, we will first develop general shaking-force-balancing relationships for multicylinder machines and then examine some specific configurations.

Figure 10.25 depicts a general arrangement in which the total number of cylinders is N. (Only three cylinders are shown in the figure.) It is assumed that all the slider-crank mechanisms have the same crank length r, connecting-rod length ℓ, and reciprocating mass m and that the crank angular velocity ω is constant. The cylinder orientations are defined by angles θ_n, $n = 1, 2, \ldots, N$, which are fixed angular positions with respect to the y-axis. The angular crank throw spacings with respect to crank 1 are represented by angles ψ_n, $n = 2, 3, \ldots, N$, which do not vary with time (i.e., each crank is rigidly attached to the same crankshaft).

Each slider-crank mechanism will generate a shaking force with a line of action along that particular cylinder's centerline (i.e., at angle θ_n with respect to the y-axis). From Eq. (10.57), the expression for the individual shaking forces is

$$F_{sn} = mr\omega^2 \cos\phi_n + mr\omega^2\left(\frac{r}{\ell}\right)\cos 2\phi_n \quad n = 1, 2, \ldots, N.$$

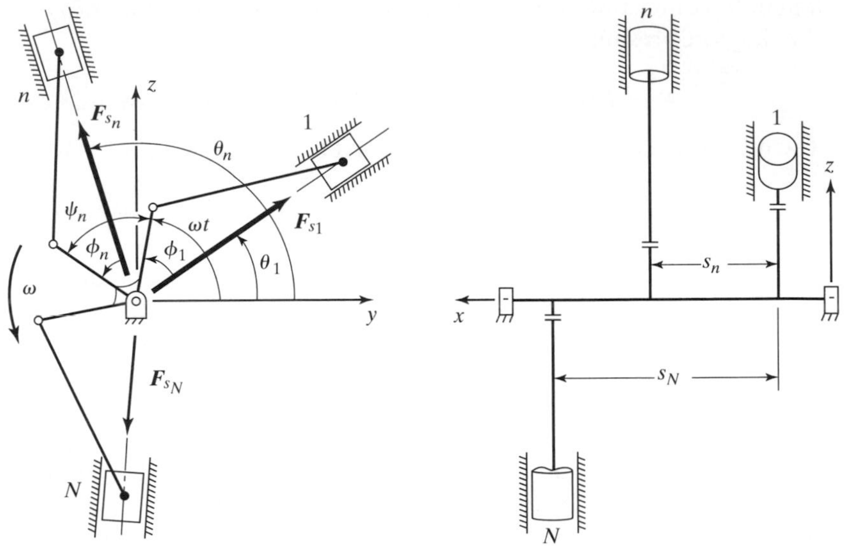

FIGURE 10.25 A multicylinder machine. The individual shaking forces combine vectorially to produce the net shaking force. Because of the axial distribution of the cylinders, the individual forces also produce a shaking moment.

Substituting the angle relationships from Figure 10.25, we can rewrite this equation as

$$F_{sn} = mr\omega^2 \cos(\omega t + \psi_n - \theta_n) + mr\omega^2\left(\frac{r}{\ell}\right)\cos[2(\omega t + \psi_n - \theta_n)]$$

$$n = 1, 2, \ldots, N, \tag{10.59}$$

where $\psi_1 = 0$ from the previous definition of angle ψ_n. The resultant shaking force will be the vector sum of all of the individual shaking forces:

$$\boldsymbol{F}_s = \sum_{n=1}^{N}(F_{sn}\cos\theta_n\boldsymbol{j} + F_{sn}\sin\theta_n\boldsymbol{k}). \tag{10.60}$$

In order for the forces to be completely balanced in the arrangement, the y and z components of Eq. (10.60) must be identically zero; that is,

$$\sum_{n=1}^{N} F_{sn}\cos\theta_n = 0 \quad \text{for all } t \tag{10.61a}$$

and

$$\sum_{n=1}^{N} F_{sn}\sin\theta_n = 0 \quad \text{for all } t. \tag{10.61b}$$

Substituting Eq. (10.59), we see that the conditions become

$$mr\omega_2 \sum_{n=1}^{N} \left[\cos(\omega t + \psi_n - \theta_n)\cos\theta_n + \left(\frac{r}{\ell}\right)\cos(2\omega t + 2\psi_n - 2\theta_n)\cos\theta_n \right] = 0$$

and

$$mr\omega_2 \sum_{n=1}^{N} \left[\cos(\omega t + \psi_n - \theta_n)\sin\theta_n + \left(\frac{r}{\ell}\right)\cos(2\omega t + 2\psi_n - 2\theta_n)\sin\theta_n \right] = 0.$$

Canceling $mr\omega^2$, which is nonzero, and factoring further, we have

$$\cos\omega t \sum_{n=1}^{N}\cos(\psi_n - \theta_n)\cos\theta_n - \sin\omega t \sum_{n=1}^{N}\sin(\psi_n - \theta_n)\cos\theta_n + \left(\frac{r}{\ell}\right)\cos 2\omega t \sum_{n=1}^{N}\cos(2\psi_n - 2\theta_n)\cos\theta_n - \left(\frac{r}{\ell}\right)\sin 2\omega t \sum_{n=1}^{N}\sin(2\psi_n - 2\theta_n)\cos\theta_n = 0 \tag{10.62a}$$

and

$$\cos\omega t \sum_{n=1}^{N}\cos(\psi_n - \theta_n)\sin\theta_n - \sin\omega t \sum_{n=1}^{N}\sin(\psi_n - \theta_n)\sin\theta_n + \left(\frac{r}{\ell}\right)\cos 2\omega t \sum_{n=1}^{N}\cos(2\psi_n - 2\theta_n)\sin\theta_n - \left(\frac{r}{\ell}\right)\sin 2\omega t \sum_{n=1}^{N}\sin(2\psi_n - 2\theta_n)\sin\theta_n = 0. \tag{10.62b}$$

The only way that these expressions can be identically zero is if the individual coefficients of the time-dependent sine and cosine functions are all zero. This yields the following eight necessary conditions for complete balance of the shaking forces:

$$\sum_{n=1}^{N}\cos(\psi_n - \theta_n)\cos\theta_n = 0; \tag{10.63a}$$

$$\sum_{n=1}^{N}\sin(\psi_n - \theta_n)\cos\theta_n = 0; \tag{10.63b}$$

$$\sum_{n=1}^{N} \cos(\psi_n - \theta_n)\sin\theta_n = 0; \tag{10.63c}$$

$$\sum_{n=1}^{N} \sin(\psi_n - \theta_n)\sin\theta_n = 0; \tag{10.63d}$$

$$\sum_{n=1}^{N} \cos(2\psi_n - 2\theta_n)\cos\theta_n = 0; \tag{10.63e}$$

$$\sum_{n=1}^{N} \sin(2\psi_n - 2\theta_n)\cos\theta_n = 0; \tag{10.63f}$$

$$\sum_{n=1}^{N} \cos(2\psi_n - 2\theta_n)\sin\theta_n = 0; \tag{10.63g}$$

$$\sum_{n=1}^{N} \sin(2\psi_n - 2\theta_n)\sin\theta_n = 0. \tag{10.63h}$$

The first four conditions account for the primary parts of the shaking forces, and if these are all satisfied, then the primary shaking forces are balanced. The last four conditions represent the secondary parts, and if those conditions are satisfied, then the secondary shaking forces are balanced. Note that the eight conditions are in terms of the cylinder orientations θ_n and the angular crank spacing ψ_n, and it follows that some arrangements of these parameters may balance the forces while other arrangements will not. Further, some arrangements may result in only primary force balancing or only secondary force balancing. Of these two possibilities, primary balancing is preferred, because it represents cancellation of the larger parts of the shaking forces.

In most multicylinder machines, the slider-crank mechanisms must be spaced axially along the crankshaft in order to avoid interference during their operation. This axial spacing is represented in Figure 10.25 by distances s_n, $n = 1, 2, \ldots, N$, measured from that cylinder designated as number 1 (therefore, $s_1 = 0$). Since the individual shaking forces will not, in general, lie in a single transverse plane, they will produce a net shaking moment, as well as a net shaking force, that will tend to cause an end-over-end rotational vibration of the crankshaft.

A set of conditions for balancing shaking moment can be established by imposing the requirement that the sum of shaking-force moments about any arbitrary axial location must be zero. Taking moments about the axial location of cylinder 1 yields

$$\sum_{n=1}^{N} s_n\boldsymbol{i} \times (F_{sn}\cos\theta_n\boldsymbol{j} + F_{sn}\sin\theta_n\boldsymbol{k}) = \boldsymbol{0}, \tag{10.64}$$

or, upon factoring,

$$\boldsymbol{i} \times \sum_{n=1}^{N} (s_nF_{sn}\cos\theta_n\boldsymbol{j} + s_nF_{sn}\sin\theta_n\boldsymbol{k}) = \boldsymbol{0}.$$

In order for this equation to be satisfied, the individual ***j*** and ***k*** components of the second factor in the cross product must be identically zero; that is,

$$\sum_{n=1}^{N} s_n F_{sn} \cos\theta_n = 0 \quad \text{for all } t \tag{10.65a}$$

and

$$\sum_{n=1}^{N} s_n F_{sn} \sin\theta_n = 0 \quad \text{for all } t. \tag{10.65b}$$

These equations are similar to Eqs. (10.61a) and (10.61b) and lead to the following similar set of conditions for balancing shaking moments.

$$\sum_{n=1}^{N} s_n \cos(\psi_n - \theta_n)\cos\theta_n = 0; \tag{10.66a}$$

$$\sum_{n=1}^{N} s_n \sin(\psi_n - \theta_n)\cos\theta_n = 0; \tag{10.66b}$$

$$\sum_{n=1}^{N} s_n \cos(\psi_n - \theta_n)\sin\theta_n = 0; \tag{10.66c}$$

$$\sum_{n=1}^{N} s_n \sin(\psi_n - \theta_n)\sin\theta_n = 0; \tag{10.66d}$$

$$\sum_{n-1}^{N} s_n \cos(2\psi_n - 2\theta_n)\cos\theta_n = 0; \tag{10.66e}$$

$$\sum_{n=1}^{N} s_n \sin(2\psi_n - 2\theta_n)\cos\theta_n = 0; \tag{10.66f}$$

$$\sum_{n=1}^{N} s_n \cos(2\psi_n - 2\theta_n)\sin\theta_n = 0; \tag{10.66g}$$

$$\sum_{n=1}^{N} s_n \sin(2\psi_n - 2\theta_n)\sin\theta_n = 0. \tag{10.66h}$$

The first four conditions guarantee primary shaking-moment balance, while the last four conditions yield secondary shaking-moment balance. Taken together, the eight equations account for the axial configuration of the cylinders, as well as for their angular orientation and the angular crank spacing. Equations (10.63a) through (10.63h) and (10.66a) through (10.66h) can be used to investigate the balancing of any piston engine or compressor.

In-Line Engines

Consider an engine, all of whose cylinders lie in a single plane and on one side of the crank axis. Suppose that these locations are given by $\theta_1 = \theta_2 = \cdots = \theta_n = \cdots$

$= \theta_N = \pi/2$. Suppose further that the cylinders are equally spaced axially with a spacing s; then, $s_n = (n - 1)s$, where the cylinders are numbered consecutively from one end of the crankshaft to the other. Substituting this information, we see that Eqs. (10.63a) through (10.63h) and (10.66a) through (10.66h) reduce to the following conditions:

$$\sum_{n=1}^{N} \sin \psi_n = 0; \quad \sum_{n=1}^{N} (n - 1)\sin \psi_n = 0;$$

$$\sum_{n=1}^{N} \cos \psi_n = 0; \quad \sum_{n=1}^{N} (n - 1)\cos \psi_n = 0;$$

$$\sum_{n=1}^{N} \cos 2\psi_n = 0; \quad \sum_{n=1}^{N} (n - 1)\cos 2\psi_n = 0;$$

$$\sum_{n=1}^{N} \sin 2\psi_n = 0; \quad \sum_{n=1}^{N} (n - 1)\sin 2\psi_n = 0.$$

Figure 10.26 shows a two-cylinder, in-line arrangement with 180° cranks; that is, $N = 2$, $\psi_1 = 0$, and $\psi_2 = \pi$. Substituting into the foregoing equations, we obtain

$$\left.\begin{aligned} \sum_{n=1}^{2} \sin \psi_n &= \sin(0) + \sin \pi = 0, \\ \sum_{n=1}^{2} \cos \psi_n &= \cos(0) + \cos \pi = 0, \end{aligned}\right\} \text{primary force}$$

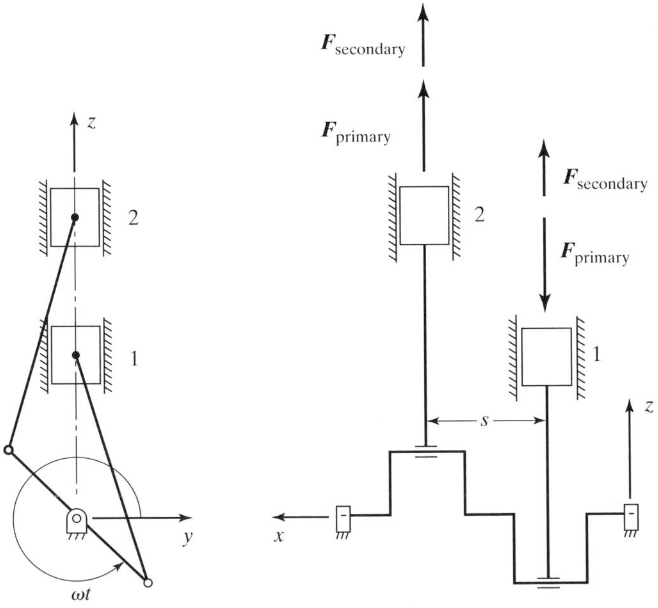

FIGURE 10.26 An in-line two-cylinder engine with 180° cranks.

$$\left.\begin{aligned}\sum_{n=1}^{2}\cos 2\psi_n &= \cos(0) + \cos 2\pi = 2,\\ \sum_{n=1}^{2}\sin 2\psi_n &= \sin(0) + \sin 2\pi = 0,\end{aligned}\right\}\ \text{secondary force}$$

$$\left.\begin{aligned}\sum_{n=1}^{2}(n-1)\sin\psi_n &= (0)\sin(0) + (1)\sin\pi = 0,\\ \sum_{n=1}^{2}(n-1)\cos\psi_n &= (0)\cos(0) + (1)\cos\pi = -1,\end{aligned}\right\}\ \text{primary moment}$$

and

$$\left.\begin{aligned}\sum_{n=1}^{2}(n-1)\cos 2\psi_n &= (0)\cos(0) + (1)\cos 2\pi = -1,\\ \sum_{n=1}^{2}(n-1)\sin 2\psi_n &= (0)\sin(0) + (1)\sin 2\pi = 0.\end{aligned}\right\}\ \text{secondary moment}$$

Thus, the primary parts of the shaking forces are always equal and opposite; therefore, they cancel, but because they are offset axially, they form a nonzero couple. This is shown in Figure 10.26. On the other hand, the secondary parts of the shaking forces are always equal with the same sense, and they therefore combine to produce a net force and also cause a net moment. From Eq. (10.60), the net shaking force is

$$\begin{aligned}\boldsymbol{F}_s &= \sum_{n=1}^{2} F_{sn}\boldsymbol{k}\\ &= \left\{mr\omega^2\cos\left(\omega t - \frac{\pi}{2}\right) + mr\omega^2\left(\frac{r}{\ell}\right)\cos\left[2\left(\omega t - \frac{\pi}{2}\right)\right]\right.\\ &\quad \left. + mr\omega^2\cos\left(\omega t + \frac{\pi}{2}\right) + mr\omega^2\left(\frac{r}{\ell}\right)\cos\left[2\left(\omega t + \frac{\pi}{2}\right)\right]\right\}\boldsymbol{k}\\ &= -2mr\omega^2\left(\frac{r}{\ell}\right)\cos 2\omega t\boldsymbol{k},\end{aligned}$$

with a maximum magnitude of $2mr\omega^2(r/\ell)$. Although this shaking force is nonzero, it nevertheless represents a significant improvement in comparison to a single-cylinder engine with respect to typical (r/ℓ) ratios. However, as noted, a shaking couple has been introduced.

Opposed Engines

In an opposed engine, all the cylinders lie in the same plane, with half on each side of the crank axis. Selecting $\theta_1 = \cdots = \theta_{N/2} = \pi/2$ and $\theta_{N/2+1} = \cdots = \theta_N = 3\pi/2$, we note that half of Eqs. (10.63a) through (10.63h) and (10.66a) through (10.66h) are automatically satisfied; these are Eqs. (10.63a), (10.63b), (10.63e), (10.63f), (10.66a), (10.66b), (10.66e), and (10.66f). This is because there will be no y-direction forces or z-direction moments in the general force and moment equations. As an example, consider the two-cylinder opposed engine of Figure 10.27a, with 180° cranks, where $N = 2, \theta_1 = \pi/2, \theta_2 = 3\pi/2, \psi_1 = 0, \psi_2 = \pi, s_1 = 0,$ and $s_2 = s$. Substituting into

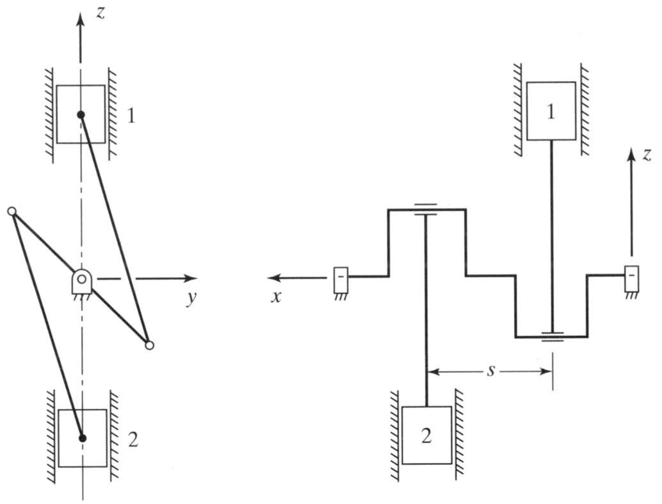

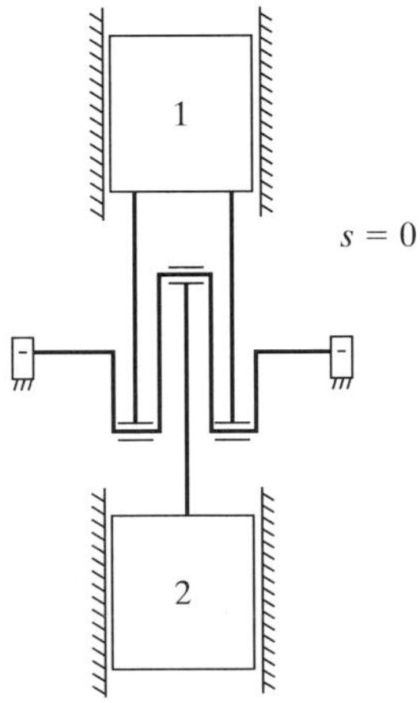

FIGURE 10.27 (a) An opposed two-cylinder engine with 180° cranks. (b) Double connecting rods for cylinder 1.

Eqs. (10.63c), (10.63d), (10.63g), (10.63h), (10.66c), (10.66d), (10.66g), and (10.66h), we have

$$\left.\begin{aligned}\sum_{n=1}^{2}\cos(\psi_n-\theta_n)\sin\theta_n &= \cos\left(-\frac{\pi}{2}\right)\sin\frac{\pi}{2}\\ &\quad+\cos\left(-\frac{\pi}{2}\right)\sin\frac{3\pi}{2}=0,\\ \sum_{n=1}^{2}\sin(\psi_n-\theta_n)\sin\theta_n &= \sin\left(-\frac{\pi}{2}\right)\sin\frac{\pi}{2}\\ &\quad+\sin\left(-\frac{\pi}{2}\right)\sin\frac{3\pi}{2}=0,\end{aligned}\right\}\ \text{primary force}$$

$$\left.\begin{aligned}\sum_{n=1}^{2}\cos(2\psi_n-2\theta_n)\sin\theta_n &= \cos(-\pi)\sin\frac{\pi}{2}\\ &\quad+\cos(-\pi)\sin\frac{3\pi}{2}=0,\\ \sum_{n=1}^{2}\sin(2\psi_n-2\theta_n)\sin\theta_n &= \sin(-\pi)\sin\frac{\pi}{2}\\ &\quad+\sin(-\pi)\sin\frac{3\pi}{2}=0,\end{aligned}\right\}\ \text{secondary force}$$

$$\left.\begin{aligned}\sum_{n=1}^{2}s_n\cos(\psi_n-\theta_n)\sin\theta_n &= (0)+s\cos\left(-\frac{\pi}{2}\right)\sin\frac{3\pi}{2}=0,\\ \sum_{n=1}^{2}s_n\sin(\psi_n-\theta_n)\sin\theta_n &= (0)+s\sin\left(-\frac{\pi}{2}\right)\sin\frac{3\pi}{2}=s,\end{aligned}\right\}\ \text{primary moment}$$

and

$$\left.\begin{aligned}\sum_{n=1}^{2}s_n\cos(2\psi_n-2\theta_n)\sin\theta_n &= (0)+s\cos(-\pi)\sin\frac{3\pi}{2}=s,\\ \sum_{n=1}^{2}s_n\sin(2\psi_n-2\theta_n)\sin\theta_n &= (0)+s\sin(-\pi)\sin\frac{3\pi}{2}=0.\end{aligned}\right\}\ \text{secondary moment}$$

The net shaking force is zero, because both parts of the individual shaking forces cancel. This is an improvement over the two-cylinder, in-line engine of Figure 10.26, but there will be a significant shaking couple (both primary and secondary) due to the staggering of the crank throws. Clearly, the smaller the spacing s, the better will be the design from the point of view of balancing. One method of reducing s to zero and thereby eliminating the shaking couple is to use double connecting rods for one of the cylinders, as shown in Figure 10.27b.

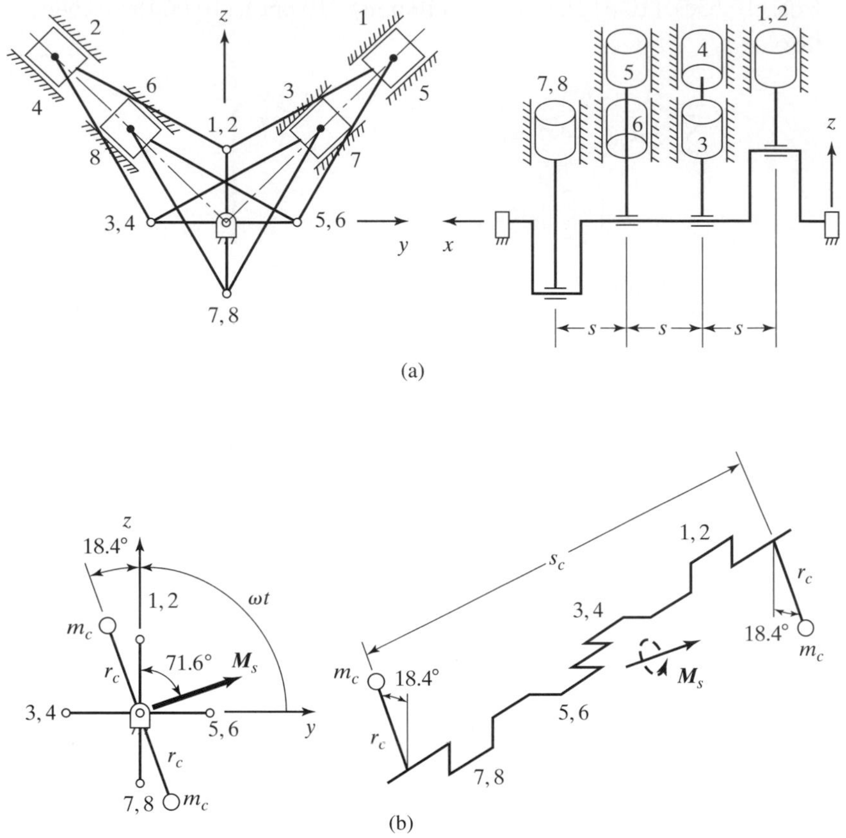

FIGURE 10.28 (a) A V-8 engine with 90° cranks. This arrangement can be completely balanced with the addition of rotating counterweights on the crankshaft. (b) Location of the counterweights.

V Engines

Due to its compact form, the V engine is common in automotive and other applications. Consider, for example, the V-8 engine of Figure 10.28a, consisting of two banks of four cylinders with an angle of 90° between banks. The four-throw crankshaft has 90° cranks, with an axial spacing s between cranks. The following quantities are determined from the figure:

$$\theta_1 = \theta_3 = \theta_5 = \theta_7 = \frac{\pi}{4};$$

$$\theta_2 = \theta_4 = \theta_6 = \theta_8 = \frac{3\pi}{4};$$

$$\psi_1 = \psi_2 = 0; \quad \psi_5 = \psi_6 = \frac{3\pi}{2};$$

$$\psi_3 = \psi_4 = \frac{\pi}{2}; \quad \psi_7 = \psi_8 = \pi;$$

$$s_1 = s_2 = 0; \quad s_5 = s_6 = 2s;$$

$$s_3 = s_4 = s; \quad s_7 = s_8 = 3s.$$

The force-balance conditions, as evaluated from Eqs. (10.63a) through (10.63h), are

$$\sum_{n=1}^{8} \cos(\psi_n - \theta_n)\cos\theta_n = \frac{1}{2} + \frac{1}{2} + \frac{1}{2} - \frac{1}{2} - \frac{1}{2} + \frac{1}{2} - \frac{1}{2} - \frac{1}{2} = 0,$$

$$\sum_{n=1}^{8} \sin(\psi_n - \theta_n)\cos\theta_n = -\frac{1}{2} + \frac{1}{2} + \frac{1}{2} + \frac{1}{2} - \frac{1}{2} - \frac{1}{2} + \frac{1}{2} - \frac{1}{2} = 0,$$

$$\sum_{n=1}^{8} \cos(\psi_n - \theta_n)\sin\theta_n = \frac{1}{2} - \frac{1}{2} + \frac{1}{2} + \frac{1}{2} - \frac{1}{2} - \frac{1}{2} - \frac{1}{2} + \frac{1}{2} = 0,$$

$$\sum_{n=1}^{8} \sin(\psi_n - \theta_n)\sin\theta_n = -\frac{1}{2} - \frac{1}{2} + \frac{1}{2} - \frac{1}{2} - \frac{1}{2} + \frac{1}{2} + \frac{1}{2} + \frac{1}{2} = 0,$$

$$\sum_{n=1}^{8} \cos(2\psi_n - 2\theta_n)\cos\theta_n = 0 + 0 + 0 + 0 + 0 + 0 + 0 + 0 = 0,$$

$$\sum_{n=1}^{8} \sin(2\psi_n - 2\theta_n)\cos\theta_n$$

$$= -\frac{1}{\sqrt{2}} - \frac{1}{\sqrt{2}} + \frac{1}{\sqrt{2}} + \frac{1}{\sqrt{2}} + \frac{1}{\sqrt{2}} + \frac{1}{\sqrt{2}} - \frac{1}{\sqrt{2}} - \frac{1}{\sqrt{2}} = 0,$$

$$\sum_{n=1}^{8} \cos(2\psi_n - 2\theta_n)\sin\theta_n = 0 + 0 + 0 + 0 + 0 + 0 + 0 + 0 = 0,$$

and

$$\sum_{n=1}^{8} \sin(2\psi_n - 2\theta_n)\sin\theta_n$$

$$= -\frac{1}{\sqrt{2}} + \frac{1}{\sqrt{2}} + \frac{1}{\sqrt{2}} - \frac{1}{\sqrt{2}} + \frac{1}{\sqrt{2}} - \frac{1}{\sqrt{2}} - \frac{1}{\sqrt{2}} + \frac{1}{\sqrt{2}} = 0.$$

Thus, the engine is completely force balanced. In fact, this configuration is force balanced for any angle between the cylinder banks, because each bank of four cylinders is force balanced independently.

Examining the shaking-moment conditions leads to the following results:

$$\sum_{n=1}^{8} s_n \cos(\psi_n - \theta_n) \cos \theta_n = 0 + 0 + \frac{s}{2} - \frac{s}{2} - s + s - \frac{3s}{2} - \frac{3s}{2} = -3s;$$

$$\sum_{n=1}^{8} s_n \sin(\psi_n - \theta_n) \cos \theta_n = 0 + 0 + \frac{s}{2} + \frac{s}{2} - s - s + \frac{3s}{2} - \frac{3s}{2} = -s;$$

$$\sum_{n=1}^{8} s_n \cos(\psi_n - \theta_n) \sin \theta_n = 0 + 0 + \frac{s}{2} + \frac{s}{2} - s - s - \frac{3s}{2} + \frac{3s}{2} = -s;$$

$$\sum_{n=1}^{8} s_n \sin(\psi_n - \theta_n) \sin \theta_n = 0 + 0 + \frac{s}{2} - \frac{s}{2} - s + s + \frac{3s}{2} + \frac{3s}{2} = 3s;$$

$$\sum_{n=1}^{8} s_n \cos(2\psi_n - 2\theta_n) \cos \theta_n = 0 + 0 + 0 + 0 + 0 + 0 + 0 + 0 = 0;$$

$$\sum_{n=1}^{8} s_n \sin(2\psi_n - 2\theta_n) \cos \theta_n$$

$$= 0 + 0 + \frac{s}{\sqrt{2}} + \frac{s}{\sqrt{2}} + \frac{2s}{\sqrt{2}} + \frac{2s}{\sqrt{2}} - \frac{3s}{\sqrt{2}} - \frac{3s}{\sqrt{2}} = 0;$$

$$\sum_{n=1}^{8} s_n \cos(2\psi_n - 2\theta_n) \sin \theta_n = 0 + 0 + 0 + 0 + 0 + 0 + 0 + 0 = 0;$$

$$\sum_{n=1}^{8} s_n \sin(2\psi_n - 2\theta_n) \sin \theta_n$$

$$= 0 + 0 + \frac{s}{\sqrt{2}} - \frac{s}{\sqrt{2}} + \frac{2s}{\sqrt{2}} - \frac{2s}{\sqrt{2}} - \frac{3s}{\sqrt{2}} + \frac{3s}{\sqrt{2}} = 0.$$

There is a primary shaking couple, but no secondary shaking couple; hence, the engine arrangement, by itself, does not yield a complete force and moment balance. However, the shaking couple has a special nature that facilitates total balancing by means of a relatively straightforward modification. To understand that nature, consider Eq. (10.64), where $\boldsymbol{M}_s$ refers to the shaking moment:

$$\begin{aligned}\boldsymbol{M}_s &= \sum_{n=1}^{8} s_n \boldsymbol{i} \times (F_{sn} \cos \theta_n \boldsymbol{j} + F_{sn} \sin \theta_n \boldsymbol{k}) \\ &= \sum_{n=1}^{8} (-F_{sn} s_n \sin \theta_n \boldsymbol{j} + F_{sn} s_n \cos \theta_n \boldsymbol{k}) \\ &= mr\omega^2 \sum_{n=1}^{8} [-s_n \cos(\omega t + \psi_n - \theta_n) \sin \theta_n \boldsymbol{j} \\ &\qquad + s_n \cos(\omega t + \psi_n - \theta_n) \cos \theta_n \boldsymbol{k}].\end{aligned}$$

In this equation, the secondary parts of the shaking forces have been disregarded, since they will cancel. Rearranging terms and substituting the results obtained earlier, we have

$$\begin{aligned} \boldsymbol{M}_s &= mrs\omega^2[(3\sin\omega t + \cos\omega t)\boldsymbol{j} + (\sin\omega t - 3\cos\omega t)\boldsymbol{k}] \\ &= mrs\omega^2\sqrt{10}[\cos(\omega t - 71.6°)\boldsymbol{j} + \sin(\omega t - 71.6°)\boldsymbol{k}]. \end{aligned}$$

The magnitude of this moment is $mrs\omega^2\sqrt{10}$, which is constant for all values of time t, and the direction of the moment is perpendicular to the crank axis and rotates with speed ω, where at any instant the angle of the moment vector with respect to the y direction is $(\omega t - 71.6°)$. This is exactly the same as the rotating, unbalanced dynamic couple discussed earlier. Thus, the net effect of this engine arrangement is what appears to be rotating dynamic unbalance. Therefore, the shaking couple can be balanced by a set of rotating counterweights that produce an equal, but opposite, rotating couple. The magnitude of this couple is given by $m_c r_c s_c \omega^2 = mrs\omega^2\sqrt{10}$, and the locations are as depicted in Figure 10.28b, where m_c is the mass, r_c is the radial position, and s_c is the axial spacing of the counterweights. Because this engine can be completely balanced in this fashion, it exhibits smooth-running performance.

SUMMARY

Newton's second law of motion is utilized in d'Alembert's principle, enabling us to portray a dynamics problem in the form of a statics problem. However, some machine components, including connecting rods and couplers, have complicated motion patterns that generate both inertial forces and inertial torques. Graphical methods are useful for illustrating principles of dynamic analysis, for analyzing a linkage in one or two positions, and for checking analytical work. But graphical methods are not recommended for dynamic analysis of a linkage over a full cycle of motion. Even analytical dynamic force analysis can be time consuming. The difficulty can be partly resolved by using approximate methods or motion simulation software.

In solving dynamics problems, we need to know component masses and mass moments of inertia. It is easy to locate the center of mass and calculate the mass moment of inertia for regular shapes, and motion simulation software can be used to find the properties of irregular shapes. Computer-generated plots and other results that look convincing sometimes contain serious errors. Accordingly, one should always verify results with simple checks. One can use common sense to spot errors; employ Newton's laws; and compare position, velocity, and acceleration curves, noting all maxima.

Sometimes the motion of a linkage component is specified. For example, crank speed in a slider crank or four-bar linkage may be a given constant value. We begin by finding all positions, velocities, and accelerations (including angular velocities and accelerations) in terms of crank position or time. Then, we can find inertial forces and torques, consider any external forces that are present, and find the required motor torque and power and forces on individual links and bearings. The problem may be long and difficult, but it *is* straightforward; we follow a step-by-step procedure. Contrast this situation with that when motion is not specified.

Now, consider the response of a linkage to a motor that has specified angular-velocity-related torque characteristics. Inertial forces on the linkage components depend on component accelerations and angular accelerations, and these accelerations in turn depend on the speed of the motor, but the inertial forces affect that speed. Velocity-related external forces might also be present, as in, say, a piston pump or compressor. This linkage motion problem is not straightforward; one might want to use numerical methods or motion simulation software to solve the problem and others like it.

When a shaft assembly is unbalanced, inertial forces cause shaking and, sometimes, catastrophic failure. If the sources of unbalance lie in a single plane perpendicular to the shaft, static balancing may be possible. If there are unbalanced rotors at multiple axial locations, dynamic balancing techniques are used. Balance is accomplished by adding or removing mass from the system. Shaking forces in single- and multicylinder machines can also be reduced by the addition of counterweights. Shaking forces can be virtually eliminated in some machine configurations.

The purpose of analysis is to serve design; that is, the results of your analysis should be used to justify and improve your tentative designs. Generate as much data as you need to understand the motion and forces on a linkage. Try changing a dimension on a linkage component. Consider changing the mass or another property of the component. If you have written a good program or if you are using motion simulation software, this can be done with a few keystrokes. Can you find ways to make the linkage better?

A Few Review Items

- Relate d'Alembert's principle to Newton's laws.
- A slender rod rotates about one end. Find the mass moment of inertia (about the center of rotation), in terms of the length and mass of the rod.
- How are the torque applied to the rod and the angular acceleration related? Suppose there is also velocity-dependent fluid resistance to the motion of the rod. Express the relationship among these variables by using d'Alembert's principle.
- Sketch a free-body diagram of a connecting rod. Include inertial effects.
- Suppose you need to check a computer program. What data will you change (temporarily) to make a connecting rod a two-force member? Then, what output data will you compare?
- The torque input to a linkage is dependent on speed. We find the variation in crank speed to be too great. Suggest one or more ways to correct the situation.
- Eccentric masses are located at several axial locations on a rotating shaft. Will static balancing be satisfactory? Explain.
- Identify some of the problems one faces when trying to balance a reciprocating engine.

PROBLEMS

10.1 Determine the magnitude and location of the equivalent offset inertial force for the connecting rod, link 2, of the slider-crank mechanism of Figure P10.1 for the position shown. The crank has a constant angular velocity of 100 rad/s counterclockwise. The mass of the connecting rod is 0.2 kg, and the moment of inertia about the center of mass, G_2, is 300 kg · mm^2.

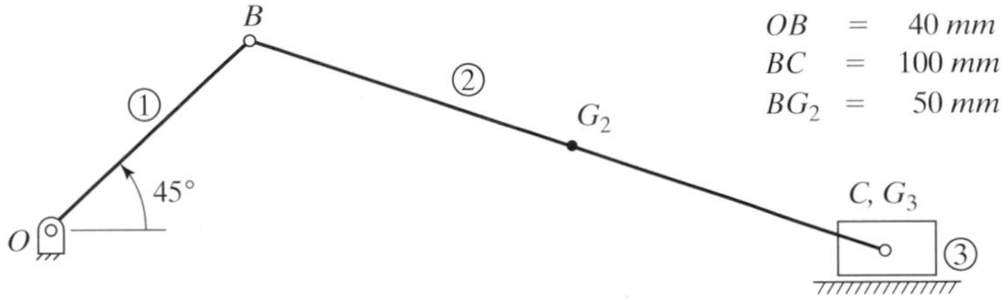

FIGURE P10.1

10.2 For the mechanism of Problem 10.1, determine the crank torque T_1 required for dynamic equilibrium at the position shown. Neglect external loads and the inertia of crank 1 and slider 3.

10.3 For the mechanism of Problems 10.1 and 10.2, determine the crank torque T_1 required for dynamic equilibrium if slider 3 has a mass of 0.3 kg. Neglect external loads and the inertia of crank 1.

10.4 Determine the magnitude and location of the equivalent offset inertial force for the coupler link 2 of the slider-crank mechanism at the position shown in Figure P10.2. The crank has a constant angular velocity of 60 rev/min clockwise. The weight of the coupler is 1000 lb, and the moment of inertia about the center of mass, G_2, is 70 lb · s^2 · ft.

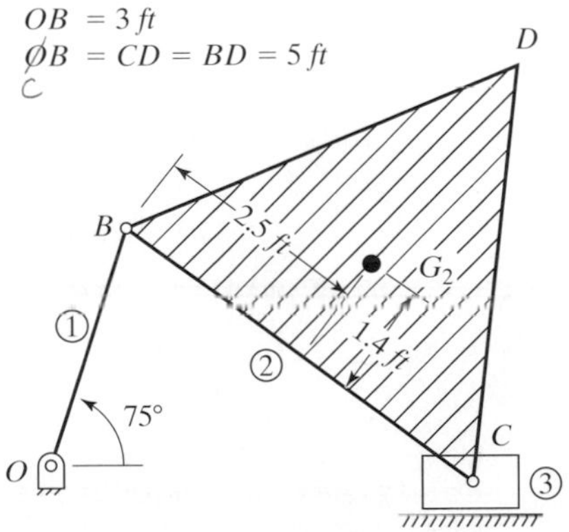

FIGURE P10.2

10.5 The four-bar linkage of Figure P10.3 has a constant crank angular velocity $\omega_1 = 60$ rad/s clockwise. Coupler link 2 has a mass of 0.3 kg and a moment of inertia about the center of mass, G_2, of 1000 kg · mm^2. Follower link 3 has a mass of 0.2 kg and a moment of inertia

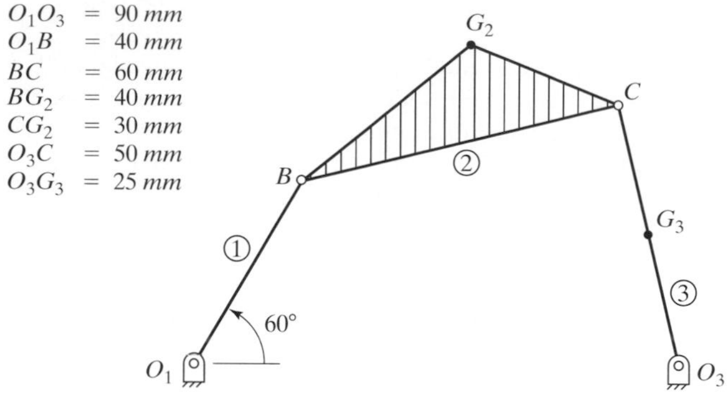

FIGURE P10.3

about its center of mass, G_3, of 150 kg · mm^2. For the position shown, determine the following:

(a) The equivalent offset inertial force for the coupler.

(b) The equivalent offset inertial force for the follower.

(c) The crank torque $\boldsymbol{T}_1$ required for dynamic equilibrium.

10.6 For the mechanism shown in Figure P10.4, member 2 has a weight of 2 lb and a moment of inertia of 0.04 lb · s^2 · in about its center of mass, G_2. Sliding block 1 has a constant velocity of 10 ft/s upward.

(a) Determine the instantaneous force $\boldsymbol{F}_1$ required to produce the motion, assuming that sliding block 3 is massless.

(b) Determine the required instantaneous force $\boldsymbol{F}_1$ if sliding block 3 has a weight of 1 lb.

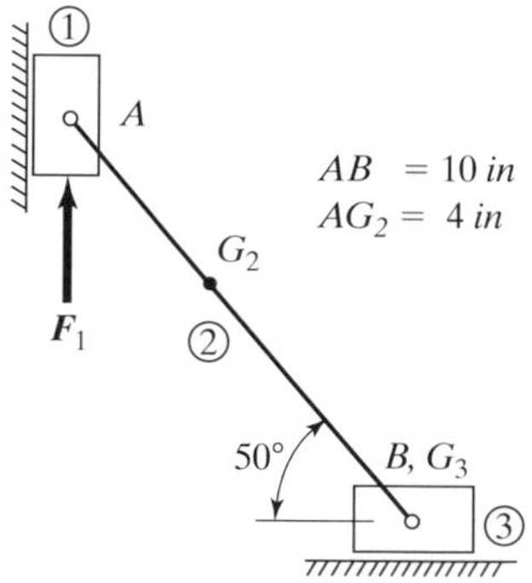

FIGURE P10.4

10.7 The slider-crank mechanism of Figure P10.5 has a constant crank angular velocity of 50 rad/s counterclockwise. The acceleration polygon is shown in the figure. The connecting rod weighs 2 lb, with a mass moment of inertia about its center of mass G_2, of 0.009 lb · s^2 · in. The piston has a weight of 1.5 lb. Determine all bearing forces and the required input torque $\boldsymbol{T}_1$ for the position shown.

$$OB = 2 \text{ in}$$
$$BC = 4 \text{ in}$$
$$BG_2 = 2 \text{ in/s}^2$$

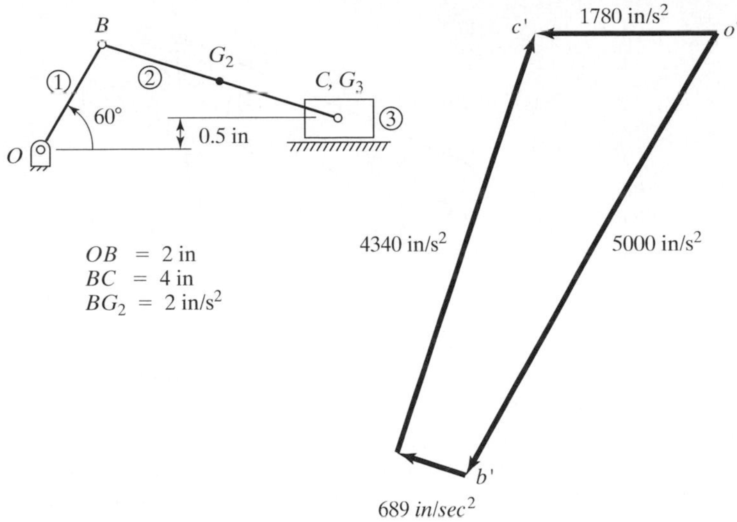

FIGURE P10.5

10.8 The crank of the slider-crank mechanism of Figure P10.6 has an instantaneous angular velocity of 10 rad/s clockwise and an angular acceleration of 200 rad/s² clockwise. Information related to acceleration is given in the figure. The connecting rod has a mass of 15 kg and a mass moment of inertia of 7500 kg · mm² about its center of mass, G_2. The slider has a mass of 8 kg. The crank has a moment of inertia of 4000 kg · mm² about its stationary center of mass, G_1. Determine all bearing forces and the input torque T_1 for the position shown.

$$OB = 30 \text{ mm}$$
$$BC = 90 \text{ mm}$$
$$a_{G_2} = 3810 \angle 325° \text{ mm/s}^2$$
$$\alpha_2 = 66.3 \text{ rad/s}^2 \text{ ccw}$$
$$a_{G_3} = 5800 \angle 0° \text{ mm/s}^2$$

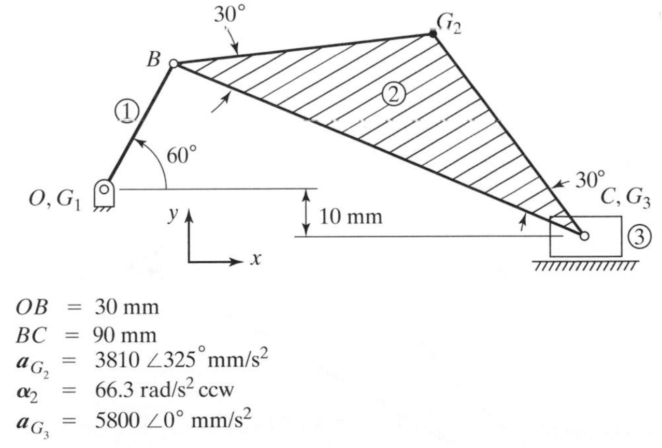

FIGURE P10.6

10.9 The four-bar linkage of Figure P10.7 has a constant input angular velocity $\omega_1 = 200$ rad/s clockwise. This results in the following accelerations: $a_{G2} = 34,500\angle298°$ in/s², $\alpha_2 = 2670$ rad/s² ccw, $a_{G3} = 14,600\angle294°$ in/s², and $\alpha_3 = 6940$ rad/s² cw. Coupler link 2 weighs 1.2 lb and has a mass moment of inertia about center of mass, G_2, equal to 0.03 lb · s² · in. Follower link 3 weighs 1.0 lb and has a mass moment of inertia about its

$$O_1B = 1 \text{ in}$$
$$BC = 4 \text{ in}$$
$$BG_2 = 2 \text{ in}$$
$$O_3C = 2 \text{ in}$$
$$O_3G_3 = 1 \text{ in}$$

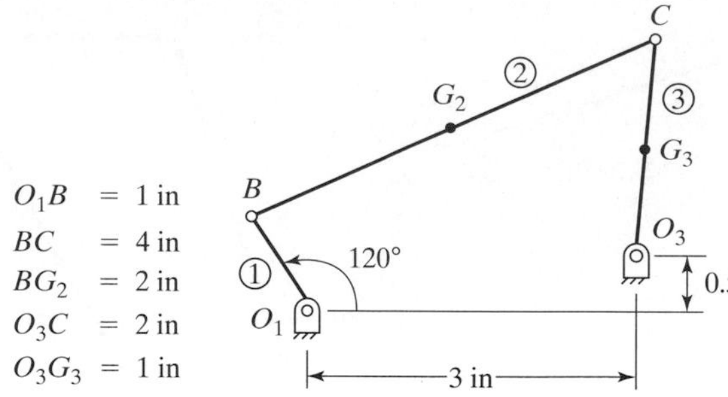

FIGURE P10.7

center of mass, G_3, equal to $0.002 \text{ lb} \cdot \text{s}^2 \cdot \text{in}$. Determine all bearing forces and instantaneous input torque T_1 by **(a)** a graphical solution and **(b)** an analytical solution.

10.10 The four-bar linkage of Figure P10.8 has a constant input angular velocity $\omega_1 = 60$ rad/s counterclockwise. The acceleration polygon is shown in the figure. The masses of coupler link 2 and follower link 3 are 1.0 kg and 0.6 kg, respectively, and the moments of inertia about the centers of mass are $700 \text{ kg} \cdot \text{mm}^2$ and $500 \text{ kg} \cdot \text{mm}^2$, respectively. Determine all bearing forces and instantaneous torque T_1 by **(a)** a graphical solution and **(b)** an analytical solution.

$$O_1B = 80 \text{ mm}$$
$$BC = 160 \text{ mm}$$
$$BG_2 = 100 \text{ mm}$$
$$O_3C = 100 \text{ mm}$$
$$O_3G_3 = 50 \text{ mm}$$
$$O_1O_3 = 200 \text{ mm}$$

168,000 mm/s^2

255,000 mm/s^2

288,000 mm/s^2

578,000 mm/s^2

4000 mm/s^2

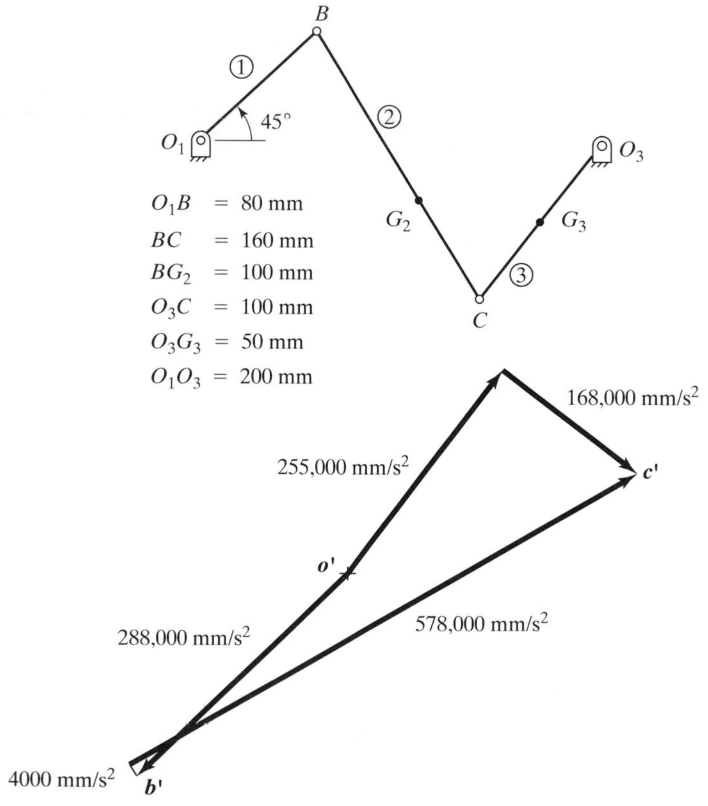

FIGURE P10.8

10.11 For the mechanism of Problem 10.10, determine the bearing forces and input torque if the constant-input angular velocity of 60 rad/s is clockwise rather than counterclockwise. How do the forces change if the input speed is doubled?

10.12 The four-bar linkage of Figure P10.9 has the following weights and moments of inertia: $w_1 = 3$ lb, and $w_2 = 5$ lb, $w_3 = 4$ lb, $I_{G_1} = 0.1\ \text{lb}\cdot\text{s}^2\cdot\text{in}$, $I_{G_2} = 0.3\ \text{lb}\cdot\text{s}^2\cdot\text{in}$, and $I_{G_3} = 0.3\ \text{lb}\cdot\text{s}^2\cdot\text{in}$. Input link 1 has instantaneous angular velocity and acceleration of 10 rad/s clockwise and 100 rad/s^2 counterclockwise, respectively, leading to the following accelerations: $\boldsymbol{a}_{G_1} = 636\angle 195°\ \text{in/s}^2$, $\boldsymbol{a}_{G_2} = 1240\angle 189°\ \text{in/s}^2$, $\boldsymbol{a}_{G_3} = 610\angle 182°\ \text{in/s}^2$, $\alpha_2 = 19$

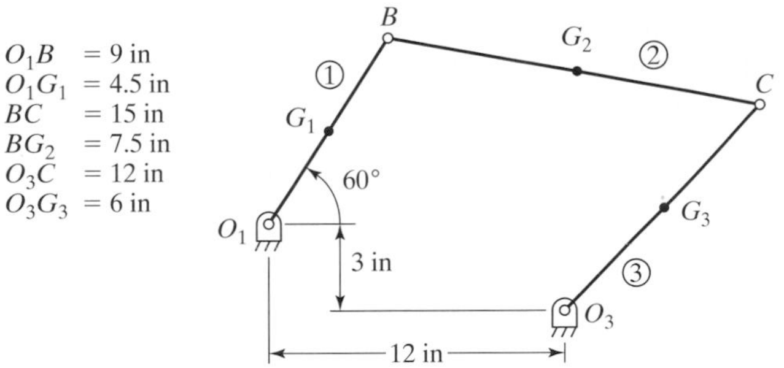

FIGURE P10.9

rad/s^2 ccw, and $\alpha_3 = 77$ rad/s^2 ccw. Determine the bearing forces and instantaneous input torque $\boldsymbol{T}_1$ by **(a)** a graphical solution and **(b)** an analytical solution.

10.13 An in-line slider-crank mechanism has a crank length of 0.1 m and a connecting-rod length of 0.5 m. The piston has a mass of 3 kg. The connecting rod has a mass of 2 kg with a center of mass located at a distance of 0.15 m from the crankpin end of the rod. Utilizing a lumped-mass approximation for the connecting rod, determine an expression for the torque required to maintain a constant crank speed of 200 rev/min. Evaluate the torque on the following crank angles ϕ: 0°, 30°, 60°, 90°, 120°, 150°, and 180°.

10.14 For the mechanism of Problem 10.13, determine the magnitudes and directions of all bearing forces for crank angle $\phi = 0, 45°, 90°, 135°$, and 180°. Assume that the center of mass of the crank is stationary.

10.15 An in-line slider-crank mechanism has a crank length of 2 ft and a connecting-rod length of 7 ft. The piston has a weight of 100 lb. The connecting rod has a weight of 75 lb with a center of gravity located at a distance of 2 ft from the crankpin end of the rod. Utilizing a lumped-mass approximation for the connecting rod, determine an expression for the torque required to maintain a constant crank speed of 60 rev/min. Evaluate the torque for the following crank angles ϕ: 0, 30°, 60°, 90°, 120°, 150°, 180°.

10.16 For the mechanism of Problem 10.15, determine the magnitudes and directions of all bearing forces for crank angle $\phi = 0, 45°, 90°, 135°$, and 180°. The crank is balanced.

10.17 Derive an expression for input torque $\boldsymbol{T}_1$ similar to Eq. (10.37) for the offset slider-crank mechanism of Figure P10.10 with constant crank angular velocity $\boldsymbol{\omega}_1$. For the parameter values of Problem 10.13 and an offset e of 0.1 m, evaluate the torque at the following crank angles ϕ: 0, 30°, 60°, 90°, 120°, 150°, and 180°.

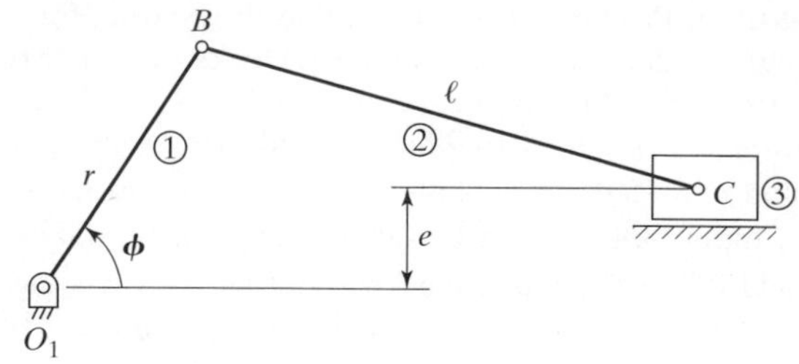

FIGURE P10.10

10.18 Determine the torque $\boldsymbol{T}_1$ required to maintain a constant crank speed of 1,000 rev/min ccw for the two-cylinder engine depicted in Figure P10.11. The individual pistons and connecting rods have masses of 1.0 kg and 0.8 kg, respectively. Consider the case where the cylinder V angle ψ is 90°, the crank spacing θ is 90°, and the crank angle ϕ is 160° · $BG_2 = DG_3 = 105$ mm.

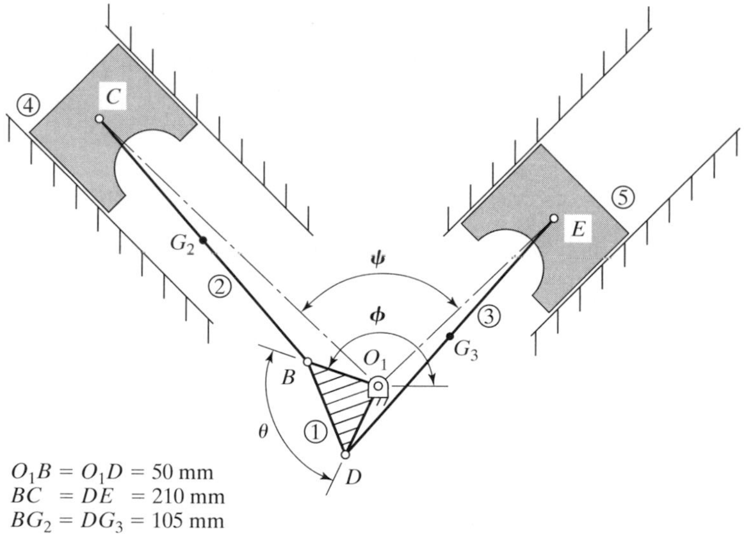

FIGURE P10.11

10.19 Determine the torque $\boldsymbol{T}_1$ required to maintain a constant crank speed of 1,000 rev/min ccw for the two-cylinder engine depicted in Figure P10.12. The individual pistons and connecting rods have masses of 1.0 kg and 0.8 kg, respectively. Consider the case where the cylinder V-angle ψ is 90° and the crank angle ϕ is 20°. $BG_2 = BG_3 = 100$ mm.

10.20 Figure P10.13 is a schematic of a three-bladed propeller. Determine the location and correction amount of the counterweight that will balance the rotor. Perform the solution by using **(a)** the graphical method and **(b)** the analytical method.

10.21 Determine the corrections needed in planes P and Q to balance the rotor shown in Figure P10.14. Carry out the solution by **(a)** the graphical method and **(b)** the analytical method.

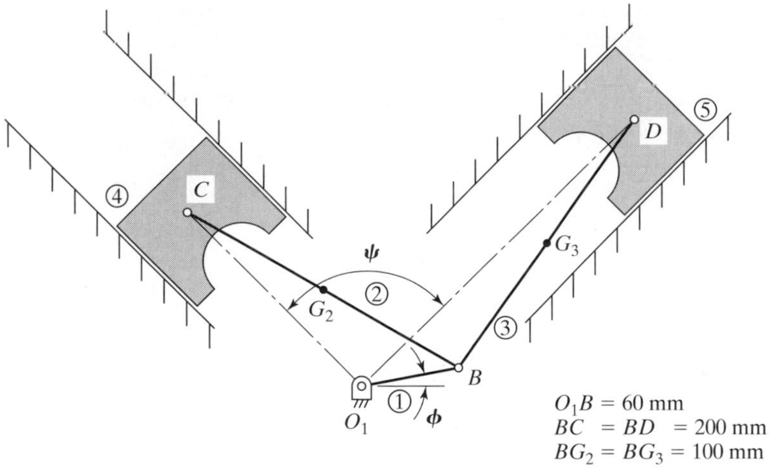

FIGURE P10.12

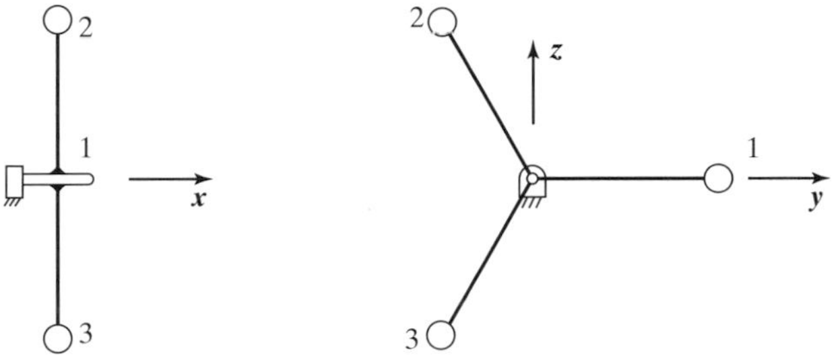

$m_1 = 1$ kg $\quad r_1 = 3$ mm $\quad \theta_1 = 0°$

$m_2 = 1$ kg $\quad r_2 = 6$ mm $\quad \theta_2 = 120°$

$m_3 = 1$ kg $\quad r_3 = 2$ mm $\quad \theta_3 = 240°$

FIGURE P10.13

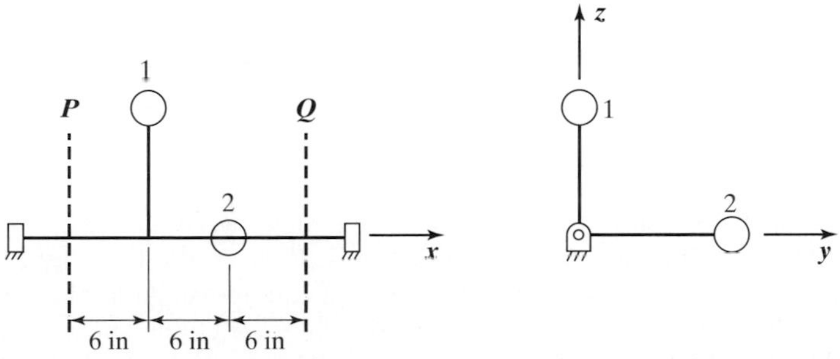

$w_1 = 4$ oz $\quad r_1 = 2$ in $\quad \theta_1 = 90°$

$w_2 = 2$ oz $\quad r_2 = 2.5$ in $\quad \theta_2 = 0$

FIGURE P10.14

10.22 For the rotor shown in Figure P10.15, unbalanced masses 1 and 2 have weights of 6 lb and 4 lb, respectively. The system is to be balanced by adding mass in the L plane at a radius of 3.0 in and removing mass in the R plane at a radius of 3.5 in. Determine the magnitudes and locations of the required corrections by **(a)** a graphical solution and **(b)** an analytical solution.

$$w_1 = 6 \text{ lb} \qquad w_2 = 4 \text{ lb}$$
$$r_1 = 4 \text{ in} \qquad r_2 = 5.25 \text{ in}$$
$$\theta_1 = 90° \qquad \theta_2 = 210°$$

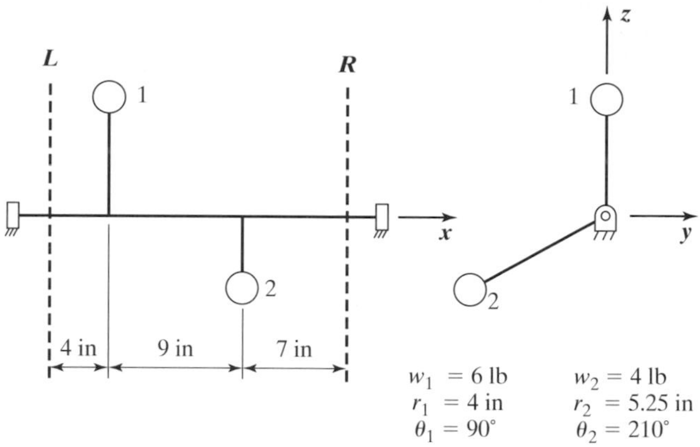

FIGURE P10.15

10.23 The rotor of Figure P10.16 has unbalanced weights $w_1 = 2$ oz, $w_2 = 3$ oz, and $w_3 = 2$ oz, at radial positions $r_1 = 3$ in, $r_2 = 3$ in, and $r_3 = 2$ in. Determine the necessary counterweight correction amounts (in · oz) and locations in balance planes at points P and Q for complete static and dynamic balance of the rotor by **(a)** a graphical solution and **(b)** an analytical solution.

$$\theta_1 = 60°$$
$$\theta_2 = 150°$$
$$\theta_3 = 270°$$

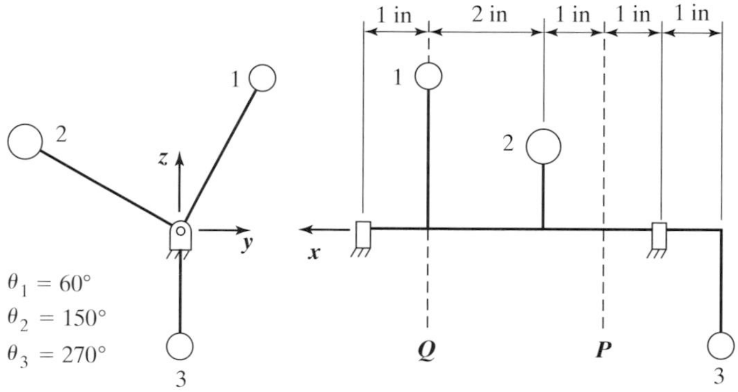

FIGURE P10.16

10.24 The rotor of Figure P10.17 has the following unbalanced amounts: $m_1 r_1 = 1500$ kg · mm, $m_2 r_2 = 2000$ kg · mm, and $m_3 r_3 = 1500$ kg · mm. Balance the rotor by determining the angular orientation and correction amount for a counterweight in plane P and the axial location and angular orientation of a second counterweight having a correction amount of 1,000 kg · mm by **(a)** a graphical solution and **(b)** an analytical solution.

10.25 Figure P10.18 depicts a four-throw crankshaft that has the following properties: $m_1 = m_2 = m_3 = m_4 = 10$ kg and $r_1 = r_2 = r_3 = r_4 = 40$ mm. Determine the balancing arrangement in correction planes P and Q.

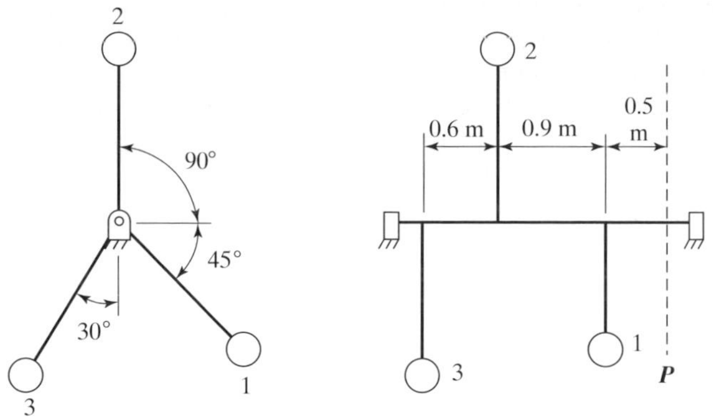

FIGURE P10.17

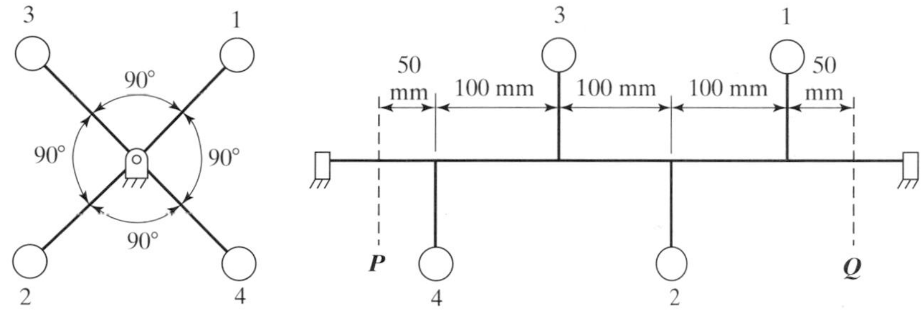

FIGURE P10.18

10.26 Figure P10.19 depicts a four-throw crankshaft that has the following properties: $m_1 = m_2 = m_3 = m_4 = 10$ kg and $r_1 = r_2 = r_3 = r_4 = 40$ mm. Determine the balancing arrangement in correction planes P and Q.

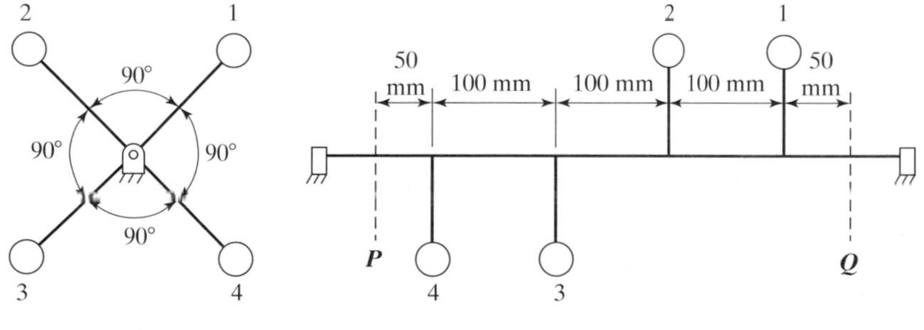

FIGURE P10.19

10.27 Repeat Problem 10.14, except include a rotating counterweight on the crankshaft at an angle of 180° relative to the crank. The counterweight has a mass of 3.5 kg and a center of mass located at a distance of 0.1 m from the crankshaft axis.

10.28 Repeat Problem 10.16, except add a rotating counterweight on the crankshaft at an angle of 180° relative to the crank. The counterweight has a weight of 100 lb and a center of mass located at a distance of 2.5 ft from the crankshaft axis.

10.29 Figure P10.20 shows an in-line, two-cylinder engine arrangement in which the cranks are spaced at 90°. Determine expressions for the net shaking force F_s and its axial position a as functions of angle ωt. For each cylinder, $mr\omega^2 = 2{,}000$ lb and $r/\ell = 0.25$, where m is the reciprocating mass, r is the crank length, and ℓ is the connecting-rod length. Axial distance $s = 4$ in. Which of the following are balanced: primary shaking force, secondary shaking force, primary shaking moment, secondary shaking moment?

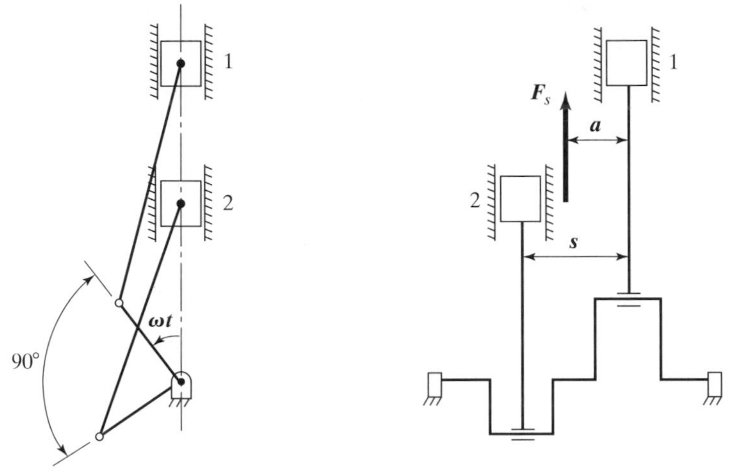

FIGURE P10.20

10.30 Figure P10.21 shows an in-line, two-cylinder engine with 180° cranks. Also shown is a gear and rotating counterweight arrangement that is driven from the crankshaft. What gear ratio, correction amount, and counterweight orientation must be used to balance the net shaking force?

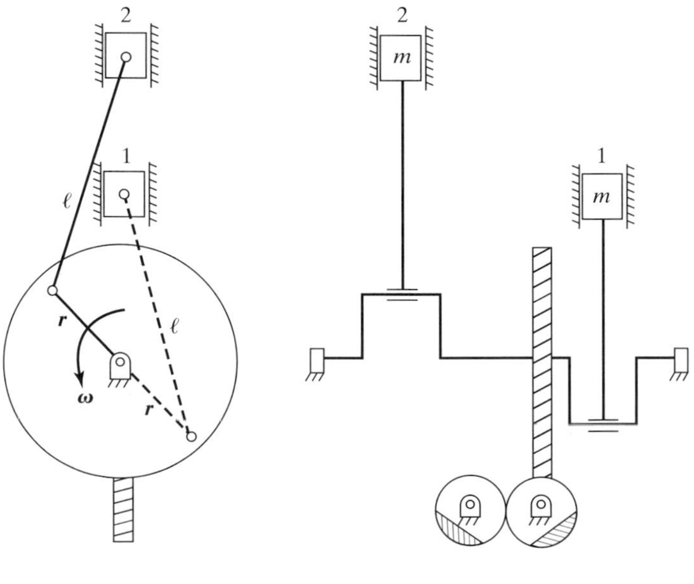

FIGURE P10.21

10.31 Examine the shaking-force balance conditions as they apply to the two-cylinder opposed engine of Figure P10.22, which has a single crank and zero distance between cylinder axes.

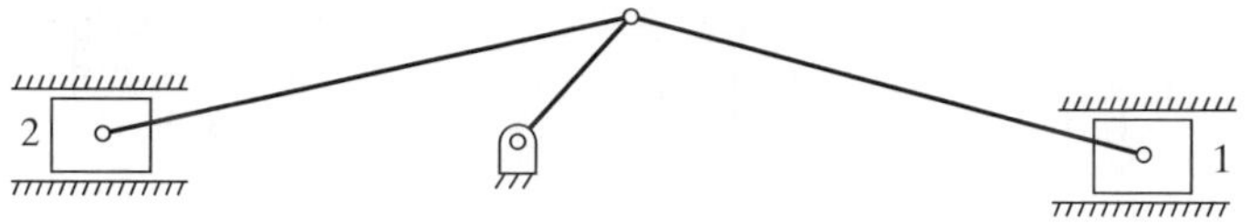

FIGURE P10.22

10.32 Examine the balance conditions, both shaking force and shaking moment, as they apply to the four-cylinder, in-line engine of Figure P10.23.

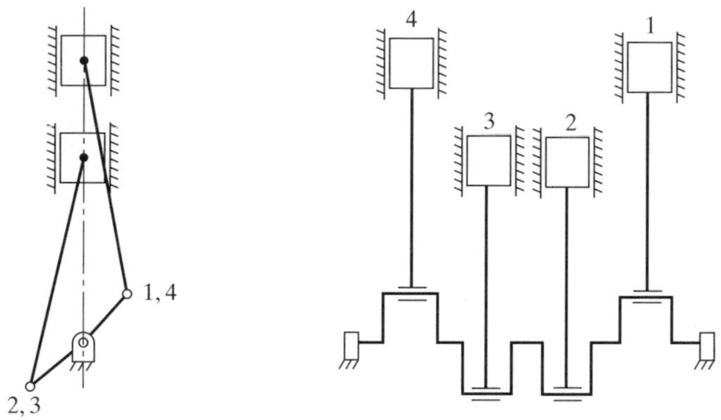

FIGURE P10.23

10.33 Examine the balance conditions, both shaking force and shaking moment, as they apply to the four-cylinder, in-line engine of Figure P10.24.

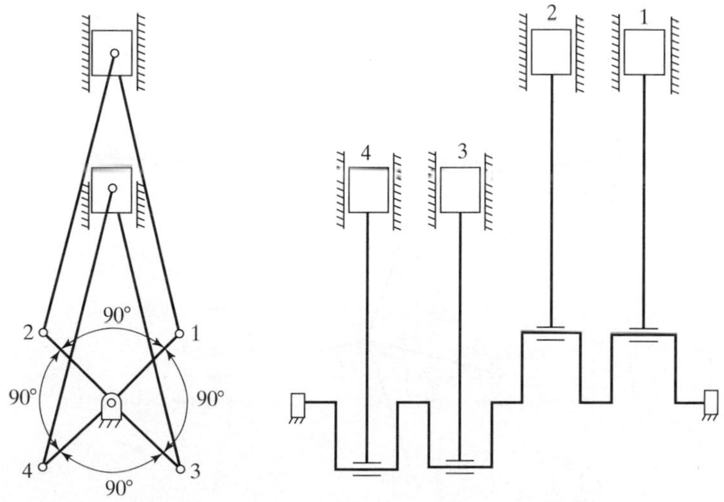

FIGURE P10.24

10.34 Examine the balance conditions, both shaking force and shaking moment, as they apply to the six-cylinder, in-line engine of Figure P10.25.

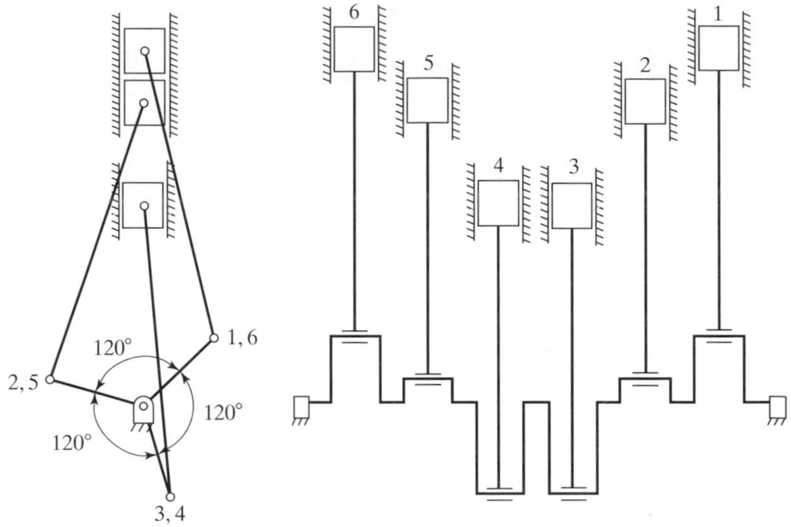

FIGURE P10.25

10.35 Examine the shaking-force balance of the two-cylinder V engine of Figure P10.11. The two cylinders lie in the same axial plane. Angle ψ is 90° and angle θ is 90°.

10.36 Examine the shaking-force balance of the two-cylinder V engine of Figure P10.12. The two cylinders lie in the same axial plane. Angle ψ is 90°.

10.37 The four-cylinder radial engine depicted in Figure P10.26 is an excellent engine from the point of view of dynamic balance. Show that the engine can be balanced by means of a single rotating counterweight mounted on the crankshaft, and determine the location and magnitude of such a counterweight. The crank length is r, the connecting rod lengths all equal ℓ, and the reciprocating masses all equal m. The rotating masses are balanced, and all four cylinders lie in a single transverse plane.

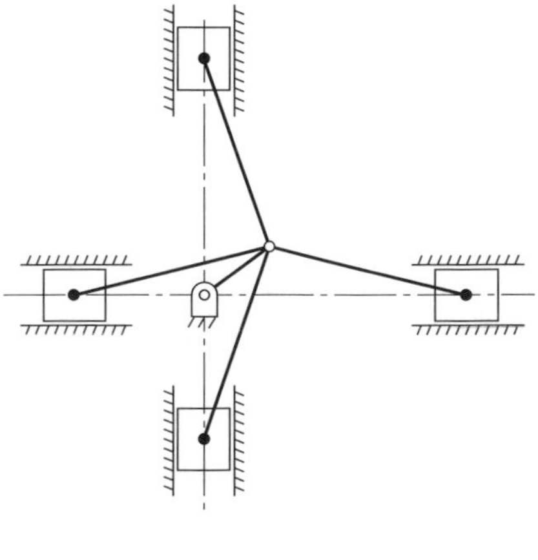

FIGURE P10.26

10.38 A 4-mm-thick steel body has the shape of an irregular polygon. The coordinates of the points describing the polygon are as shown in Figure P10.27 (in meters, where e indicates powers of 10).

(a) Sketch the body.
(b) Determine the area.
(c) Locate and mark the center of mass.
(d) Determine the mass.
(e) Determine the mass moment of inertia.

Suggestion: Use software to sketch the body and determine its properties.

	X	Y
1	0.000e+000	0.000e+000
2	0.000e+000	2.000e−002
3	2.000e−002	3.000e−002
4	3.000e−002	1.000e−002
5	4.000e−002	1.000e−002
6	4.000e−002	0.000e+000
7	2.000e−002	0.000e+000
8	1.500e−002	1.500e−002

FIGURE P10.27

10.39 A 1-mm-thick steel body has the shape of an irregular polygon. The coordinates of the points describing the polygon are as shown in Figure P10.28 (in meters, where e indicates powers of 10).

(a) Sketch the body.
(b) Determine the area.
(c) Locate and mark the center of mass.
(d) Determine the mass.
(e) Determine the mass moment of inertia.

Suggestion: Use software to sketch the body and determine its properties.

	X	Y
1	0.00000	0.00000
2	1.00000e−002	2.00000e−002
3	3.00000e−002	3.00000e−002
4	5.00000e−002	3.00000e−002
5	2.50000e−002	1.50000e−002
6	1.00000e−002	0.00000

FIGURE P10.28

10.40 A 1-mm-thick steel body has the shape of an irregular polygon. The coordinates of the points describing the polygon are as shown in Figure P10.29 (in meters, where e indicates powers of 10).

(a) Sketch the body.

(b) Determine the area.
(c) Locate and mark the center of mass.
(d) Determine the mass.
(e) Determine the mass moment of inertia.

Suggestion: Use software to sketch the body and determine its properties.

	X	Y
1	0.00000	0.00000
2	0.00000	1.00000e−002
3	5.00000e−003	2.00000e−002
4	2.00000e−002	3.00000e−002
5	4.00000e−002	3.50000e−002
6	4.50000e−002	2.00000e−002
7	4.00000e−002	1.50000e−002
8	3.00000e−002	2.00000e−002
9	2.00000e−002	1.50000e−002
10	1.00000e−002	5.00000e−003
11	1.00000e−002	0.00000

FIGURE P10.29

10.41 A 0.3 m radius by 0.04995 m thick disk with two blades is shown in Figure 10.4 in the text. The mass density of steel is about 8,000 kg/m^3.

Each blade is 1 m square by 1.25 mm thick, has a mass of 10 kg, and is subject to air resistance equal to kv^2A, where $k = 0.025$, $v =$ velocity of the blade center, and $A =$ blade area. An electric motor applying a torque of 15 N·m drives the assembly. Use mathematics software to

- Find the mass moment of inertia of the assembly.
- Find the terminal angular velocity of the assembly. Calculate the power and check the air resistance torque and angular acceleration at this speed.
- Calculate the angular acceleration and power when the rotation speed is 20 rad/s.

10.42 Use motion simulation software to examine the assembly in the previous problem. Find the power, rotation position, angular velocity, and angular acceleration at 136.1 seconds after the motor is started. Plot the following against time: air resistance force on one blade, angular velocity and acceleration of the assembly, and angular momentum of the central disk.

10.43 A rotating assembly consists of a steel disk with two blades similar to the one shown in Figure 10.4 in the text. The disk has a radius of 300 mm and is 100 mm thick. The mass density of steel is about 8000 kg/m^3. Each blade is 1 m square by 5 mm thick, has a mass of 0.5 kg, and is subject to air resistance equal to kv^2A, where $k = 0.025$, $v =$ velocity of the blade center, and $A =$ blade area. An electric motor applying a torque of 10 N·m drives the assembly. Use mathematics software to

Find the mass moment of inertia of the assembly.

Find the terminal angular velocity of the assembly. Calculate the power and check the air resistance torque and angular acceleration at this speed.

Calculate the angular acceleration and power when the rotation speed is 16 rad/s.

10.44 Use motion simulation software to examine the assembly in the previous problem. Find the rotation position 10 seconds after starting the motor. Plot the following against time: air resistance force on one blade, angular velocity and acceleration of the assembly, and angular momentum of the center disk.

10.45 A 0.3 m radius by 0.04995 m thick steel disk has two blades as shown in Figure 10.4 in the text. The mass density of steel is about 8000 kg/m^3. Each blade is 1 m square by 5 mm thick, has a mass of 10 kg, and is subject to air resistance equal to kv^2A, where $k = 0.025$, $v =$ velocity of the blade center, and $A =$ blade area. An electric motor applying a torque of 15 N · m drives the assembly.

(a) Find the following quantities 10 seconds after the motor is started: Mechanical power supplied to the assembly and angular displacement, velocity, and acceleration of the assembly.

(b) Find the same quantities one minute after the motor is started.

(c) How long does it take the motor to reach an angular velocity of 20 rad/s?

10.46 In the following problem, examine only inertial effects. Fluid forces on the piston are not considered. A pump in the form of an in-line slider-crank linkage has a 51-mm crank length and 96-mm connecting-rod length. The crank, connecting rod, and piston masses are 0.45, 0.91, and 1.36 kg, respectively. The mass moments of inertia are $9.845 \cdot 10^{-5}$ and $7.008 \cdot 10^{-4}$kg · m^2 for the crank and connecting rod, respectively (about their centers of gravity). The crank angular velocity is a constant 100 rad/s. Find the maximum crankpin force. Find the corresponding wrist-pin force, as well as the crank angle and time. (*Note*: If the problem is solved for discrete time steps, results are not likely to be exact.)

10.47 In this problem, the connecting rod is given a negligible mass so that it approximates a two-force member and we examine only inertial effects. Fluid forces on the piston are not considered. A pump in the form of an in-line slider-crank linkage has a 51-mm crank length and 96-mm connecting-rod length. Crank, connecting rod, and piston masses are 450, 1, and 1,360 grams, respectively. Mass moments of inertia are $9.845 \cdot 10^{-5}$ and $8 \cdot 10^{-7}$ kg · m^2 for the crank and connecting rod, respectively (about their centers of gravity). The crank speed is a constant 95.49 rpm. Plot the angular position, angular velocity, and angular acceleration of the connecting rod. Plot the position, velocity, and acceleration of the piston. Find the maximum motor torque and power and the corresponding time and crank angle. Find the corresponding forces at either end of the connecting rod and its position, velocity, and acceleration.

10.48 Repeat the previous problem, but consider the linkage position that results in the maximum piston acceleration magnitude. Find the piston acceleration at that instant. Find the motor torque and power and the corresponding time and crank angle. Find the corresponding forces at either end of the connecting rod and its position, velocity, and acceleration.

10.49 The piston of a pump is subject to fluid forces given by $F_d = 50\, v_C^2$ when $v_C < 0$ (i.e., when the piston is moving away from the cylinder head) and $F_d = 0$ otherwise, where $F_d =$ fluid force (N) resisting piston motion, and $v_C =$ piston velocity (m/s). The pump will have the form of an in-line slider-crank linkage with a 51-mm crank length and 96-mm connecting-rod length. The crank, connecting rod, and piston masses are 0.45, 0.91, and 1.36 kg, respectively. The mass moments of inertia are $9.845 \cdot 10^{-5}$ and $7.008 \cdot 10^{-4}$ kg · m^2 for the crank and connecting rod, respectively (about their centers of gravity). The crank angular velocity is a constant 100 rad/s. Plot the motor torque and linkage motion against time. Find the maximum motor power and the corresponding time and values of torque, linkage motion, and forces.

10.50 Repeat the previous problem, but find the magnitude of the greatest lateral force on the cylinder and the time and crank position when that force is reached. Also, find, for that instant, the motor power and torque; the piston position, velocity, and acceleration; the connecting-rod angular position, velocity, and acceleration; and the horizontal, vertical, and resultant forces on the crankpin and wrist pin.

10.51 The piston of a pump is subject to fluid forces given by $F_d = 50\,v_C^2$ when $v_C < 0$ (i.e., when the piston is moving away from the cylinder head) and $F_d = 0$ otherwise, where F_d = fluid force (N) resisting piston motion and v_C = piston velocity (m/s). The pump will have the form of an in-line slider-crank linkage with a 50-mm crank length and 90-mm connecting-rod length. The crank, connecting rod, and piston masses are 1.5, 1, and 2 kg respectively. Mass moments of inertia are $3.156 \cdot 10^{-4}$ and $6.771 \cdot 10^{-4}$ $\text{kg} \cdot \text{m}^2$ for the crank and connecting rod, respectively (about their centers of gravity). The crank angular velocity is a constant 100 rad/s. Plot the motor torque and linkage motion against time. Find the maximum motor power, the time at which the maximum power is reached, and the corresponding values of the crank position and torque. Also, find, for that instant, the piston position, velocity, and acceleration; the connecting-rod angular position, velocity, and acceleration; and the horizontal, vertical, and resultant forces on the crankpin and wrist pin.

10.52 Repeat the previous problem, but find the magnitude of the greatest lateral force on the cylinder and the time and crank position at which the maximum force is reached. Also, find, for that instant, the piston position, velocity, and acceleration; the connecting-rod angular position, velocity, and acceleration; the motor torque and power; and the remaining horizontal, vertical, and resultant forces on the crankpin and wrist pin.

10.53 A motor with torque characteristics approximated by $T = 50 - 0.005\,\omega_1^2$ drives a piston pump, where T = net torque ($\text{N} \cdot \text{m}$) and ω_1 = crank angular velocity. The piston is subject to fluid forces given by $F_d = -45\,v_C^2$ when $v_C > 0$ (i.e., when the piston is moving toward the cylinder head) and $F_d = 0$ otherwise, where F_d = fluid force (N) resisting piston motion and v_C = piston velocity (m/s). The pump will have the form of an in-line slider-crank linkage with a 50-mm crank length and 110-mm connecting-rod length. The crank, connecting rod, and piston masses are 2, 1.25, and 1.36 kg, respectively. The mass moments of inertia are 0.005 and 0.001 $\text{kg} \cdot \text{m}^2$ for the crank and connecting rod, respectively (about their centers of gravity). Show the linkage; plots of the crank angular position, velocity, and acceleration; plots of the piston position, velocity, and acceleration; and plots of the net motor torque and power, all against time. Find the maximum motor power and the corresponding values of torque, linkage motion, and forces.

10.54 A motor with torque characteristics approximated by $T = 50 - 0.005\,\omega_1^2$ drives a piston pump, where T = net torque ($\text{N} \cdot \text{m}$) and ω_1 = crank angular velocity. The piston is subject to fluid forces given by $F_d = -45\,v_C^2$ when $v_C > 0$ (i.e., when the piston is moving toward the cylinder head) and $F_d = 0$ otherwise, where F_d = fluid force (N) resisting piston motion, and v_C = piston velocity (m/s). The pump will have the form of an in-line slider-crank linkage with a 51-mm crank length and 96-mm connecting-rod length. The crank, connecting rod, and piston masses are 2, 0.91, and 1.36 kg, respectively. Mass moments of inertia are 0.0046 and 0.0007$\text{kg} \cdot \text{m}^2$ for the crank and connecting rod, respectively (about their centers of gravity). Show the linkage; plots of the crank angular position, velocity, and acceleration; plots of the piston position, velocity, and acceleration;

and plots of the net motor torque and power, all against time. Find the maximum motor power and the corresponding values of torque, linkage motion, and forces.

10.55 Repeat the previous problem, except

(a) Find the maximum crankshaft speed and the corresponding values of torque, power, linkage motion, and forces.

(b) Find the elapsed time for the first revolution of the crankshaft and the values of torque, power, linkage motion, and forces at the end of the first cycle of motion.

10.56 Refer to Figure 10.12a in the text, which shows a mechanism made up of a four-bar linkage, connecting rod, and slider. At the initial position of the mechanism, the crank angle $= 0$, and the coordinates of the points shown are given in mm as follows: motor 0_1: 0,0; fixed point 0_3: 140, 0; B: 52, 0; D: 35.5, −81.6; C: 95.4, −90.6; reentrant corner of coupler: 51.7, −71; Wrist-pin path: x, −50; E: 151.2, −50.

The bodies have the following masses (kg) and mass moments of inertia ($kg \cdot mm^2$); drive crank 0_1B: 0.08, 8.0; coupler BDC: 0.20,109; driven crank 0_3C: 0.150, 60; connecting rod DE: 0.18, 216; piston E: 0.25, 30.9. A motor drives the crank at a constant angular velocity of $100 \cdot \pi$ rad/s counterclockwise. Determine the following for the instant the magnitude of the piston velocity is maximum: time; crankshaft position; piston position, velocity, and acceleration; and forces applied by the connecting rod to the wrist pin. Sketch the linkage for this instant. Consider only inertial effects.

10.57 Repeat the previous problem, but consider the instant the piston acceleration reaches its maximum positive value (to the right). Also, consider only inertial effects.

10.58 Repeat the previous problem, but plot the motor torque against time. Find the instant in time that the torque reaches its maximum value. Sketch the linkage at that instant. Determine the following for that instant: crankshaft position; piston position, velocity, and acceleration; and forces applied by the connecting rod to the wrist pin. Consider only inertial effects.

PROJECTS

See Projects 1.1 through 1.6 and suggestions in Chapter 1.

Review the results of the study of acceleration and angular acceleration characteristics of linkages in the design. Plot representative dynamic forces and torques in the linkages. Make use of computer software wherever practical. Check your results by a graphical method for at least one linkage position. Evaluate the linkage in terms of its performance requirements.

BIBLIOGRAPHY AND REFERENCES

Bedford, A., and W. Fowler, *Engineering Mechanics: Dynamics*, Addison-Wesley, Reading, MA, 1995.

Beer, F. P., and E. R. Johnston, Jr., *Vector Mechanics for Engineers: Statics and Dynamics*, McGraw-Hill, New York, 1984.

Dimarogonas, A., *Vibration for Engineers*, 2d ed. Prentice Hall, Upper Saddle River, NJ, 1996.

James, M.L., G.M. Smith, J.C. Wolford, and P.W. Waley, *Vibration of Mechanical and Structural Systems*, Harper and Row, New York, 1989.

Knowledge Revolution, *Working Model*™ *2D User's Manual*, Knowledge Revolution, San Mateo, CA, 1996.

MathSoft, *Mathcad2000*™ *User's Guide*, MathSoft, Inc., Cambridge, MA, 1999.

Meriam, J. L., *Engineering Mechanics: Statics and Dynamics*, Wiley, New York, 1978.

Shames, I. H., *Engineering Mechanics: Statics and Dynamics*, Prentice-Hall, Upper Saddle River, NJ, 1980.

Note also the Internet references listed under the heading of **Software for calculation, design, manufacturing, motion simulation and testing** found at the end of Chapter 1. You will find more sites related to dynamics of machinery by searching the Internet yourself.

C H A P T E R 1 1

Synthesis

Methods and Concepts You Will Learn and Apply When Studying This Chapter

- What to take into account in selecting the type of mechanism and the number of links and joints in a mechanism to fill a specified need.
- Graphical design methods for guiding a link into two specified positions.
- Writing a program to do the preceding.
- Methods for guiding a link into two specified positions when the calculated pivot-point location is inaccessible.
- Analytical and graphical design methods for guiding a link into three specified positions.
- A complex-number method for proportioning a linkage to produce specified velocities and accelerations.
- Design of a function generator to produce a specified output-to-input relationship using the dot-product method.
- A complex-number matrix method for carrying out the preceding task.
- Evaluation and redesign of function generators.
- Design of linkages to produce coupler curves with specified properties.
- Design of a linkage for a specified motion requirement. Selecting and applying a coupler curve for this application.
- Utilizing motion simulation software and computational software to aid in performing the foregoing tasks, to evaluate results, and to test and evaluate redesign decisions.

If the dimensions of a mechanism are given, and we attempt to determine its motion characteristics, the process is called analysis. *Synthesis* is the inverse process: Given a set of performance requirements, we attempt to proportion a mechanism to meet those specifications. Ideally, all mechanisms would be designed by some mathematical synthesis process. However, because of the complexity of real-world machine requirements, ingenuity, judgment, and analysis are still a major part of the design process, while the application of formal synthesis is limited.

Type synthesis, number synthesis, and various forms of dimensional synthesis are described in the paragraphs that follow. Design methods involving less formal synthesis procedures are illustrated throughout the text.

In some situations, the performance requirements and our design decisions reduce a synthesis problem to a simple analysis task. In other cases, the synthesis approach yields many acceptable solutions. Whenever possible, a figure of merit is defined (e.g., minimum cost or weight), and the design is optimized. Then the proposed design is analyzed to confirm that the performance requirements are met.

11.1 TYPE SYNTHESIS

The process of deciding whether to use gears, cams, linkages, or other machine elements to transmit motion is called *type synthesis*. For example, when a precise, constant speed ratio is required, as in driving a camshaft, a gear or chain drive might be selected. A four-bar linkage might be selected to change continuous rotation to oscillation. If we need to change rotational motion into linear (rectilinear) motion, as in a compressor, a slider-crank linkage is likely to be considered. But the slider-crank linkage is also the basis for the piston engine, where the piston is driven by gas forces and the crank is the output link. We would select a cam-and-follower system to change rotational motion into linear motion when a precise input–output relationship is required.

Other considerations in type synthesis include static and inertial loading, wear, reliability, and cost. Type synthesis is largely dependent on the experience and creativity of the designer and may be followed by number synthesis, dimensional synthesis, selection of materials, and processing. As part of the adequacy assessment, analysis is an essential ingredient. Expert-system computer software packages, which are described in recent literature, may also be of interest to the designer.

11.2 NUMBER SYNTHESIS

Number synthesis is the determination of linkage configurations (the number of links and joints) that satisfy given criteria. Number synthesis allows us to consider a variety of linkages for a particular application. Suppose, for example, we wish to identify planar, one-degree-of-freedom linkages with revolute joints. Grübler's criterion (see Chapter 1) relates the number of links and pairs by the relationship

$$2n'_j - 3n_L + 4 = 0, \tag{11.1}$$

from which it follows that

$$n_L = (2n'_J + 4)/3, \tag{11.2}$$

where n_L = the number of links and n'_J = the number of revolute joints. Since the number of joints and the number of links must be integers, the following combinations satisfy the criterion:

n'_J	n_L	
1	2	
4	4	
7	6	
10	8	
13	10,	etc.

The first combination represents two links joined by a revolute (pin) joint. The second is a four-bar linkage. The Watt linkage and the Stephenson linkage (Figure 1.10) are each formed by six links and seven pin joints.

11.3 TWO-POSITION SYNTHESIS

Consider the situation in which a link must assume certain prescribed positions, as, for example, in Figure 11.1. Such requirements are common in the design of materials-handling equipment. If a link must assume three positions in a plane, the link may be guided as the coupler of a four-bar linkage. If only two planar positions are prescribed, the link may be rotated about a single pivot point, or a four-bar linkage may be used. This synthesis problem may be solved analytically or graphically.

Graphical Solution

Figure 11.2 shows a machine member in planar motion with only two positions prescribed. Two points B and C are arbitrarily selected within the member, where the initial and final positions are identified by subscripts 1 and 2, respectively. If point B moves in a circular path between the initial and final positions of the member, the center of that circular path lies on the perpendicular bisector of line B_1B_2. Likewise, if C moves in a circular path, the center lies on the perpendicular bisector of C_1C_2. The link will assume the two prescribed positions if a fixed pivot (revolute joint) is located at the intersection of the perpendicular bisectors, as shown in the figure. If translation in the z direction is required as well, a helical pair (a screw or a helical spline) could be used instead of the revolute joint.

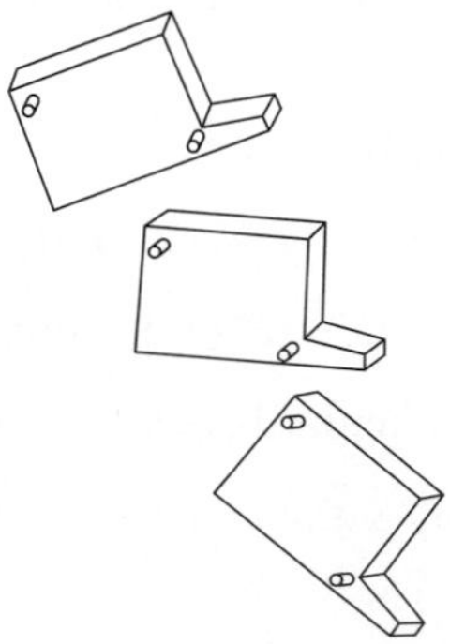

FIGURE 11.1 Design for specified positions.

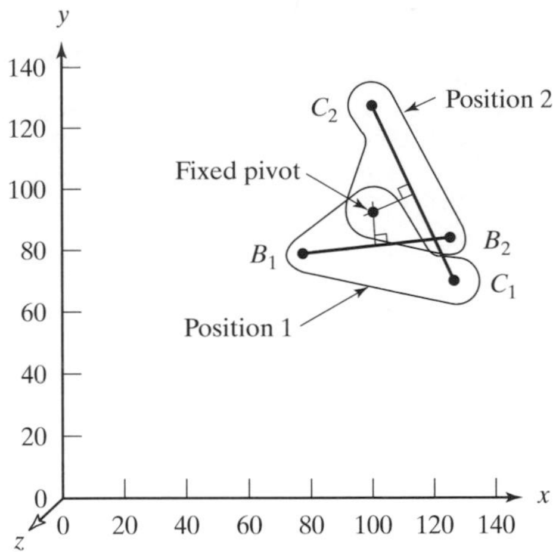

FIGURE 11.2 Two-position problem.

Computer Solution

An analytical solution of the planar two-position synthesis problem is outlined in the flowchart of Figure 11.3 Length L is the distance between points B and C, and angle β is the orientation of line BC. If the angle is input in degrees, it may be changed to radians by multiplying by $PI/180$, where $PI = 4 \times$ arctangent (1). Point C is located by the equations

$$C_x(I) = B_x(I) + L \cos \beta$$

and

$$C_y(I) = B_y(I) + L \sin \beta, \tag{11.3}$$

where the subscripts identify the x- and y-coordinates. It is suggested that input values and all relevant calculated values be printed out as an aid to debugging and to check the results graphically.

The midpoints of lines $B(1)B(2)$ and $C(1)C(2)$ are identified, respectively, by (D_x, D_y) and (E_x, E_y), where

$$D_x = [B_x(1) + B_x(2)]/2, \quad \text{etc.} \tag{11.4}$$

The slopes of the perpendicular bisectors are given by M_B and M_C, where

$$M_B = [B_x(1) - B_x(2)]/[B_y(2) - B_y(1)], \quad \text{etc.} \tag{11.5}$$

We may now write the equations of the perpendicular bisectors in the form

$$y = M_B x + K_B$$

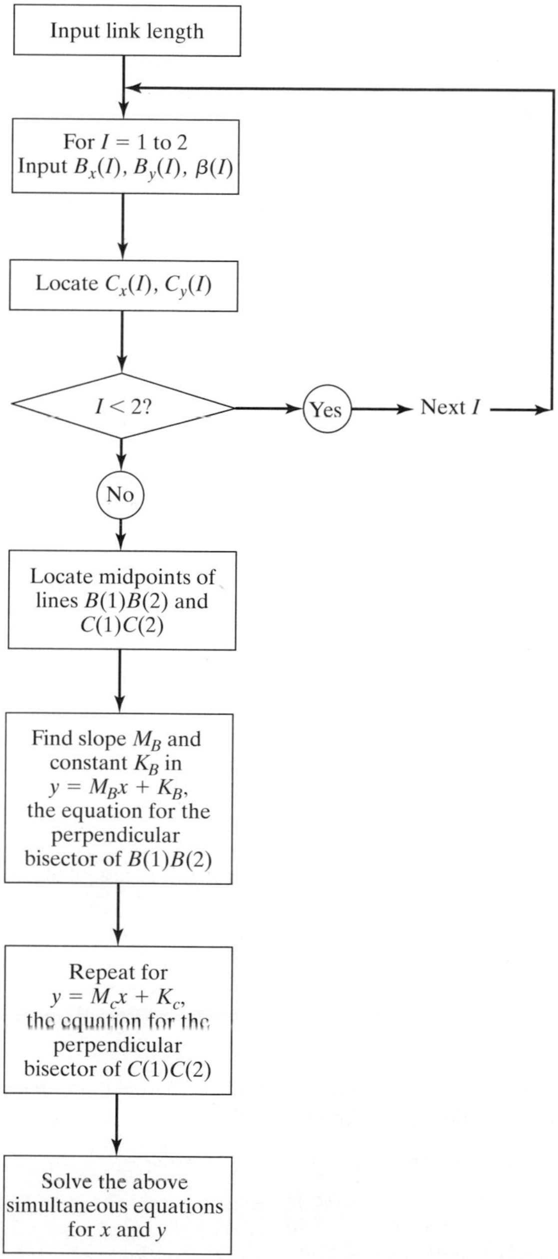

FIGURE 11.3 Flowchart for two-position problem.

and

$$y = M_C x + K_C, \tag{11.6}$$

where the constants are given by

$$K_B = D_y - M_B D_x$$

and

$$K_C = E_y - M_C E_x. \tag{11.7}$$

Solving these equations simultaneously, we find that the location of the fixed pivot point (x, y) is given by

$$x = [K_C - K_B]/[M_B - M_C]$$

and

$$y = M_B[K_C - K_B]/[M_B - M_C] + K_B. \tag{11.8}$$

If one of the selected points moves in the x direction, the perpendicular bisector will have an infinite slope. This special case can be handled by a branch in the program. A final step in the computer program may include drawing line BC in the specified positions and plotting the location of the pivot point.

SAMPLE PROBLEM 11.1

Computer-Aided Two-Position Synthesis

A link identified by points B and C, which are 50 mm apart, is to assume the following positions:

Coordinates	B_x	B_y	Angle β
Position 1	75	80	−10°
Position 2	125	85	125°

Find the location of a fixed pivot that will permit the indicated motion.

Solution. A program based on the flowchart of Figure 11.3 and equations described in the paragraph before this sample problem is used to obtain the following results:

Coordinate	x	y
Point $C(1)$	124.24	71.32
Point $C(2)$	96.32	125.96
Midpoint of $B(1)B(2)$	100	82.5
Midpoint of $C(1)C(2)$	110.28	98.64
Fixed pivot location	98.96	92.86

The results agree with the graphical solution shown in Figure 11.2.

Inaccessible Pivot Point

If the intersection of the perpendicular bisectors of the displacement is inaccessible, or if other design considerations make the use of the intersection as a pivot point impractical, a four-bar linkage may be employed. Fixed revolute joints may be located anywhere along the said perpendicular bisectors. However, transmission angles should be considered when one locates these pivots.

SAMPLE PROBLEM 11.2

Two-Position Synthesis Using a Four-Bar Linkage

A link containing points B and C, which are 75 mm apart, is to assume the following positions:

Coordinates	B_x	B_y	β
Position 1	140	180	0°
Position 2	225	185	15°

Design a linkage that will permit the indicated motion.

Solution. A program based on the flowchart in Figure 11.3 and Eqs. (11.3) through (11.8) gives the following results:

Coordinates	x	y
Midpoint of $B(1)B(2)$	182.5	182.5
Midpoint of $C(1)C(2)$	256.2	192.2
Intersection of perpendicular bisectors	163.5	505.3

	Slope	Orientation
Perpendicular bisector of $B(1)B(2)$	−17	93.37°
Perpendicular bisector of $C(1)C(2)$	−3.377	106.49°

Use of the intersection of the perpendicular bisectors as a pivot point would probably result in an unacceptable design in this case. One possible design is four-bar linkage O_BBCO_C shown in Figure 11.4, where fixed revolute joints O_B and O_C are arbitrarily located along the perpendicular bisectors. (Note that the graphical solution shown in the figure verifies the computer solution.)

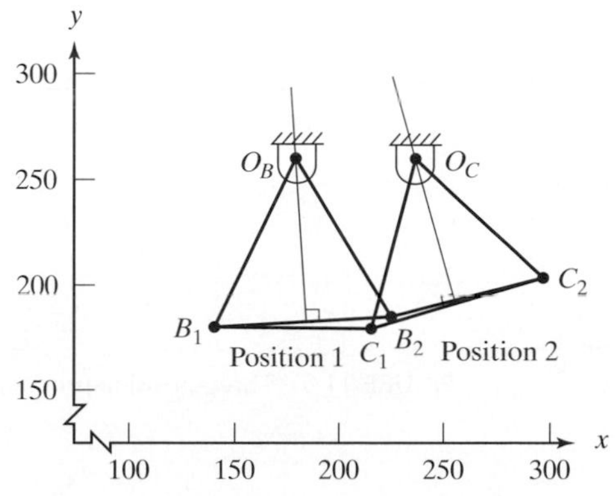

FIGURE 11.4 Two-position problem using a four-bar linkage.

11.4 THREE-POSITION SYNTHESIS USING A FOUR-BAR LINKAGE

If three positions of a link are specified, it will not generally be possible to use a single pivot point. Figure 11.5 shows a link containing points B and C in three positions in a plane, identified by subscripts 1, 2, and 3.

The concept and the design procedure are as follows:

- We will try to design a four-bar linkage with coupler BC that assumes the three required positions: B_1C_1, B_2C_2, and B_3C_3.
- The paths of points B and C will be circular arcs.
- We know from elementary geometry that three points determine a circle.
- The center of the circle lies at the intersection of the perpendicular bisectors of two chords.
- B_1B_2 is one chord of the circular path of point B; B_2B_3 is another.
- The intersection of the perpendicular bisector of B_1B_2 with the perpendicular bisector of B_2B_3 locates fixed pivot O_B of the four-bar linkage.
- Fixed pivot O_C of the four-bar linkage is located similarly, after one locates the required positions of point C.

It may be possible to drive one of the crank links (at O_B or O_C) to produce the required motion. Or the set of required positions may lead to a linkage that works best if the drive motor is located between a crank and the coupler (at point B or point C). In some cases, a four-bar linkage may not be a practical solution to the three-position synthesis problem.

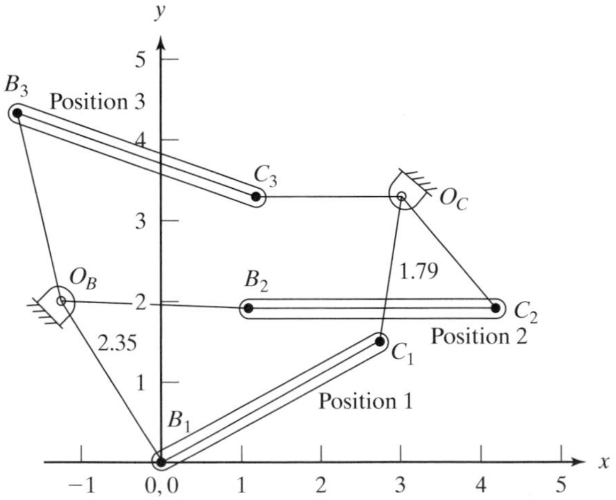

FIGURE 11.5 Three-position problem.

SAMPLE PROBLEM 11.3

Computer-Aided Three-Position Synthesis

A link containing points B and C, which are 3.1 in apart, must assume the following positions:

Coordinates	B_x	B_y	Angle β
Position 1	0	0	28°
2	1.1	1.9	0°
3	−1.7	4.3	−20°

Design a linkage that will satisfy this requirement.

Solution. (See Figure 11.5.) Fixed pivot O_B is located at the intersection of the perpendicular bisectors of chords $B(1)B(2)$ and $B(2)B(3)$. Fixed pivot O_C is similarly located with respect to the position of point C. A computer-aided solution is outlined in the flowchart given in Figure 11.6. The equations represented in the flowchart are similar to those used in two-position synthesis.

11.5 VELOCITY AND ACCELERATION SYNTHESIS BY THE COMPLEX NUMBER METHOD

A four-bar linkage may be proportioned to produce specified values of angular velocity and acceleration for each link, using a procedure developed by Block. (See Rosenauer, 1954.) Unfortunately, applications of this method of synthesis are limited, since, in general, the specified values can be produced for only an instant (i.e., for only one linkage position).

In Bloch's procedure, the links are described by complex numbers. Referring to Figure 11.7, with link 1 represented by vector $\boldsymbol{r}_1$, and so on, we see that the closed polygon formed by the four-bar linkage may be represented by the vector equation

$$\boldsymbol{r}_1 + \boldsymbol{r}_2 + \boldsymbol{r}_3 + \boldsymbol{r}_0 = \boldsymbol{0} \tag{11.9}$$

Putting the fixed-link vector on the right side of the equation and using the complex exponential form $\boldsymbol{r} = re^{j\theta}$, we have

$$r_1e^{j\theta_1} + r_2e^{j\theta_2} + r_3e^{j\theta_3} = -r_0e^{j\theta_0} \tag{11.10}$$

Differentiating with respect to time, we obtain

$$j\omega_1r_1e^{j\theta_1} + j\omega_2r_2e^{j\theta_2} + j\omega_3r_3e^{j\theta_3} = 0, \tag{11.11}$$

where angular velocity $\omega = d\theta/dt$ and $\omega_0 = 0$. Differentiating with respect to time again yields

$$(j\alpha_1 - \omega_1^2)r_1e^{j\theta_1} + (j\alpha_2 - \omega_2^2)r_2e^{j\theta_2} + (j\alpha_3 - \omega_3^2)r_3e^{j\theta_3} = 0, \tag{11.12}$$

where angular acceleration $\alpha = d\omega/dt$.

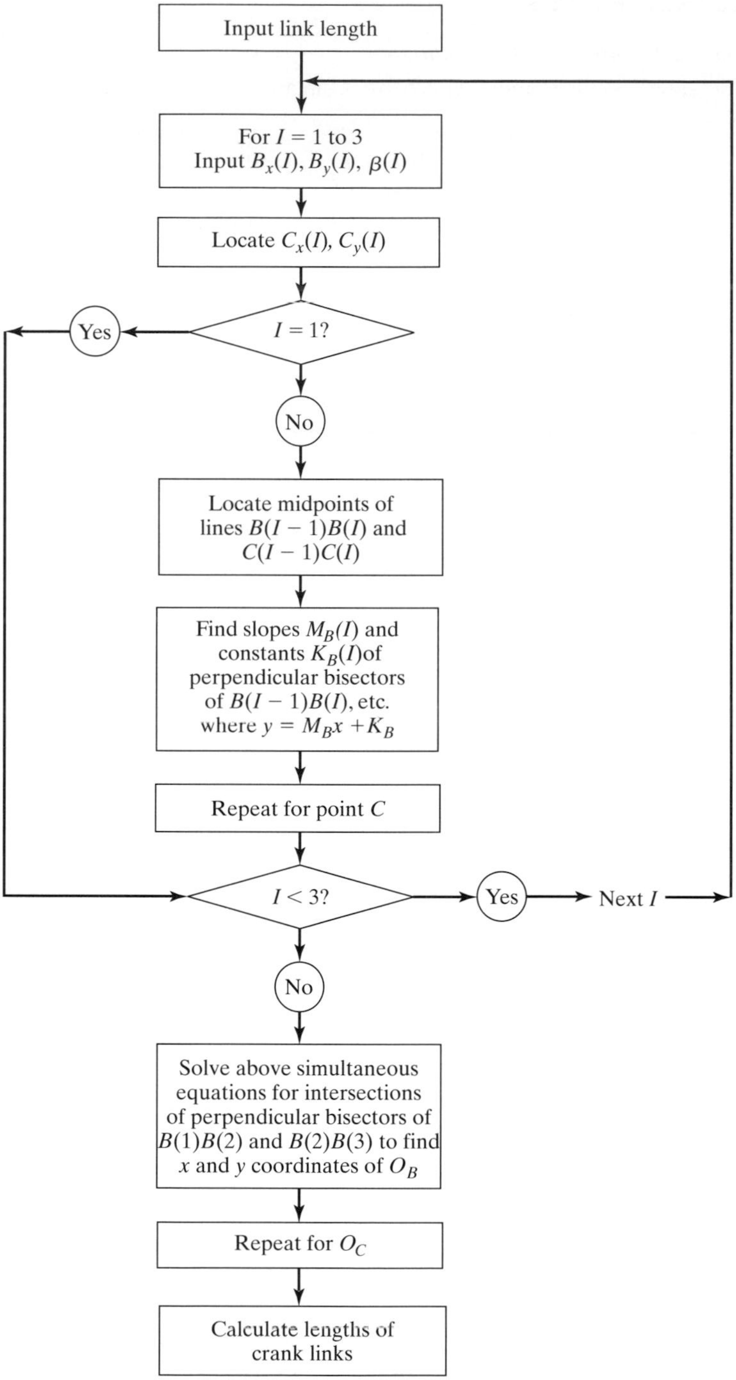

FIGURE 11.6 Flowchart for three-position problem.

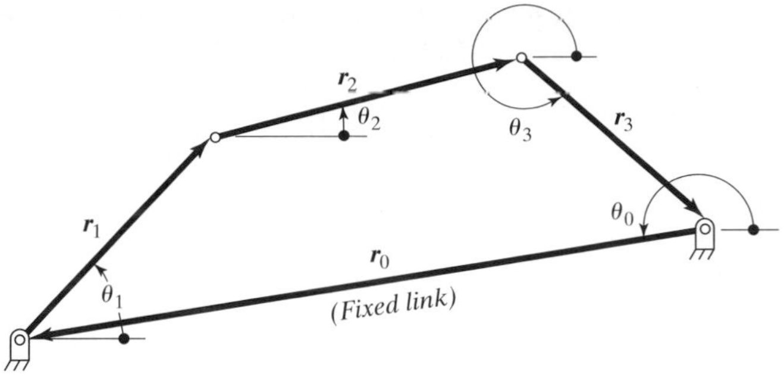

FIGURE 11.7 Mechanism synthesis for specified angular velocities and angular accelerations.

After dividing all terms of Eq. (11.11) by j, we write Eqs. (11.10) through (11.12) in vector form as

$$\mathbf{r}_1 + \mathbf{r}_2 + \mathbf{r}_3 = -\mathbf{r}_0,$$
$$\omega_1\mathbf{r}_1 + \omega_2\mathbf{r}_2 + \omega_3\mathbf{r}_3 = 0,$$

and

$$(j\alpha_1 - \omega_1^2)\mathbf{r}_1 + (j\alpha_2 - \omega_2^2)\mathbf{r}_2 + (j\alpha_3 - \omega_3^2)\mathbf{r}_3 = 0,$$

which may be expressed by the matrix equation

$$\begin{bmatrix} 1 & 1 & 1 \\ \omega_1 & \omega_2 & \omega_3 \\ j\alpha_1 - \omega_1^2 & j\alpha_2 - \omega_2^2 & j\alpha_3 - \omega_3^2 \end{bmatrix} \begin{bmatrix} \mathbf{r}_1 \\ \mathbf{r}_2 \\ \mathbf{r}_3 \end{bmatrix} = \begin{bmatrix} -\mathbf{r}_0 \\ 0 \\ 0 \end{bmatrix} \tag{11.13}$$

The solution of this equation is given in determinant form by

$$\mathbf{r}_1 = \frac{-\mathbf{r}_0}{D} \begin{vmatrix} 1 & 1 & 1 \\ 0 & \omega_2 & \omega_3 \\ 0 & j\alpha_2 - \omega_2^2 & j\alpha_3 - \omega_3^2 \end{vmatrix}$$

and so on, where

$$D = \begin{vmatrix} 1 & 1 & 1 \\ \omega_1 & \omega_2 & \omega_3 \\ j\alpha_1 - \omega_1^2 & j\alpha_2 - \omega_2^2 & j\alpha_3 - \omega_3^2 \end{vmatrix}$$

The term $-\mathbf{r}_0/D$ appears in the equation for $\mathbf{r}_2$ and $\mathbf{r}_3$ as well. For convenience, we may set $-\mathbf{r}_0/D$ equal to unity in each $\mathbf{r}$ equation and still satisfy the specified conditions (given values of ω and α). Each link will change in length by the same proportion, and each angle will change by the same value. The resulting linkage is represented by vectors $\boldsymbol{R}_0$, $\boldsymbol{R}_1$, $\boldsymbol{R}_2$, and $\boldsymbol{R}_3$ (where $\boldsymbol{R}_1 = \boldsymbol{r}_1/(-\boldsymbol{r}_0/D)$, etc.), which are given by the determinants

$$\boldsymbol{R}_1 = \begin{vmatrix} 1 & 1 & 1 \\ 0 & \omega_2 & \omega_3 \\ 0 & j\alpha_2 - \omega_2^2 & j\alpha_3 - \omega_3^2 \end{vmatrix}, \tag{11.14}$$

$$\boldsymbol{R}_2 = \begin{vmatrix} 1 & 1 & 1 \\ \omega_1 & 0 & \omega_3 \\ j\alpha_1 - \omega_1^2 & 0 & j\alpha_3 - \omega_3^2 \end{vmatrix}, \tag{11.15}$$

and

$$\boldsymbol{R}_3 = \begin{vmatrix} 1 & 1 & 1 \\ \omega_1 & \omega_2 & 0 \\ j\alpha_1 - \omega_1^2 & j\alpha_2 - \omega_2^2 & 0 \end{vmatrix}, \tag{11.16}$$

with

$$\boldsymbol{R}_0 = -\boldsymbol{R}_1 - \boldsymbol{R}_2 - \boldsymbol{R}_3. \tag{11.17}$$

SAMPLE PROBLEM 11.4

Velocity and Acceleration Synthesis

Let us examine the synthesis of a linkage for specified angular velocities and accelerations. The values are as follows:

Link:	0 (fixed)	1	2	3
ω (rad/s):	0	2	0	1
α(rad/s^2):	0	0	1	1

Specify a linkage that (instantaneously) satisfies these values.

Solution. We have

$$\boldsymbol{R}_1 = \begin{vmatrix} 1 & 1 & 1 \\ 0 & 0 & 1 \\ 0 & j & j-1 \end{vmatrix} = -j = e^{-j\pi/2},$$

$$\boldsymbol{R}_2 = \begin{vmatrix} 1 & 1 & 1 \\ 2 & 0 & 1 \\ -4 & 0 & j-1 \end{vmatrix} = -2 - j2 = 2.828e^{-j2.356},$$

$$\boldsymbol{R}_3 = \begin{vmatrix} 1 & 1 & 1 \\ 2 & 0 & 0 \\ -4 & j & 0 \end{vmatrix} = j2 = 2e^{-j\pi/2},$$

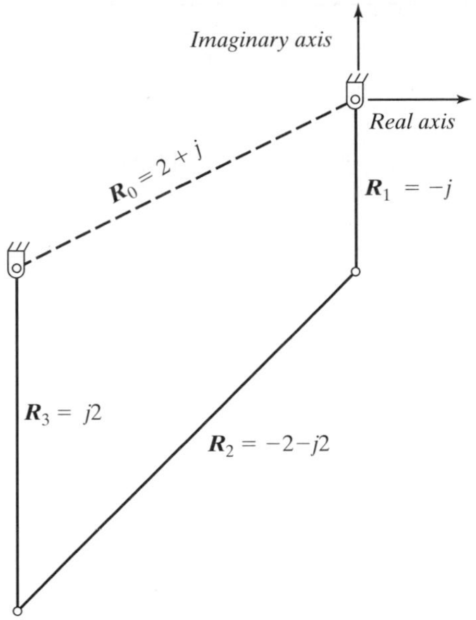

FIGURE 11.8 Synthesis solution to Sample Problem 11.4.

and

$$\boldsymbol{R}_0 = j + 2 + j2 - j2 = 2 + j = 2.236e^{j0.464}.$$

The resulting linkage is sketched in Figure 11.8. This mechanism will satisfy the required conditions only when the links pass through the relative position shown.

An Alternative Solution Using the Matrix Inverse. Let the matrix equation (11.13) be represented by

$$\boldsymbol{AX} = \boldsymbol{B}, \tag{11.18}$$

where

$$\boldsymbol{X} = \begin{bmatrix} \boldsymbol{r}_1 \\ \boldsymbol{r}_2 \\ \boldsymbol{r}_3 \end{bmatrix}, \quad \text{etc.}$$

Multiplying both sides of the equation by $\boldsymbol{A}^{-1}$, the inverse of matrix $\boldsymbol{A}$, yields

$$\boldsymbol{X} = \boldsymbol{A}^{-1}\boldsymbol{B}. \tag{11.19}$$

This form of solution is most convenient for computer calculations. A program for multiplication and inversion of complex matrices may be written by the user or purchased as part of a commercial software package.

SAMPLE PROBLEM 11.5

Computer Solution Using the Matrix Inverse

Use the matrix inverse method to solve Sample Problem 11.4.

Solution. It can be seen from the preceding results that one link vector may be chosen arbitrarily. For convenience, we let $\boldsymbol{r}_0 = 1$. Using a software package to solve Eq. (11.19), we obtain $\boldsymbol{X}$, from which it follows that $\boldsymbol{r}_1 = -0.2 - j0.4$, $\boldsymbol{r}_2 = -1.2 - j0.4$, and $\boldsymbol{r}_3 = 0.4 + j0.8$. The proportions of the linkage are the same as in the previous solution, but the size and orientation are different. If each link vector is multiplied by $2 + j$, then the results are identical to those of the earlier sample problem.

11.6 DESIGN OF A FUNCTION GENERATOR: DOT-PRODUCT METHOD

Four-bar linkages may be used as function generators. Their low friction and higher load capacity make them preferable to cams for certain applications. An important disadvantage of the four-bar-linkage function generator, however, is its inability to represent an arbitrary function exactly, except at a few points, called *precision points*. The range of input and output motion is further limited by the limiting positions of the linkage itself and sometimes by problems associated with mechanical advantage or the transmission angle.

Consider the problem of designing a four-bar linkage so that output angle ϕ is a specified function of input angle θ, where links and angles are identified as in Figure 11.9. Each of the four links of the mechanism is considered a vector, as shown in part b of the figure. The coupler link vector is equated to the vector sum of the other three link vectors, and a dot or scalar product is formed from the vectors on both sides of the equation. Thus, we have the *vector* relationship

$$-\boldsymbol{r}_2 = \boldsymbol{r}_1 + \boldsymbol{r}_0 + \boldsymbol{r}_3 \tag{11.20}$$

and

$$\boldsymbol{r}_2 \cdot \boldsymbol{r}_2 = r_2^2 = (\boldsymbol{r}_1 + \boldsymbol{r}_0 + \boldsymbol{r}_3) \cdot (\boldsymbol{r}_1 + \boldsymbol{r}_0 + \boldsymbol{r}_3), \tag{11.21}$$

where we have formed the dot product of each side of Eq. (11.20) with itself. Freudenstein (1955) developed the dot-product method to obtain linkage dimensions of a function generator by using three, four, or five precision points. Freudenstein's three-point approximation is obtained as follows:

1. Select three input angles θ_1, θ_2, and θ_3 anid the corresponding output angles ϕ_1, ϕ_2, and ϕ_3. Compute each output angle so that the required relationship between θ and ϕ is satisfactory at each of the three precision points for the prescribed function $\phi = f(\theta)$.

   Figure 11.10 shows input link 1 and output link 3 at the three precision points. The desired relationship between θ and ϕ is achieved exactly at (θ_1, ϕ_1),(θ_2, ϕ_2), and (θ_3, ϕ_3) and satisfied approximately at other values of θ and ϕ.

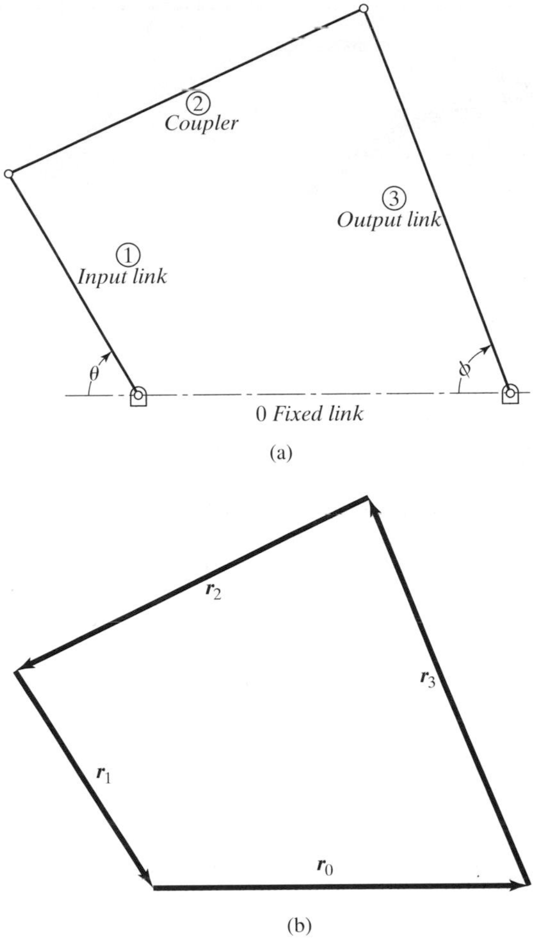

FIGURE 11.9 (a) Design of a function generator. (b) Vectors representing the linkage.

2. In order to simplify the expressions for the lengths of links 1 and 3, compute the following angle relationships, which will be used shortly.

$$A = \cos\theta_1 - \cos\theta_2;$$
$$B = \cos\theta_1 - \cos\theta_2;$$
$$C = \cos\phi_1 - \cos\phi_2;$$
$$D = \cos\phi_1 - \cos\phi_3;$$
$$E = \cos(\theta_1 - \phi_1) - \cos(\theta_2 - \phi_2);$$
$$F = \cos(\theta_1 - \phi_1) - \cos(\theta_3 - \phi_3).$$

3. If the fixed link is arbitrarily given a length r_0, the lengths of the input and output cranks are, respectively,

$$r_1 = \frac{BC - AD}{AF - BE} r_0 \quad \text{and} \quad r_3 = \frac{BC - AD}{CF - DE} r_0. \tag{11.22}$$

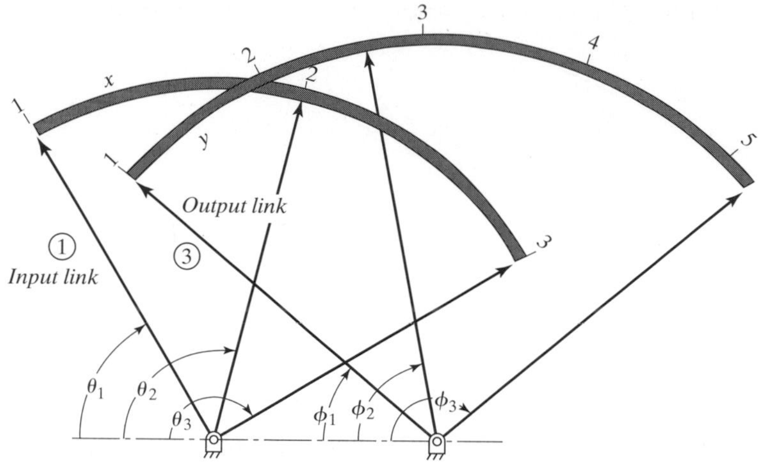

FIGURE 11.10 Proposed function generator. The input and output links of this four-bar linkage are to generate the function y (output) $= x^{1.5}$ for values of x (input) between 1 and 3. The coupler (link 2) is not shown. The sketch is not to scale, since link dimensions are yet to be determined.

If r_1 is found to be negative, link 1 is drawn in a direction opposite that shown in Figure 11.10. Similarly, if r_3 is negative, its direction is reversed.

SAMPLE PROBLEM 11.6

Design of a Function Generator

Design a four-bar linkage to generate the function $y = x^{1.5}$ for values of x between 1 and 3.

Solution.

STEP 1. Angular displacement θ of the input crank will be proportional to x, and angular displacement ϕ of the output crank will be proportional to y. At our disposal, we have the initial values of θ and ϕ, which will be taken as $\theta_0 = 60°$ and $\phi_0 = 40°$. We are also free to select the ranges of θ and ϕ, which will both be 90° for this example. (See Figure 11.10.) Measured from the initial values, the change in θ is proportional to the change in x, or, when $x_0 = 1$,

$$\theta = \theta_0 + \frac{\text{range of } \theta}{\text{range of } x}(x - x_0) = 15 + 45x. \tag{11.23}$$

Similarly,

$$\phi = \phi_0 + \frac{\text{range of } \phi}{\text{range of } y}(y - y_0). \tag{11.24}$$

As x varies from 1 to 3, y varies from $(1)^{1.5} = 1$ to $(3)^{1.5} = 5.196$, from which we obtain output crank angle

$$\phi = 18.55 + 21.45y \quad \text{(both } \theta \text{ and } \phi \text{ measured in degrees).} \tag{11.25}$$

STEP 2. For convenience, the precision points will be selected as

$$x_1 = 1, x_2 = 2, \text{and } x_3 = 3.$$

The preceding equations give the corresponding values of y, θ, and ϕ:

Point	x	y	θ	ϕ
1	1.000	1.000	60°	40°
2	2.000	2.828	105	79.22
3	3.000	5.196	150	130

STEP 3. Using these values in Eq. (11.22) and arbitrarily letting $r_0 = 2$ in, we calculate the lengths of the input and output cranks:

$$r_1 = 10.38 \text{ in} \quad \text{and} \quad r_3 = 10.06 \text{ in}$$

The coupler length $r_2 = 2.57$ in may be found analytically or simply by drawing the input and output cranks in positions θ_1 and ϕ_1, respectively. The positions for the other precision points are also sketched to check for limiting positions, as in Figure 11.11. It can be seen that the limiting positions do not fall within the range of operation of the linkage, making the design satisfactory from that standpoint. If the linkage proportions are not acceptable, the designer would try a new set of initial values or new ranges for θ and ϕ.

STEP 4. In order to transform crank angle ϕ to output y, we rewrite Eq. (11.25) to obtain

$$y = \frac{\phi - 18.55}{21.45},$$

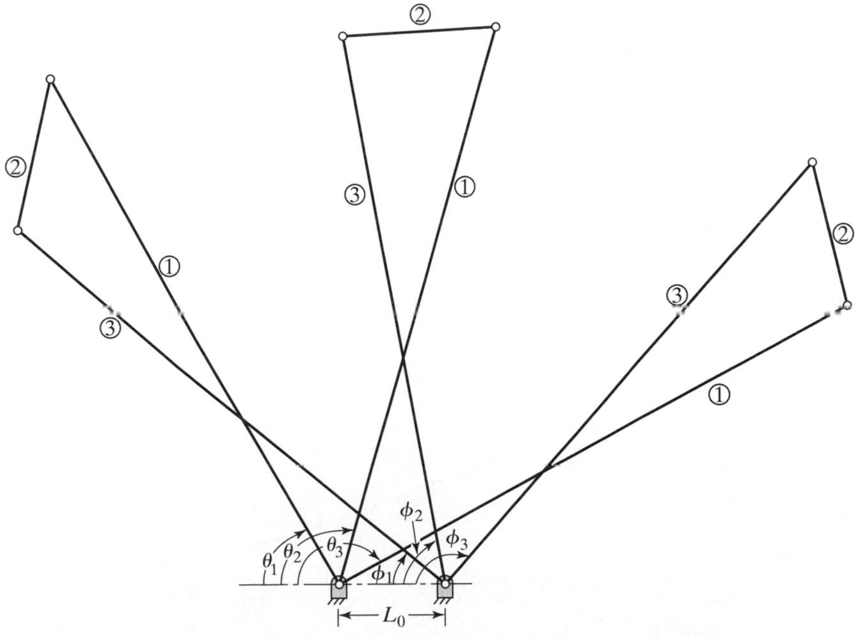

FIGURE 11.11 A linkage designed to approximate the relationship $y = x^{1.5}$.

where ϕ is measured in degrees. The linkage generates the exact function $y = x^{1.5}$ for the precision points $x = 1, 2,$ and 3 (corresponding to $\theta = 60°, 105°,$ and $150°$). An error in the generated function for any other value of x depends on the initial values and ranges of θ and ϕ, as well as on the particular value of x.

A computer solution allows the designer of a function generator to try various design changes when link ratios or other parameters are unsatisfactory. A computer solution also makes it practical to evaluate the function generator as it moves through its range of positions. The solution to this sample problem was made easier by using a spreadsheet that incorporated the position analysis equations of a four-bar linkage. Figure 11.12a shows the value of y (as generated), compared with the ideal value ($y = x^{1.5}$). The greatest error is about 1.5%. The error is, of course, zero at the precision points corresponding to $x = 1, 2,$ and 3. Figure 11.12b shows output angle ϕ and the transmission angle plotted against input angle θ. The transmission angle ranges from about 120° to 50° as the input crank moves within its 60° to 150° operating range. Transmission angles in the

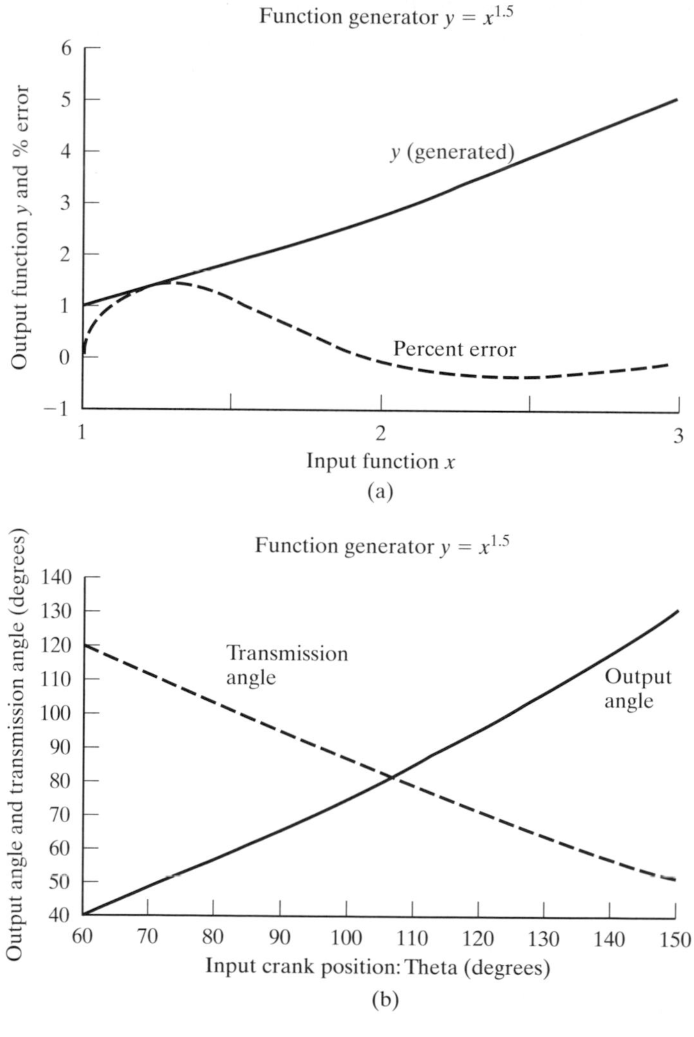

FIGURE 11.12 Function generator $x^{1.5}$. (a) Generated value of y and percent error. (b) Output angle and transmission angle.

range from about 40° or 45° to 135° or 140° are generally considered acceptable. Thus, the function generator is satisfactory from the standpoint of the transmission angle.

Selection of Precision Points

The output of a function generator will differ from the desired function, due to mechanical error and structural error. Mechanical error results from tolerances on the link length, bearing clearance, and other characteristics of the actual components of the mechanism. Structural error results from the inability of a four-bar linkage to generate an arbitrary function precisely, except at the precision points (although certain functions can be generated precisely over a continuous range of values).

In the previous example, the beginning, midpoint, and end of the range were selected as precision points. Other selections may be made in an attempt to reduce structural error (See Freudenstein 1955, 1959.) One method of selection, called Chebychev spacing, is described by Hinkle (1960). For three precision points, the values would be

$$x_1 = x_{\text{mid}} - \frac{1}{2} x_{\text{range}} \cos 30°,$$

$$x_2 = x_{\text{mid}}, \tag{11.26}$$

and

$$x_3 = x_{\text{mid}} + \frac{1}{2} x_{\text{range}} \cos 30°,$$

where x_{mid} is the mean value of x in the range of the function generator.

SAMPLE PROBLEM 11.7

Design of a Function Generator Using Chebychev Spacing

Design a four-bar linkage to generate the function $y = e^x - x$ for values of x between $x_0 = 0$ and $x_f = 1$ (corresponding to $y_0 = 1$ and $y_f = 1.7183$).

Solution. For $x_{\text{range}} = 1$ and $x_{\text{mid}} = 0.5$, we obtain

$$x_1 = 0.5 - \frac{1}{2} \cos 30° = 0.0670,$$

$$y_1 = e^{x_1} - x_1 = 1.0023,$$

and

$$x_2 = 0.5.$$

$$y_2 = 1.1487$$

$$x_3 = 0.5 + \frac{1}{2} \cos 30° = 0.9330$$

$$y_3 = 1.6091$$

We select the range of θ from $\theta_0 = 65°$ to $\theta_2 = 125°$ and the (approximate) range of ϕ from $\phi_0 = 40°$ to $\phi_2 = 80°$. For changes in θ and ϕ proportional to changes in x and y, respectively, we have

$$\theta = \theta_0 + \frac{\text{range of } \theta}{\text{range of } x}(x - x_0)$$

$$= 65° + \frac{125° - 65°}{1 - 0}(x - 0),$$

or

$$\theta(\text{degrees}) = 60x + 65°,$$

and

$$\phi = \phi_0 + \frac{\text{range of } \phi}{\text{range of } y}(y - y_0)$$

$$= 40° + \frac{80° - 40°}{1.7183 - 1}(y - 1),$$

or

$$\phi(\text{degrees}) = 51.086y - 11.086°.$$

The values of x, y, θ, and ϕ are as follows:

Position	x	$y = e^x - x$	θ	ϕ
0	0	1(ideal)	65°	40°(ideal)
1(precision point)	0.0670	1.0023	69.02°	40.13°
2(precision point)	0.5	1.1487	95°	48.28°
3(precision point)	0.9330	1.6091	120.98°	73.92°
f	1	1.7183(ideal)	125°	80°(ideal)

Equations (11.22) are used to compute link lengths (where any convenient value of r_0 may be selected). The results are $r_1/r_0 = 1.65$ and $r_3/r_0 = 1.78$, from which it follows that $r_2/r_0 = 0.45$.

We note that the sum of the lengths of the longest and shortest links is less than the sum of the other link lengths and that the coupler is shortest. Thus, on the basis of the Grashof criterion, the proposed linkage is a double rocker. Figure 11.13 shows the linkage in positions corresponding to the three precision points. Table 11.1 of input and output values tells us that something is wrong: Limiting positions of the proposed linkage do not permit an input crank angle of 65°; the software returns "ERR" instead of the desired output values.

When a function generator design fails, we have a number of alternatives. If there are to be three precision points, then the beginning and ending values of θ and ϕ are at our disposal. We can also use a different scheme to select precision points. In this example, the nature of the function

$$y = e^x - x \quad \text{for} \quad 0 \leq x \leq 1$$

is also a problem, in that

$$dy/dx = 0 \quad \text{at} \quad x = 0$$

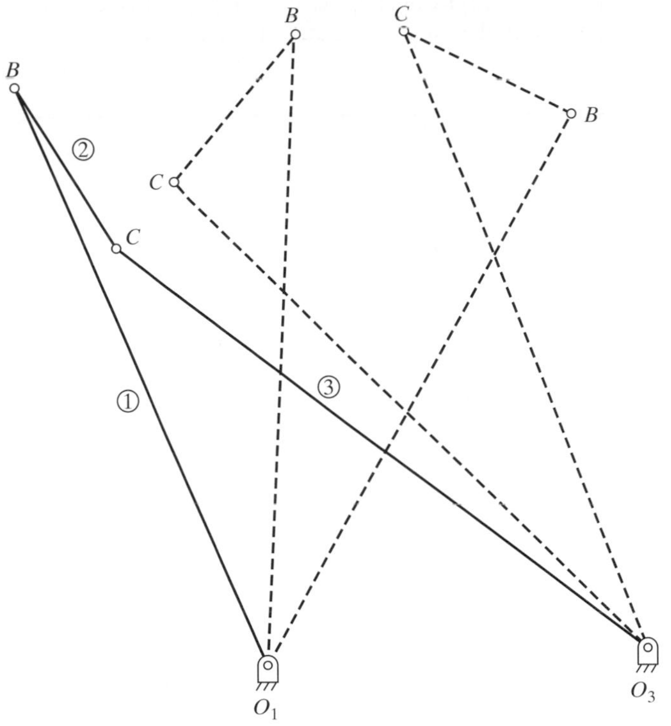

FIGURE 11.13 A linkage designed to approximate the relationship $y = e^x - x$.

TABLE 11.1 The effect of limiting positions on the performance of a function generator

Theta, deg	Phi, deg	x	y	$e^x - x$	Error %
65	ERR	0.000	ERR	1.000	ERR
70	39.811	0.083	0.997	1.004	−0.69
75	39.924	0.167	0.999	1.015	−1.58
80	41.198	0.250	1.022	1.034	−1.21
85	43.085	0.333	1.055	1.062	−0.65
90	45.455	0.417	1.098	1.100	−0.21
95	48.282	0.500	1.149	1.149	−0.00
100	51.595	0.583	1.208	1.209	−0.04
105	55.465	0.667	1.278	1.281	−0.26
110	60.018	0.750	1.359	1.367	−0.55
115	65.480	0.833	1.458	1.468	−0.69
120	72.327	0.917	1.581	1.584	−0.24
125	82.365	1.000	1.761	1.718	2.47

This condition would (ideally) require link 3 to remain stationary when link 1 begins to move. Thus, we might consider excluding $x = 0$ from the range of operation of the function generator. An alternative is to use a different type of configuration—a cam, for example—to generate the function.

11.7 DESIGNING A FUNCTION GENERATOR: COMPLEX MATRIX METHOD

In Section 11.5, a complex-number method was used to design a four-bar linkage with specified angular velocities and accelerations. Although velocity and acceleration synthesis has little practical value, complex-number synthesis is a useful and powerful tool when applied to a design of a function generator. In this application, output crank positions must be related to input crank positions by a specified function. The specifications for the function generator can be the same as in the foregoing paragraph, but the solution involves a complex matrix instead of the dot-product method. Again, three precision points will be used.

Suppose we wish to control a process with a four-bar linkage function generator. Can we design a linkage that will produce output crank positions proportional to any specified function $y(x)$, where input crank positions are proportional to x between x_0 and $x_0 + x_{\text{range}}$? Here are some steps that may lead to a satisfactory function generator:

- First, develop the complex-matrix equation. Consider a four-bar linkage like the one in Figure 11.7, except that fixed link position $\theta_0 = 0$.
- The closed vector loop is represented by Eq. (11.9), and, in complex polar form, by Eq. (11.10).
- The latter equation becomes three equations, one for each precision point.
- Since we are dealing with angles, relative link proportions matter, but we do not care about linkage scale at this step. We divide the terms in each equation by the term representing the fixed link.
- Finally, we express the result in complex-matrix form:

$$\mathbf{AR} = \mathbf{B}$$

In this equation,

$$\mathbf{A} = \begin{bmatrix} 1 & 1 & 1 \\ e^{j\theta_{1a}} & e^{j\theta_{2a}} & e^{j\theta_{3a}} \\ e^{j\theta_{1b}} & e^{j\theta_{2b}} & e^{j\theta_{3b}} \end{bmatrix},$$

$$\mathbf{R} = \begin{bmatrix} \boldsymbol{r}_1/\boldsymbol{r}_0 \\ \boldsymbol{r}_2/\boldsymbol{r}_0 \\ \boldsymbol{r}_3/\boldsymbol{r}_0 \end{bmatrix},$$

and

$$\mathbf{B} = \begin{bmatrix} -1 \\ -1 \\ -1 \end{bmatrix},$$

where $\boldsymbol{r}$ = link position vectors at first precision point (in complex form), θ = link position angles, subscripts 0, 1, 2, and 3 refer, respectively, to the fixed link, input crank, coupler, and output link, a refers to the change in angle from the first to the second precision point, and b refers to the change in angle from the first to the third precision point.

Using the matrix equation we just developed, we can try to design a function generator in accordance with the following steps:

- Select a reasonable range of motion for the input and output links.
- Select three precision points: x_1, x_2, and x_3. These points, can be the minimum, midrange, and maximum values of input variable x, or another selection procedure, such as Chebychev spacing, can be used.
- For the input crank, calculate the change in angle from the first to the second precision point:

$$\theta_{1a} = (x_2 - x_1)(\theta_{1\,\text{range}}/x_{\text{range}}).$$

- Calculate the change in angle from the first to the third precision point:

$$\theta_{1b} = (x_3 - x_1)(\theta_{1\,\text{range}}/x_{\text{range}}).$$

- Calculate the corresponding values for the output crank motion, where $y = y(x)$:

$$\theta_{3a} = [y(x_2) - y(x_1)](\theta_{3\,\text{range}}/y_{\text{range}});$$

$$\theta_{3b} = [y(x_3) - y(x_1)](\theta_{3\,\text{range}}/y_{\text{range}}).$$

- Assume values for the corresponding changes in angle for the coupler:

$$\theta_{2a} \quad \text{and} \quad \theta_{2b}.$$

- Calculate the complex link ratio vector:

$$\mathbf{R} = \mathbf{A}^{-1}\mathbf{B}$$

- Multiply all three complex components of **R** by the same real number to obtain a function generator of any desired size.

- Analyze the four-bar linkage through its range of motion. Check the motion of each link and check the transmission angle.
- Evaluate the linkage as a function generator. Compare the generated values of the output variable with exact (ideal) values of $y(x)$.
- Use the results to improve the design. You have the precision points, the angular range of input and output links, and the changes in angle of the coupler at your disposal.
- Note that the answer to the question "Can you design a linkage [to generate] *any* specified function $y(x)$ . . . " is "NO": Your function generator may be unsatisfactory because you did not try hard enough. Or it may be that *no* four-bar linkage function generator is suitable for generating the specified function over the required range.

SAMPLE PROBLEM 11.8

Using a Complex Matrix to Design a Function Generator

Design a system to produce output rotation proportional to

$$y = x^{-1/2} \cdot e^x$$

for input rotation proportional to x, where $1 \leq x \leq 5$.

Design decisions. We will try a four-bar linkage function generator with input link range $\theta_{1\text{range}} = 45°$ and output link range $\theta_{3\text{range}} = 40°$. The fixed link will be 33 mm long.

Solution summary. Precision points are selected at the beginning, the midpoint, and the end of the range of x. For the first trial, it is assumed that the coupler rotates through an angle of 0.3 radian between the first and second precision points, and 0.6 radian between the first and third precision points. Matrix $\boldsymbol{A}$ and column matrix $\boldsymbol{B}$ are formed, and the equation is solved for the complex link length ratios. Multiplying each by -33 mm, we can describe each link in complex form at the first precision point. The sum of the link vectors is zero, plus or minus a rounding error tolerance.

The cross-product method (described in Chapter 2) is then used to obtain link positions through the full range of motion of the system. We see that the counterclockwise configuration applies by comparing values of θ_3 at the first precision point. As expected, the linkage produces accurate values at the three precision points. Unfortunately, it performs very poorly as a function generator, producing an error of about -72% midway between the first two precision points.

For the second trial, we specify a coupler rotation of 0.1 radian between the first and second precision points, and 0.25 radian between the first and third precision points. The performance is improved: The error is now about -38% midway between the first two precision points. For this trial, the transmission angle could be a problem if the output link works against a significant load. Results of the second trial are shown in the detailed solution and are plotted in Figure 11.14a and 11.14b.

Note that the form of the detailed solution allows for changing the precision points. For example, x_1, the first precision point, might differ from x_0, the beginning of the range. The detailed solution also contains many redundant calculations that are used to check the linkage for closure and as an attempt to detect errors. If we were to make additional trials, both the coupler rotation and the precision points might be changed. Clearly, designing of a satisfactory function generator is likely to require many trials.

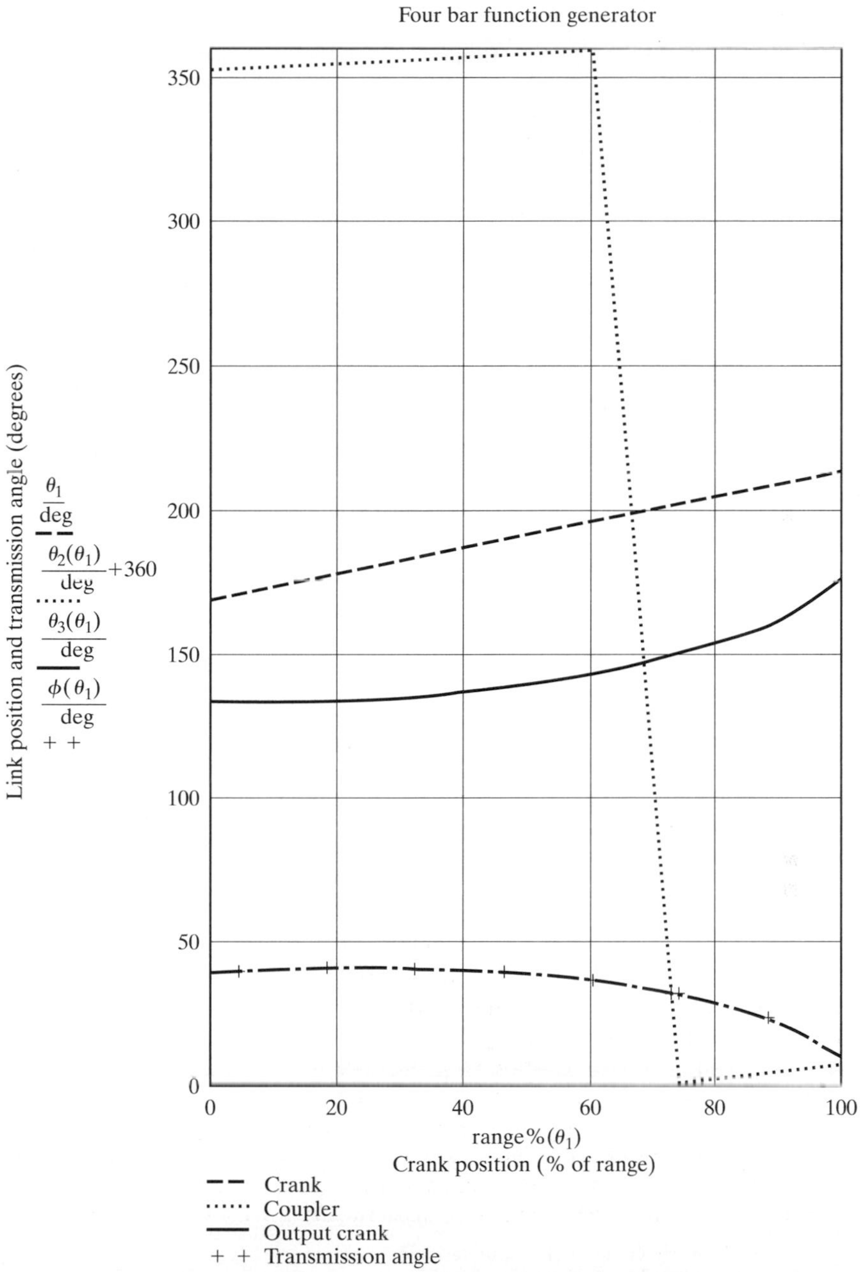

FIGURE 11.14 (a) Design and evaluation of a four-bar function generator. Link positions and transmission angle plotted against crank position (percent of crank range of motion).

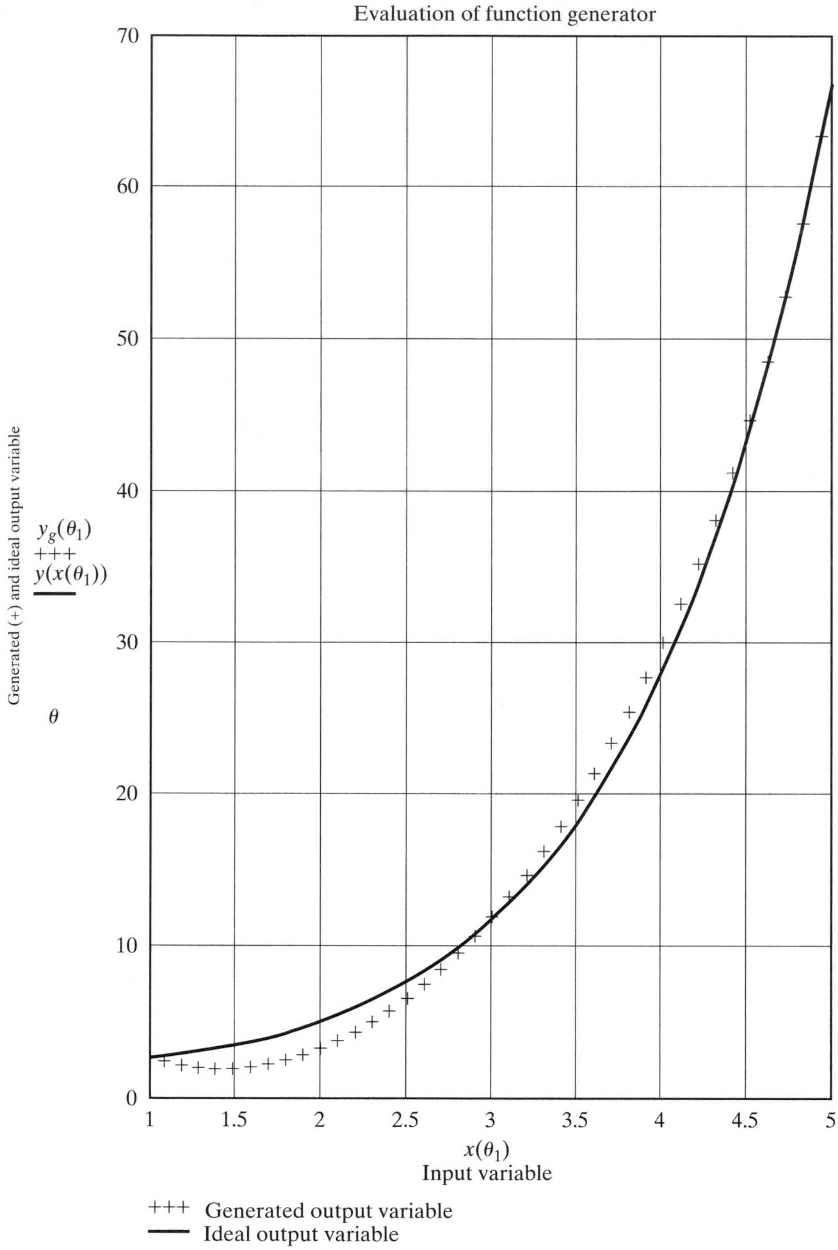

FIGURE 11.14 (b) Generated and ideal output variable plotted against input variable.

Detailed calculation. Design of a four-bar function generator using a complex-number method
$j := (-1)^{.5}$

Closed vector loop $r_0 + r_1 + r_2 + r_3 = 0$

Output function y in terms of input function x		$y(x) := x^{-0.5} \cdot \exp(x)$
Input values: start and range	$x_0 := 1$	$x_{range} := 4$
Ideal output	$y(x_0) = 2.718$	$y(x_0 + x_{range}) = 66.372$
Range of input and output links	$\theta_{1\,range} := 45 \cdot deg$	$\theta_{3\,range} := 40 \cdot deg$
Range of output variable:	$y_{range} := y(x_0 + x_{range}) - y(x_0)$	$y_{range} = 63.654$

Selection of precision points:

$$x_{mid} := \frac{x_{range}}{2} + x_0 \qquad x_1 := x_0$$

$$x_2 := x_{mid} \qquad x_3 := x_0 + x_{range}$$

$$x_1 = 1 \qquad x_2 = 3 \qquad x_3 = 5$$

Angle change from first to second precision point

Input crank $\theta 1a := \dfrac{\theta_{1\,range}}{x_{range}}(x_{mid} - x_1)$ $\quad \theta 1a = 0.393 \quad \dfrac{\theta 1a}{deg} = 22.5$

Coupler-assume value $\theta 2\,a := 1$

Output crank $\theta 3a := \dfrac{\theta_{3\,range}}{y_{range}}(y(x_{mid}) - y(x_1))$ $\quad \theta 3a = 0.097$

Angle change from first to third precision point

Input crank $\theta 1b := \dfrac{\theta_{1\,range}}{x_{range}} \cdot (x_3 - x_1)$ $\quad \theta 1b = 0.785$

Coupler-assume value $\theta 2b := .25$

Output crank $\theta 3b := \dfrac{\theta_{3\,range}}{y_{range}} \cdot (y(x_3) - y(x_1))$ $\quad \theta 3b = 0.698 \quad \dfrac{\theta 3b}{deg} = 40$

Position matrix $A := \begin{bmatrix} 1 & 1 & 1 \\ \exp(j \cdot \theta 1a) & \exp(j \cdot \theta 2a) & \exp(j \cdot \theta 3a) \\ \exp(j \cdot \theta 1b) & \exp(j \cdot \theta 2b) & \exp(j \cdot \theta 3b) \end{bmatrix} \quad B := \begin{bmatrix} -1 \\ -1 \\ -1 \end{bmatrix}$

Complex link ratio vector $R := A^{-1} \cdot B \quad R = \begin{bmatrix} 0.337 - 0.068j \\ -1.46 + 0.197j \\ 0.123 - 0.128j \end{bmatrix}$

Links in complex form at first precision point where fixed link $r_0 := -33$:

$r_1 := r_0 \cdot R_0$	$r_2 := r_0 \cdot R_1$	$r_3 := r_0 \cdot R_2$
$r_1 = -11.123 + 2.25j$	$r_2 = 48.177 - 6.485j$	$r_3 = -4.054 + 4.235j$
$\lvert r_1 \rvert = 11.348$	$\lvert r_2 \rvert = 48.612$	$\lvert r_3 \rvert = 5.863$ mm

$\theta_{11} := \arg(r_1)$ $\quad\theta_{21} := \arg(r_2)$ $\quad\theta_{31} := \arg(r_3)$ rad

$\theta_{11} := 2.942$ $\quad\theta_{21} := -0.134$ $\quad\theta_{31} := 2.334$

$$\frac{\theta_{11}}{deg} = 168{:}566 \qquad \frac{\theta_{21}}{deg} = -7.666 \qquad \frac{\theta_{31}}{deg} = 133.748$$

Locate fixed drive crank bearing 01 at 0,0; fixed driven crank bearing 03 at $-r_0$; crank pin B at r_1; and end C of coupler at $r_1 + r_2$, where

$$r_1 + r_2 = 37.054 - 4.235j$$

Check link closure: $\quad r_0 + r_1 + r_2 + r_3 = -4.441 \cdot 10^{-15} - 1.51 \cdot 10^{-14}j$

Input crank position at beginning of range

$$\theta_{10} := -\theta_{1\,range}\frac{x_1 - x_0}{x_{range}} + \theta_{11} \qquad \theta_{10} = 2.942 \qquad \frac{\theta_{10}}{deg} = 168.566$$

Position analysis: Cross-product method. Link lengths: L. Link vectors: r

$$L_0 := -r_0 \quad L_1 := |r_1| \quad L_2 := |r_2| \quad L_3 := |r_3| \quad \text{mm}$$

Rotate drive crank through range of motion $\theta_1 := \theta_{10}, \theta_{10} + \frac{\pi}{720} .. \theta_{10} + \theta_{1\,range}$

$$\text{range}\%(\theta_1) := \frac{(\theta_1 - \theta_{10})}{\theta_{1\,range}} \cdot 100$$

Redefine links as vectors

$$r_0 := \begin{bmatrix} -L_0 \\ 0 \\ 0 \end{bmatrix} \quad r_0 := \begin{bmatrix} -33 \\ 0 \\ 0 \end{bmatrix} \quad r_1(\theta_1) := \begin{bmatrix} L_1 \cdot \cos(\theta_1) \\ L_1 \cdot \sin(\theta_1) \\ 0 \end{bmatrix} \quad r_1(\theta_{10}) = \begin{bmatrix} -11.123 \\ 2.25 \\ 0 \end{bmatrix}$$

Diagonal vector

$$r_d(\theta_1) := r_0 + r_1(\theta_1) \quad r_{du}(\theta_1) := \frac{r_d(\theta_1)}{|r_d(\theta_1)|}$$

Define $a(\theta_1) := \dfrac{L_3{}^2 - L_2{}^2 + (|r_d(\theta_1)|)^2}{2|r_d(\theta_1)|}$ Rectangular unit vectors $i := \begin{bmatrix} 1 \\ 0 \\ 0 \end{bmatrix}$ $j := \begin{bmatrix} 0 \\ 1 \\ 0 \end{bmatrix}$ $k := \begin{bmatrix} 0 \\ 0 \\ 1 \end{bmatrix}$

q = 1 for assembly configuration with vector loop $r_2r_3r_d$ clockwise; −1 if counterclockwise

$q := -1$

Coupler vector

$$r_2(\theta_1) := q \cdot \sqrt{L_3{}^2 - a(\theta_1)^2} \cdot (r_{du}(\theta_1) \times k) + r_{du}(\theta_1) \cdot (a(\theta_1) - |r_d(\theta_1)|)$$

$$\theta_2(\theta_1) := \text{angle}\,(r_2(\theta_1)_0, r_2(\theta_1)_1) - 2 \cdot \pi$$

Follower crank vector $r_3(\theta_1) := -q \cdot \sqrt{L_3^2 - a(\theta_1)^2} \cdot (r_{du}(\theta_1) \times k) - r_{du}(\theta_1) \cdot a(\theta_1)$

$\theta_3(\theta_1) := \text{angle}\,(r_3(\theta_1)_0, r_3(\theta_1)_1) \quad \dfrac{\theta_3(\theta_{11})}{deg} = 133.748 \quad \dfrac{\theta_3(\theta_{10} + \theta_{1range})}{deg} = 173.748$

Compare $\dfrac{\theta_{31}}{deg} = 133.748$

$$r_0 = \begin{bmatrix} -33 \\ 0 \\ 0 \end{bmatrix} \quad r_1(\theta_{11}) = \begin{bmatrix} -11.123 \\ 2.25 \\ 0 \end{bmatrix} \quad r_2(\theta_{11}) = \begin{bmatrix} 48.177 \\ -6.485 \\ 0 \end{bmatrix} \quad r_3(\theta_{11}) = \begin{bmatrix} -4.054 \\ 4.235 \\ 0 \end{bmatrix}$$

Check closure at start $r_0 + r_1(\theta_{10}) + r_2(\theta_{10}) + r_3(\theta_{10}) = \begin{bmatrix} 9.77 \cdot 10^{-15} \\ 0 \\ 0 \end{bmatrix}$

Transmission angle: Check transmission angle at motion limits and midrange

$$\phi(\theta_1) := \text{acos}\left[\frac{L_2^2 + L_3^2 - (|r_d(\theta_1)|)^2}{2 \cdot L_2 \cdot L_3}\right]$$

$\dfrac{\phi(\theta_{10})}{deg} = 38.586 \quad \dfrac{\phi(\theta_{10} + \theta1a)}{deg} = 38.737 \quad \dfrac{\phi(\theta_{10} + \theta_{1\,range})}{deg} = 12.91$

Rotate drive crank through range of motion

$$\theta_1 := \theta_{10}, \theta_{10} + \frac{\theta_{1\,range}}{200} .. \theta_{10} + \theta_{1range}$$

Evaluate linkage as a function generator

$$x(\theta_1) := \theta_0 + (\theta_1 - \theta_{10})\frac{x_{range}}{\theta_{1\,range}} \qquad x(\theta_{10}) = 1 \quad x(\theta_{10} + \theta 1b) = 5$$

Ideal output $= y(x) \quad y(x_0) = 2.718 \quad y(x_0 + x_{range}) = 66.372$

Output function of generator

$$y_g(\theta_1) := y(x_0) + (\theta_3(\theta_1) - \theta_3(\theta_{10}))\frac{y_{range}}{\theta_{3range}}$$

$$y_g(\theta_{10}) = 2.781 \qquad y_g(\theta_{10} + \theta_{1range}) = 66.372$$

Percent error

$$y_{\%err}(\theta_1) := 100\frac{y_g(\theta_1) - y(x(\theta_1))}{y(x(\theta_1))} \qquad y_{\%err}(\theta_{10}) = 0$$

$y_{\%err}(\theta_{10} + 0.25 \cdot \theta_{1\,range}) = -38.364 \quad y_{\%err}(\theta_{10} + 0.5 \cdot \theta_{1\,range}) = 3.523 \cdot 10^{-13}$

$y_{\%err}(\theta_{10} + 0.75 \cdot \theta_{1\,range}) = 6.491 \quad y_{\%err}(\theta_{10} + \theta_{1\,range}) = 7.708 \cdot 10^{-13}$

11.8 COUPLER CURVES

We are severely limited in the flexibility of our design if we consider only the motion of the two cranks in the four-bar linkage. *Coupler curves*—curves generated by points on the coupler (connecting rod) of a four-bar linkage—provide a variety of paths to choose from. For some applications, a point on the coupler itself is utilized directly, as in a film drive mechanism or for a mixing device. In other cases, a point on the coupler is used to drive an added linkage. By using a catalog of coupler curves, we may select linkage dimensions to perform a specific function. An exhaustive catalog of coupler curves for the crank-rocker mechanism (over 7000 curves made up of about half a million plotted locations and velocities) was constructed by Hrones and Nelson (1951). We would now use motion simulation software to plot coupler curves and evaluate them for specialized design applications. A collection of coupler curves might be of help in a design task similar to the next sample problem.

SAMPLE PROBLEM 11.9

Approximate Straight-Line Motion

Design a linkage to generate approximate straight-line motion over a portion of its cycle.

Design decisions. The requirement for straight-line motion is satisfied by a piston and cylinder or a linear slide, although another configuration sometimes fits in better. A four-bar double-rocker linkage is one possibility. To satisfy the Grashof criterion, the sum of the lengths of the longest and shortest links must be less than the sum of the other two; the coupler is the shortest link. Figure 11.15a shows a tentative design with the following dimensions: links O_1B and O_3C = 110 mm; coupler BC = 65 mm; fixed link O_1O_3 = 130 mm. We extend the coupler to form a rectangle of 65 by 210.2 mm and check various coupler curves. At the instant shown, points D, E, and F lie on the line joining the fixed bearings. Point E is midway between the bearings.

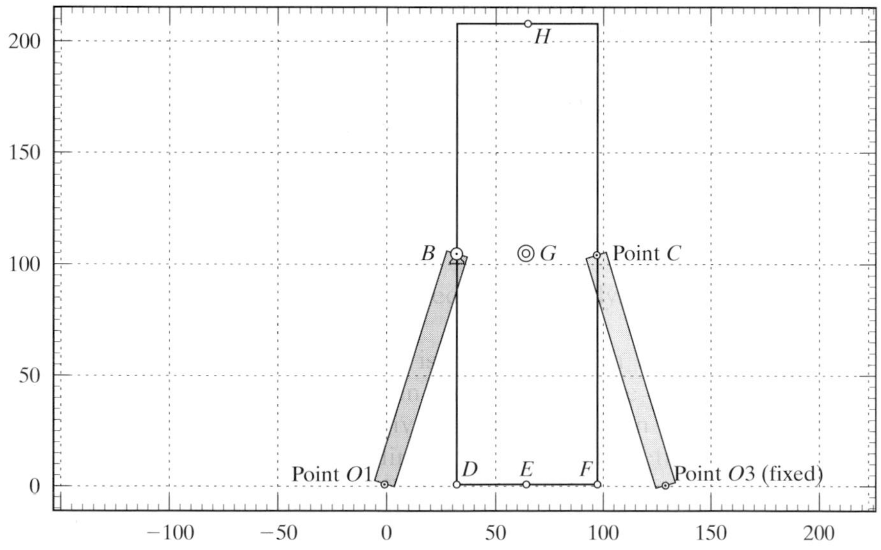

FIGURE 11.15 An attempt to generate straight-line motion. (a) A four-bar double-rocker linkage.

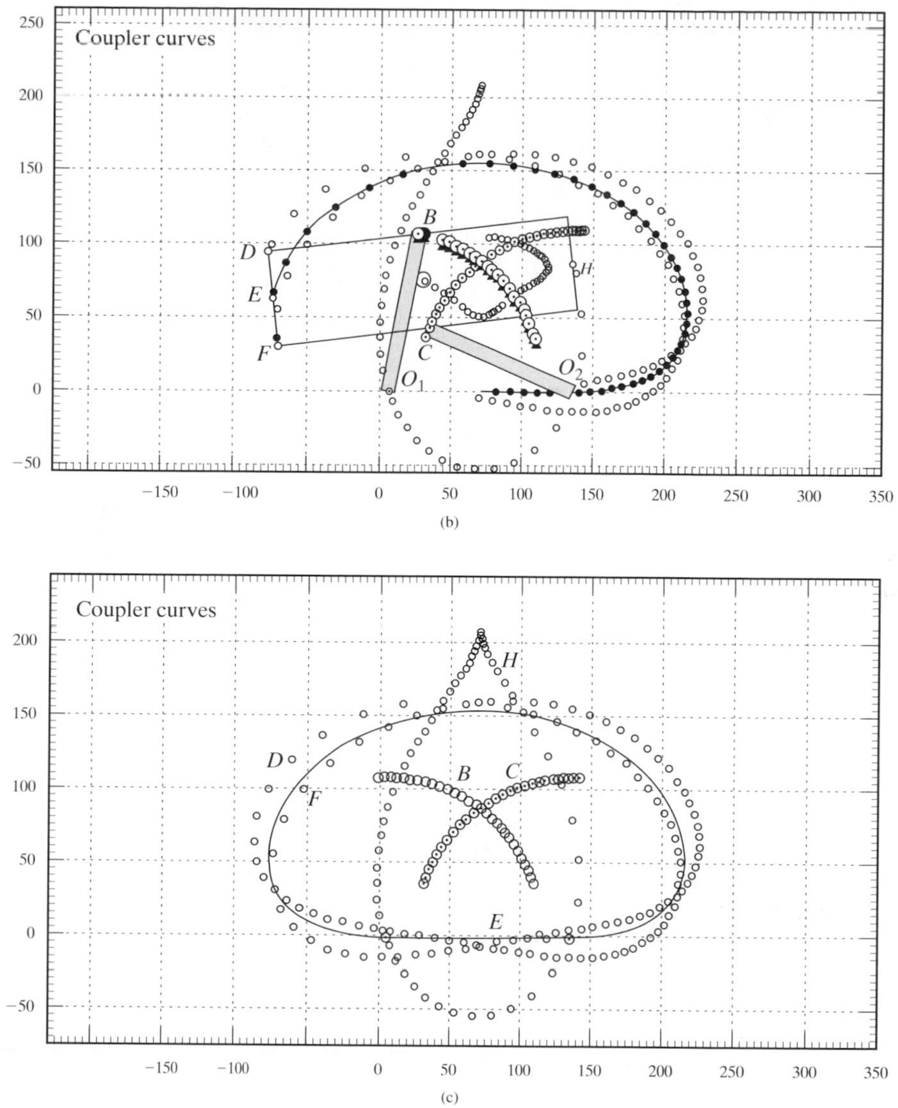

FIGURE 11.15 (b) The linkage after a partial cycle of motion, with traces of several points on the coupler. (c) Coupler curves for a full cycle.

Solution. A motor at point B drives the coupler at 1 rad/s relative to link O_1B. Figure 11.15b shows the linkage after a partial cycle of motion and traces the motion of several points on the coupler. Figure 11.15c shows coupler curves for a full cycle, with the linkage hidden. The coupler curve for point E (shown solid) looks most promising. A portion of the curve (about midway between the fixed bearings) looks like a straight line. The x- and y-coordinates of point E are plotted against time in Figure 11.15d (where the position of the linkage in part a of the figure corresponds to time $= 0$). We may want to attach another link at point E. If the result is satisfactory, we can trim down the coupler; it needs to be large enough only to accommodate bearings at points B, C, and E.

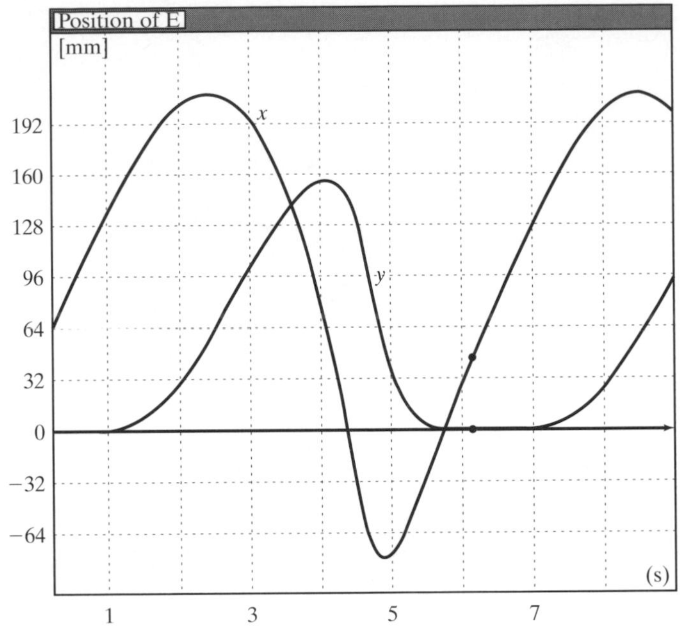

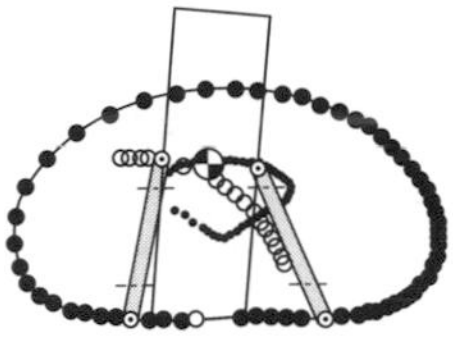

FIGURE 11.15 (d) The x- and y-coordinates of point E.

Several mechanisms for generating approximate straight-line motion were developed many years ago. They include designs equivalent to the one we reinvented in the preceding sample problem and designs that will be developed and examined in homework problems. These linkages can be evaluated by motion simulation software (e.g., Working Model™) or analyzed by vector methods (e.g., we could use Mathcad™ and the cross-product method for position analysis). In some cases, simple principles of geometry can be employed to evaluate the linkage.

SAMPLE PROBLEM 11.10

Evaluation of Coupler Curve Accuracy

Determine whether the following linkages can produce an exact straight line:

a. The linkage sketched in Figure 11.15.

b. The linkage sketched in Figure 11.16.

Solution. **a.** Noting the dimensions of the double-rocker linkage (Figure 11.15), we see that

$$O_1B = BE = EC = CO_3 \quad \text{and} \quad O_1O_3 = 2BC.$$

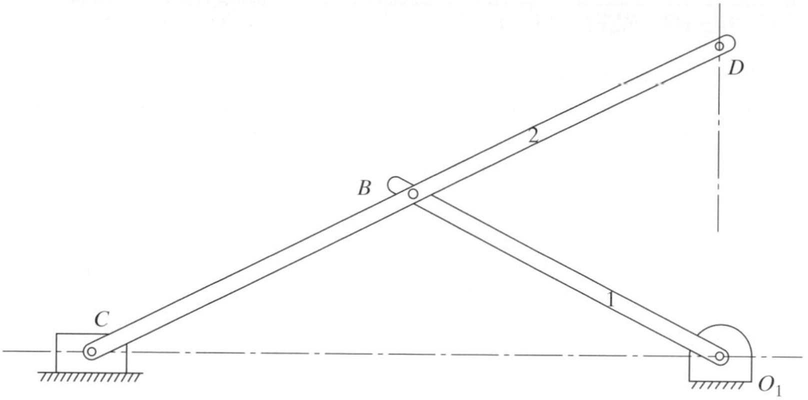

FIGURE 11.16 Exact straight-line linkage.

Assume that point E on the coupler actually moves in a straight line when it is near the midpoint between the fixed bearings. Then, a slider could be attached to point E, and the point would move horizontally. Triangles O_1BE and ECO_3 should both be isosceles. As point E moves to the right, the projection of point B on the x–axis lies midway along the base of triangle O_1BE and the projection of C lies midway along the base of triangle ECO_3. The horizontal distance between B and C is $O_1O_3/2 = BC$, but the vertical distance between B and C is not zero. Thus, if point E moves in an exact straight line, the coupler would have to stretch. Hence, our assumption is incorrect: The motion is only an approximation of a straight line.

b. Let the in-line slider-crank linkage of Figure 11.16 be proportioned so that

$$O_1B = BC = BD.$$

Then, O_1BC and O_1BD are both isosceles triangles. From the sketch, the sum of angles CBO_1 and O_1BD is 180°. Also, we know that the sum of the internal angles of a triangle equals 180°. Thus, angles are related as follows:

$$DO_1B = (180° - O_1BD)/2;$$
$$CO_1B = (180° - CBO_1)/2;$$
$$CO_1D = CO_1B + DO_1B = 180° - (CBO_1 + O_1BD)/2 = 90°.$$

Hence, we have

$$CO_1B = (180° - CBO_1)/2,$$
$$CO_1D = CO_1B + DO_1B = 180°(CBO_1 + O_1BD)/2 = 90°.$$

The last equation shows that point D moves on a line perpendicular to the path of wrist pin C. The straight-line motion for this design is exact, except for errors due to dimensional tolerances and link deflections. The trade-off for the precision is that the linkage requires a slider (e.g., a linear slide or a piston and cylinder) to produce straight-line motion of point D.

SAMPLE PROBLEM 11.11

An Application of Coupler Curves

Design a linkage to meet the following requirements: One of the links is to rotate through an angle of 20° at a rate of 1 cycle per second with a dwell of 1/6 second between oscillations. The drive link is to rotate at a constant angular velocity.

Solution. Problems of this type are common in the design of machinery. The solution almost always involves a cam, but we will attempt to use a four-bar linkage for purposes of illustrating linkage design procedures. Examining a catalog of coupler curves, we see several possibilities, one of which is sketched in Figure 11.17a. During about one-sixth of each cycle, the coupler curve described by point P in that figure approximates a straight line. By combining the four-bar linkage with a sliding contact linkage (Figure 11.17b), we may take advantage of the straight-line portion of the coupler curve to provide the required dwell.

Link 4, the output link, is located by using the coupler curve. The straight-line portion of the coupler curve defines one limiting position of link 4. The other limiting position is defined by

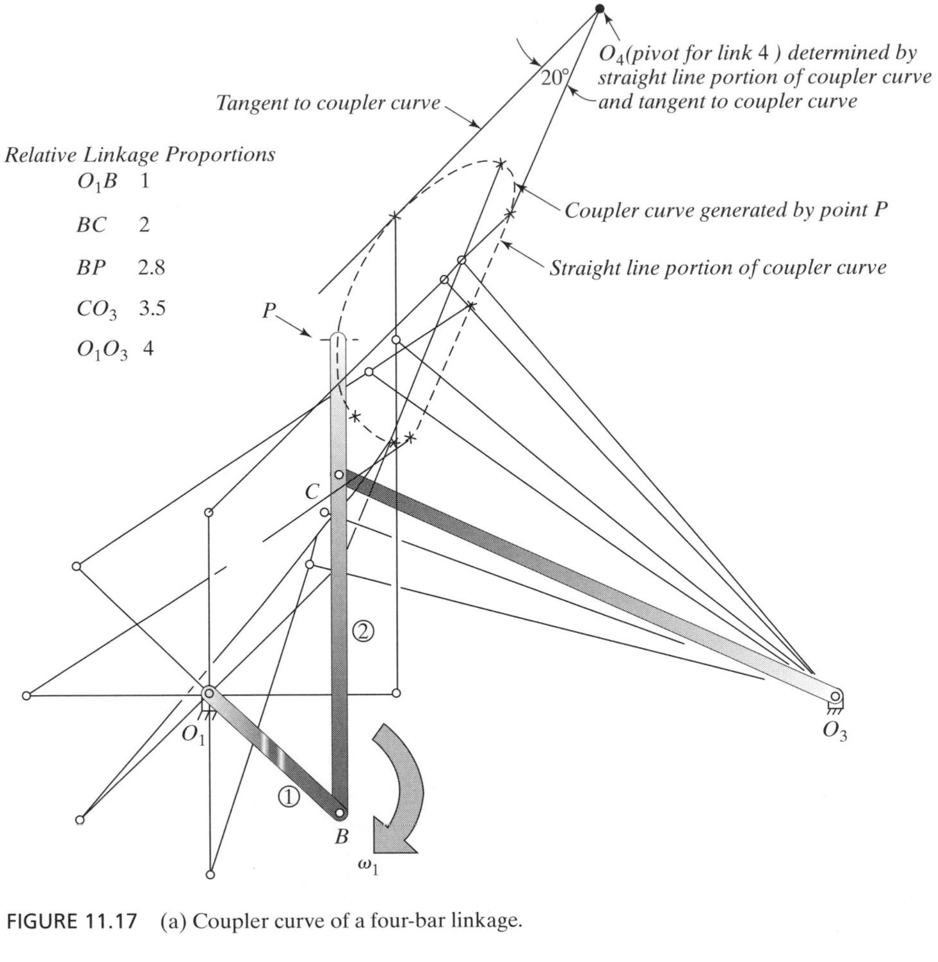

FIGURE 11.17 (a) Coupler curve of a four-bar linkage.

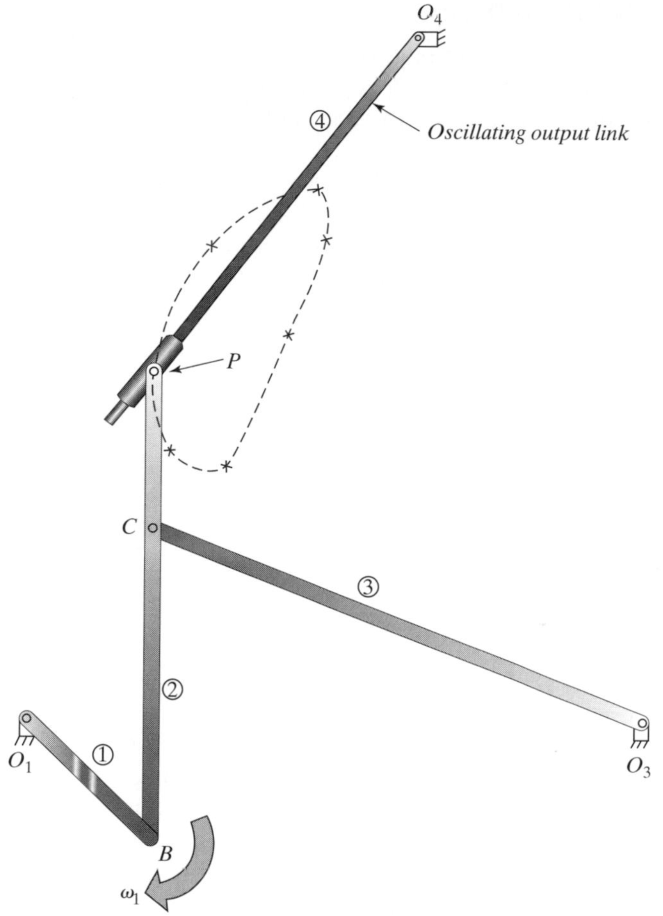

FIGURE 11.17 (b) Link 4 and a slider are added to the four-bar linkage. The output link oscillates with a dwell period.

a tangent to the coupler curve that intersects the first (limiting-position) line at an angle of 20°. Link 4 is pivoted at that intersection, and its length must be sufficient to allow the slider to assume all positions on the coupler curve. Since a complete cycle of motion corresponds to one rotation of link 1, that link will be driven at 60 rev/min.

When link 1 turns with constant angular velocity, the time required for output link 4 to rotate from one position to another is proportional to the angle between corresponding positions of link 1. On this basis, the approximate motion of link 4 is as follows: dwell, 17 percent; clockwise rotation, 30 percent; and counterclockwise rotation, 53 percent (expressed in terms of the time required for one complete cycle, i.e., one clockwise rotation of link 1).

Pattern-Matching Synthesis

The synthesis process can be further automated by pattern matching applied to coupler curves, utilizing artificial intelligence programming techniques. We are thus able to use the computer to determine the parameters of a linkage that will produce a required function.

Visual or computer-aided pattern-matching techniques may be applied to the synthesis of both planar and spatial linkages. Sodhi, et al. (1985) present a set of computer-generated curves for use in designing four-revolute spherical function generators. Their method involves matching a plot of the desired function with curves derived from computer solutions of the displacement equation.

SUMMARY

Practical mechanism design usually involves a large measure of ingenuity and judgment. Tentative designs are subject to analysis and testing, the results of which influence redesign. Formal synthesis is a direct attempt to produce a mechanism that meets specified performance requirements. Ideally, the performance specifications are fed into a computational program which outputs an acceptable linkage design. In practice, it is not that easy. However, formal synthesis techniques can be useful tools for designing mechanisms to meet certain needs.

When there is a need to guide a body through only two specified positions, a single pivot point may be located by simple graphical or analytical methods. A four-bar linkage is an alternative solution to the two-position problem if the single pivot point falls at an unacceptable location. A mechanism for guiding a body through three specified positions can also be based on a four-bar linkage.

We can form a complex matrix to design a four-bar function generator in which the positions of the input crank and the output crank must approximate a predetermined relationship. Or we can design a function generator by an alternative method based on the dot product. We can also use a complex-number method to design a four-bar linkage when angular velocities and accelerations are specified for each link.

Coupler curves offer a virtually unlimited number of design possibilities. We can look at the motion of various points on the coupler of a four-bar linkage or some other mechanism. The coupler can be extended in any direction. A selected point on the coupler can be used to guide or drive a device or a point on another linkage.

Motion simulation software (e.g., Working Model™) and computational software are useful for synthesis of mechanisms to meet specified requirements. We may trace and evaluate a large number of coupler curves in an attempt to find the best possible design for a particular application. Motion simulation software makes this task easier, leaving us more time to consider innovative design changes.

A Few Review Items

- Can you design a seven-link, planar, one-degree-of-freedom linkage with revolute joints? What is the basis for your decision?

- A body must assume two positions in a plane. A line on the body in the first position is parallel to the same line in the second position. Can the body be constrained by a single pivot?
- Can you propose a design to guide the aforementioned body through the two positions just described?
- Describe the geometric principle used to locate pivot points for three-position synthesis.
- Can you think of any application for velocity and acceleration synthesis? If you cannot, then identify a major limitation of the technique.
- In designing a function generator by the dot-product method, which parameters do you have at your disposal (i.e., which can you select)?
- In designing a function generator by the complex-matrix method, which parameters do you have at your disposal, (i.e., which can you select)?
- Compare four-bar linkage function generators with cam-and-follower systems. Note the advantages and disadvantages of each.
- Are there any applications where a mechanism incorporating the motion of a point on the coupler of a four-bar linkage might replace a cam and follower?
- Compare coupler-curve-based mechanism designs with cam-and-follower systems. Note the advantages and disadvantages of each.

PROBLEMS

11.1 A link containing points B and C, which are 150 mm apart, is to assume the following positions:

Coordinates	B_x	B_y	β (degrees)
Position 1	0	0	35
Position 2	250	0	75

Find the location of a fixed pivot that will permit the indicated motion. Usc graphical methods.

11.2 A link containing points B and C, which are 6 in apart, is to assume the following positions:

Coordinates	B_x	B_y	β (degrees)
Position 1	5	4	10
Position 2	4	7.5	−40

Find the location of a fixed pivot that will permit the indicated motion. Use graphical methods.

11.3 A link containing points B and C, which are 4 in apart, is to assume the following positions:

Coordinates	B_x	B_y	β (degrees)
Position 1	2	3	30
Position 2	7	8	60

Design a four-bar linkage that will permit the indicated motion. Use graphical methods.

11.4 Write a computer program for two-position synthesis. Include enough output data for the design of a four-bar linkage or single-pivot linkage.

11.5 A link containing points B and C, which are 150 mm apart, is to assume the following positions:

Coordinates	B_x	B_y	β (degrees)
Position 1	0	0	35
Position 2	250	0.001	75

Find the location of a fixed pivot that will permit the indicated motion. Solve the problem analytically, using a computer if available.

11.6 A link containing points B and C, which are 3.5 in apart, is to assume the following positions:

Coordinates	B_x	B_y	β (degrees)
Position 1	2	3	12
Position 2	5.5	3.2	190

Find the location of a fixed pivot that will permit the indicated motion. Solve the problem analytically, using a computer if available.

11.7 A link containing points B and C, which are 4 in apart, is to assume the following positions:

Coordinates	B_x	B_y	β (degrees)
Position 1	2	3	30
Position 2	7	8	60

Design a four-bar linkage that will permit the indicated motion. Solve the problem analytically, using a computer if available.

11.8 A link containing points B and C, which are 4.2 in apart, must assume the following three positions:

Coordinates	B_x	B_y	β (degrees)
Position 1	0	0	18
Position 2	1.3	1	8
Position 3	2.6	0.6	11

Design a linkage that will satisfy these requirements. Use graphical methods.

11.9 A link containing points B and C, which are 3 in apart, must assume the following three positions:

Coordinates	B_x	B_y	β (degrees)
Position 1	2	2	−30
Position 2	5.5	1.5	5
Position 3	7.5	0.5	40

Design a linkage that will satisfy these requirements. Use graphical methods.

11.10 Write a computer program for three-position synthesis.

11.11 A link containing points B and C, which are 60 mm apart, must assume the following three positions:

Coordinates	B_x	B_y	β (degrees)
Position 1	0	0	20
Position 2	20	40	5
Position 3	−1.5	4	−15

Design a linkage that will satisfy these requirements. Solve the problem analytically. Use a computer if available.

11.12 A link containing points B and C, which are 1.5 in apart, must assume the following three positions:

Coordinates	B_x	B_y	β (degrees)
Position 1	1	1	15
Position 2	2	3	0
Position 3	0	5	−15

Design a linkage that will satisfy these requirements. Solve the problem analytically. Use a computer if available.

11.13 A link containing points B and C, which are 100 mm apart, must assume the following three positions:

Coordinates	B_x	B_y	β (degrees)
Position 1	0	0	−10
Position 2	50	−40	5
Position 3	20	−50	30

Design a linkage that will satisfy these requirements. Solve the problem analytically. Use a computer if available.

11.14 Specify a linkage that (instantaneously) satisfies the following conditions:

Link:	0	1	2	3
ω (rad/s):	0	2	1	1
α (rad/s^2):	0	1	0	−1

11.15 **(a)** Design a four-bar linkage to satisfy the following velocity and acceleration requirements:

Link:	0	1	2	3
ω (rad/s):	0	−1	2	0
α (rad/s^2):	0	1	0	2

(b) Sketch the solution to this problem. Draw velocity and acceleration polygons. Compare your results with the given data.

11.16 **(a)** Design a four-bar linkage to satisfy the following velocity and acceleration requirements:

Link:	0	1	2	3
ω (rad/s):	0	−1	2	0
α (rad/s^2):	0	0	1	2

(b) Sketch the solution to this problem. Draw velocity and acceleration polygons. Compare your results with the given data.

11.17 Letting $r_0 = 100$, design a linkage to satisfy the following requirements:

$\omega_0 = 0$	$\omega_1 = 150$	$\omega_2 = -50$	$\omega_3 = -90$
$\alpha_0 = 0$	$\alpha_1 = 2500$	$\alpha_2 = -400$	$\alpha_3 = -1600$

(*Suggestion:* Solve this problem by the matrix inverse method, using a computer.)

11.18 The fixed link of a four-bar linkage is to be 1 in long, oriented in the positive real direction (i.e., $\boldsymbol{r}_0 = 1$). Velocities and accelerations are as given in Problem 11.14.

(a) Design a linkage to satisfy the given conditions, using the matrix inverse method.

(b) If you have already solved Problem 11.14, compare the two sets of results by multiplying the link vectors in this solution by the value of $\mathbf{r}_0$ found previously.

11.19 The fixed link of a four-bar linkage is to be 1 in long, oriented in the positive real direction (i.e., $\mathbf{r}_0 = 1$). Velocities and accelerations are as given in Problem 11.15.

(a) Design a linkage to satisfy the given conditions, using the matrix inverse method.

(b) If you have already solved Problem 11.15, compare the two sets of results by multiplying the link vectors in this solution by the value of $\mathbf{r}_0$ found previously.

11.20 The fixed link of a four-bar linkage is to be 1 in long, oriented in the positive real direction (i.e., $\mathbf{r}_0 = 1$). Velocities and accelerations are as given in Problem 11.16.

(a) Design a linkage to satisfy the given conditions, using the matrix inverse method.

(b) If you have already solved Problem 11.16, compare the two sets of results by multiplying the link vectors in this solution by the value of $\mathbf{r}_0$ found previously.

11.21 Design a four-bar linkage to generate the function $y = e^x - 0.5x^{1.1}$, where $x_0 = 0.5$, $x_{\text{range}} = 1.0$, $\theta_0 = 70°$, $\phi_0 = 40°$, $\theta_{\text{range}} = 60°$, and $\phi_{\text{range}} = 60°$. A computer solution is suggested.

(a) Determine the relative link lengths.

(b) Tabulate the input angle, the output angle, x, y (generated), y (ideal), and the percent error.

(c) Plot the output angle and the transmission angle vs. the input angle.

11.22 Design a four-bar linkage to generate the function $y = e^x - 0.2x^2$, where $x_0 = 0.5$, $x_{\text{range}} = 1.0$, $\theta_0 = 75°$, $\phi_0 = 40°$, $\theta_{\text{range}} = 60°$, and $\phi_{\text{range}} = 40°$. A computer solution is suggested.

(a) Determine the relative link lengths.

(b) Tabulate the input angle, the output angle, x, y (generated), y (ideal), and the percent error.

(c) Plot the output angle and the transmission angle vs. the input angle.

11.23 Suppose we need a mechanism that produce output rotation proportional to $y = x^3 \cdot e^{-x}$ for input rotation proportional to x, where $1 \leq x \leq 4$ As a design decision, try a four-bar linkage function generator with input link range $\theta_{1\text{range}} = 45°$ and output link range $\theta_{3\text{range}} = 40°$. The fixed link will be 55 mm long. Select precision points at the beginning, the midpoint, and the end of the range of x. Let the coupler rotate through an angle of 0.3 radian between the first and second precision point, and 0.6 radian between the first and third precision point. Use the complex-matrix method to design the linkage. Is the vector loop $\mathbf{r}_2\mathbf{r}_3\mathbf{r}_d$ clockwise or counterclockwise? Analyze the linkage through its range of motion. Check the link closure. Plot the transmission angle and the motion of each link. Plot the as-generated output variable and the ideal value. Check the error in the generated output at the precision points and between precision points.

11.24 Suppose we need a mechanism that produces output rotation proportional to $y = \text{arctangent}(x)$ for input rotation proportional to x, where $0.5 \leq x \leq 4$. As a design decision, try a four-bar linkage function generator with input link range $\theta_{1\text{range}} = 45°$, output link range $\theta_{3\text{range}} = 40°$, and a 60-mm fixed link. Select precision points at the beginning, the midpoint, and the end of the range of x. Let the coupler rotate through an angle of 0.2 radian between the first and second precision point and 0.5 radian between the first and third precision point. Use the complex-matrix method to design the linkage. Is the vector loop $\boldsymbol{r}_2\boldsymbol{r}_3\boldsymbol{r}_d$ clockwise or counterclockwise? Analyze the linkage through its range of motion. Check the link closure. Plot the transmission angle and the motion of each link. Plot the as-generated output variable and the ideal value. Check the error in the generated output at the precision points and between precision points.

11.25 A function generator designed according to the specifications listed in problem 11.24 produces an error of about 11.5% midway between the first two precision points. In addition, the transmission angle exceeds 160° during part of the range of motion. Can you reduce the error and improve the transmission angle as well? (*Suggestion:* Try increasing the range of coupler motion. There are many possible solutions.)

11.26 Design a linkage that will generate approximate straight-line motion over a portion of its cycle. As a design decision, use a four-bar double-rocker linkage. Try the following specifications: length of both rocker links = 500 mm, coupler = 200 mm between its bearings, and distance between fixed bearings = 400 mm. Extend the coupler, keeping its bearings 200 mm apart. Drive the coupler. Plot traces of various points on the coupler (between and beyond the points where the coupler joins the other links). Look for a portion of the coupler curve that approximates a straight line.

11.27 Suppose we need to guide a point on another linkage in a straight line (or an approximate straight line). There is no requirement for continuous rotation. As a design decision, try a four–bar linkage with the following dimensions: length of cranks = 200 and 250 mm, coupler = 250 mm between its bearings, and distance between fixed bearings = 539 mm.

(a) Classify this linkage, using the Grashof criterion.

(b) Extend the coupler, keeping its bearings 250 mm apart. Try to find a point to produce a coupler curve that approximates a straight line.

PROJECT

Plastic tires are used on utility carts and garden equipment. Consider the problem of forming a tire from a continuous tube. It is necessary to measure and cut the tube, form it into a circle,and join the cut ends. Determine which of these processes can be automated. It may be possible to apply number and type synthesis, three-position synthesis, or motion synthesis using coupler curves. Do not limit your study to methods discussed in this text.

BIBLIOGRAPHY AND REFERENCES

Angeles, J., *Spatial Kinematic Chains: Analysis–Synthesis–Optimization*, Springer, New York, 1982.

Freudenstein, F., "Approximate Synthesis of Four-Bar Linkages," *Transactions of the American Society of Mechanical Engineers*, vol. 77, 1955, pp. 853–861.

Freudenstein, F., "Structural Error Analysis in Plane Kinematic Synthesis," *Transactions of the American Society of Mechanical Engineers, Ser. B, Journal of Engineering for Industry*, vol. 81, 1959, pp. 15–22.

Hinkle, R., *Kinematics of Machines*, 2d ed., Prentice Hall, Upper Saddle River, NJ, 1960.

Hrones, J. A., and G. L. Nelson, *Analysis of the Four-Bar Linkage: Its Application to the Synthesis of Mechanisms*, Wiley, New York, 1951.

Kishore, A., and M. Keefe, "Synthesis of an Elastic Mechanism," *Mechanism and Machine Theory*, Penton, Cambridge, U.K. vol. 23, no. 4, 1988, pp. 305–312.

Knowledge Revolution, *Working Model™ 2D User's Manual*, Knowledge Revolution, San Mateo, CA, 1996.

Mallik, A.K., A. Ghosh, and G. Dittrich, *Kinematic Analysis and Synthesis of Mechanisms*, CRC Press, Boca Raton FL, 1994.

MathSoft, *Mathcad2000™ User's Guide*, MathSoft, Inc., Cambridge MA, 1999.

Rosenauer, N., "Complex Variable Method for Synthesis of Four-Bar Linkages," *Australian Journal of Applied Science*, vol. 5, no. 4, 1954.

Sodhi, R. S., A. J. Wilhelm, and T. E. Shoup, "Design of a Four-Revolute Spherical Function Generator with Transmission Effectiveness by Curve Matching," *Mechanism and Machine Theory*, Pergamon Press, vol. 20, no. 6, 1985, pp. 577–585.

CHAPTER 12

Introduction to Robotic Manipulators

Methods and Concepts You Will Learn and Apply When Studying This Chapter

- Robot and manipulator applications.
- Robot and manipulator analysis by matrix methods; transformation matrices; the homogeneous coordinate form.
- Forward kinematics applied to manipulators.
- Defining a transformation function for detailed analysis of spatial manipulators with the aid of mathematics software.
- Inverse kinematics.
- Practical considerations including selection among robots, manipulators, and other manufacturing methods; types of end effectors; sensors; selective compliance; part symmetry and exaggerated asymmetry for reliable assembly.
- Robot and manipulator degrees of freedom. Independent and dependent degrees of freedom; situations calling for extra degrees of freedom.

12.1 INTRODUCTION

Robots are an important class of electromechanical systems, which have applications in a diverse range of fields, including manufacturing, construction, space exploration, care for the handicapped, and many others. A robot can be defined as a programmable machine, that is, a machine capable of executing a wide variety of tasks through the specification of sets of computer software commands. These systems are useful in flexible manufacturing, because they can adapt to changes in production operations such as the introduction of new product lines. They are also useful in environments that are hazardous to humans such as undersea exploration or handling of toxic materials. The above definition of a robot suggests an almost unlimited number of potential variations.

(a) (b) (c) (d)

FIGURE 12.1 (a) A remote manipulator system lifting the Hubble Space Telescope from its berth in the space shuttle. (*Source:* NASA/Johnson Space Center.) (b) A robot arm holding a microchip. (*Source:* John Madere.) (c) An automated paint facility at an assembly plant. (*Source:* Chrysler Corp.) (d) A light-duty robot designed for educational use.

A few examples of robots and manipulators used in space, in manufacturing, and in education are shown in Figure 12.1.

There are three major components of a robot system; a manipulator mechanism, a set of actuators, and a controller. The manipulator mechanism is a multi-degree-of-freedom mechanical device, capable of performing a broad range of planar or spatial motions. Most industrial robots have from three to six degrees of freedom. Each independent degree of freedom requires separate actuation, usually applied at the joints of the robot. Electric motors are often used, but hydraulic and pneumatic actuation are also common. Some robots use a combination of actuator types. The controller provides the means for programming and executing specific motion tasks. There are different ways of programming a robot, ranging from using a teaching mode in which the robot is physically placed in desired positions with the corresponding coordinates stored in controller memory to utilizing a computer command language directly in developing motion programs. Many robots have sensing and feedback capabilities for implementing control algorithms, which enable the robot to maintain a desired motion profile within some specified degree of accuracy.

This chapter introduces concepts and methods of analysis pertaining to robotic manipulators. Background material is provided for carrying out the kinematic analysis of manipulator mechanisms. The methods presented are applicable to planar and spatial mechanisms of various types.

12.2 ROBOT TERMINOLOGY AND DEFINITIONS

A robotic manipulator consists of a multi-degree-of-freedom kinematic chain. Either closed-loop chains or open-loop chains can be used. Revolute and prismatic pairs are commonly employed as joints, because they are relatively easy to drive. Mounted at the free end of the manipulator is an end effector, which is a device for performing the specific work function of the robot. For example, the end effector may be a tool (such as a welding torch or a painting nozzle), a gripper, or another manipulator. Depending on the type of application and controller, the robot is programmed for either discrete positions (point-to-point motion) or continuous motion of the end effector.

Robots are given multiple degrees of freedom in order to produce general motion capabilities for carrying out different programmed tasks. For example, the planar manipulator in Figure 12.2a has three degrees of freedom, which is the minimum number required to produce general planar motion of a body. By means of the three rotational inputs shown, member 3, which represents the end effector in this case, can be placed at any position and angular orientation within the workspace or work envelope of the robot. This workspace is the total area or volume accessible to the end effector and is determined from the member dimensions and limits of joint motion of the robot. Similarly, the spatial robot of Figure 12.2b has six degrees of freedom, which is the minimum number required for general three-dimensional motion of a body.

The methods described in Chapter 1 can be used to determine the number of degrees of freedom of a manipulator. Repeating Eq. (1.3) here, we have

$$DF_{\text{spatial}} \geq 6(n_L - n_J - 1) + \sum_{i=1}^{n_J} f_i, \tag{12.1}$$

where DF_{spatial} is the total number of degrees of freedom of a spatial linkage, n_L is the number of links (including the fixed link), n_J is the number of joints, and f_i is the number of degrees of freedom for joint i. Consider the application of this equation to the open-chain manipulators shown in Figure 12.3. Each of these manipulators has four members and three joints, illustrating the following general condition: In an open-loop manipulator, the number of joints is one less than the number of links, and the number of degrees of freedom of the mechanism is therefore the sum of the joint degrees of freedom; that is,

$$DF_{\text{spatial}} = \sum_{i=1}^{n_J} f_i. \tag{12.2}$$

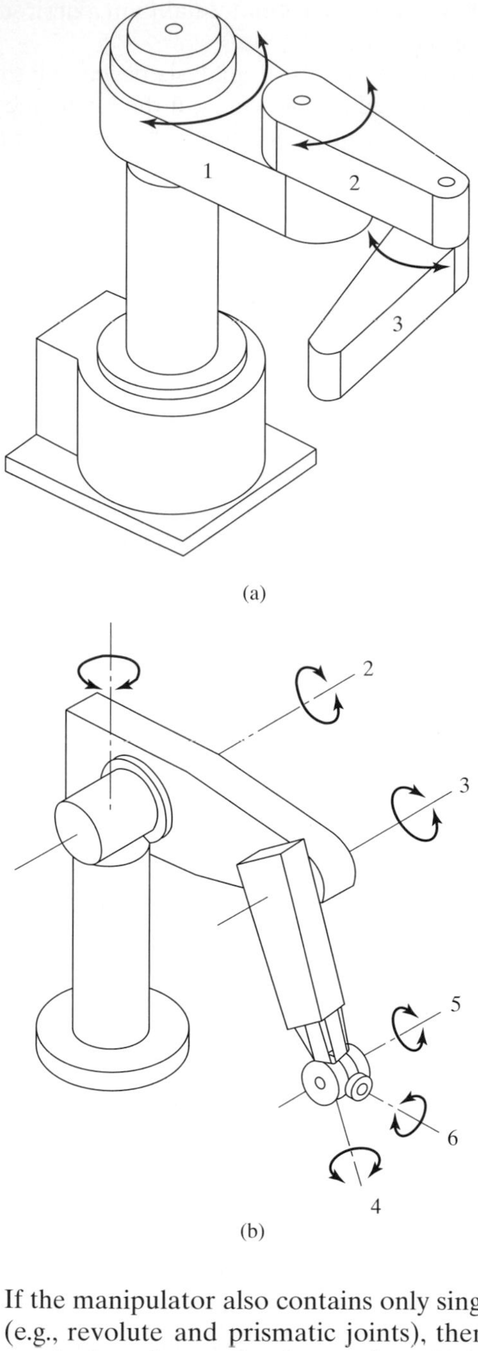

FIGURE 12.2 (a) A planar-motion robot with three degrees of freedom. (b) A spatial motion robot with six degrees of freedom. (*Source:* AEG Westinghouse, Industrial Automation Corporation.)

If the manipulator also contains only single-degree-of-freedom joints for which $f_i = 1$ (e.g., revolute and prismatic joints), then the number of degrees of freedom of the manipulator is equal to the number of joints. This is true for the manipulators shown in Figure 12.3a, b, and c, and therefore, DF $= 3$ in each of these cases. Applying Eq. (12.2) to the mechanism of Figure 12.3d results in

$$\mathrm{DF} = 1 + 3 + 2 = 6.$$

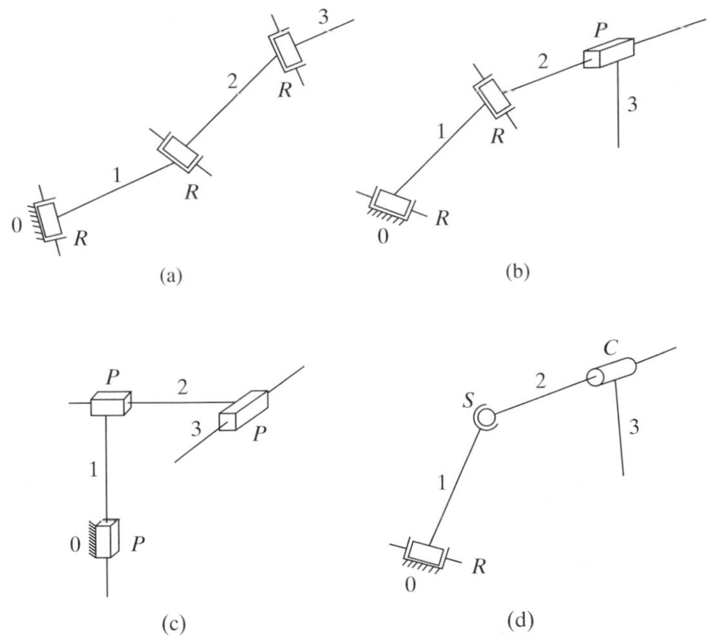

FIGURE 12.3 Examples of manipulators. (a) *RRR*. (b) *RRP*. (c) *PPP*. (d) *RSC*.

Thus, six independent actuations are required for constrained motion of this particular mechanism.

An important part of the analysis of robotic manipulators is the development of relationships between joint coordinates and end-effector coordinates. This development leads to two classes of kinematic analysis: *forward kinematics* and *inverse kinematics*. In *forward kinematics*, the end-effector coordinates are determined from knowledge of the joint coordinates. Specifically, the position and orientation of the end effector are computed from the angular positions (revolute joints) and translational positions (prismatic joints) of the joints. This, however, is not the normal procedure in programming and operating robots; we would prefer to specify prescribed end-effector coordinates and then compute the corresponding joint coordinates required for actuation. The process of going from end-effector coordinate space to joint coordinate space is referred to as *inverse kinematics*. For most typical manipulators, the inverse kinematics problem is more difficult than the forward kinematics problem. Both forms of analysis will be considered in the sections that follow.

12.3 MATRIX METHOD OF ANALYSIS

Matrix methods are very useful for the analysis of both robotic manipulators and spatial mechanisms in general. They are also applicable to the special case of planar mechanisms. With matrix methods, multiple reference frames are utilized in describing the kinematic and dynamic behavior of a mechanism. The material to be presented assumes some background in linear algebra, for which a number of references are available (e.g., Hill, 1986).

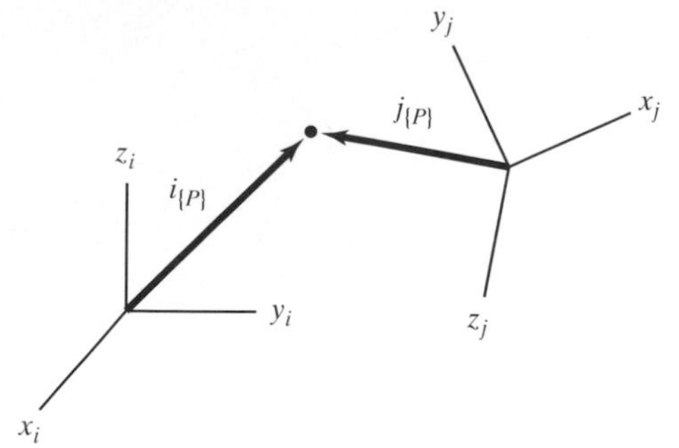

FIGURE 12.4 The vector location of a point in space with respect to two different reference frames.

Brackets will be used to identify vectors and matrices in this chapter. Consider the situation depicted in Figure 12.4, in which the position of a point in space is referred to two different frames of reference: frame i and frame j. Each of these frames is a Cartesian reference frame having a set of x-, y-, and z-coordinate axes. The position of the point, as described in the i frame, can be expressed by the vector

$$^i\{P\} = \begin{Bmatrix} P_{x_i} \\ P_{y_i} \\ P_{z_i} \end{Bmatrix}, \tag{12.3}$$

where P_{x_i}, P_{y_i}, and P_{z_i} are the x-, y-, and z-coordinates of the position vector as measured in the i frame. Similarly, the position of the point referred to the j frame is

$$^j\{P\} = \begin{Bmatrix} P_{x_j} \\ P_{y_j} \\ P_{z_j} \end{Bmatrix}, \tag{12.4}$$

where P_{x_j}, P_{y_j}, and P_{z_j} are the coordinates of the point in the j frame. There is a relationship between vectors $^i\{P\}$, and $^j\{P\}$, having the form

$$^i\{P\} = {}^i_j[T]\,{}^j\{P\}, \tag{12.5}$$

where the matrix $^i_j[T]$, which has dimensions of 3×3 in this case, relates the relative positions and orientations of the two reference frames. This matrix is called a *transformation matrix*, because it produces a coordinate mapping transformation from the j frame to the i frame. The notation convention just defined for presubscripts and presuperscripts will be used throughout the chapter. In particular, the presuperscript on a vector defines the reference frame in which the vector is based. The subscript and superscript on the transformation matrix define the frame from which, and the frame to which, a vector is mapped, respectively.

Alternatively, the transformation can take place from the i frame to the j frame, in which case

$$^j\{P\} = {}^j_i[T]\,{}^i\{P\}. \tag{12.6}$$

Comparing Eqs. (12.5) and (12.6), we observe that matrix $^j_i[T]$ is the inverse of $^i_j[T]$; that is,

$$^j_i[T] = {}^i_j[T]^{-1} \tag{12.7}$$

or

$$ {}^{i}_{j}[T]{}^{j}_{i}[T] = [I], \tag{12.8} $$

where $[I]$ is the identity matrix.

Based on the preceding concepts, multiple reference frames can be strategically located throughout a mechanism. For example, frame j in Figure 12.4 could be rigidly attached to the end effector of a manipulator, while frame i might be attached to the fixed base of the robot. Indeed, it will prove useful to have a reference frame embedded in each member of the mechanism, in which case Eq. (12.5) can be applied recursively in going successively from link to link throughout the mechanism. This use of multiple frames on adjacent links simplifies the determination of the specific forms of the transformation matrix, which is the subject of the next section. These matrices will be functions of the link dimensions and the relative joint displacements.

12.4 TRANSFORMATION MATRICES

From Figure 12.4 and Eq. (12.5), the transformation matrix ${}^{i}_{j}[T]$ depends on the three degrees of translational displacement that define the position of the origin of frame j relative to the origin of frame i, as well as on the three degrees of rotational displacement that define the angular orientation of frame j relative to that of frame i. This represents the most general case of coordinate mapping. However, we will consider some special cases before examining the general case.

Pure Rotation about the Origin

Figure 12.5 depicts a situation in which the j frame is located relative to the i frame by a rotation through angle α about the x_i-axis. From the geometry of the figure, the following equations are written relating the coordinates of any arbitrary point as

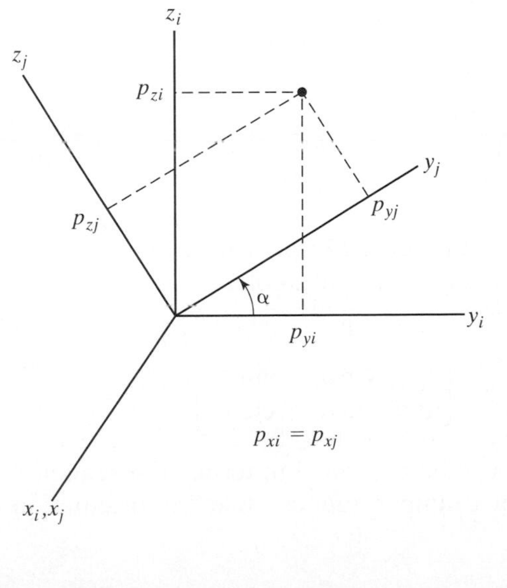

FIGURE 12.5 Rotation through angle α about the x_i axis.

described in one reference frame to coordinates of the same point as described in the other reference frame:

$$\begin{aligned} P_{x_i} &= P_{x_j}; \\ P_{y_i} &= P_{y_j}\cos\alpha - P_{z_j}\sin\alpha; \\ P_{z_i} &= P_{y_j}\sin\alpha + P_{z_j}\cos\alpha. \end{aligned}$$

In matrix form, we have

$$\begin{Bmatrix} P_{x_i} \\ P_{x_i} \\ P_{x_i} \end{Bmatrix} = \begin{bmatrix} 1 & 0 & 0 \\ 0 & \cos\alpha & -\sin\alpha \\ 0 & \sin\alpha & \cos\alpha \end{bmatrix} \begin{Bmatrix} P_{x_j} \\ P_{y_j} \\ P_{z_j} \end{Bmatrix}. \tag{12.9}$$

Expressing this equation in condensed form, we obtain

$$^{i}\{P\} = {}^{i}_{j}[R]^{j}\{P\}, \tag{12.10}$$

where the notation ${}^{i}_{j}[R]$ is used to represent a rotational transformation matrix. Specializing this notation further, we have

$$^{i}\{P\} = [R(x_i, \alpha)]^{j}\{P\} \tag{12.11}$$

where the rotational transformation

$$[R(x_i, \alpha)] = \begin{bmatrix} 1 & 0 & 0 \\ 0 & \cos\alpha & -\sin\alpha \\ 0 & \sin\alpha & \cos\alpha \end{bmatrix} \tag{12.12}$$

defines a rotation of amount α about the x_i-axis. Similar derivations lead to the determination of transformation matrices $[R(y_i, \beta)]$ for rotation about the y_i-axis through angle β, and $[R(z_i, \gamma)]$ for rotation about the z_i axis through angle γ. The result is

$$[R(y_i, \beta)] = \begin{bmatrix} \cos\beta & 0 & \sin\beta \\ 0 & 1 & 0 \\ -\sin\beta & 0 & \cos\beta \end{bmatrix} \tag{12.13}$$

and

$$[R(z_i, \gamma)] = \begin{bmatrix} \cos\gamma & -\sin\gamma & 0 \\ \sin\gamma & \cos\gamma & 0 \\ 0 & 0 & 1 \end{bmatrix}. \tag{12.14}$$

Each of the three rotation matrices defined in Eqs. (12.12) through (12.14) has the useful property that its inverse is equal to its transpose. For example,

$$[R(x_i, \alpha)]^{-1} = [R(x_i, \alpha)]^{T} = \begin{bmatrix} 1 & 0 & 0 \\ 0 & \cos\alpha & \sin\alpha \\ 0 & -\sin\alpha & \cos\alpha \end{bmatrix}. \tag{12.15}$$

Also, a general rotational transformation can be described in terms of a sequence of rotations about the x-, y-, and z-axes. For example, suppose that the orientation of frame j is obtained from the orientation of frame i as a combination of a rotation through angle α about the x_i-axis, followed by a rotation β about the y_i-axis, followed by a rotation γ about the z_i-axis, where the origins of frames i and j are coincident. Then, the transformation of the coordinates of any point from the j frame to the i frame is obtained from the relationship

$$^{i}\{P\} = {}^{i}_{j}[R]^{j}\{P\}, \tag{12.16}$$

where

$$\begin{aligned} {}^{i}_{j}[R] &= [R(z_i,\gamma][R(y_i,\beta)][R(x_i,\alpha)] \\ &= \begin{bmatrix} \cos\beta\cos\gamma & (\sin\alpha\sin\beta\cos\gamma - \cos\alpha\sin\gamma) & (\cos\alpha\sin\beta\cos\gamma + \sin\alpha\sin\gamma) \\ \cos\beta\sin\gamma & (\sin\alpha\sin\beta\sin\gamma + \cos\alpha\cos\gamma) & (\cos\alpha\sin\beta\sin\gamma - \sin\alpha\cos\gamma) \\ -\sin\beta & \sin\alpha\cos\beta & \cos\alpha\cos\beta \end{bmatrix}. \end{aligned} \tag{12.17}$$

The order of the operations in Eq. (12.17) is important, since finite rotational displacements about multiple axes are not commutative. For example, in general,

$$[R(y_i, \beta)][R(x_i, \alpha)] \neq [R(x_i, \alpha)][R(y_i, \beta)].$$

In forming the matrix ${}^{i}_{j}[R]$, and in forming transformation matrices in general, it is helpful to think in terms of the steps that are required to move frame j from its initial coincidence with frame i to its actual position and orientation. This concept is illustrated for the case of Eq. (12.17) and Figure 12.6. In Figure 12.6a, the two frames are coincident; therefore,

$${}^{i}_{j}[R] = [I].$$

Next, an intermediate frame orientation is defined in Figure 12.6b, resulting from rotating the initial frame about the x_i-axis. This yields

$${}^{i}_{j}[R] = [R(x_i, \alpha)]\,[I].$$

Now, the intermediate frame of Figure 12.6b is rotated about the y_i-axis (see Figure 12.6c), leading to

$${}^{i}_{j}[R] = [R(y_i, \beta)][R(x_i, \alpha)]\,[I].$$

Finally, this second intermediate frame is rotated about the z_i-axis into the final, actual position of frame j (Figure 12.6d):

$${}^{i}_{j}[R] = [R(z_i, \gamma)][R(y_i, \beta)][(x_i, \alpha)]\,[I].$$

This matrix is the same as the transformation matrix of Eq. (12.17).

This process we have just described will be very useful for analyzing manipulators in the sections that follow. Note that the right-hand rule applies in assigning directions of rotation; that is, a right-hand rotation about a positive axis is positive, and a left-hand rotation is negative.

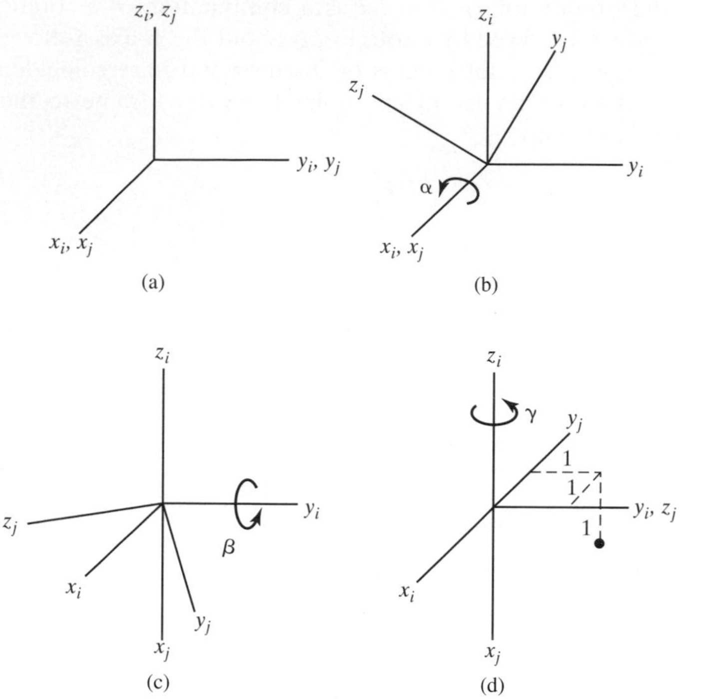

FIGURE 12.6 Rotation about the x_i-axis, (a)–(b), followed by rotation about the y_i-axis, (b)–(c), followed by rotation about the z_i-axis, (c)–(d).

SAMPLE PROBLEM 12.1

An x-y-z Rotation Sequence

Two reference frames, i and j, have the same origin. The orientation of frame j relative to frame i is given by a rotation of 60° about the x_i-axis, followed by a rotation of 90° about the y_i-axis, followed by a rotation of 150° about the z_i-axis. A point has the following coordinates in frame j:

$$^{j}\{P\} = \begin{Bmatrix} 1 \\ 1 \\ 1 \end{Bmatrix}.$$

Determine the coordinates of the point in frame i.

Solution. This situation is depicted in Figure 12.6. First, the transformation matrix is calculated as follows, using Eq. (12.17):

$$^{i}_{j}[R] = \begin{bmatrix} \cos 150^\circ & -\sin 150^\circ & 0 \\ \sin 150^\circ & \cos 150^\circ & 0 \\ 0 & 0 & 1 \end{bmatrix} \begin{bmatrix} \cos 90^\circ & 0 & \sin 90^\circ \\ 0 & 1 & 0 \\ -\sin 90^\circ & 0 & \cos 90^\circ \end{bmatrix}$$

$$\times \begin{bmatrix} 1 & 0 & 0 \\ 0 & \cos 60^\circ & -\sin 60^\circ \\ 0 & \sin 60^\circ & \cos 60^\circ \end{bmatrix}$$

$$= \begin{bmatrix} -\sqrt{3}/2 & -1/2 & 0 \\ 1/2 & -\sqrt{3}/2 & 0 \\ 0 & 0 & 1 \end{bmatrix} \begin{bmatrix} 0 & 0 & 1 \\ 0 & 1 & 0 \\ -1 & 0 & 0 \end{bmatrix} \begin{bmatrix} 1 & 0 & 0 \\ 0 & 1/2 & -\sqrt{3}/2 \\ 0 & \sqrt{3}/2 & 1/2 \end{bmatrix}$$

$$= \begin{bmatrix} -\sqrt{3}/2 & -1/2 & 0 \\ 1/2 & -\sqrt{3}/2 & 0 \\ 0 & 0 & 1 \end{bmatrix} \begin{bmatrix} 0 & \sqrt{3}/2 & 1/2 \\ 0 & 1/2 & -\sqrt{3}/2 \\ -1 & 0 & 0 \end{bmatrix}$$

$$= \begin{bmatrix} 0 & -1 & 0 \\ 0 & 0 & 1 \\ -1 & 0 & 0 \end{bmatrix}.$$

Then, from Eq. (12.16),

$${}^{i}\{P\} = \begin{bmatrix} 0 & -1 & 0 \\ 0 & 0 & 1 \\ -1 & 0 & 0 \end{bmatrix} \begin{Bmatrix} 1 \\ 1 \\ 1 \end{Bmatrix} = \begin{Bmatrix} -1 \\ 1 \\ -1 \end{Bmatrix}.$$

This result is evident from Figure 12.6d. Note that the transformation matrix can be formed from different sets of rotations from the i frame. For example, we can convert frame j from the orientation in Figure 12.6a to that in Figure 12.6d by a combination of a rotation of 90° about the z_i-axis, followed by a rotation of −90° about the x_i-axis, yielding

$${}^{i}_{j}[R] = \begin{bmatrix} 1 & 0 & 0 \\ 0 & \cos(-90^\circ) & -\sin(-90^\circ) \\ 0 & \sin(-90^\circ) & \cos(-90^\circ) \end{bmatrix} \begin{bmatrix} \cos 90^\circ & -\sin 90^\circ & 0 \\ \sin 90^\circ & \cos 90^\circ & 0 \\ 0 & 0 & 1 \end{bmatrix} = \begin{bmatrix} 0 & -1 & 0 \\ 0 & 0 & 1 \\ -1 & 0 & 0 \end{bmatrix},$$

which is the same transformation matrix as the preceding one.

Another approach to defining the rotational transformation matrix ${}^{i}_{j}[R]$ is to describe the orientation of frame j relative to frame i by means of a single rotation about a particular *non*coordinate axis, rather than by means of a sequence of rotations about the coordinate axes. Figure 12.7 illustrates this case: Frame j is obtained from

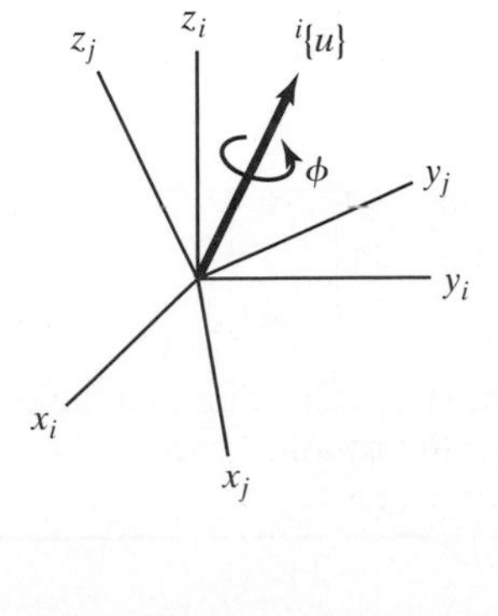

FIGURE 12.7 A general rotation about axis ${}^{i}\{u\}$ through angle ϕ.

frame i as the result of a rotation through angle ϕ about an axis defined by unit vector

$$^{i}\{u\} = \begin{Bmatrix} u_{x_i} \\ u_{y_i} \\ u_{z_i} \end{Bmatrix}, \tag{12.18}$$

where

$$u_{x_i}^2 + u_{y_i}^2 + u_{z_i}^2 = 1. \tag{12.19}$$

The derivation of the rotational transformation matrix based on these quantities can be found in various sources on robotics, including Fu, et al. (1987) and Paul (1981). Only the final result is presented here, as the matrix

$$^{i}_{j}[R] = \begin{bmatrix} (u_{x_i}^2 \text{ ver } \phi + \cos\phi) & (u_{x_i}u_{y_i} \text{ ver } \phi - u_{z_i}\sin\phi) & (u_{x_i}u_{z_i} \text{ ver } \phi + u_{y_i}\sin\phi) \\ (u_{x_i}u_{y_i} \text{ ver } \phi + u_{z_i}\sin\phi) & (u_{y_i}^2 \text{ ver } \phi + \cos\phi) & (u_{y_i}u_{z_i} \text{ ver } \phi - u_{x_i}\sin\phi) \\ (u_{x_i}u_{z_i} \text{ver } \phi - u_{y_i}\sin\phi) & (u_{y_i}u_{z_i} \text{ ver } \phi + u_{x_i}\sin\phi) & (u_{z_i}^2 \text{ ver } \phi + \cos\phi) \end{bmatrix}, \tag{12.20}$$

where

$$\text{ver } \phi = \text{versine } \phi = 1 - \cos\phi. \tag{12.21}$$

Equation (12.20) is the most general form of the rotational transformation matrix, and, in fact, Eqs. (12.12), (12.13), and (12.14) are special cases of this matrix wherein the axis of rotation $^{i}\{u\}$ is the x_i-axis, for the case of Eq. (12.12), the y_i-axis for Eq. (12.13), and the z_i-axis for Eq. (12.14).

SAMPLE PROBLEM 12.2

Rotation about a Noncoordinate Axis

Frame j has an orientation relative to frame i given by a rotation of 120° about an axis defined by the unit vector

$$^{i}\{u\} = \begin{Bmatrix} -1/\sqrt{3} \\ 1/\sqrt{3} \\ 1/\sqrt{3} \end{Bmatrix}.$$

Determine the transformation matrix $^{i}_{j}[R]$.

Solution. Noting that ver 120° $= 1 - \cos 120° = 3/2$ and $\sin 120° = \sqrt{3}/2$ and substituting into Eq. (12.20) yields

$$^{i}_{j}[R] = \begin{bmatrix} 0 & -1 & 0 \\ 0 & 0 & 1 \\ -1 & 0 & 0 \end{bmatrix}.$$

This is the same matrix as in Sample Problem 12.1, indicating that the orientation of frame j will be the same as that shown in Figure 12.6d for the rotation specified at the outset of the current sample problem.

We see that, given a sequence of x, y, and z rotations and the resulting $[R]$ matrix, we can solve for the equivalent single axis rotation (i.e., axis ${}^{i}\{u\}$ and angle ϕ). In the subsequent sections, the following representation of ${}^{i}_{j}[R]$ will be used:

$$ {}^{i}_{j}[R] = \begin{bmatrix} R_{11} & R_{12} & R_{13} \\ R_{21} & R_{22} & R_{23} \\ R_{31} & R_{32} & R_{33} \end{bmatrix} \tag{12.22} $$

Pure Translation and Homogeneous Coordinates

Figure 12.8 shows the case where frame j is located relative to frame i by a three-dimensional translation, given by vector ${}^{i}\{Q\}$, which locates the origin of frame j with respect to that of frame i; that is,

$$ {}^{i}\{Q\} = \begin{Bmatrix} Q_{x_i} \\ Q_{y_i} \\ Q_{z_i} \end{Bmatrix}. \tag{12.23} $$

The coordinate axes of the two frames are parallel. Therefore, as shown in the figure, the coordinates of any point in space, expressed in the two reference frames, are related by the equations

$$ P_{x_i} = P_{x_j} + Q_{x_i}, $$
$$ P_{y_i} = P_{y_j} + Q_{y_i}, $$

and

$$ P_{z_i} = P_{z_j} + Q_{z_i}, $$

or

$$ {}^{i}\{P\} = {}^{j}\{P\} + {}^{i}\{Q\}. \tag{12.24} $$

Equation (12.24) accurately and completely describes the transformation. However, we would like to continue the use of transformation matrices in the form of Eq. (12.5)—that is,

$$ {}^{i}\{P\} = {}^{i}_{j}[T]^{j}\{P\}, $$

where matrix ${}^{i}_{j}[T]$ would somehow be a function of translation ${}^{i}\{Q\}$ in this case.

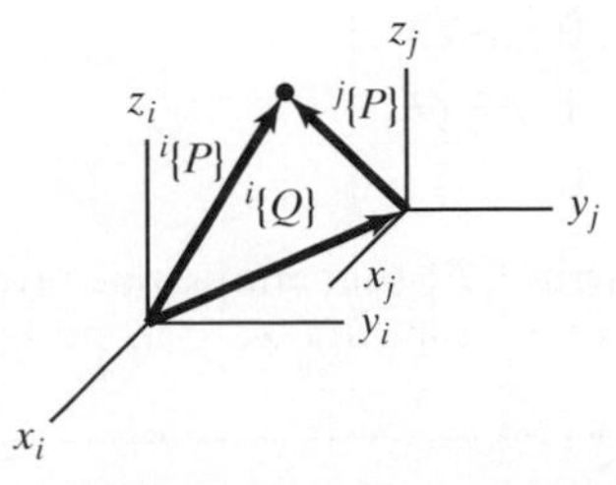

FIGURE 12.8 Frame j is located relative to frame i by means of translation vector ${}^{i}\{Q\}$.

We can accomplish this aim by revising the definition of a vector from the 3×1 dimensional form, as in Eqs. (12.3) and (12.4), to the following forms with dimensions of four rows and one column:

$$^{i}\{P\} = \begin{Bmatrix} P_{x_i} \\ P_{y_i} \\ P_{z_i} \\ 1 \end{Bmatrix} \tag{12.25}$$

and

$$^{j}\{P\} = \begin{Bmatrix} P_{x_j} \\ P_{y_j} \\ P_{z_j} \\ 1 \end{Bmatrix}. \tag{12.26}$$

Note that the fourth entry in each case is always unity. This modified vector definition is referred to as homogeneous coordinates, and from this point on, vectors will be assumed to be in homogeneous coordinate form, unless specified otherwise.

Many practical robot and manipulator designs include both revolute and prismatic pairs. Our goal is to represent rotation *and* translation by a single transformation matrix. The modified (4×1) vector is a step toward that goal. First, consider pure translation. Equation (12.14) can be written as

$$^{i}\{P\} = {}^{i}_{j}[T]^{j}\{P\}, \tag{12.27}$$

where $^{i}\{P\}$ and $^{j}\{P\}$ are given by Eqs. (12.25) and (12.26), respectively, and matrix ${}^{i}_{j}[T]$, which now has dimensions of 4×4, is defined as

$${}^{i}_{j}[T] = \begin{bmatrix} 1 & 0 & 0 & Q_{x_i} \\ 0 & 1 & 0 & Q_{y_i} \\ 0 & 0 & 1 & Q_{z_i} \\ 0 & 0 & 0 & 1 \end{bmatrix}. \tag{12.28}$$

As desired, the transformation matrix is a function of the relative position of the two reference frames and is not a function of the coordinates of specific points. In this way, the transformation can be applied to any point for the purpose of converting from coordinates in frame j to coordinates in frame i.
The inverse of matrix ${}^{i}_{j}[T]$ of Eq. (12.28) is

$${}^{i}_{j}[T]^{-1} = \begin{bmatrix} 1 & 0 & 0 & -Q_{x_i} \\ 0 & 1 & 0 & -Q_{y_i} \\ 0 & 0 & 1 & -Q_{z_i} \\ 0 & 0 & 0 & 1 \end{bmatrix}, \tag{12.29}$$

which, by definition, is the transformation matrix ${}^{j}_{i}[T]$ that maps the i frame into the j frame; that is,

$$^{j}\{P\} = {}^{j}_{i}[T]^{i}\{P\}. \tag{13.30}$$

This relationship is easily checked by carrying out the multiplication process in Eq. (12.30), which leads to the following equations:

$$P_{x_j} = P_{x_i} - Q_{x_i}; \tag{12.31a}$$

$$P_{y_j} = P_{y_i} - Q_{y_i}; \tag{12.31b}$$

$$P_{z_j} = P_{z_i} - Q_{z_i}; \tag{12.31c}$$

$$1 = 1. \tag{12.31d}$$

These equations are in agreement with Eqs. (12.24) and Figure 12.8. Note that the last equation resulting from matrix equations such as Eq. (12.30) is always an identity, because of the way in which homogeneous coordinates have been defined.

The rotational transformations of the previous section can be easily modified to incorporate the homogeneous coordinate form. In particular, the 4 × 4 transformation matrix for producing general rotation is now

$$ {}^i_j[T] = \begin{bmatrix} R_{11} & R_{12} & R_{13} & 0 \\ R_{21} & R_{22} & R_{23} & 0 \\ R_{31} & R_{32} & R_{33} & 0 \\ 0 & 0 & 0 & 1 \end{bmatrix}, \tag{12.32}$$

where the 3 × 3 $[R]$ matrix is determined as in the previous section.

Combined Rotation and Translation

We can now consider the general case where frame j has a combination of a different rotational orientation and a different translational position from that of frame i. (See Figure 12.9.) Once again, it will be helpful in deriving the transformation matrix to break down the transformation into steps required to move frame j from its initial coincidence with frame i to its actual location. For example, suppose that the translational transformation is performed first. This is a translation ${}^i\{Q\}$ to the intermediate dashed location, frame k in the figure. That translation places the origin of the coordinate

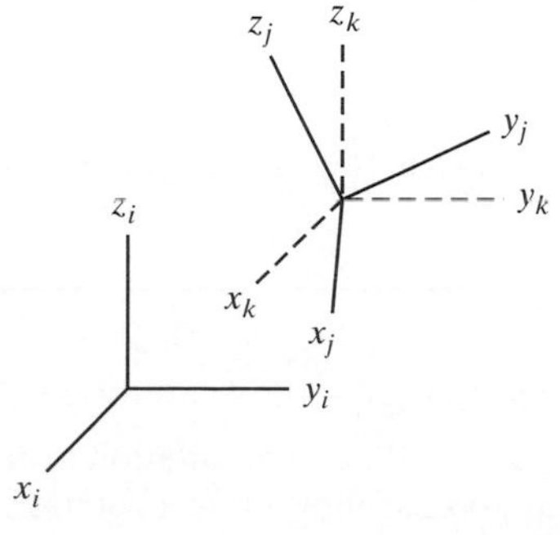

FIGURE 12.9 The location of frame j relative to frame i consists of a combination of rotation and translation.

frame at its final position; however, the coordinate axes of this intermediate frame are still parallel to frame i. Mathematically, for any point, we would have

$$^{i}\{P\} = \begin{bmatrix} 1 & 0 & 0 & Q_{x_i} \\ 0 & 1 & 0 & Q_{y_i} \\ 0 & 0 & 1 & Q_{z_i} \\ 0 & 0 & 0 & 1 \end{bmatrix} {}^{k}\{P\}. \tag{12.33}$$

Now consider the rotation of frame k about its origin until the frame reaches the orientation of frame j. Applying Eq. (12.32) to frames k and j produces

$$^{k}\{P\} = \begin{bmatrix} R_{11} & R_{12} & R_{13} & 0 \\ R_{21} & R_{22} & R_{23} & 0 \\ R_{31} & R_{32} & R_{33} & 0 \\ 0 & 0 & 0 & 1 \end{bmatrix} {}^{j}\{P\}, \tag{12.34}$$

where the 3×3 rotation matrix is ${}^{k}_{j}[R]$, which is also ${}^{i}_{j}[R]$, since the i and k frames are parallel. Substituting Eq. (12.34) into Eq. (12.33) for ${}^{k}\{P\}$ yields

$$^{i}\{P\} = \begin{bmatrix} 1 & 0 & 0 & Q_{x_i} \\ 0 & 1 & 0 & Q_{y_i} \\ 0 & 0 & 1 & Q_{z_i} \\ 0 & 0 & 0 & 1 \end{bmatrix} \begin{bmatrix} R_{11} & R_{12} & R_{13} & 0 \\ R_{21} & R_{22} & R_{23} & 0 \\ R_{31} & R_{32} & R_{33} & 0 \\ 0 & 0 & 0 & 1 \end{bmatrix} {}^{j}\{P\}. \tag{12.35}$$

Carrying out the matrix multiplication, we find that the final result is

$$^{i}\{P\} = {}^{i}_{j}[T]{}^{j}\{P\}, \tag{12.36}$$

where the general transformation matrix for combined rotation and translation is

$${}^{i}_{j}[T] = \begin{bmatrix} R_{11} & R_{12} & R_{13} & Q_{x_i} \\ R_{21} & R_{22} & R_{23} & Q_{y_i} \\ R_{31} & R_{32} & R_{33} & Q_{z_i} \\ 0 & 0 & 0 & 1 \end{bmatrix}, \tag{12.37}$$

or, in partitioned (block) form,

$${}^{i}_{j}[T] = \begin{bmatrix} {}^{i}_{j}[R] & {}^{i}\{Q\} \\ 0 & 1 \end{bmatrix}, \tag{12.38}$$

where ${}^{i}_{j}[R]$ is the 3×3 rotation matrix and ${}^{i}\{Q\}$ is the 3×1 translation vector.

SAMPLE PROBLEM 12.3

Combined Rotation and Translation

Two reference frames, 1 and 2, are shown in Figure 12.10. The orientation of frame 2 relative to frame 1 is given by a rotation of 60° about the x_1-axis, followed by a rotation of 90° about the

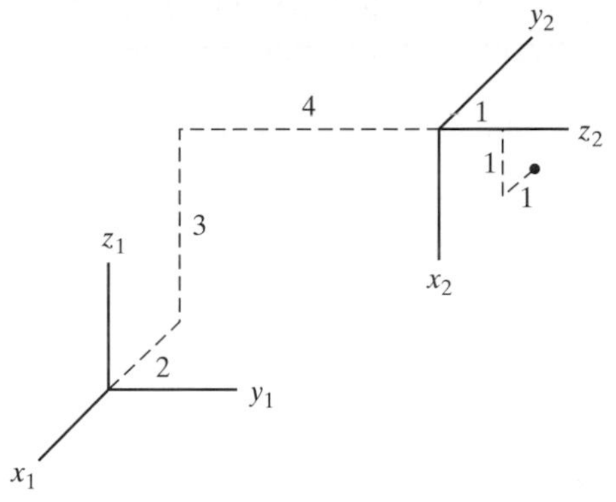

FIGURE 12.10 The situation in Sample Problem 12.3.

y_1-axis, followed by a rotation of 150° about the z_1-axis. The position of the origin of frame 2 relative to the origin of frame 1 is

$$^1\{Q\} = \begin{Bmatrix} -2 \\ 4 \\ 3 \end{Bmatrix}.$$

A point has the following coordinates in frame 2:

$$^2\{P\} = \begin{Bmatrix} 1 \\ 1 \\ 1 \end{Bmatrix}.$$

Determine the coordinates of the point in frame 1.

Solution. The rotational transformation between the two frames is the same as that considered in Sample Problems 12.1 and 12.2. Therefore, referring to either of those examples, we obtain

$$^1_2[R] = \begin{bmatrix} 0 & -1 & 0 \\ 0 & 0 & 1 \\ -1 & 0 & 0 \end{bmatrix}.$$

Substituting $^1_2[R]$, $^1\{Q\}$, and $^2\{P\}$ into Eqs. (12.36) and (12.37) yields

$$^1\{P\} = {}^1_2[T]^2\{P\} = \begin{bmatrix} 0 & -1 & 0 & -2 \\ 0 & 0 & 1 & 4 \\ -1 & 0 & 0 & 3 \\ 0 & 0 & 0 & 1 \end{bmatrix} \begin{Bmatrix} 1 \\ 1 \\ 1 \\ 1 \end{Bmatrix} = \begin{Bmatrix} -3 \\ 5 \\ 2 \\ 1 \end{Bmatrix}.$$

This result can be verified by examining Figure 12.10.

12.5 FORWARD KINEMATICS

Before applying the methods of the previous sections to robot manipulators, we need to define a consistent set of conventions and notations, that will be applicable to a broad range of robots. In particular, we define key link parameters and joint variables and establish a convention for systematically locating coordinate reference frames on each link. In this way, a single transformation matrix can be derived for transforming information between adjacent links, and this matrix can then be used repeatedly for analyzing entire mechanisms.

Definitions

Figure 12.11a shows a general member, link i, of a manipulator. It will be assumed that all members are binary members (i.e., having two joints), as shown, and that the joints are either revolute or prismatic. These assumptions are not overly restrictive, but rather, are consistent with the construction of most manipulators in common use. (See also figure 1.2, a tabulation of common linkage joints; and Figures 12.1, 12.2, and 12.3 that show examples of robots and manipulators.) The joint axes are labeled m and n in Figure 12.11a. For purposes of rigid-body kinematics, two key parameters, the link length and the link twist, totally define the link geometry. These are constant quantities that describe the relative locations of axes m and n. In particular, link length l_i is defined as the perpendicular distance between the axes, and link twist τ_i is the angle between the axes, defined as follows: Axis n is projected along the common perpendicular until it intersects axis m. (See Figure 12.11a.) Angle τ_i is the angle measured from axis m to this projected line, with the sense given by the right-hand rule with respect to the common perpendicular.

Figure 12.11b shows the coordinate frame attached to member i. Its origin is at the intersection of the common perpendicular and axis m. The x-axis of frame i is directed along the common perpendicular toward axis n. The z-axis is along axis m. Note that there are two choices for the direction of this axis (the upward direction has been selected in the figure), and once the choice has been made, all subsequent results must be consistent with it. Finally, the direction of the y-axis is obtained from axes x_i and z_i and the right-hand rule. The three axes form a right-hand coordinate system. In order to apply the analysis equations, which we will derive, each link reference frame must be defined in this manner.

In certain instances, special definitions of reference frames are required. Specifically, if the manipulator is formed from an open kinematic chain, then the

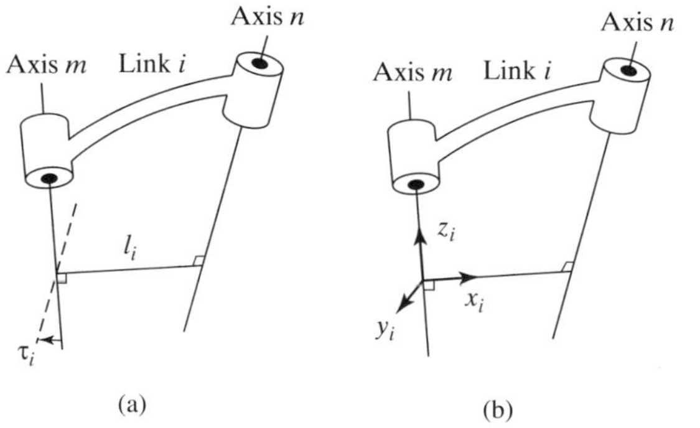

FIGURE 12.11 (a) The link parameters (l_i, τ_i) for link i of a manipulator. (b) The definition of the coordinate system for link i.

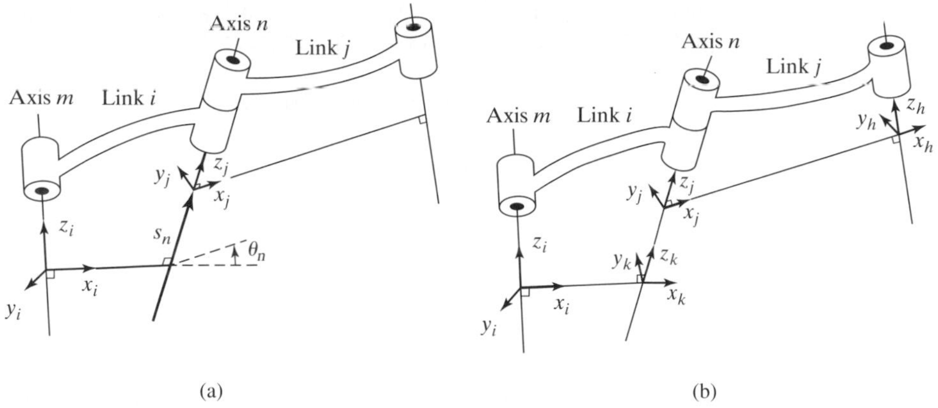

FIGURE 12.12 (a) The joint variables (s_n, θ_n) associated with joint n of a manipulator. (b) Intermediate frame k for the derivation of the transformation matrix from frame j to frame i.

coordinate systems for the first link (the fixed base) and the last link (the end effector) cannot be defined as we have done earlier, because each of these members has only one joint. For the base, it is common practice to define its frame such that it is coincident with the frame for the adjacent moving link at some predefined "home" position of the robot. In this way, the link length and link angle of the base are both zero. The frame for the last link in the chain is conveniently located such that its z-axis is along the axis of the joint that connects the link to the preceding link and its x-axis is along a key dimension of the link. Other choices of reference frames for these links are sometimes useful.

Figure 12.12a shows the joint variables associated with joint n between links i and j. The joint offset s_n is the distance along the joint axis from the common perpendicular for link i to that for link j and is positive when it is in the direction of positive z_j. The joint angle θ_n is the angle around axis n from the x_i-axis to the x_j-axis, with its sign given by the right-hand rule with respect to the z_j-axis. In the case of a revolute joint, angle θ_n is truly variable with time, while offset s_n would have a constant value fixed by the dimensions of the robot. On the other hand, for a prismatic joint, s_n is variable and θ_n is fixed. For cylindric and helical joints, both joint quantities are variable, independently of one another in the former case and dependently in the latter.

Link-to-Link Transformation Matrix

Consider again Figure 12.12 and the derivation of transformation matrix ${}^i_j[T]$, which relates coordinates measured in the frame attached to link j to those measured in the frame attached to link i. In developing this matrix, we temporarily utilize an intermediate frame, frame k in Figure 12.12b. This frame has its origin at the intersection of axis n and the common perpendicular for link i, its x-axis along axis x_i, and its z-axis along axis z_j. Now consider the relation between frames j and k. Frame j is obtained from frame k by a rotation of angle θ_n about axis z_k and a translation of distance s_n along axis z_k. (See Figure 12.12a.) Using Eq. (12.14) to form ${}^k_j[R]$, and substituting into Eq. (12.37), we find that the transformation from frame j to frame k is

$$^k\{P\} = {}^k_j[T]^j\{P\}, \tag{12.39}$$

where

$${}^k_j[T] = \begin{bmatrix} \cos\theta_n & -\sin\theta_n & 0 & 0 \\ \sin\theta_n & \cos\theta_n & 0 & 0 \\ 0 & 0 & 1 & s_n \\ 0 & 0 & 0 & 1 \end{bmatrix}. \tag{12.40}$$

Next, consider the relation between frames k and i. Frame k is obtained by a rotation τ_i about axis x_i [See Eq. (12.12)] and a translation l_i along x_i (see Figure 12.11), leading to

$$^i\{P\} = {}^i_k[T]^k\{P\}, \tag{12.41}$$

where

$${}^i_k[T] = \begin{bmatrix} 1 & 0 & 0 & l_i \\ 0 & \cos\tau_i & -\sin\tau_i & 0 \\ 0 & \sin\tau_i & \cos\tau_i & 0 \\ 0 & 0 & 0 & 1 \end{bmatrix}. \tag{12.42}$$

Substituting Eq. (12.39) into Eq. (12.41), we obtain the desired result

$$^i\{P\} = {}^i_j[T]^j\{P\}, \tag{12.43}$$

where the final transformation matrix is

$${}^i_j[T] = {}^i_k[T]{}^k_j[T] = \begin{bmatrix} \cos\theta_n & -\sin\theta_n & 0 & l_i \\ \cos\tau_i\sin\theta_n & \cos\tau_i\cos\theta_n & -\sin\tau_i & -s_n\sin\tau_i \\ \sin\tau_i\sin\theta_n & \sin\tau_i\cos\theta_n & \cos\tau_i & s_n\cos\tau_i \\ 0 & 0 & 0 & 1 \end{bmatrix}. \tag{12.44}$$

This transformation matrix will be applied to the analysis of some typical robotic manipulators in the sections that follow.

It can be shown that Eq. (12.44) is consistent with the well-known Denavit–Hartenberg representation (1955) when one considers the different coordinate systems used. These two researchers derived a transformation matrix relating frame k as the primary reference frame for link i to a similarly defined frame, frame h in Figure 12.12b, for link j. It then follows that the transformation matrix relating frames h and k is

$${}^k_h[T] = {}^k_j[T]{}^j_h[T],$$

where ${}^{k}_{j}[T]$ is given in Eq. (12.40) and ${}^{j}_{h}[T]$ is defined for link j as ${}^{i}_{k}[T]$ was defined for link i; that is,

$$
{}^{j}_{h}[T] = \begin{bmatrix} 1 & 0 & 0 & l_j \\ 0 & \cos\tau_j & -\sin\tau_j & 0 \\ 0 & \sin\tau_j & \cos\tau_j & 0 \\ 0 & 0 & 0 & 1 \end{bmatrix}.
$$

Substituting and multiplying, we get

$$
{}^{k}_{h}[T] = \begin{bmatrix} \cos\theta_n & -\sin\theta_n\cos\tau_j & \sin\theta_n\sin\tau_j & l_j\cos\theta_n \\ \sin\theta_n & \cos\theta_n\cos\tau_j & -\cos\theta_n\sin\tau_j & l_j\sin\theta_n \\ 0 & \sin\tau_j & \cos\tau_j & s_n \\ 0 & 0 & 0 & 1 \end{bmatrix}.
$$

This is the Denavit–Hartenberg transformation matrix.

2R Planar Manipulator

The manipulator of Figure 12.13 serves to illustrate the analytical approach for a case that is easily verified using traditional analysis methods. The manipulator has three members, including the fixed base (member 0), and two revolute joints whose axes are parallel. The mechanism therefore undergoes planar motion and, from Eq. (12.2), has two degrees of freedom, represented by independent rotations at joints A and B. The objective is to describe the location of point C on member 2 relative to the base of the manipulator.

The home position of the manipulator is shown in Figure 12.13a, and the link coordinate frames will be defined with reference to this figure. The frame for link 1 follows directly from the convention presented earlier. The x_1-axis is along the common normal between the joint axes at A and B, both of which are perpendicular to the page.

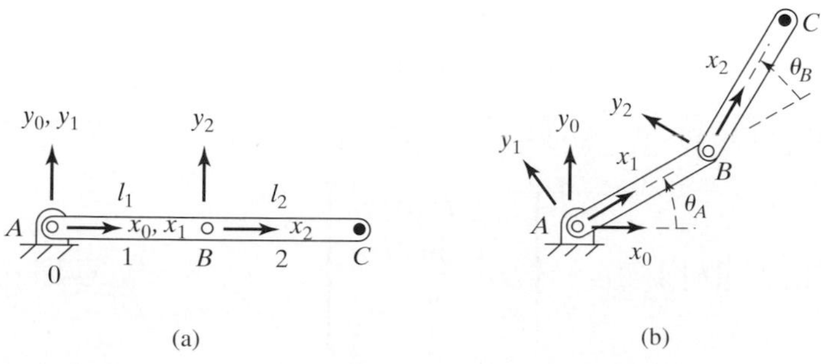

FIGURE 12.13 A two-degree-of-freedom planar manipulator.

(In this case, the x_1-axis lies on the centerline of link 1.) The z_1-axis is along the joint A axis and is chosen to point out of the page. The location of axis y_1 then follows as shown. The reference frame on the base, frame 0, is coincident with frame 1 in the home position and remains stationary as the robot moves. Frame 2 is defined with the z_2-axis, along the axis of joint B, pointing out of the page. Axis x_2 is along link 2, and axis y_2 follows for a right-hand system.

On the basis of the locations of the frames and Figure 12.13b, the following values of the link parameters and joint variables are obtained:

$$\begin{aligned} l_0 &= \tau_0 = \tau_1 = s_A = s_B = 0; \\ l_1 &= l_1; \\ \theta_A &= \theta_A; \\ \theta_B &= \theta_B. \end{aligned}$$

Recall that joint angles θ_A and θ_B are the relative rotations between the two links at each joint, as shown in Figure 12.13b. Also, the coordinates of point C in frame 2 are given by

$$^2\{P\}_C = \begin{Bmatrix} l_2 \\ 0 \\ 0 \\ 1 \end{Bmatrix}, \tag{12.45}$$

where this is a constant vector, since both frame 2 and point C are attached to link 2. Referring to Eqs. (12.39) through (12.44), we obtain, for the transformation matrices,

$$^0_1[T] = \begin{bmatrix} \cos\theta_A & -\sin\theta_A & 0 & 0 \\ \sin\theta_A & \cos\theta_A & 0 & 0 \\ 0 & 0 & 1 & 0 \\ 0 & 0 & 0 & 1 \end{bmatrix} \tag{12.46}$$

and

$$^1_2[T] = \begin{bmatrix} \cos\theta_B & -\sin\theta_B & 0 & l_1 \\ \sin\theta_B & \cos\theta_B & 0 & 0 \\ 0 & 0 & 1 & 0 \\ 0 & 0 & 0 & 1 \end{bmatrix}. \tag{12.47}$$

Applying these transformations to point C yields

$$^1\{P\}_C = {}^1_2[T]\,^2\{P\}_C = \begin{Bmatrix} l_1 + l_2\cos\theta_B \\ l_2\sin\theta_B \\ 0 \\ 1 \end{Bmatrix} \tag{12.48}$$

and

$$
{}^{0}\{P\}_C = {}^{0}_{1}[T]^{1}\{P\}_C = \begin{Bmatrix} l_1 \cos\theta_A + l_2 \cos\theta_A \cos\theta_B - l_2 \sin\theta_A \sin\theta_B \\ l_1 \sin\theta_A + l_2 \sin\theta_A \cos\theta_B + l_2 \cos\theta_A \sin\theta_B \\ 0 \\ 1 \end{Bmatrix}
$$

$$
= \begin{Bmatrix} l_1 \cos\theta_A + l_2 \cos(\theta_A + \theta_B) \\ l_1 \sin\theta_A + l_2 \sin(\theta_A + \theta_B) \\ 0 \\ 1 \end{Bmatrix}. \qquad (12.49)
$$

It can easily be observed from Figure 12.13b that the first two terms in Eq. (12.49) are the x- and y-coordinates, respectively, of point C as referred to frame 0, which is fixed to the base of the robot. Obviously, these coordinates vary with the joint rotations θ_A and θ_B as the robot moves.

RP Manipulator

Figure 12.14 shows another simple manipulator consisting of three members and two joints, in this case a revolute joint and a prismatic joint. The figure shows the locations of the three reference frames attached to the three members. Note that here the two joint axes (A and B) intersect; therefore, length $l_1 = 0$. Also, for convenience, the origin of fixed reference frame 0 has not been selected to be coincident with the origin of

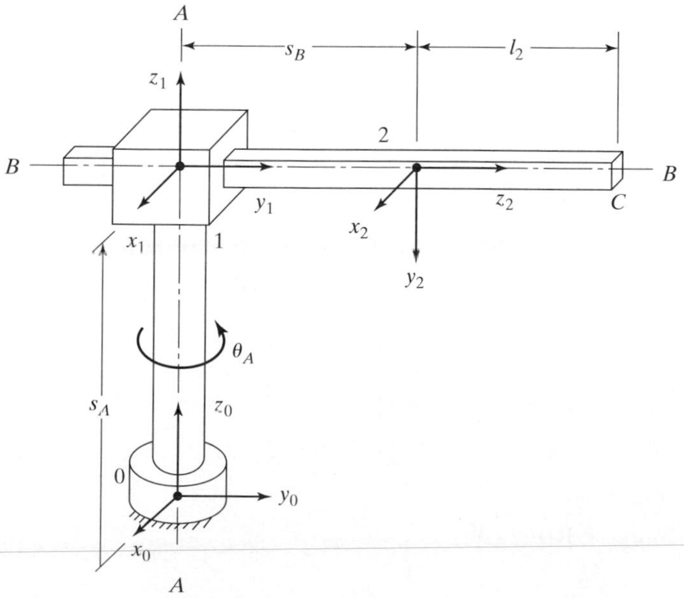

FIGURE 12.14 A manipulator with a revolute joint and a prismatic joint.

frame 1, but instead is placed at a constant offset s_A along axis A. Then, from the figure, the following values are obtained:

$$\begin{aligned} l_0 &= \tau_0 = \theta_B = 0; \\ \tau_1 &= -90°; \\ \theta_A &= \theta_A; \\ s_B &= s_B. \end{aligned}$$

From Eq. (12.44), the resulting transformation matrices are

$$ {}^0_1[T] = \begin{bmatrix} \cos\theta_A & -\sin\theta_A & 0 & 0 \\ \sin\theta_A & \cos\theta_A & 0 & 0 \\ 0 & 0 & 1 & s_A \\ 0 & 0 & 0 & 1 \end{bmatrix} \tag{12.50}$$

and

$$ {}^1_2[T] = \begin{bmatrix} 1 & 0 & 0 & 0 \\ 0 & 0 & 1 & s_B \\ 0 & -1 & 0 & 0 \\ 0 & 0 & 0 & 1 \end{bmatrix}. \tag{12.51}$$

In frame 2, point C, attached to the sliding link, has the following definition:

$$ {}^2\{P\}_C = \begin{Bmatrix} 0 \\ 0 \\ l_2 \\ 1 \end{Bmatrix}. \tag{12.52}$$

Combining Eqs. (12.50) through (12.52), we determine the coordinates of point C, as defined in reference frame 0:

$$\begin{aligned} {}^0\{P\}_C &= {}^0_1[T]{}^1_2[T]{}^2\{P\}_C \\ &= \begin{bmatrix} \cos\theta_A & -\sin\theta_A & 0 & 0 \\ \sin\theta_A & \cos\theta_A & 0 & 0 \\ 0 & 0 & 1 & s_A \\ 0 & 0 & 0 & 1 \end{bmatrix} \begin{bmatrix} 1 & 0 & 0 & 0 \\ 0 & 0 & 1 & s_B \\ 0 & -1 & 0 & 0 \\ 0 & 0 & 0 & 1 \end{bmatrix} \begin{Bmatrix} 0 \\ 0 \\ l_2 \\ 1 \end{Bmatrix} \\ &= \begin{bmatrix} \cos\theta_A & -\sin\theta_A & 0 & 0 \\ \sin\theta_A & \cos\theta_A & 0 & 0 \\ 0 & 0 & 1 & s_A \\ 0 & 0 & 0 & 1 \end{bmatrix} \begin{Bmatrix} 0 \\ l_2 + s_B \\ 0 \\ 1 \end{Bmatrix} \\ &= \begin{bmatrix} -(s_B + l_2)\sin\theta_A \\ (s_B + l_2)\cos\theta_A \\ s_A \\ 1 \end{bmatrix}. \end{aligned} \tag{12.53}$$

Once again, this result can be verified by observing Figure 12.14. In the mechanism under consideration, the time-dependent joint variables are rotation θ_A and translation s_B.

RRR Spatial Manipulator

Figure 12.15 depicts a robotic manipulator that has spatial motion with three degrees of freedom. There are three revolute axes, identified as axes A, B, and C, with corresponding joint rotations θ_A, θ_B, and θ_C, which would be produced by actuators such as electric motors, not shown in the figure. An end effector or tool would be mounted at location D on member 3. The robot is shown in its home position, and the reference frame for each of the four members is depicted. As in the previous section, it will be convenient to locate the reference frame for the base, member 0, as shown, rather than following the guideline of making it coincident with frame 1 at the home position. Instead, the two frames are parallel with frame 1, offset a constant distance s_A along the z_0-axis.

From Figure 12.15, the following constant values of the parameters listed are identified:

$$\begin{aligned} l_0 &= l_1 = 0; \\ l_2 &= l_2; \\ \tau_0 &= \tau_2 = 0; \\ \tau_1 &= 90°; \\ s_B &= s_C = 0. \end{aligned}$$

Substituting into Eq. (12.44), we find that the transformation matrices are

$$ {}^0_1[T] = \begin{bmatrix} \cos\theta_A & -\sin\theta_A & 0 & 0 \\ \sin\theta_A & \cos\theta_A & 0 & 0 \\ 0 & 0 & 1 & s_A \\ 0 & 0 & 0 & 1 \end{bmatrix}, \tag{12.54}$$

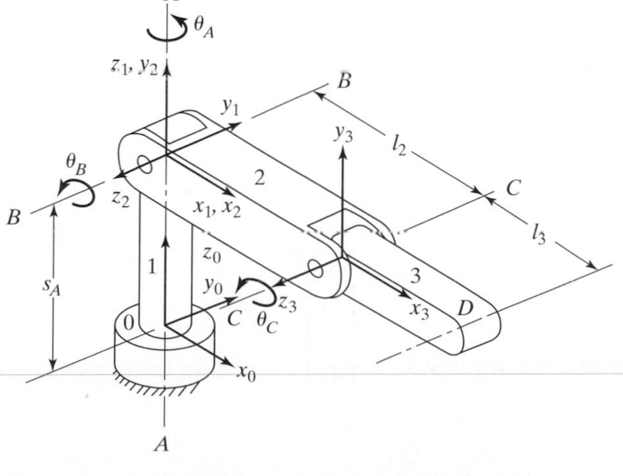

FIGURE 12.15 A three-degree-of-freedom spatial manipulator.

$$
{}^{1}_{2}[T] = \begin{bmatrix} \cos\theta_B & -\sin\theta_B & 0 & 0 \\ 0 & 0 & -1 & 0 \\ \sin\theta_B & \cos\theta_B & 0 & 0 \\ 0 & 0 & 0 & 1 \end{bmatrix}, \tag{12.55}
$$

and

$$
{}^{2}_{3}[T] = \begin{bmatrix} \cos\theta_C & -\sin\theta_C & 0 & l_2 \\ \sin\theta_C & \cos\theta_C & 0 & 0 \\ 0 & 0 & 1 & 0 \\ 0 & 0 & 0 & 1 \end{bmatrix}. \tag{12.56}
$$

Combining these transformation matrices, by means of trigonometric identities yields

$$
\begin{aligned}
{}^{0}_{3}[T] &= \begin{bmatrix} \cos\theta_A & -\sin\theta_A & 0 & 0 \\ \sin\theta_A & \cos\theta_A & 0 & 0 \\ 0 & 0 & 1 & s_A \\ 0 & 0 & 0 & 1 \end{bmatrix} \begin{bmatrix} \cos\theta_B & -\sin\theta_B & 0 & 0 \\ 0 & 0 & -1 & 0 \\ \sin\theta_B & \cos\theta_B & 0 & 0 \\ 0 & 0 & 0 & 1 \end{bmatrix} \\
&\quad \times \begin{bmatrix} \cos\theta_C & -\sin\theta_C & 0 & l_2 \\ \sin\theta_C & \cos\theta_C & 0 & 0 \\ 0 & 0 & 1 & 0 \\ 0 & 0 & 0 & 1 \end{bmatrix} \\
&= \begin{bmatrix} \cos\theta_A & -\sin\theta_A & 0 & 0 \\ \sin\theta_A & \cos\theta_A & 0 & 0 \\ 0 & 0 & 1 & s_A \\ 0 & 0 & 0 & 1 \end{bmatrix} \\
&\quad \times \begin{bmatrix} \cos(\theta_B + \theta_C) & -\sin(\theta_B + \theta_C) & 0 & l_2 \cos\theta_B \\ 0 & 0 & -1 & 0 \\ \sin(\theta_B + \theta_C) & \cos(\theta_B + \theta_C) & 0 & l_2 \sin\theta_B \\ 0 & 0 & 0 & 1 \end{bmatrix}.
\end{aligned} \tag{12.57}
$$

We now apply this transformation to point D, which has the following fixed coordinates in reference frame 3:

$$
{}^{3}\{P\}_D = \begin{Bmatrix} l_3 \\ 0 \\ 0 \\ 1 \end{Bmatrix}. \tag{12.58}
$$

The resulting coordinates of point D, as expressed in terms of fixed frame 0, are

$$^{0}\{P\}_D = {}^{0}_{3}[T]^{3}\{P\}_D = \begin{Bmatrix} l_2 \cos\theta_A \cos\theta_B + l_3 \cos\theta_A \cos(\theta_B + \theta_C) \\ l_2 \sin\theta_A \cos\theta_B + l_3 \sin\theta_A \cos(\theta_B + \theta_C) \\ s_A + l_2 \sin\theta_B + l_3 \sin(\theta_B + \theta_C) \\ 1 \end{Bmatrix}. \quad (12.59)$$

The quantities l_2, l_3, and s_A are fixed robot dimensions, and the joint angles θ_A, θ_B, and θ_C are time variant.

SAMPLE PROBLEM 12.4

Analysis of an RRR Robot

A robot of the type shown in Figure 12.15 has the following dimensions: $s_A = l_2 = l_3 = 10$ in. Determine the location of tip point D for the following joint rotations from the home position (shown in the figure): $\theta_A = 90°$, $\theta_B = 30°$, and $\theta_C = 60°$.

Solution. Substituting into Eq. (12.59), we obtain the coordinates of point D:

$$^{0}\{P\}_D = \begin{Bmatrix} 10\cos 90° \cos 30° + 10\cos 90° \cos(30° + 60°) \\ 10\sin 90° \cos 30° + 10\sin 90° \cos(30° + 60°) \\ 10 + 10\sin 30° + 10\sin(30° + 60°) \\ 1 \end{Bmatrix} = \begin{Bmatrix} 0 \\ 8.66 \\ 25.0 \\ 1 \end{Bmatrix}.$$

This position of the robot is illustrated in Figure 12.16 and confirms the foregoing results.

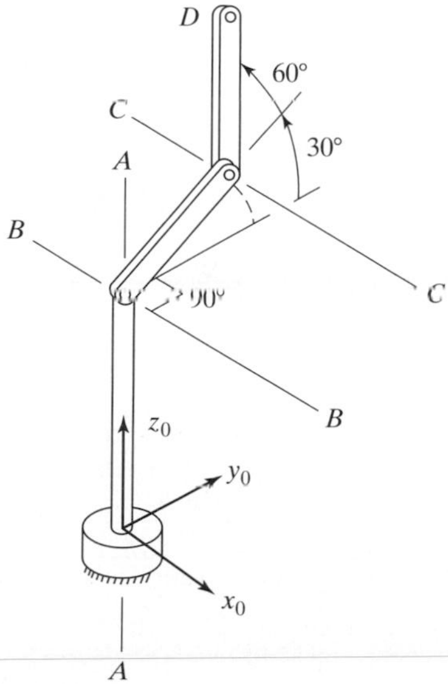

FIGURE 12.16 The manipulator of Sample Problem 12.4.

The matrix approach that has been presented here for position analysis can be extended to velocity and acceleration analysis. It also facilitates force analysis and full dynamic analysis.

Using Motion Simulation Software to Analyze Robots and Manipulators

Pencil-and-paper calculations can be quite burdensome when we need to multiply matrices and calculate a series of manipulator positions. Motion simulation software is one option for "working smart". Sometimes x-, y-, and z-direction motion components are coupled, as in general RRR, 4R, 5R, and 6R spatial manipulators and robots. Motion simulation software with a three-dimensional capability can be used to analyze those linkages. Two-dimensional motion simulation software is adequate if motion in a plane can be uncoupled from out-of-plane motion.

SAMPLE PROBLEM 12.5

Using Motion Simulation Software to Analyze Manipulator Motion in a Plane

A tentative PRRR manipulator design is shown in Figure 12.17a. Link 0, the frame, and link A (not shown) form a prismatic pair. Link A is driven by an actuator at the origin of the coordinates to provide vertical motion (along the z-axis). Links A, 1, 2, and 3 are, in turn, connected by revolute joints in the form of motors. The axes of the three revolute joints are parallel to the z-axis. Link lengths are $r_1 = 200$, $r_2 = 200$, and $r_3 = 160$ mm. The orientations of links 1, 2, and 3 are 0, 0, and π radians, respectively, at the park position.

a. Find the number of degrees of freedom of this manipulator.

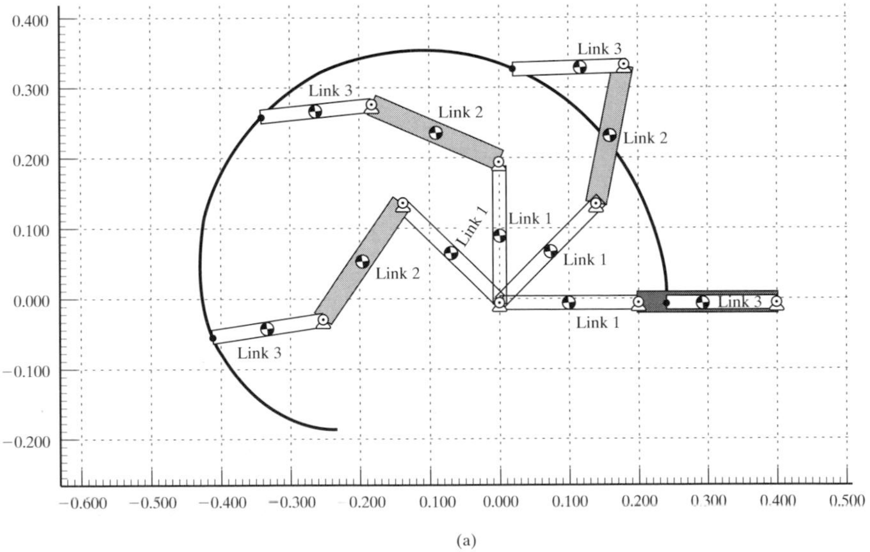

FIGURE 12.17 Using motion simulation software to analyze a manipulator: (a) Motion of the manipulator.

Time		Position of Link 2		Position of Link 3		Velocity of Link 3	
t	0.200 s	x	−0.151 m	x	−0.182 m	Vx	4.144 m/s
		y	−0.058 m	y	−0.162 m	Vy	−0.546 m/s
		rot	5.201 rad	rot	−2.942 rad	|v|	4.180 m/s
						ω	1.000 rad/s

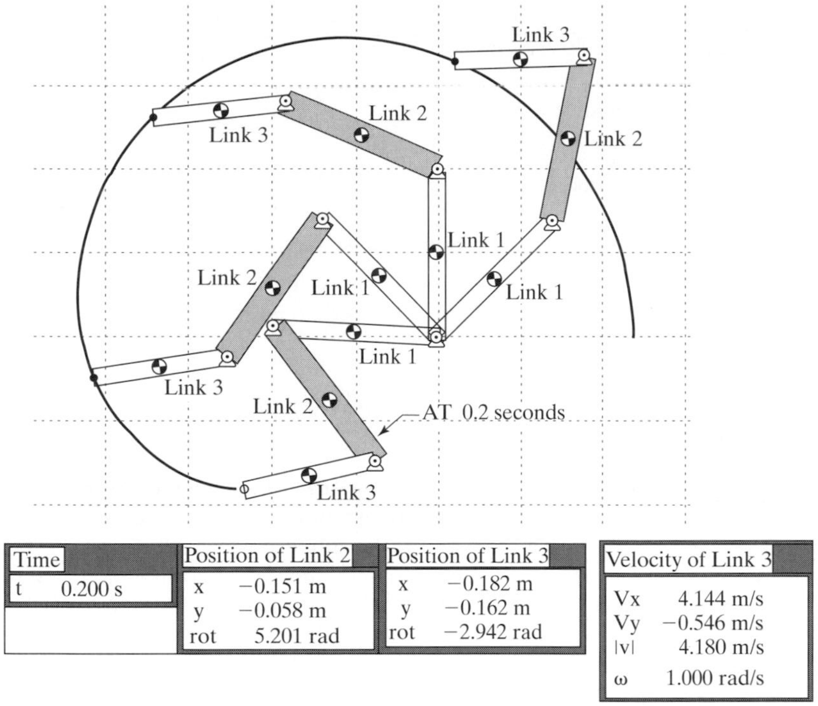

(b)

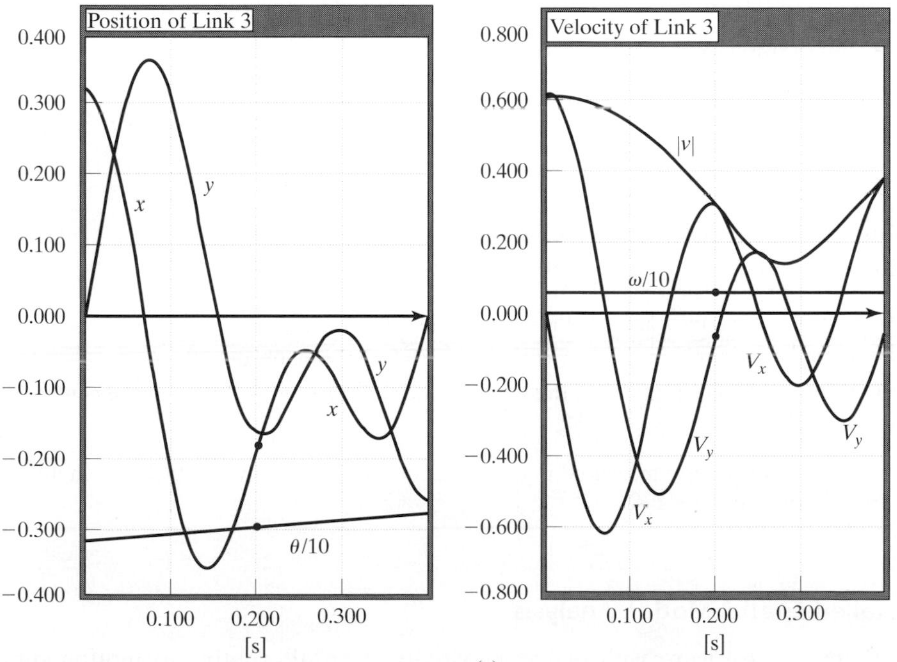

(c)

FIGURE 12.17   (b) Position and velocity components after 0.2 second. (c) Position and velocity of link 3 vs. time.

b. What are its limitations?

c. Motor angular velocities (rad/s) are

$\omega_1 = 15$ (between link A and link 1).
$\omega_2 = 11$ (link 2 relative to link 1).
$\omega_3 = -25$ (link 3 relative to link 2).

The actuator, which drives link A in the z direction, is not operating at this time. Trace the position of the extreme end of link 3 for $0 \leq t \leq 0.2$ s. Find the position of links 2 and 3 and the velocity of link 3 at $t = 0.2$ s. Plot these values against time.

Solution. **(a)** The manipulator has five links, counting the frame and link A. When the prismatic joint is active, there are four one-degree-of-freedom pairs. That is,

$$\text{DF}_{\text{spatial}} \geq 6(5 - 4 - 1) + 4 = 4.$$

The equals sign applies for this open linkage. Thus, a total of four actuators and motors is adequate for this manipulator.

(b) A general-purpose robot would ordinarily have at least six degrees of freedom. This four-degree-of-freedom manipulator can reach almost any point within a radius of $r_1 + r_2 + r_3$ from the z-axis, and any elevation within the limits of the actuator of link A. (Some positions may be inaccessible due to interference between the parts and limitations of the motors and actuator.) But the robot's motion is limited. For example, an end effector could remove a part from a horizontal conveyor belt and place it on a lower (horizontal) conveyor at a different x and y location. The missing degrees of freedom prevent the manipulator from rotating the part about the axis of link 3 or about a horizontal axis perpendicular to link 3.

(c) Working Model™ was used to model the tentative manipulator design. Figure 12.17a traces the motion of the extreme end of link 3. Part b shows the following variables at time $t = 0.2$ s:

- the x- and y-coordinates of the centers of gravity of links 2 and 3
- the angular position of links 2 and 3
- the x- and y-direction velocity components and the resultant velocity of the center of gravity of link 3
- the angular velocity of link 3

(*Note:* The linkage position at $t = 0.2$ s is the position for which $\theta_1 = 3$ rad.)
Figure 12.17c shows the following plotted against time:

- the x- and y-coordinates of the center of gravity of link 3
- the angular position of link 3 (divided by 10 for scaling purposes)
- the x- and y-coordinates of the velocity and the resultant velocity of the center of gravity of link 3.

(*Note:* All quantities are measured in the fixed coordinate system, in which the angular velocity of link 3 is constant (1 rad/s).

Detailed Spatial Motion Analysis

Mathematics software with matrix capability is an alternative to motion simulation software for dealing with long and complicated manipulator and robot problems. If you intend to "work smart," define a general transformation matrix as a function of the change in angle, the link length, translation, and time. You can use the matrix over and over, inserting the parameters for each link.

SAMPLE PROBLEM 12.6

Detailed Analysis of the Motion of a PRRR *Spatial Manipulator*

Reexamine the motion of the manipulator described in Sample Problem 12.5 and Figure 12.17a. In this case, the actuator drives link A to change the elevation of the linkage according to the equation

$$Z(t) = 0.180\,\mathrm{e}^{-1.1\,t}$$

Describe the motion of the linkage.

Solution summary. The transformation matrix can be defined as follows if the revolute joints have axes parallel to the z-axis and the prismatic pair axis is parallel to the z-axis:

$$T(\theta, r, z, t) := \begin{bmatrix} \cos\theta & -\sin\theta & 0 & r \\ \sin\theta & \cos\theta & 0 & 0 \\ 0 & 0 & 1 & z \\ 0 & 0 & 0 & 0 \end{bmatrix}.$$

The general matrix is written only once for this problem. Instead of filling in the matrix, we just need to call for it by its name and arguments. For example, to go from a coordinate system attached to link 1 to the fixed coordinate system at time = 0.2 second, we call $T(\theta_1, 0, Z, 0.2)$,

where $\theta = \theta_1 = \omega_1 \cdot t,$

$r = 0,$

$z = Z = 0.180\,\mathrm{e}^{-1.1\,t},$

and

$t = 0.2$ s.

We can describe the location of P_c, the center of link 3, in a reference frame connected to link 3 by the vector

$$P_{c3} := \begin{bmatrix} r_{3c} \\ 0 \\ 0 \\ 1 \end{bmatrix},$$

where r_{3c} = half the length of link 3. To locate the center of link 3 in the fixed reference frame, we need only type

$$P_{c0}(t) := T(\theta_1, 0, Z, t) \cdot T(\theta_2, r_1, \text{zero}, t) \cdot T(\theta_3, r_2, \text{zero}, t) \cdot P_{c3},$$

where, reading from *right to left,* the first transformation matrix takes us from link-3 to link-2 coordinates, the middle transformation matrix takes us from link-2 to link-1 coordinates, and the

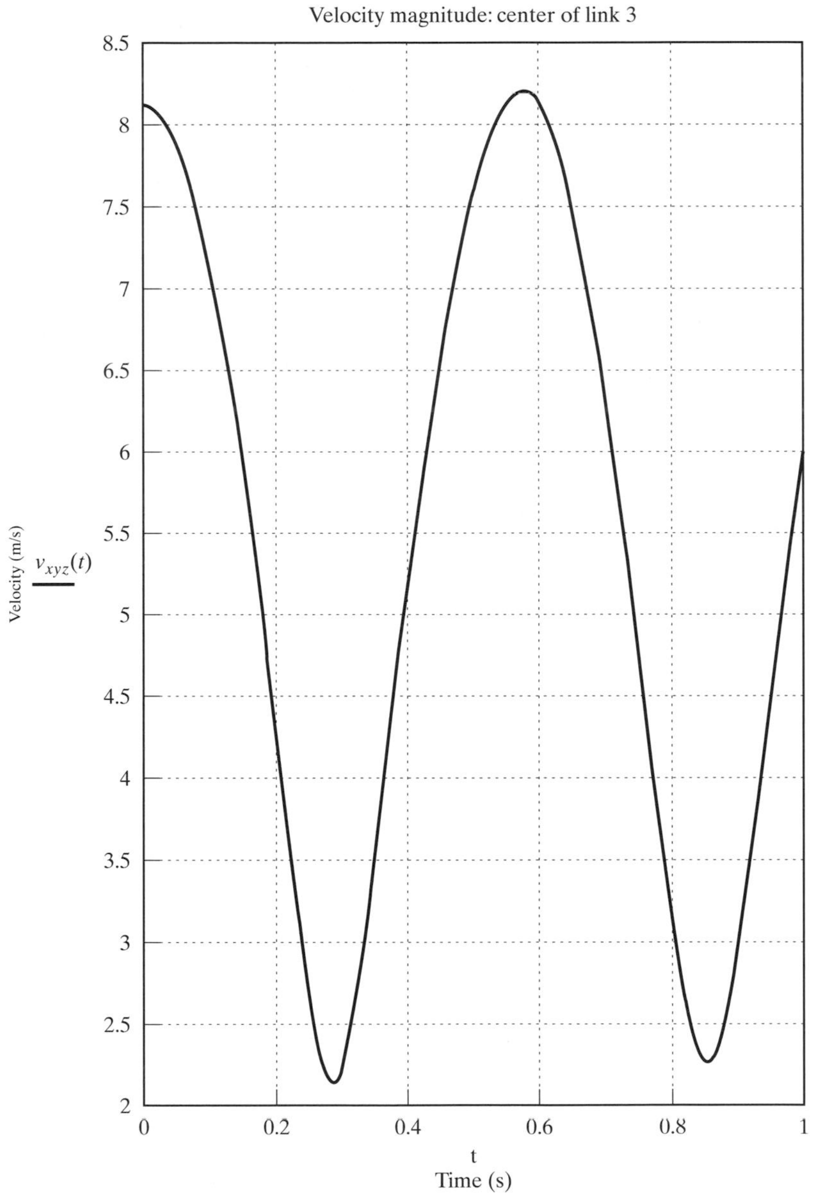

FIGURE 12.18 Resultant magnitude of velocity for the center of link 3 plotted against time, Sample Problem 12.6.

transformation matrix next to the equals sign takes us to the fixed coordinate system. The first three terms of the vector

$$P_{c0}(.2) = \begin{bmatrix} -0.183 \\ -0.164 \\ 0.144 \\ 1 \end{bmatrix}$$

are the x-, y-, and z-coordinates of the center of link 3 at time $= 0.2$ s.

The drive motor between links 2 and 3 is located in a coordinate system attached to link 2, and the transformation matrices are used to locate it in the fixed system. This point and the center of link 3 are used to find the angular position of link 3. Either the two-argument ANGLE function or the two-argument $ARCTAN_2$ function can be used. The velocity vector for the center of link 3 is found by dividing the difference in values of vector P_{c0} over a short interval by that interval. Figure 12.18 shows the resultant magnitude of the velocity for the center of link 3 plotted against time.

Detailed Calculations.

PRRR manipulator

Time $t := 0,.01..1$ seconds

Link lengths $r_1 := 0.200 \quad r_2 := 0.200 \quad r_3 := 0.160$ meters $\quad r_{3c} := \frac{r_3}{2}$

Relative angular velocities $\omega_1 := 15 \quad \omega_2 := 11 \quad \omega_3 := -25$ rad/s

Link rotation relative to adjacent link

$\theta_1(t) := \omega_1 \cdot t \qquad \theta_2(t) := \omega_2 \cdot t \qquad \theta_3(t) := \omega_3 \cdot t - \pi$

Translation: $Z(t) := 0.180 \cdot \exp(-1.1 \cdot t)$ $\text{zero}(t) := 0$ (i.e., no translation at this joint)

Define transformation matrix for z-oriented revolute joints and z-axis prismatic joint(s):

$$T(\theta, r, z, t) := \begin{bmatrix} \cos(\theta(t)) & -\sin(\theta(t)) & 0 & r \\ \sin(\theta(t)) & \cos(\theta(t)) & 0 & 0 \\ 0 & 0 & 1 & z(t) \\ 0 & 0 & 0 & 1 \end{bmatrix}$$

Transformation matrix: to fixed coordinates from coordinates on link 1 at 0 and 0.2 s:

$$T(\theta_1, 0, Z, 0) = \begin{bmatrix} 1 & 0 & 0 & 0 \\ 0 & 1 & 0 & 0 \\ 0 & 0 & 1 & 0.18 \\ 0 & 0 & 0 & 1 \end{bmatrix} \qquad T(\theta_1, 0, Z, .2) = \begin{bmatrix} -0.99 & -0.141 & 0 & 0 \\ 0.141 & -0.99 & 0 & 0 \\ 0 & 0 & 1 & 0.144 \\ 0 & 0 & 0 & 1 \end{bmatrix}$$

Locate drive motor between links 2 and 3 in reference frame connected to link 2:

$$P_{232} := \begin{bmatrix} r_2 \\ 0 \\ 0 \\ 1 \end{bmatrix} \qquad P_{232} = \begin{bmatrix} 0.2 \\ 0 \\ 0 \\ 1 \end{bmatrix}$$

Locate drive motor between links 2 and 3 in fixed reference frame 0 at time *t* and at 0.2 s:

$$P_{230}(t) := T(\theta_1, 0, Z, t) \cdot T(\theta_2, r_1, \text{zero}, t) \cdot P_{232} \qquad P_{230}(.2) = \begin{bmatrix} 0.104 \\ -0.148 \\ 0.144 \\ 1 \end{bmatrix}$$

Locate center of link 3 in reference frame 3 connected to link 3:

$$P_{c3} := \begin{bmatrix} r_{3c} \\ 0 \\ 0 \\ 1 \end{bmatrix} \qquad P_{c3} = \begin{bmatrix} 0.08 \\ 0 \\ 0 \\ 1 \end{bmatrix}$$

Locate center of link 3 in fixed reference frame 0 at time *t*, 0.001 s, and 0.2 s:

$$P_{c0}(t) := T(\theta_1, 0, Z, t) \cdot T(\theta_2, r_1, \text{zero}, t) \cdot T(\theta_3, r_2, \text{zero}, t) \cdot P_{c3}$$

$$P_{c0}(0.001) = \begin{bmatrix} 0.32 \\ 8.119 \cdot 10^{-3} \\ 0.18 \\ 1 \end{bmatrix} \qquad P_{c0}(.2) = \begin{bmatrix} -0.183 \\ -0.164 \\ 0.144 \\ 1 \end{bmatrix}$$

Orientation of link 3 in fixed reference frame 0 at time t, 0, and 0.2 s
(*Note:* Array subscripts 0, 1, 2 refer to *x*, *y*, *z* components, respectively):

$$\theta_{30}(t) := \text{angle}(P_{c0}(t)_0 - P_{230}(t)_0, P_{c0}(t)_1 - P_{230}(t)_1)$$

$$\theta_{30}(0) = 3.142 \qquad \theta_{30}(.2) - 2 \cdot \pi = -2.942$$

Velocity of center of link 3 (average over small time difference Δt) at time *t*, 0, and 0.2 s:

$$\Delta t := .0001$$

Velocity vector:

$$v(t) := \frac{P_{c0}(t + \Delta t) - P_{c0}(t)}{\Delta t} \qquad v(0) = \begin{bmatrix} -9.006 \cdot 10^{-3} \\ 8.12 \\ -0.198 \\ 0 \end{bmatrix} \qquad v(.2) = \begin{bmatrix} 4.186 \\ -0.606 \\ -0.159 \\ 0 \end{bmatrix}$$

Projection on *xy*-plane:

$$v_{xy}(t) := [(v(t)_0)^2 + (v(t)_1)^2]^{\frac{1}{2}} \qquad v_{xy}(0) = 8.12 \qquad v_{xy}(.2) = 4.229$$

Magnitude of velocity:

$$v_{xyz}(t) := [(v(t)_0)^2 + (v(t)_1)^2 + (v(t)_2)^2]^{\frac{1}{2}} \qquad v_{xyz}(0) = 8.122 \qquad v_{xyz}(.2) = 4.232 \text{ (m / s)}$$

12.6 INVERSE KINEMATICS

Inverse kinematics refers to the process of determining joint coordinates from known values of the end-effector coordinates. As mentioned earlier, this type of computation is useful for programming robots, because actuation occurs at the joints, and therefore, the joint displacements corresponding to a prescribed end-effector displacement must be known. However, as will be demonstrated in the sections that follows, inverse kinematic analysis is often more difficult than forward kinematic analysis. This is so because the mathematical relationships are nonlinear in the joint coordinates. Consequently, there is no general solution procedure that governs all cases involving robots. Indeed, in some instances it may be necessary to use numerical methods for the solution of the nonlinear equations.

Two inverse kinematic approaches will be illustrated. The first approach involves the use of forward kinematics equations, which are solved for joint variables. In the second approach, a more direct derivation of expressions for joint coordinates will be pursued. The success of the latter approach is heavily dependent upon the specific manipulator geometry.

2R Planar Manipulator

Consider the two-degree-of-freedom manipulator that was analyzed in Section 12.5 and illustrated in Figure 12.13. Expressions for the robot tip coordinates as functions of the joint coordinates were derived, yielding Eq. (12.49), which is repeated here in the form

$$x = l_1 \cos\theta_A + l_2 \cos(\theta_A + \theta_B) \tag{12.60}$$

and

$$y = l_1 \sin\theta_A + l_2 \sin(\theta_A + \theta_B), \tag{12.61}$$

where x and y refer to the tip, or end-effector, coordinates (i.e., point C in Figure 12.13) in terms of the fixed reference frame (frame 0). Subscripts and superscripts have been dropped in order to simplify the form of the equations.

Following the first approach just outlined, we will attempt to solve Eqs. (12.60) and (12.61) for θ_A and θ_B as functions of x and y. Squaring both sides of each equation and then adding the two equations results in

$$\begin{aligned} x^2 + y^2 &= l_1^2 + l_2^2 + 2l_1l_2\left[\cos\theta_A\cos(\theta_A + \theta_B) + \sin\theta_A\sin(\theta_A + \theta_B)\right] \\ &= l_1^2 + l_2^2 + 2l_1l_2\cos\theta_B. \end{aligned}$$

Rearranging terms, we obtain

$$\cos\theta_B = \frac{x^2 + y^2 - l_1^2 - l_2^2}{2l_1l_2}, \tag{12.62}$$

or

$$\theta_B = \cos^{-1}\left(\frac{x^2 + y^2 - l_1^2 - l_2^2}{2l_1l_2}\right). \tag{12.63}$$

There are two values of θ_B between $-180°$ and $+180°$ that satisfy this equation, indicating that there are multiple configurations the robot can have and still maintain the

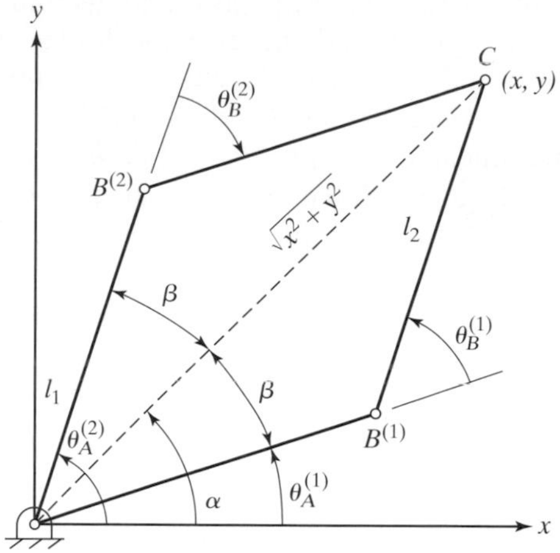

FIGURE 12.19 Multiple manipulator configurations obtained from inverse kinematics.

same tip coordinates (x, y). In fact, there are two such configurations of the robot in this example, which are illustrated in Figure 12.19. This characteristic of multiple solutions is common to most inverse kinematic analyses. The particular orientation that the robot would actually take is dictated by its previous positions and the way in which the joint motions are programmed.

Having found θ_B, we can return to Eqs. (12.60) and (12.61) in order to determine θ_A. Rewriting those equations in a modified form, we have

$$(l_1 + l_2 \cos\theta_B)\cos\theta_A - l_2 \sin\theta_B \sin\theta_A = x \tag{12.64}$$

and

$$l_2 \sin\theta_B \cos\theta_A + (l_1 + l_2 \cos\theta_B)\sin\theta_A = y, \tag{12.65}$$

where $\cos\theta_B$ is given by Eq. (12.62) and

$$\sin\theta_B = \pm\sqrt{1 - \cos^2\theta_B}, \tag{12.66}$$

where the plus and minus signs correspond to the two solutions for θ_B as noted earlier. Solving Eqs. (12.64) and (12.65) simultaneously for $\cos\theta_A$ and $\sin\theta_A$ yields

$$\cos\theta_A = \frac{x(l_1 + l_2\cos\theta_B) + yl_2\sin\theta_B}{l_1^2 + l_2^2 + 2l_1l_2\cos\theta_B} \tag{12.67}$$

and

$$\sin\theta_A = \frac{y(l_1 + l_2\cos\theta_B) - xl_2\sin\theta_B}{l_1^2 + l_2^2 + 2l_1l_2\cos\theta_B}. \tag{12.68}$$

Hence,

$$\tan\theta_A = \frac{\sin\theta_A}{\cos\theta_A} = \frac{y(l_1 + l_2\cos\theta_B) - xl_2\sin\theta_B}{x(l_1 + l_2\cos\theta_B) + yl_2\sin\theta_B}, \tag{12.69}$$

or

$$\theta_A = \tan^{-1}\left[\frac{y(l_1 + l_2\cos\theta_B) - xl_2\sin\theta_B}{x(l_1 + l_2\cos\theta_B) + yl_2\sin\theta_B}\right] \qquad (12.70)$$

The signs of the numerator and denominator in Eq. (12.70) can be used to establish the specific quadrant of angle θ_A. Thus, there are two solutions for θ_A, corresponding to the two solutions for θ_B, as represented by the two signs in Eq. (12.66). Once again, the corresponding configurations of the robot are shown in Figure 12.19. Equations (12.63) and (12.70) are the desired relationships for θ_B and θ_A as functions of x and y.

Alternatively, expressions for θ_A and θ_B can be derived directly from the geometry of Figure 12.19. From that figure, it can be seen that angle θ_A can be expressed in terms of angles α and β as

$$\theta_A = \alpha \pm \beta, \qquad (12.71)$$

where, from the x- and y-coordinates,

$$\alpha = \tan^{-1}\left(\frac{y}{x}\right). \qquad (12.72)$$

The signs of x and y will identify the proper quadrant. Angle β can be obtained from the law of cosines

$$l_2^2 = x^2 + y^2 + l_1^2 - 2l_1\sqrt{x^2 + y^2}\cos\beta.$$

Thus,

$$\beta = \cos^{-1}\left(\frac{x^2 + y^2 + l_1^2 - l_2^2}{2l_1\sqrt{x^2 + y^2}}\right), \qquad (12.73)$$

where the positive value of β between 0° and 180° would be used in Eq. (12.71). Angle θ_B can also be determined from the law of cosines. In this case,

$$x^2 + y^2 = l_1^2 + l_2^2 - 2l_1l_2\cos(\pi - \theta_B) = l_1^2 + l_2^2 + 2l_1l_2\cos\theta_B,$$

which yields

$$\theta_B = \cos^{-1}\left(\frac{x^2 + y^2 - l_1^2 - l_2^2}{2l_1l_2}\right). \qquad (12.74)$$

This is the same as Eq. (12.63) found by the first approach. Care must be exercised in combining the angles from Eqs. (12.71) and (12.74). In particular, inspection of Figure 12.19 reveals that the negative value of θ_B from Eq. (12.74) should be used in combination with the positive sign in Eq. (12.71), while the positive value of θ_B corresponds to the case of the negative sign in Eq. (12.71).

RP Manipulator

Inverse kinematics for mechanisms containing prismatic joints tends to be somewhat easier than for those with revolute joints. This is true for the manipulator of Section 12.5 and Figure 12.14, which is a two-degree-of-freedom manipulator with a combination of a revolute pair and a prismatic pair. From Eq. (12.53), the relations are

$$x = -(s_B + l_2)\sin\theta_A, \tag{12.75}$$

$$y = (s_B + l_2)\cos\theta_A, \tag{12.76}$$

and

$$z = s_A. \tag{12.77}$$

In this mechanism, s_A is a fixed dimension, and therefore, coordinate z is constant. Joint rotation variable θ_A is obtained by dividing Eq. (12.75) by Eq. (12.76), leading to

$$\theta_A = \tan^{-1}\left(\frac{-x}{y}\right), \tag{12.78}$$

where the signs of the numerator and denominator can be used to identify the quadrant. Squaring Eqs. (12.75) and (12.76) and adding yields the following relation for translational variable s_B:

$$s_B = \sqrt{x^2 + y^2} - l_2. \tag{12.79}$$

Note that multiple solutions do not exist in this case.

RRR Spatial Manipulator

A three-degree-of-freedom manipulator with three revolute pairs was also considered in Section 12.5. (See Figure 12.15.) The forward kinematics equations that were developed (see Eq. 12.59) are

$$x = l_2\cos\theta_A\cos\theta_B + l_3\cos\theta_A\cos(\theta_B + \theta_C), \tag{12.80}$$

$$y = l_2\sin\theta_A\cos\theta_B + l_3\sin\theta_A\cos(\theta_B + \theta_C), \tag{12.81}$$

and

$$z - s_A = l_2\sin\theta_B + l_3\sin(\theta_B + \theta_C). \tag{12.82}$$

The solution for angles θ_A, θ_B, and θ_C can be patterned after the approach used to solve a 2R planar manipulator. Squaring each of the preceding equations and then adding produces

$$x^2 + y^2 + (z - s_A)^2 = l_2^2 + l_3^2 + 2l_2l_3\cos\theta_C,$$

from which we obtain

$$\theta_C = \cos^{-1}\left(\frac{x^2 + y^2 + (z - s_A)^2 - l_2^2 - l_3^2}{2l_2l_3}\right). \tag{12.83}$$

As before, there are two possible solutions for θ_C.

Next, θ_A is obtained by dividing Eq. (12.81) by Eq. (12.80), yielding

$$\frac{y}{x} = \tan \theta_A,$$

or

$$\theta_A = \tan^{-1}\left(\frac{y}{x}\right). \tag{12.84}$$

Finally, rewriting Eqs. (12.80) and (12.82) yields

$$\frac{x}{\cos \theta_A} = (l_2 + l_3 \cos \theta_C) \cos \theta_B - l_3 \sin \theta_C \sin \theta_B$$

and

$$z - s_A = l_3 \sin \theta_C \cos \theta_B + (l_2 + l_3 \cos \theta_C) \sin \theta_B.$$

Solving for $\sin \theta_B$ and $\cos \theta_B$ and then combining produces

$$\theta_B = \tan^{-1}\left(\frac{(z - s_A)(l_2 + l_3 \cos \theta_C)\cos \theta_A - x l_3 \sin \theta_C}{x(l_2 + l_3 \cos \theta_C) + (z - s_A) l_3 \sin \theta_C \cos \theta_A}\right) \tag{12.85}$$

where

$$\sin \theta_C = \pm \sqrt{1 - \cos^2 \theta_C}. \tag{12.86}$$

Equations (12.83), (12.84), and (12.85) are the required equations for carrying out the inverse kinematic analysis of the RRR spatial manipulator.

12.7 PRACTICAL CONSIDERATIONS IN ROBOT AND MANIPULATOR DESIGN

Good design practice includes asking questions. For example, we might ask the following questions:

- What is it that must be done?
- Was the task always done in a particular way? Why?
- Can we automate the task?
- Will a robot perform the task best?
- Are there special environmental considerations? Is the environment hazardous to humans? Does the task require a clean room?
- Could a human operator degrade the quality of the product?
- Are we likely to change product lines? If so, will we be able to use the robot for other important tasks, or will a manipulator (with less than six degrees of freedom) work as well?

- Can a simple manipulator perform the task faster, with greater precision, or at lower cost, or is dedicated machinery the best solution?
- And, back to the original task description, Is the task really necessary? Can the task be redefined? Can parts of the task be eliminated?

When these questions are answered, we can begin to design a system to perform the required task.

End Effectors

An end effector or gripper acts as the hand of a robot. Common options include two- or three-fingered mechanical clamps (operated by electric, hydraulic, or pneumatic systems), electromagnets, and devices in which a vacuum holds the object. Users sometimes custom-design end effectors to meet special manufacturing needs, or they may mount a tool directly on the end of a robot arm.

Sensors

Despite recent advances in "robot vision," it is still difficult to design a system that can process visual information and make human-like decisions. Proximity, force, and torque sensing may be more practical at present. Force and torque sensors may be part of an end effector or may be located between an end effector and the next robot element. A strain-gage force sensor, for instance, can detect contact forces at a workpiece and feed the information back to the robot controller. Then, the robot acts accordingly to perform the task without damage to the workpiece or the robot itself.

Assembly Operations

For years, engineers have tried to design robots and manipulators for the task of assembling consumer products. Robots and manipulators have shown a competitive advantage in a number of settings, including

- clean-room applications (e.g., the manufacture of semiconductors and disk drives and the development of pharmaceuticals),
- hazardous locations (e.g., situations in which human operators would be exposed to paint spray, radiation, chemical fumes, or dust) and
- specialized tasks such as welding.

Currently, routine assembly is not one of the things that robots do best or most economically. New developments may change this, however.

Selective Compliance. Selective compliance is a special feature that can help robots perform assembly and similar tasks. In selective compliance, a robot or manipulator is compliant (flexible) in one or two directions, but not in others. For example, a manipulator that is compliant in the horizontal plane can be used to drive a pin (with a chamfer or taper) into a hole, even if the hole is not located precisely. To avoid tilting and jamming the pin, there should be no significant compliance (deviation) from the vertical.

Jigs and Fixtures. Jigs and fixtures can be used to improve the accuracy of positioning. If a jig is attached to an assembly, a selectively compliant robot can drill precisely positioned holes in the assembly.

Part Symmetry and Asymmetry. Most robots and manipulators have poor vision or no vision. We can compensate for this by making tasks easier. Sometimes it is possible to make a part symmetrical so that it can be assembled in more than one orientation. If the design *requires* a particular orientation for a given part, consider exaggerated asymmetry. Try to design the part so that it can be fed to the robot automatically with the correct orientation. Then, it is less likely that the robot will attempt an incorrect assembly.

The recommendations for symmetry or exaggerated asymmetry apply to other automated assembly operations, to in-factory manual assembly, and to products that must be assembled by the user. If you design parts so that there is more than one right way to assemble them, then you increase the probability that a robot, machine, or individual will make a correct assembly. If you exaggerate asymmetry when asymmetry is necessary, then you decrease the probability that a robot, machine, or individual will attempt an incorrect assembly.

Dependent Degrees of Freedom

Most robots are designed in the form of open-loop kinematic chains with one-degree-of-freedom joints. For these robots, the total number of degrees of freedom equals the number of joints and also equals the number of actuators. But the number of *independent* degrees of freedom—the number of degrees of freedom available to manipulate an end effector—may be less. The following robot and manipulator configurations result in *dependent* degrees of freedom:

- More than three revolute joints with axes parallel to one another.
- More than two prismatic pairs having axes parallel to the same plane.
- More than three prismatic pairs.
- More than six joints (pairs).
- Other combinations in which the motion provided by one group of joints duplicates that provided by another.

For robot and manipulator designs described by these configurations, the number of degrees of freedom available to manipulate the end effector is less than the total number of degrees of freedom. Extra (dependent) degrees of freedom do not always represent design errors; they may be an intentional feature of the design. For example, an extra revolute joint may be added to enable the end effector to reach around an obstacle.

Examples of Commercially Available Robots

Figure 12.20a shows a selective compliance assembly robot arm driven by AC-servo motors. The robot is a four-degree-of-freedom device, with two revolute joints providing motion like a shoulder and elbow held parallel to the ground. The two links comply

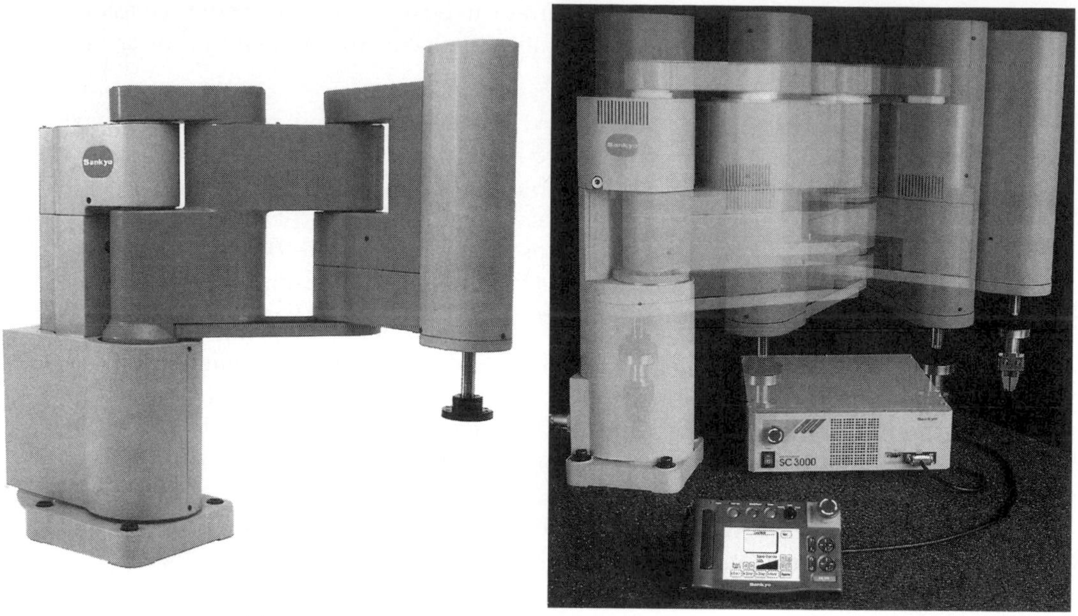

(a) (b)

FIGURE 12.20 (a) A selective compliance assembly robot arm. (*Source:* Sankyo Robotics.) (b) Time-exposure of a robot arm with a multitasking robot controller and a touch-screen pendant. (*Source:* Sankyo Robotics.)

slightly in the horizontal direction. The other two degrees of freedom are z-axis (up-and-down) motion and roll at the end of the second link. A similar robot designed for higher payload and higher rotating torque is shown in Figure 12.20b. A multitasking robot controller and touch-screen pendant are shown below the robot. The pendant is used for hand held control and to teach the robot system a sequence of motions for performing a given task.

Programmable Cartesian coordinate robots are shown in Figure 12.21. These robots incorporate AC-servo motors and precision ball screws and are used to position tools such as dispensers, drivers, cutters, and routers. The robot shown in Figure 12.21a has x, y, and z motion with a roll head. Part b of the figure shows a special configuration for load and unload operations with a special "reaching" y-axis that can be positioned side to side (in the x direction) and up and down (in the z direction).

Figure 12.22a shows a robot that is used extensively for arc welding and other industrial applications. The working envelope (working range) of a similar robot is shown in Figure 12.22b. Some industrial robots are designed for overhead mounting, a practical consideration when floor space is restricted and when a large work-processing area is required.

Rotary assembly machines utilizing index drives are sometimes used in conjunction with robots and manipulators, as sketched in Figure 12.23a. These machines are used to present parts consistently and accurately to the workstations. Rotary assembly machines similar to that in the sketch are used in automated assembly, welding,

(a) (b)

FIGURE 12.21 (a) A Cartesian coordinate robot (*Source:* Sankyo Robotics.) (b) A special configuration for load and unload operations. (*Source:* Sankyo Robotics.)

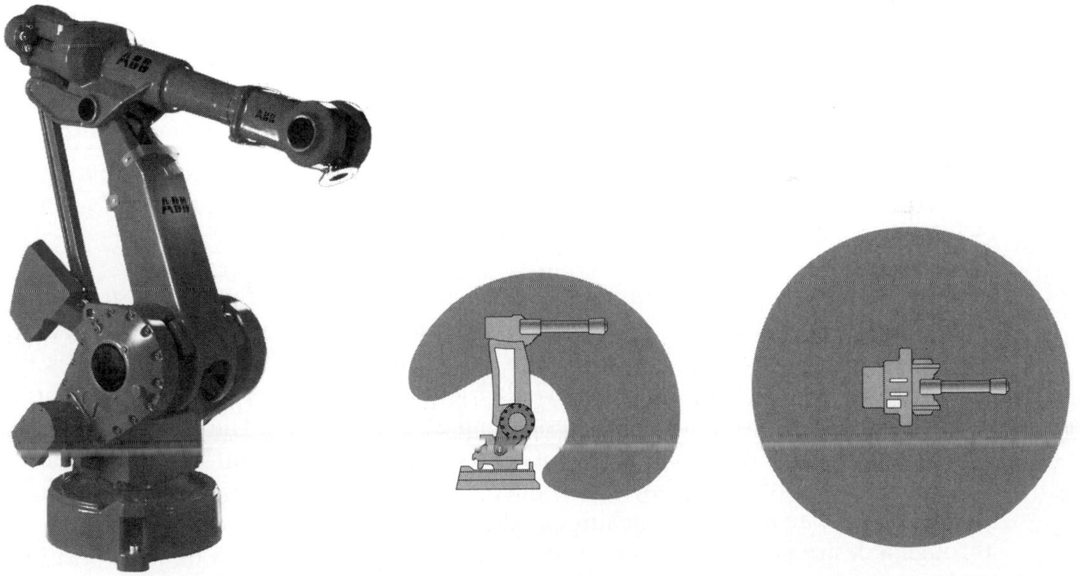

(a) (b)

FIGURE 12.22 (a) An industrial robot combining both closed- and open-loop linkages. (*Source:* ABB Flexible Automation, Inc.) (b) The working envelope of a similar robot. (*Source:* ABB Flexible Automation, Inc.)

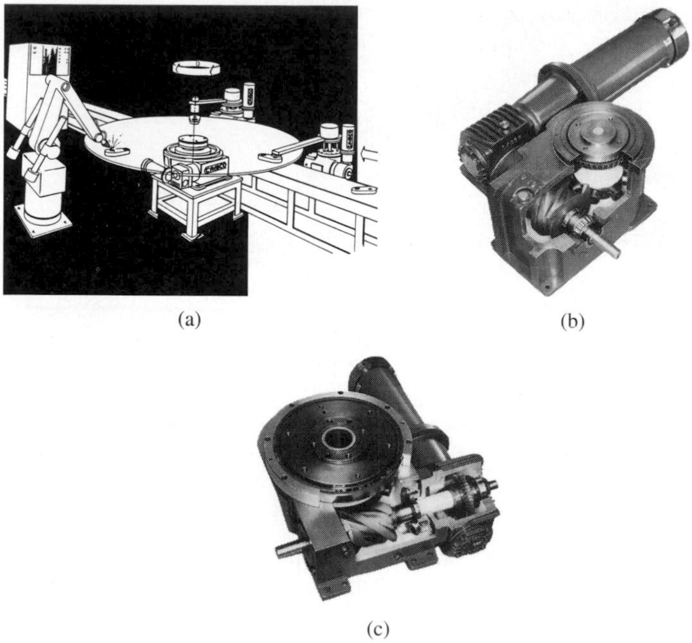

FIGURE 12.23 (a) A rotary assembly machine. (*Source:* Commercial Cam Co., Inc.) (b) and (c) Cam-driven index drives. (*Source:* Commercial Cam Co., Inc.)

and inserting applications. Figures 12.23b and c show cutaway views of cam-driven index drives.

SUMMARY

A robot system consists of a manipulator mechanism, an end effector, a set of actuators, and a controller. A typical manipulator mechanism has the form of an open kinematic chain connected by revolute joints or prismatic pairs. An end effector is the gripper or "hand" attached to the final link. Electric, hydraulic, or pneumatic actuators are located at each joint and operate the end effector as well. The controller is programmed to direct the robot through its tasks. Joint-coordinate information required for the controller program may be obtained by an analytical inverse kinematics study. If the robot system has a "teaching pendant," the operator moves the end effector through a desired cycle of motion to store the required information on joint positions in the controller.

The programmable controller, the many degrees of freedom (typically, six or more), and the task-flexibility of a robot distinguish it from machine tools dedicated to a single task. Options include vision, proximity and contact sensing, and force and torque sensing. Information from the sensors is fed back to the controller so that it can adjust the signals to the actuators. Force and torque sensing at the joints and the end effector help the robot to provide adequate force without damaging a workpiece or damaging the robot itself. If a robot is used along a moving production line, it is likely to have a preprogrammed sequence for corrective action and an escape path when it senses a malfunction.

The word "manipulator" refers to the kinematic chain of a robot; sometimes it is also used to identify a robot system with fewer than six degrees of freedom. For flexible manufacturing, a robot with six or more degrees of freedom, sensing capabilities, and other optional extras will allow for quick changes in product lines. When flexibility in performing task is unimportant, a three- or four-degree-of-freedom manipulator may provide equal or better precision, equal or greater payload capability, and higher speed and acceleration at lower cost.

Transformation matrices are the key to determining motion as a function of joint rotation and translation. Here is one way to determine end-effector motion in terms of fixed coordinates if there are n moving links:

- Attach a coordinate system to the nth link in the manipulator chain (the link with the end effector). Place the origin of those coordinates at the pair joining link n and link $n - 1$ (the final and next-to-last links).
- Locate the end effector in the nth link coordinates, using a four-element vector (homogeneous coordinate form). The first three elements are the x-, y-, and z-coordinates; the last term is unity.
- Attach a coordinate system to the $(n - 1)$th link in the manipulator chain. (The $(n - 1)$th link is next to the link with the end effector.) Place the origin of that system at the pair joining the $(n - 1)$th and $(n - 2)$th links.
- Construct a 4×4 transformation matrix to convert from n^{th} link coordinates to $(n - 1)$th-link coordinates.
- Construct another matrix to convert from $(n - 1)$th-link coordinates to $(n - 2)$th-link coordinates, and so on.
- Multiply the end-effector vector (n^{th}-link coordinates) by the transformation matrices in turn to locate the vector in the fixed coordinate system.

For detailed analysis and plotting, define the transformation matrix in function form to save time. Velocity and acceleration analysis can follow. Maximum robot speed ratings are usually based on light payloads; maximum payload ratings are based on reduced speed and acceleration.

Motion simulation software (e.g., Working Model™) offers an alternative analysis method that enables us to trace the motion of the robot arms and make tentative design changes with a few keystrokes. Inverse kinematics involves the determination of joint coordinates when the location and orientation of the end effector is specified (in a fixed coordinate system). Inverse kinematics problems often have multiple solutions; that is, two or more combinations of link orientations may produce the same location and orientation of the end effector.

A Few Review and Discussion Items.

- Identify some tasks for which robots are ideally suited.
- Identify other tasks that currently available robots cannot do competitively.

- How would you perform those tasks?
- Suppose you are assigned the task of designing a system to weld complex objects. Your assistant proposes a robot design in the form of an open kinematic chain with seven revolute joints. What is the total number of degrees of freedom? What is the maximum possible number of degrees of freedom available to manipulate an end effector?
- Will you tell the assistant to consider another line of work, or could an extra degree of freedom serve a practical purpose?
- Sketch a manipulator having two revolute joints with vertical axes and two prismatic pairs with horizontal axes. How many degrees of freedom does it have? How many degrees of freedom are available to manipulate an end effector?
- Is there an alternative to forward kinematics for programming robot motion?
- Under what circumstances is selective compliance a desirable robot attribute?
- Parts are to be handled by a robot or manipulator. What is the advantage of symmetry in parts? Why might you design a part with exaggerated asymmetry?

PROBLEMS

12.1 Determine the number of degrees of freedom of each of the spatial mechanisms shown in Figure P12.1, where, in each case, the mechanism is a closed chain in which the first and last members are part of the ground link.

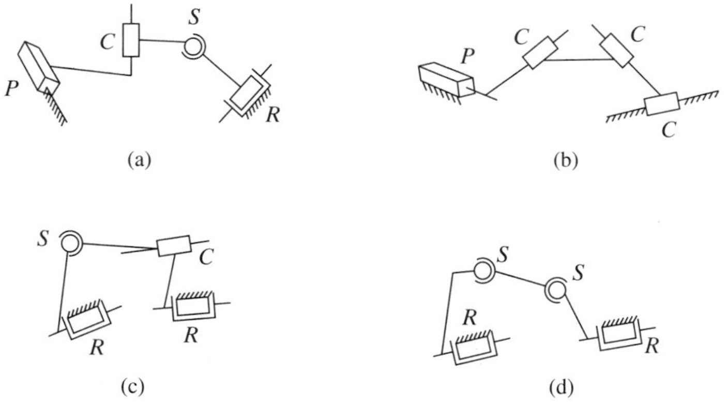

FIGURE P12.1

12.2 Determine the number of degrees of freedom of each of the spatial mechanisms shown in Figure P12.2, where, in each case, the mechanism is an open chain with the first member part of the ground and the last member free to move.

12.3 Derive the 3 × 3 rotational transformation matrix ${}^{i}_{j}[R]$ for the following sequence of rotations: rotation β about the y_i-axis, followed by rotation γ about the z_i-axis and then rotation α about the x_i-axis.

12.4 Show that finite rotations about perpendicular axes are not commutative by comparing the sequence of an x-axis rotation followed by a y-axis rotation with the sequence of a y-axis rotation followed by an x-axis rotation. Use transformation matrices.

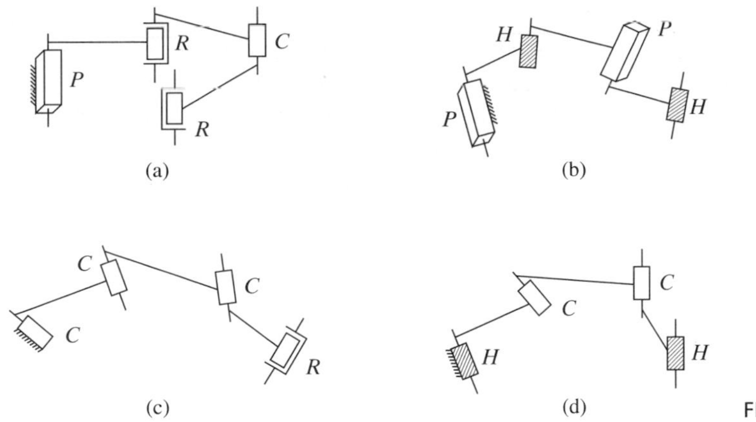

FIGURE P12.2

12.5 Reference frames i and j have the same origin. The orientation of frame j relative to frame i is given by a rotation of $-90°$ about the x_i-axis, followed by a rotation of 45° about the y_i-axis and then $-45°$ about the z_i-axis. A point has coordinates $\{1,1,1\}^T$ in the j frame. Determine the coordinates of the point in the i frame.

12.6 Reference frames i and j have the same origin. The orientation of frame j relative to frame i is given by a rotation of 90° about the y_i axis, followed by a rotation of 90° about the z_i-axis, and then 90° about the x_i-axis. Determine the rotational transformation matrices ${}^i_j[R]$ and ${}^j_i[R]$.

12.7 Reference frames i and j have the same origin. The orientation of frame j relative to frame i is given by a rotation of $-90°$ about an axis $i_{\{u\}} = (1/\sqrt{2}, 1/\sqrt{2}, 0)^T$. A point has coordinates $\{1,1,1\}^T$ in the j frame. Determine the coordinates of the point in the i frame.

12.8 Derive an expression for the rotational transformation matrix ${}^j_i[R]$, where ${}^i_j[R]$ is given by Eq. (12.17).

12.9 Derive Eq. (12.12) from Eq. (12.20).

12.10 Frame j has the same origin as frame i and an orientation obtained from the following sequence: a 90° rotation about x_i, followed by a 90° rotation about y_i, followed by a 90° rotation about z_i. Determine the axis of rotation and angle of rotation for an equivalent single-axis rotation from frame i to frame j.

12.11 The orientation of frame 2 relative to frame 1 is given by a rotation of 90° about the y_i axis, followed by a rotation of 90° about the x_i-axis. The position of the origin of frame 2 relative to the origin of frame 1 is ${}^1\{Q\} = \{2\ 3\ 0\}^T$. The coordinates of a point in frame 2 are ${}^2\{P\} = \{3\ 4\ 1\}^T$. Determine the coordinates of the point in frame 1. Draw a sketch of the frames and the point to verify your results.

12.12 The orientation of frame 2 relative to frame 1 is given by a rotation of $-45°$ about the x_i-axis, followed by a rotation of $+45°$ about the y_i-axis, followed by a rotation of $-90°$ about the z_i-axis. The position of the origin of frame 2 relative to the origin of frame 1 is ${}^1\{Q\} = \{1\ 0\ -2\}^T$. The coordinates of a point measured in frame 2 are ${}^2\{P\} = \{1\ 2\ -1\}^T$. Determine the coordinates of the point in frame 1.

12.13 An RRR planar manipulator is shown in Figure P12.3, where the axes of the three revolute joints are parallel. Assume that the link offsets are zero; that is, $s_A = s_B = s_C = 0$. Point D is a point on member 3. The home position of the robot is shown in Figure P12.3a,

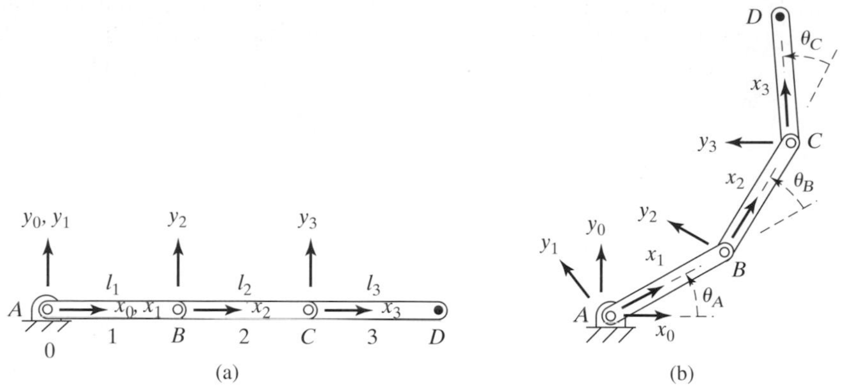

FIGURE P12.3

and joint rotations θ_A, θ_B, and θ_C are relative to this position. Derive a relationship for ${}^0\{P\}_D$ as a function of ${}^3\{P\}_D$, and evaluate these equations for the case of $l_1 = l_2 = l_3 = 10$ in and $\theta_A = \theta_B = \theta_C = 30°$.

12.14 Determine the transformation matrix ${}^0_3[T]$ for the RRR manipulator in Figure P12.4 as a function of the dimensions shown and the joint angles θ_A, θ_B, and θ_C. The position shown corresponds to $\theta_A = \theta_B = \theta_C = 0$. Compute the position vector ${}^0\{P\}_D$ for point D on member 3 when $\theta_A = \theta_B = \theta_C = 30°$ for dimensions $l_1 = l_2 = l_3 = 10$ in.

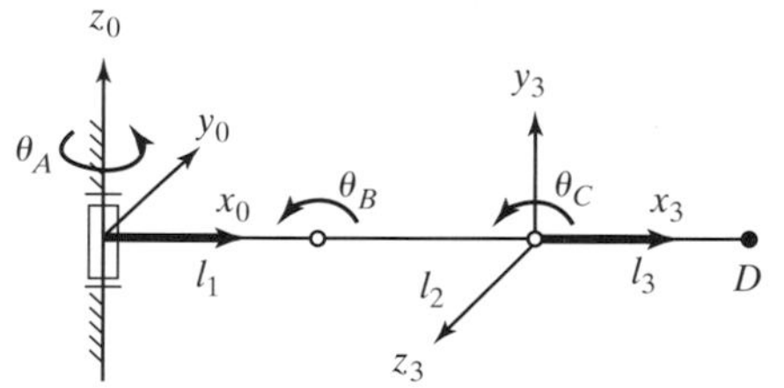

FIGURE P12.4

12.15 Determine the transformation matrix ${}^0_2[T]$ for the RR manipulator in Figure P12.5 as a function of the dimensions shown and the joint angles θ_A and θ_B. The position illustrated corresponds to $\theta_A = \theta_B = 0$. Compute the position vector ${}^0\{P\}_C$ for point C on member 2 when $\theta_A = \theta_B = 45°$ for dimensions $l_1 = l_2 = 10$ in and $b_1 = b_2 = 5$ in.

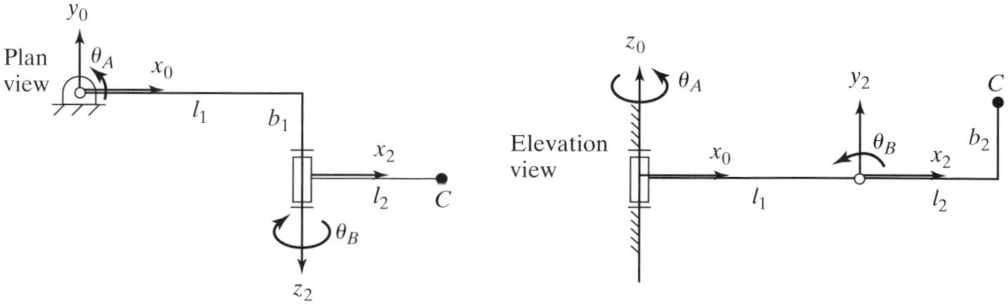

FIGURE P12.5

12.16 Determine the transformation matrix ${}^0_2[T]$ for the PP manipulator in Figure P12.6 as a function of joint translations s_A and s_B and dimensions l_2 and b_2. Compute the position vector ${}^0\{P\}_C$ for point C on member 2 when $s_A = s_B = 5$ in and $l_2 = b_2 = 10$ in.

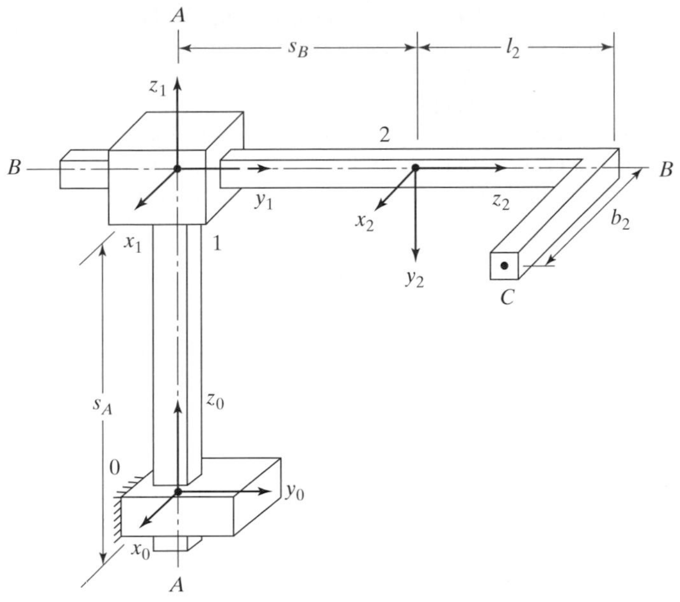

FIGURE P12.6

12.17 The RR planar manipulator of Figure 12.13 has dimensions $l_1 = l_2 = 10$ in. Carry out the inverse kinematics calculations for joint angles θ_A and θ_B when point C is at the location ${}^0\{P\}_C = \{-6\ 12\ 0\}^T$.

12.18 The manipulator of Figure 12.15 has dimensions $s_A = l_2 = l_3 = 10$ in. Carry out the inverse kinematic analysis to determine joint angles θ_A, θ_B, and θ_C when point D is at the location ${}^0\{P\}_D = \{12\ 6\ 18\}^T$.

12.19 The RP manipulator of Figure 12.14 has dimension $l_2 = 10$ in. Carry out the inverse kinematics calculations for joint coordinates θ_A, s_A, and s_B when point C is at the location ${}^0\{P\}_C = \{12\ 6\ 8\}^T$.

12.20 Derive the inverse kinematics relations for determining the joint angles θ_A, θ_B, and θ_C in terms of the end-effector coordinates ${}^0\{P\}_D$ for the manipulator of Figure P12.4.

12.21 A tentative RRR manipulator design is similar to Figure 12.17a. The frame and links 1, 2, and 3 are, in turn, connected by revolute joints in the form of motors. The axes of the three revolute joints are parallel to the z-axis. The link lengths are all 200 mm. The orientations of links 1, 2, and 3 are 0, 0, and π radians, respectively, at the park position. The angular velocities (rad/s) of the motor are $\omega_1 = 10$ (link 1 relative to the frame) and $\omega_2 = 10$ (link 2 relative to link 1). The motor between links 2 and 3 must maintain the angular position of link 3 at π radians. It is suggested that motion simulation software be used to solve this problem.

(a) Trace the position of the extreme end of link 3 for $0 \le t \le 0.2$ s.

(b) During that interval, when is link 3 farthest to the left?

(c) Show the linkage at that instant

(d) Give the x- and y-coordinates of the center of link 2 and its angular position at that instant.

(e) Give the x- and y-coordinates of the center of link 3 at that instant.

(f) Plot the x- and y-coordinates of the center of link 3 against time.

12.22 Analyze the motion of an RRR manipulator design similar to that in Figure 12.17a. Revolute joints, in the form of motors, connect the frame and links 1, 2, and 3 in turn. The axes of the three revolute joints are parallel to the z-axis. The orientations of links 1, 2, and 3 are 0, 0, and π radians, respectively, at the park position. The lengths of the links are 0.200, 0.200, and 0.250 m, respectively. Motor angular velocities (rad/s) are $\omega_1 = 12.5$ (link 1 relative to the frame), $\omega_2 = 5$ (link 2 relative to link 1), and $\omega_3 = -25$ (link 3 relative to link 2). It is suggested that motion simulation software be used to solve this problem.

(a) Trace the position of the extreme end of link 3.

(b) Show the linkage configuration at several instants in time.

(c) Give the x- and y-coordinates of the center of link 2 and its angular position at $t = 0.15$ s.

(d) Give the x- and y-coordinates of the center of link 3 at that instant.

(e) Plot the x- and y-coordinates of the center of link 3 and its angular position against time.

12.23 Analyze the motion of an RRR manipulator design similar to Figure 12.17a. Revolute joints in the form of motors connect the frame and links 1, 2, and 3 in turn. The axes of the three revolute joints are parallel to the z-axis. The orientations of links 1, 2, and 3 are 0, 0, and π radians, respectively, at the park position. The lengths of the links are 0.200, 0.200, and 0.150 m, respectively. The angular velocities (rad/s) of the motor are $\omega_1 = 8$ (link 1 relative to the frame), $\omega_2 = 12$ (link 2 relative to link 1), and $\omega_3 = -15$ (link 3 relative to link 2). It is suggested that motion simulation software be used to solve this problem.

(a) Trace the position of the extreme end of link 3.

(b) Show the linkage configuration at several instants in time.

(c) Give the x- and y-coordinates of the center of link 2 and its angular position at $t = 0.2$ s.

(d) Give the x- and y-coordinates of the center of link 3 at that instant.

(e) Plot the x- and y-coordinates of the center of link 3 and its angular position against time.

12.24 A tentative PRRR manipulator design is similar to Figure 12.17a. Fixed link 0 and link A (not shown) form a prismatic pair. Link A is driven by an actuator to move the linkage in the z direction according to the equation

$$Z(t) = 0.15 + 10 \sin(\theta_1 t),$$

where Z = change in elevation (m),

θ_1 = angular position of link 1 (rad),

and

t = time (s).

Links 1, 2, and 3 are, in turn, connected by revolute joints in the form of motors. The axes of the three revolute joints are parallel to the z-axis. The link lengths are all 200 mm. The orientations of links 1, 2, and 3 are 0, 0, and π radians, respectively, at the park position. Motor angular velocities (rad/s) are $\omega_1 = 10$ (link 1 relative to the link A) and $\omega_2 = 10$ (link 2 relative to link 1). Motor 3 (between links 2 and 3) must maintain the angular position of link 3 at π radians.

(*Suggestion*: Express the problem in transfer matrix form and use an efficient method of solution.)

(a) Find the required speed of motor 3 (the angular velocity of link 3 relative to link 2).

(b) Find the location of motor 3 at $t = 0.182$ s. Express the answer as a vector whose first three elements are the x, y, and z components.

(c) Find the location of the center of link 3 at $t = 0.001$ and 0.182 s.

(d) Find the angular position of link 3 at $t = 0$ and 0.182 s.

(e) Find the vector representing the velocity of the center of link 3 at $t = 0, 0.182$, and 0.249 s.

(f) Plot the magnitude of the resultant velocity of the center of link 3 against time.

12.25 Analyze a PRRR manipulator design similar to that in Figure 12.17a. Fixed link 0 and link A (not shown) form a prismatic pair. Link A is driven by an actuator to move the linkage in the z direction according to the equation $Z(t) = 0.350/(1 - 2t)$, where Z = change in elevation (m) and t = time (s) Links 1, 2, and 3 are, in turn, connected by revolute joints in the form of motors. The axes of the three revolute joints are parallel to the z-axis. The lengths of links 1, 2, and 3 are 200, 200, and 250 mm, respectively. The orientations of links 1, 2, and 3 are 0, 0, and π radians, respectively, at the park position. Motor angular velocities (rad/s) are $\omega_1 = 12.5$ (link 1 relative to link A), $\omega_2 = 5$ (link 2 relative to link 1) and $\omega_3 = -25$ (link 3 relative to link 2).

(*Suggestion:* Express the problem in transfer matrix form and use an efficient method of solution.)

(a) Find the location of motor 3 at $t = 0.15$ s. Express the answer as a vector whose where the first three elements are the x, y, and z components.

(b) Find the location of the center of link 3 at $t = 0.001$ and 0.15 s.

(c) Find the angular position of link 3 at $t = 0$ and 0.15 s.

(d) Find the vector representing the velocity of the center of link 3 at $t = 0, 0.15$, and 0.3 s.

(e) Plot the magnitude of the resultant velocity of the center of link 3 against time.

12.26 Analyze a PRRR manipulator design similar to Figure 12.17a. Fixed link 0 and link A (not shown) form a prismatic pair. Link A is driven by an actuator to move the linkage in the z direction according to the equation $Z(t) = 0.250\,[1 - \cos(3t)]$, where Z = change in elevation (m) and t = time (s). Links 1, 2, and 3 are, in turn, connected by revolute joints in the form of motors. The axes of the three revolute joints are parallel to the z-axis. The lengths of links 1, 2, and 3 are 200, 200, and 150 mm, respectively. The orientations of links 1, 2, and 3 are 0, 0, and π radians, respectively, at the park position. The angular velocities (rad/s) of the motor are $\omega_1 = 8$ (link 1 relative to link A), $\omega_2 = 12$ (link 2 relative to link 1), and $\omega_3 = -15$ (link 3 relative to link 2).

(*Suggestion*: Express the problem in transfer matrix form and use an efficient method of solution.)

(a) Find the location of motor 3 at $t = 0.2$ s. Express the answer as a vector whose first three elements are the x, y, and z components.

(b) Find the location of the center of link 3 at $t = 0.001$ and 0.2 s.

(c) Find the angular position of link 3 at $t = 0$ and 0.2 s.

(d) Find the vector representing the velocity of the center of link 3 at $t = 0$ and 0.2 s.

(e) Plot the magnitude of the resultant velocity of the center of link 3 against time.

PROJECT

Design a manipulator to perform some task for a disabled person. Strive for simplicity, reliability, and minimum cost.

BIBLIOGRAPHY AND REFERENCES

Craig, J. J., *Introduction to Robotics: Mechanics and Control*, Addison-Wesley, Reading, MA, 1986.

Critchlow, A. J., *Introduction to Robotics*, Macmillan, New York, 1985.

Denavit, J., and R. S. Hartenberg, "A Kinematic Notation for Lower-Pair Mechanisms Based on Matrices," *Journal of Applied Mechanics*, vol. 77, 1955, pp. 215–221.

Fu, K. S., R. C. Gonzalez, and C. S. G. Lee, *Robotics: Control, Sensing, Vision, and Intelligence*, McGraw-Hill, New York, 1987.

Gevarter, W. B., *Intelligent Machines*, Prentice Hall, Upper Saddle River, NJ, 1985.

Hill, R. O., Jr., *Elementary Linear Algebra*, Academic Press, New York, 1986.

Knowledge Revolution, *Working Model 2D User's Manual*, Knowledge Revolution, San Mateo CA, 1996.

Knowledge Revolution, *Working Model 3D User's Manual*, Knowledge Revolution, San Mateo CA, 1998.

Mathsoft, *Mathcad2000™ User's Guide*, Mathsoft, Inc., Cambridge, MA, 1999.

Paul, R. P., *Robot Manipulators: Mathematics, Programming and Control*, MIT Press, Cambridge, MA, 1981.

Note also the Internet references listed under the heading of **Robots**, **grippers**, **sensors**, **guided vehicle systems and accessories** found at the end of Chapter 1. You will find many more sites related to robots by searching the Internet yourself.

Partial Answers to Selected Problems

Note: Some of your solutions may not agree with the answers below. Perhaps the problem required design decisions, and your decisions differed from those of the authors. Do not be discouraged; your decisions may be equally valid, or they may be superior to the decisions of the authors.

CHAPTER 1

1.1 **(a)** $DF_{spatial} = 6$

(b) $DF_{spatial} \geq -5$

(c) $DF_{planar} = 2$

(d) Motion occurs in a plane or a set of parallel planes.

1.3 $v_{av} = 400$ in/s

1.5 378 and 553 in/s

1.7 For $R = 1$, piston position is given as follows:

Crank Angle T_1°	Piston Position X_2
0	2.487
20	2.433
40	2.199
60	1.844

1.9 For $R = 1$, piston position is given as follows:

Crank Angle T_1°	Piston Position X_2
0	2.755
20	2.739
40	2.550
60	2.239

1.11 **(b)** 19.99 in/s to the left

(c) 22.83 in/s to the left

1.13 Triple rocker (non-Grashof)

1.15 Change-point mechanism

1.17 Triple rocker (non-Grashof)

1.19 Change-point mechanism

1.21 Drag link

1.23 $120 \leq L_2 \leq 480$ mm

1.25 **(a)** $0 < L_1 < 20$
(b) None
(c) None
(d) $L_1 = 20$ and 60
(e) $0 < L_1 \leq 20$ and $L_1 = 60$
(f) $60 < L_1 < 140$ and $20 < L_1 < 60$

1.27 **(a)** None
(b) None
(c) $180 < L_3 < 260$
(d) $L_3 = 180$ and 260
(e) $180 \leq L_3 \leq 260$
(f) $20 < L_3 < 180$ and $260 < L_3 < 420$

1.29 **(a)** $1.5 < L_2/L_1 < 2.9$
(b) None
(c) $0 < L_2/L_1 < 0.5$
(d) $L_2/L_1 = 0.5$ or 1.5 or 2.9
(e) $0 < L_2/L_1 \leq 0.5$ and $1.5 \leq L_2/L_1 \leq 2.9$
(f) $0.5 < L_2/L_1 < 1.5$ and $2.9 < L_2/L_1 < 4.9$

1.31 Permissible values for the crank-rocker mechanism lie within the region bounded by the inequalities $L_3/L_1 < L_2/L_1 + 1$, $L_3/L_1 > 3 - L_2/L_1$, and $L_3/L_1 > L_2/L_1 - 1$

1.33 The linkage satisfies the Grashof criteria for a crank rocker, but an actual linkage with these proportions is likely to jam.

1.35 No value of L_3 satisfies the criteria.

1.37 No value of L_3 satisfies the criteria.

1.39 $L_2/L_1 = 3.045$. Other solutions are possible.

1.41 For $L_0/L_1 = 4$, the permissible region is bounded by the following points:

L_2/L_1	L_3/L_1
1.5	3.867
2	4.060
3	4.243
4	3.828
4.22	3.294
1.5	3.826
2	3.382
3	2.406
3.5	1.870

1.43 848.5 mm

1.45 494.97 mm

1.47 $0.4571L$

1.51 Set $d\phi/d\theta = R\cos\theta/(L\cos\phi) = 0$.

1.53 In a double-rocker linkage (of the first kind), the coupler must be free to rotate through 360°.

1.55 In Figure 1.29a, let $DA/DF = 2$. F is the tracing point.

1.57 In Figure 1.29a, let $AB/AC = 0.40$. Point A traces the pattern.

1.59 Many solutions are possible. Let $\phi = 30°$, $N = 6$ cylinders, $A = (d/6)^2$. Then, $d = 129.6$ mm.

1.61 $L_2 - L_1 = 0.005$ in

1.63 One possible solution: Use $L = 15$ mm, $N = 15$ teeth, $r_3/r_0 = 0.5$, and $\phi_{min} = 45°$, from which $r_{1(max)} = 0.287r_0$ and $r_2 = 1.002r_0$. A minimum crank length $r_{1(min)} = 0.090r_0$ is satisfactory.

1.65 $n_2 = 965.9$ to 1035.3 rev/min

1.67 13.86°

1.69 One possible solution: Refer to Figures 1.23 and 1.25; let $L_0 = 2$, $L_1 = 4.5$, and $L_3 = 5$. $L_2 = 7.25$ results in a ratio of 2.5 to 1 (approx).

1.71 One possible solution: $L_3 = 75$ mm. Let $L_0 = 20$ mm, $L_1 = 45$ mm, and $L_4 = 80$ mm. $L_2 = 67.5$ is satisfactory.

1.73 One possible solution: $O_1O_2 = 7$, $L_2 = 8.03$, $L_3 = 4$, and $L_{1(max)} = 4.36$.

1.75 $O_2C = 453$ mm. Selecting $O_1O_2 = 250$ mm, $O_1B_{max} = 77.25$ mm and $O_1B_{min} = 43.67$ mm.

1.77 The stroke is 3.14 in.

1.79 **(a)** 1.05 **(b)** 4.016

1.85 $100 < L_2 < 300$ mm

1.87 $95 < L_2 < 305$ mm

1.89 $\phi_{max} = 85.5°$; $L_2 = 212.13$ mm

1.91 $L_2 = 219.08$ mm; $\phi_{max} = 80.83°$

1.93 Away from crank: $v_{av} = 26{,}960$ mm/s; toward crank: $v_{av} = 21{,}940$ mm/s

1.95 $DF_{spatial} \geq 4$

1.97 $DF_{spatial} = 6$

1.101 The conditions are met with an eccentricity of 1.163 times the crank length.

1.103 Minimum transmission angle is about 39° at $R_2 = 3$ and $R_3 = 3$.

1.105 Minimum transmission angle is about 60° at $R_2 = 4$ and $R_3 = 4$.

1.107 Fixed link, drive crank, coupler and driven crank lengths of 147.8, 173.9, 189.7 and 200 mm, respectively, may be used. Minimum transmission angle is a special concern.

CHAPTER 2

2.1 **(a)** Piston, rack, reciprocating cam follower, straight edge on a drafting machine, the straight section of a belt or chain that is in tension, etc.

2.3 **(a)** 9.425 m/s
(b) 2961 m/s^2
(c) 930, 190 m/s^3

2.5 $D = S = 303$ mm

2.7 $D = S = 2.323$ in

2.9 $x = 20[1 - \cos(100t)]$; $v = 4000 \sin(200t)$; $a = 800{,}000 \cos(200t)$

2.11 44.27 rad/s

2.13 $v_{max} = 31.42$ in/s; $a_{max} = 3948$ in/s^2s

2.15 $2.909\angle 50.10°$

2.17 $1.759\angle 176.39°$

2.19 $\boldsymbol{A} \times \boldsymbol{B} = \boldsymbol{k}1$

2.21 $\boldsymbol{C} \times (\boldsymbol{D} \times \boldsymbol{E}) = \boldsymbol{k}7.954$

2.23 $\boldsymbol{A} \cdot \boldsymbol{B} = 1.732$

2.25 $\boldsymbol{A} \cdot (\boldsymbol{B} \times \boldsymbol{C}) = 0$

2.27 $\boldsymbol{C} \cdot (\boldsymbol{A} \times \boldsymbol{B}) = 0$

2.29 $-\boldsymbol{i}[3r_y + (5 + 3t)dr_y/dt] + \boldsymbol{j}[3r_x + (5 + 3t)dr_x/dt]$

2.31 In this case, reversing the order of the rotations affects only the altitude change. In general, the commutative law of vector addition does not apply to finite rotations.

2.33 $r_1 = -75.54$; $r_2 = -18.63$

2.35 $T = 3672$ N·mm cw

2.37 $\boldsymbol{r}_0 = \boldsymbol{i}0.5893r_1$ or $-\ \boldsymbol{i}2.1213r_1$; $\boldsymbol{r}_2 = r_1(-0.6428\boldsymbol{j} \mp 1.3553\boldsymbol{i})$

2.41 and 2.43 $\boldsymbol{r}_2 = \boldsymbol{i}139.6 + \boldsymbol{j}10.8$; $\boldsymbol{r}_3 = -\boldsymbol{i}49.6 - \boldsymbol{j}62.8$

2.45 $\boldsymbol{r}_2 = \boldsymbol{i}90.56 - \boldsymbol{j}106.76$; $\boldsymbol{r}_3 = -\boldsymbol{i}22.52 + \boldsymbol{j}76.77$

2.47 $d\boldsymbol{R}_1/dt = -20.78 + j12$

2.49 $\boldsymbol{R}_2 = 261.85 + j159.75 = 306.7e^{j0.548}$

2.51 $\boldsymbol{r}_0 = -0.5892r_1$ or $2.1212r_1$; $\boldsymbol{r}_2 = (\pm 1.3552 - j0.6428)r_1$

2.53 $r_3 = 180.4$ mm at $\theta = 45°$

2.55 $\boldsymbol{D} \times \boldsymbol{C} = \begin{bmatrix} 106 \\ 22 \\ 152 \end{bmatrix}$

2.57 $\boldsymbol{B} \times \boldsymbol{D} = \begin{bmatrix} -32 \\ 31 \\ -19 \end{bmatrix}$

2.59 $r_{2x} = [-r_2^2 + c_y^2 + r_3^2 - c_x^2]/[2\,c_x]$ where $c_x = -r_0 + r_1 \cos\theta$ and $c_y = r_1 \sin\theta$

2.61 $\theta_2 = 0.296$, $\theta_3 = 4.132$ rad after four iterations.

2.63 $\theta_2 = 0.293$, $\theta_3 = 4.698$, $r_D = 4$, $r_E = 4.944$

2.65 $\theta_2 = 0.16$, $\theta_3 = 5.476$, $r_D = 61.39$, $r_E = 158.79$

2.67 Transmission angle ranges from 59.4 to 104.9°.

2.69 Transmission angle ranges from 63.2 to 116.2°.

2.71 A drive crank angle of 160° results in a driven crank angle of 276.4° and a transmission metric of −0.511.

CHAPTER 3

3.1 $\boldsymbol{v} = -\boldsymbol{i}100 - \boldsymbol{j}2600 + \boldsymbol{k}2000$

3.3 $v = 1260$ in/s

3.5 $n = 12s/(\pi d)$

3.7 $\dot{R} = \boldsymbol{I}572.4 - \boldsymbol{J}82.6$

3.9 $\boldsymbol{v}_{CB} = 670.2$ in/s opposite the direction of rolling

3.11 $\boldsymbol{v}_B = 37.7\angle{-10.8°}$

3.13 $\theta_2 = 16.35°, \theta_3 = -122.19°; \omega_2 = -2.87$ rad/s, $\omega_3 = 10.86$ rad/s; $\boldsymbol{v}_C = 217.25$ mm/s $\angle 147.8°$

3.15

θ_1(degrees):	15	30	45	60	75
ω_3/ω_1:	−0.129	0.173	0.362	0.467	0.519

3.17 $\omega_2 = 5.03$ rad/s ccw

3.19 $O_2B = 79.36; \theta_2 = 5.61°; \omega_2 = 3.73$ rad/s

3.21 $O_2B = 74.31; \theta_2 = 16.59; \omega_2 = 3.55$ rad/s

3.23

θ_1(degrees):	0	15	30	45
ω_2/ω_1:	0.25	0.247	0.237	0.219

3.27 $\theta = 54.6°; v = 1.13R\omega$

3.29 $\omega_2 = 3.1$ rad/s

3.31 $\omega_2 = 60$ rad/s

3.33 **(a)** 20 in/s
(b) 20 in/s

3.35 Sliding velocity = 32 in/s

3.37 **(a)** and **(b)** $v_{CB} = \omega R$

3.39 Follower velocity = 7.1 in/s

3.41 $v_D = 17.5; v_E = 30$ in/s; $\omega_3 = 12.9$ rad/s cw

3.43 Sliding velocity = 163 in/s

3.45 $v_D = 62$ in/s

3.47 Sliding velocity = 17 in/s

3.49 $v_C = 210$ in/s

3.51 $\omega_4 = 433$ rad/s ccw

3.53 $v_C = 367$ mm/s

3.57

θ_1(Degrees)	ω_2(rad/s)	ω_3(rad/s)
30	−0.633	−4.106
60	−0.696	−3.927
90	−0.809	−3.599
120	−0.977	−3.077

3.59 $v_{B_2} = v_D = \omega_1 = 0$; $\alpha_{1(av)} = 181{,}000$ rad/s^2 cw; $a_{D(av)} = 76{,}000$ m/s^2 to the right

3.61 The angular velocities of links 2 and 3 are 1.06 rad/s cc and 32.3 rad/s cw, respectively.

3.63 and 3.65 $v_c = 7.37$ in/s to the left.

3.67 At $\theta_1 = 45°$, the angular velocities of links 2 and 3 are 0.205 and -0.238 rad/s, respectively.

3.69 At $\theta_1 = 45°$, $v_D = 490.2$ at 99.4°.

3.71 At $\theta_1 = 60°$, normalized slider velocity is -0.96.

3.73 At $\theta_1 = 30°$, the angular velocities of links 2 and 3 are 0.25 and -0.22 rad/s, respectively.

3.75 At input crank position = 70°, output crank position = 134.6°, and output crank angular velocity = 1.709 rad/s. The matrix solution agrees with values obtained by numerical differentiation. Other solutions are possible.

3.77 At input crank position = 160°, output crank position = 75.9°, and output crank angular velocity = -4.29 rad/s. The matrix solution agrees with values obtained by numerical differentiation. Other solutions are possible.

CHAPTER 4

4.1 $\boldsymbol{a} = -\boldsymbol{i}100{,}800 - \boldsymbol{j}88{,}600 - \boldsymbol{k}30{,}000$

4.3 $a_{av} = 2000$ in/s^2

4.5 **(a)** $\alpha_{av} = 5.236$ rad/s^2
(b) $\alpha_{av} - 2.09$ rad/s^2

4.7 $\boldsymbol{a}^n = 19.29$ m/s^2

4.9 $\boldsymbol{a}^t_B = 12{,}500$ mm/s^2; $\boldsymbol{a}^n_B = 72{,}000$ mm/s^2; $\boldsymbol{a}_B = 73{,}077$ mm/s^2

4.11 $\boldsymbol{R} = \boldsymbol{i}20 + \boldsymbol{j}60$; $\dot{\boldsymbol{R}} = -\boldsymbol{i}8995 + \boldsymbol{j}2985 + \boldsymbol{k}10$; $\ddot{\boldsymbol{R}} = -\boldsymbol{i}499{,}790 - \boldsymbol{j}1{,}346{,}870 + \boldsymbol{k}20$

4.13 $\boldsymbol{a}_C = 31{,}381\boldsymbol{i} - 48{,}131\boldsymbol{j} = 57{,}457$ mm/s$^2\angle -56.9°$

4.15

θ_1(Degrees)	ω_3/ω_1	α_3/ω_1^2
30	1.333	−0.246
60	1.205	−0.220
90	1.106	−0.164

4.19 $\alpha_2 = 45.7$ rad/s^2

4.21 $\boldsymbol{a}_D = 23{,}000$ in/s^2; $\alpha_2 = 2230$ rad/s^2

4.23 $\boldsymbol{a}_D = 5040$ in/s^2; $\alpha_2 = 147$ rad/s^2; $\alpha_3 = 267$ rad/s^2

4.25 $\alpha_2 = 138.5$ rad/s^2; $\alpha_3 = 440$ rad/s^2

4.27 $\alpha_1 = 63$ rad/s^2

4.29 **(b)** $a_C = 33{,}300$ in/s^2 to the left when $\theta = 0$

4.31 $\boldsymbol{a}_E = 40{,}000$ in/s^2

4.33 $\boldsymbol{a}^N_{C_1} = 100$ in/s^2; $\boldsymbol{a}^T_{C_1} = 0$; $\boldsymbol{a}_{C_2} = 70.7$ in/s^2

4.35 $\boldsymbol{a}_E = 1100$ in/s^2

4.37 $\boldsymbol{a}_D = 700 \text{ in/s}^2$

4.39 $\alpha_2 = 120 \text{ rad/s}^2$

4.41 Crank speed = 903 rev/min

4.43 $\alpha_1 = 5780 \text{ rad/s}^2$; $\alpha_3 = 5150 \text{ rad/s}^2$

4.47 $a_{max} = 1.667\ R\omega^2$

4.53 $\boldsymbol{a}_C = 625{,}000 \text{ mm/s}^2$; $\alpha_2 = 1233 \text{ rad/s}^2$

4.55 $\boldsymbol{a}_C = 1615 \text{ mm/s}^2$

4.57 $\boldsymbol{a}_C = 900 \text{ m/s}^2$

4.63 $\alpha_2/\omega_1^2 = -0.05197$ at $\theta_1 = 30°$

4.65 $\alpha_2 = 391.5 \text{ rad/s}^2$; $\alpha_3 = 541.9 \text{ rad/s}^2$

4.67 $\boldsymbol{a}_3 = \boldsymbol{a}_{s_2} = -8117 \text{ in/s}^2$

4.69

θ(Degrees)	$a_3/\omega_1^2 r_1$
0	−0.802
30	−0.727
60	−0.501

4.71 At $\theta_1 = 35°$, $\alpha_2 = 0.2938^*$ and $\alpha_3 = 0.6791^*$. *: Multiply by angular velocity squared of link 1.

4.73 At $\theta_1 = 35°$, $\alpha_2 = 0.0847^*$ and $\alpha_3 = 0.4205^*$. *: Multiply by angular velocity squared of link 1.

4.75 At $\theta_1 = 160°$, $\alpha_2 = 0.119^*$. *: Multiply by angular velocity squared of link 1.

4.77 At a crank angle of one radian, slider acceleration $= -1.264 \cdot 10^5 \text{ mm/s}^2$.

4.79 At a crank angle of one radian, angular acceleration of the connecting rod $= 4.14 \cdot 10^3 \text{ rad/s}^2$.

4.81 At a drive crank angle of one radian, angular acceleration of the driven crank $= -1265 \text{ rad/s}^2$.

CHAPTER 5

5.1 The displacement vs. time plot is linear. The cam follower cannot actually reproduce this motion at dwell, rise, and return transition points.

5.3 The displacement vs. time plot is cycloidal. Cam follower motion is smooth at dwell, rise, and return transition points.

5.5 The displacement vs. time plot is linear. The cam follower cannot actually reproduce this motion at dwell, rise, and return transition points.

5.7 The displacement vs. time plot is cycloidal. Cam follower motion is smooth at dwell, rise, and return transition points.

5.9 Although cam follower motion appears smooth, there will be acceleration jumps at dwell, rise, and return transition points.

5.11 Although cam follower motion appears smooth, there will be acceleration jumps at dwell and rise transition points.

5.13 The cam follower cannot actually produce the desired motion at dwell, rise, and return transition points.

5.15 Although the cam appears smooth, there will be acceleration jumps at dwell, rise, and return transition points.

5.19 The cam follower cannot actually produce the desired motion at dwell, rise, and return transition points.

5.23 Although the cam appears smooth, there will be acceleration jumps at dwell, rise, and return transition points.

5.25 Although the cam appears smooth, there will be acceleration jumps at dwell, rise, and return transition points.

5.27 **(b)** $v = 10.7$ in/s from $\theta = \pi/4$ to $3\pi/4$; $v = -10.7$ in/s from $\theta = 5\pi/4$ to $7\pi/4$; $a = 85.3$ in/s^2 for θ from 0 to $\pi/4$ and $7\pi/4$ to 2π; $a = -85.3$ in/s^2 for θ from $3\pi/4$ to $5\pi/4$

(c) $s = 2.89$ in, $v = 10.7$ in/s, $a = 0$, $j = 0$

(d) $s = 3.0$ in, $v = 10.7$ in/s, $a = 0$, $j = 0$

5.29 **(b)** $v = 16$ in/s at $\theta = \pi/2$, $v = -16$ in/s at $\theta = 3\pi/2$; $a = 32\pi$ in/s^2 at $\theta = \pi/4$ and $7\pi/4$, $a = -32\pi$ in/s^2 at $\theta = 3\pi/4$ and $5\pi/4$

(c) $s = 3.22$ in, $v = 12$ in/s, $a = -87.1$ in/s^2, $j = -632$ in/s^3

(d) $s = 3.0$ in, $v = 13.4$ in/s, $a = -74.7$ in/s^2, $j = -845$ in/s^3

5.31 **(b)** $v = 15$ in/s at $\theta = \pi/2$, $v = -15$ in/s at $\theta = 3\pi/2$; $a = 92.4$ in/s^2 at $\theta = 0.211\pi$ and 1.789π, $a = -92.4$ in/s^2 at $\theta = 0.789\pi$ and 1.211π

(c) $s = 3.16$ in, $v = 11.8$ in/s, $a = -71.1$ in/s^2, $j = -640$ in/s^3

(d) $s = 3.0$ in, $v = 12.7$ in/s, $a = -61.9$ in/s^2, $j = -734$ in/s^3

5.33 **(b)** $v = 75\pi$ mm/s at $\theta = 60°$, $v = -75\pi$ mm/s at $\theta = 240°$; $a = 225\pi^2$ mm/s^2 at $\theta = 0$ and $300°$, $a = -225\pi^2$ mm/s^2 at $\theta = 120$ and $180°$

5.35 **(b)** $v = 300$ mm/s at $\theta = 60°$, $v = -300$ mm/s at $\theta = 240°$; $a = 1800$ mm/s^{2}for θ from 0 to 60° and 240 to 300°, $a = -1800$ mm/s^2 for θ from 60 to 120° and 180 to 240°

5.37 $v_{max} = 9900$ mm/s

5.39 $a = 1190$ in/s^2, $v_{max} = 42.2$ in/s

5.47 **(a)** $y = (-2 + \sqrt{3})x$ and $y = (-2 - \sqrt{3})x$

(b) $x = (3/4 \pm \sqrt{3/4})\lambda$, $y = (-3/4 \pm \sqrt{3/4})\lambda$

5.49 $x = -(r_b + h\theta/\pi) \sin\theta - (h/\pi)\cos\theta$, $y = (r_b + h\theta/\pi)\cos\theta - (h/\pi)\sin\theta$; minimum follower face width $= 31.8$ mm

5.51 $x = -25 \sin\theta\{(6-\cos\theta) \pm [(6-2\cos\theta)/\sqrt{37 - 12\cos\theta}]\}, y = 25\cos\theta$ $\{(6-\cos\theta) \pm [(6 - \cos\theta + \sin\theta\tan\theta)/\sqrt{37 - 12\cos\theta}]\}$; $\phi = 8.9°$ at $\theta = 60°$

5.53 $\rho = r_b + h\theta/\pi = 100 + 15.9\theta$; neglecting possible discontinuities at ends of travel, $\rho_{min} = 100$ mm and $\rho_{max} = 150$ mm

5.55 $\rho_p = 25(37 - 12\cos\theta)^{3/2}/(38 - 18\cos\theta)$; $\rho_{p(max)} = 156.2$ mm, $\rho_{p(min)} = 147.9$ mm

5.57 Maximum acceleration $= 2.904 \cdot 10^5$ mm/s^2.

5.59 Maximum acceleration $= 3.219 \cdot 10^4$ mm/s^2.

5.61 Maximum acceleration $-$ $1.01 \cdot 10^6$ mm/s^2. This cam may cause high inertia forces, depending on the design of the follower and related components.

5.63 The displacement equation for the rise is

$$s = h\,[1 - 4x^2 + 6x^4 - 4x^6 + x^8]$$

5.65 The displacement equation for the rise is

$$s = 18[1 - 4x^2 + 6x^4 - 4x^6 + x^8]$$

5.67 Subtract wear from displacement, but note that valve will seat for negative cam follower displacement. Valve velocity $= 0$ for negative cam follower displacement except for bouncing as valve seats abruptly. Actual valve acceleration depends on mass of valve train. Velocity at mid-rise $=$ 3281 mm/s.

CHAPTER 6

6.5 $d = 20.0$ in

6.7 $n = 596.8$ rev/min

6.9 $c = 8.33$ in

6.11 $v_p = 78.5$ in/s

6.13 $N_1 = 40$, $N_2 = 100$, $v_p = 2618$ mm/s

6.15 $c = 16$ in, $r_{b2} = 11.3$ in

6.17 $N_2 = 115$, $c = 15.43$ in

6.19 Contact ratio $= 1.67$

6.21 $c = 8.0$ in

6.23 Contact ratio $= 1.66$

6.25 17.5 in/s at beginning of contact; 16.6 in/s at end of contact

6.27 $\phi = 21.9°$

6.31 Interference; remove 0.25 mm from teeth of larger gear

6.33 $N_2 = 45$

6.35 Load moves 897 in.

6.39 $N_1 = 20$, $N_2 = 100$, $v_p = 65.4$ in/s

6.41 **(a)** $c_{AB} = c_{BC} = 240$ mm

(b) $\omega_1 = 104.7$ rad/s cw; $\omega_2 = 52.4$ rad/s ccw; $\omega_3 = 20.9$ rad/s ccw

(c) $T_A = 191$ N·m, $T_B = 0$, $T_C = 955$ N·m

(d) $F_t = 2388$ N, $F_r = 869$ N

6.43 $n_A = 1200$ rev/min ccw $n_B = 480$ rev/min cw $n_C = 230$ rev/min ccw; $T_A = 239$ N · m, $T_B = 597$ N · m, $T_C = 1243$ N · m; $F_A = 3970$ N, $F_B = 12{,}280$ N, $F_C = 9144$ N

6.45 The contact ratio = 1.315 for a pair of 12 tooth, 25° pressure angle spur gears.

6.47 A full-depth 20° pressure angle, 15 tooth pinion will mesh with a 40 tooth gear without interference.

6.49 A full-depth 14.5° pressure angle 25 tooth pinion meshing with a 35 tooth gear will have a contact ratio = 1.94.

6.51 If the pinion has 16 teeth and the gear 100 teeth, a pressure angle ≥ 20° is required to avoid interference (for full-depth teeth).

6.53 If the pinion has 10 teeth and the gear 30 teeth, the minimum pressure angle required to avoid interference is approximately 25° (for full-depth teeth).

CHAPTER 7

7.1 $p^n = 0.481$ in

7.3 $N_1 = 20$, $N_2 = 38$, $\phi = 20.6°$
Other solutions are also possible.

7.5 Thrust force is to the left on gear 3. Thrust force is to the right on gear 2.

7.7 $p^n = 16.32$ mm, $p = 18.85$ mm, $N_1 = 24$, $N_2 = 60$

7.9 $c = 275$ mm,

7.11 $F_t = 5980$ N, $F_r = 2410$ N, $F_a = 2890$ N

7.13 $c = 14.2$ in

7.15 $N_1 = 24$, $N_2 = 60$

7.17 $d_w = 1.82$ in, $d_g = 3.18$ in

7.19 $c = 16.22$ in

7.21 $d_w = 2.41$ in, $d_g = 17.59$ in

7.23 $T_A = 9.55$ N · m, $T_B = 19.1$ N · m, $T_C = 764$ N · m, $F_{aA} = 87$ N, $F_{aB} = 6077$ N, $F_{aC} = 844$ N

7.25 **(a)** $\gamma_p = 16.2°$, $\Gamma_g = 43.8°$
(b) $r_{bp} = 1.56$ in, $r_{bg} = 5.18$ in

7.27 $F_{tp} = F_{tg} = 1236$ lb, $F_{rp} = F_{ag} = 402$ lb, $F_{ap} = F_{rg} = 201$ lb

7.29 If gear 2 has 72 teeth and gear 4 has 50 teeth, the output speed = 256 rpm. The helix angle of gears 3 and 4 should be 16.5° to balance thrust on the countershaft. Output torque = 44,762 N · mm (neglecting friction).

CHAPTER 8

8.1 Output speeds may range from 500 to 667 rev/min if all gears have the same module.

8.3 One possible solution: Let $N_1 = 24$. Then, $N_3 = 30$, $N_4 = 32$, $N_5 = 40$, $N_6 = 48$, $N_7 = 60$, $N_2 = 26$.

8.5 One possible solution: $N_1 = 29, N_2 = 40, N_3 = 39, N_4 = 50$

8.7 $n_R = 200(1 - N_S/N_R)$

8.9 **(a)** $n_{R_2} = 0.0915$ rev/min cw

(b) Impossible due to friction

8.11 $n_{S_2} = 31$ rev/min cw

8.13 S_4 makes one revolution in 10^8 s (about 3 years and 2 months)

8.15 $n_{S_2} = n_C + (n_{S_1} - n_C)(N_{S_1}N_{P_2})/(N_{P_1}N_{S_2})$

8.17 $n_C = (N_{S_1}N_{P_2}n_{S_1})/(N_{S_1}N_{P_2} - N_{P_1}N_{S_2})$

8.19 $n_C = (N_{R_1}N_{P_2}N_{P_4}n_{R_1})/(N_{R_1}N_{P_2}N_{P_4} + N_{P_1}N_{P_3}N_{R_2})$

8.21 $n_C = (N_{R_1}N_{P_2}n_{R_1})/(N_{P_1}N_{R_2} + N_{R_1}N_{P_2})$

8.23 $n_C = (N_SN_{P_2}n_S)/(N_SN_{P_2} - N_{P_1}N_R)$

8.25 $n_C = (n_SN_{P_2}N_{P_4}N_S)/(N_SN_{P_2}N_{P_4} - N_{P_1}N_{P_3}N_R)$

8.27 $n_C = (n_{S_1}N_{S_1}N_{P_3})/(N_{S_1}N_{P_3} + N_{P_2}N_{S_2})$

8.29 $n_C = n_{R_2}/\{1 + [N_{R_1}N_{P_2}/(N_{P_1}N_{R_2})]\}$

8.31 $n_C = n_R/\{1 - [N_SN_{P_2}/(N_{P_1}N_R)]\}$

8.33 $n_C = n_R/\{1 - [N_SN_{P_2}N_{P_4}/(N_{P_1}N_{P_3}N_R)]\}$

8.35 $n_C = (n_{S_1}N_{S_1}N_{P_2})/(N_{P_2}N_{S_1} - N_{S_2}N_{P_1})$

8.37 $n_C = (n_{R_1}N_{R_1}N_{P_2}N_{P_4})/(N_{R_1}N_{P_2}N_{P_4} + N_{R_2}N_{P_1}N_{P_3})$

8.39 $n_C = (N_{P_2}N_{R_1}n_{R_1})/(N_{P_2}N_{R_1} + N_{R_2}N_{P_1})$

8.41 $n_C = n_SN_SN_{P_2}/(N_SN_{P_2} - N_RN_{P_1})$

8.43 $n_C = (N_Sn_SN_{P_2}N_{P_4})/(N_SN_{P_2}N_{P_4} - N_RN_{P_1}N_{P_3})$

8.45 $n_C = (n_{R_2}N_{R_2}N_{P_1})/(N_{P_1}N_{R_2} + N_{R_1}N_{P_2})$

8.47 $n_C = (n_RN_RN_{P_1})/(N_RN_{P_1} - N_SN_{P_2})$

8.49 $n_C = (n_RN_RN_{P_1}N_{P_3})/(N_{P_1}N_{P_3}N_R - N_{P_2}N_{P_4}N_S)$

8.51 $N_S = 20$; $N_R = 80$; $d_R = 320$ mm; $15 \leq N_p \leq 37$

8.53 $n_{S_2}/n_{S_1} = 0.92795$

8.57 $\omega_P = 225$ rad/s cw; $\omega_C = 105.88$ rad/s cw

8.59 $\omega_{S_2} = 25.14$ rad/s

8.61 **(a)** 17.083 carrier rotations; $n_{PC} = 36.458$ rotations

(b) Carrier rotations $= (7/12)x + (5/12)y$

8.63 $5 \leq d_2 \leq 9$ in; $5 \leq d_3 \leq 7.92$ in

8.65 $r_1 = 0.3019$ in for 16 rev/min, 0.5406 for $33\frac{1}{3}$ rev/min, 0.667 for 45 rev/min, and 0.9286 for 78 rev/min

8.67 The drive may be similar to Figure 8.31 except that disk 2 drives and the diameter of disk 2 exceeds twice the diameter of disk 1.

8.71 There are 36 combinations producing 19 different ratios ranging from 0.64 to 1.5625.

8.73 $n_c/n_P = 3$

8.75 $N_{P1} = 20$

8.77 The desired reduction is produced (approximately) if the sun, planet, and ring gears have 19, 21, and 61 teeth, respectively. The train may be balanced with four planets.

8.79 The desired reduction is produced (approximately) if the sun, planet, and ring gears have 26, 18, and 62 teeth, respectively. The train may be balanced with four planets.

8.81 Torques are 30,000, 0, 82,500, and −112,500 N · mm on the sun, planet, ring, and carrier shafts, respectively.

8.83 Torques are 4375, 0, 13,125, and −17,500 N · mm on the sun, planet, ring, and carrier shafts, respectively.

8.85 The number of teeth in the planets meshing with the sun is varied from 18 to 58, and speed ratios are calculated. If those planets each have 19 teeth and the ring has 57 teeth, the output speed is 1000 rpm. If those planets each have 57 teeth and the ring has 95 teeth, the output speed is 250 rpm.

8.87 Using 19 to 88 teeth in planets #1 satisfies the approximate output speed requirement. If planets #1 each have 28 teeth and the ring 65 teeth, carrier speed is about 791 rpm, ring torque reaction is 60,981 N · mm, and carrier torque is 72,441 N · mm.

8.89 Using 19 to 90 teeth in planets #1 satisfies the approximate output speed requirement. If planets #1 each have 30 teeth and the ring 69 teeth, carrier speed is about 296 rpm, ring torque reaction is 95,329 N · mm, and carrier torque is 112,829 N · mm.

8.91 If planets #1 each have 46 teeth, the ring must have 84 teeth. For that combination, noise or vibration at the following frequencies may be significant.
Shaft error frequencies (Hz):
sun shaft = 133.3; planet carrier shaft = 11.4; planet shaft = 50.4.
Tooth meshing frequencies (Hz):
sun and planet = 2316.9; planet and ring = 957.
Tooth error frequencies (Hz):
sun = 487.8; any planet = 50.4; ring = 45.6.

CHAPTER 9

9.1 **(a)** $T_1 = 18.3$ N · m cw

9.3 $T_1 = 2.84$ N · m cw

9.5 $T_1 = 230$ N · mm cw

9.7 $T_1 = 4.92$ N · m cw

9.9 $F_s = 168$ N

9.11 $T_2 = 1340$ lb · in ccw

9.13 $T_1 = 8.22$ N · m cw

9.15 $T_1 = 121$ N · m cw

9.17 $P = 13$ lb

9.18 **(a)** $T_1 = 18.3$ N · m cw

9.20 $T_1 = 2.74$ N · m cw

9.22 $T_1 = 234$ N · mm cw

9.24 $T_1 = 5.09$ N · m cw

9.26 $F_s = 168$ N

9.28 $T_2 = 1340$ lb · in ccw

9.30 $T_1 = 8.44$ N · m cw

9.32 $T_1 = 121$ N · m cw

9.34 $P = 13$ lb

9.35 **(a)** $T_1 = 18.3$ N · m cw

9.37 $T_1 = 2.72$ N · m cw

9.39 $T_1 = 233$ N · mm cw

9.41 $T_1 = 5.09$ N · m cw

9.44 $T_2 = 1340$ lb · in ccw

9.45 $T_1 = 121$ N · m cw

9.47 **(a)** $T_1 = 22.8$ N · m cw
(b) $e = 0.80$

9.48 **(a)** $T_1 = 22.8$ N · m cw
(b) $e = 0.80$

9.51 $T_1 = 4.51$ N · m cw

9.53 $T_1 = 6.90$ lb · in cw

9.55 $T_1 = 10.7$ lb · in cw

9.57 $T_1 = 14.1$ lb · in ccw

9.59 At crank angle = 1 rad, the cylinder applies a vertical reaction force component = 15.466 N on the connecting rod. At crank angle = $\pi/4$ the torque applied by the crankshaft is −1118.4 N · mm.

9.61 When crank angle = 0, the axial piston force is −26,650 N. If crankshaft rotation is counterclockwise, it applies a small counterclockwise torque during the first half of the cycle and a significant counterclockwise torque during the second half. At crank angle = 1.75 π rad, the connecting rod is subject to a 22,513 N compressive force and crank torque = $1.311 \cdot 10^6$ N · mm.

9.63 When crank angle = 0, the axial piston force is −27,310 N. If crankshaft rotation is counterclockwise, it applies a small counterclockwise torque during the first half of the cycle and a significant counterclockwise torque during the second half. At crank angle = 1.75 π rad, the connecting rod is subjcct to a 23,063 N compressive force and crank torque = $1.673 \cdot 10^6$ N · mm.

9.65 Maximum lateral force magnitude on the piston = 132.6 N at time = 1.68 s. At that instant, spring tension = 263.5 N and motor torque magnitude = 11,761 N · mm.

CHAPTER 10

10.1 $|m_2 \boldsymbol{a}_{G2}| = 64$ N, $h_2 = 12.6$ mm

10.3 $T_1 = 4.4$ N · m ccw

10.5 **(a)** $\boldsymbol{F}_2 = 49.5\angle 45°$ N, $h_2 = 15.3$ mm

10.7 $T_1 = 26.9$ lb · in ccw

10.9 $T_1 = 34.4$ lb · in cw

10.13 $T = 8.93$ N · m when $\phi = 30°$

10.15 $T = 375$ lb · ft when $\phi = 30°$

10.17 $T = 6.94$ N · m when $\phi = 30°$

10.19 $T = 5.94$ N · m cw

10.21 $w_P r_P = 5.59$ oz · in, $\theta_P = 253°$, $w_Q r_Q = 4.27$ oz · in, $\theta_Q = 219°$

10.23 $w_P r_P = 6.36$ oz · in, $\theta_P = 35°$, $w_Q r_Q = 9.37$ oz · in, $\theta_Q = 268°$

10.25 $m_P r_P = m_Q r_Q = 141$ kg · mm, $\theta_P = 90°$, $\theta_Q = 270°$

10.27 At $\phi = 0$, $F_{32x} = 158$ N, $F_{32y} = F_{03y} = F_{12y} = F_{01y} = 0$, $F_{12x} = -251$ N, $F_{01x} = -97$ N

10.29 $F_s = 2000(\cos \omega t - \sin \omega t)$, $a = (4 \sin \omega t + \cos 2\omega t)/(\sin \omega t - \cos \omega t)$; secondary shaking forces balance

10.31 Primary forces do not balance; secondary forces balance.

10.33 Primary moments do not balance.

10.35 Neither primary nor secondary forces balance.

10.37 Correction magnitude $= 2\,mr$, counterweight location $= 180°$ from crank

10.39 The center of gravity is located at $x = 2.071 \cdot 10^{-2}$ and $y = 1.690 \cdot 10^{-2}$ m. The mass moment of inertia about a z-axis through the center of gravity $= 7.976 \cdot 10^{-7}$ kg · m^2

10.41 The mass moment of inertia about an axis through the center of rotation $= 5.084$ kg · m^2. Terminal angular velocity $= 24.2$ rad/s. At 20 rad/s, angular acceleration $= 0.243$ rad/s^2.

10.43 The mass moment of inertia about an axis through the center of rotation $= 10.9$ kg · m^2. Terminal angular velocity $= 19.766$ rad/s. At 16 rad/s, angular acceleration $= 0.316$ rad/s^2.

10.45 At time $= 10$ s: power supplied $= 111.35$ W; angular displacement $= 37.7$ rad; angular velocity $= 7.42$ rad/s; and angular acceleration $= 0.695$ rad/s^2.

10.47 Maximum motor torque $= 0.311$ N · m. Maximum power $= 3.11$ W.

10.49 Maximum motor power $= 11.11$ kW.
The corresponding torque $= 111.1$ N · m.

10.51 Maximum motor power $= 11.81$ kW. At that instant, crank angular position $= 0.9$ rad and motor torque $= 118.1$ N · m.

10.53 Maximum motor power $= 2048$ W.

10.55 Maximum crankshaft speed $= 109$ rad/s. At that instant, net motor torque $= -9.56$ N · m. Approximate elapsed time for the first revolution of the crankshaft $= 81.1$ millisec.

10.57 Maximum piston acceleration is about 3136 m/s^2. The approximate crank angle $= 1.42$ rad at that instant.

CHAPTER 11

11.1 The coordinates of the fixed pivot are 125, 344.

11.3 Fixed center O_B must lie along a line through point 4.5, 5.5 at angle 135°.

11.5 The coordinates of the fixed pivot are 124.999, 344.

11.7 Fixed center O_c must lie along a line through point 7.23, 8.23 at an angle of 151°.

11.9 Locate fixed pivot O_B at 2.5, −7.

11.11 Locate fixed pivot O_B at 25.96, 12.02.

11.13 Locate fixed pivot O_B at 26.18, −18.53.

11.15 $R_1 = j4,\ R_2 = j2,\ R_3 = 6 - j2$

11.17 $r_1 = -9.48 - j1.68,\ r_2 = -168.10 + j10.08,\ r_3 = 77.58 - j8.40$

11.19 $r_1 = -0.308 - j\,0.462,\ r_2 = -0.154 - j\,0.231,\ r_3 = -0.538 + j\,0.692$

11.21 $r_1/r_0 = 2.32,\ r_2/r_0 = 0.62,\ r_3/r_0 = 2.43$

11.23 At the first precision point, $\mathbf{r}_1 = -159.99 + j\,70.34$; $\mathbf{r}_2 = 218.94 - j\,64.23$; $\mathbf{r}_3 = -3.95 - j\,6.11$ (all mm). The error is essentially zero at the precision points. The error $= 2.37\%$ at one-quarter of the range of x.

11.25 Let the coupler rotate 0.3 radians between the first and second precision points and 0.6 radians between the first and third. At the first precision point, $\mathbf{r}_1 = -172.69 + j\,81.1$; $\mathbf{r}_2 = 240.12 - j\,66.04$; $\mathbf{r}_3 = -7.44 - j\,15.06$(all mm). The error is essentially zero at the precision points. The error $= -7.33\%$ at one-quarter of the range of x.

11.27 The linkage is a triple rocker. Extend the coupler outward from C in a straight line to a point D, where $CD = 250$ mm. Point D traces an approximately straight line over a portion of its motion.

CHAPTER 12

12.1 **(a)** $DF = 1$ **(b)** $DF = 1$ **(c)** $DF = 1$ **(d)** $DF = 2$

12.3

$$ {}^{i}_{j}[R] = \begin{bmatrix} (\cos\beta\cos\gamma) & (-\sin\gamma) & (\sin\beta\cos\gamma) \\ (\cos\alpha\cos\beta\sin\gamma + \sin\alpha\sin\beta) & (\cos\alpha\cos\gamma) & (\cos\alpha\sin\beta\sin\gamma - \sin\alpha\cos\beta) \\ (\sin\alpha\cos\beta\sin\gamma - \cos\alpha\sin\beta) & (\sin\alpha\cos\gamma) & (\sin\alpha\sin\beta\sin\gamma + \cos\alpha\cos\beta) \end{bmatrix} $$

12.5

$$ {}^{i}\{P\} = \begin{Bmatrix} \dfrac{\sqrt{2}}{2} \\ \dfrac{\sqrt{2}}{2} \\ -\sqrt{2} \end{Bmatrix} $$

12.7

$$ {}^{i}\{P\} = \begin{Bmatrix} 1 - \dfrac{1}{\sqrt{2}} \\ 1 + \dfrac{1}{\sqrt{2}} \\ 0 \end{Bmatrix} $$

12.11 $^{1}\{P\} = \begin{Bmatrix} 3 \\ 6 \\ 4 \\ 1 \end{Bmatrix}$

12.13 $^{0}\{P\}_D = \begin{Bmatrix} 5(1 + \sqrt{3}) \\ 5(3 + \sqrt{3}) \\ 0 \\ 1 \end{Bmatrix}$

12.15 $^{0}\{P\}_C = \begin{Bmatrix} 13.11 \\ 6.04 \\ 10.61 \\ 1 \end{Bmatrix}$

12.17 $\theta_A = 68.7°, 164.5°; \theta_B = 95.7°, -95.7°$

12.19 $\theta_A = -63.44°$, $s_A = 8$ in, $s_B = 3.42$ in

12.21 When the end of link 3 is farthest to the left, the center of link 2 is located at $x = -0.139$, $y = 0.146$; and the center of link 3 is located at $x = -0.327$, $y = 0.100$ (all m).

12.23 At time = 0.2 s: the center of link 2 is located at $x = -0.069$, $y = 0.124$; and the center of link 3 is located at $x = -0.175$, $y = -0.015$ (all m).

12.25 The angular position of link 3 in a fixed reference frame = 2.015 rad at time = 0.15 s. The velocity magnitude of the center of link 3 = 4.726 m/s at time = 0.15 s.

Index

G

Q

R

S

U

V